Airglow as an Indicator of Upper Atmospheric
Structure and Dynamics

Vladislav Yu. Khomich · Anatoly I. Semenov ·
Nikolay N. Shefov

Airglow as an Indicator of Upper Atmospheric Structure and Dynamics

Prof. Vladislav Yu. Khomich
Russian Foundation for Basic
Research
Leninsky prospect 32 A
Moscow
Russia 119991
khomich@rfbr.ru

Prof. Anatoly I. Semenov
Russian Academy of Sciences
Institute Atmospheric Physics
Lab. Phys. Upper Atmos.
Pyzhevsky Pereulok 3
Moscow
Russia 119017
anasemenov@yandex.ru

Prof. Nikolay N. Shefov
Russian Academy of Sciences
Institute Atmospheric Physics
Lab. Phys. Upper Atmos.
Pyzhevsky Pereulok 3
Moscow
Russia 119017
nikoshefov@yandex.ru

ISBN: 978-3-540-75832-7   e-ISBN: 978-3-540-75833-4

Library of Congress Control Number: 2007941792

© 2008 Springer-Verlag Berlin Heidelberg

This work is subject to copyright. All rights are reserved, whether the whole or part of the material is concerned, specifically the rights of translation, reprinting, reuse of illustrations, recitation, broadcasting, reproduction on microfilm or in any other way, and storage in data banks. Duplication of this publication or parts thereof is permitted only under the provisions of the German Copyright Law of September 9, 1965, in its current version, and permission for use must always be obtained from Springer. Violations are liable to prosecution under the German Copyright Law.

The use of general descriptive names, registered names, trademarks, etc. in this publication does not imply, even in the absence of a specific statement, that such names are exempt from the relevant protective laws and regulations and therefore free for general use.

Cover design: deblik, Berlin

Cover illustration: © iStockphoto.com/Roman Krochuk

Printed on acid-free paper

9 8 7 6 5 4 3 2 1

springer.com

*Devoted to the 50th anniversary of the
International Geophysical Year (1957–1959)*

Science cannot be relied on in the cases where mathematics is inapplicable or where the disciplines involved are unrelated to mathematics.

Leonardo da Vinci

Preface

Gaseous envelopes of planets have long been the object of astrophysical spectrophotometric investigations into the processes proceeding in these remote objects of the solar system. However, only the Earth's atmosphere surrounding us gives the widest opportunities for a more detailed investigation of its properties. As a rule, the study of these properties is considered as a purely geophysical problem, but in fact they are a special case of the general astronomical phenomenon.

Over the past more than half-century, atmospheric physics has become engaged in the research of the entire gaseous medium surrounding our planet extending to the interplanetary space. Despite the subdivision of the atmosphere into separate more or less homogeneous altitude layers studied separately from other atmospheric regions, these layers are in interaction, which is manifested sometimes unexpectedly for the researches of conventional problems. Of course, solar radiation whose effect in each particular case is determined by vertical profiles of gaseous atmospheric constituents and by their absorbing and scattering properties unites all these regions of the atmosphere into a single whole.

The absorbed energy of the Sun creates a certain thermal regime in each region of the atmosphere whose response is emissions arising from molecules or atoms contained in this region. These two processes participate in the establishment of thermal equilibrium in the atmosphere. The atmospheric airglow spectrum is extremely wide; it is determined by the atmospheric conditions, including photochemical processes characteristic of individual altitude layers. This allows atmospheric airglow to be used as a very informative indicator for atmospheric remote sounding.

However, the opportunity of recording the atmospheric airglow from the Earth surface is limited by the visible and near infrared ranges of the spectrum due to the specific properties of the atmospheric transparency in different ranges of the spectrum. The airglow emissions arising above 80 km refer exactly to these ranges of the spectrum. Reflecting directly the characteristics of the atmospheric region in which they arise, these emissions respond to the variations in its parameters due to the propagation of various waves generated in the ground layer of the atmosphere. This allows the airglow to be used as an indicator of very many dynamic processes occurring in the atmosphere.

Over the past more than fifty years, investigations into physics of the upper atmosphere have been extended considerably. Many of these investigations are devoted to the airglow which stimulated the foundation of the new scientific branch called aeronomy. New methods have been developed and well-known methods of measuring various emission characteristics that determine the state of the upper atmosphere have been refined to implement extended research programs. It should be emphasized that the intensities of nightglow and even twilight glow are rather low in comparison with the conventional surrounding daytime illumination. This fact imposes strong requirements to the choice of the spectral instrumentation and its sensitivity.

All the foregoing indicates that investigations of the airglow of the upper atmospheric layers call for the application of an extremely wide range of notions and data that have already been used in many other branches of natural science. Among them are problems of determining space and time coordinates, physics of atomic and molecular spectra, photochemistry, wave processes, instrumental and methodical aspects of spectrophotometric measurements, physics of solar–terrestrial relationships, etc. Therefore, brief information on these aspects of research is also given in the monograph to give physical insight into all aspects of the problem of harnessing the airglow in investigations of Earth's thermal and dynamic regimes.

Results of long-term comprehensive investigations of as though imperceptible nightglow of the upper atmosphere have given new insight into the tendency for changes in the atmospheric temperature regime and dynamic processes caused by numerous phenomena in the ground atmosphere. Large groups of researchers, who gave huge contribution to scientific knowledge of atmospheric physics, took part in accumulation and analysis of this long-term versatile material. The authors of the present monograph have also been participating in this long-term research for several decades. We make an attempt to present the invaluable experience gained from these long-term efforts which makes the monograph all the more important because many of the contributors to this work have passed away. By doing so we would like to pay tribute to the memory of those outstanding scientists.

We are aware of the fact that this first attempt to present a comprehensive review of the results obtained will suffer from certain limitations caused by incompleteness of research in some areas of interest on the one hand, and by the impossibility of a more detailed description of a wide range of problems, on the other. In connection with this, any remarks and recommendations will be appreciated.

Russia

Vladislav Yu. Khomich
Anatoly I. Semenov
Nikolay N. Shefov

Contents

Introduction 1
    References 6

1 The Radiating Atmosphere and Space 9
   1.1 Spectral Structure of the Airglow and Its Role in Atmospheric
       Research 9
   1.2 Extra-Atmospheric Radiation 15
       1.2.1 Stellar Photometric Units 17
       1.2.2 The Sun, the Moon, and Planets 19
       1.2.3 Bright Stars 23
       1.2.4 Distribution of the Starlight over the Sky 24
       1.2.5 Zodiacal Light 25
   1.3 Ultraviolet Solar Radiation 26
       1.3.1 Radiation in the 100- to 300-nm Range 26
       1.3.2 Radiation in the 20- to 100-nm Range 27
       1.3.3 Occurrence of Photoelectrons 28
   1.4 Space–Time Conditions for Detecting Radiation 30
       1.4.1 Geometric Relations for Various
             Observation Conditions 30
       1.4.2 Horizontal Coordinates of the Equator, Ecliptic,
             and Galactic Equator 42
       1.4.3 Conditions of Twilight Measurements 42
       1.4.4 Limb-Directed Measurements 46
       1.4.5 Time Coordinates 52
   1.5 Manifestation of the Variations of the Atmospheric Characteristics
       in the Properties of the Airglow 60
   1.6 The Heliogeophysical Conditions that Determine
       the Observed Airglow Characteristics 61
       1.6.1 Solar Activity and Its Variations During 11-yr Cycles ... 61
       1.6.2 Cyclic Aperiodic Variations of Solar Activity 73
       1.6.3 Long-Term Variations in Solar Activity 84

	1.6.4	Solar–Terrestrial Relationships	86

1.6.4 Solar–Terrestrial Relationships 86
1.6.5 Characteristics of the Geomagnetic Field: Their Variations in Space and Time 99
References 108

2 Processes Responsible for the Occurrence of the Airglow 119
2.1 Processes of Excitation and Deactivation of Excited Species 119
    2.1.1 Kinetics of Elementary Gaseous-Phase Photochemical Reactions 120
    2.1.2 Relaxation Processes 125
    2.1.3 Translational Relaxation 126
    2.1.4 Electron-Excited Species Relaxation 128
    2.1.5 Vibrational Relaxation 128
    2.1.6 Rotational Relaxation 134
2.2 Hydroxyl Emission 136
    2.2.1 Progress in the Studies of Hydroxyl Emission 136
    2.2.2 Principal Characteristics of Hydroxyl Emission 138
    2.2.3 Probabilities of Rotational–Vibrational Transitions 142
    2.2.4 Hydroxyl Emission Intensity 147
    2.2.5 Rotational Temperature 148
    2.2.6 The Population Distribution over Vibrational Levels 159
    2.2.7 Photochemistry 164
    2.2.8 Formation of Dissociation Products 168
    2.2.9 Recombination of Atomic Oxygen 172
2.3 Emissions from Metal Atoms 173
    2.3.1 Sodium Emission: Photochemical Excitation 173
    2.3.2 Alkali Metals: Resonant Excitation 178
2.4 Molecular Oxygen Emissions 185
    2.4.1 Ultraviolet Systems 186
    2.4.2 The Atmospheric System 189
    2.4.3 The Infrared Atmospheric System 197
2.5 557.7-nm Emission of Atomic Oxygen 202
2.6 The Atomic Oxygen 630-nm Emission 207
2.7 Helium Emission 210
2.8 Atomic Hydrogen Emissions 222
2.9 The Continuum 230
2.10 Nitric Oxide Emissions 237
2.11 Infrared Emissions from Atmospheric Atoms and Molecules 241
References 245

3 Techniques of Investigation of the Airglow 269
3.1 Photometry 270
3.2 Spectrophotometric Equipment 272
    3.2.1 Diffraction Spectrographs 273
    3.2.2 Diffraction Spectrophotometers 282

Contents xiii

	3.2.3	Photometers with Light Filters 283
	3.2.4	Photographic and Photoelectric Interferometers 292
3.3	Radiation Detectors 301	
	3.3.1	Photography 302
	3.3.2	Photomultipliers 307
	3.3.3	Image Converter Tubes 312
	3.3.4	Photoelectric Charge-Coupled Devices 317
3.4	Cooling of Radiation Detectors 327	
	3.4.1	Cooling with Solid Carbonic Acid 327
	3.4.2	Thermoelectric Cooling 328
	3.4.3	Household Refrigerators 329
3.5	Methods and Conditions of Measurements 329	
	3.5.1	Measurements at Given Directions 329
	3.5.2	Optical Recording of Wave Processes in the Upper Atmosphere 331
	3.5.3	Sky Scanning 336
	3.5.4	Scanning of a Spectrum 336
	3.5.5	Use of Fiber Glass Tubes 337
	3.5.6	Airborne Measurements 339
	3.5.7	Rocket Measurements 340
	3.5.8	Satellite Measurements 342
	3.5.9	Formation of Artificial Luminous Clouds 347
3.6	Measurement Data Processing 350	
	3.6.1	Spectrophotometry of Photographic Images of Spectra 350
	3.6.2	Electrophotometric Spectrometry 353
	3.6.3	Photographic Spectrophotometry of Interferograms 353
	3.6.4	Photoelectric Spectrophotometry of Interferograms 360
	3.6.5	Spectral Analysis of Time Series 365
	3.6.6	Spectral Characteristics of IGW Trains Detected in the Upper Atmosphere 373
3.7	Calibration of Measurements 382	
	3.7.1	Calibration of the Characteristics of Spectrophotometric Instruments 382
	3.7.2	Artificial Reference Light Sources 390
	3.7.3	Natural Reference Radiation Sources 391
3.8	Errors of Intensity and Temperature Measurements 392	
	3.8.1	Photographic Measurements 392
	3.8.2	Photoelectric Measurements 393
	3.8.3	Interferometric Measurements 394
	3.8.4	Spatial Orientation of the Instrumentation 398
	3.8.5	Statistical Processing Errors 398
References 401		

4 Regular Variations of the Airglow in the Mesopause and Thermosphere ... 413
- 4.1 Model of Hydroxyl Emission ... 415
  - 4.1.1 Factors Affecting the Variations of OH Emission Parameters ... 416
  - 4.1.2 Determination of the Characteristics of OH Emission ... 416
  - 4.1.3 Empirical Model of Variations of Hydroxyl Emission Characteristics ... 421
- 4.2 Model of the Sodium Emission ... 449
- 4.3 Model of the Molecular Oxygen Emission ... 465
  - 4.3.1 Model of the Atmospheric System of Molecular Oxygen ... 465
- 4.4 Model of the 557.7-nm Atomic Oxygen Emission ... 473
- 4.5 Model of the 630-nm Atomic Oxygen Emission ... 509
- References ... 530

5 Wave Processes in the Atmosphere ... 551
- 5.1 Internal Gravity Waves ... 551
  - 5.1.1 Detection of IGWs in the Atmosphere ... 552
  - 5.1.2 Modulation of Emission Characteristics by IGWs Propagating Through the Emission Layer ... 556
  - 5.1.3 Choice of Measurable Parameters ... 560
  - 5.1.4 Observation Conditions ... 561
  - 5.1.5 Determination of IGW Characteristics by the Variations of the OH Rotational Temperature ... 562
  - 5.1.6 Localization of IGW Sources in the Troposphere ... 567
  - 5.1.7 The Nature of IGW Sources ... 573
  - 5.1.8 Seasonal Variability of the Spectral Distribution of IGW Amplitudes ... 576
  - 5.1.9 Proportion Between the Principal Mechanisms of Hydroxyl Emission Excitation in the Mesopause ... 581
  - 5.1.10 IGW-Induced Variations of the Doppler Temperature and Intensity of the 557.7-nm Emission ... 585
  - 5.1.11 Variations of the Altitude of Hydroxyl Emission Layers with Various Vibrational Excitation ... 589
  - 5.1.12 Behavior of the Mesopause Temperature During the Propagation of IGWs in Summer and Winter ... 591
- 5.2 Orographic Disturbances ... 594
  - 5.2.1 Orographic Disturbances in the Upper Atmosphere ... 594
  - 5.2.2 Measurements ... 598
  - 5.2.3 Basic Theoretical Suppositions ... 607
  - 5.2.4 Analysis of the Input Parameters ... 610
  - 5.2.5 Calculation of the Energy Flux Spatial Distribution ... 613
  - 5.2.6 Numerical Estimates of the Energy Flux Spatial Distribution ... 617

Contents

	5.2.7	Effect of Orographic Disturbances on the Energetics and Structure of the Upper Atmosphere 619
5.3	Planetary Waves 625	
5.4	Noctilucent Clouds 629	
5.5	Spotty Structure of the Airglow Intensity 639	
	5.5.1	Recording of the Wave and Spotty Irregularities of the Airglow 640
5.6	Winds 643	
	5.6.1	Relation of the Variations of Hydroxyl Airglow Characteristics to the Tropospheric Wind 644
	5.6.2	Wind Parameters Derived from Optical Measurements 646
References 647		

6 Climatic Changes in the Upper Atmosphere 661

6.1 The Temperature Trend in the Middle Atmosphere 664
    6.1.1 Long-Term Yearly Average Temperature Trends 665
    6.1.2 Seasonal Behavior of Long-Term Temperature Trends for the Middle Atmosphere 677
    6.1.3 Latitudinal Variations of the Temperature Trend 687
6.2 Temperature Trend in the Thermosphere 695
6.3 Trends of the Atmospheric Density and Composition 696
    6.3.1 Long-Term Variations in Concentration of Neutral and Ionized Atmospheric Constituents................... 696
    6.3.2 Estimation of the Long-Term Variations of the Density of the Upper Atmosphere by the Evolution of the Parameters of Satellite Orbits 699
6.4 Long-Term Subsidence of the Middle and Upper Atmosphere 700
References 703

7 Models of Vertical Profiles of Some Characteristics of the Upper Atmosphere 711

7.1 Ozone in the Mesopause Region 711
7.2 Atomic Oxygen 718
    7.2.1 Input Experimental Data for the Model of Atomic Oxygen.. 720
    7.2.2 Photochemical Basis of the Model 720
    7.2.3 Empirical Regularities of 557.7-nm Emission Variations ... 721
    7.2.4 Model of Variations of the Atomic Oxygen Density in the Mesopause and Lower Thermosphere 722
7.3 Atomic Hydrogen 726
7.4 Temperature of the Middle Atmosphere 728
References 730

Index 735

Introduction

The Earth's atmosphere is a planetary gaseous medium studied by aeronomy. The term aeronomy comes from a Greek word meaning "air" and a Latin word meaning "law." This is a division of the astrophysics of planetary atmospheres dealing with the great diversity of physical–photochemical and hydrodynamic processes occurring in a rarefied gaseous medium exposed to UV radiation of the Sun (star) and corpuscular streams interacting with the magnetic field of the planet and the Sun.

Aeronomic investigations are aimed at the development of theories and practical methods of constructing models (algorithms) that could be used to calculate the temperatures and densities of neutral and ionized constituents and various chemically active atoms and molecules of the planetary atmosphere. Their composition determines the energy conversion by dissociation and photoionization of molecules at altitudes between 20 (km) and several thousand kilometers due to absorbed UV radiation of the Sun and high-energy electrons and protons precipitated from the magnetosphere. Galactic and solar cosmic rays (GCR and SCR) are important participants in the photochemical processes that influence the atmospheric state. Thus, aeronomy studies all the processes occurring in the planetary magnetic field that are accompanied by energy removal from the atmosphere in the form of atomic and molecular radiations under quiet and geomagnetically perturbed conditions (auroras).

To solve these problems, a wide variety of methods of mathematics, physical chemistry of elementary processes of radiation absorption, dissociation, photoionization, recombination, excitation, deactivation of atoms and molecules, radiation transfer in optically nontransparent gaseous media, condensation, hydrodynamics of rarefied gases, magnetic hydro- and electrodynamics, diffusion, and thermodynamics are used (Nicolet 1962; Banks and Kockarts 1973a,b; Whitten and Poppoff 1971; McEwan and Phillips 1975; Massey and Bates 1982; Brasseur and Solomon 2005; Krasnopolsky 1987; Rees 1989).

By analogy with subjects of astrophysical research, information on various properties of the upper atmosphere can be obtained by optical methods of measurements. Fast changes of the characteristics of photochemical and dynamic processes observed in the atmosphere with increasing altitude above the underlying surface are related to the variability of the temperature regime and composition of the atmosphere. These parameters, in turn, are determined by ultraviolet solar radiation. On average, the stationary temperature regime and composition of the atmosphere result from a stable solar energy influx whose constancy is provided by a virtually circular orbit of the Earth, small tilt angle of its rotation axis with the orbit plane, and moderate rotation velocity. These conditions provide the existence of life on our planet.

Nevertheless, variations of the atmospheric parameters about their steady average values caused by various heliogeophysical disturbances proceed permanently on various time scales. The amplitudes of these disturbances and their influence on the Earth's atmosphere at various altitudes have not gone unnoticed for the environment and people. The atmospheric response is manifested by variations of the atmospheric temperature and dynamic regime at different altitudes. In most cases, the atmospheric phenomena accompanying these disturbances occur above the troposphere and remain invisible for people from the Earth surface, except auroras repeatedly observed at high latitudes and in some rare cases at middle and low latitudes.

However, this does not mean that heliogeophysical disturbances have no impact on the organisms living on the Earth surface. The state of the Sun always affects life on the Earth, including economic human activity. Many centuries ago, optical phenomena in the atmosphere were considered as miracles of mystical origin. They were repeatedly mentioned in historical chronicles. Modern research revealed data on long-term significant changes of the solar activity for the period of several thousand years (Nagovitsyn 1997; PMAO 2004). Reliable data are available on numerous long-term periods of increased and decreased solar activity for the last 1000 years (Nagovitsyn 1997). The long Maunder period of low solar activity (1645–1715) was followed by the intensive aurora observed on March 17, 1716, that drew the attention of the researchers to the processes observed in the Earth's atmosphere once again (Schröder 2001). M. V. Lomonosov (1950) was among the first scientists who realized in 1753 that this phenomenon was connected with the upper atmospheric layers located at great altitudes. More recently, when the spectroscopic methods had already been developed in astronomy, a green line was revealed in the auroras observed at high latitudes in studies of the variability of solar activity. Detection of this emission gave impetus to investigations of its origin, since its wavelength was not present in the spectra known by that time. It should be added that even the spectral structure of the anomalous low-latitude red lights observed on September 1, 1859 (Tsurutani et al. 2003), and October 24, 1870 (Eather 1980), was not determined at that time, though the spectral equipment and photographic technique for radiation recording had already been developed.

However, the possibility itself of airglow emissions to arise at middle and low latitudes under quiet geomagnetic conditions was not recognized until the researchers tried to estimate the Earth surface illuminance at night. The first studies demonstrated that the emission of stars in the visible range makes only a part of the total night-sky emission intensity (Newcomb 1901). Some results of investigations of the star contribution to the total intensity of the night airglow were presented by Khvostikov (1937, 1948). More recent data (Sharov and Lipaeva 1973; Roach and

Gordon 1973) demonstrate that the intensity of the total star emission in the visible range is ∼10 (kilorayleigh) and that of the airglow of the upper atmosphere is ∼7 (kilorayleigh) under quiet geomagnetic conditions.

By the late 1920s, it became obvious that the Earth's atmosphere does exist at high altitudes, and moreover, processes of critical importance occur in it whose manifestation is the 557.7-nm emission of the night midlatitude airglow under quiet geomagnetic conditions. At the same time, it was quite natural that the problem of investigating the properties of this phenomenon and its relationship to the atmospheric characteristics came to the fore.

In 1933, Academician S. I. Vavilov suggested that an All Soviet Union Conference on Stratospheric Research be convened. The meeting was held in 1934. In the same year, S. I. Vavilov chaired the Stratospheric Research Commission at the USSR Academy of Sciences. Simultaneously, investigations of the Earth's airglow were started in the Soviet Union. In 1933, a spectrophotometer for investigation of the night-sky glow and Zodiacal Light was developed and subsequently used to record the first nightglow spectra (Eropkin and Kozirev 1935). In October–November 1934, V. G. Fesenkov (1937) carried out photographic measurements of the night-sky brightness with a tubular photometer in Kitabe, Tashkent, and near Moscow. Methods of analysis of the atmospheric properties based on twilight measurements were also developed (Staude 1936). Early in 1935, G. G. Sljusarev (Sljusarev and Chernjajev 1937) from the State Optical Institute (SOI) developed and manufactured a high-aperture (F: 0.58) prism spectrograph to photograph the nightglow spectrum (Khvostikov 1948). In 1936, I. A. Khvostikov took charge of the first observations of the nightglow near Elbrus, Caucasus (43.3°N, 42.5°E). At the same time, intensity variations of the atomic oxygen green emission were measured (Dobrotin et al. 1935; Lebedev and Khvostikov 1935; Vasmut et al. 1938; Ershova et al. 1939) and its optical properties were then determined (Panshin and Khvostikov 1936; Panshin et al. 1936). In 1937, V. I. Chernjajev and I. F. Vuks (1937) detected the twilight fluorescent sodium emission in the wavelength range 589.0–589.6 (nm) during twilight observations (at a sunset angle of ∼ 6°) (Khvostikov 1939). The first results were reviewed by I. A. Khvostikov (1937). In 1940, G. A. Shain and P. F. Shain (1942) measured the 557.7-nm emission absolute intensity at Crimean Observatory in Simeiz (44.4°N, 34.0°E). Electrophotometric observations of the nightglow were also carried out at Crimean Astrophysical Observatory (Chuvaev 1953a,b). N. N. Feofilov (1942) measured the night-sky energy spectrum near Elbrus, Caucasus. The first photoelectric measurements of night-sky emissions, such as the green atomic oxygen and sodium emissions and infrared emission, were carried out at this station (Osherovich et al. 1949; Rodionov and Pavlova 1949a,b; Rodionov et al. 1949). Preliminary measurements of auroral infrared radiation were also performed (Rodionov and Fishkova 1950). In 1942, regular observations of the twilight airglow were started at Abastumani Astrophysical Observatory of the Georgian Academy of Sciences (41.8°N, 42.8°E). T. G. Megrelishvili (1981) first measured the scattered radiation intensity, and then spectral measurements of emissions were started. However, at the early stage of measurements, the instrumentation was restricted to prism spectral devices having low dispersion in the red and near-infrared ranges. With the low sensitivity of the radiation detectors employed, the weak night-sky glow could be recorded only with many-hour exposure times. This limited considerably the opportunity of measuring more detailed airglow spectra. Results of Russian and foreign investigations in that period were summarized by I. A. Khvostikov in his monograph entitled "Luminescence of the night sky" (its second edition was published in 1948 (Khvostikov 1948)). By the early 1960s, this monograph had been the only manual for the researchers of the Earth's airglow in our country.

However, the actual founder of the research field dealing with night airglow and auroras in the Soviet Union and its leader for 40 years was Valerian Ivanovich Krassovsky (June 14, 1907–December 4, 1993). In the late 1940s, he first used electron-optical image converters to investigate infrared radiation of the night sky, revealed previously by the photometric method (Khvostikov 1948), to take photos of the Galaxy center which revealed a region not observed earlier, and to record with low dispersion first spectrograms in the wavelength range 600–1200 (nm) (Krassovsky 1949, 1950a,b) in March 1948. These studies demonstrated that the sky emission spectrum has the structure of molecular bands extending over the entire near-infrared range. The spectrum was identified by G. Herzberg based on spectrographic data (Meinel 1950a,b). This gave impetus to investigations of the Earth's airglow all over the world. Later on, Bates and Nicolet (1950) elaborated the basis for the photochemical theory of the hydroxyl emission arising due to the reaction between ozone and hydrogen. In 1950, V. I. Krassovsky suggested a mechanism of hydroxyl emission initiation as a result of the interaction of vibrationally excited molecular oxygen with atomic hydrogen (Krassovsky 1951a, 1963) as well as a mechanism of occurrence of the airglow continuum through the reaction between nitric oxide and atomic oxygen (Krassovsky 1951b).

The experience gained demonstrated that spectrographic instrumentation with high dispersion and aperture ratio is required to record reliably the spectra of the weak upper atmospheric emissions. It became clear that the prism spectrographs used previously would be unsuitable for these purposes. In the late 1940s, unique SP-47, SP-48, SP-49, and SP-50 diffraction spectrographs with high-aperture ratio were developed specially for recording weak light emissions in the range from UV to near-IR (Gerasimova and Yakovleva 1956; Galperin et al., 1957; Kaporsky and Nikolaeva and 1969). These instruments provided not only the opportunity of airglow investigations in the Soviet Union but also the accumulation of high-quality spectrographic data on hydroxyl, 557.7- and 630-nm atomic oxygen, 589.0–589.6-nm sodium, 656.3-nm hydrogen, and 1083.0-nm helium emissions at the national geophysical stations and astronomical observatories. These data remain uniquely significant even now.

By the International Geophysical Year (IGY) (1957–1959), the spectrographs were installed at the following observation stations: Loparskaya (68.3°N, 33.1°E), Roshchino (60.2°N, 29.6°E), Zvenigorod (55.7°N, 36.8°E) of the Institute of Atmospheric Physics of the USSR Academy of Sciences; Yakutsk (62.1°N, 129.7°E) of the Institute of Cosmophysical Research and Aeronomy of the Yakutsk Center of the USSR Academy of Sciences; Kamenskoe Plato (43.2°N, 76.9°E) of the

Astrophysical Institute of the Kazakh Academy of Sciences, Alma-Ata; Abastumani Astrophysical Observatory (41.8°N, 42.8°E) of the Georgian Academy of Sciences, and Vannovsky (38.0°N, 58.4°E) of the Physicotechnical Institute of the Turkmen Academy of Sciences, Ashkhabad. Regular photometric and spectral observations of nightglow and auroral emissions have been performed since the IGY. The organizer and supervisor of the research program, as mentioned, was V. I. Krassovsky. Radiation detected by the spectrographs was then photographed on a film. Data on the characteristics of photographic materials used by many researchers in the 20th century were reviewed by O. D. Dokuchaeva (1994). They are very valuable in analyzing photographic records taken for several decades. The electron-optical image converters (Volkov et al. 1959; Shcheglov 1963, 1980) were used in the visible and infrared ranges. This determined for many decades the regular character of observations at the majority of stations of the Soviet Union. As a result, unknown emissions were detected, such as the 656.3-nm H_α hydrogen, 1083.0-nm helium emissions and continuum in the visible and near-IR ranges, and the effect of incoherent scattering in the Fraunhofer lines of the spectrum of sunlight scattered in the Earth's atmosphere was discovered. The spectral characteristics of A- and B-type auroras and the H_α hydrogen emission with Doppler broadening were intensely investigated.

As early as 1962, "Atlas of the airglow spectrum $\lambda\lambda$ 3000–12400 Å" was published (Krassovsky et al. 1962). Later on, a set of photoelectric records of the spectrum in the visible and near-IR ranges were published (Broadfoot and Kendall 1968). In the next few years, spectrograms of individual spectral fragments were repeatedly published (Vlasyuk and Spiridonova 1993). In the early 2000s, a new edition of Atlas with emission intensities expressed in absolute units on a linear scale was compiled based on the emission spectra recorded with matrix radiation detectors (Semenov et al. 2002).

Since the beginning of the IGY, many lines of investigation of the atmospheric airglow and optical phenomena have been intensified. The main feature of the investigations performed was an attempt of regular measurements to obtain systematic data. These investigations constantly received the state support and were performed simultaneously at many stations in different spectral ranges. They solved various geophysical problems. After detection of new emissions, the key point of research was the use of the emission characteristics in atmospheric studies, since exactly these emissions allowed the researchers to establish correlations between the geophysical processes in the lower and upper atmospheric layers.

Over the past decades, the volume of research has extremely been extended. Many scientific results have radically changed our previous notion of the properties of the Earth's atmosphere and surrounding space.

The long-term (30–50 years) data on various atmospheric characteristics have been accumulated by the Russian scientific institutions as a result of regular observation at a number of stations over the last decades. Their unique significance became obvious by the end of the 20th century, when the tendency toward climatic changes in the surface layers of the Earth's atmosphere became evident. These data allowed the researchers to get an idea of the global long-term climatic changes and to describe the phenomena observed in the middle atmosphere. Thus, investigations of the Earth's airglow, including the nightglow at low and middle latitudes and auroras at high latitudes, carried out in the Soviet Union during the 20th century, are an important integral part of atmospheric astrophysics.

The present brief historical review describes the main results of investigations of the upper atmospheric emissions and atmospheric processes connected with them without full details. Therefore, references are also incomplete because their complete list would include several thousand publications. Undoubtedly, the phenomena of scientific interest to the authors are described more or less comprehensively.

This monograph presents the main results of investigations of the Earth's airglow on different, most important, lines. They involve the properties of individual emissions and their use in determining the temperature, the total atmospheric density, and the concentrations of some minor but chemically active components affecting significantly the temperature regime and dynamic processes at different altitudes. Remote spectrophotometric measurements, a conventional investigation technique for gaseous media (Earth's atmosphere in the case at hand), turned out to be an efficient tool for detecting unknown emissions and processes occurring throughout the atmospheric depth that constitute the subject of aeronomy, the field of research appeared half a century ago.

References

Banks PM, Kockarts G (1973a) Aeronomy. Pt A. Academic Press, New York
Banks PM, Kockarts G (1973b) Aeronomy. Pt B. Academic Press, New York
Bates DR, Nicolet M (1950) The photochemistry of atmospheric water vapour. J Geophys Res 55:301–327
Brasseur G, Solomon S (2005) Aeronomy of the middle atmosphere. 3rd ed. Springer-Verlag, Dordrecht, Holland
Broadfoot AL, Kendall KR (1968) The airglow spectrum 3100–10000 A. J Geophys Res 73:426–428
Chernjajev VI, Vuks MF (1937) The spectrum of the sky in the twilight. Dokl USSR Acad Sci 14:77–79
Chuvaev KK (1953a) Electrophotometric investigations of the night sky luminescence in different spectral regions. Izv Crimea astrophys observ 10:54–73
Chuvaev KK (1953b) Investigations of the night sky luminescence in different spectral regions. Astron Rep 30:472–473
Dobrotin NB, Frank I, Čerenkov P (1935) Observations of night sky luminescence by the extinction method. Dokl USSR Acad Sci 1:114–117
Dokuchaeva OD (1994) Astronomical photography. Materials and methods. Fizmatlit, Moscow
Eather RH (1980) Majestic lights. The aurora in science, history and the arts. Amer Geophys Un, Washington DC
Eropkin DI, Kozirev NA (1935) Spectrophotometry of the night sky and Zodical light. Poulkovo Observatory Circular N 13:21–25
Ershova ND, Mikhailin IM, Khvostikov IA (1939) Brightness measurement of the green line of the night sky. Izv USSR Acad Sci Ser Geograph Geophys N 2:217–221
Feofilov NN (1942) Nocturnal luminosity and energy distribution in the night sky spectrum. Dokl USSR Acad Sci 34:252–256

Fesenkov VG (1937) Night sky luminescence and the infinite of Universe. World knowledge 27:128–1B33
Galperin GI, Mironov AV, Shefov NN (1957) Spectrographs to be used for investigations of atmospheric emissions during the IGY of 1957–1958. Mém Soc Roy Sci Liège 18:68–69
Gerasimova NG, Yakovleva AV (1956) Complect of the fast spectrographs with the diffractive gratings. Instr Exper Technol N 1:83–86
Kaporsky LN, Nikolaeva II (1969) Optical instruments. Catalogue, vol 4. Nikitin VA (ed). Mashinostroenie, Moscow
Khvostikov IA (1937) Luminescence of the night sky. Vavilov SI (ed). USSR Acad Sci Publ House, Moscow-Leningrad
Khvostikov IA (1939) Twilight photoluminescence of the terrestrial atmosphere. Izv USSR Acad Sci Ser Geograph Geophys N 2:175–182
Khvostikov IA (1948) Luminescence of the night sky 2nd ed. Vavilov SI (ed). USSR Acad Sci Publ House, Moscow-Leningrad
Krasnopolsky VA (1987) Airglow physics of the planetary and comet atmospheres. Nauka, Moscow
Krassovsky VI (1949) On the night sky radiation in the infrared spectral region. Dokl USSR Acad Sci 66:53–54
Krassovsky VI (1950a) New data on the night sky radiation in the 8800–11000 A region. Dokl USSR Acad Sci 70:999–1000
Krassovsky VI (1950b) New emissions of the night sky in the 8800–11000 A region. Izv Crimea astrophys observ USSR Acad Sci 5:100–104
Krassovsky VI (1951a) On the mechanism of the night sky luminescence. Dokl USSR Acad Sci 77:395–398
Krassovsky VI (1951b) Influence of the water vapor and carbon oxides on the night sky luminescence. Dokl USSR Acad Sci 78:669–672
Krassovsky VI (1963) Chemistry of the upper atmosphere. In: Priester W (ed) Space Res. vol 3. North-Holland Publ Co, Amsterdam, pp 96–116
Krassovsky VI, Shefov NN, Yarin VI (1962) Atlas of the airglow spectrum $\lambda\lambda$ 3000–12400 Å. Planet Space Sci 9:883–915
Lebedev AA, Khvostikov IA (1935) Intensity variation of the auroral green line of the night sky. Dokl USSR Acad Sci 1:118–127
Lomonosov MV (1950) A word on the air-like phenomena caused by electrical strength. USSR Acad Sci Publ House, Moscow. 3:101–134
Massey HSW, Bates DR (1982) Atmospheric physics and chemistry. Applied atomic collision physics, vol 1. Academic Press, New York
McEwan MJ, Phillips LF (1975) Chemistry of the atmosphere. Edward Arnold, London
Megrelishvili TG (1981) Regularities of the variations of the scattered light and emission of the Earth twilight atmosphere. Khrgian AKh (ed). Metsniereba, Tbilisi
Meinel AB (1950a) Hydride emission bands in the spectrum of the night sky. Astrophys J 111:207
Meinel AB (1950b) OH emission bands in the spectrum of the night sky. I. Astrophys J 111:555–564
Nagovitsyn YuA (1997) A non-linear mathematical model of the solar cyclicity process and the possibility for the activity reconstruction in the past. Astron Lett 23:851–858
Newcomb S (1901) A rude attempt to determine the total light of all the stars. Astrophys J 14:297–312
Nicolet M (1962) Aeronomy. Preprint. Institut d'Astrophysique, Liège
Osherovich AL, Pavlova EN, Rodionov SF, Fishkova LM (1949) On the photoelectric photometry of the faint light fluxes. Exper Techn Phys 19:184–204
Panshin KB, Khvostikov IA (1936) Polarization of the luminosity of the night sky. In: Trans Elbrus expedition, USSR Acad Sci, 1934–1935. USSR Acad Sci Publ House, Moscow
Panshin KB, Khvostikov IA, Chernjajev VI (1936) Energy distribution in the night sky spectrum in different night times. In: Trans Elbrus expedition, USSR Acad Sci, 1934–1935. USSR Acad Sci Publ House, Moscow

PMAO (2004), Pulkovo Main Astronomical Observatory RAN. Data base of the mean annual Wolf numbers from 1090. http://www.gao.spb.ru/database/esai/

Rees MH (1989) Physics and chemistry of the upper atmosphere. Houghton JT, Rycroft MJ, Dessler AJ (eds). Cambridge University Press, Cambridge

Roach FE, Gordon JL (1973) The light of the night sky. D Reidel Publishing Company, Dordrecht, Holland

Rodionov SF, Fishkova LM (1950) On the infrared radiation of the polar aurorae. Dokl USSR Acad Sci 70:1001–1003

Rodionov SF, Pavlova EN (1949a) On the infrared radiation of the night sky. Dokl USSR Acad Sci 65:831–834

Rodionov SF, Pavlova EN (1949b) On the atmospheric sodium radiation. Dokl USSR Acad Sci 67:251–254

Rodionov SF, Pavlova EN, Rdultovskaya EB (1949) Measurement of the green line luminescence of the night sky with aid of the photometer with the electronic multiplier. Dokl USSR Acad Sci 66:55–57

Schröder W (2001) Vom Wunderzeichen zum Naturobjekt (Fallstudie zum Polarlicht vom 17. März 1716). Berichte Geschichte Geophys Kosm Phys 2:5–95

Semenov AI, Bakanas VV, Perminov VI, Zheleznov YuA, Khomich. VYu (2002) The near infrared spectrum of the emission of the nighttime upper atmosphere of the Earth. Geomagn Aeronomy 42:390–397

Shain GA, Shain PF (1942) Methods of the variation investigations of the emission lines in the night sky spectrum. Dokl USSR Acad Sci 35:152–156

Sharov AS, Lipaeva NA (1973) Stellar component of the night airglow. Astron Rep 50:107–114

Shcheglov PV (1963) Electronic telescopy. Fizmatgiz, Moscow

Shcheglov PV (1980) Problems of the optical astronomy. Nauka, Moscow

Sljusarev GG, Chernjajev VI (1937) Spectrographs of the large luminosity constructed in The State Optical Institute. Opt-Mechan Industry N 8:7–12

Staude NM (1936) Photometric twilight observations as a method of the upper stratospheric investigations. Vavilov SI (ed). USSR Acad Sci Publ House, Moscow-Leningrad 1:1–162

Tsurutani BT, Gonzalez WD, Lakhina GS, Alex S (2003) The extreme magnetic storm of 1–2 September 1859. J Geophys Res 108 A:1268. doi:10.1029/2002JA009504

Vasmut NA, Vertser VN, Tibilov SU, Freivert SI (1938) Observations of the green line intensity variations of the night sky. Dokl USSR Acad Sci 19:405–407

Vlasyuk VB, Spiridonova OI (1993) Night airglow spectrum in the 3100–7700 A range obtained with medium-resolution echelle-spectrometer of the 6-meter telescope. Astron Rep 70:773–791

Volkov IV, Esipov VF, Shcheglov PV (1959) The use of contact photography principle in studying of weak light fluxes. Dokl USSR Acad Sci 129:288–289

Whitten RC, Poppoff IG (1971) Fundamentals of aeronomy. John Wiley and Sons, New York

Chapter 1
The Radiating Atmosphere and Space

1.1 Spectral Structure of the Airglow and Its Role in Atmospheric Research

The upper atmosphere of the Earth (above 80 (km)) is a rather rarefied gas medium whose basic components are atomic and molecular nitrogen and oxygen, along with hydrogen and helium. The so-called small components, such as nitric oxide NO, carbon oxide CO, carbon dioxide CO_2, nitrous oxide N_2O, vapors of water H_2O, ozone O_3, nitrogen dioxide NO_2, and metastable atoms and molecules, are important for the photochemistry, energetics, and emissions of the upper atmosphere. The ionizing solar ultraviolet radiation gives rise to numerous photochemical processes in the atmosphere which induce the airglow. This glow persists both in the daytime and at night, its intensity varying within considerable limits. For more than half a century it has been the object of research for obtaining data on the state of the upper atmosphere and its structure. This is due to the fact that practically any type of emission is an indicator of the processes occurring both inside the atmosphere and in the near space, being a peculiar kind of valve for energy removal from those atmospheric regions where the atmospheric components formed as a result of absorption of solar energy are actively transformed. Therefore, various emissions, depending on their nature, can serve as energy balance controls, thermometers at various altitudes, anemometers in examining atmosphere circulation, and indicators of the intensity of ultraviolet and corpuscular radiations.

During the initial periods of research, most attention was given to the search for new emissions and to the elucidation of their nature. This lent impetus to laboratory investigations of the rates of numerous chemical reactions and of the emissive properties of various atoms and molecules. The needs of aeronomy stimulated detailed experimental investigations of the solar ultraviolet spectrum. Based on the experience gained, the airglow has been actively used as an indispensable object of remote sensing not only in ground-based, but also in satellite studies of the upper atmospheres of the Earth and other planets.

The main advantage of optical methods is that in most cases (except for artificial luminous clouds) they provide a variety of data on the medium under investigation without rendering an extraneous effect on the latter and allow one to examine the medium changes on various time scales.

The now available data give some insight into the most probable excitation processes for many atmospheric emissions which occur at night, in the twilight, and in the daytime at altitudes of 80–1000 (km). At altitudes above 80 (km), the atmosphere has low density, and the medium is optically thin for the radiations with wavelengths above 0.35 (μm). The only exception are the infrared bands of molecules in the spectral region over 2 (μm) for which radiative transfer processes are significant.

The airglow of the upper atmosphere falls almost completely within the spectral region from 0.02 to 63 (μm), more than 90(%) falling in the 0.1–20 (μm) region. It is produced in the main by 25 atmospheric components, including ionized species, 11 of which are atoms and 14 are molecules, 8 of which being diatomic.

The excitation processes, which can occur under quiet geomagnetic conditions in the nocturnal and diurnal atmosphere, give rise to 25 significant systems of bands and 130 atomic emissions. A great number of atomic emissions and, presumably, some systems of molecular bands with high excitation levels appear during auroras, which arise mainly at high latitudes due to the escape of energetic electrons and protons from the magnetosphere during geomagnetic disturbances. The airglow spectrum contains on the whole about 1000 molecular bands, each consisting of a large number (up to several hundreds or thousands) of individual spectral lines. The 0.02- to 0.1-μm ultraviolet region refers to the emissions of ionized atoms of helium, nitrogen, and oxygen. These species are present in the atmosphere mainly in the daytime due to the photoionization of atmospheric components by solar radiation and also as a result of resonance scattering. The 0.1- to 0.35-μm ultraviolet region includes 570 bands and 35 atomic lines and the 0.35- to 20-μm ultraviolet region contains 360 bands and 52 lines. However, the energetics of the long-wave spectral region is prevailing.

Even with the now available data on the airglow structure, the geophysical properties of the spectrum are well studied only for the visible and near-infrared regions from 0.35 to 2 (μm) for which long-term measurements have been conducted, mainly by ground-based means. The relevant spectrophotometric data (Krassovsky et al. 1962; Lowe 1969; Semenov et al. 2002b; Krasnopolsky 1987) are presented in Figs. 1.1–1.11. The shorter and longer wavelength regions are accessible to investigation only by rocket and satellite methods. Therefore, their use for systematic measurements involves considerable difficulties.

The altitude distribution of the airglow intensity has some features. The airglow arises mainly in the altitude range from 80 to 270 (km) since the dissociation of the molecular atmospheric components rapidly increases at altitudes above 100 (km), while their concentration decreases. Only the emissions of atomic oxygen, helium, and hydrogen cover the altitude range up to 1000 (km). Table 1.1 summarizes the characteristics of the principal emissions. The estimated emission intensities for the molecular components in the far-infrared spectral region are given for altitudes above 70 (km).

Spectroscopic data about the atoms and molecules, rates of photochemical processes, and parameters of the atmosphere at various altitudes are collected in a number of monographs and reviews. Some of them are listed in References.

1.1 Spectral Structure of the Airglow and Its Role in Atmospheric Research

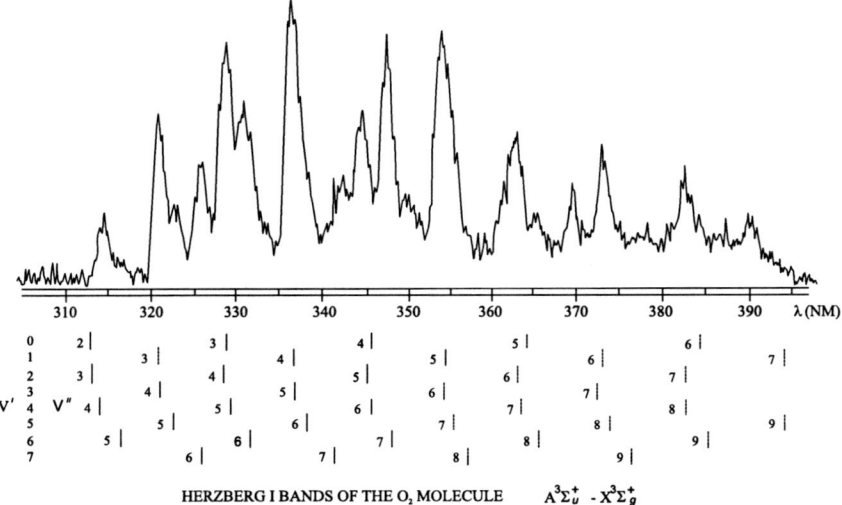

Fig. 1.1 Spectrum of the upper atmosphere nightglow, $\lambda = 305\text{--}395$ (nm) (Krassovsky et al. 1962)

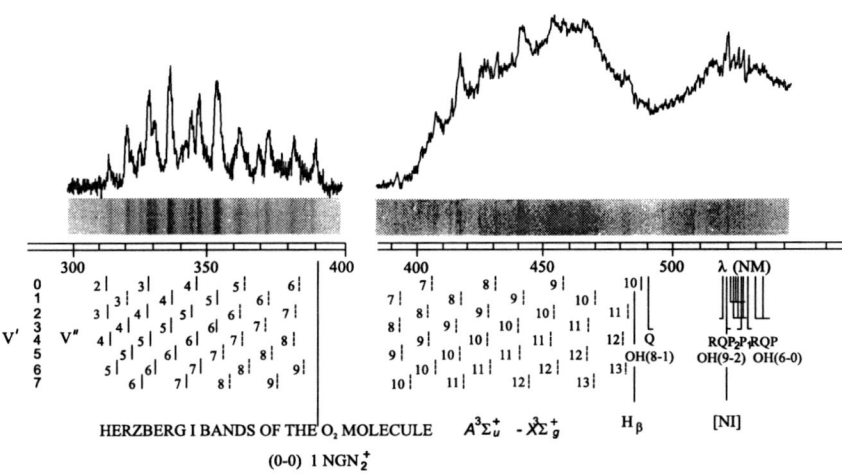

Fig. 1.2 Spectrum of the upper atmosphere nightglow, $\lambda = 300\text{--}540$ (nm) (Krassovsky et al. 1962)

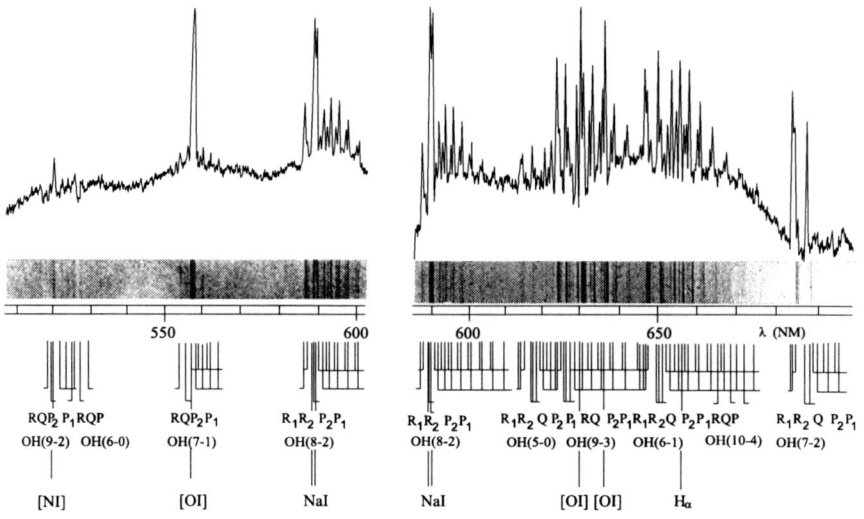

Fig. 1.3 Spectrum of the upper atmosphere nightglow, $\lambda = 510\text{–}700$ (nm) (Krassovsky et al. 1962)

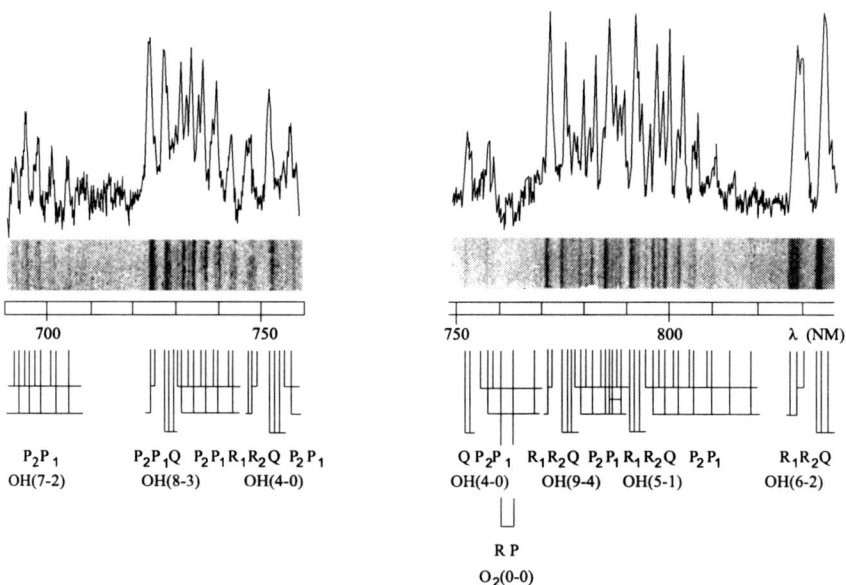

Fig. 1.4 Spectrum of the upper atmosphere nightglow, $\lambda = 690\text{–}840$ (nm) (Krassovsky et al. 1962)

1.1 Spectral Structure of the Airglow and Its Role in Atmospheric Research

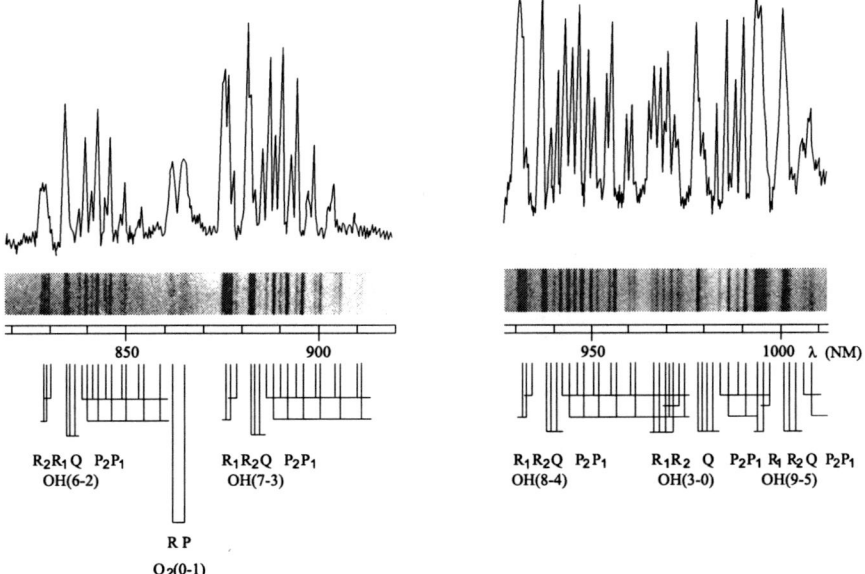

Fig. 1.5 Spectrum of the upper atmosphere nightglow, $\lambda = 820\text{--}1010\,(\text{nm})$ (Krassovsky et al. 1962)

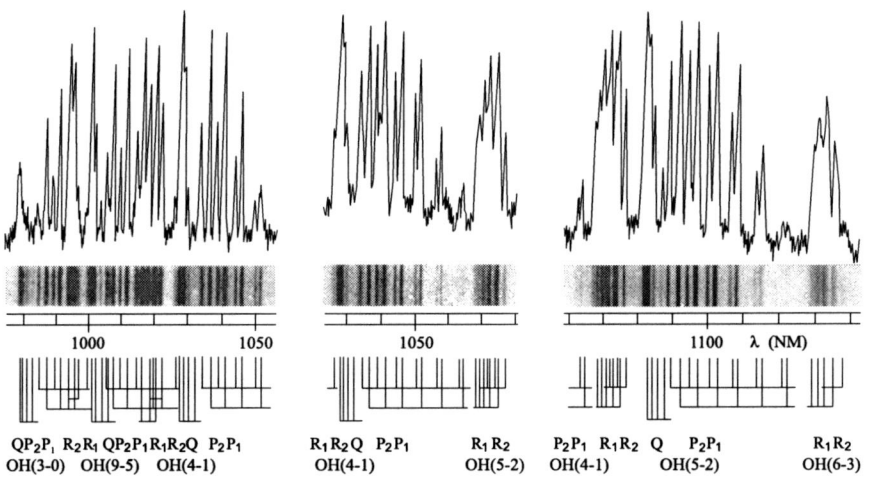

Fig. 1.6 Spectrum of the upper atmosphere nightglow, $\lambda = 980\text{--}1140\,(\text{nm})$ (Krassovsky et al. 1962)

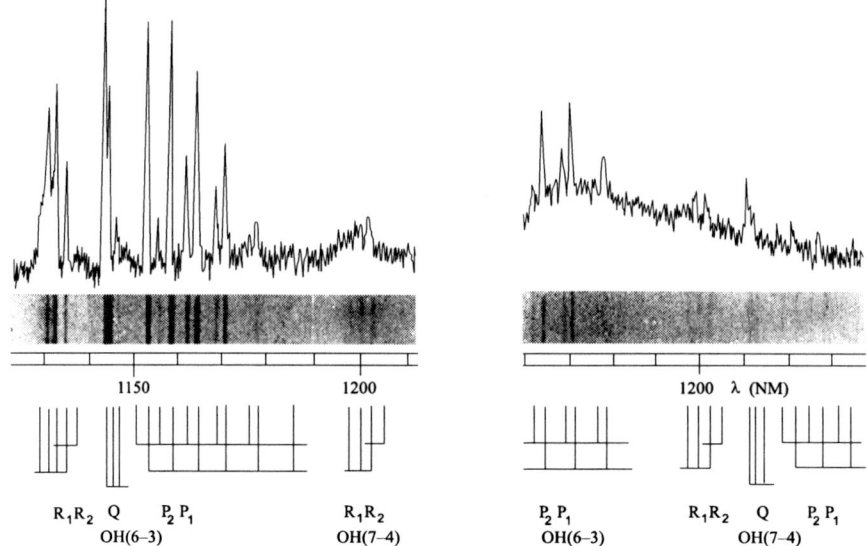

Fig. 1.7 Spectrum of the upper atmosphere nightglow, $\lambda = 1130\text{--}1240\,(\text{nm})$ (Krassovsky et al. 1962)

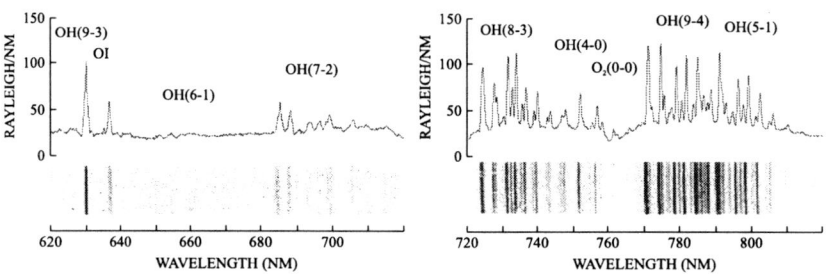

Fig. 1.8 Spectrum of the upper atmosphere nightglow, $\lambda = 620\text{--}820\,(\text{nm})$ (Semenov et al. 2002b)

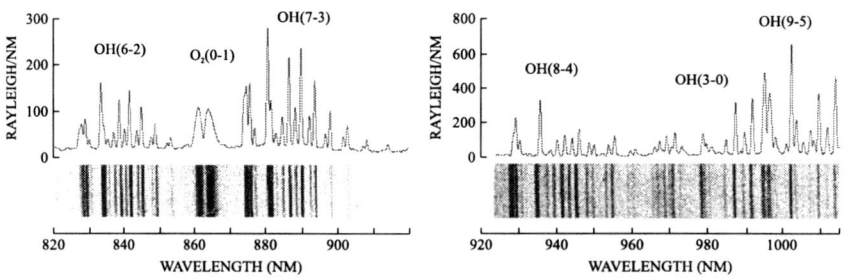

Fig. 1.9 Spectrum of the upper atmosphere nightglow, $\lambda = 820\text{--}1020\,(\text{nm})$ (Semenov et al. 2002b)

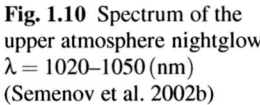

Fig. 1.10 Spectrum of the upper atmosphere nightglow, $\lambda = 1020–1050$ (nm) (Semenov et al. 2002b)

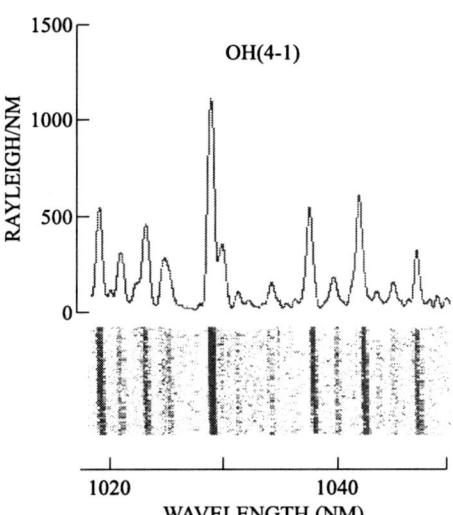

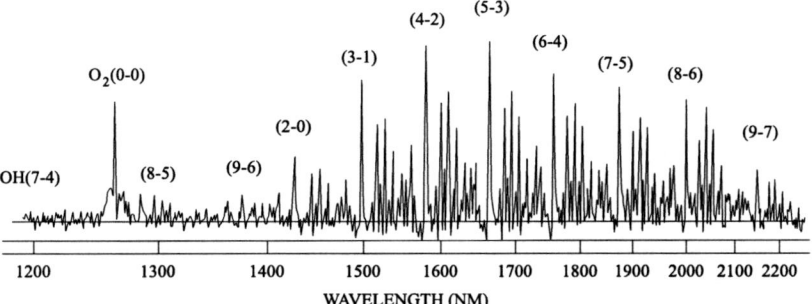

Fig. 1.11 Spectrum of the upper atmosphere nightglow, $\lambda = 1200–2200$ (nm) (Krasnopolsky 1987)

1.2 Extra-Atmospheric Radiation

The radiation of the space surrounding the Earth is a component of the night-sky radiation. It contains the light emitted by stars, nebulas, galaxies, and zodiacal constellations. They considerably vary in brightness and spectral structure and differ in these parameters from the airglow, the same as the measurement units used in astronomy.

Table 1.1 Atoms and molecules of the upper atmosphere accessible to investigation from the Earth surface

Radiating species	M	Transition	Wavelength range (μm)	Altitudes (km)	Intensity (kilorayleigh) Quiet conditions	Intensity (kilorayleigh) Auroras
1	2	3	4	5	6	7
H	1	Electronic	0.6563	100–2000	0.015	1
He	4	Electronic	1.0830	250–1500	1	10
N	14	Electronic	0.3466, 0.5200, 1.044	130–270		0.7, 0.2, 10
			0.82–0.86	130–300		3
N^+	14	Electronic	0.5755, 0.6584	130–400		0.5, 0.8
O	16	Electronic	0.5577,	90–240	0.270	100
			0.6300,	160–320	0.100	1000
			0.8446	130–280	0.013	5
O^+	16	Electronic	0.7320–0.7330	130–400	0.01	5
CH_4	16	Vibrational–rotational	7.66	(70)–140	10	
OH	17	Vibrational–rotational	0.6–4.5	70–110	1000	1000
H_2O	18	Vibrational–rotational	1.38, 1.88, 6.27	(70)–140 (70)–140	10, 10 70	
Na	23	Electronic	0.5889–0.5896	85–95	0.065	0.065
N_2	28	Electronic	VK 0.2–0.6	130–230		150
			2PG 0.25–0.5	125–230		100
			IPG 0.6–2	130–230		2000
			BW, WB 1–17	130–230		2
N_2^+	28	Electronic	1NG 0.35–0.5	115–250	0.001	100
			M 0.65–2.5	130–300		2000
CO	28	Vibrational–rotational	2.35, 4.67	(70)–200	10 1000	
NO	30	Electronic	1.22	(70)–250	0.4	
		Vibrational–rotational	2.7, 5.3	(70)–250	0.01 27 000	
NO^+	30	Vibrational–rotational	2.15, 4.27	(70)–140	0.01 2	
O_2	32	Electronic	HI 0.2–0.45	80–110	1	
			At 0.8645	60–230	0.3	5
			IR At 1.27, 1.58	50–110	100	1000
O_2^+	32	Electronic	1NG 0.5–0.73	130–300		5
CO_2	44	Vibrational–rotational	4.30, 15	(70)–200	6000 1 000 000	40 000

1.2.1 Stellar Photometric Units

In measuring the brightness of stars, the characteristics of brightness called stellar magnitudes are generally used (Albitsky et al. 1951; Albitsky and Melnikov 1973; Allen 1973; Roach and Gordon 1973; Bronshten et al. 1981; Kononovich and Moroz 2001; Kulikovsky 2002). The intensity ratio for two stars with magnitudes m_1 and m_2 has the form

$$\frac{I_2}{I_1} = 2.512^{m_1 - m_2} = 10^{0.4 \cdot (m_1 - m_2)}.$$

To denote the brightness of a star, a special notation is commonly used. Thus, for instance, the brightness of stars of magnitudes 4.3 and 0 are denoted 4.3^m and 0^m, respectively. For the Sun we have $m = -26.78$, and, hence, its brightness is -26.78^m.

Therefore, the number of stars of the 10th magnitude 10^m (S_{10}) equivalent in brightness to one star of magnitude m is

$$S_{10} = 10^{0.4(10-m)}.$$

However, they are compared in brightness based on the illuminance of the Earth surface produced by the celestial hemisphere as a whole, having an area of 20 626.5 (sq.deg). Thus, the number of 10^m stars per square degree, $S_{10}/(\text{sq.deg.})$ (m/□°) equivalent to one star of magnitude m is

$$S_{10}/(\text{sq.deg.}) = \frac{10^{0.4 \cdot (10-m)}}{20626.5}.$$

Let N_m be the number of stars of magnitude m; the equivalent number of 10^m stars is

$$I_m(S_{10}/(\text{sq.deg.})) = \frac{N_m \cdot 10^{0.4 \cdot (10-m)}}{20626.5} = N_m \cdot 10^{-0.4 \cdot (m+0.786)}.$$

The integrated radiation flux from a star at the boundary of the Earth's atmosphere near the wavelength $\lambda = 550\,(\text{nm})$ is (Allen 1973)

$$S_m(550(\text{nm})) = 3.72 \cdot 10^{-8} \cdot 10^{-0.4 \cdot m} \, (\text{erg} \cdot \text{cm}^{-2} \cdot \text{s}^{-1} \cdot \text{nm}^{-1}),$$

and near $\lambda = 440\,(\text{nm})$

$$S_m(440(\text{nm})) = 6.72 \cdot 10^{-8} \cdot 10^{-0.4 \cdot m} \, (\text{erg} \cdot \text{cm}^{-2} \cdot \text{s}^{-1} \cdot \text{nm}^{-1}).$$

These relations remain almost unchanged as the spectral type (class) of stars is changed from B to M. For the photon fluxes we have

$$S_m(550(\text{nm})) = 1.02 \cdot 10^4 \cdot 10^{-0.4 \cdot m} \, (\text{photon} \cdot \text{cm}^{-2} \cdot \text{s}^{-1} \cdot \text{nm}^{-1}),$$

$$S_m(440\,(\text{nm})) = 1.35 \cdot 10^4 \cdot 10^{-0.4 \cdot m}\,(\text{photon} \cdot \text{cm}^{-2} \cdot \text{s}^{-1} \cdot \text{nm}^{-1})\,.$$

The illuminance produced by a star is

$$E_m = 2.54 \cdot 10^{-6 \cdot 10 - 0.4 \cdot m}\,(\text{lx})\,.$$

One star of $m_v = 0$ per square degree ($\square°$) corresponds to the brightness

$$1\,S_0/\square° = 0.84 \cdot 10^{-6}\,(\text{sb})\,.$$

According to the data of Roach and Gordon (1973), we have

$$1\,S_{10}/\square° = 4.4 \cdot 10^{-3}\,(\text{Rayleigh}/\text{Å})\,.$$

The definition of the Rayleigh photometric unit is given in Sect. 3.1. Since the number of stars S_{10} corresponding to the stellar component of the luminescence of the night sky is equal to 105.20 (Roach and Gordon 1973), we have

$$105.20\,S_{10}/(\square°) = 4.63\,(\text{Rayleigh}/\text{nm})\,.$$

This is the intensity of the stellar continuum near $\lambda = 530\,(\text{nm})$.

For the spectral interval corresponding to the visibility curve for the eye, $\Delta\lambda = 100\,(\text{nm})$, we have

$$1\,S_{10}/(\square°) = 4.4\,(\text{Rayleigh})\,.$$

Thus, the airglow intensity equivalent to the intensity of light of a star or a planet of stellar magnitude m is

$$Q_m = 2.43 \cdot 10^{-0.4 \cdot m}\,(\text{Rayleigh}/\square°)\,.$$

The intensity of light of a luminary is

$$Q_m = 44 \cdot 10^{-0.4 \cdot m}\,(\text{kilorayleigh})\,.$$

In some publications, the brightness of the sky background is given as the stellar magnitude per square second ($m/\square''$). In this case, we have

$$S(m/\square'') = 27.78 - 2.5\log_{10} I_m\,(\text{erg} \cdot \text{cm}^{-2} \cdot \text{sr}^{-1} \cdot \text{s}^{-1})\,,$$
$$S(m/\square'') = 24.39 - 2.5\log_{10} Q\,(\text{Rayleigh} \cdot \text{nm}^{-1})\,,$$
$$S(m/\square'') = 29.39 - 2.5\log_{10} Q\,(\text{Rayleigh})\,.$$

In the visible spectral region, the average brightness of the sky is (Straižys 1977; Neizvestny 1982) 22–23 ($m/\square''$).

The extra-atmospheric luminescence of the sky caused by the galactic and extra-galactic components in the ultraviolet, visible, and infrared regions is described elsewhere (Abreu et al. 1982; Boulanger and Pérault 1988; Martin and Bowyer 1989).

1.2.2 The Sun, the Moon, and Planets

These luminaries can produce significant light noise for photometric instruments. At the same , they can serve as sources for calibration of spectrophotometric measurements. However, the Sun, the Moon, and planets change their positions in the sky. To determine their coordinates, it is necessary to use the methods of calculation of ephemerises described elsewhere (Abalakin et al. 1976; Abalakin 1979; Montenbruck and Pfleger 2000). Now there is a computer code (WinEphem 2002) which allows one to perform necessary calculations.

The solar flux outside the Earth's atmosphere at a distance $\rho_\odot$ from the Sun (in astronomical units, a.e. $= 1.496 \cdot 10^{13}$ (cm)) is determined as

$$S_\lambda = \frac{S_{0\lambda}}{\rho_e^2}.$$

The value of $\rho_\odot$ varies throughout a year between 0.983 (January) and 1.017 (July). The brightness of the Sun outside the atmosphere is 202 000 (sb); the apparent magnitude is -26.78^m (Allen 1973; Kononovich and Moroz 2001; Kulikovsky 2002).

According to the research data available in the literature (Makarova and Kharitonov 1972; Heath and Thekaekara 1977; Makarova et al. 1991), the average distribution of the energy of the Sun disk radiation corresponds to a color temperature of 5770 (K). However, a detailed distribution for an individual spectral region can be obtained by using the color temperature T for the given wavelength range (Makarova and Kharitonov 1972). For the visible spectral range, the temperature T is about 5800 (K); it reaches a maximum of 6360 (K) near 1.65 (μm) and then gradually decreases to 4460 (K) at about 25 (μm). For the spectral region 0.35–25 (μm), the temperature T can be approximated accurate to 50–100 (K) by the formula

$$T = \frac{6360}{1 + \frac{\sqrt{|\lambda - 1.65|}}{10.3} - \frac{0.04}{1 + 300 \cdot (\lambda - 0.46)^2}} \; (K).$$

Here, the wavelength λ is expressed in (μm).
The solar flux is given by

$$S_\lambda = \pi \left(\frac{R_e}{\rho_e}\right)^2 \cdot B(\lambda, T) \; (\text{erg} \cdot \text{cm}^{-2} \cdot \text{s}^{-1} \cdot \text{ster}^{-1}).$$

Here, $R_\odot = 6.96 \cdot 10^{10}$ (cm) is the radius of the Sun and $B(\lambda, T)$ is Planck's function:

$$B(\lambda, T) = \frac{C_1}{\lambda^5} \cdot \frac{1}{e^{\frac{C_2}{\lambda T}} - 1},$$

where $C_1 = 2h \cdot c^2 = 1.1910 \cdot 10^{-5}$ (erg $\cdot$ cm$^{-2} \cdot$ s$^{-1} \cdot$ sr^{-1}) and $C_2 = h \cdot c/k = 1.43879$ (cm $\cdot$ K). For practical use there are the following formulas:

$$S_\lambda = \frac{8100}{\lambda^5} \cdot F(\lambda T) \; (\text{erg} \cdot \text{cm}^{-2} \cdot \text{s}^{-1} \cdot \mu\text{m}^{-1}),$$

$$S_\lambda = \frac{8100}{\lambda^5} \cdot F(\lambda T) \quad 10^{-4}(\text{W} \cdot \text{cm}^{-2} \cdot \mu\text{m}^{-1}),$$

$$S_\lambda = \frac{4100}{\lambda^5} \cdot F(\lambda T) \quad 10^{12}(\text{photon} \cdot \text{cm}^{-2} \cdot \text{c}^{-1} \cdot \text{nm}^{-1})$$

where the multiplier $F(\lambda T)$ follows from Planck's formula.

For comparison, at the boundary of the Earth's atmosphere the solar flux near 550 (nm) is

$$S(550(\text{nm})) = 5.174 \cdot 10^{14} (\text{photon} \cdot \text{cm}^{-2} \cdot \text{s}^{-1} \cdot \text{nm}^{-1})$$
$$= 1870 (\text{erg} \cdot \text{cm}^{-2} \cdot \text{s}^{-1} \cdot \text{nm}^{-1})$$

and near 725 (nm)

$$S(725(\text{nm})) = 5.07 \cdot 10^{14} (\text{photon} \cdot \text{cm}^{-2} \cdot \text{s}^{-1} \cdot \text{nm}^{-1})$$
$$= 1390 (\text{erg} \cdot \text{cm}^{-2} \cdot \text{s}^{-1} \cdot \text{nm}^{-1}).$$

Outside the Earth's atmosphere, in the nearest vicinity of the Sun, the brightness is determined by the internal solar corona for angles within $\pm 2°$, by the external corona for angles up to $10°$, and by Zodiacal Light for larger angles. At an angular distance of $10°$ from the Sun, the brightness is less than that near the Sun by a factor of $6.7 \cdot 10^{10}$ (Roach and Gordon 1973).

The brightness of the full disk of the Moon is a factor 465 000 lower than that of the Sun. The apparent magnitude is $m_v = -12.71^m$. This corresponds to a mean brightness of 0.251 (sb) (Allen 1973). Ivanov (1994) points out that when the Moon is used as a source of calibrated brightness, it is necessary to consider the variations of its light flux with distance from the Earth. Detailed data on the spectrophotometric characteristics of the Moon are cited elsewhere (Pugacheva et al. 1993; Ivanov 1994).

The light flux from the Moon is

$$S_{\bullet\lambda} = 2.15 \cdot 10^{-6} \cdot S_{\odot\lambda} \cdot \eta(\Phi),$$

where the relative brightness of the visible portion of the Moon disk is determined by the phase angle Φ, which is the circumplanetary angle between the directions from the planet to the Earth and to the Sun (Fig. 1.12, Table 1.2). The ratio of the illuminated portion of the disk to its total area is

$$k = \cos^2 \frac{\Phi}{2}.$$

The relation between the Moon brightness and k is nonlinear. It has been tabulated by Kulikovsky (2002), and its approximation is presented by Hapke (1963, 1971). Within a synodical month, the phase angle is related to the Moon age

1.2 Extra-Atmospheric Radiation

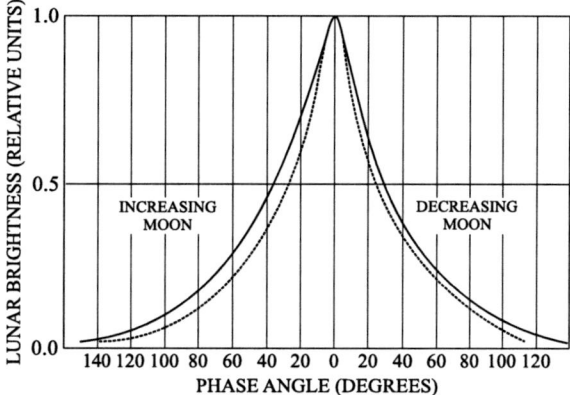

Fig. 1.12 Brightness of the Moon as a function of the phase angle. *Solid line*: Russel's data; *dotted line*: Rougier's data (Sytinskaya 1959)

t_L by the formula $\Phi = 180 \cdot (1 - t_L/14.7853)$ for an increasing Moon and $\Phi = 180 \cdot (1 + t_L/14.7853)$ for a decreasing Moon, where t_L is expressed in days.

For the full Moon we have

$$S(550\,(\text{nm})) = 1.11 \cdot 10^9 \,(\text{photon} \cdot \text{cm}^{-2} \cdot \text{s}^{-1} \cdot \text{nm}^{-1})$$
$$= 4.0 \cdot 10^{-3} \,(\text{erg} \cdot \text{cm}^{-2} \cdot \text{s}^{-1} \cdot \text{nm}^{-1}),$$
$$S(725\,(\text{nm})) = 1.09 \cdot 10^9 \,(\text{photon} \cdot \text{cm}^{-2} \cdot \text{s}^{-1} \cdot \text{nm}^{-1})$$
$$= 3.0 \cdot 10^{-3} \,(\text{erg} \cdot \text{cm}^{-2} \cdot \text{s}^{-1} \cdot \text{nm}^{-1}).$$

Table 1.2 Moon's brightness (in relative energy units E and in magnitudes m) as a function of phase angle Φ (Sytinskaya 1959)

Phase angle Φ (deg)	Age t (days)	Russel ☽ E	m	Russel ☾ E	m	Rougier ☽ E	m	Rougier ☾ E	m
0	0	1000	0.00	1000	0.00	1000	0.00	1000	0.00
10	0.8	816	0.22	816	0.22	787	0.26	759	0.30
20	1.7	666	0.44	642	0.48	603	0.55	586	0.58
30	2.5	540	0.67	505	0.74	466	0.83	425	0.86
40	3.3	436	0.90	387	1.03	365	1.12	350	1.14
50	4.2	353	1.13	299	1.31	273	1.40	273	1.41
60	5.0	283	1.37	234	1.58	210	1.69	212	1.69
70	5.8	218	1.65	180	1.86	161	1.98	157	2.02
80	6.6	161	1.98	136	2.17	120	2.30	111	2.39
90	7.5	115	2.35	100	2.50	83	2.71	78	2.77
100	8.3	77	2.78	72	2.86	56	3.13	58	3.09
110	9.2	51	3.22	49	3.27	38	3.56	41	3.48
120	10.0	31	3.77	32	3.74	25	4.01	26	3.96
130	10.8	18	4.39	19	4.30	15	4.55	16	4.50
140	11.6	9	5.14	10	4.98	–	–	9	5.09
150	12.4	4	6.09	4	5.89	–	–	5	5.86

The albedo of the Moon in the ultraviolet spectral region is given by Flynn et al. (1998). In the wavelength range 70–150 (nm), the geometric albedo is almost constant and makes 4(%); near 50, 25, and 15 (nm) it makes 1, 0.1, and 0.01(%), respectively.

The moonlight-to-sunlight intensity ratio for the infrared region at about 2.4 (μm) is about an order of magnitude greater than for the visible region (Hapke 1963, 1971). For the longer wavelength region, the radiation of the Moon becomes similar to thermal radiation with a temperature of about 400 (K) (Fig. 1.13) (Moroz 1965; Wattson and Danielson 1965; Leikin and Shvidkovskaya 1972).

The main planets – Mercury (-1.8^m), Venus (-4.2^m), Mars (-1.0^m), Jupiter (-2.05^m), and Saturn ($+0.8^m$) – possess appreciable brightness. The spectral energy distributions for the planets are shown in Fig. 1.14. The magnitudes as functions of the phase angle Φ are determined by the Müller empirical formulas (Abalakin 1979). Detailed spectrophotometric data for the planets are given elsewhere (Pugacheva et al. 1993).

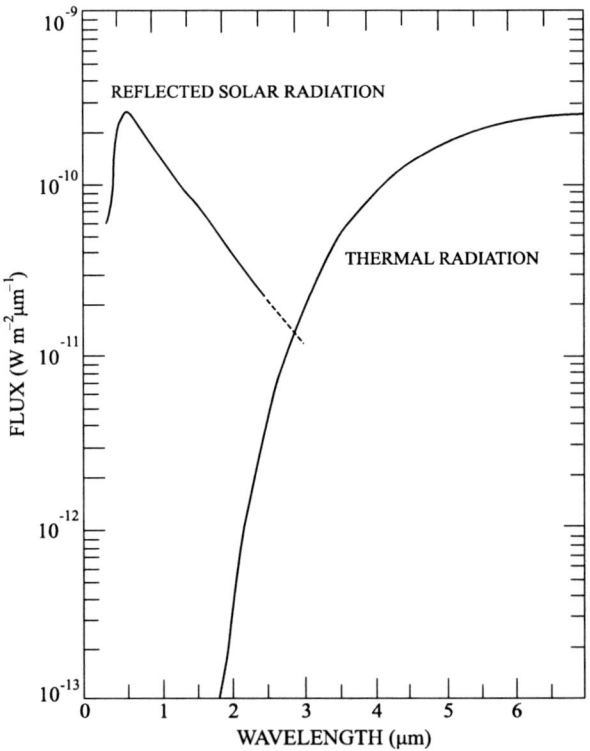

Fig. 1.13 Spectral energy distribution for the Moon in the visible and infrared regions (Moroz 1965; Leikin and Shvidkovskaya 1972)

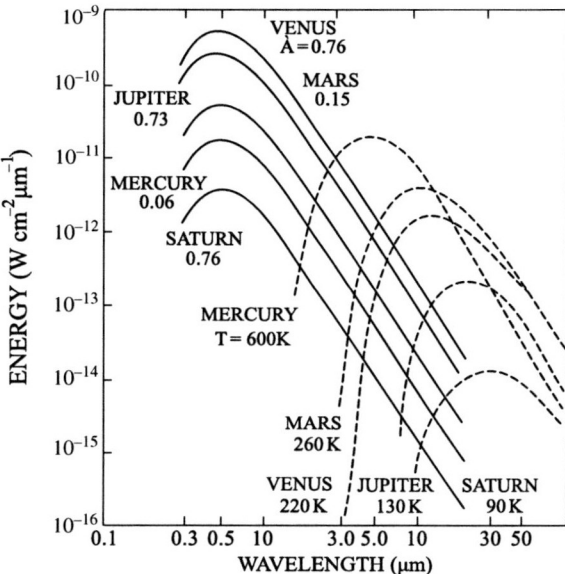

Fig. 1.14 Spectral energy distribution for the main planets (Pugacheva et al. 1993). *Solid lines* are reflected radiation. *Dashed lines* are thermal radiation. The *numbers* near the curves indicate the albedo (A) of the planets and the effective temperatures of their surfaces

It is supposed that the radiation is emitted by the "marine" surface of the Moon at a subsolar point with the phase angle zero at the mid-position between the Earth and the Moon.

1.2.3 Bright Stars

The bright stars with magnitudes corresponding to the unaided-eye visibility are numerous discrete radiation sources. The distribution of stars over the sky is presented in widely known atlases of the stellar sky (Mikhailov 1974). The photometric and spectral characteristics of many stars are collected in many publications; as to the stars brighter than 4.5^m, their characteristics are reported by Allen (1955, 1973) and Kulikovsky (2002). The magnitudes of stars are related to the solar flux as

$$S_m = 1.94 \cdot 10^{-11} \cdot 10^{-0.4m} \cdot S_\odot ,$$

or, in terms of the sky emission intensity, as

$$S_m = 44 \cdot 10^{-0.4 \cdot m} \text{ (kilorayleigh)} .$$

The distribution of stars over the sky is very nonuniform. Moreover, the number of stars of given brightness strongly varies. Throughout the sky, 4850 stars of brightness above 6^m and 1620 of brightness above 5^m are encountered. The brighter the stars, the smaller their number. The brightest star, Sirius, has a magnitude of -1.58^m. However, the basic contribution to the total brightness of the sky is due to weak stars of magnitude about 12^m (Divari 1951; Allen 1973; Kulikovsky 2002). The average distribution of the stellar component for the visible spectral region is given by Roach and Gordon (1973). This is a G2-type distribution, i.e., it is close to the solar spectral distribution.

1.2.4 Distribution of the Starlight over the Sky

The stellar component of the sky radiation concentrates toward the Galactic equator – the Milky Way. The starlight intensities have been tabulated by Roach and Gordon (1973). A spatial distribution of brightness is convenient to represent in the Aitov–Hammer–Soloviev projection (see Fig. 1.23) (Soloviev 1969). Figure 1.15 presents the distribution (in Galactic coordinates of the old system with the reference mark of galactic longitude at the point of interception of the Galaxy circle

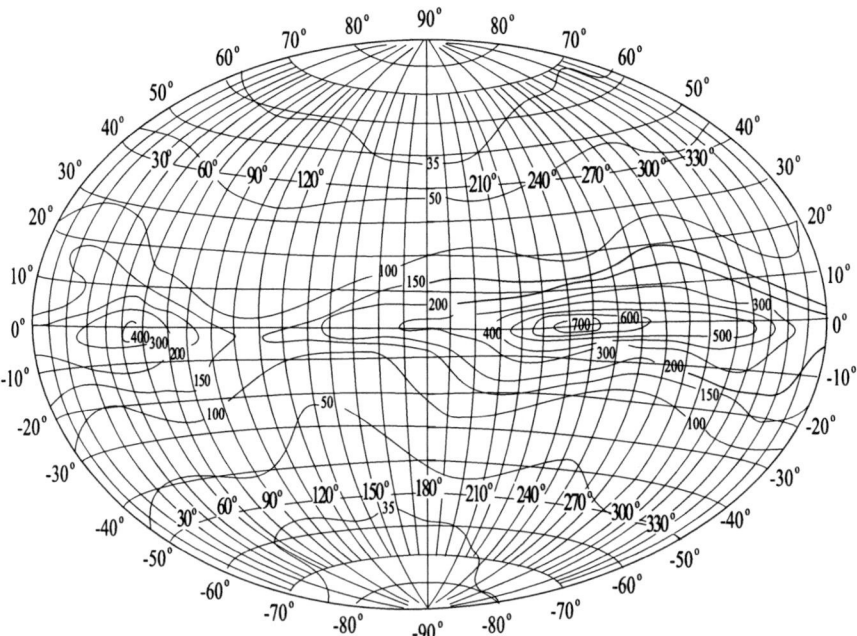

Fig. 1.15 Distribution of the total brightness of stars (number S_{10} of stars of the tenth apparent magnitude) over the sky (Roach and Megill 1961; Roach 1964; Vassy and Vassy 1976)

and the equator circle) of the total brightness of stars (number S_{10} of stars of the tenth visual value) over the sky (Roach and Megill 1961; Roach 1964; Vassy and Vassy 1976).

1.2.5 Zodiacal Light

Zodiacal Light results from the scattering of solar radiation by the interplanetary dust located near the ecliptic plane, and therefore it takes place on the background of zodiacal constellations. In ecliptic coordinates, its intensity is determined by the elongation (angular distance in longitude) of the Sun, given by $\Delta\lambda = \lambda_e - \lambda_\odot$, where λ_e is the ecliptic longitude of the line of sight and $\lambda_\odot$ the ecliptic longitude of the Sun, and by the ecliptic latitude β of the line of sight. Tables of the spatial distribution of the Zodiacal Light intensity are given by Roach and Gordon (1973). The spectral distribution for the ecliptic plane and 30° elongation, constructed based on the above measurements and a simulation performed on the assumption of the light scattering by dust of varied composition, is shown in Fig. 1.16 (Röser and Staude 1978).

The above results of the studies of extraterrestrial radiation allow some systematizing conclusions. The stellar component is stationary in character. The brightest, and, hence, the most contrast inhomogeneities (some stars, planets) are known in advance, and their contribution to the detected radiation can be calculated. The contrast ratio can reach 400 at a field of vision of one square degree. The maximum contrast ratio for the light of medium magnitude stars in the transition region from the Milky Way to Galactic Pole is more than 20. The Zodiacal Light brightness follows the location of the Sun and varies almost monotonically with elongation. Measurements

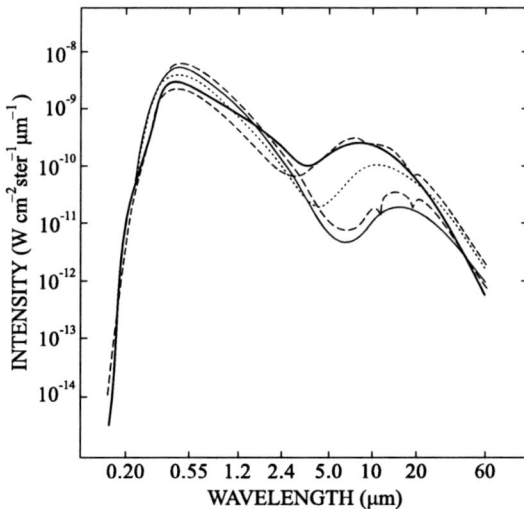

Fig. 1.16 Spectral distribution of the Zodiacal Light intensity for the ecliptic latitude $\beta = 0°$ and elongation $\lambda - \lambda_\odot = 30°$. *Thin line* is obsidian, *dotted line* is andesite, *dashed line* is olivine, *dotted dashed line* is magnetite and *thick line* is graphite (Röser and Staude, 1978)

for the visible spectral region are presented elsewhere (Karyagina 1960; Blackwell and Ingham 1961; Torr et al. 1979; Divari 2003). The ratio of the maximum intensity (at an angular distance of 10° from the Sun) to its minimum value reaches 200.

For the ultraviolet spectral region, the average intensity distribution in ecliptic latitude appears decreasing slower than for the visible region, which is due to the substantial contribution of fine dust particles (Murthy et al. 1990). According to satellite measurements, the extra-atmospheric radiation is traced in the ultraviolet region down to almost 100 (nm) as diffuse radiation consisting of Zodiacal Light and stellar radiation (125–170 (nm) (Röser and Staude 1978; Murthy et al. 1989), 165–310 (nm) (Röser and Staude 1978; Cebula and Feldman 1982; Tennyson et al. 1988; Murthy et al. 1990)) and in the short-wave region as emission lines, mainly of hydrogen (121.6 (nm) Lyman alpha and 102.6 (nm) Lyman beta) and helium (58.4 and 30.4 (nm)) (Ajello and Witt 1979).

An important feature of the spectral distribution shown in Fig. 1.16 is that the radiation in the infrared spectral region is the thermal radiation of dust. On the solar side, the temperature is about 400 (K), while on the antisolar side it is about 270 (K). In the ultraviolet region (near 180 (nm)) at small elongation angles the intensity is (Cebula and Feldman 1982)

$$I(180\,(nm)) = 0.22\,(\text{Rayleigh} \cdot \text{nm}^{-1})\,.$$

The relationship between intensity and elongation can be described as (Cebula and Feldman 1982)

$$I(\lambda - \lambda_e) \sim (\lambda - \lambda_e)^{-2.6}\,.$$

1.3 Ultraviolet Solar Radiation

The ultraviolet solar radiation is one of the basic energy sources for the upper atmosphere, and it is responsible for the dissociation and ionization of the atmospheric components. This radiation covers the wavelength range 0.1–300 (nm) and is completely absorbed in the atmosphere. By its spectral character and the role it plays in the atmosphere, this range can be divided into two: 100–300 and 0.1–100 (nm).

1.3.1 Radiation in the 100- to 300-nm Range

This spectral region contains in the main a continuum, which was approximated with Planck's function in the temperature range from 4000 to 5000 (K) (Fig. 1.17) (Makarova and Kharitonov 1972; Heath and Thekaekara 1977; Lean 1984; Makarova et al. 1991). The spectrum variability with solar activity is small, making about 2(%) near 300 (nm), 10(%) near 250 (nm), and 25(%) near 200 (nm) (Lean 1984). It becomes more pronounced in the range near 100 (nm), mainly due

1.3 Ultraviolet Solar Radiation

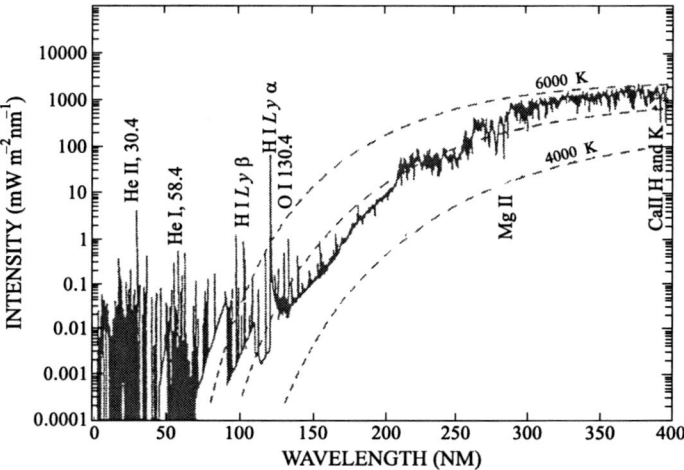

Fig. 1.17 Average spectrum of the solar radiation in the ultraviolet region (Lean 1991)

to the effect of the 121.6-nm Lyman alpha and 102.7-nm Lyman beta hydrogen and 130.4-nm atomic oxygen emissions.

Investigations of the spectral structure of solar radiation and its variability with solar activity have been performed for many years (Ivanov-Kholodny and Mikhailov 1980; Bossy and Nicolet 1981; Bossy 1983; Lean 1984).

A detailed spectrum of the Sun in the range 100–300 (nm) is published elsewhere (Nicolet 1981).

1.3.2 Radiation in the 20- to 100-nm Range

The short-wave region of the solar spectrum consists of a large number of emission lines belonging to numerous multiply ionized atoms of the atmosphere of the Sun (Bowyer et al. 1968; Ivanov-Kholodny and Mikhailov 1980; Manson 1977; Timothy 1977; Hinteregger 1981; Hinteregger et al. 1981; Fatkullin 1982).

For some regions of the spectrum, approximation formulas have been obtained; for example, for the range 0.1–104 (nm),

$$S(F_{10.7}) = S_0(144) \cdot 0.6066 \cdot \log_e \left[\frac{F_{10.7} - 40}{20} \right].$$

The photoionization of the atmosphere at night is due to the solar radiation in 58.4 and 30.4 (nm) helium lines scattered at high altitudes, the line intensities being 12 and 5 (Rayleigh), respectively (Young et al. 1968).

1.3.3 Occurrence of Photoelectrons

Photoelectrons are produced by ultraviolet solar radiation. The basic neutral components of the atmosphere at altitudes above 100 (km) are N_2 and O_2 molecules as well as O, H, and He atoms. For O_2, O, and N_2, the ionization potentials are 12, 13.6, and 15.6 (eV), respectively. Therefore, the most essential to ionization is the solar radiation with $\lambda < 110.0$ (nm). A process of the type

$$M + h\nu(\lambda < 110\,(nm)) \rightarrow M^+ + e$$

results in the occurrence of photoelectrons with energies from a few to several tens of electron-volts. This was pointed up for the first by Shklovsky (1951) who studied the solar corona. The first experimental evidence to this was obtained in the observations of the twilight emission of helium at 1083.0 (nm) (Shefov 1962; Krassovsky and Shefov 1964; Bates 1982). It is under the twilight conditions of illumination of the upper atmosphere that photoelectrons of energy about 25 (eV) appear that excite metastable orthohelium atoms. Subsequently, satellite observations were performed (Bolunova et al. 1977). Many theoretical and experimental studies (Tohmatsu et al. 1965; Stolarski and Green 1967; Nagy et al. 1977) were devoted to the electron energy distribution.

The fresh photoelectrons lose their energy in elastic and inelastic collisions with atoms and molecules, and this results in a certain superthermal energy distribution which in general depends on the altitude Z (that determines the composition of the atmosphere), time of day, season, and solar activity.

Systematization of the data on the photoelectron spectra allows the following conclusions (Krinberg 1978):

1. The energy spectrum has a maximum in the energy range $E \sim 22\text{--}27$ (eV) (due to the photoionization by solar radiation in the 30.4-nm He^+ line) and features an abrupt fall at energies over 60 (eV).
2. At altitudes above 200 (km) and at solar zenith angles $\chi_\odot < 75°$, the photoelectron energy spectrum $\Phi_0(E, Z)$ does not depend on the altitude Z, zenith angle, and geographic (or geomagnetic) coordinates, and it is a single-valued function of energy.
3. Below $Z \sim 200$ (km), the photoelectron flux Φ_0 decreases with decreasing Z, and the lower the electron energy, the faster the decrease in Φ_0.

These features of the photoelectron energy spectrum weakly depend on the altitude distribution of the concentration of neutrals and are related in the main to the variations of the solar flux $S(\lambda, Z)$ during its passage through the atmosphere. This is due to the fact that for $h\nu > 20$ (eV) the photoionization cross-sections σ_{ph} for the N_2 and O_2 molecules are almost identical, being approximately twice that for O. About the same proportion holds for the cross-sections of inelastic collisions of electrons with N_2, O_2, and O, σ_{elas}. Therefore, the photoelectron flux is determined by the relation

1.3 Ultraviolet Solar Radiation

$$\Phi_0 = S_0 \cdot \frac{\sigma_{phO} \cdot [O] + \sigma_{phO_2} \cdot [O_2] + \sigma_{phN_2} \cdot [N_2]}{\sigma_{elasO} \cdot [O] + \sigma_{elasO} \cdot [O_2] + \sigma_{elasN_2} \cdot [N_2]} \approx \frac{\sigma_{phO}}{\sigma_{elasO}} \cdot S_0 .$$

The rather abrupt decay of the energy spectrum at energies $E > 60\,(eV)$ is accounted for by both the substantial reduction in solar intensity at $h > 80\,(eV)$ ($\lambda < 15.5\,(nm)$) and the appreciable reduction of the photoionization cross-sections at the same energies.

At night, in the absence of direct insolation, photoionization occurs due to the 121.6-nm L_α and 102.6-nm L_β hydrogen radiations and the 58.4-nm neutral helium and 30.4-nm ionized helium radiations that are resonance-scattered in the geocorona. The main contribution is provided by the 30.4-nm emission. Its intensity weakly depends on the solar zenith angle at $\chi_\odot > 100°$ (Krinberg 1978). The energy spectrum of the nightglow has a maximum near 27 (eV) and is similar in shape to the dayglow spectrum, differing from it by four orders of magnitude (Krinberg 1978).

The relationship between photoelectron flux and solar activity can be represented by the empirical relation (Krinberg 1978)

$$\Phi_0 = \Phi_{00} \cdot \exp\left(\frac{F_{10.7} - 144}{72}\right) ,$$

where Φ_{00} is the "standard" spectrum for photoelectrons which corresponds to the solar radio flux $F_{10.7} = 144$ at altitudes above 250 (km). The actual relationship between the photoelectron flux and energy (at less than 1 (eV) resolution) is rather complicated because of the large number of individual solar emissions leading to the ionization of atmospheric components. However, this is of little importance for the excitation of the airglow because of both the averaging of the rates of excitation processes within the effective range of cross-sections and the local inhomogeneities of the atmospheric structure. Therefore, an average "smooth" spectral distribution of photoelectron energies suffices for the given purposes.

Based on the data reported by Krinberg (1978), the energy dependence of the photoelectron flux can be described by the following empirical relations:

for daytime conditions

$$\log_e \Phi_{00} = 6.91 + 12.43 \cdot \exp(-0.12 \cdot \log_{10}^4 E) + 2.3 \cdot \exp\left(-100 \cdot \log_{10}^2 \frac{E}{2}\right) +$$
$$+ 1.84 \cdot \exp\left(-350 \cdot \log_{10}^2 \frac{E}{25}\right) + 1.90 \cdot \exp\left(-100 \cdot \log_{10}^2 \frac{E}{50}\right)$$

and for night conditions

$$\log_e \Phi_{00n} = -\frac{E}{2.89} + 9.76 + 3 \cdot \exp\left(-350 \cdot \log_{10}^2 \frac{E}{25}\right) .$$

Here Φ_{00} is expressed in $(electron \cdot cm^{-2} \cdot s^{-1} \cdot eV^{-1} \cdot sr^{-1})$ and E in (eV).

1.4 Space–Time Conditions for Detecting Radiation

1.4.1 Geometric Relations for Various Observation Conditions

1.4.1.1 Conditions at the Earth Surface

In recording emissions in the upper atmosphere, the position of the observer over the Earth surface is determined by a set of coordinates (Fig. 1.18), namely:

x, the distance from the observer along the line of sight;
ζ, the zenith angle of the line of sight;
A, the azimuth of the line of sight which is counted from the point of the south clockwise from 0° to 360°;
Z_0, the altitude of the observer (point O) over the Earth surface;
Z, the moving altitude of a point O′ located on the line of sight;
φ_0 and λ_0, the geographic latitude and longitude of the point on the Earth surface over which the observer is situated;
Φ and λ, the same as for the point O′, and
R_E, the Earth's radius.

The intensity I (Rayleigh) of the radiation that is observed along the line of sight is determined by the relation

$$I = 10^{-6} \cdot \int_0^{x_L} Q^*(Z, \varphi, \lambda, t_d, \tau) dx,$$

where Q^* is the emission rate (photon·cm^{-3}·s^{-1}) at the altitude Z for the point on the Earth surface with coordinates φ and λ at the local solar time τ. It follows that

$$Q^* = Q \cdot \exp(-\tau_t \cdot \mathrm{Chp}\,\zeta),$$

where Q is the emission rate for the observation at zenith, τ_t is the optical thickness of the medium through which the recorded radiation propagates, and Chp ζ is the Chapman function, which determines the relationship between emission intensity and zenith angle ζ. For $\zeta \leq 70°$ we have Chp $\zeta \approx \sec \zeta$. From the above consideration it follows that

$$Z = \sqrt{x^2 + 2x(R_E + Z_0)\cos\zeta + (R_E + Z_0)^2} - R_E.$$

The geographic coordinates (φ and λ) of the point O′ are determined by the formulas of spherical trigonometry

1.4 Space–Time Conditions for Detecting Radiation

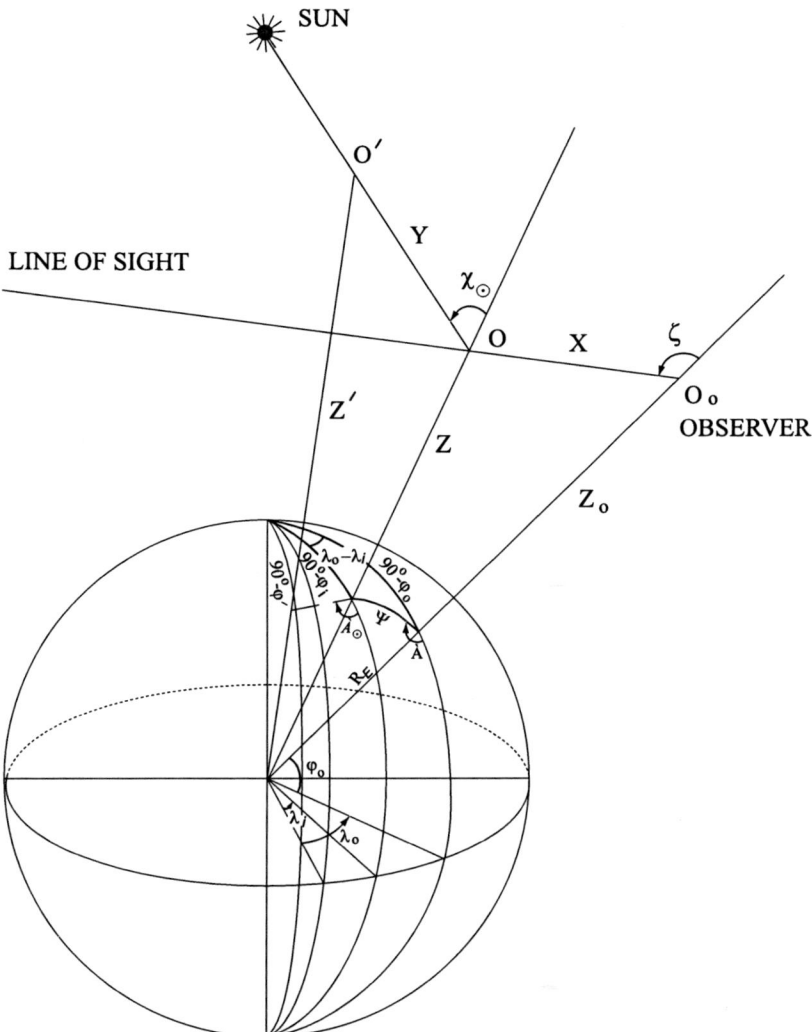

Fig. 1.18 Geometry of the orientation of the observer position over the Earth surface

$$\sin\varphi = \sin\varphi_0 \cdot \cos\psi - \cos\varphi_0 \cdot \sin\psi \cdot \cos A,$$

$$\sin(\lambda - \lambda_0) = \frac{\sin\psi \cdot \sin A}{\cos\varphi},$$

$$\cos(\lambda - \lambda_0) = \frac{-\cos\varphi_0 \cdot \cos\psi - \sin\varphi_0 \cdot \sin\psi \cdot \cos A}{\cos\varphi},$$

where

$$\mathrm{tg}\,\psi = \frac{x \cdot \sin\zeta}{R_E + Z_0 + x \cdot \cos\zeta}.$$

If $0 \leq A \leq 180°$, then $0 \leq \lambda_0 - \lambda \leq 180°$; if $180° \leq A \leq 360°$, then $180° \geq \lambda - \lambda_0 \geq 0$.

For the given altitudes, 80–100 (km), the limiting range x_L is determined by the following relations (Fig. 1.19):
for $0 \leq \zeta \leq \zeta_L$

$$x_L = -(R_E + Z_0) \cdot \cos\zeta + \sqrt{(R_E + Z_L)^2 - (R_E + Z_0)^2 \cdot \sin^2\zeta}$$

and for $\zeta_L < \zeta \leq 180°$

$$x_L = -(R_E + Z_0) \cdot \cos\zeta - \sqrt{(R_E + Z_L)^2 - (R_E + Z_0)^2 \cdot \sin^2\zeta},$$

where ζ_L corresponds to the zenith angle of the line of sight that passes at a distance $Z_L = 80$ (km) over the Earth surface and is calculated by the formula

$$\sin\zeta_L = \frac{R_E + Z_L}{R_E + Z_0}.$$

Thus, the geographic coordinates of the point O' are determined by the relations

$$\sin\varphi = \frac{(R_E + Z_0) \cdot \sin\varphi_0 + x \cdot (\cos\zeta \cdot \sin\varphi_0 - \sin\zeta \cdot \cos\varphi_0 \cdot \cos A)}{\sqrt{x^2 + 2x(R_E + Z_0)\cos\zeta + (R_E + Z_0)^2}},$$

$$\sin(\lambda - \lambda_0) = \frac{x\sin\zeta \cdot \sin A}{\sqrt{x^2 + 2x(R_E + Z_0)\cos\zeta + (R_E + Z_0)^2}},$$

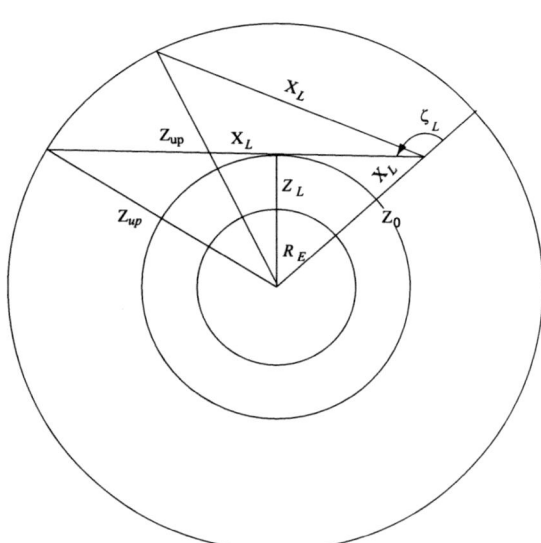

Fig. 1.19 Section in the plane of vision

1.4 Space–Time Conditions for Detecting Radiation

$$\cos(\lambda - \lambda_0) = \frac{-(R_E + Z_0)\cos\varphi_0 - x \cdot (\cos\zeta \cdot \cos\varphi_0 + \sin\zeta \cdot \sin\varphi_0 \cdot \cos A)}{\sqrt{x^2 + 2x(R_E + Z_0)\cos\zeta + (R_E + Z_0)^2}}.$$

If the observation is performed at the Earth surface in the direction with zenith angle ζ, the line of sight intersects the emission layer at an altitude Z and has at the intersection point a zenith angle ζ_Z, which is determined by the relation (Fig. 1.20)

$$\sin\zeta_Z = \frac{R_E}{R_E + Z}\sin\zeta.$$

The distance L between two points with coordinates φ_1, λ_1 and φ_2, λ_2 on the Earth surface (Fig. 1.21) is determined by the formula $L = R_E \cdot \psi$, where the angle ψ of the great circle arc is calculated by the formula

$$\cos\psi = \sin\varphi_1 \cdot \sin\varphi_2 + \cos\varphi_1 \cdot \cos\varphi_2 \cdot \cos(\lambda_1 - \lambda_2).$$

The angle ψ is related to zenith angle ζ and altitude Z. Denoting $1 + \frac{Z}{R_E} = k$, we can determine the angle ζ with the use of the relations

$$\sin\psi = \frac{1}{k} \cdot \sin\zeta \cdot \left(\sqrt{k^2 - \sin^2\zeta} - \cos\zeta\right),$$

$$\cos\psi = \frac{1}{k} \cdot \cos\zeta \cdot \left(\sqrt{k^2 - \sin^2\zeta} + \sin\zeta \cdot \text{tg}\zeta\right),$$

$$\text{tg}\psi = \text{tg}\zeta \cdot \frac{k \cdot \sqrt{1 + \left(1 - \frac{1}{k^2}\right) \cdot \text{tg}^2\zeta} - 1}{k \cdot \sqrt{1 + \left(1 - \frac{1}{k^2}\right) \cdot \text{tg}^2\zeta} + \text{tg}^2\zeta}.$$

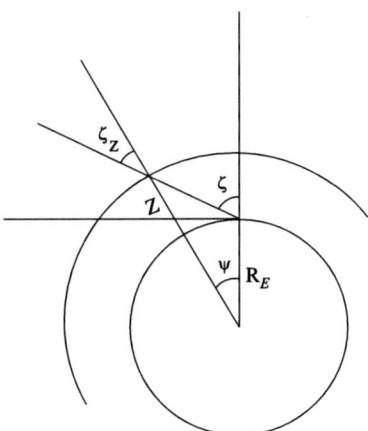

Fig. 1.20 Conditions of sighting from the Earth surface

Fig. 1.21 Spherical coordinates of two points located on the Earth surface

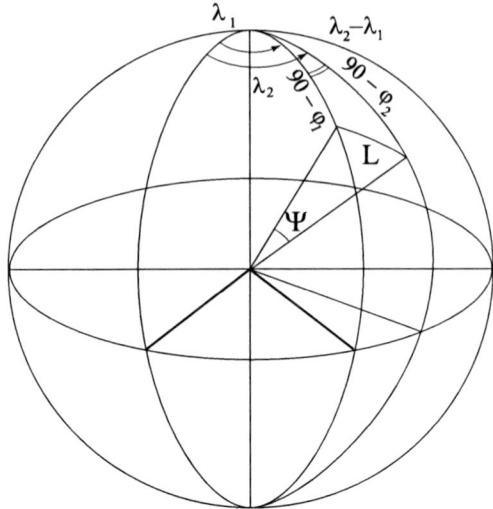

Since for the altitudes up to $Z \sim 200$ (km) we have $k \leq 1.03$, we can use the approximate relations

$$\text{tg}\psi \approx \frac{k-1}{k} \cdot \text{tg}\zeta \approx \frac{Z}{R_E} \cdot \text{tg}\zeta.$$

When considering the processes that occur in a magnetoconjugate region of the atmosphere, one should use transformations of geographic (φ, λ) and geomagnetic (Φ, Λ) coordinates. In the central dipole approximation, which well suffices to describe the processes in the middle and low latitudes, the transformation of coordinates is realized by a mere rotation of the system, which is described by relations of spherical trigonometry.

Transformation of Geographic to Geomagnetic Coordinates: $\varphi, \lambda \rightarrow \Phi, \Lambda$

The coordinates of the geomagnetic pole, φ_m, λ_m, do not remain constant (Semenov and Shefov 2003). For the epoch of 1960s, they were: $\varphi_m = 78.5°$N and $\lambda_m = 291.0°$E (Akasofu and Chapman 1972). The geomagnetic longitude is counted to the east of the meridian that passes through the geographic and geomagnetic poles. Thus, we have

$$\sin\Phi = \sin\varphi_m \cdot \sin\varphi + \cos\varphi_m \cdot \cos\varphi \cdot \cos(\lambda - \lambda_m),$$
$$\sin\Lambda = \frac{\cos\varphi \cdot \sin(\lambda - \lambda_m)}{\cos\Phi},$$
$$\cos\Lambda = \frac{-\cos\varphi_m \cdot \sin\varphi + \sin\varphi_m \cdot \cos\varphi \cdot \cos(\lambda - \lambda_m)}{\cos\Phi}.$$

1.4 Space–Time Conditions for Detecting Radiation

Transformation of Geographic to Geomagnetic Coordinates: $\Phi, \Lambda \to \varphi, \lambda$

$$\sin\varphi = \sin\varphi_m \cdot \sin\Phi - \cos\varphi_m \cdot \cos\Phi \cdot \cos\Lambda,$$
$$\sin(\lambda - \lambda_m) = \frac{\cos\Phi \cdot \sin\Lambda}{\cos\varphi},$$
$$\cos(\lambda - \lambda_m) = \frac{\cos\varphi_m \cdot \sin\Phi + \sin\varphi_m \cdot \cos\Phi \cdot \cos\Lambda}{\cos\varphi}.$$

Transformation of Geomagnetic to Geographic Coordinates: $\Phi, \Lambda \to \varphi_c, \lambda_c$

$$\sin\varphi_c = -\sin\varphi_m \cdot \sin\Phi - \cos\varphi_m \cdot \cos\Phi \cdot \cos\Lambda,$$
$$\sin(\lambda - \lambda_m) = \frac{\cos\Phi \cdot \sin\Lambda}{\cos\varphi_c},$$
$$\cos(\lambda - \lambda_m) = \frac{-\cos\varphi_m \cdot \sin\Phi + \sin\varphi_m \cdot \cos\Phi \cdot \cos\Lambda}{\cos\varphi_c}.$$

In this case, to the geographic longitudes $\lambda_m \leq \lambda \leq \lambda_m - 180°$ there correspond the geomagnetic longitudes $0 \leq \Lambda \leq 180°$, and to the range $\lambda_m - 180° \leq \lambda \leq \lambda_m$ there corresponds the range $180° \leq \Lambda \leq 360°$.

However, if we consider a process occurring at an altitude Z over the Earth surface, it is necessary to take into account the spatial behavior of the corresponding magnetic field line. In terms of the dipole description, which is well appropriate in the case of low and middle latitudes, the radius vector of a field line is given by

$$r = R_E + Z = L \cdot \cos^2\Phi.$$

Similarly, for a field line crossing the Earth surface we have

$$R_E = L \cdot \cos^2\Phi_E.$$

Thus, a point of the atmosphere at an altitude Z, having a geographic latitude φ and the corresponding geomagnetic latitude Φ_E, lies on a field line which crosses the Earth surface at a geomagnetic latitude Φ, which is determined by the formula

$$\cos^2\Phi = \left(1 + \frac{Z}{R_E}\right) \cdot \cos^2\Phi_E,$$

or

$$\cos\Phi = \sqrt{1 + \frac{Z}{R_E}} \cdot \cos\Phi_E.$$

The hour angle of the Sun (hours) at the conjugate point is determined as

$$t_{\odot c} = t_{\odot 0} + (\lambda_c - \lambda_0)/15,$$

where $t_{\odot 0}$ is the hour angle of the Sun at the given point.

To estimate the intensity of the extra-atmospheric radiation, which consists of Zodiacal Light and the total starlight, it is necessary to use the transformation of horizontal coordinates of a point on the celestial sphere to ecliptic and to galactic coordinates, respectively.

Transformation of Horizontal to Equatorial Coordinates: $\zeta, A \to \delta, t$

It is necessary to bear in mind that the right ascension α is related to the sidereal time S and hour angle t by the formula $\alpha = S - t$. The time coordinates must be used in view of the orbital motion of the Earth. The right ascension α is counted counterclockwise (to the east) from the vernal equinox point. The hour angle t is counted clockwise from the south point; S is the sidereal . These quantities are generally expressed in hours. The azimuth is counted clockwise from the south point from $0°$ to $360°$. The coordinates are transformed by the formulas

$$\sin\delta = \sin\varphi \cdot \cos\zeta - \cos\varphi \cdot \sin\zeta \cdot \cos A,$$

$$\sin t = \frac{\sin A \cdot \sin\zeta}{\cos\delta},$$

$$\cos t = \frac{\cos\varphi \cdot \cos\zeta + \sin\varphi \cdot \sin\zeta \cdot \cos A}{\cos\delta}.$$

Transformation of Equatorial to Horizontal Coordinates: $\delta, t \to \zeta, A$

$$\cos\zeta = \sin\varphi \cdot \sin\delta + \cos\varphi \cdot \cos\delta \cdot \cos t,$$

$$\sin A = \frac{\cos\delta \cdot \sin t}{\sin\zeta},$$

$$\cos A = \frac{-\cos\varphi \cdot \sin\delta + \sin\varphi \cdot \cos\delta \cdot \cos t}{\sin\zeta}.$$

Transformation of Equatorial to Ecliptic Coordinates: $\delta, \alpha \to \beta, \lambda_e$

The inclination angle of the ecliptic plane to the equator plane for the conditions of the year 1990.0 is $\varepsilon = 23.44058186°$. The secular variation of ε is given by Abalakin (1979). The ecliptic longitude λ_e is counted counterclockwise (to the east) from the vernal equinox point from $0°$ to $360°$. The coordinates are transformed by the formulas

1.4 Space–Time Conditions for Detecting Radiation

$$\sin\beta = \cos\varepsilon \cdot \sin\delta - \sin\varepsilon \cdot \cos\delta \cdot \sin\alpha,$$
$$\sin\lambda_e = \frac{\sin\varepsilon \cdot \sin\delta + \cos\varepsilon \cdot \cos\delta \cdot \sin\alpha}{\cos\beta},$$
$$\cos\lambda_e = \frac{\cos\delta \cdot \cos\alpha}{\cos\beta}.$$

Transformation of Ecliptic to Equatorial Coordinates: $\beta, \lambda_e \rightarrow \delta, \alpha$

$$\sin\delta = \cos\varepsilon \cdot \sin\beta + \sin\varepsilon \cdot \cos\beta \cdot \sin\lambda_e,$$
$$\sin\alpha = \frac{-\sin\varepsilon \cdot \sin\beta + \cos\varepsilon \cdot \cos\beta \cdot \sin\lambda_e}{\cos\delta},$$
$$\cos\alpha = \frac{\cos\beta \cdot \cos\lambda_e}{\cos\delta}.$$

Transformation of Equatorial to Galactic Coordinates: $\delta, \alpha \rightarrow b, l$

The coordinates of the galactic pole for the epoch 1950.0 are: $\alpha_g = 192.25°$, $\delta_g = 27.4°$. The transition to another reference epoch is described by Abalakin (1979). According to the new system of galactic coordinates (since 1971), the longitude of the ascending node relative to the direction to the center of the Galaxy is $l_0 = -33.0°$, longitude l being counted from the line directed to the center of the Galaxy, which is in the Sagittarius constellation. The coordinates are transformed by the formulas

$$\sin b = \sin\delta_g \cdot \sin\delta + \cos\delta_g \cdot \cos\delta \cdot \cos(\alpha - \alpha_g),$$
$$\sin(l + l_0) = \frac{\cos\delta_g \cdot \sin\delta - \sin\delta_g \cdot \cos\delta \cdot \cos(\alpha - \alpha_g)}{\cos b},$$
$$\cos(l + l_0) = \frac{\cos\delta \cdot \sin(\alpha - \alpha_g)}{\cos b}.$$

Transformation of Galactic to Equatorial Coordinates: $b, l \rightarrow \delta, \alpha$

$$\sin\delta = \sin\delta_g \cdot \sin b + \cos\delta_g \cdot \cos b \cdot \sin(l + l_0),$$
$$\sin(\alpha - \alpha_g) = \frac{\cos b \cdot \cos(l + l_0)}{\cos\delta},$$

$$\cos(\alpha - \alpha_g) = \frac{\cos\delta_g \cdot \sin b - \sin\delta_g \cdot \cos b \cdot \sin(l+l_0)}{\cos\delta}.$$

1.4.1.2 Distribution of the Extra-Atmospheric Radiation Components over the Sky

The spatial distribution of the sky brightness for the extra-atmospheric components is convenient to represent in terms of the Aitov–Hammer–Soloviev projection (Soloviev 1969), such that the area of a sphere is mapped on the area of an ellipse, i.e.,

$$\pi ab = 4\pi R^2,$$

where a and b are the semiaxes of the ellipse and R is the radius of the sphere. If we use the semiaxis ratio (any number) $h = a/b$ as a parameter, then

$$a = 2\sqrt{2} \cdot R, \qquad b = \frac{2}{\sqrt{h}} \cdot R.$$

The value $h = 2$, which corresponds to an equal-area sphere, is generally used, though other values are also sometimes applied. For $h = 2$ the above formulas become

$$a = 2\sqrt{2} \cdot R, \qquad b = \sqrt{2} \cdot R.$$

For a chosen system of spherical coordinates – geographic (φ, λ), equatorial (α, δ), ecliptic (λ, β), or galactic (l, b) – the rectangular coordinates x, y of a network in which the x-axis corresponds to longitude and the y-axis to latitude are calculated by the formulas

$$x = -\frac{2\sqrt{h} \cdot R \cdot \cos\varphi \cdot \sin\frac{\lambda}{2}}{\sqrt{1 + \cos\varphi \cdot \left|\cos\frac{\lambda}{2}\right|}} \cdot \frac{\sin\lambda}{|\sin\lambda|}, \qquad y = \frac{2R \cdot \sin\varphi}{\sqrt{h} \cdot \sqrt{1 + \cos\varphi \cdot \left|\cos\frac{\lambda}{2}\right|}}.$$

For $h = 2$ we have

$\varphi = 0°$ $\qquad\qquad\qquad \lambda = 0°$
$\lambda = 0°, x = 0$ $\qquad\qquad \varphi = 0°, y = 0$
$\lambda = 90°, x = 1.53$
$\lambda = 180°, x = 2.83 \qquad \varphi = 90°, y = 1.41$
$\lambda = 270°, x = -1.53$
$\lambda = 360°, x = 0 \qquad\qquad \varphi = -90°, y = -1.41$

1.4 Space–Time Conditions for Detecting Radiation

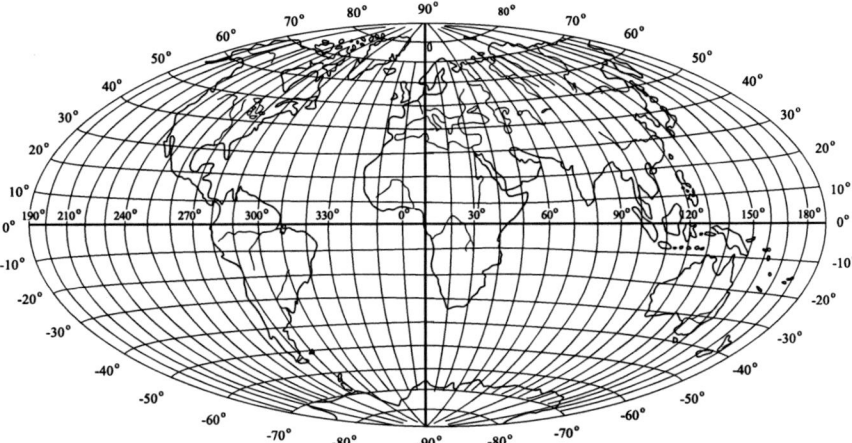

Fig. 1.22 Scale network of the Earth surface in geographic coordinates (Aitov–Hammer–Soloviev projection) (Soloviev 1969)

In Fig. 1.22, to ensure that the territory of Russia fall within the right part of the map, the axis of longitudes is shifted by $\Delta\lambda = 10°$. Therefore, in the above formulas it is necessary to use $\lambda - \Delta\lambda$ instead of λ. The scale network for the geographic coordinate system is shown in Fig. 1.22.

For the equatorial (α, δ), ecliptic (λ, β), and galactic (l, b) coordinate systems, the distribution of the glow characteristics is considered with the observer residing within a spherical system. Therefore, according to the adopted direction of counting-off of coordinates, the minus sign should be introduced in the formula for the coordinate x.

The scale network for the ecliptic coordinate system is shown in Fig. 1.23.

1.4.1.3 Refraction Effect

During the propagation of light through the atmosphere the light beam is bent; therefore, the visible zenith angle ζ of the luminary is smaller than its true value ζ_{true}. The difference between these values is called an astronomical refraction angle r (Alekseev et al. 1983; Kulikov 1969). For a spherically stratified atmosphere, a rigorous theory gives the equation

$$r = \int_1^{n_0} \frac{\frac{n_0}{n} \cdot R_E \cdot \sin\zeta_0 dn}{\sqrt{n^2 \cdot (R_E + Z)^2 - n_0^2 \cdot R_E^2 \cdot \sin^2\zeta}},$$

where n_0 is the refraction index of air at the sighting point; n is its running value at the altitude Z.

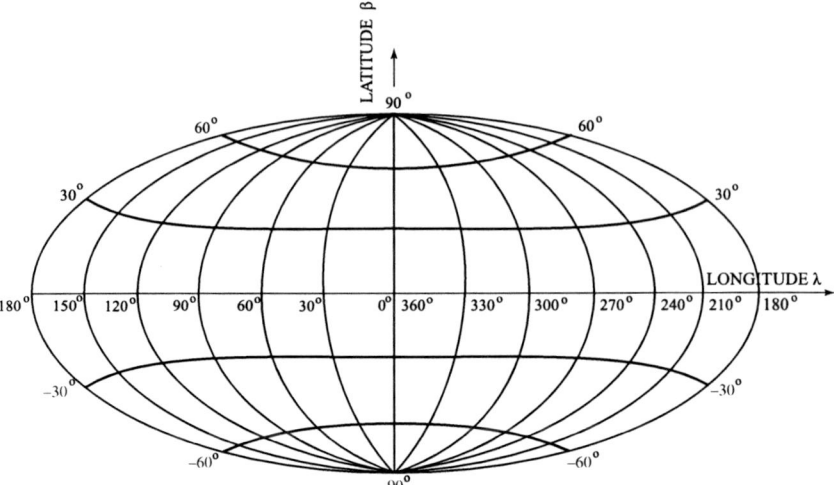

Fig. 1.23 Scale network of the projection of the celestial sphere on the area of an equal-area ellipse (Aitov–Hammer–Soloviev projection) for the ecliptic coordinate system (Soloviev 1969)

The angle between the apparent and the true line of sight directed to an object is called an Earth refraction angle:

$$r = \zeta_0 - \zeta.$$

After Alekseev et al. (1983), we have

$$r = \text{arctg}\left\{\frac{\sqrt{1-\gamma^2\cdot\sin^2\zeta}\cdot\sin\theta + \gamma\cdot\sin\zeta\cdot(1-\cos\theta)}{\sqrt{1-\gamma^2\cdot\sin^2\zeta}\cdot\cos\theta - \gamma\cdot\cos\zeta + \gamma\cdot\sin\zeta\cdot\sin\theta}\right\},$$

where

$$\theta = (R_E + Z_0)n_0\sin\zeta \int_{Z_0}^{Z}\left\{[(R_E+Z')^2n^2 - (R_E+Z_0)^2n_0^2\sin^2\zeta]^{-\frac{1}{2}}\right.$$
$$\left. - [(R_E+Z')^2n_0^2 - (R_E+Z_0)^2n_0^2\sin^2\zeta]^{-\frac{1}{2}}\right\}\cdot(R_E+Z')^{-1}dZ'$$

and

$$\gamma = \frac{R_E - Z_0}{R_E + Z_0}.$$

Here, Z_0 is the altitude of the sighting point, which is counted from the sea level; Z is the altitude of the object; $n = n(Z')$ and $n_0 = n(Z_0)$ are the refraction coefficients of air at the running altitude Z' and at the sighting point, respectively. For a radiation passing through the atmosphere, r is determined by the refraction coefficient of air, being a function of the air density, temperature (and, hence, altitude), and composition, and by the radiation wavelength (Allen 1973; Bates 1984).

In investigating the problems associated with the accuracy to which refraction is taken into account, three groups of measurements are usually considered: low-precision measurements (errors of measuring refraction angles of the order of $1''$), used in the main in optical location and communication; moderate-precision measurements ($0.1''$–$0.01''$ errors) in astronomy and geodesy, and high-precision measurements ($0.001''$–$0.0001''$ errors) in astrometry, space geodesy, and navigation.

Analysis of the refraction formulas shows that to estimate a refraction coefficient even for angles $\zeta \sim 90°$ (accurate to $0.01''$), it is necessary to have data on the refraction coefficient for altitudes below 100 (km). Therefore, to estimate the refraction in observations at high altitudes, $Z_0 \geq 100$ (km), more convenient formulas can be used that, even for near-ground conditions, would provide an accuracy no worse than $0.01''$ for $\zeta \leq 70°$, such as the Cassini formula, which was derived on the assumption of a homogeneous atmosphere:

$$r = \arcsin\left[\frac{R_E \cdot \bar{n} \cdot \sin\zeta}{R_E + \overline{H}}\right] - \arcsin\left[\frac{R_E \cdot \sin\zeta}{R_E + \overline{\overline{H}}}\right].$$

Here $\bar{n}$ and $\overline{H}$ are, respectively, the refraction coefficient and the atmospheric scale height.

The same accuracy is provided by the well-known approximate formula (Kulikov 1969; Allen 1973; Alekseev et al. 1983)

$$r = \beta \cdot \text{tg}\zeta,$$

where β is the refraction coefficient ($60.25''$ for $\zeta < 70°$) at normal pressure and temperature. For the near-ground conditions, we have

$$\beta = 60.25'' \cdot \frac{N(Z) \cdot \overline{H}(Z)}{2.15 \cdot 10^{25}} \cdot \frac{273}{T(Z)},$$

where $N(Z)$ and $\overline{H}(Z)$ are the concentration and the atmospheric scale height at the altitude Z. The value of the refraction coefficient for the near-ground conditions is discussed elsewhere (Kolchinsky 1984).

For the conditions of observation of spatial characteristics of emissions at altitudes of about 100 (km) and at greater zenith angles, it is possible to estimate the effect of refraction on the change of the distance along the emission layer. As follows from the above formulas,

$$\text{tg}\psi \approx \frac{Z}{R_E} \cdot \text{tg}\zeta,$$

whence the displacement of the observation point along the emission layer is

$$L = \Delta\psi \cdot R_E \approx Z \cdot \frac{\cos^2\psi}{\cos^2\zeta} \cdot \Delta\zeta = r \cdot Z \cdot \frac{1}{k^2} \cdot \left(\sqrt{k^2 - \sin^2\zeta} + \sin\zeta \cdot tg\zeta\right)^2.$$

For the altitude $Z \sim 100$ (km), we have $L \sim 0.7$ (km) for the zenith angle $\zeta \sim 70°$ and $L \sim 5.5$ (km) for $\zeta \sim 80°$.

Thus, at the angles $\zeta < 70°$, refraction has practically no effect on the mentioned measurements.

1.4.2 Horizontal Coordinates of the Equator, Ecliptic, and Galactic Equator

In preparing observations and in constructing the isophots of distributions of extra-atmospheric radiation intensity over the sky, it is of interest to know the position of the equator, ecliptic, and galactic equator (Milky Way) on the sky. The dependence of the zenith angle ζ on azimuth A for a point on the equator, on the ecliptic (Zodiacal Light), and on the galactic equator (Milky Way) for a given sidereal time S and a given geographic latitude ϕ are described by the formulas

$$\cos\eta = \sin\phi \cdot \cos\mu - \cos\phi \cdot \sin\mu \cdot \sin(S+\Delta)$$

$$\sin\gamma = \frac{\cos\phi \cdot \cos(S+\Delta)}{\sin\eta}; \quad \cos\gamma = \frac{\sin\mu \cdot \sin\phi + \cos\mu \cdot \cos\phi \cdot \sin(S+\Delta)}{\sin\eta}$$

$$\cos\xi = \sin(S+\Delta) \cdot \cos\gamma + \cos(S+\Delta) \cdot \sin\gamma \cdot \cos\epsilon$$

$$\sin\xi = \frac{\sin\mu \cdot \cos(A+\xi)}{\sin\eta}$$

$$\cos\zeta = \frac{\sin\eta \cdot \cos(A+\xi)}{\sqrt{1 - \sin^2(A+\xi) \cdot \sin^2\eta}}$$

where for the equator $\mu = 0$, $\Delta = 0$; for the ecliptic $\mu = \epsilon$, $\Delta = 0$; and for the galactic equator $\mu = 90 - \delta_g = 62.6°$, $\Delta = 270 - \alpha_g = 77.75°$. Here α_g and δ_g are the equatorial coordinates of the galactic pole.

1.4.3 Conditions of Twilight Measurements

The basic feature of the measurements of characteristics of the airglow of the upper atmosphere is that the shadow of the Earth substantially reduces the influence of the scattered light of the lower atmospheric layers, making possible measurements of the radiation of the upper layers lying above the boundary of the shadow. This

1.4 Space–Time Conditions for Detecting Radiation

Fig. 1.24 Geometry of twilight observations from the Earth surface

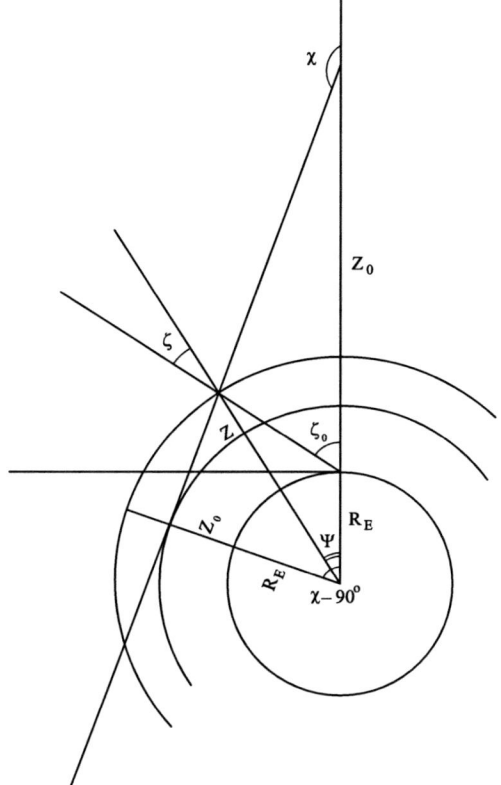

simultaneously enables one to obtain data on the altitude distribution of the radiation parameters. The typical geometry of twilight measurements is shown in Fig. 1.24.

For this case, problems arise which are associated with the propagation of solar radiation near the Earth surface, since the optical thickness is determined by the exponential variation of the density of the atmosphere with distance from the surface. For a given solar zenith angle χ, the decrease in radiation intensity along the Sun rays is determined by the optical thickness of the atmosphere:

$$I = I_0 \cdot \exp[-\tau \cdot \mathrm{Chp}\chi] \ .$$

Here τ is the optical thickness of the atmospheric layer for the vertical direction of sighting, $\mathrm{Chp}\chi$ is the Chapman function (Smith and Smith 1972; Rishbeth and Garriott 1969; Ivanov-Kholodny and Mikhailov 1980). For the solar zenith angles $\chi \leq 70°$, we have $\mathrm{Chp}\chi \approx \sec\chi$. For $\chi > 70°$, it is necessary to use an approximation of the form

$$\mathrm{Chp}(\chi) = \sqrt{\frac{\pi \cdot x \cdot \sin\chi}{2}} \left[1 \pm \mathrm{erf}\left(\sqrt{\frac{x}{2}}\cos\chi\right)\right] \cdot \exp\left(\sqrt{\frac{x}{2}}\cos\chi\right) ,$$

where $x = \frac{R_E+Z}{H}$, R_E is the Earth's radius, Z is the altitude over the Earth surface, and H is the atmospheric scale height. The "plus" and "minus" signs correspond to $\chi \geq 90°$ and $\chi \leq 90°$, respectively. The solar zenith angle χ is determined by the formulas of Sect. 1.4.1. Figure 1.25 gives examples of the χ dependence of the function Chpχ (Rishbeth and Garriott 1969).

Another problem with twilight measurements which affects the accuracy of the calculated altitude distribution of the emission intensity arises in determining the effective screen height Z_{scr}^{eff} of the ground air that is on the way of solar radiation in recording some emission, for example, of metals which has occurred in some layer of the upper atmosphere. Physically, Z_{scr}^{eff} determines the center of gravity of the emission layer that corresponds to a point on the line of sight determined by the maximum value of the product of the transmission function T(Z) by the concentration of radiating atoms, A,

$$\left[A \left(Z + Z_{scr} \cdot \frac{\cos \zeta}{\sin(\chi - \zeta_0)} \right) \right] .$$

According to Fig. 1.24, ζ_0 is the zenith angle of sight, ζ is the zenith angle of the line of sight relative to the emission layer, and Z_{scr} is the running value of the screen height. This problem was considered in detail by Toroshelidze (1968, 1970, 1972, 1991).

The main feature of the account of the screening layer is the necessity to consider a gradual decrease in transmittance of the ground atmospheric layer rather than the assumption of its sharp edge. In this case, the intensity of the radiation passed through an elementary layer is given by

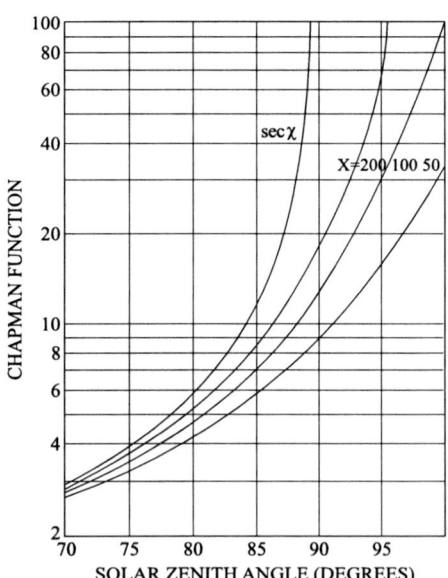

Fig. 1.25 Function Chp(χ) for three values of $x = (R_E + Z)/H$ and secχ (for comparison) versus angle χ

1.4 Space–Time Conditions for Detecting Radiation

$$dI = g \cdot T(Z) \cdot [A(Z)] \cdot \sec\zeta \cdot dZ,$$

where $T(Z)$ is the transmission function, $[A(Z)]$ is the concentration of atoms responsible for the emission under consideration which depends on altitude Z. The emission intensity is

$$I = g \cdot \int_0^\infty T(Z) \cdot [A(Z)] \cdot \sec\zeta \cdot dZ.$$

The optical thickness of the atmosphere along the ray that passes at a tangent over the Earth surface at the distance Z_{scr},

$$Z_{scr} = (Z + R_E) \cdot \sin(\chi + \zeta - \zeta_0) - R_E,$$

is determined by the expression

$$\tau = \sigma \cdot \int_{-\infty}^S [A(Z)] \cdot ds.$$

Since $S^2 = (Z + R_E)^2 - (Z_{scr} + R_E)^2$ and $Z/R_E \sim 0.05$, the above expression becomes

$$\tau = \sigma \cdot \int_{-\infty}^S \frac{[A(Z)] \cdot (Z + R_E) \cdot dZ}{\sqrt{(Z + R_E)^2 - (Z_{scr} + R_E)^2}} = \sigma \cdot \left[\int_{Z_{scr}}^\infty \frac{[A(Z)] \cdot dZ}{\sqrt{Z - Z_{scr}}} + \int_{Z_{scr}}^S \frac{[A(Z)] \cdot dZ}{\sqrt{Z - Z_{scr}}} \right].$$

The altitude distribution of the concentration of radiation-absorbing species is determined by the relevant spectral region. For the attenuation of light due to scattering we have

$$[A(Z)] = [A]_0 \cdot \exp\left(-\frac{Z}{H}\right).$$

The ray length is determined as

$$S = -(Z + R_E) \cdot \cos(\chi + \zeta - \zeta_0).$$

The relation between the zenith angle of sight, ζ_0, and the angle of the line of sight relative to the emission layer, ζ, is

$$\sin\zeta = \frac{1}{1 + \dfrac{Z}{R_E}} \cdot \sin\zeta_0.$$

Since for the conditions of twilight measurements $\chi > 100°$ and $\zeta_0 - \zeta \leq 10°$, we have $\cos(\chi + \zeta - \zeta_0) < 0$.

For $Z_{scr} < Z$, we can assume that $S \sim \infty$, and the expression for the optical thickness becomes

$$\tau = \sigma \cdot [A(Z)]_0 \cdot \sqrt{2 R_E \cdot H} \cdot \exp\left(-\frac{Z_{scr}}{H}\right) \cdot \int_0^\infty e^{-t} \cdot \frac{dt}{\sqrt{t}}.$$

Based on the data reported by Prudnikov et al. (1981), we have

$$\tau = \sigma \cdot [A(Z)]_0 \cdot \sqrt{2\pi \cdot R_E \cdot H} \cdot \exp\left(-\frac{Z_{scr}}{H}\right).$$

The transmission function is

$$T(Z_{scr}) = \exp(-\tau).$$

Thus,

$$I = g \cdot \int_0^\infty \left[A\left(Z + Z_{scr} \cdot \frac{\cos\zeta}{\sin(\chi - \zeta_0)}\right)\right] \cdot \sec\zeta \cdot \exp\left\{-\sigma \cdot [A(Z)]_0 \cdot \sqrt{2\pi \cdot R_E \cdot H} \cdot \exp\left[-\frac{Z_{scr}}{H}\right]\right\} dZ.$$

Hence, the effective screen height is determined by the formula

$$Z_{scr}^{eff} = \frac{\cos\zeta}{\sin(\chi - \zeta_0)} \cdot \frac{\int_0^\infty Z_{scr} \cdot \left[A\left(Z + Z_{scr} \cdot \frac{\cos\zeta}{\sin(\chi - \zeta_0)}\right)\right] \cdot \sec\zeta \cdot \exp\left\{-\sigma \cdot [A]_0 \cdot \sqrt{2\pi \cdot R_E \cdot H} \cdot \exp\left[-\frac{Z_{scr}}{H}\right]\right\} dZ}{\int_0^\infty \left[A\left(Z + Z_{scr} \cdot \frac{\cos\zeta}{\sin(\chi - \zeta_0)}\right)\right] \cdot \sec\zeta \cdot \exp\left\{-\sigma \cdot [A]_0 \cdot \sqrt{2\pi \cdot R_E \cdot H} \cdot \exp\left[-\frac{Z_{scr}}{H}\right]\right\} dZ}$$

The above results were used (Toroshelidze 1968, 1970, 1972, 1991) to analyze the vertical distribution of radiating sodium atoms. Based on the inferences from this analysis, the variations of the sodium layer altitude under the action of lunar tides and the coefficient of vertical turbulent diffusion, K_T, have been estimated. The value of K_T increases with altitude in the range 80–130 (km); in summertime, it increases several fold in comparison with the winter period (Toroshelidze 1991).

1.4.4 Limb-Directed Measurements

Satellite and rocket observations of the airglow of the upper atmosphere provide the possibility to perform measurements along the emission layer when it is visible as a luminous limb over the horizon. The vertical distribution of the airglow intensity inside the emission layer can be determined only with the help of rockets. However, such measurements are incidental soundings, allowing one to obtain a random sample of a great number of situations. Limb measurements with satellite-based instruments can be carried out routinely to study systematic variations of the airglow characteristics. However, this way of measuring demands the solution of an inverse problem of calculating the altitude distribution of the emission rate based on the determination of the intensity variations across the limb. This in fact implies that the measurements should be performed along the emission layer possessing spatial curvature. The geometry of these measurements is shown in Fig. 1.26.

Since the measurements are carried out above the emission layer, in general, some portion of the radiation under study can be in the illuminated region of the atmosphere and the other in the dark region. The radiation intensity can be represented as

$$I = \int_{-\infty}^{0} Q(Z) \cdot dS + \int_{0}^{\infty} Q(Z) \cdot dS.$$

Since $S^2 = (Z+R_E)^2 - (Z_0+R_E)^2$, if the atmospheric region under observation is entirely on the night side, it can be assumed with reasonable assurance that the picture is symmetric about the normal to the tangent ray. In this case, we have

$$I(Z_0, Z_m) = 2 \cdot \int_{Z_0}^{\infty} \frac{Q(Z_0, Z_m) \cdot dZ}{\sqrt{1 - \left(\frac{1+\frac{Z_0}{R_E}}{1+\frac{Z}{R_E}}\right)^2}} = \sqrt{2R_E} \cdot \int_{Z_0}^{\infty} \frac{Q(Z_0, Z_m) \cdot dZ}{\sqrt{Z-Z_0}}$$

since $Z/R_E \sim 0.05$. As can be seen in Fig. 1.26, Z_m is the maximum altitude, Z_0 is the distance between the tangent along which the measurements are performed and the Earth surface, Z is the running altitude along the tangent, Q is the emission rate at the altitude Z_0, and H is a parameter which determines the altitude distribution of the emission rate.

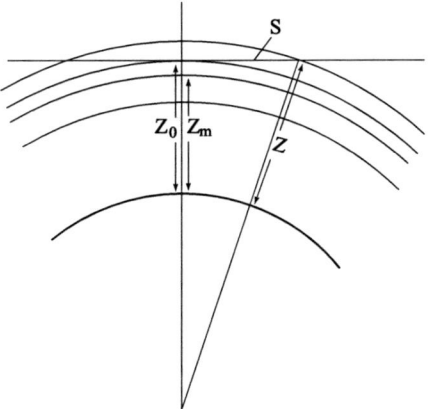

Fig. 1.26 Schematic of the observation of an emission layer along a tangent to the Earth surface

The equation obtained is an Abel equation having an explicit solution (Whittaker and Watson 1927) which can be represented as

$$Q(Z, Z_m) = \frac{1}{\sqrt{2R_E}} \cdot \frac{d}{dZ} \int_Z^\infty \frac{I(Z_0, Z_m)}{\sqrt{Z_0 - Z}} dZ_0 \,.$$

This equation can be solved with the help of numerical integration by calculating the derivative of the integral using a routine procedure (Korn and Korn 1961) and averaging over the calculated values.

However, under the conditions of the upper atmosphere, the thickness of emission layers in the region of the mesopause and lower thermosphere is comparatively small, ~ 10 (km). For the 630-nm emission of atomic oxygen the layer thickness is about 30–50 (km) at the altitudes of the ionospheric F2 region 300 (km), and for the 1083.0-nm emission of helium it is about 300–500 (km) at altitudes of 600–1000 (km). Analysis of long-term measurements has revealed the character of the altitude distributions of the emission rate for various emissions of the upper atmosphere. These distributions can be described by approximate analytic relations, which, when used in integral equations, enable an analytic calculation of the describing parameters. Thereby random fluctuations can be excluded from the distributions, which can be obtained by numerically solving the integral equations.

Rocket measurements have revealed that for many emissions the altitude distributions of the emission rate Q(Z) are well representable by asymmetric Gaussian distributions (Semenov and Shefov 1996, 1997a,b,c,d). In this case, the starting relations can be represented as (Shefov 1978)

$$Q(Z_0, Z_m) = \frac{I_0}{\sqrt{\pi} H} f(Z_0, Z_m) \,,$$

where I_0 is the emission intensity at zenith and H is the parameter of the Gaussian distribution.

If we use a Gaussian distribution in which the thickness of the layer at a half-maximum intensity is W, its upper portion above the altitude of the maximum is $P \cdot W$, and the lower portion is $(1-P) \cdot W$ (i.e., the parameter P specifies the asymmetry of the profile), it is easy to see that the area of the upper portion of the altitude distribution profile is given by $\frac{1}{2}\sqrt{\frac{\pi}{\log_e 2}} \cdot P \cdot W$, that of the lower portion by $\frac{1}{2}\sqrt{\frac{\pi}{\log_e 2}} \cdot (1-P) \cdot W$, and that of the curve as a whole by $\frac{1}{2}\sqrt{\frac{\pi}{\log_e 2}} \cdot W$.

Therefore, we have

$$Q(Z_0, Z_m) = 2\sqrt{\frac{\log_e 2}{\pi}} \cdot \frac{I_0}{W} \cdot f(Z_0, Z_m) \,,$$

and

$$f\uparrow = \exp\left(-\frac{\log_e 2 (Z_0 - Z_m)^2}{P^2 W^2}\right) \qquad f\downarrow = \exp\left(-\frac{\log_e 2 (Z_0 - Z_m)^2}{(1-P)^2 W^2}\right) \,.$$

1.4 Space–Time Conditions for Detecting Radiation

In this case, the measured radiation intensity along the limb is determined by the above two functions:

$$\psi(Z_0, Z_m) = \sqrt{\frac{2R_E}{(1-P)W_1}} \cdot \int_{Z_0}^{Z_m} \frac{\exp\left(-\frac{(Z-Z_m)^2}{(1-P)^2 W_1^2}\right) dZ}{\sqrt{\frac{Z-Z_0}{(1-P)W_1}}}$$

$$+ \sqrt{\frac{2R_E}{PW_1}} \cdot \int_{Z_m}^{\infty} \frac{\exp\left(-\frac{(Z-Z_m)^2}{P^2 W_1^2}\right) dZ}{\sqrt{\frac{Z-Z_0}{PW_1}}},$$

or $\quad \psi = \psi\downarrow + \psi\uparrow$.

Here, $W_1 = W/\sqrt{\log_e 2}$.

The arrows refer to the parameters for the lower and the upper portions of the layer, respectively. Let us introduce the following designations:

$$t_1 = \frac{Z_m - Z_0}{(1-P)W_1} \geq 0; \quad t_2 = \frac{Z_m - Z_0}{PW_1} \geq 0; \quad t_3 = \frac{Z_m - Z_0}{PW_1} \leq 0.$$

Then for the layer upper portion we have

$$\psi\uparrow = \sqrt{2R_E \cdot PW_1} \cdot \exp(-t_3^2) \cdot \int_0^{\infty} u^{-\frac{1}{2}} \cdot \exp(-u^2 + 2t_3 u) du$$

and for the lower portion

$$\psi\downarrow = \sqrt{2R_E \cdot (1-P)W_1} \cdot \exp(-t_1^2) \cdot \int_0^{t_1} u^{-\frac{1}{2}} \cdot \exp(-u^2 + 2t_1 u) du +$$

$$+ \sqrt{2R_E \cdot PW_1} \cdot \exp(-t_2^2) \cdot \int_{t_2}^{\infty} u^{-\frac{1}{2}} \cdot \exp(-u^2 + 2t_2 u) du.$$

Hence, the emission intensity along the tangent rays passing at an altitude Z_0 from the Earth surface is given for the upper portion by

$$I(Z_m, Z_0)\uparrow = I_0 \cdot \sqrt{\frac{\sqrt{2R_E}}{PW_1}} \cdot \exp\left(-\frac{t_3^2}{2}\right) \cdot D_{-\frac{1}{2}}(-\sqrt{2} \cdot t_3),$$

where D is a parabolic cylinder function (Weber function), and for the lower portion of the layer

$$I(Z_m, Z_0) \downarrow = I_0 \cdot \sqrt{\frac{\sqrt{2}R_E}{PW_1}} \cdot \exp\left(-\frac{t_2^2}{2}\right) \cdot D_{-\frac{1}{2}}(-\sqrt{2} \cdot t_2) +$$

$$+ 2I_0 \cdot \sqrt{\frac{2R_E}{PW_1}} \cdot t_2 \cdot \left[\sqrt{\left(\frac{P}{1-P}\right)^3} \cdot {}_2F_2\left(\frac{1}{2}; 1; \frac{3}{4}; \frac{1}{4}; -\left(\frac{P}{1-P}\right)^2 \cdot t_2^2\right)\right.$$

$$\left. - {}_2F_2\left(\frac{1}{2}; 1; \frac{3}{4}; \frac{1}{4}; -t_2^2\right)\right],$$

where ${}_2F_2$ is a generalized hypergeometric series. Following Bateman and Erdélyi (1953), we have

$${}_2F_2\left(\frac{1}{2}; 1; \frac{3}{4}; \frac{1}{4}; -t^2\right) = 1 - \frac{t^2}{2} \cdot \Phi\left(\frac{3}{2}; 1; -t^2\right) = 1 - \frac{t^2}{2} \cdot e^{-t^2} \cdot \Phi\left(-\frac{1}{2}; 1; t^2\right)$$

$$= 1 - \frac{t}{2} \cdot \exp\left(-\frac{t^2}{2}\right) \cdot M_{1,0}(t^2).$$

Here Φ is a confluent hypergeometric function of the first kind and $M_{1,0}$ is a Whittaker function. The function Φ has the form

$$\Phi\left(-\frac{1}{2}; 1; x\right) = 1 + \sum_{n=1}^{\infty} \frac{(2n-3)!!}{2^n \cdot (n!)^2} \cdot x^n = 1 - \sum_{n=1}^{\infty} \frac{(2n)!}{2^{2n} \cdot (n!)^3 \cdot (2n-1)} \cdot x^n$$

$$\approx 1 - \sum_{n=1}^{\infty} \frac{x^n}{\sqrt{\pi \cdot n} \cdot (2n-1) \cdot n!}.$$

The relations obtained can be used in combination with tables (Miller 1955; Slater 1960; Korobochkin and Filippova 1965; Karpov and Chistova 1968) or approximations (Abramowitz and Stegun 1964).

Thus, we have

$$I(Z_m, Z_0) \downarrow = I_0 \cdot \sqrt{\frac{\sqrt{2}R_E}{PW_1}} \cdot \exp\left(-\frac{t_2^2}{2}\right) \cdot D_{-\frac{1}{2}}(-\sqrt{2} \cdot t_2) +$$

$$+ 2I_0 \cdot \sqrt{\frac{2R_E}{PW_1}} \cdot t_2 \cdot \left\{\sqrt{\mu^3}\left[1 - \frac{1}{2}\mu^2 t_2^2 \cdot \exp(-\mu^2 t_2^2) \cdot \Phi\left(-\frac{1}{2}; 1; \mu^2 t_2^2\right)\right]\right.$$

$$\left. - \left[1 - \frac{1}{2}t_2^2 \cdot \exp(-t_2^2) \cdot \Phi\left(-\frac{1}{2}; 1; t_2^2\right)\right]\right\},$$

1.4 Space–Time Conditions for Detecting Radiation

whence

$$I(Z_m, Z_0) \downarrow = I_0 \cdot \sqrt{\frac{\sqrt{2}R_E}{PW_1}} \cdot \exp\left(-\frac{t_2^2}{2}\right) \cdot D_{-\frac{1}{2}}(-\sqrt{2} \cdot t_2) +$$

$$+ I_0 \cdot \sqrt{\frac{2R_E}{PW_1}} \cdot t_2 \left\{ t_2^2 \left[\exp(-t_2^2) \cdot \Phi\left(-\frac{1}{2}; 1; t_2^2\right) - \mu^{\frac{7}{2}} \cdot \exp(-\mu^2 t_2^2) \cdot \right.\right.$$

$$\left.\left. \Phi\left(-\frac{1}{2}; 1; \mu^2 t_2^2\right)\right] + 2\left(\mu^{\frac{3}{2}} - 1\right)\right\}.$$

Here, $\mu = \frac{P}{1-P}$. Numerical estimates show that the second term makes a small fraction of the first one.

It can be seen that if the altitude distribution of the emission rate in an emission layer is symmetric, the second term of the above expression vanishes, and the intensity distribution in the emission layer observed in the limb is described by the unified expression

$$I(Z_0, Z_m) = I_0 \cdot \sqrt{\frac{2R_E\sqrt{2\log_e 2}}{W}} \cdot \exp\left(-\frac{t^2}{2}\right) \cdot D_{-\frac{1}{2}}\left(-\sqrt{2} \cdot t\right),$$

where

$$t = 2\sqrt{\log_e 2} \cdot \frac{Z_m - Z_0}{W},$$

or

$$I(Z_0, Z_m) = I_0 \cdot \frac{122.2}{\sqrt{W}} \cdot \exp\left(-\frac{t^2}{2}\right) \cdot D_{-\frac{1}{2}}\left(-\sqrt{2} \cdot t\right).$$

Figure 1.27 gives examples of the altitude distribution $I(Z_0, Z_m)$ for three symmetric profiles. The emission intensity at zenith is taken the same for all values of W.

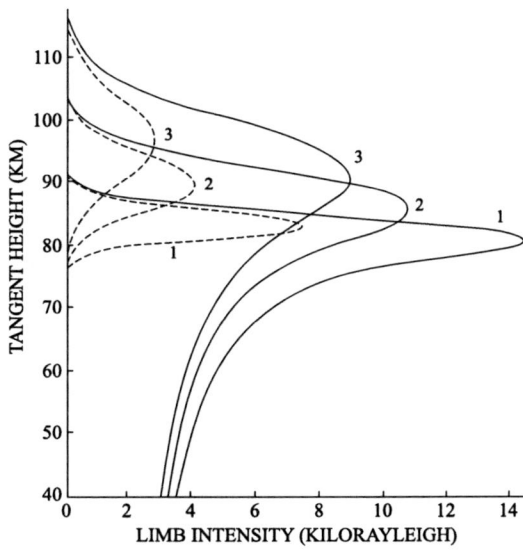

Fig. 1.27 Altitude distributions of the intensity in an emission layer for the observation along a tangent to the Earth surface (*solid lines*) and the corresponding altitude distributions of the emission rate (*dashed lines*) for the same intensity at zenith for various layer thicknesses W. W = 6 (km) (1), 11 (km) (2), and 16 (km) (3) (Shefov 1978)

As can be seen from Fig. 1.27, the altitude Z_M of the maximum of the observed intensity distribution is lower than the altitude Z_m of the emission layer maximum. As follows from the above formulas, Z_M is consistent with the requirement

$$D_{-\frac{1}{2}}\left(-\sqrt{2} \cdot t\right) = 0.$$

Thus, following Karpov and Chistova (1968), we have $t = 0.541$ and

$$Z_m = Z_M + 0.325 \cdot W.$$

The values $0.5 \cdot I(Z_0, Z_m)$ correspond to $t_{up} = 0.614$ and $t_{low} = 2.863$. Therefore, the width of the distribution observed is determined by the relation

$$W_V = \frac{W}{1.665}\left(t_{up} + t_{low}\right),$$

whence

$$W = 0.48 \cdot W_V \text{ and } Z_m = Z_M + 0.16 \cdot W_V.$$

The relations obtained allow one to estimate the length L of the emission layer under investigation at the maximum of the observed altitude distribution of the intensity I, which is given by

$$I = I_0 \cdot \frac{54}{\sqrt{\frac{W}{10}}}.$$

The length is determined by the formula

$$L = \sqrt{8 R_E P W} = 714 \cdot \sqrt{\frac{PW}{10}} \text{ (km)}.$$

This suggests that in limb measurements the irregularities of the emission layer are inevitably smoothed. These irregularities can be due to the spotty structure of the spatial distribution of the glow resulting from the propagation of waves of various scales and due to longitudinal and latitude intensity variations, which in turn are accompanied by variations of the emission layer altitude and parameters.

For the determination of the parameters of an emission layer from measurements it is reasonable to construct a set of calculated distributions for the emission under consideration.

1.4.5 Time Coordinates

The intensity of the airglow of the upper atmosphere depends not only on the geographic coordinates φ and λ, but also on the local solar τ and on the season of year, which can be defined by the number of the year day, t_d, which enters in the modern

1.4 Space–Time Conditions for Detecting Radiation

models of the upper atmosphere. In this connection, it is necessary to know the coordinates of the Sun: the solar zenith angle $\chi_\odot$; the azimuth $A_\odot$, which is counted clockwise from the south point from $0°$ to $360°$; the declination $\delta_\odot$; and the hour angle $t_\odot$, which depends on τ, t_d, and current year number, YYYY. Now there is a computer code (WinEphem 2002) which calculates the coordinates of the Sun for given geographic coordinates and points in time. The quantities used in the code are clarified below.

The local solar time τ is related to various time systems (Abalakin et al. 1976; Abalakin 1979; Bronshten et al. 1981; Kononovich and Moroz 2001; Kulikovsky 2002) as follows:

$$\tau = \tau_{UT} + \lambda/15 ,$$

where τ_{UT} is the universal time and λ is the geographic longitude (degs);

$$\tau = \tau_D - m - n - 1 + \lambda/15 ,$$

where τ_D is the decretal time of the nth zone; $m = 0$ for winter and $m = 1$ for summer.

The solar hour angle, $t_\odot$, (countered to the west from the south point) is determined by

$$t_\odot = \tau - 12^h - \eta ,$$

where the equation of time reads

$$\eta = \tau_{\text{mean}} - \tau_{\odot\text{true}} .$$

According to the data of Abalakin et al. (1976) and Bronshten et al. (1981), we have

$$\eta = 0^h.12833 \cdot \sin(\lambda_e + 78°) - 0^h.15833 \cdot \sin 2\lambda_e ,$$

where $\lambda_\odot$ is the celestial (ecliptic) solar longitude.

Thus, for a point on the line of sight, which is set by coordinates φ and λ determined by the coordinates of the observation point $(\varphi_0, \lambda_0, \tau_0)$, the solar hour angle is given by

$$t_\odot = t_{\odot 0} + (\lambda - \lambda_0)/15$$

and the local time by

$$\tau = \tau_0 + (\lambda - \lambda_0)/15 .$$

The position of the Sun in the sky is estimated (in fact, the ecliptic longitude $\lambda_\odot$, since the ecliptic latitude of the Sun (to within $1''$) $\beta_\odot = 0$) by the relations given below. The number of days since January 0.5, 1900 is determined as

$$D = [(YYYY - 1901) \cdot 365.25] + 364.5 + t_d + \frac{\tau}{24} - \frac{\lambda}{360} .$$

Here, the square brackets denote the integer part of the number and YYYY is the year number. The number of Julian centuries is found by the formula

$$T = \frac{D}{36525}; \quad T' = \frac{D - \left(\frac{\tau}{24} - \frac{\lambda}{360}\right)}{36525}.$$

Calculations of the ephemerides of the Sun and planets are described elsewhere (Abalakin et al. 1976; Bronshten et al. 1981). All coordinates are taken for the epoch of January 0.5, 1900.

The ecliptic solar longitude is determined (to within several angular seconds) by the formulas

$$\lambda_\odot = L_\odot + 1°9171 \cdot \sin l' + 0°0200 \cdot \sin 2l' + 0°0003 \cdot \sin 3l',$$
$$l' = 358°4758333 + 35999°04975 \cdot T - 0°00015 \cdot T^2 - 0°00000333 \cdot T^3,$$
$$L_\odot = 279°6966778 + 36000°76893 \cdot T + 0°000302 \cdot T^2.$$

Based on these relations, the equatorial coordinates $\alpha_\odot$ and $\delta_\odot$ are calculated by the formulas

$$\sin \delta_e = \sin \lambda_e \cdot \sin \varepsilon,$$
$$\cos \chi_{e0} = \sin \delta_e \cdot \sin \varphi_0 + \cos \delta_e \cdot \cos \varphi_0 \cdot \cos 15 t_{e0},$$
$$\sin A_{e0} = \frac{\cos \delta_e \cdot \sin 15 t_{e0}}{\sin \chi_{e0}},$$
$$\cos A_{e0} = \frac{-\cos \varphi_0 \cdot \sin \delta_e + \sin \varphi_0 \cdot \cos \delta_e \cdot \cos 15 t_{e0}}{\sin \chi_{e0}},$$

where $t_{\odot 0}$ is expressed in hours. If $0 \leq t_\odot \leq 12^h$, then $0 \leq A_\odot \leq 180°$; if $12^h \leq t_\odot \leq 24^h$, then $180° \leq A_\odot \leq 360°$.

When transforming coordinates on the celestial sphere, one has to use a sidereal time. Following Abalakin et al. (1976) and Bronshten et al. (1981), the sidereal time S (in hours) corresponding to the local time τ and to the ordinal number of day, t_d, of the year YYYY for a point having geographic longitude λ is given by

$$S = (2400.051262 + 2581 \cdot 10^{-8} \cdot T') \cdot T' + 6.6460656 - 24 \cdot (YYYY - 1900)$$
$$+ 0.000182647 \cdot \lambda + 1.002738 \cdot \tau.$$

If S turns out to be over 24 hours, it is necessary to subtract 24 hours from the value obtained by the formula, while if it is negative, 24 hours must be added.

The ordinal days of year are listed in Table 1.3 (Bronshten et al. 1981). If necessary, they can be calculated by the following formulas: If the day of month, d, and the number of month, M, of a given year YYYY are set, we have for $M = 1$ and 2

$$t_d = \left[\frac{1}{2}\left((M-1) \cdot \left(62 + \left[\frac{YYYY}{4}\right] - \left[\frac{YYYY}{4}\right] + 0.75\right)\right)\right] + d$$

and for $M = 3 - 12$

1.4 Space–Time Conditions for Detecting Radiation

Table 1.3 Ordinal days of year (Bronshten et al. 1981)

Day	Month											
	1	2	3	4	5	6	7	8	9	10	11	12
1	1	32	60	91	121	152	182	213	244	274	305	335
2	2	33	61	92	122	153	183	214	245	275	306	336
3	3	34	62	93	123	154	184	215	246	276	307	337
4	4	35	63	94	124	155	185	216	247	277	308	338
5	5	36	64	95	125	156	186	217	248	278	309	339
6	6	37	65	96	126	157	187	218	249	279	310	340
7	7	38	66	97	127	158	188	219	250	280	311	341
8	8	39	67	98	128	159	189	220	251	281	312	342
9	9	40	68	99	129	160	190	221	252	282	313	343
10	10	41	69	100	130	161	191	222	253	283	314	344
11	11	42	70	101	131	162	192	223	254	284	315	345
12	12	43	71	102	132	163	193	224	255	285	316	346
13	13	44	72	103	133	164	194	225	256	286	317	347
14	14	45	73	104	134	165	195	226	257	287	318	348
15	15	46	74	105	135	166	196	227	258	288	319	349
16	16	47	75	106	136	167	197	228	259	289	320	350
17	17	48	76	107	137	168	198	229	260	290	321	351
18	18	49	77	108	138	169	199	230	261	291	322	352
19	19	50	78	109	139	170	200	231	262	292	323	353
20	20	51	79	110	140	171	201	232	263	293	324	354
21	21	52	80	111	141	172	202	233	264	294	325	355
22	22	53	81	112	142	173	203	234	265	295	326	356
23	23	54	82	113	143	174	204	235	266	296	327	357
24	24	55	83	114	144	175	205	236	267	297	328	358
25	25	56	84	115	145	176	206	237	268	298	329	359
26	26	57	85	116	146	177	207	238	269	299	330	360
27	27	58	86	117	147	178	208	239	270	300	331	361
28	28	59	87	118	148	179	209	240	271	301	332	362
29	29	60	88	119	149	180	210	241	272	302	333	363
30	30	–	89	120	150	181	211	242	273	303	334	364
31	31	–	90	–	151	–	212	243	–	304	–	365

Note: For a leap year, after February 29, it is necessary to add a unity to all numbers of the table.

$$t_d = [(M+1) \cdot 30.6] - \left(62 + \left[\frac{YYYY}{4}\right] - \left[\frac{YYYY}{4}\right] + 0.75\right) + d,$$

where the square brackets denote the integer part of the number.

For the processes occurring in the polar zone and related to the magnetospheric phenomena, it is convenient to compare the events on the scale of local geomagnetic time counted from the local geomagnetic midnight and determined by the formula

$$\tau_G = \Lambda + \Delta\tau_G,$$

where Λ is the geomagnetic longitude of the place (degs), which is counted to the east from the meridian that passes through the geographic and geomagnetic (dipole)

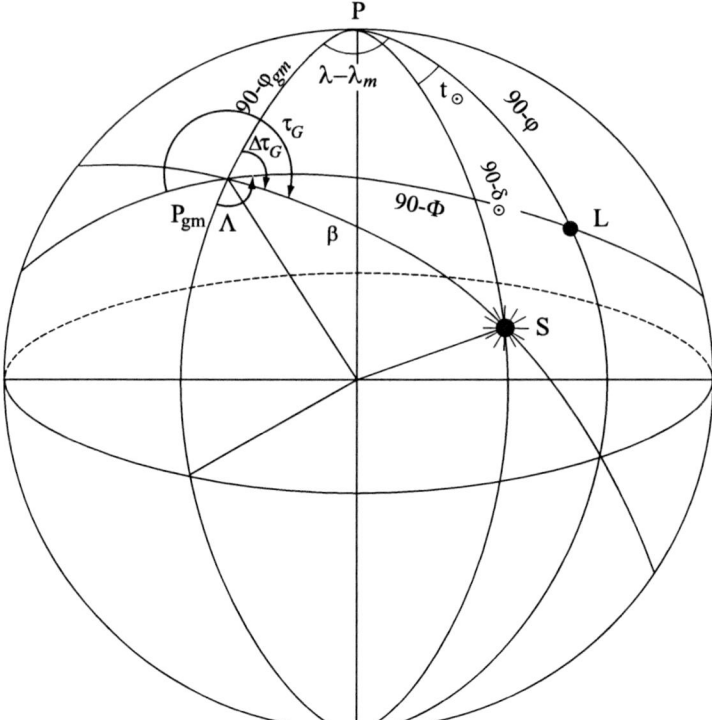

Fig. 1.28 Geometry of the determination of a local geomagnetic time

poles, and $\Delta\tau_G$ is the angle between the geomagnetic meridian that passes through the geomagnetic pole and the given point and the line directed to the Sun (Fig. 1.28). The value of $\Delta\tau_G$ is calculated by the formulas

$$\sin\Delta\tau_G = \frac{\cos\delta_\odot \cdot \sin(\lambda - \lambda_m - 15t_e)}{\sin\beta},$$

$$\cos\Delta\tau_G = \frac{\sin\delta_e \cdot \cos\varphi_m - \cos\delta_e \cdot \sin\varphi_m \cdot \cos(\lambda - \lambda_m - 15t_e)}{\sin\beta}.$$

Here, φ_m is the geographic latitude of the northern geomagnetic (dipole) pole, β is the angle between the geomagnetic pole along the geomagnetic meridian and the line directed to the Sun. Therefore, we have $\beta > 0$ and $0 < \beta < 180°$ at all times.

$$\cos\beta = \sin\varphi_m \cdot \sin\delta_e + \cos\varphi_m \cdot \cos\delta_e \cdot \cos(\lambda - \lambda_m - 15t_e).$$

Within some limits (for middle and low latitudes), the following approximation is possible:

$$\tau_G = \tau + (\lambda_m - \lambda + \Lambda)/15, \text{ hours}.$$

Nevertheless, for the polar region it is necessary to introduce a correction to the geomagnetic coordinates determined by the excentric dipole. Therefore, in view of the above formulas, the corrected geomagnetic time is $\tau_G' = \tau_G + \delta\tau_G'$, where $\delta\tau_G'$ is determined by the formulas (Simonov 1963)

$$\cos\delta\tau_G' = \frac{\cos\Phi - d\cdot\cos(\Lambda - \Lambda_0)}{\sqrt{\cos^2\Phi - 2d\cdot\cos\Phi\cdot\cos(\Lambda - \Lambda_0) + d^2}}$$

$$\sin\delta\tau_G' = \frac{d\cdot\sin(\Lambda - \Lambda_0)}{\sqrt{\cos^2\Phi - 2d\cdot\cos\Phi\cdot\cos(\Lambda - \Lambda_0) + d^2}}.$$

$$\mathrm{tg}\delta\tau_G' = \frac{d\cdot\sin(\Lambda - \Lambda_0)}{\cos\Phi - d\cdot\cos(\Lambda - \Lambda_0)},$$

where

$$d = 5.1775 - 5.587\cdot10^{-3}(\mathrm{YYYY} + t_d/365) + 1.52023\cdot10^{-6}(\mathrm{YYYY} + t_d/365)^2,$$
$$\Lambda_0 = -0.250(\mathrm{YYYY} + t_d/365) + 707.7.$$

For the year 1990 we have $d = 0.07967$ and $\Lambda_0 = 209.9°\mathrm{E}$.

To analyze the tidal influence of the Moon, it is necessary to determine the lunar time τ_o. A consideration of a lunar tide shows that its effect is determined by the tide-generating potential $W(r, \Theta)$, which depends on the angular distance of the point under consideration from the straight line connecting the Earth center and the Moon center, Θ, and on the distance from Earth to Moon, r (Chapman and Lindzen 1970). Moreover, the tide magnitude at a given point of the atmosphere strongly depends on the declination of the Moon δ_o. In the moving coordinates related to the Moon, according to the data of Fig. 1.29, we have

$$\sin\Theta = \sin\varphi\cdot\cos\delta_\mathrm{o} + \cos\varphi\cdot\sin\delta_\mathrm{o}\cdot\cos\tau_\mathrm{o},$$

$$\sin\Lambda = \frac{\cos\varphi}{\cos\Theta}\cdot\sin\tau_\mathrm{o}$$

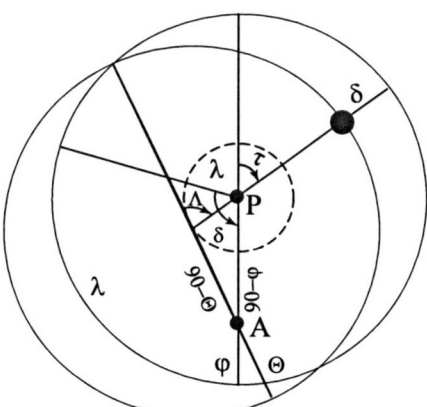

Fig. 1.29 Coordinate systems relative to the Earth and to the atmosphere reduced to the position of the Moon (Chapman and Lindzen 1970)

where φ is the latitude of the observing site; Θ and Λ are, respectively, the latitude and longitude of the measurement point in the tide-deformed atmosphere.

Figures 1.30 and 1.31 show the relations between Θ and φ and between Λ and $\tau_☉$ for various declinations of the Moon, calculated for the stations at Loparskaya (70°N), Yakutsk (64°N), Zvenigorod (58°N), and Abastumani (45°N). The latitudes are indicated for the observed regions of the atmosphere at altitudes of about 90 (km) (Kropotkina and Shefov 1977).

The exact value of lunar time can be obtained by the formulas (Chapman and Lindzen 1970)

$$\tau_☉ = \tau - \xi,$$
$$\xi = -9°26009 + 445267°12165 \cdot T + 0°00168 \cdot T^2,$$
$$\xi = -0^h6173393 + 29684^h47478 \cdot T + 0^h000112 \cdot T^2.$$

Using the data of the Annual Review of Astronomy & Astrophysics, one can determine the lunar time (to within some tenths of an hour) by the formula

$$\tau_☉ = \tau - \tau_{☉lc} + 0.8^h \cdot \lambda/360$$

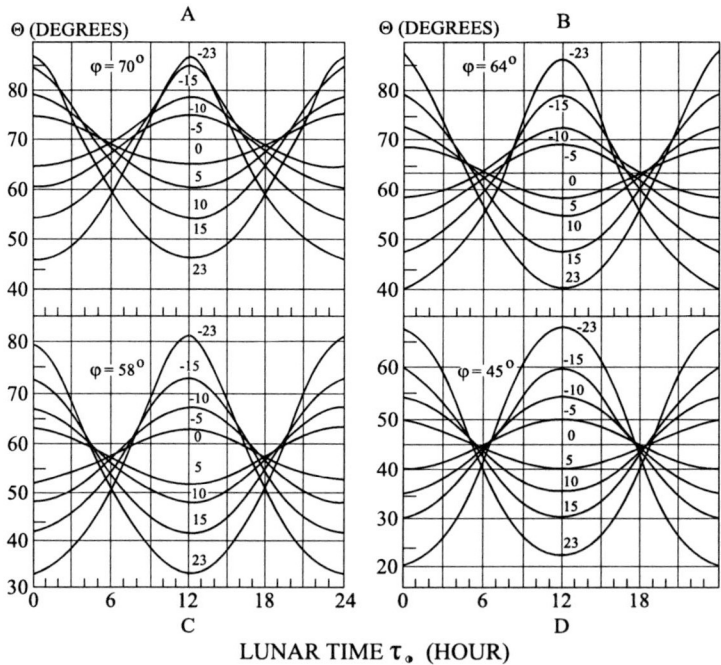

Fig. 1.30 Relations between Θ, $\tau_☉$, and φ for various declinations of the Moon, calculated for Loparskaya (70°N) (**A**), Yakutsk (64°N) (**B**), Zvenigorod (58°N) (**C**), and Abastumani (45°N) (**D**). The latitudes are indicated for the observed regions of the atmosphere at altitudes of about 90 (km) (Kropotkina and Shefov 1977)

1.4 Space–Time Conditions for Detecting Radiation

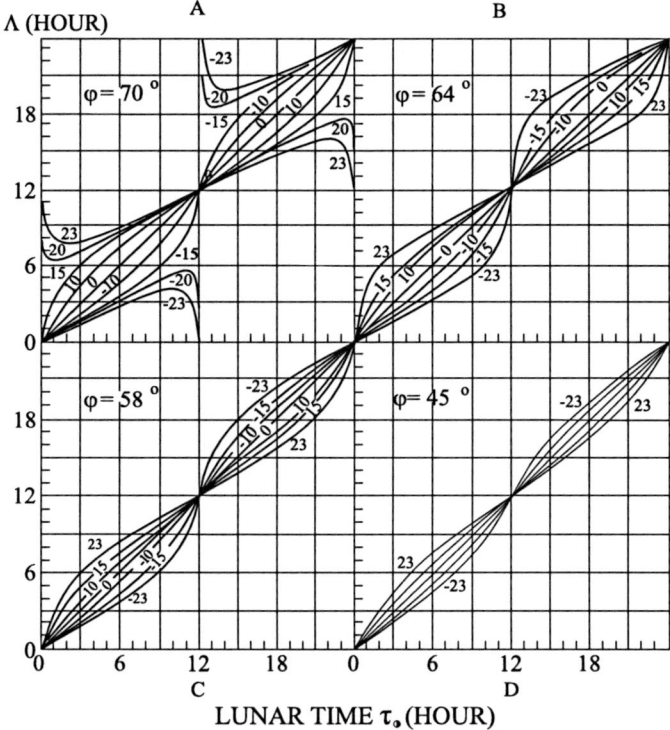

Fig. 1.31 Relations between Λ, $\tau_\mathrm{\odot}$, and φ for various declinations of the Moon, calculated for Loparskaya (70°N) (**A**), Yakutsk (64°N) (**B**), Zvenigorod (58°N) (**C**), and Abastumani (45°N) (**D**). The latitudes are indicated for the observed areas of the atmosphere at altitudes of about 90 (KM) (Kropotkina and Shefov 1977)

or
$$\tau_\mathrm{\odot} = \tau - \tau_{\mathrm{\odot}uc} + 12.4^\mathrm{h} + 0.8^\mathrm{h} \cdot \lambda/360,$$

where $\tau_{\mathrm{\odot}lc}$ is the time of the nearest previous lower culmination of the Moon on the Greenwich meridian and $\tau_{\mathrm{\odot}uc}$ is the time of the upper culmination after $\tau_{\mathrm{\odot}lc}$.

Variations of the atmospheric parameters occur with the period of the Moon's age, i.e., the synodic month equal to 29.53 days. The Moon's age t_L can be estimated to within several tenths of a day by the formula (Chapman and Lindzen 1970)

$$t_L = \left[\left(\frac{t_d}{365} + \mathrm{YYYY} - 1900 \right) \cdot 12.3685 \right] \cdot 29.53 - 1.$$

Here the square brackets denote the fractional part of the number; t_d is the number of day of the year.

1.5 Manifestation of the Variations of the Atmospheric Characteristics in the Properties of the Airglow

The occurrence of the airglow of the upper atmosphere is closely related to its chemical composition, temperature, and neutral density. These parameters determine the penetration depth of solar UV radiation in different spectral regions responsible for photoionization and photodissociation of molecules and atoms. In this connection, there occur sequences of photochemical reactions which provide energy removal from the atmosphere. The reaction products are intermediate active species whose radiation is the evidence of the existence of some mechanism of interaction between the processes involved. A knowledge of this mechanism would give us insight into the atmospheric photochemical processes.

An example of an energetically significant process is the ozone–hydrogen reaction that completes the recombination of atomic oxygen resulting from dissociation of molecular oxygen. Thus, the radiation of a hydroxyl molecule is a peculiar kind of indicator of energy balance at altitudes of 80–100 (km). Owing to the rich structure of the rotational–vibrational bands, the possibility exists of determining the temperature in the emission layer and the variations in its altitude by the vibrational temperature, which is a manifestation of the population density distribution of excited OH molecules over vibrational levels, which, in turn, depends on the concentration of neutral components. Thus, hydroxyl emission serves as a peculiar kind of thermometer of the environment.

The emission of atomic oxygen at 557.7 (nm) is a manifestation of a way of atomic oxygen recombination. However, energetically this emission is a minor sink of energy. Nevertheless, this process governs the altitude distribution of atomic oxygen near the lower thermosphere that, in turn, controls the condition of the atmosphere at the altitudes of the mesopause and lower thermosphere. Therefore, the 557.7-nm emission serves as a good indicator of dynamic processes, such as winds and horizontal diffusion.

The emission of atomic oxygen at 630.0–636.4 (nm) that arises at the altitudes of the F2 ionospheric layer allows one to determine the temperature and also the wind speed and direction in this thermospheric region based on interferometric measurements. Under quiet geomagnetic conditions, this emission results from dissociative recombination by which ionized atomic oxygen goes over in the neutral state. During the periods of geomagnetic disturbances, which show up as low-latitude red lights, the excitation is due to flows of superthermal electrons created by a ring current. This leads to intense warming of the thermosphere and to energy removal by the 630.0-nm emission.

It seems that an important part in these processes is played by the 5.3-μm emission from NO molecules, which corresponds to the (1–0) vibrational transition of the ground state, whose excitation energy is 0.233 (eV). This emission arises in various two-body reactions responsible for the formation of vibrationally excited molecules of nitrogen oxide.

The 1083-nm emission of helium arising at altitudes over 300 (km) is due to the photoionization of the neutral components by 30.4-nm solar radiation that produces photoelectrons of energy about 25 (eV). The altitude distribution of the emission intensity depends on the temperature of the thermosphere.

The 656.3-nm emission of atomic hydrogen is very low energy. However, it is a manifestation of the fluorescence of solar radiation in the Lyman beta line and arises in the main at altitudes over 1000 (km), characterizing the hydrogenous geocorona of the Earth.

1.6 The Heliogeophysical Conditions that Determine the Observed Airglow Characteristics

Heliophysical conditions constitute a set of factors which characterize solar and geomagnetic activities and are determined in the main by solar activity, which, in turn, affects the geomagnetic situation in the upper atmosphere. By geophysical conditions are implied the time–space properties of a geographic site above the Earth surface, i.e., longitude–latitude and time parameters. The idea of "solar activity" implies that the intensity of solar radiation in various spectral regions and its other characteristics do not remain constant. The time scale of these variations covers a range from several seconds to some millions of years. All these variations reflect the processes occurring in the circumplanetary space and, what is especially important, on the Earth. In this case only those types of variations of atmospheric characteristics are considered which are reflected in similar variations of the airglow of the Earth.

1.6.1 Solar Activity and Its Variations During 11-yr Cycles

1.6.1.1 Solar Activity Indices

The most pronounced feature of solar activity is its 11-yr repeatability. Investigations of solar flux variability for almost 250 years have revealed a certain law in the average behavior of solar activity indices, such as Wolf sunspot numbers and, since 1947, $F_{10.7}$ radio-frequency solar flux. Sunspots are regions of the photosphere with reduced temperature, emission intensity, and gas pressure resulting from local enhancement of the magnetic field. A Wolf number W characterizes both the number of spot groups and the number of spots in a group. It is determined by the formula (Vitinsky et al. 1986)

$$W = k(10 \cdot G + N),$$

where k is a coefficient which depends on the method of observation, instrumentation, and features of the observer and characterizes the system of a given observatory relative to the international system; G is the total number of spot groups; and N is the total number of spots in all groups. Routine observations of Wolf numbers have been performed since 1749.

The radio-frequency flux $F_{10.7}$ (expressed in $10^{-22}(W \cdot m^{-2} \cdot Hz^{-1})$ at 2800 (MHz) 10.7 (cm) wavelength) is an index which characterizes the conditions in the solar corona and has strong correlation with short-wave ultraviolet solar radiation. Routine measurements have been performed since 1947. It is of interest that the propagation of perturbations from the photosphere (W index) to the corona ($F_{10.7}$ index) occurs with a delay of $\sim$3 months, which corresponds to a velocity of $\sim$5 (m·s^{-1}) (Apostolov and Letfus 1985), i.e., to a distance of $\sim$45 000 (km) or $\sim$6(%) of the Sun radius.

On the average, there is a correlation between these indices with the correlation coefficient increasing for longer observation periods. For yearly mean values this correlation is given by (Kononovich et al. 2002)

$$F_{10.7} = 66.179 + 5.566 \cdot \left(\frac{W}{10}\right) + 0.441 \cdot \left(\frac{W}{10}\right)^2 - 0.0143 \cdot \left(\frac{W}{10}\right)^3,$$

$$W = -126.711 + 24.683 \cdot \left(\frac{F_{10.7}}{10}\right) - 0.925 \cdot \left(\frac{F_{10.7}}{10}\right)^2 + 0.0189 \cdot \left(\frac{F_{10.7}}{10}\right)^3.$$

Initially, after a minimum of solar flux, there occurs its rather rapid increase which is followed by a slower decrease, i.e., an aperiodic process is observed on the average. The relevant measurements are presented in Fig. 1.32. Attempts of analytic approximation of the dependence observed both for an individual cycle and for a sequence of cycles were made repeatedly. According to the Wolf hypothesis that there is a superposition of various variation components, many researchers attempted to use a combination of harmonic components, including the gamma distribution (Kostitsyn, 1932 (see (Waldmeier 1955, S. 155)); Stewart and Panofsky 1938; Stewart and Eggleston 1940; Waldmeier 1941, 1955; Vitinsky 1963, 1973; Bocharova and Nusinov 1983; Vitinsky et al. 1986; Ivanov-Kholodny and Nusinov 1987) and the lognormal distribution (Bothmer et al. 2002). Functions similar to Planck's function (Hathaway et al. 1994) were also used. However, the results of these investigations gave no way to describe the characteristics of the 11-yr cycle.

Ivanov-Kholodny et al. (2000a,b) reliably revealed quasi-biennial oscillation in solar activity that was first detected by Schuster (1906) and then investigated by Clough (1924, 1928), who pointed out the presence of this variability in meteorological and magnetic phenomena on the Earth. However, their presence in variations of the Earth's atmosphere was revealed independently by Clayton (1884a,b), and Woeikof (1891, 1895). Therefore, it became obvious that the observed variations in solar activity are a superposition of the aperiodic variations that characterize the 11-yr and 2- to 3-yr variations (Kononovich 1999, 2001; Khramova et al. 2002; Kononovich and Shefov 2003).

Nagovitsyn (1997) restored the yearly average Wolf numbers for a period since 1096 covering more than 900 years (see Fig. 1.32) that testify to long-term variations of solar activity.

Fig. 1.32 Variations in yearly mean values of solar activity (Wolf numbers W) during the period from 1100 to 1995 (Nagovitsyn 1997)

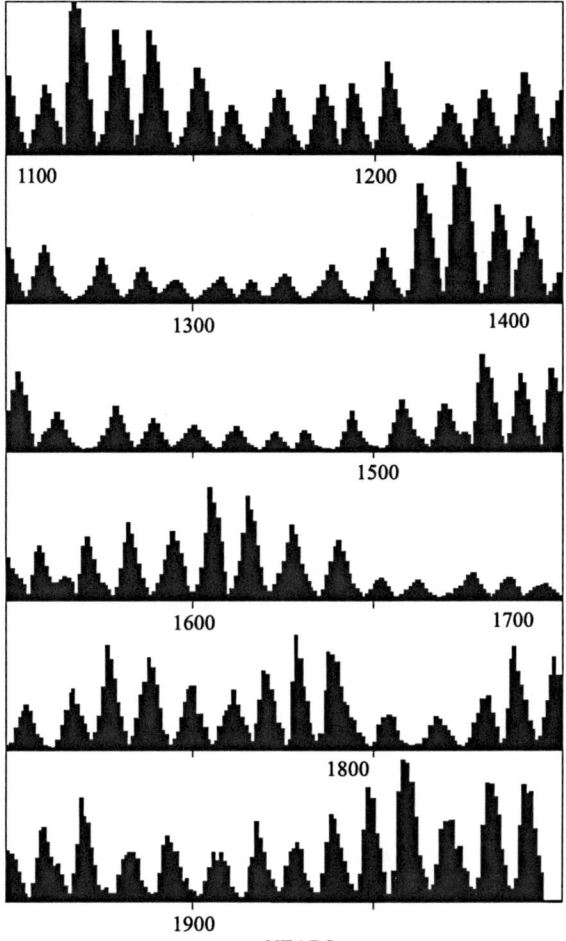

In the study of long-term variations in the activity of the stars of late spectroscopic classes F6-K7 (Baliunas et al. 1995) cyclic variations with periods from 7 to 15 years have been revealed which bear similarities to the 11-yr periodicity of solar activity. This implies the presence of general regularities in long-term cyclic activity, which appear to depend on the sizes and temperatures of stars.

1.6.1.2 Average Functional Relations for the 11-yr Cyclicity

In connection with the mentioned long-term variability in stellar and solar activities, possible approximating relations were considered which allow one to estimate the solar activity indices in different phases of the cycle (Kononovich 2004, 2005).

Among numerous functional relations suitable for describing the aperiodic variations in solar activity indices, two functions were primarily considered. First, this was the gamma distribution with a squared argument t (in a more general form, it is sometimes termed as the Nikagami distribution (Vadzinsky 2001)). It is of interest that formally it is similar in character to the Maxwellian distribution. For the first time it was used by Kostitsyn (1932) (see (Waldmeier 1955, S. 155)); however, subsequently it was overlooked by researchers. This distribution has the form

$$G(t) = B \cdot t^{2k_G} e^{-\frac{t^2}{T_G^2}},$$

where the time t (years) is counted from the cycle minimum, and the lognormal distribution is of the form

$$L(t) = \frac{A}{t} \cdot e^{-(k_L \cdot \log_e \frac{t}{T_L})^2},$$

where the argument t is counted from a point preceding the minimum of solar activity. It takes place as a statistical size distribution in the case of fragmentation of particles (Kolmogorov 1986) and as a probability density distribution of variation rates in diffusion of a passive tracers in a random velocity field (Klyatskin 1994).

The integrals S_G and S_L of these Wolf number distributions within a cycle, the argument values for the t_{MG} and t_{ML} maxima of the distributions, and the maximum values of the functions G_M and L_M are determined by the formulas

$$S_G = \frac{G_M}{2} \cdot T_G \cdot e^{k_G} \cdot k^{-k_G} \cdot \Gamma(k_G + \tfrac{1}{2}), \text{ mode } t_{MG} = \sqrt{k_G} \cdot T_G,$$

$$G_M = B \cdot k_G^{k_G} \cdot T_G^{2k_G} \cdot e^{-k_G};$$

$$S_L = \frac{\sqrt{\pi} \cdot L_M \cdot T_L}{k_L} \cdot e^{-\frac{1}{4k_L^2}}, \text{ mode } t_{ML} = T_L \cdot e^{-\frac{1}{2k_L^2}},$$

$$L_M = \frac{A}{T_L} \cdot e^{\frac{1}{4k_L^2}},$$

where Γ is the gamma function.

The width of the cyclic profile is taken at a half-maximum of the function $W_G = t_{G2} - t_{G1}$, where t_{G2} and t_{G1} are roots of the transcendental equation

$$\frac{t^2}{T_G^2} = k_G \cdot \log_e \frac{t^2}{T_G^2} - k_G \cdot \log_e k_G + k_G + \log_e 2.$$

Numerical solution of this equation for $0.50 \leq k_G \leq 1.40$ yields

$$\frac{t_{G1}}{T_G} = -0.0786 + 0.661 \cdot k_G - 0.100 \cdot k_G^2, \quad \frac{t_{G2}}{T_G} = 1.020 + 0.746 \cdot k_G - 0.130 \cdot k_G^2,$$

whence the halfwidth of the curve for an 11-yr cycle is obtained as

$$W_G = T_G \left(1.100 + 0.085 \cdot k_G - 0.030 \cdot k_G^2\right)$$

and the asymmetry as

$$P_G = \frac{t_{G2} - t_{MG}}{W_G} = \frac{1.020 - \sqrt{k_G} + 0.746 \cdot k_G - 0.13 \cdot k_G^2}{1.100 + 0.085 \cdot k_G - 0.030 \cdot k_G^2}.$$

For a lognormal function we have

$$\log_e \frac{t_{L1}}{T_L} = -\frac{1}{2k_L^2} - \frac{\sqrt{\log_e 2}}{k_L}, \quad \log_e \frac{t_{L2}}{T_L} = -\frac{1}{2k_L^2} + \frac{\sqrt{\log_e 2}}{k_L},$$

$$W_L = T_L \cdot e^{-\frac{1}{2k_L^2}} \left(e^{\frac{\sqrt{\log_e 2}}{k_L}} - e^{-\frac{\sqrt{\log_e 2}}{k_L}}\right) = 2 \cdot T_L \cdot e^{-\frac{1}{2k_L^2}} \operatorname{sh} \frac{\sqrt{\log_e 2}}{k_L},$$

$$P_L = \frac{t_{L2} - t_{ML}}{W_L} = \frac{1 - e^{-\frac{\sqrt{\log_e 2}}{k_L}}}{1 - e^{-\frac{2\sqrt{\log_e 2}}{k_L}}} = \frac{1}{1 + e^{-\frac{\sqrt{\log_e 2}}{k_L}}}.$$

With the above functions a long-term set of yearly average Wolf numbers W was analyzed (Vitinsky 1973) for all available cycles, from cycle 1 to cycle 22. For the 11-yr cycles mean annual values were used because other, short-period variations were naturally smoothed out.

The use of the above functional relations has made it possible to obtain a set of parameters to calculate an average "background" distribution of Wolf numbers for the overall period of observation of 11-yr cycles. Examples of this approximation are shown in Fig. 1.33.

1.6.1.3 The Exponential Function

Consideration of the physical mechanisms active in the convective zone of the Sun leads to an analysis of diffusion processes. Kononovich (2003) considered the combined effect of the variations caused by an 11-yr cycle and variations of smaller time scale. The solution obtained contains an exponential function of a cubic polynomial which represents an 11-yr cycle, namely,

$$E(t) = H \cdot \exp\left[-k_E r_E (t - t_0) + \frac{D_T}{3} \cdot k_E^2 (t - t_0)^3\right]$$
$$= H \cdot \exp\left[-m(t - t_0) + n(t - t_0)^3\right],$$

i.e.,

$$k_E = \sqrt{\frac{3n}{D_T}}, \quad r_E = m\sqrt{\frac{D_T}{3n}}.$$

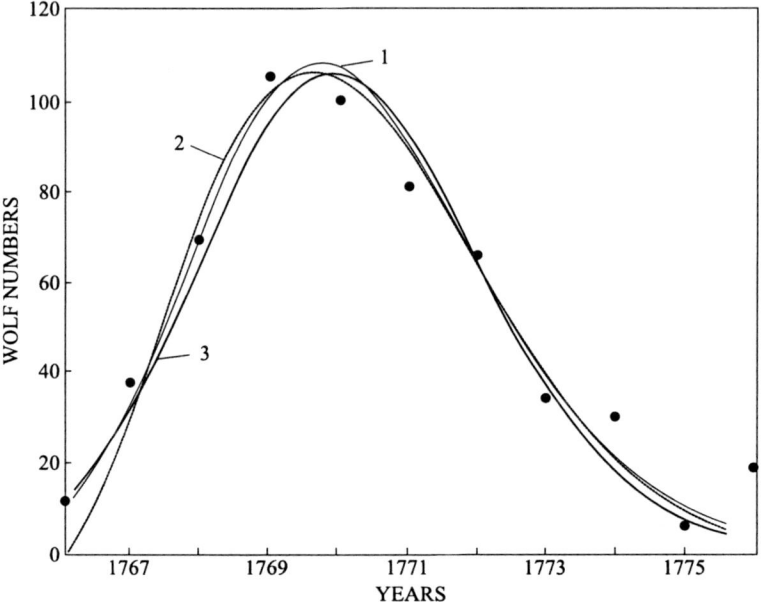

Fig. 1.33 Association between yearly average Wolf numbers for cycle 2 (1766–1775) (*points*) and approximations (*solid lines*) G(t) (*1*), L(t) (*2*), and E(t) (*3*) (Kononovich 2004, 2005)

This distribution is characterized by the mode $t_{ME} = t_0 - \sqrt{\frac{r_E}{D_T k_E}}$, and the maximum

$$E_M = H \cdot \exp\left[\frac{2 \cdot r_E}{3}\sqrt{\frac{k_E \cdot r_E}{D_T}}\right].$$

The values of the arguments for $E_M/2$ are determined by the formula

$$t_{E1} = t_0 - 2\sqrt{\frac{r_E}{D_T \cdot k_E}} \cdot \cos\left[60 - \frac{1}{3}\arccos(1 - \frac{3\log_e 2}{2 \cdot r_E\sqrt{\frac{k_E \cdot r_E}{D_T}}})\right],$$

$$t_{E2} = t_0 - 2\sqrt{\frac{r_E}{D_T k_E}} \cdot \cos\left[60 + \frac{1}{3}\arccos(1 - \frac{3\log_e 2}{2 \cdot r_E\sqrt{\frac{k_E \cdot r_E}{D_T}}})\right],$$

the halfwidth of the distribution by

$$W_E = 2\sqrt{\frac{3r_E}{D_T k_E}} \cdot \sin\left[\frac{1}{3}\arccos(1 - \frac{3\log_e 2}{2 \cdot r_E\sqrt{\frac{k_E \cdot r_E}{D_T}}})\right],$$

1.6 The Heliogeophysical Conditions that Determine the Observed Airglow Characteristics

the asymmetry by

$$P_E = \frac{t_{E2} - t_{ME}}{W_E} = \frac{1 - 2\cos\left[60 + \frac{1}{3}\arccos\left(1 - \frac{3\log_e 2}{2 \cdot r_E \sqrt{\frac{k_E \cdot r_E}{D_T}}}\right)\right]}{2\sqrt{3}\sin\left[\frac{1}{3}\arccos\left(1 - \frac{3\log_e 2}{2 \cdot r_E \sqrt{\frac{k_E \cdot r_E}{D_T}}}\right)\right]},$$

and the area by

$$S_E = E_M \cdot e^{-\frac{2 \cdot r_E}{3} \cdot \sqrt{\frac{k_E \cdot r_E}{D_T}}} \cdot \int_0^{12} \exp\left[-k_E r_E(t - t_0) + \frac{D_T}{3}k_E^2(t - t_0)^3\right] dt.$$

The total cycle duration T_E strongly depends on the choice of the minimum Wolf numbers at the cycle end, which, in real conditions, make about 5–7(%) (1/15–1/20) of E_M. In this case, we have

$$T_E = 2\sqrt{\frac{3r_E}{D_T k_E}} \cdot \sin\left[\frac{1}{3}\arccos\left(1 - \frac{3\log_e 20}{2 \cdot r_E \sqrt{\frac{k_E \cdot r_E}{D_T}}}\right)\right].$$

The duration of the rising branch is $T_{E1} = t_{EM}$ and that of the falling branch $T_{E2} = T_E - T_{E1}$.

The results of approximations of measurements of yearly average Wolf numbers are shown in Fig. 1.33 (solid line) for $D_T = 2$.

1.6.1.4 Comparison of Different Approximations

Comparison of the results of approximations using the above functions shows that the correlations between the key parameters of the distributions are strong enough and the regression lines, to within approximation errors, pass through the origin; namely, the correlation with the parameters of $G(t)$ is given by

$L_M = (1.09 \pm 0.05) G_M$, $r = 0.983 \pm 0.007$, $\sigma_L = (0.98 \pm 0.08) \sigma_G$,
$r = 0.835 \pm 0.015$,
$S_L = (1.13 \pm 0.09) S_G$, $r = 0.940 \pm 0.025$, $W_L = (0.98 \pm 0.12) W_G$,
$r = 0.884 \pm 0.048$,

with the parameters of L(t) by

$$E_M = (1.07 \pm 0.05) \, L_M, \; r = 0.977 \pm 0.010, \; \sigma_E = (0.89 \pm 0.06) \, \sigma_L,$$
$$r = 0.602 \pm 0.14,$$
$$S_E = (1.01 \pm 0.02) \, S_L, \; r = 0.996 \pm 0.002, \; W_E = (0.81 \pm 0.12) \, W_L,$$
$$r = 0.816 \pm 0.073,$$

and with the parameters of E(t) by

$$E_M = (1.11 \pm 0.06) \, G_M, \; r = 0.970 \pm 0.013, \; \sigma_E = (0.79 \pm 0.13) \, \sigma_G,$$
$$r = 0.801 \pm 0.078,$$
$$S_E = (1.16 \pm 0.09) \, S_G, \; r = 0.951 \pm 0.021, \; W_E = (0.92 \pm 0.10) \, W_G,$$
$$r = 0.911 \pm 0.037,$$
$$S_E = (1.09 \pm 0.02) \, E_M \cdot W_E, \; r = 0.996 \pm 0.002, \; \overline{S_E} = 584 \pm 195;$$
$$\overline{E_M \cdot W_E} = 537 \pm 178.$$

As can be seen, the data for all cycles have almost identical variances. However, on the average the maximum values of E_M are greater than those of L_M and G_M, the halfwidths W_E are smaller than W_L and W_G, and the distribution areas S_E are greater than S_L and S_G by about 10(%).

Considering the 22-yr Hale cyclicity (Vitinsky et al. 1986), one can notice a tendency for an increase (by $\sim$4(%) for a period of 22 years) in cycle integral ratios $S_{E(odd)}/S_{E(even)}$ in an even–odd pair (correlation factor 0.60 ± 0.26), which resembles the increase in secular solar activity (Khramova et al. 2002; Kononovich and Shefov 2003) for cycles 10 through 22, i.e., for the 1856–1996 period. It seems that there is some correlation between 22-yr cycles in their time sequence.

In studying time variations of solar activity, especially for a period of many years, the spectral composition of the observed time series is sometimes sought for. Therefore, one can try to elucidate whether the above analytic approximations of 11-yr variations of solar activity possess such properties. As already shown, all the three distributions, G(t), L(t), and E(t), provide, to within 10(%), an approximation of the observed yearly average Wolf numbers. This seems to mean that they represent various features of the process of propagation of a perturbation in the convective zone in the presence of a magnetic field of complex pattern. The best choice for analytical calculation of the Fourier coefficients is the function G(t). In this case, for the period $T_c \sim 11$ yr we have

$$G(t) = \frac{a_0}{2} + \sum_{n=1}^{\infty} A_n \cdot \cos \left[\frac{2\pi \cdot n}{T_c} (t - \varphi_n) \right],$$

where the harmonic amplitudes are given by

$$A_n = \sqrt{a_n^2 + b_n^2}$$

1.6 The Heliogeophysical Conditions that Determine the Observed Airglow Characteristics

and phases by

$$tg\varphi_n = -\frac{b_n}{a_n}.$$

The Fourier coefficients, in turn, are described by the expressions

$$a_n = \frac{2B}{T_c} \int_0^{T_c} t^{2k_G} e^{-\frac{t^2}{T_G^2}} \cos\frac{2\pi \cdot n}{T_c} t \cdot dt,$$

$$b_n = \frac{2B}{T_c} \int_0^{T_c} t^{2k_G} e^{-\frac{t^2}{T_G^2}} \sin\frac{2\pi \cdot n}{T_c} t \cdot dt.$$

The upper integration limit can be prolonged ad infinitum since the integrand rapidly approaches to zero for $t > T_c$. Therefore, according to the data of the reference book by Prudnikov et al. (1981), the solution has the form

$$a_n = \frac{\sqrt{\pi} \cdot T_G^{2k_G-1} \cdot B}{2^{k_G} \cdot T_c \cdot \cos k_G \pi} e^{-\frac{\pi^2 \cdot T_G^2 \cdot n^2}{2 \cdot T_c^2}} \left[D_{2k_G}(-X_n) + D_{2k_G}(X_n)\right],$$

$$b_n = \frac{\sqrt{\pi} \cdot T_G^{2k_G-1} \cdot B}{2^{k_G} \cdot T_c \cdot \sin k_G \pi} e^{-\frac{\pi^2 \cdot T_G^2 \cdot n^2}{2 \cdot T_c^2}} \left[D_{2k_G}(-X_n) - D_{2k_G}(X_n)\right],$$

where D_{2k_G} is a parabolic cylinder function (Weber's function) (Abramowitz and Stegun 1964), whence

$$A_n = \frac{\sqrt{\pi} \cdot T_G^{2k_G-1} \cdot B}{2^{k_G-2} \cdot T_c \cdot \sin 2k_G \pi} \cdot e^{-\frac{\pi^2 \cdot T_G^2 \cdot n^2}{2 \cdot T_c^2}} \cdot$$
$$\sqrt{D_{2k_G}^2(-X_n) - 2D_{2k_G}(-X_n)D_{2k_G}(X_n)\cos 2k_G\pi + D_{2k_G}^2(X_n)},$$

$$tg\varphi_n = -ctg k_G\pi \cdot \frac{D_{2k_G}(-X_n) - D_{2k_G}(X_n)}{D_{2k_G}(-X_n) + D_{2k_G}(X_n)},$$

where

$$X_n = \frac{\sqrt{2}\pi \cdot T_G \cdot n}{T_c}.$$

Calculations using analytic relations and numerical calculations by yearly average values of G(t) for an average profile show that the amplitudes (Wolf numbers) and phases have the following values: $A_1 = 59.4$, $A_2 = 8.7$, $A_3 = 2.5$, $A_4 = 1.3$, $\varphi_1 = 4.9$ years, $\varphi_2 = 3.7$ years, $\varphi_3 = 2.5$ years, $\varphi_4 = 1.8$ years, and the average value of $G(t) = 64.7$.

This implies that the 11-yr distribution of yearly average Wolf numbers can be approximated by the only first harmonic (cosine) whose phase practically coincides with the point of maximum of the 11-yr cycle. The contribution of the second harmonic (5.5 yr) makes $\sim 15(\%)$ and that of the subsequent harmonics is several percent. This perhaps gave impetus to some attempts to approximate an 11-yr cycle in such a way (Vitinsky et al. 1986).

1.6.1.5 Discussion of the Approximations Obtained

The cyclic variations of yearly average Wolf numbers actually reflect the solar flux variations in the latitude band on the surface of the Sun where spots appear. This evidently follows from the generalized time–space spot structure, which is represented as "Maunder's butterflies". The solar spots are manifestations of the complex processes of vertical motion of perturbations which occur in the lower-lying convection zone.

Combined analysis of the behavior of solar activity during an aperiodic cycle ($\sim$11 (yr)) and cyclic aperiodic variations (CAVs) with an average period of 2–3 years has shown that the minimum of the CAV train coincides with the common minimum of solar activity. In this case, the beginning of the CAV train is shifted from the point of minimum of the cycle by about 1–2 years. This implies that the onset of a new cycle of solar activity and the beginning of the CAV train most likely occur practically simultaneously as a result of a pulsed process, and, due to the different velocities of propagation of CAV trains from the interior of the convection zone at the Sun surface, first a CAV train is observed and then a new cycle starts. This is also evidenced by the fact that a CAV train breaks 22 years after its beginning, already in the subsequent cycle, approximately 5 years after its maximum (Kononovich and Shefov 2003).

Proceeding from physical notions, it is necessary to stress that among the functions G(t), L(t), and E(t), the last one is a solution of the parabolic differential equation

$$\frac{\partial E}{\partial t} = D_T \cdot \frac{\partial^2 E}{\partial r_E^2} - k_E \cdot r_E \cdot E,$$

which describes diffusion and heat-and-mass transfer (Polyanin 2001). The factor D_T is the diffusion coefficient (cm^2/s). The equation

$$\frac{\partial K}{\partial t} = D_T \cdot \left(\frac{\partial^2 K}{\partial r_E^2} + \frac{2}{r_E} \frac{\partial K}{\partial r_E} \right) - k_E \cdot r_E \cdot K,$$

where $E = r_E \cdot K$, describes nonstationary thermal and diffusion processes in a medium with central symmetry.

The long-term variations of solar activity within an 11-yr cycle are a consequence of the diffusion and heat-and-mass transfer caused by the conversion of the magnetic field energy in the convective zone of the Sun, which are described by the relaxation–diffusion model (Soloviev and Kirichek 2004). According to this model, each magnetic 22-yr cycle is generated by some individual "portion" of magnetic flux which has entered in the convection zone or has formed in this zone due to differential rotation and has been "processed" inside this zone by plasma eddies into some large-scale magnetic structure. The dissipation of this structure (diffusion in

1.6 The Heliogeophysical Conditions that Determine the Observed Airglow Characteristics

a spherical turbulent layer) can proceed in modes which result in the formation of magnetic structures of diffusion "wave packet" type. Once a perturbation pulse has appeared, its propagation velocity being slightly dispersed, an oscillatory process arises, which is described by the Airy function (Gill 1982). The diffusion equation is also satisfied by functions related to the Airy function.

Consideration of the development of an 11-yr cycle suggests that a toroidal perturbation arising at the base of the convection zone at heliographic midlatitudes diffuses upward and simultaneously shifts toward the equatorial plane. Therefore, the time variations of solar flux are variations of the meridional section of this floating perturbed region at the level of the photosphere. This trajectory is repeated by each subsequent cycle. However, the variations in depth of its occurrence vary the duration of the cycle and the relevant characteristics.

It is of importance that there is a significant negative correlation ($r = -0.481 \pm 0.168$) between the 11-yr cycle maximum amplitude E_M and the parameter r_E and a positive correlation between $E_M^{-1/2}$ and the time of solar maximum t_{EM} ($r = 0.748 \pm 0.094$). This gives the following regression relations:

$$E_M = (140 \pm 15) - (3.74 \pm 1.52)r_E,$$

$$\sqrt{E_M} \cdot t_{EM} = 48.6 \pm 8.5.$$

Hence, the solar maximum is obviously related to the depth of occurrence of the initial perturbation and to the velocity of its propagation toward the surface.

Besides these correlations, of interest are the relations of the mutual behavior of the parameters k_E and r_E to the time of solar maximum, t_{EM}, or, in other words, to the duration of the phase of increasing solar activity. This allows one to use average correlation relations not only for the solar cycles that occurred before 1755, invoking the database of the Pulkovo Main Astronomical Observatory at Pulkovo of the Russian Academy of Sciences (PMAO 2004), but also for the stars of the late spectroscopic classes G5–K5 (close to the solar class G2) (Baliunas et al. 1995; Bruevich et al. 2001) for which many-year cycles of duration 7–16 years have been revealed. Consideration of these correlations suggests that for both parameters k_E and r_E there is a nonlinear correlation with t_{EM} for which the data for solar cycles and 12 stars having cycles from 7 to 16 years constitute a generic set. It has been revealed that k_E decreases and r_E increases with increasing t_{EM}. In addition, there are distinct correlations of these parameters and also of t_{EM} and W_E with the star radius expressed in units of the Sun radius, which varies from 0.5 to 1.5. All this opens new possibilities for a better understanding of the nature of solar activity.

The above analysis of the long-term yearly average values of Wolf numbers allows the conclusion that the function $E(t)$, i.e., an exponential distribution of a cubic polynomial, is the preferred analytic representation of Wolf numbers, having a clear physical sense, for the 11-yr cyclicity of solar activity and also for the stars of late spectroscopic classes.

1.6.1.6 Estimation of the Parameters of the Curve of an 11-yr Cycle by Some of Its Measurable Characteristics

The importance of the above analytic expressions that describe the 11-yr variations of annual means of solar activity indices consists in that when their parameters are calculated by the least square method based on the available annual means a possibility arises to calculate the values of the indices for a given point in time within the cycle. The availability of analytic relations corresponding to a given approximation enables one to estimate the parameters of the curve based on measuring the most typical properties of the curve even before the end of the given cycle, such as the solar maximum amplitude of the cycle, its halfwidth W and asymmetry P, the time interval between the minimum and the maximum, i.e., the mode t_M, and the time interval between the minimum and the half-maximum of the amplitude, t_1, during the phase of increasing activity. Using the formulas obtained, it is possible to derive relations for the parameters of the 11-yr cycle curve.

For the gamma distribution, some relations are obtained by numerical approximations, while the other are derived analytically:

$$\log_e k_G = 20.92 - 38 \cdot P_G, \quad k_G = 3.03 - 1.74 \cdot \frac{W_G}{t_{GM}}, \quad B = G_M \cdot \left(\frac{e}{t_{GM}^2}\right)^{k_G}, \quad T_G = \frac{t_{GM}}{\sqrt{k_G}}.$$

For the lognormal distribution we have

$$k_L = \frac{\sqrt{\log_e 2}}{\log_e \frac{P_L}{1 - P_L}}, \quad T_L = \frac{W_L \cdot e^{\frac{1}{2k_L^2}}}{2 \cdot \text{sh} \frac{\sqrt{\log_e 2}}{k_L}}, \quad T_L = t_{ML} \cdot e^{\frac{1}{2k_L^2}}, \quad A = L_M \cdot T_L \cdot e^{-\frac{1}{4k_L^2}}.$$

For the exponential distribution of a cubic polynominal, to simplify the formulas, it is convenient to construct some intermediate relations, namely,

$$X = \frac{1}{3} \arccos\left(1 - \frac{3 \log_e 2}{2 \cdot r_E \sqrt{\frac{k_E \cdot r_E}{D_T}}}\right) = 2 \cdot \text{arctg}\left[2\sqrt{3}\left(P_E - \frac{1}{2}\right)\right],$$

$$Y = 2 \cdot r_E \sqrt{\frac{k_E \cdot r_E}{D_T}} = \frac{3 \ln 2}{2 \cdot \sin^2\{1.5 \cdot X\}}.$$

Based on these formulas, we have

$$H = E_M \cdot e^{-\frac{1}{3}Y}, \quad t_0 = \frac{W_E}{2\sqrt{3} \cdot \sin X} + t_{ME}, \quad k_E = \sqrt{\frac{4 \cdot Y \cdot \cos^3(60 - X)}{D_T \cdot (t_0 - t_{E1})^3}},$$

$$r_E = \sqrt{\frac{D_T \cdot Y \cdot (t_0 - t_{E1})}{4 \cdot \cos(60 - X)}}.$$

Checkout of these relations has shown that their parameters fit well to the values calculated based on approximations.

1.6.2 Cyclic Aperiodic Variations of Solar Activity

1.6.2.1 Character of the Variations

The ideas about the character of so-called quasi-biennial oscillations (QBOs) of the parameters of the mesopause and lower thermosphere have changed substantially in recent years. It has been established that the reason for these oscillations is the QBOs in solar activity. The quasi-biennial oscillations in the Earth's atmosphere were detected in 1880–1890 (Clayton 1884a,b; Woeikof 1891, 1895) and then were investigated by Clough (1924, 1928). Schuster (1906) and Clough (1924, 1928) revealed their relation to solar activity. In the early 1950s, their presence in the variations of the geomagnetic field characteristics was independently detected by Kalinin (1952) who paid attention to their relation to solar activity. Two main solar maxima with a characteristic interval of 2–3 years were investigated by Gnevyshev (1963, 1977) who observed the variability of the light intensity from the solar corona. In the early 1960s QBOs were detected again in the variations of the characteristics of stratospheric winds and since then they have been intensely investigated (Rakipova and Efimova 1975; Labitzke and van Loon 1988). It has been concluded that the QBO type depends on solar activity and on the mode of air circulation in the stratosphere.

Since the detection of QBOs in the behavior of solar activity (Schuster 1906) their periods were determined by ordinary spectral analyses, and QBOs with a period of ∼26 months have been revealed (Rakipova and Efimova 1975; Apostolov 1985; Labitzke and van Loon 1988). However, it turned out that the calculated harmonic oscillations point to the existence of periods of 2, 2.3, and 2.5 years (24, 28, and 30 months) (Gruzdev and Bezverkhnii 1999, 2003), which are not precisely constant. Temporal variability has been detected even in the effective period (Fedorov et al. 1994).

Recently new results have been obtained which testify that QBOs in the upper atmosphere are related to QBOs in solar activity (Ivanov-Kholodny et al. 2000a,b; Fadel et al. 2002), being in fact cyclic aperiodic variations (CAVs) (Kononovich and Shefov 2003). This was promoted by the application of a more appropriate method of numerical filtration of time series of observations (Ivanov-Kholodny et al. 2000a,b), which is reduced to elimination of seasonal variations of the parameters under investigation and 11- and 22-yr variations of solar activity. In a real case, the filtration is reduced to calculations of moving monthly means of an available series of monthly means of a given parameter, namely,

$$\Delta T(t_i) = 0.25 \cdot [T(t_i) - T(t_i - 12)] - [T(t_i + 12) - T(t_i)]$$
$$= 0.25 \cdot [2 \cdot T(t_i) - T(t_i + 12) - T(t_i - 12)],$$

where one month is taken for the time unit and i is the number of a term of the time series of monthly means of the parameter. This in fact is the calculation of the second differences of time series terms. In other words, this transform is the action of a linear numerical filter with the amplitude–frequency characteristic modulus

$$|G(f)| = \left|\sin^3(12\pi f)/3\pi f\right|,$$

where $f = 1/\tau$ is the frequency and τ is the period of oscillations. It has been shown (Ivanov-Kholodny and Chertoprud 1992; Antonova et al. 1996) that $|G(f)|$ reaches a maximum G_{max} at $\tau = 28.5$ months. The halfwidth boundaries of the filter are determined by the inequality $19 \leq \tau \leq 57$ months. For $\tau = 1, 2, 3, 4, 6$, and 12 months we have $G(f) = 0$.

The data processing performed for both the solar activity and the temperature of the upper atmosphere has shown that the QBOs are not harmonic oscillations (Fadel et al. 2002). This was confirmed by using the method of principal terms of time series "Caterpillar" (Danilov and Zhigljavsky 1997). It follows that the formal spectral analysis of time series used in many studies of QBOs both in solar activity and in the Earth's atmosphere failed to reveal the character of QBOs (CAV) (Shefov and Semenov 2006).

Great insight into the character of QBOs has been given by the work of Ivanov-Kholodny et al. (2000a,b) who have revealed that during an 11-yr solar cycle there are several maxima in the variations of both the index $F_{10.7}$ and the ionospheric parameters.

Earlier such a group of maxima was found by Apostolov (1985). The use of other methods of elimination of the background solar activity variations also made it possible to obtain maxima inherent in quasi-biennial oscillations (Kandaurova 1971; Baranov et al. 2001); nevertheless, in these studies no attention was paid to such a character of variations.

Actually, it turned out that the sequence of maxima in solar activity indices in an 11-yr cycle is a train of oscillations with varying period and amplitude (Fadel et al. 2002). These oscillation trains are well described by the Airy function Ai(-x), which is a solution of the second-order linear differential equation (Abramowitz and Stegun 1964)

$$y'' - xy = 0,$$

which describes the propagation of internal waves in the atmospheres and oceans of rotating planets. Here, it should be stressed that the Airy function, being a solution of hydrodynamic equations, is not an approximation of observed variations, but characterizes the processes occurring in the convection zone of the Sun. The modulation of the ultraviolet radiation emitted at various levels above the photosphere reflects the dynamics of the processes inside the Sun and, therefore, it shows up in variations of the parameters of the Earth's atmosphere (Kononovich and Shefov 2003).

For various values of the argument, the Airy function is expressed in terms of Bessel functions:

$$Ai(x) = (1/3)x^{1/2}[I_{-1/3}(2x^{3/2}/3) - I_{1/3}(2x^{3/2}/3)]$$

1.6 The Heliogeophysical Conditions that Determine the Observed Airglow Characteristics

for positive values of the argument and

$$\text{Ai}(-x) = (1/3)x^{1/2}[J_{-1/3}(2x^{3/2}/3) + J_{1/3}(2x^{3/2}/3)]$$

for its negative values.

For this case we have $x = 3 - \Delta t$, where $\Delta t = t - t_0$ is the time (years) passed from the start of the train, t_0. Since $\text{Ai}(3) \leq 0.008$, i.e., it is practically equal to zero though $\text{Ai}(x)$ asymptotically tends to zero as $x \to \infty$, it can be assumed that the beginning of the train corresponds to $x = 3$. The value of $-\text{Ai}(-1)$ is about -0.54, which seems to correspond to the yearly average $F_{10.7}$ minimum in an 11-yr cycle.

The integral representation of the Airy function has the form

$$\text{Ai}(\pm x) = \frac{1}{\pi} \int_0^\infty \cos\left(\frac{t^3}{3} \pm xt\right) dt .$$

The solar component in the index $F_{10.7}$ is

$$\Delta F_{10.7} = -22\, \text{Ai}(3 - \Delta t) .$$

Values of the Airy function can be found in tables or calculated by approximation formulas (Abramowitz and Stegun 1964):

$$\text{Ai}(x) = c_1 \cdot f(x) - c_2 \cdot g(x) ,$$

where

$$f(x) = 1 + (1/3!) \cdot x^3 + (1 \cdot 4/6!) \cdot x^6 + (1 \cdot 4 \cdot 7/9!) \cdot x^9 + \ldots,$$
$$g(x) = x + (2/4!) \cdot x^4 + (2 \cdot 5/7!) \cdot x^7 + (2 \cdot 5 \cdot 8/10!) \cdot x^{10} + \ldots.$$
$$c_1 = 3^{-2/3} \cdot \Gamma(2/3) = 0.35502, \quad c_2 = 3^{-1/3} \cdot \Gamma(1/3) = 0.25882.$$

In view of the above data, it is more correct to term oscillations of this type as cyclic aperiodic variations (CAVs). The first broad minimum has a duration of 3.8 years, the duration of the subsequent maxima is 2.8 years, decreasing to about 1.7 years during the subsequent years. The first broad minimum coincides with the minimum of an 11-yr cycle. However, the total train length is ~22 years; that is, the train covers some part of the subsequent cycle interval, and this results in interference of the maxima of the trains of two cycles. As a consequence, there is no resemblance between the CAVs observed in different 11-yr cycles (Kononovich and Shefov 2003).

The Fourier analysis of the Airy function have shown (Shefov and Semenov 2006) that if the train intervals for the arguments $3 - \Delta t$ are chosen within the limits of 12–15 years, the maximum amplitudes of harmonics correspond to periods of 2.3 and 2.8 years and depend on the position of the chosen time interval within the limits of the 11-yr solar cycle.

Figure 1.34 shows the monthly mean variations of the Wolf numbers for cycles 17 through 22, a series of quasi-biennial oscillations filtered using the method of

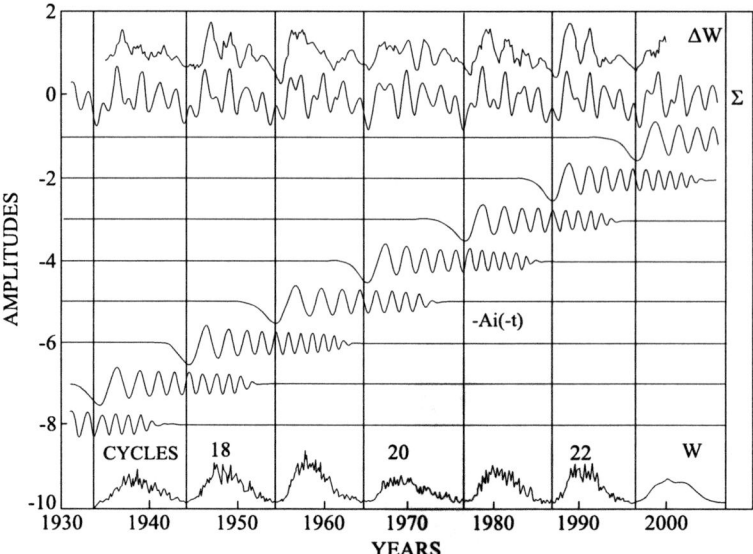

Fig. 1.34 The monthly mean Wolf numbers W for cycles 17 through 22 (*lower curve*, rel. un.). For cycle 23, the *thin line* represents forecast data (Khramova et al. 2000). The *upper curve* (ΔW) represents the series of quasi-biennial variations filtered out by the method of Ivanov-Kholodny et al. (2000a). Shown below are the superposition of eight Airy functions for cycles 17 through 22 (Σ) and, separately, the Airy function trains for each cycle, truncated within 22-yr periods (Kononovich and Shefov 2003)

Ivanov-Kholodny et al. (2000a), and the superposition of the Airy functions for each cycle, truncated within the limits of 22-yr cycles (Kononovich and Shefov 2003).

Application of the method of principal terms of time series (based on the transformation of a one-dimensional series to a many-dimensional one by means of the "Caterpillar" code) (Danilov and Zhigljavsky 1997) gave time variations of solar flux for the secular, 11-yr, and quasi-biennial components (Fig. 1.35).

Figure 1.34 presents the smoothed measured variations ΔW and the Airy functions Ai(−x) calculated for a superposition of trains. In the calculations the argument $x = 3 - K \cdot \Delta t$ was used. The coefficient K (close to unity) provided consistency between the time scale and the wave train length. When considering the interference of trains in the sequence of 11-yr cycles, the scales and shifts of individual Ai(−x) functions were adjusted by varying the coefficient K and the train duration. Thus, it can be seen that the period is ~38 months in the initial phase of the solar maximum and ~21 months in the phase of its minimum, linearly decreasing at a rate of −1.7 (month·yr^{-1}) with a correlation factor of −0.98. The best agreement between the measured and calculated variations ΔW is attained at a train duration of about two solar cycles, which is probably related to the 22-yr Hale cycle. The factor of correlation between the semiannual average and the calculated approximate values of ΔW for five cycles was 0.6 ± 0.1. The deep minimum in the beginning of the train

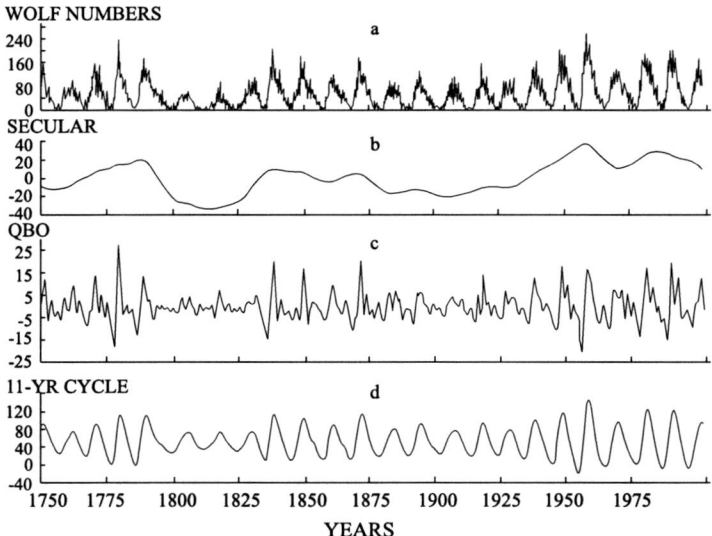

Fig. 1.35 A long-term series of monthly mean Wolf numbers W (**a**) and its decomposition into three components: secular (**b**), quasi-biennial (QBO) (**c**), and 11-yr for cycles 1 through 22 (**d**), (Kononovich and Shefov 2003)

of functions $-Ai(-x) = -0.54$ at $x = 1$ corresponds to the minimum of the yearly average W in the 11-yr cycle. This means that, according to the observations, a new cycle begins in fact before the end of the previous one, when the mentioned broad minimum starts suppressing the oscillations at the end of the previous cycle.

If all coefficients K and t_N would be identical, the full repeatability of quasi-biennial oscillations should readily be seen. In reality, because of the variability in cycle durations, the onsets of new cycles are irregular, resulting in a strong variability in time behavior of the solar activity indices represented by the Airy function, i.e., the actual behavior of the observed variations of solar activity varies from cycle to cycle. This is clearly exemplified by the 19th (1954–1964) and 22nd (1987–1996) cycles during which two maxima have been revealed in the period of maximum solar activity. Therefore, the statement about the so-called double-peak variability of solar activity and the "Gnevyshev dip" between the peaks, repeated by different researchers (Mikhailutsa and Gnevyshev 1988; Vernova et al. 1997; Obridko and Shelting 2003) is a rooted misunderstanding since these peaks do not bear witness to an independent isolated event, but represent the most pronounced part of an aperiodic process occurring during the solar cycle.

Thus, it is senseless to speak about a 2.2–2.7-yr train period. We believe that the term "quasi-biennial oscillations", though commonly used, is unhappy. It would be more correct to call variations of this type cyclic (in the sense that they are repeating and, at the same time, accompanying an 11-yr solar cycle) aperiodic variations.

1.6.2.2 Correlation Between QBOs and Variations of Solar Neutrinos

Long-term measurements of solar neutrinos are now publicated (Sakurai 1979; Lanzerotti and Raghavan 1981; Haubold and Gerth 1983; Cleveland et al. 1998; Davis 2002). Attempts have already been made to reveal various types of variations of this time series, and the presence of quasi-biennial oscillations is mentioned, though their character is not specified. In Fig. 1.36 the data about neutrinos, smoothed by means of a 1:3:5:3:1 sliding filter, are compared with the quasi-biennial component of the solar activity index, ΔW (Wolf number variation). For more convenient comparison, the neutrino flux scale is directed downward and the time scales are shifted to achieve the best mutual correspondence between the time variations (Kononovich 2004). As can be seen, the variations are concurrent in antiphase and shifted in time. For the interval 1974–1982 the correlation coefficient for these series is -0.84. Thus, CAVs in solar activity indices are observed about 1.4 years earlier than the variations in neutrino fluxes.

The effect revealed implies in fact that the CAV-generating perturbation source is in the interior of the Sun, at the base of its convective zone, and the propagation of a perturbation to the surface and to the center of the Sun takes different times because of the difference in distances.

1.6.2.3 Spectral Composition of the Cyclic Aperiodic Variations of Solar Activity and of the Earth's Atmosphere Characteristics

Spectral analysis, which was applied to the observational data on solar activity (Schuster 1906; Clough 1924, 1928; Apostolov 1985; Obridko and Shelting 2001)

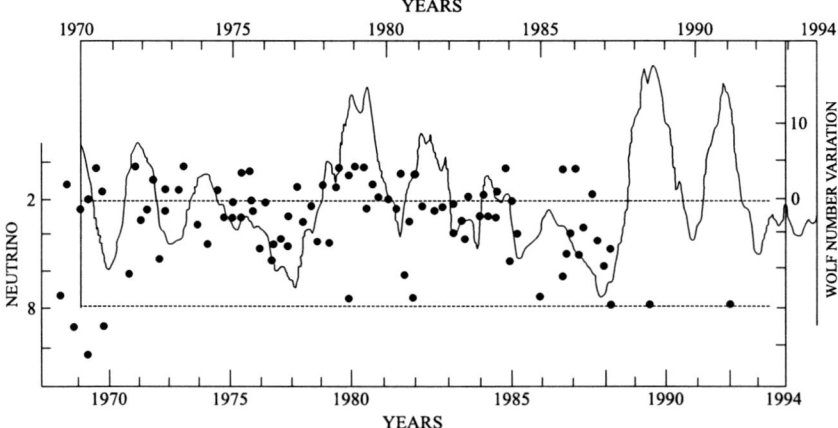

Fig. 1.36 Comparison of the cyclic aperiodic (quasi-biennial) variations in solar activity (*solid line*) (Kononovich and Shefov 2003, 2006) and in solar neutrino fluxes (*points*: data of Sakurai (1979) smoothed with a 1:3:5:3:1 filter)

and on various characteristics of the Earth's atmosphere (Clayton 1884a,b; Woeikof 1891, 1895; Clough 1924, 1928; Kalinin 1952; Rivin and Zvereva 1983; Labitzke and van Loon 1988; Gruzdev and Bezverkhnii 1999, 2003), usually led to the conclusion that there exist harmonic oscillations with periods of 2, 2.3, and 2.5 years (24, 28, and 30 months), which are not strictly constant. Time variability was noticed in QBOs (Fedorov et al. 1994; Obridko and Shelting 2001).

Investigations of this type of variations in solar activity indices by the numeral filtration method (Ivanov-Kholodny et al. 2000a; Kononovich and Shefov 2003) and by the method of principal terms ("Caterpillar" method) (Danilov and Zhigljavsky 1997) used by Kononovich and Shefov (2003) have shown that these variations are not harmonic. Analysis of the time behavior of the characteristics of the middle atmosphere (Fadel et al. 2002) has given similar results. Such cyclic aperiodic variations (CAVs), usually termed as quasi-biennial oscillations (QBOs), are well described by the Airy function (Fadel et al. 2002; Kononovich and Shefov 2003).

The long-term variations of solar activity within an 11-yr cycle are manifestations of the diffusion and heat-and-mass transfer processes resulting from the conversion of magnetic field energy in the convective zone of the Sun, which are described by the relaxation–diffusion model (Soloviev and Kirichek 2004). In hydrodynamic studies of the atmospheres of rotating planets, the occurrence of various types of oscillations propagating in them has been discussed in detail. Noteworthy is that these oscillations are described by the Airy function. Thus, a perturbation pulse propagating with a velocity having a small dispersion is accompanied by an oscillatory process described by the Airy function (Desaubies 1973; Munk 1980; Abramowitz and Stegun 1964; Gill 1982; Monin 1988). The diffusion equation is also satisfied by functions related to the Airy function.

Consideration of the development of an 11-yr cycle shows that a toroidal perturbation arising due to relaxation processes (Soloviev and Kirichek 2004) at the base of the convection zone at middle heliographic latitudes diffuses upward and simultaneously comes nearer to the equatorial plane. This seems to be due to the greatest gradients of angular velocity of rotation of the Sun taking place in the region of transition from the quasi-solid-state rotation of radiative inner layers of the Sun to the differential rotation of the convective zone (Kononovich and Shefov 2003). Thus, the time variations in solar activity represent variations in the spatial (inside the Sun) meridional section at the level of the photosphere of this perturbation moving outward, which show up as 11-yr variations. Each subsequent perturbation follows this trajectory. However, the variations in depth and, probably, in heliographic latitude of the onset of a perturbation vary the cycle duration and the corresponding atmospheric characteristics. The simultaneously propagating cyclic aperiodic variations modulate the meridional section.

This type of regular cyclic variability in solar activity shows up in some characteristics of solar magnetic activity (Ivanov-Kholodny et al. 2000a) and also in that many anomalously strong disturbances of solar activity arise near the maxima described by the Airy function. These, for example, are the intense red auroras observed on March 17, 1716 (3rd cycle), September 1/2, 1859 (10th cycle), September 24, 1870 (11th cycle), and February 11, 1958 (19th cycle) (Yevlashin 2005). The frequency of observations of red auroras also features this regularity. Within the current 23rd cycle, the red auroras of March 31, 2001, October 29 and 30, and November 20, 2003 also fell on the periods of quasi-biennial oscillation maxima.

According to measurements, the observed aperiodic variations are described by the Airy function $-Ai(-t)$. In the standard spectral analysis, the Fourier coefficients have the form

$$a_n = -\frac{2}{T} \cdot \int_{t_0}^{t_0+T} Ai(-t) \cos\left[\frac{2\pi nt}{T}\right] dt, \quad b_n = -\frac{2}{T} \cdot \int_{t_0}^{t_0+T} Ai(-t) \sin\left[\frac{2\pi nt}{T}\right] dt,$$

where T is the analyzed time interval, t_0 is the value of the Airy function argument corresponding to the beginning of the analyzed interval. In the representation of the time series as

$$\frac{a_0}{2} + \sum_{n=1}^{\infty} A_n(t) \cos\left[\frac{2\pi n}{T}(t - t_n)\right],$$

where $A_n = \sqrt{a_n^2 + b_n^2}$, the phase t_n is determined by $\cos t_n = a_n/A_n$, $\sin t_n = b_n/A_n$. The integral representation of the Airy function has the form

$$Ai(\pm t) = \frac{1}{\pi} \int_0^{\infty} \cos\left(\frac{v^3}{3} \pm tv\right) dv.$$

In this case, the Fourier coefficients for negative t for which the Airy function $-Ai(-t)$ describes the behavior of solar activity and upper atmosphere temperature at the base of the thermosphere, being damped-oscillating in character, are given by the expressions

$$a_n = -\frac{2}{\pi T} \int_{t_0}^{t_0+T} \cos\frac{2\pi nx}{T} \int_0^{\infty} \left[\cos\frac{t^3}{3}\cos xt + \sin\frac{t^3}{3}\sin xt\right] dt dx,$$

$$b_n = -\frac{2}{\pi T} \int_{t_0}^{t_0+T} \sin\frac{2\pi nx}{T} \int_0^{\infty} \left[\cos\frac{t^3}{3}\cos xt + \sin\frac{t^3}{3}\sin xt\right] dt dx,$$

whence

$$a_n = -\frac{2}{\pi T} \int_{-\pi n}^{\infty} \frac{\sin v}{v} \cos\left[\frac{8}{3T^3}(v + \pi n)^3 - v\left(\frac{2t_0}{T} + 1\right)\right] dv -$$

$$- \frac{2}{\pi T} \int_{\pi n}^{\infty} \frac{\sin v}{v} \cos\left[\frac{8}{3T^3}(v - \pi n)^3 - v\left(\frac{2t_0}{T} + 1\right)\right] dv.$$

1.6 The Heliogeophysical Conditions that Determine the Observed Airglow Characteristics

$$b_n = -\frac{2}{\pi T} \int_{-\pi n}^{\infty} \frac{\sin v}{v} \sin\left[\frac{8}{3T^3}(v+\pi n)^3 - v\left(\frac{2t_0}{T}+1\right)\right] dv +$$

$$+ \frac{2}{\pi T} \int_{\pi n}^{\infty} \frac{\sin v}{v} \sin\left[\frac{8}{3T^3}(v-\pi n)^3 - v\left(\frac{2t_0}{T}+1\right)\right] dv.$$

The use of the above formulas and forms of the Airy function has made it possible to calculate the amplitudes of harmonics for the time interval typical in considering an 11-yr solar cycle.

The harmonic amplitudes and phases were calculated first by the Fourier analysis of numerical values of the Airy function for various values of the interval T. Then a semi-analytic method was applied to calculate numerically the Fourier coefficients based on the integral representation of the Airy function.

It should be noted that in the consideration of the QBOs in solar activity and in parameters of the Earth's atmosphere, the scale of the Airy function argument practically corresponds to the time scale expressed in years. The calculations by both methods have shown that for the time interval determined by the duration of a solar cycle within which shorter periodic variations are usually analyzed the maximum harmonic amplitude corresponds to 2.3- and 2.5-yr periods, as can be seen in Fig. 1.37a,b.

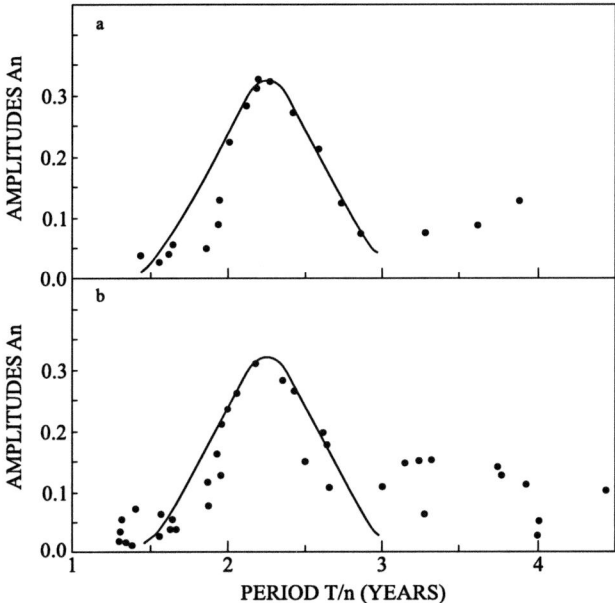

Fig. 1.37 Variations in calculated amplitudes A_n of the nth harmonics having periods T/n with the chosen time interval T of the argument of the Airy function. (**a**) Semi-analytic calculations. (**b**) Numerical analysis of the time series of the Airy function for the argument range $|-2.3| \leq x \leq |-12|$. *Solid line* is a normal distribution for the period 2.25 with a variance of 0.5

According to the properties of the Airy function, for its arguments from −2.34 to −11.94, between its every three successive zeros, there are quasi-sinusoidal oscillations with periods of 3.18, 2.70, 2.42, 2.4, 2.10, 1.98, and 1.90, having mean amplitudes of 0.400, 0.365, 0.345, 0.334, 0.322, 0.312, and 0.308, respectively. As can be seen from the calculation results (see Fig. 1.37), these periods have maximum values in a harmonic analysis. As the length of the analyzed interval is successively decreased from 12 to 2 years, the maximum amplitude of the harmonics with periods of 2.0–2.8 years increases. Successively shifting the 2.36-yr interval along the range of argument values under consideration has revealed that the maximum amplitude corresponds to the first two sine cycles whose periods are in the range 2.3–2.5 years.

The wavelet transform applied to analyze quasi-biennial oscillations in the lower and middle atmosphere testifies that the spectral distribution of the processes under investigation varies with time and with the altitude at which the oscillations occur. This transform offers the possibility to detect several simultaneously interacting oscillations (Astafieva 1996). It is of interest that the wavelet analysis of quasi-biennial oscillations of the temperature, pressure, and wind velocity in the stratosphere (Gruzdev and Bezverkhnii 1999, 2003, 2005) has shown that the oscillation periods gradually decrease within several years during the basic part of an 11-yr cycle, just as this should follow from the behavior of the periods of the Airy function. Nevertheless, the actual character of the process has not been revealed.

The presence of short-period and long-period harmonics with small amplitudes determined by the length of the analyzed interval is due to their misfit to the periodic properties of the Airy function and due to the monotonic decrease in amplitude of its maxima and minima. It is of importance that for the cases where the formal scale of the Airy function argument can be different from the mentioned time scale, the CAV periods will naturally change in proportion with the scale ratio. This situation seems to take place for the data of observations of sidereal activity.

As can be seen from Fig. 1.37, the amplitude distribution for the periods close to 2.25 years is well described by the normal distribution

$$A\left(\frac{T}{n} - 2.25\right) = 0.32 \cdot \exp\left[-\frac{\left(\frac{T}{n} - 2.25\right)^2}{(0.5)^2}\right].$$

The factor of correlation between the results of calculations and approximations is 0.9 ± 0.1.

In an actual situation of the behavior of solar activity there occurs interference of the oscillation trains of successive 11-yr cycles as a result of which the first and second maxima of a current cycle are summed up with the sixth and eighth maxima of the previous cycle, and, correspondingly, the first and second minima of a current cycle with the seventh and ninth maxima of the previous one. However, since the relative shift of trains, P (11 years), is not invariable from cycle to cycle, the actual oscillations change in character from cycle to cycle (Kononovich and Shefov 2003). Therefore, sometimes maxima are observed which are manifestations

1.6 The Heliogeophysical Conditions that Determine the Observed Airglow Characteristics

not of an independent isolated event, but of the most pronounced part of an aperiodic process occurring during a solar cycle.

Taking into account the interference of trains results in some complication of the above expressions:

$$a_n = -\frac{4}{\pi T} \int_{-\pi n}^{\infty} \frac{\sin v}{v} \cos\left[\frac{P}{T}(v+\pi n)\right] \cos\left[\frac{8}{3T^3}(v+\pi n)^3 - v\left(\frac{2t_0}{T}+1\right) - \frac{P}{T}(v+\pi n)\right] dv -$$

$$-\frac{4}{\pi T} \int_{\pi n}^{\infty} \frac{\sin v}{v} \cos\left[\frac{P}{T}(v-\pi n)\right] \cos\left[\frac{8}{3T^3}(v-\pi n)^3 - v\left(\frac{2t_0}{T}+1\right) - \frac{P}{T}(v-\pi n)\right] dv,$$

$$b_n = -\frac{4}{\pi T} \int_{-\pi n}^{\infty} \frac{\sin v}{v} \cos\left[\frac{P}{T}(v+\pi n)\right] \sin\left[\frac{8}{3T^3}(v+\pi n)^3 - v\left(\frac{2t_0}{T}+1\right) - \frac{P}{T}(v+\pi n)\right] dv +$$

$$+\frac{4}{\pi T} \int_{\pi n}^{\infty} \frac{\sin v}{v} \cos\left[\frac{P}{T}(v-\pi n)\right] \sin\left[\frac{8}{3T^3}(v-\pi n)^3 - v\left(\frac{2t_0}{T}+1\right) - \frac{P}{T}(v-\pi n)\right] dv.$$

Figure 1.38, a presents the relative position of CAV trains described by the Airy function $-Ai(x)$ within a 22-yr interval, as follows from the data by Kononovich and Shefov (2003). The circles denote the values of the parameter t_N and the horizontal arrows depict the intervals T within which the calculation of harmonics was performed. The calculation results are shown in Fig. 1.38 where the distributions of the amplitudes of 1.4- to 4.2-yr harmonics are given for the interval $T \sim 6\text{--}10$ years and for the relative shift of trains $P = 11.3$ (yr). It can be seen that the interference of two trains slightly affects the spectral distribution of the analyzed solar cycle.

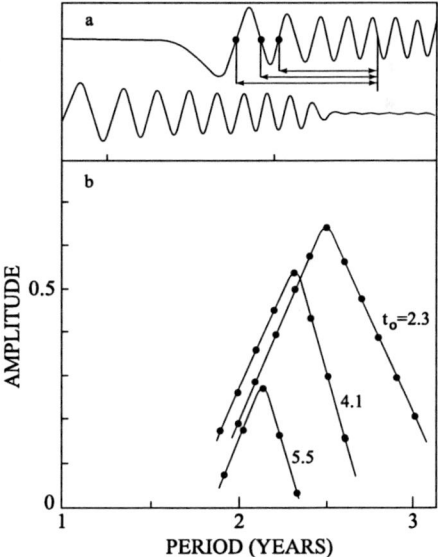

Fig. 1.38 Variations in amplitudes of the "quasi-biennial" harmonics taking into account the interference of successive CAV trains in harmonic Fourier analysis. (a) Relative positions in time of the CAV trains for the 21st and 22nd cycles of solar activity (Kononovich and Shefov 2003); *circles*: values of t_N, *horizontal arrows*: intervals T. (b) Harmonic amplitudes versus their periods for $t_0 = 2.3, 4.1,$ and 5.5

The maximum of amplitudes is about 2.1–2.5 years. Varying the shift within the limits ±1 year, which follow from the analysis of the data reported by Kononovich and Shefov (2003), with respect to the mentioned value has not revealed appreciable changes in the spectral distribution of harmonics. From the data presented in Fig. 1.38 it follows that the mentioned dependence of the amplitudes on the period reflects in fact the probability that in a harmonic analysis of a considered 11-yr interval the maximum amplitudes will always correspond to 2.3- to 2.5-yr periods. In every case, the maximum amplitude depends on the duration of the analyzed interval and on the initial value of the argument. Therefore, it is little wonder that independent spectral analyses of various characteristics of solar activity which show up in variations of the global magnetic field of the Sun also result in the conclusion about a narrow spectral interval of the periods calculated in this type of data processing (Obridko and Shelting 2001).

If the time scales of successive trains are somewhat different (by $\sim 5(\%)$), formal account of this difference substantially complicates the analysis, however improving the results insignificantly, since the properties of the first three quasi-periodic oscillations of the Airy function become dominant.

1.6.3 Long-Term Variations in Solar Activity

The problem of the long-term variability in solar activity was posed rather long ago. Various methods were applied to attack this problem and it has been found that there exist solar flux variations with periods from several tens to several thousands of years (Waldmeier 1955; Vitinsky et al. 1986; Chistyakov 1997). In doing this, conventional spectral analyses were applied to the series of Wolf numbers since 1745. Recently a series of yearly average Wolf numbers has been published which was recovered based on a nonlinear description of the data by Schove (1955, 1962, 1979, 1983) about the times at which the Wolf numbers took extreme values during the period from 1090 to 2003 (see Fig. 1.32) (Nagovitsyn 1997; PMAO 2004).

The most pronounced cycle that was revealed based on the measured solar activity indices is the 11-yr cycle. Numerous investigations have long made it obvious that the variations in annual average solar activity during an 11-yr cycle, despite that there exist variations with a shorter period, such as quasi-biennial (or cyclic aperiodic) oscillations (Apostolov 1985; Ivanov-Kholodny et al. 2000a,b; Kononovich and Shefov 2003) and other noise components, are described in general by a bell-shaped function with a shorter rising phase and a longer falling phase. Attempts were made repeatedly to explain such a behavior of solar activity (Waldmeier 1955; Vitinsky et al. 1986; Chistyakov 1997). However, all of them were aimed at constructing some approximations rather than at finding actual physical relations. In the foregoing this problem has been considered in detail from the new viewpoint based on Kononovich's work (2005) and it has been shown that the mean variations of yearly average Wolf numbers can be adequately described by an exponential of a cubic polynominal:

1.6 The Heliogeophysical Conditions that Determine the Observed Airglow Characteristics

$$E(t) = H \cdot \exp\left[-m(t-t_0) + n(t-t_0)^3\right],$$

and that this function is a solution of the diffusion equation (Polyanin 2001), which has been considered in the previous section.

It should be noted that it is more appropriate to use Wolf numbers than the index $F_{10.7}$. As noted by Apostolov (1985), Wolf numbers characterize the solar activity at the level of the photosphere whence the 200- to 300-nm UV radiation emanates which has an impact on the atmosphere at altitudes of 25–100 (km), while the index $F_{10.7}$ characterizes the activity at the altitudes of the solar corona where the UV radiation of wavelength shorter than 150 (nm) arises which acts on the atmosphere at altitudes above 100 (km) with a delay of 3 months relative to the radiation associated with Wolf numbers.

With the obtained analytic relations, which allow one to calculate 11-yr cycle parameters based on the available data about yearly average Wolf numbers (Kononovich 2004, 2005), an effort was made to analyze the long-term series (Nagovitsyn 1997) and to elucidate the behavior of the parameters of the above analytic representation of $E(t)$.

Detailed examination of different cycles has shown that the typical asymmetry ($P \sim 0.530$), which shows up most pictorially in cycles 1 through 23, is not so pronounced for the earlier period for which the reconstruction has been made by an indirect method. Though the approximation of the data of Nagovitsyn (1997) is accurate to within $\sim 30(\%)$, the distribution parameters of the function $E(t)$ vary substantially and, in some cases (~ 20 cycles of 83), take obviously anomalous values (see Fig. 1.32).

Nevertheless, examination of the long-term behavior of the mentioned parameters allows one to reveal some their typical properties. It seams that the parameter m should retain its value equal to about 0.65, and that the character of the function $E(t)$ is mainly governed by the coefficient n.

Obviously, the halfwidth W and asymmetry P of the distribution of 11-yr solar cycle parameters can be measured rather reliably. With the formulas used by Kononovich (2004, 2005) the parameter n can be calculated analytically by the asymmetry parameter P, namely,

$$n = \frac{16 \cdot m^3}{27 \cdot \ln^2 2} \cdot \sin^4\left\{3\mathrm{arctg}\left[2\sqrt{3}(P-0.5)\right]\right\}.$$

The most easily determinable characteristics of the distribution are the amplitude of the maximum, E_M, and the halfwidth W. The correlation between these quantities is negative, and, on the average, variations in E_M from 50 to 150 are accompanied by the decrease in W from 5.5 to 4 years. The parameter r, which characterizes the depth of occurrence of a perturbation, has a negative correlation with E_M. On the average, to a greater depth r there corresponds a smaller amplitude E_M and a greater halfwidth W.

Since direct measurements of solar activity are not available, to reveal its secular variations, data on various atmospheric phenomena are used which reflect the effect of solar activity, such as, for example, the frequency of observations of auroras.

These data and the relevant publications have been collected by Schröder (1997, 2000a,b, 2002). These useful collections include publications, sometimes rare and hardly accessible, in which the problems of long-term variations in solar activity are discussed. However, the visual observations of auroras at midlatitudes must be complemented with their intensity since at random observations only high-luminosity events can be detected, which are rare at midlatitudes (Schröder et al. 2004).

1.6.4 Solar–Terrestrial Relationships

The impact of the variability of solar activity on the character of the processes occurring in the circumterrestrial space and immediately at various levels of the Earth's atmosphere is the subject matter in studying solar–terrestrial relationships. For quite natural reasons, the spectrum of responses of atmospheric parameters to solar activity is extraordinarily wide. Here we shall consider the manifestations of solar activity in the properties of the lower, middle, and upper atmosphere that can be revealed by analyzing variations of the airglow of the upper atmosphere.

1.6.4.1 The Magnetosphere

The interaction of the exterior shell of the geomagnetic field of the Earth with the Sun occurs via a continuous plasma flow from the solar corona, which is called the solar wind. Under its action the geomagnetic field is deformed, being contracted on the day side of the Earth and extended as a tail on its night side for many tens of Earth radii (Fig. 1.39). This problem is discussed in numerous publications (Akasofu and Chapman 1972; Nishida 1978).

The magnetosphere is very large: it extends for about 10 Earth radii ($\sim$65 000 (km)) toward the Sun, for 15 radii ($\sim$100 000 (km)) in the direction perpendicular to the Sun, and for many hundreds and even thousands of Earth radii (some millions of kilometers) in the opposite direction from the Sun.

Quiet solar wind – solar plasma flow – continuously emanates from the corona with a velocity generally equal to about $400\,(\mathrm{km \cdot s^{-1}})$ (250–$700\,(\mathrm{km \cdot s^{-1}})$), filling up in fact the whole of the planetary Solar system. The high velocities, determined from the delays of geomagnetic storms after chromospheric flares, about $1000\,(\mathrm{km \cdot s^{-1}})$ (maximum $2000\,(\mathrm{km \cdot s^{-1}})$), are actually due to the flare-induced shock waves propagating in the solar wind plasma. These waves, interfering with the magnetosphere, practically always induce a burst of activity – a magnetospheric storm accompanied by geomagnetic disturbances and auroras.

Under the action of solar wind, a large-scale surface current layer is formed at the outer boundary of the magnetosphere. This is the so-called magnetopause whose exterior magnetic field is equal to and opposite in sign to the interior geomagnetic field, so that the net field on the outside is equal to zero, while inside it is approximately doubled. The solar wind flow is collisionless and hypersonic (since the transport

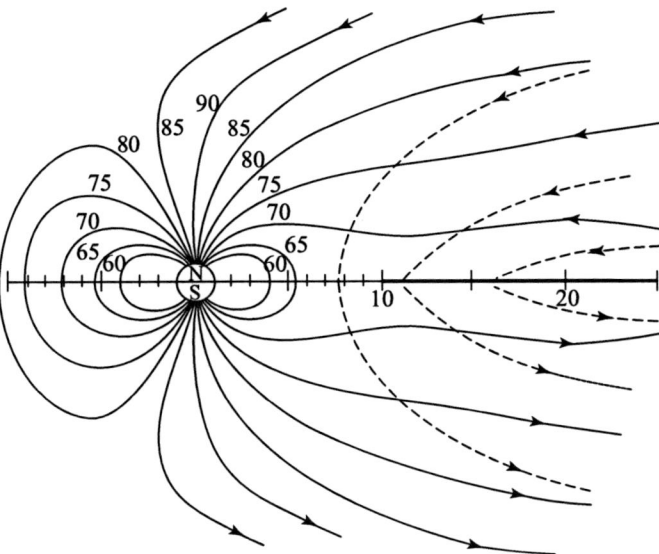

Fig. 1.39 The meridional section of the magnetospheric magnetic field. The *numbers at the field lines* indicate geomagnetic latitudes and the *numbers at the horizontal axis* indicate the distances in Earth radii (Akasofu and Chapman 1972)

velocity of the flow is several times greater than its thermal velocity). Therefore, a collisionless shock wave arises in front of an obstacle – the Earth's magnetosphere. Behind the shock wave the solar plasma flow strongly heats up, its transport velocity drops, while the plasma concentration and the magnetic field strength increase appreciably near the magnetopause. This region of hot plasma flowing round the magnetopause is called the transition layer.

The average energy of plasma particles in the transition layer is 0.1–0.2 (keV) for protons and 30–60 (eV) for electrons; the concentration is 10–30 (cm^{-3}). The magnetic field is several times greater than that in the solar wind, reaching 20–30 (γ), which is comparable to the magnetic field behind the magnetopause (Galperin 1975).

The magnetopause is not a solid obstacle invariable in shape. Its shape is determined by the balance of the plasma and magnetic field pressure forces outside and inside the magnetopause. The geomagnetic field prevails in the region between the magnetopause and the Earth. At the same time, the space outside the magnetopause is filled with solar wind particles.

The energy receipts from the solar wind to the magnetosphere are accompanied by geomagnetic disturbances and by electron and proton flows toward the polar regions of the Earth, into the auroral ovals. This gives rise to a glow in the upper atmosphere (auroras) in a wide spectral region. The physics of these processes is considered in numerous publications (Chamberlain 1961; Omholt 1971; Vallance Jones 1974; Starkov 2000; Yevlashin 2000). Most of the emissions are excited in the thermosphere and belong to nitrogen and oxygen molecules. The energetically significant emissions are the 557.7- and 630-nm emissions of atomic oxygen.

1.6.4.2 The Exosphere

At altitudes above 600–700 (km) the atmosphere is so rarefied that the collisions between the medium species are very rare. The main atmospheric constituents are hydrogen and helium with a small admixture of atomic oxygen. The altitude distribution for these light components is determined in many respects by escape processes whose rates substantially depend on the intensity of the solar short-wave ultraviolet radiation that involves the 121.6-nm Lyman alpha and 102.7-nm Lyman beta emissions of hydrogen, 58.4- and 53.7-nm emission of neutral helium, and, the most significant for these altitudes, 30.4-nm emission of ionized helium.

The interaction of these solar radiations with atmospheric components gives rise to emissions in the Earth's atmosphere. The Balmer series 656.3-nm emission of hydrogen H_α results from L_β-induced fluorescence. This makes it possible to trace the structure of the hydrogenous geocorona around the Earth up to altitudes of the order of 100 000 (km). Enhanced solar activity promotes escape processes of hydrogen, reducing its content at these altitudes. A great body of information has been obtained at the Abastumani Astrophysical Observatory of the Georgia Academy of Sciences (Fishkova 1983). These results are discussed in Sect. 2.8.

Atomic helium is less prone to escape processes. Investigating its atmospheric emissions, both ultraviolet from resonance fluorescence of solar emissions and 1083.0-nm infrared from photoelectrons produced by solar 30.4-nm emission followed by fluorescence, one can determine the neutral helium content at altitudes above 300 (km). Its altitude distribution is substantially determined by the diurnal variations in irradiance of the atmosphere and by the level of solar activity. These results are discussed in Sect. 2.7.

1.6.4.3 The Thermosphere

The upper atmosphere at altitudes above 100 (km) distinctly responds to changes in solar activity. The present notions have been formed long ago, when models of the upper atmosphere were developed based on satellite measurements of density variations (CIRA-1972; CIRA-1986; Hedin 1991).

At the thermospheric altitudes there is the basic region of the ionized component of the upper atmosphere – the F2 layer, whose participation in the photochemical processes responsible for the occurrence of the airglow is dominant. This important region of the upper atmosphere is a field for independent research for quiet helio and geophysical conditions. Also it has great interest for the investigation of its relations to magnetospheric phenomena and to the solar processes that manifest themselves in this layer. The relevant diversified problems are extensively discussed in the literature (Ivanov-Kholodny and Nikol'sky 1969; Davies 1969; Rishbeth and Garriott 1969; Kazimirovsky and Kokourov 1979; Ivanov-Kholodny and Mikhailov 1980; Chamberlain 1978; Antonova and Ivanov-Kholodny 1989).

The altitude distribution of the thermosphere temperature and composition were repeatedly considered in detail by many researchers (Nicolet 1962; Banks and Kockarts 1973a,b; Massey and Bates 1982; Rees 1989).

The thermosphere is accessible for ground-based sounding owing to the 630-nm emission of atomic oxygen and 1083-nm emission of helium present in its night airglow. These emissions allow one to determine the temperature at altitudes of 250–270 and above 300 (km), respectively, and its variations on various time scales. These methods are considered in Sects. 2.6 and 2.7.

However, the now available models that describe the temperature and density atmospheric characteristics do not consider the long-term variations in thermal conditions of the entire atmosphere that have been revealed for the thermospheric altitudes based on long-term interferometric measurements of the 630-nm emission temperature (Semenov 1996). Subsequent investigations, based on the registration of altitudes at which various types of auroras occurred (Störmer 1955) and on satellite measurements (Keating et al. 2000), have revealed a negative trend in temperatures (Evlashin et al. 1999; Starkov et al. 2000; Starkov and Shefov 2001). The temperatures at altitudes of 120–170 km determined from the data of experiments with artificial luminous clouds (Kluev 1985) also testify to the existence of a negative temperature trend.

1.6.4.4 The Middle Atmosphere

The airglow of the upper atmosphere arising in the top part of the middle atmosphere (mesopause) and in the lower thermosphere is an efficient indicator of solar activity. The dependences of the middle atmosphere characteristics on solar activity are discussed in Chap. 4. In this range of altitudes there occurs dissociation of molecular oxygen by solar ultraviolet radiation. Various processes of recombination of atomic oxygen give rise to numerous emissions and control the thermal and dynamic conditions in this atmospheric region. This also shows up in seasonal and altitudinal responses of temperature to solar activity. At the altitudes of the stratosphere and mesosphere no atmospheric emissions occur which would be accessible to ground-based observation. Therefore, there is lack of data on the temperature conditions in this range of altitudes, but they can be obtained from immediate rocket and lidar measurements and from indirect information about other phenomena, such as winds, turbulence, and waves.

Therefore, it is natural that in recent years much attention is paid to the study of the response of the temperature of the middle atmosphere to solar activity (Beig et al. 2003). This in many respects is due to the fact that for estimating the observed long-term variations in its temperature at different altitudes with the use of the measurement data obtained in different time intervals, it is necessary to reduce these data to uniform conditions to eliminate the effect of solar activity. However, this cannot be done with the now available models of the middle atmosphere since these models do not take into account the effect of solar activity at altitudes of 30–95 (km) (CIRA-1972; CIRA-1986; Hedin 1991). Some estimates of the effect of solar activity on the altitude distribution of temperature have been reviewed by Beig et al. (2003). This work presents the results of high-latitude hydroxyl emission measurements performed by the method of falling spheres (Lübken 2000; Nielsen et al. 2002) and midlatitude lidar measurements (Hauchecorne et al. 1991; She et al. 2000). However, Beig et al. (2003) point out that the results obtained are ambiguous and therefore give no way for understanding the trends observed at different latitudes, since the effect of solar activity and the long-term trend can hardly be separated because of the short observation periods used for analysis.

The response of the monthly mean temperature of the middle atmosphere to solar activity was considered based on the long-term rocket measurements and spectrophotometry of some airglow emissions during several 11-yr solar cycles (Semenov et al. 2005; Golitsyn et al. 2006).

For analysis the temperature profiles at 30–110 (km) (Semenov et al. 2002a,c) have been used which have been measured at midlatitudes for all months of the years of maximum (1980, $F_{10.7} = 198$, and 1991, $F_{10.7} = 208$) and minimum solar activity (1976, $F_{10.7} = 73$, and 1986, $F_{10.7} = 75$). These profiles show that in the mesopause region there is a pronounced winter temperature maximum and a deep summer minimum, both in the years of maximum and in the years of minimum of the 11-yr solar cycle. Based on these data and using the temperature differences at various altitudes for the profiles corresponding to years of high and low solar activity, one can use a linear approximation to find the temperature increment due to solar activity as

$$\Delta T(Z) = \delta T_F(Z) \frac{F_{10.7} - 130}{100} \ (K),$$

where $\delta T_F(Z) = dT/dF$ is the change in temperature at the altitude Z for $\Delta F_{10.7} = 100$ (sfu). Once the values of $\delta T_F(Z)$ had been found for individual levels of altitudes, the seasonal variations were constructed. Thereafter, to obtain regular relations, they were approximated by the sum of four harmonics:

$$\delta T_F(Z) = A_0 + \sum_{n=1}^{4} A_n(Z) \cos\left[\frac{2\pi n}{T_{an}}[t_d - t_n(Z)]\right] (K/100 sfu),$$

where $T_{an} = 365.2425$ days is the duration of the year and t_d is the year's day.

The amplitudes and phases were approximated by polynominals as functions of altitude Z:

$$A_n(Z) = \sum_{k=0}^{9} a_{nk}\left(\frac{Z}{100}\right)^k, \ (K/100 sfu), \ t_n = \sum_{k=0}^{9} b_{nk}\left(\frac{Z}{100}\right)^k, \text{ day of year}.$$

The approximation coefficients a_{nk} and b_{nk} are presented in Tables 1.4 and 1.5. The approximation results, $\delta T_F(Z)$, are shown in Fig. 1.40 (solid lines); the points represent measurements smoothed with a moving 3-month interval to exclude overshoots – small temperature differences (2–5 (K)) which are determined from altitude profiles corresponding to different levels of solar activity. It should be noted that for the altitudes under consideration the mean temperature is about 200 (K). The

1.6 The Heliogeophysical Conditions that Determine the Observed Airglow Characteristics

Table 1.4 Coefficients of the polynominals approximating the amplitudes of harmonics of seasonal variations and the yearly average response of atmospheric temperature, $\delta T_F(Z)$ (K/100sfu), to solar activity for various altitudes Z

k\n	A_0	A_1	A_2	A_3	A_4	C_k
a_0	2294.450351169	6663.88846900	−7907.5031625477	568.8729284113	548.0546127556	2454.7934695891
a_1	−36416.60711669	−109052.78144879	131323.35208897	−8831.32549188	−8933.15801226	−39007.66753767
a_2	247480.079171	773501.348873	−946823.373052	59134.579031	63110.767205	265641.919681
a_3	−943273.8544	−3120348.2522	3888411.9955	−224597.6547	−253824.1148	−1015805.4182
a_4	2219021.46	7889517.39	−10022316.21	534508.2063	641005.09694	2401110.6
a_5	−3335853.2564	−12965000.681	16816097.04	−828083.24252	−1054800.6484	−3634261.1665
a_6	3196084.8393	13847149.89	−18376046.695	836270.14307	1131673.9461	3515754.0994
a_7	−1872927.0739	−9269663.7182	12620055.295	−531426.25566	−763846.9292	−2089066.8139
a_8	603831.9245	3530307.7479	−4946826.1827	193010.6978	294549.65432	687634.03284
a_9	−80238.16271	−583073.12421	844034.93286	−30553.548392	−49482.427885	−94450.573208

Table 1.5 Coefficients of the polynominals approximating the phases (t, days of year) of the harmonics of seasonal variations of the response of atmospheric temperature, $\delta T_F(Z)$, to solar activity for various altitudes Z

K\n	t_1	t_2	t_3	t_4
b_0	−267504.7107	−49509.855832473	−132133.797	58919.3400526946
b_1	4370155.04989605	816671.63585655	2111539.15280742	−860359.96044322
b_2	−30795784.099081	−5761036.625868	−14533819.237505	5326464.913867
b_3	122964740.4384	22899457.6277	56612791.6267	−18198039.0593
b_4	−306759896.43	−56620705.04	−137685900.49	37386373.38
b_5	496232816	90410945.023	217149602	−46987303.5
b_6	−521180110.25	−93320143.128	−222509674.84	34720496.438
b_7	343271461.89	60095431.996	143151697.37	−13014438.1
b_8	−128906015.73	−21927495.808	−52587131.654	1140482.4465
b_9	21070244.038	3456519.315	8422943.8177	427430.72961

most substantial distinction in the behavior of altitude profiles $\delta T_F(Z)$, especially in the mesopause region, is seen in Fig. 1.41. In this figure, to demonstrate the seasonal nature of the variations in altitude profiles $\delta T_F(Z)$, they have been grouped by 3 months near solstices ((1) November–December–January and (3) May–June–July) and equinoxes ((2) February–March–April and (4) August–September–October). It can be seen that practically at all altitudes of the middle atmosphere its temperature responds to solar activity throughout the year; therefore, the response of the atmospheric temperature takes negative or positive values depending on altitude. However, the yearly average temperature response (positive or negative) to variations in solar activity becomes appreciable only at altitudes above 70 (km) (see Fig. 1.41). The calculation of $\overline{\delta T_F}(Z)$ can be performed by the formula

$$\overline{\delta T_F}(Z) = \sum_{k=1}^{9} C_k \left(\frac{Z}{100}\right)^k, \text{ (K/100 sfu)},$$

whose coefficients are given in Table 1.4.

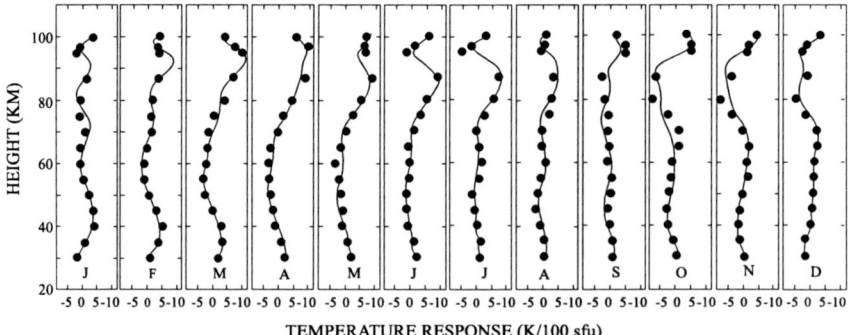

Fig. 1.40 Altitude distributions of the temperature response $\delta T_F(Z)$ to solar activity for various months of year. *Dots* are measurements; *lines* are approximations

1.6 The Heliogeophysical Conditions that Determine the Observed Airglow Characteristics 93

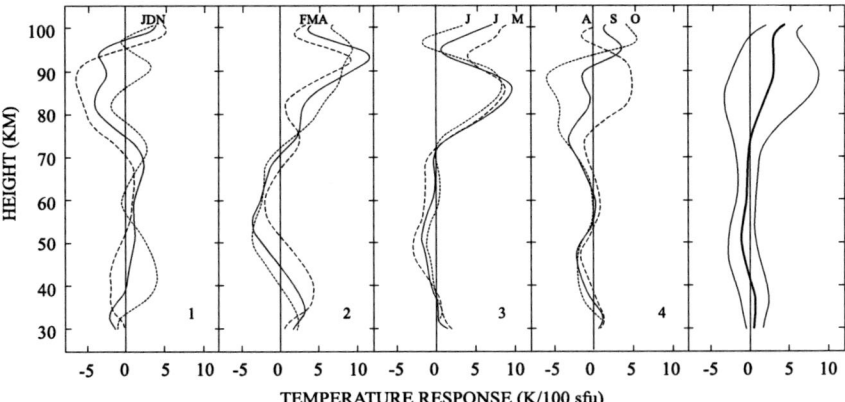

Fig. 1.41 Comparison of $\delta T_F(Z)$ model altitude profiles for different months of year combined around the winter (*1*) and the summer solstice (*3*) and around the vernal (*2*) and the autumnal equinox (*4*). The *thick line* on the right depicts the yearly average altitude distribution $\overline{\delta T}_F(Z)$. The *thin lines* show the limits of the variance due to seasonal changes

In this case, the rms error is everywhere severalfold smaller than the temperature response to solar activity, testifying to the importance of the values obtained. Thus, the greatest and the least seasonal variations are observed at 80–95 km (winter: about $-(5 \pm 1.7)\,(\text{K}/100\,\text{sfu})$, summer: about $+(8 \pm 1.7)\,(\text{K}/100\,\text{sfu}))$ and at 55–70 (km) (winter: about $+(2 \pm 0.4)\,(\text{K}/100\,\text{sfu})$, summer: about $-(1 \pm 0.4)\,(\text{K}/100\,\text{sfu}))$, respectively.

Beig et al. (2003) give values of $\delta T_F(Z)\,(\text{K}/100\,\text{sfu})$ for various latitudes without reference to seasons. This makes a comparison of these data with each other and with the results presented in Figs. 1.40 and 1.41 largely ambiguous. Thus, for example, for the high latitudes of the northern hemisphere based on the measurements at about 80 km in summer (69°N) (Lübken 2000) and at 87 (km) in winter (78°N) (Nielsen et al. 2002), the conclusion was made about no temperature response to solar activity. At the same time, a positive response of 5 (K/100 sfu) has been obtained for the high latitudes of the southern hemisphere (69°S) (French et al. 2000; French and Burns 2004). For the lower latitudes (59°N (Espy and Stegman 2002), 43°N (Offermann et al. 2002), and 43°N (Lowe 2002)), estimates of the response from 1.5 to 2.6 K/100 sfu are presented. As can be seen from Fig. 1.41, in the mesopause region the variations in the atmospheric response to the variations in solar activity during the indicated intervals are most pronounced in autumn, in winter, and in spring. For the summer months, the response practically does not vary, though it has a maximum value. Therefore, the spread in the above values probably results from the seasonal character of the observations. Nevertheless, it should be noted that the results obtained at Wuppertal (Offermann et al. 2002) practically comply with the yearly average response shown in Fig. 1.41.

The lidar measurements (1978–1989) performed at altitudes from 33 to 75 (km) (Hauchecorne et al. 1991) have also revealed a change in sign of the response

$\delta T_F(Z)$ at a certain altitude for winter and summer. It has been found that $\delta T_F(Z) \sim$ 5 (K/100 sfu) for winter and $\sim$3 (K/100 sfu) for summer at altitudes of 60–70 (km), where a maximum is observed, as well as about -6 (K/100 sfu) for winter and about -1 (K/100 sfu) for summer at about 40 (km), where a minimum is observed.

The monthly mean temperature distributions in the altitude range 83–105 (km) obtained from 1990–1999 lidar measurements are reported by She et al. (2000). The authors have noted seasonal variations in altitudes of the double-peak mesopause, but, unfortunately, they have made no attempt to relate the observed altitude profile of temperature to solar flux variations, notwithstanding that there was a maximum of solar activity in 1991 and a minimum in 1997.

Within the indicated periods, the least variations in $\delta T_F(Z)$ have been found for altitudes of 55–70 (km). At stratospheric altitudes (30–55 (km)), appreciable variations take place in winter and in spring. This seams to be related to the features of the altitude distributions of some reactive gas components and to the action of solar ultraviolet radiation that largely determine the altitude distribution of temperature. The obtained altitude profiles of the rates of temperature variations in the middle atmosphere at various altitudes as a function of solar flux are strongly nonlinear in altitude. The substantial seasonal variations of the effect of solar activity in the mesopause region seem to be due to the seasonal variations of the altitude distributions of temperature. Thus, the temperature conditions in the middle atmosphere appreciably respond to variations in solar activity, which should be taken into account in developing models.

1.6.4.5 The Lower Atmosphere

The effect of solar activity on the troposphere has been the subject of discussions for several tens of years. Much attention is paid to revealing the mechanisms by which the variable component of solar radiation affects the troposphere. This problem has active supporters who seek the phenomena through which solar activity may influence the climate of the Earth and the time scales of these phenomena. This seems to be quite natural since the Earth is in the region of the interplanetary space controlled by the Sun. The main challenge here has always been to explain by which mechanisms the variability of solar UV radiation is transferred and, in particular, why the energetics of this radiation does not compare with the energetics of the lower atmosphere of the Earth. Skeptics take into account only the long-term (some tens of thousands of years) variations in climate due to the variations in elements of the Earth's orbit.

Nevertheless, correlations between the long-term temperature variations in the ground atmosphere and the solar activity were traced repeatedly (Eddy 1976; Herman and Goldberg 1978; Lean 1991; Lean et al. 1995; Solanki and Fligge 2002; Zaitseva et al. 2003). Analysis of the studies devoted to the problem of solar–terrestrial relationships shows that all efforts of the researchers have been practically reduced to reveal possible variations in solar energy influx (Vitinsky et al. 1986; Monin 1982; Kondratiev and Nikolsky 1995; Schatten 1996). Examinations of the solar constant suggest that its variations during 11-yr cycles make no more than 0.6(%) (Lean 1991; Lean et al. 1995; Kondratyev et al. 1996; Solanki and Fligge 2002).

However, now the solution of the problem of solar activity during 11-yr and longer periods has advanced substantially. It became clear that there are several mechanisms of solar action, and they turned out radically different from those sought earlier.

First, the modulation of the flux of galactic cosmic rays by the solar wind results in variations in the degree of ionization of the atmosphere at altitudes of 10–20 (km), which varies the aerosol content at these altitudes, and, hence, the transparency of the atmosphere to visible radiation. As a result, increasing solar activity increases (by up to 6(%)) the solar influx to the Earth surface (Pudovkin and Raspopov 1992).

Nevertheless, the reason for the violation of energy balance should be sought not only in energy receipts, but also in its removal. It is well known that the basic energy removal channel is the thermal infrared radiation of the planet that is controlled by the contents of water vapor, carbon dioxide, methane, and some other small atmospheric constituents.

From the previous research it is well known that a long-term (of the order of several years) change in energy balance by $+1(\%)$ corresponds to the change of the average temperature of the Earth surface by about 1.5 (K) (Budyko et al. 1986). Therefore, it is little wonder that the long-term depression of solar activity during the $\sim$70-yr Maunder minimum period (1645–1715) was known to be accompanied by an appreciable fall of temperature (by $\sim$1 (K)) (Eddy 1976; Herman and Goldberg 1978).

However, due to active meteorological processes, the lower atmosphere generates, along with electromagnetic thermal radiation, a wide spectrum of oscillations, namely acoustic, internal gravitational, tidal, and planetary waves which transfer their energy and momentum to the upper atmosphere altitudes of $\sim$100 (km) (Chunchuzov 1978; Perminov et al. 2002; Shefov and Semenov 2004a,b; Pogoreltsev and Sukhanova 1993). Variations in the atmospheric characteristics in the region of the mesopause and lower thermosphere (80–100 (km)), which are detected by their night emissions and by the frequency of occurrence of noctilucent clouds, testify to the propagation of planetary waves with a period up to 30 days to these altitudes (Sukhanova 1996; Perminov et al. 2002; Shefov and Semenov 2004a,b).

It was supposed (Hines 1974) that the changes in the conditions of wave propagation in the middle atmosphere that result from the occurrence of an additional system of winds due to an increase in solar activity can lead to modulations of the wave flow and to its return action on the lower atmosphere, thus affecting the weather. However, this hypothesis has not been further developed (Herman and Goldberg 1978).

Kononovich and Shefov (1999a,b) pointed out that the most adequate assumption is that changes in the conditions of reflection and transmission of waves at altitudes of 50–80 (km) may result in either the removal of the energy of atmospheric waves after their dissipation in the form of UV radiation of O_3 and CO_2 molecules or the reflection of these waves back to the troposphere, thus reducing its energy losses. It has been shown (Garcia et al. 1984; Vergasova and Kazimirovsky 1994; Namboothiri et al. 1994; Petrukhin 1995) that the variations in solar UV radiation, which is absorbed at altitudes of 45–80 km, vary the wind mode (the speed changes up to twice) at these altitudes which, in turn, modulates the upward flow of atmospheric waves from the lower atmosphere. This induces variations in the additional sink of energy from the lower atmosphere.

To estimate the difference between the energy inputs at altitudes of 45–90 (km) at high and low solar activity, calculations have been performed for the yearly average conditions of the 1989 maximum ($F_{10.7} = 215$) and 1995 minimum ($F_{10.7} = 78$) (Lysenko et al. 1999; Semenov et al. 2002a).

The energy of UV solar radiation absorbed by a unit volume (1 (km) by 1 (cm^2)) of the Earth's atmosphere at altitude Z is determined as

$$E(Z) = 10^5 \cdot \sum n(Z) \cdot \int_{\lambda_1}^{\lambda_2} \sigma(\lambda) \cdot S(\lambda) \cdot \exp(-\tau \cdot \sec\chi) \cdot \cos\chi \cdot d\lambda,$$

where $\sigma(\lambda)$ is the effective absorption cross-section (cm^2); $S(\lambda)$ is the solar flux ($erg \cdot cm^{-2} \cdot s^{-1} \cdot nm^{-1}$); $n(Z)$ is the concentration of absorptive components – oxygen and ozone molecules, and χ is the solar zenith angle. The optical thickness τ is given by

$$\tau = \sum \sigma(\lambda) \cdot \int_Z^\infty n(Z')dZ'.$$

The solar energy absorbed by the illuminated hemisphere is determined by the formula

$$\overline{E}(Z) = 4 \cdot 10^5 \cdot R_E^2 \cdot \sum n(Z) \cdot \int_0^{\pi/2} \int_0^{\pi/2} \int_{\lambda_1}^{\lambda_2} \sigma(\lambda) \cdot S(\lambda) \cdot \exp(-\tau \cdot \sec\chi) \cdot \cos\chi \cdot \sin\chi \cdot d\lambda d\chi d\psi,$$

where R_E is the Earth's radius. The integration coordinates are taken relative to the subsolar point of the atmosphere, and ψ is the azimuth on the illuminated disk of the Earth.

After transformation, we have

$$\overline{E}(Z) = 2\pi \cdot R_E^2 \cdot 10^5 \cdot \sum n(z) \cdot \int \sigma(\lambda) \cdot S(\lambda) \cdot E_3(\tau) d\lambda,$$

where

$$E_3(\tau) = \int_1^\infty \frac{\exp(-\tau \cdot t)}{t^3} dt$$

1.6 The Heliogeophysical Conditions that Determine the Observed Airglow Characteristics

is the integral–exponential function calculated by its expansion into a continued fraction (Abramowitz and Stegun 1964). Thus, the global thermal energy of the atmosphere in a layer of thickness 1 km at altitude Z is given by

$$E_T(Z) = \pi \cdot 10^6 \cdot R_E^2 \cdot n^*(Z) \cdot kT(Z),$$

where $n^*(Z)$ is the concentration of the basic biatomic components.

As follows from the calculations performed, for altitudes of 45–90 km the quantity

$$\Delta = \frac{E_e(Z, \max)}{E_T(Z, \max)} - \frac{E_e(Z, \min)}{E_T(Z, \min)},$$

which describes the difference in solar energy inputs per day, $E_\odot$ ($86400\overline{E}$), related to the global thermal energy of the atmosphere, E_T, for the same atmospheric layer during the period of maximum (1989) and the period of minimum (1995), is 0.04–0.06. As already mentioned, the response of the dynamic conditions of the stratosphere and mesosphere to solar activity was pointed out in earlier publications (Garcia et al. 1984; Vergasova and Kazimirovsky 1994; Namboothiri et al. 1994; Petrukhin 1995).

It is of importance that at the equatorial latitudes from $-46°$ to $+46°$ the altitude of the turbopause and the intensity of turbulence at these altitudes (90–100 (km)), where dissipation of waves takes place) vary in antiphase with solar flux (Fig. 1.42) (Korsunova et al. 1985). This implies that an increase in intensity of solar ultraviolet radiation with solar activity produces a change in wind conditions in the mesospheric region. This hinders the upward propagation of waves and promotes their reflection back into the lower atmosphere. As a result, turbulence becomes moderate and the altitude of the turbophase decreases. It seems that this is also a manifestation of the change in spectral composition of waves.

According to the present notions, the penetration of planetary waves from the stratosphere of the winter hemisphere into the equatorial region can occur through the region of western mean-zonal winds at the mesospheric altitudes (Pogoreltsev and Sukhanova 1993; Sukhanova 1996). Long-term ionospheric observations at midlatitudes have detected planetary waves with periods of 5 and 10 days penetrating to the lower and upper ionospheric altitudes (Laštovička et al. 1994; Laštovička 1997), and the investigations of the frequency of occurrence of noctilucent clouds (82 (km)) have revealed planetary waves with a period of about 16 days (Shefov and Semenov 2004a).

In the polar range of latitudes above $60°$, the parameters of the turbophase vary in phase with solar activity. This is due to the corpuscular flows precipitated from the magnetosphere.

The solar constant is equal to $1.37 \cdot 10^6$ (erg $\cdot$ cm$^{-2} \cdot$ s^{-1}), the all-planetary daily mean solar energy is $1.75 \cdot 10^{24}$ (erg $\cdot$ s^{-1}), and the absorbed energy (at an albedo of 0.29) is $1.25 \cdot 10^{24}$ (erg $\cdot$ s^{-1}) (Monin 1982; Budyko et al. 1986). Hence, 1(%) of this value makes $1.2 \cdot 10^{22}$ (erg $\cdot$ s^{-1}) or $2.4 \cdot 10^3$ (erg $\cdot$ cm$^{-2} \cdot$ s^{-1}).

Fig. 1.42 Association of the monthly mean (*dots*) and seasonally mean values (*crosses*) of the altitude of the E$_S$ sporadic layer for winter months with solar activity (Wolf numbers W). Data of measurements performed at Heiss Island (80°N) (**a**), Moscow (56°N) (**b**), Ashkhabad (38°N) (**c**), and Djibouti (12°N) (**d**) (Korsunova et al. 1985)

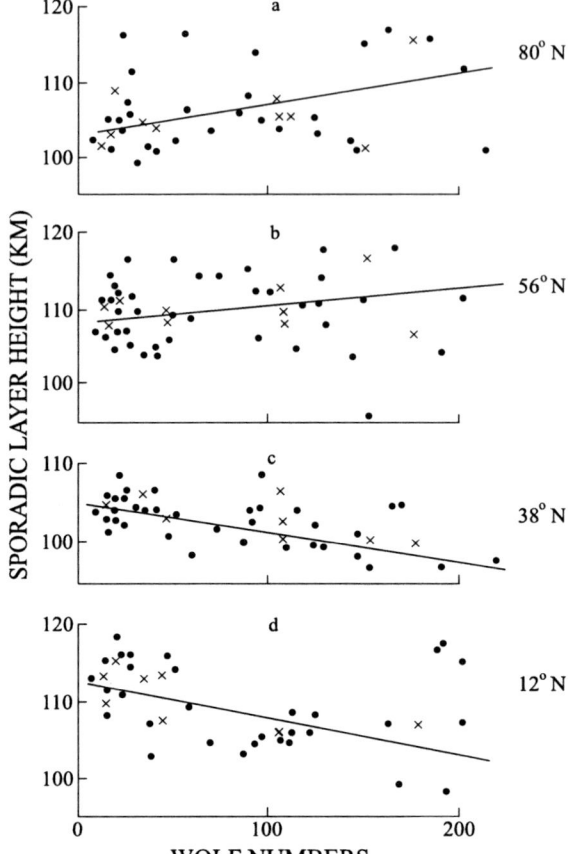

By today's data, the turbulence kinetic energy W at 100-km altitude is 700–2000 $(m^2 \cdot s^{-2})$ per unit mass (Chunchuzov 1978), which implies that on the average $W = 8 \cdot 10^3$ (erg $\cdot$ cm^{-2}) in a layer of thickness 20 (km). The maximum-to-minimum change ΔW also corresponds to this value. This characterizes mean pulsations in wind speed (30–40 $(m \cdot s^{-1})$). Hence, the turbulence energy at altitudes of 100 (km) makes 3(%) of the solar energy absorbed near the Earth surface at an area of 1 (cm^2) in 1 (s). It seems that the energy influx in the turbophase region is due to the entire spectrum of atmospheric waves. Therefore, for low and midlatitudes the turbulence at ~100 (km) is likely to be an integrated characteristic of the wave energy coming from the lower atmosphere.

Since waves may propagate to the altitudes of the ionospheric F2 region, the actual energy coming from the lower atmosphere should be greater than that obtained from the analysis of turbulence. Besides, the energy of the wind motions induced by dissipation of waves should also be considered. It seems quite admissible, as a first approximation, to double the above estimated energy. Nevertheless, it is obvious that the energetics of the turbopause is a measure of the wave energy influx.

Thus, it can be assumed that the observed minimum-to-maximum changes in turbulence kinetic energy, ΔW, are due to the changes in the conditions of wave propagation under the action of solar flux. The time it takes for a stationary state to establish at altitudes of $\sim 100\,(km)$, τ, is $\sim 3\,(h)$ (Chunchuzov 1978). Thus, the global change in lower atmosphere temperature ΔT in a time Δt can be estimated by the relation

$$\frac{5}{2} \cdot N \cdot k \cdot \Delta T = -2 \cdot \Delta W \cdot \frac{\Delta t}{\tau},$$

where N is the content of molecules in the lower atmosphere (cm^{-2}) and k is Boltzmann's constant. Hence, ΔT is about $-1\,(K)$ within $\Delta t \sim 100$ years. This is in rather good agreement with the Maunder minimum data (Eddy 1976) and with the variations in ground temperature measured within the 1960–1990 period in Equatorial Africa (Zaire), which show a distinct positive correlation with solar activity during an 11-yr cycle with an amplitude of 0.1–$0.2\,(°C)$ (Sanga-Ngoie and Fukuyama 1996).

In view of the above considerations, the removal of infrared radiation energy by means of CO_2 molecules in the mesopause region at altitudes of 80–100 (km) should have a negative correlation with solar activity. The energy removal is accompanied by the atomic oxygen layer going down during the period of high solar activity and its lower part becoming wider (Semenov and Shefov 1997c, 1999), promoting deactivation of excited CO_2 molecules by atomic oxygen and thus moderating the cooling of the atmosphere (Fomichev and Shved 1988).

The last statement is supported by the fact that the examinations of long-term variations in temperature of the middle atmosphere (Golitsyn et al. 1996, 2006) have shown that during the years of high solar activity at altitudes of 85–90 (km) a temperature maximum ($\Delta T \sim 10$–$15\,(K)$) is clearly detected below and above which there are minima, while in the period of low solar activity, preferentially in summer, there is one wide minimum. The amplitude of the maximum intimately correlates with solar activity and with the atomic oxygen content at altitudes of 85–95 (km).

Thus, the additional energy removal decreases if solar flux increases; otherwise it increases. This could produce global cooling of the lower atmosphere by 1 (K) in about 100 years if solar activity would be low for a long time. It should be stressed that both the above mechanisms act in the same way, providing an increase in temperature with increasing solar activity.

1.6.5 Characteristics of the Geomagnetic Field: Their Variations in Space and Time

The magnetic field of the Earth has a constant component – the basic field (its contribution being $\sim 99(\%)$) – and a variable component ($\sim 1(\%)$). The basic field is similar in configuration to the field of a dipole whose center is displaced relative to the Earth center and the axis is inclined to the rotation axis of the Earth by $11.5°$.

The departures from the dipole field, having a characteristic size of 10^4 (km) on the Earth surface and the maximum strength up to 10^{-5} (T), form the so-called world magnetic anomalies.

The terrestrial magnetic field and the solar wind plasma strongly affect the manifestations of solar activity in the upper atmosphere. The vector components of the field B in the Cartesian coordinate system are governed by three directions: the true north (X-component), the east (Y-component), and the vertical downward direction (Z-component). Besides, the following field elements are widely used: the declination D – the angle between the true north and the field horizontal component, countered from the direction to the north clockwise; the inclination I – the angle between the field direction and the horizontal plane, countered in the direction downward from the horizontal; and the horizontal component H (Yanovsky 1953; Akasofu and Chapman 1972; Parkinson 1983). The field elements are related as follows (Fig. 1.43):

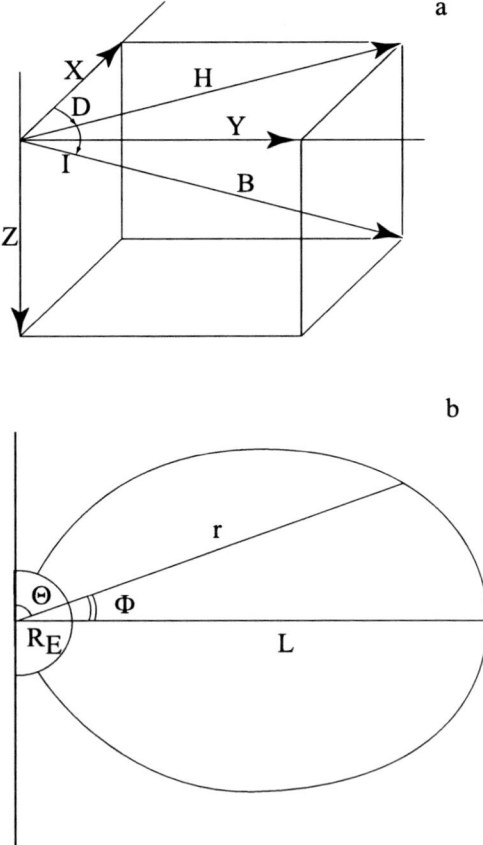

Fig. 1.43 Elements of the geomagnetic field. (a) Vectors of the geomagnetic field. (b) Meridional cross-section of a field line

1.6 The Heliogeophysical Conditions that Determine the Observed Airglow Characteristics

$$tgD = \frac{Y}{X}, \quad sinD = \frac{Y}{H}, \quad H = \sqrt{X^2 + Y^2}, \quad tgI = \frac{Z}{H},$$

$$sinI = \frac{Z}{B}, \quad B^2 = H^2 + Z^2 = X^2 + Y^2 + Z^2.$$

In the CGS system, the unit of magnetic field strength B is Gauss (G), which has dimensionality ($g^{1/2} \cdot cm^{-1/2} \cdot s^{-1}$). In geomagnetical practice, in analyzing the variations of geomagnetic field elements, the gamma unit is used which is equal to one nanotesla: $\gamma = 10^{-5}$ (G) = 1 (nT).

As a first approximation, the geomagnetic field at altitudes of the upper atmosphere can be represented by a dipole which is inside the Earth (actually, the distance from the center varies with time and now it is ~ 450 (km)). The geographic coordinates of the northern geomagnetic pole (central dipole) are: $\varphi_{gm} = 78.5°N$, $\lambda_{gm} = 291°E$. For the southern geomagnetic pole (central dipole) they are: $\varphi_{gm} = 78.5°S$, $\lambda_{gm} = 111°E$.

The radius vector of a field line is given by $r = L \cdot \cos^2$, where L is the equatorial distance of the line. If we use the geomagnetic latitude Φ_E at which the field line intersects the Earth surface, i.e., the relation $R_E = L \cdot \cos^2 \Phi_E$, we have

$$r = R_E \cdot \frac{\cos^2 \Phi}{\cos^2 \Phi_E}.$$

Therefore, when moving along a field line affixed to a reference point on the Earth surface, the altitude Z corresponds to a certain latitude Φ, i.e.,

$$Z = \left[\frac{\cos^2 \Phi}{\cos^2 \Phi_E} - 1 \right] \cdot R_E.$$

If we use the upper altitude of the auroral zone, Z_{aur}, as a starting point, we have

$$Z = (Z_{aur} + R_E) \cdot \frac{\cos^2 \Phi}{\cos^2 \Phi_E} - R_E.$$

The geomagnetic latitude of a point having an altitude Z and lying on the field line that passes through the reference point is given by

$$\cos \Phi = \sqrt{\frac{Z + R_E}{Z_{aur} + R_E}} \cdot \cos \Phi_{aur}.$$

The arc length s along the field line from the equator to the geomagnetic latitude Φ is determined by the formula (Mlodnosky and Helliwell 1962)

$$s = \frac{1}{2} \cdot \frac{R_0}{\sqrt{3} \cdot \cos \Phi_E} \cdot \left[\frac{\arcsin(\sqrt{3} \cdot \sin \Phi)}{Z} + \sqrt{3} \cdot \sin \Phi + \sqrt{1 - 3 \cdot \sin^2 \Phi} \right].$$

The field strength at some point of a field line is given by the formula (Akasofu and Chapman 1972)

$$B = \frac{M}{r^3} \cdot \sqrt{1 + 3 \cdot \sin^2 \Phi},$$

where $M = (15.77 - 0.003951 \cdot t) \cdot 10^{25} (G \cdot cm^{-3})$ is the magnetic moment of the dipole located at the center of the Earth. Here t is the time (in years) countered from the year 1900.

This expression can be represented in another form:

$$B = B_L \cdot \frac{\sqrt{1 + 3 \cdot \sin^2 \Phi}}{\cos^6 \Phi}, \quad \text{where } B_L = \frac{M}{L^3} = \frac{M}{R_E^3} \cdot \frac{1}{L_0^3} = \frac{M_0}{L_0^3}.$$

Here $M_0 = 6.078 - 0.0015228 \cdot t\,(G)$ and L_0 is expressed in Earth radii. For the year 1990 we have

$$M_0 = 5.94\,(G), \quad M = 1.54 \cdot 10^{26}\,(G \cdot cm^{-3}).$$

A widely used measure of geomagnetic disturbance is the K index that corresponds to the 3-hour interval beginning at the midnight of universal time. For each observatory the range of the field components H and D is chosen to accord integer K values from 0 to 9 (Yanovsky 1953; Akasofu and Chapman 1972; Patel 1977; Parkinson 1983). For midlatitudes, K = 9 corresponds to $\Delta H \sim 500 (\gamma)$; for the polar zone, ΔH can be 1000–2000 (γ). The lower limit for each unit of the scale is about twice the lower limit for the previous unit. Therefore, the scale of K indices appears to be nearly logarithmic. The linear index a_k is obtained by the transformation of every K index into an equivalent field variation interval. The value of K averaged over the data of chosen 12 observatories is called the planetary Kp index to which there corresponds an a_P index. The relationship between them is given by the following table:

Kp	0	1	2	3	4	5	6	7	8	9
a_P	0	2	7	15	27	48	80	140	240	400

The empirical analytic relationship between them is determined by the formulas (Jacchia 1979)

$$a_P = 6.5 \cdot \text{sh}(0.535 \cdot Kp), \quad Kp = 1.89 \cdot \text{Arsh}(0.154 \cdot a_P).$$

For the polar region, the 15-min logarithmic Q index is used which characterizes the maximum diversion of the more disturbed geomagnetic field horizontal component of the two from the undisturbed state near the midnight. The Q indices vary from 0 to 10. To describe on the average the boundaries of the auroral oval, we can use the empirical relation

$$Q = Kp + 2.4 \cdot \exp\left[-\frac{(Kp - 6)^2}{9}\right].$$

An auroral electrojet (which is responsible for fluctuations of the geomagnetic field) is characterized by the AU and AL indices which are the envelopes of

1.6 The Heliogeophysical Conditions that Determine the Observed Airglow Characteristics

deviations of the H component from the average value within a 2.5-min interval of universal time (Akasofu and Chapman 1972; Patel 1977; Parkinson 1983). The AE index is defined as AE = AU − AL and, on the average, can be described rather approximately by the expression

$$\log_{10} \text{AE} = 1.3 + 3.3 \cdot \frac{\text{Kp}}{10} - 2.7 \cdot \left(\frac{\text{Kp}}{10}\right)^2 + 1.36 \cdot \left(\frac{\text{Kp}}{10}\right)^4.$$

The magnetic field varies during a geomagnetic storm not only due to an auroral electrojet, but also due to a ring current of 10- to 120-keV protons which occurs in the magnetosphere at distances of 3–5 Earth radii. This distortion (diminution) of the field horizontal component is described by the D_{st} index (Akasofu and Chapman 1972; Patel 1977; Parkinson 1983). It is proportional to the total kinetic energy of the injected particles, E_{KT}, that are trapped in the radiation belt, provided that the particles are distributed symmetrically about the dipole axis (Akasofu and Chapman 1972). This energy is determined by the formula (Sckopke 1966; Akasofu and Chapman 1972)

$$E_{KT} = -\frac{1}{2} \cdot M \cdot \Delta B_Z,$$

where M is the magnetic moment of the Earth and ΔB_Z is the magnetic field of the ring current at the origin, or

$$E_{KT} = -7.7 \cdot 10^{20} \cdot D_{st} \text{ (erg)},$$

where D_{st} is expressed in (nT).

Since $|D_{st}|_{max} \approx |a_P|_{max}$ (Valchuk and Feldstein 1983), based on the formula given by Jacchia (1979), we have on the average

$$D_{st} = -6.5 \cdot \text{sh}\left[0.535 \cdot K_P(\tau' = \tau - 9)\right],$$

because the maximum values of Kp are $6 \div 12$ (h) ahead of the maximum values of D_{st}. The quantity D_{st} is proportional to the sum of the AE indices for the previous 10 (h). The studies devoted to geomagnetic indices are reviewed elsewhere (Lincoln 1967; Rostoker 1972; Mayaud 1980).

The interplanetary magnetic field (IMF), which is transferred by solar wind, substantially affects the intensity of the processes in the circumterrestrial space. The IMF is directed along an Archimedean spiral toward the Sun or opposite to it, following the IMF segmented structure (usually four segments). If the IMF is directed opposite to the Sun, the spiral branches run against the Earth's magnetic axis at an angle of 45° to the Sun–Earth line. In this case, in relation to the Earth's magnetic axis, the greatest southern component of the interplanetary field is observed in August, while the greatest northern component is observed in February. The equatorial plane of the Sun makes an angle of 7.2° with the ecliptic. This shows up in the shift of the season of the maximum southern component (for the IMF directed outward) to October and in the shift of the season of the maximum northern component to April. Since the occurrence of magnetic disturbances is more probable when the IMF has a southern component there are two magnetic disturbance maxima: one in October, related to the IMF directed outward, and the other in April, which corresponds to the IMF directed inward (Akasofu 1968; Akasofu and Chapman 1972; Svalgaard 1977; Ponomarev 1985; Паркинсон 1983).

Some types of disturbance of the geomagnetic field, which are characterized by the indices mentioned above, show an intimate correlation with the IMF parameters (Valchuk and Feldstein 1983). Thus, the daily mean values of the sum of K_P indices, solar wind velocity $\bar{V}$, IMF vertical component B_z, and variability of this component, σ, are related by the correlation formula

$$\sum K_P = 5.1 + 1.03 \cdot \bar{V} \cdot \bar{\sigma} - 0.51 \cdot \bar{V} \cdot B_z$$

with the correlation factor equal to 0.9.

Based on the data of long-term observations, detailed multiple correlation relations have been obtained which describe the aa indices in terms of the quantities V, σ, B_S (average value of the IMF southern component for the hours with $B_z < 0$), and B_z. With adequate accuracy (correlation coefficient is 0.8) we have

$$aa = 3.7 + 0.15 \cdot B \cdot V^2,$$

where B is expressed in nT and the solar wind speed in $100 \, (km \cdot s^{-1})$.

For the yearly average values we have

$$\overline{aa} = 1.6 \cdot \bar{V}^2 - 10.$$

The inverse relation is

$$\bar{V}^2 = 0.67 \cdot \overline{aa} + 4.78$$

with the correlation coefficient equal to 0.9.

The geomagnetic disturbances that show up in the behavior of various geomagnetic indices have a strong effect on the parameters of the neutral and ionized components of the upper atmosphere, and, hence, on its airglow.

The density of the upper atmosphere varies, depending on geomagnetic activity, is proportional to the K_p and AE indices (Truttse 1973). It has been revealed that the intensity (in Rayleighs) of the low-latitude red lights caused by the 630-nm emission of atomic oxygen can be represented by the empirical relation

$$\log_{10} I_{630} = \frac{F_{10.7} - 160}{50} + \frac{34.3 - |\Phi|}{1460} \cdot D_{st}.$$

Investigations based on the data of observations performed during the International Geophysical Year (1957–1959) have shown that in the periods of high solar activity ($F_{10.7} \geq 250$) during great geomagnetic disturbances the low-latitude red lights cover a wide range of latitudes, from 20° to 80°, and can have intensities up to 30–1000 (megarayleigh). As can be seen from the given formula at reduction of solar activity there is a sharp decrease of probability of occurrence of low latitudinal auroras.

1.6 The Heliogeophysical Conditions that Determine the Observed Airglow Characteristics

An important feature of the processes related to the geomagnetic field is the effects of conjugate fields in the northern and southern hemispheres. However, since the orientation of the geomagnetic field does not comply with geographic coordinates, the effects caused by twilight phenomena, such that the Sun starts irradiating the southern part of the upper atmosphere earlier than the northern part for the same geomagnetic longitudes, become substantial. These effects are responsible for the predawn enhancement of 630-nm emission and for the escape of photoelectrons and ions from the conjugate-illuminated region of the atmosphere into the dark region of the Earth's atmosphere.

When considering the processes in the auroral oval that are induced by charged particles moving along magnetic field lines, we see that the center of rotation of global statistical distributions is not the geographic pole, but the geomagnetic pole corrected for the shift of the dipole from the Earth center. Such distributions in corrected geomagnetic coordinates are described elsewhere (Hultqvist 1958; Sverdlov 1982; Sverdlov and Khorkova 1982). The geographic coordinates of the corrected northern geomagnetic pole for 1955 (Cole 1963) are

$$\varphi_{gm}' = 81.0°N, \quad \lambda_{gm}' = 275.3°E,$$

and those for the south pole are

$$\varphi_{gm}' = -75.0°N, \quad \lambda_{gm}' = 120.4°E.$$

To estimate the position of the auroral oval, it suffices to use its coordinates in the system of an eccentric dipole. Following Akasofu and Chapman (1972), the geographic coordinates of the projection of a displaced dipole on the Earth surface are

$$\varphi_0 = 0.1775 \cdot \left(YYYY + \frac{t_d}{365}\right) - 331.1, \quad \lambda_0 = -0.240 \cdot \left(YYYY + \frac{t_d}{365}\right) + 620.$$

The shift Δ from the center of the Earth is about 500 (km), and it determines the relationship between the geographic coordinates. Based on the data of Akasofu and Chapman (1972), we have

$$\frac{\Delta}{R_E} = 5.8747 - 6.323 \cdot 10^{-3} \cdot \left(YYYY + \frac{t_d}{365}\right) + 1.7145 \cdot 10^{-6} \cdot \left(YYYY + \frac{t_d}{365}\right)^2.$$

The corresponding distance between the axes of the central and eccentric dipoles is given by the formula

$$d = \frac{\Delta}{R_E} \cdot \frac{\cos\varphi_0 \cdot \sin(\lambda_0 - \lambda_{gm})}{\sin\Lambda_0},$$

or

$$d = 5.1775 - 5.587 \cdot 10^{-3} \cdot \left(YYYY + \frac{t_d}{365}\right) + 1.52023 \cdot 10^{-6} \cdot \left(YYYY + \frac{t_d}{365}\right)^2.$$

The geomagnetic coordinates of the point (φ_0, λ_0) in the coordinates of the central dipole are

$$\Phi_0 = 0.1496 \cdot \left(YYYY + \frac{t_d}{365}\right) - 285.5, \quad \Lambda_0 = -0.250 \cdot \left(YYYY + \frac{t_d}{365}\right) + 707.7,$$

where YYYY is the number of year and t_d is the serial number of the day of the year.

Kubova (1989) proposed to calculate geomagnetic coordinates at high latitudes as invariant coordinates; this simplifies calculations and provides an error less than 1°. In this case, the first step is the calculation of geomagnetic coordinates Φ' and Λ' in the dipole approximation (see Sect. 1.4.1).

As a first approximation, the invariant coordinates can be represented as dipole geomagnetic coordinates whose pole is shifted depending on the latitude of the observation point. These shifts are convenient to describe with interpolations in terms of geographic latitude φ:

$$\varphi_{gm} = \varphi_N \cdot \frac{1 + \sin\varphi}{2} + \varphi_S \cdot \frac{1 - \sin\varphi}{2}, \qquad \lambda_{gm} = \lambda_N \cdot \frac{1 + \sin\varphi}{2} + \lambda_S \cdot \frac{1 - \sin\varphi}{2}.$$

Here $\varphi_N = 81°$, $\lambda_N = -80°$, $\varphi_S = 74°$, and $\lambda_S = -52°$ are the reduced coordinates of the dipole pole for the northern and the southern hemispheres, respectively. After substitution of the values of φ_{gm} and λ_{gm} in the formulas for the dipole geomagnetic coordinates, we obtain, as a first approximation, the invariant coordinates that fit to those calculated by exact methods to within 5°.

In the second approximation, a correction is introduced for the invariant latitude which takes into account the asymmetry and "ellipticity" of the isolatitude lines:

$$\Phi = \Phi' + \Delta \cdot \cos\Phi'.$$

The coefficient Δ is represented as the sum of three terms:

$$\Delta = \delta_0 + \delta_1 \cdot \left\{\left[\cos(\Lambda' + \Lambda_1) + 1\right]/2\right\}^2 + \delta_2 \cdot \cos^2(\Lambda' + \Lambda_2).$$

The empirical coefficients are calculated by the formulas

$$x_i = x_i^N \cdot \frac{1 + \sin\Phi'}{2} + x_i^S \cdot \frac{1 - \sin\Phi'}{2},$$

where $x = (\delta_0, \delta_1, \delta_2, \Lambda_1, \Lambda_2)$ and

$$\delta_0^N = -5°, \; \delta_1^N = 3°, \; \delta_2^N = 9°, \; \Lambda_1^N = -170°, \; \Lambda_2^N = -180°,$$
$$\delta_0^S = -10°, \; \delta_1^S = 10°, \; \delta_2^S = 10°, \; \Lambda_1^S = -240°, \; \Lambda_2^S = -160°.$$

In the geomagnetic latitude ranges 40–90°N and 50–90°S, the calculated invariant latitudes have an error no more than 1° in comparison with their true values.

1.6 The Heliogeophysical Conditions that Determine the Observed Airglow Characteristics

The auroral ovals are ring zones of variable width over the polar latitudes of the northern and southern hemispheres within which the precipitation of electrons and protons from the magnetosphere is a maximum. As a result, wave and dynamic atmospheric disturbances of various kinds propagate from these regions to the mid and low latitudes of the atmosphere.

According to Starkov's formulas (1994b), which describe the coordinates of an auroral oval, the polar and equatorial boundaries of the oval can be described, to within some fractions of a degree, by circles whose centers are shifted from the geomagnetic pole toward the midnight part of the oval by $\Delta\Phi_P$ and $\Delta\Phi_E$ and have radii R_P and R_E, respectively. Their dependences on the level of geomagnetic disturbance can be represented by the formulas

$$R_P = 15.51 + 1.29Q - 0.365Q^2 + 0.0270Q^3,$$
$$R_E = 17.14 + 2.125Q - 0.385Q^2 + 0.0321Q^3 + (F_{10.7}240)/340,$$
$$\Delta\Phi_P = 3.05 + 0.538Q - 0.293Q^2 + 0.0247Q^3,$$
$$\Delta\Phi_E = 3.01 + 1.369Q - 0.307Q^2 + 0.0232Q^3,$$

where the quantities calculated are expressed in degrees. The effect of solar activity has been taken into account in view of the data reported elsewhere (Feldstein et al. 1968; Uspensky and Starkov 1987).

The 15-min decimal logarithmic Q index applied to the polar region characterizes the maximum deviation of the more disturbed geomagnetic field horizontal component of the two near the midnight from the undisturbed state. To describe the oval boundaries on the average, the above empirical relation can be used which, for the conditions of the problem under consideration, allows one to use an available K_P index.

The latitude Φ on the boundary of the oval is determined by the formulas

$$\sin\Phi = \frac{\cos R \cdot \cos\Delta\Phi - \sqrt{\sin^2 R - \sin^2 \Delta\Phi \cdot \sin^2 \tau_G} \cdot \sin\Delta\Phi \cdot \cos\tau_G}{1 - \sin^2 \Delta\Phi \cdot \sin^2 \tau_G}$$

$$\cos\Phi = \frac{\cos R \cdot \sin\Delta\Phi \cdot \cos\tau_G + \sqrt{\sin^2 R - \sin^2 \Delta\Phi \cdot \sin^2 \tau_G} \cdot \cos\Delta\Phi}{1 - \sin^2 \Delta\Phi \cdot \sin^2 \tau_G},$$

where τ_G is the local geomagnetic time countered from the local geomagnetic midnight.

The latitude of the oval centerline corresponding to the maximum energy of precipitating electrons is

$$\Phi_M = (\Phi_P + \Phi_E)/2.$$

The definition of geomagnetic time is given in Sect. 1.4.5.

Thus, using the geographic coordinates of the geomagnetic pole for a given year, it is necessary to calculate first the geomagnetic coordinates of the observation point and then the required coordinates of the oval that are on the geomagnetic meridian of the observation point.

References

Abalakin VK, Aksenov EP, Grebennikov EA, Demin VG, Ryabov YuA (1976) Handbook for celestial mechanics and astrodynamics, 2nd edn. Duboshin GN (ed) Nauka, Moscow

Abalakin BK (1979) Principles of ephemeride astronomy. Nauka, Moscow

Abramowitz M, Stegun IA (1964) Handbook of mathematical functions with formulas, graphs and mathematical tables. Natl Bureau Standard, Washington

Abreu VJ, Yee JH, Hays PB (1982) Galactic and zodiacal light surface brightness measurements with the Atmosphere Explorer Satellites. Appl Opt 21:2287–2290

Ajello JM, Witt N (1979) Simultaneous H(1216 A) and He(584 A) observations of the interstellar wind by Mariner 10. In: Rycroft MJ (ed) Space Research, Vol 19. Pergamon Press, Oxford, pp 417–420

Akasofu SI (1968) Polar and magnetospheric substorms. D Reidel Publ Co, Dordrecht, Holland

Akasofu SI, Chapman S (1972) Solar-terrestrial physics. The Clarendon Press, Oxford

Albitsky VA, Vyazanitsyn VP, Deutsch AN, Zeltser MS, Krat VA, Markov AB, Meiklyar PV, Melnikov OA, Nikonov VB, Sobolev VV, Shain GA, Sharonov VV (1951) Spectrophotometry. In: Mikhailov AA (ed) A course of astrophysics and celestial astronomy, Vol 1. Methods and instruments. Gostechizdat, Moscow, pp 462–489

Albitsky VA, Melnikov OA (1973) Spectral classification of the stars. In: Mikhailov AA (ed) A course of astrophysics and celestial astronomy, Vol 1. Methods and instruments, 2nd edn. Nauka, Moscow, pp 312–329

Alekseev AV, Kabanov MV, Kushtin IF, Nelyubov NF (1983) Optical refraction in the terrestrial atmosphere (slanting traces). Nauka, Novosibirsk

Allen CW (1955) Astrophysical quantities. The Athlone Press, London

Allen CW (1973) Astrophysical quantities, 3rd edn. The Athlone Press, London

Antonova LA, Ivanov-Kholodny GS (1989) Solar activity and ionosphere (at heights of 100–200 km). Nauka, Moscow

Antonova LA, Ivanov-Kholodny GS, Chertoprud VE (1996) Aeronomy of the E layer. Yanus, Moscow

Apostolov EM (1985) Quasi-biennial oscillation in sunspot activity. Bull Astron Inst Czechosl 36:97–102

Apostolov EM, Letfus V (1985) Quasi-biennial oscillations of the green corona intensity. Bull Astron Inst Czechosl 36:199–205

Astafieva NM (1996) Wavelet analysis: basic theory and some applications. Uspekhi Fiz Nauk 166:1145–1170

Baliunas SL, Donahue RA, Soon WH, Horne JH, Frazer J, Woodard-Eklund L, Bradford M, Rao LM, Wilson OC, Zhang Q, Bennett W, Briggs J, Carroll SM, Duncan DK, Figueroa D, Lanning HH, Misch A, Mueller J, Noyes RW, Poppe D, Porter AC, Robinson CR, Russell J, Shelton JC, Soyumer T, Vaughan AH, Whitney JH (1995) Chromospheric variations in main-sequence stars. II. Astrophys J 438:269–287

Banks PM, Kockarts G (1973a) Aeronomy. Pt A. Academic Press, New York

Banks PM, Kockarts G (1973b) Aeronomy. Pt B. Academic Press, New York

Baranov DG, Vernova ES, Tyasto MI, Alaniya MV (2001) Features of the time behavior of the amplitude of 27-day variations in galactic cosmic rays. Geomagn Aeronomy 41:162–167

Bateman G, Erdélyi A (1953) Higher transcendential functions. McGraw-Hill Book Co, New York

Bates DR (1982) Airglow and auroras. In: Massey HSW, Bates DR (eds) Applied atomic collision physics, Vol 1. Atmospheric physics and chemistry. Academic Press, New York, pp 149–224

Bates DR (1984) Rayleigh scattering by air. Planet Space Sci 32:785–790

Beig G, Keckhut P, Lowe RP, Roble R, Mlynczak MG, Scheer J, Fomichev V, Offermann D, French WJR, Shepherd MG, Semenov AI, Remsberg E, She CY, Luebken FJ, Bremer J, Clemesha BR, Stegman J, Sigernes F, Fadnavis S (2003) Review of mesospheric temperature trends. Rev Geophys 41:1015. doi:10.1029/2002RG000121

References

Blackwell DE, Ingham MF (1961) Observations of the Zodiacal Light from a very high altitude station. I. The average Zodiacal Light. Mon Not Roy Astron Soc 122:113–176

Bocharova NYu, Nusinov AA (1983) On the possibility of a forecast of radio emission flux from individual active regions and the entire solar disc. Solar data Bulletin N 1. Nauka, Leningrad, pp 106–110

Bolunova AD, Mulyarchik TM, Shuiskaya FK (1977) Transfer of the escaping photoelectrons in the magnetosphere. Cosmic Res 15:445–454

Bossy L (1983) Solar indices and solar u.v.-radiations. Planet Space Sci 31:977–985

Bossy L, Nicolet M (1981) On the variability of Lyman-alpha with solar activity. Planet Space Sci 29:907–914

BothmerV, Veselovsky IS, Dmitriev AV, Zhukov AN, Cargill P, Romashets EP, Yukovchuk OS (2002) Solar and heliospheric reasons for geomagnetic perturbations during the growth phase of solar cycles. Solar System Res 36:539–547

Boulanger F, Pérault M (1988) Diffuse infrared emission from the Galaxy. 1. Solar neighborhood. Astrophys J 330:964–985

Bowyer CS, Livingston PM, Price RD (1968) Upper limits to the 304 and 584 A night helium glow. J Geophys Res 73:1107–1111

Bronshten VA, Dagaev MM, Kononovich EV, Kulikovsky PG (1981) Astronomical calendar. Invariable part, 7th edn. Abalakin VK (ed) Nauka, Moscow

Bruevich EA, Katsova MM, Sokolov DD (2001) Coronal and chromospheric activity of late-type stars and types of dynamo waves. Astron Rep 78:827–838

Budyko MI, Golitsyn GS, Izrael YuA (1986) Global climatic catastrophes. Hydrometeoizdat, Moscow

Chamberlain JW (1961) Physics of the Aurora and Airglow. Academic Press, New York

Chamberlain JW (1978) Theory of planetary atmospheres. Academic Press, New York

Chapman S, Lindzen RS (1970) Atmospheric tides. D Reidel Publ Co, Dordrecht, Holland

Chistyakov VF (1997) Solar cycles and the climate variations. Trans Ussurysk phys observ, N 1. Dalnauka, Vladivostok, pp 1–156

Chunchuzov YeP (1978) On energy balance characteristics of the internal gravity waves observed from hydroxyl emission near the mesopause. Izvestiya USSR Acad Sci Atmos Oceanic Phys 14:770–772

Cebula RP, Feldman PD (1982) Ultraviolet spectroscopy of the Zodiacal light. Astrophys J 263:987–992

Chumak OV, Kononovich EV, Krasotkin SA (1998) Prognosis of the next two solar cycles. Astron Astrophys Trans 17:41–44

CIRA-1972. COSPAR International Reference Atmosphere. Stickland AC (ed) Akademie-Verlag, Berlin

CIRA-1986 (1990) COSPAR International Reference Atmosphere: 1986. Part II: Middle Atmosphere Models, Rees D, Barnett JJ, Labitzke K Adv Space Res 10:1–525

Clayton HH (1884a) A lately discovered meteorological cycle. I. Amer Meteorol J 1:130–143

Clayton HH (1884b) A lately discovered meteorological cycle. II. Amer Meteorol J 1:528–534

Cleveland BT, Daily T, Davis R, Distel JR, Lande K, Lee CK, Wildenhain PS, Ullman J (1998) Measurement of the solar neutrino flux with the Homestake chlorine detector. Astrophys J 496:505–511

Clough HW (1924) A systematically varying period with an average length of 28 months in weather and solar phenomena. Mon Weather Rev 52:421–439

Clough HW (1928) The 28-month period in solar activity and corresponding periods in magnetic and meteorological data. Mon Weather Rev 56:251–264

Cole KD (1963) Eccentric dipole coordinates. Austral J Phys 16:423–429

Danilov DL, Zhigljavsky AA (1997) Principal components of the time series: "Caterpillar" method. S. Peterburg University Press, S. Peterburg

Davies K (1969) Ionospheric radio waves. Blaisdell Publ Co, Waltham, Massachusetts

Davis R (2002) A half-century with solar neutrinos. Les Prix Nobel. Nobel Lectures 2002. http://nobelprize.org/nobelprizes/physics/laureates/2002/davis-lecture.pdf

Desaubies YJF (1973) Internal waves near the turning point. Geophys Fluid Dyn 5:143–154
Divari NB (1951) Stellar component of the night sky luminosity. Astron Rep 28:163–171
Divari NB (2003) Zodiacal light. Astroprint, Odessa
Eddy JA (1976) The Maunder minimum. Science 192:1189–1202
Espy PJ, Stegman J (2002) Trends and variability of mesospheric temperature at high–latitudes. Phys Chem Earth 27:543–553
Evlashin LS, Semenov AI, Shefov NN (1999) Long-term variations in the thermospheric temperature and density on the basis of an analysis of Störmer's aurora-height measurements. Geomagn Aeronomy 39:241–245
Fadel KhM, Semenov AI, Shefov NN, Sukhodoev VA, Martsvaladze NM (2002) Quasibiennial variations in the temperatures of the mesopause and lower thermosphere and solar activity. Geomagn Aeronomy 42:191–195
Fatkullin MN (1982) Physics of the ionosphere. In: Total results of the Science and Technique. Geomagnetism and upper layers of the atmosphere, Vol 6.VINITI, Moscow, pp 4–224
Fedorov VV, Glazkov VN, Bugaeva IV, Tarasenko DA (1994) On the connection between the quasi biennial oscillations of the equatorial circulation and the atmospheric parameter variations. Meteorol Hydrolo N 10:24–30
Feldstein YaI, Lukina LV, Shevnina NF (1968) Aurora during the minimum and maximum cycle of solar activity. In: Isaev SI, Feldstein YaI (eds) Aurorae. N 17. Nauka, Moscow, pp 50–58
Fishkova LM (1983) The night airglow of the Earth mid-latitude upper atmosphere. Shefov NN (ed) Metsniereba, Tbilisi
Flynn BC, Vallerga JV, Gladstone GR, Edelstein J (1998) Lunar reflectively from extreme ultraviolet explorer imaging and spectroscopy of the full Moon. Geophys Res Lett 25:3253–3256
Fomichev VI, Shved GM (1988) Planetary distribution of the total radiative heating effect in the middle atmosphere and sequences for the atmospheric dynamics. In: Lysenko IA (ed) Studies of the dynamic processes in the upper atmosphere. Hydrometeoizdat, Moscow, pp 40–43
French WJR, Burns GB (2004) The influence of large-scale Oscillations on long-term trend assessment in hydroxyl temperatures over Davis, Antarctica. J Atmos Solar-Terr Phys 66:493–506
French WJR, Burns GB, Finlayson K, Greet PA, Lowe RP, Williams PFB (2000) Hydroxyl (6–2) airglow emission intensity ratios for rotational temperature determination. Ann Geophys 18:1293–1303
Galperin YuI (1975) Polar auroras in the magnetosphere. Ser. Astronautics, astronomy. Znanie, Moscow
Garcia RR, Solomon S, Roble RG, Rusch DW (1984) A numerical response of the middle atmosphere to the 11-year solar cycle. Planet Space Sci 32:411–423
Gill AE (1982) Atmosphere-ocean dynamics. Academic Press, New York
Gnevyshev MN (1963) A corona and 11-year cycle of the solar activity. Astron Rep 40:401–412
Gnevyshev MN (1977) Essential features of the 11-year solar cycle. Solar Phys 51:175–183
Golitsyn GS, Semenov AI, Shefov NN, Fishkova LM, Lysenko EV, Perov SP (1996) Long-term temperature trends in the middle and upper atmosphere. Geophys Res Lett 23:1741–1744
Golitsyn GS, Semenov AI, Shefov NN, Khomich VYu (2006) The response of the middle atmosphere temperature on the solar activity during various seasons. Phys Chem Earth 31:10–15
Gruzdev AN, Bezverkhnii VA (1999) Long-term variations in the quasi-biennial oscillation of the equatorial stratospheric wind. Izvestiya Atmos Oceanic Phys 35:700–711
Gruzdev AN, Bezverkhnii VA (2003) On sources of the quasi-biennial oscillation of the atmosphere of the northern hemisphere. Dokl Earth Sci 389A:416–419
Gruzdev AN, Bezverkhnii VA (2005) Quasi biennial cyclicity in the atmosphere over Northern America according to ozone sounding data. Izvestiya Atmos Oceanic Phys 41:36–50
Hapke BW (1963) A theoretical photometric function for the lunar surface. J Geophys Res 68:4571–4586
Hapke B (1971) Optical properties of the lunar surface. In: Kopal Z (ed) Physics and astronomy of the Moon. Academic Press, pp 166–229
Hathaway DH, Wilson RM, Reichmann EJ (1994) The shape of the sunspot cycle. Solar Phys 151:177–190

Haubold YJ, Gerth E (1983) Zeitlich periodische Variationen des solaren Neutrinoflusses und das Standardmodell der Sonne. Astron Nachr 304:299–304

Hauchecorne A, Chanin ML, Keckhut P (1991) Climatology and trends of the middle atmospheric temperature (33–87) as seen by Rayleigh lidar over the south of France. J Geophys Res 96D:15297–15309

Heath DF, Thekaekara MP (1977) Solar spectrum in the 1200–3000 A region. In: White OR (ed) The solar output and its variation. University Press, Boulder, pp 212–232

Hedin AE (1991) Extension of the MSIS thermospheric model into the middle and lower atmosphere. J Geophys Res 96A:1159–1172

Herman JR, Goldberg RA (1978) Sun, weather and climate. NASA, Washington

Hines CO (1974) The upper atmosphere in motion. Heffernan Press, Worcester, Massachusetts

Hinteregger HE (1981) Representations of solar EUV fluxes for aeronomical applications. Adv Space Res 1:39–52

Hinteregger HE, Fukui K, Gilson BR (1981) Observational, references and model data on solar EUV, from measurements on AE-E. Geophys Res Lett 8:1147–1150

Hultqvist B (1958) The spherical harmonic development of the geomagnetic field, epoch 1945, transformed into rectangular geomagnetic coordinate systems. Arkiv Geofysik 3:53–64

Ivanov AV (1994) A closer definition of the lunar optical characteristics in the visible spectral region. Optical J N 4:159–160

Ivanov-Kholodny GS, Chertoprud VE (1992) Analysis of extrema of quasi-biennial variations of the solar activity. Astron Astrophys Trans 3:81–84

Ivanov-Kholodny GS, Mikhailov AV (1980) Ionospheric state forecast. Hydrometeoizdat, Leningrad

Ivanov-Kholodny GS, Nikol'sky GM (1969) Sun and ionosphere. Nauka, Moscow

Ivanov-Kholodny GS, Nusinov AA (1987) Ultraviolet radiation of the Sun and its influence on the upper atmosphere and ionosphere. In: Total results of the Science. Investigations of the cosmic processes, Vol 26. VINITI, Moscow, pp 80–154

Ivanov-Kholodny GS, Nepomnyashchaya EV, Chertoprud VE (2000a) Variability of the parameters of quasi-two-year variations in the Earth's ionosphere in the 11-year cycle. Geomagn Aeronomy 40:526–528

Ivanov-Kholodny GS, Mogilevskii EI, Chertoprud VE (2000b) Quasi-biennial oscillations in total solar irradiance and in the Earth's ionospheric parameters. Geomagn Aeronomy 40:565–569

Jacchia LG (1979) CIRA-1972, recent atmospheric models and improvements in progress. In: Rycroft MJ (ed) Space Research, Vol 19. Pergamon Press, Oxford, pp 179–192

Kalinin YuD (1952) On the certain problems of the secular variation studies of the terrestrial magnetism. In: Trans Institute Terrestrial Magnetism, Ionosphere and Radio Wave Propagations. USSR Acad Sci, N 8(18). Hydrometeoizdat, Leningrad, pp 5–11

Kandaurova KA (1971) The statistic analysis and forecast of the Zürich series of Wolf's numbers with regular part extracted. Solar data Bulletin. N 11. Nauka, Leningrad pp 80–89

Karpov KA, Chistova EA (1968) Tables of the Weber function. Ditkin VA (ed) Computer Centre USSR Acad Sci, Moscow

Karyagina ZV (1960) Energy distribution in Zodiacal Light continuum. Astrophys Bull 37:882–887

Kazimirovsky ES, Kokourov VD (1979) Ionospheric movements. Erofeev NM (ed) Nauka, Novosibirsk

Keating GM, Tolson RH, Bradford MS (2000) Evidence of long term global decline in the Earth's thermospheric densities apparently related to anthropogenic effects. Geophys Res Lett 27:1523–1526

Khramova MN, Kononovich EV, Krasotkin SA (2002) Quasi-biennial oscillations of global solar-activity indices. Solar System Res 36:548–554

Kluev OF (1985) Thermospheric temperature measurement from the emissive spectra of the AlO molecules. In: Chasovitin YuK, Portnyagin YuI (eds) Trans Institute Experimental Meteorology. Upper Atmospheric Physics, N 16(115). Hydrometeoizdat, Moscow, pp 15–25

Klyatskin VI (1994) Statistical description of the diffusion of tracers in a random velocity field. Uspekhi Phys Nauk 164:531–544

Kolchinsky IG (1984) On the problem of the changes of the "refraction constant". In: Astrometry and Astrophysics. N 52. Mauka, Kiev, pp 38–46

Kolmogorov AN (1986) On the logarithm normal law of the particle distribution during crushing. In: The probability theory and mathematical statistics. Nauka, Moscow, pp 264–267

Kondratiev KYa, Nikolsky GA (1995) Solar activity and climate. 1. Observation data. Condensation and ozone hypotheses. Earth' Investigations from Cosmos N 5:3–17

Kondratyev KYa, Nikolsky GA, Shultz EO (1996) Hourly to decadal time scale variations of the spectral and total "solar constant". Meteorol Atmos 61:119–126

Kononovich EV (1999) Quasi-biennial structure of the solar activity cycle. In: Proceedings of conference "Large-scale structure of the solar activity" (Pulkovo June 21–25, 1999). Pulkovo Main Astronomical Observatory RAS, S. Peterburg, pp 115–120

Kononovich EV (2001) Fine structure of the 11-years cycles of solar activity. In: Proceedings of international conference "Sun during epoch of the magnetic field sign changes" (S. Peterburg), pp 203–210

Kononovich EV (2003) Physics of solar and stellar activity. Abstract. Conference UIG and Baltic countries. Actual problems of the physics of the solar and stellar activity. Nizhny, Novgorod

Kononovich EV (2004) Mean variations of the solar activity cycles: Analytical representations. In: Proceedings of the XXVII annual seminar "Physics of auroral phenomena". Kola Science Center. RAS, Apatity, pp 83–86

Kononovich EV (2005) Analytical representations of mean solar activity variations during a cycle. Geomagn Aeronomy 45:295–302

Kononovich EV, Moroz VI (2001) Total course of astronomy. Ivanov VV (ed) Editorial-URSS, Moscow

Kononovich EV, Shefov NN (1999a) The influence of solar activity on long-term climatic variations. Dokl Earth Sci 367:714–717

Kononovich EV, Shefov NN (1999b) The middle atmosphere is a regulator of the solar activity influence on the long-term changes of the energy balance of the lower atmosphere. Geomagn Aeronomy 39:76–80

Kononovich EV, Shefov NN (2003) Fine structure of the 11-years cycles of solar activity. Geomagn Aeronomy 43:156–163

Kononovich EV, Shefov NN (2006) Some dependencies of the solar activity variations during 11-year cycle. Geomagn Aeromony 46:683–687

Kononovich EV, Shefov NN, Khramova MN (2002) Approximation of relationships between solar activity indices: sunspot Wolf numbers and radio emission flux at a 10.7-cm wavelength. Geomagn Aeronomy 42:430–431

Korn GA, Korn TM (1961) Mathematical handbook for scientist and engineers. McGraw-Hill Book Co, New York

Korobochkin BI, Filippova YaA (1965) Tables of the modified Whittaker function. Computer Centre USSR Acad Sci, Moscow

Korsunova LP, Gorbunova TA, Bakaldina VD (1985) Solar activity influence on the turbopause variations. In: Lysenko IA (ed) Studies of the dynamic processes in the upper atmosphere. Hydrometeoizdat, Moscow, pp 175–179

Krasnopolsky VA (1987) Airglow physics of the planetary and comet atmospheres. Nauka, Moscow

Krassovsky VI, Shefov NN (1964) Fast photoelectrons and helium emission in the upper atmosphere. Planet Space Sci 12:91–92

Krassovsky VI, Shefov NN, Yarin VI (1962) Atlas of the airglow spectrum $\lambda\lambda$ 3000–12400 Å. Planet Space Sci 9:883–915

Krinberg IA (1978) The electron kinetic in the ionosphere and plasmasphere. Nauka, Moscow

Kropotkina EP, Shefov NN (1977) Tidal emission variations in the mesopause. In: Krassovsky VI (ed) Aurorae and airglow. N 25. Soviet Radio, Moscow, pp 13–17

Kubova RM (1989) Analytical representation of invariant coordinates. Geomagn Aeronomy 29:524
Kulikov KA (1969) The course of the spherical astronomy. Nauka, Moscow
Kulikovsky PG (2002) The handbook for the astronomical amateur, 5th edn. Surdin VG (ed) Editorial URSS, Moscow
Labitzke K, van Loon H (1988) Associations between the 11-year solar cycle, the QBO and the atmosphere. P. I. The troposphere and stratosphere in the northern hemisphere in winter. J Atmos Terr Phys 50:197–206
Lanzerotti LJ, Raghavan RS (1981) Solar activity and solar neutrino flux. Nature (London) 293:122–124
Laštovička J, Fišer V, Pancheva D (1994) Long-term trends in planetary wave activity (2–15 days) at 80–100 km inferred from radio wave absorption. J Atmos Terr Phys 56:893–899
Laštovička J (1997) Observations of tides and planetary waves in the atmosphere-ionosphere system. Adv Space Res 20:1209–1222
Lean JL (1984) Estimating the variability of the solar flux between 200 and 300 nm. J Geophys Res 89A:1–9
Lean J (1991) Variations in the Sun's radiative output. Rev Geophys 29:505–535
Lean J, Beer J, Bradley R (1995) Reconstruction of solar irradiance since 1610: implications for climate change. Geophys Res Lett 22:3195–3198
Leikin GA, Shvidkovskaya TE (1972) On the infrared radiation of the Moon in the 3,5–3,9 mcm region. In: Physics of the Moon and planets. Nauka, Moscow, pp 91–95
Lincoln JV (1967) Geomagnetic indices. In: Matsushita S, Campbell WH (ed) Physics of geomagnetic phenomena, Vol 1. Academic Press, New York, pp 67–100.
Lowe RP (1969) Interferometric spectra of the Earth's airglow (1,2 to 1,6µm). Phil Trans Roy Soc London A264:163–169
Lowe RP (2002) Long-term trends in the temperature of the mesopause region at mid-latitudes as measured by the hydroxyl airglow. We-Heraeus Seminar on trends in the upper atmosphere. Kühlungsborn, Germany, p 32
Lübken FJ (2000) Nearly zero temperature trend in the polar summer mesosphere. Geophys Res Lett 27:3604–3606
Lysenko EV, Perov SP, Semenov AI, Shefov NN, Sukhodoev VA, Givishvili GV, Leshchenko LN (1999) Long-term trends of the yearly mean temperature at heights from 25 to 110 km. Izvestiya Atmos Oceanic Phys 35:393–400
Makarova EA, Kharitonov AV (1972) Energy distribution in the solar spectrum. Nauka, Moscow
Makarova EA, Kharitonov AV, Kazachevskaya TV (1991) The solar radiation flux. Nauka, Moscow
Manson JE (1977) Solar spectrum between 10 and 300 A. In: White OD (ed) The solar output and its variation. University Press, Boulder, pp 286–312
Martin C, Bowyer S (1989) Evidence for an extragalactic component of the far-ultraviolet background and constraints on Galaxy evolution for $0.1 < z < 0.6$. Astrophys J 338:677–706
Massey HSW, Bates DR (1982) Atmospheric physics and chemistry. Applied atomic collision physics, Vol 1. Academic Press, New York
Mayaud PN (1980) Derivation, meaning and use of the geomagnetic indices. Geomagnetic monograph N 22. American Geophysical Union, Washington
Mikhailov AA (1974) An atlas of the stellar sky. Nauka, Leningrad
Mikhailutsa VP, Gnevyshev MN (1988) The solar magnetic field energy, green corona emission and properties of a solar cycle. Solar data Bulletin. N 4. Nauka, Leningrad, pp 88–95
Miller JCP (1955) Tables of Weber parabolic cylinder function. H.M.S.O., London
Mlodnosky RP, Helliwell RA (1962) Graphic data on the Earth's main magnetic field in space. J Geophys Res 67:2207–2214
Monin AS (1982) An introduction to the theory of climate. Hydrometeoizdat, Leningrad
Monin AS (1988) Theoretical principles of the geophysical hydrodynamics. Hydrometeoizdat, Leningrad

Montenbruck O, Pfleger T (2000) Astronomy on the personal computer, 4th edn. Springer-Verlag, Berlin
Moroz VI (1965) Infrared spectrophotometry of the Moon and Galilean satellites of Jupiter. Astron Rep 42:1287–1295
Munk WH (1980) Internal wave spectra at the buoyant and internal frequencies. J Phys Oceanogr 10:1718–1728
Murthy J, Henry RC, Feldman PD, Tennyson PD (1989) The diffuse far-ultraviolet cosmic background radiation field observed from the Space Shuttle. Astrophys J 336:954–961
Murthy J, Henry RC, Feldman PD, Tennyson PD (1990) Observations of the diffuse near-UV radiation field. Astron Astrophys 231:187–198
Nagovitsyn YuA (1997) A non-linear mathematical model of the solar cyclycity process and the possibility for the activity reconstruction in the past. Astron Lett 23:851–858
Nagy AF, Doering JP, Peterson WK, Torr MR, Banks PM (1977) Comparison between calculated and measured photoelectron fluxes from Atmosphere Explorer C and E. J Geophys Res 82:5099–5100
Namboothiri SP, Meek CE, Manson AH (1994) Variations of mean winds and solar tides in the mesosphere and lower thermosphere over time scale ranging from 6 months to 11 yr: Saskatoon, 52°N, 107°W. J Atmos Terr Phys 56:1313–1325.
Neizvestny SI (1982) A background brightness of the night sky in SAO USSR Academy of Sciences. Astrophys Res 16:49–52
Nicolet M (1962) Aeronomy. Preprint. Institut d'Astrophysique, Liège
Nicolet M (1981) The solar spectral irradiance and its action in the atmospheric photodissociation processes. Planet Space Sci 29:951–974
Nielsen KP, Sigernes F, Raustein E, Deehr CS (2002) The 20-year change of the Svalbard OH–temperatures. Phys Chem Earth 27:555–561
Nishida A (1978) Geomagnetic diagnosis of the magnetosphere. Springer-Verlag, Berlin
Obridko VN, Shelting BD (2001) The quasi-biennial oscillations of the solar magnetic field. Astron Rep 78:1146–1152
Obridko VN, Shelting BD (2003) Global solar magnetology and reference points of the solar cycle. Astron Rep 80:1034–1045
Offermann D, Donner M, Semenov AI (2002) Hydroxyl temperatures: variability and trends. We-Heraeus Seminar on trends in the upper atmosphere. Kühlungsborn, Germany, p 38
Omholt A (1971) The optical aurora. Springer-Verlag, Berlin
Parkinson WD (1983) Introduction to geomagnetism. Scottish Academic Press, Edinburgh
Patel VL (1977) Solar-terrestrial physics. In: Bruzek A, Durrant CJ (eds) Illustrated glossary for solar and solar-terrestrial physics. D Reidel Publ Co, Dordrecht, Holland, pp 159–205
Perminov VI, Pertsev NN, Shefov NN (2002) Stationary planetary variations of the hydroxyl emission. Geomagn Aeronomy 42:610–613
Petrukhin VF (1995) Long-term variations of the dynamical regime at 90–100 km above Central Europe and Eastern Siberia. Geomagn Aeronomy 35:150–152
Pogoreltsev AI, Sukhanova SA (1993) Simulation of global structure of stationary planetary waves in the mesosphere and lower thermosphere. J Atmos Terr Phys 55:33–40
Polyanin AD (2001) Handbook. Linear equations of the mathematical physics. Fizmatgiz, Moscow
Ponomarev EA (1985) The mechanisms of the magnetospheric substorms. Nauka, Moscow
Prudnikov AP, Brychkov YuL, Marichev OI (1981) Integrals and series. Elementary functions. Nauka, Moscow
Pudovkin MI, Raspopov OM (1992) The mechanism of action of solar activity on state of the lower atmosphere and meteorological parameters (a review). Geomagn Aeronomy 32:593–608
Pugacheva SG, Novikov VV, Shevchenko VV (1993) The Moon as a natural standard for calibration of spectrophotometric under-sputnik observations. Solar System Res 27:47–63
PMAO (2004) Pulkovo Main Astronomical Observatory RAN. Data base of the mean annual Wolf numbers from 1090. 2004, http://www.gao.spb.ru/database/esai/
Rakipova LR, Efimova LK (1975) Dynamics of the upper atmospheric layers. Hydrometeoizdat, Leningrad

were the first to give information on the splitting of rotational levels. This splitting is responsible for the doublet structure of the lines of OH bands. The relative position of OH bands with different vibrationally excited states is such that at the rotational temperature corresponding to the conditions of the upper atmosphere (~ 200 (K)) some bands of the beginning of one sequence $\Delta v = v' - v''$ (i.e., transitions from lower vibrational levels) are partially overlapped by the bands of the end of another ($\Delta v + 1$) sequence (i.e., transitions from upper vibrational levels). The character of this mutual blending changes in going from large to small Δv values. As a result, for many bands from high vibrational levels their rotational structure can be recorded without noises only for lines $4 \div 6$ of the P branch (Shefov 1961a; Yarin 1961a; Krassovsky et al. 1962; Broadfoot and Kendall 1968; Perminov and Semenov 1992). In some cases, distortions of the intensity distribution in the P branch lines are observed which are due to absorptions in the bands of water vapor. Within the near-infrared spectral range there are only a few OH bands, namely the (7–2) 681.1-nm and (7–3) 882.3-nm bands, in which almost ten lines of the P branch not blended by other OH bands are visible. For the emission bands from lower vibrational levels, eight lines of the P branch can be discriminated only for the OH (5–1) 791.1-nm band.

Due to the presence of wide absorption bands of CO_2 and H_2O in the lower atmospheric layers, the radiation of hydroxyl molecules with $\lambda > 2.63$ (μm) ($\Delta v = 1$) fails to reach the Earth surface. Therefore, measurements for these bands are possible only by means of spectrometers lifted on balloons or rockets (Gush and Buijs 1964; MacDonald et al. 1968; Bunn and Gush 1972).

Hydroxyl emission has the advantage that it is characterized by three important geophysical parameters related to the conditions in the mesopause region where the emission arises, such as the radiation intensity and the rotational and vibrational temperatures, which can be determined by ground-based measurements.

The intensity of an individual band is an important parameter which characterizes the rate of photochemical processes to the emission and the complex dynamics of the upper atmosphere. Its absolute values and space–time variations provide a great deal of information to study the recombination rate of atomic oxygen, the energy removal, and the proportion between the contributions of various processes of production of excited OH molecules. The rotational and vibrational temperatures, characterizing the state of a medium and the rates of deactivation and establishment of equilibrium, at the same time give a key to finding the probabilities of rotational–vibrational transitions in an OH molecule that eventually determine all the observed parameters of the emission.

2.2.3 Probabilities of Rotational–Vibrational Transitions

Hydroxyl emission arises in the upper atmosphere at altitudes of about 87 (km). The emission characteristics are determined by observing the behavior of various OH bands. Each band is described by Einstein coefficients $A_{v',v''}$ which characterize the

2.2 Hydroxyl Emission

Table 2.2 Constants for the calculation of the energies of rotational levels for the OH molecule ground state

v	B_v (cm^{-1})	D_v (10^{-3}) (cm^{-1})	H_v (10^{-7}) (cm^{-1})	A_v (cm^{-1})	Y_v	γ_v (cm^{-1})	q_v (cm^{-1})
0	18.53104	1.9083	1.413	−139.054	−7.547	−0.1199	0.0417
1	17.82012	1.8695	1.355	−139.325	−7.876	−0.1144	0.0399
2	17.11869	1.8345	1.284	−139.593	−8.214	−0.1088	0.0377
3	16.42420	1.8045	1.200	−139.850	−8.568	−0.1031	0.0351
4	15.73331	1.7809	1.104	−140.088	−8.941	−0.0972	0.0313
5	15.04186	1.7650	0.994	−140.299	−9.368	−0.0912	0.0287
6	14.34498	1.7654	0.872	−140.439	−9.795	−0.0844	0.0269
7	13.63626	1.7838	0.734	−140.491	−10.34	−0.0770	0.0234
8	12.90513	1.7984	0.589	−140.399	−10.95	−0.0685	0.0203
9	12.14094	1.8599	0.428	−140.176	−11.58	−0.0502	0.0171
10	11.32784	1.9549	0.254	−139.518	−12.17	−0.0383	0.0140

practically not detected because of their rather low intensity. The energies of rotational levels necessary for the estimation of rotational temperatures are presented in Table 2.3.

The most complete and reliable data about the energies of rotational levels are presented elsewhere (Coxon 1980; Coxon and Foster 1982). These publications

Table 2.3 Energies of the rotational levels (cm^{-1}) relative to the vibrational levels of the ground state of the OH molecule

N′	J′	v, $^2\Pi_{3/2}$										
		0	1	2	3	4	5	6	7	8	9	10
1	3/2	−38.5	−39.7	−40.9	−42.0	−43.1	−44.2	−54.1	−46.5	−47.8	−48.7	−49.5
2	5/2	+45.3	+41.2	+36.9	+32.9	+27.9	+24.8	+19.9	+17.8	+12.1	+7.8	+4.3
3	7/2	163.9	155.2	146.7	138.4	130.2	121.8	113.8	106.8	96.0	86.9	79.7
4	9/2	317.0	303.3	288.8	275.1	261.3	246.9	233.9	221.3	204.6	188.0	171.4
5	11/2	505.7	484.9	463.3	442.8	421.9	400.9	380.4	361.3	338.1	313.1	288.0
6	13/2	729.8	700.6	670.8	641.9	613.1	583.8	555.0	527.3	494.6	460.8	426.8
7	15/2	989.4	950.3	910.7	872.3	833.8	794.5	756.4	718.8	675.6	631.2	587.0
8	17/2	1284.2	1233.9	1183.2	1133.8	1084.3	1034.5	984.8	935.8	881.6	826.3	771.0

N′	J′	v, $^2\Pi_{1/2}$										
		0	1	2	3	4	5	6	7	8	9	10
1	1/2	88.3	88.0	87.2	86.7	86.0	85.3	84.5	83.7	83.0	82.3	81.7
2	3/2	149.6	146.5	143.5	140.4	137.4	134.3	131.0	128.4	125.2	121.4	118.7
3	5/2	250.6	243.5	236.5	229.4	222.4	215.5	208.4	201.6	194.2	186.0	180.1
4	7/2	391.0	378.5	365.7	353.2	340.7	328.3	315.6	303.5	290.7	276.6	261.5
5	9/2	569.9	550.2	530.5	511.1	491.8	472.4	452.9	433.5	413.2	391.8	369.5
6	11/2	786.3	758.1	730.2	702.4	674.1	646.9	619.4	591.5	562.2	531.3	499.0
7	13/2	1039.8	1001.7	963.9	926.4	889.0	851.9	814.8	770.0	737.2	695.3	653.1
8	15/2	1329.8	1280.4	1231.3	1182.8	1134.3	1086.2	1037.4	989.5	938.2	883.3	826.0

probabilities of spontaneous transitions of an excited molecule from state v' to state v''. The geophysical importance of hydroxyl emission dictated the barest necessity to know the transition probabilities and, hence, radiative lifetimes of various vibrational states.

The hydroxyl molecule possesses appreciable anharmonicity. The determination of its potential function received unremitting attention at different times (Fallon et al. 1961; Werner et al. 1983). First calculations performed by Shklovsky (1951a) were based on the expansion of the dipole moment in a series (Scholz 1932). The absolute values were calculated by the formulas derived by Scholz (1932) for the P_1 line ($J = 3/2$) rather than for the band as a whole. Almost simultaneously, calculations of the Einstein coefficients $A_{v',v''}$ were performed with the use of two terms of the expansion of the dipole moment (Heaps and Herzberg 1952). Normalization was made with respect to the (1–0) band for which the transition probability was taken by convention equal to $100\,(s^{-1})$ in view of that for the HCl molecule $A_{1-0} = 58\,(s^{-1})$. According to the mentioned work, the band sequence $\Delta v = 1$ had the greatest transition probabilities in comparison with other bands. However, it has been shown (Ferguson and Parkinson 1963) that this conclusion is untrue, and the sequence $\Delta v = 1$ is weaker than the sequence $\Delta v = 2$. This feature was taken into account in subsequent calculations.

Unfortunately, the process of determination of the transition probabilities still remains to be completed though it lasts as long as almost 40 years. As for now, data on the transition probabilities for various OH bands are available in a great number of publications (Shklovsky 1951a, 1957; Heaps and Herzberg 1952; Cashion 1963; Phelps and Dalby 1965; Potter et al. 1971; Murphy 1971; Mies 1974; Llewellyn and Long 1978; Langhoff et al. 1986; Turnbull and Lowe 1989; Nelson et al. 1990; Dodd et al. 1991, 1993; Smith et al. 1992; Holtzclaw et al. 1993; Goldman et al. 1998). They are presented in Tables 2.4–2.13. When looking through these data, one can see that both the proportions between the transition probabilities even for the common initial higher levels and the absolute values determined in fact by the normalization with respect to the (1–0) transition have changed substantially.

Table 2.4 Einstein coefficients $A(v' \to v'')(s^{-1})$ for the radiation transitions of OH molecules ($X^2\Pi, v \leq 9$) (Shklovsky 1951a)

v''/v'	0	1	2	3	4	5	6	7	8	$\Sigma A(s^{-1})$	$\log_{10} \Sigma A$
1	580									580	2.76
2	59	1050								1110	3.05
3	6.6	160	1420							1587	3.20
4	0.43	25	300	1700						2026	3.31
5	0.17	4.3	60	440	1800					2250	3.35
6	0.03	0.43	12	105	600	1900				2618	3.42
7		0.22	3.2	29	180	750	1800			2762	3.44
8			0.87	8	45	250	860	1800		2964	3.47
9				2.5	19	86	320	930	1700	3058	3.49

Table 2.5 Einstein coefficients $A(v' \to v'')(s^{-1})$ for the radiation transitions of OH molecules $(X^2\Pi, v \le 9)$ (Heaps and Herzberg 1952)

v''/v'	0	1	2	3	4	5	6	7	8	$\Sigma A(s^{-1})$	$\log_{10} \Sigma A$
1	100									100	2.00
2	9.83	175.5								185	2.27
3	1.11	26.5	229							257	2.41
4	0.153	4.09	47.1	264						315	2.50
5	0.0252	0.718	9.31	69.3	281					360	2.56
6	0.00489	0.145	2.01	16.8	90.9	285				395	2.60
7	0.00111	0.0337	0.485	4.33	26.3	110	277			418	2.62
8	0.000286	0.0089	0.132	1.22	7.91	37.2	125	259		430	2.63
9	0.000084	0.0026	0.040	0.382	2.57	12.9	48.5	134	235	433	2.64

Table 2.6 Einstein coefficients $A(v' \to v'')(s^{-1})$ for the radiation transitions of OH molecules $(X^2\Pi, v \le 9)$ (Cashion 1963)

v''/v'	0	1	2	3	4	5	6	7	8	$\Sigma A(s^{-1})$	$\log_{10} \Sigma A$
1	390									390	2.59
2	29.5	702								731	2.86
3	2.53	82	947							1030	3.01
4	0.269	9.95	154	1125						1290	3.11
5	0.0346	1.38	24.2	240	1250					1520	3.18
6	0.00537	0.221	4.18	47	332	1320				1700	3.23
7	0.00094	0.041	0.815	9.85	78.8	421	1335			1850	3.27
8	0.00019	0.0087	0.182	2.29	19.5	119	505	1295		1940	3.29
9	0.000045	0.00212	0.0454	0.59	5.35	34.4	166	535	1230	1970	3.29

Table 2.7 Einstein coefficients $A(v' \to v'')(s^{-1})$ for the radiation transitions of OH molecules $(X^2\Pi, v \le 9)$ (Potter et al. 1971)

v''/v'	0	1	2	3	4	5	6	7	8	$\Sigma A(s^{-1})$	$\log_{10} \Sigma A$
1	3.33									3.33	0.522
2	0.217	5.88								6.10	0.785
3	0.019	0.62	8.33							8.97	0.953
4	0.0021	0.071	1.11	10						11.8	1.01
5	0.00028	0.01	0.17	1.70	10.8					12.7	1.10
6	0.000045	0.00164	0.0286	0.312	2.27	11.2				13.8	1.14
7		0.00031	0.0056	0.062	0.50	2.86	11.4			14.8	1.17
8		0.000071	0.00126	0.0147	0.120	0.714	3.45	11.2		15.5	1.19
9			0.00032	0.0038	0.0322	0.204	1.00	3.85	10.64	15.7	1.20

Rees MH (1989) Physics and chemistry of the upper atmosphere. Houghton JT, Rycroft MJ, Dessler AJ (eds) Cambridge University Press, Cambridge

Rishbeth H, Garriott OK (1969) Introduction to ionospheric physics. Academic Press, New York

Rivin YuR, Zvereva TI (1983) Frequency content of the quasi biennial variations of the geomagnetic field. In: Levitin AE (ed) Solar wind, magnetosphere and geomagnetic field. Nauka, Moscow, pp 72–90

Roach FE (1964) The light of the night sky: astronomical interplanetary and geophysical. Space Sci Rev 3:512–540

Roach FE, Gordon JL (1973) The light of the night sky. D Reidel Publ Co, Dordrecht, Holland

Roach FE, Megill LR (1961) Integrated starlight over the sky. Astrophys J 133:228–242

Röser S, Staude HJ (1978) The zodiacal light from 1500 Å to 60 micron. Astron Astrophys 67:381–384

Rostoker G (1972) Geomagnetic indices. Rev Geophys Space Phys 10:935–950

Sakurai K (1979) Quasi-biennial variation of the solar neutrino flux and solar activity. Nature (London) 278:146–148

Sanga-Ngoie K, Fukuyama K (1996) Interannual and long-term climate variability over the Zaire River during the last 30 years. J Geophys Res 101D:21351–21360

Schatten KH (1996) An atmospheric radiative-convective model: solar forcing. Astrophys J 460:69–72

Schove DJ (1955) The sunspot cycles, 649 B.C. to A.D. 2000. J Geophys Res 60:127–146

Schove DJ (1962) Auroral numbers since 500 B.C. J Brit Astron Assoc 7:30–35

Schove DJ (1979) Sunspot points and aurorae since AD 1510. Solar Phys 63:423–432

Schove DJ (1983) Sunspot cycles. Hutchinson Ross, Stroudsburg, Pennsylvania

Schröder W (1997) Geomagnetism and aeronomy. Science Edition, Bremen-Roennebeck

Schröder W (2000a) Long and short term variability in Sun's history and global change. Science Edition, Bremen-Roennebeck

Schröder W (2000b) Aurora in time. Science Edition, Bremen-Roennebeck

Schröder W (2002) Solar variability and geomagnetism. Science Edition, Bremen-Roennebeck

Schröder W, Shefov NN, Treder HJ (2004) Estimation of past solar and upper atmosphere conditions from historical and modern auroral observations. Ann Geophys 22:2273–2276

Schuster A (1906) On sunspot periodicities. Preliminary notice. Proc Roy Soc London 77:141–145

Sckopke N (1966) A general relation between the energy of trapped particles and the disturbance field near the Earth. J Geophys Res 71:3125–3130

Semenov AI (1996) A behavior of the lower thermosphere temperature inferred from emission measurements during the last decades. Geomagn Aeronomy 36:655–659

Semenov AI, Shefov NN (1996) An empirical model for the variations in the hydroxyl emission. Geomagn Aeronomy 36:468–480

Semenov AI, Shefov NN (1997a) An empirical model of nocturnal variations in the 557.7-nm emission of atomic oxygen. 1. Intensity. Geomagn Aeronomy 37:215–221

Semenov AI, Shefov NN (1997b) An empirical model of nocturnal variations in the 557.7-nm emission of atomic oxygen. 2. Temperature. Geomagn Aeronomy 37:361–364

Semenov AI, Shefov NN (1997c) An empirical model of nocturnal variations in the 557.7-nm emission of atomic oxygen. 3. Emitting layer altitude. Geomagn Aeronomy 37:470–474

Semenov AI, Shefov NN (1997d) Empirical model of the variations of atomic oxygen emission557.7 nm. In: Ivchenko VN (ed) Proceedings of SPIE (23rd European Meeting on Atmospheric Studies by Optical Methods, Kiev, September 2–6, 1997), Vol 3237. The International Society for Optical Engineering, Bellingham, pp 113–122

Semenov AI, Shefov NN (1999) Empirical model of hydroxyl emission variations. Int J Geomagn Aeronomy 1:229–242

Semenov AI, Shefov NN (2003) New knowledge of variations in the hydroxyl, sodium and atomic oxygen emissions. Geomagn Aeronomy 43:786–791

Semenov AI, Sukhodoev VA, Shefov NN (2002a) A model of the vertical temperature distribution in the atmosphere altitudes of 80–100 km that taking into account the solar activity and the long-term trend. Geomagn Aeronomy 42:239–244

Semenov AI, Bakanas VV, Perminov VI, Zheleznov YuA, Khomich. VYu (2002b) The near infrared spectrum of the emission of the nighttime upper atmosphere of the Earth. Geomagn Aeronomy 42:390–397

Semenov AI, Shefov NN, Lysenko EV, Givishvili GV, Tikhonov AV (2002c) The seasonal peculiarities of behavior of the long-term temperature trends in the middle atmosphere at the mid-latitudes. Phys Chem Earth 27:529–534

Semenov AI, Shefov NN, Perminov VI, Khomich VYu, Fadel KhM (2005) Temperature response of the middle atmosphere on the solar activity for different seasons. Geomagn Aeronomy 45:236–240

She CY, Songsheng Chen, Zhilin Hu, Sherman J, Vance LD, Vasoli V, White MA, Yu JR, Kryeger DA (2000) Eight-year climatology of nocturnal temperature and sodium density in the mesopause region (80 to 105 km) over Fort Collins, CO (41°N, 105°W). Geophys Res Lett 27:3289–3292

Shefov NN (1962) Sur l'émission de l'helium dans la haute atmosphere. Ann Géophys 18:125

Shefov NN (1978) Altitude of the hydroxyl emission layer. In: Krassovsky VI (ed) Aurorae and Airglow, N 27. Soviet Radio, Moscow, pp 45–51

Shefov NN, Semenov AI (2004) Longitudinal-temporal distribution of the occurrence frequency of noctilucent clouds. Geomagn Aeronomy 44:259–262

Shefov NN, Semenov AI (2006) Spectral composition of the cyclic aperiodic (quasi-biennial) variations in solar activity and the Earth's atmosphere. Geomagn Aeronomy 46:411–416

Shklovsky IS (1951) The solar corona. Gostekhizdat, Moscow

Simonov GV (1963) Geomagnetic time. Geophys J Roy Astron Soc 8:258–267

Solanki SK, Fligge M (2002) Solar irradiance variations and climate. J Atmos Solar-Terr Phys 64:677–686

Soloviev MD (1969) Mathematical cartography. Nedra, Moscow

Soloviev AA, Kirichek EA (2004) The diffusive theory of the solar magnetic cycle. Kalmyk State University Publishing House, Elista, S. Peterburg

Slater LJ (1960) Confluent hypergeometric functions. Cambridge University Press, Cambridge

Smith FL, Smith C (1972) Numerical evaluation of Chapman's grazing incidence integral ch(X, χ). J Geophys Res 77:3592–3597

Starkov GV (1994) Mathematical model of auroral boundaries. Geomagn Aeronomy 34:331–336

Starkov GV (2000) Planetary dynamics of auroral luminosity. In: Ivanov VE (ed) Physics of the near-Earth space. Polar Geophysical Institute, Kola Scientific Centre RAS, Apatity, pp 409–499

Starkov GV, Shefov NN (2001) Long-term changes of the auroral heights in dayside of the auroral oval. Geomagn Aeronomy 41:763–765

Starkov GV, Yevlashin LS, Semenov AI, Shefov NN (2000) A subsidence of the thermosphere during 20th century according to the measurements of the auroral heights. Phys Chem Earth Pt B 25:547–550

Stewart JQ, Eggleston FC (1940) The mathematical characteristics of sunspot variations. II. Astrophys J 91:72–84

Stewart JQ, Panofsky HAA (1938) The mathematical characteristics of sunspot variations. Astrophys J 88:385–407

Stolarski RS, Green AES (1967) Calculations of auroral intensities from electron impact. J Geophys Res 72:3967–3974

Störmer C (1955) The polar aurora. Clarendon Press, Oxford

Straižys V (1977) Multicolor stellar photometry. Mokslas, Vilnius

Sukhanova SA (1996) Propagation of the stationary planetary waves in the mesosphere and lower thermosphere altitudes. Izvestiya Atmos Oceanic Phys 32:61–68

Svalgaard L (1977) Solar wind and interplanetary medium. In: Bruzek A, Durrant CJ (eds) Illustrated glossary for solar and solar-terrestrial physics. D Reidel Publ Co, Dordrecht, Holland, pp 147–158

Sverdlov YuL (1982) The radio aurorae morphology. Nauka, Leningrad

Sverdlov YuL, Khorkova TN (1982) Coordinates of the corrected geomagnetic pole. In: Isaev SI, Starkov GV (eds) Aurorae, N 30. VINITI, Moscow, pp 98–103

Sytinskaya NN (1959) Moon nature. Fizmatgiz, Moscow

Tennyson PD, Henry RC, Feldman PD, Hartig GF (1988) Cosmic ultraviolet background radiation and Zodiacal Light. Astrophys J 330:435–444

Timothy JG (1977) Solar spectrum between 300 and 1200 A. In: White OR (ed) The solar output and its variation. University Press, Boulder, pp 257–285

Tohmatsu T, Ogawa T, Tsuruta H (1965) Photoelectronic processes in the upper atmosphere. I. Energy spectrum of primary photoelectrons. Rept Ionosph. Space Res Japan 19:482–508

Toroshelidze TI (1968) A vertical distribution of the atmospheric sodium. Bull Georgian SSR Acad Sci 51:579–584

Toroshelidze TI (1970) A study of atmospheric sodium in twilight. Indian Meteorol Geophys 21:211–218

Toroshelidze TI (1972) On the problem of the effective screen height at twilight method of the terrestrial atmosphere study. In: Kharadze EK (ed) Bull Abastumani astrophys observ, N 41. pp 105–113

Toroshelidze TI (1991) The analysis of the aeronomy problems on the upper atmosphere glow. Shefov NN (ed) Metsniereba, Tbilisi

Torr MR, Torr DG, Stengel R (1979) Zodiacal Light surface brightness measurements by Atmosphere Explorer-C. Icarus 40:49–59

Truttse YuL (1973) Upper atmosphere during geomagnetic disturbances. In: Krassovsky VI (ed) Aurorae and Airglow, N 20. Nauka, Moscow, pp 5–22

Uspensky MV, Starkov GV (1987) The auroras and the radiowave scattering. Nauka, Leningrad

Vadzinsky RN (2001) A handbook on the probability distributions. Nauka, S. Peterburg

Valchuk TE, Feldstein YaI (1983) Correlational and regressional relations between the geomagnetic activity index aa and the characteristics of the terrestrial cosmic space. In: Levitin AE (ed) Solar wind, magnetosphere and geomagnetic field. Nauka, Moscow, pp 127–145

Vallance Jones A (1974) Aurora. D Reidel Publ Co, Dordrecht

Vassy AT, Vassy E (1976) La luminescence nocturne. In: Rawer K (ed) Handbuch der Physik. Geophysik 111/5, Vol 49/5. Springer-Verlag, Berlin, pp 5–116

Vergasova GV, Kazimirovsky ES (1994) Macroscale variations of prevailing wind in the lower thermosphere. Izvestiya Atmos Oceanic Phys 30:31–38

Vernova ES, Tyasto MI, Baranov DG, Grigoryan MS (1997) Amplitude pattern of the 27-day cosmic-ray variation in the course of the solar cycle. Geomagn Aeronomy 37:234–236

Vitinsky YuI (1963) Solar activity forecasts. USSR Acad Sci Publ House, Moscow, Leningrad

Vitinsky YuI (1973) Cyclicity and solar activity forecasts. Nauka, Leningrad

Vitinsky YuI, Kopecký M, Kuklin GV (1986) Statistics of the sunspot activity of the Sun. Nauka, Moscow

Waldmeier M (1941) Ergebnisse und Probleme der Sonnenforschung. Zürich

Waldmeier M (1955) Ergebnisse und Probleme der Sonnenforschung. 2 Auflage. Leipzig

Wattson R, Danielson R (1965) The infrared spectrum of the Moon. Astrophys J 142:16–22

WinEphem (2002). http://www.geocities.com/tmarkjames/WinEphem.html 2002

Whittaker ET, Watson GN (1927) A course of modern analysis, 4th edn. Cambridge University Press, Cambridge

Woeikof A (1891) Cold and warm winter interchange. Meteorol Rep 9:409–422

Woeikof A (1895) Die Schneedecke in "paaren" und "unpaaren" Wintern. Meteorol Zeits 12: 77–78

Yanovsky BM (1953) Terrestrial magnetism, 2nd edn. Gostekhizdat, Moscow

Yevlashin LS (2000) Hydrogen emission in auroras and the precipitation of auroral protons. In: Ivanov VE (ed) Physics of the near-earth space. Polar Geophysical Institute, Kola Scientific Centre RAN, Apatity, pp 500–548, 633–665

Yevlashin LS (2005) Aperiodic variations of the observation frequency of the red type-A auroras during 11-year cycle of solar activity. Geomagn Aeronomy 45:388–391

Young JM, Carruthers GR, Holmes JC, Johnson CJ, Patterson NP (1968) Detection of Lyman-beta and helium resonance radiation in the night sky. Science 160:990–991

Zaitseva SA, Akhremtchik SN, Pudovkin MI, Galtsova YaV, Besser BP, Rijnbeek RP (2003) Long-term variations of the solar activity-lower atmosphere relationship. Int J Geomagn Aeronomy 4:167–174

Chapter 2
Processes Responsible for the Occurrence of the Airglow

The emissions of the upper atmosphere to be considered in this chapter are manifestations of the related processes giving rise to the Earth's airglow and, above all, they characterize the abundance and altitude distribution of one important component of the atmosphere. Thus, knowing the photochemical nature of a detected emission and the condition of its detection, it is possible to obtain data on the atmospheric composition and temperature conditions in a certain altitude range. If data about several emissions induced by the basic component are available, the conclusions to be made would be much more reliable. This chapter deals with the basic emissions that are accessible to ground-based observations, which make them indicators of the properties of the atmosphere.

2.1 Processes of Excitation and Deactivation of Excited Species

Emissions in a gaseous medium are produced by atoms and molecules excited as a result of various photochemical processes. These are chemical reactions and collisions of atoms and molecules with electrons and other particles possessing high kinetic energies. Collisions of the excited reactants with the environmental components strongly affect the emission process. The main thing here is the probability of a radiation transition to a lower state. This probability is characterized by the Einstein coefficient A_{ki}, where k and i are the ordinal numbers of the upper excited level and the lower level, respectively. The quantity A_{ki} determines the radiative lifetime of an excited species, $\tau = 1/A_{ki}$, during which collisions can occur as a result of which the excited species disappears, i.e., changes to another state. These phenomena are determining in all detectable emission processes.

2.1.1 Kinetics of Elementary Gaseous-Phase Photochemical Reactions

In the upper atmosphere at altitudes above 80 (km), the gaseous medium is highly rarefied: the molecular density at its lower level is $4 \cdot 10^{14}$ (cm^{-3}) and the pressure is 10 (μbar), decreasing with increasing altitude. The atoms and molecules of this gaseous medium experience collisions which result in chemical reactions giving rise to new species. According to the data available in the literature (Kondratiev 1958; Kondratiev and Nikitin 1974, 1981), the duration of an event of collision of two molecules is 10^{-13}–10^{-12} (s). The frequency of gas-kinetic collisions is determined by the relation (Bates and Nicolet 1950)

$$\beta \cdot [N] = 0.81 \cdot 10^{-10} \cdot \sqrt{\frac{T}{M}} \cdot [N] \, (s^{-1}) \,,$$

where β is the rate coefficient, T is the temperature, M is the molecular mass, and [N] is the particle concentration. As follows from this relation, the time interval between two collisions is several orders of magnitude greater than the duration of the collision event. Thus, through this time a system of two colliding particles can be considered isolated from all other particles.

According to the laws of kinetics of chemical reactions (Kondratiev 1936, 1958; Nikitin 1970; Kondratiev and Nikitin 1981), the concentration of the products of the reaction

$$AB + CD \rightarrow AC + BD + E \,,$$

where E is the energy, is given by

$$[AC] = \alpha \cdot [AB] \cdot [CD] \cdot \tau \,,$$

where the square brackets denote concentrations (cm^{-3}), α is the reaction rate (cm$^3 \cdot$ s^{-1}), and τ is the lifetime (s) of the AC molecules formed as a result of the reaction.

The lifetime of excited atoms and molecules is determined by the relation

$$\tau = \frac{1}{\sum_k A_{ik} + \sum_m \beta_{im} \cdot [M]} \, (s) \,,$$

where A_{ik} are the probabilities of transitions (s^{-1}) followed by emission from an excited state and β_{im} [M] are the rates of the processes of deactivation (s^{-1}) of excited states. The reaction rates α are functions of temperature, β is the deactivation coefficient (in a concrete case it can also depend on temperature), M designates the atoms and molecules that are responsible for deactivation of excited species both by quenching of the excited state and due to chemical transformations.

By definition, the reaction rate is given by

$$\alpha (\text{or } \beta) = \int_{E_0}^{\infty} \sigma(E) \cdot v(E) \cdot dE \,,$$

where (E) is the effective cross-section for the interaction of reactive species, v(E) is the velocity of reactive species, and E_0 is the energy at which the reaction begins. Since velocity depends on temperature, the reaction rate in typical situations under the conditions of the upper atmosphere is determined by the relation

$$\alpha = k(T) \cdot \exp\left[-\frac{E}{T}\right] (cm^3 \cdot s^{-1}) \ .$$

Here E is the activation energy which is the minimum energy that the reactive molecules should possess. The coefficient k(T) is a function of temperature and also of the sizes and masses of the reactive molecules. Besides, the reaction rate is naturally affected by the properties of the reactive particles. Thus, the Wigner rule demands that the sum of the spins of the reactive and produced species be invariable in the reaction under consideration. Violation of this rule leads either to a complete prohibition of the given reaction channel or to a substantial decrease in reaction rate. The reaction rate is affected by the so-called steric factor γ which in fact implies that molecules can react only at a certain mutual orientation.

The rate of photoionization, including that with simultaneous excitation of the ion produced, is determined by the expression

$$j = \int_0^{\lambda_i} \sigma_{ph}(\lambda) \cdot S_\lambda \cdot d\lambda \ \text{or}\ j = \int_{E_i}^{\infty} \sigma_{ph}(E) \cdot S_E \cdot dE\ ,$$

where σ_{ph} is the effective cross-section for photodissociation or photoionization, depending on wavelength or energy, respectively; λ_i and E_i are the wavelength and the ionization potential, respectively; S_λ and S_E denote the solar flux in terms of wavelengths or energies.

In an actual case, taking into account the dependence of the solar radiation intensity on the level of solar activity, we have for the photoionization rate at an altitude Z in the upper atmosphere

$$r = 0.6066 \cdot \ln\left(\frac{F_{10.7} - 40}{20}\right) \cdot \int_{E_i}^{\infty} \sigma_{ph}(E) \cdot S_{0E} \cdot \exp\left(-\tau(E) \cdot Chp\chi_e\right) \cdot dE\ ,$$

where the optical thickness of the atmospheric layer for ionizing solar radiation

$$\tau(E) = \int_Z^{\infty} \sigma_{ph}(E) \cdot n(Z) \cdot dZ\ ,$$

$Chp\chi_\odot$ is the Chapman function for the solar zenith angle $\chi_\odot$, and n(Z) is the concentration of the atmosphere. In an actual case, the optical thickness is determined as the sum of the optical thicknesses for various atmospheric components.

The contemporary data on effective photoionization cross-sections are presented in many publications (Ivanov-Kholodny and Nikol'sky 1969; Henry 1970; Hudson

1971; Stolarski and Johnson 1972; Banks and Kockarts 1973a,b; Krinberg 1978; Fatkullin 1982). For the wavelength range $10 < \lambda < 110\,(\mathrm{nm})$, these data can be approximated by the formula (Krinberg 1978)

$$\sigma_{\mathrm{ph}}(\lambda) = B_i \cdot \left(1 - \frac{\lambda}{\lambda_i}\right)^{m_i} \cdot \left[\exp\left(\frac{\lambda}{b_i}\right) - 1\right]$$

or

$$\sigma_{\mathrm{ph}}(E) = B_i \cdot \left(1 - \frac{E_i}{E + E_i}\right)^{m_i} \cdot \left[\exp\left(\frac{b_i^*}{E + E_i}\right) - 1\right],$$

whose parameters for N_2^+, O_2^+, and O^+ are reported by Krinberg (1978). For $\lambda < 10\,(\mathrm{nm})\,(E + E_i > 124\,(\mathrm{eV}))$, following Henry (1970) and Krinberg (1978), we have

$$\sigma_{\mathrm{ph}}(\lambda) = C_i \cdot \lambda^{5/2} \text{ or } \sigma_{\mathrm{ph}}(E) = C_i^* \cdot \left(\frac{E_i}{E + E_i}\right)^{5/2},$$

where E_i is the ionization potential and E is the energy of the produced photoelectrons.

It should be noted that simultaneously with photoionization there occurs excitation of the ions produced, which thus become an additional radiation source in the upper atmosphere.

Among the processes of initiation of various emissions from the basic species of the upper atmosphere in the daytime, an appreciable part is played by collisions of these species with photoelectrons:

$$M + e \rightarrow M^* + e.$$

These processes are discussed in a number of monographs (Massey and Burhop 1952; McDaniel 1964; Banks and Kockarts 1973a,b; Vainstein et al. 1979). The now available experimental and theoretical data about the effective cross-sections for electron excitation of the electronic states of nitrogen and oxygen molecules and of oxygen atoms are systematized elsewhere (Green and Dutta 1967; Jusick et al. 1967; Stolarski et al. 1967; Watson et al. 1967; Green and Sawada 1972; Green and Stolarski 1972).

An empirical representation of the effective cross-sections is

$$\sigma_i(E) = 4\pi \cdot a_0^2 \cdot A_i \cdot \left(\frac{Ry}{E_{ei}}\right)^2 \cdot \left(\frac{E_{ei}}{E}\right)^{\omega_i} \cdot \left[1 - \left(\frac{E_{ei}}{E}\right)^{\kappa_i}\right]^{\nu_i},$$

where $a_0 = 5.29 \cdot 10^{-9}\,(\mathrm{cm})$ is the Bohr radius, $Ry = 13.6\,(\mathrm{eV})$ is the Rydberg constant, and E_{ei} is the threshold excitation energy. The values of the parameters A_i, ω_i, κ_i, and ν_i are given elsewhere (Green and Stolarski 1972; Banks and Kockarts 1973a,b; Jasperse 1977; Krinberg 1978). The numerical value of the product $4\pi \cdot a_0^2 \cdot Ry^2 = 6.513 \cdot 10^{-14}\,(\mathrm{cm}^2 \cdot \mathrm{eV}^2)$.

2.1 Processes of Excitation and Deactivation of Excited Species

When atoms or molecules collide with photoelectrons, their ionization can be accompanied by excitation:

$$M + e \rightarrow M^{+*} + e + e.$$

An empirical representation of the effective cross-sections is

$$\sigma_i(E) = 4\pi \cdot a_0^2 \cdot A_i \cdot \left(\frac{Ry}{E_{ei}}\right)^2 \cdot \left(\frac{E_{ii}}{E_{ei}}\right)^{\mu_i} \cdot \left(\frac{E_{ei}}{E}\right)^{\omega_i} \cdot \left[1 - \left(\frac{E_{ei}}{E}\right)^{\kappa_i}\right]^{\nu_i}.$$

Here E_{ii} is the ionization potential equal to E_{ei} for the ground states of ions. In the upper atmosphere, a substantial part is played by the excitation of vibrational states of the N_2 and O_2 molecules by slow electrons with energies of several electron-volts. Data on the effective sections are given elsewhere (Banks and Kockarts 1973a,b; Krinberg 1978).

For the terrestrial atmosphere, the optical thickness for Rayleigh scattering is given by

$$\tau_R = \frac{0.0085}{\lambda^4} \cdot \left(1 + \frac{0.013}{\lambda^2}\right),$$

where λ is the wavelength expressed in micrometers.

For proper absorption, the absorption coefficient (effective cross-section) is given by

$$\sigma = \frac{\pi \cdot e^2}{m_e \cdot c} \cdot \sqrt{\frac{M \cdot m_H}{2\pi \cdot k \cdot T}} \cdot \lambda \cdot f,$$

where m_H and m_e are the hydrogen atom and the electron mass, respectively; M is the atomic or molecular mass; k is Boltzmann's constant; and e is an elementary charge. The oscillator strength f can be expressed in terms of known quantities (Lang 1974), for example, as

$$f = 4.2 \cdot 10^{-8} \cdot S_{273},$$

where S_{273} is the strength of the band ($cm^{-2} \cdot atm^{-1}$), which is used as an absorption characteristic (under standard atmospheric conditions, $T = 273.16\,(K)$). Therefore, we have

$$S_{273} = \frac{296}{273.16} \cdot S_{296} = 1.084 \cdot S_{296},$$

since $296\,(K) = 273\,(K) + 23\,(°C)$ is room temperature.

Besides, the oscillator strength can be expressed in terms of the transition probability A (s^{-1}) as

$$f_{12} = 1.5 \cdot 10^{-8} \cdot \lambda^2 \cdot \frac{g_2}{g_1} \cdot A_{21}.$$

Here g_2 and g_1 are the statistical weights of the upper and lower states, respectively, and λ is the wavelength expressed in micrometers.

The transition probabilities can be determined by the values of S_{273} as

$$A = \frac{2.806}{\lambda^2} \cdot S_{273}.$$

Thus, the effective cross-sections (cm^2) are determined as

$$\sigma = 1.54 \cdot 10^{-18} \cdot \sqrt{\frac{10M}{T}} \cdot \lambda \cdot S_{273} \text{ or } \sigma = 5.48 \cdot 10^{-19} \cdot \sqrt{\frac{10M}{T}} \cdot \lambda^3 \cdot \frac{g_2}{g_1} \cdot A_{21}.$$

The rate of electron excitation of atoms and molecules is given by

$$q = 4\pi \cdot \int_E^\infty \sigma(E) \cdot \Phi(E) \cdot dE,$$

where $\sigma(E)$ is the excitation cross-section (Massey and Burhop 1952; Smirnov 1968; Mott and Massey 1965); $\Phi(E) = n_e(E) = n_e(E) \cdot v(E)$ is the flux of electrons with energy E, concentration $n_{e(E)}$, and velocity $v(E)$.

In analyzing the fluorescence of atmospheric components in sunlight, it is convenient to use the photon scattering coefficient g (photon · molecule^{-1} · s^{-1}), which is determined by the relation $I = g \cdot N$, where I is the radiation intensity expressed in photons, and N is the number of unexcited atoms or molecules in the atmospheric air column of cross-section 1 (cm^2). With these suppositions, the value of g is determined by the formula (Chandrasekhar 1950; Chamberlain 1961, 1978)

$$g = \pi F_v \cdot \frac{\pi \cdot e^2}{m_e \cdot c} \cdot f_{12} \cdot \frac{A}{\Sigma A},$$

where πF_v is the solar flux at the boundary of the Earth's atmosphere (photon · cm^{-2} · s^{-1} · Hz^{-1}).

This expression can also be presented in the form

$$g_{ki} = S_\lambda \cdot \frac{\lambda^4}{8\pi \cdot c} \cdot r_\lambda \cdot \frac{g_k}{g_i} \cdot A_{ki} \cdot \frac{A_{ki}}{\sum_j A_{kj}},$$

where S_λ is the solar flux at the boundary of the Earth's atmosphere (photon · cm^{-2} · s^{-1} · cm^{-1}) (Makarova and Kharitonov 1972; Makarova et al. 1991), λ is the wavelength of the exciting radiation, r_λ is the residual intensity of the Fraunhofer line in the spectrum of the Sun at the wavelength λ.

This formula can be represented as

$$g_{ki} = 1.33 \cdot 10^{-21} \cdot S_\lambda \cdot r_\lambda \cdot \lambda^4 \cdot \frac{g_k}{g_i} \cdot A_{ki} \cdot \frac{A_{ki}}{\sum_j A_{kj}},$$

where λ is expressed in (μm) and S_λ in (photon · cm^{-2} · s^{-1} · nm^{-1}).

2.1 Processes of Excitation and Deactivation of Excited Species

In analyzing the fluorescent emissions occurring on illumination of the lower atmosphere and Earth surface, as shown by McElroy and Hunten (1966), it is necessary to consider the additional photon flux due to scattered radiation of the lower atmosphere. This can be made by introducing the relation

$$g_{\text{eff}} = (1+\Lambda) \cdot g,$$

where g_{eff} is an effective value of g, Λ is a function of the solar zenith angle, Earth surface albedo A, and optical thickness. In the limit $\tau = 0$,

$$\Lambda = 2A \cdot \cos\chi_e.$$

This, in particular, yields $g_{\text{eff}} = 3g$ for a surface which scatters light by the Lambert law and for the zenith position of the Sun. Calculations for various values of Λ, τ, and $\chi_\odot$ were performed by McElroy and Hunten (1966) based on the work of Chandrasekhar (1950).

Except for the Sun, the Earth's atmosphere, namely its thermal radiation, is also a source for the fluorescence in the infrared region at wavelengths over 3 (μm). In this case, the scattered radiation intensity depends on the temperature at the altitude that corresponds to the outgoing radiation, i.e., approximately to $\tau = 2/3$ (Makarova et al. 1973). For higher altitudes Z above the Earth surface, it is necessary to take into account the coefficient of dilution of radiation intensity, W, for Rayleigh scattering inclusive. Following Ambartsumyan et al. (1952), we have

$$W = 2 \cdot \left[1 - \sqrt{1 - \frac{1}{\left(1 + \frac{Z}{R_E}\right)^2}}\right].$$

Thus, we have $g_k^T = g_k \cdot W$.

2.1.2 Relaxation Processes

The photochemical processes in the upper atmosphere give rise to great amounts of excited atoms and molecules, whose initial energy states, both electronic (vibrational and rotational) and translational, are other than the equilibrium state in the surrounding medium. This gives rise to the process of establishment of a statistical equilibrium, i.e., relaxation, which is a multistage process since not all physical parameters of the system tend to equilibrate at the same rate. In this case, since we deal with excited atoms and molecules, the lifetime of these species is of great importance. As mentioned above, the lifetime of a particle is determined both by its radiative lifetime associated with the probability of the transition from the excited state to a lower state, which is accompanied by emission of radiation, and by the processes of deactivation in chemical reactions. Thus, the observable glow of a given atmospheric component characterizes the interactions of the radiative species with the surrounding medium. For the emissions of the upper atmosphere, the typical parameter that determines the state of the atoms and molecules under investigation is temperature.

The characteristic equation that describes such a process has the form

$$T(t) - T_0 = [T(t_0) - T_0] \cdot \exp\left[-\frac{t-t_0}{\tau}\right].$$

Here the lifetime τ depends on all interaction processes:

$$\frac{1}{\tau} = \sum_i^k A_{ik}^e + \sum A^{rot} + \sum A^{vib} + \sum \alpha \cdot [M] + \sum \beta \cdot [M].$$

2.1.3 Translational Relaxation

The reaction products possess an increased translational velocity due to the released internal energy of the interreacting components, which is greater than the thermal energy of the medium. The establishment of equilibrium over the translational degrees of freedom of molecules, resulting in a Maxwellian velocity distribution, is one of the fastest relaxation processes proceeding in gas systems. The relaxation time within which, as a first approximation, a local Maxwellian distribution is established in every volume element of a system is of the order of the mean free time (Stupochenko et al. 1965; Malkin 1971). In gas mixtures, the width of the translational relaxation zone increases with impurity gas concentration. This especially refers to mixtures of gases which differ in mass, since the temperature distribution function for thermalization depends on the mass proportion (Nikerov and Sholin 1985). As a result, for a mixture of a heavy component (M) and a small light component (m) a Maxwellian distribution is first established for the heavy component within a time τ_T. Thereafter, within a time $M \cdot \tau_T/m$, a Maxwellian distribution is established for the light component. If the concentration of the light component is not small, first an equilibrium distribution is established in each component and then a unified distribution is established for a longer time.

The process of translational relaxation is traced by measuring temperature with an interferometric technique. It is critical that the measured Doppler temperature correspond to the environment temperature. This condition is satisfied, for instance, by 557.7- and 630-nm emissions of atomic oxygen. At night the prevailing part of the 557.7-nm emission intensity is produced at altitudes of about 100 (km) and approximately 10(%) in the emission layer where 630-nm emission occurs, i.e., at altitudes of 200–400 (km).

For the 630-nm emission arising at altitudes of 200–400 (km) and having a radiative lifetime of $\sim$134 (s) (Baluja and Zeippen 1988; Link and Cogger 1988, 1989), the necessary collision frequency is provided at altitudes up to 300 (km). At higher

2.1 Processes of Excitation and Deactivation of Excited Species

altitudes, despite the smaller number of gas-kinetic collisions, an equilibrium is established due to collisions of atmospheric species with unexcited oxygen atoms, present in much greater numbers than excited atoms, which are in equilibrium with the surrounding medium:

$$O(^1D) + O(^3P) \rightarrow O(^3P) + O(^1D).$$

For the 557.7-nm emission from the $O(^1S)$ metastable state, whose radiative lifetime is 0.77 (s) (Bates 1988c), thermalization is complete at altitudes of about 100 (km). However, when measurements were performed under the conditions of the night sky and auroras, for the 557.7- and 630-nm emissions an additional portion of radiating atoms was detected at high altitudes which had not been thermalized before the onset of emission (Hernandez 1971; Frederick et al. 1976; Kopp et al. 1977; Ignatiev 1977a,b; Ignatiev and Yugov 1995). In this case, the width of the Doppler part of the line profile corresponds to an excess of translational energy with which excited metastable oxygen atoms are formed as a result of the dissociative recombination

$$O_2^+ (X^2\Pi_g) + e \rightarrow O(^3P) + O(^3P) + 6.96 \,(eV)$$

$$O_2^+ (X^2\Pi_g) + e \rightarrow O(^1D) + O(^3P) + 5.00 \,(eV)$$

$$O_2^+ (X^2\Pi_g) + e \rightarrow O(^1D) + O(^1D) + 3.04 \,(eV)$$

$$O_2^+ (X^2\Pi_g) + e \rightarrow O(^1S) + O(^3P) + 2.79 \,(eV)$$

$$O_2^+ (X^2\Pi_g) + e \rightarrow O(^1S) + O(^1D) + 0.83 \,(eV).$$

Molecular oxygen ions O_2^+ arise on the day side of the Earth and are transferred along the magnetic field to the night side, creating a layer having a maximum ion concentration near the 60° geomagnetic latitude at altitudes of about 600 (km). These ions are also produced by 0.02- to 0.4-keV electron flows which propagate in the auroral atmosphere up to altitudes of 220–300 (km) (Ignatiev and Yugov 1995).

The 589.0- to 589.6-nm sodium emission lines recorded in night and twilight airglow showed a nonthermal breadth (Hernandez 1975; Sipler and Biondi 1975, 1978) which was caused by the appearance of sodium atoms excited as a result of the reaction

$$NaO(A^2\Sigma^+) + O \rightarrow Na(^2P) + O_2 + E.$$

The excited molecules $NaO(A^2\Sigma^+)$ are produced by the reaction of sodium atoms with ozone molecules, and their radiative lifetime is 0.3 (s) (Shefov et al. 2002; Shefov and Semenov 2002). Therefore, at altitudes of 92–93 (km) the frequency of their collisions with other molecules of the surrounding medium is greater than that with oxygen atoms, providing thermalization of $NaO(A^2\Sigma^+)$ molecules. However, the excited sodium atoms have a radiative lifetime of $\sim 1.6 \cdot 10^{-8}$ (s), which is too short for their thermalization to occur (E $\sim$ 1 (eV)).

2.1.4 Electron-Excited Species Relaxation

Conversion of electron excitation energy into translational motion energy corresponds to the transfer of energy of several electron-volts. Therefore, for the temperatures inherent in the upper atmosphere this implies that there occurs deactivation of electron-excited atoms and molecules rather than establishment of an energy balance between these processes. The relaxation of an electron-excited species depends on whether it collides with an atom or a molecule (Callear and Lambert 1969). Thus, the deactivation of an electron-excited species results in a redistribution of energy for the deactivating species. The problems involved in the calculation of characteristics of processes of this type are discussed by Losev et al. (1995).

2.1.5 Vibrational Relaxation

The establishment of the equilibrium over vibrational degrees of freedom takes a much greater time than the relaxation of translational and rotational degrees of freedom. This is primarily due to the greater vibrational energy quantum. Besides, the efficiency of the energy exchange between the translational and vibrational degrees of freedom is small. Various aspects of the process of vibrational relaxation are considered in many publications (Lambert 1962; Stupochenko et al. 1965; Nikitin 1970; Callear and Lambert 1969; Capitelli et al. 1986; Losev et al. 1995).

For the conditions of the upper atmosphere, the process of vibrational relaxation is related to collisions of a vibrationally excited molecule with surrounding molecules resulting in a decrease in vibrational excitation energy. Qualitatively, a vibrational relaxation event involves two stages. During the first, fast, stage there occur intense exchange processes which do not change the total number of vibrational quanta; therefore, at the end of this stage a quasi-steady-state Boltzmann distribution of the vibrational energy is established. This distribution is characterized by a temperature which is determined only by the initial energy of the vibration and does not depend on the original distribution function of the oscillators over vibrational levels (Malkin 1971). During the second stage, as a result of the energy exchange between the translational and vibrational degrees of freedom, a new distribution is established.

In the upper atmosphere, vibrational relaxation is substantial for the OH, O_2, N_2, NO, CO_2, and O_3 molecules. The most pronounced manifestation of this process is hydroxyl emission. The vibrational relaxation of hydroxyl molecules was considered by many authors (Breig 1969; Spencer and Glass 1977a; McDade and Llewellyn 1987; Dodd et al. 1990, 1991; Shalashilin et al. 1992; Chalamala and

Copeland 1993; Shalashilin et al. 1993, 1994; Adler-Golden 1997; Grigor'eva et al. 1997).

Even early investigations of hydroxyl emission showed that the distribution of the population of the vibrational levels with v = 1–9 of the ground state $X^2\Pi_{3/2,1/2}$ fits well to the Boltzmann distribution with $T_V \sim 10000\,(K)$ (Shefov 1961g). Subsequent measurements revealed seasonal, diurnal, and other types of variations (Berg and Shefov 1963; Shefov 1969c, 1971a; Semenov and Shefov 1996, 1999a). It also turned out that the mutual correlations of the intensities and rotational temperatures of OH bands strongly depend on the compared vibrational levels (Figs. 2.1 and 2.2). For closely spaced levels, both lower and upper ones, the correlation coefficient was $\sim$0.8, while for levels distant from each other it was $\sim$0.6.

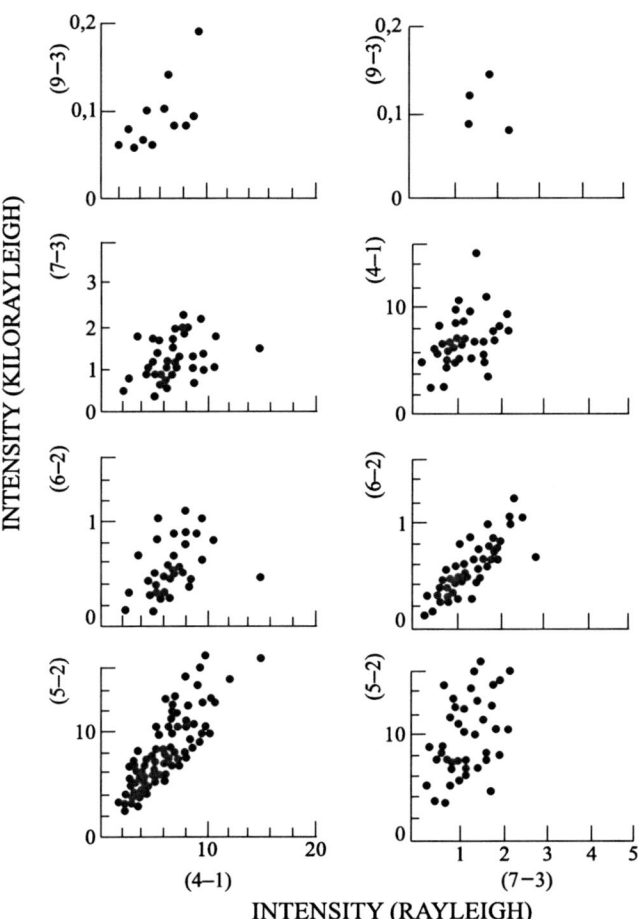

Fig. 2.1 Comparison of the intensities of OH bands from different initial lower and upper vibrational levels (Berg and Shefov 1963)

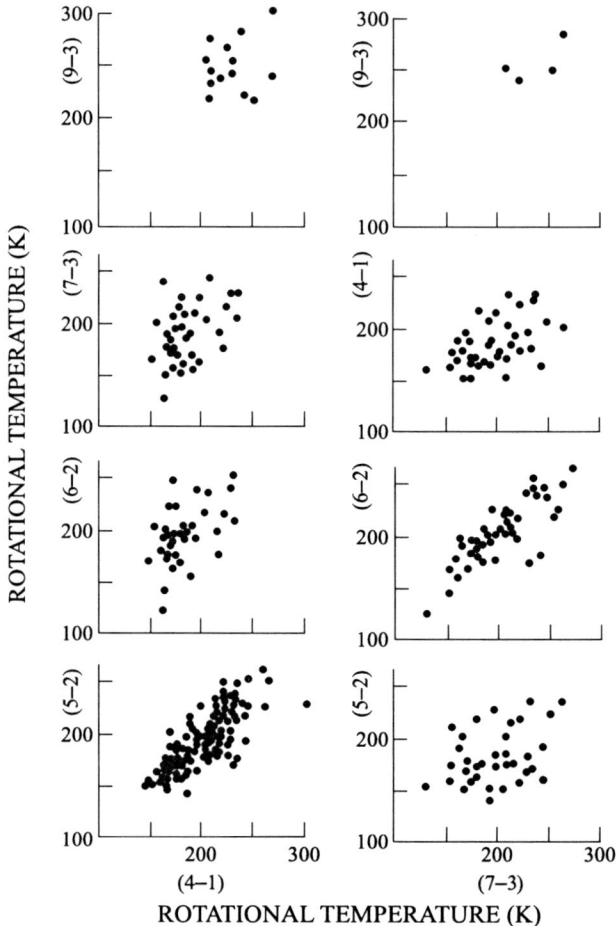

Fig. 2.2 Comparison of the rotational temperatures of OH bands from various initial lower and upper vibrational levels (Berg and Shefov 1963)

Analysis of the population distribution for lower ($v = 3–6$) and upper ($v = 6–9$) levels has revealed different vibrational temperatures T_V for these level groups: $T_{3456} \sim 7900\,(K)$ and $T_{6789} \sim 13\,100\,(K)$, respectively, the average over all levels being $T_{3456789} \sim 9800\,(K)$. In general, it is observed that the populations of the Boltzmann distributions behave as if they shake about the middle vibrational level (Fig. 2.3) (Shefov et al. 1998). Special measurements of OH emission characteristics during the night time and the choice of their near-midnight values for comparison have made it possible to compare the intensities of the OH (3–0), (4–1), (6–2), (7–3), (9–4) bands. In this case, it also turned out that the factor of correlation between closely spaced levels was 0.93 and it decreased to 0.62 for distant levels (Fig. 2.4) (Bakanas and Perminov 2003).

2.1 Processes of Excitation and Deactivation of Excited Species 131

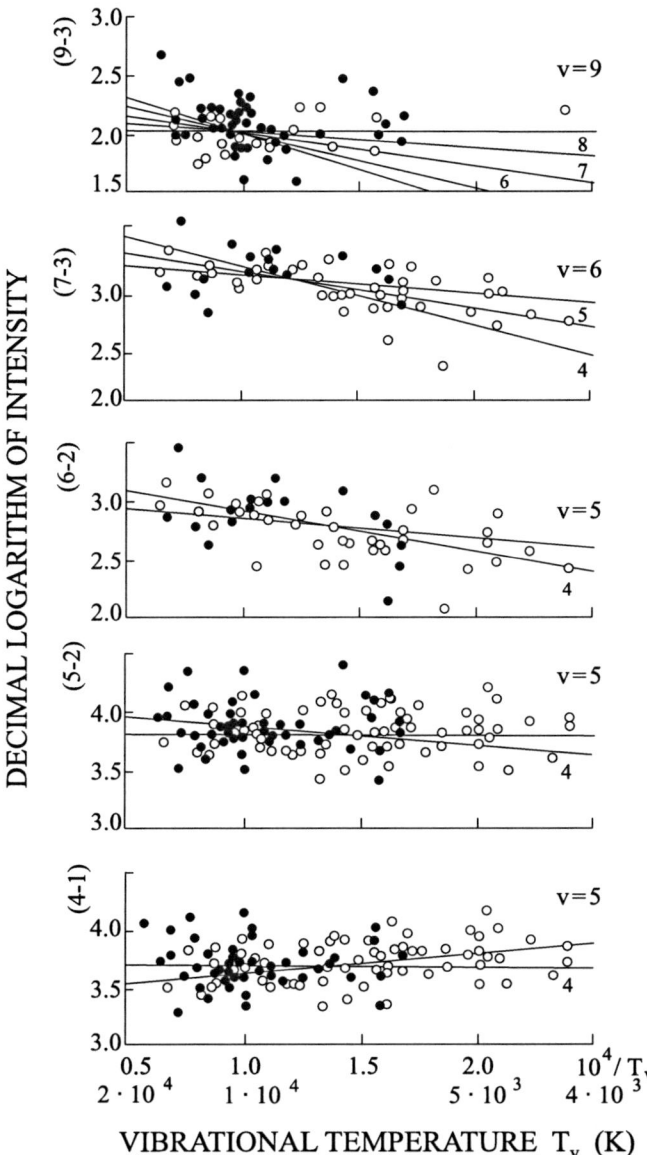

Fig. 2.3 Relation of the intensities of OH bands from variously vibrationally excited levels to vibrational temperatures. *Full circles* are winter values (October through March); *open circles* are summer values (April through September). *Straight lines* are calculated population variations for various v versus $10^4/T_V$ (Shefov et al. 1998)

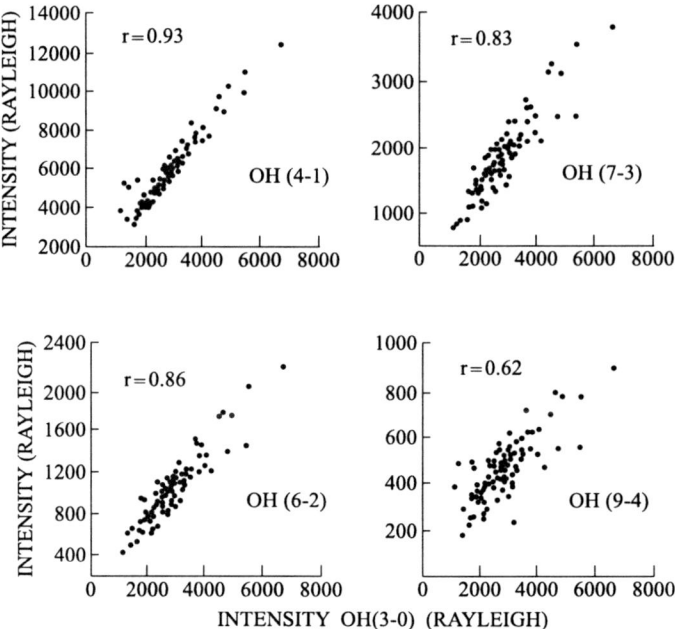

Fig. 2.4 Comparison of the intensities of OH bands from lower and upper vibrational levels (Bakanas and Perminov 2003)

Consideration of the above data in combination with rocket measurements on the altitude of the hydroxyl emission layer has shown that there is a distinct correlation between vibrational temperature and the altitude of maximum emission rate (Semenov and Shefov 1996, 1999a). The lower the altitude of the emission layer, the lower the vibrational temperature since the atmospheric density is greater. This made it possible to obtain an empirical relationship between atmospheric concentration of radiative species [M(Z)] and vibrational temperature T_V, namely,

$$[M(Z)] = 4.2 \cdot 10^{13} \cdot \left(\frac{32200}{T_V} - 1\right)^{1.25} (cm^{-3}) .$$

Thus, the measuring of vibrational temperature is a ground-based method for determining variations of the hydroxyl emission layer altitude and atmospheric density variation.

Analysis of the behavior of hydroxyl emission has allowed the conclusion that the observed variations are due to some features of vibrational relaxation, which are different for different vibrational levels. First models did not consider the variety of atmospheric components at these altitudes (Llewellyn 1978).

Subsequent investigations have revealed that the vibrational relaxation at higher and lower levels is determined by atomic and molecular oxygen. In this case, besides single-quantum transitions, there also occur multiquantum ones (Grigor'eva

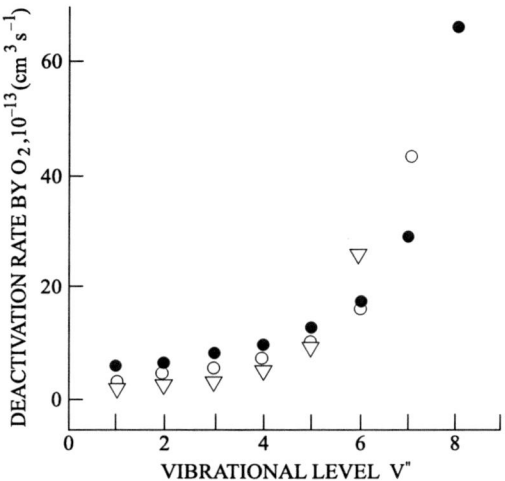

Fig. 2.5 Rates of deactivation of the higher vibrational levels of a hydroxyl molecule by oxygen molecules (Perminov et al. 1998). $k_{O_2}(9, v'')$ are *full circles*; $k_{O_2}(8, v'')$ are *open circles*, and $k_{O_2}(7, v'')$ are *inverted triangles*

et al. 1997; Perminov et al. 1998) (Figs. 2.5 and 2.6). Atomic oxygen is an active component of the mesopause region which determines the energy regime of the latter. The altitude distribution of atomic oxygen varies during the seasons of year and depends on solar activity (Semenov and Shefov 1999b, 2005; Shefov et al. 2000). This results in variability of the conditions for vibrational relaxation of excited hydroxyl molecules.

As a result of photochemical reactions, oxygen molecules vibrationally excited to metastable electronic states $A^3\Sigma_u^+$, $A'^3\Delta_u$, $c^1\Sigma_u^-$, $b^1\Sigma_g^+$, $a^1\Delta_g$ appear in the upper atmosphere. The radiative lifetimes of these states are rather long: 0.07, 1.1, 1, 7.1, and 5260 (s), respectively. Therefore, they can exist only at the mesopause altitudes.

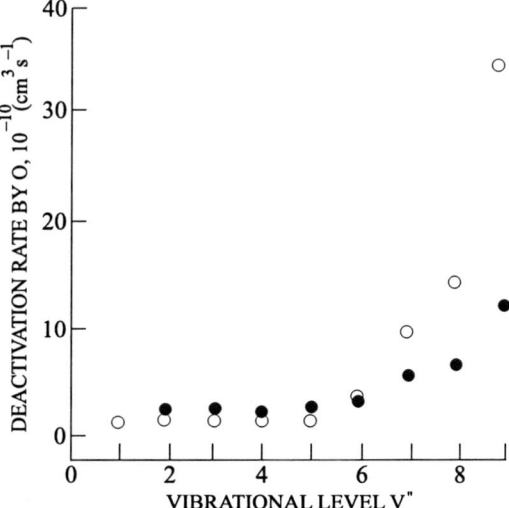

Fig. 2.6 Rates of deactivation of the vibrational levels of a hydroxyl molecule by atomic oxygen (Perminov et al. 1998). $k_O(v', v'')$ are *full circles*, transitions for all Δv are equally probable (n = 0); transitions accompanied by loss of all vibrational energy (n = −5) are *open circles*

Vibrationally excited oxygen molecules in the ground state $X^3\Sigma_g^-$ can be produced only by radiative transitions from the higher states. Since radiative transitions between vibrational levels of one electronic state for biatomic molecules with identical atoms are forbidden, the vibrational relaxation of excited oxygen molecules can occur only on their collisions with other oxygen molecules and atoms (Jones et al. 1965; Breig 1969; Breen et al. 1973; Slanger et al. 2000). As a consequence of these exchange processes, the distribution of the populations of vibrational levels of excited O_2 molecules is not a Boltzmann, but a nonuniform distribution.

2.1.6 Rotational Relaxation

Under the conditions of the upper atmosphere, rotationally and vibrationally excited molecules are formed as a result of chemical reactions. The distribution of the reaction products over rotational and vibrational states differs from the equilibrium distribution, which is determined by the environment temperature. Therefore, the rotational excitation of molecules is relaxed due to the conversion of energy of rotation of atoms of a molecule into their translational motion. The small energy of rotational quanta is responsible for the rather easy energy exchange between these degrees of freedom. For the lower rotational levels, several collisions are usually required (Gordiets et al. 1980; Bogdanov et al. 1991). However, for the higher rotational levels, having high rotational numbers, the rotational quanta are rather great, and therefore the necessary number of collisions is greater and the rotational relaxation time is longer.

The process of relaxation of rotationally excited molecules, and, in the overwhelming majority of cases, of rotational temperature T_r, is generally described by the equation

$$\frac{dT_r}{dt} = -\frac{T_r - T_k}{\tau_r},$$

where T_k is the kinetic temperature of the environment and τ_r is the relaxation time constant.

The solution of this simple equation,

$$T_r = T_k + (T_0 - T_k) \cdot \exp\left(-\frac{t}{\tau_r}\right),$$

determines the quantity τ_r or the related number of collisions that provide the transition to an equilibrium state

$$Z_r = \tau_r \cdot [M] \cdot \pi\sigma_k^2 \cdot \sqrt{\frac{8kT}{\pi \cdot \mu}}.$$

Here [M] is the concentration of the medium molecules; $\pi\sigma_k^2$ is the cross-section for elastic scattering; k is the Boltzmann's constant; $\mu = m \cdot m_M/(m + m_M)$ is the reduced mass; m and m_M are the masses of the relaxing molecule and the surrounding molecules, respectively.

The characteristic relaxation time τ_r is $\sim 10^2 \cdot \tau_0$ for H_2 and D_2 molecules and $\sim (1-10) \cdot \tau_0$ for other molecules; here τ_0 is the mean free time of the molecules (Losev et al. 1995).

It is of importance that the number of collisions for rotational relaxation refers to the time interval that determines the excited state lifetime, which, in turn, depends both on the radiative lifetime,

$$\tau_{rad} = 1 \bigg/ \sum_{v''=0}^{v'-1} A_{v',v''},$$

and on the frequency of collisions resulting in deactivation of the excited state. Experimental data show that for molecules which do not make radiative vibrational transitions in the ground electronic state, the quantity Z_r strongly depends on the energy of rotational quanta. Thus, for the temperatures $\sim 200-250$ (K), which are typical of the conditions of the middle atmosphere, Z_r has the following values: 300–500 for H_2, 200 for D_2, 15 for HD, 5 for N_2, and 4.5 for O_2 (Osipov 1985). For the hydroxyl molecule OH, according to the data reported by Kistiakowsky

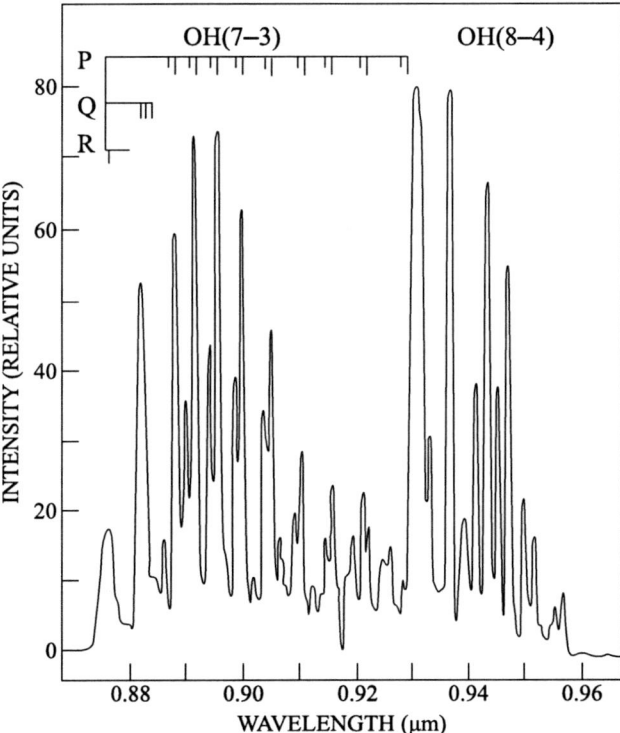

Fig. 2.7 Spectrogram of the OH (7–3) band. Zvenigorod, 16.03.1991, 20:12–20:32 (Perminov and Semenov 1992)

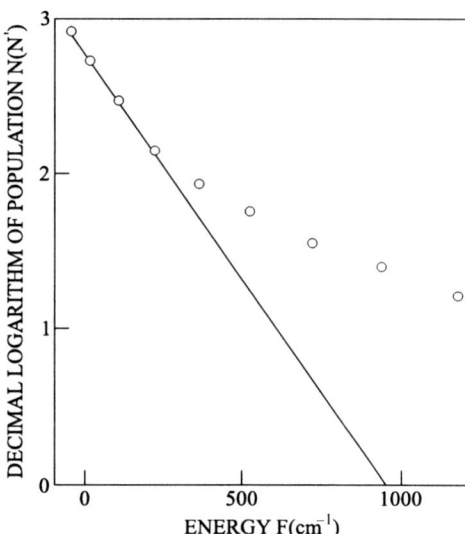

Fig. 2.8 Populations of the rotational levels of the OH (7–3) band as a function of their energy (Perminov and Semenov 1992). *Circles* are measurements; *straight line* is the regression line drawn through the first four P lines of the OH band by which rotational temperature is usually determined. For the P lines with $N' = 6/9$, the rotational temperature is $\sim 1000\,(K)$

and Tabbutt (1959), Z_r is about 10. The problem of rotational relaxation of atmospheric molecules, the hydroxyl molecule included, is considered elsewhere (Nicholls et al. 1972; Hotlzclaw et al. 1997; Strekalov 2003; Lazarev et al. 2003).

For the hydroxyl molecule the radiative lifetime of the seventh vibrational level of the ground state is 0.007 (s) (Semenov and Shefov 1996). The vibrational relaxation of this molecule in collisions with atomic and molecular oxygen reduces this lifetime (Perminov et al. 1998). Therefore, while at the altitudes of the maximum emission rate in the emission layer the necessary number of collisions is provided for the rotational temperature to correspond to the medium temperature, at the altitudes near 100 (km) the number of collisions becomes ~ 1. Though the part of the emission layer on the line of sight above 100 (km) is small, the intensity of rotational lines with $J' > 13/2\,(N' > 6)$ that arise in this part of the layer becomes prevailing (Fig. 2.7). This results in a nonequilibrium rotational temperature of $\sim 1000\,(K)$ (Fig. 2.8) (Perminov and Semenov 1992). This also explains why the emissions due to high ($N' = 33$) pure rotational transitions from the vibrational levels 0, 1, and 2 were observed along the limb at altitudes of 80–100 (km) on the Shattle spacecraft (Smith et al. 1992; Dodd et al. 1994).

2.2 Hydroxyl Emission

2.2.1 *Progress in the Studies of Hydroxyl Emission*

First observations of night-sky hydroxyl emission were performed long before the nature of this emission had been elucidated. In 1928, a series of emission peaks at 653, 687, and 727 (nm) were detected by Slipher (1929) who photographed

2.2 Hydroxyl Emission

nightglow spectra with a small dispersion in the range 580–770 (nm). Somewhat later, when studying the radiation spectra of planets, he published photos of spectra for wavelengths up to 860 (nm) that showed four emission peaks (Slipher 1933). However, it was impossible to identify the peaks observed because of poor resolution of the spectra. The work performed in the subsequent years gave similar results (Sommer 1932). In 1938, night airglow spectra were obtained with a greater dispersion at the Mount Wilson observatory (Babcock 1939). In these spectra, some structures of unknown molecular bands were seen. Unfortunately, the spectroscopic resolution achieved in this case also gave no way of revealing the nature of the emission observed.

Besides spectrographic observations, photometric methods for recording the nightglow of the upper atmosphere came into use in the early 1940s. The first observation with the use of photomultipliers detected intense night-sky infrared emissions in the 700-nm range and near 1000 and 1050 (nm) (Rodionov 1940; Herman et al. 1942; Elvey 1943; Stebbins et al. 1945) whose intensity was almost two orders of magnitude greater than that of the well-known green emission of atomic oxygen. This circumstance inclined researchers to suppose that the radiation observed is emitted by nitrogen atoms and molecules (NI – 1040 (nm), (0–0) band 1051 (nm) 1PG N_2) (Khvostikov 1937, 1948; Bates 1948; Elvey 1942; Nicolet 1948). It is curious that the presence of OH bands in the visible spectrum of night airglow was mentioned and discussed (Déjardin and Bernard 1938; Nicolet 1948); however, the presence of OH molecules in the atmosphere was considered as extremely doubtful.

Only in 1948, when Meinel (1948) and Krassovsky (1949) took photos of spectra (in the range 700–800 (nm) and 700–1100 (nm), respectively) with a dispersion of about $25\,(\text{nm} \cdot \text{mm}^{-1})$, the rotational structure was clearly revealed for emission bands not identified earlier for which R, Q, and P branches were easily seen. Analysis of the spectra, performed by Krassovsky, has shown that in the 1044 (nm) range there is no emission peak, and thus the idea that the observed infrared radiation belongs to atomic and molecular nitrogen has been rejected (Krassovsky 1950a,b,c). Subsequently this result was confirmed by photometric observations performed by Kron (1950). The identity of the observed spectra to the vibrational–rotational bands of hydroxyl molecules was proposed by Herzberg and experimentally confirmed by Meinel (1950a,b,c). Subsequently Shklovsky (1950a,b, 1951a) identified hydroxyl bands by the spectra measured by Meinel (for the range 700–900 (nm) where Meinel was not able to do this) and Krassovsky (900–1200 (nm)). He was also the first to calculate the transition probabilities for the rotational–vibrational bands of hydroxyl (Table 2.4) and, based on them, to estimate the emission intensity for other, virtual, OH bands (Shklovsky 1951a, 1957). It turned out that the OH emission is associated in the main with the infrared range up to 4500 (nm), its total intensity reaching about 5 (megarayleigh). Here it should be stressed that this intensity, if it would pertain to green emission, should be equivalent to the brightness of IBC IV aurora entirely covering the sky. Later, more exact and detailed calculations of the transition probabilities were performed and the intensity distribution in the rotational–vibrational spectrum of the hydroxyl molecule was calculated (Table 2.5) (Heaps and Herzberg 1952).

Almost immediately after the identification of hydroxyl emission and estimation of its total energy (2–4 (erg·cm^{-2}·s^{-1})), suppositions about the photochemical nature of its occurrence were made (Bates and Nicolet 1950; Herzberg 1951; Krassovsky 1951a). These ideas were developed and perfected (Krassovsky 1963a; Breig 1970) for more than 40 years. The bibliography on hydroxyl emission encounters a great number of publications. Nevertheless, many questions related to the processes underlying the occurrence of hydroxyl emission still remain to be elucidated. These questions arise in analyzing the data on the time variations of the intensity of emission bands from various vibrationally excited levels that have been considered in Sect. 2.1.5.

2.2.2 Principal Characteristics of Hydroxyl Emission

Present-day data allow the statement that hydroxyl emission arises due to the rotational–vibrational transitions of the ground state $X^2\Pi$. The potential curve of the ground state is presented in Fig. 2.9 (Fallon et al. 1961; Krassovsky et al. 1962). The energies of vibrational levels, $G(v)$, are determined by the formula (Herzberg 1945, 1950, 1971)

$$G(v) = \omega_e \cdot \left(v + \frac{1}{2}\right) - \omega_e x_e \cdot \left(v + \frac{1}{2}\right)^2 + \omega_e y_e \cdot \left(v + \frac{1}{2}\right)^3 - \omega_e z_e \cdot \left(v + \frac{1}{2}\right)^4 + \ldots$$

With the constants published by Huber and Herzberg (1979), this equation has the form

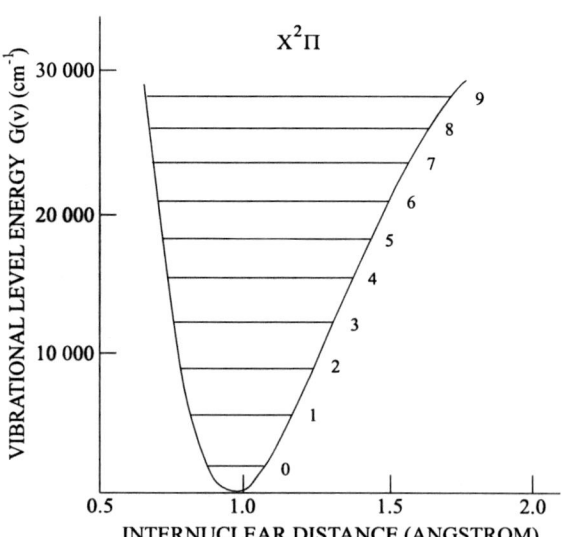

Fig. 2.9 Potential curve of the ground state of the hydroxyl molecule (Fallon et al. 1961; Krassovsky et al. 1962)

2.2 Hydroxyl Emission

$$G(v) = 3737.76 \cdot \left(v + \frac{1}{2}\right) - 84.881 \cdot \left(v + \frac{1}{2}\right)^2 + 0.540 \cdot \left(v + \frac{1}{2}\right)^3$$
$$- 0.0213 \cdot \left(v + \frac{1}{2}\right)^4 - 0.0011 \cdot \left(v + \frac{1}{2}\right)^5 + \ldots .$$

Data on the potential curve and energy of vibrational levels G(v) (Fallon et al. 1961) are given in Table 2.1. The ground state includes two systems of sublevels, $X^2\Pi_{3/2}$ and $X^2\Pi_{1/2}$. This is an inverse state, and the $X^2\Pi_{1/2}$ levels are located higher than $X^2\Pi_{3/2}$ (Fig. 2.10).

The energies of rotational levels are determined with the help of the quantum number J by the formulas

$$^2\Pi_{3/2} : F_1(J) = B_v \cdot \left[\left(J + \frac{1}{2}\right)^2 - 1 - \frac{1}{2} \cdot \sqrt{4\left(J + \frac{1}{2}\right)^2 + Y_v(Y_v - 4)} \right] - D_v \cdot J^4,$$

$$^2\Pi_{1/2} : F_2(J) = B_v \cdot \left[\left(J + \frac{1}{2}\right)^2 - 1 + \frac{1}{2} \cdot \sqrt{4\left(J + \frac{1}{2}\right)^2 + Y_v(Y_v - 4)} \right] - D_v \cdot J^4$$

or with the help of the quantum number N by the formulas

$$^2\Pi_{3/2} : F_1(N) = B_v \cdot \left[(N+1)^2 - 1 - \frac{1}{2} \cdot \sqrt{4(N+1)^2 + Y_v(Y_v - 4)}\right] - D_v \cdot N^2 \cdot (N+1)^2,$$

$$^2\Pi_{1/2} : F_2(N) = B_v \cdot \left[N^2 - 1 + \frac{1}{2} \cdot \sqrt{4N^2 + Y_v(Y_v - 4)}\right] - D_v \cdot N^2 \cdot (N+1)^2,$$

$$^2\Pi_{3/2} : N = J - \frac{1}{2}; \qquad ^2\Pi_{1/2} : N = J + \frac{1}{2}; \qquad Y_v = \frac{A_v}{B_v}.$$

Table 2.1 Structure of the vibrational levels of the ground state of the OH molecule

v	r(A) min	r(A) max	G(v) (cm^{-1})	$G_0(v) = G(v) - G(0)$ (cm^{-1})	(eV)	(erg) 10^{-12}	(kcal)
0	0.884	1.081	1847.78	0.00	0.000	0.000	0.00
1	0.831	1.179	5417.41	3569.64	0.443	0.709	10.21
2	0.800	1.257	8821.36	6973.68	0.865	1.385	19.94
3	0.777	1.329	12061.61	10214.05	1.266	2.029	29.19
4	0.758	1.399	15139.28	13291.82	1.648	2.640	37.99
5	0.743	1.468	18054.52	16207.12	2.009	3.219	46.31
6	0.731	1.538	20805.95	18958.82	2.350	3.765	54.17
7	0.720	1.610	23392.1	21544.30	2.671	4.279	61.57
8	0.710	1.683	25806.7	23958.99	2.970	4.759	68.47
9	0.702	1.760	28043.5	26196.06	3.248	5.203	74.88
10			30095.1	28245.36	3.502	5.610	80.73

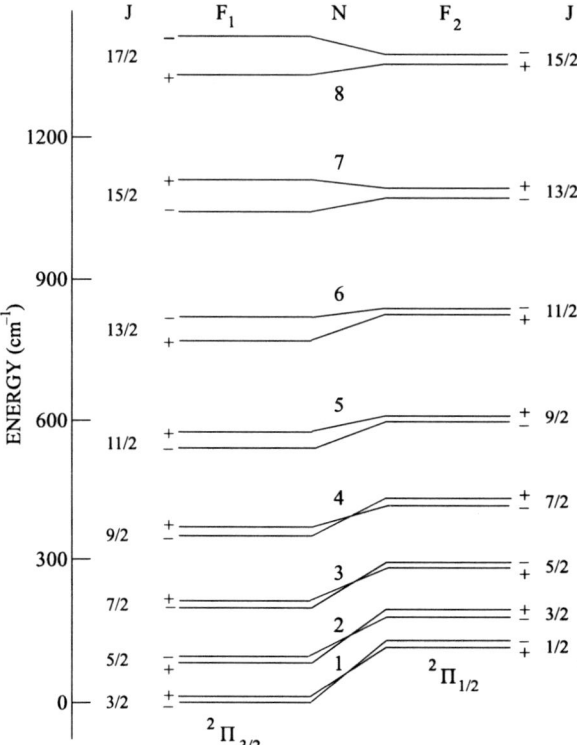

Fig. 2.10 Rotational structure of the vibrational levels of the ground state of the OH molecule (Herzberg 1974)

The constants for the rotational levels are given in Table 2.2 (Krassovsky et al. 1962; Coxon 1980; Coxon and Foster 1982).

In view of the Λ doubling, the energies of rotational levels are calculated by the formulas

$$F_{vJ}^{\pm} = B_v \cdot f^{\pm}(J) - D_v \cdot [f^{\pm}(J)]^2 + H_v \cdot [f^{\pm}(J)]^3 + \ldots \pm \gamma_{vJ}\left(J + \frac{1}{2} \pm 1\right),$$

where

$$f^{\pm}(J) = \left(J + \frac{1}{2}\right) \cdot \left(J + \frac{1}{2} \pm 1\right); \quad \gamma_{vJ} = \gamma_v + \gamma_{Dv} \cdot J(J+1) + \gamma_{Hv} \cdot J^2(J+1)^2 + \ldots.$$

The Λ doubling is given by $F_d - F_c = q_v N(N+1)$.

The fine structure of the lowest rotational level is responsible for the radio emission at a wavelength of 18 (cm) that is observed in the interstellar space. The rotational–vibrational transitions corresponding to $X^2\Pi_{3/2} R_1$, Q_1, and P_1 lines and to $X^2\Pi_{1/2} R_2$, Q_2, and P_2 branches occur within each subsystem of levels. The wavelengths of the OH bands of the entire system are given elsewhere (Shefov and Piterskaya 1984). The intercombination $X^2\Pi_{1/2} \rightarrow X^2\Pi_{3/2}$ transitions are

2.2 Hydroxyl Emission

Table 2.8 Einstein coefficients $A(v' \to v'')(s^{-1})$ for the radiation transitions of OH molecules $(X^2\Pi, v \leq 9)$ (Murphy 1971)

v''/v'	0	1	2	3	4	5	6	7	8	$\sum A(s^{-1})$	$\log_{10} \sum A$
1	1									1.00	0.00
2	0.563	1.36								1.92	0.283
3	0.0443	1.58	1.26							2.88	0.459
4	0.00543	0.167	2.94	0.878						4.04	0.606
5	0.0010	0.0247	0.396	4.55	0.390					5.36	0.729
6	0.00025	0.00536	0.0676	0.747	6.26	0.0474				7.12	0.852
7	0.000071	0.00148	0.0167	0.154	1.25	7.94	0.0753			9.44	0.975
8	$2.2 \cdot 10^{-5}$	0.00048	0.0051	0.0384	0.272	1.89	9.38	0.700		12.28	1.09
9	$7.7 \cdot 10^{-6}$	0.00017	0.0018	0.0128	0.0766	0.563	2.56	10.5	2.12	15.82	1.20

Table 2.9 Einstein coefficients $A(v' \to v'')(s^{-1})$ for the radiation transitions of OH molecules $(X^2\Pi, v \leq 9)$ (Mies 1974)

v''/v'	0	1	2	3	4	5	6	7	8	$\sum A(s^{-1})$	$\log_{10} \sum A$
1	20									20	1.3
2	14	25								39	1.59
3	0.92	40	21							63	1.80
4	0.079	4.3	73	12						89	1.95
5	0.05	0.39	11	108	4.5					124	2.09
6		0.053	1.3	21	141	2,3				166	2.22
7			0.18	2.9	37	163	9,1			212	2.33
8			0.030	0.57	5.7	61	167	26		260	2.4
9				0.13	1.2	11	90	147	51	300	2.48

Table 2.10 Einstein coefficients $A(v' \to v'')(s^{-1})$ for the radiation transitions of OH molecules $(X^2\Pi, v \leq 9)$ (Llewellyn and Solheim 1978)

v''/v'	0	1	2	3	4	5	6	7	8	$\sum A(s^{-1})$	$\log_{10} \sum A$
1	20.2									20	1.30
2	14.1	25.2								39	1.54
3	0.92	39.9	20.9							62	1.74
4	0.079	4.29	72.6	12.3						89	1.95
5	0.05	0.392	10.6	108	4,47					123	5.09
6	0.0045	0.053	1.27	21.0	142	2,35				167	2.22
7	0.00128	0.0265	0.182	2.91	37.3	163	9.14			213	2.33
8	$4.04 \cdot 10^{-4}$	$8.63 \cdot 10^{-3}$	0.030	0.564	5.67	60.8	169	25.8		262	2.42
9	$1.38 \cdot 10^{-4}$	$3.07 \cdot 10^{-3}$	0.0108	0.130	1.19	10.8	90.3	147	50.7	300	2.48

Table 2.11 Einstein coefficients $A(v' \to v'')(s^{-1})$ for the radiation transitions of OH molecules $(X^2\Pi, v \leq 9)$ (Langhoff et al. 1986)

v″ / v′	0	1	2	3	4	5	6	7	8	$\Sigma A(s^{-1})$	$\log_{10} \Sigma A$
1	15									15	1.18
2	8.5	21								29.5	1.48
3	0.69	23.5	19							43.19	1.63
4	0.065	2.7	43	14						59.765	1.78
5	0.011	0.34	6.4	65	7					78.751	1.90
6		0.055	0.91	12.6	85	3				101.565	2.01
7			0.19	2.1	21	103	3			129.29	2.11
8				0.42	4.0	33	111	8		156.42	2.19
9					0.96	6.9	48	108	21	184.86	2.27

Table 2.12 Einstein coefficients $A(v' \to v'')(s^{-1})$ for the radiation transitions of OH molecules $(X^2\Pi, v \leq 9)$ (Turnbull and Lowe 1989)

v″ / v′	0	1	2	3	4	5	6	7	8	$\Sigma A(s^{-1})$	$\log_{10} \Sigma A$
1	22.7									22.7	1.36
2	15.4	30.4								45.8	1.66
3	2.03	40.3	28.1							70.43	1.85
4	0.30	7.19	69.8	20.3						97.59	1.99
5	0.051	1.32	15.9	99.4	11.0					127.67	2.11
6	0.010	0.27	34.8	27.9	126	4.0				161	2.21
7		0.063	0.85	7.16	42.9	145	2.3			198.27	2.30
8			0.23	2.01	12.7	60.0	154	8.6		237.54	2.38
9				0.62	4.05	19.9	78.6	149	23.7	275.87	2.44

Table 2.13 Einstein coefficients $A(v' \to v'')(s^{-1})$ for the radiation transitions of OH molecules $(X^2\Pi, v \leq 9)$ (Goldman et al. 1998)

v″ / v′	0	1	2	3	4	5	6	7	8	$\Sigma A(s^{-1})$	$\log_{10} \Sigma A$
1	17.33									17.33	1.24
2	10.31	23.58								33.89	1.53
3	1.12	27.55	22.20							50.87	1.71
4	0.13	4.12	48.68	16.45						69.38	1.84
5	0.019	0.63	9.43	70.70	9.31					90.09	1.95
6	0.003	0.108	1.74	17.14	91.44	3.78				114.21	2.06
7		0.023	0.37	3.78	27.19	107.25	2.27			140.88	2.15
8			0.088	0.92	7.04	39.09	116.58	7.25		170.97	2.23
9				0.25	1.96	11.72	51.92	117.6	20.41	203.86	2.31

However, some authors considered the calculated transition probabilities in relation to the early intensity measurements (Krassovsky et al. 1962). For these measurements, the seasonal intensity variations and the data used for the calculation of transition probabilities were not reduced to identical heliogeophysical conditions. This should have a detrimental effect on the determination of the coefficients at the terms of the expansion of the dipole moment of the OH molecule in a series. Measurements of the intensities of various OH bands reduced to identical yearly average conditions were published (Shefov 1976; Shefov and Piterskaya 1984). However, it seems that they have not yet been used in calculations of the transition probabilities. Since the calculations of the transition probabilities demanded a knowledge of absolute values of $A_{v',v''}$, experimental investigations of the absorption coefficients for some resonance transitions were undertaken to determine these values (Benedict et al. 1953; Phelps and Dalby 1965; Potter et al. 1971; Murphy 1971; Worley et al. 1971, 1972; Roux et al. 1973; Langhoff et al. 1986). However, in some cases, these investigations gave a rather large spread in values of A(1–0): from 395 (s^{-1}) (Phelps and Dalby 1965) to 3.3 (s^{-1}) (Potter et al. 1971). Today it is accepted that the most correct data are those obtained based on the results of Goldman et al. (1998), according to which $A_{(1-0)} = 17.33\,(s^{-1})$. Nevertheless, the complexity of laboratory absorption measurements still casts some doubt upon the correctness of the last version of the absolute values of the transition probabilities.

2.2.4 Hydroxyl Emission Intensity

The OH emission intensity, as already mentioned, is extraordinarily high under the conditions of the night upper atmosphere. Its value characterizes the rates of photochemical processes, which, for the conditions of the mesopause, provide a major channel for recombination of atomic oxygen. However, as can be seen from the discussion in the previous section, the total hydroxyl emission intensity is difficult to estimate because of the large spectral range covered by this emission. Under the conditions of the ground-based observations of the upper atmosphere it is possible to detect only the bands with $\Delta v \geq 2$, which correspond to $\lambda \leq 2.1\,(\mu m)$. According to the initial data on transition probabilities, the transitions with $\Delta v = 1 (\lambda = 2.8\text{–}5\,(\mu m)$, which are inaccessible to ground-based measurements) should have the highest intensity. However, since Ferguson and Parkinson (1963) stated that the transition probabilities for the bands $\Delta v = 1$ should be lower, including in comparison with those for the bands $\Delta v = 2$, the total power of hydroxyl emission was re-estimated. It is well known that the intensity of the bands of the sequence $\Delta v = 1$ was never measured from the Earth surface, but were estimated from ground measurements of the band intensities in other transitions. Therefore, the total emission power decreased from 2.5–3 $(erg \cdot cm^{-2} \cdot s^{-1})$ to 1 $(erg \cdot cm^{-2} \cdot s^{-1})$ (1 megarayleigh) (Shefov 1976; Shefov and Piterskaya 1984; Semenov and Shefov 1996, 1999a). This quantity is essential to the understanding of the role of hydroxyl molecules as an energy sink in the atmospheric energy balance in the mesopause region. Based on the systematization of the measurements of OH band intensities in various spectral regions (Bakanas and Perminov 2003) and transition probabilities (Goldman et al. 1998), the yearly average intensities of the OH bands for the year 2000 are presented in Table 2.14.

Based on long-term observations, various regular time variations of the OH emission intensity at night have been studied. These data allow one to get a notion about the behavior of hydroxyl emission (Fishkova 1983; Semenov and Shefov 1996, 1999a) (see Sect. 4.1).

In the previous sections, the behavior of the intensities of OH bands from different vibrationally excited levels has been considered. Special investigations of the seasonal intensity variations have shown that this behavior is related to the variations of the vibrational temperature that reflect the variations of the altitude of the emission layer.

2.2.5 Rotational Temperature

The rotational temperature is determined by the distribution of the emission intensity over the rotational–vibrational bands, mainly of the P branch (Shefov 1961c). This problem was also considered by Prokudina (1959b) and Kvifte (1961). Besides, the temperature can be determined by the relative total intensities of groups of lines, more often, of the R_1, Q_1, and P_1 branches. The data necessary to do this are available (Piterskaya and Shefov 1975).

The starting expression (for the intensity of a line in (Rayleighs)) is

$$I(J',J'') = C \cdot \lambda^{-3}(J',J'') \cdot i(J') \cdot \exp\left[-\frac{hc}{k} \cdot \frac{F(J')}{T_r}\right].$$

Here C is a constant, $\lambda(J',J'')$ is the wavelength of the rotational–vibrational line, $i(J')$ is the intensity factor for the line, $F(J')$ is the energy of the rotational level J' relative to the energy $G_{v'}$ of the vibrational level v' (see Table 2.3), and T_r is the rotational temperature.

The procedure of determination of the rotational temperature typically consists in constructing a correlation relation:

$$\log_e\left[\frac{I(J',J'') \cdot \lambda^3}{i(J')}\right] = -\frac{hc}{k} \cdot \frac{F(J')}{T_r} + C = -1.4388 \cdot \frac{F(J')}{T_r} + C.$$

An example of this relation is presented in Fig. 2.11 (Shefov 1961c). As can be seen, the temperature is determined by the coefficient of regression.

Under the conditions of observations in the upper atmosphere at temperatures of 170–230 (K) it is usually possible to record with confidence 4–5 lines of P_1 branches in the radiation spectrum which are best spectroscopically resolved in the

2.2 Hydroxyl Emission

Table 2.14 Wavelengths (nm) of the OH bands and the related yearly average intensities (Rayleigh) for the period of the maximum solar activity of the year 2000

v″ / v′	0	1	2	3	4	5	6	7	8	9	$\sum_{v''=0}^{v'-1} I_{v'-v''}$
1	2800.7 170000										170000
2	1433.6 54000	2936.9 124000									178000
3	978.8 3000	1504.7 74500	3085.4 60100								137600
4	752.2 200	1027.3 6200	1582.4 73300	3248.3 24800							104500
5	616.9 20	791.1 560	1082.8 8400	1668.2 63000	3429.4 8300						80280
6	527.3 2	649.6 65	834.2 1070	1143.3 10500	1764.2 56200	3633.4 2300					701400
7	464.0 0.5	556.2 10	686.1 180	882.3 1900	1211.4 13500	1873.0 53100	3865.8 1100				69790
8	417.3 0.2	490.3 4	588.6 40	727.3 390	937.2 3000	1289.6 16400	1999.2 49000	4140.3 3000			71830
9	381.6 0.04	441.8 0.8	520.1 8	625.6 60	774.6 480	1010.8 2900	1381.3 12700	2149.3 28800	4469.5 5000		49940
10	353.9	405.1	467.0	554.4	668.5	830.2	1076.3	1491.6	2331.5	4873.4	

$\Sigma = 932080$

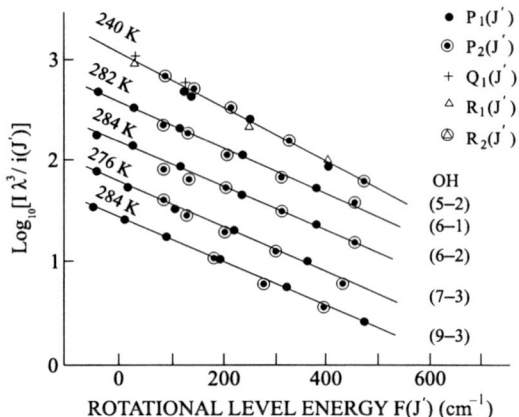

Fig. 2.11 Plots for determining the rotational temperatures of OH bands from different vibrationally excited levels (measurement date: 6.12.1959) (Shefov 1961c)

band structure. According to the above equation, the intensities of the lines in the P branch with $N' \leq 5$ are well described by a Boltzmann distribution with temperature T_r (Shefov 1961c).

For many years the values of $i(J')$ published by Benedict et al. (1953) (Table 2.15) were used. When looking at the procedure of determination of the rotational temperature of hydroxyl emission, one can see that the main problem is that an unambiguous complete set of intensity factors $i(J)$ is not available. Now various theoretical approaches are used which consider intricate relationships between the rotational and vibrational transitions in the ground state of the OH molecule. The original data of Benedict et al. (1953) were repeatedly revised (Roux et al. 1973; Mies 1974; Turnbull 1987; Langhoff et al. 1986; Turnbull and Lowe 1989; Goldman et al. 1998; French et al. 2000; Pendleton and Taylor 2002).

Based on the data reported by Kovács (1969) and Whiting et al. (1973), Turnbull (1987) proposed formulas for calculating the intensity factors (Table 2.16). They were used to calculate $i(J')$ for some lines of OH vibrational–rotational bands (Table 2.17). The new intensity factors calculated for the temperatures of the upper atmosphere ($J' \leq 8.5$) turned out practically the same for all investigated bands.

These and all values of $i(J)$ calculated later both by direct formulas and from transition probabilities differ from the values given by Benedict et al. (1953). Nevertheless, comparing $i(J)$ values for various bands, one can see that for the usually used OH bands the values of $i(J)$ for the $P_1(J')$ branch ($J' \leq 5.5$) differ by no more than 0.01–0.02. The differences are more substantial for the $P_2(J' \geq 3.5)$, $Q_2(J' \geq 3.5)$, and $R_2(J' \geq 3.5)$ branches, which are practically not used in ground-based measurements to determine rotational temperatures below 300 (K). However, according to many studies (French et al. 2000), the use of the values of $i(J)$ for P_1, Q_1, and R_1 branches gives different temperature values for different branches. As can readily be seen from Table 2.18, the values of intensity factors obtained by different authors, normalized to the value for P_1 ($J' = 1.5$) equal to 1.62 (Benedict et al. 1953), give lower temperature values compared to those obtained with the use of the data of Benedict et al. (BPH) (1953). If we put $\Delta T = T$ (BPH) $- T$ (new), then, based on

2.2 Hydroxyl Emission

Table 2.15 Intensity factors of the OH vibrational–rotational bands of the ground state $X^2\Pi$ (Benedict et al. 1953)

N'	J'	$P_1(N')$	$i(J')$	$Q_1(N')$	$i(J')$	$R_1(N')$	$i(J')$	N'	J'	$P_2(N')$	$i(J')$	$Q_2(N')$	$i(J')$	$R_2(N')$	$i(J')$
1	3/2	$P_1(1)$	1.62	$Q_1(1)$	2.30	–	–	1	1/2	$P_2(1)$	1.29	$Q_2(1)$	0.667	–	–
2	5/2	$P_1(2)$	2.90	$Q_1(2)$	1.40	$R_1(2)$	1.62	2	3/2	$P_2(2)$	2.34	$Q_2(2)$	0.300	$R_2(2)$	1.29
3	7/2	$P_1(3)$	4.04	$Q_1(3)$	0.98	$R_1(3)$	2.90	3	5/2	$P_2(3)$	3.37	$Q_2(3)$	0.223	$R_2(3)$	2.34
4	9/2	$P_1(4)$	5.14	$Q_1(4)$	0.73	$R_1(4)$	4.04	4	7/2	$P_2(4)$	4.37	$Q_2(4)$	0.188	$R_2(4)$	3.37
5	11/2	$P_1(5)$	6.21	$Q_1(5)$	0.57	$R_1(5)$	5.14	5	9/2	$P_2(5)$	5.38	$Q_2(5)$	0.17	$R_2(5)$	4.37
6	13/2	$P_1(6)$	7.26	$Q_1(6)$	0.46	$R_1(6)$	6.21	6	11/2	$P_2(6)$	6.39	$Q_2(6)$	0.16	$R_2(6)$	5.38
7	15/2	$P_1(7)$	8.30	$Q_1(7)$	0.37	$R_1(7)$	7.26	7	13/2	$P_2(7)$	7.39	$Q_2(7)$	0.15	$R_2(7)$	6.39
8	17/2	$P_1(8)$	9.32	$Q_1(8)$	0.33	$R_1(8)$	8.30	8	15/2	$P_2(8)$	8.40	$Q_2(8)$	0.15	$R_2(8)$	7.39
9	19/2	$P_1(9)$	10.34	$Q_1(9)$	0.29	$R_1(9)$	9.32	9	17/2	$P_2(9)$	9.41	$Q_2(9)$	0.14	$R_2(9)$	8.40
10	21/2	$P_1(10)$	11.36	$Q_1(10)$	0.26	$R_1(10)$	10.34	10	19/2	$P_2(10)$	10.42	$Q_2(10)$	0.14	$R_2(10)$	9.41
11	23/2	$P_1(11)$	12.37	$Q_1(11)$	0.24	$R_1(11)$	11.36	11	21/2	$P_2(11)$	11.42	$Q_2(11)$	0.12	$R_2(11)$	10.42
12	25/2	$P_1(12)$	13.38	$Q_1(12)$	0.22	$R_1(12)$	12.37	12	23/2	$P_2(12)$	12.43	$Q_2(12)$	0.11	$R_2(12)$	11.42
13	27/2	$P_1(13)$	14.39	$Q_1(13)$	0.20	$R_1(13)$	13.38	13	25/2	$P_2(13)$	13.43	$Q_2(13)$	0.10	$R_2(13)$	12.43
14	29/2	$P_1(14)$	15.40	$Q_1(14)$	0.18	$R_1(14)$	14.39	14	27/2	$P_2(14)$	14.43	$Q_2(14)$	0.10	$R_2(14)$	13.43
15	31/2	$P_1(15)$	16.41	$Q_1(15)$	0.16	$R_1(15)$	15.40	15	29/2	$P_2(15)$	15.44	$Q_2(15)$	0.10	$R_2(15)$	14.43
16	33/2	$P_1(16)$	17.42	$Q_1(16)$	0.15	$R_1(16)$	16.41	16	31/2	$P_2(16)$	16.44	$Q_2(16)$	0.10	$R_2(16)$	15.44
17	35/2	$P_1(17)$	18.42	$Q_1(17)$	0.14	$R_1(17)$	17.42	17	33/2	$P_2(17)$	17.44	$Q_2(17)$	0.10	$R_2(17)$	16.44
18	37/2	$P_1(18)$	19.43	$Q_1(18)$	0.13	$R_1(18)$	18.42	18	35/2	$P_2(18)$	18.44	$Q_2(18)$	0.09	$R_2(18)$	17.44
19	39/2	$P_1(19)$	20.43	$Q_1(19)$	0.12	$R_1(19)$	19.43	19	37/2	$P_2(19)$	19.44	$Q_2(19)$	0.09	$R_2(19)$	18.44
20	41/2	$P_1(20)$	21.44	$Q_1(20)$	0.11	$R_1(20)$	20.43	20	39/2	$P_2(20)$	20.44	$Q_2(20)$	0.09	$R_2(20)$	19.44

Table 2.16 Formulas for the intensity factors of the OH vibrational–rotational bands of the ground state $X^2\Pi$ (Turnbull 1987)

Branch	Intensity factors
$P_1(J')$	$\dfrac{(J'+1)\cdot(J'+1.5)}{4(J'+1)\cdot C'^+(J')\cdot C''^+(J'+1)} \cdot [u'^+(J')\cdot u''^+(J'+1) + 4(J'-0.5)\cdot(J'+2.5)]^2$
$Q_1(J')$	$\dfrac{(J'+1)}{2J'\cdot(J'+1)\cdot C'^+(J')\cdot C''^+(J')} \cdot [0.5\cdot u'^+(J')\cdot u''^+(J') + 6(J'-0.5)\cdot(J'+1.5)]^2$
$R_1(J')$	$\dfrac{(J'-0.5)\cdot(J'+0.5)}{4(J')\cdot C'^+(J')\cdot C''^+(J'-1)} \cdot [u'^+(J')\cdot u''^+(J'-1) + 4(J'-1.5)\cdot(J'+1.5)]^2$
$P_2(J')$	$\dfrac{(J'+1)\cdot(J'+1.5)}{4(J'+1)\cdot C'^-(J')\cdot C''^-(J'+1)} \cdot [u'^-(J')\cdot u''^-(J'+1) + 4(J'-0.5)\cdot(J'+2.5)]^2$
$Q_2(J')$	$\dfrac{(J'+0.5)}{2J'\cdot(J'+1)\cdot C'^-(J')\cdot C''^-(J')} \cdot [u'^-(J')\cdot u''^-(J') + 6(J'-0.5)\cdot(J'+1.5)]^2$
$R_2(J')$	$\dfrac{(J'+0.5)\cdot(J'+0.5)}{4J'\cdot C'^-(J')\cdot C''^-(J'-1)} \cdot [u'^-(J')\bullet u''^-(J'-1) + 4(J'-1.5)\cdot(J'+1.5)]^2$

$$C'^+(J') = 0.5\cdot\{[u'^+(J')]^2 + 4\cdot[(J'+0.5)^2 - 1]\}$$

$$C'^-(J') = 0.5\cdot\{[u'^-(J')]^2 + 4\cdot[(J'+0.5)^2 - 1]\}$$

$$C''^+(J') = 0.5\cdot\{[u''^+(J')]^2 + 4\cdot[(J'+0.5)^2 - 1]\}$$

$$C''^-(J') = 0.5\langle\{[u''^-(J')]^2 + 4\cdot[(J'+0.5)^2 - 1]\}$$

$$u'^+(J') = [Y'\cdot(Y'-4) + 4(J'+0.5)]^{0.5} + (Y'-2) \qquad Y' = A_{v'}/B_{v'}$$

$$u'^-(J') = [Y'\cdot(Y'-4) + 4(J'+0.5)]^{0.5} - (Y'-2) \qquad Y' = A_{v'}/B_{v'}$$

$$u''^+(J') = [Y''\cdot(Y''-4) + 4(J'+0.5)]^{0.5} + (Y''-2) \qquad Y'' = A_{v''}/B_{v''}$$

$$u''^-(J') = [Y''\cdot(Y''-4) + 4(J'+0.5)]^{0.5} - (Y''-2) \qquad Y'' = A_{v''}/B_{v''}$$

Table 2.17 Intensity factors for the OH vibrational–rotational bands of the ground state $X^2\Pi$, calculated by the formulas of Kovács (1969), Turnbull (1987)

N'	J'	$P_1(N')$	$i(J')$	$Q_1(N')$	$i(J')$	$R_1(N')$	$i(J')$	N'	J'	$P_2(N')$	$i(J')$	$Q_2(N')$	$i(J')$	$R_2(N')$	$i(J')$
1	3/2	$P_1(1)$	2.00	$Q_1(1)$	2.99	–	–	1	1/2	$P_2(1)$	1.95	$Q_2(1)$	0.667	–	–
2	5/2	$P_1(2)$	3.34	$Q_1(2)$	1.79	$R_1(2)$	1.60	2	3/2	$P_2(2)$	2.92	$Q_2(2)$	0.294	$R_2(2)$	2.61
3	7/2	$P_1(3)$	4.50	$Q_1(3)$	1.28	$R_1(3)$	2.86	3	5/2	$P_2(3)$	3.89	$Q_2(3)$	0.217	$R_2(3)$	3.53
4	9/2	$P_1(4)$	5.60	$Q_1(4)$	0.99	$R_1(4)$	4.00	4	7/2	$P_2(4)$	4.87	$Q_2(4)$	0.188	$R_2(4)$	4.49
5	11/2	$P_1(5)$	6.67	$Q_1(5)$	0.81	$R_1(5)$	5.09	5	9/2	$P_2(5)$	5.85	$Q_2(5)$	0.177	$R_2(5)$	5.47
6	13/2	$P_1(6)$	7.72	$Q_1(6)$	0.69	$R_1(6)$	6.16	6	11/2	$P_2(6)$	6.84	$Q_2(6)$	0.172	$R_2(6)$	6.46
7	15/2	$P_1(7)$	8.75	$Q_1(7)$	0.59	$R_1(7)$	7.20	7	13/2	$P_2(7)$	7.83	$Q_2(7)$	0.171	$R_2(7)$	7.45
8	17/2	$P_1(8)$	9.78	$Q_1(8)$	0.53	$R_1(8)$	8.24	8	15/2	$P_2(8)$	8.85	$Q_2(8)$	0.171	$R_2(8)$	8.43

2.2 Hydroxyl Emission

Table 2.18 Comparison of the intensity factors for the OH vibrational–rotational bands of the ground state $X^2\Pi$ (Benedict et al. 1953) to those calculated by the data of Mies (1974), Turnbull (1987), Langhoff et al. (1986), Turnbull and Lowe (1989) and normalized to $i(P_1(1.5))$ (Benedict et al. 1953)

N′	J′	P branch lines	i(J′)				
			BPH	T	TL	LWR	M
1	3/2	$P_1(1)$	1.62	1.62	1.62	1.62	1.62
2	5/2	$P_1(2)$	2.90	2.71	2.92	3.03	3.02
3	7/2	$P_1(3)$	4.04	3.64	4.11	4.44	4.44
4	9/2	$P_1(4)$	5.14	4.54	5.85	5.90	5.91
5	11/2	$P_1(5)$	6.21	5.40	7.38	7.45	7.47

the regression equation

$$\Delta \log_e \left[\frac{i(BPH)}{i(new)}\right] = \rho \cdot \Delta F; \quad \Delta \log_e \left[\frac{I(J') \cdot \lambda^3}{i(J')}\right] = C - 1.4388 \cdot \frac{\Delta F}{T},$$

where ρ is the coefficient of regression and ΔF is the difference of the energies of the rotational levels the transitions between which are used to calculate the temperature T, we obtain the relation

$$\Delta T = -\frac{0.695 \cdot \rho \cdot T^2}{1 + 0.695 \cdot \rho \cdot T} \approx -0.695 \cdot \rho \cdot T^2.$$

From the preceding it can be deduced that the coefficient ρ is negative because the new intensity factors i(new) are systematically greater than i(BPH).

This allows one to estimate a systematic correction ΔT for the temperatures that were calculated with the use of different intensity factors. As follows from comparison of various data, ΔT is 4–8 (K) for $T \sim 200$ (K).

Here it should be noted that the use of the intensity factors calculated following Benedict et al. (1953) and of all possible rotational lines, i.e., P_1 (1.5), P_1 (2.5), P_1 (3.5), P_1 (4.5), P_1 (5.5), Q_1 (1.5), R_1 (2.5), P_2 (0.5), P_2 (1.5), P_2 (2.5), P_2 (3.5), and R_2 (1.5), on the condition that the spectrogram was of high quality, had the result (Shefov 1961c) that the points on the plots used to determine the temperature by various OH bands (see Fig. 2.11) were, as a rule, rather close to the line of regression and there were no systematic deviations (high values of the correlation coefficients). Similar results for the temperature of the OH (4–1) band measured for assurance simultaneously by two spectrographs (Shefov 1961c) are shown in Fig. 2.12. The use of the new intensity factors results in a more appreciable spread of points in similar plots (French et al. 2000).

The independent determination of the OH temperature by the Doppler broadening of the P_1 ($J' = 1.5$) rotational line of the 839.93-nm OH (6–2) emission with the use of a Fabry–Perot interferometer (Hernandez et al. 1993; Conner et al. 1993; Greet et al. 1994) gives an error of order $\sqrt{T}$, reaching 40 (K), which is substantially

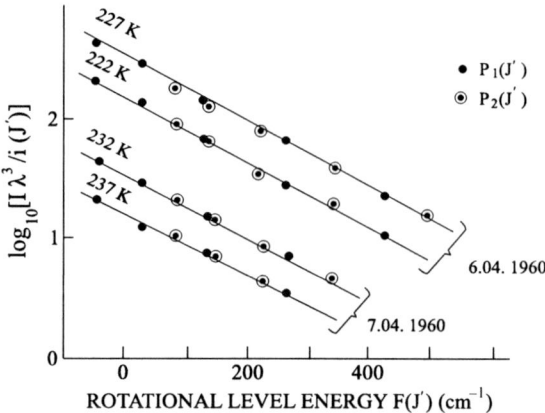

Fig. 2.12 Plots for determining the rotational temperature of the OH (4–1) band by the spectra obtained simultaneously on two spectrographs of the same type (Shefov 1961c)

greater than the error in the calculated rotational temperature (Choi et al. 1998). Lidar measurements of the temperature performed simultaneously with ground-based and satellite spectrophotometric measurements have made it possible to obtain, based on seasonal variations, mean temperature differences within 6–7 (K) (She and Lowe 1998).

Thus, from the data considered it can be seen that the small difference in absolute values of the temperature obtained by various methods affects the absolute values of the mesopause temperature insignificantly since its natural variations are substantially greater. To estimate the mesopause temperature for summertime, when there are conditions for the occurrence of noctilucent clouds, it is necessary to know more exact absolute values of the rotational temperature.

Since different authors use different values of i(J) and, the more so, different sets of rotational lines, it is necessary to bear in mind this systematic difference when comparing the data of different researchers. This difference is most substantial in analyzing long-term temperature variations since different methods (values of i(J)) used for different portions of a long-term data set will inevitably distort the character of a long-term trend.

As mentioned in Sect. 2.1.6, the rotational temperature is close in value to the temperature of the medium in which hydroxyl emissions occur. Therefore, measuring this parameter allows one to investigate the temperature conditions in the mesopause and to study the effect of complex dynamic processes resulting from various geophysical phenomena in the lower atmosphere on these conditions. An important problem in the determination of the rotational temperature is its conformity to the environmental temperature. This problem received the attention of many researchers (Suzuki and Tohmatsu 1976; Dick 1977; Krassovsky et al. 1977; Hotlzclaw et al. 1993, 1997).

As already stressed above, the absolute values of transition probabilities are of great importance since they determine the conditions of rotational relaxation and the lifetime of the excited molecules formed in the reactions discussed above.

2.2 Hydroxyl Emission

The absolute values of transition probabilities are used in determining the mesopause temperature based on measurements of the rotational temperature of OH. From the formula given above it follows that at an altitude of about 90 (km) the number of collisions in a second is about $2 \cdot 10^4$ (s^{-1}). Hence, even for the transitions from the ninth vibrational level for which $A_9 = 204$ (s^{-1}) (Goldman et al. 1998), n_0 is about 70, which, apparently, satisfies quite well the above requirements.

It should be noted that the hydroxyl emission layer is located in the upper atmosphere near the temperature minimum and, therefore, within the thickness of the layer there exists a certain temperature gradient, and the number of collisions which provide relaxation of OH molecules to equilibrium conditions can vary. Therefore, the determined rotational temperature of the layer is a weighted mean (Shefov 1961c). If we suppose that each elementary layer has its own temperature and the populations over rotational levels are described by a Boltzmann distribution corresponding to this temperature, the net population distribution corresponding to the observed distribution of the rotational line intensities will not be, strictly speaking, a Boltzmann distribution. This distinction, however, will be appreciable only for lines with high rotational numbers J'. As already mentioned, in practice, rotational temperature can be determined only based on five rotational lines, and, therefore, this effect is imperceptible.

An important feature of the behavior of the rotational temperature and intensities of OH bands was revealed by comparing data corresponding to different vibrationally excited levels. Analysis of even first data (Shefov 1961c) showed that the nightly mean temperature for higher vibrational levels is sometimes noticeably greater than that for lower levels (see Fig. 2.11). Comparison of the band intensities and rotational temperatures for higher and lower levels has shown that the coefficient of correlation between close levels were ~ 0.8 and between higher and lower levels were much less than ~ 0.4 (Fig. 2.2) (Berg and Shefov 1963). Takahashi and Batista (1981) also noted that the correlation between the intensities of bands from higher levels is appreciably stronger and the dispersion of values about the line of regression is less than in the case where the intensities of bands from higher and lower vibrational levels are compared.

Differences in rotational temperatures for bands from upper and lower vibrational levels were also noted by other authors (Karyagina 1962; Yarin 1962a; Sivjee and Hamwey 1987). It is of interest that the factors of correlation both between the temperatures and between the intensities of OH bands (nightly means) for vibrational levels with identical Δv turned out to be rather close. Moreover, the dependence of the correlation coefficient r(T, I) on Δv is linear (the correlation coefficient is -0.952 ± 0.023):

$$r(T,I) = (0.955 \pm 0.034) - (0.175 \pm 0.014)\Delta v.$$

Subsequently special investigations of this phenomenon were performed based on measurements of the OH intensities and rotational temperatures carried out with ~ 10-min exposures (see Fig. 2.13) (Bakanas and Perminov 2003). It has been established that the earlier revealed difference in temperatures also shows up in seasonal

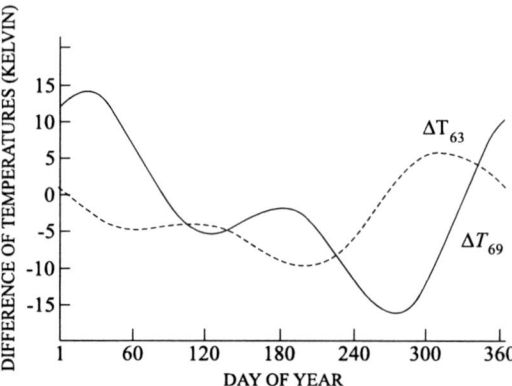

Fig. 2.13 Seasonal variations of the rotational temperature differences ΔT_{63} and ΔT_{69} (Bakanas and Perminov 2003)

variations for the near-midnight values of these parameters for both upper and lower levels (see Fig. 2.2) (Bakanas and Perminov 2003), and this is a consequence of the use of temperature values which refer to different seasons. In this case, the dependence of the correlation coefficient r(T, I) on Δv is also linear (correlation coefficient is -0.965 ± 0.034):

$$r(T,I) = (1.000 \pm 0.032) - (0.058 \pm 0.009)\Delta v.$$

The temperature difference between different levels is described by the empirical relation

$$T_V - T_6 = \Delta T_{V6} = a_0 + a_1 \cdot \cos\left[\frac{2\pi}{365.25}(t_d - t_1)\right] + a_2 \cdot \cos\left[\frac{4\pi}{365.25}(t_d - t_2)\right],$$

$$a_0 = 0.24 \cdot v^2 - 3.02 \cdot v + 9.55; \quad a_1 = -0.89 \cdot v^2 + 10 \cdot v - 27.81;$$

$$a_2 = -0.62 \cdot v^2 + 6.81 \cdot v - 18.51,$$

$$t_1 = 12.8 \cdot v - 68.5; \quad t_2 = -19 \cdot v + 182.$$

Since simultaneous variations in temperature and intensity occur synchronously and are caused by the variations in altitude of the emission layer, the difference in rate of variation of the correlation coefficients revealed between the nightly mean and the near-midnight values seems to reflect the difference in scales of variations of the emission layer altitude for OH bands with different vibrationally excited levels.

The data considered were used as a basis for studying the rotational relaxation and the possibility of its completion (Nicholls et al. 1972; Suzuki and Tohmatsu 1976). It was found again that the vibrational temperatures that correspond to the population distribution for lower ($v = 3-6$) and upper ($v = 7-9$) vibrational levels are different and show different seasonal variations (Shefov 1976; Semenov and Shefov 1996, 1999a).

In view of the problem of the difference in temperatures between different OH bands, Krassovsky et al. (1977), based on the data of more than 150 many-hour observations at Zvenigorod scientific station of the Institute of Atmospheric Physics of the Russian Academy of Sciences, have analyzed the nature of the equilibrium of the rotational temperature of hydroxyl emission. The observed difference in rotational temperatures calculated for bands from different vibrational levels (Shefov 1961c; Berg and Shefov 1963; Shagaev 1977) was accounted for by the fact that the emissions in bands with high vibrational excitation occur in the main at higher altitudes, in regions of positive temperature gradients. Besides, in the opinion of Krassovsky et al. (1977), the observed difference in temperatures can be in part due to the incomplete thermodynamic equilibrium of the excited hydroxyl molecules with the surrounding medium in the top region of the emission layer, i.e., at altitudes of 95–100 (km). By the way, hydroxyl emission spectra for which the distribution of the intensity of lines in the P branch was other than the Boltzmann distribution were studied in laboratory experiments (Charters and Polanyi 1960; Charters et al. 1971).

Here some publications should be mentioned in which it was reported about anomalously high relative intensities of a line of the branch R of the OH (3–1) band which should belong to the line $P_1(7)$ of the OH (2–0) band (spectral range near 1.53 (μm)) (Gattinger and Vallance 1973; Pendleton et al. 1989) and also of the line P_1 (N = 12) of the OH (7–4) band among the lines of the OH (8–5) band (spectral range near 1.36 (μm)).

Nevertheless, the observations performed near Zvenigorod have revealed a clear nonequilibrium intensity distribution over high rotational levels with $N' > 6$ for OH molecules (Perminov and Semenov 1992) which results from the contribution of the upper region of the emission layer (Figs. 2.7 and 2.8). It should be noted that the effective temperature corresponding to the population distribution over the levels with N = 7–9 varies during the night. The mean variations point to a maximum of 1000 (K) in the midnight and to 750 (K) in the beginning and at the end of the night. The vibrational temperature calculated for the groups of levels with v = 3–5 and v = 7–9 also varies during the night and, as mentioned, correlates with solar zenith angle. For the example under consideration, this temperature is ∼8500 (K) in twilight and ∼13 500 (K) in the midnight. Since vibrational temperature is an indicator of the altitude of the emission layer (Semenov and Shefov 1996, 1999a), the observed variations of the nonequilibrium rotational temperature reflect the variations of the altitude. The nonequilibrium is due to the contribution of the upper region of the emission layer near 95–100 (km). This is confirmed by direct observations in the afternoon, in twilight, and at night from a spacecraft at an altitude of 260 (km) along the limb. At limb distances from the Earth surface of 81–83 and 95–99 (km), rotational lines with $N' = 33$ have been detected for the vibrational levels with v = 0–6 (Smith et al. 1992; Dodd et al. 1993).

In practice a need arises to estimate the rotational temperature in relation to individual components of a band and the total intensity of a band by the intensity of any of its components. The results of such calculations for various vibrational levels are presented elsewhere (Piterskaya and Shefov 1975). An example for the sixth vibrational level is given in Table 2.19.

Table 2.19 Relative intensities of the lines and branches* of the bands of the sixth vibrational level for the hydroxyl molecule (Piterskaya and Shefov 1975)

Line (N')	T (K)															
	150	160	170	180	190	200	210	220	230	240	250	260	270	280	290	300
$R_2(2)$	1.96	2.04	2.09	2.14	2.18	2.20	2.22	2.23	2.23	2.24	2.23	2.23	2.22	2.21	2.20	2.18
$Q_2(1)$	1.60	1.61	1.62	1.62	1.61	1.60	1.59	1.57	1.55	1.54	1.52	1.50	1.48	1.46	1.44	1.42
$Q_1(1)$	18.86	17.71	16.64	15.62	14.74	13.94	13.21	12.59	11.99	11.46	10.98	10.53	10.11	9.73	9.38	9.02
$Q_2(2)$	0.45	0.47	0.48	0.49	0.50	0.50	0.51	0.51	0.51	0.51	0.51	0.51	0.51	0.51	0.50	0.50
$Q_1(2)$	6.05	5.87	5.69	5.52	5.35	5.19	5.04	4.89	4.75	4.61	4.49	4.37	4.25	4.15	4.04	3.94
$Q_1(3)$	1.71	1.75	1.78	1.80	1.82	1.83	1.83	1.83	1.82	1.81	1.80	1.79	1.78	1.76	1.75	1.73
$Q_1(4)$	0.39	0.43	0.47	0.50	0.53	0.56	0.58	0.60	0.62	0.64	0.66	0.67	0.68	0.69	0.70	0.71
$Q_1(5)$	0.072	0.088	0.10	0.12	0.13	0.15	0.16	0.18	0.19	0.20	0.21	0.22	0.24	0.25	0.26	0.27
$P_2(1)$	3.03	3.06	3.08	3.07	3.06	3.04	3.02	2.99	2.95	2.92	2.88	2.84	2.81	2.77	2.73	2.69
$P_1(1)$	12.95	12.16	11.42	10.72	10.12	9.57	9.07	8.64	8.23	7.87	7.54	7.23	6.94	6.68	6.44	6.19
$P_2(2)$	3.40	3.53	3.63	3.71	3.77	3.81	3.85	3.86	3.87	3.88	3.87	3.87	3.85	3.83	3.81	3.78
$P_1(2)$	12.08	11.72	11.36	11.02	10.69	10.37	10.07	9.76	9.48	9.21	8.96	8.72	8.49	8.28	8.07	7.86
$P_2(3)$	2.27	2.46	2.63	2.79	2.93	3.05	3.16	3.25	3.33	3.40	3.45	3.51	3.55	3.59	3.62	3.64
$P_1(3)$	6.71	6.87	6.99	7.09	7.15	7.20	7.21	7.19	7.17	7.13	7.09	7.05	7.00	6.94	6.88	6.81
$P_2(4)$	1.01	1.17	1.32	1.47	1.63	1.77	1.90	2.02	2.13	2.23	2.33	2.42	2.51	2.59	2.66	2.72
$P_1(4)$	2.60	2.86	3.10	3.32	3.52	3.71	3.87	4.02	4.15	4.26	4.36	4.45	4.53	4.60	4.66	4.70
$P_2(5)$	0.32	0.40	0.49	0.58	0.68	0.78	0.88	0.97	1.06	1.16	1.25	1.34	1.43	1.51	1.59	1.66
$P_1(5)$	0.73	0.89	1.05	1.20	1.36	1.51	1.65	1.79	1.92	2.05	2.17	2.29	2.40	2.51	2.61	2.70
$P_2(6)$	0.07	0.10	0.14	0.15	0.22	0.27	0.32	0.38	0.43	0.49	0.55	0.61	0.67	0.74	0.80	0.86
$P_1(6)$	0.16	0.21	0.27	0.33	0.40	0.48	0.56	0.64	0.72	0.81	0.90	0.98	1.07	1.15	1.23	1.31
ΣR	20.46	21.28	22.00	22.73	23.41	23.94	24.66	25.20	25.66	26.15	26.60	27.02	27.43	27.82	28.17	28.43
ΣQ	29.36	28.18	27.06	25.98	25.02	24.13	23.31	22.58	21.87	21.23	20.65	20.09	19.57	19.10	18.63	18.17
ΣP	45.36	45.90	45.56	45.56	45.69	45.75	45.81	45.82	45.81	45.85	45.86	45.90	45.91	45.91	45.93	45.83

* in percents of the total intensity expressed in (Rayleighs)

2.2 Hydroxyl Emission

Consideration of pure rotational transitions inside the rotational level system has shown that the transition probabilities may have an effect only for high values of N′ for which, under the conditions of the upper atmosphere, in measuring the intensity of the corresponding lines at the Earth surface, vibrational–rotational transitions are not observed (Goorvitch et al. 1992).

According to numerous investigations of the behavior of the rotational temperature, seasonal and long-term variations have been detected. This is discussed in detail in Chap. 4.

2.2.6 The Population Distribution over Vibrational Levels

As a result of photochemical production of vibrationally excited hydroxyl molecules followed by their vibrational relaxation, a quasi-Boltzmann distribution of molecules is established which corresponds on the average to a vibrational temperature of about 10 000 (K).

The radiation intensity of a hydroxyl band for a given vibrational level v is determined by the expression

$$I_{v',v''} = A_{v',v''} \cdot N_{v'},$$

where $N_{v'}$ is the population of the vibrational level v' and $A_{v',v''}$ is the probability of the radiation transition from the upper level v' to the lower level v''. In the general form, the relative population of the vibrational levels N_v is described by the Boltzmann distribution

$$N_v = N \cdot \frac{\exp(-G_v \cdot hc/kT_V)}{\sum_{v=1}^{9} \exp(-G_v \cdot hc/kT_V)},$$

where N is the total population of all vibrational levels, G_v is the energy of the vibrational level above the zero electronic state (cm^{-1}) (Table 2.1), and T_V is the distribution parameter which is termed as vibrational temperature. Thus, knowing the band intensity ratio $I_{v'_1,v''_1}/I_{v'_2,v''_2}$, we can estimate the vibrational temperature by the formula

$$T_V = \frac{hc}{k} \cdot \frac{G_{v'_1} - G_{v'_2}}{\log_e \left(\frac{I_{v'_1,v''_1}}{A_{v'_1,v''_1}}\right) - \log_e \left(\frac{I_{v'_2,v''_2}}{A_{v'_2,v''_2}}\right)} \quad (K).$$

The vibrational temperature T_V, which characterizes on the average the Boltzmann distribution of the vibrational level populations, is in the range 5000–15 000 (K) depending on the season and time of day and reflects the processes of excitation of molecules and deactivation of the excited molecules. Examples of the distribution of populations over vibrational levels and the corresponding values of vibrational temperatures are presented in Fig. 2.14.

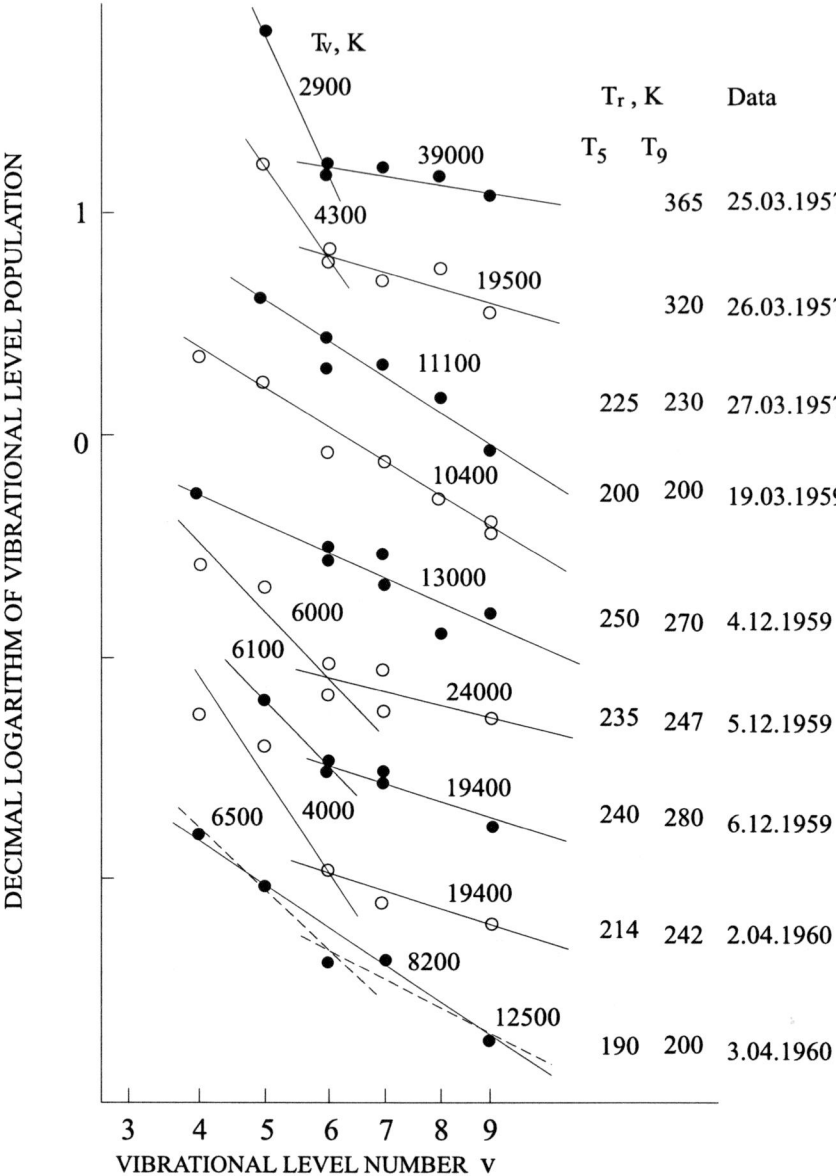

Fig. 2.14 Examples of the distribution of populations over vibrational levels and the corresponding vibrational and rotational temperatures, according to observations in Zvenigorod

If the value of T_V is known, the total intensity of hydroxyl emission can be calculated from measurements for individual bands. Using the above expression and measurements of the absolute intensity of OH bands, it is possible to find the populations of the corresponding vibrational levels. Analysis of the distribution of OH molecules over vibrational levels allows one to estimate the yield of excited molecules. First attempts to estimate this yield, which did not take into account possible processes of deactivation of excited OH molecules due to their collisions with molecules of the surrounding medium, have led the researchers to the conclusion that the rate of production of excited molecules is nearly invariable (Chamberlain and Smith 1959). However, subsequent calculations with various values of vibrational temperature have revealed its variations (Shefov 1961e; Yarin 1962b). Some authors who used more precise values of transition probabilities (Llewellyn and Long 1978) have concluded that the ninth, eighth, and seventh vibrational levels are populated preferentially. Analysis of atmospheric and laboratory measurements shows that the effect of vibrational deactivation is rather substantial (Breig 1969; Velculescu 1970; Worley et al. 1971, 1972; Streit and Johnston 1976; Nagy et al. 1976; Llewellyn and Long 1978; Llewellyn et al. 1978; Finlayson-Pitts and Kleindienst 1981; Turnbull and Lowe 1983; Smith and Williams 1985; Khayar and Bonamy 1987; McDade and Llewellyn 1987; López-Moreno et al. 1987; Grigor'eva et al. 1994, 1997; Perminov et al. 1998).

To analyze this process, the following deactivation reactions were considered:

$$OH(v') + O_2 \rightarrow OH(v'' < v') + O_2 \quad k_{O_2}(v', v''),$$
$$OH(v') + N_2 \rightarrow OH(v' - 1) + N_2 \quad k_{N_2}(v'),$$
$$OH(v') + O \rightarrow OH(v'' < v') + O \quad k_O(v', v''),$$
$$OH(v') + O \rightarrow H + O_2 \quad k_d.$$

Based on the data of observations of various OH bands in the airglow of the upper atmosphere, seasonal variations were chosen for which substantial changes in altitude distributions of atomic oxygen are characteristic, as was shown by analyzing the model of intensity variations for 557.7-nm emission (Semenov and Shefov 1997a,b,c,d). (This is exemplified by Fig. 2.15 where the data for the

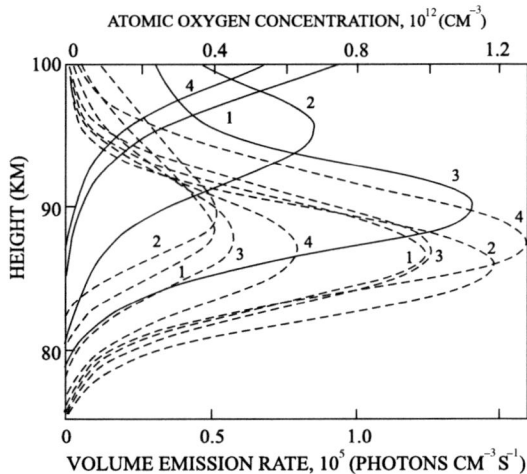

Fig. 2.15 Altitude distribution of the atomic oxygen concentration (*solid lines*) and volume emission rate of OH for $v = 4$ (*dot-and-dash lines*) and $v = 9$ (*dashed lines*) for four days of the year 1972. (1) t_d is 100 (April 9); (2) t_d is 173 (June 21); (3) t_d is 290 (October 16); and (4) t_d is 355 (December 21)

conditions of the year 1972 are presented (Perminov et al. 1998)). This was made purposely in order to have conditions of predominance of molecular oxygen (winter and spring) or atomic oxygen (summer and autumn).

Theoretical investigations have shown that the vibrational deactivation of OH(v) molecules by nitrogen molecules is a single-quantum process (Shalashilin et al. 1992, 1993). When the deactivation is produced by oxygen molecules, this is a multiquantum process for higher vibrational levels (v ≥ 7) and a nearly single-quantum process for lower vibrational levels (Shalashilin et al. 1992, 1994; Grigor'eva et al. 1994). For this case, the coefficients of deactivation of OH by molecular nitrogen were taken from the work by Makhlauf et al. (1995) in which, by means of interpolation between earlier measured values k_{N_2} (2) and k_{N_2} (12) (Sappey and Copeland 1990), their values for v = 3–9 and their extrapolated value k_{N_2} (1) were determined. They fall in the range of values theoretically obtained by Shalashilin et al. (1993). The rates of deactivation of the vibrational levels of OH (v) by molecular oxygen have been determined for all v (Chalamala and Copeland 1993; Knutsen and Copeland 1993). Their dependences on Δv (number of deactivated quanta) turned out to be impossible to estimate. Such an estimation was made based on the data about the distributions of [O_2], [N_2], and hydroxyl emission at altitudes of 80–87 (km) (i.e., in the main below the maximum of OH emission intensity) for $t_d = 100$ (spring) and 355 (winter) for which the deactivation by atomic oxygen could be neglected. It was supposed that the probability of the $v' \to v''$ transition at collisions of OH (v) with O_2 is proportional to $(v' - v'')^{-n}$, where for large positive n the deactivation is a stepped single-quantum process, for large negative n it occurs with a preferred transition to the zero vibrational level, and for n = 0 all transitions are equally probable (Dodd et al. 1994). In this case, the procedure of estimation of $k_{O_2}(v', v'')$ involved the determination of the exponent n for all v. The criterion for an optimum estimate was the achievement of a maximum for the correlation coefficient (no less than 0.95) when seeking consistence between the distributions of the relative populations of vibrational levels obtained by an empirical model (Semenov and Shefov 1996, 1999a) and by a photochemical model of OH emission in the altitude range 80–87 (km) (Semenov 1997). The results of determination of $k_{O_2}(v', v'')$ for v = 7–9 are shown in Fig. 2.5. It can be seen that about one half of the transitions between level v' and level v'' of a hydroxyl molecule in its collisions with oxygen molecules occur with $\Delta v = 1$. Analysis also shows that for v'<7 transitions of this type prevail.

The determination of the character of the deactivation of vibrational levels of hydroxyl by atomic oxygen and estimation of the rates $k_O(v', v'')$ were performed in a similar manner. The relative vibrational distributions of the populations were considered for the altitude range above 87 (km) where the deactivation by oxygen atoms plays a significant part. The coefficient k_d was taken equal to $1.05 \cdot 10^{-10} (cm^3 \cdot s^{-1})$. Though this value was obtained in laboratory conditions for v' = 1 (Spencer and Glass 1977b), it was taken identical for all v'. In this study the rate of deactivation of OH (1) by atomic oxygen was also found: $k_O(1,0) = 1.45 \cdot 10^{-10} (cm^3 \cdot s^{-1})$ (Spencer and Glass 1977a). Therefore, when seeking consistence between the model

2.2 Hydroxyl Emission

population distributions over vibrational levels [OH (v, Z)], the rates $k_O(v', v'')$ were estimated only for higher vibrational levels.

Thus, it turned out that for large positive n, i.e., for single-quantum deactivation, $k_O(v', v'')$ can be taken equal to $2.2 \cdot 10^{-10}(\text{cm}^3 \cdot \text{s}^{-1})$ for all $v' > 1$. For the lower n a dependence of this coefficient on v' arises in which its maximum values correspond to $v' = 9$. As an example, Fig. 2.6 presents the v' dependence of n for $n = 0$ and -5. Makhlauf et al. (1995) point out that the upper limit for the value of $k_O(v', v'')$ in the mesopause region is $2.5 \cdot 10^{-10}(\text{cm}^3 \cdot \text{s}^{-1})$. In the simulation under consideration, this criterion is satisfied only in the case of single-quantum deactivation. The coefficient of correlation between the population distributions over vibrational levels obtained by the empirical model of hydroxyl emission variations (Semenov and Shefov 1996, 1999a) and by the photochemical model (Semenov 1997) is 0.86 (Perminov et al. 1998). Thus, the above consideration shows a substantial part played by atomic oxygen in vibrational deactivation of hydroxyl.

Investigations of the variations of the vibrational temperature and its correlations with other parameters of hydroxyl emission enable one to analyze various features of its behavior. In the first years of intense accumulation of data about hydroxyl emission an attempt was made to investigate the parameters of the photochemical processes giving rise to excited OH molecules. The parameter of interest was the activation energy that determines the rate of the ozone–hydrogen reaction (Krassovsky 1961; Krassovsky et al. 1961). This parameter was determined based on the correlation relation between the logarithm of the intensity of the hydroxyl emission band and the inverse rotational temperature. However, in constructing such a relation by data obtained at various stations it was found that the calculated coefficient of regression (earlier treated as the activation energy) shows precise proportionality to the vibrational level energy. This result implied that the obtained correlation relations arise due to the variations of vibrational level populations that result from variations of vibrational temperature.

The complexity of the variations of the parameters of hydroxyl emission is determined not only by photochemical processes, but also by the dynamic modulation of the entire structure of the atmosphere. When relating the intensity logarithms for bands with different vibrationally excited levels to inverse vibrational temperature, it has been revealed that they depend on both the absolute populations and on their variations caused by the variations of vibrational temperature (Shefov et al. 1998). As follows from Fig. 2.3, for the vibrational levels with $v = 4, 5, 6$, and 7 the points are grouped rather systematically about the average curve corresponding to middle (4–6) vibrational levels. For the ninth level, the distribution of points corresponds to the average curves related to the upper levels. A formal regression relation between the effective vibrational level number and a given v for $v = 4 \div 9$ can be given by

$$v_{\text{eff}} = 2.3 \times \exp(v/7.15), r = 0.956.$$

Thus, the populations described by Boltzmann distributions behave as if they shake about the population of level v_{eff}. Examples of distributions for some nights have already been presented in Fig. 2.14.

It is interesting that there is a similar correlation between the intensity of 864.5-nm (0–1) O_2 emission and the vibrational temperature of OH (Shefov 1970c, 1971b) for which the mean altitude of the emission layer is ~ 95 (km). The coefficient of regression ρ is -0.49, and it is statistically significant ($r = -0.745$).

Now it is quite clear that the variations of vibrational level populations caused by altitude variations are different in character for upper and lower levels. The seasonal altitude variations for $v = 5$ within a year are small (Semenov and Shefov 1996, 1999a). The predictions of the model of seasonal variations of the altitude distribution of atomic oxygen (Semenov and Shefov 2005) point to its substantial variations within the hydroxyl emission (Fig. 2.15). All this testifies to a considerable effect of dynamic processes on the behavior of hydroxyl emission.

The variations of vibrational temperature reflect the variations of the emission layer altitude, and, hence, the variations of atmosphere density (Semenov and Shefov 1996, 1999a).

2.2.7 Photochemistry

The theory of the occurrence of hydroxyl emission was actively developed once its spectra had been identified. Since the radiation power is rather high, it became obvious at once that this emission results from a series of reactions of recombination of the atomic oxygen produced in the atmosphere by photodissociation of molecular oxygen and that it is a sink of absorbed solar energy. Initially it was hypothesized that hydroxyl emission arises due to the ozone–hydrogen reaction (Bates and Nicolet 1950; Herzberg 1951; Heaps and Herzberg 1952; Kaufman 1969; Nicolet 1989b). According to this hypothesis, ozone molecules are produced by the reactions

$$O + O_2 + O_2 \rightarrow O_3 + O_2 \quad \alpha_{O_3}(O_2) = 5.96 \cdot 10^{-34} \cdot (300/T)^{2.37} \left(cm^6 \cdot s^{-1}\right),$$

$$O + O_2 + N_2 \rightarrow O_3 + N_2 \quad \alpha_{O_3}(N_2) = 5.7 \cdot 10^{-34} (300/T)^{2.62} \left(cm^6 \cdot s^{-1}\right),$$

$$H + O_3 \rightarrow OH(v \leq 9) + O_2 \quad \alpha_{HO_3} = 1.4 \cdot 10^{-10} \cdot \exp[-480/T] \left(cm^3 \cdot s^{-1}\right).$$

The oxygen–hydrogen reaction (Krassovsky 1951a)

$$H + O_2^* \rightarrow OH^* + O$$

and the ozone–hydrogen reaction were first considered as possible paths of recombination processes in the early 1950s.

Krassovsky (1963a) proposed one more mechanism of the vibrational excitation of hydroxyl molecules – the reaction of perhydroxyl with atomic oxygen, whose rates are given elsewhere (Nicovich and Wine 1987; Nicolet 1989b):

2.2 Hydroxyl Emission

$$H + O_2 + M \rightarrow HO_2 + M \qquad \alpha_{HO_2} = 3.3 \cdot 10^{-33} \cdot \exp(800/T) \left(cm^6 \cdot s^{-1}\right),$$

$$O + HO_2 \rightarrow OH(v \leq 6) + O_2 \qquad \alpha_{OHO_2} = 2.9 \cdot 10^{-11} \cdot \exp(200/T) \left(cm^3 \cdot s^{-1}\right).$$

The vibrationally excited OH molecules ($v \leq 9$) radiate in the spectral range 0.5–5 (μm) (Semenov and Shefov 1996, 1999a) with effective transition probability:

$$OH(v \leq 9) \rightarrow OH + h\nu(0.5 \leq \lambda \leq 5(\mu m)) \qquad \bar{A} = 48\,(s^{-1}).$$

Data of different authors on transition probabilities $A_{v',v''}$ for various vibrational–rotational bands are presented in Tables 2.4–2.13. Many calculations of transition probabilities had the disadvantage that they used measurements of the intensities of individual OH bands (Krassovsky et al. 1962) not reduced to some reference conditions. Therefore, further refinement of $A_{v',v''}$ values is necessary.

The effective transition probability is determined by the relation

$$I(OH) = \sum_{v'=1}^{9} \sum_{v''=0}^{8} I_{v'-v''} = \sum_{v'=1}^{9} \sum_{v''=0}^{8} A_{v',v''} \cdot N_{v'} = \frac{\sum_{v'=1}^{9} A_{v'} \cdot \exp(-G_{v'}/kT_V)}{\sum_{v'=1}^{9} \exp(-G_{v'}/kT_V)} \cdot N_{eff} = \bar{A} \cdot N_{eff}.$$

The energies of vibrational levels, $G_v(cm^{-1})$, are given by

$$G_v = 3653v - 85v^2 + 0.54v^3.$$

Both hypotheses assume that atomic hydrogen is regenerated by reactions of deactivated hydroxyl molecules with atomic oxygen (Nicolet 1989b):

$$OH + O \rightarrow O_2 + H; \qquad \beta_{OHO} = 2.3 \cdot 10^{-11} \cdot \exp(40/T)(cm^3 \cdot s^{-1}).$$

In view of the processes considered, the emission rate (photon·cm^{-3}·s^{-1}) at an altitude Z is determined by the formula

$$Q_{OH}(Z) = \bar{A} \cdot \alpha_{HO_3}[H] \cdot [O_3]/\{\bar{A} + k_{O_2} \cdot [O_2] + k_{N_2} \cdot [N_2] + (k_O + k_d) \cdot [O]\}.$$

If the parameters of hydroxyl emission and the reaction rates are known, one can determine the concentrations of ozone

$$[O_3] = \frac{Q_{OH}(Z) \cdot \{\bar{A} + k_{O_2} \cdot [O_2] + k_{N_2} \cdot [N_2]\}}{\bar{A} \cdot \alpha_{HO_3} \cdot [H] \cdot B} \, (cm^{-3})$$

and atomic oxygen

$$[O] = \frac{Q_{OH}(Z) \cdot \{\bar{A} + k_{O_2} \cdot [O_2] + k_{N_2} \cdot [N_2]\}}{\bar{A} \cdot \{\alpha_{O_3}(O_2) \cdot [O_2] + \alpha_{N_2}(N_2) \cdot [N_2]\} \cdot [O_2] \cdot B} \, (cm^{-3}).$$

Here
$$B = 1 - \frac{Q_{OH}(Z) \cdot (k_O + k_d)}{\bar{A} \cdot \{\alpha_{O_3}(O_2) \cdot [O_2] + \alpha_{O_3}(N_2) \cdot [N_2]\} \cdot [O_2]}.$$

Initially it seemed that the problem of choice between the proposed mechanisms could be solved if the altitude of the hydroxyl emission layer was measured precisely. However, the situation turned out much more complicated. Despite some success in experimental investigation of the characteristics of hydroxyl emission, there were additional difficulties in understanding the kinetics of a great many photochemical processes accompanying the terminating event of excitation of a hydroxyl molecule. First, the part played by excited molecular reactants in these processes still remains unclear. Second, the rates of many reactions participating in the excitation process have been determined rather approximately. Various measuring techniques yield values which can differ by an order of magnitude. The rates of many reactions proceeding with participation of excited components are sometimes absolutely unknown. Besides, for some reactions there exist probabilistic alternatives of the formation of end products. The quantitative relations between the rates of such processes are known only for a restricted number of reactions. Of great importance is the excitation energy distribution among the reaction products. In this connection, special attention should be paid to the formation of metastable electron-excited or vibrationally excited molecules which can be active participants in subsequent series of reactions.

In recent years, in a description of the processes of excitation of hydroxyl the reaction with participation of the perhydroxyl molecule is often used as a mechanism of excitation of OH (v = 0–3), though energetically it is capable of providing excitation up to v = 6. In the literature there is a significant diversity of opinion regarding the role of this reaction. Some authors (Llewellyn 1978; McDade and Llewellyn 1987) believe that oxygen–perhydroxyl reaction does not result in excitation of OH molecules. On the other hand, Lunt et al. (1988) report on laboratory data about the possibility of excitation of OH in this reaction; however, they could not measure the coefficient of branching of the distribution of OH molecules over vibrational levels. Some authors (López-Moreno et al. 1987) supposed that the perhydroxyl reaction provides excitation of OH (v) to the (3–6) vibrational levels, while others (Kaye 1988) assumed that OH can be excited by this reaction only to the (0–3) levels. Semenov (1989b), having analyzed the variations of the hydroxyl radiation intensity and rotational temperature caused by internal gravity waves, has shown that the contribution of the ozone–hydrogen and perhydroxyl (alternative) reactions to the excitation of vibrational levels with $v \leq 5$ is seasonal in character.

Irrespective of the solution of the question on the contribution of the perhydroxyl process, it should be borne in mind that its involvement in the scheme of photochemical reactions of excitation of hydroxyl was dictated by the need to interpret experimental data rather than by the search for an alternative mechanism. The matter is that, as already mentioned, all available measurements which characterize the behavior of the intensity and temperature of the emission bands from upper and lower vibrational levels testify to lack of close correlation between radiation intensity and temperature, which, however, should exist if the excitation is to occur only

2.2 Hydroxyl Emission

by the ozone–hydrogen mechanism. Actually, as found in a number of studies and shown above, the population distributions for the (1–6) and (6–9) vibrational levels are associated with different values of vibrational temperature T_V which have quasi-independent variations. The characteristics of the processes that show up under the action of IGWs (η and θ) (see Sect. 5.1.2) have different values for the lower and upper vibrational levels.

In view of the fact that one has to consider alternative processes of formation of vibrationally excited molecules of hydroxyl because of intricate variability of its parameters, laboratory experiments have been carried out to investigate the perhydroxyl mechanism with the ^{18}O isotope of atomic oxygen participating in the process

$$^{18}O + H^{16}O_2 \rightarrow {}^{18}O - H^{16}O_2 \rightarrow {}^{18}OH + {}^{16}O_2$$
$$^{18}O + {}^{16}O_2H \rightarrow {}^{18}O - {}^{16}O_2H \rightarrow {}^{16}OH + {}^{18}O^{16}O$$
$$\rightarrow {}^{18}OH + {}^{16}O_2.$$

The formation of hydroxyl molecules was detected by the fluorescence of the $P_1(5)$ line of the (0–0) band of the $A^2\Sigma^+ - X^2\Pi$ electronic transition. The wavelength of this line for ^{18}OH and ^{16}OH is equal to 310.1975 and 310.2129 (nm), respectively, i.e., the shift is 0.0154 (nm), which is three times greater than the line width of the laser used (0.0050 (nm)). It turned out that only ^{16}OH molecules were detected, i.e., the complex HO_3^* arising in the reaction dissociated into $^{18}O^{16}O$ and ^{16}OH. Based on these observations, the conclusion has been made that as a result of the reaction the excitation energy is delivered to O_2 rather than to OH molecules (Sridharan et al. 1985; Kaye 1988; Kukuy et al. 1996).

Other processes which could be responsible for the appearance of excited hydroxyl molecules were also proposed (Reynard and Donaldson 2001) in which the reaction of metastable oxygen molecules with hydrogen molecules gives rise to vibrationally excited hydrogen molecules (Hohmann et al. 1994). These excited molecules enter into a reaction with atomic oxygen resulting in the formation of hydroxyl molecules (Johnson and Winter 1977):

$$H_2 + O_2(b^1\Sigma_g^+) \rightarrow H_2(v=1) + O_2(a^1\Delta_g) \quad 6.3 \cdot 10^{-12} \cdot \exp(-574/T)\,(cm^3 \cdot s^{-1}),$$

$$H_2(v=1) + O(^3P) \rightarrow OH + H \quad 4.65 \cdot 10^{-14} \cdot \exp(-1868/T)\,(cm^3 \cdot s^{-1}).$$

According to the estimates of Reynard and Donaldson (2001), the rate of production of OH molecules is a maximum ($100\,cm^3 \cdot s^{-1}$) at altitudes near 85 (km); the thickness of the layer is ~ 12 (km). However, this process can provide excitation of hydroxyl only to the first vibrational level.

Besides the processes considered, some other reactions can proceed in the upper atmosphere in conditions different from those typical of the nocturnal mesopause.

Under daytime conditions (below 55 (km)) the following process is possible (Nicolet 1971; Banks and Kockarts 1973a,b):

$$H_2 + O(^1D) \rightarrow OH(v \leq 2) + H \quad \alpha_{H_2O(^1D)} = 3.1 \cdot 10^{-11} \, (cm^3 \cdot s^{-1}) \ .$$

If water vapors are dispersed at high altitudes, the excitation of hydroxyl can occur as a result of the reaction chain (Virin et al. 1979)

$$H_2O + O^+ \rightarrow H_2O^+ + O \quad \alpha_{H_2OO^+} = 3.2 \cdot 10^{-9} \, (cm^3 \cdot s^{-1}),$$

$$H_2O^+ + e \rightarrow OH(^2\Pi, v \leq 9) + H \quad \alpha_{H_2O^+} = 4.1 \cdot 10^{-6} \cdot \left(\frac{292}{T_e}\right)^{0.5} (cm^3 \cdot s^{-1}).$$

In anomalous conditions, the following processes (Virin et al. 1979) can occur:

$$NO_2(X^2A_1) + H(^2S) \rightarrow NO(^2\Pi) + OH(^2\Pi, v \leq 3) \quad \alpha_{NO_2H} = 2 \cdot 10^{-12} \cdot \sqrt{T} \, (cm^3 \cdot s^{-1})$$

and

$$N_2O(X^1\Sigma^+) + H(^2S) \rightarrow N_2(X^1\Sigma) + OH(^2\Pi, v \leq 6) \quad \alpha_{N_2OH} = 8.4 \cdot 10^{-11} \, (cm^3 \cdot s^{-1}) \ .$$

Thus, despite the common viewpoint that the mechanism of the ozone–hydrogen reaction is the basic and preferred process, the existing relations for the variations of hydroxyl emission parameters demand a careful consideration of other processes of formation of excited hydroxyl molecules and the processes of vibrational deactivation and relaxation of these molecules. Therefore, systematic and comprehensive investigations of the variations of the emission characteristics of hydroxyl molecules both in laboratory conditions and under the conditions of the upper atmosphere are necessary.

2.2.8 Formation of Dissociation Products

According to the present-day notions, the photodissociation of molecular oxygen occurs under the action of the ultraviolet radiation of the Sun in the Herzberg continuum ($204 \leq \lambda \leq 242.4 \, (nm)$) and in the Schumann–Runge bands ($175 \leq \lambda \leq 204 \, (nm)$). The least dissociation energy for oxygen molecules is 5.11 (eV), which corresponds to the Herzberg continuum whose long-wave limit is 242.4 (nm). In this case, the following process takes place:

$$O_2 + h\nu(204 \leq \lambda \leq 242.4 \, (nm)) \rightarrow O(^3P) + O(^3P) \ ,$$

for which the photodissociation factor $j_{Herz} = 5.87 \cdot 10^{-10} (s^{-1})$ (Nicolet and Kennes 1988).

With the molecular oxygen column content at an altitude Z denoted by $N(O_2)$ we have $j_{Herz} = 3.2 \cdot 10^4 \, (N(O_2))^{-0.7567} (s^{-1})$ (Nicolet and Kennes 1988). The cross-sections corresponding to this process range from $1.5 \cdot 10^{-24} (cm^2)$ for $\lambda = 242.4 \, (nm)$ to $1.5 \cdot 10^{-23} (cm^2)$ near the beginning of the Herzberg continuum at

2.2 Hydroxyl Emission

204 (nm) (Ditchburn and Young 1962; Nicolet 1981; Cheung et al. 1984). In the mesopause the rate of this process makes 2–4(%) of the total rate of photodissociation.

The dissociation by the radiation of shorter wavelength, $175 \leq \lambda \leq 204$ (nm) (Schumann–Runge bands), also results in the formation of oxygen atoms in the ground state:

$$O_2 + h\nu(175 \leq \lambda \leq 204\,\text{nm}) \rightarrow O(^3P) + O(^3P).$$

The photodissociation rate j_{SR} for this case is $1.25 \cdot 10^{-7}(s^{-1})$ (Nicolet et al. 1989).

For a varying altitude Z and, hence, the column density $N(O_2)$, the photodissociation rate can be estimated by the expression (Nicolet et al. 1989)

$$j_{SR} = 6.5 \cdot 10^6 (N(O_2))^{-0.7567} (s^{-1}).$$

The measurement accuracy of the cross-section for absorption in Schumann–Runge bands is largely related to the capabilities of the spectroscopic instrumentation used to resolve the structure of these bands (Ackerman and Biaume 1970). The measurements reported elsewhere (Watanabe et al. 1953; Hudson and Carter 1969; Ackerman et al. 1970; Nicolet 1981) show that the absorption cross-sections vary from $4 \cdot 10^{-23} (cm^2)$ for 200 (nm) to $1.7 \cdot 10^{-19} (cm^2)$ near 175 (nm).

The main reaction of absorption of solar radiation by molecular oxygen at thermospheric altitudes occurs in the Schumann–Runge continuum. As a result of dissociation, metastable oxygen atoms are formed:

$$O_2 + h\nu(130 \leq \lambda \leq 175\,(\text{nm})) \rightarrow O(^3P) + O(^1D).$$

The rate j_{SRc} of this process is $4.1 \cdot 10^{-6} (s^{-1})$ (Hinteregger 1976).

It should be noted that the oxygen atoms formed in the thermosphere as a result of this process are transported down to the mesopause, where they recombine. In this case, it is important to know the number of oxygen atoms formed in this photodissociation process in a column of the thermosphere. This number strongly depends on the number of photons $(cm^{-2} \cdot s^{-1})$ radiated by the Sun in the wavelength range $130 \leq \lambda \leq 175$ (nm). The estimates made based on experimental investigations of the solar ultraviolet energy fluxes (Nicolet 1971; Rottman 1981) show that the total number of oxygen atoms produced by the absorption of solar radiation in the Schumann–Runge continuum lies between $(1.5 \pm 0.5) \cdot 10^{12} (cm^{-2} \cdot s^{-1})$ (quiet Sun) and $(2.5 \pm 0.5) \pm 10^{12} (cm^{-2} \cdot s^{-1})$ (maximum of solar activity) (Nicolet 1981). The cross-sections for absorption in the Schumann–Runge continuum are substantially greater than those for absorption in the Schumann–Runge bands and Herzberg continuum and they are equal to $2 \cdot 10^{-19} (cm^2)$ for 175 (nm), $1.5 \cdot 10^{-17} (cm^2)$ for 140 (nm), and $3 \cdot 10^{-18} (cm^2)$ for 130 (nm) (Watanabe et al. 1953).

In addition to these three spectroscopic intervals involved in the photodissociation of oxygen molecules, it is necessary to consider the solar radiation taking part in the emission of Lyman alpha (L_α) radiation at 121.6 (nm) that dissociates molecular oxygen giving rise to metastable oxygen atoms

$$O_2 + h\nu(\lambda\ 121.6\,(nm)) \rightarrow O(^3P) + O(^1D)$$

at a rate $j_L = 3 \cdot 10^{-9}(s^{-1})$ (Nicolet 1971).

The absorption cross-section for this process is very small, being equal to about 10–$20\,(cm^2)$ (Nicolet 1962). Owing to this, the solar radiation in the Lyman alpha can penetrate to altitudes of 60–70 (km). This seems to be a reason for the solar activity effect at mesospheric altitudes (Chanin et al. 1989; Givishvili et al. 1996; Semenov et al. 2005; Golitsyn et al. 2006).

Thus, the total rate of photodissociation of molecular oxygen is

$$j_{O_2} = 4.4 \cdot 10^{-6}(s^{-1}).$$

In the spectral range under consideration ($\lambda < 300\,(nm)$), depending on the level of solar activity, variations of radiation intensity become pronounced whose degree increases with decreasing wavelength. Thus, according to Nicolet (1989a), the variations of solar flux $S(\lambda)$ make about 2(%) in the wavelength range near 300 (nm), 10(%) near 250 (nm), and 25(%) near 200 (nm). These variations can be approximately described by the empirical relation

$$S(\lambda, F_{10.7}) = S(\lambda, 144) \cdot \frac{1 + \frac{F_{10.7} - 70}{26} \cdot \exp(-18 \cdot \lambda)}{1 + 2.85 \cdot \exp(-18 \cdot \lambda)},$$

where λ is the wavelength expressed in μm and $F_{10.7}$ is the flux of solar radio radiation ($10^{-22}\,W \cdot m^{-2} \cdot Hz^{-1}$) that characterizes the solar radiation in the ultraviolet spectral region.

The variations of solar radiation intensity affect the rates of dissociation of molecular oxygen in the spectral ranges under consideration. Though the accuracy of the calculation of photodissociation rates makes only $10 \pm 50(\%)$, the tendencies of their variations depending on solar activity can be considered (Nicolet 1989a).

Some other atmospheric components, which are formed in the mesopause as a result of photodissociation of water vapors, are also essential to the occurrence of hydroxyl emission. According to the present-day notions (Brasseur and Solomon 2005), the density of water vapors in the middle atmosphere is determined by the density of molecular methane (CH_4) produced by a sequence of reactions. Therefore, methane, as one of the greenhouse gases, is of significance to the problem of hydroxyl emission since its energetics is a manifestation of not only the recombination rate of methane, but also its concentration in the Earth's atmosphere. For the absorption range of the Herzberg and Schumann–Runge continua (Nicolet 1984) we have

$$H_2O + h\nu(\lambda < 246\,(nm)) \rightarrow H_2(X^1\Sigma_g^+) + O(^3P),$$

$$H_2O + h\nu(\lambda < 242\,(nm)) \rightarrow OH(X^2\Pi) + H(^2S),$$

$$H_2O + h\nu(\lambda < 176\,(nm)) \rightarrow H_2(X^1\Sigma_g^+) + O(^1D),$$

2.2 Hydroxyl Emission

$$H_2O + h\nu(\lambda < 136\,(nm)) \rightarrow OH(A^2\Sigma^+) + H(^2S),$$

$$H_2O + h\nu(\lambda < 130\,(nm)) \rightarrow O(^3P) + H(^2S) + H(^2S),$$

$$H_2O + h\nu(\lambda < 130\,(nm)) \rightarrow H_2(X^1\Sigma_g^+),$$

and also, due to the Lyman alpha (Nicolet 1984),

$$\begin{aligned}H_2O + h\nu(\lambda 121.6\,(nm)) &\rightarrow OH(X^2\Pi) + H(^2S) \\ &\rightarrow OH(A^2\Sigma^+) + H(^2S) \\ &\rightarrow H_2(X^1\Sigma_g^+) + O(^1D) \\ &\rightarrow O(^3P) + H(^2S) + H(^2S)\end{aligned}$$

The first reaction of the branch, which provides 70(%) of the photodissociation yield, is most efficient. The subsequent reactions provide 8, 10, and 12(%), respectively (Nicolet 1984). The photodissociation rates of these processes are different: $j_{176} = 4.3 \cdot 10^{-6}(s^{-1})$ for the range $\lambda \leq 176\,(nm)$, $j_{SRlsa} = 1.2 \cdot 10^{-6}(s^{-1})$ for low solar activity and $j_{SRhsa} = 1.4 \cdot 10^{-6}(s^{-1})$ for high solar activity for the range $175 \leq \lambda \leq 200\,(nm)$ and the conditions of the upper atmosphere.

The attenuation of solar radiation in the atmosphere at an altitude Z can be taken into account by the relation

$$j_{242}(Z) = j_{SRc} \cdot \exp[-10^{-7} \cdot N(O_2)^{0.35}],$$

where $N(O_2)$ is the column density of oxygen molecules at the altitude Z. The rate of photodissociation of water vapors due to absorption of Lyman alpha radiation is estimated in the main by the rate of the first reaction, which, for the upper atmosphere, is given by

$$j_{L\alpha} = 4.5 \cdot 10^{-6} \cdot \left[1 + 0.2 \cdot \frac{F_{10.7} - 65}{100}\right](s^{-1}).$$

The attenuation of Lyman alpha radiation by the Earth's atmosphere at an altitude Z is taken into account by the expression (Nicolet 1984)

$$j_{L\alpha}(Z) = j_{L\alpha} \cdot \exp\left[-4.4 \cdot 10^{-19} \cdot (N(O_2))^{0.917}\right].$$

At the mesopause altitudes, the strong attenuation of solar radiation in several spectral regions is due to ozone. Absorption of the radiation in the visible and ultraviolet spectral regions results in the following dissociation processes (Nicolet 1989b):

$$O_3 + h\nu(400 \leq \lambda \leq 800(nm)) \rightarrow O_2(^3\Sigma_g^-) + O(^3P) \qquad j_{Chp} = 3 \cdot 10^{-4}\,(s^{-1})$$

for the visible region (Chappuis band) and

$$O_3 + h\nu(310 \leq \lambda \leq 400(nm)) \rightarrow O_2(^1\Delta_g) + O(^1D) \qquad j_{Hug} = 2 \cdot 10^{-4} \, (s^{-1})$$

for the ultraviolet region (Huggins band).

The most substantial absorption occurs in the Hartley bands ($210 \leq \lambda \leq 310$ (nm)) where the absorption cross-section reaches $10^{-17} (cm^2)$ (Nicolet 1971, 1984):

$$O_3 + h\nu(210 \leq \lambda \leq 310(nm)) \rightarrow O_2(^1\Delta_g, ^1\Sigma_g^+) + O(^1D) \qquad j_{Hart} = 1 \cdot 10^{-2} \, (s^{-1}).$$

This reaction determines in the main the total rate of photodissociation of ozone molecules (about 90(%)).

2.2.9 Recombination of Atomic Oxygen

The atomic oxygen produced by dissociation of molecular oxygen, due to atmospheric mixing and vertical transport, is transferred downward, and its active recombination occurs with participation of various components, atomic hydrogen included. Such efficient channels of production of hydroxyl molecules are related to three-body collision-induced reactions which result in the formation of important intermediate products such as ozone (O_3) and perhydroxyl (HO_2) molecules (Nicolet 1989b):

$$O + O_2 + M \rightarrow O_3 + M \qquad \alpha_{O_3} = 6 \cdot 10^{-34} \cdot \left(\frac{300}{T}\right)^{2.3} \left(cm^6 \cdot s^{-1}\right)$$

and

$$H + O_2 + M \rightarrow HO_2 + M \qquad \alpha_{HO_2} = 3.3 \cdot 10^{-33} \cdot \exp(800/T) \left(cm^6 \cdot s^{-1}\right).$$

The excited species O_3^* and HO_2^* resulting from these reactions are of importance for further recombination of atomic oxygen and for the efficiency of hydroxyl emission as an energy sink, as stressed by Krassovsky (1963b). However, the data on these important branches of reactions are rather scanty and uncertain (Gershenzon et al. 1985).

The sequence of hydroxyl reactions actually ends with a process in which (with participation of a third component) two oxygen molecules are formed from ozone and atomic oxygen. The reaction of immediate collision of these two particles is also possible (Nicolet 1971):

$$O + O_3 \rightarrow O_2 + O_2 \qquad \alpha_{OO_3} = 2.4 \cdot 10^{-11} \cdot \exp(-2350/T) \left(cm^3 \cdot s^{-1}\right).$$

This forward reaction corresponds to 94.1 (kcal) of released energy, which is almost equal to the energy of the excited state $O_2(c^1\Sigma_u^-)$. For a temperature of about 200 (K) this yields the reaction rate

$$\alpha_{OO_3} = 1.9 \cdot 10^{-16} \, (\text{cm}^3 \cdot \text{s}^{-1}) \, .$$

As can readily be shown, the ratio of the yields of oxygen molecules produced by these two recombination channels is

$$\frac{[O_2]_{OH}}{[O_2]_{OO_3}} = \frac{\alpha_{HO_3} \cdot \beta_{OHO}}{\alpha_{OO_3} \cdot \bar{A}_{OH}} \cdot [H] \cdot m \, ,$$

where $m = [OH]/[OH^*] > 10$ (McEwan and Phillips 1975), whence $[O_2]_{OH}/[O]_{O3} \sim 10^3$. Thus, the recombination proceeding with emission of hydroxyl radiation is an essentially dominant process.

2.3 Emissions from Metal Atoms

In the Earth's upper atmosphere, metals are small nongaseous components. Their presence is related both to their transport from oceans together with reek (sodium) and to the intrusion of meteoric matter. They are indicators of dynamic phenomena in the upper atmosphere.

2.3.1 Sodium Emission: Photochemical Excitation

The emission from sodium at wavelengths of 589.0–589.6 (nm) was first detected in the nightglow of the upper atmosphere by Slipher (1929). However, because of the very low dispersion of the spectrographs used, this emission could be identified only in 1936 upon its observation in twilight (Chernjajev and Vuks 1937; Khvostikov 1948). The presence of atomic sodium in the upper atmosphere is caused by its transport upward with water evaporated from oceans (Chamberlain 1961). It seems that the meteors intruding into the atmosphere and burning down at the mesopause altitudes also make a certain contribution to the abundance of atomic sodium (Fiocco et al. 1974). Lidar measurements at the wavelength of sodium emissions very often detect narrow (~ 1 (km)) intense (up to tenfold) maxima in the altitude distribution of sodium concentration (von Zahn et al. 1987). The nightly mean intensity of sodium emission is 70 (Rayleigh), to which there corresponds $\sim 7 \cdot 10^9 (\text{cm}^{-2})$ sodium atoms in a column. The emission layer is at an altitude of ~ 92 (km); its thickness is 10 (km). Therefore, the maximum concentration of sodium is $[Na] \sim 10^4 (\text{cm}^{-3})$.

Sodium emissions arise due to the following transitions between the first excited state and the ground state (the transition probabilities are taken from the work by Bates (1982)):

$$\text{Na}(3^2P^o_{3/2}) \rightarrow \text{Na}(3^2S_{1/2}) + h\nu(589.0\,(\text{nm})) \quad A_{D_2} = 6.26 \cdot 10^7 \, (\text{s}^{-1}) ,$$
$$\text{Na}(3^2P^o_{1/2}) \rightarrow \text{Na}(3^2S_{1/2}) + h\nu(589.6\,(\text{nm})) \quad A_{D_1} = 6.26 \cdot 10^7 \, (\text{s}^{-1}) ,$$

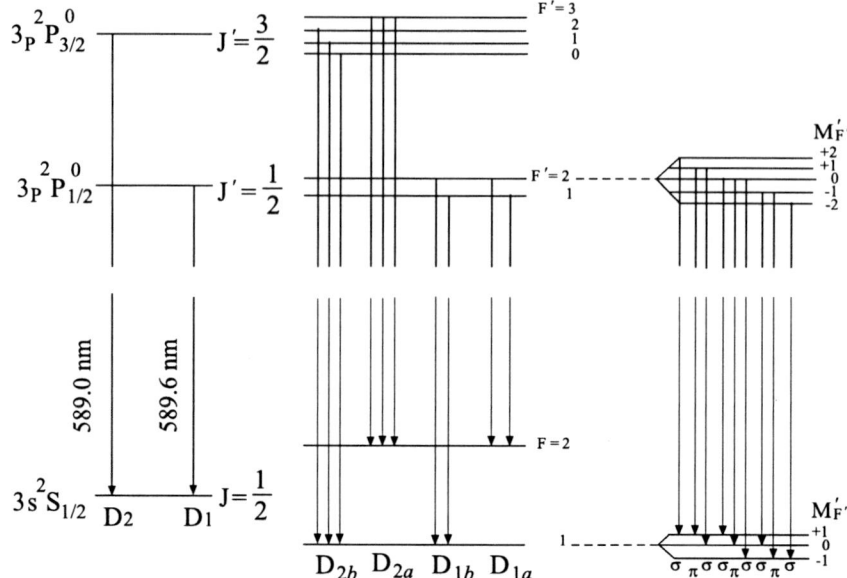

Fig. 2.16 Structure of the energy levels of the ground (3s) and the first excited state (3p) of the Na atom. Hyperfine levels for the nuclear spin I = 3/2 are scaled-up. The Zeeman levels and the Zeeman component lines are shown on the *right* (Chamberlain 1978)

which actually have a complex Zeeman structure of the upper and lower levels due to the nuclear spin I = 3/2 (Fig. 2.16) (Chamberlain 1978). Therefore, the observed doublet structure of lines D_2 and D_1 (589.0–589.6 (nm)) actually consists of several components, resulting in the complex spectral profile measured for the emission (Fig. 2.17) (Chamberlain 1961).

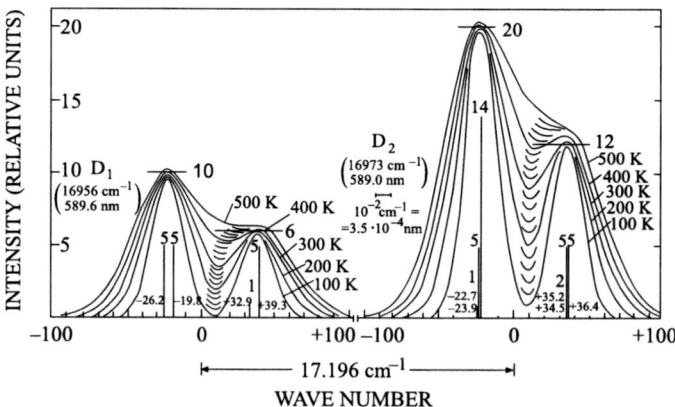

Fig. 2.17 Resonance line profiles for the 589.0- to 589.6-nm sodium emission at different temperatures (Chamberlain 1961)

Since the radiative lifetimes of excited states are $\sim 2 \cdot 10^{-8}$ (s), collisional deactivation at altitudes above 80 (km) is insignificant.

The mechanism of the photochemical excitation of sodium emission under the conditions of a night atmosphere was proposed by Chapman (1939). The present-day studies of the photochemical processes initiating the emission from sodium (Kirchhoff 1986a,b; Chikashi et al. 1989; Plane 1991; Herschbach et al. 1992; Helmer and Plane 1993; Clemesha et al. 1995; McNeil et al. 1995; Shefov and Semenov 2002; Shefov et al. 2002) consider the most essential processes with participation of atomic sodium:

$$Na + O_3 \rightarrow NaO(A^2\Sigma^+) + O_2(X^3\Sigma_g^-) \; f_1 \cdot \alpha_{NaO},$$

$$\alpha_{NaO} = 1.1 \cdot 10^{-9} \exp(-116/T), (cm^3 \cdot s^{-1}),$$

$$Na + O_3 \rightarrow NaO(X^2\Pi, v) + O_2(a^1\Delta_g) \quad (1-f_1) \cdot \alpha_{NaO}, \; f_1 = 0.67,$$

$$NaO(A^2\Sigma^+) \rightarrow NaO(X^2\Pi) + h\nu(6.67\,(\mu m)) \; A_{NaO} \sim 3, (s^{-1}),$$

$$NaO + O \rightarrow Na(^2P) + O_2 \quad f_2 \cdot \beta_{NaO}, \beta_{NaO} = 2.2 \cdot 10^{-10}(T/200)^{1/2}, (cm^3 \cdot s^{-1}),$$

$$NaO + O \rightarrow Na(^2S) + O_2 \quad (1-f_2) \cdot \beta_{NaO}, \quad f_1 \cdot f_2 = 0.13, \quad f_2 = 0.19,$$

$$Na(^2P) \rightarrow Na(^2S) + h\nu(589.0 - 589.6\,(nm)) \; A = 6.2 \cdot 10^7, (s^{-1}),$$

$$NaO + O_3 \rightarrow NaO_2 + O_2 \quad \beta_{O_3}^{(1)} = 1.1 \cdot 10^{-9} \exp(-568/T), (cm^3 \cdot s^{-1}),$$

$$NaO + O_3 \rightarrow Na + O_2 + O_2 \quad \beta_{O_3}^{(2)} = 3.2 \cdot 10^{-10} \exp(-550/T), (cm^3 \cdot s^{-1}),$$

$$NaO + O_2 + N_2 \rightarrow NaO_3 + N_2 \quad \beta_M = 5.3 \cdot 10^{-30}(200/T), (cm^6 \cdot s^{-1}),$$

$$NaO_3 + O \rightarrow Na + O_2 + O_2 \quad \beta_O = 2.5 \cdot 10^{-10}(T/200)^{1/2}, (cm^3 \cdot s^{-1}),$$

$$Na + O_2 + N_2 \rightarrow NaO_2 + N_2 \quad \alpha_M = 5.0 \cdot 10^{-30}(200/T)^{1.22}, (cm^6 \cdot s^{-1}),$$

where α and β are the reaction rates; A_{NaO} and A are the transition probabilities for NaO molecules and Na atoms, respectively; f_1 and f_2 are the coefficients of branching of the reactions producing excited NaO molecules ($A^2\Sigma^+$) and Na atoms (^2P), respectively; and T is the temperature.

Based on these reactions, it is possible to find an equation for the emission rate $Q_{Na}(Z)$ (photon·cm$^{-3} \cdot s^{-1}$). However, under the actual atmospheric conditions at altitudes of 80–100 (km), the contribution of ozone, oxygen, and nitrogen molecules in their reactions with Na, affecting the lifetime of NaO, is small compared to the contribution of the reaction of Na with atomic oxygen. Therefore, the sodium emission rate is determined by the relation

$$Q_{Na}(Z) = \frac{f_1 \cdot f_2 \cdot \alpha_{NaO} \cdot [O_3] \cdot [Na]}{1 + \frac{A_{NaO}}{\beta_{NaO} \cdot [O]}}.$$

In this equation, it is naturally supposed that the reaction rates and the concentrations of the reactants depend on altitude Z.

Thus, the ozone concentration is determined by the relation

$$[O_3] = \frac{Q_{Na}}{f_1 \cdot f_2 \cdot \alpha_{NaO} \cdot [Na]} \cdot \left[1 + \frac{A_{NaO}}{\beta_{NaO} \cdot [O]}\right] \approx \frac{Q_{Na}}{f_1 \cdot f_2 \bullet \alpha_{NaO} \cdot [Na]}.$$

In describing the photochemical process of excitation of the sodium nightglow a problem arises which is related to the indeterminacy of the coefficients of branching, f_1 and f_2, for the reactions that produce NaO ($A^2\Sigma^+$) and ($X^2\Pi$) molecules and also sodium atoms in an excited Na(^{2}P) and a nonexcited state Na(^{2}S), respectively. The radiative lifetime of NaO molecules ($A^2\Sigma^+$) should correspond to the magneto-dipole transition. According to the available rough estimates, the radiative lifetime is considerably longer than the lifetime determined by photochemical reactions (0.04 (s)) at the altitudes of the emission layer (Chikashi et al. 1989; Herschbach et al. 1992). The coefficients of branching were estimated to be $f_1 \sim 0.67$ and f_2 from 0.01 to 0.15–0.20 (Clemesha et al. 1995). Based on the analysis of data on variations of the sodium emission intensity, the radiation transition probability has been estimated for NaO ($A^2\Sigma^+$) molecules (6.67 (μm), $A_{NaO} \sim 3\,(s^{-1})$) and the coefficient of branching ($f_2 \sim 0.19$) (Shefov and Semenov 2002; Shefov et al. 2002).

In view of a feature of the excitation process (see Sect. 2.1.3) and blending of complex profiles, the interferometric method was not used to determine the temperature. Besides, the sodium emission lines are in the close neighborhood with the $Q_1(1)$(588.9 (nm)), $Q_2(2)$(589.4 (nm)), $Q_1(2)$(589.5 (nm)), and $Q_2(3)$(590.1 (nm)) lines of the OH (8–2) band, whose intensities are on the average equal to 4, 0.2, 1.5, and 0.1 Rayleigh, respectively (Shefov and Piterskaya 1984). The total intensity of this band is about 40 Rayleigh and at 5.5 (nm) about 20 Rayleigh. The continuum intensity in this spectral region is 20–40 (Rayleigh·nm^{-1}) (Shefov 1959, 1960, 1961d). This gives rise to difficulties in electrophotometer measurements using interference filters.

Nevertheless, the variations of the intensity of sodium nightglow were investigated by many researchers (Nasyrov 1967, 2003; Pal 1973; Fishkova 1976, 1979, 1983; States and Gardner 1999, 2000a,b) and, based on the data obtained, the diurnal and seasonal variations have been estimated. The seasonal variations of the intensity of sodium emission have an important feature that the intensity decreases in summer. This is related to the photoionization (Swider 1970; McEwan and Phillips 1975) of sodium and the formation of its molecular ions in the daytime, which considerably elongates in summer:

$$Na + h\nu(\lambda \leq 241.6\,(nm)) \rightarrow Na^+ + e \qquad j_{Na} = 2 \cdot 10^{-5}\,(s^{-1}).$$

Sodium ions can be produced not only by direct photoionization, but also in charge exchange reactions (McEwan and Phillips 1975):

$$Na + O_2^+ \rightarrow Na^+ + O_2 \beta_{O_2^+} = 6.7 \cdot 10^{-10} \, (cm^3 \cdot s^{-1}),$$

$$Na + N_2^+ \rightarrow Na^+ + N_2 \beta_{N_2^+} = 5.8 \cdot 10^{-10} \, (cm^3 \cdot s^{-1}),$$

$$Na + NO^+ \rightarrow Na^+ + NO \beta_{NO^+} = 7.0 \cdot 10^{-11} \, (cm^3 \cdot s^{-1}),$$

$$Na + H_2O^+ \rightarrow Na^+ + H_2O \beta_{H_2O^+} = 2.7 \cdot 10^{-9} \, (cm^3 \cdot s^{-1}),$$

$$Na + N_2O^+ \rightarrow Na^+ + N_2O \beta_{N_2O^+} = 2.0 \cdot 10^{-9} \, (cm^3 \cdot s^{-1}).$$

The loss of ions due to subsequent photoionization occurs with low rates ($<10^{-10}$ (s^{-1})) since for the Mg^+, Ca^+, Sr^+, Ba^+, and Eu^+ ions the ionization potential is greater than 10 eV, i.e., the photoionization wavelengths are shorter than 100 (nm).

Recombination of metal ions with electrons, for example,

$$Na^+ + e \rightarrow Na + h\nu \quad \beta_e = 10^{-12} \, (cm^3 \cdot s^{-1}),$$

occurs with low rates, providing the effect of increased electron density on the density of metal ions in the ionospheric E region. A more efficient channel of recombination of metal ions is their transformation into molecular ions due to the greater O_2 and NO densities, for example, by the reactions

$$Na^+ + O_2 \rightarrow NaO^+ + O\beta_{O_2} = 1 \cdot 10^{-13} \, (cm^3 \cdot s^{-1}),$$

$$Na^+ + NO \rightarrow NaO^+ + N\beta_{NO} = 1 \cdot 10^{-13} \, (cm^3 \cdot s^{-1}),$$

$$Na^+ + O_3 \rightarrow NaO^+ + O_2 \beta_{O_3} = 1 \cdot 10^{-11} \, (см^3 \cdot с^{-1}).$$

Thus, the ionized-to-neutral sodium density ratio is $\sim 10^{-3}$ and can be greater.

The use of lidars operating at the wavelength of sodium (589.0 (nm)) gave impetus to investigations of the sodium layer as an indicator of atmospheric processes (Richter et al. 1981; Gardner et al. 1986; States and Gardner 1999, 2000a,b). Along with the detection of sporadic perturbations, which manifest themselves in the altitude distribution of sodium, internal gravity waves propagating to the sodium layer altitudes were also investigated (Molina 1983; Gardner and Voelz 1985).

Besides, based on lidar measurements of the sodium layer, a close correlation between the sodium column content N_{Na} and the temperature in this altitude range has been found for the nocturnal and seasonal variations (Qian and Gardner 1995; States and Gardner 1999). The existence of this correlation for all latitudes from South Pole to the high latitudes of the northern hemisphere was proposed by (Plane et al. 1998). Since the radiation intensity is proportional to the sodium column density, based on the data on the night variations of the emission intensity, a close correlation between emission intensity and temperature has been revealed, which is described by the following regression equation (correlation coefficient $r = 0.952 \pm 0.020$) (Shefov and Semenov 2001):

$$T_{Na} = (185 \pm 0.8) + (0.20 \pm 0.01) \cdot I_{Na}, (K),$$

where the intensity I_{Na} is expressed in (Rayleighs).

This equation allows one to estimate the temperature at the altitudes of the emission layer ($\sim$92 (km)) for both the nocturnal and the seasonal variations from emission intensity measurements.

2.3.2 Alkali Metals: Resonant Excitation

The presence of metals in small concentrations in the Earth's upper atmosphere is mainly the result of the intrusion of meteoric bodies, which vaporize at altitudes of 80–100 (km) (Jégou et al. 1985a,b). The column densities of metals in the atmosphere are insignificant (Table 2.20). Their radiation can be observed only when the upper atmosphere is illuminated by the Sun, since it is only in this case that the excitation of their resonant emissions due to fluorescence can be provided.

The emissions of sodium have the greatest intensity in twilight, and it reaches $\sim$1000 (Rayleigh). The 769.9-nm emission of potassium and the 280.0-nm emission of ionized magnesium have intensities of $\sim$400 and $\sim$300 (Rayleigh), respectively. The intensities of the natural emissions of lithium (670.8 (nm)), ionized calcium (393.5 (nm)), magnesium (285.2 (nm)), and iron (386.0 (nm)) are no more than several tens of (Rayleighs).

For the fluorescence process, the intensity I (photon·cm^{-2}·s^{-1}) of fluorescent emission (at zenith) is determined by the relation

Table 2.20 Typical mean intensities of metal emissions in the upper atmosphere

Atom	λ(nm)	I (Rayleigh)			N(cm^{-2})	Notes
		night	twilight	day		
LiI	670.8	3	25	–	7(6)	Natural content
NaI	589.3	70	1000	5000	7(9)	
MgI	285.2	–	10	–	3(8)	
MgII	280.0	–	300	1500	5(8)	
KI	769.9	20	400	–	4(7)	
CaI	422.7	–	10	–	3(7)	
CaII	393.5	–	50	–	2(8)	
FeI	386.0	–	10	–	7(8)	
SrI	460.7	–	–	–	–	released
SrII	407.8	–	–	–	–	
CsI	455.5	–	–	–	–	
CsI	852.1	–	–	–	–	
BaI	553.5	–	–	–	–	
BaII	455.4	–	–	–	–	

2.3 Emissions from Metal Atoms

$$I = Q \cdot N(M) = g \cdot [M] \cdot W,$$

where Q (photon·cm^{-3}·s^{-1}) is the emission rate, W is the thickness of the emission layer, N(M) is the total density of atoms in an atmospheric column, [M] is the concentration of M-type atoms, and g (s^{-1}) is the coefficient of photon scattering in the solar line (defined in Sect. 2.1.1).

The parameters of various metal atoms that are given in Table 2.21 have the following notation: E_u denotes the excitation energy; g_u and g_l the statistical weights of the upper and lower states, respectively; σ the coefficient of absorption at the line center (cm^2); S(λ) the solar radiation flux (photon·cm^{-2}·s^{-1}·nm^{-1}); and r the residual intensity of the Fraunhofer lines in the solar spectrum.

Under the conditions of solar illumination of the upper atmosphere in twilight and in the daytime there occurs photoionization of metal atoms, resulting in variations in equilibrium concentration of neutral atoms (Table 2.22). These processes have been considered in the previous section.

In twilight, the ionizing radiation passes through the atmospheric layers lying beneath the horizon. In this case, the photoionization rate is determined by the relation

$$r = j \cdot \exp(-\tau \cdot \text{Chp}\chi).$$

Here j is the photoionization rate (s^{-1}), τ is the optical thickness of the higher-lying atmospheric layer, and Chpχ is the Chapman function of the solar zenith angle χ (Smith and Smith 1972; Rishbeth and Garriott 1969; Ivanov-Kholodny and Mikhailov 1980).

The twilight emissions have been studied most comprehensively for sodium (Galperin 1956a,b; Hunten 1956, 1967; Lytle and Hunten 1959; Gadsden 1969; Graham et al. 1971; Toroshelidze and Chilingarashvili 1975; Jégou et al. 1985a,b; Toroshelidze 1991). When observations are carried out at large zenith angles, an optically thick medium arises along the line of sight for which it is necessary to solve the problem of radiation transfer. For the sodium emission, this problem is complicated because of the necessity of considering the anisotropic dissipation of the radiation by the fine structure of atomic levels (Chamberlain 1961, 1978; Chamberlain and Hunten 1987).

Once these problems are solved, it becomes possible to use measurements of the intensity of sodium emission during twilight for the determination of various characteristics of the upper atmosphere. Toroshelidze (1991) considered the solution of the problem of finding the altitude distribution of the coefficient of eddy diffusion K_T by analyzing the altitude distribution of the sodium emission rate obtained from twilight measurements. The possibility of doing this is related to the concept that the altitude of the Earth's shadow varying in the course of successive measurements in twilight allows one to obtain data on the altitude to which there corresponds the measured radiation intensity. The principles underlying this procedure were stated in Sect. 1.4.3.

Due to the increased coefficient of eddy diffusion and the higher altitude of the turbopause (the altitude at which the eddy diffusion coefficient K_T is comparable to

Table 2.21 Parameters of the resonant lines of metals

Atom	M	λ(nm)	Transition	E_u (eV)	g_u / g_l	$A(s^{-1})$	$\sigma(cm^2)$	$S(\lambda)$	r	$g(s^{-1})$
1	2	3	4	5	6	7	8	9	10	11
Li	6	670.792	$2^2P^0_{3/2} - 2^2S_{1/2}$	1.85	4/2	3.66(7)	7.32(−12)	5.22(14)	1.0	10.3
		670.807	$2^2P^0_{1/2} - 2^2S_{1/2}$	1.85	2/2	3.66(7)	3.60(−12)	5.22(14)	1.0	5.15
Li	7	323.263	$3^2P_{3/2} - 2^2S_{1/2}$	3.83	4/2	1.00(6)	2.20(−14)	1.40(14)	0.7	0.00284
			$3^2P_{1/2} - 2^2S_{1/2}$	3.83	2/2	1.00(6)	1.10(−14)	1.40(14)	0.7	0.00142
Li	7	670.776	$2^2P^0_{3/2} - 2^2S_{1/2}$	1.85	4/2	3.66(7)	7.32(−12)	5.22(14)	1.0	10.3
		670.791	$2^2P^0_{1/2} - 2^2S_{1/2}$	1.85	2/2	3.66(7)	3.60(−12)	5.22(14)	1.0	5.15
Na	23	330.232	$4^2P^0_{3/2} - 3^2S_{1/2}$	3.75	4/2	2.75(6)	1.17(−13)	1.61(14)	0.056	0.000784
		330.299	$4^2P^0_{1/2} - 3^2S_{1/2}$	3.75	2/2	2.75	5.85(−14)	1.61(14)	0.083	0.000583
Na	23	342.686	$4^2D_{3/2} - 3^2S_{1/2}$	3.61	4/2	1.60(6)	7.60(−14)	1.66(14)	0.80	0.00782
Na	23	588.995	$3^2P^0_{3/2} - 3^2S_{1/2}$	2.10	4/2	6.26(7)	1.52(−11)	5.31(14)	0.05	0.54
Na	23	589.592	$3^2P^0_{1/2} - 3^2S_{1/2}$	2.10	2/2	6.26(7)	7.57(−12)	5.31(14)	0.06	0.33
Mg	24	285.213	$3^1P^0_1 - 3^1S_0$	4.35	3/1	4.60(8)	1.93(−11)	4.74(13)	0.0256	0.015
Mg⁺	24	279.553	$3^2P^0_{3/2} - 3^2S_{1/2}$	4.43	4/2	2.60(8)	6.87(−12)	4.75(13)	0.2	0.119
Mg⁺	24	280.270	$3^2P^0_{1/2} - 3^2S_{1/2}$	4.42	2/2	2.60(8)	3.45(−12)	4.75(13)	0.2	0.0179
Al	27	394.403	$4^2S_{1/2} - 3^2P^0_{1/2}$	3.14	2/2	3.30(7)	1.20(−12)			0.62
K	39	404.414	$5^2P^0_{3/2} - 4^2S_{1/2}$	3.065	4/2	1.20(6)	1.22(−13)	3.35(14)	0.80	0.0229
		404.720	$5^2P^0_{1/2} - 4^2S_{1/2}$	3.063	2/2	1.20(6)	6.10(−14)	3.35(14)	0.80	0.0115
K	39	766.490	$4^2P^0_{3/2} - 4^2S_{1/2}$	1.62	4/2	4.14(7)	2.87(−11)	4.85(14)	0.08	1.48
K	39	769.896	$4^2P^0_{1/2} - 4^2S_{1/2}$	1.61	2/2	4.05(7)	1.42(−11)	4.82(14)	0.2	2.70
Ca	40	422.673	$4^1P^0_1 - 4^1S_0$	2.93	3/1	2.11(8)	3.72(−11)	3.69(14)	0.07	0.694
Ca	40	657.278	$4^3P^0_1 - 4^1S_0$	1.89	3/1	2.06(3)	1.37(−15)	5.19(14)	0.80	0.000637
Ca⁺	40	393.478	$4^2P^0_{3/2} - 4^2S_{1/2}$	3.15	4/2	1.50(8)	1.43(−11)	2.18(14)	0.10	0.193
Ca⁺	40	396.959	$4^2P^0_{1/2} - 4^2S_{1/2}$	3.12	2/2	1.40(8)	6.81(−12)	2.45(14)	0.10	0.103
Ca⁺	40	849.802	$4^2P^0_{3/2} - 3^2D_{3/2}$	3.15	4/4	1.40(6)	6.70(−13)	4.36(14)	0.28	0.0018
Ca⁺	40	854.209	$4^2P^0_{3/2} - 3^2D_{5/2}$	3.15	4/6	1.20(7)	3.88(−12)	4.33(14)	0.20	0.015
Ca⁺	40	866.214	$4^2P^0_{1/2} - 3^2D_{3/2}$	3.12	2/4	1.30(7)	3.27(−12)	4.27(14)	0.24	0.0096
Fe	56	385.991	$4^5D^0_4 - 4^5D_4$	3.20	9/9	3.60(6)	1.91(−13)	1.99(14)	0.10	0.014

2.3 Emissions from Metal Atoms

Table 2.21 (continued)

Atom	M	λ(nm)	Transition	E_u (eV)	g_u / g_l	$A(s^{-1})$	$\sigma(cm^2)$	$S(\lambda)$	r	$g(s^{-1})$
Rb	85	780.023	$5^2P^0_{3/2} - 5^2S_{1/2}$	1.59	3 / 1	1.00(8)	1.50(−10)			18.42
Sr	88	460.733	$5^1P^0_1 - 5^1S_0$	2.69	3 / 1	1.90(8)	6.10(−11)	4.79(14)	0.60	13.39
Sr	88	689.259	$5^3P^0_1 - 5^1S_0$	1.80	3 / 1	4.23(4)	4.80(−14)	5.21(14)	0.85	0.0169
Sr^+	88	407.771	$5^2P^0_{3/2} - 5^2S_{1/2}$	3.04	4 / 2	1.30(8)	2.03(−11)	3.52(14)	0.15	0.505
Sr^+	88	421.552	$5^2P^0_{1/2} - 5^2S_{1/2}$	2.94	2 / 2	1.20(8)	1.04(−11)	3.70(14)	0.10	0.186
Cs	133	455.528	$7^2P^0_{3/2} - 6^2S_{1/2}$	2.72	4 / 2	1.93(6)	5.20(−13)	4.72(14)	0.97	0.100
Cs	133	459.317	$7^2P^0_{1/2} - 6^2S_{1/2}$	2.70	2 / 2	8.85(5)	1.22(−13)	4.78(14)	0.92	0.023
Cs	133	852.112	$6^2P^0_{3/2} - 6^2S_{1/2}$	1.46	4 / 2	3.31(7)	5.80(−11)	4.35(14)	1.0	22.06
Cs	133	854.346	$6^2P^0_{1/2} - 6^2S_{1/2}$	1.39	2 / 2	3.20(7)	2.83(−11)	4.35(14)	0.60	5.92
Ba	138	307.158	$7^1P^0_1 - 6^1S_0$	4.04	3 / 1	4.01(7)	5.05(−12)	9.60(13)	0.61	0.084
Ba	138	350.111	$6^1P^0_1 - 6^1S_0$	3.54	3 / 1	1.81(7)	3.37(−12)	1.77(14)	0.95	0.183
Ba	138	553.548	$6^1P^0_1 - 6^1S_0$	2.40	3 / 1	1.16(8)	8.55(−11)	5.16(14)	0.50	22.33
Ba^+	138	455.403	$6^2P^0_{3/2} - 6^2S_{1/2}$	2.72	4 / 2	1.10(8)	3.00(−11)	4.72(14)	0.20	12.55
Ba^+	138	493.409	$6^2P^0_{1/2} - 6^2S_{1/2}$	2.51	2 / 2	9.00(7)	1.57(−11)	4.95(14)	0.13	0.457
La	139	357.443		3.47		2.60(8)*	1.60(−11)*			0.97
La^+	139	408.672		3.03		1.00(8)*	9.19(−12)*			1.24
Zr	140	351.960	$5^3P^0_2 - 5^3F_2$	3.52	5 / 5	1.00(8)	4.26(−12)			0.36
Zr^+	140	357.247	$5^4G^0_{5/2} - 5^4F_{3/2}$	3.47	6 / 4	6.20(7)	5.73(−12)			1.41
Nd	144	492.453		2.52		2.00(8)*	3.28(−11)			7.3
Eu	153	311.143	$6^8P - 6^8S^0_{7/2}$	3.98	8	1.10(9)*	6.30(−12)	1.07(14)	0.86	0.157
Eu	153	321.281	$6^8P - 6^8S^0_{7/2}$	3.86	8	9.60(8)*	6.05(−12)	1.31(14)	0.56	0.125
Eu	153	333.433	$6^8P - 6^8S^0_{7/2}$	3.72	8	6.60(8)*	4.65(−12)	1.62(14)	0.67	0.147
Eu	153	459.403	$6^8P - 6^8S^0_{7/2}$	2.70	8	6.70(8)*	1.23(−11)	4.79(14)	0.80	1.90
Eu	153	462.722	$6^8P - 6^8S^0_{7/2}$	2.68	8	5.60(8)*	1.05(−11)	4.80(14)	0.97	1.99
Eu	153	466.188	$6^8P - 6^8S^0_{7/2}$	2.66	8	4.60(8)*	8.90(−12)	4.82(14)	0.62	1.08
Eu	153	564.580	$6^8P_{5.2} - 6^8S^0_{7/2}$	2.20	6 / 8	8.30(5)	1.71(−13)	5.17(14)	0.92	0.04
Eu	153	576.520	$6^8P_{7.2} - 6^8S^0_{7/2}$	2.15	8 / 8	9.00(5)	2.62(−13)	5.27(14)	1.0	0.07
Eu	153	601.815	$6^8P_{9.2} - 6^8S^0_{7/2}$	2.05	10 / 8	7.40(5)	3.06(−13)	5.29(14)	1.0	0.085
Eu^+	153	368.842	$6^9P - 6^9S^0_4$	3.36	9	9.70(7)*	8.22(−13)	1.99(14)	0.43	0.023
Eu^+	153	372.494	$6^9P - 6^9S^0_4$	3.32	9	2.80(8)*	2.45(−12)	2.00(14)	0.75	0.120
Eu^+	153	381.967	$6^9P_5 - 6^9S^0_4$	3.24	11 / 9	4.36(7)	3.67(−12)	2.00(14)	0.30	0.091
Eu^+	153	412.970	$6^9P - 6^9S^0_4$	3.00	9	1.90(8)*	2.24(−12)	3.65(14)	0.65	0.194
Eu^+	153	420.505	$6^9P_3 - 6^9S^0_4$	2.95	7 / 9	4.57(7)	4.3(−12)	3.72(14)	0.53	0.291

* $= A_{ul} \cdot g_u$

Table 2.22 Rates of photoionization of metal atoms in the sunlight and effective cross-sections for absorption

Atom	M	E_{ion} (eV)	λ_{ion} (nm)	$S(\lambda)$	σ_{ion} (cm^2)	$j(s^{-1})$	σ(cm^2)
LiI	7	5.390	230.3	7.93(12)	1.7(−18)	2.4(−4)	4(−24)
NaI	23	5.138	241.6	7.78(12)	1.2(−19)	2.0(−5)	2(−24)
MgI	24	7.644	162.4	1.70(10)	1.2(−18)	4.0(−7)	4(−18)
MgII	24	15.03	82.6	1.50(9)	−	−	3(−17)
KI	39	4.339	286.0	2.64(13)	1.2(−20)	3.3(−5)	1(−25)
CaI	40	6.111	203.1	8.50(11)	4.5(−19)	3.4(−5)	1(−23)
CaII	40	11.87	104.6	4.17(9)	1.7(−19)	2.0(−8)	2(−18)
FeI	56	7.896	157.2	1.30(10)	6.0(−18)	5.3(−7)	5(−18)
SrI	88	5.692	218.0	5.36(12)	−	2.2(−4)	7(−24)
SrII	88	11.03	112.5	6.50(8)	−	−	5(−19)
CsI	133	3.893	318.8	1.20(14)	2.2(−19)	3.0(−4)	1(−25)
BaI	138	5.210	238.2	7.43(12)	−	5.1(−4)	3(−24)
BaII	138	10.01	124.0	3.10(11)	−	−	2(−19)
EuI	153	5.67	218.9	6.48(12)	−	−	7(−24)
EuII	153	11.24	109.8	7.20(8)	−	−	7(−19)

the coefficient of molecular diffusion D), the diffusion separation begins at higher altitudes, resulting in smaller densities of atmospheric components at high altitudes. The altitude of the layer of maximum intensity of sodium emission is very sensitive to the value of K_T, and this value affects the transport of sodium from the region of intrusion of meteoric matter to the altitude of the sodium layer in the Earth's atmosphere (Kirchhoff and Clemesha 1983).

As a first approximation, K_T can be estimated by the data on stratification of the sodium layer (Hunten and Wallace 1967):

$$K_T = \frac{L^2}{t}.$$

Here L is the distance between the maxima of the layer vertical structure and t is the time during which this structure persists. According to the measurements performed at Abastumani, the multilayered structure of the sodium layer repeatedly remained rather stable at least for 60–70 (h) (Toroshelidze 1991). According to these measurements, L was about 6 (km) and t varied from 24 to 70 (h). This gives the estimate $K_T \sim (1.5 - 4) \cdot 10^6 (\text{cm}^2 \cdot \text{s}^{-1})$, which agrees with the results obtained by other methods (Hocking 1985). The use of the altitude distribution of radiation intensity also allows one to obtain data about altitude variations of the eddy diffusion coefficient (Fig. 2.18) (Toroshelidze 1991).

The resonance lines of metals, preferentially of alkali metals, such as Li, K, Ca, Ca^+, Mg, Mg^+, Ba, and Ba^+, have a special place in weak emissions. Some of these species are of natural origin, entering the upper atmosphere with meteors, while others are dispersed artificially to investigate diffusion, winds, etc. Owing to the high efficiency of the fluorescent mechanism of excitation, many metals are used to study the dynamic characteristics of the upper atmosphere. For this purpose they are

2.3 Emissions from Metal Atoms

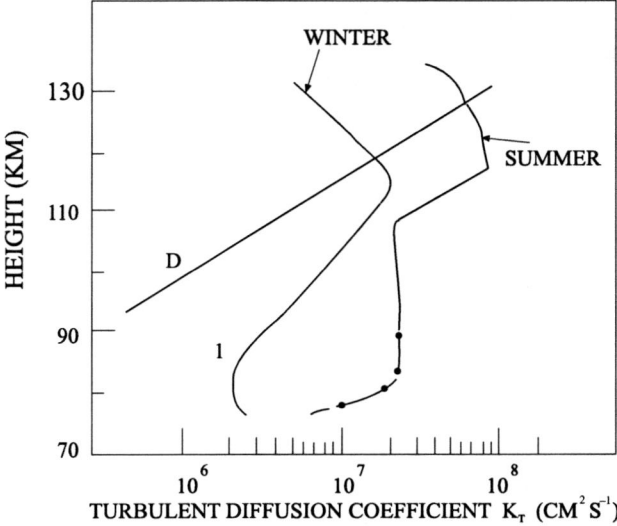

Fig. 2.18 Altitude distribution of the eddy diffusion coefficient K_T for winter and summer, derived from the data obtained at Abastumani from 1962 to 1972. The variations in molecular diffusion (Nicolet 1962) are shown by a *straight line* (Toroshelidze 1991)

released to form artificial luminous clouds (Filipp et al. 1986; Andreeva et al. 1991). These problems are considered in Sect. 3.5.9.

Sporadic emissions were also the subject of active research (Vallance Jones 1958; Link 1966; Gadsden 1969; Mitra 1974a,b; Granier et al. 1985, 1989; Grebowsky and Reese 1989; Toroshelidze 1991; IAPh 1994).

The greatest attention was paid to the data on the 670.8-nm emission of lithium obtained under twilight conditions since its occurrence accompanied the nuclear tests conducted in the atmosphere from the late 1950s to the early 1960s (Gadsden and Salmon 1958; Delannoy 1960). Besides, Li was always present in small amounts in the artificial clouds produced by releasing sodium. In contrast to the emissions of Na, for which the residual intensities in the Fraunhofer spectrum of the Sun make about 6(%), for the emission of Li the solar spectrum is impaired insignificantly. Therefore, the scattering factor for Li is much greater than that for Na. This feature made it possible to estimate the wind speed in the upper atmosphere at the altitudes where Li was nebulized.

Since the emission of lithium testified that a nuclear explosion had been effected, special interferometric measurements were performed to determine the isotopic composition for the lithium emission. As mentioned earlier, the nuclear charge shell made of hard material, D^6Li, contains the isotope ^6Li whose participation in a nuclear explosion results in the formation of tritium, which is necessary for the subsequent nuclear reaction to occur. Figure 2.19 shows interferograms of the 670.8-nm emission of lithium (IAPh 1994). As can be seen, each group of rings in the interferograms contains three emission lines. The wavelengths of the doublet

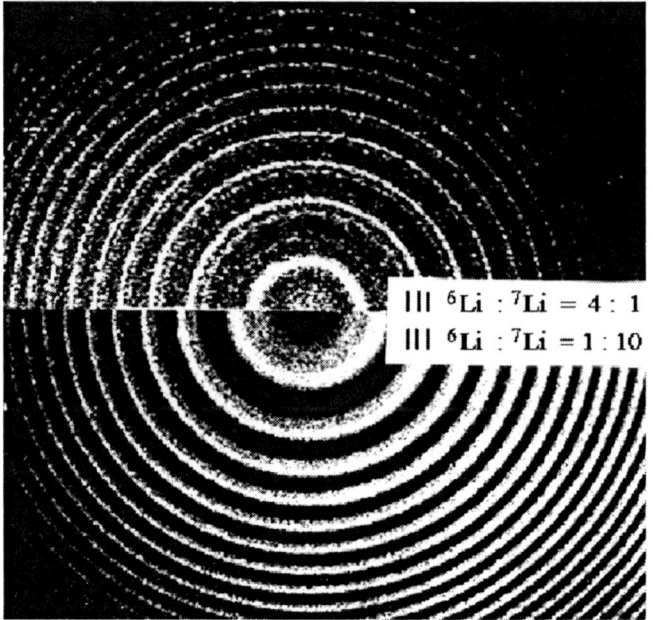

Fig. 2.19 Comparison of two interferograms for emissions of the lithium isotopes ^{7}Li and ^{6}Li

of the isotope ^{6}Li are shifted relative to those of the isotope ^{7}Li to the long-wave spectral range by 0.0154 (nm), i.e., practically by the width of the doublet interval (0.0158 (nm)) for both isotopes (Kaliteevsky and Chaika 1970). As a consequence, three rings are seen, the inner belonging to ^{6}Li. Therefore, in the interferogram corresponding to the atmospheric tests, the inner ring (smaller component) refers to ^{6}Li and the outer to ^{7}Li (greater component), while the middle ring is associated with the total of ^{7}Li and ^{6}Li. A natural sample of lithium contains both isotopes in the proportion ^{7}Li:^{6}Li $= 10 : 1$. In the atmosphere, within several days after a nuclear explosion (e.g., in 1962 over the Johnston island in Pacific ocean), a different intensity distribution was observed which corresponded to the proportion ^{7}Li:^{6}Li $= 1 : 4$. The intensity of the lithium emission observed in this case was several kilorayleighs.

An exotic radiation which sporadically appears in the atmosphere at altitudes of 80–100 (km) results from iron-resonant emissions at $\lambda = 385.991$ (nm) [($z^5D_4^o - a^5D_4$), $A = 1.6 \cdot 10^7 (s^{-1})$] and 371.994 (nm) [($z^5F_4^o - a^5D_4$), $A = 2.5 \cdot 10^7 (s^{-1})$]. The first of them was first detected by Broadfoot and Johanson (1976) and later it was investigated by other researchers (Tepley et al. 1981). The intensity of these emissions varied from 2 to 20 (Rayleigh). In the subsequent years, the altitude distribution of the atomic iron column density was analyzed with the help of laser sounding (Granier et al. 1989; Alpers et al. 1990). It turned out that the column content of iron atoms varies, usually being $(3-4) \cdot 10^9 (cm^{-2})$; the iron concentration is about $3 \cdot 10^3 (cm^{-3})$. However, it can increase 5–8 times, and layers of thickness about 2–3 (km) with a maximum concentration of $8 \cdot 10^4 (cm^{-3})$ can be formed. As a result of observations it was repeatedly noted that the layer of iron atoms was located several kilometers below the sodium layer, which, probably, is related to photochemical processes in the atmosphere.

2.4 Molecular Oxygen Emissions

Molecular oxygen is undoubtedly the major constituent of the Earth's atmosphere. It not only absorbs the ultraviolet radiation of the Sun of wavelength shorter than 300 (nm), but also participates in reactions giving rise to numerous chemically active components. The oxygen molecule has only one electronic state from which an emission not forbidden by usual rules is possible. The other six electronic states are metastable, and the transitions between them are responsible for nine systems of bands. Six of them are in the ultraviolet spectral region and three in the infrared one. Figure 2.20 presents potential curves plotted by the data of Gilmore (1965). Since the potential curves of the upper excited states are considerably shifted relative to those of the lower excited states, in the Delandres table containing vibrational transitions the bands with maximum probabilities are placed, according to the Franck–Condon principle, along the Condon parabola expanded with respect to the main diagonal. Figure 2.21 shows systems of transitions between various states.

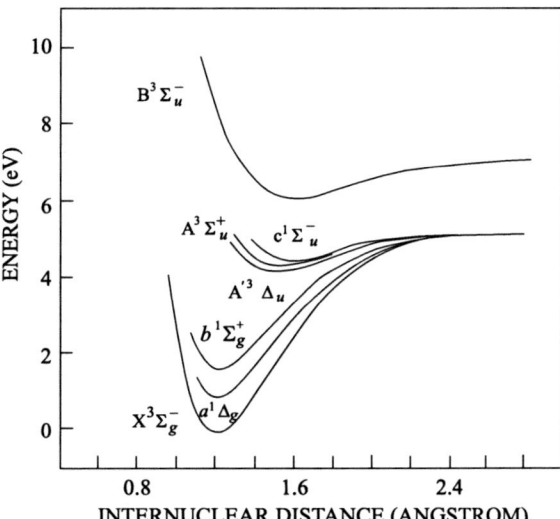

Fig. 2.20 Potential curves of the oxygen molecule

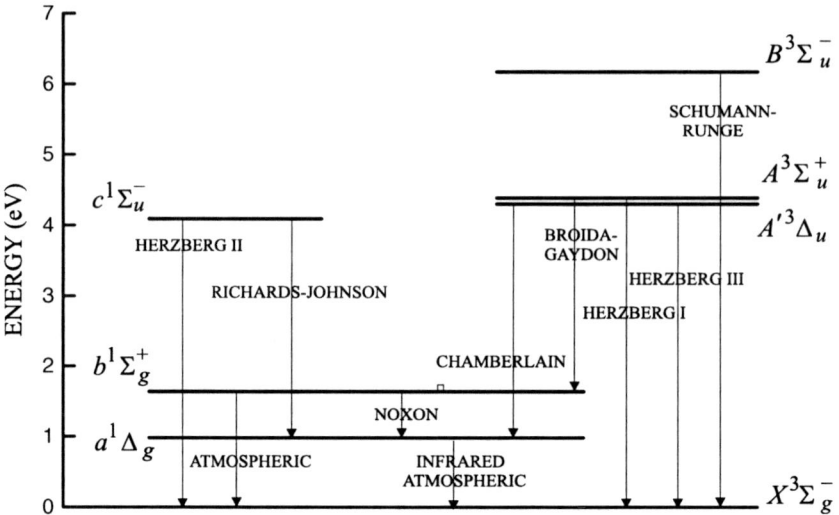

Fig. 2.21 Schematic of transitions between various electronic states

2.4.1 Ultraviolet Systems

The oxygen molecule has some electronic transitions responsible for the ultraviolet emission bands in the spectral range 200–550 (nm). They have the following names and designations:

Herzberg I system: $A^3\Sigma_u^+ - X^3\Sigma_g^-$, 240–520 (nm), $A(AX) = 15\,(s^{-1})$
Herzberg II system: $c^1\Sigma_u^+ - X^3\Sigma_g^-$, 250–530 (nm), $A(cX) = 1.0\,(s^{-1})$
Herzberg III system: $A'^3\Delta_u - X^3\Sigma_g^-$, 260–600 (nm), $A(A'X) = 1.2\,(s^{-1})$
Chamberlain system: $A'^3\Delta_u - a^1\Delta_g$, 300–870 (nm), $A(A'a) = 0.89\,(s^{-1})$
Broida–Gaydon system: $A^3\Sigma_u^+ - b^1\Sigma_g^+$, 300–1100 (nm), $A(Ab) =$
Richards–Johnson system: $c^1\Sigma_u^+ - a^1\Delta_g$, 280–1000 (nm), $A(ca) =$
Schumann–Runge system: $B^3\Sigma_u^- X^3\Sigma_g^-$, 137–200 (nm), $A(BX) \approx 2 \cdot 10^7\,(s^{-1})$

All of them correspond to initial metastable states (Bates 1988a).

Investigations of the molecular oxygen emissions in the ultraviolet spectral region have demonstrated that they occur as a result of recombination of atomic oxygen (Barth 1964):

$$O + O + M \rightarrow O_2^* + M, \quad \alpha_{O_2} = 5.5 \cdot 10^{-33} \cdot (200/T)^2 (cm^3 \cdot s^{-1}).$$

In the reaction of production of O_2^* excited molecules, several metastable electron states, such as $^5\Pi_g$, $A^3\Sigma_u^+$, $A'^3\Delta_u$, $c^1\Sigma_u^-$, $b^1\Sigma_g^+$, $a^1\Delta_g$, $X^3\Sigma_g^-$, are realized in the proportion 0.5:0.06:0.18:0.04:0.03:0.07:0.12, respectively (Bates 1988a). The lifetime of the molecules in these metastable states is determined by the collisions (Bates 1988a; Johnston and Broadfoot 1993)

2.4 Molecular Oxygen Emissions

$$O_2(A^3\Sigma_u^+) + O_2 \to O_2(X^3\Sigma_g^-) + O_2 \quad \beta_{O_2}(A^3\Sigma_u^+) = 1.5 \cdot 10^{-10} \, (cm^3 \cdot s^{-1}),$$

$$O_2(A^3\Sigma_u^+) + N_2 \to O_2(X^3\Sigma_g^-) + N_2 \quad \beta_{N_2}(A^3\Sigma_u^+) = 3.8 \cdot 10^{-11} \, (cm^3 \cdot s^{-1}),$$

$$O_2(A^3\Sigma_u^+) + O \to O_2(X^3\Sigma_g^-) + O \quad \beta_O(A^3\Sigma_u^+) = 1.5 \cdot 10^{-11} \, (cm^3 \cdot s^{-1}),$$

$$O_2(A^3\Sigma_u^+) \to O_2(X^3\Sigma_g^-) + h\nu(\lambda 240 - 520\,(nm)) \quad A(AX) = 15\,(s^{-1}) \text{ Herzberg I};$$

$$O_2(A'^3\Delta_u) + O_2 \to O_2(X^3\Sigma_g^-) + O_2 \quad \beta_{O_2}(A'^3\Delta_u) = 7.4 \cdot 10^{-11} \, (cm^3 \cdot s^{-1}),$$

$$O_2(A'^3\Delta_u) + N_2 \to O_2(X^3\Sigma_g^-) + N_2 \quad \beta_{N_2}(A'^3\Delta_u) = 2.0 \cdot 10^{-11} \, (cm^3 \cdot s^{-1}),$$

$$O_2(A'^3\Delta_u) + O \to O_2(X^3\Sigma_g^-) + O \quad \beta_O(A'^3\Delta_u) = 1.5 \cdot 10^{-11} \, (cm^3 \cdot s^{-1}),$$

$$O_2(A'^3\Delta_g) \to O_2(X^3\Sigma_g^-) + h\nu(\lambda 240 - 520\,(nm)) \quad A(A'X) = 1.2\,(s^{-1}) \text{ Herzberg III};$$

$$O_2(A'^3\Delta_g) \to O_2(a^1\Delta_g) + h\nu(\lambda 300 - 870\,(nm)) \quad A(A'a) = 0.89\,(s^{-1}) \text{ Chamberlain};$$

$$O_2(c^1\Sigma_u^-) + O_2 \to O_2(X^3\Sigma_g^-) + O_2 \quad \beta_{O_2}(c^1\Sigma_u^-) = 1.6 \cdot 10^{-11} \, (cm^3 \cdot s^{-1}),$$

$$O_2(c^1\Sigma_u^-) + N_2 \to O_2(X^3\Sigma_g^-) + N_2 \quad \beta_{N_2}(c^1\Sigma_u^-) = 4.3 \cdot 10^{-12} \, (cm^3 \cdot s^{-1}),$$

$$O_2(c^1\Sigma_u^-) + O \to O_2(X^3\Sigma_g^-) + O \quad \beta_O(c^1\Sigma_u^-) = 2.7 \cdot 10^{-11} \, (cm^3 \cdot s^{-1}),$$

$$O_2(c^1\Sigma_u^-) \to O_2(X^3\Sigma_g^-) + h\nu(\lambda 250 - 530\,(nm)) \quad A(cX) = 1.0\,(s^{-1}) \text{ Herzberg II}.$$

In these terms, the emission rate (photon $\cdot$ cm$^{-3} \cdot$ s^{-1}) of any system is determined as

$$Q(ki) = \frac{\alpha_{O_2} \cdot [O]^2 \cdot [M]}{1 + \{\beta_{O_2}(k) \cdot [O_2] + \beta_{N_2}(k) \cdot [N_2] + \beta_O(k) \cdot [O]\}/A(ki)}.$$

In actual conditions, the intensities of these band systems are rather insignificant. Their absolute values were determined by a number of researchers (Yarin 1961b, 1962c; Hennes 1966; Degen 1968, 1969, 1977; Degen and Nicholls 1969; Hasson et al. 1970; Slanger and Huestis 1981; Torr et al. 1985; Sharp 1986; McDade and Llewellyn 1986; Saxon and Slanger 1986; Ogawa et al. 1987; Swenson et al. 1989). As follows from the measurements for the spectral range 310–390 (nm), the average intensity is 0.09 (Rayleigh$\cdot$nm^{-1}). The intensity of the Herzberg I system as a whole is ~600 (Rayleigh). The Herzberg II and III systems and the Chamberlain system are weaker, their intensities being about 100, 70, and 200 (Rayleigh), respectively (Krasnopolsky 1987; Sharp and Siskind 1989). Tables 2.23–2.25 list the wavelengths of vibrational–rotational bands of the Herzberg I and II systems and of the Chamberlain system. Figure 2.22 presents spectrograms of these systems which are shown in Figs. 1.1 and 1.2 in integrated form. In Table 2.26, the probabilities of transitions are given for the bands of the Schumann–Runge system.

The distributions of the populations of the excited molecules over vibrational levels for the states $A^3\Sigma_u^+$, $A'^3\Delta_u$, $c^1\Sigma_g^-$ have maxima for $v \sim 5 \div 7$ and show an abrupt fall in going toward the lower ($v \sim 0$–1) and upper levels ($v \sim 10$) (Stegman 1991). The population distributions based on these data are presented in Fig. 2.23. The altitude distributions of the integrated emission rate have maxima near 97 (km) and at layer thickness $W \sim 10$ (km). These data, based on the results reported elsewhere (Krasnopolsky 1987), are presented in Fig. 2.24.

Table 2.23 Wavelengths (nm) and intensities (Rayleigh) of the Herzberg I system bands for O_2 (the total intensity is taken equal to 1000 (Rayleigh)) (Stegman 1991)

v''/v'	0	1	2	3	4	5	6	7	8	9	10	11
0	285.6	298.8	313.2	328.8	345.7	364.1	384.2	406.4	430.7	457.7	487.6	521.1
	0.0	0.0	0.0	0.1	0.4	0.9	1.8	2.7	3.5	3.5	2.9	2.1
1	279.4	292.1	305.8	320.6	336.7	354.7	373.1	394.0	416.8	442.0	469.9	500.8
	0.0	0.0	0.3	1.3	3.3	6.1	8.4	8.1	5.4	1.7	0.1	0.5
2	273.7	285.9	299.0	313.2	328.4	345.0	363.1	382.7	404.3	427.9	454.0	482.9
	0.0	0.3	1.9	6.0	12.0	16.7	14.7	6.5	0.5	1.1	4.5	4.9
3	268.5	280.2	292.8	306.3	320.9	336.8	353.9	372.6	393.0	415.3	439.8	466.8
	0.2	1.6	7.3	19.1	29.1	26.9	11.2	0.3	4.0	10.3	7.1	0.8
4	263.7	275.0	287.1	300.1	314.1	329.3	345.7	363.4	382.8	404.0	427.1	452.5
	0.4	3.9	15.4	30.8	36.0	19.3	1.4	4.2	12.8	8.1	0.3	3.1
5	259.4	270.3	282.0	294.5	308.0	322.5	338.2	355.2	373.7	393.8	415.8	439.9
	0.8	6.6	20.7	35.4	28.0	6.1	1.2	12.0	10.1	0.7	3.0	7.0
6	255.4	266.0	277.3	289.4	302.5	316.5	331.6	347.9	365.6	384.9	405.8	428.7
	2.2	15.3	42.8	53.9	27.9	0.5	11.7	20.5	4.7	2.2	11.3	5.2
7	251.9	262.2	273.2	285.0	297.6	311.1	325.7	341.5	358.5	377.0	397.1	419.0
	2.8	17.4	39.6	39.6	10.8	1.4	15.8	11.1	0.0	7.8	7.6	0.1
8	248.9	258.9	269.6	281.1	293.3	306.5	320.6	335.9	352.4	370.2	389.6	410.6
	1.6	8.7	17.4	12.7	1.3	2.6	7.4	1.7	1.0	4.1	1.1	0.6
9	246.3	256.2	266.6	277.8	289.8	302.6	316.4	331.2	347.3	364.6	383.3	403.7
	1.7	8.2	13.5	7.6	0.1	3.6	4.6	0.2	2.0	2.6	0.0	1.5
10	244.3	254.0	264.3	275.3	287.0	299.6	313.1	327.6	343.3	360.2	378/5	398.3
	1.3	6.0	8.9	3.7	0.0	3.0	2.3	0.0	1.9	1.1	0.1	1.3
11	242.9	252.5	262.7	273.5	285.1	297.5	310.8	325.1	340.5	357.2	375.1	294.6
	0.2	1.0	1.3	0.5	0.0	0.5	0.3	0.0	0.3	0.1	0.0	0.2

2.4 Molecular Oxygen Emissions

Table 2.24 Wavelengths (nm) and intensities (Rayleigh) of the Herzberg II system bands for O_2 (the total intensity is taken equal to 1000 (Rayleigh)) (Stegman 1991)

v''/v'	0	1	2	3	4	5	6	7	8	9	10	11
0	286.6	299.9	314.4	330.1	347.1	365.6	386.0	408.3	432.9	460.1	490.4	524.3
	0.0	0.2	0.7	1.6	2.2	1.6	0.5	0.0	0.3	0.4	0.2	0.0
1	281.1	293.9	307.8	322.8	339.1	356.8	376.1	397.3	420.5	446.2	474.6	506.2
	0.1	0.8	2.7	4.7	4.5	1.7	0.0	0.7	1.2	0.4	0.0	0.3
2	276.0	288.4	301.8	316.2	331.8	348.7	367.1	387.3	409.3	433.6	460.4	490.1
	0.4	3.0	8.3	11.3	7.1	0.8	0.9	3.0	1.5	0.0	0.7	0.9
3	271.4	283.4	296.3	310.1	325.1	341.4	359.0	378.3	399.3	422.3	447.7	475.8
	1.3	8.2	18.7	18.7	6.5	0.1	5.0	4.9	0.3	1.1	2.2	0.4
4	267.2	278.8	291.3	304.7	319.1	334.7	351.7	370.1	290.2	412.2	436.4	463.0
	3.8	19.2	36.7	26.2	3.7	3.5	11.7	3.8	0.6	4.5	1.9	0.1
5	263.4	274.6	286.7	299.7	313.7	328.8	345.1	362.8	382.1	403.2	426.3	451.6
	6.2	28.2	44.3	22.1	0.2	10.1	11.9	0.5	4.0	5.0	0.1	1.9
6	259.9	270.9	282.6	295.2	308.8	323.4	339.1	356.3	374.9	395.1	417.2	441.5
	8.3	32.7	42.3	13.1	0.9	13.6	6.9	0.4	6.5	2.3	0.6	2.9
7	256.8	267.5	278.9	291.2	304.3	318.5	333.8	350.4	368.4	387.9	409.2	432.5
	16.2	56.1	59.1	9.7	6.2	22.1	4.1	4.1	10.0	0.5	3.5	3.5
8	254.0	264.5	275.6	287.6	300.4	314.2	329.1	345.2	362.7	381.6	402.2	424.7
	16.5	50.7	43.1	3.0	10.6	16.7	0.5	7.1	5.8	0.2	4.6	1.1
9	251.5	261.8	272.7	284.4	297.0	310.5	325.0	340.7	357.6	376.0	396.0	417.8
	8.2	22.8	15.2	0.2	6.1	5.2	0.0	3.8	1.3	0.6	1.8	0.0
10	249.4	249.4	270.2	281.7	294.0	307.2	321.4	336.7	353.3	371.2	390.7	411.9
	7.7	19.3	10.6	0.0	6.0	3.0	0.5	3.3	0.4	1.2	1.2	0.0

Thus, it has been shown that the characteristics of ultraviolet emissions, irrespective of detailed features of each of them, are determined by the concentration of atomic oxygen because they are manifestations of the process of its recombination. Therefore, joint measurements of the time variations and altitude distributions of the emission rate enable one to study the properties of atomic oxygen.

2.4.2 The Atmospheric System

The emission from the $O_2(b^1\Sigma_g^+) \rightarrow O_2(X^3\Sigma_g^-)$ Atmospheric system of molecular oxygen was identified in the nightglow simultaneously with hydroxyl emission (Meinel 1950d). The bands of this system are presented in Table 2.27. The most intense transition occurs in the 761.9-nm (0–0) band. The 864.5-nm (0–1) band is the second in significance. Other bands are practically not observed in the nightglow atmosphere since the populations of the levels with $v' > 0$ are almost two orders of magnitude lower than those of the level with $v' = 0$; hence, their intensities are small (Slanger et al. 2000).

Table 2.25 Wavelengths (nm) and intensities (Rayleigh) of the bands of the Chamberlain system for O_2 (the total intensity is taken equal to 1000 (Rayleigh)) (Stegman 1991)

v''/v'	0	1	2	3	4	5	6	7	8	9	10	11
0	379.5	402.1	427.2	455.0	486.1	521.0	560.4	605.3	656.8	716.5	786.2	868.8
	0.0	0.0	0.0	0.1	0.2	0.5	0.8	1.1	1.1	0.9	0.5	0.3
1	368.4	389.7	41.32	439.2	468.1	500.4	536.6	577.7	624.4	678.1	740.2	813.0
	0.0	0.1	0.3	1.0	2.1	3.2	3.2	2.0	0.7	0.0	0.2	0.5
2	358.4	378.5	400.6	425.0	452.0	482.0	515.6	553.3	596.1	644.8	700.8	765.7
	0.0	0.5	2.1	5.1	8.3	7.4	3.6	0.4	0.4	1.7	1.6	0.4
3	349.3	368.4	389.3	412.2	437.6	465.6	496.9	531.9	571.3	615.8	666.7	725.2
	0.3	2.3	8.0	16.1	16.9	8.0	0.5	1.6	4.5	2.6	0.1	0.8
4	341.0	359.2	379.0	400.7	424.7	541.0	480.3	512.9	549.4	590.6	637.2	690.4
	1.1	7.7	23.0	32.5	21.7	2.8	2.3	9.3	5.1	0.0	2.6	3.3
5	333.5	350.9	369.8	390.5	413.1	438.1	465.6	496.2	530.3	568.5	611.6	660.4
	2.5	15.4	37.4	39.0	12.6	0.2	10.9	9.5	0.3	3.4	5.0	0.5
6	326.8	343.4	361.5	381.3	402.8	426.5	452.6	481.5	513.5	549.2	589.3	634.5
	4.0	20.9	40.2	28.3	2.8	5.2	13.1	2.8	1.8	6.1	1.3	0.9
7	320.8	336.8	354.2	373.1	393.8	416.4	441.2	468.6	498.8	532.5	570.1	612.3
	9.4	41.2	64.1	29.7	0.0	17.8	14.9	0.0	9.2	5.9	0.2	5.0
8	315.5	331.0	347.8	366.0	385.9	407.5	431.3	457.4	486.2	518.2	553.7	593.4
	11.0	43.2	51.1	13.0	2.8	18.5	5.5	2.4	9.2	1.1	2.8	3.9
9	311.0	326.0	342.3	359.9	379.1	400.0	422.9	447.9	475.5	506.0	539.9	577.6
	5.7	19.4	17.7	2.1	3.4	7.2	0.4	2.7	2.8	0.0	2.1	0.6
10	30.72	321.9	337.7	354.9	373.5	393.8	415.9	440.1	466.8	496.1	528.6	564.7
	5.5	16.5	12.0	0.4	4.2	4.5	0.0	3.0	1.2	0.5	1.6	0.0
11	304.2	318.6	334.1	350.9	369.1	388.9	410.5	434.0	459.9	488.4	519.8	554.7
	4.8	12.9	8.1	0.0	3.9	2.5	0.2	2.4	0.3	0.9	0.9	0.1

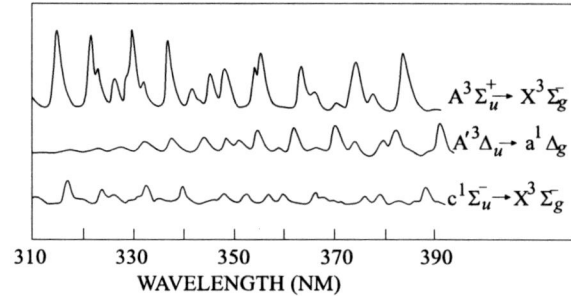

Fig. 2.22 Calculated spectral distributions of the intensities of the ultraviolet bands of the Herzberg I and II systems and of the Chamberlain system for molecular oxygen (Krasnopolsky 1987)

2.4 Molecular Oxygen Emissions

Table 2.26 Transitions probabilities (s^{-1}) for the bands of the Schumann–Runge system of molecular oxygen (Allison et al. 1971)

v'/v''	0	1	2	3	4	5	6	7	8	9	10
0	4.51(−1)	1.15(1)	1.41(2)	1.09(3)	6.02(3)	2.52(4)	8.29(4)	2.20(5)	4.81(5)	8.71(5)	1.32(6)
1	5.52(0)	1.29(2)	1.41(3)	9.75(3)	4.72(4)	1.70(5)	4.74(5)	1.04(6)	1.79(6)	2.45(6)	2.60(6)
2	3.48(1)	7.38(2)	7.30(3)	4.47(4)	1.90(5)	5.87(5)	1.36(6)	2.40(6)	3.16(6)	3.02(6)	1.88(6)
3	1.52(2)	2.93(3)	2.61(4)	1.42(5)	5.25(5)	1.38(6)	2.64(6)	3.63(6)	3.41(6)	1.92(6)	3.71(5)
4	5.16(2)	9.07(3)	7.29(4)	3.52(5)	1.13(6)	2.52(6)	3.88(6)	3.96(6)	2.34(6)	4.46(5)	8.45(4)
5	1.45(3)	2.34(4)	1.69(5)	7.26(5)	2.02(6)	3.75(6)	4.53(6)	3.19(6)	8.49(5)	2.49(4)	9.51(5)
6	3.48(3)	5.12(4)	3.36(5)	1.28(6)	3.07(6)	4.70(6)	4.29(6)	1.78(6)	3.48(4)	7.22(5)	1.57(6)
7	7.20(3)	9.75(4)	5.79(4)	1.96(6)	4.04(6)	5.02(6)	3.25(6)	5.40(5)	2.51(5)	1.56(6)	1.24(6)
8	1.31(4)	1.64(5)	8.83(5)	2.65(6)	4.69(6)	4.64(6)	1.92(6)	1.23(4)	1.02(6)	1.75(6)	4.80(5)
9	2.14(4)	2.48(5)	1.22(6)	3.25(6)	4.90(6)	3.75(6)	8.10(5)	1.88(5)	1.64(6)	1.27(6)	3.03(4)
10	3.19(4)	3.43(5)	1.54(6)	3.67(6)	4.70(6)	2.69(6)	1.78(5)	7.03(5)	1.79(6)	6.13(5)	8.72(4)
11	4.38(4)	4.39(5)	1.82(6)	3.88(6)	4.22(6)	1.72(6)	1.37(1)	1.19(6)	1.52(6)	1.58(5)	4.02(5)
12	5.54(4)	5.22(5)	2.00(6)	3.84(6)	3.55(6)	9.73(5)	1.06(5)	1.45(6)	1.07(6)	2.28(3)	6.76(5)
13	6.44(4)	5.74(5)	2.06(6)	3.60(6)	2.84(6)	4.79(5)	3.09(5)	1.45(6)	6.37(5)	4.98(4)	7.75(5)
14	6.89(4)	5.86(5)	1.98(6)	3.18(6)	2.16(6)	1.99(5)	4.78(5)	1.27(6)	3.26(5)	1.61(5)	7.14(5)
15	6.78(4)	5.55(5)	1.78(6)	2.66(6)	1.57(6)	6.62(4)	5.54(5)	1.02(6)	1.44(5)	2.45(5)	5.73(5)
16	6.29(4)	4.99(5)	1.53(6)	2.16(6)	1.14(6)	1.54(4)	5.53(5)	7.78(5)	5.55(4)	2.78(5)	4.29(5)
17	5.52(4)	4.27(5)	1.27(6)	1.70(6)	8.07(5)	1.21(3)	5.00(5)	5.74(5)	1.76(4)	2.69(5)	3.06(5)
18	4.52(4)	3.43(5)	9.93(5)	1.28(6)	5.59(5)	3.78(2)	4.12(5)	4.06(5)	4.12(3)	2.31(5)	2.09(5)
19	3.46(4)	2.58(5)	7.35(5)	9.20(5)	3.76(5)	2.27(3)	3.15(5)	2.77(5)	4.91(2)	1.80(5)	1.38(5)
20	2.31(4)	1.71(5)	4.79(5)	5.87(5)	2.29(5)	3.03(3)	2.09(5)	1.70(5)	2.59(−3)	1.21(5)	8.25(4)
21	1.47(4)	1.08(5)	3.00(5)	3.64(5)	1.38(5)	2.65(3)	1.32(5)	1.03(5)	7.16(1)	7.72(4)	4.89(4)
Σ	6.15(5)	5.42(6)	1.98(7)	3.82(7)	4.29(7)	3.27(7)	2.70(7)	2.64(7)	2.06(7)	1.64(7)	1.50(7)

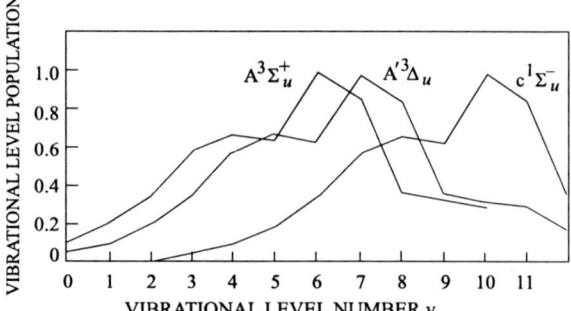

Fig. 2.23 Distributions of the populations of the excited molecules over vibrational levels for the states $A^3\Sigma_u^+$, $A'^3\Delta_u$, $c^1\Sigma_g^-$ according to the data of Stegman (1991)

The studies of the emission from the bands of this system in the upper atmosphere have shown that the range of altitudes of occurrence of this emission is 90–100 (km) (Hunten 1967; Izod and Wayne 1968; Llewellyn and Evans 1971; Bates 1982). The altitude distribution of the emission rate is shown in Fig. 2.25 (Tarasova 1962; Krasnopolsky 1987).

According to the present-day theoretical notions, several mechanisms of the formation of metastable molecules $O_2\left(b^1\Sigma_g^+\right)$, whose radiative lifetime is about 12 (s), are considered (Bates 1954, 1982; Wallace and Hunten 1968; Greer et al. 1981). Under the conditions of the night atmosphere (Bates 1954), the reaction considered above,

$$O + O + M \rightarrow O_2^* + M,$$

produces excited molecules, including $O_2\left(c^1\Sigma_u^-\right)$ (Greer et al. 1981; Bates 1982), which radiate the Herzberg II system bands.

As a result of deactivation of these molecules by the processes

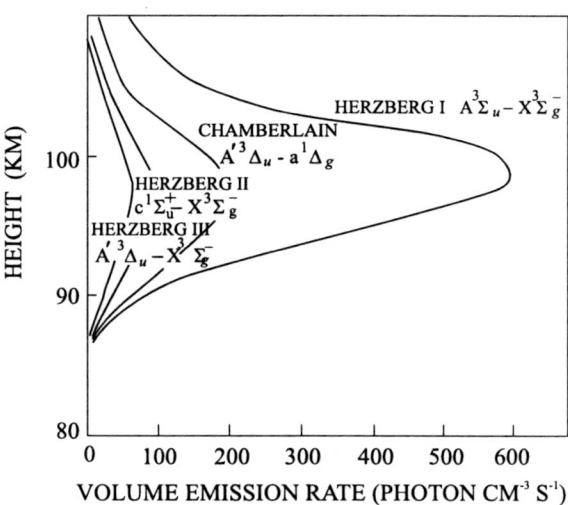

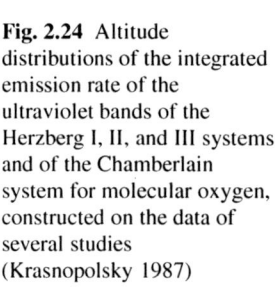

Fig. 2.24 Altitude distributions of the integrated emission rate of the ultraviolet bands of the Herzberg I, II, and III systems and of the Chamberlain system for molecular oxygen, constructed on the data of several studies (Krasnopolsky 1987)

2.4 Molecular Oxygen Emissions

Table 2.27 The wavelengths λ (nm), absolute transition probabilities A (s^{-1}), coefficients of scattering g (s^{-1}), and relative intensities I(%) of the Atmospheric system bands of molecular oxygen

v''/v'	0	1	2
0	759–773 7.93(−2) 8.5(−9) 88.54	862–870 3.91(−3) 4.2(−10) 4.38	994–1002 9.71(−5) 1.0(−11) 0.104
1	687–697 7.99(−3) 6.0(−11) 0.625	768–777 6.73(−2) 5.0(−10) 5.21	871–880 7.75(−3) 5.8(−11) 0.604
2	628–635 4.34(−4) 1.4(−13) 0.0015	670–700 1.57(−2) 5.2(−12) 0.054	777–785 5.52(−2) 1.8(−11) 0.188

Transition probability $A_{bX} = 0.087 (s^{-1})$

$$O_2\left(c^1\Sigma_u^-\right) + O_2 \rightarrow O_2\left(b^1\Sigma_g^+\right) + O_2 \quad \alpha_{O_2}(b^1\Sigma_g^+) = 5 \cdot 10^{-13}\ (cm^3 \cdot s^{-1}),$$

$$O_2\left(c^1\Sigma_u^-\right) + O \rightarrow O_2\left(b^1\Sigma_g^+\right) + O \quad \alpha_O(b^1\Sigma_g^+) = 3 \cdot 10^{-11}\ (cm^3 \cdot s^{-1}),$$

$O_2\left(b^1\Sigma_g^+\right)$ molecules can be formed. The subsequent deactivation of the excited states of the O_2 Atmospheric system is also essential (Izod and Wayne 1968):

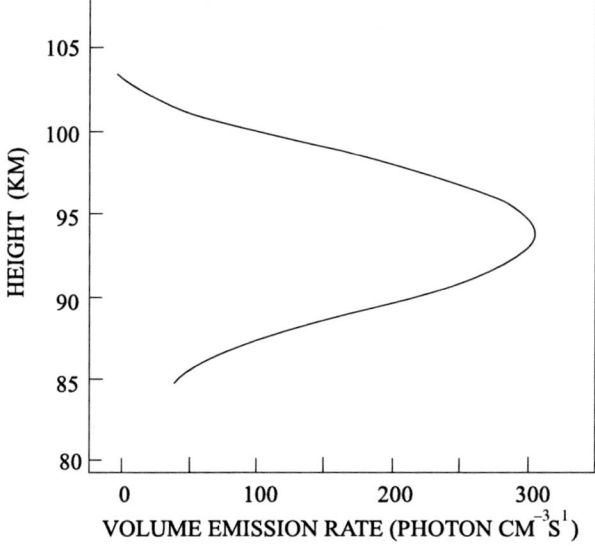

Fig. 2.25 Example of the altitude distribution of the emission rate for the band (0–1) of the O_2 Atmospheric system, constructed based on various measurement sets (Krasnopolsky 1987)

$$O_2(b^1\Sigma_g^+) + O \to O_2(X^3\Sigma_g^-) + O \quad \beta_O(b^1\Sigma_g^+) = 8 \cdot 10^{-14} \, (cm^3 \cdot s^{-1}),$$

$$O_2(b^1\Sigma_g^+) + N_2 \to O_2(X^3\Sigma_g^-) + N_2 \beta_{N_2}(b^1\Sigma_g^+) = 2.2 \cdot 10^{-15} \, (cm^3 \cdot s^{-1}).$$

These processes are responsible for the formation of the Atmospheric system bands at altitudes below 100 (km). At higher altitudes, the following process can occur (Bates 1982):

$$O(^1D) + O_2 \to O + O_2(b^1\Sigma_g^+) \quad \alpha_{O^1D}(b^1\Sigma_g^+) = 6 \cdot 10^{-11} \, (cm^3 \cdot s^{-1}),$$

which considerably intensifies in the daytime due to $O(^1D)$ atoms produced additionally as a result of the photoionization of oxygen molecules in the Schumann–Runge continuum.

The direct fluorescence

$$O_2(X^3\Sigma_g^-) + h\nu(761.9 \,(nm)) \to O_2(b^1\Sigma_g^+) \quad g(b^1\Sigma_g^+) = 8.5 \cdot 10^{-9} \, (s^{-1})$$

has a small coefficient of photon scattering; therefore, the contribution of this process is most considerable at altitudes of about 50 (km).

At altitudes above 80 (km), $O_2(b^1\Sigma_g^+)$ molecules can be produced in the daytime by the reactions of photolysis of ozone

$$O_3 + h\nu(\lambda \leq 310(nm)) \to O_2(a^1\Delta_g) + O(^1D) \quad j_{O_3}(a^1\Delta_g) = 1.0 \cdot 10^{-2} \, (s^{-1}),$$

giving rise to metastable oxygen atoms, and

$$O_3 + h\nu(\lambda \leq 463(nm)) \to O_2(b^1\Sigma_g^+) \quad j_{O_3}(b^1\Sigma_g^+) = 1.0 \cdot 10^{-3} \, (s^{-1}).$$

At lower altitudes, the reaction

$$O_3 + O \to O_2(b^1\Sigma_g^+) + O_2 \quad \alpha_{O_3}(b^1\Sigma_g^+) \approx 1.3 \cdot 10^{-21} \, (cm^3 \cdot s^{-1})$$

can make some contribution.

In the range of altitudes where hydroxyl emission occurs, the reaction

$$O_3 + H \to O_2(b^1\Sigma_g^+) + OH(v' \leq 4) \quad \alpha_{O_3H}(b^1\Sigma_g^+) = 1 \cdot 10^{-10} \cdot \exp\left(-\frac{470}{T}\right) (cm^3 \cdot s^{-1})$$

can take place.

During auroras, oxygen molecules can be excited by electron impact, for which the effective cross-section is a maximum at energies of 7–8 (eV) (Vallance Jones 1974):

$$O_2 + e \to O_2(b^1\Sigma_g^+) + e \quad \sigma_e(b^1\Sigma_g^+) = 2 \cdot 10^{-18} \, (cm^2),$$

and also by the process (Vallance Jones and Gattinger 1974; Bates 1982)

$$O_2^+ + NO \to O_2(b^1\Sigma_g^+) + NO^+ \quad \alpha_{O_2^+}(b^1\Sigma_g^+) = 8 \cdot 10^{-10} \, (cm^3 \cdot s^{-1}).$$

2.4 Molecular Oxygen Emissions

The terminating emission process is

$$O_2(b^1\Sigma_g^+) \rightarrow O_2(X^3\Sigma_g^-) + h\nu(\lambda 760\,(nm)) \quad A(bX) = 8.7 \cdot 10^{-2}\,(s^{-1})\,.$$

Under the conditions of ground measurements the 864.5-nm band of O_2 (0–1) is detected:

$$O_2(b^1\Sigma_g^+) \rightarrow O_2(X^3\Sigma_g^-) + h\nu(\lambda 864.5\,(nm))\ A(bX) = 3.7 \cdot 10^{-3}\,(s^{-1})\,.$$

The spectral structure of this band is shown in Fig. 2.26; the energies of the vibrational levels are given in Table 2.28; the wavelengths of the 864.5-nm (0–1) band are listed in Table 2.29 (Krassovsky et al. 1962; Berg and Shefov 1962b, 1963).

The results of systematization of rocket measurements of the emission layer altitude (Shefov 1975b) show that it varies substantially within a day; at night the emission intensity maximum is higher, at about 90 (km), and in the daytime the

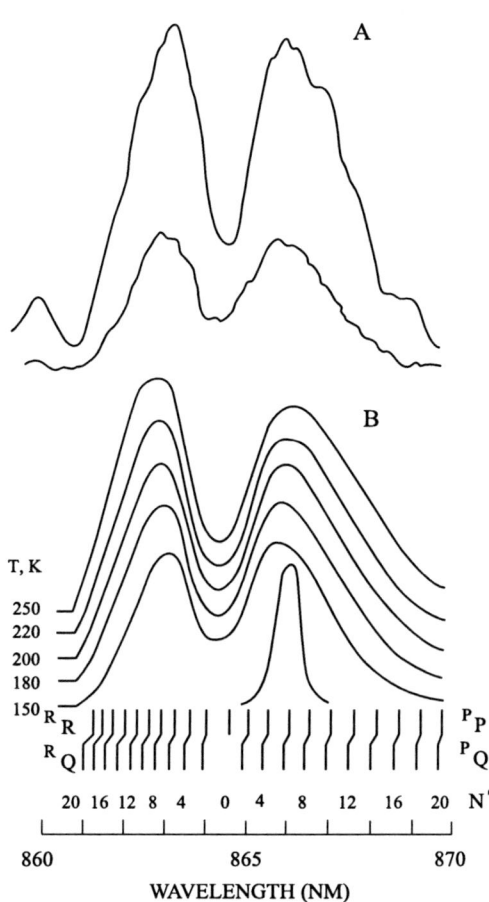

Fig. 2.26 Spectral structure of the 864.5-nm band of O_2 (0–1). (**A**) Examples of measured profiles; (**B**) calculated synthetic profiles for the given temperatures and width of the instrumental profile (Berg and Shefov 1962a)

Table 2.28 Energies of the rotational levels of the O_2 states $b^1\Sigma_g^+$ (v = 0) and $X^3\Sigma_g^-$ (v = 1)

$b^1\Sigma_g^+$ (v = 0) $\nu_{00} = 13120.9080\,(cm^{-1})$		$X^3\Sigma_g^-$ (v = 1) $\nu_{10} = 1556.3856\,(cm^{-1})$			
N	F(cm^{-1})	N	F_1(cm^{-1})	F_2(cm^{-1})	F_3(cm^{-1})
0	0.0000	1	0.9511	2.8436	−1.1340
2	8.3478	3	15.1030	17.0631	+14.8749
4	27.8244	5	40.6578	42.6551	40.6375
6	58.4263	7	77.5908	79.6158	77.6355
8	100.1477	9	125.8908	127.9391	125.9863
10	152.9808	11	185.5477	187.6173	185.6880
12	216.9158	13	256.5508	258.6405	256.7330
14	291.9408	15	338.8884	340.9973	339.1093
16	378.0416	17	432.5465	434.6741	432.8054
18	475.2022	19	537.5098	539.6557	537.8057
20	583.4044	21	653.7610	655.9249	654.0932

layer lowers to an altitude of about 50 (km) because of the change of the excitation processes.

Systematic ground measurements of the nightglow characteristics, such as the intensity and rotational temperature of the 864.5-nm band (0–1), have revealed their seasonal variations; maxima are observed in winter and minima in summer. Nevertheless, like for many other emissions, the yearly average intensity shows long-term variations.

The rotational temperature is an important parameter of the emission of O_2 (0–1). The spectral band of this emission consists of ^RR, ^RQ, ^PQ, and ^PP lines. Their intensity factors i(J), which determine the probabilities of transitions within the band, were calculated in several studies (Miller et al. 1969). Their differences from values calculated by Schlapp (1937) formulas and presented in Table 2.30 do not exceed a few percent.

Table 2.29 Wavelengths (nm) of the lines of the O_2 band (0–1) branches

N′	^PP	^PQ	^RQ	^RR
0	864.689	–	–	–
2	865.128	864.964	863.924	864.065
4	865.587	865.436	863.527	863.677
6	866.056	865.915	863.150	863.300
8	866.559	866.412	862.793	862.945
10	867.077	866.929	862.456	862.609
12	867.607	867.463	862.138	862.292
14	868.159	868.017	861.838	861.995
16	868.732	868.590	861.559	861.716
18	869.333	869.182	861.300	861.457
20	869.931	869.793	861.059	861.218

2.4 Molecular Oxygen Emissions

Table 2.30 Intensity factors $i(N')$ of the rotational lines of the O_2 band (0–1) (Krassovsky et al. 1962)

N'	^PP	^PQ	^RQ	^RR
0	1	–	–	–
2	2	1.38	1.12	0.5
4	3	2.38	2.12	1.5
6	4	3.38	3.12	2.5
8	5	4.38	4.12	3.5
10	6	5.38	5.12	4.5
12	7	6.38	6.12	5.5
14	8	7.38	7.12	6.5
16	9	8.38	8.12	7.5
18	10	9.38	9.12	8.5
20	11	10.38	10.12	9.5

Due to the large radiative lifetime, rotational relaxation has time to complete and the rotational temperature reflects the temperature of the atmosphere. However, with the dispersion of the spectroscopic instruments used, the spectral structure of the band was not resolved. Therefore, the temperature was determined by comparing the measured and theoretical synthetic profiles calculated for various temperatures, as can be seen from Fig. 2.26. The yearly average temperature turned out equal to 190 (K) (Berg and Shefov 1962a, 1963; Shefov 1975b). The error of individual values was ≤ 20 (K). The intensities of OH and O_2 emissions and their temperatures, according to Zvenigorod data, are compared in Figs. 2.27 and 2.28, respectively. It can be seen that there is a correlation between the intensities, which, however, depends on the vibrational level number of the hydroxyl bands, and this, in turn, is determined by the different character of seasonal variations for the bands of lower and upper levels. This is especially pronounced in comparing monthly mean intensities. In this case, the correlation coefficient increases in going from the lower (0.2–0.5) to the upper levels (0.88). When comparing individual temperature values for the OH and O_2 bands, we see that the correlation coefficients have insignificant values (~ 0.3). The monthly mean temperatures show a strong correlation (0.9), which seems to be due to seasonal variability of the temperature regime in the mesopause. Eventually, all the basic variations of the intensity of molecular oxygen emission, which strongly depend on the vertical distribution of atomic oxygen, are determined by the variations in vertical eddy diffusion (Perminov et al. 2004).

2.4.3 The Infrared Atmospheric System

The Infrared Atmospheric system of molecular oxygen was first observed in dayglow with the use of instruments elevated on balloons (Gopstein and Kushpil 1964);

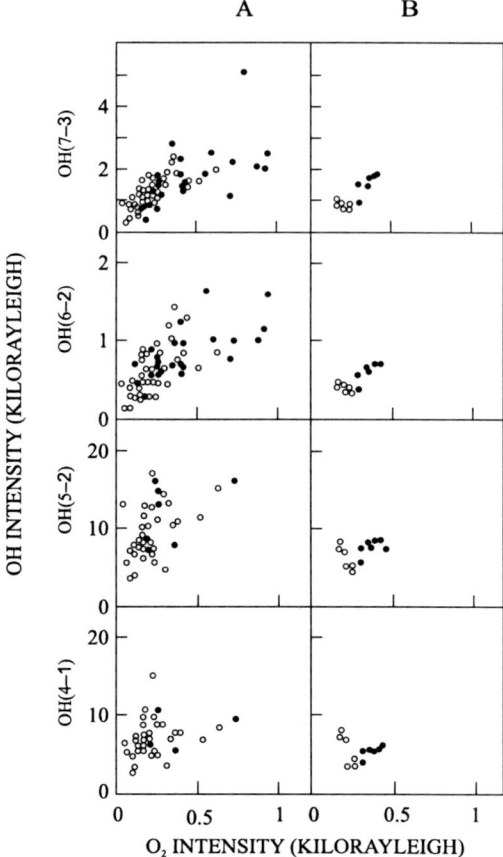

Fig. 2.27 Comparison of the measured (**A**) and monthly average (**B**) intensities of bands from different vibrationally excited levels with the intensity of the 864.5-nm O_2 (0–1) band, according to Zvenigorod measurements. *Full circles* are period from October to March. *Open circles* are period from April to September

however, it was not identified at that time. Noxon and Vallance Jones (1962) described first observations of this system in twilight. This system is produced by the following transition with the probability determined by Mlynczak and Nesbitt (1995):

$$O_2(a^1\Delta_g) \rightarrow O_2(X^3\Sigma_g^-) + h\nu(1.27(\mu m)) \quad A(aX) = 1.47 \cdot 10^{-4} \, (s^{-1}) \, .$$

Table 2.31 gives the wavelengths (nm), absolute values of transition probabilities A (s^{-1}), scattering coefficients g (s^{-1}), and relative intensities I(%) of the bands of the O_2 Infrared Atmospheric system.

Only the 1.27-μm (0–0) and 1.58-μm (0–1) bands are actually observed. Despite the low transition probability, the absorption coefficient $\sigma_{1.27} = 6 \cdot 10^{-22}(cm^2)$ appears sufficient for the optical thickness of the atmospheric layer from the surface to the altitudes of occurrence of emission (~ 80–$100\,(km)$) to be high, $\tau_{1.27} \approx 10^4$. Therefore, the radiation in the 1.27-μm (0–0) band is strongly attenuated (to 4(%)) in its propagation toward the Earth surface. Only part of the band produced by

2.4 Molecular Oxygen Emissions

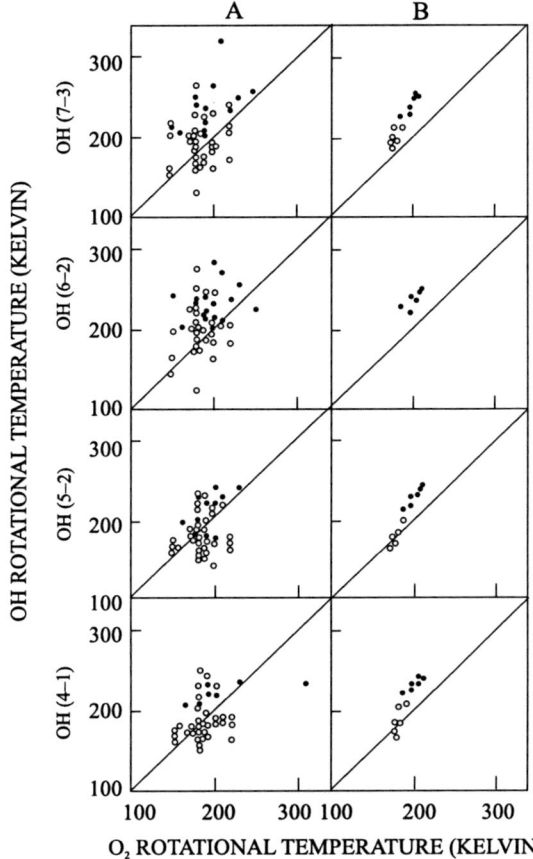

Fig. 2.28 Relation of the measured (**A**) and yearly average (**B**) rotational temperatures of bands from different vibrationally excited levels to the rotational temperature of the 864.5-nm O_2 (0–1) band, according to measurements performed at Zvenigorod. *Full circles* are period from October to March. *Open circles* are period from April to September

Table 2.31 Wavelengths (nm), absolute values of transition probabilities A (s^{-1}), scattering coefficients g (s^{-1}), and relative intensities I(%) of the bands of the O_2 Infrared Atmospheric system

v''/v'	0	1	2
0	1268.7	1580.8	2086.1
	1.47(–4)	1.86(–6)	
	4.4(–10)	3.0(–12)	
	99.32	0.67	
1	1067.7	1280.4	1593.0
	6.25(–6)	2.67(–4)	4.03(–6)
	1.4(–13) –	6.0(–12) –	9.0(–14) –

Transition probability $A_{aX} = 1.47 \cdot 10^{-4}$ (s^{-1})

transitions between higher rotational levels is recorded whose spectral distribution is strongly distorted. Under night conditions, the intensity is about 100 (kilorayleigh) and in the daytime it reaches 30 (megarayleigh). The transition probability for the 1.58-μm (0–1) band is $A_{1.58}(aX) = 3.5 \cdot 10^{-6}(s^{-1})$ (Haslett and Fehsenfeld 1969). This value agrees with the measurements (Findlay 1969).

A mechanism of the production of $O_2(a^1\Delta_g)$ excited molecules at night is related to the transitions of the Chamberlain band system considered above. Some other reactions can also be involved (Gattinger 1971; Ali et al. 1986; Howell et al. 1990; Vlasov et al. 1997):

$$O(^3P) + O(^3P) + M \leftarrow O_2(a^1\Delta_g) + M \quad \alpha_{OOM}(a^1\Delta_g) = 2.2 \cdot 10^{-34} \cdot (200/T)^2 \left(cm^3 \cdot s^{-1}\right),$$

$$O(^3P) + OH(v) \rightarrow O_2(a^1\Delta_g) + H \quad \alpha_{OOH}(a^1\Delta_g) = 3 \cdot 10^{-12} \cdot \sqrt{T} \left(cm^3 \cdot s^{-1}\right).$$

In the daytime, $O_2(a^1\Delta_g)$ molecules are produced due to photolysis of ozone (Nicolet 1971):

$$O_3 + h\nu(\lambda \leq 611(nm)) \rightarrow O(^3P) + O_2(a^1\Delta_g) \quad j_{O_3}(a^1\Delta_g) = 1 \cdot 10^{-2} \left(s^{-1}\right).$$

However, the actual process of photolysis of ozone occurs due to the solar ultraviolet radiation emitted in the Hartley bands:

$$O_3 + h\nu(\lambda \leq 310(nm)) \rightarrow O(^1D) + O_2(a^1\Delta_g) \quad j_{O_3}(a^1\Delta_g) = 1 \cdot 10^{-2}(s^{-1}).$$

This process is used to determine the ozone density by the observed intensity of 1.27-μm emission. Based on this process, a model for the atomic oxygen density has been developed at CIRA–1996 (Llewellyn and McDade 1996). However, the complex many-stage procedure of reconstruction of the ozone density by the intensity of 1.27-μm emission followed by the reconstruction of the atomic oxygen density from the reaction of ozone production with the use of standard data on the densities of other atmospheric components and on the temperature made this model incapable of predicting the actual variations of the atomic oxygen density (Semenov and Shefov 2005).

Measurements of the altitude distribution of the emission rate of the O_2 Infrared Atmospheric system have revealed that the emission layer has two maxima of nearly the same emission rate at about 87 and 96 (km). The minimum, which takes place at altitudes about 91–92 (km), makes 25(%) of the amplitudes of the maxima (Fig. 2.29) (Evans et al. 1972).

The nocturnal intensity variations of the 1.27-μm emission in most cases closely correlate with those of hydroxyl emission (Vallance Jones 1973). In twilight periods, the intensity abruptly decreases from 20 (megarayleigh) to 250 (kilorayleigh) in the evening as the solar zenith angle χ increases from 80° to 100°, and the rate of this decrease in summer is greater than in winter. In the seasonal behavior of the twilight intensity, a deep minimum in summer and a maximum in the middle of winter are observed. The daytime intensity slightly varies during a year.

2.4 Molecular Oxygen Emissions

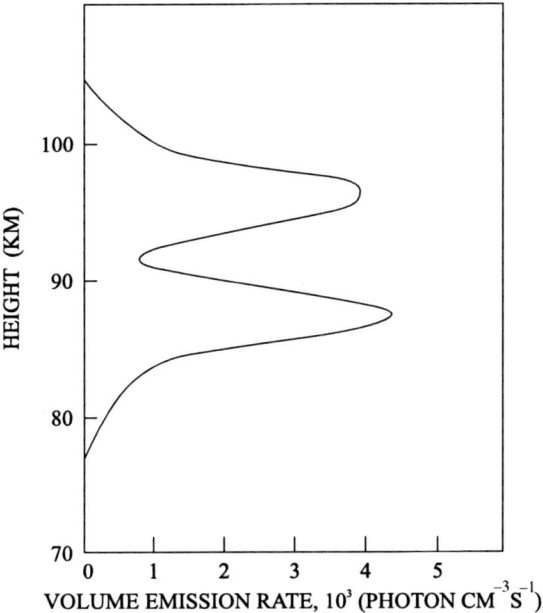

Fig. 2.29 Altitude distribution of the emission rate of the O_2 Infrared Atmospheric system, constructed based on the data of Evans et al. (1972)

The deactivation of $O_2(a^1\Delta_g)$ excited molecules occurs as a result of the reactions (Llewellyn and Long 1978)

$$O_2(a^1\Delta_g) + O_2 \rightarrow O_2 + O_2 \quad \beta_{O_2}(a^1\Delta_g) = 2.2 \cdot 10^{-18} \cdot (T/300)^{0.78} \; (cm^3 \cdot s^{-1}),$$

$$O_2(a^1\Delta_g) + N_2 \rightarrow O_2 + N_2 \quad \beta_{N_2}(a^1\Delta_g) \leq 10^{-20} \; (cm^3 \cdot s^{-1}),$$

$$O_2(a^1\Delta_g) + O \rightarrow O_2 + O \quad \beta_O(a^1\Delta_g) = 1.3 \cdot 10^{-16} \; (cm^3 \cdot s^{-1}),$$

$$O_2(a^1\Delta_g) + H \rightarrow OH + O \quad \beta_H(a^1\Delta_g) \leq 1 \cdot 10^{-13} \; (cm^3 \cdot s^{-1}).$$

The state $O_2(b^1\Sigma_g^+)$ is the starting for the Noxon electronic transition (Noxon 1961)

$$O_2(b^1\Sigma_g^+) \leftrightarrow O_2(a^1\Delta_g) + h\nu(1.908(\mu m)) \quad A(ba) = 2.5 \cdot 10^{-3} \; (s^{-1}).$$

The intensity of nightglow in the upper atmosphere is estimated to be 600 (Rayleigh), but it can be observed only at altitudes above 20–30 (km) because of the strong atmospheric absorption in this spectral range by CO_2 and H_2O molecules. Since the 1.908-μm emission corresponds to the initial state for Atmospheric system, different variations of this emission should be proportional to the variations of the Atmospheric system.

2.5 557.7-nm Emission of Atomic Oxygen

The 557.7-nm emission was the first discrete emission in the nightglow of the upper atmosphere, which was detected under the conditions of middle latitudes. It had been known earlier as one of the "nebular" emissions shown up in spectra of gaseous nebulas – relics of supernovas and high-latitude auroras. Nevertheless, its nature was not elucidated at once. Details of the relevant investigations are presented in monographs by Khvostikov (1948) and Chamberlain (1961). It turned out that this emission belongs to atomic oxygen and corresponds to a transition from one lower metastable state to another. Therefore, it can occur only in rarefied gaseous media. For the Earth's atmosphere such conditions are provided at altitudes of about 100 (km).

The energy level structure of the lower metastable states of the oxygen atom is shown in Fig. 2.30. Table 2.32 presents the parameters of transitions from the initial level ^{1}S.

The first mechanism of the production of O(^{1}S) metastable oxygen atoms under the conditions of the Earth's atmosphere was proposed by Chapman (1931) who considered it as a consequence of the recombination of atomic oxygen produced by dissociation of molecular oxygen under the action of the solar ultraviolet radiation. This process goes as follows:

$$O + O + O \rightarrow O_2^* + O(^1S_0) \,.$$

Numerous investigations of the variations in green emission intensity, including those based on rocket measurements, and the use of various laboratory data on photochemical reaction rates have revealed no correlation with direct measurements of atomic oxygen concentrations at the altitudes of the 557.7-nm emission.

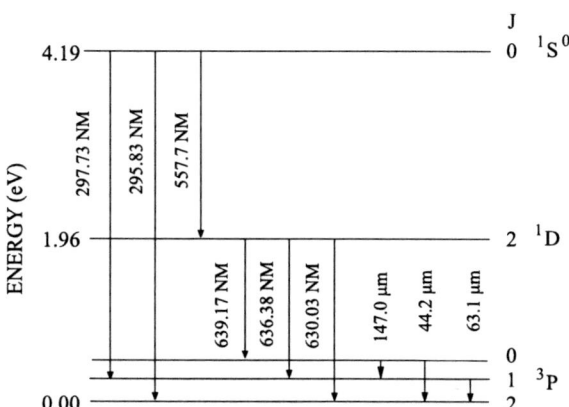

Fig. 2.30 Structure of the metastable levels and the wavelengths of the transitions between these levels for a neutral oxygen atom

2.5 557.7-nm Emission of Atomic Oxygen

Table 2.32 Parameters of the transitions from the level ^1S of atomic oxygen

λ(nm)	Transition	J	A(s^{-1})
295.83	^1S–^3P	0–2	3.7(–4)
297.2325	^1S–^3P	0–1	7.6(–2)
557.7345	^1S–^1D	0–2	1.215

Another process was considered by Barth and Hildebrandt (1961):

$$O + O + M \rightarrow O_2^* + M \quad \alpha_{O_2},$$

$$O_2^* + N_2 \rightarrow O_2 + N_2 \quad \beta_{N_2}^*,$$

$$O_2^* + O_2 \rightarrow O_2 + O_2 \quad \beta_{O_2}^*,$$

$$O_2^* + O \rightarrow O_2 + O \quad \beta_O^*,$$

$$O_2^* + O \rightarrow O_2 + O(^1S_0) \quad \alpha_O,$$

$$O_2^* \rightarrow O_2 + h\nu \quad A^*,$$

$$O(^1S_0) \rightarrow O(^1D_2) + h\nu(557.7\,(\text{nm})) \quad A_{557.7},$$

$$O(^1S_0) \rightarrow O(^3P_1) + h\nu(297.2\,(\text{nm})) \quad A_{297.2},$$

$$O(^1S_0) + O_2 \rightarrow O + O_2 \quad \beta_{O_2},$$

$$O(^1S_0) + O \rightarrow O + O \quad \beta_O.$$

Here α are the rates of production of the excited components, β are their deactivation rates in collisions with atoms and molecules, and A are the Einstein probabilities of the radiation transitions.

As already considered in Sect. 2.4, in the reaction producing O_2^* excited molecules, several metastable electronic states, namely $A^3\Sigma_u^+$, $A'^3\Delta_u$, $c^1\Sigma_u^-$, are realized. It was argued (Witt et al. 1979; McDade et al. 1986a; McDade and Llewellyn 1988) that the most probable state for the initiation of the 557.7-nm emission is $c^1\Sigma_u^-$. Subsequently it was discussed (Wraight 1982; Bates 1988a) whether the molecules in the state $^5\Pi_g$ contribute substantially to the production of $O(^1S)$ atoms because such molecules can make only ~0.5 of all O_2^* molecules.

Investigations of the variations of the Doppler temperature under the action of IGWs have made it possible to determine the numbers η and θ, which characterize the relationship between the relative variations in intensity and temperature as well as the mean time delay of the emission intensity oscillations caused by the propagation of IGWs relative to the temperature variations (Krassovsky 1972). The value of θ turned out to be ~90 (s). The obtained values of η and θ made possible the conclusion that excited oxygen molecules – preferentially in the $^5\Pi_g$ state – are formed in the Barth process (Semenov 1989a).

Thus, the emission intensity is determined by the relation

$$I_{557.7} = \int_0^\infty Q(Z)dZ, (\text{Rayleigh}),$$

where the emission rate depends on the atmospheric constituents and reaction rates:

$$Q_{557.7} = \frac{A_{557.7} \cdot \alpha_O \cdot \alpha_{O_2}[O]^3[M]}{\left\{A_{557.7} + A_{297.2} + \beta_{O_2}[O_2] + \beta_O[O]\right\}\left\{A^* + \beta_{O_2}^*[O_2] + \beta_{N_2}^*[N_2] + (\alpha_O + \beta_O^*)[O]\right\}} \cdot (\text{photon} \cdot \text{cm}^{-3} \cdot \text{s}^{-1}).$$

In this case, if we take into account the deactivation of excited atoms $O(^1S_0)$ and molecules O_2^* by the surrounding unexcited oxygen atoms, which is neglected in many studies (McDade et al. 1986a), we obtain the well-known cubic equation. Its solution consists of two parts, which transform into one another on varying the parameters of the emission layer:

$$[O] = D\left\{\left[0.5 + 0.5 \cdot \sqrt{1-E}\right]^{\frac{1}{3}} + \left[0.5 - 0.5 \cdot \sqrt{1-E}\right]^{\frac{1}{3}}\right\}, (\text{cm}^{-3}),$$

for $E \leq 1$ and

$$[O] = D \cdot 4^{\frac{1}{3}} \cdot E^{\frac{1}{6}} \cdot \cos\left[\frac{1}{3}\arccos\frac{1}{\sqrt{E}}\right], (\text{cm}^{-3}),$$

for $E \geq 1$.

Here

$$E = \frac{4}{27}D^3 d^3,$$

where

$$D = \left\{\frac{Q_{557.7}\left(A^* + \beta_{O_2}^*[O_2] + \beta_{N_2}^*[N_2]\right)\left(A_{557.7} + A_{297.2} + \beta_{O_2}[O_2]\right)}{A_{557.7} \cdot \alpha_O \cdot \alpha_{O_2}[M]}\right\}^{\frac{1}{3}},$$

$$d = \frac{\beta_O}{A_{557.7} + A_{297.2} + \beta_{O_2}[O_2]} + \frac{\alpha_O + \beta_O^*}{A^* + \beta_{O_2}^*[O_2] + \beta_{N_2}^*[N_2]}.$$

It should be noted that if other excited states of O_2^* are also taken into account, the degree of the equation increases and its analytic solution becomes impossible. Therefore, in solving the cubic equation, effective values of reaction rates are used for some metastable states of the oxygen molecule for which the effective emission probability is equal to A^*.

The emission rate $Q_{557.7}$ (photon $\cdot$ cm$^{-3} \cdot$ s^{-1}) for an altitude profile function $f(Z)$ is most adequately described by an asymmetric Gauss distribution function (Sect. 4.4):

2.5 557.7-nm Emission of Atomic Oxygen

$$Q_{557.7} = Q_m(Z_m) \cdot f(Z) ,$$

where

$$Q_m(Z_m) = 2\sqrt{\frac{\log_e 2}{\pi}} \cdot \frac{I_{557.7}}{W_{557.7}} , \quad I_{557.7} = Q_m \cdot W_{557.7} \cdot \sqrt{\frac{\pi}{4\log_e 2}} ,$$

and $I_{557.7}$ is the emission intensity at zenith (photon $\cdot$ cm$^{-3} \cdot$ s^{-1}). The parameters of the emission layer, such as its thickness $W_{557.7}$, i.e., the difference between the altitudes corresponding to the values $0.5\,Q_m(Z_m)$, or

$$W = Z_{up}[0.5 \cdot Q_m(Z_m)] - Z_{low}[0.5 \cdot Q_m(Z_m)] ,$$

and the asymmetry

$$P = \frac{Z_{up}[0.5 \cdot Q_m(Z_m)] - Z_m}{W} ,$$

depend on the altitude Z_m (Semenov and Shefov 1997c), or, more precisely, on atmospheric density n.

Thus, we have

$$W_{557.7} = 1.89 \cdot \log_e\left(\frac{2.44 \cdot 10^{15}}{n}\right) , (km)$$

and

$$n = 2.44 \cdot 10^{15} \exp(-0.53 \cdot W_{557.7}), (cm^{-3}) .$$

The rates of the above processes have the following values (Witt et al. 1979; Mc-Dade et al. 1986a; McDade and Llewellyn 1988; Bates 1988b,c; Shefov et al. 2000, 2002):

$$A_{557.7} = 1.215(s^{-1}),$$

$$A_{297.2} = 0.076(s^{-1}),$$

$$A_{295.8} = 0.00037(s^{-1}),$$

$$A^* = 1.0 s^{-1}(c^1\Sigma_u^-), 1.75 s^{-1}(A'^3\Delta_u), 15 s^{-1}(A^3\Sigma_u^+), A^*_{eff} = 3 s^{-1},$$

$$\alpha_O = 1 \cdot 10^{-12}(cm^3 \cdot s^{-1}),$$

$$\alpha_{O_2} = 5.5 \cdot 10^{-33}(200/T) \cdot 2(cm^3 \cdot s^{-1}),$$

$$\beta_O = 5.0 \cdot 10^{-11}\exp(-305/T)(cm^3 \cdot s^{-1}),$$

$$\beta_{O_2} = 4.3 \cdot 10^{-12}\exp(-865/T)(cm^3 \cdot s^{-1}),$$

$$\alpha_O + \beta_O^* = 5.9 \cdot 10^{-12}(cm^3 \cdot s^{-1}),$$

$$\beta^*_{O_2} = 3 \cdot 10^{-14} (\text{cm}^3 \cdot \text{s}^{-1}),$$

$$\beta^*_{N_2} = 4.7 \cdot 10^{-9} (200/T)^2 \exp(-1506/T)(\text{cm}^3 \cdot \text{s}^{-1}).$$

The Bart mechanism was used to calculate the atomic oxygen density under given heliogeophysical conditions for which, based on the empirical model of variations of the 557.7-nm emission of atomic oxygen (Semenov, Shefov 1997a,b,c,d), the intensity, temperature, altitude, and thickness of the emission layer have been determined. Comparison of the rocket measurements of altitude distributions of atomic oxygen density in the ETON experiment (Greer et al. 1986) (dashed line) with the altitude distribution calculated by the empirical model (Semenov and Shefov 1997a,b,c,d) (solid line) for the same heliogeophysical conditions (Semenov 1997) (Fig. 2.31) has shown a disagreement between experiment and calculation within 10(%).

Another mechanism of the excitation of the 557.7-nm emission is realized at the altitudes of the ionospheric F2 layer simultaneously with the 630-nm emission (Sect. 2.5).

In this case, the mechanism of excitation of metastable oxygen atoms at night in quiet geomagnetical conditions is related to the dissociative recombination (Bates 1982)

$$O^+ + O_2 \rightarrow O_2^+ + O \qquad \alpha_{O+O_2} = 2.8 \cdot 10^{-11} \left(\text{cm}^3 \cdot \text{s}^{-1}\right).$$

This is followed by the dissociative recombination (Frederick et al. 1976; Bates 1982)

$$O_2^+ + e \rightarrow O + O(^1S) \qquad \alpha_{O_2^+ e} = 1.7 \cdot 10^{-8} \cdot (300/T)^{0.5} \left(\text{cm}^3 \cdot \text{s}^{-1}\right).$$

As a result, the intensity of the 557.7-nm emission in the F2 layer makes about 30(%) of that of the 630-nm emission.

The excitation of oxygen atoms at night occurs due to their collisions with electrons:

$$O + e \rightarrow O(^1S) + e \quad q_{1Se} = 4.0 \cdot 10^{-9} (\text{s}^{-1}).$$

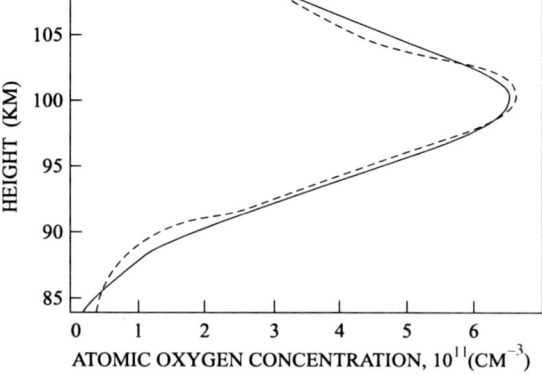

Fig. 2.31 Comparison of the rocket measurements of the atomic oxygen density in the ETON experiment (Greer et al. 1986) (*dashed line*) with the altitude distribution calculated for the same helio and geophysical conditions (Semenov 1997) by an empirical model (Semenov and Shefov 1997c) (*solid line*)

The contribution of photoelectrons is determined by the rate $2 \cdot 10^{-12} \, (\text{s}^{-1})$.

A great number of studies was devoted to the intensity and temperature variations of the 557.7-nm emission. Systematization of these results, which made it possible to gain an impression of the behavior of the green emission of atomic oxygen, has offered a way of developing a model of the variations of atomic oxygen concentration. The relevant considerations are given in Sects. 4.4 and 7.2.

2.6 The Atomic Oxygen 630-nm Emission

The atomic oxygen 630-nm emission is an important representative of the family of "nebular" emissions which were first detected by the spectra of nebulas. Subsequently they were attributed to the forbidden transitions from the metastable states ^{1}D of oxygen atoms. For quiet geomagnetic conditions, the intensity of this atmospheric emission at night is 50–100 (Rayleigh). The anomalous intensity of this emission in the Sun-illuminated aurora that took place on February 11, 1958 was detected at Zvenigorod (Mironov et al. 1959; Shefov and Yurchenko 1970), Loparskaya (Yevlashin 1962), and Abastumani (Fishkova 1983); it reached 30 (megarayleigh) at the midlatitudes. According to the observations performed in this period in the western hemisphere, the intensity of this emission reached 100 (megarayleigh) (Manring and Pettit 1959). A somewhat lower value was estimated for the low-latitude aurora observed on January 21, 1957 (Shefov and Yurchenko 1970). A detailed analysis of the behavior of the intensity of this emission in low-latitude red auroras was made by Truttse (1968a,b, 1969, 1972a,b, 1973), Truttse and Gogoshev (1977). A close correlation of the intensity logarithm with the solar radio flux (index $F_{10.7}$) and with the level of geomagnetic disturbance D_{st} has been shown. It was revealed that the intensity variations in the predawn time are related to the transport of superthermal ions from the conjugate region (Krassovsky et al. 1976). The mechanisms of the occurrence of red low-latitude arcs were analyzed by Pavlov (1998).

The atomic oxygen red emission is generated by the transitions from the O(^{1}D) metastable state (see Fig. 2.30):

$$O(^1D) \rightarrow O(^3P_2) + h\nu(630.0\,(\text{nm})), \quad A_{630.0} = 5.63 \cdot 10^{-3}\,(\text{s}^{-1}),$$

$$O(^1D) \rightarrow O(^3P_1) + h\nu(636.4\,(\text{nm})) A_{636.4} = 1.82 \cdot 10^{-3}\,(\text{s}^{-1}),$$

$$O(^1D) \rightarrow O(^3P_0) + h\nu(639.2\,(\text{nm})) A_{639.2} = 8.92 \cdot 10^{-7}\,(\text{s}^{-1}).$$

The effective transition probability is $A_{1_D} = 7.45 \cdot 10^{-3}\,(\text{s}^{-1})$ (Baluja and Zeippen 1988; Link and Cogger 1988, 1989). This corresponds to an excited state lifetime of $\sim 134\,(\text{s})$ and this, naturally, suggests that these atoms can radiate only at the

Table 2.33 Parameters of the transition levels related to the 630-nm emission

λ(nm)	Transition	J	$A(s^{-1})$
630.0308	$^1D-^3P$	2–2	5.63(−3)
636.3790	$^1D-^3P$	2–1	1.82(−3)
639.17	$^1D-^3P$	2–0	8.92(−7)
44.2 (μm)	$^3P-^3P$	0–2	1.8(−9)
63.1 (μm)	$^3P-^3P$	1–2	8.9(−5)
147.0 (μm)	$^3P-^3P$	0–1	1.7(−5)

altitudes of the ionospheric F2 layer (∼270 km). The parameters of the transition levels that correspond to the 630-nm emission are presented in Table 2.33.

The mechanism of the excitation of metastable oxygen atoms at night in quiet geomagnetical conditions is related to the dissociative recombination (Bates 1982)

$$O^+ + O_2 \rightarrow O_2^+ + O \qquad \alpha_{O^+O_2} = 2.8 \cdot 10^{-11} \, (\text{cm}^3 \cdot \text{s}^{-1}) \, .$$

A more detailed formula for the rate of this reaction is given elsewhere (Torr and Torr 1982). Besides, the following processes take place (Bates 1982):

$$O^+ + N_2 \rightarrow NO^+ + N \qquad \alpha_{O^+N_2} = 1 \cdot 10^{-12} \, (\text{cm}^3 \cdot \text{s}^{-1}),$$

$$O_2^+ + N \rightarrow NO^+ + O \qquad \alpha_{O_2^+N} = 1.2 \cdot 10^{-10} \, (\text{cm}^3 \cdot \text{s}^{-1}).$$

The subsequent reactions realize dissociative recombination:

$$O_2^+ + e \rightarrow O + O(^1D) \qquad \alpha_{O_2^+e} = 1.9 \cdot 10^{-7} \cdot (300/T_e)^{0.5} \, (\text{cm}^3 \cdot \text{s}^{-1}),$$

$$NO^+ + e \rightarrow N(^4S) + O(^1D) \qquad \alpha_{NO^+eS} = 1 \cdot 10^{-7} \cdot (300/T_e)^{0.7} \, (\text{cm}^3 \cdot \text{s}^{-1}),$$

$$NO^+ + e \rightarrow N(^2D) + O(^3P) \qquad \alpha_{NO^+eD} = 3 \cdot 10^{-7} \cdot (300/T_e)^{0.7} \, (\text{cm}^3 \cdot \text{s}^{-1}),$$

where T_e is the temperature of electrons.

However, for the first reaction in which the nitric oxide ion participates, the Wigner rule of conservation of spin does not hold, and therefore the actual contribution of this process is insignificant (Bates 1982). The second reaction gives rise to metastable nitrogen atoms (76%) which radiate at 520 (nm). It can be followed by the reactions

2.6 The Atomic Oxygen 630-nm Emission

$$N(^2D) + O_2 \rightarrow NO + O(^1D) \qquad \alpha_{NDO_2} = 6 \cdot 10^{-12} \, (cm^3 \cdot s^{-1}),$$

$$N(^2D) + O \rightarrow N(^4S) + O \qquad \alpha_{NDO} = 4 \cdot 10^{-13} \, (cm^3 \cdot s^{-1}),$$

$$N(^2D) + e \rightarrow N(^4S) + e \qquad \alpha_{NDe} = 1.35 \cdot 10^{-10} \cdot (T_e - 220)^{0.5} \, (cm^3 \cdot s^{-1}),$$

$$N(^2D) \rightarrow N(^4S) + h\nu(\lambda 520 \, (nm)) \qquad A_{520} = 2.3 \cdot 10^{-5} \, (s^{-1}).$$

At altitudes over 300 (km) its contribution becomes insignificant.

The processes of formation of excited oxygen atoms are accompanied by the processes of their deactivation in collisions with atmospheric constituents (Pavlov et al. 1999):

$$O(^1D) + O \rightarrow O(^3P) + O(^3P) \qquad \beta_O = 2.5 \cdot 10^{-12} \, (cm^3 \cdot s^{-1}),$$

$$O(^1D) + O_2 \rightarrow O(^3P) + O_2 \qquad \beta_{O_2} = 2.9 \cdot 10^{-11} \cdot \exp(67.5/T) \, (cm^3 \cdot s^{-1}),$$

$$O(^1D) + N_2 \rightarrow O(^3P) + N_2 \qquad \beta_{N_2} = 2.0 \cdot 10^{-11} \cdot \exp(107.8/T) \, (cm^3 \cdot s^{-1}),$$

$$O(^1D) + e \rightarrow O(^3P) + e \qquad \beta_e = 8.3 \cdot 10^{-10} \cdot (T_e/1000)^{0.86} \, (cm^3 \cdot s^{-1}).$$

The participation of ionized species in these basic photochemical processes determines the importance of the solar ultraviolet irradiation of the atmosphere at altitudes of occurrence of the 630-nm emission. During twilight the intensity of the 630-nm emission decreases, as the Sun sets beneath the horizon, due to the lesser effect of the solar UV radiation and the decrease in reactant concentrations resulting from dissociative recombination. The influx of photoelectrons and O^+, He^+, and H^+ ions from the magnetically conjugate regions of the upper atmosphere has an effect as well. This feature is related to the fact that the magnetic dipole is not arranged along the rotation axis of the Earth. Therefore, there is a certain angle between the geographic and geomagnetic meridians because of which the magnetically conjugate points in one hemisphere appear illuminated earlier than the corresponding region in the other hemisphere. When the magnetically conjugate region of the atmosphere is irradiated with solar ultraviolet, the photoelectrons and ions produced provide an additional excitation of atomic oxygen in the evening in winter and an enhancement of the emission at down.

During geomagnetic disturbances in low and middle latitudes, red auroras occur whose intensity is substantially greater than that of the 630-nm emission in quiet geomagnetic conditions. In the recovery phase of geomagnetic storms at midlatitudes, subauroral red (SAR) arcs arise which are induced by collisions of oxygen atoms with thermal electrons (Pavlov 1996, 1997):

$$O(^3P) + e \rightarrow O(^1D) + e.$$

The energy exchange between the thermal plasma and the ring current increases the electron temperature at the altitudes of the F region due to the intensification of the heat flux from the plasmasphere (Pavlov et al. 1999). In this case, the effective cross-section of the excitation process is $3 \cdot 10^{-17}$ (cm^2) for electrons of energy ~ 5 (eV) (Rees 1989).

The electron excitation rate for the 630-nm emission is determined by the formula (Pavlov et al. 1999)

$$P_{630} = 4.73 \cdot 10^{-12} \cdot T_e^{0.7} \cdot \exp(-22829/T_e), \, (\text{cm}^3 \cdot \text{s}^{-1}) \, ,$$

where T_e is the electron temperature.

In high-altitude auroras, the excitation of atomic oxygen is also provided by direct electron impact of the secondary electrons formed upon ionization of atmospheric neutrals by precipitating high-energy electrons.

The excitation of the emission due to fluorescence is essentially possible:

$$O(^3P_2) + h\nu(\lambda 630 \, (\text{nm})) \rightarrow O(^1D_2) \qquad g_{630} = 4.5 \cdot 10^{-10} \, (\text{s}^{-1}) \, .$$

However, by virtue of the smallness of the scattering factor (Chamberlain 1978), this process has no effect on the emission intensity.

The chain of reactions, as mentioned, is completed with the radiative process

$$O(^1D) \rightarrow O(^3P) + h\nu(\lambda 630 \, (\text{nm})) \qquad A_D = 7.45 \cdot 10^{-3} (\text{s}^{-1}) \, .$$

Thus, the rate of the 630-nm emission is determined by the relation

$$Q_{630}(Z) = \frac{\alpha_{O^+O_2} \cdot [O^+] \cdot [O_2] + \alpha_{NO^+eS} \cdot [NO^+] \cdot n_e + \alpha_{NO^+eD} \cdot \alpha_{NDO_2} \cdot [NO^+] \cdot [O_2] \cdot n_e}{\{A_{520} + \alpha_{NDO_2} \cdot [O_2] + \beta_{NDO} \cdot [O] + \beta_{NDe} \cdot n_e\} \cdot \left\{1 + \frac{\beta_{N_2} \cdot [N_2] + \beta_{O_2} \cdot [O_2] + \beta_O \bullet [O] + \beta_e \cdot n_e}{A_{630} + A_{636.4}}\right\}}.$$

The emission intensity near midnight is generally 50–100 (Rayleigh). In evening and morning twilights, it equals 1000 (Rayleigh). In the daytime, the average intensity is ~ 10 (kilorayleigh).

Much research was devoted to the behavior of the emission intensity, Doppler temperature, and characteristics of the altitude distribution of the emission rate. Systematization of the data obtained is considered in Sect. 4.5.

2.7 Helium Emission

Helium is one of the major constituents of the upper atmosphere at altitudes above 300 (km). The helium concentration in the upper atmosphere was generally estimated only theoretically by its concentration in the lower atmosphere. For excitation of helium atoms high energies are required (Fig. 2.32) (Telegin and Yatsenko 2000), therefore the helium emission registration is difficult, even in auroras.

2.7 Helium Emission

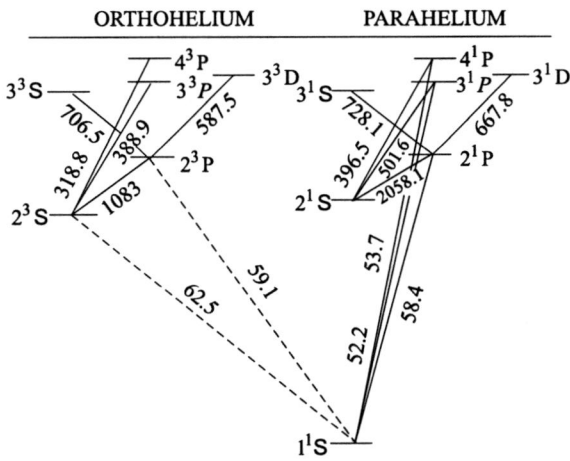

Fig. 2.32 The energy level structure of the helium atom and the wavelengths of the basic transitions

Radiation from atmospheric helium was first detected at midlatitudes during the sunlight-irradiated aurora on February 11, 1958 (Mironov et al. 1959; Krassovsky and Galperin 1960). This was possible owing to the fact that the 1083.0-nm helium emission, which, under usual conditions, is blended by the Q_1 line of the (5–2) band of the hydroxyl molecule, was much more intense (68 (kilorayleigh)) than in usual conditions (1 (kilorayleigh)).

As can be seen from the energy level structure of the helium atom (Fig. 2.32; Tables 2.34 and 2.35), the 1083.0-nm line corresponds to the 2^3P–2^3S resonant transition of orthohelium whose lower level is metastable and requires high energy (19.73 (eV)) for its excitation from the ground state 1^1S. Thus, the emission occurs due to the reaction

$$He(2^3P) \rightarrow He(2^3S) + h\nu(1083.0\,(nm))\,.$$

This is an allowed transition with the probability $A_{1083} = 1.05 \cdot 10^7\,(s^{-1})$ (Allen 1973).

Analysis of the process initiating this helium emission, in view of the fact that only the 1083.0-nm emission is detectable and the other well-known helium emissions do not show up in the visible spectral region, shows that the detectable emission occurs due to the fluorescence of the metastable orthohelium atoms formed by collisions of neutral helium atoms with electrons that takes place in sunlight.

The electron excitation of helium was extensively studied both theoretically and experimentally (Vriens et al. 1968; Moussa et al. 1968).

For triplet states, in particular for the 2^3S level, the excitation follows the reaction

$$He(1^1S) + e \rightarrow He(2^3S) + e\sigma_{2^3S} = 5 \cdot 10^{-18}\,(cm^2)$$

in which electrons with energies of about 25 (eV) participate. The effective cross-sections for this type of excitation are close to 20–25 (eV) (Vriens et al. 1968; Moussa et al. 1968). In the upper atmosphere, electrons with these energies can be

Table 2.34 Wavelengths, level energies, transition probabilities, and photon scattering factors for the emissions from orthohelium atoms

λ (nm)	Transition	Level energy (eV)		A (s^{-1})	g (s^{-1})
1083.0341	2^3P–2^3S	20.97	19.82	1.05(7)	16.8
1083.0250		20.97	19.82		
1082.9081		20.97	19.82		
388.8648	3^3P–2^3S	23.01	19.82	9.4(6)	0.0485
318.7747	4^3P–2^3S	23.71	19.82	6.1(6)	0.0287
706.5719	3^3S–2^3P	22.72	20.97	1.6(7)	0.89
706.5188		22.72	20.97		
587.5989	3^3D–2^3P	23.08	20.97	7.2(7)	9.6
587.5650		23.08	20.97		
587.5618		23.08	20.97		
471.3373	4^3S–2^3P	23.60	20.97	3.3(6)	0.0335
471.3143		23.60	20.97		
447.1688	4^3D–2^3P	23.74	20.97	2.0(7)	0.715
447.1477		23.74	20.97		
412.0993	5^3S–2^3P	23.97	20.97	2.0(7)	0.008
412.0812		23.97	20.97		
402.6362	5^3D–2^3P	24.05	20.97	1.2(7)	0.143
402.6189		24.05	20.97		
62.5585	2^3S–1^1S	19.82	0.00	2.2(-5)	
59.1406	2^3P–1^1S	20.97	0.00	<1(3)	

produced due to auroral precipitation of electrons with energies of about 10 (keV) (Shefov 1961a,b,e,f).

Under quiet geomagnetic conditions, such electrons appear in the upper atmosphere as a result of photoionization of helium by solar ultraviolet radiation

Table 2.35 Wavelengths, level energies, transition probabilities, and photon scattering factors for the emissions from parahelium atoms

λ (nm)	Transition	Level energy (eV)		A (s^{-1})	g (s^{-1})
2058.130	2^1P–2^1S	21.22	20.62	2.0(6)	13.5
501.5675	3^1P–2^1S	23.09	20.62	1.4(7)	1.73
396.4727	4^1P–2^1S	23.74	20.62	8.7(6)	0.0463
361.3641	5^1S–2^1P	24.05	20.62	4.3(6)	0.0255
728.1349	3^1S–2^1P	22.92	21.22	1.9(7)	1.13
667.8149	3^1D–2^1P	23.08	21.22	6.6(7)	13.0
504.7736	4^1S–2^1P	23.68	21.22	5.5(6)	0.079
492.1929	4^1D–2^1P	23.74	22.22	2.0(7)	1.25
443.7549	5^1S–2^1P	24.01	21.22	3.4(6)	0.0186
438.7928	5^1D–2^1P	24.05	21.22	8.6(6)	0.12
58.4328	2^1P–1^1S	21.22	0.00	2.33(9)	1.9(−7)
53.7014	3^1P–1^1S	23.09	0.00	5.6(8)	1.0(−8)
52.2186	4^1P–1^1S	23.74	0.00	2.3(8)	1.8(−9)

2.7 Helium Emission

whose most intense component is emitted in the 30.4-nm line of ionized helium He$^+$. The detection of the 1083.0-nm helium emission was the first immediate evidence of the existence of significant photoelectron fluxes in the Earth's upper atmosphere (Shefov 1962a,b; Shcheglov 1962a,b), which was earlier suggested by Shklovsky (1951b).

A helium atom excited due to the above reaction absorbs a photon:

$$\text{He}(2^3\text{S}) + h\nu(1083.0\,\text{nm}) \rightarrow \text{He}(2^3\text{P})\,.$$

The photon scattering factor for this emission, g_{1083}, is 16.8 (s^{-1}) (Shefov 1961b, 1962a,b). Subsequently other estimates were obtained, e.g., $g_{1083} = 17.7$ (s^{-1}) (Bishop and Link 1993). The intensity distribution in the solar spectrum (Minnaert et al. 1940; Molher et al. 1950) near the wavelengths corresponding to the orthohelium and parahelium lines is presented in Fig. 2.33.

Besides the above process of production of He(2^3S) metastable atoms, some other processes may occur, such as the recombination of helium ions

$$\text{He}^+(1^2\text{S}) + e \rightarrow \text{He}(2^3\text{S}) + h\nu(\lambda < 267.0\,(\text{nm}))$$

$$\alpha_{\text{He}^+,2^3\text{S}} = 1 \cdot 10^{-12} \cdot \left(\frac{1000}{T_e}\right)^{0.5} \quad (\text{cm}^3 \cdot \text{s}^{-1})\,,$$

where T_e is the electron temperature.

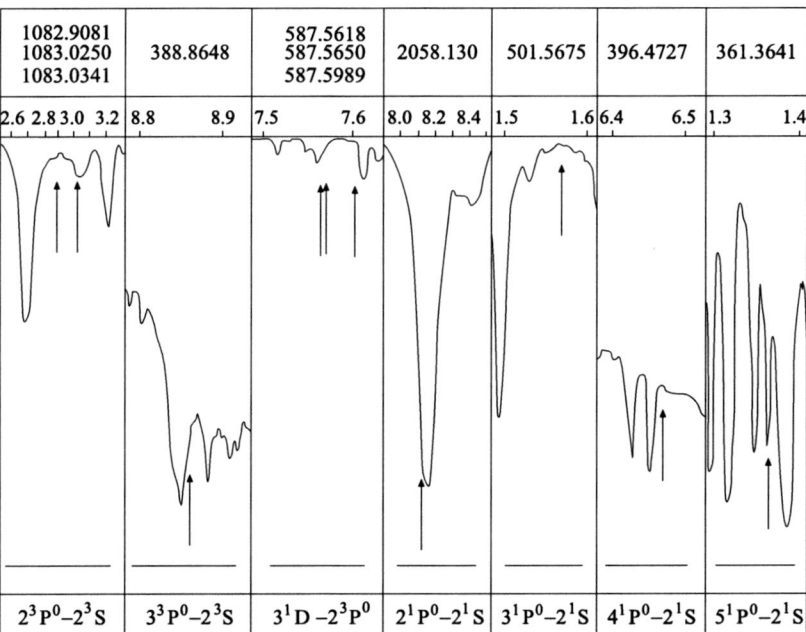

Fig. 2.33 Emission intensity distribution in the solar spectrum (Minnaert et al. 1940; Molher et al. 1950) near the wavelengths (nm) corresponding to the orthohelium and parahelium lines

However, with the mentioned photorecombination rate, the contribution of this process at altitudes above 1000 (km) can be substantial because of the low helium ion density (10^4 (cm^{-3})).

Metastable helium atoms can be produced by the following sequence of reactions (Shefov 1963a, b). The resonant absorption of the 58.4- and 53.7-nm solar emissions gives rise to He(2^1P) and He(3^1P) atoms:

$$He(1^1S) + h\nu(58.4\,(nm)) \rightarrow He(2^1P) \quad g_{58.4} = 1.9 \cdot 10^{-7} \quad (s^{-1}),$$

$$He(1^1S) + h\nu(53.7\,(nm)) \rightarrow He(3^1P) \quad g_{53.7} = 1.0 \cdot 10^{-8} \quad (s^{-1}).$$

The scattering factors were calculated by the data on the intensity of UV solar radiation (Ivanov-Kholodny and Mikhailov 1980).

The above excited states are responsible both for the emissions that result in scattered resonant radiation in the Earth's atmosphere, including the nightglow at high altitudes over the Earth's shadow, and for the 2058.1- and 501.6-nm emissions, respectively:

$$He(2^1P) \rightarrow He(1^1S) + h\nu(58.4\,(nm)) \qquad A_{58.4} = 2.33 \cdot 10^9\,(s^{-1}),$$

$$He(2^1P) \rightarrow He(2^1S) + h\nu(2058.1\,(nm)) \quad A_{2058.1} = 2.0 \cdot 10^6\,(s^{-1})$$

$$g_{2058.1} = 13.5\,(s^{-1}),$$

$$He(3^1P) \rightarrow He(1^1S) + h\nu(53.7(nm)) \qquad A_{53.7} = 5.6 \cdot 10^8\,(s^{-1}),$$

$$He(3^1P) \rightarrow He(2^1S) + h\nu(501.6\,(nm)) \quad A_{501.6} = 1.4 \cdot 10^7\,(s^{-1}) \quad g_{501.6} = 1.73\,(s^{-1}).$$

The photon scattering factors for the 2058.1-nm emission are smaller than for the 1083.0-nm emission, since the solar spectrum contains a strong Fraunhofer line, and they are greater for the 501.6-nm emission than for the 388.9-nm (3^3P–2^3S) emission, since the latter occurs in the range of the strong Fraunhofer lines of ionized calcium. However, in contrast to orthohelium atoms, absorption of 2058.1- and 501.6-nm photons results in the main in 58.4- and 53.7-nm emissions because their transition probabilities are much higher. Therefore, the intensities of the 2058.1- and 501.6-nm parahelium emissions are considerably lower compared to those of the corresponding orthohelium emissions.

Metastable parahelium atoms are produced, similar to orthohelium atoms, by electron impact:

$$He(1^1S) + e \rightarrow He(2^1S) + e \qquad \sigma_{2^1S} = 6 \cdot 10^{-18}\,(cm^2)\,.$$

However, the cross-sections have wider peaks depending on the electron energy (Massey and Burhop 1952). Besides, the formation of He(2^1S) metastable parahelium atoms occurs due to photorecombination:

2.7 Helium Emission

$$He^+(2^2S) + e \rightarrow He(2'^1S) + h\nu (\lambda < 321 nm)$$

$$\alpha_{He^+,2^1S} = 10^{-12} \cdot \left(\frac{1000}{T_e}\right)^{0.5} \quad (cm^3 \cdot s^{-1}).$$

The processes that give rise to metastable parahelium atoms are of interest because the reaction of electron exchange is possible. The excitation energy for the 2^1S level is greater by 0.78 (eV) than for the 2^3S level. Therefore, collisions of $He(2^1S)$ with thermal electrons occur as follows:

$$He(2^1S) + e \rightarrow He(2^3S) + e, \sigma_{2^1S,2^3S} = 3 \cdot 10^{-14} (cm^2),$$

$$\alpha_{2^1S,2^3S} = 5.4 \cdot 10^{-7} \cdot \left(\frac{T_e}{1000}\right)^{0.5} (cm^3 \cdot s^{-1}).$$

However, the contribution of this process to helium airglow is insignificant.

Besides the excitation of the 2^3S metastable state, there occur deactivation of this state in collisions with atoms, molecules, and electrons, photoionization, and the transition to the ground state. The level 2^3S is purely metastable. According to the available estimates, the two-photon radiation transition to the state 1^1S is possible. As this takes place, the transition energy, which corresponds to the wavelength 62.6 (nm), is divided continuously between two photons (the transition probability is taken from the work by Mathis (1957)):

$$He(2^3S) \rightarrow He(1^1S) + h\nu + h\nu \quad A_{62.6} \approx 2.2 \cdot 10^{-5} (s^{-1}).$$

Later studies also considered one-photon transitions (Woodworth and Moos 1975):

$$He(2^3S) \rightarrow He(1^1S) + h\nu \quad A_{62.6} \approx 1.10 \cdot 10^{-4} (s^{-1}).$$

The most contributory deactivation processes are Penning reactions [whose rates were taken from other publications (Ferguson et al. 1964; Moussa et al. 1968; Lindinger et al. 1974; Cook et al. 1974) for temperatures of ~900 (K) corresponding to the altitudes of the thermosphere]:

$$He(2^3S) + N_2 \rightarrow He(1^1S) + N_2^+ + e \quad \beta_{N_2} = 3.5 \cdot 10^{-10} (cm^3 \cdot s^{-1}),$$

$$He(2^3S) + O_2 \rightarrow He(1^1S) + O_2^+ + e \quad \beta_{O_2} = 6.5 \cdot 10^{-10} (cm^3 \cdot s^{-1}),$$

$$He(2^3S) + O \rightarrow He(1^1S) + O^+ + e \quad \beta_O = 1.25 \cdot 10^{-10} (cm^3 \cdot s^{-1}).$$

Besides, the density of helium atoms in the 2^3S state decreases due to photoionization from this excited level, whose rate is given elsewhere (Rundle 1960; McElroy 1965; Patterson 1967):

$$He(2^3S) + h\nu(\lambda < 267.0(nm)) \rightarrow He^+ + e \quad j_{2^3S} = 1.6 \cdot 10^{-3} (s^{-1}).$$

Woodworth and Moos (1975), based on the data on photoionization cross-sections (Stebbings et al. 1973), have obtained $j_{2^3S} = 1.9 \cdot 10^{-3}\,(s^{-1})$.

The reverse reaction of deactivation by thermal electrons

$$He(2^3S) + e \rightarrow He(2^1S) + e$$

$$\beta_{2^3S,2^1S} = 5.4 \cdot 10^{-7} \cdot \left(\frac{T_e}{1000}\right)^{0.5} \cdot \exp\left(-\frac{9100}{T_e}\right) \quad (cm^3 \cdot s^{-1})$$

proceeds at a considerably lower rate. The cross-section for the reverse process is $\sigma_{2^3S,2^1S} = (1.5 \div 8) \cdot 10^{-16}\,(cm^2)$ (Ivanov et al. 1991).

The presence of twilight helium emission with an intensity of 1 (kilorayleigh) was confirmed experimentally by direct measurements (Shefov 1961f; Shcheglov 1962a, b; Fedorova 1962, 1967). Figure 2.34 shows an interferogram of 1083.0-nm helium emission, taken in twilight (Shcheglov 1962a, b).

Subsequently measurements were performed in usual twilight at middle latitudes (Shefov 1963a, b, 1967, 1968, 1969a, d, 1973; Taranova 1967; Tinsley 1968a; Shefov and Yurchenko 1970; Toroshelidze 1970, 1971, 1976, 1991; Christensen et al. 1971; Teixeira et al. 1976; Tinsley and Christensen 1976; Suzuki 1983), during a solar eclipse (Shouiskaya 1963), and during auroras illuminated by the Sun (Fedorova 1967; Harrison and Cairns 1969; Sukhoivanenko and Fedorova 1976). During auroras, the emission intensity was as high as 10–12 (kilorayleigh), but it did not reach 68 (kilorayleigh), the maximum observed value measured in the red aurora on February 11, 1958.

According to the observations at high latitudes in the Tiksi bay ($\Phi = 65.6°N$, $\Lambda = 195.2°E$), on the Spitsbergen island ($\Phi = 73.1°N$, $\Lambda = 129.4°E$), and in Norway ($\Phi = 64.3°N$, $\Lambda = 104.9°E$), the 587.6-nm helium emission with a shifted wavelength profile was regularly observed during intense auroras which was caused by precipitating helium ions (Stoffregen 1969; Henriksen and Sukhoivanenko 1982). Its intensity reached 120 (Rayleigh).

Fig. 2.34 Interferogram of the 1083.0-nm helium emission taken in twilight (Shcheglov 1962a,b)

2.7 Helium Emission

Besides the neutral helium emission, the airglow spectrum of the Earth's atmosphere shows the presence of the 30.4-nm emission from He^+ helium ions. Satellite measurements (Meier 1974) have shown that the intensity of the atmospheric nightglow is ~ 12 (Rayleigh) and it is largely determined by the conditions of solar irradiation of the upper atmosphere, reflecting the structure of its illuminated part. The penetration of this emission into the atmosphere is restricted to altitudes of 130–140 (km).

The emission from helium ions calls for the highest excitation energy among all atmospheric constituents. Since the helium ion is a hydrogen-like atom, it can make a series of transitions similar to the Lyman and Balmer series of hydrogen. Table 2.36 presents the wavelengths, transition probabilities, and absorption cross-sections for helium emissions.

The emissions from helium ions are provided mainly due to fluorescence of solar radiation:

$$He^+(2^2P^o) \rightarrow He^+(1^2S) + h\nu(\lambda 30.4\,(nm)) \qquad A_{30.4} = 1.00 \cdot 10^{10}\,(s^{-1}),$$

$$He^+(3^2P^o) \rightarrow He^+(1^2S) + h\nu(\lambda 25.6\,(nm)) \qquad A_{25.6} = 2.68 \cdot 10^{9}\,(s^{-1}),$$

$$He^+(4^2P^o) \rightarrow He^+(1^2S) + h\nu(\lambda 24.3\,(nm)) \qquad A_{24.3} = 1.09 \cdot 10^{9}\,(s^{-1}).$$

The solar emission intensities at these wavelengths (Ivanov-Kholodny and Mikhailov 1980; Ivanov-Kholodny and Nusinov 1987),

$$S(30.4\,(nm)) = 1.3 \cdot 10^{10}\,(photon \cdot cm^{-2} \cdot s^{-1}),$$

$$S(25.6\,(nm)) = 6.6 \cdot 10^{8}\,(photon \cdot cm^{-2} \cdot s^{-1}),$$

$$S(24.3\,(nm)) = 3.0 \cdot 10^{7}\,(photon \cdot cm^{-2} \cdot s^{-1}),$$

Table 2.36 Wavelengths (Striganov and Odintsova 1982) and transition probabilities for the helium ion (Allen 1973; Wiese et al. 1966)

λ (nm)	Transition	E (eV)	E (eV)	g_u	g_l	$A\,(s^{-1})$	$\sigma\,(cm^2)$
23.73307	5^2P^o–1^2S	52.24	0.0	6	2	5.50(8)	2.42(−15)
24.30266	4^2P^o–1^2S	51.02	0.0	6	2	1.09(9)	5.14(−15)
25.63170	3^2P^o–1^2S	48.37	0.0	6	2	2.68(9)	1.48(−14)
30.37822	2^2P^o–1^2S	40.81	0.0	6	2	1.00(10)	9.22(−14)
99.2364	7^2D–2^2P^o	53.31	40.81	10	6	8.23(7)	1.47(−14)
102.5273	5^2D–2^2P^o	52.90	40.81	10	6	1.51(8)	2.97(−14)
108.4944	4^2D–2^2P^o	52.24	40.81	10	6	3.30(8)	7.7)(−14)
121.5088	4^2D–2^2P^o	51.02	40.81	4	2	1.55(8)	6.10(−14)
121.5171	3^2D–2^2P^o	51.02	40.81	10	4	1.07(9)	5.26(−13)
164.0332		48.37	40.81	4	2	3.59(8)	3.47(−13)
164.0474	3^2D–2^2P^o	48.37	40.81	6	4	1.03(9)	7.48(−13)
164.0490	3^2D–2^2P^o	48.37	40.81	4	4	1.00(8)	4.84(−14)

correspond to the 1-a. u. distance from the Sun and to the solar activity $F_{10.7} = 144$.

Following Ivanov-Kholodny and Mikhailov (1980), the dependence of the emission intensity on solar activity can be represented as

$$S_{30.4}(F_{10.7}) = S_{30.4}(144) \cdot 0.56 \cdot \log_e \left(\frac{F_{10.7} - 17}{23} \right)$$

(Ivanov-Kholodny and Nusinov (1987) give more cumbersome relations), which is somewhat different from the average dependence for the entire ultraviolet radiation flux.

For the excitation of helium due to resonant fluorescence we have

$$He^+(1^2S) + h\nu(\lambda 30.4\,(nm)) \to He^+(2^2P^o) \qquad g_{30.4} = 4.1 \cdot 10^{-5}\,(s^{-1}),$$

$$He^+(1^2S) + h\nu(\lambda 25.6\,(nm)) \to He^+(3^2P^o) \qquad g_{25.6} = 3.4 \cdot 10^{-7}\,(s^{-1}),$$

$$He^+(1^2S) + h\nu(\lambda 24.3\,(nm)) \to He^+(4^2P^o) \qquad g_{24.3} = 5.5 \cdot 10^{-9}\,(s^{-1}).$$

However, in this case, when performing calculations, one has to take into account the difference between the Doppler profiles of solar emission lines and the profiles of their absorption in the Earth's atmosphere. According to the available measurements (Doschek et al. 1974; Dere 1977; Ivanov-Kholodny and Mikhailov 1980), the profiles of solar emission lines are very well described by the Doppler distribution (Fig. 2.34)

$$I = I_0 \cdot \exp\left[-\frac{4 \cdot \log_e 2 \cdot (\lambda - \lambda_0)^2}{(\Delta\lambda)^2} \right] = 2\sqrt{\frac{\log_e 2}{\pi}} \cdot \frac{S}{\Delta\lambda} \cdot \exp\left[-\frac{4 \cdot \log_e 2 \cdot (\lambda - \lambda_0)^2}{(\Delta\lambda)^2} \right],$$

where the profile halfwidth (nm) is given by the formula

$$\Delta\lambda = 2 \cdot \frac{\lambda_0}{c} \cdot \sqrt{\frac{2 \cdot \log_e 2 \cdot k \cdot N \cdot T}{M}} = 7.16 \cdot 10^{-7} \cdot \lambda_0 \cdot \sqrt{\frac{T}{M}},$$

where T is the temperature and M is the atomic mass of helium.

For the lines of helium in the solar spectrum we have $\Delta\lambda(30.4\,(nm)) = 0.010\,(nm)$, $\Delta\lambda(25.6\,(nm)) = 0.0084\,(nm)$, and $\Delta\lambda(24.3\,(nm)) = 0.0080\,(nm)$ (Doschek et al. 1974; Dere 1977; Ivanov-Kholodny and Mikhailov 1980). These values correspond to a temperature of 845 000 (K).

At the same time, for the helium lines ($M = 4$) and atmospheric altitudes above 500 (km) ($T \sim 1000\,(K)$) we have $\Delta\lambda(30.4\,(nm)) = 0.00034\,(nm)$, $\Delta\lambda(25.6\,(nm)) = 0.00029\,(nm)$, and $\Delta\lambda(24.3\,(nm)) = 0.00028\,(nm)$. The effective solar fluxes for these emissions (near the peaks of the profiles) are

2.7 Helium Emission

$$S_m(30.4\,(\text{nm})) = 1.2 \cdot 10^{12} \quad (\text{photon} \cdot \text{cm}^{-2} \cdot \text{s}^{-1} \cdot \text{nm}^{-1}),$$
$$S_m(25.6\,(\text{nm})) = 7.5 \cdot 10^{10} \quad (\text{photon} \cdot \text{cm}^{-2} \cdot \text{s}^{-1} \cdot \text{nm}^{-1}),$$
$$S_m(24.3\,(\text{nm})) = 3.6 \cdot 10^{9} \quad (\text{photon} \cdot \text{cm}^{-2} \cdot \text{s}^{-1} \cdot \text{nm}^{-1}).$$

For these values of solar fluxes the scattering factors for the mentioned emissions were presented above. It should be noted that these values are several fold lower than those obtained by Chamberlain (1978) based on theoretical profiles for the solar emission lines.

Figure 2.35 shows, for the 30.4-nm emission profile, the positions of the line triplet of the doubly ionized oxygen atom O^{2+} whose wavelength, 30.3799 (nm), practically coincides with the wavelength of the helium ion. Owing to the latter circumstance, fluorescence occurs (Bowen 1934, 1947) due to which the solar radiation is additionally scattered by oxygen ions.

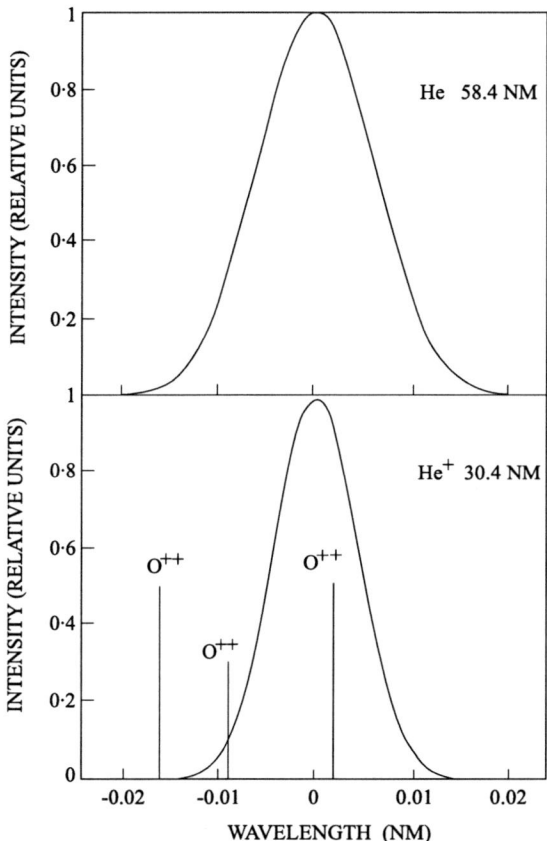

Fig. 2.35 Profiles of the 30.4-nm (He+) and 58.4-nm (He) emissions in solar radiation (Doschek et al. 1974). For the 30.4-nm emission the positions of the multiplet components of the ionized oxygen atom are shown

Excitation of He$^+$ states can occur due to photoionization of neutral helium atoms:

$$\text{He} + h\nu(\lambda \leq 19\,(\text{nm})) \rightarrow \text{He}^+(2^2\text{P}^\text{o}) + e \quad j_{30.4} = 1.1 \cdot 10^{-8}(\text{s}^{-1}),$$

$$\text{He} + h\nu(\lambda \leq 17\,(\text{nm})) \rightarrow \text{He}^+(3^2\text{P}^\text{o}) + e \quad j_{25.6} = 2.3 \cdot 10^{-9}(\text{s}^{-1}),$$

$$\text{He} + h\nu(\lambda \leq 16.5\,(\text{nm})) \rightarrow \text{He}^+(4^2\text{P}^\text{o}) + e \quad j_{24.3} = 1.1 \cdot 10^{-9}(\text{s}^{-1}).$$

The altitude variations of the ultraviolet radiation spectral distribution due to different coefficients of absorption at different wavelengths are presented in Table 2.37.

Collisions of helium atoms with photoelectrons may result in excitation of the atoms followed by their ionization:

$$\text{He} + e(\text{E} \geq 65\,(\text{eV})) \rightarrow \text{He}^+(2^2\text{P}^\text{o}) + e + e \quad q_{30.4} = 3.2 \cdot 10^{-11}(\text{s}^{-1}),$$

$$\text{He} + e(\text{E} \geq 73\,(\text{eV})) \rightarrow \text{He}^+(3^2\text{P}^\text{o}) + e + e \quad q_{25.6} = 3.0 \cdot 10^{-11}(\text{s}^{-1}),$$

$$\text{He} + e(\text{E} \geq 75\,(\text{eV})) \rightarrow \text{He}^+(4^2\text{P}^\text{o}) + e + e \quad q_{24.3} = 2.8 = 10^{-11}(\text{s}^{-1}).$$

The radiation of the Sun contains, besides the radiation of ionized helium atoms, the radiation emitted by neutral helium. Rocket measurements have established that the intensity of the 58.4-nm He emission is about 1000 (Rayleigh) in the daytime (measured at $\sim$180 (km)) and about 4–5 (Rayleigh) at night, and it is closely related to the illumination of the atmosphere by the Sun. The intensity of the interplanetary component of the 58.4-nm helium emission is 7–10 (Rayleigh) (Ajello and Witt 1979). The parahelium emission arises due to the transitions

$$\text{He}(2^1\text{P}) \rightarrow \text{He}(1^1\text{S}) + h\nu(\lambda 58.4\,(\text{nm})) \quad A_{58.4} = 2.33 \cdot 10^9\,(\text{s}^{-1}),$$

$$\text{He}(3^1\text{P}) \rightarrow \text{He}(1^1\text{S}) + h\nu(\lambda 53.7\,(\text{nm})) \quad A_{53.7} = 5.66 \cdot 10^8\,(\text{s}^{-1}),$$

$$\text{He}(4^1\text{P}) \rightarrow \text{He}(1^1\text{S}) + h\nu(\lambda 52.2\,(\text{nm})) \quad A_{52.2} = 2.43 \cdot 10^8\,(\text{s}^{-1}).$$

Table 2.37 Coefficients of absorption of helium ion emissions by atmospheric components (Banks and Kockarts 1973a,b)

Component	$\sigma\,(\text{cm}^2)$		
	30.4 (nm)	25.6 (nm)	24.3 (nm)
He	1.0(−18)	8.0(−19)	7.0(−19)
N$_2$	4.0(−18)	4.0(−18)	4.0(−18)
O$_2$	6.8(−18)	5.2(−18)	5.2(−18)
O	3.4(−18)	2.6(−19)	2.6(−18)

2.7 Helium Emission

The excitation of helium due to fluorescence in solar radiation occurs as follows:

$$He(1^1S) + h\nu(\lambda 58.4\,(nm)) \rightarrow He(2^1P) \qquad g_{58.4} = 1.8 \cdot 10^{-5}\,(s^{-1}),$$

$$He(1^1S) + h\nu(\lambda 53.7\,(nm)) \rightarrow He(3^1P) \qquad g_{53.7} = 5.4 \cdot 10^{-7}\,(s^{-1}),$$

$$He(1^1S) + h\nu(\lambda 52.2\,(nm)) \rightarrow He(4^1P) \qquad g_{52.2} = 1.0 \cdot 10^{-7}\,(s^{-1}).$$

Here, as for the helium ion, the differences in Doppler profiles of the solar and atmospheric emissions that are presented in Fig. 2.34 have been taken into account.

The coefficients of absorption of the neutral helium emissions by atmospheric species are presented in Table 2.38.

The excitation of helium atoms by photoelectrons occurs due to the reactions

$$He(1^1S) + e(E \geq 22\,(eV)) \rightarrow He(2^1P) + e \qquad q_{58.4} = 4.6 \cdot 10^{-9}\,(s^{-1}),$$

$$He(1^1S) + e(E \geq 23\,(eV)) \rightarrow He(3^1P) + e \qquad q_{53.7} = 1.2 \cdot 10^{-9}\,(s^{-1}),$$

$$He(1^1S) + e(E \geq 24\,(eV)) \rightarrow He(4^1P) + e \qquad q_{52.2} = 4.4 \cdot 10^{-10}\,(s^{-1}).$$

For photoelectrons with energies of about 30 (eV) the optical thickness to ultraviolet radiation is practically the same for all helium emissions. The coefficients of absorption are

$$\sigma(He) = 3.0 \cdot 10^{-18}\,(cm^2), \sigma(N_2) = 1.0 \cdot 10^{-17}\,(cm^2), \sigma(O_2) = 1.7 \cdot 10^{-17}\,(cm^2),$$

$$\sigma(O) = 1.0 \cdot 10^{-17}\,(cm^2).$$

Excited states can arise due to recombination (Allen 1973):

$$He^+ + e \rightarrow He^* + h\nu \qquad \alpha_{He^*} = 1.3 \cdot 10^{-12} \cdot \sqrt{\frac{1000}{T_e}}\,(cm^3 \cdot s^{-1}).$$

Table 2.38 Coefficients of absorption of the neutral helium emissions by atmospheric species (Banks and Kockarts 1973a,b)

Component	$\sigma\,(cm^2)$			
	58.4 (nm)	53.7 (nm)	52.2 (nm)	Factor
He	1.5(–13)	2.9(–14)	1.2(–14)	$\sqrt{1000/T}$
N_2	2.5(–17)	2.5(–17)	2.5(–17)	
O_2	2.7(–17)	2.7(–17)	2.7(–17)	
O	1.1(–17)	1.1(–17)	1.1(–17)	

However, the contribution of this process to helium emissions is insignificant because of the small helium ion density.

For the helium density in the upper atmosphere, of importance are the processes responsible for its spatial variability resulting from diurnal variations of the exospheric temperature and some escape. All this provides intra-atmospheric migration of helium (Shefov 1970a,b). To account for the necessary escape of helium, various chemical reactions were considered, such as the reaction of He^+ with O_2 and N_2 as well as the reaction of $He(2^3S)$ with O, O_2, and N_2 (Bates and Patterson 1962). However, Patterson (1967) showed that the reactions related to metastable helium atoms can provide no more than 1(%) of the necessary escape of helium. Moreover, Krassovsky (1969) noted that though in these reactions an energy greater than the eluding energy is released, these processes fail to provide the required escape of helium since the necessary quantities of oxygen and nitrogen molecules are far below the boundary of the exosphere from which helium atoms can escape. Therefore, the helium atoms that possess excessive energy will not be able to conserve it in their motion to the base of the exosphere because of their collisions with species of the dense atmosphere.

At the same time, the transport of ionospheric plasma from one hemisphere into another can occur due to a process which was proposed by Krassovsky (1959) and later considered by Rishbeth (1967). Therefore, the transport of helium ions from the summer to the winter hemisphere followed by their neutralization can produce an excess of neutral helium.

2.8 Atomic Hydrogen Emissions

Atomic hydrogen is the lightest constituent of the atmosphere and, therefore, the hydrogenous atmosphere extends up to altitudes of some tens of thousands of kilometers and forms a geocorona, which is sensitive to solar activity.

The hydrogen airglow is characterized by a series of emission lines. The structure of transitions is shown in Fig. 2.36; the parameters of the transitions are given in Table 2.39. Atomic hydrogen produces H_α (656.3 (nm)) and H_β (486.1 (nm)) Balmer series emissions in the visible spectral region and L_α (121.6 (nm)) and L_β (102.7 (nm)) Lyman series emissions in the ultraviolet region. In the Earth's atmosphere, the Lyman series radiation was detected first outside the auroral zone (1955) (Byram et al. 1957; Kupperian et al. 1959) and then the H_α radiation was detected at Zvenigorod (1957) (Shklovsky 1958, 1959; Prokudina 1959a) and at Abastumani (1958) (Fishkova and Markova 1958).

The explanation of the origin of the H_α emission under the conditions of the upper atmosphere was given by Shklovsky (1958, 1959) who showed that it is caused by the dissipation of the L_β (102.7 (nm)) solar radiation by the atomic hydrogen of the Earth's atmosphere; that is, of the three possible transitions (Fig. 2.37)

2.8 Atomic Hydrogen Emissions

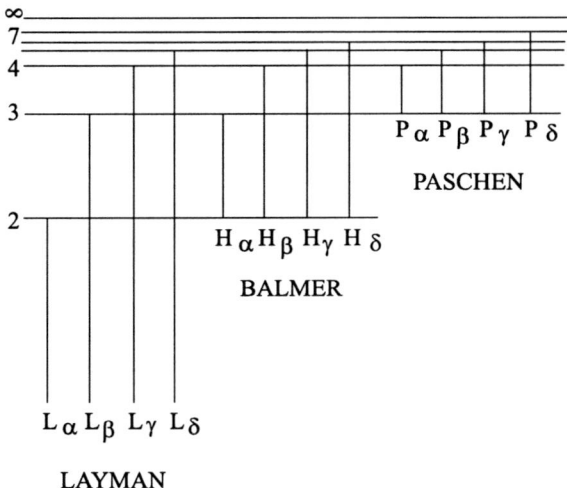

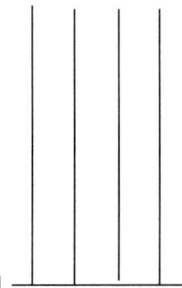

Fig. 2.36 Energy levels structure of the hydrogen atom and the wavelengths of the basic transitions

Table 2.39 Wavelengths (Striganov and Odintsova 1982), level energies (Moore 1949), transition probabilities (Allen 1973; Condon and Shortley 1935), and photon scattering factors (Donahue 1964) for the transitions of the hydrogen atom

λ(nm)	Symbol	Transition	E_u(eV)	E_l(eV)	$A(s^{-1})$	$g(s^{-1})$
121.56683	L_α	$2p^2P^o_{3/2} - 1s^2S_{1/2}$	10.20	0.00	6.23(8)	2.1(−3)
121.56737	L_α	$2p^2P^o_{1/2} - 1s^2S_{1/2}$	10.20	0.00	4.68(8)	–
102.57219	L_β	$3p^2P^o_{3/2} - 1s^2S_{1/2}$	12.09	0.00	1.65(8)	2.3(−6)
102.57230	L_β	$3p^2P^o_{1/2} - 1s^2S_{1/2}$	12.09	0.00	5.54(7)	–
656.27256	H_α	$3p^2P^o_{3/2} - 2s^2S_{1/2}$	12.09	10.20	2.23(7)	3.4(−7)
656.27720	H_α	$3p^2P^o_{1/2} - 2s^2S_{1/2}$	12.09	10.20	2.23(7)	–
656.28520	H_α	$3d^2D_{5/2} - 2p^2P^o_{3/2}$	12.09	10.20	6.43(7)	–
656.28520	H_α	$3d^2D_{3/2} - 2p^2P^o_{3/2}$	12.09	10.20	1.07(7)	–
656.27101	H_α	$3d^2D_{3/2} - 2p^2P^o_{1/2}$	12.09	10.20	5.36(7)	–
656.29099	H_α	$3s^2D_{1/2} - 2p^2P^o_{3/2}$	12.09	10.20	4.20(6)	–
656.27520	H_α	$3s^2D_{1/2} - 2p^2P^o_{1/2}$	12.09	10.20	2.10(6)	–

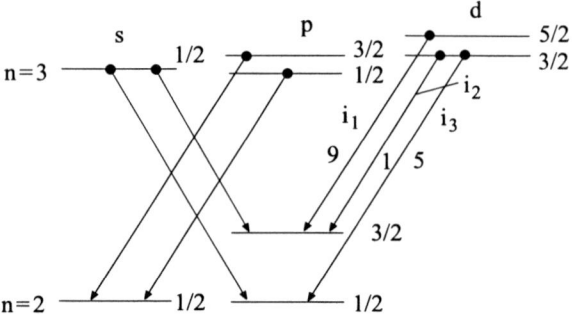

Fig. 2.37 Schematic of the transitions responsible for the 656.3-nm H_α emission. The intensities of the 2p–3d multiplet lines are indicated (Krasnopolsky 1987)

$$H(3p^2P) \rightarrow H(2s^2S) + h\nu(\lambda 656.3\,(nm)),$$

$$H(3s^2S) \rightarrow H(2p^2P) + h\nu(\lambda 656.3\,(nm)),$$

$$H(3d^2D) \rightarrow H(2p^2P) + h\nu(\lambda 656.3\,(nm)),$$

only the first one is due to absorption of ultraviolet radiation:

$$H(1s^2S) + h\nu(\lambda 102.7\,(nm)) \rightarrow H(3p^2P) \qquad \sigma_{L_\beta} = 2.8 \cdot 10^{-14}\,(cm^2),$$

and then reverse transitions with emission

$$H(3p^2P) \rightarrow H(1s^2S) + h\nu(\lambda 102.7\,(nm)),$$

$$H(3p^2P) \rightarrow H(2s^2S) + h\nu(\lambda 656.3\,(nm)) \qquad g_{656.3} = 3.4 \cdot 10^{-7}(s^{-1}).$$

The scattering factor was estimated by Donahue (1964). It should be stressed once again that the H_α emission arises due to several transitions ($3p^2P - 2s^2S$; $3s^2S - 2p^2P$; $3d^2D - 2p^2P$) of which only the $3p^2P - 2s^2S$ transition participates in the fluorescence process. In these calculations, the L_β emission intensity was taken equal to 0.06 (erg·cm^{-2}·s^{-1}), which corresponds to $3.1 \cdot 10^9$ (photon·cm^{-2}·s^{-1}), and to 0.75 (erg·cm^{-2}·s^{-1}·nm^{-1}) at the center of the line, and $F_{10.7} \sim 210$ was set. The profile of the L_α line in solar radiation is shown in Fig. 2.38. (Nicolet 1989a) and that of the L_β line is presented in Fig. 2.39 (Meier et al. 1987). The dependence of the L_β emission intensity on solar activity can be described, following Timothy (1977), by the formula (correlation coefficient is 0.88)

$$S(L_\beta) = \frac{F_{10.7} + 236}{4880}(erg \cdot cm^{-2} \cdot s^{-1})\ or\ S(L_\beta)$$
$$= 1.06 \cdot \frac{F_{10.7} + 236}{100} \cdot 10^9\,(photon \cdot cm^{-2} \cdot s^{-1}).$$

2.8 Atomic Hydrogen Emissions

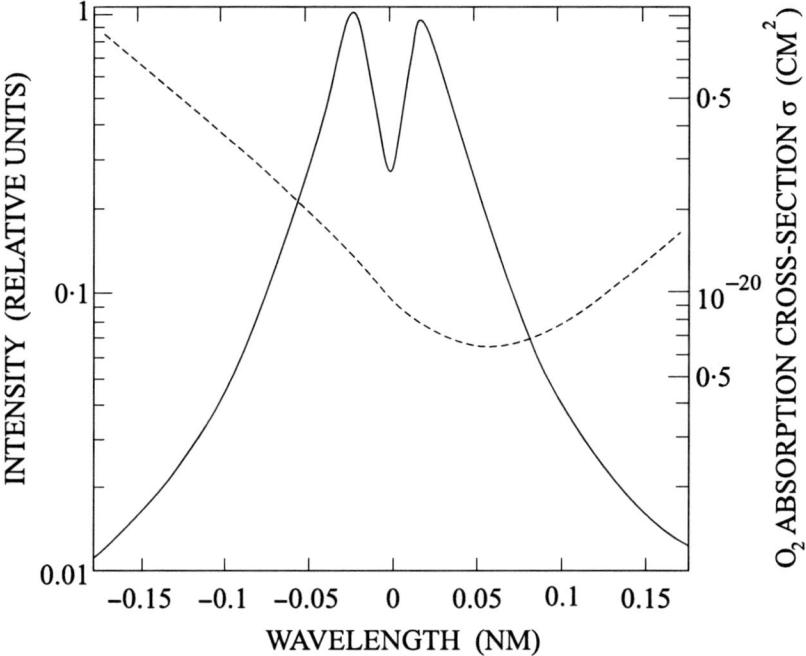

Fig. 2.38 Average profile of the L_α solar hydrogen emission (Nicolet 1989a)

These relations can be used to correct the coefficient $g_{656.3}$, though the correlation between the intensity and solar activity may be nonlinear.

The intensity of the H_α emission turned out equal to 10–15 (Rayleigh) (Shefov 1959); other hydrogen lines are much less intense since the intensity of the solar emissions producing these lines is lower compared to the L_β emission. Though

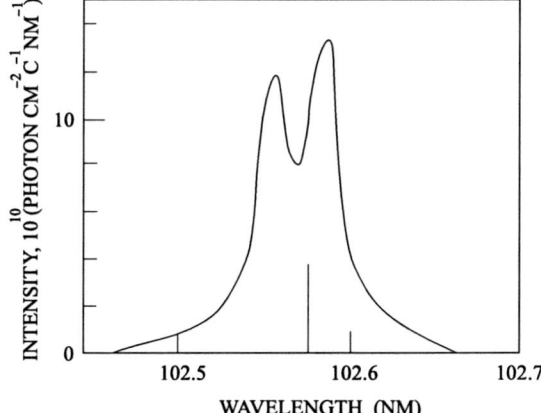

Fig. 2.39 Average profile of the L_β solar hydrogen emission (Nicolet 1989a). The *vertical line* indicates the position of the 102.576-nm line of atomic oxygen (Meier et al. 1987)

the intensity of hydrogen airglow is insignificant compared to other emissions of the upper atmosphere, its altitude distribution, as already mentioned, covers a very large altitude range; therefore, this airglow can become one of the basic atmospheric glows detected at altitudes above 700–800 (km) in sighting at 0°–90° zenith angles.

Observations performed at Abastumani and at other observatories for many years made it possible to obtain data on diurnal, seasonal, and long-term variations of the intensity of the H_α emission (Fishkova 1963, 1972, 1983; Krassovsky et al. 1966; Fishkova and Martsvaladze 1967; Armstrong 1967; Tinsley 1968b, 1969; Shefov and Truttse 1969; Shefov 1969a; Krassovsky 1971b; Tinsley and Meier 1971; Weller et al. 1971; Martsvaladze et al. 1971; Martsvaladze 1972; Fishkova 1972; Mange 1973; Martsvaladze and Fishkova 1982).

The H_α emission intensity varies significantly during a day, a season, and a solar cycle. The mean intensity is about 10 (Rayleigh). In the morning the intensity is higher than in the evening for the same solar zenith angles, and this is accounted for by the enhanced escape of hydrogen atoms in the daytime. The variations vary in character from night to night during different seasons (the solstice and the equinox). This seems to be related to the change of circulation systems in the upper atmosphere which provide intermixing of atmospheric constituents and transport of molecular and atomic hydrogen upward and downward, respectively. This is especially pronounced after artificial release of water vapor into the atmosphere resulting from launches of heavy satellites and rockets (Krassovsky et al. 1982; Martsvaladze and Fishkova 1982). An increase in intensity of hydrogen airglow by up to 30(%) was systematically detected 2–6 (days) after launches of space vehicles with apogees above 800 (km) in the western hemisphere.

The H_α emission intensity in the years of maximum solar activity is about half that in the years of its minimum. After geomagnetic disturbances, several-day variations occur which are similar to the variations of OH emission (Martsvaladze et al. 1971). A distinct asymmetry of the intensity in the northern and southern hemispheres is observed. The key feature of the variations is that the hydrogen density in the upper atmosphere is strongly affected by the escape of hydrogen atoms resulting from temperature. Therefore, the intensity of H_α emission does not increase, as is the case with other emissions, but decreases with temperature. The degree of elongation of the geocorona on the night side is considered in a number of publications (Kurt 1963; Shcheglov 1963, 1967; Meier 1969, 1974; Meier and Prinz 1970; Meier et al. 1977).

Consideration of the mechanism of occurrence of H_α emission shows that the optical thickness of the Earth's atmosphere to L_β radiation is determined by the atomic hydrogen and molecular oxygen densities:

$$\tau_{L_\beta}(Z) = \sigma_{L_\beta}(H) \cdot [H] \cdot H_H(Z) + \sigma_{L_\beta}(O_2) \cdot [O_2] \cdot H_{O_2}(Z)$$

and is equal to unity for altitudes of about 105 (km) due to absorption by molecular oxygen. The effective cross-section $\sigma_{L_\beta}(O_2)$ corresponds to the process

$$O_2 + h\nu(\lambda 102.7(nm)) \rightarrow O_2^+ + e \quad \sigma_{L_\beta}(O_2) = 1.6 \cdot 10^{-18} \, (cm^2) \ .$$

2.8 Atomic Hydrogen Emissions

The absorption of the radiation by hydrogen provides an optical thickness equal to unity at altitudes of about 120 (km). Thus, for altitudes above 300 (km) (for which the hydrogen emission should be considered) there is no need to solve a complicated problem of radiation transfer, and the rate of the process is determined only by fluorescence. The contribution of the radiation back-scattered in the lower thermosphere is due to the central part of the solar emission profile and makes ∼42(%), which is in good agreement with the data reported elsewhere (Brandt and Chamberlain 1959; Kurt 1963).

The H_α emission can also arise due to photoelectron excitation of hydrogen atoms in the daytime and in the auroral zone. An increase in the intensity of the narrow-band H_α emission to 400 (Rayleigh) was also observed (Harang and Pettersen 1967) which was accompanied by emissions from metastable ions of atomic nitrogen [NII] at $\lambda = 658.4$ (nm). The excitation of H_α emission in this case proceeds as follows:

$$H(1s^2S) + e \rightarrow H(3^2P, 3^2S, 3^2D) + e; \quad \sigma_{H_\alpha}(e) = 6 \cdot 10^{-18} (cm^2);$$

$$q_{H_\alpha}(e) = 6.5 \cdot 10^{-8} (s^{-1}).$$

Here it should be noted that the L_α emission that occurred due to electron excitation (Kondo and Kupperian 1967) cannot have an intensity (which is three times greater than that of the L_β emission) comparable to the observed intensity (∼2500 (Rayleigh)) (Byram et al. 1957; Kupperian et al. 1959; Kurt 1963). This was the subject of detailed consideration (Eather 1968; Donahue 1968; Krassovsky 1968b; Shefov 1969b,d, 1970a, 1973).

The H_α emission can also arise due to a process (Bowen 1934, 1947) which is related to the wavelength of L_β emission being different by 0.004 (nm) from the wavelength of atomic oxygen emission which can be excited by photoelectrons (Table 2.40):

$$O(2^3P) + e \rightarrow O(3^3D) + e; \quad \sigma_O(e) = 8 \cdot 10^{-18} (cm^2); \quad q_O(e) = 6 \cdot 10^{-9} (s^{-1}),$$

$$O(3^3D) \rightarrow O(2^3P) + h\nu(\lambda 102.7 (nm)) \quad A_{102.7}(O) = 1.5 \cdot 10^7 (s^{-1}) \quad (Allen\ 1973).$$

Table 2.40 Wavelengths of emissions from hydrogen and oxygen atoms (Striganov and Odintsova 1982)

Atom	Transition	λ (nm)
H	$3d^2D_{5/2} - 1s^2S_{1/2}$	102.57219
H	$3d^2D_{3/2} - 1s^2S_{1/2}$	102.57219
H	$3p^2P_{3/2} - 1s^2S_{1/2}$	102.57219
H	$3s^2S_{1/2} - 1s^2S_{1/2}$	102.57230
H	$3p^2P_{1/2} - 1s^2S_{1/2}$	102.57230
O	$3d^3D_{3,2,1} - 1p^{4^3}P_2$	102.57618

Thereafter, the state of hydrogen that is initial for the L$_\beta$ and H$_\alpha$ emissions is excited:

$$H(1s^2S) + h\nu[\lambda 102.7\,(nm); O(3^3D)] \to H(3p^2P) \,.$$

At altitudes above 120 (km) the optical thickness to the 102.7-nm atomic oxygen emission is less than unity. Therefore, the intensity of the H$_\alpha$ emission is determined by the relation

$$I(H_\alpha) = \frac{1}{3} \cdot \frac{A_{656.3}}{A_{656.3} + A_{102.7}} \cdot [H] \cdot H_H(Z) \cdot I(O) \cdot \int_0^\infty K_H(\lambda) \cdot f_O(\lambda) \cdot d\lambda \,,$$

where I(O) is the intensity of the atomic oxygen emission; f_O is the spectral profile function for this emission; $K_H(\lambda)$ is the absorption factor of atomic hydrogen; [H] and $H_H(Z)$ are, respectively, its density and scale height. In deriving this formula the assumption was made that only one-third of the radiation of atomic oxygen is directed upward to the geocorona.

The oxygen emission profile can be considered as a Doppler profile:

$$f_O(\lambda) = \frac{1}{\sqrt{\pi}} \cdot \exp\left[-\frac{(\lambda - \lambda_0)^2}{D_O \cdot T_O}\right] \,,$$

where λ_0 is the wavelength of atomic oxygen emission and

$$D_O = \frac{2k \cdot N \cdot \lambda_0^2}{M_O \cdot c^2} = 1.22 \cdot 10^{-8} \,.$$

Here, $k = 1.38 \cdot 10^{-16}$ (erg $\cdot$ K^{-1}) is the Boltzmann constant, $N = 6.025 \cdot 10^{23}$ (molecule $\cdot$ g^{-1} $\cdot$ mol^{-1}) is the Avogadro number, $c = 3 \cdot 10^{10}$ (cm $\cdot$ s^{-1}) is the velocity of light, and $M_O = 16$ is the atomic mass of oxygen.

The absorption constant, in view of damping and Doppler broadening, can be represented by a Voigt profile (Mitchell and Zemansky 1934; Unsöld 1938; Ambartsumyan et al. 1952):

$$K(x) = K_0 \cdot \frac{a}{2\pi} \cdot \int_{-\infty}^\infty \frac{e^{-y^2} \cdot dy}{\left(\frac{a}{2}\right)^2 + (x-y)^2} \,,$$

where

$$x = \frac{\lambda_H - \lambda}{\sqrt{D_H \cdot T_H}}; \qquad a = \frac{\gamma}{\sqrt{D_H \cdot T_H}};$$

and the coefficient of absorption at the center of the line

$$K_0 = \frac{\sqrt{\pi} \cdot e^2}{m_e \cdot c^2} \cdot \frac{\lambda_H^2}{\sqrt{D_H \cdot T_H}} \,.$$

2.8 Atomic Hydrogen Emissions

Here,

$$D_H = \frac{2k \cdot N \cdot \lambda_H^2}{M_H \cdot c^2} = 1.95 \cdot 10^{-7},$$

and also $e = 4.77 \cdot 10^{-10}$ is the elementary charge, $m_e = 9.0 \cdot 10^{-28}$ (g) is the electronic mass, $M_H = 1$ is the atomic mass of hydrogen, and $\gamma = 1.18 \cdot 10^{-5}$ (nm) is the damping constant.

The emission wavelength for a hydrogen atom is $\lambda_H = \lambda_0 + \delta\lambda$, where $\delta\lambda = 3.88 \cdot 10^{-3}$ (nm). Approximation of the absorption factor yields (Unsöld 1938)

$$K(\lambda) = K_0 \cdot \frac{\gamma}{2\pi} \cdot \frac{\sqrt{D_H \cdot T_H}}{(\lambda_H - \lambda)^2},$$

whence the effective cross-section for absorption is

$$\sigma_{102.7}(H) = K_0 \cdot \gamma \cdot \frac{\sqrt{D_H \cdot T_H} \cdot \sqrt{D_0 \cdot T_0}}{2\pi \cdot (\delta\lambda)^2} = \frac{\sqrt{\pi} \cdot e^2}{m_e \cdot c^2} \cdot \frac{\lambda_H^2}{2\pi \cdot (\delta\lambda)^2} \cdot \sqrt{D_0 \cdot T_0}$$

$$= 2.3 \cdot 10^{-11} \cdot \sqrt{\frac{T_0}{1000}}, (cm^2).$$

Assuming that the hydrogen column density at altitudes above 100 (km) is $\sim 10^{13}$ (cm^{-2}), we obtain the intensity of the H$_\alpha$ emission as

$$I(H_\alpha) = I(O) \cdot \sqrt{\frac{T_0}{1000}}.$$

According to calculations (Green and Barth 1967), the intensity of the 102.7-nm atomic oxygen emission is ~ 100 (Rayleigh). Estimates (Tinsley 1969) show that the contribution of the interplanetary medium is about 4 (Rayleigh).

Excited hydrogen atoms can also arise due to photorecombination (Allen 1973):

$$H^+ + e \rightarrow H(3^2P, 3^2S, 3^2D) + h\nu(\lambda < 814\,(nm))$$

$$\alpha_{H_\alpha} = 1.52 \cdot 10^{-13} \cdot \sqrt{\frac{1000}{T_e}}\,(cm^3 \cdot s^{-1}).$$

For the L$_\beta$ emission, the recombination rate is

$$\alpha_{L_\beta} = 6.32 \cdot 10^{-14} \cdot \sqrt{\frac{1000}{T_e}}, (cm^3 \cdot s^{-1}).$$

However, since the electron and proton densities are small ($n_e \sim 10^5$ (cm^{-3}), [H$^+$] $\sim 2 \cdot 10^4$ (cm^{-3})) at altitudes above 700 (km), the contribution of this process is of the order of 0.01 (Rayleigh).

The atomic hydrogen density at the altitudes of the lower thermosphere depends on solar activity and shows a long-term trend. Based on rocket measurements of the L$_\alpha$ emission intensity, the atomic hydrogen density has been estimated

(Semenov 1997; Shefov and Semenov 2002) and its proportionality to the level of solar activity and long-term increase has also been revealed. These results are presented in Sect. 7.3.

Of importance for the density of hydrogen atoms, like for the helium density in the upper atmosphere, are its spatial variations caused by the diurnal variations of exospheric temperature and by some escape of hydrogen. All this gives rise to intra-atmospheric hydrogen migration (Shefov 1970a,b).

At the same time, the transport of ionospheric plasma from one hemisphere into another can occur due to a process which was proposed by Krassovsky (1959) and later considered by Rishbeth (1967). Therefore, the transport of hydrogen ions from the summer to the winter hemisphere, followed by their neutralization, can produce an excess of neutral hydrogen.

2.9 The Continuum

The continuum present in the nightglow of the upper atmosphere has long attracted the attention of researchers who observed the net stellar light and the time and space variations of the sky glow in those spectral regions where no discrete emissions were detected (Straižys 1977; McDade et al. 1986b). The true spectral structure of this quasi-continuous radiation is still not exactly known. Therefore, the notion of continuum implies the total of the radiation giving an actually continuous spectrum and a possible radiation producing a series of weak diffuse molecular bands which cannot be resolved spectroscopically.

Most of the investigations of the space–time characteristics of the continuum were performed in the visible range, including near 530 (nm) (Chuvaev 1952, 1961; Heppner and Meredith 1958; Shefov 1959, 1960, 1961d; Karyagina and Tulenkova 1959; Yarin 1961b; Taranova 1962; Tarasova and Slepova 1964; Gindilis 1965; Davis and Smith 1965; Greer and Best 1967; Huruhata et al. 1967; Baker and Waddoups 1967, 1968; Sparrow et al. 1968; Fishkova 1969, 1970, 1983; Robley and Vilkki 1970; Robley 1973; Gadsden and Marovich 1973; Straižys 1977; Sobolev 1978a,b, 1979; Witt et al. 1981; Misawa and Takeuchi 1982; McDade et al. 1984, 1986b). The data of this research have shown that the intensity of the atmospheric component is 10 (Rayleigh·nm^{-1}).

The airglow has a minimum intensity near the equator ($\sim$1–2 (Rayleigh·nm^{-1})); at the polar latitudes its intensity reaches 10–20 (Rayleigh·nm^{-1}) (Taranova 1962; Davis and Smith 1965). The intensity of the extraterrestrial radiation component consisting of stellar and zodiacal light, investigated by a number of researchers (Megill and Roach 1961; Roach and Megill 1961; Chuvaev 1961; Sternberg and Ingham 1972; Sharov and Lipaeva 1973; Roach 1964; Roach and Gordon 1973), is on the average 10 (Rayleigh·nm^{-1}). From these measurements it could be suggested that there is some dependence of the continuum intensity on geomagnetic coordinates.

2.9 The Continuum

Midlatitude measurements show that the continuum intensity varies from night to night and during a night, and, on the average, is a minimum after local midnight. According to data obtained at midlatitudes (Sobolev 1978a,b; Fishkova 1969, 1970, 1983), there are seasonal intensity variations.

Measurements of the continuum spectral distribution have revealed the existence of a wide peak near 600 (nm) and a decrease in intensity in the ultraviolet and infrared regions (Fig. 2.40) (Shefov 1959, 1961d; Yarin 1961b; Gindilis 1965; Fishkova 1969, 1970, 1983; Noxon 1978; Sobolev 1978a,b). The continuum intensity is on the average 20–30 (Rayleigh·nm^{-1}).

The spectral distribution was estimated based on ground-based electrophotometric measurements. From these data it follows that the spectral composition of the continuum varies during seasons on the average within about 20(%), though some larger diversions have also been detected that testify to several processes responsible for the continuum formation (Sobolev 1978a,b, 1979). The correlations between the continuum and the emissions arising in the mesopause region were repeatedly

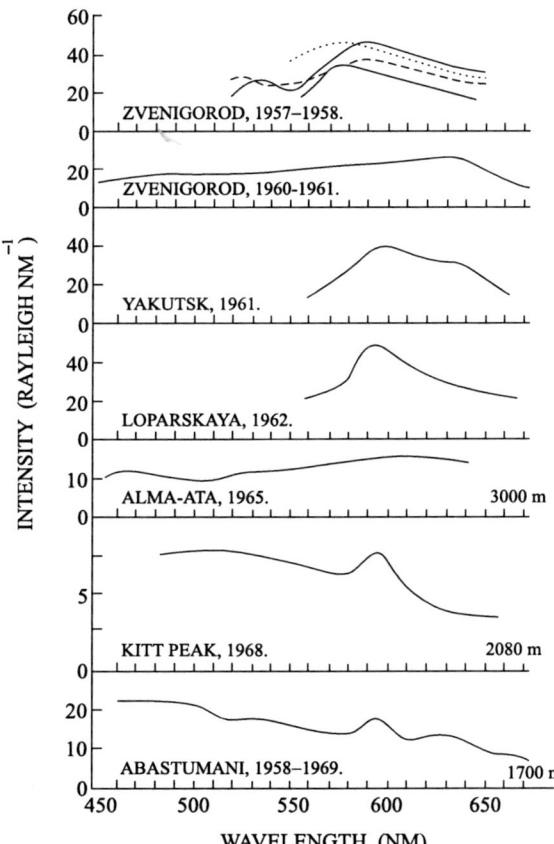

Fig. 2.40 Intensity distributions in the nightglow continuum spectrum, according to observations at various stations (for the mountain stations the altitude above sea level is indicated) (Fishkova 1970, 1983)

investigated (Shefov 1961d; Yarin 1961b; Fishkova 1969, 1970, 1983). Many researchers noted both noncorrelating and correlating continuum components.

To elucidate the nature of the continuum, rocket measurements of altitude distributions were repeatedly performed. They have shown that the maximum of the continuum emission rate is located at altitudes of ~ 100 (km) and below, and the entire altitude range of the emission layer is from 70 to 130 (km). However, it should be noted that alongside the main maximum in the altitude distribution of the emission rate, in some cases secondary maxima are detected in the lower part (and sometimes in the upper one) of the main emission layer. This is confirmed by recent measurements (Fig. 2.41) (Gurvich et al. 2002).

It should be stressed that notwithstanding the spectral emission rate of the continuum being small (though in some cases it was measured to be as high as 100–150 (Rayleigh·nm^{-1}) (Yarin 1961b), the integrated intensity appears rather significant over a wide spectral range. Thus, for the range 0.35–1.2 (μm) the intensity is about 10 (kilorayleigh) and can reach 20–30 (kilorayleigh); that is, the observed nightglow is in fact the continuum glow. However, for the more long-wave range practically no data about the continuum are available. A theoretical study where an increase in intensity (to 150 (kilorayleigh)) in the long-wave spectral range was proposed (Wraight 1975, 1977, 1986) was based on incorrect use of diverse measurements of the spectra of the nightglow of the upper atmosphere (Shefov 1978).

The first theoretical notions about the nature of the atmospheric continuum (Krassovsky 1951b) proposed a chemical process where nitrogen dioxide molecules radiate due to the sequence of reactions

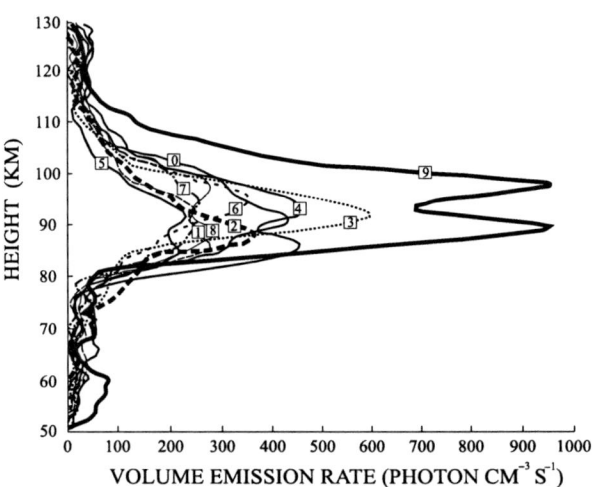

Fig. 2.41 Altitude distribution of the emission rate in the range 420–530 (nm) for the upper atmosphere (Gurvich et al. 2002). The *numerals on the curves* correspond to the numbers of measuring sessions on the "Mir" spacecraft

2.9 The Continuum

$$NO + O \rightarrow NO_2^* \quad \alpha_{NOO} = 2.9 \cdot 10^{-17} \cdot \exp(-530/T) \; (cm^3 \cdot s^{-1}),$$

$$NO_2^* \rightarrow NO_2 + h\nu(\lambda 0.4 - 2 \, (\mu m)) \quad A_{NO_2^*} = 2.2 \cdot 10^4 \; (s^{-1}).$$

This process is well known in laboratory investigations (Karmilova and Kondratiev 1951; Kondratiev 1958; Kondratiev and Nikitin 1974, 1981). Subsequently this viewpoint received support (Becker et al. 1971; Gadsden and Marovich 1973; Whitten and Poppoff 1971; Noxon 1978; Krassovsky 1978; Krassovsky et al. 1980; Bates 1982).

Simultaneously with the above reactions, a three-body reaction proceeds (Machael et al. 1976; Bates 1982):

$$NO + O + M \rightarrow NO_2^* + M \quad \alpha_{NOOM} = 1.55 \cdot 10^{-32} \cdot \exp(584/T) \; (cm^6 \cdot s^{-1}).$$

The spectral distributions of the emissions resulting from these reactions cannot be separated since the reactions proceed simultaneously. Therefore, it is possible to consider a unified distribution, as this was done by Kenner and Ogrizlo (1984). Following these authors, emission was detected in the range 0.4–1.6 (μm), while Golde et al. (1973) observed it up to 3 (μm).

The spectral distribution can be approximated by the relation (photons, relative units)

$$I_{NOO} = 100 \cdot \exp\left[-\left(2.78 \cdot \log_e \frac{\lambda}{0.66}\right)^2\right],$$

where the wavelength λ is expressed in μm.

Another possible process is also related to nitric oxide (Clough and Thrush 1967):

$$NO + O_3 \rightarrow NO_2^* + O_2 \quad \alpha_{NOO_3} = 1.26 \cdot 10^{-12} \cdot \exp(-2100/T) \; (cm^3 \cdot s^{-1}).$$

The spectral distribution of the radiation intensity (0.6–3.0 (μm)) for this reaction is given elsewhere (Kenner and Ogrizlo 1984). It has a maximum at about 1.2–1.25 (μm) and can be approximated as (photons, relative units)

$$I_{NOO_3} = 100 \cdot \exp\left[-\left(2.38 \cdot \log_e \frac{\lambda}{1.25}\right)^2\right].$$

Here, as in the previous formula, the wavelength λ is expressed in μm.

The previous reaction can occur with participation of an excited ozone molecule (Kenner and Ogrizlo 1984)

$$NO + O_3^* \rightarrow NO_2^* + O_2 \quad \alpha_{NOO_3^*} = 1.8 \cdot 10^{-15} \; (cm^3 \cdot s^{-1}),$$

whose formation is provided by the process in which an excited oxygen molecule takes part which radiates the Herzberg I system bands:

$$O_2(A^3\Sigma_u^+) + O_2 \to O_3^* + O \quad \alpha_{O_2^*O_2} = 1.8 \cdot 10^{-15} \, (cm^3 \cdot s^{-1}) \, .$$

The lifetime of excited ozone molecules is determined both by photodissociation (Banks and Kockarts 1973a)

$$O_3^* + h\nu(\lambda<1180\,(nm)) \to O_2 + O \quad j_{O_3^*} = 1 \cdot 10^{-2} \, (s^{-1})$$

and by the emissions

$$O_3^* \to O_3 + h\nu(\lambda 9.065\,(\mu m)) \quad A_{9.065} = 0.667 \, (s^{-1})$$

and

$$O_3^* \to O_3 + h\nu(\lambda 14.4\,(\mu m)) \quad A_{14.4} = 0.252 \, (s^{-1}) \, ,$$

and also by the reaction

$$O_3^* + M \to O_3 + M \quad \beta_{O_3^*M} = 2 \cdot 10^{-14} \, (cm^3 \cdot s^{-1}) \, .$$

The spectral distribution of the emission $(0.5-1.4\,(\mu m))$ that arises due to the reaction of nitric oxide with an excited ozone molecule was determined by Kenner and Ogrizlo (1984). The long-wave portion of the distribution was estimated up to 2 (μm); the distribution peaks near $0.85-0.87$ (μm). Based on experimental data (Fontijn and Schiff 1961; Fontijn et al. 1964; Clough and Thrush 1967; Sobolev 1978), an approximation formula (photons, relative units) was constructed:

$$I_{NOO_3^*} = 100 \cdot \exp\left[-\left(2.53 \cdot \log_e \frac{\lambda}{0.87}\right)^2\right] \, ,$$

where the wavelength λ is expressed in μm. The results of calculations for various processes are presented in Table 2.41 and Fig. 2.42.

Using ground-based measurements (Sobolev 1978a,b, 1979), it is possible to estimate the intensity of the continuum near 1.2 (μm) produced due to the reaction of nitric oxide and ozone to be about 15 (Rayleigh·nm^{-1}). Thus, the integrated intensity of the continuum in the infrared region is about 15 (kilorayleigh). However, it should be stressed once again that this estimate refers to ground-based measurements rather than to data obtained at altitudes above 100 (km).

The ozone density at altitudes above 100 (km) is low and, moreover, it decreases in the daytime due to dissociation. Therefore, the reactions with participation of ozone can have an appreciable effect only at altitudes below 100 (km).

Thus, the continuum airglow is related mainly to the pre-dissociative emission from nitrogen dioxide molecules. The emission intensity seems to be also affected by the deactivation processes (practically at altitudes below 100 (km)) (Becker et al. 1971)

$$NO_2^* + M \to NO_2 + M \quad \beta_{NO_2^*M} = 2.44 \cdot 10^{-11} \, (cm^3 \cdot s^{-1}) \, .$$

2.9 The Continuum

Table 2.41 Relative spectral distributions of the intensity (photons) of continua calculated based on laboratory measurements by approximation formulas

λ(μm)	NO+O	NO+O$_3^*$	NO+O$_3$	λ(μm)	NO+O	NO+O$_3^*$	NO+O$_3$
0.40	5	2	–	1.20	6	52	99
0.45	33	6	–	1.25	–	43	100
0.50	56	14	–	1.30	–	36	98
0.55	78	26	–	1.35	–	29	96
0.60	93	42	5	1.40	–	24	92
0.65	100	58	9	1.5	–	15	82
0.70	96	74	15	1.6	–	10	70
0.75	86	87	23	1.7	–	–	58
0.80	74	96	33	1.8	–	–	46
0.85	59	100	44	1.9	–	–	36
0.90	47	99	55	2.0	–	–	28
0.95	35	95	66	2.1	–	–	21
1.00	26	88	76	2.2	–	–	16
1.05	19	80	85	2.3	–	–	12
1.10	14	70	92	2.4	–	–	9
1.15	9	61	96	2.5	–	–	–

Theoretical analysis of the problems relevant to the nitric oxide density in the thermosphere shows that almost 40 reactions contribute to the NO density, and an important role is played by atomic nitrogen, including that in the metastable state ^{2}D.

Some processes responsible for the formation of NO (Gordiets and Markov 1983) in the thermosphere during intense auroras under the conditions of high temperatures $(T \sim 3500\,(K))$, such as

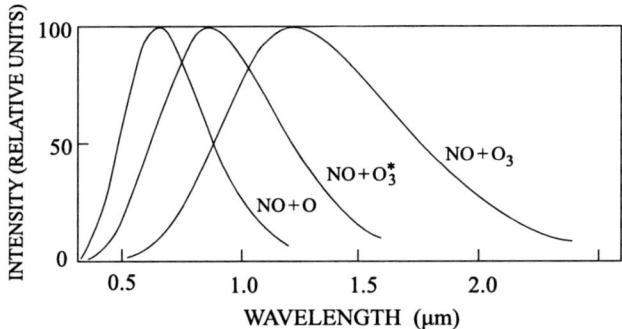

Fig. 2.42 Approximate spectral distributions of the intensity (photon·cm^{-2}·s^{-1}, relative units) of a continuum for the reactions of different species indicated near each curve

$$N_2 + O \to NO + N \qquad \alpha_{N_2O} = 1.2 \cdot 10^{-10} \cdot \exp\left[-\frac{38000}{T}\right] (cm^3 \cdot s^{-1}),$$

$$N + O_2 \to NO + O \qquad \alpha_{NO_2} = 2.2 \cdot 10^{-14} \cdot \exp\left[-\frac{3570}{T}\right] (cm^3 \cdot s^{-1}),$$

which are accompanied by the deactivation

$$NO + N \to N_2 + O \qquad \beta_{NON} = 2.6 \cdot 10^{-11} (cm^3 \cdot s^{-1}),$$

$$NO + O \to N + O_2 \qquad \beta_{NOO} = 6 \cdot 10^{-12} \cdot \exp\left[-\frac{19860}{T}\right] (cm^3 \cdot s^{-1}),$$

can provide a considerable increase both in nitric oxide density and in continuum intensity (Zipf et al. 1970; Truttse 1973).

An important part in the formation of nitric oxide is played, besides the above processes, by reactions with participation of metastable nitrogen atoms, such as

$$N(^2D) + O_2 \to NO + O \qquad \alpha_{N^2DO_2} = 7.4 \cdot 10^{-12} \cdot \left(\frac{T}{300}\right)^{0.5} (cm^3 \cdot s^{-1}).$$

These atoms are produced due to the reactions

$$NO^+ + e \to N(^2D) + O \qquad \alpha_{NO^+e} = 3 \cdot 10^{-7} \cdot \left(\frac{300}{T}\right)^{0.5} (cm^3 \cdot s^{-1}),$$

$$N_2^+ + e \to N(^2D) + N \qquad \alpha_{N_2^+e} \ll 3 \cdot 10^{-7} (cm^3 \cdot s^{-1}),$$

$$N_2^+ + O \to NO^+ + N(^2D) \qquad \alpha_{N_2^+O} = 1.4 \cdot 10^{-10} \cdot \left(\frac{300}{T}\right)^{0.44} (cm^3 \cdot s^{-1}).$$

The set of reactions involved can be extended further.

The photochemical equations for estimating the NO density should be complemented with an equation describing the diffusion flow of NO molecules downward which would compensate the losses due to photodissociation. The loss flux can be estimated by the formula (Banks and Kockarts 1973a)

$$F(NO) = \frac{1}{2} \cdot [NO] \cdot j_{NOh\nu}(NO) \cdot H_{NO}$$

for the process of photodissociation

$$NO + h\nu(\lambda < 190 (nm)) \to N + O \qquad j_{NOh\nu} = 1.34 \cdot 10^{-5} (s^{-1}).$$

The NO influx due to diffusion can be approximately described by the formula

$$F(NO) = [NO] \cdot \frac{D(NO)}{H_c} \cdot \left[\frac{H_c}{H_{NO}} - \frac{M(NO)}{M_c}\right].$$

The diffusion coefficient of the component M_i can be determined by the formula

$$D(i) = 3 \cdot 10^{17} \cdot \frac{T^{0.5}}{[i]}.$$

Here H_c and H_{NO} are the scale heights of neutral atmosphere and NO molecules, respectively; M_c and $M(NO)$ are the corresponding molecular masses.

At altitudes below 120 (km) the molecular diffusion coefficient is

$$D(NO) \sim 2.5 \cdot 10^7 \; (cm^2 \cdot s^{-1}).$$

For altitudes above 120 (km), instead of this quantity, the coefficient of eddy diffusion should be used:

$$K(NO) \sim 5 \cdot 10^6 \; (cm^2 \cdot s^{-1}).$$

Thus, it is obvious that the airglow continuum of the upper atmosphere, providing a significant portion of the nightglow in the visible spectral region, reflects the density of nitric oxide and its interaction with atomic oxygen.

2.10 Nitric Oxide Emissions

Nitric oxide is an important constituent of the upper atmosphere. The structure of its molecule allows many electronic states with like potential functions most of which having practically no shift relative to one another (Gilmore 1965). In a molecule of this type, according to the Franck–Condon principle, only transitions with $\Delta v = 0$ can occur.

Emission bands of the $\beta(B^2\Pi - X^2\Pi)$, $\gamma(A^2\Sigma^+ - X^2\Pi)$, $\delta(C^2\Pi - X^2\Pi)$, and $\varepsilon(D^2\Sigma^+ - X^2\Pi)$ systems were identified in the ultraviolet region (170–250 (nm)) of dayglow spectra of the upper atmosphere. However, they were not investigated comprehensively.

The infrared emissions that arise due to vibrational transitions from the ground state (1–0) at 5.3 (μm) and (2–0) at 2.7 (μm) represent the greatest interest:

$$NO(X^2\Pi, v = 1) \rightarrow NO(X^2\Pi, v = 0) + h\nu(5.3\,(\mu m)), A_{1-0} = 13.38\,(s^{-1}),$$

$$NO(X^2\Pi, v = 2) \rightarrow NO(X^2\Pi, v = 0) + h\nu(2.7\,(\mu m)), A_{2-0} = 0.852\,(s^{-1}).$$

For the 1.8-μm (3–0) band, $A_{3-0} = 0.067\,(s^{-1})$ (Schurin and Ellis 1966; Rothman et al. 1983).

Measurements of the intensity of the 5.3-μm (1–0) band under the conditions of the upper atmosphere are reported elsewhere (Stair et al. 1985; Baker et al. 1977; Ulwick et al. 1985). At altitudes of about 140 (km) the thickness of the emission layer was 60–70 (km). The conditions of the occurrence of this emission were considered theoretically (Degges 1971; Ogawa 1976; Ogawa and Kondo 1977; Baker et al. 1977; Witt et al. 1981; Caledonia and Kennealy 1982).

The greatest contribution to the formation of vibrationally excited NO molecules is provided by the reaction (Ogawa 1976)

$$NO(v=0) + O \rightarrow NO(v=1) + O \quad \alpha_{NO,O} = 6.5 \cdot 10^{-11} \cdot \exp\left(-\frac{2900}{T}\right) (cm^3 \cdot s^{-1}).$$

The fast oxygen atoms that appear during auroras considerably increase the yield of this reaction (Krassovsky 1971a).

At altitudes below 100 (km), the transfer of vibrational energy from oxygen and nitrogen molecules to NO molecules is of importance (Ogawa 1976):

$$NO(v=0) + O_2(v=1) \rightarrow NO(v=1) + O_2$$

$$\alpha_{NO,O_2} = 2 \cdot 10^{-14} \cdot \exp\left(-\frac{460}{T}\right) (cm^3 \cdot s^{-1}),$$

$$NO(v=0) + N_2(v=1) \rightarrow NO(v=1) + N_2(v=0)$$

$$\alpha_{NO,N_2} = 4.22 \cdot 10^{-10} \cdot \exp\left(-\frac{86.35}{T^{1/3}}\right) (cm^3 \cdot s^{-1}).$$

The O_2 and N_2 molecules are vibrationally excited in collisions with thermal electrons (Gordiets et al. 1978; Rusanov and Fridman 1984):

$$O_2 + e \rightarrow O_2(v) + e, \quad \sigma_{O_2,e} = 1 \cdot 10^{-17} cm^2,$$

$$\alpha_{O_2,e} = 1.1 \cdot 10^{-10} \cdot \left(\frac{T_e}{1000}\right)^{0.5} \cdot \exp\left(-\frac{2240}{T_e}\right) (cm^3 \cdot s^{-1}),$$

$$N_2 + e \rightarrow N_2(v) + e, \quad \sigma_{N_2,e} = 3 \cdot 10^{-16} cm^2,$$

$$\alpha_{N_2,e} = 3.3 \cdot 10^{-11} \cdot \left(\frac{T_e}{1000}\right)^{0.5} \cdot \exp\left(-\frac{3380}{T_e}\right) (cm^3 \cdot s^{-1}).$$

At altitudes above 100 (km) there occur the reactions (Malkin 1971; Kennealy et al. 1978)

2.10 Nitric Oxide Emissions

$$N(^4S) + O_2 \to NO(v = 0 \div 10) + O,$$

$$\alpha_{N(S),O_2} = 1.1 \cdot 10^{-14} \cdot T \cdot \exp\left(-\frac{3151}{T}\right) (cm^3 \cdot s^{-1}),$$

$$N(^2D) + O_2 \to NO + O, \quad \alpha_{N(D),O_2} = 7.4 \cdot 10^{-12} \cdot \left(\frac{T}{300}\right)^{0.5} (cm^3 \cdot s^{-1}),$$

$$N(^2P) + O_2 \to NO + O, \quad \alpha_{N(P),O_2} = 2.6 \cdot 10^{-12} (cm^3 \cdot s^{-1}).$$

These processes increase in importance during auroras.

At altitudes of 80–110 (km), the prevailing contribution is provided by fluorescence:

$$NO(v = 0) + h\nu(5.3\,(\mu m)) \to NO(v = 1)$$

and a severalfold smaller contribution is made by the re-emission from the level with $v = 2$:

$$NO(v = 0) + h\nu(2.7\,(\mu m)) \to NO(v = 2),$$

$$NO(v = 2) \to NO(v = 1) + h\nu(5.4\,(\mu m)).$$

In this case, fluorescence occurs both due to solar radiation, whose excitation rate is given by (Ogawa 1976)

$$S = g_{S5.3} \cdot \exp(-\tau \cdot Ch p\chi)\,(s^{-1}), \quad g_{S5.3} = 1.0 \cdot 10^{-4}\,(s^{-1}) \quad \text{(Ogawa 1976)},$$

and due to the thermal radiation of the lower atmosphere for which $g_{T5.3} = 3.8 \cdot 10^{-4}\,(s^{-1})$ (Ogawa 1976). For this case it was assumed that the absorption factor is $\sigma_{NO,5.3} = 1.0 \cdot 10^{-15}\,(cm^2)$. The optical thickness $\tau_{5.3}$ is practically determined by the absorption of NO molecules, including at altitudes below 80 (km), since the attenuation of radiation due to Rayleigh scattering ($\sigma_R = 5 \cdot 10^{-31}\,(cm^2)$) is insignificant. Therefore, we have $\tau_{5.3} = \sigma_{NO} n(NO) \cdot H_{NO}(Z)$. Absolutely the same relation takes place for the 2.7-μm emission. For this case, we have $g_{S2.7} = 3.1 \cdot 10^{-4}\,(s^{-1})$, $g_{T2.7} = 2.3 \cdot 10^{-9}\,(s^{-1})$, and $\sigma_{NO,5.3} = 1.0 \cdot 10^{-15}\,(cm^2)$ (Degges 1971). The contribution of the (2–1) transition is small and it can be neglected.

In the daytime, the formation of NO ($v = 1$ and 2) due to fluorescence in ultraviolet transitions is possible:

$$NO(v = 0) + h\nu(\gamma; 226\,(nm)) \to NO(A^2\Sigma^+, v)$$
$$\to NO(X^2\Pi, v = 1) g_{\gamma 1} = 3 \cdot 10^{-5}\,(s^{-1}),$$

$$NO(v = 0) + h\nu(\delta; 190\,(nm)) \to NO(C^2\Pi, v)$$
$$\to NO(X^2\Pi, v = 1)\ g_{\delta 1} = 2.5 \cdot 10^{-5}\,(s^{-1}),$$

$$NO(v = 0) + h\nu(\varepsilon; 190\,(nm)) \to NO(D^2\Sigma^+, v) \to NO(X^2\Pi, v = 1).$$

The excitation of the vibrational levels of the ground state can occur not only due to the above processes, but also in collision of NO molecules with electrons:

$$NO(v=0) + e \rightarrow NO(v=1) + e,$$

$$\alpha_{NO,1,e} = 1.8 \cdot 10^{-10} \cdot \left(\frac{T_e}{1000}\right)^{0.5} \exp\left(-\frac{2680}{T_e}\right) \, (cm^3 \cdot s^{-1}),$$

$$NO(v=0) + e \rightarrow NO(v=2) + e,$$

$$\alpha_{NO,2,e} = 1.8 \cdot 10^{-10} \cdot \left(\frac{T_e}{1000}\right)^{0.5} \exp\left(-\frac{5360}{T_e}\right) \, (cm^3 \cdot s^{-1}).$$

It can be seen that under usual conditions the contribution of the electron excitation of NO molecules is much less than that of the excitation by other processes.

Alongside the excitation there occur deactivation of the molecules:

$$NO(v=1) + M \rightarrow NO(v=0), \quad \beta_{NO,10} = 6.5 \cdot 10^{-11} \, (cm^3 \cdot s^{-1}),$$

$$NO(v=2) + M \rightarrow NO(v=1), \quad \beta_{NO,21} = 3.0 \cdot 10^{-11} \, (cm^3 \cdot s^{-1}),$$

$$NO(v=2) + M \rightarrow NO(v=0), \quad \beta_{NO,20} = 4.1 \cdot 10^{-11} \, (cm^3 \cdot s^{-1}),$$

and their photodissociation:

$$NO + h\nu(\lambda \leq 190\,(nm)) \rightarrow N + O \quad j_{NO,i} = 1.34 \cdot 10^{-5} \, (s^{-1}),$$

resulting in destruction and excitation of the molecules (Nicolet 1981). This process is more efficient than the photoionization (Rothman et al. 1978)

$$NO + h\nu(\lambda \leq 134\,(nm)) \rightarrow NO^+ + e \quad j_{NO,ph} = 5 \cdot 10^{-7} \, (s^{-1}).$$

It should, however, be noted that Ogawa and Kondo (1977) took the photoionization rate equal to $1.38 \cdot 10^{-6} (s^{-1})$.

In addition, the following ion-exchange reaction can occur:

$$NO + O_2^+ \rightarrow NO^+ + O_2 \quad \beta_{NO,O_2^+} = 4.4 \cdot 10^{-10} \, (cm^3 \cdot s^{-1}).$$

The collisions of NO molecules with other species have much lower rates.

In view of the above reactions, the 5.3-μm emission rate for the NO molecule is determined by the relation

$$Q_{5.3} = \left\{ [\text{NO}] \cdot \left[\alpha_{\text{NO},\text{O}} \cdot [\text{O}] + \alpha_{\text{NO},\text{O}_2^*} \cdot [\text{O}_2^*] + \alpha_{\text{NO},\text{N}_2^*} \cdot [\text{N}_2^*] + S + g_{S5.3} \right] \right\}$$

$$\left/ \left\{ 1 + \frac{\beta_{\text{NO},10} \cdot \left[[\text{O}] + [\text{O}_2] + [\text{N}_2]\right] + \beta_{\text{NO},\text{O}_2^+} \cdot [\text{O}_2^+]}{A_{5.3}} \right\} \right. .$$

The importance of the 5.3-μm emission from nitric oxide molecules is related to the participation of NO in the excitation of the 630-nm emission from oxygen atoms during intense red auroras, resulting in energy removal from the upper atmosphere. Krassovsky (1969, 1971a) pointed out that the intensity of the continuum increases during intense geomagnetic disturbances and red auroras. This implies, as mentioned in Sect. 2.9, that the density of NO molecules increases. The increase in NO density testifies to the fact that the red emission intensifies due to its stimulation in the upper part of the ionospheric F2 layer which rather strongly heats up due to absorption of magnetohydrodynamic waves (Krassovsky 1968a). On collisions of oxygen atoms with nitric oxide molecules, favorable conditions for the excitation of the $O(^1D)$ state are created. The cross-section for the charge exchange between atomic ions of atmospheric constituents with their neutral atoms is rather large ($\sim 10^{-15}\,(\text{cm}^2)$). Therefore, magnetohydrodynamic oscillations of frequency more than 1 (Hz) can be efficiently absorbed in the upper ionosphere. Besides, the infrared radiation of NO cools the upper atmosphere (Dalgarno 1963). All this provides the removal of energy from the atmosphere.

2.11 Infrared Emissions from Atmospheric Atoms and Molecules

The infrared radiation emitted by atmospheric atoms and molecules in the spectral region beyond 2 (μm) can be detected only by instruments placed on rockets and satellites since these species radiate predominantly at altitudes over 90 (km). The above-considered hydroxyl, nitric oxide, and continuum emissions cover the spectral region approximately up to 5.5 (μm). More long-wave emissions of terrestrial origin were also identified which arise from various molecular, so-called minor, atmospheric constituents, such as NO^+, CO, CO^+, CH_4, NO_2, O_3, and CO_2.

Of greatest energetic and photochemical importance are the emissions from ozone molecules at 9.065 (μm) and 9.6 (μm) and from carbon dioxide molecules at 4.26 (μm) and 14.98 (μm). They cover an altitude range of several tens of kilometers in the middle atmosphere. However, for the problems relevant to the upper atmosphere, the region above 80 (km) is of interest.

The lowest vibrational transitions of the ozone molecule correspond to the emissions (McClatchey et al. 1973)

$$O_3(100) \rightarrow O_3(000) + h\nu(\lambda 9.065\,(\mu m)) \quad A_{O_3}(100) = 0.667\,(s^{-1}),$$

$$O_3(001) \rightarrow O_3(000) + h\nu(\lambda 9.596\,(\mu m)) \quad A_{O_3}(001) = 12.37\,(s^{-1}),$$

$$O_3(010) \rightarrow O_3(000) + h\nu(\lambda 14.267\,(\mu m)) \quad A_{O_3}(010) = 0.252\,(s^{-1}).$$

Vibrationally excited ozone molecules are produced first of all due to three-body recombination collisions (von and Trainor 1973):

$$O + O_2 + M \rightarrow O_3^* + M \quad \alpha_{OO_2M} = 1.1 \cdot 10^{-34} \cdot \exp\left(\frac{510}{T}\right) (cm^6 \cdot s^{-1}),$$

and due to thermal excitation (Ogawa 1976):

$$O_3 + M \rightarrow O_3(100) + M \quad \alpha_{O_3M}(100) = 2 \cdot 10^{-14} \cdot \exp\left(-\frac{1590}{T}\right) (cm^3 \cdot s^{-1}),$$

$$O_3 + M \rightarrow O_3(001) + M \quad \alpha_{O_3M}(001) = 2 \cdot 10^{-14} \cdot \exp\left(-\frac{1500}{T}\right) (cm^3 \cdot s^{-1}),$$

$$O_3 + M \rightarrow O_3(010) + M \quad \alpha_{O_3M}(010) = 2 \cdot 10^{-14} \cdot \exp\left(-\frac{1010}{T}\right) (cm^3 \cdot s^{-1}).$$

Absorption of the thermal radiation of the lower atmosphere by ozone molecules results in their fluorescence in various bands:

$$O_3 + h\nu(\lambda 9.065(\mu m)) \rightarrow O_3(100) \quad g_{O_3}^T(9.065) = 1.2 \cdot 10^{-3}\,(s^{-1})$$
$$g_{O_3}^S = 1.1 \cdot 10^{-5}\,(s^{-1}),$$

$$O_3 + h\nu(\lambda 9.596(\mu m)) \rightarrow O_3(001) \quad g_{O_3}^T(9.596) = 3.2 \cdot 10^{-2}\,(s^{-1})$$
$$g_{O_3}^S = 2.0 \cdot 10^{-4}\,(s^{-1}),$$

$$O_3 + h\nu(\lambda 14.267(\mu m)) \rightarrow O_3(010) \quad g_{O_3}^T(14.267) = 3.6 \cdot 10^{-3}\,(s^{-1})$$
$$g_{O_3}^S = 5.7 \cdot 10^{-6}\,(s^{-1}).$$

From these relations it can be seen that the thermal excitation rate ($g_{O_3}^T$) is about two orders of magnitude greater than the solar excitation rate ($g_{O_3}^S$).

Consideration of the processes resulting in excitation of higher vibrational levels has shown that the vibrational temperature corresponding to their population distribution can reach 300–500 (K) at altitudes of 80–85 (km) depending on the altitude profiles of the atomic oxygen density and atmospheric temperature (Manuilova and Shved 1989).

The reactions of deactivation of excited ozone molecules give rise to a variety of other species, including those producing emissions in the Atmospheric Infrared system bands of molecular oxygen (Dvornikov and Kulagina 1984)

2.11 Infrared Emissions from Atmospheric Atoms and Molecules

$$O_3^* + O \rightarrow O_2(a^1\Delta_g) + O_2(X^3\Sigma_g^-) \quad \beta_{O_3^*O} = 1.7 \cdot 10^{-12} \, (cm^3 \cdot s^{-1}) ,$$

hydroxyl emission (Bevan and Johnson 1973)

$$O_3^* + H \rightarrow OH^* + O_2 \quad \beta_{O_3^*H} = 2.6 \cdot 10^{-11} \, (cm^3 \cdot s^{-1}) ,$$

and continuum (Kurilo et al. 1974; Hui and Cool 1978)

$$O_3^* + NO \rightarrow NO_2^* + O_2 \quad \beta_{O_3^*NO} = 1.8 \cdot 10^{-15} \, (cm^3 \cdot s^{-1}) .$$

The 9.6-μm (001) band is, naturally, most intense. The maximum of the emission rate at night is $1 \cdot 10^7$ (photon·cm^{-3}·s^{-1}) at altitudes near 90 (km); the thickness of the emission layer is 10 (km). This corresponds to an emission intensity of ~10 (megarayleigh).

The most important atmospheric constituent is carbon dioxide CO_2. Its infrared radiation diffuses upward from the lower atmosphere due to transport processes which proceed in the optically thick medium.

The CO_2 molecule is linear. Therefore, the mode $v_2(100)$ corresponds to symmetric vibrations and has no dipole moment for the CO_2 molecule, which consists of identical isotopes. Thereof the rotational–vibrational spectrum of this mode is absent, as in the case of biatomic molecules with identical nuclei.

Above the 79- to 75-km level, the radiation of CO_2 is an efficient sink of energy from the upper atmosphere (Kutepov and Shved 1978, 1981). The most intense processes are (Rothman and Benedict 1978; Caledonia et al. 1985)

$$CO_2(00^01^1) \rightarrow CO_2(00^0\,0^1) + h\nu(\lambda 4.26\,(\mu m)) \quad A_{CO_2}(001) = 432.7\,(s^{-1}),$$

$$CO_2(01^10^1) \rightarrow CO_2(00^0\,0^1) + h\nu(\lambda 14.98\,(\mu m)) \quad A_{CO_2}(010) = 1.28\,(s^{-1}).$$

The excitation occurs due to fluorescence caused by the thermal emission of the lower atmosphere and by sunlight:

$$CO_2(00^0\,0^1) + h\nu(\lambda 4.26\,(\mu m)) \rightarrow CO_2(00^01^1) \quad g^T_{CO_2}(001) = 1.3 \cdot 10^{-5}\,(s^{-1})$$

$$g^S_{CO_2} = 2.8 \cdot 10^{-3}\,(s^{-1}),$$

$$CO_2(00^00^1) + h\nu(\lambda 14.98\,(\mu m)) \rightarrow CO_2(01^10^1)$$

$$g^T_{CO_2}(010) = 2.1 \cdot 10^{-3}\,(s^{-1}) \quad g^S_{CO_2} = 3.1 \cdot 10^{-5}\,(s^{-1}).$$

The effective cross-sections for radiation absorption are $\sigma(4.26\,(\mu m)) = 2.5 \cdot 10^{-14}\,(cm^2)$ and $\sigma(14.98\,(\mu m)) = 3.2 \cdot 10^{-15}\,(cm^2)$.

An important process is the reaction of CO_2 molecules with atomic oxygen (Khvorostovskaya et al. 2002)

$$CO_2(00^0\,0^1) + O \rightarrow CO_2(01^1 0^1) + O$$

$$\alpha_{CO_2O}(010) = (1.56 \div 1.40) \cdot 10^{-12} \cdot \exp\left(-\frac{960}{T}\right) (cm^3 \cdot s^{-1}),$$

$$CO_2(01^1 0^1) + O \rightarrow CO_2(00^0\,0^1) + O$$

$$\beta_{CO_2O}(010) = (1.56 \div 1.40) \cdot 10^{-12} \, (cm^3 \cdot s^{-1}).$$

Laboratory investigations (Khvorostovskaya et al. 2002) have shown that the rate of these reactions has different values for the temperatures below 260 (K) and above 300 (K): $k^O_{01^1 0, 00^0\, 0} = 1.56 \cdot 10^{-12}\,(cm^3 \cdot s^{-1})$ and $k^O_{01^1 0, 00^0\, 0} = 1.40 \cdot 10^{-12}\,(cm^3 \cdot s^{-1})$, respectively. For intermediate temperatures, the reaction rate varies practically linearly. Therefore, it can be well approximated (Semenov and Shefov 2005) by a unified formula for the temperature range characteristic of both the upper mesosphere and the lower thermosphere:

$$k^O_{01^1 0, 00^0\, 0} = \left\{ 1.40 + 0.16 \bigg/ \left[1 + \exp\left(\frac{T-280}{10}\right)\right] \right\} \cdot 10^{-12}\,(cm^3 \cdot s^{-1}).$$

Comparison of the measured temperature profiles at the mentioned altitudes with the altitude distributions of the atomic oxygen density calculated for the 557.7-nm emission has shown a close quantitative relationship between them (Semenov and Shefov 1999b), which testifies to an important part played by the interaction of oxygen atoms with CO_2 molecules. Obviously, the reverse process of excitation of CO_2 molecules by collisions with oxygen atoms also takes place. It plays a significant part in the cooling of the medium at altitudes above 100 (km) where the optical thickness of the atmosphere to the 15-μm emission is less than unity (Kutepov and Shved 1985).

Ulwick et al. (1985) performed measurements of the emission intensity for O_3 and CO_2 molecules. At altitudes of 85–90 (km) the intensity was 6.1 and 95 (megarayleigh), respectively. For the 15-μm emission that occurs at altitudes of 85–90 (km) there is a plateau which testifies to the presence of an emission layer. With the ∼10-km thickness of this layer, the emission rate should be ∼10^8 (photon·$cm^{-3} \cdot s^{-1}$). However, the variations of these emissions were not measured regularly.

The suggestion of an important part of the O_2 infrared emission was first made by Bates (1951). The ground state of atomic oxygen has three sublevels (Table 2.42) the transitions between which correspond to the 63- and 147-μm lines. The first results of rocket measurements of the altitude distribution of the emission intensity were obtained on March 16, 1977 at 23:04 at the rocket range of the Andoya island (69°N, 16°E) (Grossmann and Offermann 1978; Offermann and Grossmann 1978). Subsequently measurements were performed at the Esrange rocket range, Sweden, on December 9, 1981 at 21:25 (Grossmann et al. 1983).

Table 2.42 The far-infrared lines of atmospheric atoms and molecules

Atom	Wavelength	Transition	$A\,(s^{-1})$
H	21.1 (cm)	$^2S \to {}^2S$	2.87(−15)
N^+	203.6 (μm)	$^3P_1 \to {}^3P_0$	2.13(−6)
	121.6 (μm)	$^3P_2 \to {}^3P_1$	7.48(−6)
	76.14 (μm)	$^3P_2 \to {}^3P_0$	1.30(−12)
N^{2+}	57.29 (μm)	$^2P_2^0 \to {}^2P_1^0$	4.77(−5)
O	147.0 (μm)	$^3P_0 \to {}^3P_1$	1.70(−5)
	63.07 (μm)	$^3P_1 \to {}^3P_2$	8.95(−5)
	44.14 (μm)	$^3P_0 \to {}^3P_2$	1.00(−10)
O^{2+}	88.16 (μm)	$^3P_1 \to {}^3P_0$	2.62(−5)
	51.69 (μm)	$^3P_2 \to {}^3P_1$	9.75(−5)
	32.59 (μm)	$^3P_2 \to {}^3P_0$	3.50(−11)
OH	17.4 (cm)	$F_{2+} \to F_{1-}$	0.940(−11)
	18.0 (cm)	$F_{2+} \to F_{2-}$	7.698(−11)
	18.0 (cm)	$F_{1+} \to F_{1-}$	7.103(−11)
	18.6 (cm)	$F_{1+} \to F_{2-}$	1.289(−11)

The emission intensity at zenith at an altitude of 100 (km) was $7 \cdot 10^{-10}$ (W · cm^{-2} · sr^{-1}) or 2.8 (megarayleigh). As follows from the character of the altitude variations of the intensity, the atomic oxygen layer becomes optically thick at altitudes below 100 (km). At the measuring time the index K_p was 2 and the solar activity corresponded to $F_{10.7} = 74$. It turned out that the measured intensity was much lower than that predicted by theoretical calculations and, hence, it had no significance for the cooling of the thermosphere.

The infrared transitions of oxygen and nitrogen atoms and ions in the microwave region and of hydrogen atoms and hydroxyl molecules in the centimeter region, presented in Table 2.42, because of the low densities of these species, can serve only as indicators of the altitude distributions of the relevant processes.

References

Ackerman M, Biaume F (1970) Structure of the Schumann-Runge bands from the (0–0) to the (13–0) band. J Mol Spectrosc 35:73–82

Ackerman M, Biaume F, Kockarts G (1970) Absorption cross-sections of the Schumann-Runge bands of molecular oxygen. Planet Space Sci 18:1639–1651

Adler-Golden S (1997) Kinetic parameters for OH nightglow modeling consistent with recent laboratory measurements. J Geophys Res 102A:19969–19976

Ajello JM, Witt N (1979) Simultaneous H(1216 A) and He(584 A) observations of the interstellar wind by Mariner 10. In: Rycroft MJ (ed) Space Research, Vol 19. Pergamon Press, Oxford, pp 417–420

Ali AA, Ogrizlo EA, Shen YQ, Wassel PT (1986) The formation of $O_2(a^1\Delta_g)$ in homogeneous and heterogeneous atom recombination. Can J Phys 64:1614–1620

Allen CW (1973) Astrophysical quantities, 3rd edn. The Athlone Press, London
Allison AC, Dalgarno A, Pasachoff NW (1971) Absorption by vibrationally excited molecular oxygen in the Schumann-Runge continuum. Planet Space Sci 19:1463–1473
Alpers M, Höffner J, von Zahn U (1990) Iron atom densities in the polar mesosphere from lidar observations. Geophys Res Lett 17:2345–2348
Ambartsumyan VA, Mustel ER, Severny AB, Sobolev VV (1952) Theoretical astrophysics. Ambartsumyan VA (ed) Gostekhizdat, Moscow
Andreeva LA, Kluev OF, Portnyagin YuI, Khananiyan AA (1991) Studies of the upper atmospheric processes by the artificial luminous cloud method. Hydrometeoizdat, Leningrad
Armstrong EB (1967) Observation of the airglow H_α emission. Planet Space Sci 15:407–425
Babcock HW (1939) Radiations of the night sky photographed with a grating. Publ Astron Soc Pac 51:47–50
Bakanas VV, Perminov VI (2003) Some features in the seasonal behavior of the hydroxyl emission characteristics in the upper atmosphere. Geomagn Aeronomy 43:363–369
Baker DJ, Waddoups RO (1967) Rocket measurements of midlatitude night airglow emissions. J Geophys Res 72:4881–4883
Baker DJ, Waddoups RO (1968) Correction to paper by D. Baker, R. Waddoups. Rocket measurements of midlatitude night airglow emissions. J Geophys Res 73:2546–2547
Baker KD, Baker DJ, Ulwick JC, Stair AT (1977) Measurements of 1.5–5.3 µm infrared enhancements associated with a bright aurorae. J Geophys Res 82:3518–3528
Baluja KL, Zeippen CJ (1988) M1 and E2 transition probabilities for states within the $2p^4$ configuration of the OI isoelectronic sequence. J Phys B 21:1455–1471
Banks PM, Kockarts G (1973a) Aeronomy. Pt A. Academic Press, New York
Banks PM, Kockarts G (1973b) Aeronomy. Pt B. Academic Press, New York
Barth CA (1964) Three-body reaction. Ann Géophys 20:182–196
Barth CA, Hildebrandt AF (1961) The 5577 Å airglow emission mechanism. J Geophys Res 66:985–986
Bates DR (1948) Theoretical considerations regarding the night sky emission. In: The emission spectra of the night sky and aurorae (Reports of the Gassiot Committee). The Phys Soc London, pp 21–33
Bates DR (1951) The temperature of the upper atmosphere. Proc Roy Soc 64B:805–821
Bates DR (1954) The physics of the upper atmosphere. In: Kuiper GP (ed) The Earth as a planet. University Chicago Press, Chicago, pp 576–643
Bates DR (1982) Airglow and auroras. In: Massey HSW, Bates DR (eds) Applied atomic collision physics. Atmospheric physics and chemistry, Vol 1. Academic Press, New York, pp 149–224
Bates DR (1988a) Transition probabilities of the bands of the oxygen systems of the nightglow. Planet Space Sci 36:869–873
Bates DR (1988b) Excitation and quenching of the oxygen bands in the nightglow. Planet Space Sci 36:875–881
Bates DR (1988c) Excitation of 557.7 nm OI line in nightglow. Planet Space Sci 36:883–889
Bates DR, Nicolet M (1950) The photochemistry of atmospheric water vapour. J Geophys Res 55:301–327
Bates DR, Patterson TNL (1962) Helium ions in the upper atmosphere. Planet Space Sci 9:599–605
Becker KH, Groth W, Thran D (1971) The airglow reaction $NO + O + (M) \rightarrow NO_2^* + (M)$ at low pressure. In: Fiocco G (ed) Mesospheric models and related experiments. D Reidel Publishing Company, Dordrecht, pp 261–265
Benedict WS, Plyler EK, Humphreys CJ (1953) The emission spectrum of OH from 1.4 to 1.7 µ. J Chem Phys 21:398–402
Berg MA, Shefov NN (1962a) Emission of the hydroxyl bands and the (0,1) λ 8645 Å atmospheric band of oxygen in the nightglow. Planet Space Sci 9:167–171
Berg MA, Shefov NN (1962b) OH emission and atmospheric O_2 band λ 8645 A. In: Krassovsky VI (ed) Aurorae and Airglow, N 9. USSR Academic Science Publishing House, Moscow, pp 46–52

References

Berg MA, Shefov NN (1963) The hydroxyl emission with different vibrational excitation. In: Krassovsky VI (ed) Aurorae and Airglow, N 10. USSR Academic Science Publishing House, Moscow, pp 19–23

Bevan PLT, Johnson GRA (1973) Kinetics of ozone formation in the pulse radiolysis of oxygen gas. J Phys Chem 69:216–217

Bishop J, Link R (1993) Metastable He 1083 nm intensities in the twilight: a reconsideration. Geophys Res Lett 20:1027–1030

Bogdanov AV, Dubrovsky GV, Osipov AI, Strelchenya VM (1991) Rotational relaxation in gases and plasma. Energoatomizdat, Moscow

Bowen IS (1934) The excitation of the permitted OIII nebular lines. Publ Astron Soc Pac 46:146–148

Bowen IS (1947) Excitation by line coincidence. Publ Astron Soc Pac 59:196–198

Brandt JC, Chamberlain JW (1959) Interplanetary gas. I. Hydrogen radiation in the night sky. Astrophys J 130:670–682

Breen JE, Quy RB, Glass GP (1973) Vibrational relaxation of O_2 in the presence of atomic oxygen. J Chem Phys 59:556–557

Breig EL (1969) Statistical model for the vibrational deactivation of molecular by atomic oxygen. J Chem Phys 51:4539–4547

Breig EL (1970) Secondary production processes for the hydroxyl atmospheric airglow. Planet Space Sci 18:1271–1274

Broadfoot AL, Johanson AE (1976) Fe (3860 A) emission in the twilight. J Geophys Res 81:1331–1334

Broadfoot AL, Kendall KR (1968) The airglow spectrum 3100–10000 A. J Geophys Res 73:426–428

Bunn FE, Gush HP (1972) Spectrum of the night airglow between 3 and 4 microns. Can J Phys 50:213–215

Byram ET, Chubb TA, Friedman H, Kupperian JE (1957) Far ultraviolet radiation in the night sky. In: Zelikoff M (ed) The threshold of space. Pergamon Press, London, pp 203–210

Caledonia GE, Kennealy JP (1982) NO infrared radiation in the upper atmosphere. Planet Space Sci 30:1043–1056

Caledonia GE, Green BD, Nadile RM (1985) The analysis of SPIRE measurements of atmospheric limb $CO_2(v_2)$ fluorescence. J Geophys Res 90A:9783–9788

Callear AB, Lambert JD (1969) Energy exchange between the chemical species. In: Bamford CH, Tipper CFH (eds) Comprehensive chemical kinetics. The formation and decay of excited species, Vol 3. Elsevier Publishing Company, Amsterdam, pp 214–317

Capitelli M, Cacciatore M, De Benedictis S, Dilonardo M, Gorse C, Gordiets BF, Zhdanok SA, Billing GD, Smith IWM, Aquilanti V, Laganà A, Wadehra JM, Bréchignac P, Taran JP, Rich JW, Bergman RC, Rusanov VD, Fridman AA, Sholin GV, Ricard A (1986) Nonequilibrium vibrational kinetics. Capitelli M (ed) Springer-Verlag, Berlin

Cashion K (1963) A method for calculating vibrational transition probabilities. J Mol Spectrosc 10:182–231

Chalamala BR, Copeland RA (1993) Collision dynamics of $OH(X^2\Pi, v = 9)$. J Chem Phys 99:5807–5811

Chamberlain JW (1961) Physics of the Aurora and Airglow. Academic Press, New York

Chamberlain JW (1978) Theory of planetary atmospheres. Academic Press, New York

Chamberlain JW, Hunten DM (1987) Theory of planetary atmospheres, 2nd edn. Academic Press, San Diego

Chamberlain JW, Smith CA (1959) On the excitation rates and intensities of OH in the airglow. J Geophys Res 64:611–614

Chandrasekhar S (1950) Radiative transfer. Clarendon Press, Oxford

Chanin ML, Keckhut P, Hauchecorne A, Labitzke K (1989) The solar activity – Q.B.O. effect in the lower thermosphere. Ann Geophys 7:463–470

Chapman S (1931) Absorption and ionizing effect of monochromatic radiation in an atmosphere of a rotating Earth. Proc Phys Soc London 43(26):483–501

Chapman S (1939) Notes on atmospheric sodium. Astrophys J 90:309–316

Charters PE, Polanyi JC (1960) In improved technique for the observation of infrared chemiluminescence: resolved emission of the OH arising from $H + O_3$. Can J Chem 38:1742–1755

Charters PE, Macdonald RG, Polanyi JC (1971) Formation of vibrationally excited OH by the reaction $H + O_3$. Appl Opt 10:1747–1754

Chernjajev VI, Vuks MF (1937) The spectrum of the sky in the twilight. Dokl USSR Acad Sci 14:77–79

Cheung ASC, Yoshino K, Parkinson WH, Freeman DE (1984) Herzberg continuum cross section of oxygen in the wave length region 193.5–204.0 nm and band oscillator strength of the (0–0) and (1–0) Schumann–Runge bands. Can J Phys 62:1752–1764

Chikashi Y, Masahara F, Hirota E (1989) Detection of the NaO radical by microwave spectroscopy. J Chem Phys 90:3033–3037

Chuvaev KK (1952) On luminescence of the terrestrial atmosphere in continuum spectrum. Dokl USSR Acad Sci 87:551–554

Chuvaev KK (1961) On the separation of the night sky luminescence on components. Astron Rep 38:692–705

Choi GH, Monson IK, Wickwar VB, Rees D (1998) Seasonal variations of temperature near the mesopause from Fabry-Perot interferometer observations of OH Meinel emissions. Adv Space Res 21:843–846

Clemesha BR, Simonich DM, Takahashi H, Melo SML, Plane JMC (1995) Experimental evidence for photochemical control of the atmospheric sodium layer. J Geophys Res 100D:18909–18916

Clough PN, Thrush BA (1967) Mechanism of chemiluminescent reaction between nitric oxide and ozone. Trans Faraday Soc 63:915–925

Condon EU, Shortley G (1935) The theory of atomic spectra. Cambridge University Press, London

Conner JF, Smith RW, Hernandez G (1993) Techniques for deriving Doppler temperatures from multiple-line Fabry-Perot profiles: an analysis. Appl Opt 32:4437–4444

Cook TB, West WP, Dunning FB, Rundel RD, Stebbings RF (1974) Absolute cross sections for Penning ionization of atomic oxygen by helium metastable atoms. J Geophys Res 79:678–680

Coxon JA (1980) Optimum molecular constants and term values for the $X^2\Pi(v \leq 5)$ and $A^2\Sigma^+(v \leq 3)$ states of OH. Can J Phys 58:933–949

Coxon JA, Foster SC (1982) Rotational analysis of hydroxyl vibration rotation emission bands: molecular constants for OH $X^2\Pi$, $6 \leq v \leq 10$. Can J Phys 60:41–48

Dalgarno A (1963) Vibrationally excited molecules in atmospheric reactions. Planet Space Sci 10:19–28

Davis TN, Smith LL (1965) Latitudinal and seasonal variations in the night airglow. J Geophys Res 70:1127–1138

Degen V (1968) The Herzberg II $(c^1\Sigma_u^- - X^3\Sigma_g^-)$ system of O_2 in emission in the oxygen-argon afterglow. Can J Phys 46:783–787

Degen V (1969) Vibrational populations of $O_2(A^3\Sigma_u^+)$ and synthetic spectra of the Herzberg bands in the night airglow. J Geophys Res 74:5145–5154

Degen V (1977) Nightglow emission rates in the O_2 Herzberg bands. J Geophys Res 82:2437–2438

Degen V, Nicholls RW (1969) Intensity measurements on the $A^3\Sigma_u^+ - X^3\Sigma_g^-$ Herzberg I band system of O_2. J Phys B Ser 2 2: 1240–1250

Degges TC (1971) Vibrationally excited nitric oxide in the upper atmosphere. Appl Opt 10:1856–1860

Déjardin G, Bernard R (1938) Les bandes de la molécule OH dans le spectre du ciel nocturne. Comptes Rendus Acad Sci 206:1747–1749

Delannoy J (1960) Sur les observations crépusculaires de la raie du lithium au cours de l' Année Géophysiques Internationale. Ann Géophys 16:236–252

Dere KP (1977) Extreme ultraviolet spectra of solar active region and their analysis. Solar Phys 82:77–93

Dick KA (1977) On the rotational temperature of the airglow hydroxyl emissions. Planet Space Sci 25:595–596

Ditchburn RW, Young PA (1962) The absorption of molecular oxygen between 1850 and 2500 Å. J Atmos Terr Phys 24:127–139
Dodd JA, Lipson SJ, Blumberg WAM (1990) Vibrational relaxation of OH ($X^2\Pi_i$, v = 1 – 3) by O_2. J Chem Phys 92:3387–3393
Dodd JA, Lipson SJ, Blumberg WAM (1991) Formation and vibrational relaxation of OH($X^2\Pi_i$, v) by O_2 and CO_2. J Chem Phys 95:5752–5762
Dodd JA, Blumberg WAM, Lipson SJ, Lowell JR, Armstrong PS, Smith DR, Nadile RM, Wheeler NB, Huppi ER (1993) OH(v, N) column densities from high-resolution Earthlimb spectra. Geophys Res Lett 20:305–308
Dodd JA, Lipson SJ, Armstrong PS, Blumberg WAM, Nadile RM, Adler-Golden SM, Marinelli WJ, Holtzclaw KW, Green BD (1994) Analysis of hydroxyl earthlimb air glow emissions: kinetic model for state-to-state dynamics of OH (v, N). J Geophys Res 99D:3559–3585
Donahue TM (1964) H_α excitation in the hydrogen near the Earth. Planet Space Sci 12:149–159
Donahue TM (1968) Discussion of paper by Y. Kondo and J.E. Kupperian, Jr., Interaction of neutral hydrogen and charged particles in the radiation belts: the consequent Lyman-alpha emission. J Geophys Res 73: 4455–4457
Doschek GA, Behring WE, Feldman U (1974) The profiles of the solar HeI and HeII lines at 584, 537 and 304 Å. Astrophys J 190:L141–L142
Dvornikov IV, Kulagina LV (1984) The quenching mechanism of the $O_2(b^1\Sigma_g^+)$ by atomic oxygen. Optics Spectroscopy 57:1015–1021
Eather RH (1968) Discussion of paper by Yoji Kondo and James E. Kupperian, Jr., Interaction of neutral hydrogen and charged particles in the radiation belts: the consequent Lyman-alpha emission. J Geophys Res 73:3599–3600
Elvey CT (1942) Light of the night sky. Rev Modern Phys 14:140–150
Elvey CT (1943) Observations of the light of the night sky with photoelectric photometer. Astrophys J 97:65–71
Evans WFJ, Llewellyn EJ, Vallance Jones A (1972) Altitude distribution of the $O_2(^1\Delta)$ nightglow emission. J Geophys Res 77:4899–4901
Fallon RJ, Tobias I, Vanderlice JT (1961) Potential energy curve for OH. J Chem Phys 34:167–168
Fatkullin MN (1982) Physics of the ionosphere. In: Total results of the Science and Technique. Geomagnetism and upper layers of the atmosphere, Vol 6. VINITI, Moscow, pp 4–224
Fedorova NI (1962) Twilight fluorescence of the 10830 A helium emission. Izvestiya USSR Acad Sci Geophys Ser N 4:538–547
Fedorova NI (1967) Twilight emission of helium at high latitudes. In: Krassovsky VI (ed) Aurorae and Airglow N 13. USSR Academic Science Publishing House, Moscow, pp 53–63
Ferguson AF, Parkinson D (1963) The hydroxyl bands in the nightglow. Planet Space Sci 11:149–159
Ferguson EE, Fehsenfeld FC, Dunkin DB, Schmeltekopf AL, Schiff HI (1964) Laboratory studies of helium ion loss processes of interest in the ionosphere. Planet Space Sci 12:1169–1171
Filipp ND, Oraevsky VN, Blaunstein NSh, Ruzhin YuYa (1986) Evolution of the artificial plasma inhomogeneities in the Earth' ionosphere. Gusev VD (ed) Stiintsa, Kishinev
Findlay FD (1969) Relative band intensities in the atmospheric and infrared atmospheric systems of molecular oxygen. Can J Phys 47:687–691
Finlayson-Pitts BJ, Kleindienst TE (1981) The reaction of hydrogen atoms with ozone as a source of vibrationally excited $OH(X^2\Pi_i)_{v=9}$ for kinetic studies. J Chem Phys 74:5643–5658
Fiocco G, Fua D, Visconti G (1974) Origin of the upper atmospheric Na from sublimating dust: a model. Ann Géophys 30:517–528
Fishkova LM (1963) On the spatial distribution and variations of the H_α emission in the night sky. In: Krassovsky VI (ed) Aurorae and Airglow, N 10. USSR Academic Science Publishing House, Moscow, pp 35–39
Fishkova LM (1969) On distribution of intensity in a complete spectrum of the night sky luminescence in the region 5500–6700 A. Geomagn Aeronomy 9:568–570
Fishkova LM (1970) Nightglow continuum in the visual region. In: Kharadze EK (ed) Bull Abastumani astrophys observ. N 39. pp 117–150

Fishkova LM (1972) The investigation of the upper atmosphere and geocorona hydrogen by observations of the H_α emission line in the airglow spectrum. In: Fishkova LM, Kharadze EK (eds) Bull Abastumani astrophys observ N 42. pp 131–181

Fishkova LM (1976) Regular nocturnal and seasonal variations of the emission intensity of OH, NaD, 5577 A of the upper atmosphere. In: Krassovsky VI (ed) Aurorae and Airglow. N 24. Soviet Radio, Moscow, pp 5–15

Fishkova LM (1979) Nocturnal sodium emission in the Earth' upper atmosphere. In: Problems of the atmospheric optics. Leningrad State University, Leningrad, pp 154–172

Fishkova LM (1983) The night airglow of the Earth mid-latitude upper atmosphere. Shefov NN (ed) Metsniereba, Tbilisi

Fishkova LM, Markova GV (1958) On the H_α line in the night sky spectrum. Astron Circ USSR Academic Science 196:8–9

Fishkova LM, Martsvaladze NM (1967) On variations H_α emission and distribution of hydrogen in the upper atmosphere in Abastumani. In: Krassovsky VI (ed) Aurorae and Airglow. N 13. USSR Acad Science Publishing House, Moscow, pp 69–72

Fishkova LM, Shcheglov PV (1972) The diurnal variations of the night airglow H_α emission. In: Fishkova LM, Kharadze EK (eds) Bull Abastumani astrophys observ. N 42. pp 29–36

Fontijn A, Schiff HI (1961) Absolute rate constant for light emission of the air afterglow reaction for the wavelength region 3875–6200 Å. In: Cadle RD (ed) Chemical reactions in the lower and upper atmosphere. Interscience, New York, pp 239–254

Fontijn A, Meyer CB, Schiff HI (1964) Absolute quantum yield measurements of the NO + O reaction and its use as a standard for chemiluminescent reactions. J Chem Phys 40:64–70

Frederick JE, Rusch DW, Victor GA, Sharp WE, Hays PB, Brinton HC (1976) The OI (λ 5577 Å) airglow: observations and excitation mechanisms. J Geophys Res 81:3923–3930

French WJR, Burns GB, Finlayson K, Greet PA, Lowe RP, Williams PFB (2000) Hydroxyl (6–2) airglow emission intensity ratios for rotational temperature determination. Ann Geophys 18:1293–1303

Galperin GI (1956a) Ration of the intensity components of the sodium yellow doublet in the twilight spectrum. Astron Reports 33:173–181

Galperin GI (1956b) The ratio of the intensities of the components of the sodium doublet in the twilight spectrum. In: Armstrong EB, Dalgarno A (eds) The airglow and the aurorae. Pergamon Press, London, pp 91–94

Gadsden M (1969) Antarctic twilight observations. 1. Search for metallic emission lines. Ann Géophys 25:119–126

Gadsden M, Marovich E (1973) The nightglow continuum. J Atmos Terr Phys 35:1601–1614

Gadsden M, Salmon K (1958) Presence of 6707 Å radiation in the twilight sky. Nature 182:1598

Gardner CS, Voelz DG (1985) Lidar measurements of gravity wave saturation effects in the sodium layer. Geophys Res Lett 12:765–768

Gardner CS, Voelz DG, Sechrist CF, Segal AC (1986) Lidar studies of the nighttime sodium layer over Urbana, Illinois. 1. Seasonal and nocturnal variations. J Geophys Res 91A:13659–13673

Gattinger RL (1971) Interpretation of airglow in terms of excitation mechanisms. In: McCormac BM (ed) The radiating atmosphere. D Reidel Publishing Company, Dordrecht-Holland, pp 51–63

Gattinger RL, Vallance Jones A (1973) Observation and interpretation of hydroxyl airglow emissions. In: McCormac BM (ed) Physics and chemistry of upper atmospheres. D Reidel Publishing Company, Dordrecht-Holland, pp 184–192

Gershenzon YuM, Grigor'eva VM, Konoplev AV, Rosenstein VB (1985) An analysis of the influence of the vibrationally excited ozone on the ozone and atomic oxygen concentrations in the terrestrial upper atmosphere. Russian J Chem Phys 4:544–550

Gilmore FR (1965) Potential energy curves for N_2, NO, O_2 and corresponding ions. J Quant Spectrosc Radiat Transfer 5:369–390

Gindilis LM (1965) The absolute measurements of the nightglow continuum. In: Krassovsky VI (ed) Aurorae and Airglow. N 11. Nauka, Moscow, pp 26–34

Givishvili GV, Leshchenko LN, Lysenko EV, Perov SP, Semenov AI, Sergeenko NP, Fishkova LM, Shefov NN (1996) Long-term trends of some characteristics of the Earth's atmosphere. I. Experimental results. Izvestiya Atmos Oceanic Phys 32:303–312

Golde MF, Roche AE, Kaufman F (1973) Absolute rate constant for the O + NO chemiluminescence in the near infrared. J Chem Phys 59:3953–3959

Goldman A, Schoenfeld WG, Goorvitch D, Chackerian C, Dothe H, Mélen F, Abrams MC, Selby JEA (1998) Updated line parameters for OH $X^2\Pi - X^2\Pi(v'', v')$ transitions. J Quant Spectrosc Radiat Transfer 59:453–469

Golitsyn GS, Semenov AI, Shefov NN, Khomich VYu (2006) The response of the middle atmosphere temperature on the solar activity during various seasons. Phys Chem Earth 31:10–15

Goorvitch D, Goldman A, Dothe H, Tipping RH, Chackerian C (1992) Hydroxyl $X^2\Pi$ pure rotational transitions J Geophys Res 97D:20771–20786

Gopstein NM, Kushpil BI (1964) Dayglow of the upper atmosphere of the Earth in the 1.25 mcm. Cosmic Res 2:619–622

Gordiets BF, Markov MN (1983) IR-radiarion and NO concentration in the essentially heated upper atmosphere. Geomagn Aeronomy 23:446–450

Gordiets BF, Markov MN, Shelepin LA (1978) The theory of the infrared radiation of the near-earth space. In: Trans Physical Institute Acad Sci. Infrared radiation in the Earth' atmosphere and in space, Vol 105. Nauka, Moscow, pp 7–71

Gordiets BF, Osipov AI, Shelepin LA (1980) Kinetic processes in gases and molecular lasers. Nauka, Moscow

Graham DA, Ichikawa T, Kim JS (1971) Observations of sodium, lithium and potassium twilight glow at Moscow, Idaho, USA. Ann Géophys 27:483–491

Granier C, Jégou JP, Megie G (1985) Resonant lidar detection of Ca and Ca^+ in the upper atmosphere. Geophys Res Lett 12:655–658

Granier C, Jégou JP, Megie G (1989) Iron atoms and metallic species in the Earth's upper atmosphere. Geophys Res Lett 16:243–246

Grebowsky JM, Reese N (1989) Another look at equatorial metallic ions in the F region. J Geophys Res 94A:5427–5440

Green AES, Barth CA (1967) Calculations of the photoelectron excitation of the dayglow. J Geophys Res 72:3975–3986

Green AES, Dutta SK (1967) Semi-empirical cross sections for electron impacts. J Geophys Res 72:3933–3941

Green AES, Sawada T (1972) Ionization cross sections and secondary electron distribution. J Atmos Terr Phys 34:1719–1728

Green AES, Stolarski RS (1972) Analytical models of electron impact excitation cross sections. J Atmos Terr Phys 34:1703–1717

Greer RGH, Best GT (1967) A rocket-borne photometric investigation of the oxygen lines at 5577 Å and 6300 Å, the sodium D-lines and the continuum at 5300 Å in the night airglow. Planet Space Sci 15:1857–1881

Greer RGH, Llewellyn EJ, Solheim BH, Witt G (1981) The excitation of $O_2(b^1\Sigma_g^+)$ in the nightglow. Planet Space Sci 29:383–389

Greer RGH, Murtagh DP, McDade IC, Dickinson PHG, Thomas L, Jenkins DB, Stegman J, Llewellyn EJ, Witt G, Mackinnon DJ, Williams ER (1986) ETON 1: A data base pertinent to the study of energy transfer in the oxygen nightglow. Planet Space Sci 34:771–788

Greet PA, Innis J, Dyson PL (1994) High-resolution Fabry-Perot observations of mesospheric OH (6–2) emissions. Geophys Res Lett 21:1153–1156

Grigor'eva VM, Gershenzon YuM, Shalashilin DV, Umanskii SYa (1994) A new kinetic mechanism of the hydroxyl emission of the night sky and the peculiarities of the vibrational relaxation of the OH upper levels (v = 7–9) by O_2. Russian J Chem Phys 13:3–25

Grigor'eva VM, Gershenzon YuM, Semenov AI, Umanskii SYa, Shalashilin DV, Shefov NN (1997) Excitation kinetics of the mesopause hydroxyl emission and the role of vibrational relaxation of upper vibrational levels. Geomagn Aeronomy 37:94–100

Grossmann KU, Offermann D (1978) Atomic oxygen emission at 63 μm as a cooling mechanism in the thermosphere and ionosphere. Nature (London) 276:594–595

Grossmann KU, Barthol P, Frings W, Hennig R, Offermann D (1983) A new spectrometric measurement of atmospheric 63 μm emission. Adv Space Res 2:111–114

Gurvich AS, Vorobiev VV, Savchenko SA, Pakhomov AI, Padalka GI, Shefov NN, Semenov AI (2002) The 420–530 nm region nightglow of the upper atmosphere as measured onboard Mir research platform in 1999. Geomagn Aeronomy 42:514–519

Gush HP, Buijs HL (1964) The near infrared spectrum of the night airglow observed from high altitude. Can J Phys 42:1037–1045

Harang O, Pettersen H (1967) Variation in width of the H_α – line in aurora. Planet Space Sci 15:1599–1603

Harrison AW, Cairns CD (1969) Helium emission (1.083 μ) in sunlit aurora. Planet Space Sci 17:1213–1219

Haslett JC, Fehsenfeld FC (1969) Ratio of the $O_2(^1\Delta_g - ^3\Sigma_g^-)$ (0,0), (0,1) transitions. J Geophys Res 74:1878–1879

Hasson V, Nicholls RW, Degen V (1970) Absolute intensity measurements on the $A^3\Sigma_u^+ - X^3\Sigma_g^-$ Herzberg I band system of molecular oxygen. J Phys B 3:1192–1194

Heaps HS, Herzberg G (1952) Intensity distribution in the rotation-vibration spectrum of the OH molecule. Zeits Phys 133:48–64

Helmer M, Plane JMC (1993) A study of the reaction $NaO_2 + O \rightarrow NaO + O_2$: implications for the chemistry of sodium in the upper atmosphere. J Geophys Res 98D:23207–23222

Hennes JP (1966) Measurement of the ultraviolet nightglow spectrum. J Geophys Res 71:763–770

Henriksen K, Sukhoivanenko PYa (1982) The detection and interpretation of the orthohelium emission at 5876 Å in aurora. Planet Space Sci 30:695–699

Henry RJW (1970) Photoionization cross-sections for atoms and ions of carbon, nitrogen, oxygen and neon. Astrophys J 161:1153–1155

Heppner JP, Meredith LH (1958) Nightglow emission altitudes from rocket measurements. J Geophys Res 63:51–65

Herman R, Herman L, Gauzit J (1942) Spectre du ciel nocturne dans le proche infrarouge. Cahiers Phys 12:46–48

Hernandez G (1971) The signature profiles of $O(^1S)$ in the airglow. Planet Space Sci 19:467–476

Hernandez G (1975) Reaction broadening of the line profiles of atomic sodium in the night airglow. Geophys Res Lett 2:103–105

Hernandez G, Fraser J, Smith RW (1993) Mesospheric 12-hour oscillation near South Pole, Antarctica. Geophys Res Lett 20:1787–1790

Herschbach DR, Kolb CE, Worsnop DR, Shi X (1992) Excitation mechanism of the mesospheric sodium nightglow. Nature (London) 356:414–416

Herzberg G (1945) Infrared and Raman spectra of polyatomic molecules. Van Nostrand Reinhold Co, New York

Herzberg G (1950) Molecular spectra and molecular structure. I. Spectra of diatomic molecules, 2nd edn. Van Nostrand Reinhold Co, New York

Herzberg G (1951) The atmospheres of the planets. J Roy Astron Soc Canada 45:100–123

Herzberg G (1971) The spectra and structures of simple free radicals. Cornell University Press, Ithaca and London

Hinteregger HE (1976) EUV fluxes in the solar spectrum below 2000 Å. J Atmos Terr Phys 38:791–806

Hocking WK (1985) Turbulence in the altitude region 80–120 km. In: Labitzke K, Barnett JJ, Edwards B (eds) Handbook for MAP, Vol 16. SCOSTEP, Urbana, pp 290–304

Hohmann J, Müller G, Schönnenbeck G, Stuhl F (1994) Temperature-dependent quenching of $O_2(b^1\Sigma_g)$ by H_2, D_2, CO_2, HN_3, DN_3, HNCO, and DNCO. Chem Phys Lett 217:577–581

Hotlzclaw KW, Person JC, Green BD (1993) Einstein coefficients for emission from high rotational states of the $OH(X^2\Pi)$ radical. J Quant Spectrosc Radiat Transfer 49:223–235

Holtzclaw KW, Upschulte BL, Caledonia GE, Cronin JF, Green BD, Lipson SJ, Blumberg WAM, Dodd JA (1997) Rotational relaxation of high-N states of $OH(X^2\Pi, v = 1-3)$ by O_2. J Geophys Res 102A:4521–4528

Howell CD, Michelangeli DV, Allen M, Yung YI, Thomas RJ (1990) SME observation of $O_2(a^1\Delta_g)$ nightglow: an assessment of the chemical production mechanism. Planet Space Sci 38:529–537

Huber KP, Herzberg G (1979) Molecular spectra and molecular structure. IV. Constants of diatomic molecules. Van Nostrand Reinhold Co, New York

Hudson RD (1971) Critical review of ultraviolet photoabsorption cross sections for molecules of astrophysical and aeronomic interest. Rev Geophys Space Phys 9:305–406

Hudson KD, Carter VL (1969) Absorption in the spectral range of the Schumann-Runge bands. Can J Chem 47:1840–1846

Hui KK, Cool TA (1978) Experiments concerning the laser-enhanced reaction between vibrationally excited O_3 and NO. J Chem Phys 68:1022–1037

Hunten DM (1956) Seasonal variations of twilight sodium emission. In: Armstrong EB, Dalgarno A (eds) The airglow and the aurorae. Pergamon Press, London, pp 114–121

Hunten DM (1967) Spectroscopic studies of the twilight airglow. Space Sci Rev 6:493–573

Hunten DM, Wallace L (1967) Rocker measurements of the sodium dayglow. J Geophys Res 72:69–79

Huruhata M, Nakamura T, Steiger WR (1967) A rocket observation of (OI) 5577 Å emission and continuum at 5300 Å in night airglow. Rep Ionosph Space Res Japan 21:229–232

IAPh (1994) Obukhov Institute of Atmospheric Physics of the Russian Academy of Sciences. Booklet. N 4462M. VneshTorgIzdat, Moscow

Ignatiev VM (1977a) Unusual profiles of the 5577 Å and 6300 Å emissions in aurorae. Astron Circ USSR Acad Sci 940:2–4

Ignatiev VM (1977b) Peculiarities of contours of 5577 Å and 6300 Å lines in auroras. Geomagn Aeronomy 17:153–154

Ignatiev VM, Yugov VA (1995) Interferometry of the large-scale dynamics of the high-latitudinal thermosphere. Shefov NN (ed) Yakut Scientific Centre Siberian Branch RAN, Yakutsk

Ivanov VA, Prikhod'ko AS, Skoblo YuA (1991) Deactivation of the 2^1S state of atomic helium by low-velocity electrons. Optics Spectroscopy 70:507–510

Ivanov-Kholodny GS, Mikhailov AV (1980) Ionospheric state forecast. Hydrometeoizdat, Leningrad

Ivanov-Kholodny GS, Nikol'sky GM (1969) Sun and ionosphere. Nauka, Moscow

Ivanov-Kholodny GS, Nusinov AA (1987) Ultraviolet radiation of the Sun and its influence on the upper atmosphere and ionosphere. In: Total results of the Science. Investigations of the cosmic processes, Vol 26. VINITI, Moscow, pp 80–154

Izod TPJ, Wayne RP (1968) The formation, reaction and deactivation of $O_2(b^1\Sigma_g^+)$. Proc Roy Soc London A 308:81–94

Jasperse JR (1977) Electron distribution function and ion concentrations in the Earth's lower ionosphere from Boltzmann–Fokker–Planck theory. Planet Space Sci 25:743–756

Jégou JP, Granier C, Chanin ML, Megie G (1985a) General theory of the alkali metals present in the earth's upper atmosphere. I. Flux model: chemical and dynamical processes. Ann Geophys 3:163–175

Jégou JP, Granier C, Chanin ML, Megie G (1985b) General theory of the alkali metals present in the earth's upper atmosphere. II. Seasonal and meridional variations. Ann Geophys 3:299–312

Johnson BR, Winter NW (1977) Classical trajectory study of the effect of vibrational energy on the reaction of molecular hydrogen with atomic oxygen. J Chem Phys 66:4116–4120

Johnston JE, Broadfoot AL (1993) Midlatitude observations of the night airglow: implications to quenching near the mesopause. J Geophys Res 98A:21593–21603

Jones DG, Lambert JD, Stretton JL (1965) Vibrational relaxation in mixtures containing oxygen. Proc Phys Soc London 86:857–860

Jusick AT, Watson CE, Peterson LR, Green AES (1967) Electron impact cross sections for atmospheric species. 1. Helium. J Geophys Res 72:3943–3951

Kaliteevsky NI, Chaika MP (1970) Fabry-Perot interferometer and its some applications in the spectroscopy. In: Frish SE (ed) Spectroscopy of the gaseous discharge plasma. Nauka, Leningrad, pp 160–200

Karmilova LB, Kondratiev VN (1951) Measurements of the atomic oxygen concentration in the flames with aid of NO. J Phys Chem 25:312–322

Karyagina ZV (1962) The hydroxyl emission in the airglow spectrum according to observations in Alma-Ata. In: Krassovsky VI (ed) Aurorae and Airglow. N 8. USSR Academic Science Publishing House, Moscow, pp 6–8

Karyagina ZV, Tulenkova LN (1959) Spectrophotometric investigation of the continuum and emission spectrum of the night sky in the visual spectral region. Izvestiya Astrophys. Institute Kazakh. SSR Acad Sci 9:86–95

Kaufman F (1969) Neutral reactions involving H and other minor constituents. Can J Chem 47:1917–1926

Kaye JA (1988) On the possible role of the reaction $O + HO_2 \rightarrow OH + O_2$ in OH airglow. J Geophys Res 93:285–288

Kennealy JP, Del Greco FP, Caledonia GE, Green BD (1978) Nitric oxide chemiexcitation occurring in the reaction between metastable nitrogen atoms and oxygen molecules. J Chem Phys 69:1574–1584

Kenner RD, Ogrizlo EA (1984) Orange chemiluminescence from NO_2. J Chem Phys 80:1–6

Khayar A, Bonamy J (1987) Calculation of mean collision cross sections of free radical OH with foreign gases. J Quant Spectrosc Radiat Transfer 28:199–212

Khvorostovskaya LE, Potekhin IYu, Shved GM, Ogibalov BP, Uzyukova TV (2002) Measurement of the rate constant for quenching $CO_2(01^10)$ by atomic oxygen at low temperatures: reassessment of the rate of cooling by the CO_2 15µm emission in the lower thermosphere. Izvestiya Atmos Oceanic Phys 38:613–624

Khvostikov IA (1937) Luminescence of the night sky, Vavilov SI (ed) USSR Academic Science Publishing House, Moscow, Leningrad

Khvostikov IA (1948) Luminescence of the night sky, 2nd edn. Vavilov SI (ed) USSR Acad Sci Publ House, Moscow, Leningrad

Kirchhoff VWJH (1986a) Comment on General theory of the alkali metals present in the Earth's upper atmosphere. Ann Geophys 4:413–418

Kirchhoff VWJH (1986b) Theory of the atmospheric sodium layer: a review. Can J Phys 64:1664–1672

Kirchhoff VWJH, Clemesha BR (1983) Eddy diffusion coefficients in the lower thermosphere. J Geophys Res 88:5765–5768

Kistiakowsky GB, Tabbutt FD (1959) Gaseous detonations. XII. Rotational temperatures of the hydroxyl free radicals. J Chem Phys 30:577–581

Knutsen K, Copeland RA (1993) Vibrational relaxation of $OH(X^2\Pi, v = 7, 8)$ by O_2, N_2, N_2O and CO_2. Abstract. EOS, Trans. AGU, Vol 43. Fall Meeting Suppl. p 472

Kondo Y, Kupperian JE (1967) Interaction of the neutral hydrogen and charged particles in the radiation belts: the consequent Lyman-alpha emission. J Geophys Res 72:6091–6097

Kondratiev VN (1936) The elementary chemical processes. ChimTheoret, Leningrad

Kondratiev VN (1958) Kinetic of the chemical gaseous reactions. USSR Academic Science Publishing House, Moscow

Kondratiev VN, Nikitin EE (1974) Kinetic and mechanism of the gaseous phase reactions. Nauka, Moscow

Kondratiev VN, Nikitin EE (1981) Chemical processes in gases. Nauka, Moscow

Kovács I (1969) Rotational structure in the spectra of diatomic molecules. Akadémiai Kiadó, Budapest

Krasnopolsky VA (1987) Airglow physics of the planetary and comet atmospheres. Nauka, Moscow

Krassovsky VI (1949) On the night sky radiation in the infrared spectral region. Dokl USSR Acad Sci 66:53–54

Krassovsky VI (1950a) New data on the night sky radiation in the 8800–11000 A region. Dokl USSR Acad Sci 70:999–1000
Krassovsky VI (1950b) New emissions of the night sky in the 8800–11000 A region. Izvestiya Crimea astrophys. observ. USSR Acad Sci 5:100–104
Krassovsky VI (1950c) Nature of the infrared luminescence of the night sky. Dokl USSR Acad Sci 73:679–682
Krassovsky VI (1951a) On the mechanism of the night sky luminescence. Dokl USSR Acad Sci 77:395–398
Krassovsky VI (1951b) Influence of the water vapor and carbon oxides on the night sky luminescence. Dokl USSR Acad Sci 78:669–672
Krassovsky VI (1959) Energy sources of the upper atmosphere. Planet Space Sci 1:14–19
Krassovsky VI (1961) On the nature of the OH emission in the upper atmosphere. In: Krassovsky VI (ed) Spectral, electrophotometrical and radar researches of aurorae and airglow. N 5. USSR Academic Science Publishing House, Moscow, pp 29–31
Krassovsky VI (1963a) Chemistry of the upper atmosphere. In: Priester W (ed) Space Research, Vol 3. North-Holland Publ Co, Amsterdam, pp 96–116
Krassovsky VI (1963b) The hydroxyl emission in the upper atmosphere. In: Krassovsky VI (ed) Aurorae and Airglow. N 10. USSR Academic Science Publishing House, Moscow, pp 24–34
Krassovsky VI (1968a) Heating of the upper atmosphere during geomagnetic disturbances. Nature (London) 217:1136
Krassovsky VI (1978b) Discussion of paper by Y. Kondo and J.E. Kupperian, Jr., Interaction of neutral hydrogen and charged particles in the radiation belts: the consequent Lyman-alpha emission. J Geophys Res 73:6402–6403
Krassovsky VI (1969) The upper atmosphere as a regulator of geomagnetic storms, substorms and aurorae. Geomagn Aeronomy 9:29–40
Krassovsky VI (1971a) The calms and the storms in the upper atmosphere (Physics of the upper atmosphere and near-Earth space). Nauka, Moscow
Krassovsky VI (1971b) Atmospheric H_α emission of the atomic hydrogen by observations in Zvenigorod, Abastumani, Alma-Ata. Cosmic Res 9:418–429
Krassovsky VI (1972) Infrasonic variations of the OH emission in the upper atmosphere. Ann Géophys 28:739–746
Krassovsky VI (1978) NO_x dissociation of H_2O and winter anomaly of ionospheric absorption. Geomagn Aeronomy 18:151–153
Krassovsky VI, Galperin YuI (1960) Review of observational results on the airglow and aurorae. In: Trans Intern Astron Union (Moscow 1958), Vol 10A. Cambridge University Press, Cambridge, pp 327–328
Krassovsky VI, Shefov NN, Yarin VI (1961) On the OH airglow. J Atmos Terr Phys 21:46–53
Krassovsky VI, Shefov NN, Yarin VI (1962) Atlas of the airglow spectrum $\lambda\lambda$ 3000–12400 Å. Planet Space Sci 9:883–915
Krassovsky VI, Shefov NN, Vaisberg OL (1966) Atomic hydrogen and helium in airglow. Ann Géophys 22:208–216
Krassovsky VI, Semenov AI, Shefov NN, Yurchenko OT (1976) Predawn emission at 6300 Å and super-thermal ions from conjugate points. J Atmos Terr Phys 38:999–1001
Krassovsky VI, Potapov BP, Semenov AI, Shagaev MV, Shefov NN, Sobolev VG (1977) On the equilibrium nature of the rotational temperature of hydroxyl airglow. Planet Space Sci 25:596–597
Krassovsky VI, Rapoport ZTs, Semenov AI, Sobolev VG, Shefov NN (1980) Nitric oxide, water vapor, noctilucent clouds, emissions and radiowave absorption near the mesopause. Geomagn Aeronomy 20:657–663
Krassovsky VI, Rapoport ZTs, Semenov AI (1982) New emissions of the upper atmosphere as a sequence of the anthropogenic influence on the ionosphere. Cosmic Res 20:237–243
Krinberg IA (1978) The electron kinetic in the ionosphere and plasmasphere. Nauka, Moscow
Kron GE (1950) Photoelectric measurements of night-sky radiation beyond 9000 Angstroms. Publ Astron Soc Pac 62:264–266

Kukuy AS, Zelenov VV, Dodonov AF, Grigor'eva VM, Gershenzon YuM (1996) Reaction of $OH(v = 7 \div 9) + O_2 = OH_2 + O$ and its role in the kinetic mechanism of the hydroxyl emission in nightglow. Russian J Chem Phys 15:76–87

Kupperian JE, Byram ET, Chubb TA, Friedman H (1959) Far ultraviolet radiation in the night sky. Planet Space Sci 1:3–6

Kurilo MJ, Braun W, Kaldor A, Freund SM, Wayne RP (1974) Infra-red laser enhanced reactions: chemistry of vibrationally excited O_3 with NO and $O_2(^1\Delta)$. J Photochem 3:71–87

Kurt VG (1963) Neutral hydrogen in the near-earth neighbourhood and interplanetary space. Uspekhi Phys Nauk 81:249–270

Kutepov AA, Shved GM (1978) Radiative transfer of the 15-µm CO_2 band with the breakdown of local thermodynamic equilibrium in the Earth's atmosphere. Izvestiya USSR Acad Sci Atmos Oceanic Phys 14:28–43

Kutepov AA, Shved GM (1981) Radiation intensities of the 4, 3 and 15 mcm of CO_2 in the Earth' upper atmosphere for quiet conditions. Cosmic Res 19:483–486

Kutepov AA, Shved GM (1985) On the cooling of the lower thermosphere by radiation in the 15-µm CO_2 band. Izvestiya USSR Acad Sci Atmos Oceanic Phys 21:421–423

Kvifte G (1961) Temperature measurements from OH bands. Planet Space Sci 5:153–157

Lambert JD (1962) Relaxation in gases. In: Bates DR (ed) Atomic and molecular processes. Academic Press, New York, pp 679–699

Lang KR (1974) Astrophysical formulae. Springer-Verlag, Berlin

Langhoff SR, Werner HJ, Rosmus P (1986) Theoretical transition probabilities for the OH Meinel system. J Mol Spectrosc 118:507–529

Lazarev AV, Zastenker NN, Trubnikov DN (2003) Rotational energy relaxation in the free azot stream. Russian J Chem Phys 22:10–15

Lindinger W, Schmeltekopf AL, Fehsenfeld FC (1974) Temperature dependence of de-excitation rate constants of $He(2^3S)$ by Ne, Ar, Xe, H_2, N_2, O_2, NH_3 and CO_2. J Chem Phys 61:2890–2895

Link JK (1966) Measurement of the radiative lifetimes of the first excited states of Na, K, Rb, and Cs by means of the phase – shift method. J Opt Soc Amer 56:1195–1199

Link R, Cogger LL (1988) A reexamination of the OI 6300 Å nightglow. J Geophys Res 93A:9883–9892

Link R, Cogger LL (1989) Correction to "A reexamination of the OI 6300 Å nightglow" by R. Link and L.L. Cogger. J Geophys Res 94A:1556

Llewellyn EJ, Evans WFJ (1971) The dayglow. In: McCormac BM (ed) The radiating atmosphere. D Reidel Publishing Company, Dordrecht, pp 17–33

Llewellyn EJ, Long BH (1978) The OH Meinel bands in the airglow. The radiative lifetime. Can J Phys 56:581–586

Llewellyn EJ, McDade IC (1996) A reference model for atomic oxygen in the terrestrial atmosphere. Adv Space Res 18:209–226

Llewellyn EJ, Solheim BH (1978) The excitation of the Infrared Atmospheric oxygen bands in nightglow. Planet Space Sci 26:533–538

Llewellyn EJ, Long BH, Solheim BH (1978) The quenching of OH* in the atmosphere. Planet Space Sci 25:525–531

López-Moreno JJ, Rodrigo R, Moreno F, Lopez-Puertaz M, Molina A (1987) Altitude distribution of vibrationally excited states of atmospheric hydroxyl at levels $v = 2$ to $v = 7$. Planet Space Sci 35:1029–1038

Losev SA, Umansky SYa, Yakubov IT (1995) Physical-chemical processes in the gaseous dynamics. In: Cherny GG, Losev SA (eds) Dynamics of the physical-chemical processes in the gas and plasma, Vol 1. Moscow State University Press, Moscow

Lunt ST, Marston G, Wayne RP (1988) Formation of $O_2(a^1\Delta_g)$ and vibrationally excited OH in the reaction between O atoms and HO_x species. J Chem Soc Faraday Trans 2 84:899–912

Lytle EA, Hunten DM (1959) The ratio of sodium to potassium in the upper atmosphere. J Atmos Terr Phys 16:236–245

MacDonald RG, Buijs HL, Gush HP (1968) Spectrum of the night airglow between 3 and 4 microns. Can J Phys 46:2575–2578
Machael JV, Payne WA, Whytock DA (1976) Absolute rate constants for $O + NO + M(=He, Ne, Ar, Kr) \rightarrow NO_2 + M$ from 217–500 K. J Chem Phys 65:4830–4834
Makarova EA, Kharitonov AV (1972) Energy distribution in the solar spectrum. Nauka, Moscow
Makarova EA, Kharitonov AV, Kazachevskaya TV (1991) The solar radiation flux. Nauka, Moscow
Makarova NM, Mikirov AE, Smerkalov VA (1973) Generalized dependence of the terrestrial and water-surface albedo on the solar height over horizon. In: Trans Institute Applied Geophysics. Certain problems of upper atmospheric physics. N 17. Hydrometeoizdat, Moscow, pp 203–210
Makhlauf UB, Picard RH, Winick JR (1995) Photochemical-dynamical modeling of the measured response of airglow to gravity waves. 1. Basic model for OH airglow. J Geophys Res 100D:11289–11311
Malkin OA (1971) Relaxation processes in the gas. Atomizdat, Moscow
Mange P (1973) Hydrogen and helium emissions. In: McCormac BM (ed) Physics and chemistry of upper atmospheres. D Reidel Publishing Company, Dordrecht, pp 248–259
Manuilova RO, Shved GM (1989) The origin of the glow of vibrationally excited ozone in the atmosphere. In: Feldstein YaI, Shefov NN (eds) Aurorae and Airglow. N 33. VINITI, Moscow, pp 43–47
Martsvaladze NM (1972) Spatial distribution of the upper atmosphere H_α emission. Its variations during the solar cycle and dependence on geomagnetic disturbances. In: Fishkova LM, Kharadze EK (eds) Bull Abastumani astrophys observ. N 42. pp 39–45
Martsvaladze NM, Fishkova LM (1982) On the one possible reason of the irregular variations of the hydrogen emission of the upper atmosphere. Cosmic Res 20:773–775
Martsvaladze NM, Fishkova LM, Shefov NN (1971) Disturbed variations of the hydrogen emission. Astron Circ USSR Acad Sci 619:5–6
Massey HSW, Burhop EHS (1952) Electronic and ionic impact phenomena. Clarendon Press, Oxford
Mathis JS (1957) Statistical equilibrium of triplet levels of neutral helium. Astrophys J 125:318–327
McClatchey RA, Benedict WS, Clough SA, Burch DE, Calfee RF, Fox K, Rothman LS, Garing JS (1973) AFCGL atmospheric absorption line parameters compilation. AFCRL-TR-73-0096. L.G. Hanscom Field. Ma., 01730, Begford N 434
McDade IC, Llewellyn EJ (1986) The excitation of $O(^1S)$ and O_2 bands in the nightglow: a brief review and preview. Can J Phys 64:1626–1630
McDade IC, Llewellyn EJ (1987) Kinetic parameters related to sources and sinks of vibrationally excited OH in the night. J Geophys Res 92A:7643–7650
McDade IC, Llewellyn EJ (1988) Mesospheric oxygen atom densities inferred from night-time OH Meinel band emission rates. Planet Space Sci 36:897–905
McDade IC, Greer RGH, Murtagh DP (1984a) Thermospheric nitric oxide concentrations derived from a measurement of the altitude profile of the green nightglow continuum. Ann Geophys 2:487–494
McDade IC, Murtagh DP, Greer RGH, Dickinson PHG, Witt G, Stegman J, Llewellyn EJ, Thomas L, Jenkins DB (1986a) ETON 2: Quenching parameters for the proposed precursors of $O_2(b^1\Sigma_g^+)$ and $O(^1S)$ in the terrestrial nightglow. Planet Space Sci 34:789–800
McDade IC, Llewellyn EJ, Greer RGH, Murtagh DP (1986b) ETON 3: altitude profile of the nightglow continuum at green and near infrared wavelengths. Planet Space Sci 34:801–810
McDaniel EW (1964) Collision phenomena in ionized gases. John Wiley and Son Inc, New York
Mcelroy MB (1965) Excitation of atmospheric helium. Planet Space Sci 13:403–433
McElroy MB, Hunten DM (1966) A method of estimating the Earth albedo for dayglow measurements. J Geophys Res 71:3635–3638
McEwan MJ, Phillips LF (1975) Chemistry of the atmosphere. Edward Arnold, London
McNeil WJ, Murad E, Lai ST (1995) Comprehensive model for the atmospheric sodium layer. J Geophys Res 100D:16847–16855

Megill AB, Roach FE (1961) The integrated star-light over the sky. Nat Bur Stand, Washington DC 106:1–76
Meier RR (1969) Balmer Alpha and Balmer Beta in the hydrogen geocorona. J Geophys Res 74:3561–3574
Meier RR (1974) A survey of the ultraviolet airglow from 1216 to 304 Å. Ann Géophys 31:91–104
Meier RR, Prinz DK (1970) Absorption of the solar Lyman alpha line by geocoronal atomic hydrogen. J Geophys Res 75:6969–6979
Meier RR, Carruther GR, Page TL, Lavasseur-Regourd AC (1977) Geocoronal Lyman β and Balmer α emissions measured during the Apollo 16 mission. J Geophys Res 82:737–739
Meier RR, Anderson DE, Paxton LJ, McCoy RP (1987) The OI 3d $^3D^o - 2p^{43}P$ transition at 1026 A in the day airglow. J Geophys Res 92A:8767–8773
Meinel AB (1948) The near-infrared spectrum of the night sky and aurorae. Publ Astron Soc Pac 60:373–378
Meinel AB (1950a) Hydride emission bands in the spectrum of the night sky. Astrophys J 111:207
Meinel AB (1950b) OH emission bands in the spectrum of the night sky. I. Astrophys J 111:555–564
Meinel AB (1950c) OH emission bands in the spectrum of the night sky. II. Astrophys J 112:120–130
Meinel AB (1950d) O_2 emission band in the infrared spectrum of the night sky. Astrophys J 112:464–468
Mies FH (1974) Calculated vibrational transition probabilities of $OH(X^2\Pi)$. J Mol Spectrosc 53:150–188
Miller JH, Boese RW, Giver LP (1969) Intensity measurements and rotational intensity distribution for the oxygen A-band. J Quant Spectrosc Radiat Transfer 9:1507–1517
Minnaert M, Mulders GFW, Houtgast J (1940) Photometric atlas of the solar spectrum from λ 3612 to λ 8771 Å. D Schnabel Amsterdam Kampert Helm, Amsterdam
Mironov AV, Prokudina VS, Shefov NN (1959) Auroral observations on 10–11 February, 1958, Moscow. In: Krassovsky VI (ed) Spectral, electrophotometrical and radar researches of aurorae and airglow. N 1. USSR Acad Sci Publ House, Moscow, pp 20–24
Misawa K, Takeuchi I (1982) Nightglow intensity variations in the O_2(0–1) atmospheric band, the Na D lines, the OH (6–2) band, the yellow-green continuum at 5750 Å and the oxygen green line. Ann Géophys 38:781–788
Mitchell ACG, Zemansky MW (1934) Resonance radiation and excited atoms. Cambridge University Press, Cambridge
Mitra V (1974a) Origin of alkali metals in the Earth's atmosphere. Ann Géophys 30:421–427
Mitra V (1974b) Deposition of twilight lithium by a high altitude thermonuclear explosion. Ann Géophys 30:497–502
Mlynczak MG, Nesbitt DJ (1995) The Einstein coefficient for spontaneous emission of the $O_2(a^1\Delta_g)$ state. Geophys Res Lett 22:1381–1384
Molher OC, Pierce AM, McMath RR, Goldberg L (1950) Photometric atlas of the infra-red solar spectrum λ 8465 to λ 25242 Å. Michigan University Press, Ann Arbor
Molina A (1983) Sodium nightglow and gravity waves. J Atmos Sci 40:2444–2450
Moore CE (1949) Atomic energy levels, Vol 1. N 467. Nat Bur Stand, Washington DC
Mott NF, Massey HSW (1965) The theory of atomic collisions, 3rd edn. Clarendon Press, Oxford
Moussa HRM, de Heer FJ, Schutten J (1968) Excitation of helium by 0.05–6 keV electrons and polarization of the resulting radiation. Physica 40:517–549
Murphy RE (1971) Infrared emission of the OH in the fundamental and first overtone bands. J Chem Phys 54:4852–4859
Nagy AF, Liu SC, Baker DJ (1976) Vibrationally excited hydroxyl molecules in the lower atmosphere. Geophys Res Lett 3:731–734
Nasyrov GA (1967) Spatial variations of nightglow in the region λ 5893 A. In: Krassovsky VI (ed) Aurorae and Airglow. N 13. Nauka, Moscow, pp 10–12
Nasyrov GA (2003) Statistical regularities of variations in the sodium nightglow observed in Ashkhabad during the solar activity minimum. Geomagn Aeronomy 43:402–404

Nelson DD, Schiffman A, Nesbit DJ, Orlando JJ, Burkholder JB (1990) H + O_3 Fourier-transform infrared emission and laser absorption studies of OH($X^2\Pi$) radical: an experimental dipole moment function and state-to-state Einstein A coefficients. J Chem Phys 93:7003–7019

Nicholls DC, Evans WFJ, Llewellyn EJ (1972) Collisional relaxation and rotational intensity distributions in spectra of aeronomic interest. J Quant Spectrosc Radiat Transfer 12:549–558

Nicolet M (1948) Deduction regarding the state of the high atmosphere. In: The emission spectra of the night sky and aurorae (Reports of the Gassiot Committee). Phys Soc London, pp 36–48

Nicolet M (1962) Aeronomy. Preprint. Institut d'Astrophysique, Liège

Nicolet M (1971) Aeronomic reactions of hydrogen and ozone. In: Fioccho G (ed) Mesospheric model and related experiments. D Reidel Publishing Company, Dordrecht, pp 1–51

Nicolet M (1981) The solar spectral irradiance and its action in the atmospheric photodissociation processes. Planet Space Sci 29:951–974

Nicolet M (1984) On the photodissociation of water vapour in the mesopause. Planet Space Sci 32:871–880

Nicolet M (1989a) Solar spectral irradiances with their diversity between 120 and 900 nm. Planet Space Sci 37:1249–1289

Nicolet M (1989b) Aeronomic chemistry of ozone. Planet Space Sci 37:1621–1652

Nicolet M, Kennes R (1988) Aeronomic problems of molecular oxygen photodissociation–IV. The various parameters for the Herzberg continuum. Planet Space Sci 36:1069–1076

Nicolet M, Cieslik KS, Kennes R (1989) Aeronomic problems of molecular oxygen photodissociation–V. Predissociation in the Schumann-Runge bands of oxygen. Planet Space Sci 37:427–458

Nicovich JM, Wine PH (1987) Temperature dependence of the O + HO_2 rate coefficient. J Phys Chem 91:5118–5123

Nikerov VA, Sholin GV (1985) A kinetic of the degradational processes. Energoatomizdat, Moscow

Nikitin EE (1970) A theory of the elementary atomic-molecular processes in the gases. Chemistry, Moscow

Noxon JF (1961) Observation of the $(b^1\Sigma_g^+ - a^1\Delta_g)$ transition in O_2. Can J Phys 39:1110–1119

Noxon JF (1978) The near infrared nightglow continuum. Planet Space Sci 26:105–115

Noxon JF, Vallance Jones A (1962) Observation of the (0,0) band of the $(^1\Delta_g - ^3\Sigma_g^-)$ system of oxygen in the day and twilight airglow. Nature (London) 196:157–158

Ogawa T (1976) Excitation processes of infrared atmospheric emissions. Planet Space Sci 24:749–756

Ogawa T, Kondo Y (1977) Diurnal variability of thermospheric N and NO. Planet Space Sci 25:735–742

Ogawa T, Iwagami N, Nakamura M, Takano M, Tanabe H, Takeuchi A, Miyashita A, Suzuki K (1987) A simultaneous observation of the height profiles of the night airglow OI 5577 Å, O_2 Herzberg and Atmospheric bands. J Geomag Geoelectr 39:211–228

Offermann D, Grossmann KU (1978) Spectrometric measurement of atomic oxygen 63 μm emission in the thermosphere. Geophys Res Lett 5:387–390

Osipov AI (1985) Rotational relaxation in gases. Ingeneering-Physical J 49:154–170

Pal SR (1973) Features of sodium emission in nightglow. Tellus 25:69–79

Patterson TNL (1967) Metastable helium in the upper atmosphere. Planet Space Sci 15:1219–1222

Pavlov AV (1996) Mechanism of the electron density depletion in the SAR arc region. Ann Geophys 14:211–212

Pavlov AV (1997) Subauroral red arcs as a conjugate phenomenon: comparison of OV1-10 satellite data with numerical calculations. Ann Geophys 15:984–998

Pavlov AB (1998) Interpreting the observations of auroral red arcs in magnetically conjugate regions. Geomagn Aeronomy 38:803–807

Pavlov AV, Pavlova NM, Drozdov AB (1999) Production rate of $O(^1D)$, $O(^1S)$ and $N(^2D)$ in the subauroral red arc region. Geomagn Aeronomy 39:201–205

Pendleton W, Espy P, Baker D, Steed A, Fetrow M, Henriksen K (1989) Observation of OH Meinel (7,4) P($N'' = 13$) transitions in the night airglow. J Geophys Res 94:505–510

Pendleton WR, Taylor MJ (2002) The impact of L-uncoupling on Einstein coefficients for the OH Meinel (6,2) band: implications for Q-branch rotational temperatures. J Atmos Solar-Terr Phys 64:971–983

Perminov VI, Semenov AI (1992) The nonequilibrium of the rotational temperature of OH bands under high level rotational excitation. Geomagn Aeronomy 32:306–308

Perminov VI, Semenov AI, Shefov NN (1998) Deactivation of hydroxyl molecule vibrational states by atomic and molecular oxygen in the mesopause region. Geomagn Aeronomy 38:761–764

Perminov VI, Semenov AI, Bakanas VV, Zheleznov YuA, Khomich VYu (2004) Regular variations in the (0–1) band intensity of the oxygen emission Atmospheric system. Geomagn Aeronomy 44:498–501

Phelps DH, Dalby FW (1965) Optical observations of the Stark effect of OH. Can J Phys 43:144–154

Piterskaya NA, Shefov NN (1975) Intensity distribution of the OH rotation-vibration bands. In: Krassovsky VI (ed) Aurorae and Airglow. N 23. Nauka, Moscow, pp 69–122

Plane JMC (1991) The chemistry of meteoric metals in the Earth's upper atmosphere. Int Rev Phys Chem 10:55–106

Plane JMC, Cox RM, Qian J, Pfenninger WM, Papen GC, Gardner CS, Espy PJ (1998) Mesospheric Na layer at extreme high latitudes in summer. J Geophys Res 103D:6381–6389

Potter AE, Coltharp RN, Worley SD (1971) Mean radiative lifetime of vibrationally excited ($v = 9$) hydroxyl. Rate of the reaction of vibrationally excited hydroxyl ($v = 9$) with ozone. J Chem Phys 54:992–996

Prokudina VS (1959a) Observations of the line λ 6562 A in the night airglow spectrum. In: Krassovsky VI (ed) Spectral, electrophotometrical and radar researches of aurorae and airglow. N 1. USSR Academic Science Publishing House, Moscow, pp 43–44

Prokudina VS (1959b) Determination of the hydroxyl rotational temperature in the upper atmosphere. Izvestiya USSR Acad Sci Ser Geophys 4:629–631

Qian J, Gardner CS (1995) Simultaneous lidar measurements of mesospheric Ca, Na, and temperature profiles at Urbana, Illinois. J Geophys Res 100D:7453–7461

Rees MH (1989) Physics and chemistry of the upper atmosphere. Houghton JT, Rycroft MJ, Dessler AJ (eds) Cambridge University Press, Cambridge

Reynard LM, Donaldson DJ (2001) OH production from the reaction of vibrationally excited H_2 in the mesosphere. Geophys Res Lett 28:2157–2160

Richter ES, Rowlett JR, Gardner CS, Sechrist CF (1981) Lidar observations of the mesospheric sodium layer over Urbana, Illinois. J Atmos Terr Phys 43:327–337

Rishbeth H (1967) Transequatorial diffusion in the topside ionosphere. Planet Space Sci 15:1261–1265

Rishbeth H, Garriott OK (1969) Introduction to ionospheric physics. Academic Press, New York

Roach FE (1964) The light of the night sky: astronomical interplanetary and geophysical. Space Sci Rev 3:512–540

Roach FE, Gordon JL (1973) The light of the night sky. D Reidel Publishing Company, Dordrecht, Holland

Roach FE, Megill LR (1961) Integrated starlight over the sky. Astrophys J 133:228–242

Robley R (1973) Variation annuelle des luminances de l'émission continue atmosphérique et de la lumière zodiacalle du pole célestre. Ann Géophys 29:321–328

Robley R, Vilkki E (1970) Le continuum dans la lumière du ciel nocturne. Ann Géophys 26:195–199

Rodionov SF (1940) Light counter. J Exp Theor Phys 10:294–304

Rothman LS, Benedict WS (1978) Infrared energy levels and intensities of carbon dioxide. Appl Opt 17:2605–2611

Rothman LS, Clough SA, McClatchey RA, Young LG, Snider DE, Goldman A (1978) AFGL trace gas compilation. Appl Opt 17:507

Rothman LS, Goldman A, Gillis JR, Gamache RR, Pickett HM, Poynter RL, Husson N, Chedin A (1983) AFGL trace gas compilation: 1982 version. Appl Opt 22:1616–1627

Rottman GJ (1981) Rocket measurements of the solar spectral irradiance during solar minimum, 1972–1977. J Geophys Res 86A:6697–6705

Roux F, d'Incan J, Cerny D (1973) Experimental oscillator strengths in the infrared vibration-rotation spectrum of the hydroxyl radical. Astrophys J 186:1141–1156

Rundle HN (1960) Ionization of a static interplanetary gas and expected emission lines from this gas. Planet Space Sci 2:86–98

Rusanov VD, Fridman AA (1984) Physics of the chemically active plasma. Nauka, Moscow

Sappey AD, Copeland RA (1990) Collision dynamics of $OH(X^2\Pi_i, v = 12)$. J Chem Phys 93:5741–5746

Saxon RP, Slanger TG (1986) Molecular oxygen absorption continua at 195–300 nm and O_2 radiative lifetimes. J Geophys Res 91D:9877–9879

Schlapp RJ (1937) Fine structure in the $^3\Sigma$ ground state of the oxygen molecule, and the rotational intensity distribution in the atmospheric band. Phys Rev 51:343–345

Scholz K (1932) Zur quantenmechanischen Berechnung von Intensitäten ultrarotes Banden. Zeits Phys B 78:751–770

Schurin B, Ellis RE (1966) First and second – overtone intensity measurements for CO and NO. J Chem Phys 45:2528–2532

Semenov AI (1989a) The specific features of the green emission excitation process in the nocturnal atmosphere. In: Feldstein YaI, Shefov NN (eds) Aurorae and Airglow. N 33. VINITI, Moscow, pp 74–80

Semenov AI (1989b) Relation between the ozone-hydrogen and superhydroxyl excitation mechanism of hydroxyl emission. Geomagn Aeronomy 29:687–689

Semenov AI (1997) Long-term changes in the height profiles of ozone and atomic oxygen in the lower thermosphere. Geomagn Aeronomy 37:354–360

Semenov AI, Shefov NN (1996) An empirical model for the variations in the hydroxyl emission. Geomagn Aeronomy 36:468–480

Semenov AI, Shefov NN (1997a) An empirical model of nocturnal variations in the 557.7-nm emission of atomic oxygen. 1. Intensity. Geomagn Aeronomy 37:215–221

Semenov AI, Shefov NN (1997b) An empirical model of nocturnal variations in the 557.7-nm emission of atomic oxygen. 2. Temperature. Geomagn Aeronomy 37:361–364

Semenov AI, Shefov NN (1997c) An empirical model of nocturnal variations in the 557.7-nm emission of atomic oxygen. 3. Emitting layer altitude. Geomagn Aeronomy 37:470–474

Semenov AI, Shefov NN (1997d) Empirical model of the variations of atomic oxygen emission 557.7 nm. In: Ivchenko VN (ed) Proceedings of SPIE (23rd European Meeting on Atmospheric Studies by Optical Methods, Kiev, September 2–6, 1997), Vol 3237. The International Society for Optical Engineering, Bellingham, pp 113–122

Semenov AI, Shefov NN (1999a) Empirical model of hydroxyl emission variations. Int J Geomagn Aeronomy 1:229–242

Semenov AI, Shefov NN (1999b) Variations of the temperature and the atomic oxygen content in the mesopause and lower thermosphere region during change of the solar activity. Geomagn Aeronomy 39:484–487

Semenov AI, Shefov NN (2005) Model of the vertical profile of the atomic oxygen concentration in the mesopause and lower ionosphere region. Geomagn Aeronomy 45:797–808

Semenov AI, Shefov NN, Perminov VI, Khomich VYu, Fadel KhM (2005) Temperature response of the middle atmosphere on the solar activity for different seasons. Geomagn Aeronomy 45:236–240

Shagaev MV (1977) The nightglow OH rotational temperatures with different vibrational excitation Astron Circ USSR Acad Sci 936:3–4

Shalashilin DV, Umanskii SYa, Gershenzon YuM (1992) Dynamics of vibrational energy exchange in collisions of OH and OD radicals with N_2. Application to the kinetics of OH-vibrational deactivation in the upper atmosphere. Chem Phys 168:315–325

Shalashilin DV, Umansky SYa, Gershenzon YuM, Grigor'eva VM (1993) Vibrational energy exchange dynamics during the OH and OD radical collisions with N_2. Application to the vibrational kinetics of the OH quenching in the upper atmosphere. Russian J Chem Phys 12:435–445

Shalashilin DV, Umansky SYa, Gershenzon YuM, Grigor'eva VM, Lara-Ochoa F, Mishchenko AV (1994) Trajectory investigation of the effective VT-exchange of the vibrationally excited hydroxyl collision with the oxygen molecule. Russian J Chem Phys 13:9–21

Sharov AS, Lipaeva NA (1973) Stellar component of the night airglow. Astron Rep 50:107–114

Sharp WE (1986) Sources of the emission features between 2000 and 8000 Å in the thermosphere. Can J Phys 64:1594–1607

Sharp WE, Siskind DE (1989) Atomic emission in the ultraviolet nightglow. Geophys Res Lett 16:1453–1456

Shcheglov PV (1962a) Twilight enhancement of the infrared helium line 10830 A. Astron Rep 39:158–159

Shcheglov PV (1962b) Observation of the twilight helium emission λ 10830 A with Fabry-Perot interferometer. In: Krassovsky VI (ed) Aurorae and Airglow. N 9. USSR Acad Sci Publ House, Moscow, pp 59–60

Shcheglov PV (1963) Electronic telescopy. Fizmatgiz, Moscow

Shcheglov PV (1967) The neutral hydrogen distribution in the terrestrial atmosphere by observations of the H_α nightglow. Astron Circ USSR Acad Sci 427:5

She CY, Lowe RP (1998) Seasonal temperature variations in the mesopause region at mid–latitude: comparison of lidar and hydroxyl rotational temperatures using WINDII/UARS OH height profiles. J Atmos Solar-Terr Phys 60:1573–1583

Shefov NN (1959) Intensities of some twilight and night airglow emissions. In: Krassovsky VI (ed) Spectral, electrophotometrical and radar researches of aurorae and airglow. N 1. USSR Acad Sci Publ House, Moscow, pp 25–29

Shefov NN (1960) Intensities of some night sky emissions. In: Krassovsky VI (ed) Spectral, electrophotometrical and radar researches of aurorae and airglow. N 2–3. USSR Acad Sci Publ House, Moscow, pp 57–59

Shefov NN (1961a) On the nature of helium emission λ 10830 Å in aurorae. Planet Space Sci 5:75–76

Shefov NN (1961b) Émission de l'helium dans la haute atmosphère. Ann Géophys 17:395–402

Shefov NN (1961c) On determination of the rotational temperature of the OH bands. In: Krassovsky VI (ed) Spectral, electrophotometrical and radar researches of aurorae and airglow. N 5. USSR Acad Sci Publ House, pp 5–9

Shefov NN (1961d) Continuous emission in the night airglow. In: Krassovsky VI (ed) Spectral, electrophotometrical and radar researches of aurorae and airglow. N 5. USSR Acad Sci Publ House, Moscow, pp 39–41

Shefov NN (1961e) On the nature of helium emission λ 10830 Å in aurorae. In: Krassovsky VI (ed) Spectral, electrophotometrical and radar researches of aurorae and airglow. N 5. USSR Acad Sci Publ House, pp 47–48.

Shefov NN (1961f) Twilight enhancement of the λ 10830 Å helium emission. Astron Circ USSR Acad Sci 222:11–12

Shefov NN (1961g) On the vibrational population of OH molecules. In: Krassovsky VI (ed) Spectral, electrophotometrical and radar researches of aurorae and airglow. N 6. USSR Academic Science Publishing House, pp 21–27

Shefov NN (1962a) Sur l'émission de l'helium dans la haute atmosphère. Ann Géophys 18:125

Shefov NN (1962b) The helium emission in the upper atmosphere. In: Krassovsky VI (ed) Aurorae and Airglow. N 8. USSR Academic Science Publishing House, pp 50–65

Shefov NN (1963a) The behaviour of the helium λ 10830 A emission in twilight. In: Krassovsky VI (ed) Aurorae and Airglow. N 10. USSR Academic Science Publishing House, pp 56–64

Shefov NN (1963b) Helium in the upper atmosphere. Planet Space Sci 10:73–77

Shefov NN (1967) Statistical properties of the helium emission. In: Krassovsky VI (ed) Aurorae and Airglow. N 13. USSR Academic Science Publishing House, pp 64–68

Shefov NN (1968) Twilight helium emission during low and high geomagnetic activity. Planet Space Sci 16:1103–1107

Shefov NN (1969a) Concentration of hydrogen and helium in the outer atmosphere: geocorona. Ann IQSY. The M.I.T. Press, Cambridge Mass, 5:215–228

Shefov NN (1969b) Discussion of paper by Y. Kondo and J.E. Kupperian, Jr., Interaction of neutral hydrogen and charged particles in the radiation belts: the consequent Lyman-alpha emission. J Geophys Res 74:922–924

Shefov NN (1969c) Hydroxyl emission of the upper atmosphere. II. Effect of a sunlit atmosphere. Planet Space Sci 17:1629–1639

Shefov NN (1969d) Hydrogen and helium emissions in the upper atmosphere. Geomagn Aeronomy 9:1048–1052

Shefov NN (1970a) Migration of the H and He inside the atmosphere and their escape. Geomagn Aeronomy 10:278–282

Shefov NN (1970b) Migration of the H and He inside the atmosphere and their escape. In: Donahue TM, Smith PA, Thomas L (eds) Space Research, Vol 10. North-Holland Publishing Company, Amsterdam, pp 623–632

Shefov NN (1970c) On the correlation between the intensity emission of the atmospheric system of O_2 and the vibrational temperature of the OH bands. Astron Circ USSR Acad Sci 589:7–8

Shefov NN (1971a) Hydroxyl emissions of the upper atmosphere. III. Diurnal variations. Planet Space Sci 19:129–136

Shefov NN (1971b) Hydroxyl emission of the upper atmosphere. IV Correlation with the molecular oxygen emission. Planet Space Sci 19:795–796

Shefov NN (1973) Hydrogen and helium emissions and concentrations in the upper atmosphere. In: Krassovsky VI (ed) Aurorae and Airglow. N 20. Nauka, Moscow, pp 40–56

Shefov NN (1975a) Results of studies of the hydroxyl emission. In: Krassovsky VI (ed) Aurorae and Airglow. N 22. Nauka, Moscow, pp 71–76

Shefov NN (1975b) Emissive layer altitude of the atmospheric system of molecular oxygen. In: Krassovsky VI (ed) Aurorae and Airglow. N 23. Nauka, Moscow, pp 54–58

Shefov NN (1976) Seasonal variations of the hydroxyl emission. In: Krassovsky VI (ed) Aurorae and Airglow. N 24. Nauka, Moscow, pp 32–36

Shefov NN (1978) Airglow. In: Total results of the Science and Technique. Geomagnetism and upper layers of the atmosphere, Vol 4. VINITI, Moscow, pp 199–230

Shefov NN, Truttse YuL (1969) Hydrogen and hydroxyl emissions in the nightglow. Ann IQSY. The M.I.T. Press, Cambridge Mass 4:400–406

Shefov NN, Yurchenko OT (1970) Absolute intensities of the auroral emissions in Zvenigorod. In: Krassovsky VI (ed) Aurorae and Airglow. N 18. Nauka, Moscow, pp 50–96

Shefov NN, Piterskaya NA (1984) Spectral and space-time characteristics of the background luminosity of the upper atmosphere. Hydroxyl emission. In: Galperin YuI (ed) Aurorae and Airglow. N 31. VINITI, Moscow, pp 23–123

Shefov NN, Semenov AI (2001) An empirical model for nighttime variations in atomic sodium emission: 2. Emitting layer height. Geomagn Aeronomy 41:257–261

Shefov NN, Semenov AI (2002) The long-term trend of ozone at heights from 80 to 100 km at the mid-latitude mesopause for the nocturnal conditions. Phys Chem Earth 27:535–542

Shefov NN, Semenov AI, Tikhonova VV, Yurchenko OT, Novikov NN (1998) Variations in the distribution of vibrational-level populations of hydroxyl molecules. Geomagn Aeronomy 38:823–826

Shefov NN, Semenov AI, Pertsev NN (2000) Dependencies of the amplitude of the temperature enhancement maximum and atomic oxygen concentrations in the mesopause region on seasons and solar activity level. Phys Chem Earth Pt B 25:537–539

Shefov NN, Semenov AI, Yurchenko OT (2002) Empirical model of the ozone vertical distribution at the nighttime mid-latitude mesopause. Geomagn Aeronomy 42:383–389

Shklovsky IS (1950a) Identification of the infrared luminescence of the night sky with the rotation-vibration bands of the OH hydroxyl molecules. Dokl USSR Acad Sci 75:371–374

Shklovsky IS (1950b) Quantitative analysis of the hydroxyl emission intensity of the night sky. Dokl USSR Acad Sci 75:789–792

Shklovsky IS (1951a) On the nature of the infrared radiation of the night sky. Izvestiya Crimea astrophys observ. 7:34–58

Shklovsky IS (1951b) The solar corona. Gostekhizdat, Moscow

Shklovsky IS (1957) The intensity of the rotation-vibration bands of the OH molecule. Mém Soc Roy Sci Liège 18:420–425

Shklovsky IS (1958) Elementary processes in the upper atmosphere and their manifestation in emissions. Ann Géophys 14:414–424

Shklovsky IS (1959) On the hydrogen emission in the night sky. Planet Space Sci 1:63–65

Shouiskaya FK (1963) An attempt to detect the proper glow of atmosphere during the solar eclipse on February 15, 1961. In: Krassovsky VI (ed) Aurorae and Airglow. N 10. USSR Acad Sci Publ House, Moscow, pp 44–53

Sipler DP, Biondi MA (1975) Evidence for chemiexcitation as the source of the sodium nigh glow. Geophys Res Lett 2:106–108

Sipler DP, Biondi MA (1978) Interferometric studies of the twilight and nightglow sodium D-line profiles. Planet Space Sci 26:65–73

Sivjee GG, Hamwey RM (1987) Temperature and chemistry of the polar mesopause OH. J Geophys Res. 92A:4663–4672

Slanger TG, Huestis DL (1981) $O_2(c^1\Sigma_u^- \rightarrow X^3\Sigma_g^-)$ emission in the terrestrial nightglow. J Geophys Res 86A:3551–3554

Slanger TG, Cosby PC, Huestis DL, Osterbrock DE (2000) Vibrational level distribution of $O_2(b^1\Sigma_g^+, v = 0$–$15)$ in the mesosphere and lower thermosphere region. J Geophys Res 105D:20557–20564

Slipher VM (1929) Emissions in the spectrum of the light of the night sky. Publ Astron Soc Pac 41:262–263

Slipher VM (1933) Spectrographic studies of the planets. Mon Not Roy Astron Soc 93:657–668

Smirnov BM (1968) The atomic collisions and the elementary processes in the plasma. Atomizdat, Moscow

Smith FL, Smith C (1972) Numerical evaluation of Chapman's grazing incidence integral $ch(X, \chi)$. J Geophys Res 77:3592–3597

Smith IWM, Williams MD (1985) Vibrational relaxation of OH ($v = 1$) and OD ($v = 1$) by HNO_3, DNO_3, H_2O, NO and NO_2. J Chem Soc Faraday Trans 2 81:1849–1860

Smith DR, Blumberg WAM, Nadile RM, Lipson SJ, Huppi ER, Wheeler NB (1992) Observation of high-N hydroxyl pure rotation lines in atmospheric emission spectra by the CIRRIS 1A Space Shuttle experiment. Geophys Res Lett 19:593–596

Sobolev VG (1978a) Continuum in night airglow between 8000 and 11000 Å. Planet Space Sci 26:703–704

Sobolev VG (1978b) Continuum of the near infrared range of nightglow spectrum. In: Krassovsky VI (ed) Aurorae and Airglow. N 27. Soviet Radio, Moscow, pp 30–35

Sobolev VG (1979) Correlation between nightglow continuum and ionospheric absorption. Astron Circ USSR Acad Sci 1083:7–8

Sommer LA (1932) Über den langwelligen Teil des sichtbaren Spektrums des Nachthimmellichtes. Zeits Phys 77:374–390

Sparrow JG, Ney EP, Burnett GB, Stoddart JW (1968) Airglow observations from OSO-B2 satellite. J Geophys Res 73:857–866

Spencer JE, Glass GP (1977a) The production and subsequent relaxation of vibrationally excited OH in the reaction of atomic oxygen with HBr. Int J Chem Kin 9:97–109

Spencer JE, Glass GP (1977b) Some reactions of OH($v = 1$). Int J Chem Kin 9:111–122

Sridharan UC, Klein FS, Kaufman F (1985) Detailed course of the $O + HO_2$ reaction. J Chem Phys 82:592–593

Stair AT, Sharma RD, Nadile RM, Baker DJ, Grieder WF (1985) Observations of limb radiance with cryogenic spectral infrared rocket experiment. J Geophys Res 90A:9763–9775

States RJ, Gardner CS (1999) Structure of the mesospheric Na layer at $40°$ N latitude: seasonal and diurnal variations. J Geophys Res 104D:11783–11898

States RJ, Gardner CS (2000a) Thermal structure of the mesopause region (80–105 km) at 40°N latitude. Part I: seasonal variations. J Atmos Sci 57:66–77

States RJ, Gardner CS (2000b) Thermal structure of the mesopause region (80–105 km) at 40°N latitude. Part II: diurnal variations. J Atmos Sci 57:78–92

Stebbins J, Whitford AE, Swings P (1945) A strong infra-red radiation from molecular nitrogen in the night sky. Astrophys J 101:39–46

Stebbings RF, Dunning FB, Tittel FK, Rundel RD (1973) Photoionization of helium metastable atoms near threshold. Phys Rev Lett 30:815–817

Stegman J (1991) Spectroscopic and kinetic studies of atmospheric oxygen emissions. Stockholm University, Stockholm

Sternberg JR, Ingham MT (1972) Observations of the airglow continuum. Mon Not Astron Roy Soc 159:1–20

Stoffregen W (1969) Transient emissions on the wavelength of helium I, 5876 Å recorded during auroral break–up. Planet Space Sci 17:1927–1935

Stolarski RS, Dulock VA, Watson CE, Green AES (1967) Electron impact cross sections for atmospheric species. 2. Molecular nitrogen. J Geophys Res 72:3953–3960

Stolarski RS, Johnson NP (1972) Photoionization and photoabsorption cross sections for ionospheric calculations. J Atmos Terr Phys 34:1691–1701

Straižys V (1977) Multicolor stellar photometry. Mokslas, Vilnius

Streit GE, Johnston HS (1976) Reaction and quenching of vibrationally excited hydroxyl radicals. J Chem Phys 64:95–103

Strekalov ML (2003) The rotational relaxation of the diatomic molecules. A model of the angle moment transfer. Russian J Chem Phys 22:3–9

Striganov AR, Odintsova GA (1982) Tables of the spectral lines of the atoms and ions. Handbook. Energoizdat, Moscow

Stupochenko EV, Losev SA, Osipov AI (1965) The relaxation processes in the shock waves. Nauka, Moscow

Sukhoivanenko PYa, Fedorova NI (1976) Fast registration of the λ 10830 Å helium emission. In: Shefov NN, Savrukhin AP (eds) Studies of the upper atmospheric emission. Ylym, Ashkhabad, pp 12–16

Suzuki K (1983) Observation of helium 10830 Å airglow emission in midlatitude. J Geomagn Geoelectr 35:321–330

Suzuki K, Tohmatsu T (1976) An interpretation of the rotational temperature of the airglow hydroxyl emissions. Planet Space Sci 24:665–671

Swenson GR, Mende SB, Llewellyn EJ (1989) Imaging observations of lower thermospheric O(^{1}S) and O$_2$ airglow emissions from STS 9: implications of height variations. J Geophys Res 94A:1417–1429

Swider W (1970) Ionic reactions for meteoric elements. Ann Géophys 26:595–599

Takahashi H, Batista PP (1981) Simultaneous measurements of OH(9,4), (8,3), (7,2), (6,2) and (5,1) bands in the airglow. J Geophys Res 86A:5632–5642

Taranova OG (1962) Continuum emission in airglow and aurorae. In: Krassovsky VI (ed) Aurorae and Airglow. N 8. USSR Academic Science Publishing House, Moscow, pp 21–23

Taranova OG (1967) On diurnal variations of helium emission. In: Krassovsky VI (ed) Aurorae and Airglow. N 13. USSR Academic Science Publishing House, Moscow, pp 50–52

Tarasova TM (1962) Direct measurements of the night sky in the λ 8640 A spectral region. In: Artificial satellites of the Earth. N 13. USSR Academic Science Publishing House, Moscow, pp 107–109

Tarasova TM, Slepova VA (1964) Height distribution of radiation intensity of the night sky main emission lines Geomagn Aeronomy 4:321–327

Teixeira NR, Angreji PD, Sahai Y, Tinsley BA, Christensen AB (1976) Tropical twilight HeI 10830 emission. Planet Space Sci 24:303–312

Telegin GG, Yatsenko AS (2000) The optical spectra of the atmospheric gases. Rautian SG (ed) Nauka, Novosibirsk

Tepley CA, Meriwether JW, Walker JCG, Mathews JD (1981) Observations of neutral iron emission in twilight spectra. J Geophys Res 86:4831–4835

Timothy JG (1977) Solar spectrum between 300 and 1200 A. In: White OR (ed) The solar output and its variation. University Press, Boulder, pp 257–285

Tinsley BA (1968a) Measurements of twilight helium 10830 Å emission. Planet Space Sci 16:91–99

Tinsley BA (1968b) Temporal variations in geocoronal Balmer Alpha. J Geophys Res 73:4139–4149

Tinsley BA (1969) Reinterpretation of geocoronal observations with increased high/low altitude hydrogen ratio. Planet Space Sci 17:769–771

Tinsley BA, Meier RR (1971) Balmer Alpha distribution over a solar cycle: comparison of observations with theory. J Geophys Res 76:1006–1016

Tinsley BA, Christensen AB (1976) Twilight helium 10, 830 Å calculations and observations. J Geophys Res 81:1253–1263

Toroshelidze TI (1970) Twilight emission of the helium by the observations of Abastumani. Geomagn Aeronomy 10:1037–1042

Toroshelidze TI (1971) The emission of atmospheric helium 10830 A at the predawn period. Astron Circ USSR Acad Sci 652:1–3

Toroshelidze TI (1976) On the certain particularities of the 10830 A helium emission in twilight. In: Shefov NN, Savrukhin AP (eds) Studies of the upper atmospheric emission. Ylym, Ashkhabad, pp 22–32

Toroshelidze TI (1991) The analysis of the aeronomy problems on the upper atmosphere glow. Shefov NN (ed) Metsniereba, Tbilisi

Toroshelidze TI, Chilingarashvili SP (1975) Study of the sodium layer variations according to twilight observations of D emission. In: Kharadze EK (ed) Bull Abastumani astrophys observ. N 46, pp 235–250

Torr MR, Torr DG (1982) The role of the metastable species in the thermosphere. Rev Geophys Space Phys 20:91–144

Torr MR, Torr DG, Laher RR (1985) The O_2 Atmospheric 0–0 band and related emissions at night from Spacelab 1. J Geophys Res 90A:8525–8538

Truttse YuL (1968a) Upper atmosphere during geomagnetic disturbances. I. Some regular features of low-latitude auroral emissions. Planet Space Sci 16:981–992

Truttse YuL (1968b) Upper atmosphere during geomagnetic disturbances. II. Geomagnetic storms oxygen emission at 6300 Å and heating of the upper atmosphere. Planet Space Sci 16:1201–1208

Truttse YuL (1969) Upper atmosphere during geomagnetic disturbances. III. Some regularities in density variations. Planet Space Sci 17:181–187

Truttse YuL (1972a) Oxygen emission at 6300 Å. Ann Géophys 28:169–177

Truttse YuL (1972b) Night variations of intensity of emission 6300 A in quiet geomagnetic conditions. Geomagn Aeronomy 12:561–564

Truttse YuL (1973) Upper atmosphere during geomagnetic disturbances. In: Krassovsky VI (ed) Aurorae and Airglow. N 20. Nauka, Moscow, pp 5–22

Truttse YuL, Gogoshev MM (1977) Red oxygen line 6300 Å and electron content in night F-region. Dokl Bulg Acad Sci 30:45–48

Turnbull DN (1987) An empirical determination of the electric dipole moment function and transition probabilities of $OH(X^2\Pi)$. University of West Ontario, London

Turnbull DN, Lowe RP (1983) Vibrational population distribution in the hydroxyl night airglow. Can J Phys 61:244–250

Turnbull DN, Lowe RP (1989) New hydroxyl transition probabilities and their importance in airglow studies. Planet Space Sci 37:723–738

Ulwick JC, Baker KD, Stair AT, Frings W, Hennig R, Grossmann KU, Hegblom ER (1985) Rocket-borne measurements of atmospheric fluxes. J Atmos Terr Phys 47:123–131

Unsöld A (1938) Physik der Sternatmosphären mit besonderer Derücksichtigung der Sonne. Springer-Verlag, Berlin

Vainstein LA, Sobelman II, Yukov EA (1979) An atom excitation and the spectral line broadening. Nauka, Moscow
Vallance Jones A (1958) Calcium and oxygen in the twilight airglow. Ann Géophys 14:179–185
Vallance Jones A (1973) The infrared spectrum of the airglow. Space Sci Rev 15:355–400
Vallance Jones A (1974) Aurora. D Reidel Publishing Company, Dordrecht
Vallance Jones A, Gattinger RL (1974) The $O_2(b^1\Sigma_g^+ \to X^3\Sigma_g^-)$ system in aurora. J Geophys Res 79:4821–4822
Velculescu VG (1970) On the production of excited hydroxyl radicals in the $H+O_3$ – atomic flame. Zeits Phys 237:69–74
Virin LI, Dzhagatspanyan RV, Karachevtsev GV, Potapov VK, Talrose VL (1979) Ion-molecular reactions in gases. Nauka, Moscow
Vlasov MN, Klopovsky KS, Lopaev DV, Popov NA, Rakhimov AT, Rakhimova TB (1997) The mechanism of singlet oxygen emission in the upper atmosphere. Cosmic Res 35:235–242
von Rosenberg CW, Trainor DW (1973) Observations of vibrationally excited O_3 formed by recombination. J Chem Phys 59:2142
von Zahn U, von der Gathen P, Hansen G (1987) Forced release of sodium upper atmospheric dust particles. Geophys Res Lett 14:76–79
Vriens L, Bonsen TFM, Smith JA (1968) Excitation to the metastable states and ionization from ground and metastable states in helium. Physica 40:229–252
Wallace L, Hunten DM (1968) Dayglow of the oxygen A band. J Geophys Res 73:4813–4834
Watanabe K, Inn ECY, Zelikoff M (1953) Absorption coefficient of oxygen in the vacuum ultraviolet. J Chem Phys 21:1026–1030
Watson CE, Dulock VA, Stolarski RS, Green AES (1967) Electron impact cross sections for atmospheric species. 3. Molecular oxygen. J Geophys Res 72:3961–3966
Weller CS, Meier RR, Tinsley BA (1971) Simultaneous measurements of the hydrogen airglow emissions of Lyman alpha, Lyman beta and Balmer alpha. J Geophys Res 76:7734–7744
Werner HJ, Rosmus P, Reinsch EA (1983) Molecular properties from MCSCF-SCEP wave function. I. Accurate dipole moment functions of OH, OH^- and OH^+. J Chem Phys 79:905–916
Whiting EE, Paterson JA, Kovács I, Nicholls R (1973) Computer checking of rotational line intensity factors for diatomic transitions. J Mol Spectrosc 47:84–98
Whitten RC, Poppoff IG (1971) Fundamentals of aeronomy. John Wiley and Sons, New York
Wiese WL, Smith MW, Glennon BM (1966) Atomic transition probabilities H through Ne, Vol 1. NSRDS-NBS 4, Washington
Witt G, Stegman J, Solheim BH, Llewellyn EJ (1979) A measurement of the $O_2(b^1\Sigma_g^+ \to X^3\Sigma_g^-)$ atmospheric band and the $O(^1S)$ green line in the nightglow. Planet Space Sci 27:341–350
Witt G, Rose J, Llewellyn EJ (1981) The airglow continuum at high latitudes – an estimate of the NO concentration. J Geophys Res 86A:623–628
Woodworth JR, Moos HW (1975) Experimental determination of the single-photon transition rate between the 2^3S_1 and 1^1S_1 states of He. Phys Rev A 12:2455–2463
Worley SD, Coltharp RN, Potter AE (1971) Quenching of vibrationally excited hydroxyl (v = 9) with oxygen. J Chem Phys 55:2608–2609
Worley SD, Coltharp RN, Potter AE (1972) Rates of interaction of vibrationally excited hydroxyl (v = 9) with diatomic and small polyatomic molecules. J Chem Phys 56:1511–1514
Wraight PC (1975) Is there a continuum near infra-red dayglow? J Atmos Terr Phys 37:731–737
Wraight PC (1977) The near infrared nightglow continuum. Planet Space Sci 25:787–794
Wraight PC (1982) Association of atomic oxygen and airglow excitation mechanism. Planet Space Sci 30:251–259
Wraight PC (1986) Theory of the nightglow continuum. Planet Space Sci 34:1373
Yarin VI (1961a) The OH emission according to observations in Yakutsk. In: Krassovsky VI (ed) Spectral, electrophotometrical and radar researches of aurorae and airglow. N 5. USSR Academic Science Publishing House, Moscow, pp 10–17
Yarin VI (1961b) Continuous emission and the Herzberg O_2 bands in the night airglow. In: Krassovsky VI (ed) Spectral, electrophotometrical and radar researches of aurorae and airglow. N 5. USSR Academic Science Publishing House, Moscow, pp 35–38

Yarin VI (1962a) On the dependence of intensity of OH bands on the rotational temperature. In: Krassovsky VI (ed) Aurorae and Airglow. N 8. USSR Academic Science Publishing House, Moscow, pp 9–10

Yarin VI (1962b) Variations of the vibrational population rates of OH molecules. In: Krassovsky VI (ed) Aurorae and Airglow. N 9. USSR Academic Science Publishing House, Moscow, pp 10–18

Yevlashin LS (1962) Prominent aurora of February 11, 1958. Geomagn Aeronomy 2:74–78

Zipf EC, Borst WL, Donahue TM (1970) A mass spectrometer observation of NO in auroral arc. J Geophys Res 75:6371–6376

Chapter 3
Techniques of Investigation of the Airglow

The airglow that arises in the rarefied gaseous medium of the higher atmospheric layers is a sensitive indicator of the conditions under which it is formed, characterizing the parameters of the atmosphere, such as its density and temperature and their variations in space and time. All this information is practically contained in the airglow intensity and spectral distribution measurements. In airglow research it is necessary to use spectroscopic equipment which would provide a required spectral resolution to measure the spectral distribution of the airglow intensity and its variations on various time scales. This requirement can be fulfilled by using various types of spectroscopic equipment depending on a task in view.

For recording weak optical radiations of the upper atmosphere, four types of equipment are used: spectrographs, spectrometers, photometers, and interferometers. All optical measurement techniques using these devices constitute an integral part of the physical research of the Earth's upper atmosphere and circumterrestrial space. Any theoretical models which describe the properties and behavior of the atmospheric characteristics should be based on experimental data or be checked up with the help of these data. For the last half-century the methods and equipment used have undergone significant improvements and changes. The prevailed photographic methods of recording radiation spectra and analog photoelectric processing methods have been gradually replaced by matrix radiation detectors allowing digital image processing. Nevertheless, since numerous data have been obtained by former methods, it is natural that these methods should be described with all necessary technological details. This should provide the succession of experimental technology in order that the subsequent researchers would comprehend all features of the observation data that have enabled researchers to obtain the up-to-date notions about the physics of the Earth's upper atmosphere.

3.1 Photometry

In astrophysical investigations of radiation sources, it is critically important to measure quantitatively the luminous flux, i.e., the light energy that passes in a unit time through the cross-section of a cone originating from the radiation source and having a certain cone angle. The luminous flux can serve as a measure of the luminous power of a source. The area S that is cut out by a cone on a sphere of radius r, whose center is at the vertex of the cone, corresponds to a solid angle $\omega = S \cdot r^{-2}$. If $S = r^2$, then $\omega = 1$ (sr). To measure light intensity, the candela unit (cd), formerly known as international candle, is generally used. According to the definition, one candela is the intensity of light radiated perpendicularly from the surface of a black emitter $1.6 \cdot 10^{-5}$ (m^2) in size at the solidification temperature of platinum at a pressure of 101,325 (Pa) (Kamke and Krämer 1977). The luminous flux corresponding to a solid angle equal to 1 (sr) and having the light intensity equal to 1 (cd) is called the lumen (lm).

The luminous flux dF that passes within a solid angle $d\omega$ through an area $d\sigma$ perpendicular to the flux direction in a unit time dt within a unit wavelength interval $d\lambda$ (or frequency interval $d\nu$) is termed as the intensity I (Ambartsumyan et al. 1952; Martynov 1977), i.e.,

$$dF = I \cdot d\sigma \cdot dt \cdot d\omega \cdot d\lambda.$$

If the area $d\sigma$ makes an angle θ with the incident ray direction, we have

$$dF = I \cdot d\sigma \cdot \cos\theta \cdot dt \cdot d\omega \cdot d\lambda.$$

The luminous flux incident on a unit area determines the area illuminance E. For perpendicular incidence we have

$$E = \frac{F}{S} = \frac{I \cdot \omega}{S} = \frac{I}{r^2}.$$

For the unit of illuminance, one lux (lx), the illuminance is taken as that produced by a flux of 1 (lm) passing through a 1-m^2 area perpendicular to the light flux or the illuminance produced by 1 (cd) at a distance of 1 (m). If a 1-lm flux passes through an area of 1 (cm^2), the illuminance is equal to 1 (phot). Thus, 1 (phot) = 10,000 (lx).

The brightness of a spatial light source is defined as the light intensity which is provided by a 1-(cm^2) area along a normal to this area. The brightness at which the radiation of a surface of area 1 (cm^2) yields, in the direction normal to the surface, the light intensity equal to 1 (cd) or a flux of 1 (lm) in a steradian is taken as a unit brightness and called the stilb (sb). The unit of brightness being one 10,000th of a stilb is termed the nit (nt). It corresponds to 1 (cd) per 1 (m^2).

All these units are not used in practical spectrophotometric investigations, but can be useful in comparing different photometric data. Therefore, the values of photometric units for some light sources (Kulikovsky 2002) are given below:

3.1 Photometry

The illuminance produced by a star of magnitude m is

$$E_m = 2.77 \cdot 10^{-6} \cdot 10^{-0.4m} \text{ (lx)}.$$

The illuminance produced by the full Moon at zenith on the Earth surface perpendicular to the direction of incident rays is equal to 0.25 (lx). The mean brightness of the full Moon is 0.251 (sb).

The illuminance produced by the Earth on the Moon at a new moon is 15 (lx).

The moonless night sky has a brightness of 10^{-7} (sb) or 10^{-3} (nt) in the visible region of the spectrum.

The illuminance produced by the Sun at a distance of 1 (a.u.) (astronomical unit is $1.496 \cdot 10^{13}$ (cm), mean distance between the Sun and the Earth) outside the atmosphere is $1.35 \cdot 10^5$ (lx).

The luminous flux at a wavelength of 550 (nm) is 688 (lm) = 1 (W).

It is common astronomical practice to measure the brightness of extra-atmospheric objects in magnitudes m. The scale of magnitudes is determined by the relation

$$m_2 - m_1 = 2.5 \log_{10}(I_2/I_1);$$

that is, the difference of five magnitudes implies the intensity ratio equal to 100.

The radiation flux produced by a star of magnitude m belonging to the spectroscopic class of the Sun type is (Allen 1955, 1973)

$$S_m = 2.54 \cdot 10^{-5} \cdot 10^{-0.4m} \text{ (erg} \cdot \text{cm}^{-2} \cdot \text{s}^{-1})$$

or

$$S_m = 1.17 \cdot 10^7 \cdot 10^{-0.4m} \text{ (photon} \cdot \text{cm}^{-2} \cdot \text{s}^{-1}).$$

For the boundary of the Earth's atmosphere at a wavelength of about 550 (nm) this corresponds to

$$S_m(550 \text{ (nm)}) = 3.72 \cdot 10^{-8} \cdot 10^{-0.4m} \text{ (erg} \cdot \text{cm}^{-2} \cdot \text{s}^{-1} \cdot \text{nm}^{-1})$$

and at about 725 (nm) to

$$S_m(725 \text{ (nm)}) = 2.77 \cdot 10^{-8} \cdot 10^{-0.4m} \text{ (erg} \cdot \text{cm}^{-2} \cdot \text{s}^{-1} \cdot \text{nm}^{-1})$$

or, in terms of photons,

$$S_m(550 \text{ (nm)}) = 1.02 \cdot 10^4 \cdot 10^{-0.4m} \text{ (photon} \cdot \text{cm}^{-2} \cdot \text{s}^{-1} \cdot \text{nm}^{-1}),$$
$$S_m(725 \text{ (nm)}) = 1.00 \cdot 10^4 \cdot 10^{-0.4m} \text{ (photon} \cdot \text{cm}^{-2} \cdot \text{s}^{-1} \cdot \text{nm}^{-1}).$$

Within the limits of the spectral sensitivity of an eye (Kulikovsky 2002), we have

$$S_m(\text{vis}) = 3.97 \cdot 10^{-6} \cdot 10^{-0.4m} \text{ (erg} \cdot \text{cm}^{-2} \cdot \text{s}^{-1}),$$
$$S_m(\text{vis}) = 1.09 \cdot 10^6 \cdot 10^{-0.4m} \text{ (photon} \cdot \text{cm}^{-2} \cdot \text{s}^{-1}).$$

The function of the relative spectroscopic luminous efficacy of eyesight can be approximated by an asymmetric Gaussian function (Budin 1997):

$$V(\lambda) = \exp[-(\lambda - \lambda_0)^2 \cdot K],$$

where λ is the wavelength; $\lambda_{0D} = 0.555$ (μm), $K_{Dblue} = 337$ (μm^{-2}) (the error σ is no worse than 15(%)) and $K_{Dread} = 235$ (μm^{-2}) ($< 10(\%)$) for the daytime function $V_D(\lambda)$; $\lambda_{0N} = 0.507$ (μm), $K_{Nblue} = 270$ (μm^{-2}) ($< 10(\%)$) and $K_{Nread} = 395$ (μm^{-2}) ($< 0.5(\%)$) for the night function $V_N(\lambda)$.

The intensity of the airglow of the upper atmosphere is usually measured in Rayleigh. By definition 1 (Rayleigh) = 10^6 (photon·cm^{-2}·s^{-1}), i.e., this is the flux of omnidirectional radiation of an atmospheric column of cross-section 1 (cm^2). This is related to the concept that the Rayleigh characterizes the emission rate, i.e., the number of photochemical processes.

When comparing the units of surface brightness following Roach and Gordon (1973), we obtain that the brightness of one star of m = 10 is equivalent to $4.4 \cdot 10^{-2}$ (Rayleigh·nm^{-1}) = $1.31 \cdot 10^{-8}$ (erg·cm^{-2}·s^{-1}·sr^{-1}·nm^{-1}), whence the intensity

$$I_\lambda (\text{erg} \cdot \text{cm}^{-2} \cdot \text{s}^{-1}) = 1.98 \cdot 10^{-3} \cdot \frac{I_\lambda (\text{Rayleigh})}{\lambda (\text{nm})},$$

$$I_\lambda (\text{Rayleigh}) = 505 \cdot \lambda(\text{nm}) \cdot I_\lambda (\text{erg} \cdot \text{cm}^{-2} \cdot \text{s}^{-1}).$$

3.2 Spectrophotometric Equipment

During the last years of airglow observations there was a highly important stage at which the optical technique was mastered for monitoring IGWs in the upper atmosphere. This technique is based on determining the variations of the emissions of the upper atmosphere, which are characterized by a very low intensity (10–100 (Rayleigh)) and the temperature of the medium where they arise.

To determine the temperature from the spectra of radiating species, spectrophotometric equipment providing high spectroscopic resolution (0.0001–0.1 (nm)) is required. At the same time, to detect the time variations of the emission intensity caused by the passage of IGWs through emission layers (for the mesopause altitudes the Brunt–Väisälä period is about 5.2 (min)), it is necessary to use high-aperture equipment which would be capable of recording radiation spectra with exposures no more than 2–3 (min). To combine these two requirements, which seemingly are mutually exclusive for optical spectroscopic equipment, special spectrophotometric complexes should be developed and built in which these requirements would be balanced to solve specific geophysical problems. The issue is aggravated by the fact that in geophysical investigations, measurements are often performed in field conditions which poses additional difficulties. The special complexes designed to solve all these problems and the technology of their operation in ground conditions are described in detail elsewhere (Potapov 1974; Potapov et al. 1978).

It should be stressed that the new technologies and methods developed for the past half a century for measuring weak luminous fluxes have increased significantly the capabilities of investigation of their space and time distributions and this has considerably extended the range and improved the quality of geophysical investigations. Nevertheless, the fundamental geophysical results obtained by former methods remain obviously topical and important.

3.2.1 Diffraction Spectrographs

Spectroscopic instruments are devices in which ultraviolet, visible, or infrared electromagnetic radiation is decomposed into monochromatic components by means of dispersive elements (such as a prism or a diffraction grating). Now diffraction gratings are most widely used in spectroscopy as dispersive elements. In spectroscopic equipment, gratings of various types are used: flat and concave, having grooves of varied shapes and profiles. As to the purpose, the spectroscopic equipment can be classed into several types:

- monochromators, which isolate a narrow spectral range or a spectral line;
- polychromators, which isolate simultaneously several narrow spectral ranges or several spectral lines;
- spectrographs, which photograph simultaneously a wide spectral range;
- spectrometers, which scan a recorded spectrum.

The classical design of a diffraction (flat reflection grating) slit spectroscopic device, presented schematically in Fig. 3.1, contains a collimating lens which forms a parallel beam of light incident on a dispersive element. In the focus of the collimating lens are the entrance slit, the dispersive element (diffraction grating), the camera lens focusing the image of the entrance slit on the radiation detector at various wavelengths of the decomposed light, and the radiation detector.

Numerous special publications (Zaidel et al. 1976; Lebedeva 1977, 1986; Tarasov 1968, 1977; Meaburn 1976) are devoted to the theory and practical issues of the development and building of spectroscopic equipment. We shall not discuss the relevant problems in detail, but only give a description of the devices specially intended for recording the weak airglow of the upper atmospheric layers. It should be noted that the spectroscopic devices have found wide use. The range of problems which can be solved by means of spectroscopic equipment is extremely wide. They are employed successfully in pure and applied research. In particular, they have many applications in astrophysical and geophysical investigations in solving problems of spectral analysis of the radiation of remote astronomical objects and of the airglow of the upper layers of the Earth's atmosphere.

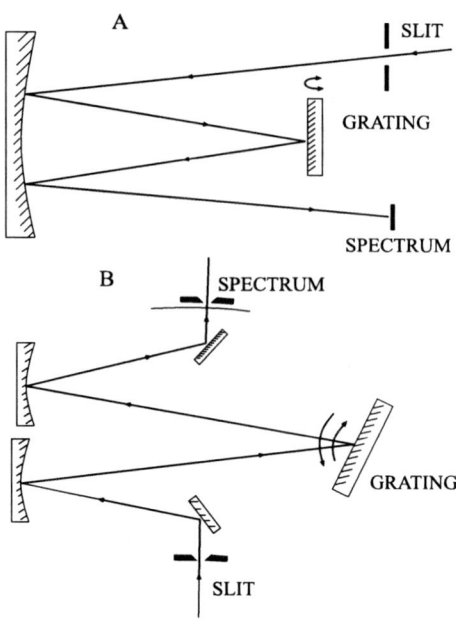

Fig. 3.1 Optical arrangement of devices with flat diffraction gratings: the Ebert (**A**) and the Cherny–Terner scheme (**B**) (Lebedeva, 1986)

3.2.1.1 The SP-48, SP-49, and SP-50 Spectrographs

Astrophysical and geophysical investigations involve the need of recording extremely weak radiations with high resolution. Successful realization of these requirements in one device is practically impossible since increasing its optical efficiency results in considerable loss of resolving ability. Therefore, in developing spectroscopic equipment one has to make a reasonable compromise depending on the purposes of the equipment and the conditions of its operation. For the conditions of recording the airglow of the upper atmosphere, the optimum linear dispersion of spectrographs is 8–10 (nm·mm^{-1}), which provides the necessary high optical efficiency. In this regard, the complete set of diffraction spectrographs intended for recording weak radiations of small gas constituents of the Earth's upper atmosphere, developed on V. I. Krassovsky's initiative at S. I. Vavilov State Optical Institute, turned out very successful. The complete set consisted of three devices of type SP-48, SP-49, and SP-50 (Figs. 3.2–3.5) intended for recording the spectra of weak glows in the visible, the ultraviolet, and the infrared region, respectively. The devices possessed high optical efficiency and made it possible to obtain high-quality spectra with 0.2–0.5 (nm) resolution. Subsequently, prior to the beginning of the International Geophysical Year (1957), a series of such devices was produced at the State Optical-Mechanical Plant in St. Petersburg (former Leningrad) for equipment of the stations and observatories that conducted spectrophotometric investigations. These devices were successfully used both in stationary and in field conditions. Tables 3.1–3.3 list the characteristics of these spectrographs.

3.2 Spectrophotometric Equipment

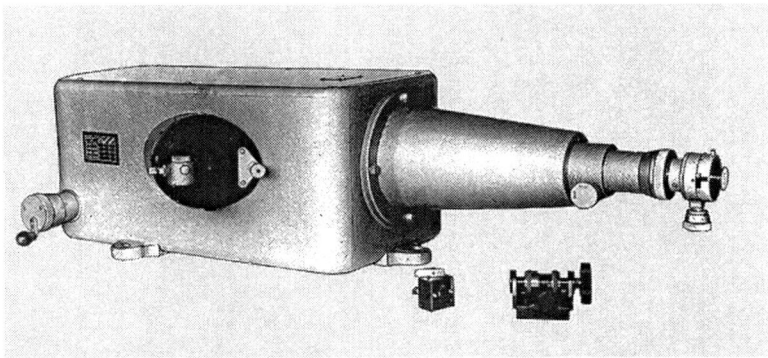

Fig. 3.2 The SP-48 spectrograph

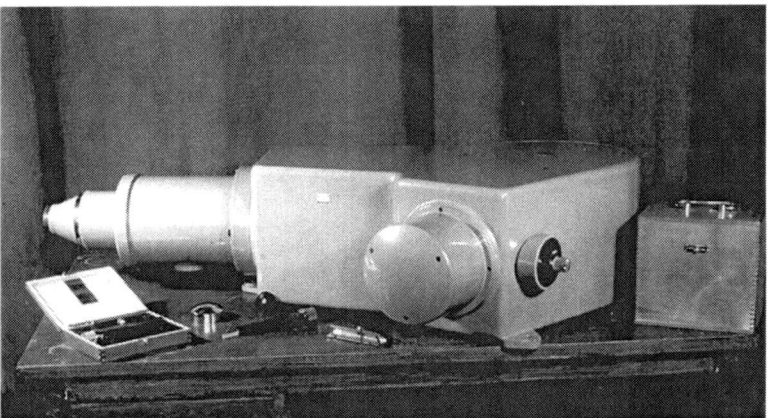

Fig. 3.3 The SP-49 spectrograph

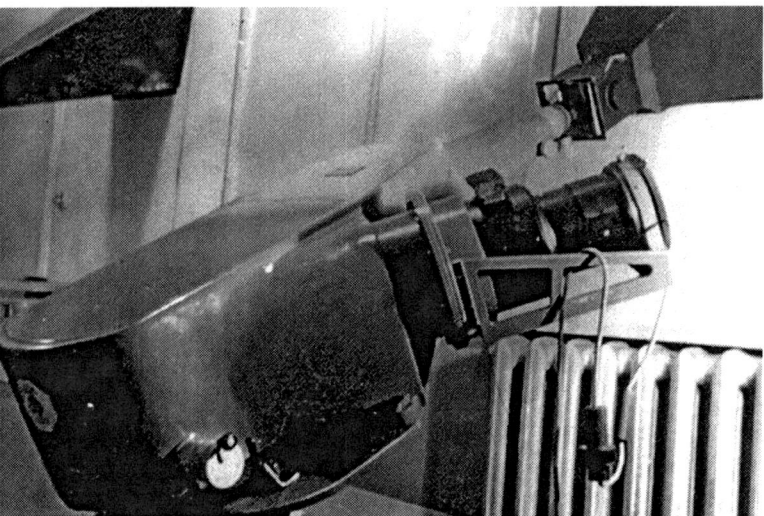

Fig. 3.4 The SP-50 spectrograph with the FKT-1 image converter tube

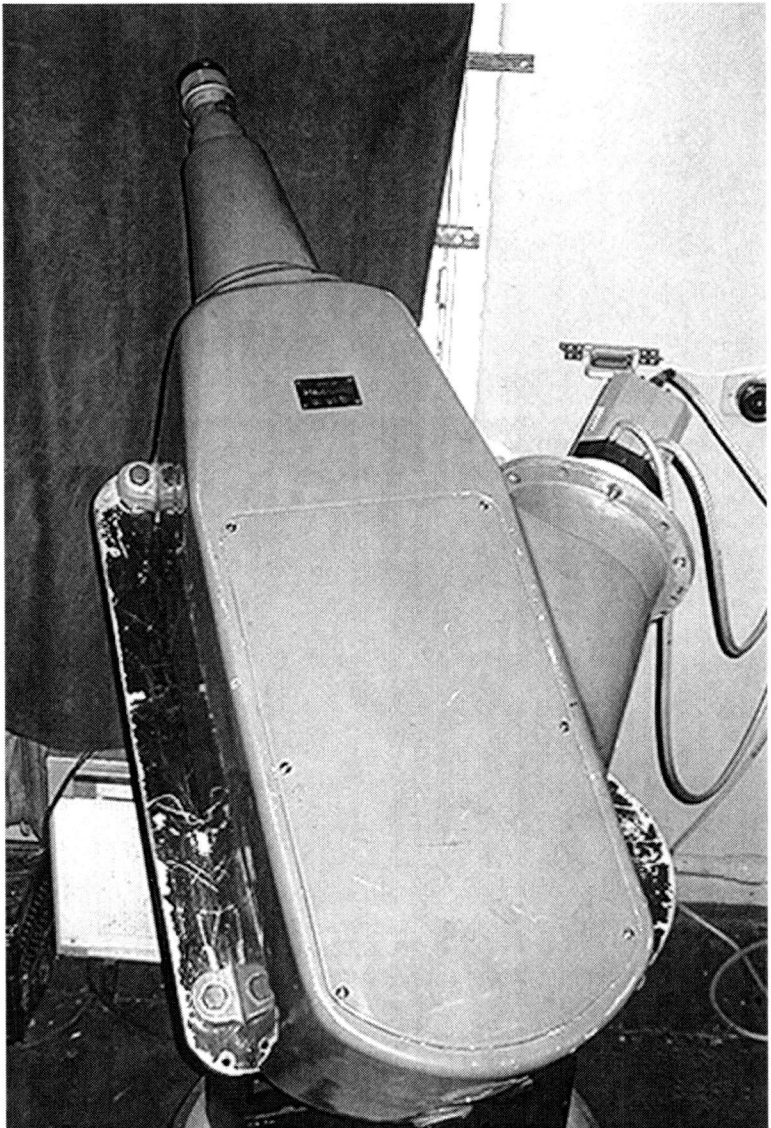

Fig. 3.5 The SP-50 spectrograph with the FPZS INSTASPEC IV unit

3.2.1.2 Spectrography

Spectrography of the variations of characteristics of the upper atmosphere airglow was carried out by the SP-48, SP-49, and SP-50 high-aperture diffraction spectrographs providing high spectral resolution (several tenths of a nanometer)

3.2 Spectrophotometric Equipment

Table 3.1 Characteristics of the SP-48 spectrograph

Lens of collimator:	Working spectral range (nm) 380–800
Focal distance (mm) 630	Reciprocal linear dispersion (nm/mm) 10
Relative aperture 1:4.7	Full length of spectrum (mm) 48
Lens of camera:	Measured length of spectrum (mm) 18
Focal distance (mm) 70	Limiting resolution (nm) 0.3
Relative aperture 1:0.8	
Magnification of optical system $0.1\times$	Method of spectrum recording: photography
Replica of diffraction grating: flat	Overall dimensions (mm):
Number of grooves per 1 (mm) 1200	Length 1045
Size of shaded part (mm) 136×90	Width 450
Concentration of light: order 1	Height 270
	Mass (kg) 60

(Gerasimova and Yakovleva 1956; Galperin et al. 1957; Kaporsky and Nikolaeva 1969). The optical schemes of these spectrographs are presented in Figs. 3.6–3.8.

The SP-48 spectrograph was used for recording radiation in the spectral range 700–800 (nm). Initially, in compliance with the manufacturer specifications, spectra were recorded on a 16-(mm) film which was moved out from a special cartridge to the focal region of the spectrograph, which had a cylindrical surface with a 100-mm radius of curvature. Spectra were photographed at night with exposures from 3 to 10 (h).

The SP-49 spectrograph differed in design from the SP-48 and SP-50 devices. Its lenses were mirror since they were intended for use in the ultraviolet spectral

Table 3.2 Characteristics of the SP-49 spectrograph

Lens of collimator:	Working spectral region (nm) 275–395
Focal distance (mm) 2370	Reciprocal linear dispersion
Relative aperture 1:15	(nm/mm) 8
	Full length of spectrum (mm) 17
Lens of camera:	Limiting resolution (nm) 0.2
Focal distance (mm) 170	Focal surface: cylindrical with radius (mm)
Relative aperture 1:1.25	264
Magnification of optical system: $0.07\times$	Method of spectrum recording: photography
Replica of diffraction grating: flat	Overall dimensions (mm):
Number of grooves per 1 (mm) 600	Length 1195
Size of shaded part (mm) 150×140	Width 620
Concentration of light: order 1	Height 290
	Mass (kg) 95

Table 3.3 Characteristics of the SP-50 spectrograph

Lens of collimator:	Working spectral region (nm) 800–1000
Focal distance (mm) 820	Reciprocal linear dispersion (nm/mm) 10
Relative aperture 1: 6.8	Limiting resolution (nm) 0.5
Lens of camera:	Measured length of spectrum (mm) 15
Focal distance (mm) 135	Method of spectrum recording:
Relative aperture 1:1.5	photography
Magnification of optical system: $0.16\times$	Overall dimensions (mm):
Replica of diffraction grating: flat	Length 1250
Number of grooves per 1 (mm) 600	Width 550
Size of shaded part (mm) 120×95	Height 270
Concentration of light: order 1	Mass (kg) 57

range. The Schmidt lens of the camera had a quartz plate with the surface of the fourth order of curvature. The focal surface was inside the spectrograph. Therefore, the 8-(mm) film was moved in the focal surface with a special cartridge. With this design it was impossible to use image amplifiers.

To provide recording spectra with exposures of several minutes, multistage image converter tubes (ICTs) were used; the spectrum image was photographed from the ICT screen. For the SP-48 spectrograph, three-stage ICTs with multi-alkali cathodes were employed. The working spectral range for the SP-50 spectrograph

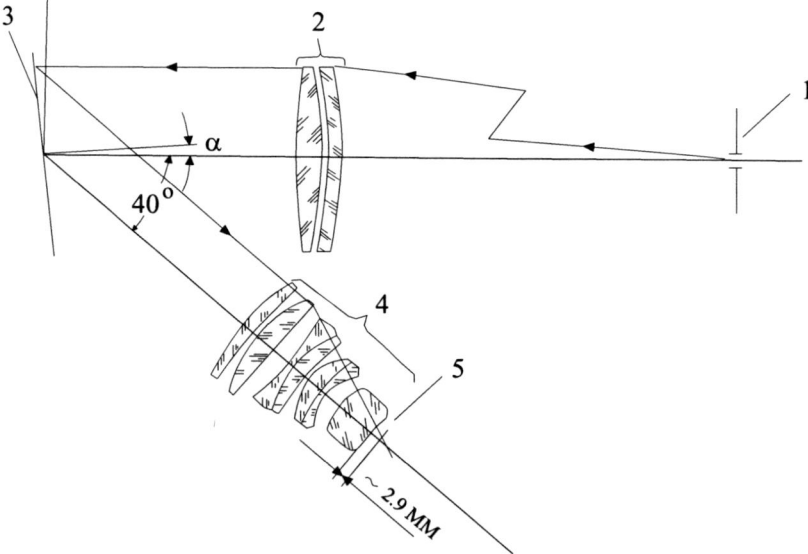

Fig. 3.6 Optical arrangement of the SP-48 diffraction spectrograph. 1 – Slit; 2 – lens of the collimator; 3 – grating; 4 – lens of the camera; and 5 – focal surface

3.2 Spectrophotometric Equipment

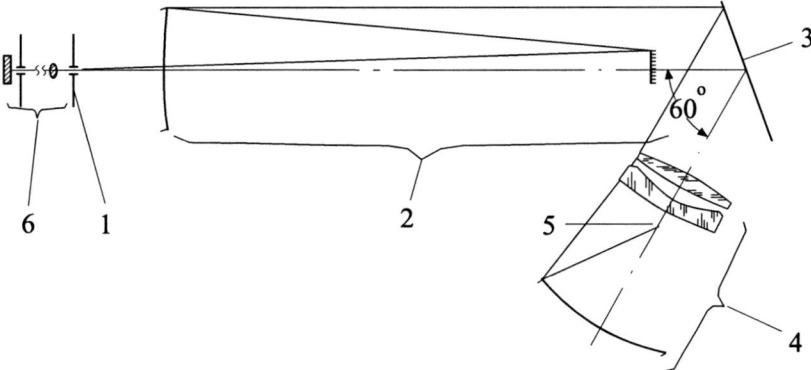

Fig. 3.7 Optical arrangement of the SP-48 diffraction spectrograph. 1 – Slit; 2 – lens of the collimator; 3– grating; 4– lens of the camera; 5 – focal surface; 6 – attachment for reducing the field of vision

was 1000–1100 (nm). Therefore, this device incorporated a multistage ICT with an oxide-silver-cesium cathode, which must be cooled because of a considerable dark noise at room temperature. Thus, to operate the SP-50 spectrograph, a cooler was required. This was the only difference between the two recording systems. The rest were absolutely identical; therefore, they are described in common below, and the operation of the cooler is discussed separately.

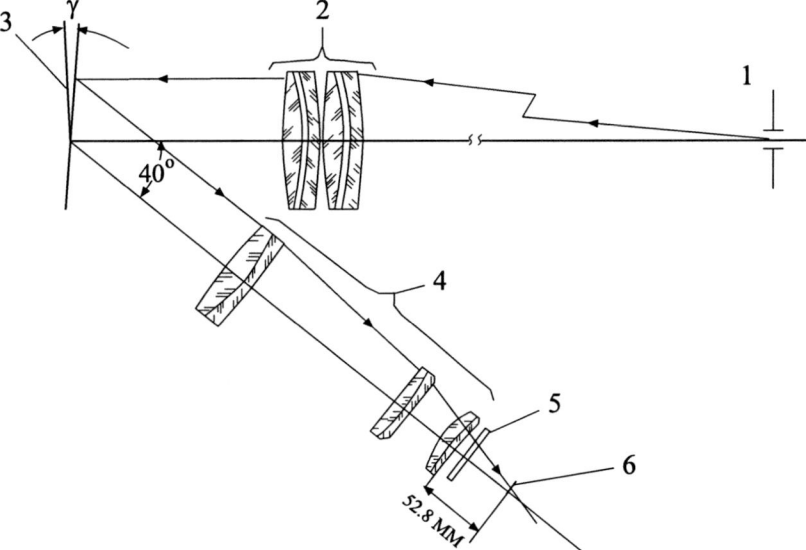

Fig. 3.8 Optical arrangement of the diffraction spectrograph of the SP-50. 1 – Slit; 2 – lens of the collimator; 3 – grating; 4 – lens of the camera; 5 – glass filter; 6 – detector

The optical arrangement of the setup was as follows (see Fig. 3.8): Light from the sky arrives at the entrance slit of the device, passes through the lens of the collimator, and then, as a parallel beam, falls on a flat diffraction grating. The light decomposed in wavelengths is collected by the lens of the camera on the ICT cathode.

Initially single-camera ICTs of the FKT type (see Fig. 3.33) (Volkov et al. 1962; Shcheglov 1963) with contact transfer of the image appearing on the screen onto a film were used. This made it possible to obtain high-quality spectrograms regularly and compose the first atlas of the sky nightglow (Krassovsky et al. 1962). The spectrum in the range 300–400 (nm) was obtained with the SP-49 spectrograph, in the range 400–700 (nm) with the SP-48 spectrograph, and in the range 700–1240 (nm) with the SP-50 spectrograph.

Subsequently, to record hydroxyl emissions with exposures of several minutes, three-camera ICTs were used. In this case, the image of the spectrum, converted into visible form and amplified in brightness, was transferred from the screen onto a film with the help of two high-aperture lenses (the screen was in the focal plane of the first lens and the film was in the focal plane of the second lens). For the optics transferring the image from the screen onto the film, type RO-109 high-aperture lenses (relative aperture 1:1.2) were used. When the SP-50 spectrograph was operated, a type KS-13 red filter was placed in front of the slit to exclude superimposition of the second-order spectra on the spectral region to be recorded.

In assembling an ICT it was necessary to meet special requirements (Shcheglov 1963). The ICT was placed in an insulating plexiglas block in which elastic sealing gaskets were mounted on the cathode and the screen side. This was necessary to prevent moisture penetration to the high-voltage contacts of the ICT which would immediately result in breakdowns and upset the normal operation of the device. The assembled converter was enclosed in a tube-shaped screen, made of soft magnetic steel to exclude the defocusing action of induced magnetic fields, and fastened to the housing of the spectrograph. On the side of the ICT screen, the recording camera providing automatic film transport was placed.

The arrangement of the setup with the SP-50 spectrograph was complicated by the necessity to cool the oxide-silver-cesium cathode of the ICT to -60 to -70 (°C). Cooling was performed with a mixture of solid carbonic acid with alcohol which was put in the cooler surrounding the ICT cathode. The cooler was carefully heat-insulated. To protect the cathode against fogging, a vacuum window was used and the space between the window and the spectrograph lens was sealed off.

Structurally, the ICT attachment to spectrographs was designed so that it enabled one to move the ICT unit without loss of sealing, which made it possible to focus the spectrum image onto the coolable ICT photocathode keeping the cathode unfogged.

As already mentioned, type U-32 (3312) (Fig. 3.34) and U-72 (3308) commercial three-stage image converter tubes with electrostatic focusing were used in the setups. To maintain their normal operation, it was necessary to apply a high voltage, $\sim (10–15)$ (kV), between the cathode and the screen of each stage. Besides, to provide electrostatic focusing of the image, an additional voltage (tens of volts) should be applied to some electrodes which could be varied within the limits ± 100 (V) relative to the cathode of a given stage (Zaidel and Kurenkov 1970; Berkovsky

3.2 Spectrophotometric Equipment

et al. 1976, 1988). Thus, a three-cascade image converter tube of this type is a multi-electrode electron-optical system to the electrodes of which voltages of the order of 45 (kV) are applied. Of all possible ways of voltage application to the ICT leads, the power supply from one high-voltage source with a high-resistance divider was chosen. Since the current flowing directly through the ICT is negligible ($\sim 10^{-10}$ (A)), the current consumed by the system was dictated by the resistance of the voltage divider. The resistance was chosen so that, on the one hand, it was much lower than the internal resistance of the ICT and, on the other hand, it should be much greater than the internal resistance of the power supply in order that the latter not be shunted. In these setups, the total resistance of the voltage divider was 10^{10} (Ω).

For the power supply, commercially produced air ionizers were used which are capable of producing an output voltage discretely (in 5 (kV)) controllable from 5 to 50 (kV). The devices were modernized to allow smooth control of the output voltage. The operating voltage was monitored by the current flowing through the voltage divider of the ICT or by means of a static kilovoltmeter.

Subsequently the recording unit of the SP-48 spectrograph was re-designed. This made it possible to employ an ICT of later modification—an EP-16-type modular ICT with static electron focusing in which a fiber optics was used which enabled the spectrum image to be transferred from the cathode on the ICT screen practically without distortion. Moreover, this ICT had a high-voltage transducer built in the housing which required an operating voltage less than 1 (kV). This circumstance substantially simplified its service.

To take photos of nightglow spectra from the ICT screen, automatic cameras were mounted which were operated with the help of an electronic timer. The timer provided commutation of the equipment control circuits in given time intervals. These time intervals were determined by the required exposure time and film transport time for one frame. For the usual range of exposures from 1 to 5 (min) the exposure time control accuracy was within 1 (s) and the one-frame transport time control accuracy was within 0.1 (s). The operation of the shutter was synchronized with film transport with the help of an electronic timer. The timer also supplied signals to the frame counter.

To provide three-azimuthal recording of radiation, a system of mirrors mounted on the entrance slit of the device was used which projected the light from three sites of the sky simultaneously onto three sites of the slit. Immediately on the slit, two mirrors were installed (N1 and N3, Fig. 3.43) that projected light from two regions of the firmament (N1 and N3, respectively) cut out from the emission layer by the field of view of the spectrograph onto the top and bottom parts of the slit. The mirror N2, also installed on the slit, was pushed a little forward so that the overhanging parts of the device did not close the field of view. With this mirror the region N2 is projected onto the middle portion of the slit.

The position of the regions on the firmament to be observed depends on the orientation of the device (its optical axis) in azimuth and zenith angles and also on the angles of rotation of the mirrors and can be varied over a wide range. The area of the sky regions under observation at the altitude of hydroxyl airglow with the used zenith angles of sighting is about 30×50 (km^2).

In the setup with the SP-50 spectrograph, the viewed regions were chosen to be arranged so that they formed an almost equilateral triangle with a side (base) of about 250 (km). The spectrograph axis was oriented to a point with an azimuth of 210°. With this arrangement, the azimuths of the regions N1, N2, and N3 were 90°, 210°, and 330° and the zenith distances 63°, 23°, and 63°, respectively. All azimuths were counted clockwise from the south point.

The dimensions of the mirrors N1 and N3, equal to $10 \times 20 \,(\mathrm{mm}^2)$, were chosen so that the required sky regions could be viewed even for a base of 50 (km). The mirror dimensions were dictated by its distance from the slit and by the field of view of the spectrograph.

In the setup with the SP-48 spectrograph, prisms were used. The principle of three-azimuthal photography was also applied in this case. Because of the design features of the system, the geometry of the base triangle was changed. The zenith was chosen to be the location of one point; the azimuths and zenith angles of the other two points were 180°, 270° and 30°, 30°, respectively.

The spectrographic method described possesses a doubtless advantage from the viewpoint of the spectral resolution of the emissions observed. This enables one to select very precisely the necessary spectral ranges and, therefore, to use simultaneously several emission bands related to different vibrationally excited electron levels. Moreover, measuring emissions simultaneously from three regions of the sky in many spectral lines considerably reduces the noises from random small clouds during observations. Nevertheless, the disadvantage of the method is, certainly, the laborious photometric processing of observation data.

3.2.2 Diffraction Spectrophotometers

Spectrophotometers were used both for obtaining data about the fast variations of the Earth's airglow characteristics (occurring within several minutes and some tens of minutes) caused by the changes of the excitation processes under the conditions of auroras (Vaisberg 1960) and for finding regularities in these variations. At the Institute of Atmospheric Physics a type DFS-14 high-aperture spectrometer (Fig. 3.9) was employed which was specially developed and manufactured at the State Optical-Mechanical Plant (Vaisberg 1960). Its optical arrangement was of Ebert type (Fig. 3.10); the specifications of the spectrometer are given in Table 3.4.

The most comprehensive measurements were conducted at Zvenigorod to investigate the variations of the hydroxyl emission temperature and intensity in twilight (Taranova 1967; Taranova and Toroshelidze 1970) and at Yakutsk (Ammosov et al. 1992) to examine the spectral characteristics of the continuum (Sobolev 1978a,b, 1979).

The main advantage of the spectrometer is the higher dispersion allowing one to detect fast temperature variations by measurements of closely located rotational lines of hydroxyl emission.

3.2 Spectrophotometric Equipment 283

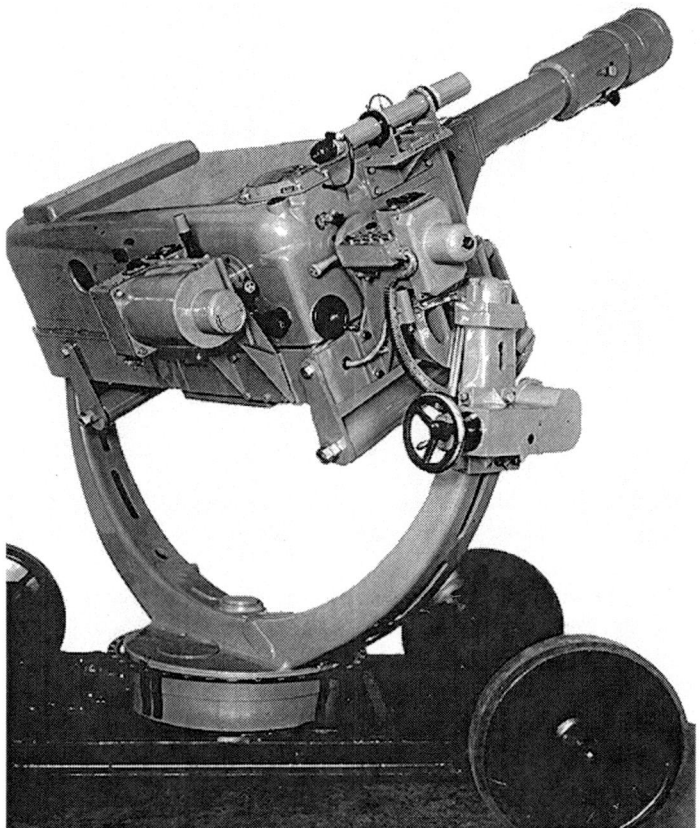

Fig. 3.9 The DFS-14 spectrometer

3.2.3 Photometers with Light Filters

3.2.3.1 Arrangement of an Electrophotometer

When studying the nightglow, which includes the airglow of the upper atmosphere and the extraterrestrial light, one has to measure extremely weak luminous fluxes. While the photographic method of recording enables one to obtain a required signal from the emission source under observation due to a long exposure (of the order of several hours), photoelectric detectors can be used if there is a need to record emission variations within substantially shorter time intervals. Conceptually, photometers have long been developed; their description is given in a number of publications (Nikonov 1973a), and their peculiarities are dictated by the tasks for which they are used. Therefore, finalized commercial photoelectric instruments which would be applicable to measurements of any type are not available. Such equipment is generally

Fig. 3.10 Optical arrangement of the DFS-14 spectrometer. 1 – Input lens for constructing the image of the object of photometry (e.g., an aurora) on the entrance slit; 2 – entrance slit; 3 – deflecting mirror; 4 – spherical collimating mirror; 5 – diffraction grating; 6 – camera spherical mirror; 7 and 10 – exit slits; 9 – deflecting mirror; 8 and 11 – Fabry lenses transferring light to the cathodes of the photomultipliers

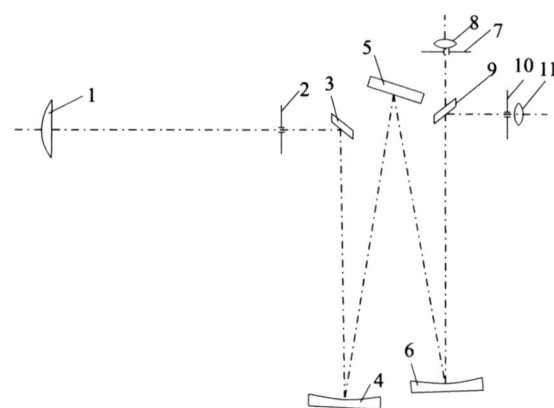

developed by researchers as individual devices (or small series) for solving concrete measuring problems. This concerns in full measure to the investigations of the Earth's airglow.

The basic optical unit of a photometer includes an entrance lens O_1; a diaphragm D, which determines the field of vision of the photometer; a Fabry lens O_2, which is behind the diaphragm and projects the entrance pupil onto the cathode of a radiation detector; and a photomultiplier S (Fig. 3.11). Since the entrance pupil of the photometer is illuminated practically uniformly, its image on the photocathode remains unchanged and does not depend on the sensitivity distribution over the cathode.

Table 3.4 Specifications of the DFS-14 spectrometer

Collimator lens:	Working spectral range (nm)
Focal distance (mm) 1200	310–670 (2)
Relative aperture 1:5.9	580–1660 (1)
	Reciprocal linear dispersion (nm/mm) 0.6
	Limiting resolution (nm) 0.3
Camera lens:	Maximum sweep speed (nm/s) 30
Focal distance (mm) 1200	Number of speeds 8
Relative aperture 1:5.9	
Inlet tube:	Method of spectrum recording: photoelectric
Focal distance (mm) 1000	Slit dimensions (mm) 0.42; 0.75; 1.5; 2.3; 3; 6
Relative aperture 1:7.3	Slit height (mm) 60
Magnification of the optical system: 1	Overall dimensions (mm):
	Length 3500
Replica of the diffraction grating: flat	Width 2100
Number of grooves per 1 (mm) 600	Height 900
Dimensions of the shaded part (mm) 150 × 140	Mass (kg) 700
	Guiding system (kg) 400
Concentration of light: order 2	

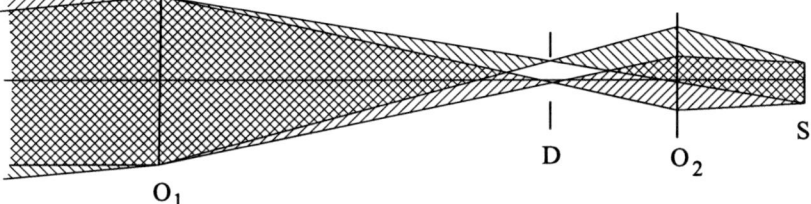

Fig. 3.11 Optical arrangement of an electrophotometer. O_1, O_2 lenses; D diaphragm; S focal surface

A natural component of the photometer is a light filter, which allows one to select a required spectral range.

3.2.3.2 Electrophotometry

As applied to the study of the airglow emissions, electrophotometry possesses some advantages over photography, allowing quicker and more convenient obtainment of initial data. Thus, it offers the capability of obtaining measurements in numeral form, so that their subsequent interpretation can be readily computerized. Below we give examples of photometers designed for goal-directed investigations.

For recording IGWs, three absolutely identical photometers had been developed and built, subsequently each operated at its own azimuth. Each photometer had a unit for changing the interference filters, a lens, a PMT unit with a cooler, and a preamplifier unit. For measurements the hydroxyl emission (4–1) band was chosen which is located in the spectral range near 1030 (nm) where there are practically no absorption bands of the Earth's atmosphere. The wavelengths of the filters were chosen so that the P (1034 (nm)) and Q (1029 (nm)) branches could be recorded. The spectral halfwidths of the filters were about 5 (nm).

The photometer was of conventional arrangement. The lens had a focal distance of 300 (mm) and a 1:4.5 relative aperture. The diaphragm located in the focal plane of the lens provided a 5° field of vision. Behind the diaphragm, a Fabry lens was placed which constructed the image of the photometer entrance pupil, uniformly illuminated by the incident light, on the photocathode. This made it possible to exclude the effect of nonuniform sensitivity distribution over the cathode surface.

Since measurements were performed in the infrared spectral region, PMTs with an oxide-silver-cesium photocathode which demanded a rather deep cooling (to about $-70(°C)$) to reduce the basic noise or with a gallium arsenide cathode (Reisse et al. 1974; Berkovsky et al. 1976, 1988) were used. Used for cooling was solid carbonic acid mixed with alcohol, as was done in spectrography.

In front of the lens the filter-changing unit was placed whose main component was a rotating disk. On this disk, in special mounts, 80-mm diameter interference filters were lodged. The inclination of filters could vary gradually during their rotation from 0° to 15°. The principle of operation of the filter rotator is described elsewhere

(Lyakhov and Managadze 1975; Potapov et al. 1978) (Fig. 3.12). Its advantage over a mere swinging of the filter is that it ensures a fine adjustment and reproducibility of a required inclination angle and also possesses opacity to stray light.

For this case, the inclination angle of the filter to the optical axis, ψ, depending on the angle of rotation of the filter mount about its axis, φ, making an angle α with the line of sight is determined by the relation (Fig. 3.13)

$$\sin\frac{\psi}{2} = \sin\alpha \cdot \sin\frac{\varphi}{2}.$$

The mounted light filter was at the angle α to its plane of rotation, i.e., in one of its positions it was perpendicular to the optical axis.

Change of filters was executed within several seconds by a command of a digital printing device (DPD). The construction of the disk allowed for the installation of four light filters for recording hydroxyl emissions from different vibrational levels. The operating capacity of the photometers was tested in triangulation measurements of the hydroxyl emission altitude (Potapov 1975a,b, 1976).

For automatic operation of the device as a whole it was necessary to match the change of the photometer filters and the operation of the frequency meters and DPD. For this purpose a matching unit and a filter-changing unit were made. The operational cycle of each photometer was as follows: recording of the signal of the P branch of the OH (4–1) band during 10 (s) followed by data input in the DPD during 5 (s); recording of the signal of the Q branch of the OH (4–1) band during 10 (s) followed by data input in the DPD during 5 (s); recording of the dark current of the PMT during 10 (s) followed by data input in the DPD during 5 (s). Thus, the complete recording cycle lasted 45 (s).

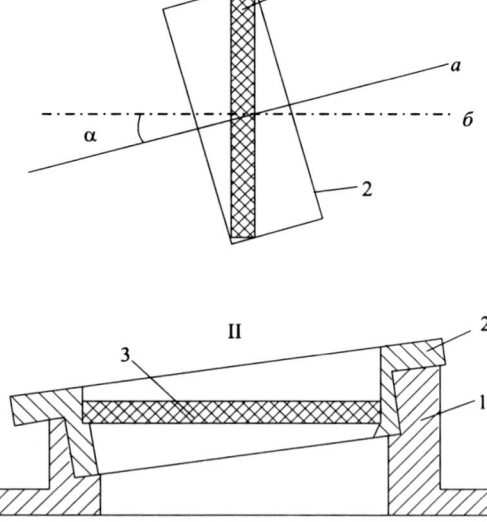

Fig. 3.12 Rotator intended to vary the inclination of an interference light filter to the optical axis (Lyakhov and Managadze 1975; Potapov et al. 1978). **I** – Sketch of the orientation of a mounted light filter; **II** – sketch of the filter rotator. 1 – Stator; 2 – rotor; 3 – light filter. a – Rotation axis of the filter; b – optical axis of the system

3.2 Spectrophotometric Equipment

Fig. 3.13 Scheme showing rotary changes of the inclination of an light filter to the photometer optical axis

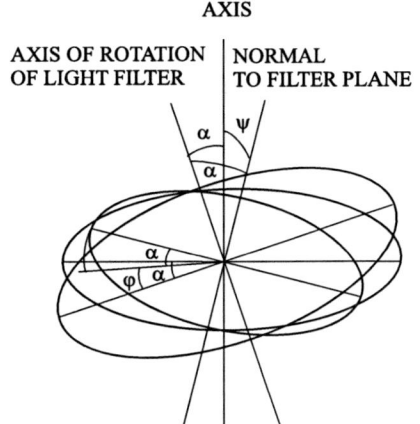

For recording the variations of the hydroxyl intensity and rotational temperature related to IGWs the three identical photometers were arranged so that their optical axes were co-oriented at a zenith distance of 45° to the following azimuths: 60°, 180°, and 300°.

Figure 3.14 presents an example of records of the rotational temperature variations during the night of March 15–16, 1975, at three azimuths.

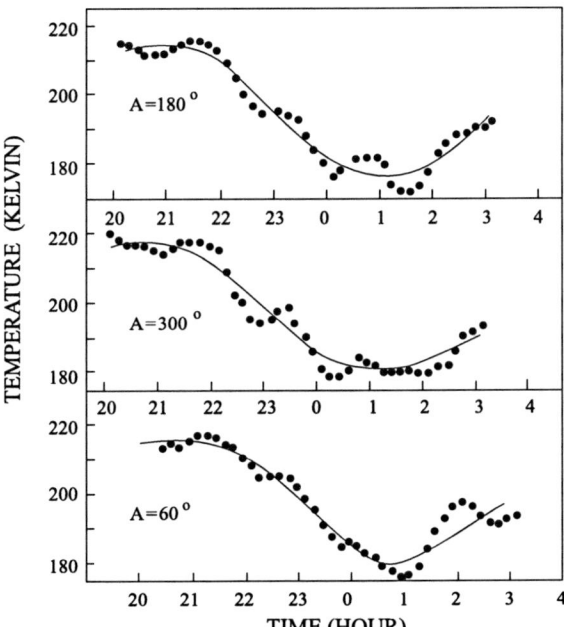

Fig. 3.14 Example of the rotational temperature records with a photometer

The general properties of the photoelectric equipment used in investigations of orographic perturbations also met the above requirements. As mentioned in Sect. 5.2, two photometers were used: a double-channel photometer for measuring the temperature along the direction perpendicular to the median line of the Caucasian edge and an eight-channel one for investigations of wave processes.

The two-channel photometer (Fig. 3.15) had two parallel channels for measuring simultaneously two spectral ranges of the OH (8–3) band with the use of light filters: the Q branch (at 727.55 (nm) in vertical position and at 725.7 (nm) in inclined position for measuring the noise) and the R branch (at 724.75 (nm) in vertical position and at 722.7 (nm) in inclined position for measuring the noise). The interference light filter unit was mounted in a special casing for providing inclination (up to $15°$) and placed in a thermostat with a temperature of $30(°C)$. The light filters had halfwidths of 1.1 (nm) (Q branch) and 2.15 (nm) (R branch) in vertical position and 1.6 (nm) (Q branch) and 2.34 (nm) (R branch) in inclined position. The measuring cycle for each channel at each zenith angle (Fig. 5.21) included four readings: the light intensity measured with the filter in vertical and in inclined position and with a luminophor, and the dark current. The multi-alkali photocathodes of the FEU-79 photomultiplier were cooled by three-stage thermoelectric coolers which allowed to reduce the temperature of the photocathodes up to $-40(°C)$. Accumulation of photons in each channel was performed for 8 (s); the transition to another state took 2 (s). The cycle of operation at one orientation of the line of sight lasted 40 (s); the complete cycle of obtaining a profile in the plane perpendicular to the edge lasted 14 (min) 40 (s).

In the eight-channel photometer, each channel operated as an individual photometer with its own optical system and light filter. All of them were combined into one unit placed in the cold store of a compression cooler. The optical axes of the photometers were arranged in given directions by means of identical periscopic

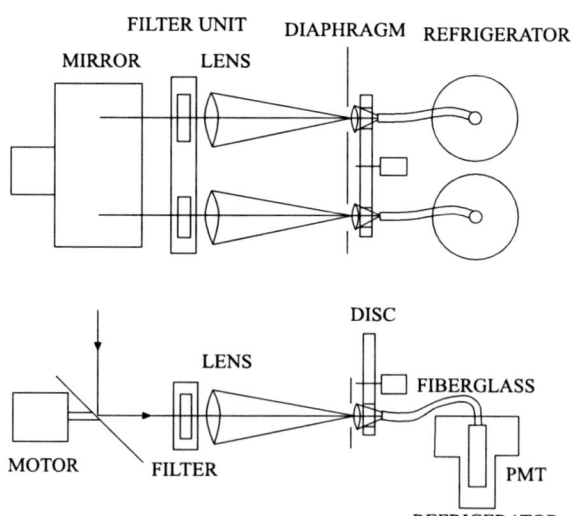

Fig. 3.15 Arrangement of a two-channel photometer designed for determination of the temperature of hydroxyl emission

3.2 Spectrophotometric Equipment

mirror systems. The halfwidths of the light filters were ~1.8–2.0 (nm) at a wavelength of 724.7–724.9 (nm). The configuration of the channels is shown in Figs. 5.22 and 5.23.

Under the conditions of photometric measurements at altitudes over 100 (km) above the Earth surface in the presence of bright light sources, such as the Sun, in the sky, careful protection against the light scattered at the entrance parts of the photometer is necessary. For this purpose, specially and carefully developed blinds are used which substantially reduce the level of possible noises (Fig. 3.16) (Davydov et al. 1975; Davydov 1986). As can be seen, the blind unit consists of an exterior complex blind (1), a lens (2), an opaque damper (3), an interior blind (4), a field diaphragm (5), lenses (6 and 8), mirrors (7), condensers (10 and 12), focal surface (13), a PMT (14), interference filters (9), and a modulator (11) in the immediate proximity of which a light gate is located.

The length of the exterior blind, L, and the diameter of its inlet, D, are calculated by the formulas

$$D = D_0 + 2a + 2L \cdot tg\beta; \qquad L = \frac{D_0 + 2a}{tg\varphi_{min} - tg\beta},$$

where D_0 is the diameter of the entrance pupil of the lens; 2β is the angle of vision of the lens, and φ_{min} is the minimum light-striking angle at which direct rays from

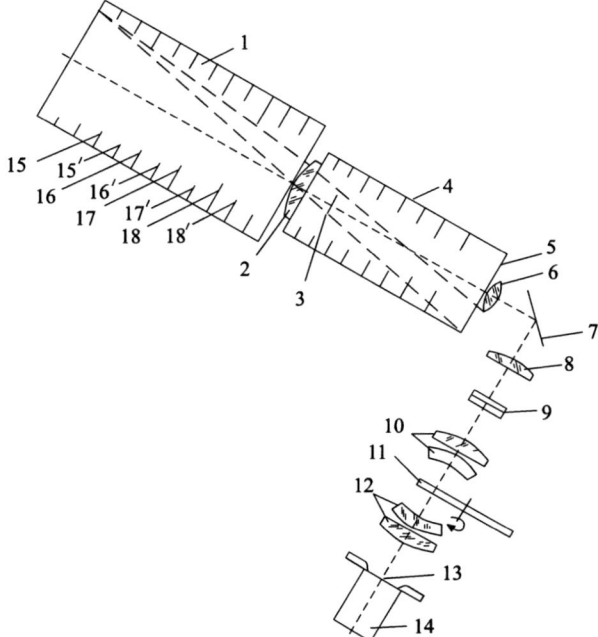

Fig. 3.16 Schematic diagram of a photometer with a blind attenuating light noises from bright light sources. See explanatory notes in the text

a lateral light source fall on the lens; a is the distance between the extreme field ray and the line on which the interior edges of diaphragms of smaller diameters lie.

If the angles of light-striking from a lateral light source are greater than φ_{min}, direct light does not fall on the lens, and it is quenched in the blind. To increase the degree of attenuation of light in the blind, besides basic diaphragms 15, 16, 17, and 18, additional diaphragms 15′, 16′, 17′, 18′ can be set up.

The blind of this design substantially reduces the light scattered by its interior surfaces and makes it possible to decrease the number of diaphragms whose interior edges reflect the first-order light into the lens. Moreover, the level of noise in the range of scattering angles from 15° to 95° is reduced by about four orders of magnitude.

3.2.3.3 Calibration of Photometric Measurements

The photoelectric observations using narrow-band interference light filters feature low spectral resolution. Therefore, the problem is to obtain information on a discrete emission from measurements in several spectral regions. In the previous section it was pointed out that the vibrational temperature of hydroxyl can be determined from measurements in two spectral regions. However, in this case, the possibility should exist to transform the measured intensity ratios into absolute values of temperature. To do this when operating with interference filters, it is necessary to know the transmission curves of the filters for various angles of their inclination relative to the optical axis of the photometer (the inclination of the filter changes the spectral characteristics of the device) and the theoretical intensity distributions for various band lines depending on temperature.

However, because of some practical difficulties, it appears inconvenient to construct the transmission curve of a filter in the working state with a required accuracy. In this connection, to determine the temperature of hydroxyl emission, another method of the transformation of intensity ratios into temperature was used, namely, comparison of the results of measurements performed on a photometer with those obtained with a DFS-14 spectrometer. This stationary device, with the spectral width of the entrance slit equal to 0.9 (nm), allowed measurements of the emission intensity of individual rotational lines of the P branch of the hydroxyl bands located in the range 700–1010 (nm) based on which the rotational temperature was determined.

As a radiation detector, a PMT with a cooled oxide-coated cathode was used which operated as a photon counting receiver. The dark noise of the PMT was 1–2 (pulse · s^{-1}). To increase the response speed of the spectrometer, a system for stepwise rotation of the diffraction grating was developed that minimized the time it took to switch from one line of the OH band to another during measurements.

An interference filter with a maximum transmission at a wavelength of 846 (nm) ($P_1(3)$ OH (6–2) line) was used in the photometer. An SP-50 spectrograph was used to photograph the transmission spectral regions of the filter at various inclination angles. At an angle of 15° the transmission band displaced to the R and Q branches of the OH (6–2) band. The radiation intensities of hydroxyl emission rated in these two branches show different temperature dependences. Therefore, their ratio is a sensitive function of temperature.

For calibration of photometric measurements with the DFS-14 spectrometer, the intensities of the $P_1(1)$ and $P_1(3)$ lines of the OH (6–2) bands were measured. The temperature dependence of their intensity ratio was obtained as described elsewhere (Piterskaya and Shefov 1975).

When performing calibration, the photometer and spectrometer were oriented to the same region of the sky. Measurements were conducted in clear moonless nights with a good atmospheric transparence to minimize the contribution of the continuous background. Comparison was carried out by the corresponding harmonic amplitudes obtained by a harmonic analysis of observation data sequences. Both measurement sequences were of identical duration and consisted of data obtained for identical time intervals. After smoothing, elimination of a trend, determination of the average value and variance, centering, and construction of periodograms, maxima were revealed in both periodograms. When the periods of oscillations corresponding to these maxima coincided on the photometer and on the spectrometer, their amplitudes were compared.

Since the spectrometer measured the emission intensities of the $P_1(1)$ and $P_1(3)$ lines of the OH band, the temperature was determined by the formula (Sect. 2.2.5)

$$T = 1.44 \cdot \Delta F_{1,3} / \left(\log_e \frac{P_1(1)}{P_1(3)} - \log_e \frac{i_1}{i_3} \right),$$

where $\Delta F_{1,3} = F_1 - F_3$ is the difference between the energies of levels for $P_1(1)$ and $P_1(3)$; i_1 and i_3 are the intensity factors.

The nightly mean temperature and the oscillation amplitudes were determined in absolute values directly from the spectrometer measurements. For the photometer, these quantities were determined in relative units. To reduce them to absolute temperature values, plots were constructed to determine the harmonic amplitudes and the mean temperature from the photometer data (Fig. 3.17a, b). Separate calibrations for the mean temperature and the amplitude are dictated by the need to investigate the effect of the background on each of these quantities. As can be seen in Fig. 3.17a, the slope of the regression line is equal to 20(K) per 1(%) of the intensity ratio for individual lines or P branches, which corresponds to an ideal relationship in the absence of a background. At small temperature variations some background effect is possible. However, for usually measured amplitudes ($> 0.5(\%)$) this effect can be neglected. Figure 3.17b illustrates the relation between the mean temperatures obtained from spectrometer data and the mean intensity ratios obtained from photometer data for the same observation intervals as in Fig. 3.17a. In this case, the effect of the background shows up in untrue values of mean temperatures: the temperatures lower than 250(K) are overestimated, while those higher than 250(K) are underestimated, and the error increases with background intensity. The spread in experimental points is also explained by the background effect.

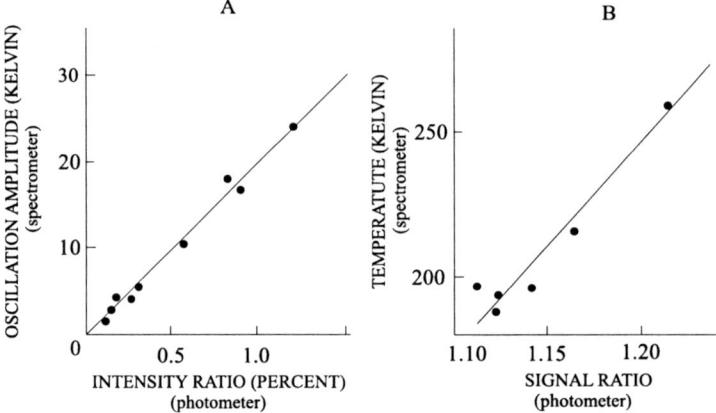

Fig. 3.17 Calibration of photometer measurements of the rotational temperature by spectrometric measurements. (**A**) Relation of temperature oscillation amplitudes A_s to oscillation amplitudes of intensity ratio A_{ph} (photometer). (**B**) Relation of mean rotational temperatures T_s (spectrometer) to readout ratios T_{ph} (photometer)

Subsequently these plots were used to determine oscillation amplitudes and mean temperatures in processing photometric data.

3.2.4 Photographic and Photoelectric Interferometers

Interferometers are widely used in physical research and for industrial inspection of characteristics of various products. However, all these applications are preferentially aimed at measuring wavelengths and linear displacements of objects. Temperature measurements are carried out under the conditions of high temperatures and radiation intensities, allowing rather fast times of signal recording. As a rule, for these purposes interferometers of special design were commercially produced. The conditions of using interferometers in investigations of weak luminous fluxes in astronomy and in observations of the night airglow of the upper atmosphere impose much heavier demands on their stability during long-term (lasting many hours) periods and noise immunity to stray light sources, as well as on the sensitivity of the radiation detectors and recording systems.

The optical arrangement of a Fabry–Perot interferometer includes a system of two parallel high-reflectivity flat mirrors (Fig. 3.18). The main advantage of the interferometer over other spectroscopic instruments is its considerably greater aperture ratio, L, which is defined as the product of the entrance pupil area S and the solid angle Ω (Meaburn 1976):

$$L = S \cdot \Omega, \text{ where } \quad S = \frac{\pi D^2}{4} \quad \text{and} \quad \Omega = \frac{\pi d^2}{4F^2}.$$

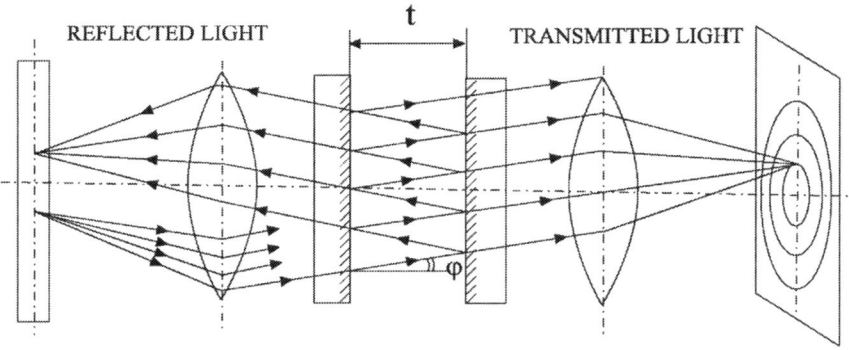

Fig. 3.18 Optical arrangement of a Fabry–Perot interferometer

For a spectrometer the entrance area is determined as the product of the width of the slit (~ 0.02 (cm)) and its height (~ 1.8 (cm)), i.e., $S = 0.036$ (cm^2); for an interferometer with a ~ 5-cm diameter of the plates $S = 19.6$ (cm^2). The solid angle of the collimator lens of diameter $d \sim 15$ (cm) with focal distance $F \sim 70$ (cm) is $\Omega = 0.036$ (sr) and for an interferometer with the size of the first ring $d \sim 0.2$ (cm) and focal distance $F \sim 10$ (cm) is $\Omega = 0.00031$ (sr). Thus, we have for the spectrometer $L_{SP} = 0.0013$ and for the interferometer, allowing much higher resolution, $L_{FP} = 0.0022$.

The use of Fabry–Perot interferometers began soon after the detection of the Earth's airglow. Being high-aperture instruments with high resolving ability (Tolansky 1947; Korolev 1953; Frish 1980), they were used for exact determination of the wavelength of the night-sky green emission whose identification was an important problem at the beginning of the 20th century (Khvostikov 1937, 1948). Another example of the use of an interferometer of this type for the determination of the spectral structure of emissions was the detection of the isotopic composition of the 670.8-nm lithium emission during nuclear explosions realized in the upper atmosphere (IAPh 1994). The matter is that the nuclear charge structurally contains a solid constituent, lithium deuteride D^{6}Li, which participates in the nuclear reaction giving rise to tritium necessary for the subsequent explosion. Figure 2.19 shows a photo of an interferogram on which components of the ^{7}Li and ^{6}Li lines of the lithium atom are clearly seen. For comparison the isotope ratio for natural conditions (^{7}Li :^{6} Li $= 10 : 1$) and that for the atmospheric conditions after an explosion (^{7}Li :^{6} Li $= 1 : 4$) are given.

The main goal of the use of a Fabry–Perot interferometer in atmospheric investigations was to determine the Doppler temperature of the upper atmosphere by the emission lines of atomic oxygen at 557.7 (nm) (Hernandez 1971, 1976, 1977; Ignatiev 1977a,b; Semenov 1989; Ignatiev and Yugov 1995) and 630.0 (nm) (Mulyarchik 1960a,b; Semenov 1975a,b, 1976, 1978; Ignatiev 1977a,b; Hernandez and Killeen 1988; Hernandez et al. 1992; Ignatiev and Yugov 1995), by the anomalous broadening of the nightglow emission lines of sodium (Hernandez 1975), and by the $Q_1(2)$ and $P_1(3)$ lines of the vibration–rotation bands of the hydroxyl molecule

(Hernandez et al. 1992; Greet et al. 1994; Choi et al. 1998). Besides, the interferometer was used as a narrow-band light filter for the determination of the intensity of both the H_α hydrogen emission in the nightglow (Shcheglov 1967a,b) and of the H_β emission from extra-atmospheric radiants (Aitova and Shcheglov 1977; Aitova 1977). An important application of these interferometers was the determination of wind velocities at the altitudes of emission layers (Ignatiev and Yugov 1995; Roble and Shepherd 1997).

To determine the Doppler temperature of emission lines under real conditions, two techniques for recording the interference pattern were used: (1) photographic recording of an immobile image (Mulyarchik 1959, 1960a,b; Shcheglov 1967a; Semenov 1975a,b, 1976, 1978; Ignatiev and Yugov 1995) and (2) spectrophotometric recording with scanning the spectrum by varying the pressure inside the chamber (Hernandez and Turtle 1965) containing an interferometer or by varying the distance between the plates with the help of piezoelements (Hernandez 1966, 1970, 1974). Typical interferograms of the 557.7- and 630-nm emissions photographed under the conditions of auroras are shown in Fig. 3.19.

During investigations of the upper atmosphere temperature the radiating medium is usually optically thin, and therefore the profile of the emission line produced by a gas parcel at a constant temperature can be described by the Doppler distribution

$$I(\lambda) = I_0 \cdot \exp\left[-\frac{1}{2} \cdot \frac{(\delta\lambda)^2}{\left(\lambda_0\sqrt{kNT/Mc^2}\right)^2}\right],$$

where I_0 is the peak intensity, λ_0 is the wavelength of the line under investigation, $\delta\lambda = \lambda - \lambda_0$, k is Boltzmann's constant $(1.380 \cdot 10^{-16}\,(\text{erg} \cdot \text{K}^{-1}))$, T is the temperature, M is the atomic weight, c is the velocity of light $(2.9979 \cdot 10^{10}\,(\text{cm} \cdot \text{s}^{-1}))$, and $N = 6.025 \cdot 10^{23}$ is the Avogadro number.

The halfwidth $\Delta\lambda_{HW}$ of the line profile, i.e., its width at the intensity equal to half the maximum, and the Doppler dispersion $\Delta\lambda_D$ are given (expressed in wavelengths), respectively, as

$$\Delta\lambda_{HW} = 7.18 \cdot 10^{-7}\lambda_0\sqrt{\frac{T}{M}} \text{ and } \Delta\lambda_D = 3.84 \cdot 10^{-7}\lambda_0\sqrt{\frac{T}{M}}.$$

However, a recording optical system distorts the actual profile because of the finite width of its instrument function. It is well known that the ideal instrumental profile of the Fabry–Perot etalon is described by the Airy function (Tolansky 1947; Tarasov 1968; Born and Wolf 1964; Hays and Roble 1971; Zaidel et al. 1972; Lebedeva 1977; Ignatiev and Yugov 1995)

$$I_{Airy}(\theta) = \frac{(1-\mu)^2}{1 - 2\mu \cdot \cos\theta + \mu^2}.$$

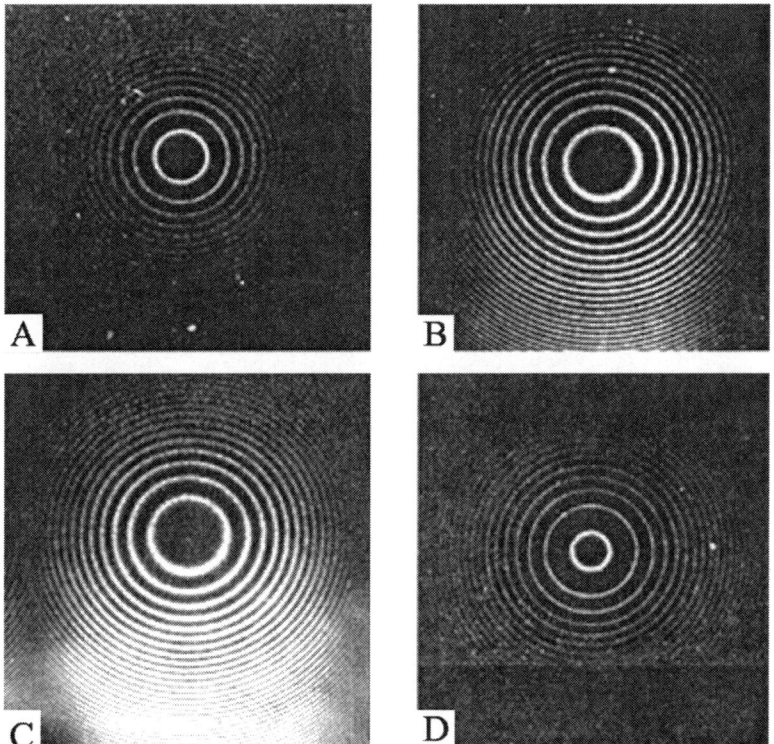

Fig. 3.19 Photos of interferograms of the emissions. (**A**) At 557.7 (nm) in an aurora (t = 200 (mm)). (**B**) At 630 (nm) in an A-type aurora of moderate intensity. (**C**) At 630 (nm) in a bright A-type aurora. (**D**) At 587.092 (nm) (yellow line of krypton) from a laboratory source (Mulyarchik and Shcheglov 1963)

Here $\mu = \sqrt{\mu_1 \cdot \mu_2}$, where μ_1 and μ_2 are the reflectivities of the two mirrors of the interferometer, $\theta = 4\pi \cdot n \cdot t \cdot \left(\frac{1}{\lambda_\angle} - \frac{1}{\lambda_\perp} \right) \cdot \cos\varphi$ is the phase difference between two neighboring interfering rays, $\lambda_\angle$ and $\lambda_\perp$ are, respectively, the wavelengths of the inclined and normal rays, incident on the plates of the interferometer, n is the refraction index of the medium between the plates, t is the distance between the plates, and φ is the angle of incidence of light on the mirrors.

The above formula of the instrument function of an ideal interferometer is normed to its maximum. The transmittance of such an interferometer is

$$\tau_{max} = \left(\frac{1 - \mu - \varepsilon}{1 - \mu} \right)^2,$$

where ε is the absorption coefficient of the mirror coatings of the plates. Thus, the full halfwidth of an ideal interferometer is

$$\delta\lambda_{\text{Airy}} = \frac{1-\mu}{2\pi \cdot \sqrt{\mu}} \cdot \frac{\lambda^2}{t}.$$

However, the actual instrumental profile is a convolution of the Airy function and the following functions (Hays and Roble 1971): (1) the function describing microscopic irregularities of the interference plate surfaces, which is a normal distribution for the argument $2\pi \cdot t \cdot \cos\varphi / \lambda$; (2) the quadratic function describing spherical defects in the interference plates; and (3) the aperture function similar in form to the previous one, with the only difference that for this function the halfwidth is the instrument geometry factor $a = d^2/16\lambda \cdot f^2$, where d is the diameter of the circular aperture, f is the focal distance of the lens, and this factor does not depend on the distance t between the plates; φ is the angle of incidence of rays (Zhiglinsky and Kuchinsky 1983).

Thus, the main constituents of an actual instrumental profile are the Airy and the Gauss functions (Ivanov and Fishman 1973). If the instrumental profile is described by the Lorentz function and the profile under investigation by the Doppler function,

$$I_L = \frac{1}{\pi} \cdot \frac{\Delta\lambda_L}{(\Delta\lambda_L)^2 + (\Delta\lambda)^2} \qquad I = \frac{1}{\sqrt{\pi} \cdot \Delta\lambda_D} \cdot \exp\left[-\frac{(\Delta\lambda)^2}{(\Delta\lambda_D)^2}\right],$$

then the recorded profile will be described by the Voigt function

$$V(t,a) = \frac{1}{\sqrt{\pi} \cdot \Delta\lambda_D} \cdot \frac{a}{\pi} \cdot \int_{-\infty}^{\infty} \frac{e^{-x} \cdot dx}{(t-x)^2 + a^2},$$

where $t = \frac{\Delta\lambda}{\Delta\lambda_D}$ and $a = \frac{\Delta\lambda_L}{\Delta\lambda_D}$. The solution of this equation is considered, for instance, by Zhiglinsky and Kuchinsky (1983).

However, all these approaches take into account only the factors that influence the image of an interference figure. If the image is recorded by photographing there occurs radiation scattering in the photographic layer or imposition of the discrete structure of the matrix detector. This implies that the actual instrumental profile should be determined experimentally by recording narrow emission lines, such as the lines of a krypton laser.

If the Doppler profile of a line is observed, one more Gauss function is imposed on the instrumental profile, and therefore there is a complex convolution of functions at the output of the device. Under the actual conditions of measuring weak emission lines in the spectral transmission band of an interferometer interferences occur due to a possible blinding by close spectral components and also due to the noise of the radiation detector and fluctuations of the background. Therefore, the true profile of the line under investigation can be revealed only by Fourier analysis, and thereafter the least square method can be used to determine the Doppler temperature (Hays and Roble 1971; Cooper 1971; Ivanov and Fishman 1973).

In investigations of low-intensity emission lines of the upper atmosphere, the recording time required for signal accumulation inevitably increases. In this case, of great importance are the intensity time variations that can distort the line profile

3.2 Spectrophotometric Equipment

if recorded by the method of successive scanning of the spectrum, as this occurs on varying the pressure in the interferometer chamber. To eliminate errors of this type, the method of recording of the entire interferometric pattern is preferred. Initially photographic recording with electrooptical image converters was used for this purpose (Trunkovsky and Semenov 1978). Interferograms were obtained with the use of a three-camera ICT with a multi-alkali photocathode which was cooled to $-18(°C)$ with the help of an absorption refrigerator (Fig. 3.20). This can best be done by using modern matrix radiation detectors.

A

B

Fig. 3.20 The Fabry–Perot interferometer with an image converter tube, used for the determination of the temperature of the 630-nm emission (Semenov 1975a) at Zvenigorod. (**A**) View from the laser light source; (**B**) view from the refrigerator

Interferometry allows not only to determine the emission layer temperature by broadening the Doppler profile, but also to measure small wavelength variations caused by the Doppler shift resulting from horizontal wind motions in the upper atmosphere.

To determine the horizontal wind velocity V in an emission layer by interferometric measurements in two opposite azimuthal directions, the change ΔD_1 of the diameter D_1 of the interior ring resulting from the Doppler shift of the wavelength λ relative to λ_0 is directly measured. An increase in diameter D_1 results in a decrease in wavelength. This implies that the wind velocity vector is directed to the observer. Since the radial velocity V_r is determined immediately, in view of Fig. 3.21, we have

$$V = \frac{V_r}{\sin \zeta_0},$$

where ζ_0 is the zenith angle of the line of sight at the emission layer point at which the measurement is performed. Here

$$\sin \zeta_0 = \frac{\sin \zeta}{1 + \frac{Z}{R_E}},$$

where ζ_0 is the zenith angle of the sky region under observation (the emission layer altitude region) at the observation place, Z is the altitude of the emission layer, and $R_E = 6356 \, (\text{km})$ is the Earth's radius.

For the wavelength Doppler shift $\delta \lambda$ we have

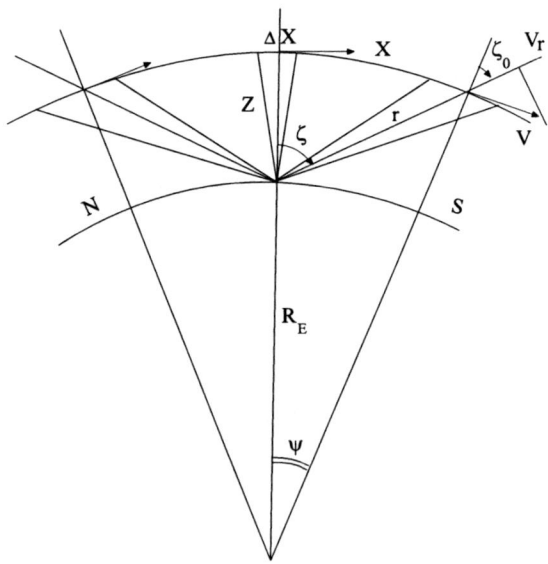

Fig. 3.21 Scheme illustrating the determination of the velocity of a wind

3.2 Spectrophotometric Equipment

$$V_r = \frac{c}{\lambda_0} \cdot \delta\lambda,$$

where c is the velocity of light. Thus,

$$V = \frac{c}{\lambda_0} \cdot \frac{1 + \frac{Z}{R_E}}{\sin\zeta} \cdot \delta\lambda.$$

According to the dispersion formula for an interferometer, for instance, for the northern (N) and southern (S) directions,

$$(\delta\lambda)_N = -\Delta\lambda \cdot \frac{2D_N \cdot \delta D_N}{D_{2N}^2 - D_{1N}^2}, \qquad (\delta\lambda)_S = -\Delta\lambda \cdot \frac{2D_S \cdot \delta D_S}{D_{2S}^2 - D_{1S}^2}.$$

The variations in ring diameters in the northern (D_N) and southern (D_S) directions relative to the direction toward the zenith (D_0) have different signs; that is, if we assume that D_N is greater than D_S, we have

$$\delta D_N = D_N - D_0 \text{ and } \delta D_S = D_0 - D_S$$

and then

$$2\delta D_1 = \delta D_N + \delta D_S = D_N - D_S.$$

Since the actual variations of the diameter of a ring are small ($\Delta D/D \sim 0.05$), the diameters (not the diameter variations!) of the rings in the northern and southern directions can be considered identical (where they are factors in the formulas above). Therefore, we have

$$2\delta\lambda = -\Delta\lambda \cdot \frac{2D_1 \cdot \delta D_1}{D_2^2 - D_1^2},$$

whence the wind velocity in the direction from north to south (+) is

$$V = c \cdot \frac{\Delta\lambda}{\lambda_0} \cdot \frac{1 + \frac{Z}{R_E}}{\sin\zeta} \cdot \frac{\frac{\delta D_1}{D_1}}{\frac{D_2^2}{D_1^2} - 1}.$$

From the viewpoint of the actual observation conditions, the optimum zenith angle of the direction of the line of sight, ζ, is 60°. Therefore, for the mentioned characteristics of interferometers we have

$$V(557.7\,(\text{nm})) = 3267 \cdot \Phi(\Delta D_1), \,(\text{m}\cdot\text{s}^{-1}),$$
$$V(630.0\,(\text{nm})) = 16{,}230 \cdot \Phi(\Delta D_1), \,(\text{m}\cdot\text{s}^{-1}),$$
$$V(630.0\,(\text{nm})) = 14{,}173 \cdot \Phi(\Delta D_1), \,(\text{m}\cdot\text{s}^{-1}),$$

where

$$\Phi(\Delta D_1) = \frac{\frac{\Delta D_1}{D_1}}{\frac{D_2^2}{D_1^2} - 1}.$$

The distance X along the emission layer from the zenith to the sighting point is determined by the relation

$$\sin\zeta = \left(1 + \frac{Z}{R_E}\right) \cdot \sin\left(\zeta - \frac{X}{R_E} \cdot \frac{180}{\pi} \cdot \frac{1}{1 + \frac{Z}{R_E}}\right),$$

whence we have for the zenith angle ζ

$$\text{tg}\,\zeta = \frac{\left(1 + \frac{Z}{R_E}\right) \cdot \sin\left(\frac{180}{\pi} \cdot \frac{\frac{X}{R_E}}{1 + \frac{Z}{R_E}}\right)}{\left(1 + \frac{Z}{R_E}\right) \cdot \cos\left(\frac{180}{\pi} \cdot \frac{\frac{X}{R_E}}{1 + \frac{Z}{R_E}}\right) - 1},$$

for the vector radius r

$$r = (R_E + Z) \cdot \left[\sqrt{1 - \frac{\sin^2\zeta}{\left(1 + \frac{Z}{R_E}\right)^2}} - \frac{\cos\zeta}{1 + \frac{Z}{R_E}}\right],$$

for the distance X

$$\text{tg}\left(\frac{180}{\pi} \cdot \frac{\frac{X}{R_E}}{1 + \frac{Z}{R_E}}\right) = \frac{\left(1 + \frac{Z}{R_E}\right) \cdot \sqrt{1 + \left[1 - \frac{1}{\left(1 + \frac{Z}{R_E}\right)^2}\right] \cdot \text{tg}^2\zeta} - 1}{\left(1 + \frac{Z}{R_E}\right) \cdot \sqrt{1 + \left[1 - \frac{1}{\left(1 + \frac{Z}{R_E}\right)^2}\right] \cdot \text{tg}^2\zeta} + \text{tg}^2\zeta} \cdot \text{tg}\,\zeta,$$

and for the angle Ψ subtended by the arc of the great circle

$$\Psi = \frac{180}{\pi} \cdot \frac{X}{R_E} \cdot \frac{1}{1 + \frac{Z}{R_E}}.$$

For the emissions and observation conditions considered above we have

X (557.7 (nm)) = 166 (km), r (557.7 (nm)) = 192 (km), Ψ (557.7 (nm)) = 1.47°;
X (630.0 (nm)) = 411 (km), r (630.0 (nm)) = 476 (km), Ψ (630.0 (nm)) = 3.57°.

Assuming that for the central ring the angle of sight of the interferometers is about 1°, we obtain for the dimension ΔX of the emission layer region at $\zeta = 60°$

$$\Delta X_{60} (557.7 \,(nm)) = 6.4 \,(km), \quad \Delta X_{60} (630.0 \,(nm)) = 15.0 \,(km)$$

and at zenith

$$\Delta X_0 (557.7 \,(nm)) = 1.7 \,(km), \quad \Delta X_0 (630.0 \,(nm)) = 4.4 \,(km).$$

To fit the actual conditions under which the temperature of the upper atmosphere is determined, the emissions of atomic oxygen at 557.7 (nm) (altitude ~97 (km) and temperature ~200 (K)) and 630 (nm) (altitude ~270 (km) and temperature ~700–1500 (K)) are mainly used. Therefore, the following parameters are commonly chosen for the interferometers:

$t(557.7\,(nm)) = 30\,(mm), \Delta\lambda(557.7\,(nm)) = 0.00518\,(nm), \gamma(557.7\,(nm)) =$
$\qquad 0.52845150\,(K^{-1})$;
$t(630.0\,(nm)) = 7\,(mm), \Delta\lambda(630.0\,(nm)) = 0.0284\,(nm), \gamma(630.0\,(nm)) =$
$\qquad 0.22434056\,(K^{-1})$;
$t(630.0\,(nm)) = 8\,(mm), \Delta\lambda(630.0\,(nm)) = 0.0248\,(nm), \gamma(630.0\,(nm)) =$
$\qquad 0.29419895\,(K^{-1})$.

3.3 Radiation Detectors

The first detector of light capable of quantitatively evaluating its intensity was, certainly, the naked eye. Therefore, all photometric notions arose initially as a result of visual determination of the luminosity of optical radiation. However, this turned out to be absolutely insufficient to obtain detailed quantitative data. Therefore, technological progress gave rise to new instrumental methods for research of light, and measurement of its quantitative characteristics appeared.

It is quite clear that the ideal radiation detector does not exist. Nevertheless, it is possible to specify the necessary properties of such a detector. It should be an efficient and precise instrument, i.e., be capable of detecting most of the incident photons and determining reliably and precisely the input signal from measurements of the output signal. Such a detector should operate over a broad range of values of the input signal without nonlinear distortion or saturation. It should ensure accumulation, storage, and release of the output signal on any required time scale. For one- or two-dimensional image detectors, high spatial stability and homogeneity, good resolution, and a large size of the light detection zone are required. Moreover, the detector should be simple to service, reliable in operation, and durable (Eccles et al. 1983).

Naturally, none of the existing radiation detectors meets the above requirements. Therefore, when specifying the goal of a study, one must choose instruments which would best suit the necessary requirements, bearing in mind the optimum modes of their operation.

The primary characteristic of a detector is its sensitivity. It can be defined as the output-to-input signal ratio. However, it speaks nothing about the efficiency of detection of photons. The quantum yield of a detector is defined as the number of detected photons related to the number of photons which would be detected by an ideal detector; it is measured in percent. Nevertheless, this characteristic is more convenient when different detectors are compared by sensitivity to radiation of different wavelength, since it does not take into account the photon losses at subsequent stages of recording. It is more appropriate to define a generalized quantum yield DQE as the ratio of the squared signal/noise ratios at the inlet and at the outlet of the detector (Eccles et al. 1983):

$$\mathrm{DQE} = \left(\frac{S_{out}}{N_{out}}\right)^2 \bigg/ \left(\frac{S_{in}}{N_{in}}\right)^2,$$

which is usually measured in percent.

The spatial resolution of a detector is generally expressed by the number of discernible stripes per millimeter. It is closely related to the image element of the smallest region of the detector-sensitive surface for which the signal level is noticeably different from that for the neighboring region.

3.3.1 Photography

Photography was the first method of light recording in the form of an image of both the light source and the light spectrum (Dokuchaeva 1994). The main constituent of the modern black-and-white photographic materials used in scientific research is the photosensitive emulsion coating consisting of a solidified mixture of microcrystals of silver halides (silver bromide AgBr, silver iodide AgI) and gelatine. The microcrystals are usually about 1 μm in size; their number reaches 10^9–10^{10} per 1 (cm^2) of the coating surface. The emulsion coating thickness is different for different photographic materials. For slow films it is 4–8 (μm) and for fast films and plates it varies between 10 and 20 (μm) (Mees 1942; Mees and James 1966). Glass, film, or paper can serve as substrates for photographic emulsions, but only photographic materials applied on film were used for spectrophotometric investigations of the Earth's airglow. Used for fastening the photolayer is hardened gelatine with admixtures of organosilicon substances and some other ingredients. The subbing layer has a thickness no more than 1 μm. The substrate thickness is 0.07, 0.10, or 0.18 (mm). Some photographic materials have an antihalation coating.

When a photomaterial is exposed to light, a photochemical reaction takes place which results in the formation of microcrystals of silver halide producing a latent image. The most important property of a photographic emulsion is its ability to accumulate incident light photons.

Since to record weak optical radiations calls for high sensitivity, which allows one to reduce the duration of exposure, the airglow investigations were carried out with the use of films of special types, such as Kodak 103aF and also A-700U- and A-660-type films produced at Kazan Chemicals Plant. For the best emulsions, the generalized quantum yield DQE is only 4(%).

Development of Photographic Materials

An obligatory primary treatment of an exposed photographic material is chemical development in a developer solution. The developer is highly selective in its action. However long term the development, the difference between strongly irradiated and not irradiated layer regions remains appreciable.

In the course of development, particles of metallic silver are pushed out from the crystal since a silver atom is larger in volume than the ion. When halides are removed from a microcrystal, it is transformed to a tangled ball of filaments. Fine-grained emulsions contain short filaments, while coarse-grained emulsions include larger balls of silver filaments. Centers of a latent image are formed both on the surface and in the bulk of microcrystals.

Photographic developers conventionally have in their composition the following components: a developing substance, a compound providing an alkaline reaction of the solution, an antioxidant preventing fast oxidation of the solution by the air oxygen, an antifogging agent, and water as solvent. The most typical developer components are the following (Zhurba 1990):

Metol M-143 $(C_6H_4OHCH_3)_2 \cdot H_2SO_4$ is an easily water-soluble, softly working developing agent.

Hydroquinone H-142 $C_6H_4(OH)_2$ is a developing agent whose activity strongly increases when an alkali is added to the developer. Hydroquinone is generally used together with Metol. In mixed developers, Metol begins development and hydroquinone prolongs this process and promotes enhancement of contrast. The activity of developing agents strongly depends on the alkali, which is necessary for neutralization of hydrobromide acid HBr formed on reduction of silver bromide. The most commonly used are sodium carbonate and potash. In alkali-free Metol developers, the alkaline medium is produced by sodium sulfite Na_2SO_3.

For the developer antioxidant, only sodium sulfite Na_2SO_3 is used. It is of importance that this component prevents the developed photographic image and gelatine from coloration by the products of decomposition of developing agents.

Potassium bromide KBr is commonly used as an antifogging agent.

For dissolution of the developer components, it is absolutely necessary to use only distilled water and to dissolve them in the sequence indicated in manuals.

Table 3.5 gives the compositions of developers used in processing of photographic materials at different times.

A developer can be used not earlier than 12 (h) after preparation. The temperature of development should be observed carefully for providing stable photometric characteristics of the results obtained in long-term systematic investigations. Routinely this is 20(°C). During development it is necessary to ensure intermixing of the bath with the developed photographic material. A developer should be stored in a densely closed vessel with the air volume above its surface as small as possible. The duration of use should be restricted.

After development, the photographic material is dipped for several seconds in a water bath for rinsing not only to interrupt the development, but also to prevent ingress of the developer in the fixing agent solution.

Fixation is performed with hyposulfite $Na_2S_2O_3$ and is accompanied by mixing. As a result of this process, unexposed silver halide which has not decomposed under the action of light is removed from the photosensitive layer. The total time of fixation is determined as the double time during which there occurs clarification of the layer. It is recommended to use sequentially two fixing baths. In this case, in the first bath, 95(%) of silver halide goes from the layer into solution, and in the second bath, within the same time, the residual is washed out with fresh fixing solution. Nevertheless, it is not allowed to considerably increase the time of fixation.

After fixation, it is necessary to wash the negatives in running water for no less than an hour to remove the heels of hyposulfite and other impurities. After washing, the negatives are rinsed in distilled water. Careful observance of the technology ensures a long shelf life of negatives.

Negatives are dried at room temperature in a housing protected from ingress of dust.

Table 3.5 Compositions of developers for different films

Builder	UP-2	D-19	D-76
Metol (g)	5	2.2	2
Anhydrous sodium sulfite (g)	40	96	100
Hydroquinone (g)	6	8.8	5
Anhydrous soda (g)	31	48	
Crystalline borax (g)			2
Potassium bromide (g)	4	5	2
Water (l)	Upto 1	Upto 1	Upto 1
Duration of development (min)	6	5−10	12−16

Photometric Processing of Negatives

A photographic negative is the basic material for further investigations. The light having been incident on a photosensitive layer during the exposure time produces a latent image in the emulsion which, after development, gets a certain blackening that characterizes the light source. The blackening density of a photographic image is presented as a record (Fig. 3.22), which is obtained with a special instrument—a microphotometer.

An important property of a photographic emulsion is the nonlinear dependence of blackening density on exposure time. In photography, the "signal" amplitude, i.e., the amount of light which has acted on the photosensitive layer, is estimated by the quantity termed an exposure dose, which is defined as the product of the illuminance at a given layer region by the time for which this illuminance persists, which is termed an exposure time, i.e., $H = Et$ (Martynov 1977; Avgustinovich 1990). The choice of this quantity is related to the fact that the blackening density of a photosensitive layer is determined by the total amount of absorbed energy. As follows from Fig. 3.22, the blackening density produced by a measured signal is higher than the fog density. Thus, the measure of blackening on a record is the ratio of the useful signal density measured from the fog background level h to the deviation of the background level L from the photometer signal that corresponds to its dark level, i.e., h/L.

The curve described by the relation

$$\lg \frac{h}{L} = f(\log_{10} H)$$

is called a characteristic curve (Fig. 3.23). In this curve, three sections can be distinguished: the section of underexposures, the linear section, and the section of overexposures. The tangent of the slope angle γ of the linear section is called the contrast coefficient of the photographic material. The characteristic curve of a photographic

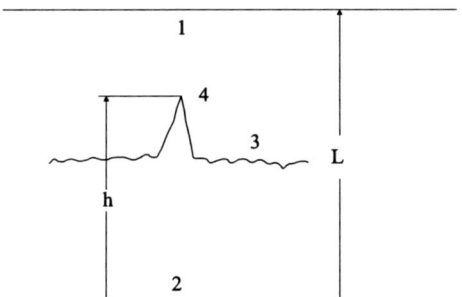

Fig. 3.22 Record of the blackening of a photographic image, obtained with a microphotometer. 1 – the level of the dark signal of the microphotometer; 2 – the level of the for blackening background of the photographic image; 3 – blackening from the continuous spectrum background; 4 – blackening from an emission line

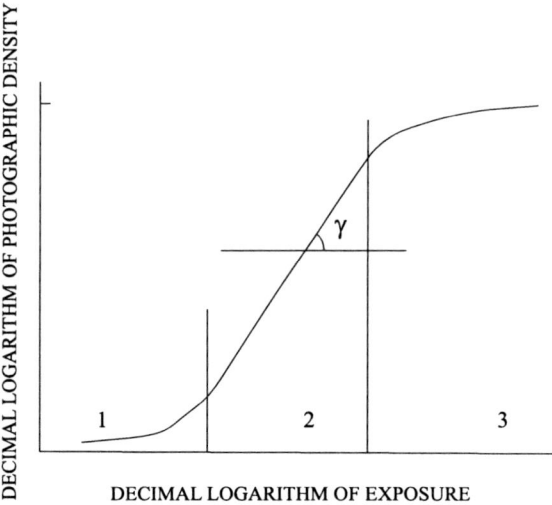

Fig. 3.23 Characteristic curve of a photosensitive layer in logarithmic coordinates (h/L blackening, H exposure). 1 – section of underexposures; 2 – linear section; 3 – section of overexposures; γ is the slope angle of the linear section

material is usually constructed by taking the image of a spectrally neutral step attenuator on this material. Such an attenuator has a set of areas with specified optical densities allowing one to determine the proportion between the amounts of light energy that act on the photolayer during the same time interval. Usually no less than two exposures at different initial source strengths are required to cover the entire blackening section of the characteristic curve. Since the curve sections are constructed in logarithmical coordinates, they are stitched by shifting one relative to another along the axis of exposure doses.

In photometric processing of spectrograms, blackening records were first obtained with the use of an MF-4 microphotometer. Subsequently, to automate and speed up the processing, various versions of analog and digital conversion of blackening into light intensity were used (Kononovich 1962, 1963). In this case, the characteristic curve initially constructed in logarithmic coordinates was reconstructed in linear coordinates. An example of such a characteristic curve is shown in Fig. 3.24.

As mentioned above, the blackening density depends on the exposure dose determined by the illuminance and exposure time. For actual conditions, they are not interchangeable. According to the Schwarzschild equation, the relation $(E \cdot t^p)_{D=\text{const}} = \text{const}$ is valid. Nevertheless, it was found that the value of p is different not only for different photographic materials, but also for the same material developed in different developers, for the same developer and different development times, and even for the same material developed during the same time in the same developer, but with different relations between E and t determining the same exposure (Schwarzschild effect). Because of the complexity of the Schwarzschild effect, it is impossible to characterize it only by the Schwarzschild equation (Avgustinovich 1990). For long exposure times inherent in recording weak light fluxes, we have $p < 1$ or, more specifically, $p \sim 0.7$–0.9.

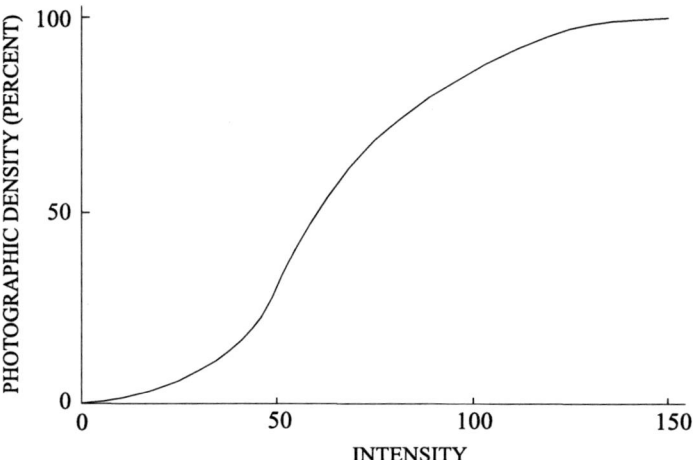

Fig. 3.24 The characteristic curve of a photosensitive layer in linear coordinates (h/L is blackening, I is intensity)

To eliminate the Schwarzschild effect, all photographic measurements of radiation intensity should be performed with the same exposure time.

When recording spectra of the nightglow of the upper atmosphere, special astronomical films having high sensitivity at exposure times were used. Nevertheless, the necessity of hypersensitization of the photographic materials used inevitably arose. In the literature, various methods of hypersensitization were repeatedly described (Mulyarchik and Petrova 1957; Breido 1973; Martynov 1977). More often, preliminary illumination with an exposure time of ~ 0.01 (s) before taking photos was used.

The inherent property of photographic materials is the wavelength dependence of the sensitivity. Therefore, the spectral calibration for the purposes of photometric processing should be performed on the same spectroscopic instrument that was used in recording the nightglow spectra and under the same conditions at which the spectra were photographed. For this purpose it is possible to use secondary light sources, such as incandescent lamps and luminophors, whose radiation spectral distributions have been determined by using other standard radiation sources.

3.3.2 Photomultipliers

Photoelectric recording was developed to record weak light fluxes. A photomultiplier tube (PMT) is an electrovacuum instrument which converts radiation into an electric signal followed by amplification of the signal. PMTs are conditionally distinguished by the type of cathodes which are operative in various spectral ranges: cesium-antimonide photocathodes in the visible spectral range (300–650 (nm)), multi-alkali photocathodes in the range 600–800 (nm), and oxide–silver–cesium photocathodes in 700–1250 (nm) (see Fig. 3.35). It is natural that various practical applications of PMTs dictated their various parameters such as overall dimensions, the photocathode size, etc.

The conditions under which small light fluxes are measured pose strong requirements on the quality of the PMTs used. Since the principle of operation of a PMT consists in consecutive amplification of the electron flux developed at the cathode by dynodes, an important part is played by the voltage distribution between the dynodes (Fig. 3.25).

The operating experience of photometers has shown that the instrument sensitivity can be increased substantially by matching the voltage divider resistors for each dynode. This matching is performed on a special bench.

A PMT can operate in one of the three modes: the mode with a dc amplifier, the mode with an ac amplifier where the input light flux is modulated, and the photon counting mode. In any mode, a dark background signal is present, which should be weakened. Therefore, the mode of high-voltage supply is also matched to the properties of the PMT. While in the mode with a dc amplifier all electrons from the photocathode are recorded, in the photon counting mode the amplifier circuit should include a pulse amplitude discriminator downstream of the PMT to cut off weak background pulses and restrict the signal amplitudes in order that the pulse counting device operated with pulses of the same amplitude.

In Poisson statistics, the variance D of a measurable quantity is numerically equal to the average number of readouts N for a chosen time t, i.e., $D = N$. Then for a useful signal $N_S = N - N_n$, and D_S can be expressed in terms of directly measured quantities as

$$D_S = D + D_n = N + N_n = (n + n_n) \cdot t,$$

where n is the average rate of counting of one-electron pulses and n_n is the rate of counting of the dark current pulses.

In photometric measurements, of great importance is the ratio of the measured signal amplitude to the root square of its variance, i.e., the signal/noise ratio:

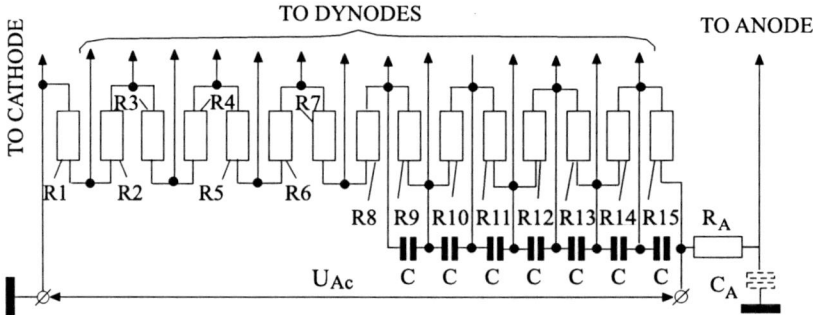

Fig. 3.25 Circuit diagram of the voltage divider of a PMT power supply

3.3 Radiation Detectors

$$\rho = \frac{N - N_t}{\sqrt{N + N_t}} = \frac{N_S}{\sqrt{N_S + 2N_t}} = (n - n_t) \cdot \sqrt{\frac{t}{n + n_t}} = n_S \cdot \sqrt{\frac{t}{n_S + 2n_t}}.$$

One-electron pulses are counted if their amplitude exceeds some threshold value or the number of electrons in a pulse is over a threshold value M_{thr}. Then the probability for a pulse to be recorded is given by the formula

$$a_1 = \alpha \cdot \frac{n_S + n_D}{n} + \beta \cdot \frac{n_n}{n},$$

where n_n is the noise component whose average amplitude differs from that of the one-electron pulse,

$$\alpha = 1 - \exp(-M_0) \cdot \sum_{M=0}^{M_{thr}-1} \frac{M_0^M}{M!}; \qquad \beta = 1 - \exp(-M_1) \cdot \sum_{M=0}^{M_{thr}-1} \frac{M_1^M}{M!},$$

where M_1 is the number of electrons in the pulse, M_0 is the same for the signal coming from the photocathode, and M is the number of electrons in the anode pulse.

The coefficients α and β can be reduced to a unified Poisson scale if we replace M_1 by M_0 in the expression for β and simultaneously change the limit of summation by setting $M_0/M_1 = \sigma$:

$$\beta = 1 - \exp(-M_1) \cdot \sum_{M=0}^{M_{thr} \cdot \sigma - 1} \frac{M_0^M}{M!}.$$

For the case of no signal, we obtain a similar expression $a_2 = a_1$ ($n_S = 0$).

Amplitude selection of this type does not violate the Poisson statistics of the number of readouts at the PMT output, since we consider the case of recording of independent events. Denoting $n_1 = n_D + n_n$, in view of the threshold, we can write

$$P_N = \frac{(a_1 \cdot n \cdot t)^N}{N!} \cdot \exp(-a_1 \cdot n \cdot t) \qquad P_{N_1} = \frac{(a_2 \cdot n_1 \cdot t)^{N_1}}{N_1!} \cdot \exp(-a_2 \cdot n_1 \cdot t)$$

with averages

$$\overline{N} = (n_S + n_D) \cdot \alpha \cdot t + n_n \cdot \beta \cdot t; \qquad \overline{N_1} = n_D \cdot \alpha \cdot t + n_n \cdot \beta \cdot t.$$

Then the signal/noise ratio is

$$\rho = n_S \cdot \alpha \cdot \sqrt{\frac{t}{n_S \cdot \alpha + 2(n_D \cdot \alpha + n_n \cdot \beta)}}.$$

If background light pulses with counting rate n_{bg} are present in the output signal, the equation for the signal/noise ratio becomes

$$\rho = n_S \cdot \alpha \cdot \sqrt{\frac{t}{n_S \cdot \alpha + 2[n_D \cdot \alpha + n_n \cdot \beta + n_{bg} \cdot (\alpha + \gamma \cdot \beta)]}},$$

where γ is a coefficient taking into account the amplitude distribution of the background pulses.

It seems that optimization of the discrimination threshold cannot be expected for $n_S \gg n_D$ and $n_S \gg n_{bg}$. If we consider the signal variance to be small compared to the variance of the background, for $n_D \gg n_{bg}$ or $n_{bg} \gg n_D$ we have

$$\rho = \varepsilon \cdot n_S \cdot \sqrt{\frac{t}{2 n_D}}.$$

For the cases $n_D \gg n_{bg}$ and $n_{bg} \gg n_D$ there correspond $\varepsilon_1 = \frac{\alpha}{\sqrt{\alpha+\beta}}$ and $\varepsilon_2 = \frac{\alpha}{\sqrt{\alpha+\gamma\cdot\beta}}$ and, obviously, $\varepsilon_1 = \varepsilon_2$ for $\gamma = 1$.

Figure 3.26 gives the family of curves $\varepsilon(M_{thr})$ (Vetokhin et al. 1986). The presence of a maximum at $\gamma \neq 0$ points to the existence of an optimum threshold for amplitude discrimination. Since ε increases with σ, the location of this threshold is stable.

The sources of noise can be subdivided into primary and secondary, interior and exterior, essentially eliminable and inherent. The thermal emission from the photocathode and first dynodes and the leakage current through the PMT anode are assigned to internal primary and essentially inherent noise sources. The optical and ionic feedbacks are considered internal secondary and essentially eliminable noise sources; they are collectively called gas-discharge phenomena. Internal primary essentially eliminable noise sources include autoelectronic emission and the ionizing radiation of the nuclei of the constituent species of the PMT constructional materials. The existence of external primary essentially eliminable sources of noise is

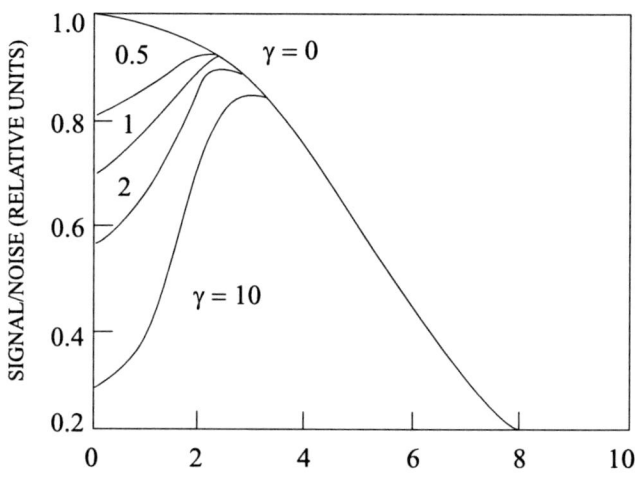

Fig. 3.26 The effect of the amplitude discrimination threshold on the signal/noise ratio in one-electron recording (Vetokhin et al. 1986)

associated with the background action of hard and optical radiations; these radiations are also capable of producing secondary effects, such as the long-term glow of the luminescent materials present in a PMT.

An important characteristic of a PMT which is used to analyze the amplitude distribution of pulses is its counting-rate curve. For a PMT acceptable for operation the counting-rate curve has three sections showing different dependences of the number of pulses on the power supply voltage.

The first section is characterized by a fast rise in the number of pulses with voltage, which is accounted for by the fact that the system amplification capability is insufficient to produce recordable pulses from each electron which has arrived at the first dynode and produced secondary electrons.

The second section of the counting-rate curve—a plateau—corresponds to the recording of each electron multiplied by the dynode system. Some rise is due to the following reasons: (1) the recording of some portion of thermionic electrons from the first dynodes; (2) the decrease of the probability of the secondary emission coefficients becoming zero; and (3) the gas-discharge phenomena gradually developing with increasing power supply voltage.

The third section, which demonstrates an abrupt increase in the number of pulses at high power supply voltages, is associated with intensely developing gas-discharge phenomena.

For the majority of photomultipliers for which the amplitude distribution of pulses follows approximately the Poisson law, the one-electron peak is accompanied by a non-Poisson additive in the region of small amplitudes; the intensity of this additive falls exponentially with increasing amplitude (Fig. 3.27). In most cases, it is accepted to choose the optimum discrimination threshold on a level A corresponding to the transition region between the exponential branch and the one-electron peak. This choice turned out to be rather successful (Vetokhin et al. 1986).

The FEU-79 photomultiplier turned out to be the most acceptable for recording weak signals by its sensitivity and noise levels and also by the amplitude distribution of the output pulses, which is shown in Fig. 3.27.

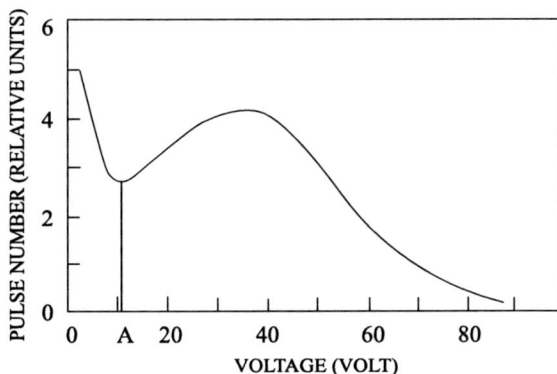

Fig. 3.27 Typical amplitude distribution of the output pulses of a one-electron PMT. A is the amplitude discrimination level (Vetokhin et al. 1986)

The dark background can be reduced by cooling the photocathode to −30 to −40(°C) (Nikonov 1973a; Vetokhin et al. 1986). Nevertheless, there is evidence that the sensitivity decreases on cooling; however, this is not always the case, including for the FEU-79.

In investigations of weak light fluxes, PMTs with a photocathode of small working area, such as FEU-64 and FEU-79, are preferred. This is essential because the image of the photometer entrance pupil, produced by the Fabry lens on the photocathode, is 5–8 (mm) in size, and this allows one to use completely the photocathode sensitive part at which the noise signal occurs. On cooling (FEU-79, FEU-83) the amplitude distribution of noise pulses changes in character: the one-electron peak disappears (Figs. 3.28–3.30). At the same time, a decrease in temperature practically does not change the spectrum of the useful signal at a weak light flux (Fig. 3.31) (Vetokhin et al. 1986).

The photomultipliers considered above were used in various photometric instruments for investigations of the airglow of the upper atmosphere.

3.3.3 Image Converter Tubes

Image converter tubes (ICTs) are vacuum radiation detectors operating in various spectral ranges. They are used to enhance the brightness of an image or to convert an image obtained in x-ray, ultraviolet, or infrared radiation into a visible image.

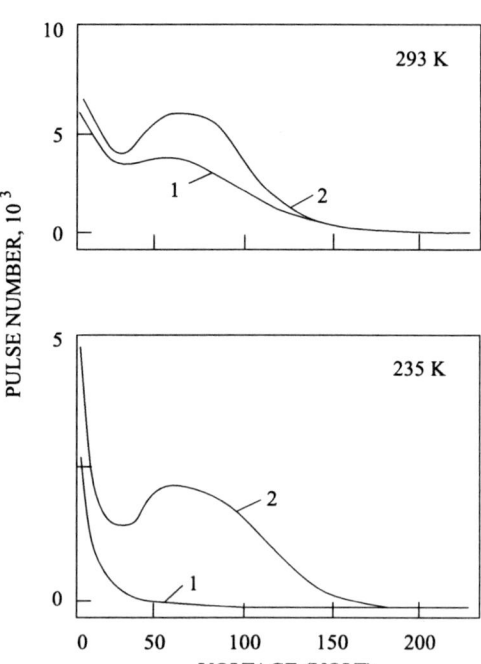

Fig. 3.28 Temperature effect on the amplitude spectrum of the dark current (1) and on the amplitude spectrum in the weak light flux mode (2) for the FEU-79 photomultiplier (Vetokhin et al. 1986)

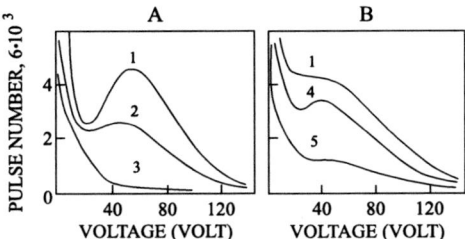

Fig. 3.29 Temperature effect on the amplitude spectrum of the dark current for the FEU-79 (**A**), FEU-62 and FEU-83 photomultipliers (**B**) at temperatures of +20 (1), +5 (2), −15 (3), −20 (4), and 40(°C) (5) (Vetokhin et al. 1986)

The idea to transfer an electron image from a photocathode on a screen by means of a uniform electric field was proposed and realized by Hulst in the 1930s. In the 1940s, a series of single-stage ICTs were developed with participation of Krassovsky, Butslov and co-workers (Butslov et al. 1978).

The possibility of visual observation of objects in infrared rays gave rise to extensive use of ICT in night-vision instruments for the observation of objects illuminated with infrared light or objects which radiate in this spectral range. In the late 1940s, ICTs were successfully employed in astronomy for photographic recording of the infrared radiation emitted from the center of the Galaxy and also for spectrographic investigations of the Earth's airglow (Krassovsky 1949, 1950a,b; Kalinyak et al. 1950). The results of these experiments stimulated the development of new ICT design versions, which substantially extended the field of their scientific applications in geophysics, nuclear physics, medicine, etc. (Berkovsky et al. 1976, 1988). The use of image converter tubes in the Soviet Union in investigations of the Earth's airglow and various astronomical objects in the near-infrared range, initiated by Krassovsky, gave essentially new scientific results. Owing to the regular spectrophotometric measurements with the use of ICTs at various stations carried out for several tens of years, unique material has been accumulated.

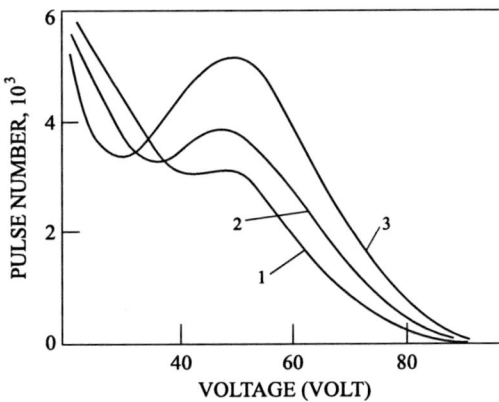

Fig. 3.30 Amplitude distribution of one-electron pulses for the FEU-62 and FEU-83 photomultipliers at temperatures of +20 (1), 0 (2), and 40(°C) (3) (Vetokhin et al. 1986)

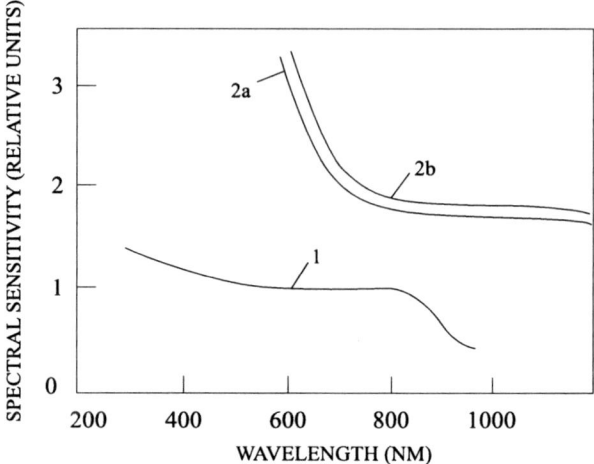

Fig. 3.31 Temperature effect on the spectral sensitivity (relative units) for FEU-79 (1), FEU-62 and FEU-83 (2a,b): -20 (1), -10 (2a), and $-40(°C)$ (2b) (Vetokhin et al. 1986)

Principle of Operation

The simplest ICT consists of a vacuum cylindrical flask; at its end faces a semitransparent photocathode and a luminescent screen are disposed (Zaidel and Kurenkov 1970; Iznar 1977). Inside the flask, between the photocathode and the screen, a unit for electron-optical focusing of electrons is placed (Fig. 3.32). The image of the object under observation (1) is projected with a lens (2) onto the photocathode (3). When the photocathode is illuminated due to photoelectric emission, an electron image of object 1 is produced in which the distribution density of photoelectrons corresponds to the light intensity distribution in the optical image. Since an accelerating voltage is applied between the photocathode and the screen, the electron image starts moving toward the screen. The (electrical or magnetic) focusing system (4) forces electrons to move from the photocathode in such a manner that all electrons from any point of the photocathode are focused, irrespective of the initial velocities and escape direction, at a certain point of a luminescent screen (5) and excite it. Thus, the electron optics forms an electron image of the object on the photocathode in the plane of the screen; the image can be detected visually or recorded photographically with an eye lens (6). In this case, the image brightness increases several tens of times. Initially single-stage ICTs were used. The type FKT-1 contact ICT (Fig. 3.33) turned out to be most efficient. In these devices, the screen was a thin mica film to which a photographic film was attached by immersion, allowing efficient use of the light radiated by the screen (Volkov et al. 1962; Shcheglov 1963). Later three-stage ICT of the U-32 type (Fig. 3.34) came into use for investigations of internal gravity waves and for interferometric temperature measurements (Semenov 1975a). In the 1970s, ICTs were equipped with built-in electron image intensifiers based on microchannel plates (MCPs). An MCP is designed as a spacer of certain diameter and

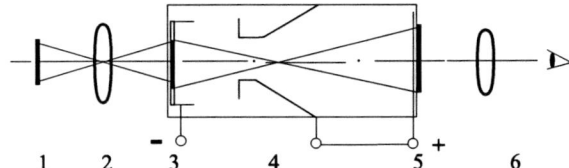

Fig. 3.32 Schematic diagram of an image converter tube. 1 – Object under observation; 2 – lens; 3 – photocathode; 4 – focusing system; 5 – screen; and 6 – eye at an eyepiece

thickness 0.4–0.5 (mm) consisting of several millions of glass tubules (channels) of diameter 10–40 (μm). Each tubule is a miniature secondary-electron multiplier with a dynode system of distribution type. Due to the potential difference between the ends of a channel, once an electron hits on its internal surface, a significant (10^3–10^4 times) multiplication of the number of electrons takes place. The use of MCP has made feasible a planar ICT with no focusing of the electron flow. This considerably reduced the ICT dimensions and therefore appreciably extended the field of their use.

Characteristics of Image Converter Tubes

Any ICT is characterized by certain parameters, a knowledge of which enables one to choose an instrument which would meet specific goals. Let us consider these parameters (Zaidel and Kurenkov 1970).

The integral sensitivity of a photocathode, φ, is the ratio of the photoelectric current I_{ph} to the incident light flux F from an incandescent lamp whose filament has a color temperature T = 2854 (K):

$$\varphi = I_{ph}/F.$$

The integral sensitivity of a photocathode to a light flux is measured in amperes per lumen.

The spectral sensitivity of a photocathode, $\varphi(\lambda)$, is the ratio of the photoelectric current I_{ph} to the incident light flux $F(\lambda)$ from a monochromatic radiation source:

Fig. 3.33 FKT-1 contact image converter tube (Volkov et al. 1962)

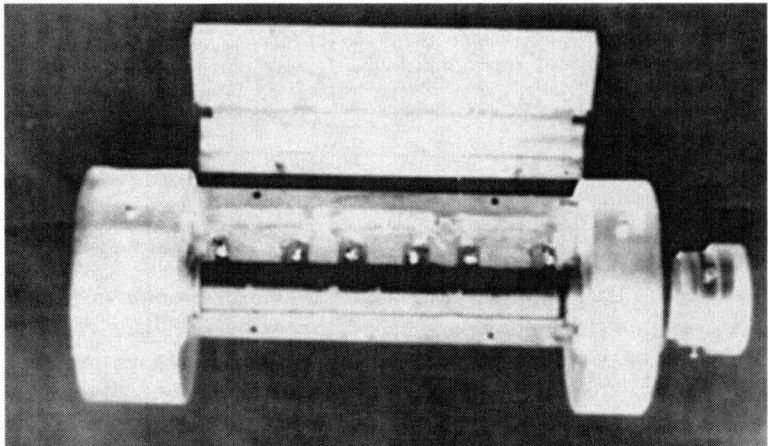

Fig. 3.34 U-32 three-stage image converter tube in the assembly unit, used in an interferometer

$$\varphi(\lambda) = I_{ph}/F(\lambda).$$

Figure 3.35 presents the spectral sensitivities $\varphi(\lambda)$ of cesium–antimony SbCs, multi-alkali SbKNa-Cs, and oxide–silver–cesium O-AgCs photocathodes. For comparison the curve of eye visibility $K(\lambda)$, the radiation intensity distribution $F(\lambda)$ for a standard radiation source of type A (black body at 2856(K)), and the transmission curves of filters for red, $\tau_r(\lambda)$, and infrared, $\tau_{ir}(\lambda)$, radiations are given.

The conversion coefficient η is the ratio of the light flux radiated by the ICT screen to the light flux incident on the ICT photocathode from an incandescent lamp whose filament has a color temperature T = 2856(K):

$$\eta = \frac{\Phi_{SCR.RAD}}{\Phi_{FL.INC}} = \pi \cdot \gamma \cdot \varphi \cdot U \cdot K_G^n,$$

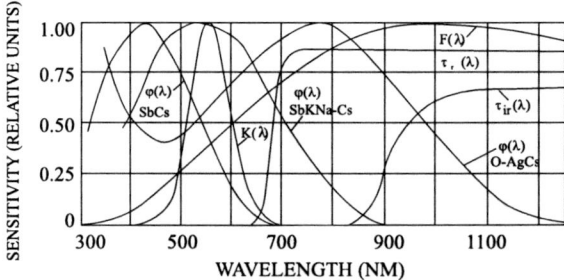

Fig. 3.35 Spectral distributions of the sensitivity $\varphi(\lambda)$ of cesium–antimony SbCs, multi-alkali SbKNa-Cs, and oxide–silver–cesium O-AgCs photocathodes. For comparison the eye visibility curve $K(\lambda)$, the radiation intensity distribution $F(\lambda)$ for a standard source of type A (black body at 2856(K)), and the transmission curves of filters for red, $\tau_r(\lambda)$, and infrared, $\tau_{ir}(\lambda)$, radiations are given

where γ is the luminous efficacy of the screen (cd · W^{-1}); φ is the integral sensitivity of the input photocathode (A · lm^{-1}); V is the operating voltage (V), K$_G$ is the stage current gain, and n is the number of amplification stages.

The maximum resolution is the number of stripe pairs per millimeter of the image of a black-and-white test pattern on a photocathode which are discernible in the image on the screen in four directions at the optimum screen brightness and good magnification of the eye lens.

The brightness coefficient η_B is the ratio of the brightness of the ICT screen, B, to the corresponding illuminance of the photocathode, E:

$$\eta_B = \frac{B}{E} = \frac{\eta}{M_{ICT}^2},$$

where η is the conversion coefficient and M_{ICT}^2 is the electron-optical magnification. This parameter allows one to estimate the efficiency of operation of an ICT in visual observation of the image.

The brightness of the dark background is the brightness of the ICT screen luminescence in the absence of illumination of the photocathode. This luminescence is caused by the photocathode thermoemission, back light-striking inside the device, and scintillations arising on bombardment of the accessory and walls inside the ICT by scattered electrons. The temperature dependence of the photocathode thermoemission current is described in general form by the expression

$$I_T = S_{PhC} \cdot A \cdot T^2 \cdot e^{\frac{-\varepsilon_A}{kT}},$$

where S_{PhC} is the area of the photocathode, ε_A is the work function of the photosensitive layer of the photocathode, k is Boltzmann's constant, T is the temperature of the photocathode, and A is a factor depending on the photocathode type.

To perform photographic recording of radiation spectra in the 700- to 1200-nm near-infrared region, image converter tubes have been used since the mid-1950s. The arrangement of these image amplifiers is described in a number of publications (Volkov et al. 1959, 1962; Shcheglov 1958, 1963; Zaidel and Kurenkov 1970; Soule 1968; Iznar 1977; Butslov et al. 1978).

3.3.4 Photoelectric Charge-Coupled Devices

Recently the radiation detectors based on semiconductor systems of image reception and transmission have been actively introduced in the practice of recording and investigation of weak emissions. These devices, combining advantages of both photographic and photoelectric methods, have extended substantially the scope of investigations of the characteristics of the Earth's airglow. They do not demand vacuum, have small dimensions, and possess high quantum yield (up to 80%). In the late 1960s, V. Boyle and H. Smith at the Bell Laboratory, USA, discovered the principle of operation of solid-state photoelectric converters based on the transfer of a localized charge. In 1970, this principle was used to develop commercial charge-coupled devices (CCDs) (Séquin and Tompsett 1975). They received wide application in electronic systems for memorization of great bodies of data, signal analysis, recording of weak light signals, etc. CCDs are produced as single line arrays (one-dimensional arrangement of light-sensitive cells) or matrices (two-dimensional arrangement of cells) (Press 1981). The use of one or another type of detector in the spectrometry of the Earth's airglow depends on the way of scanning of radiation spectra in the focal plane of the spectrometer.

These matrix devices, capable of producing images much as this is done in photography and also in digital form applicable for real-time processing, in many cases have successfully replaced the earlier used instruments. However, to operate CCD efficiently, the experimenter must understand many features of these devices. In this section we briefly outline the available information about CCD.

Practically, a charge-coupling device consists of a set of miniature MOS (metal-oxide-semiconductor) capacitors (Abramenko et al. 1984). An MOS capacitor is a metal electrode deposited on an oxidized surface of p-silicon (Fig. 3.36). If a potential positive with respect to the substrate is applied to the metal electrode, the majority carriers ("holes") in the silicon, which appear in p-silicon under the action of light, will be repelled from the electrode and leave the adjacent semiconductor layer. As a result, a potential well is formed at the oxide–semiconductor (silicon) interface in which electrons (minority carriers of p-silicon) pile up. The charge accumulated in this hole is proportional to the intensity of the radiation incident on the detector and to the recording time.

In charge-coupled devices, MOS capacitors are mounted on the common layer of oxide and semiconductor so close to each other that their potential wells appear practically "coupled". In this case, the charge of minority carriers builds up at the place where the potential is greater, i.e., the charge behaves as if it flows in the deepest part of the potential well. This effect allows one to control the charge transfer along the oxide–semiconductor interface from one metal electrode to another. The charge movement along the transport channel can be realized due to a certain potential distribution at the metal electrodes. This charge movement reminds "a fire

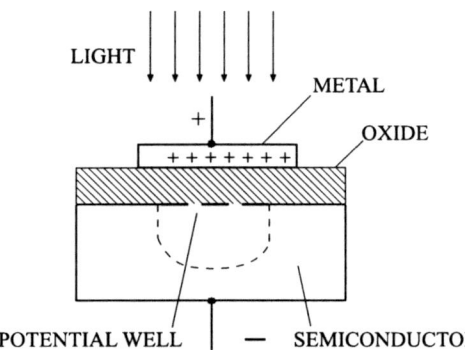

Fig. 3.36 Block diagram of a MOS capacitor (cross-section) (Lebedeva 1986)

chain"—handing over of a bucket filled with water for fire suppression, i.e., charges transfer from one potential well into the next one throughout the channel.

Let us consider the charge transfer by the example of three-phase voltage application to electrodes. Figure 3.37 shows three storage cells of a CCD. Each of them has three electrodes, and the third electrodes of the cells are connected to the same busbar. The dashed line beneath the semiconductor characterizes the potential distribution for different voltages applied to the electrodes. In the mode of charge storage beneath electrodes 1, 4, and 7 (see Fig. 3.36a) a potential V_2 is applied to these electrodes which is greater in amplitude than the potential V_1 applied to other electrodes. When a greater potential V_3 ($V_3 > V_2 > V_1$) is applied to electrodes 2, 5, and 8, the charges move in deeper potential wells beneath these electrodes (see Fig. 3.36b). After their transport, the storage mode is actuated again (see Fig. 3.36c, d), but charges have already moved for the dimension of one electrode compared to the situation shown in Fig. 3.36a. Thereafter all processes repeat, and the charges reach the edge of the detector from which they are extracted.

Like any radiation detector, a CCD has a light-signal and a frequency-contrast characteristic which relate the output signal to the illuminance in its plane. The light-signal characteristic is linear over a wide range of light fluxes (Fig. 3.38).

The best CCD detectors have a dynamic range of $\sim 5 \cdot 10^3$. In the light-signal characteristic, three points are selected for which the detector parameters are measured. For the middle part, it is convenient to define an integrated voltage sensitivity as the ratio of the variation of the light-signal amplitude to the variations of the illuminance ($V \cdot lx^{-1}$), irradiance ($V \cdot W^{-1} \cdot cm^{-2}$), and light exposure ($V \cdot J^{-1} \cdot m^{-2}$)

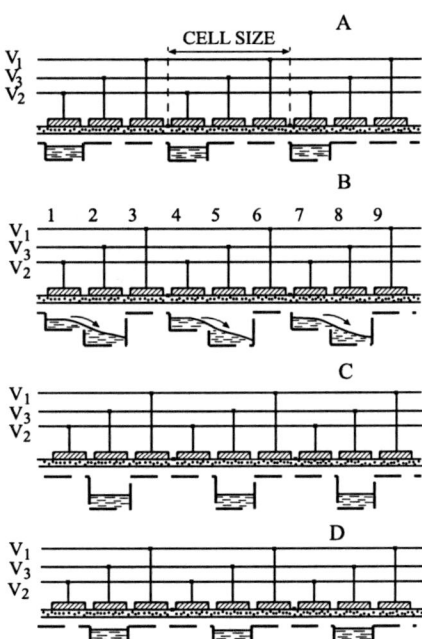

Fig. 3.37 Sketch of the charge transfer in a CCD with three-phase control. (**A**) Mode of charge storage in cells 1, 4, and 7; (**B**) mode of charge transfer; (**C**) mode of charge storage in cells 2, 5, and 8; (**D**) mode of charge storage in cells 2, 5, and 8 at potential V_2. Preparation for the next charge transfer event; $V_1 < V_2 < V_3$ (Lebedeva 1986)

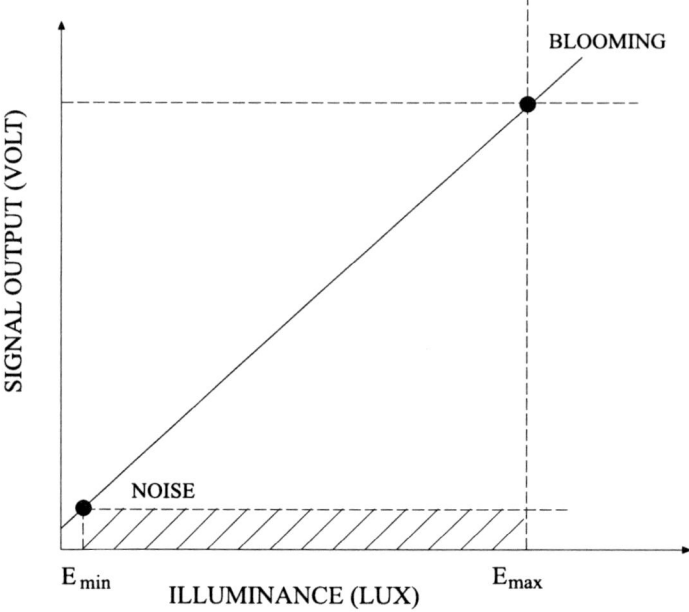

Fig. 3.38 Light-signal characteristic of a CCD (Press 1991)

in a given spectral range (Press 1991). Since the light-signal characteristic is linear, it is possible to measure merely the ratio of the light-signal amplitude to the illuminance or to the exposure. The integral sensitivity is 10^{-2}–10^{-3} ($V \cdot lx^{-1} \cdot s^{-1}$), and it is determined by design and technological factors.

The bottom portion of the light-signal characteristic gives an idea of the operation of the CCD in the threshold region, i.e., at extremely low values of illuminance. If the incident light is completely cut off, some dark signal will remain at the output which is characterized by the voltage and the uniformity of the voltage distribution. The dark signal is formed by the deterministic and fluctuation noises of the CCD. The deterministic noise, in turn, consists of the clock noise (as a rule, from the pulses applied to the gate of the reset transistor) and geometric dark noise. The geometric dark noise represents the nonuniformity of the dark current distribution. It arises due to thermal carrier generation giving rise to a dark current and associated with surface and volumetric generation–recombination centers. The occurrence of these centers is related in the main to the deposition of metals at places of disruption of the crystal lattice.

The pattern of a dark signal taken from the CCD output shows that the signal is nonuniform, the local outbursts in overwhelming majority (80(%)) coincide with the locations of metal deposition places. The dark signal parameters characterize the technological level of manufacture of the CCD and the temperature of the device during operation. On going from the dark mode to low levels of illuminance, the geometric imperfection of the signal often increases since new carrier generation centers—recombinations—are actuated and the existing centers become more active under the action of light. Local luminescence, which is most pronounced at low illuminance levels, should also be attributed as geometric imperfection. Four types of local luminescence are possible: the first one is caused by clocking the output register, the second is related to the prebreakdown state of p–n junctions in the input–output unit, the third arises on puncturing MOS transistors in the output devices, and the fourth is associated with luminous spots.

The first type of luminescence is observed in matrix CCDs in modes of low illuminance and long charge accumulation time. The output register operates during accumulation, and the clocking of its phase electrons can give rise to long-wave photons absorbed by the elements of vertical shift registers located most closely to the output register. The luminosity falls rapidly with distance from the register output (10–20 cells) and depends on the phase shift, frequency, and risetime of the clock pulses of the output register, and also on temperature. The generation of photons by p–n junctions in a prebreakdown state is a phenomenon well known for a long time. Its observation in a CCD has shown that in this case a near-infrared radiation is emitted to which the silicon CCDs are rather sensitive. In a TV image taken from a light-protected CCD, a luminous region can be seen which extends from the output device to which a too high voltage has been applied. A similar situation arises as a channel of the output MOS transistor is punctured. When the voltage at the drain becomes much greater than the voltage at the gate, conditions arise for avalanche generation of electron–hole pairs and emission of photons in the near-infrared region. The most serious concern is luminous spots which appear as a result of local leakages between the phases and between a phase and the substrate. Since the area of a matrix CCD is great, the probability of the appearance of spots is significant. One can distinguish luminous spots from white dots on the image, which are caused by an excess dark current, by cooling the CCD. A way for preventing the appearance of luminous spots is to reduce the phase voltages and to control (reduce) the substrate bias during charge accumulation.

The fluctuation noise of a CCD includes two basic components: the carrier transport noise and the noise of the output device. The transport noise results from the inefficient transport of charge carriers and is directly proportional to the degree of inefficiency. For highly efficient modern CCD this noise is of minor importance. The main problem is the noise of the output device. It is proportional to the root square of the capacitance of the reading unit and is essentially unavoidable. A way of reducing this noise is to optimize the design of output devices. The fluctuation noise is characterized by the voltage of the noise component, which is the root-mean-square value of the time-dependent signal fluctuation at a fixed load resistance in the given frequency range and in the absence of radiation. By measuring the voltage of the noise component and using the signal-to-noise ratio one can determine the parameters that characterize the threshold behavior of a CCD: the threshold irradiance, the threshold flux, and the threshold exposure. The threshold illuminance of a photosensitive surface corresponds to the voltage of the signal equal to the voltage of the noise time component. The threshold flux is defined similarly. The threshold exposure dose is the product of the threshold illuminance by the charge accumulation time. The noises and the threshold parameters of a CCD are measured for one cell of the device, obtaining a time series of measurements (the longer the series, the more accurate the resulting parameter value).

The top portion of the light-signal characteristic reflects two parameters: the maximum output signal, which is characterized by the saturation voltage of the output signal and corresponds to filled potential wells, and the response of the CCD to local overillumination, i.e., the overfilling of potential wells at sites of increased illuminance. Uniform overillumination of a CCD, causing overfilling of all potential wells in the photosensitive region, is not so harmful. It is suppressed with the use of neutral optical filters or by varying the charge accumulation time. Local overilluminations are much more difficult to suppress, and the struggle against them is a major problem. At local illuminations, blooming spots and stripes extended in the direction of transport of charge packets are observed on a TV image. The degree of blooming can be estimated by various methods, for instance, by measuring the width of the bloomed image and relating it to the width of the projected stripe or by determining the ratio of the excess local illuminance to the illuminance that causes the filling of the potential well.

Blooming can be eliminated by preventing the propagation of excess carriers from the site of local overillumination by modifying the circuit or design of the device. The circuit-based antiblooming is reduced to intensification of the recombination of excess carriers at surface local centers. This method is used in matrix CCD with frame transfer; in this case, the CCD design remains unchanged and it suffices to change the sequence of performing pulses. Some electrodes in the accumulation cells are switched from a depleting to an inverting bias during the reverse motion along the line array. The majority carriers from the nearest stop-channel regions emerge beneath these electrodes. As the device is switched to a depleting bias, excess carriers fill the surface traps and recombine in them. Switching occurs several times during the reverse motion along the array. In each cell where the charge exceeds the maximum storage capability of the cell, the surface centers are filled alternately with electrons and holes. This results in more intense recombination of the excess charge at the surface traps.

A design-based solution of the problem calls for complication of both the design and the technique of manufacturing of CCD.

Blooming has the appearance of smearing: an image spreads vertically downward as though potential wells were slightly overfilled. However, the origin of this phenomenon is other than that of blooming. In matrix CCDs with frame transfer, smearing occurs due to the flare light during the transfer of charge packets from the photosensitive section to the memory section. Though the accumulation time during charge transfer is small, the addition of carriers of the subsequent frame to the carriers of the preceding one is possible at high illuminances, and this results in smearing of the image. This effect can be inhibited by increasing the transfer frequency. At a transport frequency of 4 (MHz), which is attained in the best matrices, the phenomenon of image smearing is suppressed. In matrix CCD with line-frame transfer, smearing is caused by that the carriers generated in the bulk by long-wave photons get in the vertical shift registers. In matrices of this type, as distinct from matrices with line-frame transfer, smearing shows up at low illuminances.

The frequency-contrast characteristic (FCC) of a CCD determines the resolving ability of the device (Fig. 3.39). It describes the response of the CCD to an optical signal in the form of a system of stripes (meander) with a certain spatial frequency. To obtain the FCC of a CCD, a stripe test chart, whose step sets the spatial frequency of CCD operation, is projected on the CCD. It is convenient to use the ratio of the given spatial frequency f to the maximum frequency f_{max} at which the step of the test chart is equal to the step of the photosensitive cells of the CCD. In an ideal CCD, upon projecting a test chart, rectangular signals should appear at the output; in actuality, the shape of signals is flattened, and at high spatial frequencies it resembles a sine curve. The ratio of the total amplitude of the output signal during the transfer of a mira of given spatial frequency or of an electrically introduced signal of given frequency to the total amplitude of the signal from a large detail of the image or of an electrically introduced signal (which fills no less than five potential wells) determines the modulation transfer coefficient K_M. Practically, to characterize a CCD, it suffices to measure the modulation transfer ratio at any fixed frequency, more often at a half of the maximum spatial frequency, because, according to the Nyquist theorem, a radiation detector transfers without distortion spatial frequencies not exceeding a half of the maximum frequency. In projecting higher spatial frequencies, the Moire effect shows up in the form of false images.

The form of the FCC of any CCD is determined by three factors: the geometry of the active part of the device, i.e., the dimensions and step of photosensitive cells (integration FCC, expressed by the modulation transfer ratio K_{M1}); the efficiency of carrier transfer (transfer FCC, expressed by the modulation transfer ratio K_{M2}), and

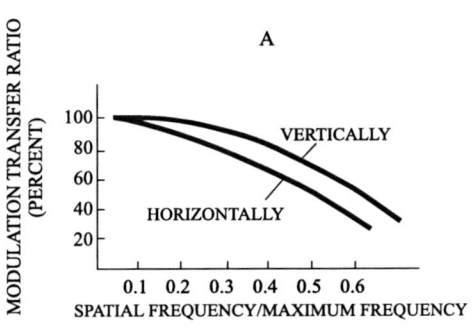

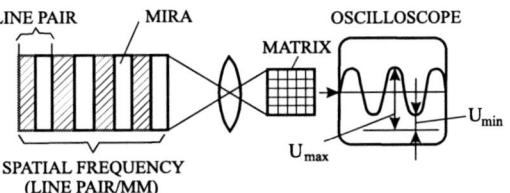

Fig. 3.39 The frequency-contrast characteristic of a CCD (**A**) and the method of measuring the modulation factor (**B**) (Press 1991)

the diffusion of carriers generated by long-wave photons (diffusion FCC, expressed by the modulation transfer ratio K_{M3}).

For rectangular photosensitive cells of size s, arranged in line with a step p, the dependence of K_{M1} on spatial frequency f is described by the expression

$$K_{M1} = \frac{\sin\left(\frac{f}{f_{max}} \cdot \frac{\pi \cdot s}{2p}\right)}{\frac{f}{f_{max}} \cdot \frac{\pi \cdot s}{2p}},$$

where $f_{max} = 1/2p$. If the cells are densely spaced (s = p), then $K_M = 0$ for $f = 2f_{max}$. For the maximum spatial frequency we have $K_{M1} = 64(\%)$. Such an FCC is inherent, for example, in a CCD with frame transfer. It should be borne in mind that an FCC along the vertical is different from that along the horizontal. If the step of cells differs from their size (as in matrices with line-frame transfer where photosensitive columns alternate with shift registers along the horizontal), the form of the FCC is different. Here, the modulation transfer ratio for the maximum frequency makes ~85(%). It should be stressed that the expression for K_{M1} refers not to the FCC phase component, but to the best case where the test chart stripes and the CCD cells coincide in phase. If the phases do not coincide, the resolving ability of the device decreases.

The fall of an FCC depending on the number of transfer events, n, and on the inefficiency of one transfer event, ε, is determined by the expression

$$K_{M2} = \exp\left\{-n \cdot \varepsilon \cdot \left[1 - \cos\left(n \cdot \frac{f}{f_{max}}\right)\right]\right\}.$$

For the modulation transfer ratio measured at a half-maximum of the spatial frequency this expression takes the form

$$K_{M2} = \exp(-n \cdot \varepsilon).$$

For a CCD with a greater number of cells and high carrier transfer efficiency, the diffusion of carriers generated by long-wave photons is a serious factor restricting the resolving ability. The long-wave photons (0.66–1.1 (μm)) penetrate deep in the bulk of the semiconductor since they are weakly absorbed by silicon. The generated carriers diffuse in all directions. Most of them get in a proper photocell, but a significant amount of them appear in the adjacent photocells. Thus a crosstalk noise arises, and the resolving ability decreases. The closer the photocells are located, the larger is the photon penetration depth, and the larger the diffusion length of the carriers, the steeper the fall of the FCC and the lower the modulation factor K_{M3}. It can be determined by the relation

$$K_{M3} = \frac{\operatorname{ch}\left(\frac{d}{L_0}\right)}{\operatorname{ch}\left(\frac{d}{L}\right)}, \quad \frac{1}{L^2} = \frac{1}{L_0^2} + (2\pi f)^2,$$

where L_0 is the diffusion length of carriers in silicon and d is the distance from the place of light absorption to the depleted layer.

If the input optical image contains spatial frequencies above the Nyquist limit and if the FCC of the detector beyond this limit has a rather great value of K_M, the output signal will contain stray components giving rise to false images. Since in this case converted signals of higher spatial frequencies fall in the working band, this effect is referred to as a reduction of spatial frequency. At a frequency f much greater than the Nyquist frequency f_N, distinct false components appear. If the optical image is moved relative to the detector, the moving false images are perceived as Moire effects. To exclude the reduction and, at the same time, to retain good resolution in the transmission band ($0 < f < f_N$), it is necessary to prefilter the higher spatial frequencies with an optical filter with a near-rectangular characteristic (Fig. 3.40). However, it is well known that the FCC of a light detector having a rectangular point spread function is represented by a curve which, at high frequencies, gets to the region of negative values of the contrast transfer ratio, and the effect of "false resolution" is observed at these frequencies. However, such a filter is physically unrealizable. Usually, in each concrete case, it is possible to find a more or less successful compromise between the requirements of sufficient sharpness and a minimum of false signals.

One more important property of a CCD, along with the characteristics discussed above, is the spectral sensitivity S_λ. For a CCD with the MOS storage photocells having polysilicon electrodes, the spectral sensitivity maximum is near 800–900 (nm) (Fig. 3.41) (Abramenko et al. 1984; Khomich et al. 1999, 2002; Khomich and Zheleznov 2000; Khomich and Schutter 2000). Photodiode matrices show the widest spectral sensitivity band. They provide for recording radiation in the optical range 0.3–1 (µm).

To reduce the noise and attain the highest possible sensitivity, it is necessary to cool the CCD with thermoelectric microcoolers. At temperatures from −20 to −40(°C) the dark current can be reduced by two or three orders of magnitude. In this case, it is necessary to take into account that on cooling the transmission band of a CCD is appreciably shifted toward the short-wave spectral range.

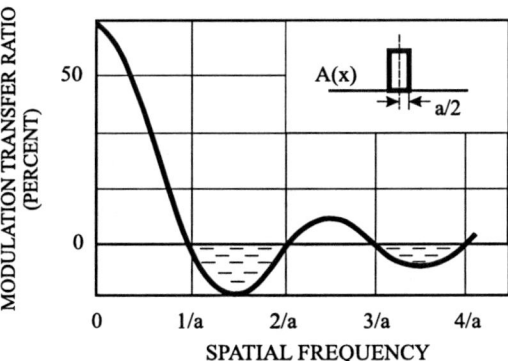

Fig. 3.40 Frequency-contrast characteristic K_M for a rectangular function of scattering $A(x)$. f is the spatial frequency

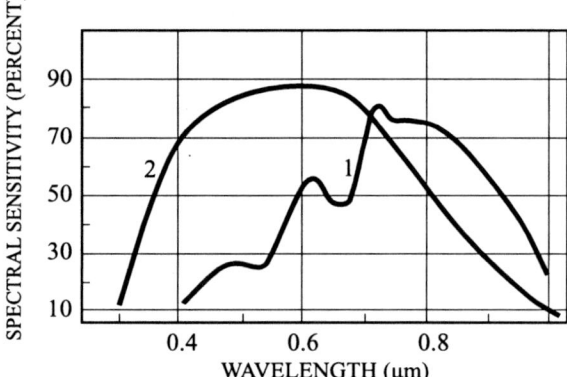

Fig. 3.41 Spectral dependence of the quantum efficiency. 1 – Illumination of the device on the side of the electrode structure; 2 – illumination of the device from behind

When recording the airglow of the higher atmospheric layers with the purpose of studying various nonstationary processes related both to photochemical mechanisms of initiation of various emissions and to internal gravity waves propagating through emission layers, two-dimensional image detectors (matrices) are most often used (Nakamura et al. 1999; Ammosov and Gavrilieva 2000; Gavrilieva and Ammosov 2001; Semenov et al. 2002; Bakanas et al. 2003). Figure 3.42 shows a radiation recording unit with an INSTASPEC IV CCD manufactured by the ORIEL firm (1999) which was mounted on an SP-50 spectrograph. The CCD matrix consists of 1024×256 cells. The pixel size is $26 \times 26 (\mu m)$. To attain the best signal/noise ratio, the CCD matrix is cooled to $-40(°C)$ with a thermoelectric cooler.

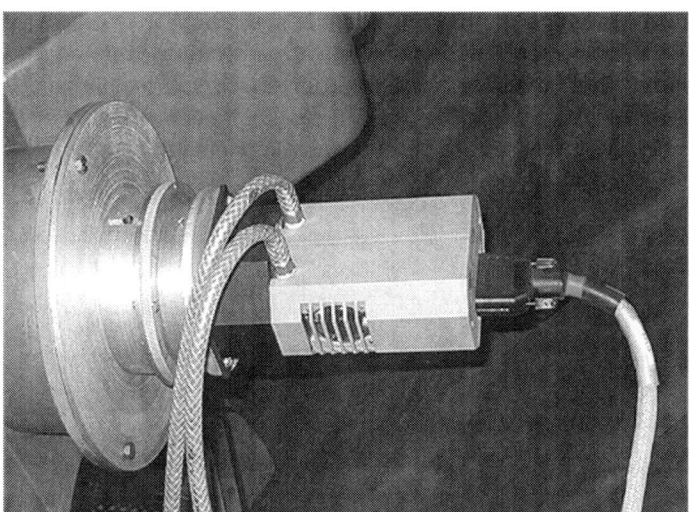

Fig. 3.42 The radiation recording unit with a CCD, mounted on an SP-50 spectrograph

Figures 1.8–1.10 give examples of spectrograms. An example of recorded spatial wave disturbances of hydroxyl emission is shown in Fig. 5.1.

Devices of this type are capable of recording radiation from all elements of the image projected on their photosensitive surface. They ensure high quantum efficiency and good time resolution and are capable of recording low-contrast images and, most importantly, to perform real-time computer processing of the image. The small dimensions and mass of the matrix allow its easy cooling aimed at reducing noise signals. As a rule, the thermoelectric refrigerators harnessing the Peltier effect are used for these purposes. The cooling temperature of the matrix is set and controlled with high precision by means of a computer.

3.4 Cooling of Radiation Detectors

In recording weak light fluxes it is necessary to provide the greatest possible signal-to-noise ratio. The dark background of a photoelectric detector can be efficiently reduced by decreasing the temperature of the photosensitive part of the detector. For the conditions under which the nightglow of the upper atmosphere is recorded only a limited number of photoelectric detector types are usable. These are photomultipliers operating in various spectral regions, image converters, and charge-coupled photoelectric devices. For devices of these types, which operate at wavelengths shorter than $1.2\,(\mu m)$, a significant reduction of the dark background is attained at temperatures below $-20(°C)$. The photoresistors used in the more long-wave region demand deeper cooling, which can be provided by using liquid nitrogen ($-196(°C)$). In this case, the specificity of the photoresistors consists only in special designs of liquid-nitrogen cryostats with which they are furnished. Here we do not consider the photoelectric detectors that are used for measurements of the far-infrared spectral region with use of rockets.

3.4.1 Cooling with Solid Carbonic Acid

The first and simplest cooling agent for the cathodes of photoelectric detectors was solid carbonic acid formed on abrupt expansion of a condensed gas. The temperature of the resulting "dry ice" is $-78(°C)$. For cooling, a detector is placed in a cold store. To improve the contact between the external surface of the photocathode and the cooler walls, a thin rubber sheet is placed in the space between them. Alcohol is added to maintain a permanent thermal contact of the cooling dry ice with the metal construction. The cooler is enclosed in a heat insulator, usually made of cellular plastic. The inlet window near which the photocathode is placed is made as a degasified cylindrical glass flask $\sim 50\,(mm)$ in diameter with two plane-parallel sides. The distance between these parallel glasses is $\sim 20\,(mm)$. This vacuum window prevents moisture penetration to the cooled photocathode. Nevertheless, to reliably protect the photocathode against fogging during long-term measurements, the space between the exit part of the camera lens and the vacuum window was encapsulated with a thin-walled metal cylinder. Its one end was rigidly and hermetically fastened on the foam envelope of the cooler, while the other could slide by its interior surface, greased with lubricant, over a specially made exterior casing mounted on the lens of the spectrograph camera. This design made it possible to properly focus the spectrum image onto the photocathode of the image converter tube. With a carefully manufactured cold storage, about 400–500 (cm^3) of dry ice suffices for 1–2 (h) of work at room temperature of the ambient air. This method was employed for image converter tubes used to photograph nightglow spectra and for photomultipliers with silver–oxygen–cesium photocathodes operated in the near-infrared range.

To use photoresistors for recording radiation in the wavelength range over 1.3 (μm), liquid nitrogen providing the temperature $-196(°C)$ is required.

3.4.2 Thermoelectric Cooling

The radiation detectors that demand low temperatures and small powers can be successfully cooled with thermoelectric coolers which depend for their operation on the Peltier effect (Antonov et al. 1969). A cooling device of this type is based on an ordinary thermal cell which consists of two series-connected semiconductor arms, one possessing electron (n) and the other hole conductance (p). When the thermal cell carries a dc electric current in the n-to-p direction, a temperature difference arises between the switching plates that connect the thermal cells and power supply leads. Heat is absorbed at one junction and released at the other junction. If the temperature at the hot junction is kept constant by heat removal, the temperature of the cold junction will decrease. At a given current, the temperature decrement depends on the thermal load of the junction. This load includes the heat input from the surrounding medium, the heat coming from hot junction due to the thermal conduction of the thermal cell arms, and the Joule heat released in the thermal cell arms as they carry a current.

The operation of a thermal cell is substantially rendered by Joule heat. Actually, if the absorbed Peltier heat P is proportional to the first power of current, i.e., $Q_1 = P \cdot I \cdot t$, the Joule heat released in a thermal cell is proportional to the squared current: $Q_2 = I^2 \cdot R \cdot t$. Thus, as a first approximation, about half the Joule heat falls on the cold junction of the thermal cell, and this reduces the effect of cooling. Therefore, the maximum temperature decrement at the cold junction is attained at a certain optimum current. In real thermoelectric coolers, as the current is changed by $\pm 10(\%)$ of its optimum value, the degree of cooling remains practically unchanged.

For the operation of a real thermoelectric cooler, forced cooling of the hot junction is necessary. In the investigations of orographic perturbations of the upper atmosphere emissions performed at the Institute of Atmospheric Physics of the Russian

Academy of Sciences, air cooling was used for the radiators of the thermoelectric coolers that were employed in the two-stage photometer considered in Sect. 3.2.3 (see Fig. 3.15) (Voronin et al. 1984). When IGW investigations were carried out by means of a set of photometers, water-cooled thermoelectric coolers were used. The INSTASPEC IV photosensitive charge-coupled detectors (see Fig. 3.42) manufactured by the ORIEL company (1999) operate with water cooling. The necessary temperature is set and controlled with a computer.

3.4.3 Household Refrigerators

The absorptive and compressive refrigerating units used in household refrigerators are convenient coolers for radiation detectors (Veinberg and Vain 1974). For this purpose, the cooled detector (or several detectors, since the cold store is generally rather large in volume and rather powerful) is put in a specially made metal case and placed in the cold store. This facility as a whole demands careful heat insulation and airtight packing (it is desirable that it would meet the requirements placed on vacuum operating conditions) to preclude moisture penetration inside the device, since the system is in permanent operation because of the long time (6–8 (h)) it takes for steady-state conditions ($-18(°C)$) to establish inside of the cold store. To reduce the pressure difference between the air inside the refrigerator and the surrounding air, a simple device in the form of a rubber boot was used which was put on the tube outgoing from the interior of the refrigerator and varied in volume as the air pressure decreased on cooling. Devices of this type were successfully employed both for multistage photometers in recording wave processes by means of hydroxyl emission (Potapov et al. 1976, 1978) and orographic perturbations (Sukhodoev et al. 1989a,b) and for Fabry–Perot interferometers (see Fig. 3.20) in recording 557.7- and 630-nm emissions from atomic oxygen (Semenov 1975a). The long-term experience of these devices has shown that the humidity inside the cold store starts exceeding the normal level only after continuous operation of the refrigerating unit for several months.

3.5 Methods and Conditions of Measurements

3.5.1 Measurements at Given Directions

The zenith angle ζ and horizontal range X corresponding to an emission layer under consideration which is located at an altitude Z above the Earth surface are determined by the formulas

$$\text{tg}\zeta = \frac{(1+Z/R_E)\cdot\sin\left(\dfrac{X/R_E}{1+Z/R_E}\right)}{(1=Z/R_E)\cdot\cos\left(\dfrac{X/R_E}{1+Z/R_E}\right)-1},$$

$$\text{tg}\left(\frac{X/R_E}{1+Z/R_E}\right) = \frac{(1+Z/R_E)\cdot\sqrt{1+\left[1-\dfrac{1}{(1+Z/R_E)^2}\right]\cdot\text{tg}^2\zeta}-1}{(1+Z/R_E)\cdot\sqrt{1+\left[1-\dfrac{1}{(1+Z/R_E)^2}\right]\cdot\text{tg}^2\zeta}+\text{tg}^2\zeta},$$

where R_E is the Earth's radius.

When performing measurements aimed at location of IGW sources, the azimuths of the lines of sight are determined by the formulas (Fig. 3.43)

$$\sin(A_{vis}-A_s) = \frac{R_0}{\rho}\cdot\sin\zeta,\ \cos(A_{vis}-A_s) = \frac{1}{2}\left(\frac{r}{\rho}+\frac{\rho}{r}-\frac{R_0^2}{r\cdot\rho}\right),$$

$$\rho = \sqrt{R_0^2+r^2-2R_0\cdot r\cdot\cos\zeta},\ \rho = r\cdot\cos(A_{vis}-A_s)+\sqrt{R_0^2-r^2\cdot\sin^2(A_{vis}-A_s)},$$

$$\sin\zeta = \frac{\rho}{R_0}\cdot\sin(A_{vis}-A_s),\ \cos\zeta = \frac{1}{2}\left(\frac{R_0}{r}+\frac{r}{R_0}-\frac{\rho^2}{r\cdot R_0}\right),\ R_0 = Z\cdot\sqrt{\left(\frac{\tau}{\tau_g}\right)^2-1},$$

where R_0 is the radius of the circle in which the emission layer regions at an altitude of about 90 (km) viewed by the photometer are arranged; r is the distance from the observation point to the IGW source; ρ is the distance from the observation point to the sighting region (along the layer); ζ is the zenith angle counted from the line between the wave source and the observation point; τ is the measured wave period;

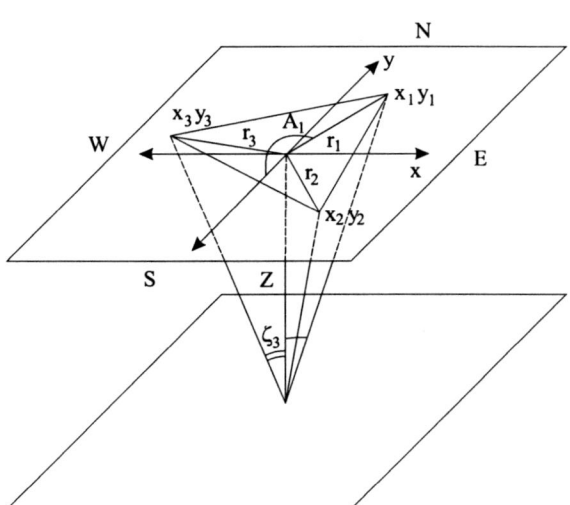

Fig. 3.43 Scheme of three-azimuthal measurements (Potapov et al. 1978). See explanations in the text

$\tau_g = 5.2\,(\text{min})$ is the Brunt–Väisälä period; A_{vis} is the azimuth of the line of sight at the observation point, and A_s is the azimuth of the source relative to the observation point.

3.5.2 Optical Recording of Wave Processes in the Upper Atmosphere

The results of investigations of the fast variations of airglow emissions, including hydroxyl emission, at some one line of sight, have long indicated the apparently oscillating character of the intensity and rotational temperature variations of the emissions in OH bands (Taranova 1967). This type of variations was earlier presumed to be caused by internal gravity waves (Krassovsky 1957a,b; Krassovsky and Shagaev 1974a,b). Therefore, to detect perturbations of this type which propagate in the upper atmosphere at altitudes of about 90 (km) and to determine their velocity and direction of propagation, it is necessary to perform measurements simultaneously at several sites of the emission layer that are spaced along the Earth surface at some hundreds of kilometers, the order of the IGW length. This measuring technique using three observation points has already been known in ionospheric (Mitra 1952) and seismic investigations (Mathesn 1966; Eiby 1980).

For measuring the parameters of the airglow emissions arising at altitudes of about 90 (km), the side of the basic triangle was chosen to be about 200–500 (km) (see Fig. 3.43). This corresponded to about $55°–60°$ zenith angles of the directions of the optical axes of the instruments used (Shagaev 1975).

Based on Fig. 3.43, one can derive the necessary relations for the velocity and azimuth of propagation of IGWs. The distance r_i from the origin of coordinates (observation point) to the ith vertex of the triangle is given by

$$r_i = Z \cdot \mathrm{tg}\zeta_i.$$

If the optical axis of the device is at an angle α relative to the line directed to the zenith, the center of the ellipsoidal region of the horizontal layer under observation is off the optical axis. Therefore, r_i is determined as

$$r_i = \frac{Z \cdot \mathrm{tg}\zeta_i}{1 - \sec^2\zeta_i \cdot \sin^2\frac{\alpha}{2}}$$

or, in view of the curvature of the Earth surface, as

$$r_i = Z \cdot \mathrm{tg}\zeta_i \cdot \left[1 + \frac{Z}{R_E} \cdot \left(1 - \frac{Z}{R_E}\right) \cdot \mathrm{tg}^2\zeta_i\right],$$

where R_E is the Earth's radius. Therefore, the coordinates of the triangle vertices or, in general, of any observation point are given by

$$x_i = -r_i \cdot \sin A_i = -Z \cdot \mathrm{tg}\zeta_i \cdot \sin A_i,$$
$$y_i = -r_i \cdot \cos A_i = -Z \cdot \mathrm{tg}\zeta_i \cdot \cos A_i,$$

where A_i is the azimuth of the point (x_i, y_i). The triangle side length l_{ij} is determined by the formula

$$l_{ij}^2 = r_i^2 + r_j^2 - 2r_i \cdot r_j \cdot \cos(A_i - A_j).$$

Here the x-axis is directed to the east and the y-axis to the north; the azimuth A_i is counted clockwise from the south point S from 0 to 360°.

Spectral analysis of the observed variations of emission parameters at each point (x_i, y_i) for the pth component of frequency ω_p (period $\tau_p = 2\pi/\omega_p$) yields the amplitude

$$B(\tau_p) = (\mathrm{Re}^2(p) + \mathrm{Im}^2(p\varphi))^{1/2}$$

and the phase

$$\varphi(\tau_p) = \mathrm{arctg}(-\mathrm{Im}/\mathrm{Re}),$$

where Re and Im are the real and imaginary parts of the Fourier transform. For each point (x_i, y_i), the phase $\varphi(\tau_p)$ at a given point in time can be represented in the form

$$\varphi_i = \varphi_o + k_x \cdot x_i + k_y \cdot y_i,$$

where φ_o is the phase at a point O; k_x and k_y are the components of the wave vector. In view of the relation

$$|k| = \sqrt{k_x^2 + k_y^2},$$

the wave velocity and azimuth are $C = \dfrac{\omega}{|k|} = \dfrac{2\pi}{\tau |k|}$ and $A^* = \mathrm{arctg}\left(-\dfrac{k_y}{k_x}\right)$, respectively.

However, the direction of counting of the angle A^* differs from the astronomical azimuth $\overline{A}$ and is related to this quantity as

$$\overline{A} - A^* = 270.$$

Therefore, we have

$$\overline{A} = \mathrm{arcctg}(-k_y/k_x) = \mathrm{arctg}(k_x/k_y).$$

Thus, the quantities k_x and k_y for the basic triangle are determined by the system of equations

$$k_x \cdot x_1 + k_y \cdot y_1 + \varphi_o = \varphi_1;$$
$$k_x \cdot x_2 + k_y \cdot y_2 + \varphi_o = \varphi_2;$$
$$k_x \cdot x_3 + k_y \cdot y_3 + \varphi_o = \varphi_3.$$

3.5 Methods and Conditions of Measurements

For the number of observation points n more than three, using the least square method, one can determine k_x and k_y by the system of conditional equations

$$k_x \cdot \Sigma x_i^2 + k_y \cdot \Sigma x_i y_i + \varphi_o \cdot \Sigma x_i = \Sigma x_i \varphi_i;$$
$$k_x \cdot \Sigma x_i y_i + k_y \cdot \Sigma y_i^2 + \varphi_o \cdot \Sigma y_i = \Sigma y_i \varphi_i;$$
$$k_x \cdot \Sigma x_i + k_y \cdot \Sigma y_i + \varphi_o \cdot n = \Sigma \varphi_i.$$

Hence,

$$k_x = D_x/D; k_y = D_y/D,$$

where D_x, D_y, and D are the determinants of the system:

$$D_x = \begin{vmatrix} \varphi_1 y_1 1 \\ \varphi_2 y_2 1 \\ \varphi_3 y_3 1 \end{vmatrix}, \qquad D_y = \begin{vmatrix} x_1 \varphi_1 1 \\ x_2 \varphi_2 1 \\ x_3 \varphi_3 1 \end{vmatrix}, \qquad D = \begin{vmatrix} x_1 y_1 1 \\ x_2 y_2 1 \\ x_3 y_3 1 \end{vmatrix}.$$

These relations allow one to derive simple expressions for the azimuth $\overline{A}$ and wave velocity C if the basic triangle is equilateral, its center being at the origin and one of its vertices on a coordinate axis. The side of the triangle is equal to $r \cdot \sqrt{3}$. If the first vertex lies on the y-axis, the coordinates of the vertices are

$$x_1 = 0, \qquad y_1 = r;$$
$$x_2 = r \cdot \frac{\sqrt{3}}{2}, \qquad y_2 = -0,57 r;$$
$$x_3 = -r \cdot \sqrt{3}, \qquad y_3 = -0,57 r.$$

The wave phase is given by

$$\varphi = t \cdot 2\pi/t_i,$$

where t_i is the point in time at which the wave passes through the ith vertex of the triangle. Thus,

$$tg\overline{A} = \frac{\sqrt{3}}{2} \cdot \frac{t_2 - t_3}{t_1 - 0.5 \cdot (t_2 - t_3)}, \qquad C = \frac{3 \cdot r \cdot \cos \overline{A}}{2 \cdot [t_1 - 0.5 \cdot (t_2 - t_3)]}.$$

When vertex 3 lies on the x-axis, we have

$$x_1 = 0.5 \cdot r, \ y_1 = r \cdot \frac{\sqrt{3}}{2};$$
$$x_2 = 0.5 \cdot r, \ y_2 = -r \cdot \frac{\sqrt{3}}{2};$$
$$x_3 = -r, \ y_3 = 0.$$

In this case,

$$\mathrm{tg}\overline{A} = \frac{2\sqrt{3}}{3} \cdot \frac{0.5 \cdot (t_1 + t_2) - t_3}{t_1 - t_2}, \quad C = \frac{\sqrt{3} \cdot r \cdot \cos\overline{A}}{t_1 - t_2}.$$

These relations can be easily transformed for a given orientation of the basic triangle.

Let us now consider the effect of the variations of the initial quantities Z, ζ_i, and A_i on the calculated IGW parameters: the velocity C and azimuth A. As follows from the above relations,

$$\frac{dx}{x} = \frac{dZ}{Z} + \frac{d\zeta}{\sin\zeta \cdot \cos\zeta} - \mathrm{ctg}A \cdot dA,$$

$$\frac{dy}{y} = \frac{dZ}{Z} + \frac{d\zeta}{\sin\zeta \cdot \cos\zeta} + \mathrm{tg}A \cdot dA,$$

$$\frac{dC}{c} = \frac{d\tau}{\tau} + \sin^2\overline{A} \cdot \frac{dk_x}{k_x} + \cos^2\overline{A} \cdot \frac{dk_y}{k_y},$$

$$d\overline{A} = 0.5 \cdot \sin 2\overline{A} \cdot \left(\frac{dk_x}{k_x} - \frac{dk_y}{k_y}\right).$$

From the expressions for k_x and k_y we have

$$\frac{dC}{C} = \frac{d\tau}{\tau} - \frac{dD}{D} + \sin^2\overline{A}\frac{dD_x}{D_x} + \cos^2\overline{A}\frac{dD_y}{D_y},$$

$$d\overline{A} = 0.5 \cdot \sin 2\overline{A} \cdot \left(\frac{dD_x}{D_x} - \frac{dD_y}{D_y}\right).$$

For actual observation conditions, the errors in various quantities have some features. The effect of the error of determination of the coordinates of the basic triangle vertices on the error in the phase is given by the formula

$$\Delta\varphi_i = k_x \cdot \Delta x_i + k_y \cdot \Delta y_i.$$

The inaccuracy of the determination of $\Delta\zeta_i$ is practically determined by the method of positioning of the optical axes of the photometers or mirrors. Therefore, if measurements are performed at the same zenith distances ζ_i, the corresponding measurement errors will be practically identical. This is equally valid for the determination of the azimuths A_i; that is, we have

$$|\Delta\zeta_1| = |\Delta\zeta_2| = |\Delta\zeta_3| \text{ and } |\Delta A_1| = |\Delta A_2| = |\Delta A_3|.$$

The error in the altitude of the emission layer is due to the average altitude $Z = 87$ (km) that has to be taken because of lack of data about the true altitude of the layer during observations. However, this error is insignificant, since the data of rocket measurements (Shefov and Toroshelidze 1974, 1975) show a random spread on the average within ± 5 (km).

3.5 Methods and Conditions of Measurements

Thus, if we use an equilateral basic triangle, we have

$$\frac{\Delta C}{c} = \frac{\Delta \tau}{\tau} + \left(\frac{\Delta Z}{Z} + \frac{\Delta \zeta}{\sin\zeta \cdot \cos\zeta}\right)\left[\cos^2 A_3 + \frac{1}{3}\cos(\overline{A} + A_3) \cdot \cos(\overline{A} - A_3)\right]$$
$$+ \frac{4}{3} \cdot \sin A_3 \cdot \sin \overline{A} \cdot \sin(\overline{A} - A_3) \cdot \Delta A,$$

$$\Delta \overline{A} = \frac{4}{3}\left(\frac{\Delta Z}{Z} + \frac{\Delta \zeta}{\sin\zeta \cdot \cos\zeta}\right) \cdot \sin(\overline{A} + A_3) \cdot \cos(\overline{A} - A_3)$$
$$+ \left[1 + \frac{8}{3}\sin A_3 \cdot \cos\overline{A} \cdot \sin(\overline{A} + A_3)\right] \cdot \Delta A,$$

where A_3 is the azimuth of the third vertex of the basic triangle. Setting $A_3 = 60°$, we obtain

$$\frac{\Delta C}{c} = \frac{\Delta \tau}{\tau} + \cos^2 \overline{A} \cdot \left(\frac{\Delta Z}{Z} + \frac{\Delta \zeta}{\sin\zeta \cdot \cos\zeta}\right) + \frac{2}{\sqrt{3}} \sin \overline{A} \cdot \sin(\overline{A} - 60) \cdot \Delta A,$$

$$\Delta \overline{A} = \frac{4}{3}\cos(\overline{A} - 30) \cdot \sin(\overline{A} + 30) \cdot \left(\frac{\Delta Z}{Z} + \frac{\Delta \zeta}{\sin\zeta \cdot \cos\zeta}\right)$$
$$+ 2\left[1 + \frac{1}{\sqrt{3}}\sin 2(\overline{A} + 30)\right] \cdot \Delta A,$$

whence, on the average, the errors are given by

$$\frac{\Delta C}{c} = \frac{\Delta \tau}{\tau} + 0.5 \cdot \left(\frac{\Delta Z}{Z} + \frac{\Delta \zeta}{\sin\zeta \cdot \cos\zeta}\right) + 0.4 \cdot \Delta A,$$

$$\Delta \overline{A} = 0.6 \cdot \left(\frac{\Delta Z}{Z} + \frac{\Delta \zeta}{\sin\zeta \cdot \cos\zeta}\right) + 2 \cdot \Delta A.$$

Thus, if the error of the determination of the angles of positioning of the instrument mirrors (i.e., $\Delta \zeta_i$ and ΔA_i) is about $1°$, for the error in measuring an altitude of about 5 (km), $\zeta_i \sim 60°$, and $Z \sim 87$ (km), we have

$$\frac{\Delta C}{C} = \frac{\Delta \tau}{\tau} + 5(\%).$$

Estimation of the maximum values of the errors yields

$$\frac{\Delta C}{C} = \frac{\Delta \tau}{\tau} + 9(\%), \quad \Delta \overline{A} = 6°.$$

The above reasoning and conclusions show the ways of increasing the accuracy of determination of IGW parameters and suggest that they can be determined reliably.

3.5.3 Sky Scanning

This measuring technique can involve terrestrial and satellite measurements.

Measurements of the former type were performed with a photometer, multichannel in some cases, whose optical axis was moved, following a certain program, along almucantarats at given zenith angles. Roach et al. (1953) were among the first to perform such measurements for studying the spotty structure of the 557.7-nm emission. To study the behavior of the 630-nm red oxygen emission at the Haute Provence observatory, a multichannel photometer was developed (Barbier 1955).

An example of this type of device developed and built at the Institute of Atmospheric Physics of the USSR Academy of Sciences is described elsewhere (Jordjio 1961). In this case, the zenith angles were $10°$, $20°$, $30°$, $40°$, $50°$, $60°$, and $70°$, and switching from one almucantarat to another occurred when the photometer was oriented to the south. The time of complete survey of the sky was about 16 (min). In this photometer, provision was made to record the light from a standard luminophor source every 30 (min). With this device mounted at the Vannovsky station of the Physicotechnical Institute of the Turkmen Academy of Sciences (near Ashkhabad), a great body of data were obtained about the spatial distribution of the intensities of the 557.7- and 630-nm emissions from atomic oxygen (Korobeynikova et al. 1966, 1968, 1972; Korobeynikova and Nasyrov 1972).

In carrying out investigations of orographic perturbations in the mesopause near the Caucasian ridge, the mesopause region was scanned with a two-channel photometer over the vertical plane in 11 discrete directions (see Sect. 5.2).

When the airglow of the upper atmosphere is recorded with instruments erected on a satellite, the scanning of the firmament occurs due to the motion of the satellite.

3.5.4 Scanning of a Spectrum

In optical radiation measurements a need arises to consecutively record a spectrum at the entrance of the spectrophotometer by moving the spectrum image along the exit slit of the instrument. This scanning of a spectrum demands that some conditions should be fulfilled in order that the resulting intensity distribution over the spectrum not be distorted during the measuring process by inherent variations of some emissions.

The first condition is that the resolution should be high enough to detect important details of the spectrum under investigation which are used to determine characteristics of the upper atmosphere.

This imposes the requirement on the speed of scanning of the given section of the spectrum to minimize the distortions of the spectral intensity distribution that are caused by time variations of the intensity during the scanning of the given spectrum.

The scanning speed, in turn, is determined by the sensitivity of the radiation detector and, hence, by the time constant of the recording device.

The observance of these requirements is of particular importance in recording interferograms of emissive line profiles which are used to determine the Doppler temperature.

At the Zvenigorod station, to provide for rapid (some minutes) recording of hydroxyl emission spectra, a DFS-14 spectrometer was employed.

3.5.5 Use of Fiber Glass Tubes

The fiber glass tubes are convenient to transfer a light flux from an optical device to a detector when these units are difficult to interface immediately. This problem also arises when the focal surface of the spectrometer is curved, while the surface of the radiation detector is flat.

The modules of image converter tubes that have fiber glass tubes on the photocathode and on the screen efficiently correct the curvature and the distortion of the transferred image. This enables one to increase considerably the effective area of the screen and to obtain an image of greater dimensions (about 30 (mm)) on the screen. However, the sensitivity of an ICT as a whole depends on the structure of the fiber glass tubes and on their matching to one another. For quite obvious reasons, the cross-sections of various fiber glass tubes consisting of numerous fibers cannot be entirely identical in structure. But even if some fiber glass tubes are not too different from each other, their imperfect matching in series-connected modules can affect substantially the resulting amplification of the system (Kapani 1967; Trofimova et al. 1972; Sattarov 1973; Veinberg and Sattarov 1977; Avdoshin 1990; Kravchuk et al. 1990).

Fiber glass tubes are usually regular in structure. The structure quality is determined by the technology of manufacturing of fiber glass bundles. Typical fiber diameters are 7–15 (μm). The thickness of the fiber envelope makes 0.15–0.25 of its outer radius.

The propagation of light in fibers is determined by the Straubel invariant, according to which

$$n_1 \cdot \sin u_1 = n_c \cdot \sin u_c = n_2 \cdot \sin u_2 = \sqrt{(n_c^2 - n_u^2)} = A_0,$$

where n_1, n_2, n_c, and n_i are the refraction indices of the medium at the inlet and outlet of the fiber, the material of the core and the insulating envelope, respectively; u_1, u_c, and u_2 are the inclination angles of the ray to the axis prior to the fiber glass tube inlet, inside the fiber glass tube, and at its outlet, respectively.

If for the medium outside a fiber glass tube $n_1 = 1$, the nominal numerical aperture $A_0 = \sin u_0$, where u_0 is the nominal aperture angle of a ray bundle in air (relative to the fiber).

This implies that if for the medium at the outlet $n_2 < n_c$, complete internal reflection of a ray bundle from the outlet end face is possible. From the Straubel invariant it follows that $\sin u_2 = 1$ for $n_2 = A_0$. If $n_2 < A_0$, there occurs vignetting of the light flux due to the return of some bundles reflected from the outlet end face. This, naturally, results in a decrease of the flux passed through the fiber glass tube.

If a conical bundle of rays with an aperture angle over u_0 is incident on the inlet of a fiber glass tube, the geometric transmittance of the fiber glass tube is less than unity, since some rays experience vignetting caused when they hit the fiber glass tube lateral surface at an angle j smaller than the critical angle of complete internal reflection

$$j_k = \arcsin \frac{n_u}{n_c}.$$

Skew rays penetrating in a fiber glass tube propagate so that they pass at a distance b from the fiber glass tube. Thus,

$$b = \sqrt{1 - (\sin u_0 / \sin u_1)^2}.$$

Therefore, the greatest inclination angle of a ray passing through a fiber glass tube without vignetting depends on the distance b between the ray and the fiber axis and is determined by the relation

$$\sin u = \sqrt{[(n_c^2 - n_u^2)/(1 - b^2)]} = A_0/\sqrt{1 - b^2}.$$

Thus, for the TF5 glass of the fiber core ($n_c = 1.755$) and K17 glass of the envelope ($n_u = 1.516$) the critical angle providing complete internal reflection $j_k = 60°$. At a numerical aperture in air $A_1 = 1$, light passes through the fiber glass tube, but only the circumferential part of the fiber glass tube manufactured of TF5 and K17 glasses, which makes only 81(%) of the fiber glass tube cross-section, participates in passing rays with $u = 90°$. The working cross-section of the fiber glass tube increases with decreasing inclination angle, and if $u \leq u_0$, all rays undergo complete internal reflection.

The nominal numerical aperture A_o of a fiber glass tube depends on the wavelength of the light incident on the inlet end face. Thus, for a wavelength of 400, 550, 650, and 850 (nm) we have $A_o = 0.57, 0.54, 0.53$, and 0.52 (nm), respectively. The angle $2u_o$ is the cone angle of the rays entering the fiber glass tube.

In butt-jointed fiber glass tube, the propagation of light will be determined by the degree of coincidence of the cross-sections of individual fibers. Increasing the number of butt-joints appreciably attenuates the transmitted radiation.

Assuming that the interior part of a fiber has a round cross-section, δ is the thickness of the fiber envelope related to its diameter, and Δ is the mutual displacement of two fibers, and that the fiber transmission function is rectangular, we can estimate the transmittance P by the relation

$$P = \frac{2}{\pi} \cdot \left[\frac{\pi}{3} - \sqrt{3} \cdot \frac{\Delta}{2(1-\delta)} + \arccos \frac{\Delta}{2(1-\delta)} + \arccos \frac{1-\Delta/2}{1-\delta} - 2\arccos \frac{1-\Delta}{\sqrt{3}} \right.$$
$$\left. - \left[\frac{\Delta}{2(1-\delta)} \cdot \sin \left(\arccos \frac{\Delta}{2(1-\delta)} \right) + \frac{1-\Delta/2}{1-\delta} \cdot \sin \left(\arccos \frac{1-\Delta/2}{1-\delta} \right) \right] \right].$$

3.5 Methods and Conditions of Measurements

As follows from calculations, a significant effect is rendered by the thickness of the fiber envelope in which stray light scattering takes place. This is of particular concern in recording weak radiation.

When the luminophor of the ICT screen is butt-jointed with a fiber glass tube, light is emitted from the luminophor within an angle $\omega_0 \sim 180°$ and the fiber aperture ω is $\sim 60°$. The solid angle ratio for this case is

$$\eta = \frac{1 - \cos(\omega/2)}{1 - \cos(\omega_0/2)},$$

whence for $\omega \sim 60°$ and $\omega_0 \sim 180°$ we have $\eta \sim 0.13$, while for $\omega \sim 60°$ and $\omega_0 \sim 120°$ $\eta \sim 0.27$. Thus, for butt-jointed n modules of the same transmittance P the effective transmittance is given by

$$P_{eff} = P^{2n-1}.$$

As a result, for a three-stage ICT of the EP-16 type, the transmittance can vary from 0.05 to 0.5 for $\delta \sim 0.1$–0.2.

Adjustment of the image of the spectrum of the recorded radiation source having small linear dimensions on the ICT screen to the photosensitive elements of the CCD should also affect the transfer of fine weak details of the image. When matching the ICT screen and the matrix pixels (which are rectangular in shape and 15–20 (μm) in size and are spaced at about 5 (μm)) either by placing them in contact or by transferring the image with a lens not changing the image scale, a weak image of small objects can be attenuated substantially on its reproduction with the CCD matrix.

When fiber units are butt-jointed, the resolving ability of the instrument also changes. There is an empirical formula for the resolving ability v_{rm} of n fiber glass tube units, each having a resolving ability v_{mi} (Trofimova et al. 1972; Veinberg and Sattarov 1977):

$$v_{nm} = \left(\sum_1^n v_{mi}^{-7}\right)^{-0.143}.$$

For the typical values of envelope thickness $\delta \sim 0.1$–0.4 the resolving ability is not over 30.

Should the need arise to perform measurements in directions which differ from the direction of the optical axis of the spectroscopic device, in particular simultaneously in several different directions, a fiber optics can be used. When doing this, however, it is impermissible to bend the fiber glass tube when a need arises to change the line of sight by rotating its inlet end face.

3.5.6 Airborne Measurements

Airborne measurements offer certain opportunities. Investigations of this type were conducted during the solar eclipse of February 15, 1961, in the region of Rostov-on-Don. At an altitude of flight of 10 (km), the sky brightness on the opposite side of the Sun at the zenith viewing angle during the maximum eclipse phase was approximately an order of magnitude lower than that with no eclipse (Shouyskaya 1963). This was equivalent to the conditions of twilight observations on the Earth surface when the depression angle of the Sun beneath the horizon was $\sim 6°$. Considerable difficulties in this type of measurements arise due to inevitable vibrations of the airplane.

Besides, airborne measurements were performed to determine the spatial variations of the characteristics of airglow emissions over a small area of the Earth, for instance, in the region of a mountain ridge in studying the effect of orographic disturbances (Semenov et al. 1981; Shefov et al. 1983; Shefov and Pertsev 1984).

3.5.7 Rocket Measurements

Rockets were the first space vehicles used in the study of the Earth's airglow. With their help altitude distributions of emission intensity were measured at altitudes above the rocket flight altitudes. The main difficulty in these measurements was related to the use of photometers with interference light filters whose spectral widths, in particular at the initial stages of investigation, were responsible for a considerable contribution of the continuum to the measured intensities of discrete emissions. Another difficulty was encountered in monitoring the orientation of the rocket in flight. An important problem was associated with calibration of the measured data.

Rocket measurements were carried out on many testing grounds (Fig. 3.44) by means of various types of rockets (Grinberg et al. 1979). Table 3.6 presents the geographic (φ, λ) and geomagnetic (Φ, Λ) coordinates of the places of rocket launches, and also the geographic coordinates (φ_{con}, λ_{con}) of the conjugate region of the atmosphere for the altitude $Z = 270\,(km)$.

Notwithstanding the above problems, the published results of all rocket measurements of various emissions have been used for statistical data processing. When a certain body of information of rocket investigations was accumulated, it was initially thought that there was a large spread in the measured parameters. However, subsequently it was found that the available data involved a broad spectrum of variations of the intensity, temperature, altitude, and thickness of emission layers caused by diurnal, seasonal, tidal (lunar), and long-term variations among which there were those associated with solar cycles. A great many measurements of altitude distributions of many emissions from molecular and atomic oxygen were performed under the ETON program (Greer et al. 1986; McDade et al. 1986a,b). Some results were involved in empirical models of the variations of emissions from hydroxyl (87 (km)), sodium (92 (km)), molecular oxygen (95 (km)), and atomic oxygen at 557.7 (nm) (97 (km)) and 630 (nm) (270 (km)) (Semenov and Shefov 1996, 1997a,b,c,d, 1999; Fishkova et al. 2000, 2001a,b). They are presented in Chap. 4.

3.5 Methods and Conditions of Measurements

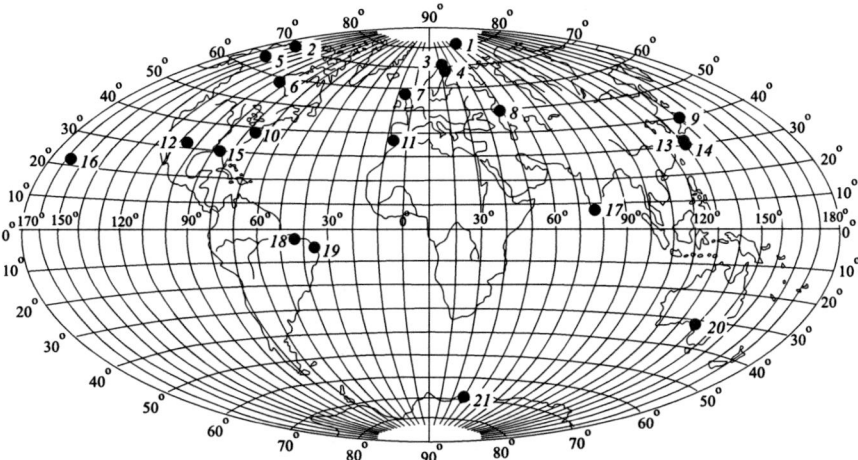

Fig. 3.44 Rocket ranges from which rockets were launched to study the characteristics of the Earth's airglow emissions. *1* through *21* are the numbers of stations. Their names and coordinates are given in Table 3.6

Table 3.6 Rocket range on which investigations of altitude distributions of the airglow emissions were performed

No[a]	Station	Country	φ (deg)	λ (deg)	Φ (deg)	Λ (deg)	Z = 270 (km)	
							φ$_{con}$ (deg)	λ$_{con}$ (deg)
1	Heiss Island	USSR	80.6°N	58.4°E	71.3°N	156.2°E	60.8°S	95.9°E
2	Fort Barrow	USA	71.3°N	203.4°E	68.6°N	241.3°E	61.7°S	152.5°E
3	Andoya	Norway	69.8°N	10.8°E	68.7°N	110.5°E	63.0°S	63.8°E
4	Esrange	Sweden	68.6°N	21.0°E	65.8°N	117.0°E	59.5°S	66.3°E
5	Poker Flat	USA	64.9°N	212.0°E	64.6°N	256.3°E	60.4°S	166.7°E
6	Fort Churchill	Canada	58.8°N	265.8°E	68.7°N	322.6°E	76.4°S	223.8°E
7	South Uist	England	57.4°N	252.6°E	61.2°N	80.0°E	61.7°S	32.7°E
8	Volgograd	USSR	48.7°N	45.8°E	42.9°N	125.1°E	36.9°S	63.8°E
9	Uchinoura	Japan	39.6°N	139.7°E	29.3°N	205.1°E	20.8°S	133.8°E
10	Wallops Island	USA	37.8°N	284.5°E	49.5°N	352.1°E	61.6°S	280.4°E
11	El Arenocilo	Spain	37.0°N	352.7°E	41.7°N	74.4°E	45.8°S	12.3°E
12	White Sands	USA	32.4°N	254.0°E	41.3°N	317.5°E	50.5°S	239.6°E
13	Kagoshima	Japan	31.5°N	131.1°E	20.6°N	198.2°E	12.6°S	128.1°E
14	Utinuara	Japan	30.3°N	131.3°E	19.4°N	198.5°E	11.6°S	128.4°E
15	Eglin	USA	30.0°N	273.0°E	40.9°N	239.3°E	52.8°S	265.3°E
16	Barking Sands	USA	22.0°N	200.9°E	21.5°N	265.3°E	22.9°S	191.3°E
17	Thumba	India	8.5°N	76.8°E	1.0°S	146.2°E	2.1°S	78.0°E
18	Natal	Brazil	5.9°S	324.8°E	3.8°N	32.5°E	13.5°S	324.5°E
19	Alcantara	Brazil	2.5°S	44.4°E	8.0°N	24.8°E	18.4°S	317.8°E
20	Woomera	Australia	31.0°S	137.0°E	41.2°S	209.9°E	50.9°N	147.4°E
21	Syowa	Japan	69.0°S	39.6°E	69.7°S	77.7°E	64.7°N	343.6°E

[a] See Fig. 3.44

3.5.8 Satellite Measurements

Satellite measurements of characteristics of the airglow emissions have both advantages and disadvantages. On the one hand, they allow one to obtain global distribution of the different characteristics of the atmosphere which indicate latitudinal and longitudinal variations for rather short time intervals. In this case, in sighting from a spacecraft there is an opportunity of seeing a panorama of the luminous atmosphere. On the other hand, to obtain statistically regular variations demands long-term measurements which encounters certain difficulties. Besides, in sighting along the horizon, the integrated intensity along the limb is measured which is an average for a large range of latitudes and longitudes and, hence, for a large interval of local times. Therefore, in twilight periods, the line of sight inevitably passes both through atmospheric regions still being in a shadow and through regions illuminated by the Sun. To determine the vertical distribution of the emission rate, it is necessary to solve an inverse problem using a number of assumptions (Sect. 1.4.4).

The procedure of satellite measurements also depends on whether they are performed by an automatically operated instrument or by an astronaut.

When an observer (instrument) moves over the Earth surface, its position, specified by coordinates (geographic latitude, geographic longitude, local time) and an altitude, continuously varies according to the law of motion. For an orbital motion, the time–space coordinates should be calculated based on the orbit elements and the running time. Thus, if we consider an unperturbed elliptical orbit, it is specified by the following elements (Abalakin et al. 1976; Abalakin 1979): the major semiaxis of the orbit, a; the eccentricity e; the inclination of the orbital plane to the equatorial plane, i; the longitude of the ascending node, Ω; the angular distance from the ascending node to the perigee, ω; and the epoch mean anomaly M_0. The quantity Ω usually refers to the vernal equinoctial point.

These elements are determined by the initial conditions of the satellite motion: the velocity V_0 and the distance from the Earth center, r_0. The circular velocity is given by

$$V_{circ} = \sqrt{\frac{k^2 \cdot m_E}{r_0}} = 7905.36 \cdot \sqrt{\frac{R_E}{r_0}}, (m \cdot s^{-1}).$$

Here $k^2 \cdot m_E = 3.986005 \cdot 10^{14}$ $(m^3 \cdot s^{-2})$ is the geocentric gravitation constant equal to the product of the gravitation constant k^2 by the Earth's mass, m_E. The equatorial radius of the Earth, R_E, is 6378.140 (km).

If the initial velocity V_0 is directed horizontally, we have

$$a = \frac{r_0}{2 - \left(\frac{V_0}{V_{circ}}\right)^2} = \frac{R_E}{\frac{2}{1 + \frac{Z}{R_E}} - \left(\frac{V_0}{7.9}\right)^2},$$

3.5 Methods and Conditions of Measurements

where V_0 is expressed in kilometers per second. The initial eccentricity is determined by the formula

$$e_0 = \left(\frac{V_0}{V_{circ}}\right)^2 - 1.$$

In this case, the initial point is the perigee and the perigee orbital altitude is given by $Z_{per} = r_0 - R_E$ and the apogean altitude by $Z_{apog} = a_0 \cdot (1 + e_0) - R_E$.

The motion period depends only on the semiaxis a_0 and is determined by the formula

$$P_0 = \frac{2\pi \cdot a_0^{3/2}}{\sqrt{k^2 \cdot m_E}} = 84.489 \cdot \left(\frac{a_0}{R_E}\right)^{\frac{3}{2}} \text{ (minute)}.$$

If the zenith angle ζ_0 of the direction of the initial velocity is not equal to 90°, then the satellite orbit is elliptical with the same major semiaxis a_0 (it does not depend on the direction of velocity), but the eccentricity of the orbit, e_ζ, is greater and the perigee is displaced by the angle $\delta\omega_\zeta$ relative to the initial position of the satellite. For this case, the orbit elements are determined by the following formulas (Bronshten et al. 1981):

$$e_\zeta^2 = e_0^2 + (1 - e_0^2) \cdot \cos^2\zeta; \quad \text{tg}(\delta\omega_\zeta) = (1 + e_0) \cdot \frac{\text{tg}\zeta}{e_0 \cdot \text{tg}^2\zeta - 1}; \quad r_{per} = r_0 \cdot \frac{1 - e}{1 - e_0};$$

$$r_{apog} = r_0 \cdot \frac{1 + e}{1 - e},$$

where e_0 is the eccentricity for $\zeta = 90°$. In this case, the perigee altitude Z_{per} is less than the initial altitude. We have an approximate equality $Z_0 - Z_{per} \approx r_0 \cdot (e - e_0)$. For example, for $Z_0 \sim 200$–300 (km) and $\zeta \sim 89°$ we obtain that $Z_0 - Z_{per} \sim 100$ (km), and the orbital motion of a satellite becomes impossible because of its drag in the lower atmosphere.

If we set the azimuth A_{init} (from the south point) of the initial direction of the horizontal velocity and the geographic coordinates—latitude φ_{init} and longitude λ_{init}—of the initial zero point, the orbit elements will be determined by the relations

$$\cos i_0 = -\cos\varphi_{init} \cdot \sin A_{init}; \quad \sin\omega_0 = \frac{\sin\varphi_{init}}{\sin i_0};$$

$$\cos\omega_0 = -\frac{\sin\varphi_{init} \cdot \cos A_{init}}{\sin i_0}$$

$$\sin(\Omega_{ginit}^0 - \lambda_{init}) = \frac{\sin\varphi_{init} \cdot \sin A_{init}}{\sin i_0}; \quad \cos(\Omega_{ginit}^0 - \lambda_{init}) = -\frac{\cos A_{init}}{\sin i_0}.$$

Here the longitude of the ascending node, Ω_{ginit}^0, at the zero point is counted, like the longitude λ_{init}, from the Greenwich meridian. Since the orbital plane retains its orientation in space, we have $\Omega_{ginit}^0 = \Omega_0 - S_0$, where Ω_0 is the longitude of the ascending node relative to the vernal equinoctial point, and S_0 is the sidereal time at the Greenwich meridian at the time zero (see Sect. 1.4.5).

The epoch mean anomaly is determined as

$$M_0 = n \cdot (\tau_{init} - \tau_{per}),$$

where τ_{init} is the initial time (epoch), τ_{per} is the point in time at which the satellite passes through the perigee, and the mean motion is given by

$$n_0 = \sqrt{\frac{k^2 \cdot m_E}{a_0^3}} = \frac{631.3482}{a_0^{3/2}}, (s^{-1}),$$

where a_0 is expressed in kilometers, or by

$$n_0 = \frac{1.2394 \cdot 10^{-3}}{a_0^{3/2}}, (s^{-1}),$$

where a_0 is expressed in units of the Earth's equatorial radius.

Thus, the mean anomaly is determined as

$$M = n \cdot (\tau - \tau_{init}) + M_0.$$

The eccentric anomaly E is described by Kepler's equation

$$E - e \cdot \sin E = M.$$

This equation is solved by the iteration method

$$E_1 = E_0 + \frac{M + e \cdot \sin E_0 - E_0}{1 - e \cdot \cos E_0},$$

until a required accuracy is achieved. Since the quantities M and E are expressed in degrees, we have

$$E_1 = E_0 + \frac{M + \frac{180}{\pi} \cdot e \cdot \sin E_0 - E_0}{1 - e \cdot \cos E_0}.$$

Thereafter the true anomaly v is calculated:

$$\operatorname{tg}\frac{v}{2} = \sqrt{\frac{1+e}{1-e}} \cdot \operatorname{tg}\frac{E}{2},$$

whence the longitude in the orbit (countered from the ascending node) is determined as $l = v + \omega_0$.

Based on these parameters, one can determine the geographic coordinates of a satellite along the orbit:

$$\sin \varphi = \sin i_0 \cdot \sin l; \quad \sin(\lambda - \Omega^0_{gpeg}) = \frac{\cos i_0 \cdot \sin l}{\cos \varphi}; \quad \cos(\lambda - \Omega^0_{ginit}) = \frac{\cos l}{\cos \varphi}.$$

3.5 Methods and Conditions of Measurements

The distance from the center of the Earth is determined as $r = Z + R = a_0 \cdot (1 - e_0 \cdot \cos E)$. The dependence of the Earth's radius on latitude is described by the formula (Bronshten et al. 1981)

$$R = R_E \cdot (1 - 0.003325 \cdot \sin^2 \varphi - 0.000028 \cdot \sin^4 \varphi),$$

whence $Z = a_0 \cdot (1 - e_0 \cdot \cos E) - R$.

From the given formulas it follows (Skrebushevsky 1990) that a satellite shifts in longitude from the east to the west. This shift per one circuit, in a first approximation, is determined as (Skrebushevsky 1990)

$$\Delta \lambda_{orb} = \omega_E \cdot T_{sat},$$

where ω_E is the angular velocity of the Earth and T_{sat} is the satellite period.

The ratio

$$n_\lambda = \frac{2\pi}{\Delta \lambda_{orb}} = \frac{T_E}{T_{sat}}$$

specifies the number of circuits per day. In the general case, if n_λ is an irrational number, the satellite does not return to its initial position and, hence, sequentially passes over all points of the Earth surface in the range of latitudes

$$-i \leq \varphi \leq i \quad \text{for} \quad 0 \leq i \leq \frac{\pi}{2},$$

$$i - \pi \leq \varphi \leq \pi - i \quad \text{for} \quad \frac{\pi}{2} < i < \pi.$$

Here i is an orbit element—the inclination of the orbit to the equatorial plane.

In the real atmosphere there occurs a drag of a satellite, and therefore its orbit elements change. Approximate formulas which describe the decrements of some orbit elements per one circuit have the form (Bronshten et al. 1981)

$$\frac{\delta a}{a} = K \cdot (1 + \frac{1}{8v} + 2e - \frac{3}{4} \cdot \frac{e}{v}) \cdot \sqrt{\frac{2\pi}{v}}; \quad \delta e = K \cdot (1 - \frac{3}{8v} + e) \cdot \sqrt{\frac{2\pi}{v}};$$

$$K = 2.2 \cdot \rho_{per} \cdot \frac{\sigma}{m} \cdot a;$$

$$\frac{\delta P}{P} = \frac{3}{2} \cdot \frac{\delta a}{a};$$

$$\frac{\delta Z_{per}}{a} = \frac{\delta a}{a} \cdot (1 - e) - \delta e; \quad \frac{\delta Z_{apog}}{a} = \frac{\delta a}{a} \cdot (1 + e) + \delta e; \quad v = \frac{a \cdot e}{H}.$$

Here H is the scale height at the altitude Z_{per}, P is the satellite period, σ is the cross-sectional area of the satellite (cm^2), m is its mass (g), and ρ_{per} is the atmospheric density (g$\cdot$cm^{-3}) at the perigee altitude:

$$\rho_{per} = 1.673 \cdot 10^{-24} \cdot \{28 \cdot [N_2] + 32 \cdot [O_2] + 16 \cdot [O] + 4 \cdot [He]\},$$

where [N_2], [O_2], [O], and [He] are the concentrations of the atmospheric constituents.

The disturbing acceleration of a satellite at the thermospheric altitudes, caused by the compression of the Earth, is greater than that resulting from the drag in the atmosphere. However, it manifests itself in permanent shifts of the orbit perigee and ascending node. For one circuit we have

$$\delta\Omega = -0.58° \cdot \left(\frac{R_E}{a}\right)^2 \cdot \frac{\cos i}{(1-e^2)^2}; \quad \delta\omega = 0.29° \cdot \left(\frac{R_E}{a}\right)^2 \cdot \frac{5\cdot\cos^2-1}{(1-e^2)^2}.$$

These approximate formulas are too rough to yield accurate values of the satellite orbit elements and position for a long time, but they are applicable to estimate the coordinates of an observer moving over the Earth surface with the purpose of estimating the parameters of an expected airglow.

A satellite intended for investigations of the characteristics of various airglow emissions was WINDII/UARS launched into an orbit in 1991. Its description is given elsewhere (Shepherd et al. 1993a,b).

With the instrumentation erected on this satellite, the rotational temperature of OH (8–3) emission, the emission intensity, and the wind speed were determined from the 557.7-nm emission measurements performed by means of a Fabry–Perot interferometer (Shepherd et al. 1993a,b).

The Explorer satellites have made it possible to reveal the distribution of the airglow over the Earth surface and, in particular, to obtain the image of the auroral oval in the ultraviolet spectral range by the measured parameters of the 130.4-nm emission from atomic oxygen (Frank et al. 1986; Craven and Frank 1987).

The optical observations performed by astronauts on piloted spacecrafts were visual in the prevailing number of cases and gave no quantitative information which could be used in a joint analysis together with ground-based, rocket, and satellite measurements.

It was managed to extract interesting results from satellite measurements performed voluntarily by the astronauts on the "Mir" spacecraft for pale stars during their set beyond the horizon (Gurvich et al. 2002). The immediate goal of these measurements was to investigate the space–time distribution of the air density in the stratosphere by observations of stellar scintillations. For this purpose bright stars were observed which substantially reduced background noises. Besides, the spectral range of emissions under investigation (420–530 (nm)) was chosen outside that of the discrete emissions of the upper atmosphere. However, in recording pale stars, a considerable contribution to the recorded radiation was made by the airglow continuum. Since the measurements for various types of stars were performed during successive periods of motion of the "Mir" station at various dates, it appeared that the local time was the same for all cases. This made it possible to exclude the possible effect of the variations of the continuum emission intensity during a night. Processing of these data has yielded the altitude distribution of the airglow at altitudes of 75–120 (km) and revealed some features of the emissions (see Fig. 2.41). The line of sight over the Earth surface for different measurements appeared over different longitudes at latitudes of 46–52°N. This made it possible to obtain the longitudinal distribution of the irregularities in the airglow intensity (Fig. 5.45). An important advantage of this measurement technique, which involved the sighting of stars in their set beyond the horizon, was the opportunity of exact coordinate fixation of the line of sight and also the possibility of absolute calibration of the airglow intensity by the radiation spectra of sighted stars when they were at large zenith angles. This ensured the absence of noises from the airglow in the field of vision of the instrument.

3.5.9 Formation of Artificial Luminous Clouds

Bates (1950) proposed to use rockets to release various chemicals into the upper atmosphere to create artificial luminous clouds. The goal of experiments of this type is not only to investigate the action of various substances on the gaseous medium of the upper atmosphere, but also to determine its temperature, the wind velocity, and the turbulent and molecular diffusion coefficients.

Systematic rocket experiments with release of alkali metal atoms, whose fluorescence could be observed by ground-based instruments, in the upper atmosphere illuminated by the Sun have been conducted since 1954.

First measurements were performed with sodium as the released glow indicator. They have shown the possibility to determine the velocity of a wind in the middle atmosphere and the atmospheric density at altitudes up to 400 (km) (Shklovsky and Kurt 1959), and also as an indicator of the position of a spacecraft launched toward the Moon (Shklovsky 1960). Subsequently various substances were released to produce plasma, luminescent, and smoke clouds. Used for this purpose were lithium (Li), whose photon scattering factor is much greater than that of sodium (see Table 2.21), other alkali metals, such as potassium (K), cesium (Cs), barium (Ba), and strontium (Sr), and also trimethyl aluminum ($Al(CH_4)_3$) which forms AlO molecules upon interaction with atomic oxygen in the thermosphere (Andreeva et al. 1991). Long-term investigations of the upper atmosphere have been carried out by the Institute of Experimental Meteorology (Kluev 1979, 1985; Andreeva and Katasev 1983; Andreeva 1985; Andreeva et al. 1991).

An important line of research was the determination of the thermospheric temperature. For this purpose, chemical agents were released which form fluorescent molecules of alumina (Johnson 1965; Cole and Kolb 1981; Kluev 1985). Though irregular, the measurements performed gave results that could be compared with model predictions for the altitudes of the upper atmosphere (120–170 (km)) that are inaccessible to study by other methods (Guseva and Kluev 1977). Interesting results were obtained in observations of luminous tracks of released substances over a wide range of altitudes. This made it possible not only to determine the altitude distribution of the atmospheric temperature, but also to find out its oscillations at altitudes from 110 to 170 (km) (Kluev 1985; Andreeva et al. 1991). As mentioned in the cited publications, in a number of experiments conducted in the western Europe

(El Arenocilo, Spain, 37.1°N, 353.3°E) oscillations of the temperature profiles with periods of about 10 (km) were observed at altitudes of 20–80 (km) of the middle atmosphere and near the turbopause at altitudes of 90–135 (km), which corresponded in periods (10–200 (min)) to internal gravity waves (Friedrich et al. 1977; Offermann 1977, 1979; Offermann et al. 1979; Schmidlin et al. 1982). This suggests that the oscillations revealed in the thermosphere are not local processes, but manifestations of IGWs propagating from the troposphere.

For the mentioned altitude range, the damping coefficient of IGWs can be approximated by the relation (correlation coefficient r = 0.998) (Kluev 1985)

$$k(Z) = \exp\left(-\frac{Z-90}{20}\right).$$

A similar relation was obtained on investigations of IGWs in the altitude range 135–170 (km) (Justus and Woodrum 1973).

Comparison of thermospheric temperature measurements with the model predictions by Barlier et al. (1978) and Köhnlein (1980) has shown that even for these models, more perfect compared to the CIRA-1972, the measurement data exceed the predictions by some tens of degrees (Kelvin) (Kluev 1985). This was considered as a good reason for refining empirical models.

However, a possible explanation of the observed discrepancies between the temperature measurements and predictions is that the model values of temperatures corresponded to higher altitudes. Relation of these model altitudes and the differences between them to the measured altitudes has shown high correlation (r = 0.750 ± 0.081). The average altitude difference for altitudes of 135–140 (km) was ~9 (km). Since the data on which the above models are based were obtained in the 1960s, while the midlatitude measurements (Kluev 1985) were performed in the period from 1976 to 1979, the yearly average rate of subsidence of the atmosphere, estimated for a period of ~10 years, is several hundreds of meters per year. It does not contradict to the data reported elsewhere (Evlashin et al. 1999; Semenov et al. 2000; Starkov et al. 2000).

When measuring the wind velocity, temperature, and diffusion coefficients, one has to determine the absolute brightness of artificial luminous clouds. In the case of spherically symmetric release of agents, as the cloud reaches a critical radius (within 0.1–1 (s) after release of chemicals), its spatial evolution occurs in accordance with the solution of the diffusion equation (Dorokhova et al. 1990)

$$n(r,t) = \frac{N}{\pi^{3/2} \cdot (R_{cr}^2 + 4D \cdot t)^{3/2}} \cdot \exp\left(-\frac{r^2}{R_{cr}^2 + 4D \cdot t}\right),$$

where $n(r, t)$ is the concentration of released particles, D is the diffusion coefficient, R_{cr} is the critical radius of the cloud, r is the running radius of the cloud, and t is the time.

The emission intensity corresponding to the given spatial distribution of the particle concentration in the cloud is given by the formula

3.5 Methods and Conditions of Measurements

$$I = \int_{-\infty}^{\infty} g \cdot n \cdot dl,$$

where g is the coefficient of photon scattering due to resonance (for alkali metals) or diffuse scattering (for an aerosol cloud) and l is the distance along the line of sight.

The brightness distribution over a cloud, provided that the optical thickness of the emission layer is less than unity, is described by the formula

$$I = \frac{\sqrt{\pi} \cdot g \cdot R_{cr} \cdot n(0, t_{cr})}{1 + 4D \cdot t/R_{cr}^2} \cdot \exp\left[-\frac{r^2/R_{cr}^2}{1 + 4D \cdot t/R_{cr}^2}\right],$$

where R_{cr} is the critical radius of the cloud at the maximum of its brightness.

Measurements have shown that the emission intensity from a cloud is generally 10–100 (kilorayleigh).

Experiments in which ethylene was released showed an interesting feature. In this case, during the formation and expansion of a cloud, fast photochemical reactions, whose rates are greater than the diffusion rates, took place. The basic chemical process in such a cloud at altitudes of 100–150 (km) was the reaction of ethylene with atomic oxygen (Konoplev et al. 1985). The fast expansion of the cloud at the initial stage was the result of spontaneous ignition of ethylene (Gershenzon et al. 1982, 1984, 1990). An interesting inference from this experiment was the detection of an increased intensity of hydroxyl emission in the cloud within several minutes at altitudes of about 100 (km) which was due to the reaction

$$HCO + O \rightarrow OH(v \leq 9) + CO, \quad \alpha_{HCO,O} \sim 10^{-11}\,(cm^3 \cdot s^{-1}).$$

It should be noted that in this case the formation of OH molecules excited up to the ninth vibrational level is provided. However, the observed radiation intensity in the OH (6–2) band was ∼900 (kilorayleigh), which is about three orders of magnitude greater than the intensity of the night airglow.

A peculiar kind of artificial clouds are those containing alkali metals, such as barium, released at high altitudes in the thermosphere. When the atmosphere is illuminated by the Sun, ionization of barium takes place. Therefore, alongside the green glow of barium at the 553.5-nm wavelength, the blue glow of the barium ion at wavelengths of 455.4 and 493.4 (nm) arises which moves along the magnetic field lines (Föppl et al. 1967; Rieger 1974; Filipp et al. 1986).

Observations have shown that a neutral cloud of barium vapors released at high altitudes (930 (km)) expands as a thin spherical sheath with a velocity of about 1.2 (km · s^{-1}). The total brightness of a cloud of barium atoms—a neutral (not ionized) cloud—falls rapidly due to photoionization, whose rate is $j_{Ba} = 5.1 \cdot 10^{-4}\,(s^{-1})$ (Filipp et al. 1986). As a result of photoionization, a cloud of ionized atoms appears which is readily distinguished by color and shape. Soon after the release of barium, the ionic cloud becomes visible in the form of stripes extended along the geomagnetic field lines (Milinevsky et al. 1990). Within 2 (min) after barium release at the mentioned altitude the diameter of the neutral cloud was about 200 (km), while the diameter of the ionized clouds reached ~1000 (km) within about 10 (min). By tracking the motion of such a cloud, one can determine the characteristics of the Earth's electric field at high altitudes.

The clouds formed by release of chemical agents make it possible not only to investigate the processes occurring in the upper atmosphere, but also to actively affect the thermal regime and state of various atmospheric constituents, both neutral and ionized. These problems have received much attention (Zinn et al. 1982; Biondi and Sipler 1984; Vlasov and Pokhunkov 1986; Kachurin 1990).

3.6 Measurement Data Processing

3.6.1 Spectrophotometry of Photographic Images of Spectra

To provide high accuracy of photometric processing of photographic spectrograms, observance of all rules of spectrophotometry is necessary. With this purpose, for a spectrograph supplied with an ICT a special device was developed which superimposed the spectrum of a standard lamp and the image of a nine-stage attenuator on the film on which the night airglow was exposed in the beginning and at the end of the cycle of night observations. The exposure was the same for the superimposition and the recording of airglow spectra. In illuminating the spectrograph slit by a standard lamp, a spreading screen was used. The stability of the lamp operation was monitored by the readings of 0.1-grade instruments connected in the power supply circuit and by the signal from a selenium photocell that was recorded with a microammeter. To obtain reference spectral lines from a standard source on a frame, light from a neon lamp was introduced in a part of the slit. The reference and the neon lamp were enclosed in a light-tight housing.

The absolute and spectral calibration of the reference lamp was performed in reference to the Sun; therefore, the lamp was a secondary standard source (Krat 1973).

To obtain the characteristic curve of a film that would allow one to translate the film darkening into the intensity of the nightglow emission that caused this darkening, a nine-stage attenuator with measured transmission coefficients of each stage was used. The procedure of its superimposition on the film was the following: The attenuator, through an opal glass and a neutral and a green filter (as this was the color of the luminescence of the ICT screen), was illuminated with light from an incandescent lamp. The image of the attenuator in green light (whose wavelength corresponded to the wavelength of the luminescence of the ICT screen) was transferred on the film by a rotary prism with the help of a lens whose optical axis was perpendicular to the optical axis of the spectrograph. The luminance of the illuminating lamp and, hence, the intensity of light passed through the attenuator was controlled with a photoelectric cell. This device was also placed in a light-tight housing.

A special benchboard was fabricated to control the operation of the reference lamp and the optical unit of the nine-stage attenuator by varying the filament voltage

3.6 Measurement Data Processing

of the lamps. The luminance of the lamps was controlled with a photoelectric cell connected with a microampermeter. All this made it possible to reliably control the light conditions of the standard sources.

The effective spectral sensitivity curve obtained by photographing the reference lamp and the characteristic curve obtained by photographing the nine-stage attenuator were used subsequently in processing spectrograms. These curves usually remained practically unchanged from night to night; however, when a new type of film was used, spectrograms were processed with proper corrections.

In photometric processing of spectrograms of the Earth's airglow, photographs obtained with SP-48 and SP-50 spectrographs were processed. The images of spectra on the film were about 15×2 (mm) in size; in a photograph the width of the instrumental profile of individual lines, whose spectral width was 0.3–0.4 (nm), was ~ 0.05 (mm). For microphotometry, MF-4 microphotometers were used which made it possible to obtain spectrum records of blackenings. In this case, the dark background corresponding to the closed shutter was recorded in the beginning and at the end of the procedure. Initially spectrum records were obtained by a standard way on photoplates, and then the needed parameters were measured. Subsequently the spectrum records were reproduced on a paper ribbon with an EPP-09 self-recording unit.

The processing automation was refined with the help of an additional rheochord, which had ten uniform sections, built in the EPP-09 self-recording unit (Kononovich 1962, 1963). The rheochord tuning for obtaining a characteristic curve on a linear scale was performed with wire-wound resistors connected in parallel to each of its section. In this case, blackenings were automatically converted into intensities, and the spectrum records required only a correction for the spectral sensitivity which would take into account the absolute calibration.

When it became possible to use computers, spectrum records, which were read out from a microphotometer, by using the characteristic curve and the spectral sensitivity memorized in numerical form in the computer, were printed as intensities of individual lines of the spectrum under investigation. Since for the most part these were spectrograms of hydroxyl bands, the temperature and its error were calculated by the method of least squares with the use of the known relations between line intensities and rotational temperature.

At the start of investigations of the variations of intensity and rotational temperature resulting from the modulation of the upper atmosphere by internal gravity waves, the exposure time of photographic recording of spectra was substantially reduced (to 5 (min)) owing to the use of three-stage ICTs. This substantially increased the amount of spectrophotometric processing and called for methods which would allow one to determine short-term variations of relative intensities and temperatures.

The rotational temperature measured by hydroxyl bands could be determined in several ways. All of them are based on the supposition that the population distribution over the rotational levels of the OH molecule is described by a Boltzmann relation corresponding to a temperature T (see Sect. 2.2.5). Therefore, the intensities I(J) (in photons) of the lines of the OH vibration–rotation bands will be determined by the corresponding formula.

The simplest way of determining the rotational temperature is to use its dependence on the relation between the intensities of two or three most intense lines of the P branch, for example,

$$T = f\left\{\frac{I[P_1(1)] - I[P_2(1)]}{I[P_1(2)] - I[P_2(2)]}\right\}.$$

Application of this method enabled one not to take into account the background, assuming that the background intensity remains practically unchanged within the spectral range taken by three P lines (no more than 5 (nm)). Then the photometry procedure is reduced to measuring intensities at the maxima of three P lines. Besides, for rapid processing, the temperature was evaluated by the difference in blackening from two P lines in the hydroxyl band. In this case, in usual operation within the linear portion of the characteristic curve of the photographic emulsion it was possible to readily determine the temperature, not performing intermediate calculations, by the formula

$$T = [(F_2 - F_1)/k]/[\alpha(D_1 - D_2)/\gamma - \log_e(i_1/i_2)],$$

where F_1 and F_2 are the energies of the levels, α is a coefficient taking into account the spectral sensitivity of the system as a whole, D_1 and D_2 are the blackenings from the P-line emissions, γ is the contrast factor, i_1 and i_2 are the respective intensity factors.

However, the most widespread, though somewhat complicated, method is the determination of temperatures simultaneously by the first three, most intense, P lines of OH bands, which yields the least error. With this method, three temperatures, $T_{1,2}$, $T_{2,3}$, and $T_{1,3}$, were calculated from the intensities of P-line pairs by the formulas

$$T_{ml} = 1.4388 \cdot (F_m - F_1) / \left(\log_e \frac{I_m}{I_1} - \log_e \frac{i_m}{i_1}\right)$$

or

$$T_{ml} = 0.625 \cdot (F_m - F_1) / \left(\log_{10} \frac{I_m}{I_1} - \log_{10} \frac{i_m}{i_1}\right),$$

where the sense of the designations is the same as above. Afterward the rotational temperature was determined as an average:

$$T = \frac{1}{3} \cdot (T_{1,2} + T_{2,3} + T_{1,3}).$$

Let us consider the errors that can occur in determining the temperature by the above methods. First, they can arise on conversion of blackenings from individual lines into intensities because of an inexact knowledge of the characteristic curve. As revealed by a special check, a characteristic curve obtained in different nights remains practically unchanged from night to night. Therefore, for a photographic emulsion of the same coating the characteristic curve was constructed by photographs of the nine-stage attenuator obtained for no less than ten nights. In this case, the great number of experimental points ensured high accuracy of the curve. For the usually used range of radiation intensities of hydroxyl bands, which corresponds to the linear portion of the characteristic curve, the accuracy of the conversion of blackenings into intensities of individual lines was estimated to be within 0.5(%). This agrees with the data available in the literature (Sawyer 1951; Malyshev 1979).

The effective spectral sensitivity curve obtained by photographing a reference lamp also showed high reproducibility from night to night and, therefore, it was constructed by spectra obtained for several nights. As to the errors in its determination, which could arise because of the instability of power supply voltages applied to the reference lamp, they were evaluated experimentally. The spectrum of the reference lamp was photographed under usual conditions of night operation (i.e., at the same power supply voltage and exposure during 3 (h)). Processing of the spectrum records has shown that the intensities in the three chosen ranges of the spectrum of the reference lamp corresponding to the first three P lines of the OH band have a variance of about 1(%), and the proportion between them is retained to within 0.5(%).

Besides, the spectrum of a neon lamp was photographed and the proportion between the intensities of its lines produced by the transitions from the common state $3D_1$ to the states $3P_2^\circ$ and $3P_1^\circ$ at wavelengths of 621.7 and 638.3 (nm), respectively, was determined. In this case, the root-mean-square deviation from the average value for the intensity ratio of these lines was not over 1(%), implying that the temperature determined by this method was accurate to within 5 (K).

3.6.2 Electrophotometric Spectrometry

Since the electric signal of a radiation detector is proportional to the measured radiation flux, the primary problem in measurements of this type is to take into account the spectral sensitivity of the whole of the instrument complex. In this case, the spectral calibration procedure is absolutely the same as that used in spectrographic measurements. The only difference is that the scanning of the airglow and standard light source spectra should be performed with electronics operating in the same mode.

3.6.3 Photographic Spectrophotometry of Interferograms

An essential peculiarity of the processing of interference images used to determine a radiation temperature is that the dispersion of the interferograms obtained, and, hence, of the spectrum records, is nonlinear, becoming infinite at the center of the interference pattern. This presents some difficulties in analyzing the measured profiles since their shape should be corrected for this effect.

This correction is performed as follows: Assume that an emission arises at two close wavelengths, λ_0 and $\lambda_0 + \delta\lambda$, belonging to the Doppler profile of a line for which λ_0 is the central wavelength. To these wavelengths there correspond certain radii of two sequential interference rings. Since in practice the spectrum record of the first ring is most often processed as being most intense, we shall consider the first and second rings. In Fig. 3.45 to the wavelengths λ_0 and $\lambda_0 + \delta\lambda$ there correspond the radii R_1 and $R_1 + \delta R_1$ in the first ring and R_2 and $R_2 + \delta R_2$ in the second one.

Let the above wavelengths at the center of the interference pattern be associated with interference orders $n(\lambda_0)$ and $n(\lambda_0 + \delta\lambda)$.

Assume that for all wavelengths of the Doppler profile within the limits of the first ring the integer parts of the interference order n_i are identical, while the fractional parts ε are different.

Then, following Tolansky (1947), we can write

$$n(\lambda_0) = n_1 + \varepsilon(\lambda_0),$$
$$n(\lambda_0 + \delta\lambda) = n_1 + \varepsilon(\lambda_0 + \delta\lambda),$$
$$n(\lambda_0) - n(\lambda_0 + \delta\lambda) = \varepsilon(\lambda_0) - \varepsilon(\lambda_0 + \delta\lambda).$$

The interference order at the center of the interference pattern for a standard light source with distance t between the plates is given by

$$n(\lambda_0) = \frac{2t}{\lambda_0} \quad \text{and} \quad n(\lambda_0 + \delta\lambda) = \frac{2t}{\lambda_0 + \delta\lambda};$$

therefore, following Tolansky (1947), we have

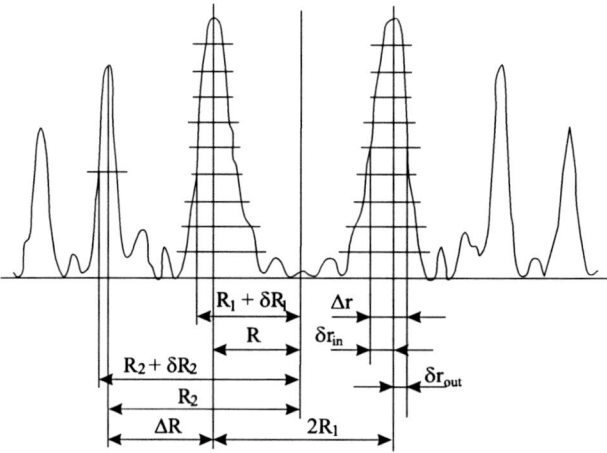

Fig. 3.45 Curve obtained by the photoelectric scanning of interferogram of atomic oxygen emission (630 (nm)) registered by the photographic method (Trunkovsky and Semenov 1978). See explanations in the text

3.6 Measurement Data Processing

$$\frac{1}{\lambda_0} - \frac{1}{\lambda_0 + \delta\lambda} = \frac{\varepsilon(\lambda_0) - \varepsilon(\lambda_0 + \delta\lambda)}{2t},$$

or, approximately,

$$\frac{\delta\lambda}{\lambda_0^2} = \frac{1}{2t} \cdot [\varepsilon(\lambda_0) - \varepsilon(\lambda_0 + \delta\lambda)],$$

whence

$$\delta\lambda = \frac{\lambda_0^2}{2t} \cdot [\varepsilon(\lambda_0) - \varepsilon(\lambda_0 + \delta\lambda)] = \Delta\lambda \cdot [\varepsilon(\lambda_0) - \varepsilon(\lambda_0 + \delta\lambda)],$$

where $\Delta\lambda = \lambda_0^2/2t$ is the free spectral interval. For example, for a line with $\lambda_0 = 630$ (nm) and for a laser line with $\lambda_0 = 632.8$ (nm) we have $\Delta\lambda = 0.0248$ (nm) and $\Delta\lambda = 0.0250$ (nm), respectively.

Let us write down the expressions relating $\varepsilon(\lambda_0)$ and $\varepsilon(\lambda_0 + \delta\lambda)$ with the corresponding radii in the first and the second ring:

$$\varepsilon(\lambda_0) = \frac{R_1^2}{R_2^2 - R_1^2} = \frac{R_2^2}{R_2^2 - R_1^2} - 1;$$

$$\varepsilon(\lambda_0 + \delta\lambda) = \frac{(R_1 + \delta R_1)^2}{(R_2 + \delta R_2)^2 - (R_1 + \delta R_1)^2} = \frac{(R_2 + \delta R_2)^2}{(R_2 + \delta R_2)^2 - (R_1 + \delta R_1)^2} - 1.$$

In view of the well-known formula (Tolansky 1947)

$$R_2^2 - R_1^2 = \frac{\lambda_0 \cdot f^2}{t} = \frac{\lambda_0^2}{2t} \cdot \frac{2f^2}{\lambda_0} = \Delta\lambda \cdot \frac{2f^2}{\lambda_0},$$

we can write

$$(R_2 + \delta R_2)^2 - (R_1 + \delta R_1)^2 = \frac{(\lambda_0 + \delta\lambda)^2}{2t} \cdot \frac{2f^2}{(\lambda_0 + \delta\lambda)}$$

$$= \left[\frac{\lambda_0^2}{2t} + \frac{2\lambda_0 \cdot \delta\lambda + (\delta\lambda)^2}{2t}\right] \cdot \frac{2f^2}{(\lambda_0 + \delta\lambda)}.$$

Since the width of the Doppler profile of a line, for example, the 630-nm line, is such that $\delta\lambda/\lambda_0 \sim 10^{-6} - 10^{-5}$, we obtain

$$\frac{2\lambda_0 \cdot \delta\lambda}{2t} \leq 10^{-5} \cdot \frac{\lambda_0^2}{2t} \quad \text{and} \quad \frac{(\delta\lambda)^2}{2t} \leq 10^{-10} \cdot \frac{\lambda_0^2}{2t}.$$

Neglecting these terms in the above equation and taking into account that the deletion of $\delta\lambda$ in the denominator of the multiplier $\frac{2f^2}{\lambda_0 + \delta\lambda}$ results in the relative error

$$\left(\frac{2f^2}{\lambda_0} - \frac{2f^2}{\lambda_0 + \delta\lambda}\right) \bigg/ \frac{2f^2}{\lambda_0 + \delta\lambda} = \frac{\delta\lambda}{\lambda_0} \leq 10^{-5},$$

we have
$$(R_2 + \delta R_2)^2 - (R_1 + \delta R_1)^2 = \frac{\lambda_0^2}{2t} \cdot \frac{2f^2}{\lambda_0} = \Delta\lambda \cdot \frac{2f^2}{\lambda_0}.$$

From the above formulas we obtain
$$\delta\lambda = \Delta\lambda \cdot \left[\frac{R_1^2}{R_2^2 - R_1^2} - \frac{(R_1 + \delta R_1)^2}{(R_2 + \delta R_2)^2 - (R_1 + \delta R_1)^2}\right] = \frac{\Delta\lambda \cdot [R_1^2 - (R_1 + \delta R_1)^2]}{\Delta\lambda \cdot \frac{2f^2}{\lambda_0}},$$

whence
$$(R_1 + \delta R_1)^2 - R_1^2 = -\frac{2f^2}{\lambda_0} \cdot \delta\lambda.$$

Now, introducing new designations by setting $\Delta R = R_2 - \delta R_1$ and $\delta r = \delta R_1$ (see Fig. 3.45), we obtain

$$-\frac{\delta\lambda}{\Delta\lambda} = \frac{(R_1 + \delta r)^2 - R_1^2}{R_2^2 - R_1^2} = \frac{\delta r}{\Delta R}\left[\frac{2R_1 + \delta r}{2R_1 + \Delta R}\right] = \frac{\delta r}{\Delta R} \cdot \frac{2R_1 + \delta r}{\Delta R} \cdot \frac{1}{1 + \frac{2R_1}{\Delta R}},$$

whence
$$\delta\lambda = -\Delta\lambda \cdot \frac{\delta r}{\Delta R} \cdot \frac{1}{1 + \frac{2R_1}{\Delta R}} \cdot \frac{2R_1 + \delta r}{\Delta R}.$$

When analyzing the last formula, it is necessary to take into account that the radii increase toward the periphery, while the wavelengths increase toward the center. The radius R_1 corresponds to the intensity maximum of the first ring and to the wavelength λ_0. Let us agree that δr is a positive increment if the corresponding point of the profile is farther from the center than the maximum. The corresponding wavelength increment $\delta\lambda$ will be of opposite sign to δr.

Since the expression $(2R_1 + \delta r)/\Delta R$ is always positive irrespective of the sign of δr, we have $2R > \delta r$. Therefore, the sign of $\delta\lambda$ is determined only by the sign of δr. However, the absolute value of the expression $(2R_1 + \delta r)/\Delta R$ depends on the sign of δr. To find the full width of a profile at any level, it is necessary to add up the absolute values of the wavelength increments corresponding to the displacements in both directions from the radius R_1 to the profile points lying at this level.

Let us designate the radius increments corresponding to the displacement outward and inward in relation to the center by δr_{out} and δr_{in}, respectively, and introduce similar designations for the respective wavelengths.

Taking into account that $\delta r_{out} > 0$ and $|\delta r_{out}| = \delta r_{out}$, and also that $\delta r_{in} < 0$ and $|\delta r_{in}| = -\delta r_{in}$, we have

$$|\delta\lambda_{out}| = \Delta\lambda \cdot \frac{|\delta r_{out}|}{\Delta R} \cdot \frac{1}{1 + \frac{2R_1}{\Delta R}} \cdot \frac{|2R_1 + \delta r_{out}|}{\Delta R}$$

$$= \Delta\lambda \cdot \frac{|\delta r_{out}|}{\Delta R} \cdot \frac{1}{1 + \frac{2R_1}{\Delta R}} \cdot \left[\frac{2R_1}{\Delta R} + \frac{|\delta r_{out}|}{\Delta R}\right],$$

3.6 Measurement Data Processing

$$|\delta\lambda_{in}| = \Delta\lambda \cdot \frac{|\delta r_{in}|}{\Delta R} \cdot \frac{1}{1 + \frac{2R_1}{\Delta R}} \cdot \left[\frac{2R_1}{\Delta R} - \frac{|\delta r_{in}|}{\Delta R}\right].$$

Introducing new designations by setting

$$\Delta r = |\delta r_{out}| + |\delta r_{in}| \text{ and } \delta = |\delta r_{out}| - |\delta r_{in}|,$$

we obtain the full width of the profile

$$\delta\lambda = |\delta\lambda_{out}| + |\delta\lambda_{in}| = \Delta\lambda \cdot \frac{\Delta r}{\Delta R} \cdot \frac{1}{1 + 2R_1/\Delta R} \cdot \left[\frac{2R_1}{\Delta R} - \frac{\delta}{\Delta R}\right]$$

$$= \Delta\lambda \cdot \frac{\Delta r}{\Delta R} \cdot \frac{2R_1/\Delta R}{1 + 2R_1/\Delta R} \cdot \left[1 - \frac{\delta}{2R_1}\right];$$

or, setting

$$z = \frac{2R_1}{\Delta R},$$

we obtain the formula

$$\delta\lambda = \Delta\lambda \cdot \frac{\Delta r}{\Delta R} \cdot \frac{z}{1+z} \cdot \left[1 - \frac{\delta}{2R_1}\right] = \Delta\lambda \cdot \frac{\Delta r}{\Delta R} \cdot \frac{z}{1+z} \cdot \left[1 - \frac{1}{z} \cdot \frac{\Delta r}{\Delta R} \cdot \frac{\delta}{\Delta r}\right].$$

The quantity $\delta/\Delta r$ characterizes the degree of asymmetry of the profile in relation to its width. The asymmetry is caused by the nonlinear dispersion, and because of this nonlinearity we have $|\delta r_{in}| > |\delta r_{out}|$ and $\delta > 0$. The factor $z/(1+z)$ characterizes the total broadening of the profile in a spectrum record in relation to the width of the actually observed profile due to the nonlinearity of dispersion. Both factors, δ and $z/(1+z)$, depend on the diameter of the ring under consideration. Therefore, in practice, to reduce the asymmetry of the intensity distribution in the recorded profile, it is necessary to adjust the interferometer so that the diameter of the first ring be as large as possible.

The second part of the last formula can be used with tabulated data (Trunkovsky and Semenov 1978). The values of $\delta/\Delta r$ can vary in the range 0.01–0.1, $\Delta r/\Delta R$ in the range 0.1–1, and z in the range 1.5–3.

Based on the values of $\delta\lambda_i$, calculated by the formula

$$\delta\lambda_i = \rho \cdot \sqrt{\log_e \frac{I_0}{I_i}} = 7.18 \cdot 10^{-7} \cdot \lambda_0 \cdot \sqrt{\frac{T}{M}} \cdot \sqrt{\log_e \frac{I_0}{I_i}},$$

the correlation plot is constructed (Fig. 3.46) and the line of regression is drawn through the points with greatest "weights", which, if the profile is of Gaussian type, should pass through the origin of coordinates (Semenov 1975a). This procedure, which is used in statistics, allows one to check whether the line profile is Gaussian and whether the background intensity level from which running intensities of the line are counted is drawn correctly, which is of particular importance for the profile

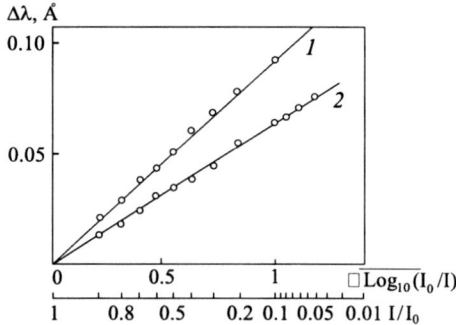

Fig. 3.46 Relation between the measured profile width (at the levels of the normalized intensity I/I_0) and $\sqrt{\log_{10}(I_0/I)}$. 1 – 630-nm emission of the upper atmosphere; 2 – 632.8-nm emission of the calibration source (Semenov 1975a)

wings. This gives the important advantage of using not only the halfwidth, but also many points of the profile. It can be seen that the squared regression coefficient is proportional to the estimated temperature T_{meas}.

Thereafter a correction for instrumental broadening should be made. If the instrumental profile is well described by a Gaussian distribution, which was verified by photographic recording (Trunkovsky and Semenov 1978), the relation

$$|\delta\lambda_{meas}|^2_{HW} = |\delta\lambda_{atm}|^2_{HW} + |\delta\lambda_{instr}|^2_{HW},$$

holds and the atmospheric temperature is equal to the difference between the measured temperature T_{meas} and the temperature corresponding to the instrumental profile of the device, T_{instr}:

$$T_{atm} = T_{meas} - T_{instr}.$$

The error introduced in determining the temperature is calculated by using the conventional formulas (Taylor 1982)

$$\sigma^2_{\delta\lambda} = \frac{1}{N-2} \cdot \sum_{i=1}^{N} \left(\delta\lambda_i - \rho \cdot \sqrt{\frac{I_0}{I_i}}\right)^2$$

$$= \frac{1}{N-2} \cdot \left[\sum_{i=1}^{N}(\delta\lambda_i)^2 - 2\rho \cdot \sum_{i=1}^{N}\left(\delta\lambda_i \cdot \sqrt{\frac{I_0}{I_i}}\right) + \rho^2 \cdot \sum_{i=1}^{N}\left(\sqrt{\frac{I_0}{I_i}}\right)^2\right],$$

$$\sigma^2_\rho = \frac{N \cdot \sigma^2_{\delta\lambda}}{N \cdot \left[\sum_{i=1}^{N}\left(\sqrt{\frac{I_0}{I_i}}\right)^2\right] - \left[\sum_{i=1}^{N}\sqrt{\frac{I_0}{I_i}}\right]^2},$$

which yield

$$\sigma_T = 2T \cdot \frac{\sigma_\rho}{\rho}.$$

As considered above, in recording an immobile interference image, for example, by photographing it or by using a matrix image detector, the directly obtained

3.6 Measurement Data Processing

wavelength increment depends nonlinearly on the ring radius:

$$\delta\lambda = -\Delta\lambda \cdot \frac{r^2 - R_1^2}{R_2^2 - R_1^2}.$$

Since the Fourier analysis deals with the interval $\Delta\lambda$, it is necessary to choose interval boundaries in the spectrum record near the analyzed profile such that for

$$r = r_{in}, = \Delta\lambda/2 \text{ and } r = r_{out}, = -\Delta\lambda/2$$

the condition $0 \leq \delta\lambda \leq \Delta\lambda$ be satisfied, and the dependence of the wavelength increment $\delta\lambda$ on radius should be determined by the relation

$$\delta\lambda = \frac{\Delta\lambda}{2} - \Delta\lambda \cdot \frac{r^2 - R_1^2}{R_2^2 - R_1^2}.$$

In photometry, the interval under investigation is partitioned into equal intervals δr, i.e., we have $r = \delta r(j-1) + r_{in}$, where j is the number of a point of partition, or

$$\frac{\delta\lambda_j}{\Delta\lambda} = 1 - \frac{\delta r(j-1)}{\sqrt{R_2^2 - R_1^2}} \cdot \left[\frac{\delta r(j-1)}{\sqrt{R_2^2 - R_1^2}} + 2\frac{r_{in}}{\sqrt{R_2^2 - R_1^2}} \right], \text{ where } r_{ins} = \sqrt{\frac{3R_1^2 - R_2^2}{2}}.$$

A similar formula holds for $r_{out} = \sqrt{\frac{R_1^2 + R_2^2}{2}}$. The previous formula demands that the interferometer be adjusted so that the condition $\frac{R_2}{R_1} < \sqrt{3}$ is satisfied. This condition implies that the interval $\Delta\lambda/2$ should fall within the radius of the first ring, and it is desirable, with some margin. Under real conditions, the partition step δr is set by the electron-mechanical device producing the spectrum record. Therefore, subject to the condition that

$$j = 1 \to \delta\lambda = \Delta\lambda \text{ and } j = p \to \delta\lambda = 0,$$

the number of points will be determined by the expression

$$p = \left[1.5 + \frac{\sqrt{\frac{R_1^2 + R_2^2}{2}} - \sqrt{\frac{3R_1^2 - R_2^2}{2}}}{\delta r} \right],$$

where the square brackets denote the integer part of the number and 0.5 is added for its correct rounding. Since the ring dimensions in emission profiles are different for the sky and a laser (in obtaining characteristics of the instrumental profile), p and Δr for them can be different.

Thus, a spectrum record is discretized linearly in r and appears nonlinearly partitioned in $\delta\lambda$. Therefore, the Fourier coefficients should be calculated by formulas which would take into account the step irregularity. If trapezoid formulas are used to evaluate the integrals, we have

$$a_n = \frac{1}{2} \sum_{j=1}^{p-1} \left[S(\delta\lambda_j) \cos \frac{2\pi n}{\Delta\lambda} \delta\lambda_j + S(\delta\lambda_{j+1}) \cos \left(\frac{2\pi n}{\Delta\lambda} \cdot \delta\lambda_{j+1} \right) \right] \cdot (\delta\lambda_{j+1} - \delta\lambda_j),$$

$$b_n = \frac{1}{2} \sum_{j=1}^{p-1} \left[S(\delta\lambda_j) \cdot \sin \left(\frac{2\pi n}{\Delta\lambda} \cdot \delta\lambda_j \right) + S(\delta\lambda_{j+1}) \cdot \sin \left(\frac{2\pi n}{\Delta\lambda} \cdot \delta\lambda_{j+1} \right) \right] \cdot (\delta\lambda_{j+1} - \delta\lambda_j),$$

$$c_n = \frac{1}{2} \sum_{j=1}^{p-1} \left[W(\delta\lambda_j) \cdot \cos \left(\frac{2\pi n}{\Delta\lambda} \cdot \delta\lambda_j \right) + W(\delta\lambda_{j+1}) \cdot \cos \left(\frac{2\pi n}{\Delta\lambda} \cdot \delta\lambda_{j+1} \right) \right] \cdot (\delta\lambda_{j+1} - \delta\lambda_j),$$

$$d_n = \frac{1}{2} \sum_{j=1}^{p-1} \left[W(\delta\lambda_j) \cdot \sin \left(\frac{2\pi n}{\Delta\lambda} \cdot \delta\lambda_j \right) + W(\delta\lambda_{j+1}) \cdot \sin \left(\frac{2\pi n}{\Delta\lambda} \cdot \delta\lambda_{j+1} \right) \right] \cdot (\delta\lambda_{j+1} - \delta\lambda_j).$$

Naturally, other numerical integration formulas can also be used.

The minimum of the function K(T) that corresponds to the sought-for Doppler temperature T of an airglow emission can be found by a computer step-by-step evaluation. The initial value of T is generally chosen at the level of the lower limit.

3.6.4 Photoelectric Spectrophotometry of Interferograms

Photoelectric interferograms are obtained on scanning the interference pattern image by varying the optical thickness of the medium between the interference plates. This can be realized by varying the pressure inside the interferometer or the distance between the plates. In the latter case, piezoelectric cells changing their dimensions on application of a dc voltage are used. In either case, the spectrum record obtained shows a linear distribution of the wavelength along the abscissa (Fig. 3.47) (Hernandez and Turtle 1965).

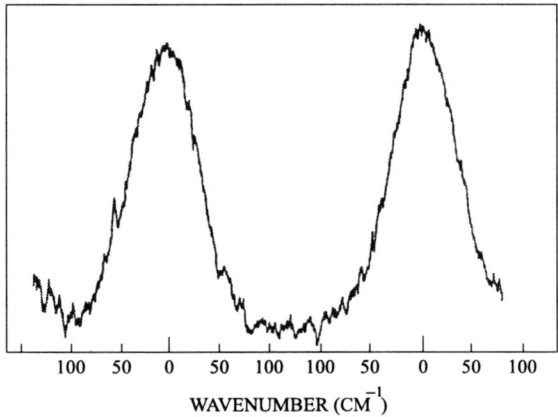

Fig. 3.47 An interferogram of the 557.7 (nm) emission obtained by a photoelectric method at scanning with the help of change of pressure within 15 min (Hernandez and Turtle 1965)

3.6 Measurement Data Processing

When the image of an interference pattern is projected on the plane of a round diaphragm, since the path-length difference depends on the mirror spacing of the standard source, t, on the index of refraction of the medium between the mirrors, n, and on the inclination angle φ of interfering rays, the spectrum can be scanned by varying any of these parameters.

The scanning by varying the mirror spacing demands high-precision mechanical displacements. For these purposes, the best choice is to use barium titanate piezoelectric cells which change their dimensions under the action of dc voltage. In this case, however, there are difficulties with selection of cells having identical properties.

Scanning by varying the gas pressure in photoelectric recording of signals was actively used by Hernandez and Turtle (1965). It is best applicable if the spacing t is large. The index of refraction of a gas, n, is related to the pressure p linearly (Tarasov 1977; Lebedeva 1977; Ignatiev and Yugov 1995):

$$n - 1 = (n_0 - 1) \cdot \frac{p}{p_0},$$

where n_0 is the index of refraction at normal pressure p_0. For the angle of incidence of rays on the interferometer close to zero, the interference order k is related to n and t as

$$k = \frac{2nt}{\lambda}.$$

Estimates obtained with the above relations show that the scanning of a spectrum by varying the gas pressure within 1 (atm) is appropriate at t over 10 (mm). One of the most convenient way of providing a uniform pressure rise is to supply the gas through a narrow capillary tube (∼0.01 (mm)), since the amount of gas flowing through a capillary tube with a supersonic speed does not depend on the pressure difference between the inlet and the outlet. For example, at an inlet pressure of about 2.5 (atm), the pressure rise in the chamber was uniform to within 1.5(%).

Under the conditions of measurements of the nightglow emissions, whose intensities were some tens to hundreds of Rayleighs, the time required for scanning one or two rings was about 15 (min) even at plate diameters of ∼120 (mm). To eliminate the time variations in emission intensity, it was necessary to perform simultaneous intensity measurements with a photometer.

The method of varying the angle of inclination of the interferometer plates with a small spacing (t ∼ 60 (μm)) was successfully used in measuring the intensity of the 656.3-nm H_α emission (Shcheglov 1967a; Fishkova 1983). In this case, a Fabry–Perot interferometer was used as a narrow-band filter with a spectral bandpass of 3.6 (nm), naturally, in combination with a conventional interference light filter with a bandpass of 2–3 (nm).

The inclination angle can be varied periodically with the use of any swinging device. In a device of this type, a standard light source and an interference light filter (Lyakhov and Managadze 1975; Potapov et al. 1978) are placed in a rotating disk inclined at an angle α/2 to a plane perpendicular to the optical axis of the device as a whole in which the standard source is also at an angle α/2 to the disk plane. In this case, as the disk is rotated, the inclination angle of the standard source periodically varies from zero to α.

In photoelectric recording of interference images by means of scanning, the recorded profiles of emission lines have a linear scale wavelength. In this case, naturally, alongside with the desired signal, a noise component is present. To eliminate the noise and, simultaneously, the instrument function of the standard source, it was proposed to apply the Fourier analysis method to the obtained spectrum records (Hays and Roble 1971; Killeen and Hays 1984). To do this, it is necessary to represent the measured ($S(\delta\lambda)$) and the instrumental profile ($W(\delta\lambda)$) as Fourier series:

$$S(\delta\lambda) = \frac{a_0}{2} + \sum_{n=1}^{\infty} \left[a_n \cdot \cos\left(\frac{2\pi n}{\Delta\lambda} \cdot \delta\lambda\right) + b_n \cdot \sin\left(\frac{2\pi n}{\Delta\lambda} \cdot \delta\lambda\right) \right]$$

$$= \frac{a_0}{2} + \sum_{n=1}^{\infty} S_n \cdot \cos\left(\frac{2\pi n}{\Delta\lambda} \cdot \delta\lambda - \psi_n\right),$$

$$W(\delta\lambda) = \frac{c_0}{2} + \sum_{n=1}^{\infty} \left[c_n \cdot \cos\left(\frac{2\pi n}{\Delta\lambda} \cdot \delta\lambda\right) + d_n \sin\left(\frac{2\pi n}{\Delta\lambda} \cdot \delta\lambda\right) \right]$$

$$= \frac{c_0}{2} + \sum_{n=1}^{\infty} W_n \cdot \cos\left(\frac{2\pi n}{\Delta\lambda} \cdot \delta\lambda - \psi_n\right).$$

Here $\Delta\lambda$ is the spectral interval of the interferometer, i.e., the distance between the maxima on the wavelength scale:

$$\Delta\lambda = \frac{\lambda^2}{2t}.$$

The Fourier coefficients are given by

$$a_n = \int_{-\Delta\lambda/2}^{\Delta\lambda/2} S(\delta\lambda) \cdot \cos\left(\frac{2\pi n}{\Delta\lambda} \cdot \delta\lambda\right) d\delta\lambda, \quad b_n = \int_{-\Delta\lambda/2}^{\Delta\lambda/2} S(\delta\lambda) \cdot \sin\left(\frac{2\pi n}{\Delta\lambda} \cdot \delta\lambda\right) d\delta\lambda,$$

$$c_n = \int_{-\Delta\lambda/2}^{\Delta\lambda/2} W(\delta\lambda) \cdot \cos\left(\frac{2\pi n}{\Delta\lambda} \cdot \delta\lambda\right) d\delta\lambda, \quad d_n = \int_{-\Delta\lambda/2}^{\Delta\lambda/2} W(\delta\lambda) \cdot \sin\left(\frac{2\pi n}{\Delta\lambda} \cdot \delta\lambda\right) d\delta\lambda,$$

the harmonic amplitudes by

$$S_n = \frac{2}{\Delta\lambda} \cdot \sqrt{a_n^2 + b_n^2}, \qquad W_n = \frac{2}{\Delta\lambda} \cdot \sqrt{c_n^2 + d_n^2},$$

and the phases, in units of $\Delta\lambda$, by

$$\psi_n = \frac{\Delta\lambda}{2\pi n} \cdot \text{arctg} \frac{b_n}{a_n}, \qquad \psi'_n = \frac{\Delta\lambda}{2\pi n} \cdot \text{arctg} \frac{d_n}{c_n}.$$

3.6 Measurement Data Processing

Since

$$I(\delta\lambda) = I_0 \cdot \exp\left[-\frac{(\delta\lambda)^2}{(\delta\lambda_D)^2}\right],$$

where

$$\delta\lambda_D = \frac{\lambda_0}{c} \cdot \sqrt{\frac{2kNT}{M}}$$

and

$$S(\delta\lambda) = \int_{-\infty}^{\infty} I(\delta\lambda) \cdot W(\delta\lambda - \delta\lambda') \cdot d\delta\lambda',$$

we have

$$S_T(\delta\lambda) = A_0 + \sum_{n=1}^{\infty} W_n \cdot e^{-\frac{n^2}{4} \cdot \gamma \cdot T} \cdot \cos\left(\frac{2\pi n}{\Delta\lambda} \cdot \delta\lambda\right),$$

where

$$\gamma = \frac{4\pi^2}{(\Delta\lambda)^2} \cdot \beta \quad \text{and} \quad \beta = \frac{2kN \cdot \lambda_0^2}{M \cdot c^2}.$$

For an instance, $\beta\,(557.7\,(\text{nm})) = 3.5417403 \cdot 10^{-8}$ and $\beta\,(630.0\,(\text{nm})) = 4.5833681 \cdot 10^{-8}$.

The functions $S(\delta\lambda)$ and $S_T(\delta\lambda)$ can be compared by the method of least squares, according to which the function

$$K(T) = \int [S - S_T]^2 \, dx \text{ or } K(T) = \sum_{n=1}^{N} S_n^2 - \frac{\left[\sum_{n=1}^{N} W_n \cdot S_n \cdot e^{-\frac{n^2}{4} \cdot \gamma \cdot T}\right]^2}{\sum_{n=1}^{N} W_n^2 \cdot e^{-\frac{n^2}{2} \cdot \gamma \cdot T}}$$

should have a minimum for the sought-for temperature. As a rule, the number of harmonics $N \leq 8$ (Hays and Roble 1971).

The function $K(T)$ is cuspidal in character and has one extremum. It can be approximated well enough by a polynomial of the fourth degree. The use of a second-order parabola (Hays and Roble 1971) is possible only within a narrow range of T values, and this is inconvenient in practice. Thus, if we assume that

$$K(T) = A + B \cdot T + C \cdot T^2 + D \cdot T^3 + E \cdot T^4$$

and introduce $T = T_0 \cdot m$, then, setting $m = 1, 2, 3, 4$, and 5, we can obtain K_1, K_2, K_3, K_4, and K_5 and construct the system of equations

$$A + B \cdot T_0 + C \cdot T_0^2 + D \cdot T_0^3 + E \cdot T_0^4 = K_1,$$

$$A + 2B \cdot T_0 + 4C \cdot T_0^2 + 8D \cdot T_0^3 + 16E \cdot T_0^4 = K_2,$$

$$A + 3B \cdot T_0 + 9C \cdot T_0^2 + 27D \cdot T_0^3 + 81E \cdot T_0^4 = K_3,$$

$$A + 4B \cdot T_0 + 16C \cdot T_0^2 + 64D \cdot T_0^3 + 256E \cdot T_0^4 = K_4,$$

$$A + 5B \cdot T_0 + 25C \cdot T_0^2 + 125D \cdot T_0^3 + 625E \cdot T_0^4 = K_5.$$

The solution of this system yields the coefficients of the equation:

$$Q_4 = 24E \cdot T_0^4 = K_5 - 4K_4 + 6K_3 - 4K_2 + K_1,$$

$$Q_3 = 24D \cdot T_0^4 = -10K_5 + 44K_4 - 72K_3 + 52K_2 - 14K_1,$$

$$Q_2 = 24C \cdot T_0^4 = 35K_5 - 164K_4 + 294K_3 - 236K_2 + 71K_1,$$

$$Q_1 = 24B \cdot T_0^4 = -50K_5 + 244K_4 - 468K_3 + 428K_2 - 154K_1.$$

To obtain the value of T from the minimum K(T), it is necessary to solve the equation

$$4E \cdot T^3 + 3D \cdot T^2 + 2C \cdot T + B = 0$$

or

$$T^3 + 3d \cdot T^2 + 2c \cdot T + b = 0,$$

where

$$d = \frac{Q_3}{4Q_4} \cdot T_0, \quad c = \frac{Q_2}{4Q_4} \cdot T_0^2, \quad b = \frac{Q_1}{4Q_4} \cdot T_0^3.$$

Introducing

$$p = \left[\frac{2}{3} \cdot \frac{Q_2}{4Q_4} - \left(\frac{Q_3}{4Q_4}\right)^2\right] \cdot T_0^2, \quad q = \left[\left(\frac{Q_3}{4Q_4}\right)^3 - \frac{Q_3}{4Q_4} \cdot \frac{Q_2}{4Q_4} + \frac{1}{2} \cdot \frac{Q_1}{4Q_4}\right] \cdot T_0^3,$$

which are coefficients of the equation

$$y^3 + 3py^2 + 2q = 0,$$

we obtain

$$\frac{p^3}{q^2} = \frac{\left[\frac{2}{3} \cdot \frac{Q_2}{4Q_4} - \left(\frac{Q_3}{4Q_4}\right)^2\right]^3}{\left[\left(\frac{Q_3}{4Q_4}\right)^3 - \frac{Q_3}{4Q_4} \cdot \frac{Q_2}{4Q_4} + \frac{1}{2} \cdot \frac{Q_1}{4Q_4}\right]^2},$$

whence

$$T = T_0 \left[\sqrt[3]{\sqrt{p^3 + q^2} - q} - \sqrt[3]{\sqrt{p^3 + q^2} + q} - \frac{Q_3}{4Q_4} \right],$$

$$T = T_0 \cdot \left[\sqrt[3]{q} \cdot \left(\sqrt[3]{\sqrt{1 + \frac{p^3}{q^2}} - q} + \sqrt[3]{\sqrt{1 + \frac{p^3}{q^2}} + q} \right) - \frac{Q_3}{4Q_4} \right],$$

$$T = T_0 \cdot \left\{ \sqrt{p} \cdot \left[\frac{\left(1 + \sqrt{1 + \frac{p^3}{q^2}} \right)^{1/3}}{\left(\frac{p^3}{q^2} \right)^{1/6}} \right] \cdot \left[1 - \frac{\left(\frac{p^3}{q^2} \right)^{1/3}}{\left(1 + \sqrt{1 + \frac{p^3}{q^2}} \right)^{2/3}} \right] - \frac{Q_3}{4Q_4} \right\}.$$

Since the actual temperature range is 200–1800(K), we can use in calculations the T_0 values that correspond to a particular type of emission. For example, for the 630-nm emission we have $T_0 \sim 350$–400(K) and for the 557.7-nm emission $T_0 \sim 100$–150(K).

As considered above, in photoelectric recording of an immobile interference image, for example, with matrix image detectors, a directly obtained spectrum record shows a nonlinear dependence on the ring radius and its processing is carried out as described in Sect. 3.6.3.

3.6.5 Spectral Analysis of Time Series

In investigations of time variations of the intensity and temperature of airglow emissions a need arises to reveal regular variations with various time scales: daily, lunar phase, seasonal, and long term. The last, in turn, include quasi-biennial, 5.5-yr, and 11-yr solar cycles, as well as secular scales. Some variations can be superimposed by disturbed changes caused by geomagnetic storms. It is quite natural that to separate these components, it is necessary not only to analyze carefully the formal spectral composition of the variations, but also to consider the physical nature of the involved phenomena.

In this connection, it is necessary to consider various analyses of the data constituting the time sequence. The first and most natural and customary method is Fourier analysis.

It is quite natural that the time series of the characteristics of the Earth's airglow, by virtue of its optical specificity, are not ideally continuous. Hence, time intervals for which there are no data are inevitable. Therefore, an understanding of the laws of the variations and a certain skill to reveal their properties are required. Anyway, in analyzing the behavior of processes which occur on smaller time intervals, regularities for more long-term variations should be revealed and eliminated from the analyzed data.

The spectrum analysis of time series is most applied in investigations of internal gravity waves since, in this case, the variation phases are not known in advance.

In the presence of random outliers of the high-frequency component type, it is practical in some cases to perform smoothing by means of numerical windows whose character affects the results of the spectral analysis (Harris 1978). An important feature of the spectral analysis is that each signal under processing should have a finite duration. The signal duration can be varied, which, in some cases, is necessary for obtaining reliable information (see Fig. 3.51), but it should necessarily be finite. In processing signals of finite duration, interesting and interdependent problems arise which must be considered in the course of analysis. The finiteness of the interval of observation affects the resolvability of periods with small amplitudes in the presence of close periods with strong amplitudes, the resolvability of oscillations of varying frequency, and the accuracy of estimates of the parameters of all mentioned signals. Therefore, smearing of a signal or spectral leakage from the continuum may occur.

The reason for the smearing is that signals with frequencies other than basic ones are aperiodic in the observation window. If the recorded signal is incommensurable in duration with the observation interval, the periodic extension of the signal will have discontinuities at the interval boundaries. These discontinuities make spectral contributions at all basic frequencies, thereby smearing the signal. The windows represent weighting functions, which are used to reduce the smearing of spectral components resulting from the finiteness of the observation interval. Therefore, the action of a window on a data array (as a multiplicative weighting function) consists in reducing the order of a discontinuity at the boundary of the periodic extension. To provide this, the weighed data should smoothly tend to zero at the boundaries of the interval, so that the periodic extension of the signal be continuous up to the derivatives of higher orders.

In Fourier analysis, windows are used to diminish undesirable spectral leakage effects. Windows affect many characteristics of a harmonic processor, such as the detection ability, resolving ability, dynamic range, reliability, and realizability of computing operations (Harris 1978).

One of these features is considered in Sect. 3.6.6.

The time series of emission intensities and temperatures measured in various regions of the sky, even in visual examination, sometimes testify to the presence of perturbations propagating in the upper atmosphere. Examples of such data were presented in a number of publications. In these series it is possible to distinguish, even visually, perturbed sections (trains), to evaluate their delay relative to each other, and thus to solve the problem of finding the azimuth and propagation velocity of IGWs. For doing this, it is convenient to apply the method of mutual correlation functions, $R_{lm}(r)$, which are determined for a delay r:

$$R_{lm}(r) = \frac{1}{(N-r) \cdot \sqrt{\sigma_l \cdot \sigma_m}} \cdot \sum_{s=1}^{N-1} |T_l(s) - \overline{T}_l| |T_m \cdot (s+r) - \overline{T}_m|.$$

3.6 Measurement Data Processing

Here T_l and T_m are the time series under investigation with variances σ_l and σ_m and averages $\overline{T}_l$ and $\overline{T}_m$. The presence of a maximum $R_{lm}(r)$ for a delay r can be interpreted as if the series $T_l(s)$ is shifted relative to the series $T_m(s)$ by r corresponding to the time interval $\delta t_{lm} = r \cdot \Delta$, where Δ is the sampling interval. When measurements are performed at three points (1, 2, and 3), for IGWs the obvious condition should be fulfilled:

$$\delta t_{1,2} + \delta t_{2,3} + \delta t_{3,1} = 0.$$

For all maxima of the correlation functions that satisfy this condition, the IGW azimuth and propagation velocity can be determined from the known coordinates of the observation points.

However, this method has a number of major disadvantages. First, it gives no information about important IGW parameters, such as amplitude and period. Moreover, in analyzing the mutual correlation functions, a significant number of maxima are almost always detected for the delays that satisfy the above condition, complicating the interpretation of results and sometimes even making it impossible.

This is due to many disturbances hampering the recording of the effect and the superposition of IGWs with different periods. These are instrumental noises, various distorting factors in the surface air (variations of the transparency, highlights from urban lights), and the variations of the structural parameters of the atmosphere related to the circulation at the altitude of the emission layer. Therefore, to determine reliably IGW parameters (amplitude, period, azimuth, and propagation velocity), statistical methods should be applied to select a signal on the noise background. Based on these methods, widely described in the literature (see, e.g., Serebryannikov and Pervozvansky 1965; Bendat and Piersol 1966; Jenkins and Watts 1969), a procedure for processing time series of temperatures obtained as a result of observations of hydroxyl emission variations at three points of the firmament has been developed.

The procedure of processing spectrographic and photometric data consists of two stages. First of all, it is necessary to determine temperatures at three observation points and then select a desired signal (IGW, in our case) from the obtained time series by spectral analysis and determine its characteristics. As to the first stage, it has some features related to the peculiarities of the operation of spectrographs and photometers. The second stage is absolutely identical for spectrographic and photometric data. The procedure of primary processing of spectrographic observations has already been described. A description of the primary processing of photometric observations is given below.

As described above, the information coming from a photometer consisted of six readings of the dark background in the beginning of observations, x_{db}^o, an array of skyglow readings, and six readings of the background at the end of observations x_{db}^N. The array of skyglow readings was subdivided into groups of six numbers, each group corresponding to one cycle of sequential observation of the sky at three points and each number corresponding to a certain arrangement of the filter and mirrors, i.e., each number of this array is represented as x_{ijn}, where $i = 1, 2$ is the number of the filter position; $j = 1, 2, 3$ is the number of the mirror position, and $n = 1, 2, 3\ldots$ N is the cycle number.

The temperature at each of the three points of the sky was determined by the ratio of the readings corresponding to two different positions of the filter. To take into account a possible variation of the dark background during a night, its arithmetic mean values were determined at the beginning of observations, $\overline{x}_{db}^o = \frac{1}{6} \cdot \sum_1^6 x_{db}^N$, and at their end, $\overline{x}_{db}^N = \frac{1}{6} \cdot \sum_1^6 x_{db}^N$, and then the value of the dark background corresponding to the nth cycle was found by the formula

$$x_{db}(n) = \overline{x}_{db}^o + \frac{n}{N}(\overline{x}_{db}^N - \overline{x}_{db}^o).$$

The temperature at the jth point of the sky for the nth cycle was determined as

$$T_{jn} = \frac{x_{1j}(n) - x_{db}(n)}{x_{2j}(n) - x_{db}(n)} \cdot K,$$

where K is the calibration factor. These operations yield three series of length N for the temperature T_n that correspond to the three observation points, and each value within this series is taken with a time interval Δ. Each series can be represented as the sum of three components: a desired signal which is present in the form of one or several periodic components corresponding to IGWs of different periods (1); random noises (2); and a slowly varying low-frequency component which can be superimposed on the desired signal (3). To analyze the desired signal, it is necessary to select it from the measurement data.

The effect of noises is that they increase the variance of spectral estimates, and therefore spectral density maxima appear at high frequencies, which cannot be interpreted as IGWs. To reduce the effect of noises, it is necessary to smooth the data. Usually a smoothing filter has the form

$$\overline{T}_n = \sum_{k=-m}^{m} \omega_k \cdot T_{n+k},$$

where $\overline{T}_n$ are smoothed values of temperature and ω_k is the so-called weighting function.

For smoothing with equal weights $\omega_k = 1/(2m+1)$, the filtering procedure is equivalent to the smoothing of a time series by a rectangular window of width $(2m+1) \cdot \Delta$. The frequency characteristic of such a filter is given by

$$H(f) = \frac{\sin|(2m+1)\pi f \cdot \Delta|}{(2m+1) \cdot \sin \pi f \Delta}.$$

The value of m was chosen proceeding from the form of this frequency characteristic, and it was also taken into account that the IGW periods are longer than the Brunt–Väisälä period $\tau_g \sim 5.2\,(\text{min})$. For the photometer, $m = 3$ was chosen. In

3.6 Measurement Data Processing

processing the data of spectral observations, m was varied from 1 to 3 depending on exposure time. This corresponded to an averaging time interval of 10.5 (min) for the photometric data and 7–12 (min) for the spectral data. Moreover, smoothing of initial data reduces the variance for the process under investigation, which is especially important in processing photometer readings whose distribution is statistical in character. In this case, the statistical error of an individual measurement decreases on smoothing by a factor of $\sqrt{2m+1}$ because of the increase in the total number of photons making a contribution to this measurement.

If the observation period is less than the period of the low-frequency component, the time series under investigation represents a fluctuating process with a slowly varying mean (trend). Without going into the physical nature of the processes responsible for the occurrence of this trend, it should be noted that it distorts the spectrum of the desired signal because of low-frequency power leakage (Jenkins and Watts 1969). Therefore, to reduce the shift of spectral estimates, the trend should be filtered. A filter of first differences, $\widetilde{T}_n = \overline{T}_n - \overline{T}_{n-1}$, is generally used. However, for the case under consideration, such a filter is unacceptable since its frequency characteristic (Jenkins and Watts 1969) can distort the signal spectrum at low frequencies. Therefore, to eliminate the trend, a procedure based on selection of a trend of known functional form by the method of least squares was used. It is well known (Taranova 1967; Takahashi et al. 1974) that the overall smoothed run of the hydroxyl emission rotational temperature within a night (6–7 (h)) shows no more than one extremum (see, e.g., Fig. 3.14). Besides, the geophysical conditions of the processes responsible for the generation of IGWs in active meteorological structures and propagation of IGWs in the region of their recording indicate that it would be practical to investigate the obtained time series within periods of about 3–4 (h). All this gives sufficient grounds to approximate the trend by a parabola:

$$\overline{T}_n = a_0 + a_1 \cdot n + a_2 \cdot n^2,$$

though other analytic representations can also be used (Chetyrkin 1975).

To determine the coefficients α_o, α_1, and α_2, a system of normal equations was composed:

$$\alpha_o \cdot N + \alpha_2 \cdot \sum n + \alpha_2 \cdot \sum n^2 = \sum \overline{T}_n,$$
$$\alpha_o \cdot \sum n + \alpha_1 \cdot \sum n^2 + \alpha_2 \cdot \sum n^3 = \sum n \overline{T}_n,$$
$$\alpha_o \cdot \sum n^2 + \alpha_1 \cdot \sum n^3 + \alpha_2 \cdot \sum n^4 = \sum n^2 \overline{T}_n..$$

The determinant of this system of equations is

$$D = \begin{vmatrix} N & \sum n & \sum n^2 \\ \sum n & \sum n^2 & \sum n^3 \\ \sum n^2 & \sum n^3 & \sum n^4 \end{vmatrix} = \frac{N^3 \cdot (N^2 - 1)^2 \cdot (N^2 - 4)}{2160}.$$

With a conventional notation for the minors of this determinant: $A_{1,1}$, $A_{1,2}$, etc., the solutions are written as

$$\alpha_0 = \frac{A_{1,1} \cdot \sum \overline{T}_n + A_{2,1} \cdot \sum n \overline{T}_n + A_{3,1} \cdot \sum n^2 \overline{T}_n}{D},$$

$$\alpha_1 = \frac{A_{1,2} \cdot \sum \overline{T}_n + A_{2,2} \cdot \sum n \overline{T}_n + A_{3,2} \cdot \sum n^2 \overline{T}_n}{D},$$

$$\alpha_2 = \frac{A_{1,3} \cdot \sum \overline{T}_n + A_{2,3} \cdot \sum n \overline{T}_n + A3 \cdot \sum n^2 \overline{T}_n}{D},$$

or

$$a_0 = \frac{30}{N \cdot (N-1) \cdot (N^2-2)} \cdot \sum_{n=1}^{N} \overline{T}_n \cdot \left[n^2 - \frac{3}{5}(2N+1) \cdot n + \frac{1}{10}(3N^2 + 3N + 2) \right],$$

$$a_1 = \frac{-180}{N \cdot (N-1) \cdot (N^2-4)} \cdot \sum_{n=1}^{N} \overline{T}_n \cdot \left[n^2 - \frac{(2N+1) \cdot (8N+11)}{15(N+1)} \cdot n + \frac{1}{10}(N+2) \cdot (2N+1) \right],$$

$$a_2 = \frac{180}{N \cdot (N^2-1) \cdot (N^2-4)} \cdot \sum_{n=1}^{N} \overline{T}_n \cdot \left[n^2 - (N+1) \cdot n + \frac{1}{6}(N+1) \cdot (N+2) \right].$$

The coefficients of the parabola completely determined the character of the trend. The low-frequency component determined in such a way was then subtracted from the initial series T_n:

$$\tilde{T}_n = \overline{T}_n - (\alpha_0 + \alpha_1 \cdot n + \alpha_2 \cdot n^2).$$

All subsequent operations were performed with filtered series $\tilde{T}_n$. In many cases of data processing, the square term of the trend approximation was small enough.

The results of trial calculations have shown that the use of polynomials of higher orders for selection of the trend form is inexpedient.

The next stage in processing time series is the construction of a so-called periodogram. A time series is represented as the sum of N/2 cosine functions whose frequencies are divisible by the base frequency $f_1 = 1/N \cdot \Delta$:

$$\tilde{T}_n = \frac{2\Delta}{N} \cdot \sum_{p=1}^{N/2} B_p \cdot \cos(2\pi \cdot n \cdot f_1 + \varphi_p).$$

Here B_p is the amplitude and φ_p is the phase of the pth harmonic. The amplitude and phase can be expressed in terms of Fourier coefficients:

$$B_p(\tau_p) = \sqrt{\text{Re}^2(p) + \text{Im}^2(p)}; \quad \varphi_p(\tau_p) = \text{arctg}\left[-\frac{\text{Im}(p)}{\text{Re}(p)} \right],$$

$$\text{Re}(p) = \sum_{n=1}^{N} \tilde{T}_n \cdot \cos(2\pi \cdot p \cdot n \cdot f_1), \quad \text{Im}(p) = \sum_{n=1}^{N} \tilde{T}_n \sin(2\pi \cdot p \cdot n \cdot f_1).$$

It is more practical to use a spectrum of phases restricted by the interval $[-\pi, \pi]$. To obtain a variation of $\varphi_p(\tau_p)$ on this interval, it suffices to analyze the signs of the real and imaginary parts of the Fourier transform for the frequency f_p and to find the value of φ'_p. Thus, for finding the phase we have the following conditions:

3.6 Measurement Data Processing

$$\text{Re} = 0, \text{Im} = 0, \varphi' = 0,$$
$$\text{Re} < 0, \text{Im} = 0, \varphi' = -\pi,$$
$$\text{Re} > 0, \text{Im} = 0, \varphi' = \pi,$$
$$\text{Re} = 0, \text{Im} < 0, \varphi' = -\pi/2,$$
$$\text{Re} < 0, \text{Im} < 0, \varphi' = \varphi - \pi,$$
$$\text{Re} > 0, \text{Im} < 0, \varphi' = \varphi,$$
$$\text{Re} = 0, \text{Im} > 0, \varphi' = \pi/2,$$
$$\text{Re} < 0, \text{Im} > 0, \varphi' = \varphi + \pi,$$
$$\text{Re} > 0, \text{Im} > 0, \varphi' = \varphi.$$

If the length of the observation data series is small, the frequency step $\Delta f = 1/N \cdot \Delta$ can be too large and there is a risk to miss a maximum of interest in the periodogram. Therefore, to obtain a more detailed plot of the spectral density, the latter should be calculated with a step smaller than $1/N \cdot \Delta$ by a factor of two or three. Though in this case the spectral density estimates at the next frequencies become correlated, the positions of maxima in periodograms can be determined with high accuracy.

Subsequently the spectra of the phases of temperature time series measured at three points of the sky are used to determine the azimuth and velocity of propagation of IGWs. For a wave of frequency $\omega = 2\pi f$ which propagates in the xy-plane with velocity C (see Fig. 3.43), the wave vector is determined as $K = \frac{\omega}{|C|} \cdot C$. The oscillation phase φ_i, measured at a point with coordinates (x_i, y_i), is related to the wave vector components k_x and k_y as

$$\varphi_i = \varphi_0 + k_x \cdot x_i + k_y \cdot y_i,$$

where φ_0 is the phase at the point zero. It is well known that

$|k| = \sqrt{k_x^2 + k_y^2}$; the velocity of a wave $C = \frac{\omega}{|k|} = \frac{2\pi}{|k|}$; and its azimuth $A^* = \text{arctg}\left(-\frac{k_y}{k_x}\right)$.

However, the direction of counting of the angle A^* differs from the astronomical azimuth A and is related to it as $A = A^* + 270°$. Therefore, we have

$$A = \text{arcctg}\left(-\frac{k_y}{k_x}\right) = \text{arctg}\frac{k_x}{k_y}.$$

Thus, the values of k_x and k_y are determined for the base triangle from the system of equations

$$k_x \cdot x_1 + k_y \cdot y_1 + \varphi_0 = \varphi_1;$$
$$k_x \cdot x_2 + k_y \cdot y_2 + \varphi_0 = \varphi_2;$$
$$k_x \cdot x_3 + k_y \cdot y_3 + \varphi_0 = \varphi_3.$$

In our case, it turned out that a simpler way was to determine the wave azimuth and velocity for all calculated frequency values and detect a wave by the spectral density maximum at one or another frequency at all three vertices of the measuring triangle.

To validate this way of determination of the wave parameters at a given frequency, the correlation between the temperature variations at three points was estimated by means of the coefficient of coherence (Potapov et al. 1978).

When all three coefficients of coherence at a given frequency f were greater than some threshold value depending on the data series length, it was considered that a wave with a period $\tau = 1/f$ passing through three points of the firmament had been detected. For usually investigated data series the threshold value of the coefficient of coherence was chosen in the range from 0.6 to 0.9, depending on the data series length, following a well-known procedure (Jenkins and Watts 1969). It should be noted that practically in all cases the presence of a spectral density maximum was confirmed by a high value of the coefficient of coherence.

The above algorithm served as a basis in developing data processing codes.

The operation of a code was first mustered on artificial series. Three series were formed, each consisting of two sinusoids of different periods shifted in phase relative to each other. Then these series were clogged with random numbers distributed by a normal law. The variance of these random numbers, i.e., the noise level, varied over a wide range. As a result of the operation of the code it turned out that the parameters of a wave (amplitude, period, azimuth, and velocity of propagation) can be determined confidently for the noise level approaching 100(%) of the wave amplitude.

The error of the determination of the wave phase, $\Delta\varphi$, was ~ 0.3 (rad), while the errors of the period, $\Delta\tau$, was of the order of the sampling interval.

In particular cases, in addition to harmonic Fourier analysis, other methods should be used. In a spectral analysis of oscillatory processes within obvious time intervals, such as a day, the phase age of the Moon, and a year, harmonics are calculated whose periods are divisible by these time intervals. For oscillatory processes with fuzzy time limits, such as biennial, formal procedures of Fourier analysis or wavelet transformation (Astafieva 1996), which have been widely used for many years, are inefficient. Repeated analyses by these methods gave no insight into the physical nature of such a process.

It turned out that to elucidate the character of the quasi-biennial variations of both solar activity and atmospheric phenomena, numerical filtering of data is necessary for eliminating both seasonal and 11-yr variations, which was proposed by Ivanov-Kholodnyi et al. 2000a,b). This was considered in detail in Sect. 1.6.2. Here it can be mentioned that the filtering procedure involved the calculation of the second differences of the time series aimed at elimination of the seasonal variations and solar activity effects during the cycle. It turned out that this process is not harmonic (Fadel et al. 2002; Kononovich and Shefov 2003) and it is described by the Airy function. Therefore, the results obtained by harmonic methods (Labitzke and van Loon 1988; Gruzdev and Bezverkhnii 1999, 2000, 2003) formally yield a mean period and give no idea of the physical process (Shefov and Semenov 2006).

For cases like this, the "Caterpillar" method of principal terms of time series (Danilov and Zhiglyavsky 1997) can be successfully used for which a developed processing code is given in the cited publication.

Attempts to analyze the long-term series of 11-yr solar cycles with the use of Fourier transforms do not consider the circumstance that the variations in each cycles are determined by aperiodic relaxation processes which cannot be "squeezed" in any harmonic sequence.

3.6.6 Spectral Characteristics of IGW Trains Detected in the Upper Atmosphere

The observed variance of the amplitudes of internal gravity waves (IGWs) detected in the mesopause region in their spectral dependence on period has a normal distribution and it is substantially greater than possible errors.

Spectral analysis of the wave trains of short duration generated by tropospheric gusts shows that this is due to the fact that the amplitudes of detected spectral components depend on the degree of attenuation of the oscillations in the train. The degree of attenuation, in turn, depends on the conditions of IGW generation which are determined by the interaction of the wind flow with an obstacle. To reveal these conditions, the harmonics were calculated analytically for various durations of the interval of the detected train portion and its position relative to the beginning of the train.

Long-term observations of IGWs in the upper atmosphere, in particular, measurements of the characteristics of hydroxyl emission (Krassovsky et al. 1977, 1978; Novikov 1981; Spizzichino 1971), have revealed that the dependence of the wave amplitude δT on period τ is described by the universal formula

$$(\delta T/T)^2 \sim \tau^{5/3},$$

where T is the atmospheric temperature.

Records of wave disturbances in the downwind regions of the Urals (Semenov et al. 1981; Shefov et al. 1983; Shefov and Pertsev 1984) and Caucasus Mountains (Sukhodoev et al. 1989a,b; Sukhodoev and Yarov 1998; Shefov et al. 1999, 2000) have shown that mountain obstacles generate various wave trains whose phases depend on the speed and fluctuations of the slip wind.

The observed variance of some values of wave amplitudes relative to the average regression dependence on period is greater than possible measurement errors and, therefore, it has a real physical content. Besides the seasonal variations of atmospheric characteristics (Semenov 1988), the finite duration of the generated wave trains and the random relative time position of a train and the time period of its recording should also have an effect. This inevitably should result in underestimated IGW amplitudes.

The spectral dependences of IGW amplitudes δT on periods τ reported in a number of publications (Krassovsky et al. 1977, 1978; Novikov 1981; Spizzichino 1971) were obtained by spectral analysis of time series of length from two to several hours. It is obvious, however, that this time interval was arbitrarily superimposed on the wave train.

The measurement data obtained near the Caucasian ridge with an eight-channel photometer, whose visual fields were distributed in a certain manner in the region of the hydroxyl emission layer over the downwind region of the mountains, have revealed a substantial nonuniformity of the wave field in the horizontal region covered by the eight-channel device (∼50 (km)) (Sukhodoev et al. 1989a,b). According to the measurements, appreciable fluctuations of the wave amplitudes and periods were observed in some sections of the emission layer (at ∼87 (km)) which seem to result from the irregularity of the underlying surface and wind field and from fluctuations of gusts responsible for the appearance of local wave sources. Undoubtedly, for sighting regions corresponding to large zenith angles, additional spatial smoothing of the recorded waves by the apertures of spectroscopic instruments can take place (Shefov 1989).

Analysis of the dependences of the calculated amplitudes of IGWs on their periods (Novikov 1981) shows (Fig. 3.48) that there is some difference from the experimental, reliably established $\tau^{5/3}$ dependence (Krassovsky et al. 1977, 1978) and also from a theoretically derived relation (Obukhov 1988). Figure 3.49 presents statistical distributions of the deviations of measured amplitudes, $\Delta \log_{10}(\delta T/T)$, from the relation $\delta T/T \propto \tau^{5/6}$ (Shefov and Semenov 2004). The sampling was made for several intervals of periods ($\Delta \log_{10} \tau = 0.4$). As follows from these data, the dis-

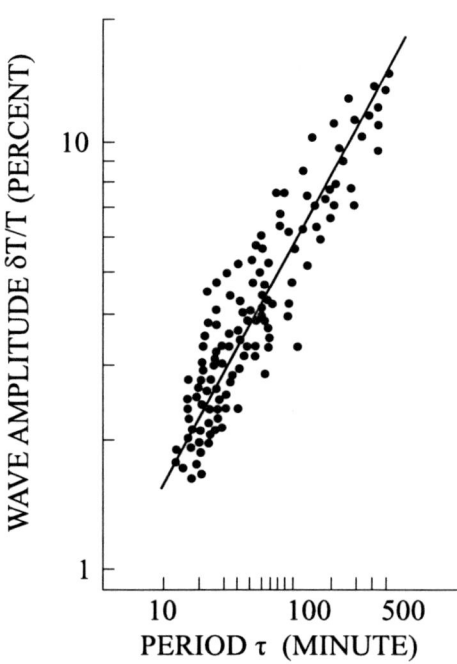

Fig. 3.48 Dependence of the IGW amplitudes on their period (Novikov 1981)

3.6 Measurement Data Processing

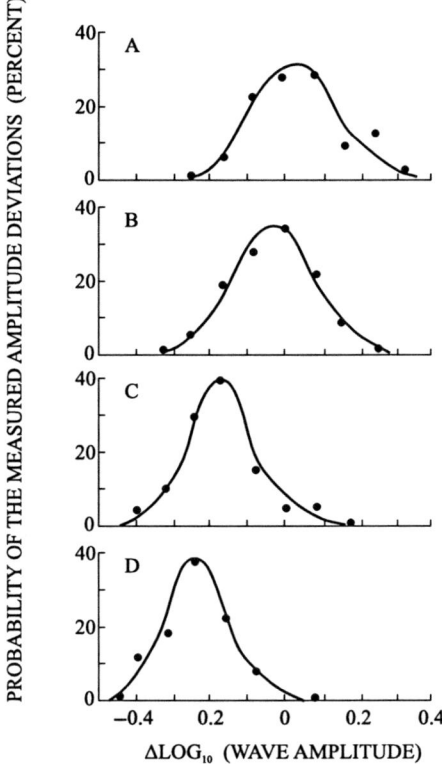

Fig. 3.49 Mean distributions of the number of deviations (N) of the measured IGW amplitudes, $\Delta\log_{10}(\delta T/T)$ from the relation $\delta T/T \sim \tau^{5/6}$ for various period intervals τ ($\Delta\log_{10}\tau = 0.4$). Time intervals: 12–32 (**A**), 32–80 (**B**), 80–200 (**C**), and 200–500 (**D**) (Shefov and Semenov 2004)

tributions are well described by the normal law (correlation factor r = 0.999, 0.997, 0.995, and 0.999, respectively), suggesting that the observed deviations were random in character. The variances σ of these distributions decrease with increasing periods τ according to the empirical relation

$$\sigma\left[\Delta\log_{10}\frac{\delta T}{T}\right] = (0.248 \pm 0.012) - (0.054 \pm 0.0058)\log_{10}\frac{\tau_m}{\tau_0},$$

where $\tau_m = 22, 56, 140$, and 350 (min) and, for the chosen intervals, $\tau_0 = 1$ (min).

It can clearly be seen that the calculated amplitudes decrease with increasing periods. It should be stressed that, according to the requirements of the standard spectral analysis, the condition $T_0/\tau \geq 3$, where T_0 is the analyzed time series interval, was satisfied. The decrease in IGW amplitudes for the periods $\sim$7–15 (min) is most likely due to the fact that not all IGWs pass through the middle atmosphere (Gossard and Hooke 1975).

The variations of the revealed (measured) amplitude deviations (Novikov 1981) from the theoretical relation can be approximated by a square polynomial in $\log_{10}\tau$:

$$\log_{10}\left[\frac{\left(\frac{\delta T}{T}\right)_{meas}}{\left(\frac{\delta T}{T}\right)_{theor}}\right] = -0.2856 + 0.4221 \cdot \log_{10}\frac{\tau}{\tau_0} - 0.1642 \cdot \left(\log_{10}\frac{\tau}{\tau_0}\right)^2,$$

where τ is expressed in minutes and $\tau_0 = 1$ (min).

Theoretical estimates of the possible oscillations propagating upward from the regions of their origin testify to the attenuation character of the trains including several periodic oscillations. The basic time dependence of the oscillation amplitudes has the form (Chunchuzov 1988, 1994)

$$\delta T = \delta T_{00}\left(\frac{2\pi t}{\tau}\right)^{1/2}\exp\left(-\frac{2\pi t}{\alpha\tau}\right)\cos\left(\frac{2\pi t}{\tau}+\frac{\pi}{4}\right).$$

Here τ and δT_{00} are the wave period and amplitude, respectively; t is the time elapsed from the beginning of the wave train; α is the attenuation parameter determined by the width a of the obstacle of height Z to which the wave propagates and by the zenith angle of its propagation θ, namely,

$$\alpha = \frac{Z}{a \cdot \sin\theta \cdot \cos\theta}$$

or

$$\alpha = \frac{\frac{Z}{a}\left(\frac{\tau}{\tau_g}\right)^2}{\left[\left(\frac{\tau}{\tau_g}\right)^2 - 1\right]^{1/2}} \sim \frac{Z}{a}\cdot\frac{\tau}{\tau_g}$$

since the Brunt–Väisälä period $\tau_g \sim 5.2$ (min) for altitudes up to 100 (km). For the initial time t_0 of the measurement interval from 0 to T_0 the formula for the train oscillation amplitude can be represented as

$$\delta T = \delta T_{max}\sqrt{2e\frac{2\pi(t-t_0)}{\alpha\tau}}\exp\left[-\frac{2\pi(t-t_0)}{\alpha\tau}\right]\cos\left[\frac{2\pi(t-t_0)}{\tau}+\frac{\pi}{4}\right],$$

since the maximum amplitude δT_{max} of the curve enveloping the wave train is given by

$$\delta T_{max} = \delta T_{00}\cdot\sqrt{\frac{\alpha}{2e}}.$$

This formula takes into account the circumstance that the oscillation could appear earlier or later than the starting of the procedure of recording of the radiation modulated by this oscillation.

Fig. 3.50 Examples of wave trains for various values of the parameter α (Shefov and Semenov 2004)

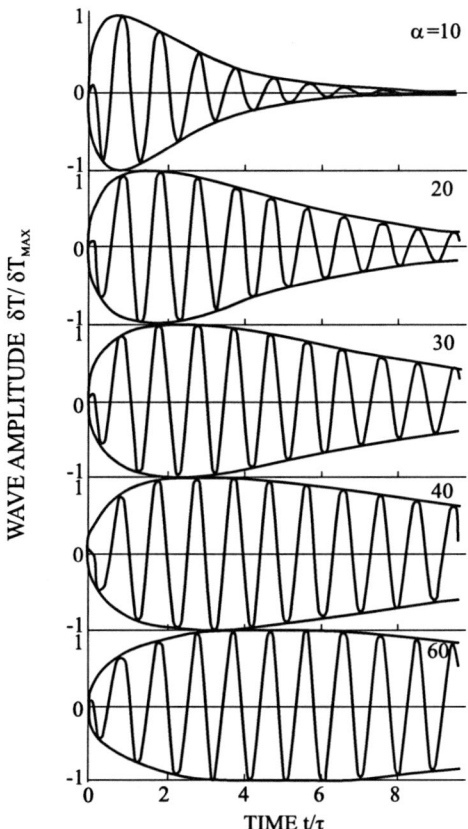

Examples of trains calculated by this formula for various values of α are shown in Fig. 3.50.

In a Fourier analysis of the oscillations observed during a time period T_0 for the above train of damped oscillations, the Fourier coefficients are determined by the formulas

$$a_n = 2\sqrt{\frac{2e}{\alpha}} \cdot \frac{\delta T_{max}}{T_0} \int_0^{T_0} \left[\frac{2\pi(t-t_0)}{\tau}\right]^{1/2} \cdot \exp\left[\frac{2\pi(t-t_0)}{\alpha\tau}\right] \cdot \cos\left[\frac{2\pi(t-t_0)}{\tau} + \frac{\pi}{4}\right] \cdot \cos\frac{2\pi nt}{T_0} dt,$$

$$b_n = 2\sqrt{\frac{2e}{\alpha}} \cdot \frac{\delta T_{max}}{T_0} \int_0^{T_0} \left[\frac{2\pi(t-t_0)}{\tau}\right]^{1/2} \cdot \exp\left[\frac{2\pi(t-t_0)}{\alpha\tau}\right] \cdot \cos\left[\frac{2\pi(t-t_0)}{\tau} + \frac{\pi}{4}\right] \cdot \sin\frac{2\pi nt}{T_0} dt.$$

Introducing new designations by the relations

$$X = t/T_0, \ X_0 = t_0/T_0, \ 2\pi T_0/\tau = 2\pi k,$$

we obtain

$$a_n = \delta T_{max}\sqrt{\frac{2e}{\alpha}}\sqrt{2\pi k}\int_{-X_0}^{1-X_0}\sqrt{Z}\exp\left(-\frac{2\pi k(X-X_0)}{\alpha}\right)$$
$$\times\left\{\cos\left[2\pi(k-n)Z-(2\pi nX_0-\frac{\pi}{4})\right]+\cos\left[2\pi(k+n)Z+(2\pi nX_0+(2\pi nX_0)+\frac{\pi}{4})\right]\right\}dZ,$$

$$b_n = \delta T_{max}\sqrt{\frac{2e}{\alpha}}\sqrt{2\pi k}\int_{-X_0}^{1-X_0}\sqrt{Z}\exp\left(-\frac{2\pi k(X-X_0)}{\alpha}\right)$$
$$\times\left\{-\sin\left[2\pi(k-n)Z-(2\pi nX_0-\frac{\pi}{4})\right]+\sin\left[2\pi(k+n)Z+(2\pi nX_0+\frac{\pi}{4})\right]\right\}dZ,$$

whence

$$a_n = \delta T_{max}\sqrt{\frac{2e}{\alpha}}\sqrt{2\pi k}\left[C_{1n}\cos(2\pi nX_0-\frac{\pi}{4})+C_{2n}\cos(2\pi nX_0+\frac{\pi}{4})\right.$$
$$\left.+S_{1n}\sin(2\pi nX_0-\frac{\pi}{4})-S_{2n}\sin(2\pi nX_0+\frac{\pi}{4})\right],$$

$$b_n = \delta T_{max}\sqrt{\frac{2e}{\alpha}}\sqrt{2\pi k}\left[C_{1n}\sin(2\pi nX_0-\frac{\pi}{4})+C_{2n}\sin(2\pi nX_0+\frac{\pi}{4})\right.$$
$$\left.-S_{1n}\cos(2\pi nX_0-\frac{\pi}{4})+S_{2n}\cos(2\pi nX_0+\frac{\pi}{4})\right],$$

where

$$C_{1n} = \int_{-X_0}^{1-X_0}\sqrt{Y}\cdot\exp\left(-\frac{2\pi kY}{\alpha}\right)\cdot\cos\left[2\pi(k-n)Y\right]dY,$$

$$C_{2n} = \int_{-X_0}^{1-X_0}\sqrt{Y}\cdot\exp\left(-\frac{2\pi kY}{\alpha}\right)\cdot\cos\left[2\pi(k+n)Y\right]dY,$$

$$S_{1n} = \int_{-X_0}^{1-X_0}\sqrt{Y}\cdot\exp\left(-\frac{2\pi kY}{\alpha}\right)\cdot\sin\left[2\pi(k-n)Y\right]dY,$$

$$S_{2n} = \int_{-X_0}^{1-X_0}\sqrt{Y}\cdot\exp\left(-\frac{2\pi kY}{\alpha}\right)\cdot\sin\left[2\pi(k+n)Y\right]dY,$$

where $Y = X - X_0$. Calculation of the functions C and S based on the data by Prudnikov et al. (1981) yields the relations

3.6 Measurement Data Processing

$$CC(W) = \frac{\sqrt{\pi}(U^2+V^2)}{2(\beta^2+\xi^2)^{3/4}} \cdot \cos(\frac{3}{2}\mu - \delta) - \frac{\sqrt{W}}{\beta^2+\xi^2} \cdot \exp(-\beta W) \cdot \cos(\mu + \xi W),$$

$$SS(W) = \frac{\sqrt{\pi}(U^2+V^2)}{2(\beta^2+\xi^2)^{3/4}} \cdot \sin(\frac{3}{2}\mu - \delta) - \frac{\sqrt{W}}{\beta^2+\xi^2} \cdot \exp(-\beta W) \cdot \sin(\mu + \xi W),$$

where

$$\sin\mu = \frac{\xi}{\sqrt{\beta^2+\xi^2}}; \qquad \cos\mu = \frac{\beta}{\sqrt{\beta^2+\xi^2}}.$$

Here $\beta = 2\pi k/\alpha$, $\xi = 2\pi (k \pm n)$,

$$\cos\delta = \frac{U}{\sqrt{U^2+V^2}}, \qquad \sin\delta = \frac{V}{\sqrt{U^2+V^2}}.$$

The functions U and V correspond to the representation of the function

$$\mathrm{erf}\, z = \frac{2}{\sqrt{\pi}} \cdot \int_0^z e^{-x^2} dx$$

of a complex argument in the form

$$\mathrm{erf}\,(\rho + i\eta) = U(\rho,\eta) + iV(\rho,\eta),$$

where the argument is determined by the relation

$$\rho + i\eta = \sqrt{(\beta + i\xi)W} = (\beta^2+\xi^2)^{1/4} \cdot \sqrt{W} \cdot \cos\frac{\mu}{2} + i \cdot (\beta^2+\xi^2)^{1/4} \cdot \sqrt{W} \cdot \sin\frac{\mu}{2}.$$

As follows from the above formulas,

$$\rho = 2\sqrt{\frac{\pi k}{\alpha}} \cdot \left(\sqrt{\alpha^2 \cdot \left(1 \pm \frac{n}{k}\right)^2 + 1} + 1\right)^{1/2} \cdot \sqrt{W},$$

$$\eta = 2\sqrt{\frac{\pi k}{\alpha}} \cdot \left(\sqrt{\alpha^2 \cdot \left(1 \pm \frac{n}{k}\right)^2 + 1} - 1\right)^{1/2} \cdot \sqrt{W},$$

and $\rho\eta = 4\pi(k \pm n)W$. The sign "+" refers to the functions C_{2n} and S_{2n} and the sign "−" to the functions C_{1n} and S_{1n}.

Following Abramowitz and Stegun (1964), we have

$$U(\rho,\eta) = \mathrm{erf}\rho + \frac{2e^{-\rho}}{\pi} \cdot \left[\frac{1-\cos 2\rho\eta}{4\rho} + \sum_{m=1}^{\infty} \frac{e^{-\frac{m}{2}}}{m^2+4\rho^2} \cdot f_m(\rho,\eta)\right],$$

$$V(\rho,\eta) = \frac{2e^{-\rho}}{\pi} \cdot \left[\frac{\sin 2\rho\eta}{4\rho} + \sum_{m=1}^{\infty} \frac{e^{-\frac{m}{2}}}{m^2+4\rho^2} \cdot g_m(\rho,\eta)\right],$$

where

$$f_m(\rho, \eta) = 2\rho - 2\rho\,\text{ch}\,m\eta \cdot \cos 2\rho\eta + m\,\text{sh}\,m\eta \cdot \sin 2\rho\eta,$$
$$g_m(\rho, \eta) = 2\rho \cdot \text{ch}\,m\eta \sin 2\rho\eta + m\,\text{sh}\,m\eta \cdot \cos 2\rho\eta.$$

For actual calculations $m = 5$ is sufficient.

For a real variable the function erf x has an approximation (Abramowitz and Stegun 1964)

$$\text{erf}\,x = 1 - (a_1 t + a_2 t^2 + a_3 t^3)\exp(-x^2),$$
$$t = 1/(1+rx),\ r = 0.47047,\ a_1 = 0.3480242,\ a_2 = -0.0958798,$$
$$a_3 = 0.7478556.$$

Thus,

$$\begin{aligned}
C_{1n} &= CC(1-X_0) - CC(-X_0) & \xi &= 2\pi(k-n),\\
C_{2n} &= CC(1-X_0) - CC(-X_0) & \xi &= 2\pi(k+n),\\
S_{1n} &= SS(1-X_0) - SS(-X_0) & \xi &= 2\pi(k-n),\\
S_{2n} &= SS(1-X_0) - SS(-X_0) & \xi &= 2\pi(k+n),
\end{aligned}$$

and in all relations

$$\beta = \frac{2\pi k}{\alpha}.$$

The wave amplitude is $\delta T_n = \sqrt{a_n^2 + b_n^2}$ or

$$\delta T_n = \delta T_{\max} \cdot \sqrt{\frac{4\pi ek}{\alpha}} \cdot \sqrt{(C_{1n} - S_{2n})^2 + (C_{2n} - S_{1n})^2}.$$

The oscillation phase for the representation of the components in the form $\cos(2\pi n Y - \varphi_n)$ is defined as

$$\varphi_n = \text{arctg}(b_n/a_n)$$

or

$$\varphi_n = (\kappa_n - \nu_n)/2 + 2\pi n X_0 - \pi.$$

For the beginning of the train the phase is given by $\varphi_n^0 = 2\pi n X_0$.
By virtue of the definition of the train waveform, we have $\pi - (\kappa_n - \nu_n)/2 = \pi/4$.
Here,

$$\cos\kappa_n = \frac{C_{2n}}{\sqrt{C_{1n}^2 + S_{1n}^2}};\qquad \sin\kappa_n = \frac{S_{2n}}{\sqrt{C_{1n}^2 + S_{1n}^2}};$$

$$\cos\nu_n = \frac{C_{1n}}{\sqrt{C_{1n}^2 + S_{1n}^2}};\qquad \sin\nu_n = \frac{S_{1n}}{\sqrt{C_{1n}^2 + S_{1n}^2}}.$$

3.6 Measurement Data Processing

This eventually allows one to determine which part of the wave train was measured.

Based on the relations obtained, the functions of attenuation of the calculated wave amplitude δT_n related to its real value δT_{max} were constructed: $\delta T_n/\delta T_{max} = f(\alpha, t/\tau, T_0/\tau)$. The results are presented in Fig. 3.51. It can be seen that in all cases the calculated amplitudes of IGW harmonics are less than the maximum amplitude of the train. The optimum measurement conditions corresponded to the situation where the analyzed time interval was five times greater than the wave period, instead of three times as usual. Besides, it is necessary that the oscillation under investigation completely fill the analyzed time interval. In this case, the value of α should be within the range 25–60. Naturally, if the analyzed time interval is taken only by a portion of the train, the calculated amplitudes will by considerably lower than the real ones.

As follows from the calculations, the calculated wave amplitude strongly depends on the ratio of the integer number of train periods to the duration of the analyzed time interval, namely,

$$\delta T_n/\delta T_{max} \sim \left| \frac{n\tau}{T_0} - 1 \right|^{15}.$$

Thus, as this ratio deviates from unity by 0.04, the amplitude halves. Therefore, it is necessary to vary sequentially the analyzed interval to detect (n) periods divisible by this interval since the sought-for IGW period is the nth harmonic of the analyzed interval T_0.

The generation of IGWs is usually related to the region of interaction of a jet stream and a baric structure of cold front (Krassovsky et al. 1977, 1978). The jet stream of moderate latitudes (45–65°N) is usually at altitudes of 8–12 (km) (Zverev 1977); the width of the region of IGW generation is 50–200 (km). For these conditions we have $\alpha \sim (0.1$–$0.5)\tau$. Thus, for waves with periods of 15–30 (min), the amplitudes decrease by a factor of $\sim$0.9, which agrees with the measurement data (Novikov 1981) based on which the formula approximating the deviations of the measured amplitudes from their theoretical values was obtained.

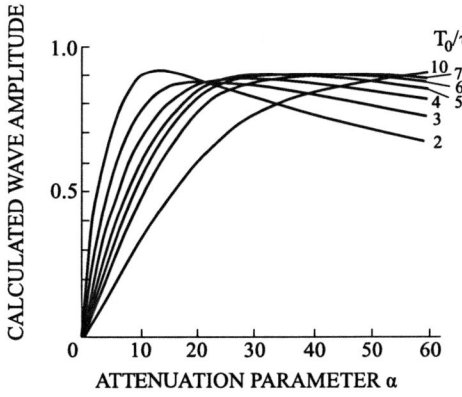

Fig. 3.51 Effective relative wave amplitudes obtained by spectral analysis of the interval of a time series having different characteristics at different values of T_0/τ (Shefov and Semenov 2004)

The above analysis of the expected amplitudes of recordable IGWs has shown that their values make no more than 90(%) of the maximum amplitude of oscillations in the train. The optimum measurement conditions correspond to the situation where the analyzed time interval is five times greater than the wave period, and it is completely filled with the oscillation under investigation. In this case, the value of α should be in the range 25–60, which corresponds to periods of 15–50 (min). The proportion between the calculated harmonic amplitudes can serve as an indicator of the character of the IGW train under investigation (Shefov and Semenov 2004).

3.7 Calibration of Measurements

3.7.1 Calibration of the Characteristics of Spectrophotometric Instruments

The spectral properties peculiar to all spectrophotometric instruments are determined by the sensitivity of the radiation detectors and the transparency of the optics. Therefore, these instruments should be calibrated to give absolute values of emission intensities. For various types of light measurements various systems of units are used depending on the properties of the phenomena under investigation. For optical investigations of the Earth's airglow, typical parameters are the emission wavelength (nm) and intensity (Rayleigh = 10^6 (photon $\cdot$ cm$^{-2} \cdot$ s^{-1})). This is conditioned by the fact that the airglow emissions arise due to photochemical processes with participation of atoms and molecules which convert the solar energy absorbed by the atmosphere into radiation. The rate of this conversion can be estimated by measuring the intensities of the emissions under observation.

Notwithstanding that the airglow is spatially nonuniform in structure (rather fine scale: from several kilometers to several tens of kilometers, i.e., $1°–10°$), most of the spectrophotometric instruments perceive the sky glow as a spatial radiation source which fills up the instrument aperture. Therefore, the calibration procedure demands that the instrument aperture be filled by the light from a reference source. Used for the reference source was a screen whose scattered radiation filled up the instrument aperture.

3.7.1.1 Optical Characteristics of Spectrographs and Spectrometers

As already mentioned, the key feature of spectrophotometric measurements of the night airglow is that the optical object is a spatial radiation source. Therefore, a luminous flux is measured which is perceived by the entrance optical system within its visual angle. This determines the conditions for which absolute intensities of emissions are calculated.

Calibration of spectrophotometric devices implies determination of a variety of their parameters. For spectrographs, these are the linear dispersion and the shape and width of the instrumental profile on the photographic image of a narrow spectral line at a given width of the entrance slit. Thus, for SP-48 and SP-50 spectrographs, the focal distance ratio of the collimating and the camera lens is 9 and 6.1, respectively.

Minimum image size on the photofilm was determined by light scattering in photosensitive layer. It was 0.03 (mm). Therefore, the projection of this size on the entrance slit was ~0.2 (mm). This was the optimum width of the slit, since a reduction of the slit width could not improve resolution, but only resulted in light flux losses. The spectral line width was 0.35 (nm) for the SP-48 and 0.4 (nm) for the SP-50 spectrograph.

For the spectral ranges inherent in these spectrographs, the dispersion curves were slightly nonlinear.

The spectral sensitivity should be determined within the spectral range used. A shift of the spectral range called for a new determination of the wavelength dependence of emission intensity because of the vignetting effects that occurred in the optics at the edges of the spectral range. These effects were most pronounced when image converter tubes were used because of the distortion of the spectrum image on the screen, such that the spectral line width was the narrowest at the screen center and increased toward the edges. This, naturally, resulted in a gradual broadening of the instrumental profile of the spectral lines from the center of the spectrogram to its edges. ICT with fiber glass tube at the cathode and screen moderated this effect, but the curved focal surface of the camera lens also distorted the spectrum image on the flat outside surface of the fiber spacer with a photocathode.

As to spectrophotometers, certain requirements should also be fulfilled. In the case under consideration, the optical system, similar to that of the DFS-14 spectrometer (Vaisberg 1960), had the collimating and camera lenses with the same focal distance. To provide a greater luminous flux, not only mirrors and a diffraction grating of larger dimensions were used, but also the heights of the slits were increased. In doing this, the curvature of the spectral lines at the outlet slit was taken into account.

3.7.1.2 Characteristics of Electrophotometers

In contrast to the spectrographs and spectrophotometers considered above, the photometers with light filters were specially fabricated for realization of concrete research programs.

Prior to performing spectral measurements for an airglow emission one should know its intensity and space–time characteristics. This necessitates calibration of the photometer.

An important stage in photometer calibration is the investigation of the function of the photometer visual field, ω. Knowledge of this function is necessary to estimate the effective solid angle and the contribution of light from individual stars, which may significantly increase the signal intensity when these stars pass through the visual field. This can serve as an additional check of the absolute calibration if the characteristics of the passing star are known.

Experimental investigations show that the field function ω can be rather well approximated by the expression (Miroshnikov 1977)

$$\omega = \cos^k\left(\frac{\pi}{2} \cdot \frac{r_X}{r_0}\right),$$

where r_0 is the radius of the field diaphragm aperture and r_X is the distance from the optical axis (Fig. 3.52).

Since the luminous flux measured by a photometer is determined by the solid angle, the existence of the field function, playing the role of a weighting function η, makes the effective solid angle smaller than the geometric angle

$$\omega_G = \frac{\pi \cdot r_0^2}{f^2},$$

where f is the focal distance of the lens, i.e.,

$$\omega_{\text{eff}} = \eta \cdot \omega_G,$$

where

$$\eta = \frac{8}{\pi^2} \cdot \int_0^{\pi/2} x \cdot \cos^k x \cdot dx.$$

Numerical evaluation of the integral yields the following analytic approximation:

$$\eta = 0.99 - 0.366 \cdot k + 0.613 \cdot k^2 - 0.201 \cdot k^3 + 0.025 \cdot k^4.$$

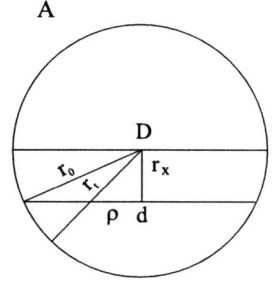

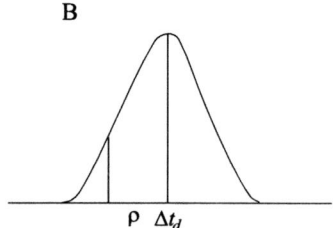

Fig. 3.52 Geometry of measuring the field function of a photometer with a star passing through its field of vision. (**A**) Field of vision; (**B**) record of the signal from a star passing along a chord d at a distance r_x from the field center

3.7 Calibration of Measurements

The plots of ω and η are presented in Fig. 3.53. For k = 2, which closely corresponds to the real geometry of a vignetted field, the analytic representation yields

$$\eta(2) = \frac{1}{2} - \frac{2}{\pi^2} = 0.297.$$

The function ω_{eff} was determined experimentally by recording the signals of a photometer as its field of vision was crossed by bright stars. For this purpose, the photometer was fixed at a given zenith angle. Since a star moves in the photometer field with constant velocity, the signal waveform will depend on the distance r_X from the field center. When a star crosses the field along its diameter D, the passage time Δt_D depends on the declination of the star:

$$\Delta t_D = \frac{4D(\text{degree})}{\cos\delta},$$

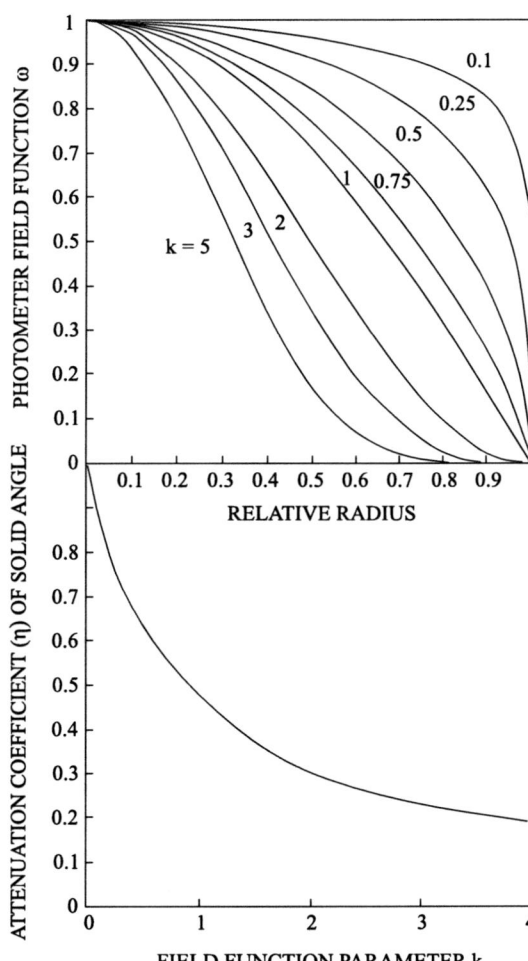

Fig. 3.53 The field function ω of a photometer and the effective solid angle η versus parameter k

where

$$\sin\delta = \sin\varphi \cdot \cos\zeta - \cos\varphi \cdot \sin\zeta \cdot \cos A,$$

where φ is the geographic latitude, ζ is the zenith angle of sight, A is the azimuth which is counted from the south point to the west. The length d of the chord along which a star passes (Fig. 3.54) is determined as

$$d = \sqrt{D^2 - 4r_X^2},$$

whence the distance r_X (deg) is

$$r_X = \frac{\cos\delta}{8} \cdot \sqrt{\Delta t_D^2 - \Delta t_d^2},$$

where the times Δt_D and Δt_d are expressed in minutes. In analyzing the waveform of the signal from a star passing through the photometer field the relative distance from the center is determined by the formula (Fig. 3.54)

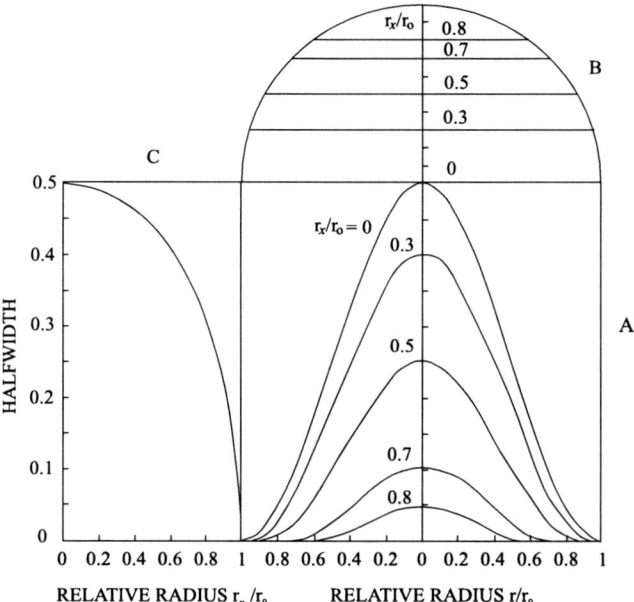

Fig. 3.54 Calculated relative values of the signal from a star (**A**) passing through the photometer field along chords at various distances r_X/r_0 from the field center (**B**) on the assumption that the field function is $\omega = \cos^2\left(\frac{\pi}{2} \cdot \frac{r_X}{r_0}\right)$. The halfwidth of the profile in fractions of the field geometric diameter as a function of the relative distance from the field center (**C**)

3.7 Calibration of Measurements

$$\frac{r_X}{r_0} = \sqrt{1 + \left[\left(\frac{2\rho}{\Delta t_d}\right)^2 - 1\right] \cdot \left(\frac{\Delta t_d}{\Delta t_D}\right)^2},$$

where $D = 2r_0$, ρ is the distance of the star along a chord from the chord middle point (mm), and r_X is the distance from the field center to the star moving along the chord.

Measurements performed for the photometer used for orographic measurements in Kislovodsk have shown that the field function can be rather well described by (Fig. 3.55)

$$\omega = \cos^{1.7}\left(\frac{\pi}{2} \cdot \frac{r}{r_0}\right),$$

which gives $\eta = 0.341$. The dependence of η on the field function exponent k is shown in Fig. 3.53.

For this case, Fig. 3.54 presents the relative values of the signal from a star passing through the photometer field along chords at various distances r_X/r_0 from the field center. The halfwidth of the signal profile (in fractions of the angular geometric diameter D) is given by

$$\Delta = \sqrt{\left\{\frac{2}{\pi} \cdot \arccos\left[\frac{1}{\sqrt{2}} \cdot \cos\left(\frac{\pi}{2} \cdot \frac{r_X}{r_0}\right)\right]\right\}^2 - \left(\frac{r_X}{r_0}\right)^2}.$$

Many difficulties emerged when interference light filters were used whose spectral characteristics were not always suited to the research goals. Moreover, the actual properties of filters of this type depend substantially on the conditions under which they are used.

Before using an interference light filter, its characteristics should be checked. The matter is that in manufacturing a light filter for a certain spectral range, where it should have a narrow transmission band, $\Delta\lambda \sim 1\text{--}2\,(\text{nm})$ near the wavelength λ_0, an additional transmission band of width $\sim 50\,(\text{nm})$ always appears in a more

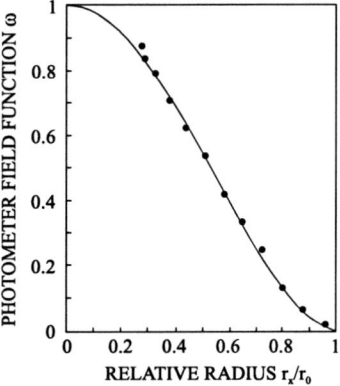

Fig. 3.55 The measured field function of a two-channel photometer. Points are measurement data, solid line is calculation by the formula $\omega = \cos^{1.7}\left(\frac{\pi}{2} \cdot \frac{r_X}{r_0}\right)$

long-wave spectral range near $\lambda \sim 1.2\lambda_0$. If the spectral sensitivity range of a radiation detector covers this spectral range, it is necessary to use additional light filters.

The operating conditions of a light filter in the light beam of a photometer are of great importance. If the filter is in a converging beam either in front of the lens or inside the photometer before the field diaphragm, the transmission band will be expanded and shifted toward the short-wave spectral range. Therefore, the real transmission curve of a light filter should be measured either immediately on placing it in a photometer or in conditions corresponding to the given place in the photometer. Since the transmission band of a filter shifts to the short-wave spectral range and expands on varying the slope of the filter relative to the optical axis, this should be a subject of investigation.

The transmission band of a light filter is sensitive to ambient temperature. Therefore, it is necessary to stabilize the temperature to within $\sim 1(^\circ C)$. Moreover, the ageing of a light filter results in a shift of the transmission band to the short-wave range. Therefore, it would be expedient that narrow-band filters intended for some special use be manufactured to pass a greater wavelength, so that a required spectral range be obtained by varying the slope of the filter. Figure 3.56 presents the transmission curves of the interference light filters that were used in a two-channel photometer to measure the rotational temperature of the OH (8–3) band in relation to the intensities of the Q and R branches in investigations of orographic perturbations (Sect. 5.2). It can be seen that when the slope of the light filter unit was changed by 15° relative to the optical axis, the transmission band shifted noticeably to the short-wave spectral range.

3.7.1.3 Characteristics of Interferometers

In investigations of the upper atmosphere, Fabry–Perot interferometers were employed in the main to determine Doppler temperatures and wind speeds. For this purpose, the atomic oxygen 557.7- and 630-nm emission lines of the P branch of the OH (8–3) band were used. Originally, interference plates of diameter 50 (mm) with dielectric coatings were used; the accuracy of the surfaces was $\sim\lambda/50$. Only the interference plates, the plate separators, and the housing in which the interference unit was assembled were commercially produced. The remaining parts of the device construction were manufactured to suit particular goals of the study of airglow emissions. Therefore, to realize the necessary capabilities of the plates, the latter were coated with dielectric multilayers. An example of such a device is shown in Fig. 3.20. Photographic recording was carried out with the help of image amplifiers (Semenov 1975a).

The piezoelectric method was used for scanning interference images (Hernandez 1966, 1970, 1974, 1978; Hernandez and Mills 1973; Ignatiev and Yugov 1995).

To determine the instrumental profile of the interferometer, the 632.8-nm laser radiation was used. With the recording tools employed—an image converter tube and a photographic film—the instrumental profile was determined by the properties of the ICT luminescent screen and by the scattering coefficient of the photographic

3.7 Calibration of Measurements

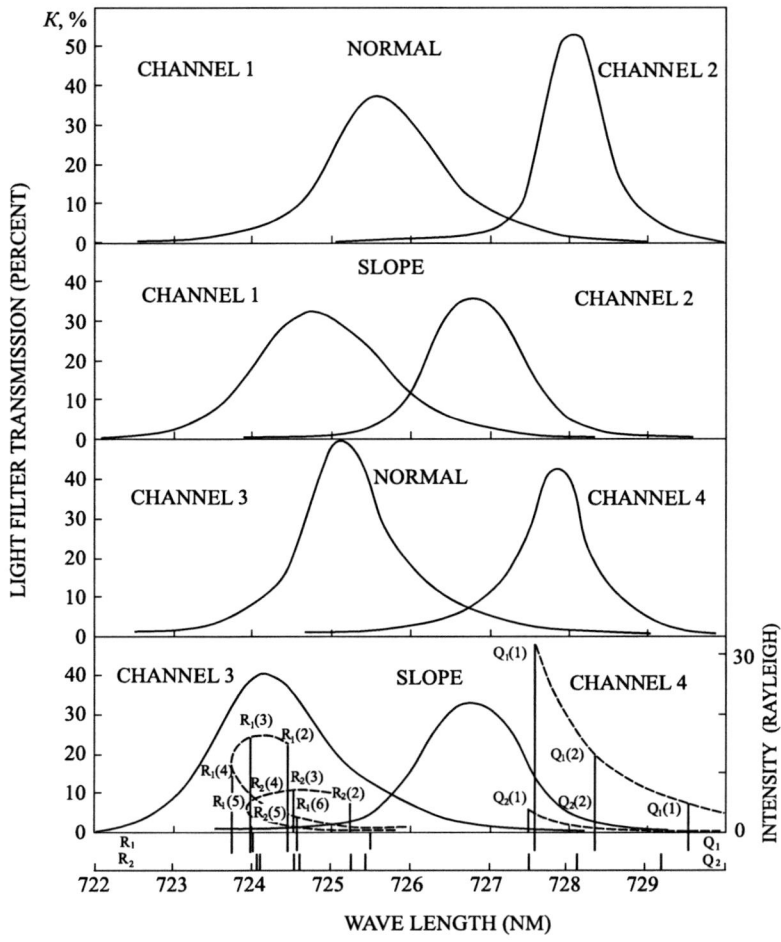

Fig. 3.56 Transmission curves of the interference light filters of different channels of a two-channel photometer in upright and in inclined position. The positions of the lines of the R and Q branches of the OH (8–3) bands are shown and their yearly average intensities are given in the *bottom*

emulsion and was rather well described by a Gaussian distribution. An example of such a distribution is given in Fig. 3.56.

For absolute calibration many requirements should be fulfilled to ensure a necessary accuracy of the radiation intensity measurements because all subsequent calculations are based on these primary data. It should be noted that the real absolute calibration error is unlikely to be less than 10(%). However, in view of a substantial variability of the parameters of the emissions under consideration, this accuracy is quite sufficient for absolute intensity estimates.

3.7.2 Artificial Reference Light Sources

Secondary reference light sources used for calibration of spectrophotometric data are standard lamps and phosphor screens (Nikonov 1973b). The conventional incandescent lamps applied as standard light sources possess a radiation spectrum similar to that of a black body at a temperature of 2856 (K), termed as an A-type light source. Nevertheless, since actual emitters differ from a black body, it is necessary to introduce a grayness coefficient into the spectral distribution (Ribaud 1931; Harrison 1960). However, for the operating conditions of the spectrophotometric instrumentation used for measuring the nightglow intensity, the radiation intensity of such a lamp should be substantially attenuated, and this, naturally, should affect the final spectral distribution of the light source under observation. Besides, powerful lamps of this type carry high currents. To provide emission stability, it is necessary to control the voltage and current by instruments of high accuracy rating, since the brightness of a lamp is proportional to the seventh power of current (Ribaud 1931). Therefore, in regular operation, one has to use secondary reference sources of substantially lower intensity, so that the radiation signals from the sky and the calibration source are comparable. Secondary radiation sources, in turn, should be calibrated against natural radiation sources, such as stars and the Moon. A secondary radiation source of this type is shown schematically in Fig. 3.57. Several small incandescent lamps, operated at a reduced voltage in order that there was no vaporization of the lamp filament and, hence, the radiation characteristics did not vary, were placed in a closed housing. The outgoing radiation was attenuated with neutral filters. The size of the output window of the radiator was made equal to the size of the photometer lens, so that the calibration source had the properties of a spatial radiator. On the opposite inner wall of the housing, a selenium photocell was placed that recorded the total radiation of the lamps. The photocell current was measured with a microammeter.

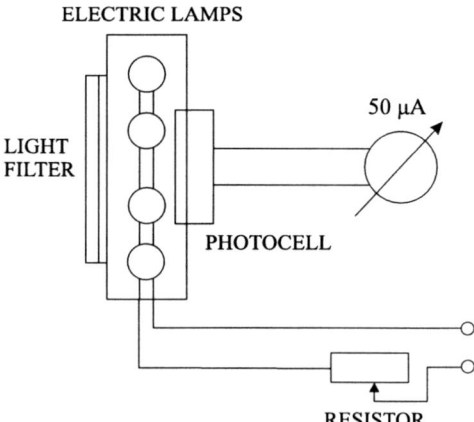

Fig. 3.57 Scheme of a secondary calibration source for photometers

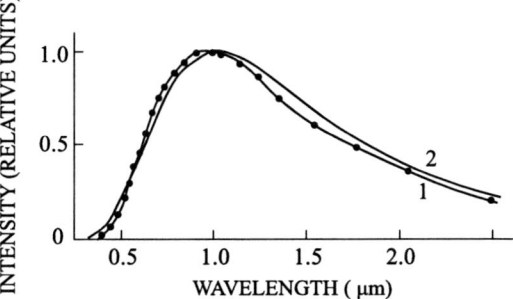

Fig. 3.58 Radiation spectra of a reverse-shifted silicon p–n transition at room temperature (1) and of a black body at 2856 (K) (2) (Kosyachenko et al. 1990)

For the spectrographs, a special secondary source was manufactured which imitated a spatial radiation source.

Now low-intensity semiconductor radiation sources are available which are closely similar to A-type sources in the radiation spectral distribution (Kosyachenko et al. 1990). An example of such a spectral distribution is shown in Fig. 3.58.

3.7.3 Natural Reference Radiation Sources

Extraterrestrial luminaries, such as stars, the Moon, and the Sun, are the most convenient reference radiation sources whose absolute intensities have been measured rather reliably. Possible variations of their spectral and time characteristics are well studied. They have been discussed in Sect. 1.2.

When calibration is performed against the Moon and the Sun, it is most appropriate to illuminate a white screen placed before the entrance pupil of the spectrograph, spectrometer, or photometer to make their aperture to be filled with light to exclude the vignetting of the device field. Besides, it is necessary that the light flux directly entering the device be comparable in intensity to that corresponding to the conditions of night measurements. This should be provided quantitatively by attenuating the light of the Moon and, particularly, of the Sun. When doing this, it is necessary to bear in mind that the spectra of these radiation sources have broad ranges distorted due to the absorption of radiation by the terrestrial atmosphere.

In ground-based photoelectric observations, the most reliable calibration method involves the use of stars with observance of necessary requirements during measurements. Nevertheless, the use of stars calls for feeding optics. Thus, a recent absolute calibration of a lamp for spectrophotometric measurements has been performed against the spectrum of the Chapella star (α Aur, $\alpha = 5^h15^m$, $\delta = 45°58'$) of class G8III (magnitude 0.03^m) with a spectrograph erected on a telescope. However, for stationary instrumentation a special watching system is required to keep the light source in the instrument field, which considerably complicates the calibration task.

A good case in point is the use of stars of magnitude m ~ 5 in the measurements of the altitude distributions of airglow emissions performed on the Mir orbiting spacecraft. The advantage of pale stars is that their intensity is a small fraction of that of the atmospheric radiation. Therefore, in sighting and guiding them during measurements above the emission layer of the upper atmosphere, absolute calibration of the photometer was performed. Subsequently, for the period during which a star crossed the emission layer until its set beneath the horizon, a series of data precisely referred to an altitude above the Earth surface was obtained. Mathematical processing of the measured intensities has made it possible to determine the altitude distribution of the emission rate in the spectral range selected by the interference light filter (Gurvich et al. 2002).

3.8 Errors of Intensity and Temperature Measurements

3.8.1 Photographic Measurements

Photography errors are mainly caused by the nonlinear dependence of the blackening density on incident intensity. In obtaining a spectrum record, inevitable errors arise in drawing the fog background level which serves as a reference in evaluating the blackening density. For instance, in a spectrum record obtained on the paper chart of an EPP-09 self-recorder the background level was about 200 (mm) apart from the level corresponding to the closed microphotometer shutter. The noise recording track of the fog background had a swing no more than 10 (mm), and its average level was drawn accurate to within 1 (mm). Thus, the error of this record was no more than 1(%). The error of the determination of the level of the upper atmosphere radiation background relative to the fog background was the sum of the error aroused in drawing the average level of the spectrum background in the record and the error caused by the variations of film blackening associated with the presence of weak atmospheric absorption bands. On the average, it seemed to be not over 2–3(%). Thus, the error of the determination of the intensity of a hydroxyl band line was no more than 5(%). Since the background level near the rotational lines of the OH band that were used for the determination of rotational temperature was practically the same for all lines, the errors were also nearly the same and practically of the same sign. Special check has shown that in these cases the relative error in the determination of rotational temperature was not over half the error of the determination of line intensity, i.e., it was less than 2–3(%), which corresponds to 4–6 (K) at an average temperature of ~200 (K). This can well be seen from Figs. 2.11 and 2.12 (Shefov 1961).

The conditions of processing of spectrograms corresponded to photographing both immediately on SP-48 spectrographs and with the use of image converter tubes on SP-50 spectrographs. It is natural that in the latter case distorting factors might come into play, such as the nonuniformity of the ICT luminescence field.

Fluctuations in the transparency of the atmosphere, both during the exposure time and throughout the observation period, and the background noise from artificial radiants and the Moon could also produce false effects in the time variations of the OH rotational temperature in three regions of the firmament.

Possible transparency fluctuations could distort true temperature values only if the radiation attenuation factor would vary highly nonuniformly along the spectrum. However, for the spectral ranges used neither these effects nor indications of selective absorption for individual lines of OH bands have been revealed (except for the Q branch of the (8–3) band). Trial observations under the conditions of a strong background (the Moon, twilight, highlight from urban lights) have shown that the continuum of such a background distorts the intensity distributions of individual lines of the hydroxyl emission bands under investigation and, hence, the temperature, inappreciably. Probably, the background affects radiation records in those spectral ranges where it is superimposed by the absorption bands of water vapor. Nevertheless, the observations were carried out only in clear moonless nights with good air transparency. The results of special investigations were in agreement with inferences from other works that the rotational temperature of OH can be determined with high accuracy (Shefov 1961, 1967; Taranova 1967; Sivjee and Hamwey 1987).

Obviously, each temperature value obtained during the exposure time can be referred to the middle of the time interval only if the air transparency and the instrumentation sensitivity were stable. If these parameters vary gradually during the exposure time, the effective times at which these temperatures are measured may shift within the exposure time, perhaps, by no more than 20–25(%) of the time interval.

A similar situation occurs in processing photographic images of interferograms. The correctness of the drawing of the background level near the emission lines (see Fig. 3.45) is checked once the correlation plot has been constructed (see Fig. 3.46).

To conclude the discussion of the effect of various errors on the determination of temperature, it should be noted that the statistical processing methods applied to IGWs are rather efficient. These methods will be described in detail below. They allow one to determine reliably the IGW parameters (amplitude, propagation azimuth, and velocity) even when the noise level reaches 100(%) of the IGW amplitude. The noises caused by instrumental errors and by the errors encountered in the method of determination of temperature are much lower, as follows from the discussion.

3.8.2 Photoelectric Measurements

Photoelectric measurements have the feature that the signal accumulation time is either determined by the time constant of the amplifying unit for the analog measuring method or set by the operating mode of the recording device if the photon counting method is used. In view of the natural variability of emission intensity, the obtained intensity value is a mean value for a given time interval.

3.8.3 Interferometric Measurements

The errors of primary photometric processing of interferograms were discussed in Sects. 3.8.1 and 3.8.2. Below the effect of atmospheric factors is considered.

3.8.3.1 The Errors Caused by a Possible Broadening of the Doppler Profile of an Emission due to Tropospheric Scattering

In interferometric investigations of the atomic oxygen 630.0-nm red emission that arises at the altitudes of the F_2 ionospheric region, the temperature of the emission layer is determined by the Doppler broadening of the emission profile and the wind speed in this altitude range by the Doppler shift of the emission wavelength. However, during measurements the instrument records not only the radiation directly from the region of the sky under observation, but also some portion of the radiation of the upper atmosphere scattered by the lower atmosphere. In this case, the radiation intensity for the Doppler profile of the emission from the sighted region of the sky is summed up with that for the profiles having other Doppler shifts.

Thus, some "smearing" of the recorded profile should be expected if the transparency of the lower atmosphere is not high enough. The estimation of the increase in temperature due to a decrease in transparency was considered by Akmamedov (1991).

Sighting of an emission layer is usually performed at a zenith distance ζ_0. Therefore, the Doppler shift of the emission intensity profile at an altitude Z above the Earth surface for the horizontal wind velocity V lying in the vertical sighting plane (see Fig. 3.21) will be determined by the relation

$$\delta\lambda = \frac{\lambda_0}{c} \cdot \frac{V}{1 + \frac{Z}{R_E}} \cdot \cos\omega \cdot \sin\zeta_0,$$

where ω is the angle counted from the vertical sighting plane.

The transparency p of the lower atmosphere, measured by a conventional method with reference to stars (Rosenberg 1963), is determined in the main by the aerosol component. Therefore, the profile portion associated with the directly sighted radiation corresponds to the intensity determined by the ratio

3.8 Errors of Intensity and Temperature Measurements

$$\frac{\exp\left[-\left(\lambda-\lambda_0-\dfrac{\lambda_0}{c}\cdot\dfrac{V}{1+\dfrac{Z}{R_E}}\cdot\sin\zeta_0\right)^2\Big/(\delta\lambda_D)^2\right]\cdot p^{\sec\zeta_0}}{\sqrt{1-\dfrac{\sin^2\zeta_0}{\left(1+\dfrac{Z}{R}\right)^2}}},$$

where the denominator is the van Rhijn factor (Chamberlain 1978), which takes into account the variation of the emission layer thickness with increasing zenith angle ζ_0.

The scattered portion of the radiation from a spatial source is less attenuated, and the degree of attenuation can be represented as (Rosenberg 1963)

$$\frac{1}{1+0.5\cdot\tau}=\frac{1}{1-\dfrac{\ln p}{2}\cdot\sec\zeta},$$

where τ is the optical thickness corresponding to the transparency $p=e^{-\tau}$. Besides, radiation scattering occurs with a certain indicatrix, which can be determined by the Henyey–Greenstein formula (Chamberlain 1978)

$$\rho(\theta)=\beta\cdot\frac{1-g^2}{(1-2g\cdot\cos\theta+g^2)^{3/2}},$$

where g is a parameter, which, in fact, is the mean cosine of the scattering angle, β is a normalizing factor, and θ is the scattering angle calculated by the formula

$$\cos\theta=\cos\zeta_0\cdot\cos\zeta+\sin\zeta_0\cdot\sin\zeta\cdot\cos\omega.$$

The value of g determines the elongation of the indicatrix. On the average, for the real case of single scattering by a mist (Gorchakov et al. 1976a,b), the ratio of the forward/backward scattering energies is about three, which corresponds to an approximately tenfold forward/backward elongation of the indicatrix. For this case we have $g\sim 0.35$.

For a more turbid scattering medium, multiple scattering predominates, and the indicatrix becomes practically spherical (Gorchakov et al. 1976a,b). This corresponds to $g\sim 0$.

Thus, for practical estimation of the degree of scattering, it can be assumed that $g\approx 0.4\cdot p$. In view of the indicatrix shape, the normalizing factor is (Abramowitz and Stegun 1964; Prudnikov et al. 1979)

$$\beta=\frac{1}{4\pi}.$$

Moreover, when scattered, light is partially absorbed by the atmosphere. For the albedo we can take a value approximately equal to $1-p$, where p is the transparency of the atmosphere.

Thus, based on the approach considered, a recorded emission intensity profile, approximated as a Doppler profile, will be determined by the relation

$$= \frac{I_0 \cdot \exp\left[-\left(\lambda - \lambda_0 - \frac{\lambda_0}{c} \cdot \frac{V}{1 + \frac{Z}{R_E}} \cdot \sin\zeta_0\right)^2 \bigg/ (\delta\lambda_D)^2\right]}{}$$

$$= \frac{\exp\left[-\left(\lambda - \lambda_0 - \frac{\lambda_0}{c} \cdot \frac{V}{1+\frac{Z}{R_E}} \cdot \sin\zeta_0\right)^2 \bigg/ (\delta\lambda_D)^2\right] \cdot p^{\sec\zeta_0}}{\sqrt{1 - \frac{\sin^2\zeta_0}{\left(1+\frac{Z}{R}\right)^2}}}$$

$$+ 2(1-p) \cdot \beta \cdot \int_0^\pi \int_0^{\frac{\pi}{2}} \frac{p(\theta) \cdot}{\left(1 - \frac{\ln p}{2} \cdot \sec\zeta\right)}$$

$$\frac{\exp\left[-\left(\lambda - \lambda_0 - \frac{\lambda_0}{c} \cdot \frac{V}{1+\frac{Z}{R_E}} \cdot \cos\omega \cdot \sin\zeta_0\right)^2 \bigg/ (\delta\lambda_D)^2\right] \cdot \sin\xi \cdot d\zeta d\omega}{\sqrt{1 - \frac{\sin^2\zeta_0}{\left(1+\frac{Z}{R}\right)^2}}},$$

where

$$\sin\xi = \sin\zeta \cdot \left[\sqrt{1 - \frac{\sin^2\zeta}{\left(1+\frac{Z}{R_E}\right)^2}} - \frac{\cos\zeta}{1+\frac{Z}{R_E}}\right].$$

The effective Doppler width $\delta\lambda_{\text{eff}}$ of a profile under observation is convenient to determine by the method considered above (Semenov 1975a) in which the regression equation (see Fig. 3.46)

$$\delta\lambda_i = \delta\lambda_{\text{eff}} \cdot \sqrt{\log_e \frac{I_0}{I_i}}$$

3.8 Errors of Intensity and Temperature Measurements

is used to calculate the regression coefficient $\delta\lambda_{\text{eff}}$. Then the temperature and its error are determined as

$$T_{\text{eff}} = T \cdot \left(\frac{\delta\lambda_{\text{eff}}}{\delta\lambda_D}\right)^2 \text{ and } \sigma(T) = 2T \cdot \frac{\sigma(\delta\lambda_{\text{eff}})}{\delta\lambda_{\text{eff}}}.$$

The value of $\sigma(\delta\lambda_{\text{eff}})$ is determined by the above Taylor formula (Taylor 1982), and it serves as a criterion for the accuracy of calculation of the temperature.

Numerical calculations of the effect of the profile broadening due to the transparency of the atmosphere by the formulas presented above were performed for the atomic oxygen 630.0-nm red emission at $T = 1000\,(K)$, $Z = 250\,(km)$, $\zeta_0 = 60°$, and various values of the transparency p and wind speed V. For this case, $\sigma(T)$ was 5–20 (K), which is much lower than the possible experimental errors.

Based on these calculations, an empirical relation has been constructed which approximates the obtained equation for the transparency values from 0.2 to 1, which are reasonable both from the viewpoint of the feasibility of measurements and from the viewpoint of the validity of the theoretical approximations made. As a result, we have

$$\frac{T_{\text{eff}}}{T} = 1 + \frac{V}{100} \cdot (1-p)^{1.5},$$

where the velocity V is expressed in meters per second.

As can be seen, measurements performed under low transparency conditions can give overestimated temperature values. Thus, for the above conditions, the temperature increment ΔT is not over 20 (K) at $p \sim 0.8$ and can reach almost 100 (K) at $p \sim 0.5$ for $V = 300\,(m \cdot s^{-1})$. For $V = 100\,(m \cdot s^{-1})$ at $p \sim 0.5$ we have $\Delta T \sim 35\,(K)$.

Obviously, the limiting value T_{eff} at $p = 1$ corresponds to the initial value of T. An estimate of T_{eff} by the above relation at another limiting value of transparency, $t \sim 0$, seems to be incorrect since the radiation scattering should be taken into account in a more complicated way. However, in practice this is not necessary because it is obvious that the temperature measured under these conditions is strongly distorted due to the Doppler "smearing" of the emission profile caused by the wind in the emission layer.

The errors in the determination of temperature and wind speed for the obtained material of observations can be estimated based on the above formulas with the help of the relations

$$\Delta T = T \cdot \left[\left(\frac{2\Delta(\delta\lambda)}{\delta\lambda}\right)^2 + \left(\frac{\Delta\left(\log_e \frac{I_0}{I(\lambda)}\right)}{\log_e \frac{I_0}{I(\lambda)}}\right)^2\right]^{1/2};$$

$$\Delta V = V \cdot \left[\left(\frac{2\Delta D_1}{D_{N;W} - D_{S;E}}\right)^2 + \left(\frac{3\Delta D_1}{D_1}\right)^2 + \left(\frac{2\Delta D_1}{D_2 - D_1}\right)^2\right].$$

In most cases, the error for temperature was 70–100 (K) and for wind speed 15–30 $(m \cdot s^{-1})$.

The effect of the atmosphere transparency on the determined values of wind speed was considered theoretically (Abreu et al. 1983; Abreu 1985). However, actual estimates of the variations of the measured wind speed values were not given.

3.8.4 Spatial Orientation of the Instrumentation

In Sect. 3.5.1, when the calculations of the orientation of lines of sight for three-azimuthal measurements were considered, the errors of the determination of the zenith angles of optical axes and of the dimensions of a basic triangles have been estimated.

3.8.5 Statistical Processing Errors

Results of any measurements, irrespective of their carefulness, are subject to some errors. Theoretical relations for the estimation of errors are given in many works on statistical processing of measurements (Shchigolev 1960; Taylor 1982). It should be borne in mind that errors may occur as random deviations from a true value realized as an average of repeated measurements and they may be systematic errors occurring for some latent reasons, including personal errors introduced by the operator who performed measurements. In analyzing direct measurements and calculations of quantities of interest, all intermediate errors encountered in the calculations of the components used are inevitably accumulated. Therefore, in some cases, the relative errors of initial data can be much less than the absolute errors. However, this may have no effect on the resulting correlation relations. It should also be borne in mind that the observed variance of the values of measured quantities is not always caused by errors. In many cases, the spread in values about an average value may result from time variations of the measured quantities on various time scales. Therefore, an obvious rule of any analysis of data is to preclude the influence of distorting processes. Sometimes repeated iterations are required for revealing a more reliable relation to describe the behavior of a quantity under investigation.

In some cases, errors can be associated with too rough approximations used and with the simplifying assumptions applied in solving equations which describe some processes.

The obtaining of statistical relationship between mutual variations of two compared characteristics is the important task. The typical procedure consists in constructing a correlation field for obtaining parameters of the regression function, more often linear:

$$y = a + \rho \cdot x.$$

This problem is solved by the method of least squares, since the sum of squared deviations of individual data in an n-dimensional space determines the squared

3.8 Errors of Intensity and Temperature Measurements

distance from the origin of coordinates. The requirement of its minimum determines the sense of the method of least squares. The system of conditional equations

$$a \cdot N + \rho \cdot \Sigma x_i = \Sigma y_i,$$
$$a \cdot \Sigma x_i + \rho \cdot \Sigma x_i^2 = \Sigma x_i \cdot y_i,$$

has the obvious solution

$$a = \frac{(\Sigma x_i^2) \cdot (\Sigma y_i) - (\Sigma x_i) \cdot (\Sigma x_i \cdot y_i)}{N \cdot (\Sigma x_i^2) - (\Sigma x_i)^2}; \qquad \rho = \frac{N \cdot (\Sigma x_i \cdot y_i) - (\Sigma x_i) \cdot (\Sigma y_i)}{N \cdot (\Sigma x_i^2) - (\Sigma x_i)^2}.$$

These formulas are used in computer codes. However, the errors of the regression equation coefficients are not always found by specially derived formulas. Sometimes they are determined in terms of the sums entering in the conditional equations:

$$\sigma_\rho = \sqrt{\frac{N \cdot [\Sigma y_i^2 + \rho \cdot (\rho \cdot \Sigma x_i^2 - 2 \cdot \Sigma x_i \cdot y_i) - a^2 \cdot N]}{(N-2) \cdot [N \cdot (\Sigma x_i^2) - (\Sigma x_i)^2]}}; \qquad \sigma_a = \sigma_\rho \cdot \sqrt{\frac{\Sigma x_i^2}{N}}.$$

The correlation coefficient r is determined by the well-known formula

$$r = \frac{N \cdot \Sigma x_i \cdot y_i - \Sigma x_i \cdot \Sigma y_i}{\sqrt{[N \cdot \Sigma x_i^2 - (\Sigma x_i)^2] \cdot [N \cdot \Sigma y_i^2 - (\Sigma y_i)^2]}}; \qquad \sigma_r = \frac{1 - r^2}{\sqrt{N-1}}.$$

It is well known that anomalous values present among a data set subject to analysis can substantially distort the statistical analysis result, i.e., the values of the regression coefficient ρ and correlation coefficient r (Taranova 1965). Therefore, visual examination of the correlation field is necessary to detect anomalous situations.

The significance of correlation factors is evaluated by calculating the Student coefficient (Gmurman 1975)

$$t_{St} = \frac{r \cdot \sqrt{N-2}}{\sqrt{1-r^2}}.$$

For a two-sided critical region, a significance level of 0.05, and $N > 10$ we have $t_{St} \leq 2.20$. Thus, the following condition should be satisfied:

$$r > \left(1 + \frac{N-2}{t_{St}^2}\right)^{-0.5}.$$

In practice, the significance of calculated values of ρ and r is determined by the ratios

$$\frac{\rho}{\sigma_\rho} \geq 3 \quad \text{and} \quad \frac{r}{\sigma_r} \geq 3.$$

In any case of construction of a linear correlation field for a given data set there are two usual regression lines corresponding to the equations

$$y = a + \rho_x \cdot x \quad \text{and} \quad x = b + \rho_y \cdot y,$$

which form a certain angle ψ. If we normalize these equations with respect to average $\bar{y}$ and $\bar{x}$, i.e., represent them in the form

$$\frac{y}{\bar{y}} = \frac{a}{\bar{y}} + \rho_x \cdot \frac{\bar{x}}{\bar{y}} \cdot \frac{x}{\bar{x}} \quad \text{and} \quad \frac{x}{\bar{x}} = \frac{b}{\bar{x}} + \rho_y \cdot \frac{\bar{y}}{\bar{x}} \cdot \frac{y}{\bar{y}}$$

or

$$Y = A + \rho_{nx} \cdot X \quad \text{and} \quad X = B + \rho_{ny} \cdot Y,$$

then the angle ψ is determined by the formula

$$\text{tg}\,\psi = \frac{\rho_{nx} \cdot (1 - r^2)}{r^2 + \rho_{nx}^2},$$

since $\rho_x \cdot \rho_y = \rho_{nx} \cdot \rho_{ny} = r^2$.

When the spread of points in the correlation field is great enough, i.e., the significant correlation coefficient is small, a straight line

$$Y = C + \bar{\rho}_{nx} \cdot X$$

can be constructed, such that the sum of squared distances from this line to all measurement data points is a minimum. The angle φ this line makes with the x-axis is determined by the relation (Franklin 1955; Shefov 1965)

$$\text{tg}\,2\varphi = \frac{2\rho_{nx} \cdot r^2}{r^2 - \rho_{nx}^2},$$

whence the effective regression coefficient is given by

$$\text{tg}\,\varphi = \bar{\rho}_{nx} = \frac{\rho_{nx}^2 - r^2 + \sqrt{(r^2 - \rho_{nx}^2)^2 + 4\rho_{nx}^2 \cdot r^4}}{2\rho_{nx} \cdot r^2}.$$

The bisectrix of the angle ψ, which also characterizes the mean trend of the regression relation, makes an angle θ with the x-axis, which is determined by the formula

$$\text{tg}\,2\theta = \frac{\rho_{nx} \cdot (1 + r^2)}{r^2 - \rho_{nx}^2},$$

whence

$$\text{tg}\,\theta = \frac{\rho_{nx}^2 - r^2 + \sqrt{(1 + \rho_{nx}^2) \cdot (r^4 + \rho_{nx}^2)}}{\rho_{nx} \cdot (1 + r^2)}.$$

As follows from these formulas, for r = 1 we have tgφ = tgθ = ρ_{nx}. For another limiting case, r → 0, we have tgφ = ρ_{nx} → 0 and, at the same time, ρ_{nx} → $1/\rho_{ny}$, i.e., it becomes indefinite as ρ_{ny} → 0. As r → 0, we have tgθ → 1, i.e., the bisectrix remains in position.

References

Abalakin VK (1979) Principles of ephemeride astronomy. Nauka, Moscow
Abalakin VK, Aksenov EP, Grebennikov EA, Demin VG, Ryabov YuA (1976). In: Duboshin GN (ed) Handbook for celestial mechanics and astrodynamics, 2nd edn. Nauka, Moscow
Abramenko AN, Agapov ES, Anisimov VF, Galinsky ND, Prokofieva VV, Sinenok SM (1984). In: Nikonov VB (ed) Television astronomy. Nauka, Moscow
Abramowitz M, Stegun IA (1964) Handbook of mathematical functions with formulas, graphs and mathematical tables. Nat Bur Stand, Washington, DC
Abreu VJ (1985) Atmospheric scattering effects on ground-based Fabry–Perot measurements of thermospheric winds: on inversion technique. Planet Space Sci 33:1049–1055
Abreu VJ, Schmitt GA, Hays PB, Meriwether JW, Tepley CA, Cogger LL (1983) Atmospheric scattering effects on ground-based measurements of thermospheric winds. Planet Space Sci 31:303–310
Aitova GA (1977) Observations of Hβ emission in high velocity cloud OUH 308 direction (α = 18^h03^m, δ = $+36°5$) with a Fabry–Perot spectrometer. Astron Circ USSR Acad Sci N 936:7–8
Aitova GA, Shcheglov PV (1977) Observation of Hβ emission in M31 direction with a Fabry–Perot spectrometer. Astron Circ USSR Acad Sci N 936:5–7
Akmamedov Kh (1991) Broadening of the recorded Doppler profile of emissions of the upper atmosphere due to scattering in the troposphere of light from different sections of the sky with a wind field present in the radiating layer. Geomagn Aeron 31:698–701
Allen CW (1955) Astrophysical quantities. The Athlone Press, London
Allen CW (1973) Astrophysical quantities, 3rd edn. The Athlone Press, London
Ambartsumyan VA, Mustel ER, Severny AB, Sobolev VV (1952). In: Ambartsumyan VA (ed) Theoretical astrophysics. Gostekhizdat, Moscow
Ammosov PP, Gavrilieva GA (2000) Infrared digital spectrograph for the hydroxyl rotational temperature measurement. Instr Exp Techn N 6:73–78
Ammosov PP, Gavrilieva GA, Ignatiev VM (1992) Registration of wave disturbances over Yakutia. Adv Space Res 12:145–150
Andreeva LA (1985) A wind velocity in the polar thermosphere in the height interval of 150–180 km. In: Portnyagin YuI, Chasovitin YuK (eds) Trans Inst Exp Meteorol. N16(115). Hydrometeoizdat, Moscow, pp 3–14
Andreeva LA, Katasev LA (1983) Studies of the upper atmospheric dynamics with aid of artificial luminous clouds at Volgograd and Heiss Island. In: Teslenko VP (ed) Trans Inst Exp Meteorol. N 15(111). Hydrometeoizdat, Moscow, pp 6–35
Andreeva LA, Kluev OF, Portnyagin YuI, Khananiyan AA (1991) Studies of the upper atmospheric processes by the artificial luminous cloud method. Hydrometeoizdat, Leningrad
Antonov EI, Il'in VE, Kolenko EA, Petrovsky YuV, Smirnov AI (1969). In: Epifanova VI (ed) Devices for cooling of the radiation detectors. Mashinostroenie, Leningrad
Astafieva NM (1996) Wavelet analysis: basic theory and some applications. Uspekhi Fiz Nauk 166:1145–1170
Avdoshin ES (1990) Components of the fiber technique. Opt Mech Ind N 3:3–15
Avgustinovich KA (1990) Principles of photographic metrology. Legprombytizdat, Moscow
Bakanas VV, Perminov VI, Semenov AI (2003) Seasonal variations of emission characteristics of the mesopause hydroxyl with different vibrational excitation. Adv Space Res 32:765–770

Barbier D (1955) Les photomètre automatiques pour l'étude de la lumière du ciel nocturne. Rev d'optique 34:353–360

Barlier F, Berger C, Falin JL, Kockarts G, Thuillier G (1978) A thermospheric model based on satellite drag data. Ann Geophys 34:9–24

Bates DR (1950) Suggestion regarding the use rockets to vary the amount of atmosphere sodium. J Geophys Res 55:347–349

Bendat JS, Piersol AG (1966) Random data: analysis and measurement procedures. Wiley-Interscience, New York

Berkovsky AG, Gavanin BA, Zaidel IN (1976) Vacuum photoelectronic devices. Energy, Moscow

Berkovsky AG, Gavanin BA, Zaidel IN (1988) Vacuum photoelectronic devices, 2nd edn. Radio and Svyaz, Moscow

Biondi MA, Sipler DP (1984) Studies of equatorial 630.0 nm airglow enhancements produced by chemical release in F-region. Planet Space Sci 32:1605–1610

Born M, Wolf E (1964) Principles of optics, 2nd edn. Pergamon Press, Oxford

Brasseur G, Solomon S (2005) Aeronomy of the middle atmosphere. 3rd ed. Springer-Verlag, Dordrecht Holland

Bronshten VA, Dagaev MM, Kononovich EV, Kulikovsky PG (1981). In: Abalakin VK (ed) Astronomical calendar. Invariable part, 7th edn. Nauka, Moscow

Breido II (1973) Principal properties of the photographic materials. In: Mikhailov AA (ed) A course of astrophysics and celestial astronomy, vol 1. Methods and instruments, 2nd edn. Nauka, Moscow, pp 107–133

Budin VP (1997) A function approximation of the relative spectral light effectiveness of the eyesight by two different branches of the Gaussian distribution. Opt Mech Ind N 1:55–56

Butslov MM, Stepanov BM, Fanchenko SD (1978). In: Zavoisky EK (ed) Electronic–optical converters and their application in the scientific studies. Nauka, Moscow

Cole JL, Kolb CE (1981) On the upper atmospheric chemiluminescent emission observed upon release of aluminium vapour and its compounds. J Geophys Res 86:9125–9135

Chamberlain JW (1978) Theory of planetary atmospheres. Academic Press, New York

Chetyrkin EM (1975) Statistical methods of the forecasting. Statistica, Moscow

Choi GH, Monson IK, Wickwar VB, Rees D (1998) Seasonal variations of temperature near the mesopause from Fabry–Perot interferometer observations of OH Meinel emissions. Adv Space Res 21:843–846

Christensen AB, Patterson TNL, Tinsley BA (1971) Observation and computation of twilight helium 10830 Å emission. J Geophys Res 76:1764–1777

Chunchuzov IP (1988) Orographic waves in the atmosphere produced by a varying wind. Izv USSR Acad Sci Atmos Oceanic Phys 24:5–12

Chunchuzov IP (1994) On possible generation mechanism for nonstationary mountain waves in the atmosphere. J Atmos Sci 15:2196–2206

Cooper VG (1971) Analysis of Fabry–Perot interferograms by means of their Fourier transforms. Appl Opt 10:525–530

Craven JD, Frank LA (1987) Latitudinal motions of the aurora during substorms. J Geophys Res 92A:4565–4573

Davydov VS (1986) Study of the upper atmospheric emissions of the Earth in the visual spectral region. Mashinostroenie, Moscow

Davydov VS, Medvedev VE, Mikheev AS (1975) The optical scheme of the scanning radiometer. Opt Mech Ind N 3:37–39

Dokuchaeva OD (1994) Astronomical photography. Materials and methods. Fizmatlit, Moscow

Dorokhova IV, Esin VP, Khananayan AA (1990) Estimations of the brightness characteristics of the artificial luminous cloud in the upper atmosphere. In: Khananayan AA (ed) Trans Inst Exp Meteorol. N 21(143). Hydrometeoizdat, Moscow, pp 52–57

Eccles MJ, Sim ME, Tritton KP (1983) Low light level detectors in astronomy. Cambridge Univ Press, Cambridge

Eiby GA (1980) Earthquakes. Van Nostrand Reinhold, New York

Evlashin LS, Semenov AI, Shefov NN (1999) Long-term variations in the thermospheric temperature and density on the basis of an analysis of Störmer's aurora–height measurements. Geomagn Aeron 39:241–245

Fadel KhM, Semenov AI, Shefov NN, Sukhodoev VA, Martsvaladze NM (2002) Quasibiennial variations in the temperatures of the mesopause and lower thermosphere and solar activity. Geomagn Aeron 42:191–195

Filipp ND, Oraevsky VN, Blaunstein NSh, Ruzhin YuYa (1986). In: Gusev VD (ed) Evolution of the artificial plasma inhomogeneities in the Earth' ionosphere. Stiintsa, Kishinev

Fishkova LM (1983). In: Shefov NN (ed) The night airglow of the Earth mid-latitude upper atmosphere. Metsniereba, Tbilisi

Fishkova LM, Martsvaladze NM, Shefov NN (2000) Patterns of variations in the OI 557.7-nm. Geomagn Aeron 40:782–786

Fishkova LM, Martsvaladze NM, Shefov NN (2001a) Long-term variations of the nighttime upper-atmosphere sodium emission. Geomagn Aeron 41:528–532

Fishkova LM, Martsvaladze NM, Shefov NN (2001b) Seasonal variations in the correlation of atomic oxygen 557.7-nm emission with solar activity and in long-term trend. Geomagn Aeron 41:533–539

Föppl H, Haerendel G, Haser L, Loidi J, Lütjens P, Lüst R, Melzner F, Meyer B, Neuss H, Rieger E (1967) Artificial strontium and barium clouds in the upper atmosphere. Planet Space Sci 15:357–372

Frank LA, Signarth JB, Craven JD (1986) On the flux of small comets in to the Earth's upper atmosphere. I. Observations. Geophys Res Lett 13:303–306

Franklin P (1955) Fundamental mathematical formulas. In: Menzel DH (ed) Fundamental formulas of physics. Prentice-Hall, New York, pp 9–96

Friedrich VH, Offermann D, Trinks H, von Zahn U (1977) The cryo mass spectrometer in the winter anomaly campaign. J Geophys 44:139–146

Frish SE (1980) Optical methods of the measurements. Pt.2. Leningrad State Univ, Leningrad

Galperin GI, Mironov AV, Shefov NN (1957) Spectrographs to be used for investigations of atmospheric emissions during the IGY of 1957–1958. Mém Soc R Sci Liège 18:68–69

Gavrilieva GA, Ammosov PP (2001) Observations of gravity wave propagation in the all-sky infrared airglow. Geomagn Aeron 41:363–369

Gerasimova NG, Yakovleva AV (1956) Complect of the fast spectrographs with the diffractive gratings. Instr Exp Techn N 1:83–86

Gershenzon YuM, Konoplev AV, Chekin SK (1982) Thermal self-ignition in a man-made cloud in the upper atmosphere. Izv USSR Acad Sci Atmos Oceanic Phys 18:270–274

Gershenzon YuM, Konoplev AV, Maksutov ShSh (1984) Evolution dynamics of the fast reaction artificial cloud in the upper atmosphere. Russ J Chem Phys 3:1196–1199

Gershenzon YuM, Konoplev AV, Maksutov ShSh (1990) Chemistry of the artificial gaseous clouds in the terrestrial upper atmosphere. In: Khananayan AA (ed) Trans Inst Exp Meteorol. N 21(143). Hydrometeoizdat, Moscow, pp 3–31

Gmurman VE (1975) Manual to the task solution on theory of probability and mathematic statistics, 2nd edn. High School, Moscow

Gorchakov GI, Yemilenko AS, Izakov AA, Sviridenkov MA (1976a) The directional light scattering coefficient in the angle region of 0.5–170°. Izv USSR Acad Sci Atmos Oceanic Phys 12:637–642

Gorchakov GI, Izakov AA, Sviridenkov MA (1976b) Statistical relations between the scattering coefficient and the scattering directively coefficient at the angle range 0.5–165°. Izv USSR Acad Sci Atmos Oceanic Phys 12:861–868

Gossard EE, Hooke WH (1975) Waves in the atmosphere. Elsevier Scientific, Amsterdam

Grinberg VN, Pozin AA, Soboleva VV, Khvostov VG, Shidlovsky AA (1979) Research and meteorological rockets in the world. Hydrometeoizdat, Leningrad

Greet PA, Innis J, Dyson PL (1994) High-resolution Fabry–Perot observations of mesospheric OH (6–2) emissions. Geophys Res Lett 21:1153–1156

Greer RGH, Murtagh DP, McDade IC, Dickinson PHG, Thomas L, Jenkins DB, Stegman J, Llewellyn EJ, Witt G, Mackinnon DJ, Williams ER (1986) ETON 1: a data base pertinent to the study of energy transfer in the oxygen nightglow. Planet Space Sci 34:771–788

Gruzdev AN, Bezverkhnii VA (1999) Long-term variations in the quasi-biennial oscillation of the equatorial stratospheric wind. Izv Atmos Oceanic Phys 35:700–711

Gruzdev AN, Bezverkhnii VA (2000) Corrections to the paper "Long-term variations of the quasi-biennial oscillation in the equatorial stratospheric wind". Izv Atmos Oceanic Phys 36:678–679

Gruzdev AN, Bezverkhnii VA (2003) On sources of the quasi-biennial oscillation of the atmosphere of the Northern Hemisphere. Dokl Earth Sci 389A:416–419

Gurvich AS, Vorobiev VV, Savchenko SA, Pakhomov AI, Padalka GI, Shefov NN, Semenov AI (2002) The 420–530 nm region nightglow of the upper atmosphere as measured onboard Mir research platform in 1999. Geomagn Aeron 42:514–519

Guseva NN, Kluev OF (1977) Calculation of the emission spectrum of the AlO molecule transition $A^2\Sigma^+ \leftrightarrow X^2\Sigma^+$ for the determination of the upper atmospheric temperature. In: Katasev LA (ed) Trans Inst Exp Meteorol. N 6(74). Hydrometeoizdat, Moscow, pp 52–63

Harris FJ (1978) On the use of windows for harmonic analysis with the discrete Fourier transform. Proc IEEE 66:51–83

Harrison TR (1960) Radiation pyrometry and its underlying. Wiley, New York

Hays PB, Roble RG (1971) A technique for recovering Doppler line profiles from Fabry–Perot interferometer fringes of very low intensity. Appl Opt 10:192–200

Hernandez G (1966) Analytical description of a Fabry–Perot photoelectric spectrometer. Appl Opt 5:1745–1748

Hernandez G (1970) Analytical description of a Fabry–Perot photoelectric spectrometer. 2: Numerical results. Appl Opt 9:1591–1596

Hernandez G (1971) The signature profiles of $O(^1S)$ in the airglow. Planet Space Sci 19:467–476

Hernandez G, Mills OA (1973) Feedback stabilized Fabry–Perot interferometer. Appl Opt 12:126–130

Hernandez G (1974) Analytical description of a Fabry–Perot photoelectric spectrometer. 3: Off axis behavior and interference filters. Appl Opt 13:2654–2661

Hernandez G (1975) Reaction broadening of the line profiles of atomic sodium in the night airglow. Geophys Res Lett 2:103–105

Hernandez G (1976) Lower thermosphere temperatures determined from the line profiles of the OI 17,924-K (5577Å) emission in the night sky. 1. Long-term behaviour. J Geophys Res 81:5165–5172

Hernandez G (1977) Lower thermosphere temperatures determined from the line profiles of the OI 17,924-K (5577Å) emission in the night sky. 2. Interaction with the lower atmosphere during stratospheric warming. J Geophys Res 82:2127–2131

Hernandez G (1978) Analytical description of a Fabry–Perot photoelectric spectrometer. 4: Signal noise limitations in data retrieval; winds, temperature, and emission rate. Appl Opt 17:2967–2972

Hernandez GJ, Turtle JP (1965) Nightglow 5577 A [OI] line kinetic temperatures. Planet Space Sci 13:901–904

Hernandez G, Mills OA (1973) Feedback stabilized Fabry–Perot interferometer. Appl Opt 12:126–130

Hernandez G, Killeen TL (1988) Optical measurements of winds and kinetic temperatures in the upper atmosphere. Adv Space Res 8:149–213

Hernandez G, Smith RW, Conner JF (1992) Neutral wind and temperature in the upper mesosphere above South Pole, Antarctica. Geophys Res Lett 19:53–56

IAPh (1994) Obukhov Institute of Atmospheric Physics of the Russian Academy of Sciences. Booklet. VneshTorgIzdat, Moscow, N 4462M

Ignatiev VM (1977a) Unusual profiles of the 5577 A and 6300 A emissions in aurorae. Astron Circ USSR Acad Sci N 940:2–4

Ignatiev VM (1977b) Peculiarities of contours of 5577 A and 6300 A lines in auroras. Geomagn Aeron 17:153–154

Ignatiev VM, Yugov VA (1995). In: Shefov NN (ed) Interferometry of the large-scale dynamics of the high-latitudinal thermosphere. Yakut Sci Centre Siberian Branch RAN, Yakutsk

Ivanov VM, Fishman IS (1973) Use of the Fourier-transform for the component determination of the Voigt profile. Opt Spectrosc 25:1175–1177

Ivanov-Kholodnyi GS, Nepomnyashchaya EV, Chertoprud VE (2000a) Variability of the parameters of quasi-two-year variations in the Earth's ionosphere in the 11-year cycle. Geomagn Aeron 40:526–528

Ivanov-Kholodnyi GS, Mogilevskii EI, Chertoprud VE (2000b) Quasi-biennial oscillations in total solar irradiance and in the Earth's ionospheric parameters. Geomagn Aeron 40:565–569

Iznar AN (1977) Electronic–optical instruments. Mashinostroenie, Moscow

Jenkins GM, Watts DG (1969) Spectral analysis and its applications. Honden-Day, San Francisco

Johnson ER (1965) Twilight resonance radiation of AlO in the upper atmosphere. J Geophys Res 70:1275–1277

Jordjio NV (1961) Automatic scanning photometer. Geomagn Aeron 1:1005–1008

Justus GG, Woodrum A (1973) Upper atmosphere planetary-wave and gravity-wave observations. J Atmos Sci 30:1267–1275

Kachurin LG (1990) Physical principles of the influence on the atmospheric processes. Hydrometeoizdat, Leningrad

Kalinyak AA, Krassovsky VI, Nikonov VB (1950) Observations of the Galactic center region in the radiation near 10000 A. Izv Crimea Astrophys Observ USSR Acad Sci 6:119–129

Kamke D, Krämer K (1977) Physikalische Grundlagen der Masseinheiten. BG Teubner, Stuttgart

Kapani NS (1967) Fiber optics. Principles and applications. Academic Press, New York

Kaporsky LN, Nikolaeva II (1969). In: Nikitin VA (ed) Optical instruments. Catalogue, vol 4. Mashinostroenie, Moscow

Khomich VYu, Schutter J (2000) Investigation of high temperature plasma radiation using multichannel spectrometry system and CCD detectors. Proc Intern Conf Appl CCD Spectrosc, Belfast

Khomich VYu, Zheleznov YuA (2000) Registration of visible and X-ray radiation spectra excited in plasma channels. Proc Intern Conf Appl CCD Spectrosc, Belfast

Khomich VYu, Moshkunov SI, Danilov VA, Zheleznov YuA, Dakhnovsky VO (1999) The elaboration of the registration system of the emission spectra in the optical and X-ray regions excited in plasma canals of the Besselian beams. Report. Scientific-Technological Centre of the Unique Instrument-Making RAS, Moscow

Khomich VYu, Zheleznov YuA, Moshkunov SI (2002) An experimental method and the instrument complex for the plasma research caused by plasmatrons in the electrophysical installations for the defense of the surrounding medium. Report. Scientific-Technological Centre of the Unique Instrument-Making RAS, Moscow

Khvostikov IA (1937). In: Vavilov SI (ed) Luminescence of the night sky. USSR Acad Sci, Moscow-Leningrad

Khvostikov IA (1948). In: Vavilov SI (ed) Luminescence of the night sky, 2nd edn. USSR Acad Sci, Moscow-Leningrad

Killeen TL, Hays PB (1984) Doppler line profile analysis for a multichannel Fabry–Perot interferometer. Appl Opt 23:612–627

Kluev OF (1979) Thermospheric temperature measurement over Heiss Island and Volgograd from the emissive spectra of the AlO molecules in the artificial luminous clouds. In: Nadubovich YuA (ed) Investigations of the optical nightglow. Yakut Depart Siberian Branch USSR Acad Sci, Yakutsk, pp 56–60

Kluev OF (1985) Thermospheric temperature measurement from the emissive spectra of the AlO molecules. In: Portnyagin YuI, Chasovitin YuK (eds) Trans Inst Exp Meteorol. N 16(115). Upper atmospheric physics. Hydrometeoizdat, Moscow, pp 15–25

Köhnlein W (1980) A model of thermospheric temperature and composition. Planet Space Sci 28:225–243

Konoplev AV, Korpusov VN, Novikov NN (1985) Experimental study of the artificial luminous ethylene cloud evolution in the night atmosphere at heights of 103 and 122 km. In: Portnyagin

YuI, Chasovitin YuK (eds) Trans Inst Exp Meteorol. N 16(115). Hydrometeoizdat, Moscow, pp 25–31

Kononovich EV (1962) The intensity microphotometer. Astron Circ USSR Acad Sci N 227:13–15

Kononovich EV (1963) The intensity microphotometer on the base of MF-4. In: Rozhkovsky DA (ed) New technique in astronomy. USSR Acad Sci, Moscow-Leningrad, pp 158–164

Kononovich EV, Shefov NN (2003) Fine structure of the 11-years cycles of solar activity. Geomagn Aeron 43:156–163

Kopp JP, Frederick JE, Rusch DW, Victor GA (1977) Morning and evening behavior of the F-region green line emission: evidence concerning the sources of $O(^1S)$. J Geophys Res 82:4715–4719

Korobeynikova MP, Nasyrov GA (1972) Study of the nightglow emission λ 5577 A for 1958–1967 in Ashkhabad. Ylym (Nauka), Ashkhabad

Korobeynikova MP, Nasyrov GA, Khamidulina VG (1966) The nightglow emission λ 5577 Å. Tables and maps of isophotes. Ashkhabad, 1964. In: Kalchaev KK, Shefov NN (eds). VINITI, Moscow

Korobeynikova MP, Nasyrov GA, Khamidulina VG (1968). In: Kalchaev KK, Shefov NN (eds) The nightglow emission λ 5577 Å. Tables and maps of isophotes. Ashkhabad, 1965–1966. VINITI, Moscow

Korobeynikova MP, Nasyrov GA, Khamidulina VG (1972). In: Savrukhin AP (ed) The nightglow emission λ 5577 Å, λ 6300 Å. Tables and maps of isophotes. Ashkhabad, 1967. Ylym (Nauka), Ashkhabad

Korolev FA (1953) High resolution spectroscopy. Gostekhizdat, Moscow

Kosyachenko LA, Kukhto EF, Sklyarchuk VM, Shemyakin VA (1990) Semiconductor emitter with the spectrum of the A source. Opt Mech Ind N 6:64–66

Krassovsky VI (1949) On the night sky radiation in the infrared spectral region. Dokl USSR Acad Sci 66:53–54

Krassovsky VI (1950a) New data on the night sky radiation in the 8800–11000 A region. Dokl USSR Acad Sci 70:999–1000

Krassovsky VI (1950b) New emissions of the night sky in the 8800–11000 A region. Izv Crimea Astrophys Observ USSR Acad Sci 5:100–104

Krassovsky VI (1957a) Nature of the intensity variations of the terrestrial atmosphere emission. Mém Soc R Sci Liège 18:58–67

Krassovsky VI (1957b) Nature of the intensity changes of the terrestrial atmosphere. Izv USSR Acad Sci Ser Geophys 664–669

Krassovsky VI, Shagaev MV (1974a) Optical method of recording of acoustic gravity waves in the upper atmosphere. J Atmos Terr Phys 36:373–375

Krassovsky VI, Shagaev MV (1974b) Inhomogeneities and wavelike variations of the rotational temperature of atmospheric hydroxyl. Planet Space Sci 22:1334–1337

Krassovsky VI, Shefov NN, Yarin VI (1962) Atlas of the airglow spectrum $\lambda\lambda$ 3000–12400 Å. Planet Space Sci 9:883–915

Krassovsky VI, Potapov BP, Semenov AI, Shagaev MV, Shefov NN, Sobolev VG, Toroshelidze TI (1977) The internal gravity waves near mesopause and hydroxyl emission. Ann Geophys 33:347–356

Krassovsky VI, Potapov BP, Semenov AI, Sobolev VG, Shagaev MV, Shefov NN (1978) Internal gravity waves near mesopause. I. Results of studies of hydroxyl emission. In: Galperin YuI (ed) Aurorae and airglow. N 26. Soviet Radio, Moscow, pp 5–29

Krat AV (1973) Spectrophotometry. In: Mikhailov AA (ed) A course of astrophysics and celestial astronomy, vol 1. Methods and instruments, 2nd edn. Nauka, Moscow, pp 507–530

Kravchuk GS, Sen' YuV, Tyutikov AM, Ivanov VN (1990) Improvement of the resolution capability of the microcanal plate–screen system. Opt Mech Ind N 5:66–69

Kulikovsky PG (2002). In: Surdin VG (ed) The handbook for the astronomical amateur, 5th edn. Editorial URSS, Moscow

Labitzke K, van Loon H (1988) Associations between the 11-year solar cycle, the QBO and the atmosphere. P. I. The troposphere and stratosphere in the Northern Hemisphere in winter. J Atmos Terr Phys 50:197–206

Lebedeva VV (1977) Optical spectroscopy technique. Moscow State Univ Press, Moscow
Lebedeva VV (1986) Optical spectroscopy technique, 2nd edn. Moscow State Univ Press, Moscow
Lyakhov SB, Managadze GG (1975) Photometer with the rotating interference filter. Instr Exp Techn N 3:200–201
Malyshev VI (1979) An introduction to the experimental spectroscopy. Nauka, Moscow
Manring ER, Pettit HB (1959) Photometric observations of the 5577 A and 6300 A emissions made during the aurora of February 10–11, 1958. J Geophys Res 64:149–153
Martynov DYa (1977) A course of the practical astrophysics, 3rd edn. Nauka, Moscow
Mathesn HA (1966) A nomogram for determining azimuth and horizontal trace velocity from tripartite measurements. Earthquake Notes 37:33–37
McDade IC, Murtagh DP, Greer RGH, Dickinson PHG, Witt G, Stegman J, Llewellyn EJ, Thomas L, Jenkins DB (1986a) ETON 2: quenching parameters for the proposed precursors of $O_2(b^1\Sigma_g^+)$ and $O(^1S)$ in the terrestrial nightglow. Planet Space Sci 34:789–800
McDade IC, Llewellyn EJ, Greer RGH, Murtagh DP (1986b) ETON 3: altitude profile of the nightglow continuum at green and near infrared wavelengths. Planet Space Sci 34:801–810
Meaburn J (1976) Detection and spectrophotometry of faint light. Reidel, Dordrecht
Mees CEK (1942) The theory of the photographic process. Macmillan, New York
Mees CEK, James TH (1966) The theory of the photographic process, 3rd edn. Macmillan, New York
Milinevsky GP, Romanovsky YuA, Evtushevsky AM, Savchenko BA, Alpatov VV, Gurvich AV, Lifshits AI (1990) Optical observations in the active experiments on the investigation of the upper atmosphere and ionosphere of the Earth. Cosmic Res 28:418–429
Miroshnikov MM (1977) The theoretical principles of the optical–electronic instruments. Mashinostroenie, Leningrad
Mitra SK (1952) The upper atmosphere, 2nd edn. The Asiatic Soc, Calcutta
Mulyarchik TM (1959) Interferometric measurements of the line width of the λ 5577 Å [OI] in the aurorae. Izv USSR Acad Sci Ser Geophys 1902–1903
Mulyarchik TM (1960a) Interferometric measurement of λ 6300 Å [OI] and λ 5198–5200 Å [NI] emissions from auroras. Dokl USSR Acad Sci 130:303–306
Mulyarchik TM (1960b) Interferometric measurements of the upper atmospheric temperature from the widths of the some emission lines. Izv USSR Acad Sci Ser Geophys 449–458
Mulyarchik TM, Petrova KI (1957) Increasing the sensitivity of some photographic emulsions by a preliminary short exposure. Astron Rep 34:102–104
Mulyarchik TM, Shcheglov PV (1963) Temperature and corpuscular heating in the auroral zone. Planet Space Sci 10:215–218
Nakamura T, Higashikawa A, Tsuda T, Matsushita Y (1999) Seasonal variations of gravity wave structures in OH airglow with a CCD imager at Shigaraki. Earth Planets Space 51:897–906
Nikonov VB (1973a) Photoelectric astrophotometry. In: Mikhailov AA (ed) A course of astrophysics and celestial astronomy, vol. 1. Methods and instruments, 2nd edn. Nauka, Moscow, pp 392–433
Nikonov VB (1973b) Photometric systems, standards and catalogues. In: Mikhailov AA (ed) A course of astrophysics and celestial astronomy, vol 1. Methods and instruments, 2nd edn. Nauka, Moscow, pp 434–459
Novikov NN (1981) Internal gravity waves near mesopause. VI. Polar region. In: Galperin YuI (ed) Aurorae and airglow. N 29. Soviet Radio, Moscow, pp 59–67
Obukhov AM (1988) Turbulence and atmospheric dynamics. Hydrometeoizdat, Leningrad
Offermann D (1977) Some results from the European winter anomaly campaign 1975/76. In: Grandal B, Holtet JA (eds) Dynamical and chemical coupling between the neutral and ionized atmosphere. Reidel, Dordrecht, pp 17–33
Offermann D (1979) Recent advances in the study of the D-region winter anomaly. J Atmos Terr Phys 41:735–752
Offermann D, Curtis P, Cisneros IM, Satrustegui I, Lauche H, Rose G, Petzoldt K (1979) Atmospheric temperature structure during the Western European Winter Anomaly Campaign 1975/76. J Atmos Terr Phys 41:1051–1062

ORIEL (1999). The book of photon tools. Oriel Instruments, Stratford
Piterskaya NA, Shefov NN (1975) Intensity distribution of the OH rotation–vibration bands. In: Krassovsky VI (ed) Aurorae and airglow. N 23. Nauka, Moscow, pp 69–122
Potapov BP (1974) Dependence between variations of intensity and rotational temperature of hydroxyl emission of night time-sky. Geomagn Aeron 14:1056–1060
Potapov BP (1975a) Determination of the effective height of fluctuations of hydroxyl emissions. Planet Space Sci 23:1346–1347
Potapov BP (1975b) Triangular measurements of the hydroxyl emission height. Astron Circ USSR Acad Sci N 856:5–7
Potapov BP (1976) Measurement of the altitude of the OH emission by the fluctuations of the rotational temperature. In: Krassovsky VI (ed) Aurorae and airglow. N 24. Soviet Radio, Moscow, pp 21–26
Potapov BP, Sobolev VG, Chubukov VP (1976) Using of the cooled photomultipliers with the oxygen–caesium cathodes for the photon counters. In: Krassovsky VI (ed) Aurorae and airglow. N 24. Soviet Radio, Moscow, pp 75–78
Potapov BP, Semenov AI, Sobolev VG, Shagaev MV (1978) Internal gravity waves near mesopause. II. Instruments and optical methods of measurements. In: Galperin YuI (ed) Aurorae and airglow. N 26. Soviet Radio, Moscow, pp 30–65
Press FP (1981) Videosignal generators in the charge-coupled devices. Radio and Svayz, Moscow
Press FP (1991). Photosensitive charge-coupled devices. Radio and Svayz, Moscow
Prudnikov AP, Brychkov YuL, Marichev OI (1979) Integrals and series. Nauka, Moscow
Prudnikov AP, Brychkov YuL, Marichev OI (1981) Integrals and series. Elementary functions. Nauka, Moscow
Ribaud G (1931) La traité de pyrométrie optique. Edition de la Revue d'Optique théorique et instrumentale, Paris
Rieger E (1974) Neutral air motions deduced from barium releases experiments. Vertical winds. J Atmos Terr Phys 36:1377–1385
Roach FE, Gordon JL (1973) The light of the night sky. Reidel, Dordrecht
Roach FE, Pettit HB, Williams DR, St Amand P, Davis DN (1953) A four-year study of OI 5577 Å in the nightglow. Ann d'Astrophys 16:185–205
Roble RG, Shepherd GG (1997) An analysis of wind imaging interferometer observations of $O(^1S)$ equatorial emission rates using the thermosphere–ionosphere–mesosphere–electrodynamics general circulation model. J Geophys Res 102A:2467–2474
Reisse R, Creecy R, Poulthney SK (1974) Single photon detection and subnanosecond timing resolution with the RCA C31034 photomultiplier. Rev Sci Instrum 44:1666–1668
Rosenberg GV (1963) The twilight. Nauka, Moscow
Sattarov DK (1973) Fiber optics. Mashinostroenie, Leningrad
Sawyer RA (1951) Experimental spectroscopy, 2nd edn. Academic Press, New York
Schmidlin FJ, Carlson M, Offermann D, Philbrick CR, Rees D, Widdel HU (1982) Wind structure and small-scale wind variability in the stratosphere and mesosphere during the November 1980 energy budget campaign. Adv Space Res 2:125–128
Semenov AI (1975a) Interferometric measurements of the upper atmosphere temperature. I. Application of the cooled image converters. In: Krassovsky VI (ed) Aurorae and airglow. N 23. Nauka, Moscow, pp 64–65
Semenov AI (1975b) Doppler temperature and intensity of emission 6300 A. Geomagn Aeron 15:876–880
Semenov AI (1976) Interferometric measurements of the upper atmosphere temperature. III. 6300 A emission and characteristics of the atmosphere and ionosphere. In: Krassovsky VI (ed) Aurorae and airglow. N 24. Soviet Radio, Moscow, pp 44–51
Semenov AI (1978) Interferometric measurements of the upper atmospheric temperature. V. Semiannual temperature variations according to 6300 emission. In: Krassovsky VI (ed) Aurorae and airglow. N 27. Soviet Radio, Moscow, pp 85–86
Semenov AI (1988) Seasonal variations of the hydroxyl rotational temperature. Geomagn Aeronomy 28:333–334

Semenov AI (1989) The specific features of the green emission excitation process in the nocturnal atmosphere. In: Feldstein YaI, Shefov NN (eds) Aurorae and airglow. N 33. VINITI, Moscow, pp 74–80
Semenov AI, Shefov NN (1996) An empirical model for the variations in the hydroxyl emission. Geomagn Aeron 36:468–480
Semenov AI, Shefov NN (1997a) An empirical model of nocturnal variations in the 557.7-nm emission of atomic oxygen. 1. Intensity. Geomagn Aeron 37:215–221
Semenov AI, Shefov NN (1997b) An empirical model of nocturnal variations in the 557.7-nm emission of atomic oxygen. 2. Temperature. Geomagn Aeron 37:361–364
Semenov AI, Shefov NN (1997c) An empirical model of nocturnal variations in the 557.7-nm emission of atomic oxygen. 3. Emitting layer altitude. Geomagn Aeron 37:470–474
Semenov AI, Shefov NN (1997d) Empirical model of the variations of atomic oxygen emission 557.7 nm. In: Ivchenko VN (ed) Proc SPIE (23rd European Meeting on Atmospheric Studies by Optical Methods, Kiev, September 2–6, 1997), vol 3237. The Intern Soc Opt Engin, Bellingham, pp 113–122
Semenov AI, Shefov NN (1999) Empirical model of hydroxyl emission variations. Int J Geomagn Aeron 1:229–242
Semenov AI, Shagaev MV, Shefov NN (1981) On the effect of orographic waves on the upper atmosphere. Izv USSR Acad Sci Atmos Oceanic Phys 17:982–984
Semenov AI, Shefov NN, Givishvili GV, Leshchenko LN, Lysenko EV, Rusina VYa, Fishkova LM, Martsvaladze NM, Toroshelidze TI, Kashcheev BL, Oleynikov AN (2000) Seasonal peculiarities of long-term temperature trends of the middle atmosphere. Dokl Earth Sci 375:1286–1289
Semenov AI, Bakanas VV, Perminov VI, Zheleznov YuA, Khomich VYu (2002) The near infrared spectrum of the emission of the nighttime upper atmosphere of the Earth. Geomagn Aeron 42:390–397
Séquin CH, Tompsett MF (1975) Charge-coupled device. Academic Press, New York
Serebryannikov MG, Pervozvansky AA (1965) Discovery of the latent periodicities. Nauka, Moscow
Shagaev MV (1975) Study of the hydroxyl emission with time resolution of some minutes. In: Krassovsky VI (ed) Aurorae and airglow. N 23. Nauka, Moscow, pp 22–27
Shcheglov PV (1958) Some methodical problems in applying image converters. Astron Rep 35:651–655
Shcheglov PV (1963) Electronic telescopy. Fizmatgiz, Moscow
Shcheglov PV (1967a) Photoelectric interferometric installation with Fabry–Perot etalon. Astron Circ USSR Acad Sci N 411:1–4
Shcheglov PV (1967b) The neutral hydrogen distribution in the terrestrial atmosphere by observations of the T_α nightglow. Astron Circ USSR Acad Sci N 427:5
Shchigolev BM (1960) Mathematical treatment of the observations. Fizmatgiz, Moscow
Shefov NN (1961) On determination of the rotational temperature of the OH bands. In: Krassovsky VI (ed) Spectral, electrophotometrical and radar researches of aurorae and airglow. N 5. USSR Acad Sci, Moscow, pp 5–9
Shefov NN (1965) Correlation between the upper atmosphere emissions. In: Krassovsky VI (ed) Aurorae and airglow. N 11. USSR Acad Sci, Moscow, pp 43–47
Shefov NN (1967) Some properties of the hydroxyl emission. In: Krassovsky VI (ed) Aurorae and airglow. N 13. USSR Acad Sci, Moscow, pp 37–43
Shefov NN (1989) The recording of wave and spotted inhomogeneities of upper atmospheric emission. In: Feldstein YaI, Shefov NN (eds) Aurorae and airglow. N 33. VINITI, Moscow, pp 81–84
Shefov NN, Toroshelidze TI (1974) Dynamics of minor constituent emissions. Ann Geophys 30:79–83
Shefov NN, Toroshelidze TI (1975) Upper atmosphere emission as an indicator of the dynamic processes. In: Krassovsky VI (ed) Aurorae and airglow. N 23. USSR Acad Sci, Moscow, pp 42–53

Shefov NN, Pertsev NN (1984) Orographic disturbances of upper atmosphere emissions. In: Taubenheim J (ed) Handbook for Middle Atmosphere Program, vol 10. SCOSTEP, Urbana, pp 171–175

Shefov NN, Semenov AI (2004) Spectral characteristics of the IGW trains registered in the upper atmosphere. Geomagn Aeron 44:763–768

Shefov NN, Semenov AI (2006) Spectral composition of the cyclic aperiodic (quasi-biennial) variations in solar activity and the Earth's atmosphere. Geomagn Aeron 46:411–416

Shefov NN, Pertsev NN, Shagaev MV, Yarov VN (1983) Orographically caused variations of upper atmospheric emissions. Izv USSR Acad Sci Atmos Oceanic Phys 19:694–698

Shefov NN, Semenov AI, Pertsev NN, Sukhodoev VA, Perminov VI (1999) Spatial distribution of IGW energy inflow into the mesopause over the lee of a mountain ridge. Geomagn Aeron 39:620–627

Shefov NN, Semenov AI, Pertsev NN, Sukhodoev VA (2000) The spatial distribution of the gravity wave energy influx into the mesopause over a mountain lee. Phys Chem Earth Pt B 25:541–545

Shepherd GG, Thuillier G, Gault WA, Solheim BH, Hersom C, Alunni M, Brun JF, Brune S, Chalot P, Cogger LL, Desaulniers DL, Evans WFJ, Gattinger RL, Girod F, Harvie D, Henn RH, Kendall DJW, Llewellyn EJ, Lowe RP, Ohrt J, Pasternak F, Peillet O, Powell I, Rochon Y, Ward WE, Wiens RH, Wimperis J (1993a) WINDII, the wind imaging interferometer on the upper atmosphere research satellite. J Geophys Res 98D:10725–10750

Shepherd GG, Thuillier G, Solheim BH, Chandra S, Cogger LL, Duboin ML, Evans WFJ, Gattinger RL, Gault WA, Hersé M, Hauchecorne A, Lathuilliere C, Llewellyn EJ, Lowe RP, Teitelbaum H, Vial F (1993b) Longitudinal structure in atomic oxygen concentrations observed with WINDII on UARS. Geophys Res Lett 20:1303–1306

Shklovsky IS (1960) Artificial comet as a method of optical observations of the cosmical rockets. In: Sedov LI (ed) Artificial satellites of the Earth. N 4. USSR Acad Sci, Moscow, pp 195–204

Shklovsky IS, Kurt VG (1959) Atmospheric density determination at height of 403 km by method of sodium vapor diffusion. In: Sedov LI (ed) Artificial satellites of the Earth. N 3. USSR Acad Sci, Moscow, pp 66–77

Shouyskaya FK (1963) An attempt to detect the proper glow of atmosphere during the solar eclipse on February 15, 1961. In: Krassovsky VI (ed) Aurorae and airglow. N 10. USSR Acad Sci, Moscow, pp 44–53

Sivjee GG, Hamwey RM (1987) Temperature and chemistry of the polar mesopause OH. J Geophys Res 92A:4663–4672

Skrebushevsky BS (1990) Development of the spacecraft orbits. Mashinostroenie, Moscow

Sobolev VG (1978a) Continuum in night airglow between 8000 and 11000 Å. Planet Space Sci 26:703–704

Sobolev VG (1978b) Continuum of the near infrared range of nightglow spectrum. In: Krassovsky VI (ed) Aurorae and airglow. N 27. Soviet Radio, Moscow, pp 30–35

Sobolev VG (1979) Correlation between nightglow continuum and ionospheric absorption. Astron Circ USSR Acad Sci N 1083:7–8

Soule HV (1968) Electro-optical photography at low illumination levels. Wiley, New York

Spizzichino A (1971) Meteor trail radar winds over Europe. In: Webb WL (ed) Thermospheric circulation. The MIT Press, Cambridge, pp 117–180

Starkov GV, Yevlashin LS, Semenov AI, Shefov NN (2000) A subsidence of the thermosphere during 20th century according to the measurements of the auroral heights. Phys Chem Earth Pt B 25:547–550

Sukhodoev VA, Yarov VI (1998) Temperature variations of the mesopause in the leeward region of the Caucasus ridge. Geomagn Aeron 38:545–548

Sukhodoev VA, Pertsev NN, Reshetov LM (1989a) Variations of characteristics of hydroxyl emission caused by orographic perturbations. In: Feldstein YaI, Shefov NN (eds) Aurorae and airglow. N 33. VINITI, Moscow, pp 61–66

Sukhodoev VA, Perminov VI, Reshetov LM, Shefov NN, Yarov VN, Smirnov AS, Nesterova TS (1989b) The orographic effect in the upper atmosphere. Izv USSR Acad Sci Atmos Oceanic Phys 25:681–685

Takahashi H, Clemesha BR, Sahai Y (1974) Nightglow OH (8,3) band intensities and rotational temperature at 23° S. Planet Space Sci 22:1323–1329

Taranova OG (1965) On the statistical treatment of the photoelectric measurements with the filters. In: Krassovsky VI (ed) Aurorae and airglow. N 11. USSR Acad Sci, Moscow, pp 35–42

Taranova OG (1967) Study of space-time properties of the hydroxyl emission. In: Krassovsky VI (ed) Aurorae and Airglow. N 13. USSR Acad Sci, Moscow, pp 13–21

Taranova OG, Toroshelidze TI (1970) On the measurements of the hydroxyl emission in twilight. In: Krassovsky VI (ed) Aurorae and airglow. N 18. Nauka, Moscow, pp 26–32

Tarasov KI (1968) Spectral instruments. Mashinostroenie, Leningrad

Tarasov KI (1977) Spectral instruments, 2nd edn. Mashinostroenie, Leningrad

Taylor JR (1982) An introduction to error analysis. Univ Science Book, Mill Valley, CA

Tolansky S (1947) High resolution spectroscopy. Methuen and Co, London

Trofimova LS, Sattarov DK, Konaeva GYa (1972) Once more on the optical characteristics of the pile of the fiber optical elements. Opt Spectrosc 32:1211–1215

Trunkovsky EM, Semenov AI (1978) Interferometric measurements of the upper atmosphere temperature. IV. Analysis of photographic interferograms. In: Krassovsky VI (ed) Aurorae and airglow. N 27. Soviet Radio, Moscow, pp 66–84

Vaisberg OL (1960) First observations of the auroral spectra with the photoelectric spectrometer. Izv USSR Acad Sci Ser Geophys 1277–1278

Veinberg BS, Vain LN (1974) Household compressive refrigerators. Food Industry, Moscow

Veinberg VB, Sattarov DK (1977) Fiber optics, 2nd edn. Mashinostroenie, Moscow

Vetokhin SS, Gulakov IR, Pertsev AN, Reznikov IV (1986) One-electron photodetectors, 2nd edn. Energyatomizdat, Moscow

Vlasov MN, Pokhunkov SA (1986) The effect on the upper atmosphere produced by release of some chemically active gases. Geomagn Aeron 26:756–761

Volkov IV, Esipov VF, Shcheglov PV (1959) The use of contact photography principle in studying of weak light fluxes. Dokl USSR Acad Sci 129:288–289

Volkov IV, Esipov VF, Shcheglov PV (1962) The contact photography of weak light objects. Astron Rep 39:323–329

Voronin AN, Galperin VL, Kudacov AS, Semenov AI, Shefov NN (1984) Thermoelectric refrigerator. Author certificate N 1084560 from 8 December, 1983. Official Bull Govern Com Invention N 13

Zaidel IN, Kurenkov GI (1970) Image converter tubes. In: Balashov VP (ed) Soviet Radio, Moscow

Zaidel AN, Ostrovskaya GV, Ostrovsky YuN (1972) Technique and practices of the spectroscopy. Ser Phys Techn Spectral Analysis. Nauka, Moscow

Zaidel AN, Ostrovskaya GV, Ostrovsky YuN (1976) Technique and practices of the spectroscopy. Ser Phys Techn Spectral Analysis, 2nd edn. Nauka, Moscow

Zhiglinsky AG, Kuchinsky VV (1983) Real Fabry–Perot interferometer. Mashinostroenie, Leningrad

Zhurba YuI (1990) A brief handbook on the photographic processes and materials, 2nd edn. Art, Moscow

Zinn J, Sutherland CD, Stone SN, Duncan LM (1982) Ionospheric effects of rocket exhaust products. HEAO-C, Skylab. J Atmos Terr Phys 44:1143–1171

Zverev FS (1977) Synoptic meteorology, 2nd edn. Hydrometeoizdat, Leningrad

Chapter 4
Regular Variations of the Airglow in the Mesopause and Thermosphere

Each emission in the upper atmosphere is confined to a certain layer which is characterized by the emission intensity I (Rayleigh), temperature T (K) inside the layer, represented by its value near the maximum of the emission rate $Q_m(Z_m)$ (photon·cm^{-3}·s^{-1}), vertical distribution of the emission rate Q(Z), layer thickness W at the level $Q_m(Z_m)/2$, as well as by the thickness W_u above the altitude Z_m and thickness W_l below Z_m, which define the layer asymmetry P equal to the ratio W_u/W. According to this definition, the total emission intensity, which characterizes the rate of energy conversion and removal, is given by

$$I = \int_0^\infty Q(z)dz .$$

An important characteristic in this definition is the shape of the emission rate altitude profile. Theoretical investigations of a number of emissions point to an asymmetric profile such that the layer fraction above the altitude of maximum emission rate Z_m is greater than that below Z_m (Moreels et al. 1977), i.e., the asymmetry P > 0.5.

Originally, it was a common practice to describe an altitude distribution of emission rate by the function proposed by Chapman (Rao et al. 1982)

$$Q(Z) = Q_m(Z_m) \exp\left\{ S\left[1 - \frac{Z - Z_m}{\sigma \cdot W} - \exp\left(-\frac{Z - Z_m}{\sigma \cdot W} \right) \right] \right\} ,$$

where the emission rate at the layer maximum is given by

$$Q_m(Z_m) = e^{-S} \cdot S^S \cdot I_{v'v''}/\Gamma(S)\sigma \cdot W ,$$

where S and σ are parameters, W is the layer thickness at the level $Q_m(Z_m)/2$, and Γ is the gamma function, or by

$$Q_m(Z_m) = 12S\sqrt{\frac{S}{2\pi}} \cdot \frac{I_{v'v''}}{\sigma \cdot W(12S+1)} .$$

From the description of the emission rate altitude profile by the Chapman formula it follows that

$$P = \sigma \cdot \ln\left[\sigma\left(\exp\frac{1}{\sigma} - 1\right)\right] \text{ and } S = \frac{\ln 2}{\frac{P}{\sigma} + \exp\frac{P}{\sigma} - 1},$$

whence approximate solutions can be obtained:

$$\ln\sigma = -3.04 + 2.01\left(\frac{1}{P} - 1\right) + 1.78\left(\frac{1}{P} - 1\right)^2$$

$$\ln S = -2.95 + 0.747\left(\frac{1}{P} - 1\right) + 5.464\left(\frac{1}{P} - 1\right)^2.$$

However, analysis of rocket data has shown that the distribution of emission rate $Q(Z)$ can be represented more adequately by a Gaussian asymmetric distribution such that its top part is wider than the bottom one. In this case, we have

$$Q(Z) = Q_m(Z_m)\exp\left[-\frac{\ln 2(Z - Z_m)^2}{P^2 W^2}\right]$$

for $Z \geq Z_m$ and

$$Q(Z) = Q_m(Z_m)\exp\left[-\frac{\ln 2(Z - Z_m)^2}{(1 - P)^2 W^2}\right],$$

for $Z \leq Z_m$. The emission rate at the layer maximum is determined by the formula

$$Q_m(Z_m) = 2\sqrt{\frac{\ln 2}{\pi}} \cdot \frac{I}{W},$$

where P is the asymmetry. The investigations performed show that the variations of the layer altitude Z_m are reflected in variations of the emission intensity I, temperature T, and layer thickness W and asymmetry P.

When constructing a model of the behavior of characteristics of any atmospheric emission, one has to deal with a rather wide spectrum of variations which contains both regular and irregular components. In this case, it is meant that the values of regular variations for a given place and a given time can be calculated based on given heliogeophysical conditions and that irregular variations arise due to some random phenomena, but subsequently their behavior is certainly regular.

Since variations of any type are in fact modulations of some average values of the emission characteristics, the following representation was taken as a first approximation for the emission parameters T_r, T_V, Z_m, W, and P:

$$f = f_0 + \sum_i^N \Delta f_i,$$

and for the emission intensity I,

$$f = f_0 \cdot \prod_i^N (1 + \Delta f_i),$$

where f_0 designates some global mean for given heliogeophysical conditions: geographic latitude $\varphi = 45°N$; geographic longitude $\lambda = 40°E$; time of day: local solar midnight $\tau = 0$; season of year: equinox; day of year $t_d = 80$; solar flux $F_{10.7} = 130$; geomagnetic activity $K_p = 0$, and year 1972.5.

Thus, at the given stage of research, because of lack of data, it is supposed that in most cases the behavior of an emission is the same in the northern and southern hemispheres.

The quantities Δf_i describe variations of different types, namely (1) regular variations: diurnal variations during a night, $\Delta f_d(\chi_\odot, \varphi)$ ($\chi_\odot$ is the zenith angle of the Sun and φ is the geographic latitude); terminator variations in the evening and morning, $\Delta f_{te}(\tau)$ (τ is the local time); lunar variations during a day, $\Delta f_L(\tau_L)$ (τ_L is the lunar time); lunar variations during a synodic month of 29.53 days, $\Delta f_L(t_{L\varphi})$ ($t_{L\varphi}$ is the phase age of the Moon); seasonal variations, $\Delta f_s(t_d)$ (t_d is the day of year); cyclic aperiodic (quasi-biennial) variations with a variable period of 32 to 22 months, $\Delta f_{SAO}(t_{SAO})$ [t_{SAO} is the year of the beginning of a cycle (the minimum of a 11-yr solar cycle)]; 5.5-yr cycles, $\Delta f_{5.5}(t_{5.5})$ ($t_{5.5}$ is the year of the beginning of a cycle); variations during a solar cycle, $\Delta f_F(F_{10.7})$ ($F_{10.7}$ is the Sun radioradiation flux; 22-yr variations are also taken into account by the index $F_{10.7}$); a long-term trend $\Delta f_{tr}(t_{tr})$ (t_{tr} is the year of the trend origin); latitudinal, $\Delta f_\varphi(\varphi)$, longitudinal, $\Delta f_\lambda(\lambda)$, and orographic variations, $\Delta f_{or}(V, L)$ (V is the velocity of a tropospheric wind and L is the distance from the mountain), and (2) irregular variations: 27-day variations caused by variations of solar flux in ultraviolet radiation, $\Delta f_{27}(t_{27})$ (t_{27} is the date of the beginning of the nearest Carrington cycle); variations after stratospheric warming, $\Delta f_{sw}(t_{sw})$ (t_{sw} is the date of the beginning of the process); variations after a geomagnetic disturbance, $\Delta f_{gm}(t_{gm}, K_p, \Phi)$ (t_{gm} is the date of the beginning of the geomagnetic storm, K_p is the planetary index of the geomagnetic disturbance, and Φ is the geomagnetic latitude); variations after intrusion of meteoric flows, $\Delta f_{mf}(t_{mfd})$ (t_{mf} is the date of the beginning of intrusion), and variations caused by internal gravity waves, $\Delta f_{gw}(\tau_w)$ (τ_w is the wave period).

However, such a list of variations could be compiled for not all emissions of the upper atmosphere, since the available measurement series are of varied length and measurements were performed not for all parameters which characterize the properties of an emission layer.

4.1 Model of Hydroxyl Emission

The hydroxyl emission is one of the most energetically significant emissions occurring in the upper atmosphere. It arises due to recombination of atomic oxygen. An essential feature of this emission is that its spectral and aeronomic parameters, which characterize not only the processes of excitation of hydroxyl radicals but also the properties and state of the atmospheric layer in which it arises, can be measured reliably.

These parameters are the intensity (Rayleigh) of individual molecular bands $(v' - v)$, $I_{v',v''}$, the equilibrium, T_r, and nonequilibrium, T_{nr}, rotational temperatures that characterize the distribution of excited OH molecules over rotational states (J, N), the vibrational temperature T_V that characterizes the distribution of OH molecules over vibrational states v and the related photochemical processes, the emission layer altitude Z_m, and the parameters of its altitude distribution: thickness W and asymmetry P.

Investigations of the hydroxyl emission parameters have been conducted for 55 years though rather nonuniformly from the viewpoint of accumulation of measurement data during time and their planetary distribution. This causes difficulties in revealing particular types of variations. Nevertheless, the need in such data is obvious, since the observed time–space variations are manifestations of a wide spectrum of photochemical and dynamic processes in the atmosphere as a whole.

The photochemical processes resulting in the excitation of hydroxyl emission in the atmosphere was considered in detail in Sect. 2.2 where it has been shown that the basic atmospheric constituents responsible for its occurrence are ozone, atomic oxygen, and hydrogen, and also the ultraviolet radiation of the Sun. Their space–time behavior in the mesopause can be studied based on the data about the hydroxyl emission. Therefore, an attempt was made to systematize the material accumulated for many years (Semenov and Shefov 1996a, 1999a). It should be stressed that the available data refer only to the night interval of day. For daytime conditions, there are ad hoc published measurement data (Le Texier et al. 1989) and theoretical model calculations (López-González et al. 1996). Nevertheless, they are insufficient for any systematization.

4.1.1 Factors Affecting the Variations of OH Emission Parameters

Attempts to systematize some measurements of the emission intensity I, rotational, T_r, and vibrational, T_V, temperatures and of the layer altitude Z_m at which the emission rate Q(Z) is a maximum to gain a general idea of the behavior of hydroxyl emission were made repeatedly (Shefov 1969a,b, 1971a,b, 1973a, 1974a,b, 1975a, 1976, 1978b; Semenov and Shefov 1979a,b; Fishkova 1955, 1976, 1981, 1983; Shefov and Piterskaya 1984; Toroshelidze 1991; Semenov and Shefov 1996a, 1999a).

Subsequently, some improvement of the relations obtained have been made (Semenov and Shefov 2003). They have already been considered in the previous sections.

4.1.2 Determination of the Characteristics of OH Emission

As already mentioned, the hydroxyl emission arises due to rotational–vibrational electronic transitions $(v' = 9 \div 1, v'' = (v' - 1) \div 0)$ of the ground state $X^2\Pi$. Its

4.1 Model of Hydroxyl Emission

total intensity I is the sum of the intensities $I_{v'v''}$ of bands distributed in the 0.5–5.0 (μm) spectral range. The greatest intensity (photons) is produced by the sequence of vibrational transitions $\Delta v = 2$ in the spectral range 1.4–2.2 (μm) (Shefov and Piterskaya 1984).

The population distribution over vibrational levels can be on the average represented by a Boltzmann distribution with vibrational temperature $T_V \sim 10000$ (K), though there is evidence (see Fig. 2.15) of the existence of two sections in the distributions for $v = 1 - 5$ and $v = 6 - 9$ for which $T_{9876} \geq T_{54321}$ (Shefov 1976; Fishkova 1981, 1983).

Thus, the total radiation intensity (photons) is determined as

$$I = \sum_{v'=1}^{9} \sum_{v''=0}^{8} I_{v'v''} = \sum_{v'=1}^{9} \sum_{v''=0}^{8} A_{v'v''} \cdot N_{v'} = \frac{\sum_{v'=1}^{9} A_{v'} \cdot \exp(-G_{v'}/kT_V)}{\sum_{v'=1}^{9} \exp(-G_{v'}/kT_V)} N_{\text{eff}} = \bar{A} \cdot N_{\text{eff}},$$

where $A_{v'v''}$ are the probabilities of the transitions; $A_{v'} = \sum_{v''=0}^{v'} A_{v'v''}$; $N_{v'}$ is the population of the v'th vibrational level; k is Boltzmann's constant; and N_{eff} is the effective number (cm^{-2}) of excited OH molecules. The values of $A_{v'v''}$ obtained by calculations based on various data about the intensities of OH bands are given in Tables 2.1–2.10. It should be stressed that many calculations of transition probabilities had the disadvantage that they used the data of nonsimultaneous measurements of intensities of some OH bands presented by Krassovsky et al. (1962) that were not reduced to some standard conditions because many laws of variations have not been established yet at that time. Notwithstanding that the calculation techniques have been improved, the values of $A_{v'v''}$ call for further refinement. The energies of vibrational levels, $G_v(\text{cm}^{-1})$, are determined by the relation

$$G_v = 3653\,v - 85\,v^2 + 0.54\,v^3 \,;$$

$\bar{A}(T_V)$ ($\sim 48(\text{s}^{-1})$ for $T_V = 10000$ (K)) is the effective transition probability. It can be found from the approximate relation

$$\ln \bar{A}(T_V) = 2.3265 + 2.7832 \cdot T'_V - 1.6776 \cdot T'^2_V + 0.5001 \cdot T'^3_V - 0.0587 \cdot T'^4_V \,,$$

where $T'_V = T_V/10000$. Thus, the intensity (photons) of an OH band is determined by the formula

$$I_{v'v''} = I \cdot \frac{A_{v'v''}}{\bar{A} \cdot Q_v} \exp(-G_{v'}/kT_V) \,,$$

where the sum over the vibrational states

$$Q_{v'} = \sum_{v'=1}^{9} \exp(-G_{v'}/kT_V)$$

can be approximated by the relation

$$\ln Q_{v'} = -2.97 + 7.14 \cdot T'_V - 5.547 \cdot T'^2_V + 2.20 \cdot T'^3_V - 0.345 \cdot T'^4_V.$$

The total hydroxyl emission energy is determined by the formula

$$E = N_{eff} \cdot \sum_{v'=1}^{9} \left(\sum_{v''=0}^{8} \frac{hc}{\lambda_{v'v''}} A_{v'v''} \right) / [Q_{v'} \exp(-G_{v'}/kT_V)]$$

or

$$E = \frac{2 \cdot 10^{-12}}{\lambda(T_V)} \cdot I (\text{erg} \cdot \text{cm}^{-2} \cdot \text{s}^{-1}),$$

where $\lambda_{v'v''}$ are the wavelengths of the OH bands, and the effective wavelength (micrometers) is approximated as

$$\lambda(T_V) = 3.095 - 2.862 \cdot T'_V + 2.549 \cdot T'^2_V - 1.051 \cdot T'^3_V + 0.165 \cdot T'^4_V.$$

The OH rotational temperature T_r (equilibrium) is determined by the intensity distribution for the first four to five lines of the P branch of the rotational band (Shefov 1961). Transition probabilities for various lines (Benedict et al. 1953) can be found in the Atlas composed by Krassovsky et al. (1962). The measured temperature T_r is the weighted average of the atmospheric temperature T inside the emission layer and practically corresponds to the altitude of maximum emission rate Q_m. However, there exists a dependence of T_r on vibrational level number v (Berg and Shefov 1963; Bakanas and Perminov 2003; Bakanas et al. 2003), which is related to seasonal variations of the respective emission layer altitude Z_{mv}. Special investigations have revealed that the difference between rotational temperatures corresponding to different vibrational levels depends on the season and is determined by the harmonic amplitudes and phases. It can be described by the relation

$$\Delta T_{v6} = a_0 + a_1 \cdot \cos[2\pi(t_d - t_1)/365.25] + a_2 \cdot \cos[4\pi(t_d - t_2)/365.25],$$

where $\Delta T_{v6} = T_v - T_6$ is the difference between the temperatures corresponding to the fifth and sixth vibrational levels and t_d is the day of year,

$$a_0 = 9.55 - 3.02 \cdot v + 0.24 \cdot v^2,$$
$$a_1 = -27.81 + 10 \cdot v - 0.89 \cdot v^2,$$
$$a_2 = -18.51 + 6.81 \cdot v - 0.62 \cdot v^2,$$
$$t_1 = -68.5 + 12.8 \cdot v,$$
$$t_2 = 182 - 19 \cdot v.$$

The calculated seasonal variations of the rotational temperature differences for vibrational levels 2–9 and the distribution of these differences over the levels for different months of year are presented in Fig. 4.1. The data for the second

4.1 Model of Hydroxyl Emission

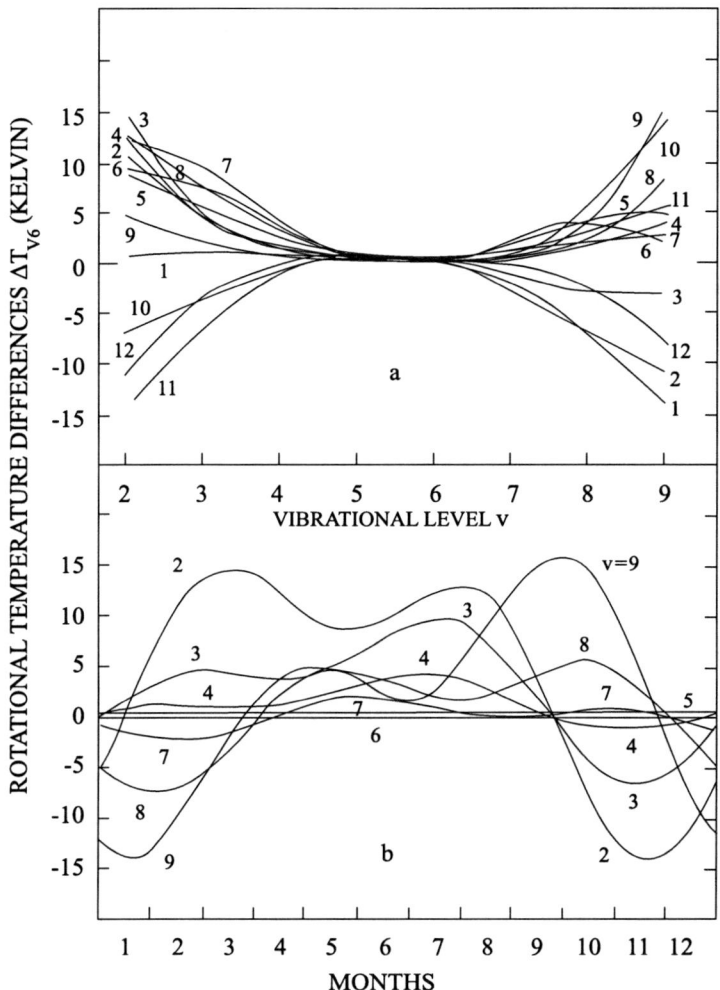

Fig. 4.1 Calculated variations of the differences between rotational temperatures for the OH bands related to the temperature for the bands with $v' = 6$. (**a**) Variations depending on the vibrational level number v for different months of the year ($1 \div 12$); (**b**) seasonal variations for various vibrational levels ($v = 2$–9)

vibrational level have been obtained based on an empirical formula, and they call for refinement.

The nonequilibrium rotational temperature T_{nr} (~ 1100 (K)) has been revealed by analyzing the intensity distribution for the lines of the 882.4-nm (7–3) band of the P branch with rotational numbers $N' = 6$–12 (Perminov and Semenov 1992). The nonequilibrium and, as a consequence, high rotational temperatures are due to the incompleteness of rotational relaxation for large N in the top part of the emission layer. Satellite measurements have detected transitions for $N' \sim 33$

(Dodd et al. 1993, 1994) and pure rotational transitions between high rotational levels in the spectral range 10–25 (μm).

The diurnal variations of T_{nr} are synchronous with the variations of the vibrational temperature and reflect the altitude variations of the upper part of the OH emission layer. Based on a small series of data obtained at Zvenigorod under night conditions near the equinox, the variations can be described by the relation

$$T_{nr} = 250 - 1580 \cdot \cos\chi_\odot \, .$$

The emission rate altitude distribution was investigated until recently only by rocket methods. At present, only about 50 rocket measurements carried out between 1956 and 1992 are known (Shefov and Toroshelidze 1975; Baker and Stair 1988), 30–35 of which are suitable for analysis.

In these data, the most uncertain characteristic is the shape of the altitude profile. Theoretical investigations suggest an asymmetric distribution in which the share of the emission layer above the altitude of emission rate maximum Z_m is greater than that below Z_m (Moreels et al. 1977), i.e., the asymmetry $P > 0.5$.

Special investigations carried out on the UARS satellite testify to the asymmetry $P \sim 0.65$ (Lowe and Le Blanc 1993). In measuring the intensity vertical distribution for the emission layer sighted at a tangent to the limb, the solution of the inverse problem of reconstruction of the shape of the bottom part of a layer presents some difficulties since this demands that the form of the sought-for function be preset. Comparison of simultaneously measured values of Z_m and W for various rocket launches has revealed a nonlinear correlation between them and a distinct separation into two subgroups of vibrational levels: 1–5 and 6–9. For these levels, it has been obtained that, on the average, for $v = 4$,

$$W_4 = 2.7 + \frac{(Z_m - 65)^2}{76} \, ,$$

and for $v = 8$,

$$W_8 = 3.7 + \frac{(Z_m - 65)^2}{76} \, .$$

Dependences of the altitude Z_m and layer thickness variations ΔW on v have also been revealed:

$$Z_m = 87 + \frac{v-5}{4.3} \text{ and } \Delta W = -2.44 + 0.477 \, v \, .$$

Therefore, on analyzing the given measurement data with the purpose of revealing variations of different types, all of them were reduced to $v = 5$. Thus, the following relation has been obtained:

$$W = 2.63 + \frac{(Z_m - 65)^2}{76} \, .$$

By analogy to the variations of other emission layer parameters, we can write

$$Z_m = Z_m^0 + \sum_i \Delta Z_{mi}; W = W_0 + \sum_i \Delta W_i; P = P_0 + \sum_i \Delta P_i .$$

For Z_m^0, W_0, and P, their values corresponding to standard heliogeophysical conditions should be taken.

4.1.3 Empirical Model of Variations of Hydroxyl Emission Characteristics

To reveal the above variations, statistical systematization of the ground-based measurements of the emission intensity and rotational (equilibrium) and vibrational temperatures carried out at Zvenigorod and at other stations has been performed (Berthier 1955, 1956; Shefov 1967a, 1968b, 1969a,b,c, 1970a,b, 1971a,b, 1972a,b, 1973a,b, 1974a,b, 1975a, 1976, 1978a,b; Taranova 1967; Taranova and Toroshelidze 1970; Lowe and Lytle 1973; Takahashi et al. 1974, 1984, 1990, 1995, 1996; Toroshelidze 1975; Shefov and Toroshelidze 1975; Kropotkina 1976; Kropotkina and Shefov 1977; Moreels et al. 1977; Fishkova 1955, 1976, 1978, 1981, 1983; Semenov and Shefov 1979a,b; Takahashi and Batista 1981; Potapov et al. 1983, 1985; Myrabø et al. 1983; Myrabø 1948; Shefov and Piterskaya 1984; Baker and Stair 1988; Agashe et al. 1989; Scheer and Reisin 1990, 2000, 2002; Toroshelidze 1991; Turnbull and Lowe 1991; Perminov et al. 1993) and of the rocket and satellite measurements of the emission layer altitude (Yee et al. 1997; Skinner et al. 1998). The total list of publications is not given because of its great length.

Based on the systematization of the above emission characteristics for the earlier specified heliogeophysical conditions, the following values have been obtained:

Rotational temperature $T_r^0 = 195$ (K)
Total intensity $I^0 = 0.95$ (megarayleigh)
Maximum emission rate $Q_m^0 = 1.1 \cdot 10^6 (cm^{-3} \cdot s^{-1})$
Total energy $E^0 = 1 (erg \cdot cm^{-2} \cdot s^{-1})$
Vibrational temperature $T_v^0 = 10000$ (K)
Altitude of the emission layer maximum $Z_m^0 = 87$ (km)
Thickness of the emission layer $W^0 = 9$ (km)
Asymmetry $P = 0.65$
Mean effective transition probability $\bar{A} = 48$ (s^{-1})
Effective wavelength $\bar{\lambda} = 1.9$ (μm)
Geographic latitude $\varphi = 45°N$
Geographic longitude $\lambda = 40°E$
Year YYYY $= 1972.5$
Season of year: equinox, day of year $t_d = 80$
Time of day (local solar midnight) $\tau = 0$
Average solar flux $F_{10.7} = 130$
Geomagnetic activity $K_p = 0$

422   4 Regular Variations of the Airglow in the Mesopause and Thermosphere

Notwithstanding that a large body of observation data is available, not all types of variations can be investigated in detail. This especially concerns the data on the emission layer altitude and vibrational temperature. Therefore, in revealing the character of variations based on scanty data, the knowledge of the behavior of other parameters was helpful since all variations are mutually related.

Various types of variations of the considered parameters are graphically presented in Figs. 4.2, 4.3, 4.4, and 4.5. On the plots for the variations of T_r, I, and T_V, the points represent mean values for given arguments. For the variations of Z_m, the points refer to individual values.

Empirical approximations have been obtained to calculate the variations of the parameters considered relative to their standard values under given heliogeophysical conditions.

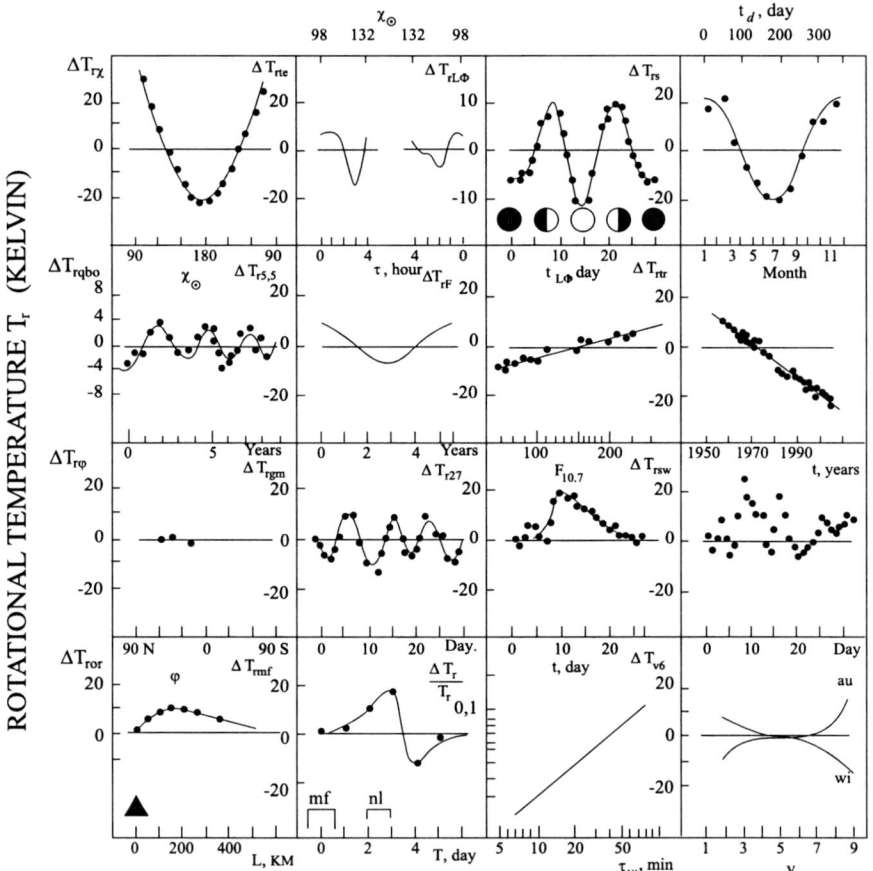

Fig. 4.2 Variations of the rotational temperature of hydroxyl emission about its mean value for standard conditions versus different parameters (Semenov and Shefov 1996a, 1999). The designations are given in the preface to Chap. 4. MF – meteoric flows, NC – noctilucent clouds

4.1 Model of Hydroxyl Emission 423

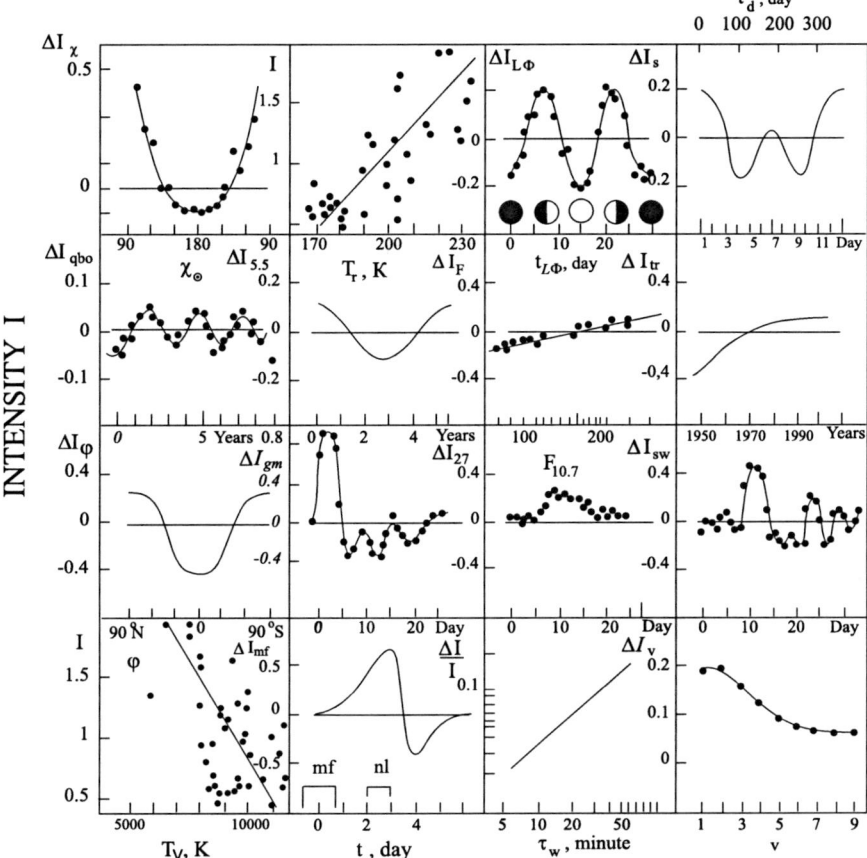

Fig. 4.3 Variations of the hydroxyl emission intensity about its mean value for standard conditions versus different parameters (Semenov and Shefov 1996a, 1999). Designations are given in the preface to Chap. 4

As it became obvious in recent years, the variations of the vibrational temperature T_V (and the nonequilibrium rotational temperature T_{nr}) reflect the variations of the atmospheric density inside the emission layer on varying altitude Z_m (Shefov 1978b; Perminov and Semenov 1992; Perminov et al. 1993). The emission intensity I and rotational temperature T_r also vary with the altitude of the emission layer. Therefore, the validity of the relations obtained can be checked to some extent with the use of empirical approximations of the OH emission parameters by calculating them for the conditions of rocket measurements (Shefov and Toroshelidze 1975; Baker and Stair 1988) and comparing the results with the data of rocket measurements of Z_m reduced to the parameter values for $v = 5$. These results are also presented in Figs. 4.2–4.5. The regression relations have the form

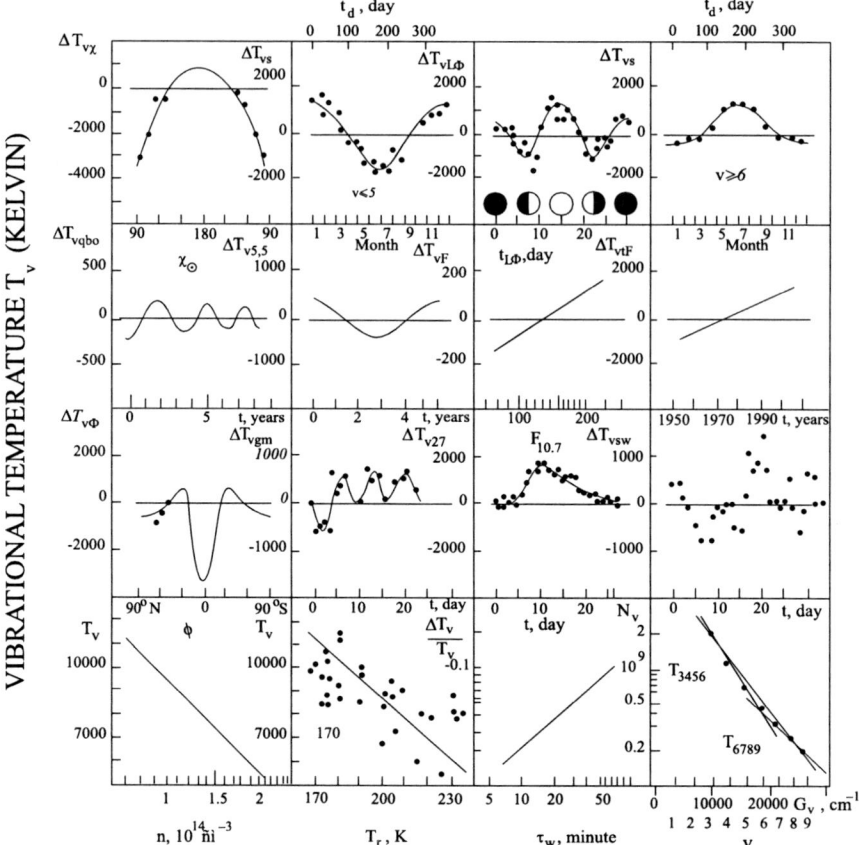

Fig. 4.4 Variations of the vibrational temperature of hydroxyl emission about its mean value for standard conditions versus different parameters (Semenov and Shefov 1996a, 1999). Designations are given in the preface to Chap. 4

$$Z_m = 78 + T_V/1000 \text{ (km)}, (r \sim 0.74);$$

$$Z_m = 91 - 4I \text{ (km)}, (r \sim -0.64);$$

$$Z_m = 105 - 0.09 T_r \text{ (km)}, (r \sim -0.63);$$

$$I = -3.3 + T_r/45 \text{ (megarayleigh)}, (r \sim 0.77);$$

$$I = 4.3 - T_V/2850 \text{ (megarayleigh)}, (r \sim -0.51);$$

$$T_V = 25800 - 86 T_r \text{ (K)}, (r \sim -0.58);$$

$$T_V = 32200/(1 + 1.26 \cdot 10^{-11} n^{0.8}) \text{ (K)}, (r \sim -0.7),$$

where n is the concentration of neutral constituents (cm^{-3}).

It should be stressed that the correlations between the hydroxyl emission parameters obtained for a wide spectrum of heliogeophysical conditions during rocket

4.1 Model of Hydroxyl Emission

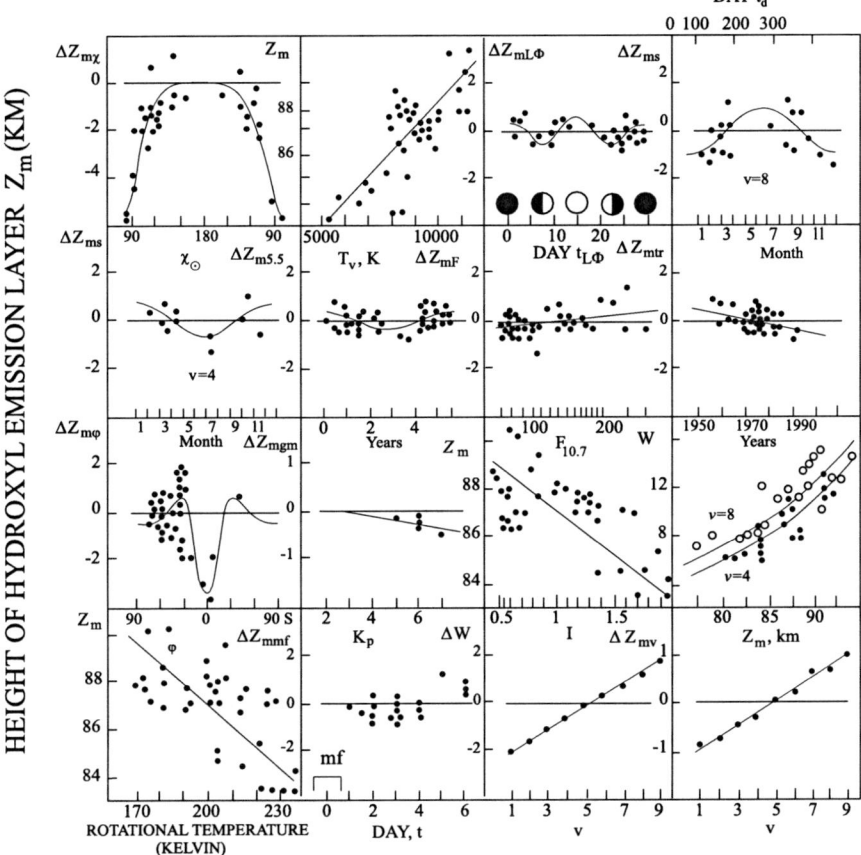

Fig. 4.5 Variations of the maximum altitude of the OH emission layer about the mean value for standard conditions versus different parameters (Semenov and Shefov 1996a, 1999). Designations are given in the preface to Chap. 4

measurements are in good agreement with the results of measurements carried out at different stations (Takahashi et al. 1974, 1984, 1990; Shefov 1975a; Fishkova 1983; Agashe et al. 1989; Scheer and Reisin 1990, 2000, 2002; Toroshelidze 1991). From these relations, we have for regular nocturnal variations

$$\Delta T_V / \Delta Z_m \sim 1000 \ (K \cdot km^{-1}),$$
$$\Delta T_r / \Delta Z_m \sim -10 \ (K \cdot km^{-1}),$$
$$\Delta I / \Delta Z_m \sim -25\% \ (km^{-1}),$$
$$\Delta n / \Delta T_V \sim -19\% (1000 \ (K))^{-1}.$$

Below we give empirical relations for different types of variations of hydroxyl emission parameters.

Nocturnal Variations

$$\Delta T_{r\chi} = 44 - 62 \cdot \cos(\varphi + \delta_\odot) + \frac{50 \cdot [\cos\chi_\odot + \cos(\varphi + \delta_\odot)]}{1 + \exp[(28 - |\varphi|)/4]} (K),$$

$$\cos\chi_\odot = \sin\varphi \cdot \sin\delta_\odot - \cos\varphi \cos\delta_\odot \cos\tau ,$$

where τ is the local mean solar time and $\delta_\odot$ is the declination of the Sun.

$$\Delta I_\chi = 0.89 \cdot \left[|\cos\chi_\odot|^{-0.33} - |\cos(\varphi + \delta_\odot)|^{-0.33} \right] ;$$

$$\Delta T_{V\chi} = 12700 \cdot \left[|\cos\chi_\odot|^{0.2} - |\cos(\varphi + \delta_\odot)|^{0.2} \right] (K);$$

$$\Delta Z_{m\chi} = -\frac{5.5}{1 + \exp[-8 \cdot (\cos\chi_\odot + 0.26)]} (км) .$$

The nightly mean variations have been derived from the data of numerous observations (Fishkova 1955; Shefov 1971a, 1972a,b; Takahashi et al. 1974, 1984, 1990; Shefov and Toroshelidze 1975; Toroshelidze 1975, 1991; Shefov 1978a; Fishkova 1981, 1983; Myrabø et al. 1983; Potapov et al. 1983; Baker and Stair 1988; Agashe et al. 1989; Scheer and Reisin 1990, 2000, 2002; Turnbull and Lowe 1983, 1991). Undoubtedly, the nocturnal variations of hydroxyl emission parameters are affected by the phase shifts of semidiurnal thermal tides resulting from the emission layer altitude variations (Petitdidier and Teitelbaum 1977; Takahashi et al. 1984) and by the perturbations caused by internal gravity waves.

Special attempts have been made to measure the variations of OH emission parameters at evening and morning twilights for $\chi_\odot = 98° - 108°$ (Toroshelidze 1968a, 1991; Taranova and Toroshelidze 1970; Moreels et al. 1977; Scheer and Reisin 1990; Takahashi et al. 1990; Turnbull and Lowe 1991). It seems that the observed maxima were related to the effect of the motion of the terminator (Toroshelidze 1991). Some features of the variations can be studied based on the measurements performed at summer nights at the latitudes where $\chi_\odot \leq 112°$ (Shefov 1971a). For these conditions, periodic temperature variations (4-h and 2-h harmonics) have been revealed (Taranova and Toroshelidze 1970; Toroshelidze 1975, 1991). The temperature variations in the evening (ev) at latitudes near 45°N can be determined, on the average, by the formula

$$\Delta T_{rte}^{ev} = 10 \cdot \sin\frac{\pi}{2}(\tau - \tau_\chi) + 6 \cdot \cos\pi(\tau - \tau_\chi)(K) ,$$

where τ_χ is the mean solar time (h) for $\chi_\odot = 98°$. For this zenith angle, the given atmospheric region is at a distance of 900 (km) from the terminator line.

The temperature variations in the morning (mo) are given by

$$\Delta T_{rte}^{mo} = 6 \cdot \cos\frac{\pi}{2}(\tau_\chi - \tau) - 3 \cdot \sin\pi(\tau_\chi - \tau)(K) .$$

4.1 Model of Hydroxyl Emission

However, the available data are insufficient for constructing relations which would describe the behavior of the hydroxyl emission parameters for solar zenith angles between 80° and 100°. Therefore, the above approximate formulas are inapplicable in this range of the argument.

Lunar Variations

The lunar-tidal variations of the OH emission observable at a given latitude depend substantially on the position of the lunar orbit plane relative to the plane of the Earth's equator, i.e., on the declination of the Moon, δ_L, as can be seen from Fig. 4.6. The angle of inclination of the Moon orbit plane to the ecliptic plane ranges about 5.15°. Therefore, the angle of inclination of the Moon orbit plane to the Earth's equator plane varies between 18.3 and 28.63°. The latitude Θ and longitude Λ of the observation point in the tide-deformed atmosphere relative to the straight line connecting the Earth and Moon centers are determined by the formulas (Kropotkina and Shefov 1977)

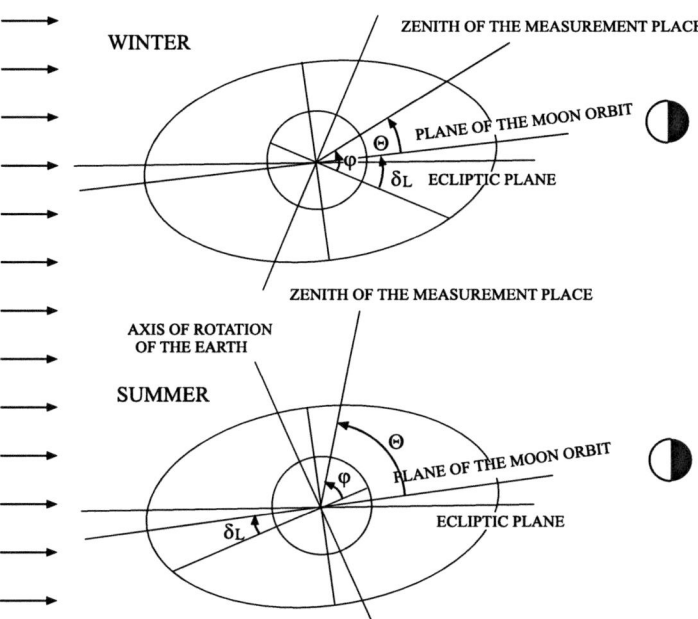

Fig. 4.6 Schema of the orientation of the tide-deformed Earth's atmosphere under the influence of the Moon. The scale of the Moon-related longitudinal-to-transverse deformation ratio is considerably increased

$$\sin\Theta = \sin\varphi \cdot \cos\delta_L + \cos\varphi \cdot \sin\delta_L \cdot \cos t_L,$$

$$\sin\Lambda = \frac{\cos\varphi}{\cos\Theta} \cdot \sin t_L,$$

$$\cos\Lambda = \frac{-\cos\varphi \cdot \cos\delta_L + \sin\varphi \cdot \sin\delta_L \cdot \cos t_L}{\cos\Theta},$$

where t_L is the hour angle and δ_L is the declination of the Moon. The latitudinal distribution of the tidal disturbance amplitude is determined by the relation

$$A_L = 0.5(3\cos^2\Theta - 1).$$

In the equatorial coordinate system, this distribution has the form

$$A_L = \frac{1}{3}(1 - 3\cdot\sin^2\delta_L)\cdot(1 - 3\cdot\sin^2\varphi) + \sin 2\delta_L \cdot \sin 2\varphi \cdot \cos t_L$$
$$+ \cos^2\varphi \cdot \cos^2\delta_L \cdot \cos 2t_L.$$

The relationships between lunar coordinates and time for various geographic latitudes are presented in Figs. 1.29–1.31.

However, the position of the Moon orbit plane varies due to precession, making a turn in 18.61 years. This means that the orbit plane turns by 19.3° in a year. Therefore, the seasonal character of lunar tides will change from year to year. Therefore, the seasonal variations of lunar tides should be constructed for the same values of the Moon declination δ_L for a given latitude φ of the observation place. The published data do not always give a clear indication whether the relevant observations met these requirements.

According to the available data, the diurnal lunar-tidal variations of the OH emission parameters have small amplitudes and cannot be revealed reliably. Based on the data of long-term measurements (Shefov and Toroshelidze 1975; Scheer and Reisin 1990), the following estimates have been obtained:

$$\Delta T_{rL} = 0.5 \cdot \cos\frac{\pi}{12}\tau_L - 1.5 \cdot \cos\frac{\pi}{6}\tau_L (K);$$

$$\Delta I_L = 0.01 \cdot \cos\frac{\pi}{12}\tau_L - 0.02 \cdot \cos\frac{\pi}{6}\tau_L;$$

$$\Delta T_{VL} = -60 \cdot \cos\frac{\pi}{12}\tau_L + 150 \cdot \cos\frac{\pi}{6}\tau_L (K);$$

$$\Delta Z_{mL} = -3.5 \cdot \cos\frac{\pi}{12}\tau_L + 4.5 \cdot \cos\frac{\pi}{6}\tau_L (km) \text{ (Fig. 4.27)}$$

Here τ_L is the lunar time (h). Following Chapman and Lindsen (1970), it can be estimated as

$$\tau_L = \tau - \xi,$$

where $\xi = -0.6173393 + 0.8127167 \cdot D$ and $D = [(YYYY - 1901) \cdot 365.25] + 364.5 + t_d$.

In this formula, the square brackets denote the integer part of the number and YYYY designates the year number.

4.1 Model of Hydroxyl Emission

The variations of the nonequilibrium rotational temperature in the Moon orbit plane are determined by the formula

$$\Delta T_{nrL} = -70 \cdot \cos\frac{\pi\tau_L}{12} + 200 \cdot \cos\frac{\pi\tau_L}{6}.$$

The variations associated with the Moon age period, i.e., the synodic month equal to 29.53 days, have greater amplitudes than the variations associated with a lunar day:

$$\Delta T_{rL\Phi} = 3 \cdot \left(2.2 - 1.5 \cdot \cos\frac{2\pi}{365.25}t_d\right) \cdot \cos\frac{2\pi}{29.53}t_{L\Phi} - 9 \cdot \cos\frac{4\pi}{29.53}t_{L\Phi}(K);$$

$$\Delta I_{L\Phi} = 0.05 \cdot \cos\frac{2\pi}{29.53}t_{L\Phi} - 0.18 \cdot \cos\frac{4\pi}{29.53}t_{L\Phi};$$

$$\Delta T_{VL\Phi} = -400 \cdot \cos\frac{2\pi}{29.53}t_{L\Phi} + 1000 \cdot \cos\frac{4\pi}{29.53}t_{L\Phi}(K);$$

$$\Delta Z_{mL\Phi} = -0.3 \cdot \cos\frac{2\pi}{29.53}t_{L\Phi} + 0.6 \cdot \cos\frac{4\pi}{29.53}t_{L\Phi}(KM).$$

The data for ΔT_{rL}, ΔI_L, and ΔT_{VL} have been obtained in the latitude 57°N (Shefov 1974a,b), for ΔZ_{mL} on the average at latitudes 30–40°N, and all of them have been reduced to the Moon orbit plane. A rough estimate of the Moon age to within several tenth of day can be obtained by the following formula derived from Meeus' formulas (Meeus 1982):

$$t_{L\Phi} = [(t_d/365.25 + YYYY - 1900) \cdot 12.3685] \cdot 29.53 - 1, \text{ days}.$$

Here, the square brackets denote the fractional part of the number.

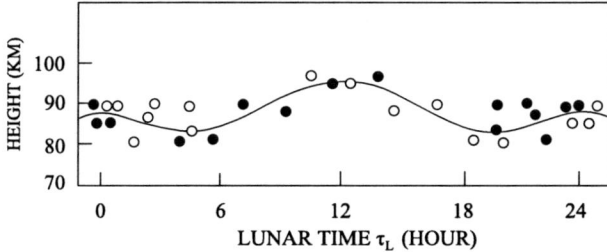

Fig. 4.7 Lunar variations of the hydroxyl emission layer altitude according to rocket measurements (Shefov and Toroshelidze 1975). *Full circles* are measurement data; *hollow circles* are mirror transfer of points to the other half of the lunar day relative to the upper culmination

The data obtained have been used to construct the relations of the tide-deformed atmosphere parameters to the coordinates, latitude Θ and longitude Λ, of the observation point relative to the straight line connecting the Earth and the Moon centers. Analysis of the temperature and altitude variations (Fig. 4.8a) (Kropotkina and Shefov 1977) shows, as expected, that the quantity ΔT changes its sign as the latitude increases from 0 to 90°. Of interest is to estimate the magnitude of the tide-induced altitude variations. For this purpose, one can use the relations between the altitude and temperature variations obtained by Forbes and Geller (1972). The dependence of the tide-induced altitude variations ΔZ on latitude Θ constructed based on these relations is given in Fig. 4.8. The solid line depicts the function $2A_L = 3 \cdot \cos^2 \Theta - 1$ in relative units. The values of ΔZ agree with the data obtained from rocket measurements of hydroxyl emission altitudes (Shefov and Toroshelidze 1975).

The nature of these oscillations in the upper atmosphere is accounted for by the planetary waves with periods of about 15 and 30 days that occur in the lower atmosphere (Reshetov 1973) and are caused by parametrically excited atmospheric circulation.

Seasonal Variations

The seasonal variations of hydroxyl emission characteristics have the greatest amplitudes among other types of regular changes of these characteristics. Based on the available data of measurements performed at different latitudes, the following approximations have been obtained:

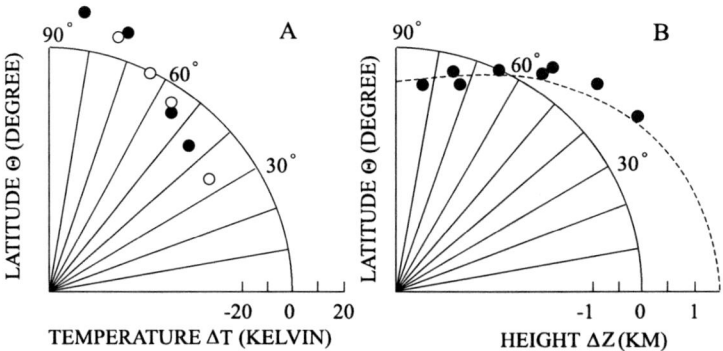

Fig. 4.8 Dependence of the atmospheric parameters variations on latitude Θ (Kropotkina and Shefov 1977). (**A**) Rotational temperature variations, *open circles* for $\Lambda = 0$, *full circles* for $\Lambda = 12$ (h); (**B**) altitude variations ΔZ (*full circles*). The *dashed line* corresponds to the theoretical tidal potential distribution

4.1 Model of Hydroxyl Emission

$$\Delta T_{rs} = 38 \cdot \frac{1 - \exp[-1.43 \cdot \varphi/(92 - |\varphi|)]}{1 + \exp[-1.43 \cdot \varphi/(92 - |\varphi|)]} \cdot \cos \frac{2\pi}{365.25} t_d (K);$$

$$\Delta I_s = 0.09 \cdot \cos \frac{2\pi}{365.25} t_d + 0.12 \cdot \cos \frac{4\pi}{365.25} t_d;$$

$$\Delta T_{Vs}^{36} = 11800 \cdot \left[(\cos(\varphi + \delta_\odot))^{0.33} - (\cos\varphi)^{0.33} \right] (K);$$

$$\Delta T_{Vs}^{69} = \left[(\cos(\varphi + \delta_\odot))^{-0.16} - (\cos\varphi)^{-0.16} \right] (K);$$

$$\Delta T_{Vs}^{39} = 19000 \cdot \left[\frac{(\cos(\varphi + \delta_\odot))^{0.33}}{(\cos(\varphi + \delta_\odot))^{0.49} + 0.8} - \frac{(\cos\varphi)^{0.33}}{(\cos\varphi)^{0.49} + 0.8} \right] (K);$$

$$\Delta Z_{ms} = \frac{1 - \exp(2v - 10)}{1 + \exp(2v - 10)} \cdot \cos \frac{2\pi}{365.25} t_d, (km) .$$

Nevertheless, the harmonic amplitudes and phases contain components depending both on the solar flux and on the long-term trend (Semenov and Shefov 2003). For the latitude interval 45–55°N, the relations for the rotational temperature variations have the form

$\Delta T_{r1} = -(0.071 \pm 0.013)(F_{10.7} - 130)(K)$, correlation coefficient r
$\quad = -0.960 \pm 0.030$,

$t_1 = (0.152 \pm 0.017)(F_{10.7} - 130)$ (days of year), correlation coefficient r
$\quad = -0.970 \pm 0.030$,

$\Delta T_{r2} = -(0.027 \pm 0.01)(F_{10.7} - 130)(K)$, correlation coefficient r $= -0.82 \pm 0.15$,

$t_2 = 94 \pm 8$ (days of year) .

The seasonal variations, as well as diurnal ones, have the greatest amplitude among other types of variations (Shefov 1969a; Gavrilieva and Ammosov 2002a,b). Their amplitude increases with latitude, and in the early works (Shefov and Yarin 1961, 1962; Kvifte 1967) this effect was attributed to the increase in T_r with altitude. The harmonic amplitude ratio will vary with latitude, since the semiannual harmonic prevails near the equator (Golitsyn et al. 2000).

It seems that the difference in the relations of seasonal and yearly mean temperatures to solar flux was the reason for the negative correlation and large variance obtained by Neumann (1990). It should be noted that this character of the effect of solar activity on the hydroxyl emission temperature and intensity is due to the features of the variations of altitude temperature distributions in the mesopause (Semenov et al. 2005; Golitsyn et al. 2006). This shows up in that the yearly average temperature increases with solar flux while the difference between the winter and the summer temperatures decreases with increasing solar flux. This behavior is related in the main to the more substantial decrease in temperature in the mesopause during the summer period of solar flux minima (Semenov et al. 2002a,c).

In some studies, the response of the mesopause temperature determined by the hydroxyl emission to solar activity was taken into account only by using yearly average variations (Espy and Stegman 2002; Reisin and Scheer 2002). As mentioned above, this is insufficient since the effect of solar activity is different in different periods of the year; therefore, the contribution of seasonal variations should be taken into consideration.

The altitude variations ΔZ_{ms} agree in character with the data of satellite measurements (Hernandez et al. 1995; Yee et al. 1997; Skinner et al. 1998).

Cyclic Aperiodic (Quasi-biennial) Variations (CAVs)

Recently the notions about the character of the so-called quasi-biennial oscillations (QBOs) of the mesopause and lower thermosphere parameters have changed substantially. It has been established that the reason for these oscillations is QBOs of solar flux (Sect. 1.6.2). Quasi-biennial oscillations in the Earth's atmosphere were revealed in 1880–1890 (Clayton 1884a,b; Woeikof 1891, 1895) and were investigated later (Clough 1924, 1928). Their relation to solar activity was found (Schuster 1906; Clough 1924, 1928). In the beginning of 1950, Kalinin (1952) independently detected their presence in the variations of the geomagnetic field characteristics and paid attention to their relation to solar activity. In the early 1960s, QBOs were detected again in the variations of the stratospheric wind characteristics, and since then they have been intensely investigated (Rakipova and Efimova 1975; Labitzke and van Loon 1988). It has been concluded that the QBO type depends on solar activity and on the mode of circulation in the stratosphere.

Processing of data on solar fluxes and temperatures in the upper atmosphere has shown that QBOs are not harmonic variations.

It turned out that the sequence of solar flux maxima during an 11-yr cycle is a train of oscillations with varying period and amplitude (Fadel et al. 2002). Their variations are well described by the Airy function $-Ai(-x)$, which is a solution of the second-order linear differential equation (Abramowitz and Stegun 1964)

$$y'' - xy = 0,$$

which describes internal waves propagating in the atmospheres and oceans of rotating planets. This suggests that QBOs are manifestations of the dynamic processes occurring in the interior of the Sun (Kononovich and Shefov 2003).

The results of the Fourier analysis of the Airy function (Shefov and Semenov 2006) for arguments $3 - \Delta t$ for 12- to 15-yr intervals have shown that the maximum amplitudes of the harmonics correspond to periods of $\sim$2.3 and 2.8 years and depend on the position of the chosen time interval within an 11-yr solar cycle (Sect. 5.1).

These CAVs for the hydroxyl emission can be described by the formulas

$$\Delta I_{SAO}(OH) = -0.08 \cdot Ai(3 - \Delta t),$$

$$\Delta T_{rSAO}(OH) = -5 \cdot Ai(3 - \Delta t)(K),$$

$$\Delta T_{VSAO}(OH) = -430 \cdot Ai(3 - \Delta t)(K),$$

$$\Delta Z_{mSAO}(OH) = -0.5 \cdot Ai(3 - \Delta t)(km).$$

Data of measurements and their approximations are presented in Figs. 4.2–4.5.

5.5.5-yr Variations

These variations occur practically in phase with an 11-yr solar cycle, and therefore they have not been adequately investigated. They are mentioned in a number of publications (Fishkova 1983; Shefov and Piterskaya 1984; Megrelishvili and Fishkova 1986). The vibrational temperatures T_V have been estimated by the correlation between the yearly average values of T_r and T_V. Based on the measurements for $t_{5.5} = 1959.0, 1970.0, 1981.0$, and 1992.0, the following formulas for the variations have been obtained:

$$\Delta T_{r5.5} = 8 \cdot \cos \frac{2\pi}{5.5}(t - t_{5.5})(K);$$

$$\Delta I_{5.5} = 0.12 \cdot \cos \frac{2\pi}{5.5}(t - t_{5.5});$$

$$\Delta T_{V5.5} = 400 \cdot \cos \frac{2\pi}{5.5}(t - t_{5.5})(K);$$

$$\Delta Z_{m5.5} = 0.4 \cdot \cos \frac{2\pi}{5.5}(t - t_{5.5})(km).$$

According to the analysis of 11-yr solar flux data (Kononovich 2005), this period seems to be the second harmonic of an 11-yr cycle, making ~15(%) of the first one. The phase of the second harmonic is equal to 3.7 years around the solar flux minimum.

Variations Depending on the 11-yr Variability of Solar Flux

The mean variations of the hydroxyl emission parameters can be determined by the formulas

$$\Delta T_{rF} = 25\log_{10}[F_{10.7}(t-0.42)/150],$$

$$\Delta I_F = 0.40\log_{10}[F_{10.7}(t-0.42)/150],$$

$$\Delta T_{VF} = 600\log_{10}[F_{10.7}/150],$$

$$\Delta Z_{mF} = (F_{10.7} - 130)/230.$$

The yearly average values of the solar radio flux $F_{10.7}$ with a shift by the 0.42 part of year give the greatest correlation (~0.98) (Shefov and Piterskaya 1984; Semenov and Shefov 1979a,b). They are mentioned elsewhere (Shefov 1969a; Wiens and Weill 1973; Fishkova 1983).

As shown in Sect. 7.4, the effect of solar activity on the temperature of the middle atmosphere varies with season. This effect can be taken into account based on the data presented in Figs. 7.10–7.11.

It should be borne in mind that the ultraviolet radiation of the Sun, which causes dissociation of molecular oxygen at the altitudes of the mesopause and lower thermosphere, is emitted at the level of the photosphere where the level of solar activity is evaluated by solar spots (Wolf numbers W). The solar radio-frequency emission, whose index $F_{10.7}$ determines the solar activity in the corona, indicates the activity level which is delayed by about 3 months relative to the solar activity determined by Wolf numbers (Apostolov and Letfus 1985). Hence, the global response of the upper atmosphere seems to be delayed by almost half a year.

Long-Term Trend

During the last century, the Earth's atmosphere experienced considerable changes whose rate (trend) did not remain constant. In the surface layer the temperature variation rate was about +0.007 (K/yr), while in the middle atmosphere it changed its sign depending on altitude. In the mesopause, the long-term yearly average trend was about −0.68 (K/yr).

In our calculations, the long-term temperature trend was estimated in reference to the year 1972.5, since for this point in time the yearly average $F_{10.7}$ was about 130, which corresponds to the average smoothed (with a 22-yr sliding interval) value for the 19th–22nd periods of solar activity:

$$\Delta T_{rtr} = -0.68(t - 1972.5),$$

$$\Delta I_{tr} = 0.97t' - 3.67t'^2 + 2.77t'^3 + 27.8t'^4, \text{ where } t' = (t - 1972.5)/100,$$

$$\Delta T_{vtr} = 40(t - 1972.5),$$

$$\Delta Z_{mtr} = -0.002(t - 1972.5).$$

4.1 Model of Hydroxyl Emission

The trends are based on the data of a number of studies (Shefov 1969a; Wiens and Weill 1973; Fishkova 1983; Shefov and Piterskaya 1984; Semenov and Shefov 1979a,b; Givishvili et al. 1996; Golitsyn et al. 1996, 2000, 2001; Evlashin et al. 1999; Lysenko et al. 1999; Semenov 2000; Semenov et al. 2000, 2002a). The nature of a trend seems to be complicated in character, being a manifestation of both man's impact and long-period variations of solar flux.

However, the trend shows seasonal variability (Golitsyn et al. 2000; Semenov 2000; Semenov et al. 2000, 2002a). It turned out that at the hydroxyl emission altitudes ($\sim$87 (km)) in winter (December/January) the trend is negative, and according to a linear approximation for the period 1955–1995, it is -0.92 ± 0.08. During the summer period (June/July), there is no trend (0.02 ± 0.08), and the yearly average value is -0.64 ± 0.05.

The data obtained allow the conclusion that the found constancy of the mesopause temperature during the summer period and the observed long-term increase in the frequency of occurrence of noctilucent clouds (Gadsden 1990, 1998), though questionable (Gadsden 2002), testify to an increase in humidity in the upper atmosphere. This is confirmed by the increase in concentrations of atomic hydrogen, methane, and water vapors in the atmosphere (Semenov 1997; Chandra et al. 1997; Shefov and Semenov 2002).

Independent evidence for the cooling of the atmosphere at the mesopause altitudes is the detection of a long-term reduction of the meridional wind speed (Merzlyakov and Portnyagin 1999) as a result of the decrease in temperature difference between the winter and the summer hemispheres (Semenov 2000).

The data on long-term temperature variations at the hydroxyl altitudes seem to testify to a possible decrease in trend since the 1990s, indicating its nonlinear behavior. This nonlinearity was already mentioned (Lysenko et al. 1997a,b, 2003; Lysenko and Rusina 2002a,b, 2003). It may be related, in particular, to long-term variations of solar flux. However, to gain more adequate notions about long-term variations of the trend, further investigations are necessary. A possible character of these temperature variations for winter conditions can be described by the relation

$$T_w = 234 - 1.3 \cdot (t - 1972.5) + 0.014 \cdot (t - 1972.5)^2 (K)$$

and for yearly average conditions by

$$T_{ma} = 230 - 0.9 \cdot (t - 1972.5) + 0.010 \cdot (t - 1972.5)^2 (K).$$

The decrease in temperature of the middle atmosphere during several decades is naturally responsible for the overall subsidence of the atmosphere on the average, which, in separately taken time intervals, is compensated by seasonal temperature variations and solar flux variations. Therefore, the rate of subsidence depends on the seasonal rearrangement of the temperature altitude distribution. At the altitudes of the formation of noctilucent clouds in summertime (82–83 (km)), the subsidence rate is about -50 (m · yr^{-1}). This corresponds to the decrease in atmospheric density by about $-1.5(\% \cdot \text{yr}^{-1})$. No wonder noctilucent clouds arise practically at the same altitude (1–1.5 (km)) for many years (Semenov et al. 2000).

Latitudinal Variations

For nightly mean rotational temperatures, we have

$$\Delta T_{r\varphi} = 0,$$

and for midnight values,

$$\Delta T_{r\varphi} = 44 - 62 \cdot \cos(\varphi + \delta_\odot) \text{ (K)},$$

$$\Delta I_\varphi = \frac{0.68}{1 + \exp[(41 - \varphi)/6.1]} - 0.45,$$

$$\Delta T_{V\varphi} = 3500 \cdot \exp\{-[(\varphi + \delta_\odot)/50]^4\} - 6500 \cdot \exp\{-[(\varphi + \delta_\odot)/28]^2\} - 500 \text{ (K)},$$

$$\Delta Z_{m\varphi} = 3.5 \cdot \exp\{-[(\varphi + \delta_\odot)/50]^4\} - 6.5 \cdot \exp\{-[(\varphi + \delta_\odot)/28]^2\} - 0.5 \text{ (km)}.$$

All previous statements about the increase in T_r with latitude φ (Shefov and Yarin 1961, 1962; Kvifte 1967) were based on data obtained in winter, which therefore reflected the latitudinal dependence of the amplitude of seasonal variations rather than of the behavior of the yearly average values of T_r. The latitudinal variations ΔT_V have been derived from the relation of their correlation with ΔZ_m.

Disturbed Variations After Geomagnetic Storms

These variations have been described in detail based on the data obtained at Zvenigorod ($\Phi = 51°$N) and traced down to the equator using the data of measurements performed at a number of stations (Figs. 4.9, 4.10, and 4.11) (Shefov 1968b, 1969a,c, 1972a,b, 1973a, 1975a; Truttse and Shefov 1971). The measurements have shown that geomagnetic disturbances are followed by intensity and temperature oscillations which propagate from the auroral zone toward the equator. This effect is traced not only on the average for many storms, but also from the data of individual observations for a rather long period (10 nights) (Figs. 4.12 and 4.13) (Truttse and Shefov 1970a,b).

4.1 Model of Hydroxyl Emission

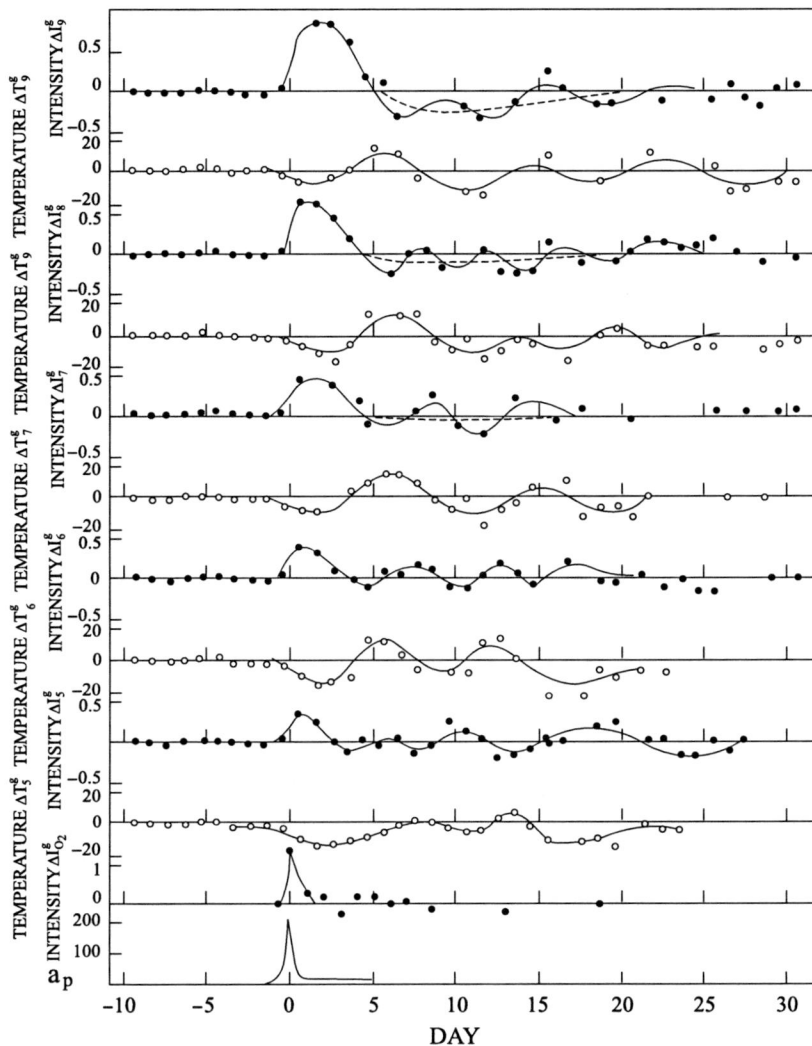

Fig. 4.9 The average variations of relative intensity increments ΔI^g_{OH} and $\Delta I^g_{O_2}$ (*full circles*) and temperature increments ΔT^g_{OH} (K) (*open circles*) after geomagnetic disturbances with various indices K_P derived from the data of measurements performed at Zvenigorod (Shefov 1969a, 1973a). The *solid lines* are approximations

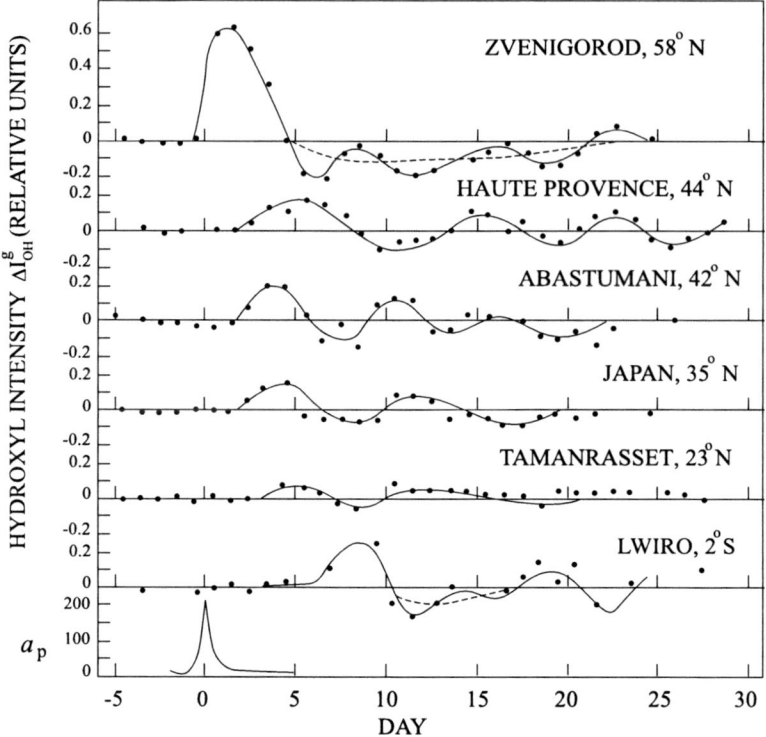

Fig. 4.10 The average variations of relative intensity increments ΔI^g_{OH} after the geomagnetic disturbances with indices K_P varied between 6_- and 9_0 that occurred at different latitudes (Shefov 1969a, 1973a). The *solid lines* are approximations

The variations associated with the given process for the disturbance maximum at the latitude of Zvenigorod have been approximated by the relations

$$\Delta T_{rgm} = -14 + 2 \cdot (K_P - 5)(K), \quad r = 0.76 \pm 0.21,$$
$$\Delta I_{gm} = 0.30 + 0.10 \cdot (K_P - 5), \quad r = 0.98 \pm 0.01,$$
$$\Delta T_{Vgm} = 650 - 44 \cdot (K_P - 5)(K), \quad r = 0.90 \pm 0.1,$$
$$\Delta Z_{mgm} = -0.17 - 0.1 \cdot (K_P - 5)(km), \quad r = 0.6 \pm 0.3.$$

The velocity of composition waves is 5–10 $(m \cdot s^{-1})$; these waves transfer additional amounts of water vapor and nitric oxide (Shefov 1968b, 1969a, 1973a, 1978a). This follows from the revealed correlation of hydroxyl emission variations with ionospheric absorption (Fig. 4.14) (Rapoport and Shefov 1974, 1976; Shefov 1978a). For the emission intensity, the correlation formula is

$$\Delta I/I = 0.97 \cdot \overline{\Delta f_{min}} + 0.13, \text{correlation coefficient } r = 0.80$$

4.1 Model of Hydroxyl Emission

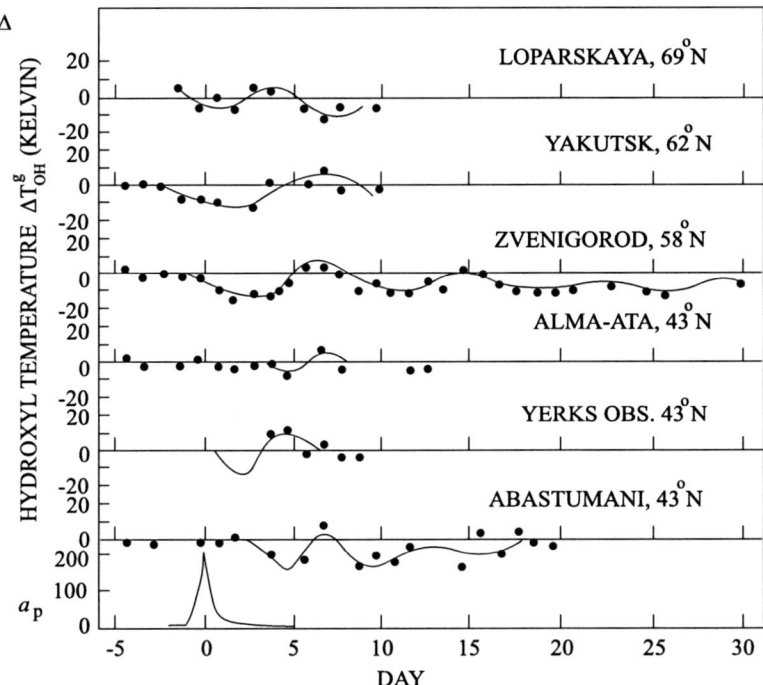

Fig. 4.11 The average variations of relative temperature increments ΔT^g_{OH} (K) after geomagnetic disturbances with K_P indices between 6_- and 9_0 at various latitudes (Shefov 1969a, 1973a). The *solid lines* are approximations

and for the temperature

$$\Delta T = -21.2 \cdot \overline{\Delta f_{min}} + 4.1 (K), \text{correlation coefficient } r = -0.84 \, .$$

On the average, the maximum disturbance amplitudes are 80–90(%) for the intensity, 10–15 (K) for the rotational temperature, and 1000 (K) for the vibrational temperature. Rocket measurements of the altitude Z_m allow one to estimate only the decrease in layer altitude in relation to the index K_p at latitudes near 40–45°N within 5 or 6 days after a geomagnetic disturbance.

The latitude dependence of the magnitude of the aftereffect of a magnetic storm gives an idea of the rates of dynamic processes occurring in the upper atmosphere, such as dynamic and photochemical relaxation. As follows from measurement data (Shefov 1973a), the beginning of the aftereffect is traced down to the equator within 10 days after a storm, and the periodic oscillation persists from 25 days at latitudes ~58°N to ~9 days near the equator (see Figs. 4.10 and 4.11). The latitude dependence of the duration of variations of this type can be described by a linear approximate relation (Fig. 4.15) ($r = 0.992 \pm 0.007$):

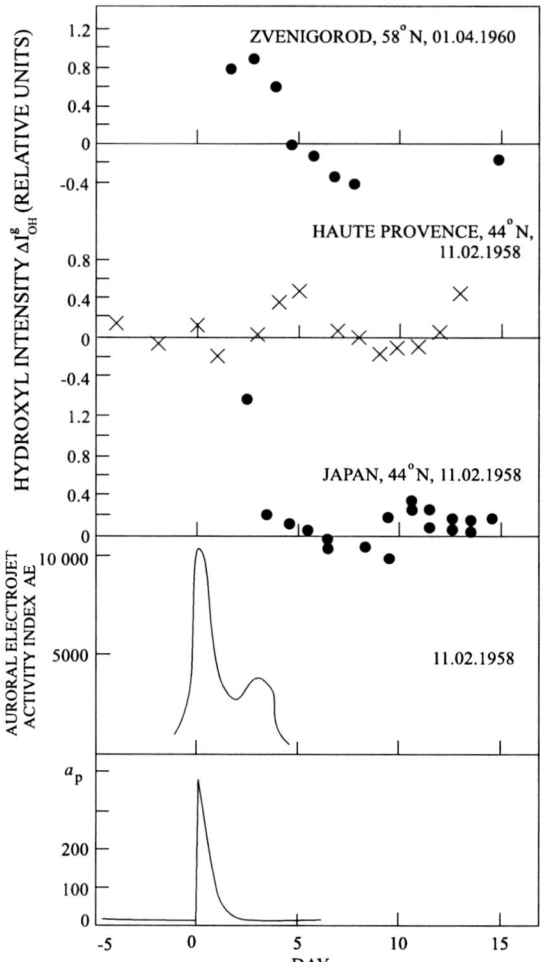

Fig. 4.12 Examples of the variations of ΔI_{OH}^g after strong geomagnetic storms by the data for Zvenigorod (April 1, 1960), Haute Provence, and Japan, and the 10-h mean $\overline{AE}_{10}$ (February 11, 1958) (Shefov 1969a, 1973a)

$$\Delta t_{osc} = (8.11 \pm 0.70) + (0.30 \pm 0.02) \cdot \varphi \, (\text{days}) \, .$$

This testifies to a slow relaxation process, i.e., the relaxation time constant is several days.

Since the ring-shaped disturbed atmospheric region containing active constituents extends toward the low latitudes, its area should increase proportional to cos Φ, where Φ is the geomagnetic latitude, provided its latitudinal span remains unchanged (Shefov 1978a). Therefore, the rate of attenuation of the aftereffect process

4.1 Model of Hydroxyl Emission

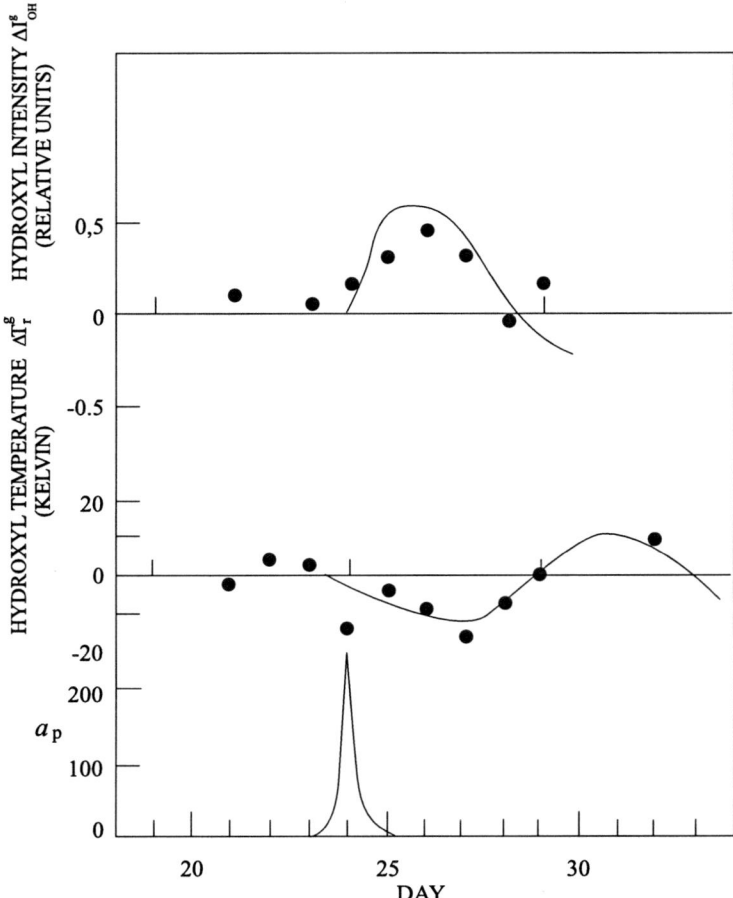

Fig. 4.13 Variations of the nightly mean relative intensity increments ΔI^g_{OH} and temperature increments ΔT^g_{OH} (K) (*dots*) after the geomagnetic storm of March 24, 1969, by Zvenigorod data. The *solid lines* are the mean variations for a given level of disturbance (Truttse and Shefov 1970a, b)

should increase. In view of the fact that the ring-shaped disturbance area is smeared in altitude and latitude, the attenuation will proceed even faster.

The relation of the values of $\lg \Delta I$ and $\Delta t = t - t_{gm}$ presented in Fig. 4.15 shows that the attenuation of the aftereffect can be approximated by the relation

$$\Delta I = \Delta I_0 \cdot \cos\Phi \cdot \exp(-\Delta t/\tau) ,$$

where $\Phi = 68°N$, $\tau = 2.3 \pm 0.5$ (days) is the attenuation time constant, which, probably, varies with latitude, $\Delta I_0 = 3.2 \pm 0.5$; and the correlation factor $r = -0.923 \pm 0.066$.

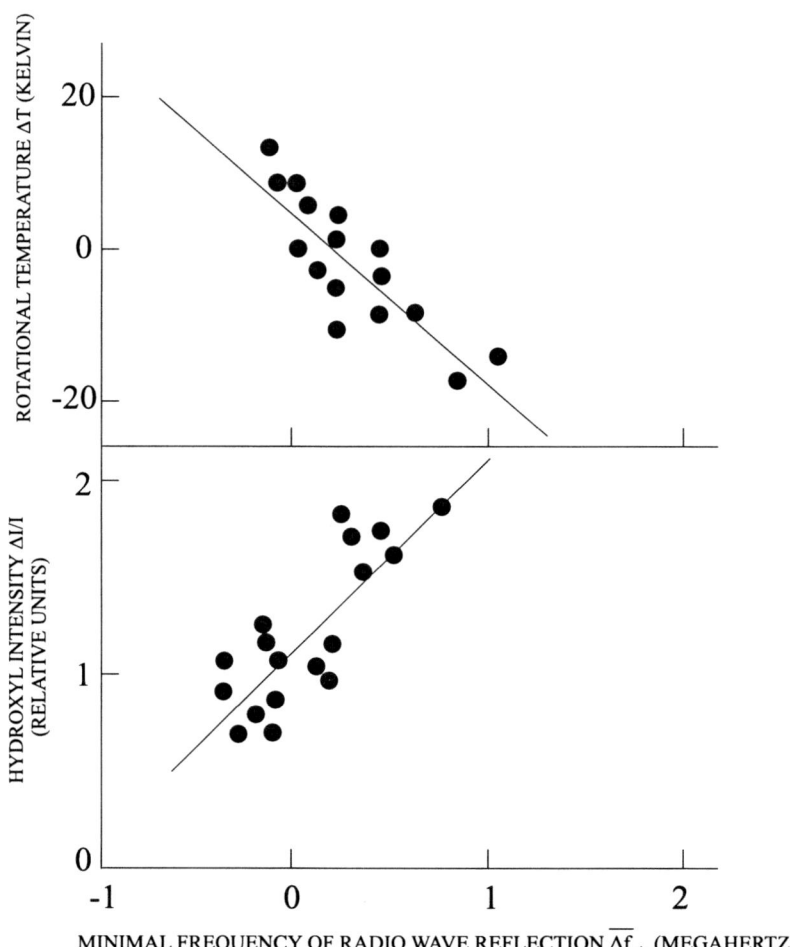

Fig. 4.14 Relation of the hydroxyl emission intensity increment $\Delta I/I$ and rotational temperature increment ΔT to ionospheric parameters $\overline{\Delta f}_{min}$ (Rapoport and Shefov 1974, 1976; Shefov 1978a). The *straight lines* are regression lines

As shown by Shefov (1978a), relating the hydroxyl emission intensity increment, ΔI, to the cosmic radiowave absorbed by the ionosphere (18–25 (MHz)) L ($\cos\chi = 0.2$) at various latitudes (from the equator to 60°N) (Rishbeth and Garriot 1969), we obtain a close correlation between these quantities (Fig. 4.16):

$$L(dB) = 22 \cdot \Delta I + 15.$$

This suggests that the observed moving disturbance is caused by the inflow of chemically active atmospheric constituents from high latitudes.

4.1 Model of Hydroxyl Emission

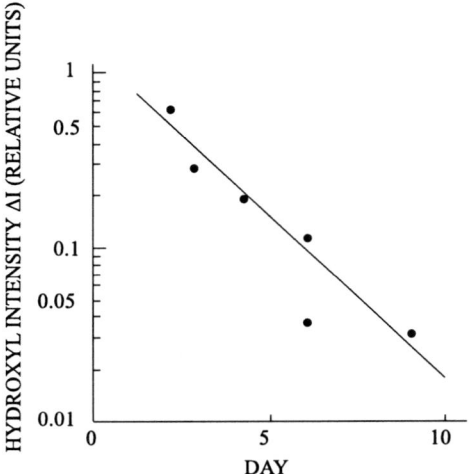

Fig. 4.15 Time variations of the mean relative increments of hydroxyl emission intensity caused by a geomagnetic-storm-induced disturbance which propagated from the auroral zone toward the equator (Shefov 1978a)

Analysis of the character of the obtained variations allows the assumption that the observed wavelike process propagating from the auroral zone is similar to an oscillation arising in the meridional flow on a rotating planet. This process can be described by the Airy function (Munk 1980; Monin 1988) since it is a solution of the linearized Korteweg–de Vries wave equation (Miropolsky 1981) that considers

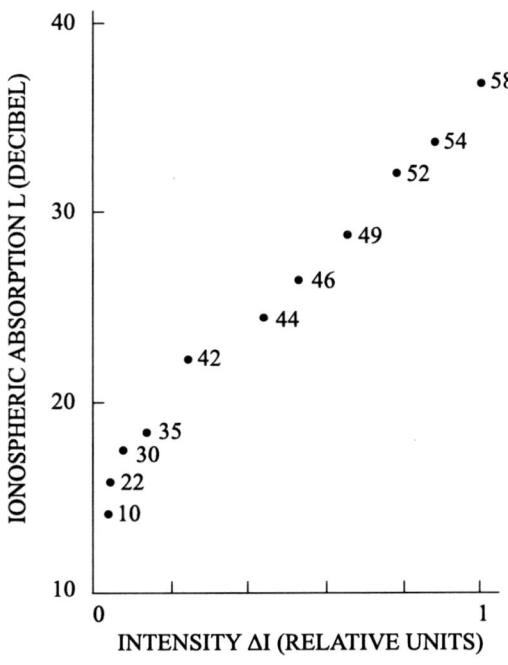

Fig. 4.16 Relation of the mean values of the ionospheric absorption L to the hydroxyl emission intensity increment ΔI (Shefov 1978a) for different geomagnetic latitudes. The latitude is indicated at the respective points

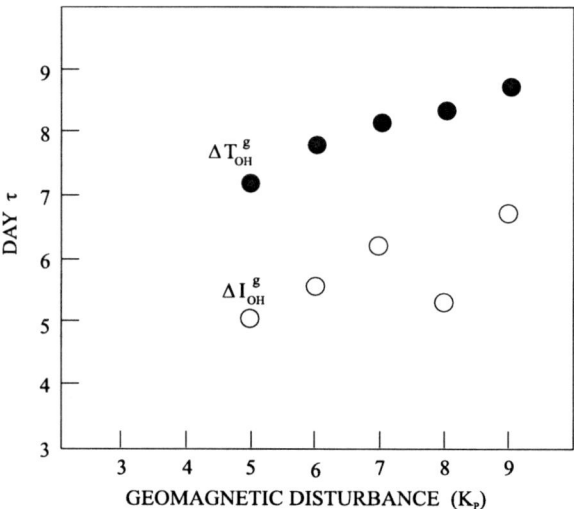

Fig. 4.17 Average dependences of the periods of the variations of intensity increments, $\tau(\Delta I^g_{OH})$ (*open circles*), and temperature increments, $\tau(\Delta T^g_{OH})$ (*full circles*) on the geomagnetic disturbance magnitude K_P (Shefov 1969a, 1973a)

the effect of dispersion and nonlinear effects. For a weakly dispersing medium, the Airy function describes an attenuating oscillatory process.

In the case under consideration, the observed intensity and temperature variations are attenuating processes described by the Airy function. The attenuation seems to be related to the transfer of active chemical agents: nitric oxide and water vapors.

The data presented in Fig. 4.8 indicate that there is a dependence of the period of the intensity increment variations on the magnitude of the geomagnetic disturbance (Fig. 4.17). Approximation gives the following expressions:

$$\tau(\Delta T_{rgm}) = 7.3 + 0.3 \cdot (K_P - 5) \text{ (day)} \qquad r = 0.95 \pm 0.05,$$

$$\tau(\Delta I_{gm}) = 5 + 0.3 \cdot (K_P - 5) \text{ (day)} \qquad r = 0.68 \pm 0.25.$$

27-Day Variations

This type of variations is associated with the Carrington cycles of rotation of the Sun which manifest themselves in the 27-day cyclicity of solar activity that is determined by the active regions on the Sun persisting for several turns. This cyclicity is responsible for the corresponding variations of the UV radiation of the Sun. However, since solar spots appear and disappear in a sporadic manner, the cyclicity is not strictly regular, and a new group of spots gives rise to a new point of counting for a 27-day cycle. Therefore, the point of onset of the initial disturbance in a cycle is not fixed and persists on the average only during no more than several turns of the Sun. Variations of this type are described elsewhere (Shefov 1967; Yarin 1970). They can be estimated by the formulas

$$\Delta T_{r27} = 45 \cdot \left(\frac{t-t_{27}}{3}\right)^{1.7} \cdot \exp\left[-\frac{t-t_{27}}{3}\right] (K),$$

$$\Delta I_{27} = 0.45 \cdot \left(\frac{t-t_{27}}{3}\right)^{1.7} \cdot \exp\left[-\frac{t-t_{27}}{3}\right],$$

$$\Delta T_{V27} = 3500 \cdot \left(\frac{t-t_{27}}{3}\right)^{1.7} \cdot \exp\left[-\frac{t-t_{27}}{3}\right] (K).$$

The available rocket data are insufficient to describe the behavior of ΔZ_{m27}.

The increase in intensity and rotational temperature at a simultaneous increase in vibrational temperature that is associated with the displacement of the emission layer upward testifies to a warming of this mesopause region with increasing solar activity, and this agrees with the results obtained by Golitsyn et al. (2005) and Semenov et al. (2005).

Variations Induced by Stratospheric Warmings

The large-scale dynamic processes developing at the stratospheric altitudes are reflected in variations of the characteristics of emission layers in the mesopause. They were detected by various methods in various geographic regions (Shefov 1973b, 1975a; Kropotkina 1976; Fishkova 1978, 1983; Matveeva and Semenov 1985; Hauchecorne et al. 1987). These variations are described by the formulas

$$\Delta T_{rsw} = 4.5 \cdot \left(\frac{t'}{2.3}\right)^3 \cdot \exp\left(-\frac{t'}{2.3}\right) + 10 \cdot \sin\frac{2\pi}{9}t' \cdot \exp\left(-\frac{t'}{30}\right) (K),$$

$$\Delta I_{sw} = 0.11 \cdot \left(\frac{t'}{2.3}\right)^3 \cdot \exp\left(-\frac{t'}{2.3}\right) - 0.25 \cdot \sin\frac{2\pi}{9}t' \cdot \exp\left(-\frac{t'}{30}\right) (K),$$

$$\Delta T_{Vsw} = -220 \cdot \left(\frac{t'}{2.3}\right)^3 \cdot \exp\left(-\frac{t'}{2.3}\right) - 500 \cdot \sin\frac{2\pi}{9}t' \cdot \exp\left(-\frac{t'}{30}\right) (K),$$

where $t' = t_{SW} - 5$.

The shift in the response time in the behavior of hydroxyl emission depends on the relative position of the warming source and observation place. In some cases, systematization was carried out by the maxima of the disturbance observed. According to some measurements, the temperature was observed not only to increase but also to decrease (Shefov 1973b, 1975a; Kropotkina 1976; Fishkova 1978, 1983; Offermann et al. 1983; Matveeva and Semenov 1985; Hauchecorne et al. 1987; Nikolashkin et al. 2001). The data of rocket measurements are insufficient to reveal the behavior of ΔZ_{msw}.

Nevertheless, it should be noted that the construction of mean variations after stratospheric warmings depends substantially on the choice of the initial moment of a disturbance if the superimposed epoch method is applied, for example, to variation maxima. It was noted (Taubenheim 1969; Taubenheim and Feister 1975) that the use of the method of autosynchronization by variation maxima can lead to wrong conclusions.

Orographic Variations

These spatial variations of hydroxyl emission parameters are related to the IGWs generated near a mountain relief as an air flow whose speed is the speed V_{600} of the prevailing wind at the 600-mbar altitude interacts with this relief.

The spatial temperature variations for the region of the Caucasian Mountains have been estimated by the measurement data reported in a number of publications (Shefov and Pertsev 1984; Sukhodoev et al. 1989a,b; Sukhodoev and Yarov 1998; Shefov et al. 1999, 2000b). The maximum amplitude of a disturbance, ΔT_{roro}, its distance from the ridge, X_M (schematically, its position is shown by the full triangle in Fig. 5.49), and horizontal width of this phenomenon, ΔX, correlate with wind velocity and vary with wind azimuth A_V^*:

$$\Delta T_{roro} = 5 \cdot \left(\frac{V_{600}}{V_0}\right)^{1/2} (K), r = 0.587 \pm 0.128, \qquad V_0 = 1 (m \cdot s^{-1}),$$

where V_{600} is the speed of the wind along the 600-mbar isobaric surface ($\sim$4 (km)),

$$X_M = (120 \pm 13) + (0.80 \pm 0.24) \cdot A_V^* \text{ (km)} \qquad r = 0.470 \pm 0.125,$$
$$\Delta X = (180 \pm 23) + (0.90 \pm 0.44) \cdot A_V^* \text{ (km)} \qquad r = 0.400 \pm 0.160.$$

Theoretical calculations of the spatial distribution of a temperature disturbance in the leeward region of the mesopause were performed based on the mechanism of generation of internal gravity waves by gusts arising on the interaction of an air flow with relief irregularities (Chunchuzov 1988). It was found that the spatial distribution of the fluctuations of the prevailing wind and gust azimuths is described by the Henyey–Greenstein function (Shefov et al. 1999, 2000b)

$$p(\cos A) = \frac{1}{4}(1 - g^2) \cdot (1 - 2g \cdot \cos A + g^2)^{-3/2},$$

where the parameter g, being the average cosine of the distribution, is equal to 0.5.

As follows from measurement data, an increase in rotational temperature ($\sim$10 (K)) in the region of an orographic disturbance corresponds to a decrease in vibrational temperature (~ -1500 (K)). This points to a decrease of the altitude of the emission layer.

The temperature variations in the direction perpendicular to the ridge line can be described by the empirical formulas

4.1 Model of Hydroxyl Emission

$$\Delta T_{roro} = 2.7 \cdot V_{600} \cdot \left(\frac{X+7}{147}\right)^{1.14} \cdot \exp\left(-\frac{X+7}{147}\right) \text{ (K)},$$

$$\Delta T_{Voro} = -400 \cdot V_{600} \cdot \left(\frac{X+7}{147}\right)^{1.14} \cdot \exp\left(-\frac{X+7}{147}\right) \text{ (K)},$$

$$\Delta Z_{moro} = -0.4 \cdot V_{600} \cdot \left(\frac{X+7}{147}\right)^{1.14} \cdot \exp\left(-\frac{X+7}{147}\right) \text{ (km)}.$$

Calculations of the energy afflux in the leeward region of the mesopause have shown that the vertical flux at the disturbance maximum is $[F_z] = 3 \cdot 10^{-3}$ (W·m^{-2}) = 3 (erg·cm^{-2}·s^{-1}). Based on these data, in view of the average length of the mountains in the northern and southern hemispheres, the planetary average energy flux is estimated to be ~ 4 (erg·cm^{-2}·s^{-1}) (Shefov 1985) (Sect. 5.2).

Variations Induced by Meteoric Flows and Noctilucent Clouds

These variations have been revealed from the data of observations at Zvenigorod in summertime. The results testify to intensification of the OH emission after intrusion of meteoric flows (MF in Figs. 4.2 and 4.3) and on occurrence of noctilucent clouds (NLC in Figs. 4.2, 4.3) (Shefov 1968a, 1970b), whose average altitude is about 82 (km) (Bronshten and Grishin 1970). From the data of measurements performed on Alaska, Taylor et al. (1995) have estimated the temperatures during the periods of observation of noctilucent clouds to be 150–155 (K). However, the temperature variations before the occurrence of noctilucent clouds were different in character compared to those observed at Zvenigorod. The distribution of meteoric flows during a year is given elsewhere (Bronshten et al. 1981). These variations have been described by the relations

$$\Delta T_{rmf} = 18 \cdot \frac{\exp[10 \cdot (3.5-t)] - 1}{\exp[10 \cdot (3.5-t)] + 1} \cdot \exp\left[-\frac{(t-3)^2}{2}\right] \text{ (K)},$$

$$\Delta I_{mf} = 0.7 \cdot \frac{\exp[10 \cdot (3.5-t)] - 1}{\exp[10 \cdot (3.5-t)] + 1} \cdot \exp\left[-\frac{(t-3)^2}{2}\right],$$

$$\Delta T_{Vmf} = -500 \cdot \frac{\exp[10 \cdot (3.5-t)] - 1}{\exp[10 \cdot (3.5-t)] + 1} \cdot \exp\left[-\frac{(t-3)^2}{2}\right] \text{ (K)},$$

$$\Delta Z_{mmf} = -0.5 \cdot \frac{\exp[10 \cdot (3.5-t)] - 1}{\exp[10 \cdot (3.5-t)] + 1} \cdot \exp\left[-\frac{(t-3)^2}{2}\right] \text{ (km)}.$$

IGW-Induced Variations

Variations of this type cannot be predicted within a night since they are caused by IGWs propagating in the mesopause. IGWs are generated in the main by meteorological sources in the lower atmosphere (Krassovsky et al. 1977; Semenov and Shefov 1989), whose arrangement relative to the observation place and, hence, the wave period and the time at which IGWs are generated are random in character. The rotational temperature variations can be estimated by the relation

$$\Delta T_{rgw} = T_r \cdot 10^A \cdot \tau_w^k \cdot \sin \frac{2\pi}{\tau_w} t,$$

where τ_w is the wave period (min).

$$A = -2.45 + 0.48 \cdot \cos \frac{2\pi}{365} t_d; \qquad k = 0.78 - 0.25 \cdot \cos \frac{2\pi}{365} t_d.$$

Intensity and vibrational temperature variations can be estimated by the relations

$$\Delta I_{gw} = I \cdot \eta \cdot \frac{\Delta T_r}{T_r}; \Delta T_{Vgw} = -T_V \cdot \frac{\Delta T_r}{T_r}.$$

The parameter η, which defines the ratio of the relative amplitudes of the OH emission intensity and temperature variations that are induced by IGWs propagating through the emission layer, has different values for the upper and the lower vibrational levels (Shagaev 1978). Based on these data, they vary during a year according to the formulas

$$\eta_9 = 2.4 + 1.11 \cdot \cos^{0.5} \frac{2\pi}{365} t_d; \qquad \eta_5 = 1.7 + 0.75 \cdot \cos^{0.5} \frac{2\pi}{365} t_d.$$

The time of delay of the intensity variations relative to the temperature variations, θ (min), also depends on vibrational level number v and on season:

$$\theta_9 = 1.70 + 1.4 \cdot \cos^{0.5} \frac{2\pi}{365} t_d; \qquad \theta_5 = 0.65 + 0.55 \cdot \cos^{0.5} \frac{2\pi}{365} t_d.$$

The data for other vibrational levels have not been adequately investigated. On the assumption that η and θ linearly vary with v, the following general expressions have been obtained:

$$\bar{\eta}_v = 0.825 + 0.175 \cdot v; \qquad \bar{\theta}_v = -0.66 + 0.25 \cdot v.$$

In developing the presented empirical global model of the space–time variations for the OH emission, the attention was focused on obtaining average relations and absolute values of variations. In doing this, the measurement data were reduced to standard heliogeophysical conditions as fully as possible. This provided a way of performing theoretical calculations for particular situations and comparing different types of variations. The presented set of variations is a refined version of that obtained initially (Semenov and Shefov 1996a, 1999a), and doubtlessly, it will be further refined based on goal-directed measurements. All this will promote the construction of models which would describe the behavior of the concentrations of small atmospheric constituents that participate in photochemical processes initiating hydroxyl emission, which practically cannot be determined by other methods.

The presented empirical model has been verified based on the materials of the international joint measurements performed under the CRISTA/MAHRSI program during the period from October 1 to December 31, 1994, and on the data of measurements performed at Zvenigorod in cooperation with the Canadian scientists from the University of Western Ontario. An important inference of the analysis performed is that the values of rotational temperature T_r measured at various stations are in rather good agreement and that 16- and 30-day variations, global in character and induced by planetary waves, have been revealed though the random conditions of their occurrence could not be taken into consideration by the given model.

The problems of determination of the characteristics of IGWs are discussed in Sects. 3.6.4, 3.6.5, and 5.1.

4.2 Model of the Sodium Emission

When constructing a model of the emission from sodium atoms, one should use first of all the emission intensity and emission rate altitude distribution data. Direct interferometric measurements of the atmospheric sodium temperature have shown that for altitudes ~ 90 (km) a near-equilibrium temperature (215 ± 15 (K)) was obtained only during twilight measurements (Sipler and Biondi 1975, 1978), since the emission is caused by resonance fluorescence. At night the Doppler temperature turned out to be far above the equilibrium temperature (600 ± 50 (K)) because of the formation of excited sodium atoms of high kinetic energy from $(NaO_2)^*$-excited molecules (Hernandez 1975; Sipler and Biondi 1975, 1978). Laboratory investigations have also revealed this phenomenon (Gann et al. 1972).

The sodium emission, as well as other airglow emissions, responds to a variety of regular and irregular variations which take place in the mesopause. In the studies based on lidar measurements, the content of sodium atom is estimated by the number of atoms in a vertical column of cross-section 1 (cm^2), N (cm^{-2}).

It should be noted that for the atomic sodium emission, much less rocket measurement data on the emission rate altitude distribution Q(Z) are available than for the emissions of OH and O (557.7 (nm)). The relevant measurements were carried out at White Sands (32.4°N, 106.4°W) in the mid-1950s and at Alcantara (Brazil) (2.5°S, 44.4°W) in the 1990s. The measurement data clearly point to that Q(Z) for Na is on the average asymmetric relative to the maximum Q altitude. The asymmetry is P ~ 0.55 (Packer 1961; Clemesha and Takahashi 1995; Takahashi et al. 1996).

The much more numerous lidar measurements of the altitude profile of sodium atom column density testify that these profiles are, on the average, symmetric

(P = 0.5) (States and Gardner 1999, 2000a,b), and the altitude of maximum sodium density is the same as that of maximum emission rate.

The reason for this seems to be that, according to the mechanism of excitation of sodium emission at night (Chapman 1939; Plane 1991; Helmer and Plane 1993), the emission rate is proportional to the sodium and ozone concentrations, i.e.,

$$Q(Na) \sim [Na] \cdot [O_3]$$

or, finally,

$$Q(Na) \sim [Na] \cdot \frac{[O] \cdot [O_2] \cdot \{[O_2] + [N_2]\}}{[H]},$$

since the formation of ozone is related to recombination of atomic oxygen. The increase in atomic oxygen concentration with altitude is the reason for the asymmetry of the sodium emission rate altitude distribution.

The most numerous measurements of the Na emission intensity have been performed by the photometric method in 1958–1968 at Haute Provence (43.9°N, 5.7°E) (Yao 1962; Wiens and Weill 1973; Vassy and Vassy 1976), in 1957–1968 at Abastumani (41.8°N, 42.8°E) (Fishkova 1983; Toroshelidze 1991), in 1957–1990 in Japan (38.1°N, 140.6°E) (Yao 1962; Smith et al. 1968; Fukuyama 1976, 1977a,b; Takeuchi et al. 1986, 1989a,b), in 1964–1965 at Kitt Peak (32°N, 111.6°W) (Smith et al. 1968; Fishkova 1983), in 1961–1966 at Haleakala (20.7°N, 156.3°W) (Fukuyama 1977a), in 1964–1973 at Poone (18.5°N, 73.9°E) (Smith et al. 1968; Toroshelidze 1991), in 1968–1970 at Adi Ugri (14.9°N, 38.8°N) (Wiens and Weill 1973), in 1986–1991 at Fortaleza (3.9°S, 38.4°W) (Batista et al. 1990, 1994), in 1977–1991 at Cachaeira Paulista (22.7°S, 45.0°W) (Takahashi et al. 1984; Batista et al. 1994), and in 1967–1971 at Zeekoegat (33.1°S, 22.5°E) (Wiens and Weill 1973). Spectrographic measurements were performed at Abastumani (Fishkova 1983) and in 1957–1993 at Zvenigorod (55.7°N, 36.8°E).

Nocturnal Variations

The lidar observations of the sodium layer allow the conclusion that the variations of its parameters during daytime hours and at night are different types of regular variations. This conclusion is based on the nocturnal behavior of the emission intensity whose spatial distribution shows a pronounced spotty structure (Nasyrov 1967a,b) and the occurrence of the intensity nightly minimum depends on season (Shefov et al. 2000a). This effect also takes place for the 557.7-nm atomic oxygen emission. Its reason is the displacement of the atmospheric tidal fluctuations phase to the region above the mesopause, where emission layers are localized, due to the variations of the emission layer altitudes (Batista et al. 1990). In the daytime, the sodium density and the layer altitude are determined in the main by other processes because of the action of solar radiation.

Therefore, as follows from many studies (Yao 1962; Wiens and Weill 1973; Vassy and Vassy 1976; Smith et al. 1968; Fukuyama 1976; Takahashi et al. 1984;

4.2 Model of the Sodium Emission

Takeuchi et al. 1986, 1989a,b), the diurnal variations reflect the effect of the thermal tide that arises in the stratospheric ozone layer. Therefore, to systematize measurement data for various seasonal and latitudinal conditions, it is necessary to change from time to space coordinates, namely to solar zenith angles, $\chi_\odot$. Thus, we have

$$\Delta I_{\chi\odot} = 0.12 \cdot \sin 2(135 - \chi_\odot - \theta).$$

Figure 4.18 presents the dependence of $\Delta I_{\chi\odot}$ on $\chi_\odot$ for $\theta = 180°$. The solar zenith angle $\chi_\odot$ is determined by the relation

$$\cos \chi_\odot = \sin \varphi \cdot \sin \delta - \cos \varphi \cdot \cos \delta \cdot \cos \tau.$$

Here δ is the declination of the Sun and τ is the local solar time. When calculating $\Delta I_{\chi\odot}$ for the points in time after midnight, it is necessary to use $-\chi_\odot$. The variations presented in Fig. 4.18 correspond to the equinox conditions. However, the phase θ associated with a given emission intensity depends on the season due to the wind variations in the global circulation of the atmosphere (Vincent et al. 1998) that result in variations of the emission layer altitude. Figure 4.18 presents these variations, which are described by the relation

$$\theta = 150 + 24.7 \cdot \cos\frac{2\pi}{365}(t_d - 355) + 7.0 \cdot \cos\frac{4\pi}{365}(t_d - 6).$$

As follows from these data, the time of nightly minimum emission intensity depends on season and corresponds to a solar zenith angle $\chi_\odot^{min}$. It is determined by the relation

$$\cos \chi_\odot^{min} = \sin(45° - \theta).$$

The minimum I is determined by the condition $\chi_\odot(I_{min}) = \min[\chi_\odot^{min}; 180° - \varphi - \delta]$.

In the calculations of the night ($\chi_\odot > 100°$) values of ΔN_χ, $\Delta Z_{m\chi}$, and ΔW_χ for a local time τ after midnight, the value of $\chi_\odot$ is taken with the minus sign:

$$\Delta N_\chi = 0.13\cos(\chi_\odot - \theta - 275) + 0.088\cos 2(\chi_\odot - \theta - 78) + 0.042\cos 3(\chi_\odot - \theta - 11),$$
$$\Delta Z_{m\chi} = 0.23\cos(\chi_\odot - \theta - 217) + 0.20\cos 2(\chi_\odot - \theta - 74) + 0.12\cos 3(\chi_\odot - \theta - 7),$$
$$\Delta W_\chi = 0.15 \cdot \cos(\chi_\odot - \theta - 116) + 0.48 \cdot \cos 2(\chi_\odot - \theta - 80) + 0.06 \cdot \cos 3(\chi_\odot - \theta - 103),$$

where $\theta = 5 + 24.7\cos[2(t_d - 355)/365] + 7.0 \cdot \cos[4(t_d - -6)/365]$ (t_d is the day of year).

Consideration of the combined variations of the layer altitude, its thickness, and the atomic sodium density based on the data reported by States and Gardner (1999) has shown that their mutual diurnal correlations are complex and nonlinear in character. The nocturnal variations ($\chi_\odot > 100°$) are also nonlinear:

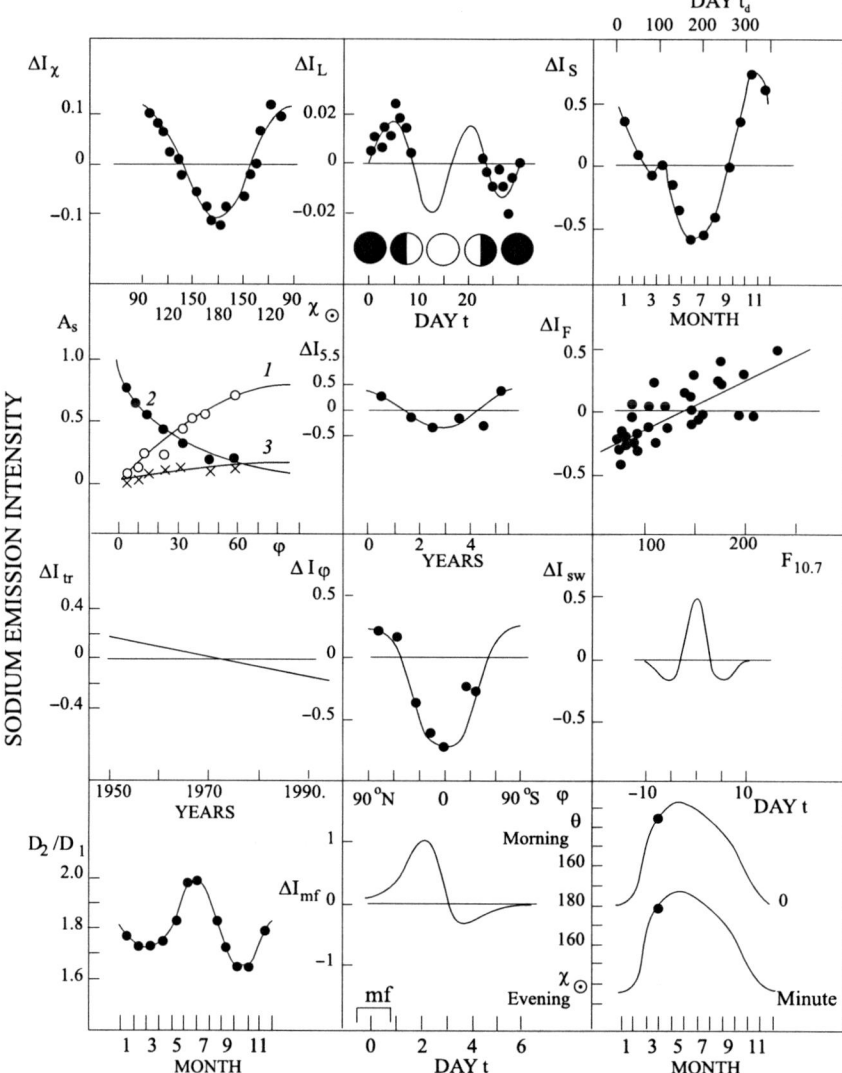

Fig. 4.18 Nocturnal variations of the intensity of the 589.0–589.6 (nm) atomic sodium emission about its mean value for standard heliogeophysical conditions in relation to different parameters. The designations are described in the preface to Chap. 4

4.2 Model of the Sodium Emission

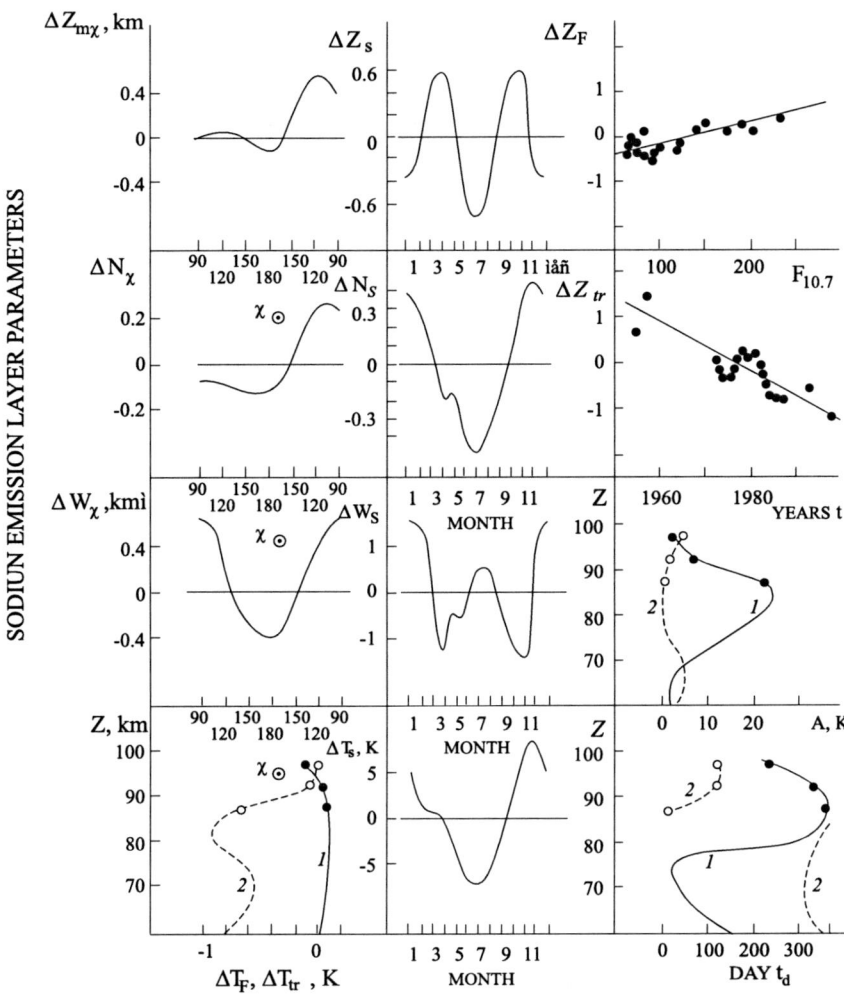

Fig. 4.19 Nocturnal variations of the sodium layer parameters about their mean values corresponding to standard heliogeophysical conditions. 1 – Altitude distributions of temperature response to solar activity δT_F; 2 – long-term trend ΔT_{tr}, determined according to emission measurement data. The altitude variations of amplitudes A and phases t_d of the annual (*full circles*) and semiannual (*open circles*) harmonics derived from emission data. The *thick solid dashed lines* represent rocket measurement data for all cases (Lysenko et al. 1997, 1999) and the *thin solid dashed lines* represent approximations of emission and rocket data

$$\Delta N_\chi = -0.13 + 0.106 \cdot \Delta Z_{m\chi} - 1.1 \cdot (\Delta Z_{m\chi})^2,$$
$$\Delta W_\chi = -0.49 - 4.86 \cdot \Delta Z_{m\chi} + 16.14 \cdot (\Delta Z_{m\chi})^2 + 0.71 \cdot (\Delta Z_{m\chi})^3 - 24.1 \cdot (\Delta Z_{m\chi})^4.$$

Here $\Delta Z_{m\chi}$ and ΔW_χ are expressed in kilometers (Fig. 4.19).

Lunar Variations

The emission intensity variations depending on the phase age of the Moon have been described for the intensity at midlatitudes varying with the period of a synodical month equal to 29.53 days (Fishkova 1983):

$$\Delta I_L = 0.042 \cdot \cos\frac{2\pi}{29.53}(t_{L\Phi}+1) - 0.174 \cdot \cos\frac{4\pi}{29.53}(t_{L\Phi}+2.5).$$

These variations seem to be associated with yearly average conditions. Here, the above notes regarding the lunar variations of hydroxyl emission are applicable.

The Moon's age can be estimated to within several tenth of day by the formula (Semenov and Shefov 1996a)

$$t_{L\Phi} = [(t_d/365 + YYYY - 1900) \cdot 12.3685] \cdot 29.53 - 1.$$

Here YYYY is the year number; the square brackets denote the fractional part of the number.

Seasonal Variations

The seasonal variations, as usual, have the greatest amplitudes compared to other periodic changes.

The seasonal intensity variations can be described by the relation

$$\Delta I_S = A_1(\varphi) \cdot B_1(F_{10.7},t) \cdot \cos\frac{2\pi}{365}(t_d - 355)$$
$$+ A_2(\varphi) \cdot B_2(F_{10.7},t) \cdot \cos\frac{4\pi}{365}(t_d - 110)$$
$$+ A_3(\varphi) \cdot B_3(F_{10.7},t) \cdot \cos\frac{6\pi}{365}(t_d - 75).$$

The seasonal variations have been derived from the data of a number of stations such as Zvenigorod, Abastumani, Kiso, Adi Ugri, Cachoeira Paulista, Zeekoegat, and Fortaleza (Wiens and Weill 1973; Fishkova 1983; Fukuyama 1977a; Takeuchi et al. 1986, 1989a,b; Takahashi et al. 1984, 1995; Batista et al. 1994). As follows from these data, the seasonal variability depends on geographic latitude, solar flux, and year's number. The data for $\varphi = 45°N$ are presented in Fig. 4.18. The harmonic amplitudes A_S are given by the relations

4.2 Model of the Sodium Emission

$$A_1(\varphi) = 0.82 \cdot \sin^{0.88} \varphi, \text{correlation factor } r = 0.998 \pm 0.002,$$
$$A_2(\varphi) = 1 - 0.91 \cdot \sin^{0.5} \varphi, r = 0.995 \pm 0.004,$$
$$A_3(\varphi) = 0.18 \cdot \sin^{0.64} \varphi, r = 0.993 \pm 0.006.$$

Fukuyama (1977a) pointed to the latitude dependence of the harmonic phases. The analysis performed by this author has shown that the possible dependence of the phases on latitude is weak. The multipliers depending on solar flux and long-term trend are determined by the formulas

$$B_1(F_{10.7}, t) = 1 + (0.00287 \pm 0.00146)$$
$$(F_{10.7} - 130) - (0.135 \pm 0.045)(t - 1972);$$
$$B_2(F_{10.7}, t) = 1 + (0.00089 \pm 0.00025)$$
$$(F_{10.7} - 130) - (0.0242 \pm 0.0087)(t - 1972);$$
$$B_3(F_{10.7}, t) = 1 + (0.00122 \pm 0.0005)$$
$$(F_{10.7} - 130) - (0.0173 \pm 0.0072)(t - 1972).$$

Probably, the regression coefficients depend on latitude.

An important characteristic of the sodium emission is the ratio of the doublet components $D_2(589.0(nm))/D_1(589.6(nm))$, which also varies during a year (Fishkova 1983). For measurements at the zenith (at latitude $45°N$), we have

$$D_2/D_1 = 1.78 + 0.110 \cdot \cos \frac{2\pi}{365}(t_d - 173) + 0.103 \cdot \cos \frac{4\pi}{365}(t_d - 6).$$

This ratio characterizes the optical thickness of the sodium layer (Chamberlain 1961). However, based on the measurements performed at the Laurel Ridge Observatory ($40.1°N, 79.2°W$) in October, 1974 (solar flux minimum), it was obtained that at night $D_2/D_1 = 2$ (Sipler and Biondi 1978). Estimation of the intensity of sodium emission for the given conditions using the model under consideration gives 65 (Rayleigh).

Based on the data of lidar observations (Clemesha et al. 1992; Plane et al. 1999; States and Gardner 1999), it was obtained that the variations of the sodium content, layer altitude (km), and layer thickness (km) are determined by the sums of harmonics:

$$\Delta N_S = 0.41 \cdot \cos[2\pi(t_d - 355)/365]$$
$$+ 0.041 \cdot \cos[4\pi(t_d - 118)/365] + 0.021 \cdot \cos[6\pi(t_d - 35)/365],$$
$$\Delta Z_{mS} = 0.23 \cdot \cos[2\pi(t_d - 10)/365]$$
$$+ 0.59 \cdot \cos[4\pi(t_d - 100)/365] + 0.03 \cdot \cos[6\pi(t_d - 70)/365],$$
$$\Delta W_S = 0.31 \cdot \cos[2\pi(t_d - 27)/365]$$
$$+ 1.15 \cdot \cos[4\pi(t_d - 11)/365] + 0.41 \cdot \cos[6\pi(t_d - 3)/365].$$

Noteworthy is that the harmonic phases for the sodium density and the sodium emission intensity (Shefov et al. 2000a), obtained in fact from independent data, are close to each other.

Close correlation has been revealed (Qian and Gardner 1995; States and Gardner 1999) between the sodium density in the layer, N_{Na}, and the atmospheric temperature, T_{Na}, at the layer maximum altitude for both the nocturnal and the seasonal variations. This correlation holds for all latitudes, from the South Pole to the high latitudes of the northern hemisphere (Plane et al. 1998). Naturally, the sodium emission intensity I_{Na} is proportional to N_{Na}. Based on the data by Shefov et al. (2000a), the seasonal variations of the intensity I_{Na} (Rayleigh) have been calculated for the conditions of lidar measurements performed in 1997 (States and Gardner 1999) and compared to the temperature seasonal variations. As a result, the following relation has been obtained:

$$T_{Na} = (185 \pm 0.8) + (0.20 \pm 0.01) \cdot I_{Na}, (K),$$

with the correlation coefficient $r = 0.952 \pm 0.020$.

The revealed close relation between the sodium emission intensity and the atmospheric temperature makes it an important means for the estimation of the seasonal variations of the atmospheric temperature at altitudes of $\sim$92 (km) that can be presented by the formula

$$\Delta T_{Na} = A_1 \cdot \cos[2\pi(t_d - t_1)/365] \\ + A_2 \cdot \cos[4\pi(t_d - t_2)/365] + A_3 \cdot \cos[6\pi(t_d - t_3)/365],$$

where, for the conditions of the year 1997, the amplitudes of the annual, semiannual, and 4-month harmonics are, respectively, $A_1 = 6.0$ (K), $A_2 = 1.7$ (K), and $A_3 = 0.9$ (K); the respective phases are $t_1 = 345$ (days), $t_2 = 120$ (days), and $t_3 = 86$ (days).

Comparison of the yearly average temperatures T_{Na} at $\sim$92 (km) calculated by the above formula for the period 1955–1995 and the T_{92} values taken from yearly average temperature profiles (Semenov and Shefov 1999a; Lysenko et al. 1999) has shown their good agreement, namely $T_{Na} = T_{92} \pm 4$ (K), i.e., the error is $\sim$2(%). Thus, based on measurements of the sodium emission intensity, it is possible to evaluate the behavior of the temperature at altitudes of about 92 (km).

Cyclic Aperiodic (Quasi-biennial) Variations

As mentioned in Sect. 4.1, the present-day notions about the character of the quasi-biennial variations (QBO) of the mesopause and lower thermosphere parameters are essentially other than the former ones. It turned out that the original cause of these variations is solar flux QBOs. All researchers who used the standard Fourier analysis to determine the periodicity parameters arrived at average periods of $\sim$2.3 years.

4.2 Model of the Sodium Emission

The problem of these variations was considered in detail in Sect. 1.6.2. The solar activity component in the index $F_{10.7}$ makes

$$\Delta F_{10.7} = -22 \cdot \text{Ai}(3 - \Delta t_{SAO}) ,$$

where Δt_{SAO} (years) is the time interval counted from the point of solar flux minimum.

The Airy function values can be obtained from tabulated data or by approximate formulas (Abramowitz and Stegun 1964).

The quasi-biennial oscillations are presented in Fig. 4.20. The data for the sodium emission are based on the measurements performed at Abastumani (Fishkova et al. 2001a) and Ashkhabad (Nasyrov 2003). These quasi-biennial oscillations can be described by the formula (Semenov and Shefov 2003)

$$\Delta T_{92}(589.3\,\text{nm}) = 5\,\text{Ai}(3 - \Delta t)(K) .$$

This formula is confirmed by the intensity and temperature data obtained at Zvenigorod, Abastumani, and Yakutsk (Fadel et al. 2002) and at Ashkhabad (Nasyrov 2003).

5.5-yr Variations

The 5.5-yr variations of the sodium emission intensity are described by the formula

$$\Delta I_{5.5} = 0.04 \cdot \cos \frac{2\pi}{5.5}(t - t_{5.5}) .$$

Variations of this type have been revealed by analyzing the data of long-term measurements performed at Zvenigorod, Abastumani, Haute Provence, and Kiso. The $\Delta I_{5.5}$ maxima seem to occur synchronously with solar flux maxima since the approximation of an 11-yr solar cycle with harmonics shows that the phase of the first harmonic is near the cycle maximum and the second (5.5-yr) harmonic has a phase of 3.7 (yr) relative to the minimum and an amplitude of $\sim 15(\%)$ of the first harmonic amplitude (Kononovich 2005). Therefore, for the years of solar flux minima the observed emission intensities should inevitably contain a small contribution of 5.5-yr variations. The corresponding data for the sodium layer altitude are not available.

Variations Associated with Solar Activity

In earlier studies of the seasonal variations of sodium emission intensity, it has been established that the typical annual variations show a maximum at the end of the year and a minimum in summertime. As well as for all characteristics of the mesopause region, there is a dependence on solar flux. Analysis of long-term measurement data

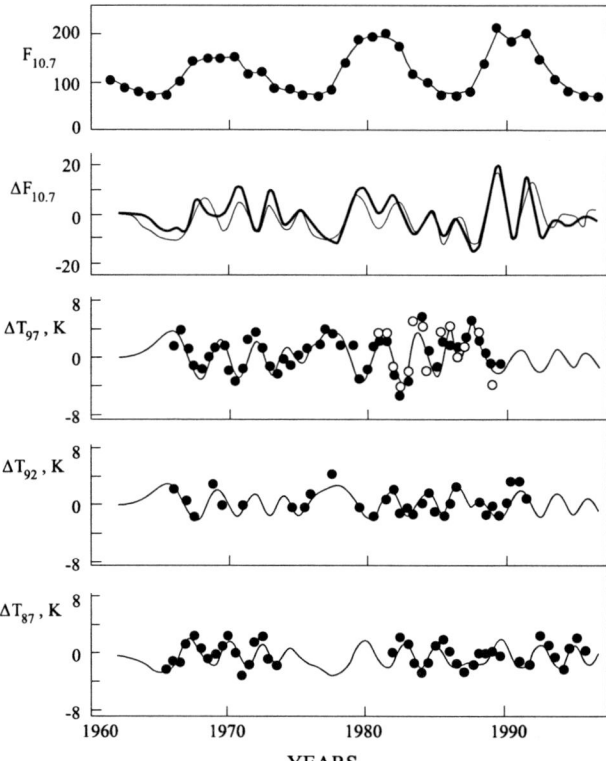

Fig. 4.20 Cyclic aperiodic (quasi-biennial) temperature variations ΔT at altitudes of 97 ([OI] 557.7 (nm)), 92 (Na), and 87 (km) (OH) derived from spectrophotometric measurements (*dots*). The *circles* represent the results obtained from interferometric measurements (Yugov et al. 1997). The *solid line* represents the approximation by the Airy function. In the top of the figure, the variations of the yearly average values of $F_{10.7}$ and its quasi-biennial oscillations $\Delta F_{10.7}$ (*thick line*) are presented. The *thin line* represents the approximation by the Airy function (Fadel et al. 2002)

has shown that the dependence of the emission intensity on solar flux changes in character during a year (Fishkova et al. 2001a). These data are presented in Fig. 4.21. It can be seen that the correlation coefficient changes its sign during the autumn–winter season when an intensity maximum is observed. Thus, the seasonal variations can be approximated by the relations

$$I_S(F,i) = [I_{MS}(i) \pm \sigma_{FS}] + [\delta I_F(i) \pm \sigma_F] \cdot (F_{10.7} - 130) \text{ (Rayleigh)}$$

for which the correlation coefficient is equal to $r_F \pm \sigma_{rF}$, and the values of the Student factor t_{SF} are given in Table 4.1. Here i is the month's number. In turn, the variations of monthly mean intensities reduced to the solar flux $F_{10.7} = 130$ (1972) can be represented by the sum of three harmonics with periods of 12, 6, and 4 months:

$$I_{MS} = I_M \cdot (1 + \sum A_{Ik}) \text{ and } \delta I_F = \delta I_{MF} \cdot (1 + \sum A_{Fk}),$$

4.2 Model of the Sodium Emission

where

$$\sum A_{Ik} = 0.58 \cdot \cos[2\pi(t_d - 343)/365]$$
$$+ 0.12 \cdot \cos[4\pi(t_d - 134)/365] + 0.06 \cdot \cos[6\pi(t_d - 70)/365].$$

The straight lines are regression lines. The seasonal mean variations of intensity I_{MS} and temperature response to solar activity δI_F are shown below. The solid lines depict the sums of the three harmonics: annual, semiannual, and 4-monthly.

$$\sum A_{Fk} = 1.07 \cdot \cos[2\pi(t_d - 142)/365] + 0.83 \cdot \cos[4\pi(t_d - 56)/365]$$
$$+ 0.26 \cdot \cos[6\pi(t_d - 27)/365].$$

Here $I_M = 65$ (Rayleigh) and t_d is the day of the year.

There is satisfactory agreement of these data with the results obtained by Shefov et al. (2000a), in particular, for the annual harmonic, which makes the basic contribution to seasonal variations. The yearly average intensity shows a significant positive correlation with solar flux ($r = 0.520 \pm 0.143$):

$$I_{MA} = I_{MA}^0 \cdot [1 \pm 0.077 + (0.0015 \pm 0.0005) \cdot (F_{10.7} - 130)] \text{ (Rayleigh)}.$$

To reveal the dependence of the sodium layer altitude on solar flux, lidar measurement data on the layer altitude for a 14-yr period (Clemesha et al. 1992) and for shorter periods (States and Gardner 1999) and also rocket measurement data (Koomen et al. 1956; Tousey 1958; Packer 1961; Clemesha et al. 1993) were used. For individual rocket measurements, preliminary corrections were made for diurnal

Table 4.1 Statistical characteristics of the seasonal dependence of the sodium emission intensity on solar flux

Month	I_{MS} (Rayleigh)	σ_{FS} (Rayleigh)	δI_F (Rayleigh · $(\Delta F = 1)^{-1}$)	σ_F (Rayleigh · $(\Delta F = 1)^{-1}$)	r_F	σ_{rF}	t_{SF}
1	88	5	0.073	0.029	0.466	0.160	2.5
2	76	6	0.090	0.039	0.446	0.171	2.3
3	60	7	0.112	0.044	0.432	0.151	2.5
4	48	7	0.105	0.045	0.406	0.158	2.3
5	36	6	0.082	0.042	0.351	0.163	2.0
6	30	5	0.078	0.032	0.431	0.156	2.5
7	33	5	0.104	0.035	0.508	0.145	3.0
8	42	5	0.093	0.029	0.546	0.140	3.2
9	65	6	0.072	0.039	0.342	0.170	1.9
10	93	6	0.026	0.041	0.128	0.197	0.6
11	113	7	−0.072	0.043	−0.317	0.176	1.7
12	107	6	−0.073	0.040	−0.338	0.171	1.9
Yearly average	65	5	0.058	0.033	0.361	0.161	2.1

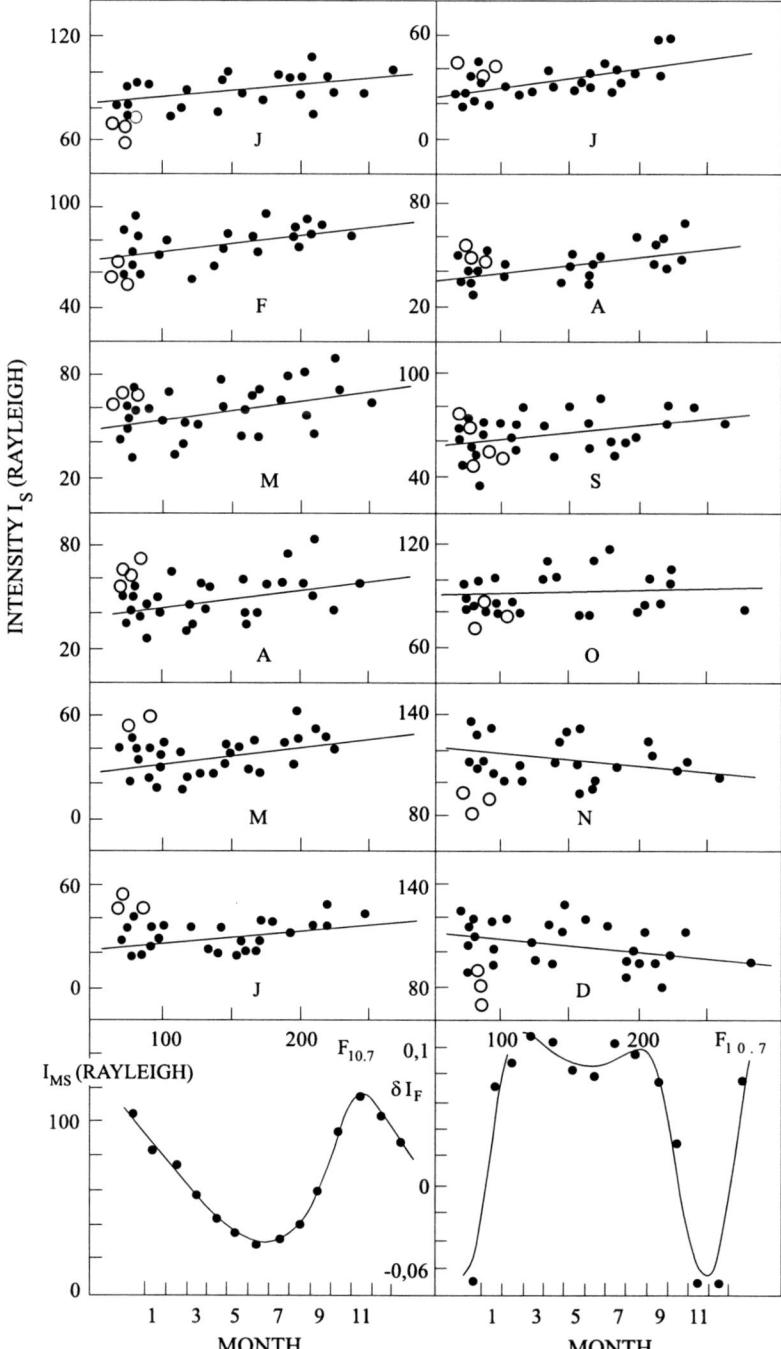

Fig. 4.21 Long-term variations of the monthly mean sodium emission intensities [*full circles*: data by Fishkova et al. (2001a) and *open circles*: data by Nasyrov (2003)] after elimination of the dependence of the intensity on solar flux for various months of the year (1–12). The *straight lines* are regression lines. The seasonal mean variations of the intensity I_{MS} and temperature response to solar activity δI_F are shown below. The *solid lines* depict the sums of the three harmonics

4.2 Model of the Sodium Emission

and seasonal variations based on the data presented in Fig. 4.19. As a result, it has been obtained that

$$\Delta Z_{mF} = (0.0046 \pm 0.0003)(F_{10.7} - 130)$$

and the correlation coefficient $r = 0.840 \pm 0.069$.

The dependence of the sodium layer temperature on solar flux is given by the relation

$$\Delta T_F(Na) = 0.04(F_{10.7} - 130)(K) .$$

The Long-Term Trend

After elimination of the dependence of the sodium emission intensity on solar flux, long-term intensity variations have been obtained for different months of the year (Fig. 4.22) (Fishkova et al. 2001a). From Fig. 4.22, it can be seen that the trend strongly varies from month to month and even changes its sign. The long-term intensity trend can be estimated by the approximate formula

$$I_{trS}(t,i) = [I_{MS}(i) \pm \sigma_{trS}] + [\delta I_{tr}(i) \pm \sigma_{tr}] \cdot \{t - [1972 + t_d(i)/365]\} \text{ (Rayleigh)} ,$$

for which the correlation coefficients are equal to $r_{tr} \pm \sigma_{rtr}$ and the Student factors t_{Str} are given in Table 4.2. Here t is the year's number and i is the month's number.

Table 4.2 Statistical characteristics of the seasonal dependence of the sodium emission intensity on the long-term trend

Month	I_{MS} (Rayleigh)	σ_{trS} (Rayleigh)	δI_{tr} (Rayleigh · yr^{-1})	σ_{tr} (Rayleigh · yr^{-1})	r_{tr}	σ_{rtr}	t_{Str}
1	93	12	−0.56	0.15	−0.577	0.136	3.4
2	81	12	−0.60	0.15	−0.649	0.118	4.1
3	65	14	−0.50	0.18	−0.345	0.176	1.9
4	53	14	−0.32	0.18	−0.415	0.169	2.2
5	43	12	−0.18	0.16	−0.008	0.200	0.1
6	32	10	−0.02	0.13	−0.103	0.200	0.5
7	30	13	0.10	0.17	0.122	0.193	0.6
8	43	12	0.16	0.17	0.191	0.197	0.9
9	60	16	0.27	0.21	0.291	0.183	1.5
10	88	17	0.26	0.23	0.156	0.195	0.8
11	109	15	−0.01	0.20	−0.296	0.182	1.5
12	110	10	−0.45	0.14	−0.703	0.103	4.7
Yearly average	65	7	−0.24	0.10	−0.489	0.166	2.5
Winter	102	11	−0.44	0.14	−0.550	0.142	3.2
Summer	30	8	0.10	0.11	0.187	0.211	0.9

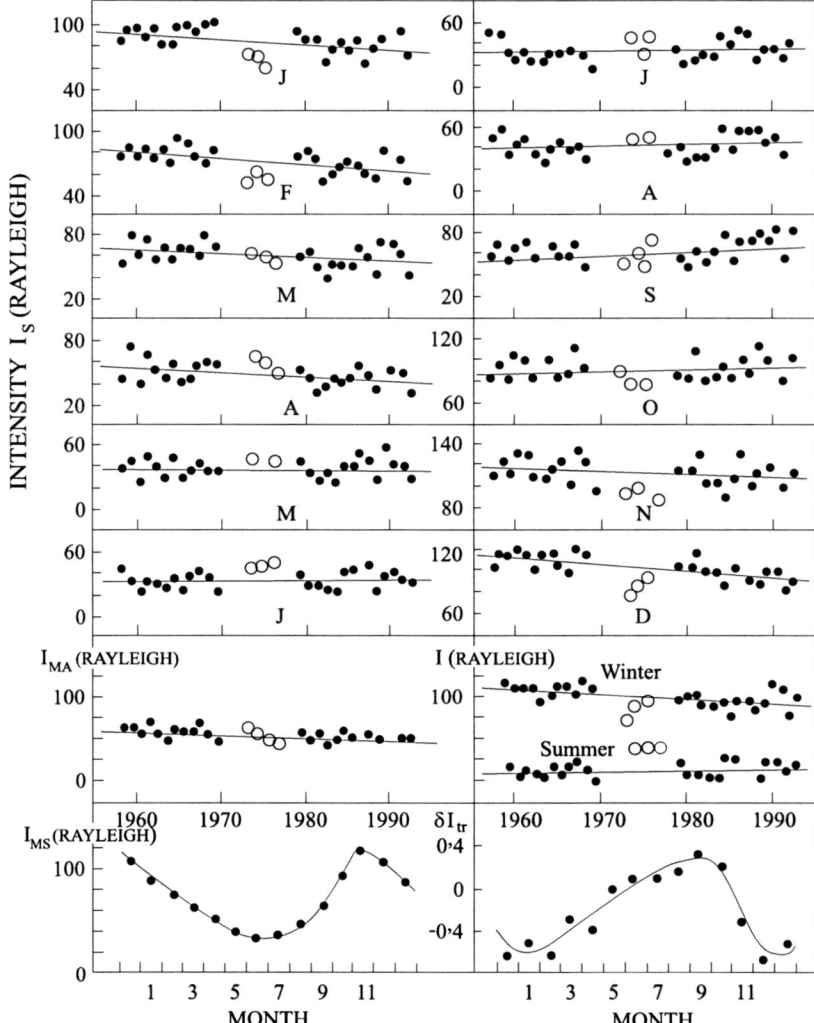

Fig. 4.22 Dependence of the monthly mean intensities of sodium emission (*full circles* (Fishkova et al. 2001a) and *open circles* (Nasyrov 2003)) on solar flux ($F_{10.7}$) for different months of the year (1 ÷ 12). The *straight lines* are regression lines. Yearly average intensities are denoted by I_{MA}; the subscripts wi and su refer to winter and summer conditions, respectively. The seasonal mean variations of intensity I_{MS} and trend δI_{tr} are shown below. The *solid lines* depict the sums of the three harmonics: annual, semiannual, and four-monthly

4.2 Model of the Sodium Emission

As follows from these data, the maximum negative trend takes place in winter. In summer, the trend is practically absent. The yearly average of the normalized intensity trend is 0.0040 ± 0.0016.

The seasonal mean variations of the normalized intensity and its long-term trend, reduced to standard heliogeophysical conditions (the year 1972.5 and $F_{10.7} = 130$), can be represented as the sums of three harmonics:

$$\Delta I_{MS} = 0.58 \cdot \cos[2\pi(t_d - 344)/365] + 0.13 \cdot \cos[4\pi(t_d - 133(/365]$$
$$+ 0.06 \cdot \cos[6\pi(t_d - 77)/365],$$
$$\delta I_{tr} = 2.7 \cdot \cos[2\pi(t_d - 218)/365] + 0.69 \cdot \cos[4\pi(t_d - 90)/365]$$
$$+ 0.07 \cdot \cos[6\pi(t_d - 35)/365].$$

For this case, we have $I_0 = 64$ (Rayleigh) and $\delta I_0 = -0.16$ (Rayleigh $\cdot$ yr^{-1}).

It can be seen that the approximations of the seasonal mean intensity variations for the above heliogeophysical conditions appear identical, to within estimation errors, for the dependences on solar flux and on long-term trend.

The long-term trend for the sodium layer altitude has been revealed based on the data of lidar and rocket measurements (Koomen et al. 1956; Tousey 1958; Packer 1961; Clemesha et al. 1992, 1993):

$$\Delta Z_{mtr} = -(0.0500 \pm 0.0067)(t - 1972.5),$$

where t is the time measured in years; $r_{tr} = -0.875 \pm 0.055$.

The mean trend for the period 1955–1996 is close to that obtained for the period 1972–1986 (Clemesha et al. 1992).

For the yearly average temperature, it was obtained that

$$\Delta T_{tr}(Na) = -0.04 \cdot (t - 1972.5)(K).$$

The seasonal temperature variations at altitudes of about 92 (km) are presented in Table 4.3 (Fishkova et al. 2001a).

Latitudinal Variations

Based on data of a number of observations, it has been obtained that

$$\Delta I_\varphi = -0.25 - 0.46 \cos 2\varphi$$

and the correlation coefficient $r_\varphi = 0.927 \pm 0.050$.

To construct this relation, the results of the analysis of seasonal variations at various stations in the northern and southern hemispheres have been used. The seasonal variations in the regions located closer to the polar areas have been revealed from satellite measurement data (Walker and Reed 1976) from which it follows that the intensity at latitudes 80–90°N is on the average 15(%) greater than that near

Table 4.3 Statistical characteristics of the seasonal dependence of the atmospheric temperature at altitudes of ~92 km on the long-term trend

Month	T_{MS} (K)	σ_T (K)	δT_{tr} (K)	σ_{tr} (K)	r_{tr}	σ_{rtr}	t_{Str}
1	202	3	−0.10	0.03	−0.577	0.136	3.4
2	199	3	−0.12	0.03	−0.649	0.118	4.1
3	197	4	−0.07	0.04	−0.345	0.176	1.9
4	195	4	−0.08	0.04	−0.415	0.169	2.2
5	192	3	−0.001	0.03	−0.010	0.200	0.1
6	191	3	0.013	0.03	0.103	0.200	0.5
7	192	3	0.021	0.03	0.122	0.193	0.6
8	194	3	0.031	0.03	0.191	0.197	1.0
9	197	4	0.06	0.04	0.291	0.183	1.5
10	203	4	0.04	0.05	0.156	0.195	0.8
11	208	4	−0.06	0.04	0.296	0.182	1.5
12	206	3	−0.13	0.03	−0.703	0.103	4.7
Yearly average	198	2	−0.04	0.01	−0.408	0.181	2.4
Winter	205	2	−0.12	0.03	−0.602	0.123	3.9
Summer	191	2	0.01	0.03	0.112	0.187	0.6

60°N. This well fits to the behavior of the amplitude of the first harmonic of the seasonal variations. There is good agreement with the data of ground-based measurements performed in the winter periods of 1991–1997 at the Eureka station (80.1°N, 82.7°W) (McEven et al. 1998).

The latitudinal variations of the sodium emission intensity revealed by Davis and Smith (1965) cannot be directly used as latitudinal variations because the measurements were conducted on a ship during the period from March to November (9 months) 1962, and covered the altitude range 67°N–62°S along the meridian ~70°W. Therefore, the data obtained were distorted by seasonal variations, which, moreover, were different in character at different latitudes. Also, the effect of the OH emission contribution within the passband of the interference optical filter could not be excluded.

Variations After Stratospheric Warmings

These intensity variations can be approximated by the formula

$$\Delta I_{sw} = 0.16 \cos \frac{\pi}{10}(t - t_{sw}) + 0.21 \cos \frac{\pi}{5}(t - t_{sw}),$$

which was obtained based on the measurements reported by Fukuyama (1977b). The time of the onset of stratospheric warming at the 10-mbar isobaric level was taken for the zero date. Here the notes made in Sect. 4.1 about the application of the superimposed epoch method to the data of Taubenheim (1969) and Taubenheim and Feister (1975) are of importance.

Variations During the Periods of Meteoric Activity

These variations are described by the formula

$$\Delta I_{mf} = e^{-\frac{(t-3)^2}{2}} \cdot \frac{\exp[10(3.5-t)] - 1}{\exp[10(3.5-t)] + 1},$$

which was obtained based on the data reported by Shefov (1968a, 1970b). In Fig. 4.18, the letters "MF" denote the time of the onset of intrusion of meteoric flows.

4.3 Model of the Molecular Oxygen Emission

In Sect. 2.4, the processes of excitation of the oxygen molecule in the upper atmosphere have been considered for all realizable band systems of the molecule. However, the ultraviolet systems were recorded in the main for the definition of their spectral characteristics. Regular measurements necessary to gain an idea of their behavior were not performed. Some conclusions can be made based on the available correlations of this emission with other emissions, such as the 557.7-nm atomic oxygen emission, that arise due to the same sequence of reactions.

Regular investigations were carried out in fact only for the O_2 (0–1) 864.5-nm emission from the Atmospheric system and the O_2 (0–0) 1268.7-nm and (0–1) 1580.8-nm emissions from the Infrared Atmospheric system. However, the Infrared Atmospheric system was investigated practically only during twilight periods.

4.3.1 Model of the Atmospheric System of Molecular Oxygen

In quiet geomagnetic conditions, the airglow of the O_2 Atmospheric system in the upper atmosphere is presented by the 760-nm (0–0) band and the 864.5-nm (0–1) band emissions arising at altitudes of 94 (km) (Greer et al. 1981). However, owing to the large optical thickness of the atmospheric oxygen layer from the Earth surface to the altitudes of emission layers, the 760-nm emission cannot be detected in ground-based observations even with the small coefficient of absorption at these forbidden transitions. Therefore, the unique detectable emission is that at 864.5 (nm) from the (0–1) band.

However, since the early 1950s, investigations of this band were performed on a regular basis only at low latitudes (Takahashi et al. 1986, 1998) and incidentally at middle and high latitudes (Berthier 1955, 1956; Dufay 1959; Berg and Shefov 1962a,b; Shefov 1971b, 1975b; Myrabø et al. 1984; Myrabø 1987; Zhang et al. 1993). Meanwhile, the emission from $O_2(b^1\Sigma_g^+)$ is a sensitive indicator of the recombination of oxygen atoms in the lower thermosphere. Thus, the variations of the characteristics of this emission can give information on the variations of the atomic oxygen density on different time scales at the base of the thermosphere and on various dynamic atmospheric processes which affect the oxygen density. The occurrence of this emission is related to the general process of excitation of the molecular oxygen ultraviolet emissions and atomic oxygen green emission.

Similar to other atmospheric emissions, the Atmospheric system airglow is characterized by intensity I, rotational temperature T_r, emission layer altitude Z_m, layer thickness W, and its asymmetry P. However, in contrast to other emissions, the parameters of the O_2 (0–1) 864.5-nm emission were measured in most cases to obtain data on their nocturnal variations. Practically no long-term regular measurement data are available. Rocket measurements of the parameters of the emission layer were rather sparse. All this restricts the possibilities of revealing various types of variations.

The parameters of variations and their types are described in the preface to Chap. 4. The mean characteristics of the 864.5-nm emission from O_2 (0–1) for the conditions of the year 2000 obtained by analyzing the data presented below are $I_0 = 320$ (Rayleigh) $T_r^0 = 190$ (K), $Z_m^0 = 93$ (km), $W_0 = 15$ (km), and $P = 0.52$. Various types of variations are presented in Figs. 4.23, 4.24, and 4.25.

Correlations Between the Emission Parameters

Mutually relating the variations of some characteristics of the 864.5-nm molecular oxygen emission, one has an important opportunity to analyze their behavior. Of equal importance are the correlations with other emissions arising in the same range of altitudes. There exists a correlation ($r = 0.886 \pm 0.039$) between intensity (Rayleigh) and temperature (K):

$$T = (160 \pm 4) + (0.077 \pm 0.007) \cdot I.$$

The data obtained by Yee et al. (1997) testify to a negative correlation between the emission intensity and the emission maximum altitude. In reality, this correlation is nonlinear and can be represented as ($r = -0.983 \pm 0.010$)

$$I = 360 \cdot \exp\left(-\frac{Z_m - 93}{2}\right) \text{ (Rayleigh)}.$$

Based on the relations obtained, it is possible to derive a relation between the emission temperature and the emission layer altitude:

$$T = 160 + 28 \cdot \exp\left(-\frac{Z_m - 93}{2}\right) \text{ (K)}.$$

Rocket measurements of the altitude profiles of the emission point to some tendencies in the behavior of the emission layer thickness W and asymmetry P, which can be described by the relations

4.3 Model of the Molecular Oxygen Emission

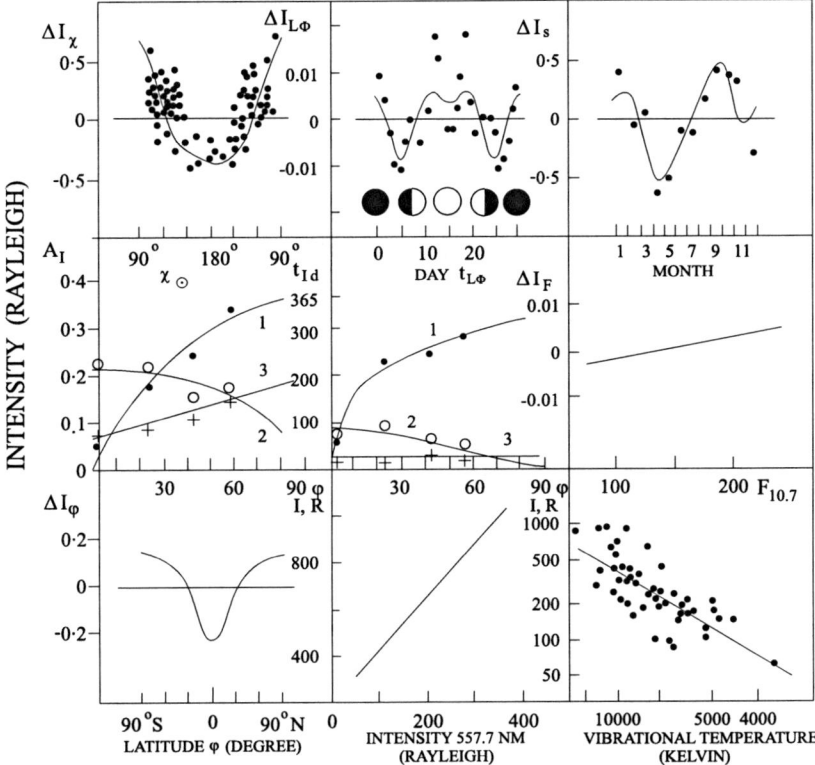

Fig. 4.23 Variations of the intensity of the 864.5-nm emission from the O_2 (0–1) atmospheric system about the mean value for standard conditions in relation to different parameters. The designations are described in the preface to Chap. 4. The *full circles* are individual measurements for diurnal variations and for comparison to the vibrational temperature of hydroxyl emission and average values in the other plots. The seasonal variations refer to measurements performed at Zvenigorod. The numbers 1, 2, and 3 are the latitudinal variations of the amplitudes and phases of the seasonal harmonics

$$W = 15 + 0.93 \cdot (Z_m - 93) \text{ (km)}, \qquad r = 0.790 \pm 0.133,$$

$$P = 0.51 \pm 0.012 \cdot (Z_m - 93), \qquad r = -0.589 \pm 0.231.$$

In addition, correlations have been revealed between the intensity of the O_2 (0–1) emission and the intensity of the atomic oxygen 557.7-nm emission (Berthier 1955; Dufay 1959; Berg and Shefov 1962a,b; Misawa and Takeuchi 1976, 1977; Takahashi et al. 1986). According to the data of the last cited work, we have

$$I_{O_2} = 192 + 2.3 \cdot I_{557.7} \text{ (Rayleigh)}, \qquad r \sim 0.9,$$

and the relation of I_{O_2} to the vibrational temperature of hydroxyl emission (Shefov 1970a; Shefov 1971b)

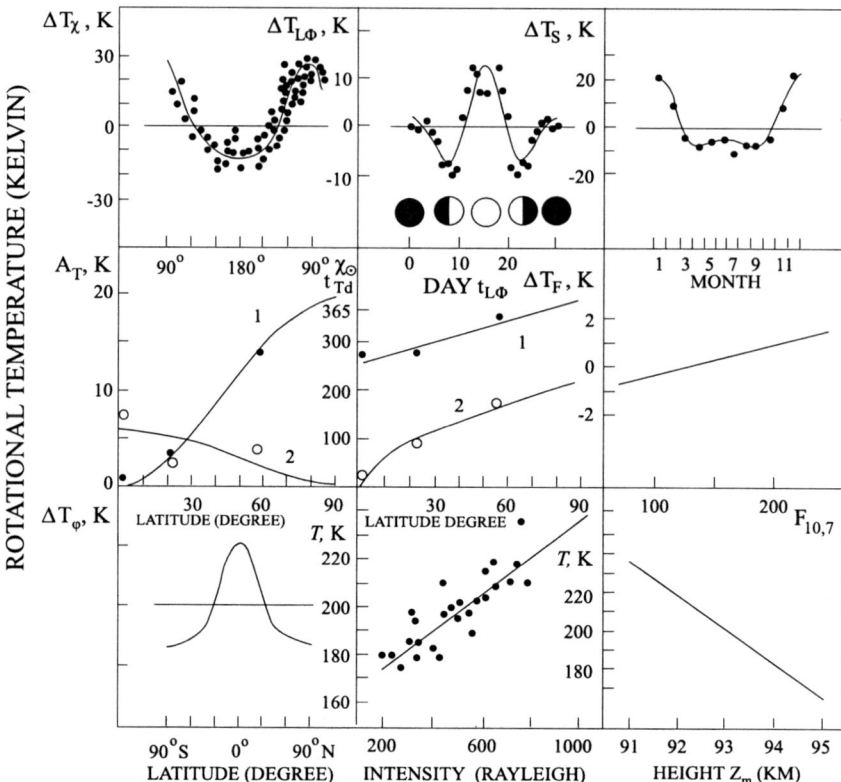

Fig. 4.24 Variations of the rotational temperature of the 864.5-nm emission from the O_2 (0–1) Atmospheric system about the mean value for standard conditions in relation to different parameters. The designations are described in the preface to Chap. 4. The *full circles* are individual measurements for diurnal variations and for comparison to the emission intensity and mean values in the other plots. The seasonal variations refer to measurements performed at Zvenigorod. The numbers 1, 2, and 3 are the latitudinal variations of the amplitudes and phases of the seasonal harmonics

$$I_{O_2} = 1200 \cdot \exp\left(-\frac{11000}{T_V}\right), (\text{Rayleigh}) \qquad r = -0.742 \pm 0.065.$$

The variations of the vibrational temperature T_V of hydroxyl emission reflect the variations of the atmospheric density, i.e., the variations of the altitude of the hydroxyl emission layer. Therefore, an increase in vibrational temperature implies an elevation of the hydroxyl layer. This leads to a greater contribution of photochemical processes to the initiation of molecular oxygen emission due to the deactivation of hydroxyl molecules vibrationally excited to over the fourth vibrational level by atomic oxygen (Witt et al. 1979).

4.3 Model of the Molecular Oxygen Emission   469

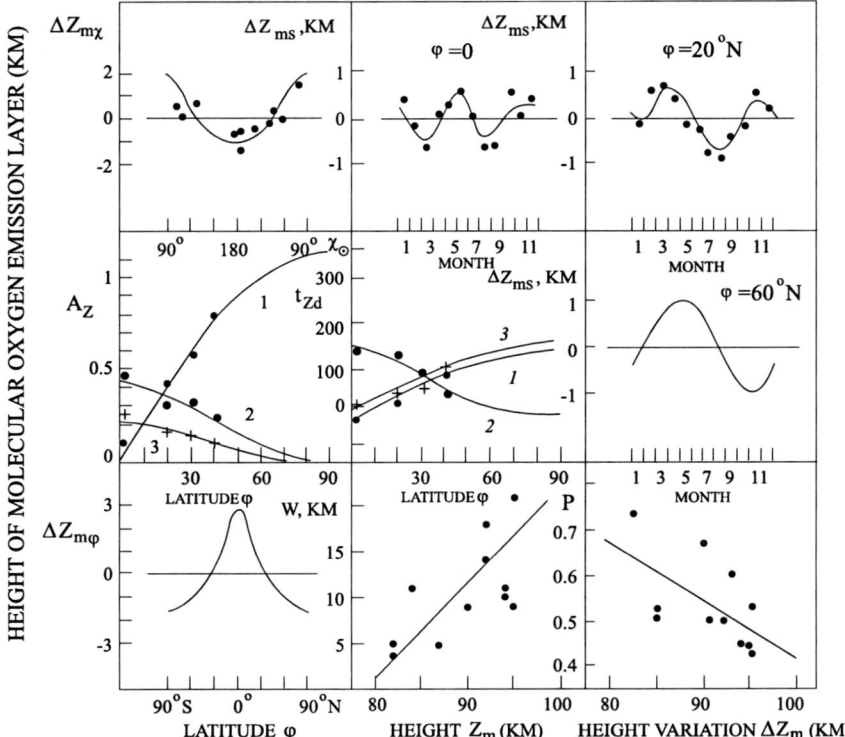

Fig. 4.25 Variations of the altitude of the maximum of the 864.5-nm emission from the O_2 (0–1) Atmospheric system about its mean value for standard conditions in relation to different parameters. The designations are described in the preface to Chap. 4. The *full circles* are individual measurements for diurnal variations and for comparison to the layer thickness and asymmetry of the altitude profile and mean values in the other plots. The seasonal variations refer to measurements performed at Zvenigorod. The numbers 1, 2, and 3 are the latitudinal variations of the amplitudes and phases of the seasonal harmonics

Nocturnal Variations

The relations for the nocturnal variations as a function of the solar zenith angle $\chi_\odot$ have been constructed based on the data of intensity and temperature measurements performed at Zvenigorod (Perminov et al. 2004, 2007) and the data reported elsewhere (Myrabø et al. 1984; Takahashi et al. 1986, 1998; Scheer and Reisin 1998, 2002). The variations of the emission layer altitude have been derived from the data of rocket (Packer 1961; Tarasova 1961, 1962; Tarasova 1963, 1967; Tarasova and Slepova 1964; Makino and Hagiwara 1971; Deans et al. 1976; Watanabe et al. 1976, 1981; Witt et al. 1979, 1984; Harris 1983; Ogawa et al. 1987; Murtagh et al. 1990; Siskind and Sharp 1991; López-González et al. 1992) and satellite measurements (Skinner et al. 1998).

The measurement data reported by Takahashi et al. (1998) show that the minimum values of the intensity and temperature of the molecular oxygen emission correspond to different points in local time. This suggests that the nocturnal variations of the emission parameters are affected by the phase shifts of the semidiurnal thermal tide resulting from the variations of the emission layer altitude (Petitdidier and Teitelbaum 1977; Takahashi et al. 1984) and by the disturbances caused by internal gravity waves.

Analysis of the data of Takahashi et al. (1998) shows that the position of the minimum in the diurnal distribution of both intensity and temperature, which takes place in the first half of a night, can be represented by the solar zenith angle $\chi_\odot^{min}$ for this point in time as a function of the day of year, t_d,

$$\chi_\odot^{min} = 135 + 35 \times \cos\frac{2\pi}{365}t_d + 10 \times \cos\frac{4\pi}{365}t_d .$$

In view of this, the variations of the emission parameters will be determined by the relations

$$\Delta I_\chi = -0.47 \cdot \cos 2(\chi_\odot - \chi_\odot^{min}) + 0.16 \cdot \cos 4(\chi_\odot - \chi_\odot^{min}) - 0.07 \cdot \cos 6(\chi_\odot - \chi_\odot^{min}),$$

$$\Delta T_\chi = -17 \cdot \cos 2(\chi_\odot - \chi_\odot^{min}) + 6 \cdot \cos 4(\chi_\odot - \chi_\odot^{min}) - 3 \cdot \cos 6(\chi_\odot - \chi_\odot^{min})(K),$$

$$\Delta Z_{m\chi} = -1.4 \cdot \cos 2(\chi_\odot - \chi_\odot^{min}) + 0.32 \cdot \cos 4(\chi_\odot - \chi_\odot^{min})$$
$$-0.16 \cdot \cos 6(\chi_\odot - \chi_\odot^{min})(km) .$$

The variations of the layer altitude have been estimated based on satellite measurements (Skinner et al. 1998) and approximated by average relations. Figures 4.23–4.25 present the data reduced to the winter solstice conditions.

For the local time before midnight, the solar zenith angle $\chi_\odot$ is calculated by the formula

$$\cos\chi_\odot = \sin\varphi \cdot \sin\delta_\odot - \cos\varphi \cdot \cos\delta_\odot \cdot \cos\tau ,$$

where τ is the local mean solar time, φ is the geographic latitude, and $\delta_\odot$ is the declination of the Sun. For the day period after midnight, the value $360 - \chi_\odot$ is used in the above formulas.

The solar zenith angle for the emission maximum, which is observed in the second half of a night, is determined by the formula

$$\chi_\odot^{max} = 130 - 35 \cdot \cos\frac{2\pi}{365}t_d - 10 \cdot \cos\frac{4\pi}{365}t_d .$$

Naturally, neither the minimum nor the maximum is detected at solar zenith angles corresponding to twilight conditions.

4.3 Model of the Molecular Oxygen Emission

Lunar Variations

The variations depending on the phase age of the Moon have been derived from measurements performed in Zvenigorod. They are described by the formulas

$$\Delta I_{L\Phi} = 0.040 \cdot \cos \frac{2\pi}{29.6}(t_{L\Phi} - 14.8)$$

$$+ 0.036 \cdot \cos \frac{4\pi}{29.6}(t_{L\Phi} - 14.8) + 0.048 \cdot \cos \frac{6\pi}{29.6} t_{L\Phi},$$

$$\Delta T_{L\Phi} = 4.1 \cdot \cos \frac{2\pi}{29.6}(t_{L\Phi} - 14.8)$$

$$+ 7.4 \cdot \cos \frac{4\pi}{29.6}(t_{L\Phi} - 14.8) + 1.7 \cdot \cos \frac{6\pi}{29.6}(t_{L\Phi} - 5) \; (K).$$

Here $t_{L\Phi}$ is the phase age of the Moon. Its rough estimate to within several tenths of day can be obtained by the formula (Meeus 1982)

$$t_{L\Phi} = 29.53 \cdot \left[\left(\frac{t_d}{365} + YYYY - 1900 \right) \cdot 12.3685 \right] - 1.$$

The square brackets denote the fractional part of the number and YYYY the year's number.

The available rocket data are insufficient to obtain information on the variations of the emission layer altitude.

Seasonal Variations

The seasonal variations have been derived in the main from the data of a few stations: Zvenigorod (55.7°N) (Berg and Shefov 1962a,b; Shefov 1975b; Perminov et al. 2004, 2007), Haute Provence (43.9°N) (Berthier 1956), El Leoncito (32°S) (Scheer and Reisin 2000), Cachoeira Paulista (22.7°S), and Fortaleza (3.9°S) (Takahashi et al. 1998). Their behavior is determined mainly by the amplitudes A and phases t of annual and semiannual harmonics:

$$\Delta f = A_1 \cdot \cos \frac{2\pi}{365}(t_d - t_1) + A_2 \cdot \cos \frac{4\pi}{365}(t_d - t_2) + A_3 \cdot \cos \frac{6\pi}{365}(t_d - t_3).$$

where for the intensity

$$A_{I1} = 0.4 \cdot \sin^{0.84}|\varphi|, \quad A_{I2} = 0.21 \cdot \cos^{0.47}\varphi, \quad A_{I3} = 0.071 + 0.133 \cdot \left(\frac{|\varphi|}{90} \right),$$

$$t_{I1} = 324 \cdot \left(\frac{|\varphi|}{90} \right)^{0.3} (\text{day}), \quad t_{I2} = 80 \cdot \cos^{1.2}\varphi(\text{day}), \quad t_{I3} = 25 \; (\text{day});$$

for the temperature

$$A_{T1} = 20 \cdot \sin^2|\varphi| \text{ (K)}, \quad A_{T2} = 6 \cdot \cos^{1.5}\varphi \text{ (K)},$$

$$t_{T1} = 267 + 126 \cdot \left(\frac{|\varphi|}{90}\right) \text{ (day)}, \quad t_{T2} = 223 \cdot \left(\frac{|\varphi|}{90}\right)^{0.6} \text{ (day)};$$

and for the altitude of the emission layer maximum

$$A_{Z1} = 1.15 \cdot \sin^{0.88}|\varphi| \text{ (km)}, \quad A_{Z2} = 0.43 \cdot \cos^{2.4}\varphi \text{ (km)},$$

$$A_{Z3} = 0.18 \cdot \cos^{2.2}\varphi \text{ (km)};$$

$$t_{Z1} = 140 \cdot \sin|\varphi| \text{ (day)}, \quad t_{Z2} = 148 \cdot \cos^{3.5}\varphi \text{ (day)},$$

$$t_{Z3} = 140 \cdot \sin|\varphi| + 20 \text{ (day)}.$$

Solar Activity

The relations of the emission parameter variations to solar flux were estimated based on the data obtained at Zvenigorod (Perminov et al. 2007) and El Leoncito (Scheer et al. 2005):

$$\Delta I_F = 0.05 \cdot \frac{F_{10.7} - 130}{100},$$

$$\Delta T_F = 1.3 \cdot \frac{F_{10.7} - 130}{100} \text{(K)}.$$

Latitudinal Variations

The yearly average dependence of the variations on latitude can be estimated from the yearly average emission intensities derived from the data obtained at the mentioned stations and from the UARS satellite data (Skinner et al. 1998). Since these data are available only as isophots for the period of measurements from March 1994 to April 1995, the obtained relations for the latitudinal variations are approximate in character:

$$\Delta I_\varphi = 0.14 \cdot \cos 2\varphi + 0.055 \cdot \cos 4\varphi + 0.023 \cdot \cos 6\varphi,$$

$$\Delta T_\varphi = 3.4 \cdot \cos 2\varphi + 1.4 \cdot \cos 4\varphi + 0.6 \cdot \cos 6\varphi \text{ (K)},$$

$$\Delta Z_{m\varphi} = 1.84 \cdot \cos 2\varphi + 0.45 \cdot \cos 4\varphi + 0.42 \cdot \cos 6\varphi \text{ (km)}.$$

The temperature variations have been estimated based on the correlation formula presented in the previous sections.

Long-Term Variations

The available measurement data cover small time intervals of continuous measurements. The estimates obtained for the period 1998–2002 (Scheer et al. 2005) give the relations for the intensity trend

$$\Delta I_{tr} = -0.015 \cdot (t - 1998)$$

and for the temperature trend

$$\Delta T_{tr} = 0.5 \cdot (t - 1998), (K \cdot yr^{-1}) \ .$$

Estimates based on the measurements performed at Zvenigorod in the years of practically identical solar activity give the intensity trend value ∼0.002.

The results of the investigations of the altitude distributions of temperature trends (Semenov et al. 2002a) show that at altitudes of 90–95 (km) the trend is practically equal to zero throughout the year. The data on the sodium emission that arise at these altitudes also point to the intensity and temperature trends close to zero.

4.4 Model of the 557.7-nm Atomic Oxygen Emission

The 557.7-nm atomic oxygen emission – the nightglow green emission – was historically the first detected airglow emission. It arises at altitudes of ∼100 (km) due to recombination of atomic oxygen. The half-thickness of its emission layer is ∼10 (km). The dissipation of internal gravity waves in this altitude range significantly adds dynamism to this emission. The photochemical processes responsible for its initiation were considered in Chap. 2.

The results of investigations of the 557.7-nm emission carried out during several tens of years at various stations have been systematized to reveal regular variations of some its characteristics such as intensity I, Doppler temperature T, and emission layer maximum altitude Z_m. This is necessary to analyze the fast variations occurring on the background of the regular behavior of airglow characteristics, to reveal the irregular variations caused by various sporadic phenomena, and to construct a model of the behavior of atomic oxygen at these altitudes.

The first attempt to do this was undertaken by Shefov and Kropotkina (1975) who have analyzed the data available at that time. Regular variations of the emission layer maximum altitude and the emission intensity depending on the time of day, solar zenith angle $\chi_\odot$, season t_d, and solar flux $F_{10.7}$ have been revealed. Subsequently, the spectrum of types of variations of the 557.7-nm emission parameters (intensity I, Doppler temperature T, and emission layer maximum altitude Z_m) was extended substantially and empirical relations describing the behavior of the newly detected variations were constructed.

The data of ground-based photometric measurements presented in a number of publications refers in the main to the behavior of the emission intensity and, mainly, to middle latitudes. Based on this material, data on various types of regular variations have been obtained. The behavior of the emission parameters associated with various sporadic disturbances was much less investigated.

The variations of the temperature inside the emission layer have been considered mainly based on interferometric measurements of the Doppler temperature (Hernandez 1976, 1977; Hernandez and Killeen 1988) and on lidar measurements of the atmosphere temperatures at altitudes ~97 (km) (She et al. 1993). From the investigations performed, it follows (Shefov and Kropotkina 1975) that the variations of the layer altitude are reflected in the variations of emission intensity I, temperature T, layer thickness W, and its asymmetry P.

Because of the limited body of data, it is supposed that the behavior of the emission characteristics in the northern and southern hemispheres is identical for a given season, and this is supported by the UARS satellite measurements (Ward 1999). The longitudinal variations should be taken into account in going from average heliogeophysical conditions to concrete situations in which measurements were performed. The similarity of seasonal variations follows from the early work by Lord and Spencer (1935). Though some difference in data for the northern and the southern hemispheres was mentioned in some publications (Birnside and Tepley 1990), it seems that their account will be possible upon improvement of the given model.

There is some uncertainty in the data available for the transition equatorial zone where seasonal variations in one hemisphere should change into seasonal variations in the other with a half-year shift. The measurement data for the near-polar region are few in number, and they correspond only to the winter time. At these latitudes, auroral processes distort the nightglow, which arises only due to recombination of atomic oxygen.

It has long been known that the 557.7-nm emission is a manifestation of the behavior of atomic oxygen in the upper atmosphere. Recombination of atomic oxygen gives rise to the airglow (Bates 1981). According to numerous investigations, irrespective of the process responsible for the occurrence of the emission by the Chapman or Bart mechanism (Bates 1981), the emission intensity $I_{557.7}$ is proportional to the cubed atomic oxygen density (Reed and Chandra 1975; Bates 1981; Rao et al. 1982; Greer et al. 1986; McDade and Llewellyn 1986). This is also supported by occurrence probability distribution of the cubic root (and also the logarithm) of the observed emission intensity that is close to a normal distribution (Fig. 4.26) (Korobeynikova and Nasyrov 1972).

The most numerous measurements of the emission intensity were performed between 1923 and 1934 at Terling (52°N, 1°W) (Lord and Spencer 1935; Hernandez and Silverman 1964) and between 1948 and 1992 at Abastumani (41.8°N, 42.8°E) (Fishkova 1983; Fishkova et al. 2000, 2001b; Toroshelidze 1991), Ashkhabad (38.0° N, 58.4°E) (Korobeynikova and Nasyrov 1972), Haute Provence (43.9°N, 5.7°E) (Dufay and Tcheng Mao-Lin 1946, 1947a,b; Barbier 1959b; Christophe-Glaume 1965), and in Japan (38.1°N, 140.6°E) (Takahashi and Okuda 1974). Some measurement data are mentioned in other publications (Christophe-Glaume 1965; Korobeynikova and Nasyrov 1972; Fishkova 1983; Toroshelidze 1991; Takahashi

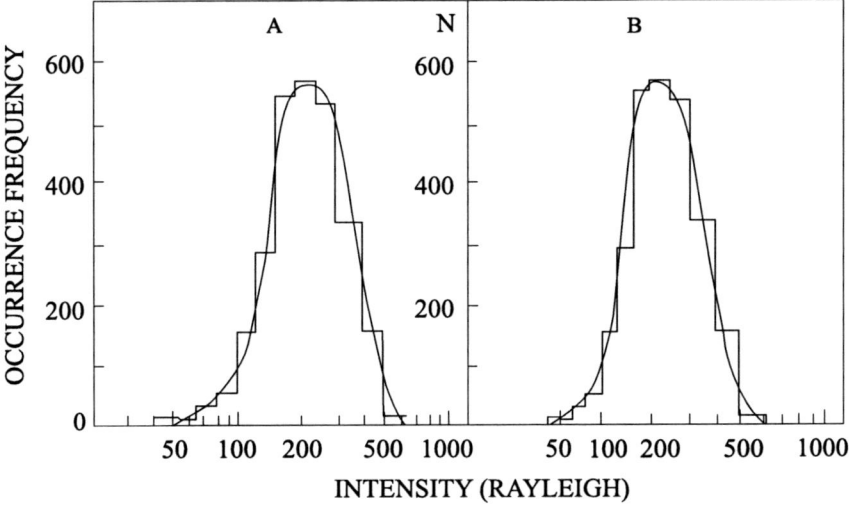

Fig. 4.26 Frequency of distribution of occurrence of intensity of emission 555.7 (nm) depending on the decimal logarithm of the intensity (Rayleigh) (**A**) and of the cubic root of the intensity (**B**) (Korobeynikova and Nasyrov 1972)

and Okuda 1974; Fukuyama 1976, 1977a,b). A review of the relevant investigations was made by Silverman (1970).

Interferometric measurements of the Doppler temperature based on the 557.7-nm emission records were carried out rather incidentally and for short periods. The longest data series (1965–1974) are available from measurements performed at the Fritz Peak Observatory (39.9°N, 105.5°W) (Hernandez 1976, 1977; Hernandez and Killeen 1988). Their total duration is much shorter than that of the intensity data. The UARS satellite measurements are presented in the literature only for a few cases (Shepherd et al. 1993a). The lidar data on the temperature for these altitudes are also not numerous (She et al. 1993).

According to the available publications, rocket measurements of the altitude profiles of the 557.7-nm emission were carried out from 1955 to the late 1990s. For this period it was informed about more than 40 rocket launches and also on satellite measurements. However, for seven of them, no data of measurements near the mesopause were reported. The majority of measurements (17) were performed in the latitude range 30–40°N. For these latitudes, it has been shown (Shefov and Kropotkina 1975) that the mean altitude of the emission layer maximum, without reference to concrete heliogeophysical conditions, was 98 (km) and the layer thickness was 8 (km).

Ten measurements were made in the latitude range 45–70°N. However, two of them gave the layer maximum altitude below 90 (km), which, in view of the available data on the nocturnal altitude distribution of atomic oxygen (in quiet geomagnetic conditions at the midlatitudes), seems to be improbable. Thus, the data of only about 30 measurements can be used for analysis.

However, as already mentioned (Semenov and Shefov 1997a), the dynamic structure of the emission layer on small time scales is rather disturbed. The vertical and the horizontal structures of the emission layer have been revealed (Ross et al. 1992) by the records of the Herzberg I bands of the 557.7-nm O_2 emission arising due to the general photochemical process. These data explain why the correlation coefficient is 0.6 when the layer parameters W, P, and I are related to Z_m.

Based on the most reliable rocket data, significant correlations have been obtained:

$$\Delta W/\Delta Z_m = 0.52 \ (r = 0.54),$$

$$\Delta P/\Delta Z_m = 0.016 (km^{-1}) \ (r = 0.57),$$

$$\Delta I/\Delta Z_m = -4.2 (\% \cdot km^{-1}) \ (r = -0.56),$$

$$\Delta T/\Delta Z_m = -2 (K \cdot km^{-1}) \ (r = -0.5),$$

$$\Delta T/\Delta I = 0.48 (K \cdot (\%)^{-1}) \ (r = 0.95).$$

It should be borne in mind that in relating the values of Z_m to the values of I and T, the latter were calculated by the empirical approximate formulas considered below that were obtained for concrete heliogeophysical conditions under which rocket measurements of the emission layer altitudes were performed.

Thus, from the aforesaid it follows that all characteristics of the emission for the mentioned heliogeophysical conditions have the following planetary average values:

$$I^0_{557.7} = 270 \text{ Rayleigh}, \ T = 196 \text{ (K)}, \ Z^0_m = 97 \text{ (km)},$$
$$W_0 = 9 \text{ (km)}, \ P_0 = 0.61, \sigma_0 = 0.35, \ S_0 = 0.78.$$

Horizontal Eddy Diffusion

The IGWs propagating in the mesopause region and their dissipation at altitudes of about 100 (km) form a turbulence zone which clearly manifests itself in irregularities of the spatial distribution of airglow parameters. This spotty structure of the airglow was studied based on the intensity of the 557.7-nm atomic oxygen green emission measured during IGY (1957) and in subsequent years at a number of mid-latitude stations, including Ashkhabad (Fig. 4.27) (Truttse 1965; Korobeynikova et al. 1966, 1968, 1970, 1972; Korobeynikova and Nasyrov 1972). Subsequent investigations gave interesting information on the statistical characteristics of this atmospheric region. It has been found (Nasyrov 1967a,b; Korobeynikova and Nasyrov 1972) that the spots vary in size from 10 to 1000 (km) and that the mean size is about 90 (km). The mean size of large-scale irregularities is about 1000 (km). Once appeared, the spots develop, increasing in size, and, finally, break up into smaller irregularities. The most probable mean size of a spot is about 75 (km). However, it has been shown (Korobeynikova et al. 1984) that the variations of the layer maximum altitude between 92 and 103 (km), depending on season and time of day, are accompanied by variations of the layer thickness W between 5 and 12 (km) and

4.4 Model of the 557.7-nm Atomic Oxygen Emission

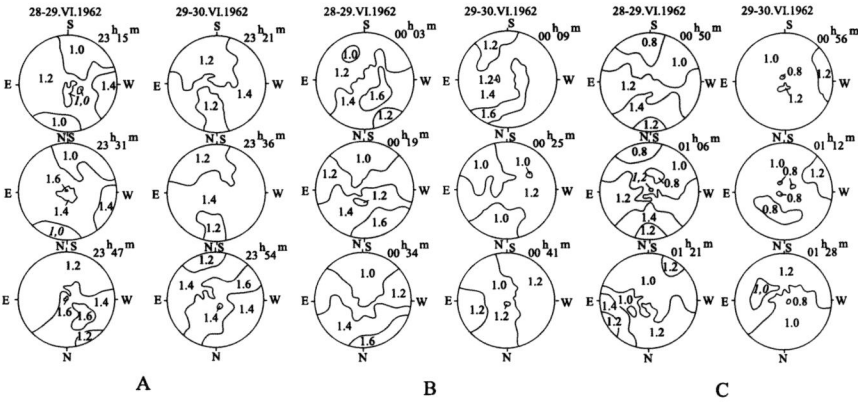

Fig. 4.27 Maps of the spatial distribution of the 557.7-nm emission intensity spots (relative units) during at close points in sidereal time for two neighboring nights, plotted by observations at Ashkhabad (Truttse 1965). The diameter of the observation region of the emission layer is 600 (km)

by an increase in spot mean size from 75 to 110 (km). For this case, the correlation coefficient is about 0.7. Since the medium-scale irregularities appear due to wave processes, the spotty structure is also observed in the lower hydroxyl emission layers. An interesting illustration of the spotty structure of the upper atmosphere is provided by the DE-1 satellite observations of the 130.4-nm atomic oxygen dayglow emission (Frank et al. 1986). The observed emission corresponds to altitudes of about 200 (km). Evaluation of the sizes of the spotty image cells gives a mean spot size of about 200 (km). Krassovsky and Semenov (1987) suggested that this structure might be caused by IGWs whose presence at these altitudes was detected by investigations of artificial luminous clouds (Kluev 1985).

The behavior of the spot luminance possesses some features. The amplitude of intensity of a luminescence of spots is 20–30% of the mean value of a background, but sometimes it is greater. The spots move simultaneously varying in sizes. These properties are accounted for by eddy diffusion. As this takes place, the turbopause region appreciably varies in altitude (Yee and Abreu 1987; Shashilova 1983) and experiences different variations on varying solar flux at different latitudes (Korsunova et al. 1985). Therefore, the 557.7-nm emission layer, which is in the bottom of the turbopause, offers the possibility of identifying the mixing processes occurring in this important atmospheric region. According to the diffusion law (Marchuk 1982), the area of a spot is given by the formula

$$S = 2\pi \cdot K_H \cdot t,$$

where K_H is the coefficient of horizontal eddy diffusion (~ 1 $(km^2 \cdot s^{-1})$) and t is the time. This law is confirmed by measurements of the spot area variations (Fig. 4.28) (Korobeynikova et al. 1984). As follows from Fig. 4.28, the horizontal and the vertical scales of fluctuations is 10–1000 (km) and 1–5 (km), respectively, and the

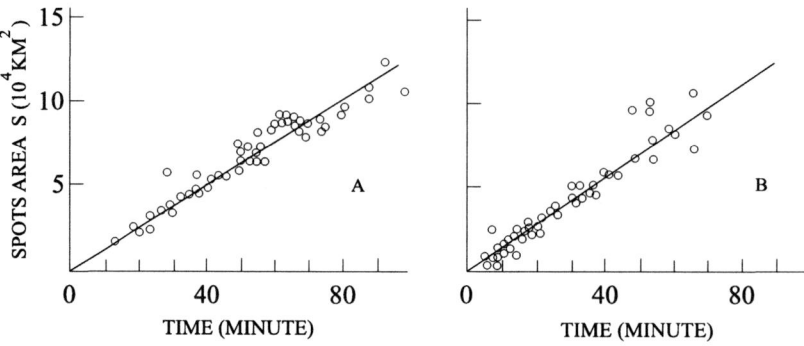

Fig. 4.28 Time dependence of the effective area of disturbance spots in the emission layer at the level 1/e derived from average profiles of $\sqrt[3]{I_{557.7}}$ ($K_H = 1.7 \cdot 10^6$ (m$^2 \cdot$ s^{-1})) (**A**) and from isophot maps ($K_H = 2.6 \cdot 10^6$ (m$^2 \cdot$ s^{-1})) (**B**) (Korobeynikova et al. 1984)

time scale is 1000–10 000 (s) (Korobeynikova et al. 1984). This variability is due to the peculiarity of the atmosphere near the turbopause: the presence of a turbulized layer of thickness 10–20 (km) at altitudes of about 100 (km).

Relating the nightly mean emission intensity variations, the altitude Z correlating with emission intensity, and the mean spot size, L, one can obtain the empirical formula

$$L = 132 \cdot \left(\frac{Z}{56} - 1\right) \text{ (km)}.$$

According to the available data, the average value $K_H^0 = 2.5 \cdot 10^6$ (m$^2 \cdot$ s^{-1}) corresponds to $Z_0 \sim 97$ (km) and $L_0 \sim 97$ (km). Therefore,

$$K_H = 4.64 \cdot \left(\frac{Z}{56} - 1\right)^2 \cdot 10^6 \text{ (m}^2 \cdot \text{s}^{-1}\text{)}.$$

Figure 4.29 presents K_H as a function of altitude for some months. The faltering line depicts calculations by the formula obtained; the solid line corresponds to data of Ebel (1980) for the summer period (Korobeynikova et al. 1984).

The vertical component of the mean temperature gradient at these altitudes is positive, resulting in suppression of the vertical motion. Based on measurement data, the rate of rise of the spot area is on the average

$$\frac{dS}{dt} = 400 \cdot \text{(km}^2 \cdot \text{min}^{-1}\text{)}.$$

The time interval in which a spot develops is several tens of minutes, and the spot size reaches 200–300 (km).

The coefficient of vertical eddy diffusion K_Z, according to the available estimates, is about three orders of magnitude lower than the coefficient of horizontal diffusion (Chunchuzov and Shagaev 1983).

4.4 Model of the 557.7-nm Atomic Oxygen Emission

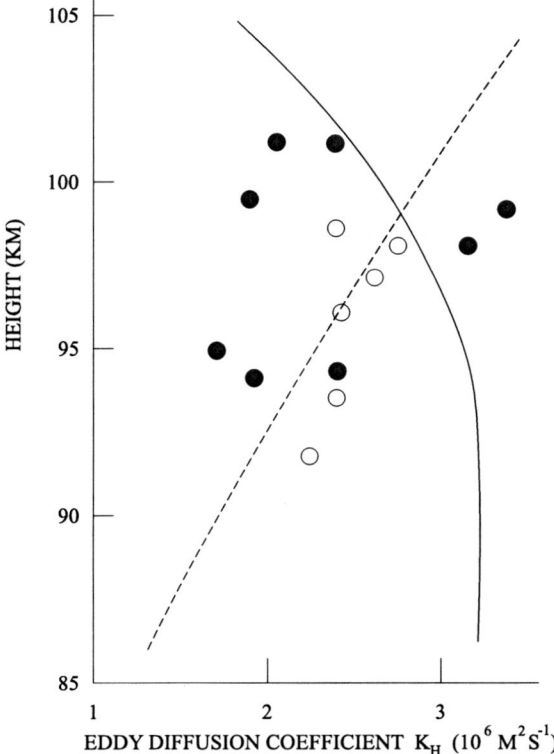

Fig. 4.29 Altitude dependence of K_H (Korobeynikova et al. 1984). *Full circles* are July–August; *open circles* are September–November; *dashed line* is average empirical dependence, and *solid line* is data of Ebel (1980)

A tentative statistical systematization of the frequency distributions of the spot intensities, sizes, and velocities of motion has shown that they are asymmetric and can be well approximated by lognormal distributions (Semenov and Shefov 1989). In other words, normal distributions involve logarithms of the spot intensity, size, and velocity and of the thickness of the emission layer.

$$P(\Delta I) = \frac{0.42}{1+\Delta I} \exp\left[-(4 \cdot \log_{10}(1+\Delta I))^2\right]$$

$$= \frac{0.42}{1+\Delta I} \exp\left[-(1.74 \cdot \log_e(1+\Delta I))^2\right] (\%/\%),$$

$$P(L) = \frac{90}{L} \exp\left[-\left(3.8 \cdot \log_{10}\frac{L}{90}\right)^2\right]$$

$$= \frac{90}{L} \exp\left[-\left(1.65 \cdot \log_e\frac{L}{90}\right)^2\right] (\%/km),$$

$$P(V) = \frac{62}{V} \exp\left[-\left(2.5 \cdot \log_{10} \frac{V}{150}\right)^2\right]$$

$$= \frac{62}{V} \exp\left[-\left(1.09 \cdot \log_e \frac{V}{150}\right)^2\right] \ (\%/(m/s)),$$

$$P(W) = \frac{9}{W} \exp\left[-\left(4 \cdot \log_{10} \frac{W}{9}\right)^2\right]$$

$$= \frac{9}{W} \exp\left[-\left(1.74 \cdot \log_e \frac{W}{9}\right)^2\right] \ (\%/km).$$

Here the change of the spot luminance ΔI is expressed in percentage, the spot size L in (km), the spot velocity V in (km/s), and the layer thickness W in (km).

It should be stressed that theoretical considerations of the diffusion of a passive impurity in a random field of velocities lead to the conclusion that the frequency functions of the rates of change of the observable atmospheric characteristics are described by lognormal distributions (Klyatskin 1994).

The actual small-scale space–time structure of the emission layer is smoothed by the angular aperture of the measuring device (Shefov 1989). Nevertheless, it inevitably differs from its average representation for which an analytic approximation is used.

All relations presented below have been obtained based on the statistical analysis of a given type of variations by eliminating other types of variations. In the figures that show various types of variations, the points, as a rule, represent mean values for given arguments.

Nocturnal Variations

The values of the emission parameters obtained during the night time of day have been systematized not on the local time scale, but as functions of the solar zenith angle. This has made it possible to combine the data of measurements performed during different seasons.

The results are shown in Figs. 4.30, 4.31, and 4.32. The nocturnal variations can be described by the relations

$$\Delta I \chi_\odot = -0.22 - 0.39 \cdot \cos(\chi_\odot - 20) + 0.1 \cdot \cos(2\chi_\odot + 25),$$

$$\Delta T \chi_\odot = -11 - 19 \cdot \cos(\chi_\odot - 20) + 5 \cdot \cos(2\chi_\odot + 25)(K),$$

$$\Delta Z \chi_\odot = 5.5 + 9.5 \cdot \cos(\chi_\odot - 20) - 2.5 \cdot \cos(2\chi_\odot + 25)(km),$$

$$\Delta W \chi_\odot = 2.8 + 4.9 \cdot \cos(\chi_\odot - 20) - 1.3 \cdot \cos(2\chi_\odot + 25)(km).$$

4.4 Model of the 557.7-nm Atomic Oxygen Emission

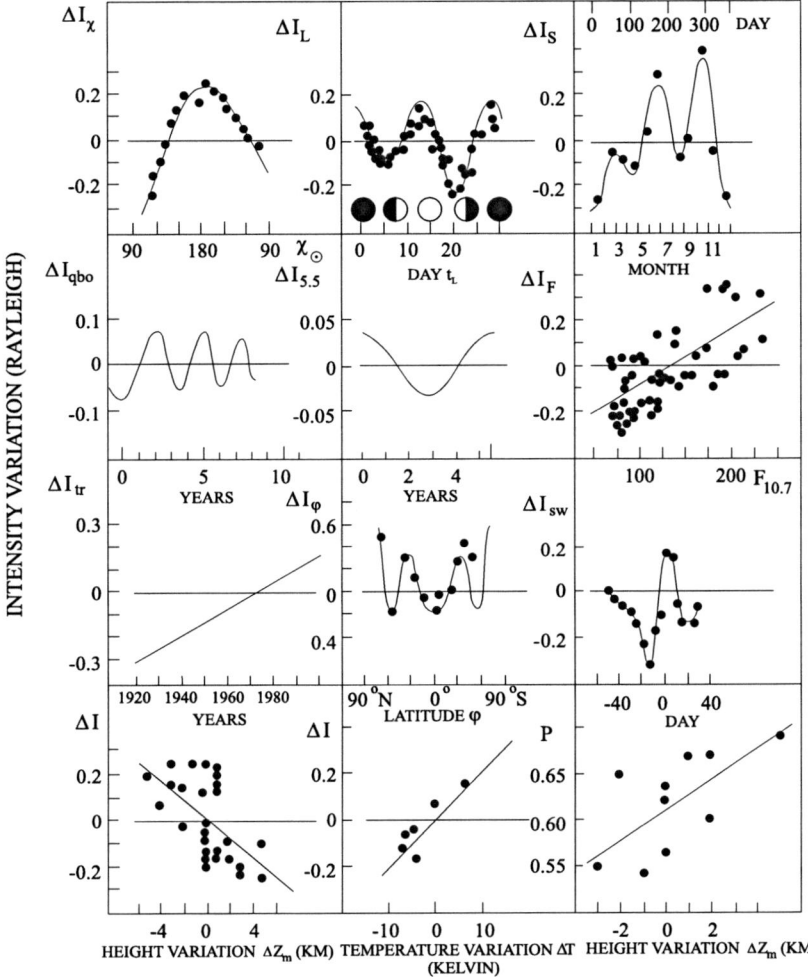

Fig. 4.30 Intensity variations of the 557.7-nm atomic oxygen emission about its mean value for standard conditions in relation to different parameters (Semenov and Shefov 1997a,d). The designations are described in the preface to Chap. 4

The solar zenith angle $\chi_\odot$ is determined by the formula

$$\cos\chi_\odot = \sin\varphi \cdot \sin\delta_\odot - \cos\varphi \cdot \cos\delta_\odot \cdot \cos\tau ,$$

where $\delta_\odot$ is the declination of the Sun and τ is the local solar time. To describe some asymmetry of the emission variations in relation to $\chi_\odot$ when calculating their values for the evening time ($\tau \leq 24$ (h)), $\chi_\odot$ is estimated using the above formula. For the morning part of day ($\tau \geq 0$ (h)), the value of $\chi_\odot$ is taken with the minus sign.

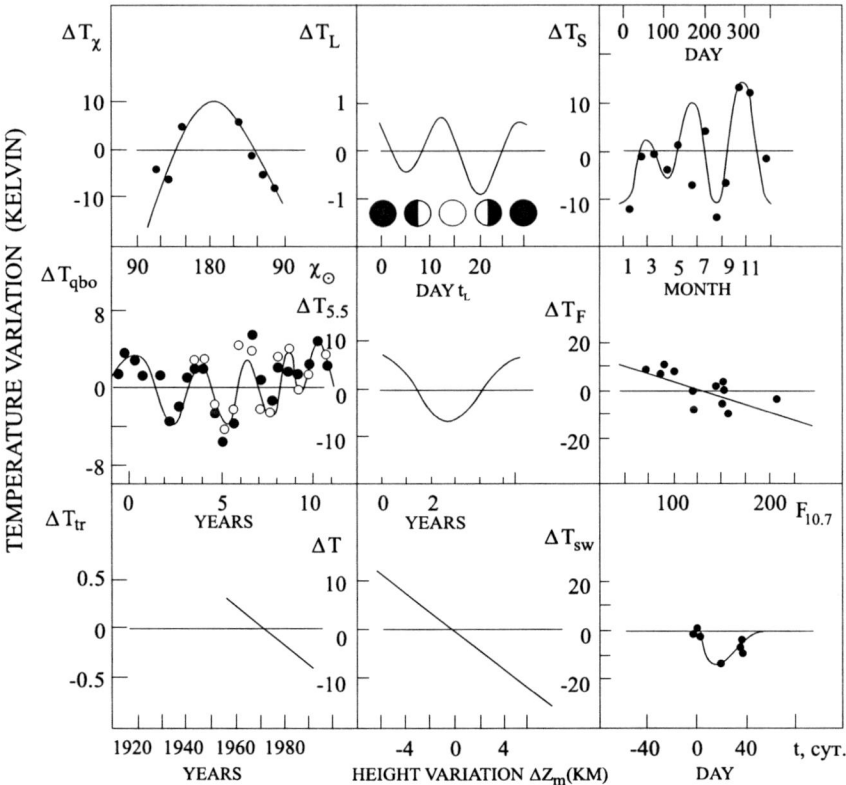

Fig. 4.31 Variations of the 557.7-nm atomic oxygen emission layer temperature (K) about its mean value for standard conditions in relation to different parameters (Semenov and Shefov 1997b,d). The designations are described in the preface to Chap. 4. The *open circles* in the plot for ΔT_{qbo} are Doppler temperature measurements (Yugov et al. 1997)

From measurement data obtained at various latitudes, it has been revealed that on the average there is a nightly maximum whose position in time varies during a year (Christophe-Glaume 1965; Fukuyama 1976; Rao et al. 1982; Birnside and Tepley 1990). It has been suggested (Petitdidier and Teitelbaum 1977) that this character of variations is related to the variation of the phase of the solar thermal semidiurnal tide, responsible for the character of the nocturnal variations, with the emission layer altitude. The relations of the parameters to $\chi_\odot$ have been constructed based on the data reported in the above publications.

From the data presented by Takahashi et al. (1984), it follows that the maximum of the semidiurnal component falls to various points in local time τ_0 during a year. An approximation yields

$$\tau_0 = 3 - 3\cos\frac{2\pi}{365}(t_d - 15).$$

4.4 Model of the 557.7-nm Atomic Oxygen Emission

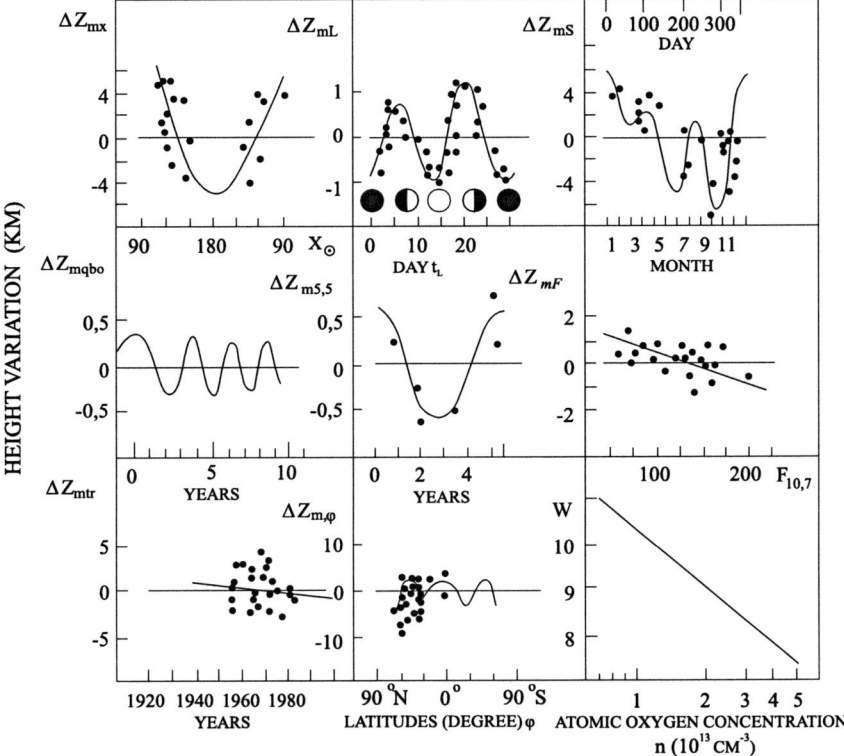

Fig. 4.32 Variations of the altitude (km) of the 557.7-nm atomic oxygen emission maximum about its mean value for standard conditions in relation to different parameters (Semenov and Shefov 1997c,d). The designations are described in the preface to Chap. 4

However, the seasonal variations of the emission layer altitude that should reflect the seasonal variations of τ_0 are much more intricate in character. Therefore, these data need to be refined.

Lunar Variations

These variations are described by the formulas

$$\Delta I_{L\Phi} = 0.007 \cdot \cos \frac{2\pi}{29.53}(t_{L\Phi} - 5.5) - 0.018 \cdot \cos \frac{4\pi}{29.53}(t_{L\Phi} - 5.5),$$

$$\Delta T_{L\Phi} = 0.26 \cdot \cos \frac{2\pi}{29.53}(t_{L\Phi} - 5.5) - 0.7 \cdot \cos \frac{4\pi}{29.53}(t_{L\Phi} - 5.5) \text{ (K)},$$

$$\Delta Z_{mL\Phi} = -0.3 \cdot \cos \frac{2\pi}{29.53}(t_{L\Phi} - 5.5) + 1 \cdot \cos \frac{4\pi}{29.53}(t_{L\Phi} - 5.5) \text{ (km)},$$

$$\Delta W_{L\Phi} = -0.17 \cdot \cos\frac{2\pi}{29.53}(t_{L\Phi} - 5.5) + 0.52 \cdot \cos\frac{4\pi}{29.53}(t_{L\Phi} - 5.5) \text{ (km)},$$

where $t_{L\Phi}$ is the phase age of the Moon (days). The Moon's age can be roughly estimated to be within several tenths of day by the formula (Semenov and Shefov 1996a, 1999a)

$$t_{L\Phi} = \left[\left(\frac{t_d}{365} + YYYY - 1900\right) \cdot 12.3685\right] \cdot 29.53 - 1.$$

Here YYYY is the year's number, the square brackets imply the fractional part of the number, and t_d is the day of year.

Variations with a period equal to the phase age of the Moon, i.e., the synodical month of 29.53 days, have been revealed for the midlatitudes (Fishkova 1983), and they seem to be manifestations of yearly average conditions (see Figs. 4.30–4.32).

The origin of these fluctuations in the upper atmosphere is accounted for by planetary waves (with periods ~15 and 30 days) propagating in the lower atmosphere (Reshetov 1973). These waves arise due to parametrically excited atmospheric circulation.

According to some investigations (Fishkova 1983), the diurnal lunar variations are small, making only 1–2(%) of the mean intensity of the 557.7-nm emission. Therefore, they are difficult to reliably reveal on the background of other variations. Here it should be noted that in constructing the lunar-tidal variations mentioned in Sect. 4.1, it is necessary to take into account the declination of the Moon.

Seasonal Variations

The seasonal variations of the parameters of the atomic oxygen green emission depend not only on season but also on geographic latitude, long-term trend, and solar activity.

Based on the results of long-term measurements performed at Abastumani (Fishkova 1983; Fishkova et al. 2000, 2001b), the following formula has been obtained for the seasonal variations of the emission intensity at midlatitudes (see Fig. 4.30):

$$\Delta I_S = 0.176 \cdot \cos\frac{2\pi}{365}(t_d - 230) + 0.094 \cdot \cos\frac{4\pi}{365}(t_d - 120) + 0.184 \cdot \cos\frac{6\pi}{365}(t_d - 50).$$

The formula for the seasonal temperature variations has been obtained based on the data of Hernandez (1976,1977) after elimination of the dependence on solar flux. The correlation coefficient between ΔI and Δt is ~0.89. It should be noted that the mentioned character of the seasonal variations and the approximations presented (see Fig. 4.31) have been derived from the data on the corresponding seasonal intensity variations by the above formula, but not from the data reported by Hernandez and Killeen (1988). For the temperature variations, the following relation has been obtained:

4.4 Model of the 557.7-nm Atomic Oxygen Emission

$$\Delta T_S = 2.74 \cdot \cos\frac{2\pi}{365}(t_d - 230) + 4.67 \cdot \cos\frac{4\pi}{365}(t_d - 120) + 9.94 \cdot \cos\frac{6\pi}{365}(t_d - 50).$$

The seasonal variations of the emission layer altitude have been derived by analyzing rocket measurements upon elimination of the diurnal variations and solar flux dependence. It should also be stressed that the approximations of the seasonal variations of the layer altitude and thickness (see Fig. 4.32) have been made by analogy with the seasonal variations of the emission intensity. As can be noticed, the greater the altitude of the layer, the lower the emission intensity and the temperature in the layer (Shefov et al. 2000a). For these variations, the following relations have been obtained:

$$\Delta W_S = -2.17 \cdot \cos\frac{2\pi}{365}(t_d - 230) - 1.16 \cdot \cos\frac{4\pi}{365}(t_d - 120)$$

$$-2.26 \cdot \cos\frac{6\pi}{365}(t_d - 50),$$

$$\Delta Z_{mS} = -3.52 \cdot \cos\frac{2\pi}{365}(t_d - 230) - 1.88 \cdot \cos\frac{4\pi}{365}(t_d - 120)$$

$$-3.68 \cdot \cos\frac{6\pi}{365}(t_d - 50),$$

$$\Delta P = -0.056 \cdot \cos\frac{2\pi}{365}(t_d - 230) - 0.030 \cdot \cos\frac{4\pi}{365}(t_d - 120)$$

$$-0.059 \cdot \cos\frac{6\pi}{365}(t_d - 50).$$

The use of data for other latitudes has made it possible to reveal the dependences of the seasonal variations on latitude (Fig. 4.33), solar flux (Fig. 4.34), long-term trend (Fig. 4.35) (Fishkova et al. 2000, 2001). In view of these dependences, the seasonal variations for the mentioned parameters can be presented as

$$I_S = I_{S0}(\varphi) \cdot I_{SA} \cdot I_{Str} \cdot I_{SS},$$

$$I_{S0}(\varphi) = 60 + 220 \cdot [\cos 1.8(\varphi - 40)]^{4.5} \text{ (Rayleigh)},$$

$$I_{SA} = \left[1 + 0.22 A_1(\varphi) \cos\frac{2\pi}{365}(t_d - t_\varphi) + 0.11 A_2(\varphi) \cos\frac{4\pi}{365}(t_d - 120)\right.$$

$$\left. + 0.23 A_3(\varphi) \cos\frac{6\pi}{365}(t_d - 50)\right],$$

where the amplitudes of the latitudinal harmonics, normalized for the harmonic amplitude at latitude 40°N, are given by the relations

$$A_1(\varphi) = (1.3 \pm 0.13) \cdot [\sin\varphi]^{0.6 \pm 0.10}, A_2(\varphi) = (2.60 \pm 0.21) \cdot \left[1 - (\sin\varphi)^{1.1 \pm 0.08}\right],$$

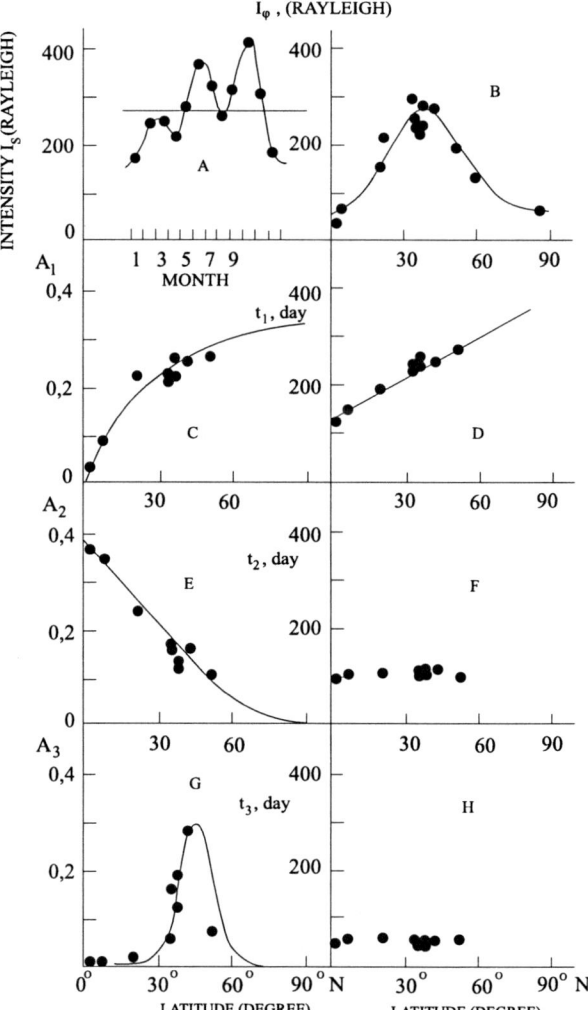

Fig. 4.33 Average variations of the 557.7-nm emission intensity for the conditions of the mean solar flux $F_{10.7} = 130$. Seasonal variations of the absolute intensity by Abastumani data (**A**); latitudinal variations by data of different stations: the yearly average absolute intensity (**B**); the amplitude (**C**) and phase (**D**) of the annual harmonic; the amplitude (**E**) and phase (**F**) of the semiannual harmonic; and the amplitude (**G**) and phase (**H**) of the four-month harmonic. *Dots* are data of different stations; *solid lines* are approximations (Fishkova et al. 2000)

$$A_3(\varphi) = (1.28 \pm 0.04) \cdot [\cos 2(\varphi - 45)]^{16 \pm 1}, t_\varphi = (133 \pm 7) + (2.8 \pm 0.16) \cdot \varphi.$$

The phase t_φ is determined as $t_\varphi = (133 \pm 7) + (2.80 \pm 0.16)$, where φ is the geographic latitude (degree), and the correlation coefficient is estimated as $r = 0.986 \pm 0.009$. The phases t_{S2} and t_{S3} do not depend on latitude:

4.4 Model of the 557.7-nm Atomic Oxygen Emission

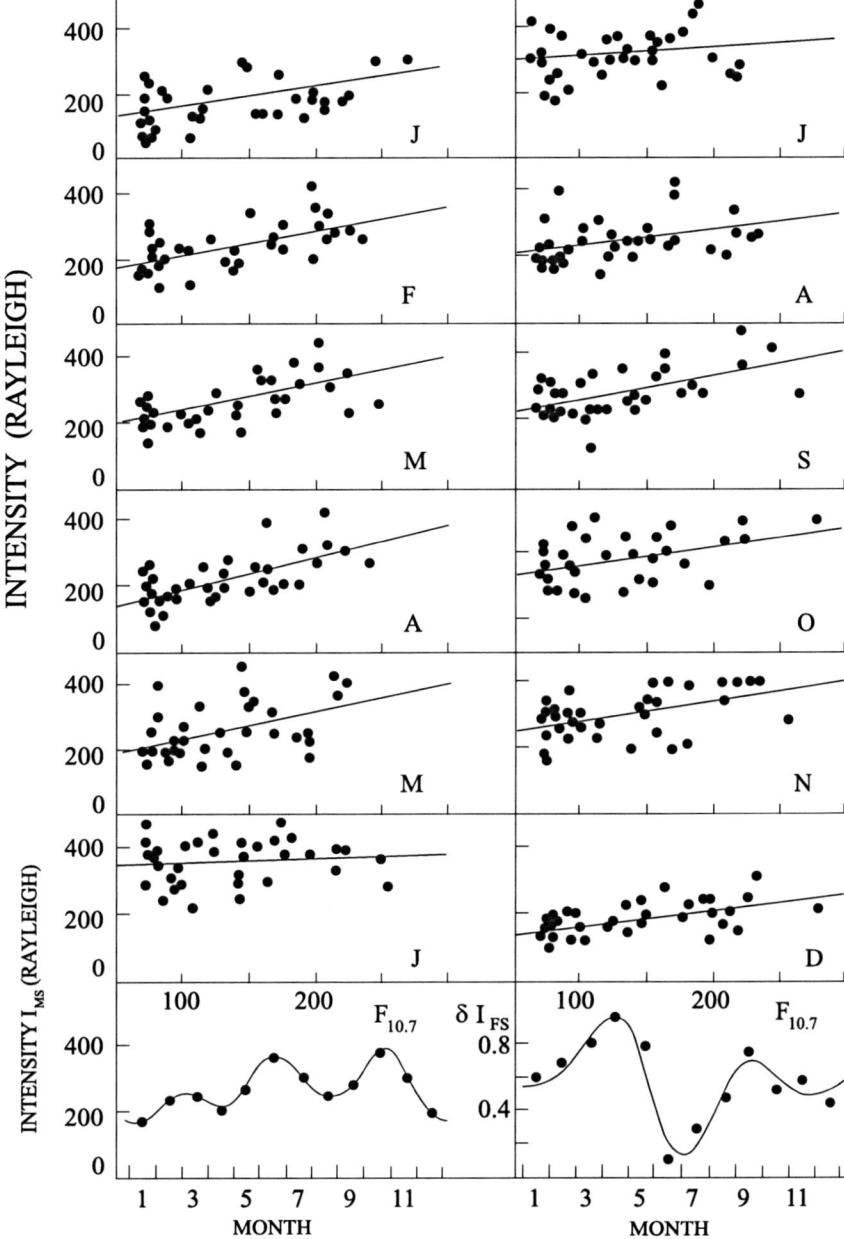

Fig. 4.34 Dependence of the monthly mean intensity of the 557.7-nm atomic oxygen emission (*dots*) on solar flux ($F_{10.7}$) for different months. The *straight lines* are regression lines. The seasonal mean variations of intensity I_{MS} and intensity response to solar activity δI_{FS} (Rayleigh/($F_{10.7} = 1$)) are shown at the bottom. The *solid lines* are the sums of three harmonics with periods of 12, 6, and 4 months (Fishkova et al. 2001)

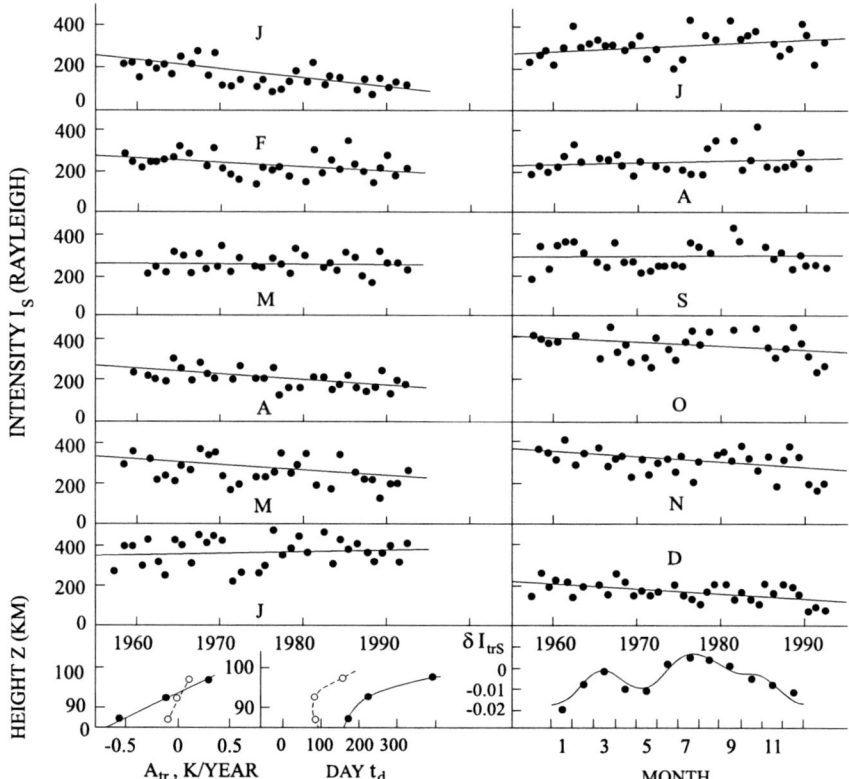

Fig. 4.35 Long-term variations of the monthly average intensity of the emission of atomic oxygen after elimination of the effect of solar activity (*dots*) for different months of the year. The *straight lines* are regression lines. The average seasonal variations of trend δI_{trS} (Rayleigh·yr^{-1}) are shown on the lower right. The *solid lines* are the sums of three harmonics with periods of 12, 6, and 4 months. The altitude variations of the amplitudes and phases of the annual (*dots* and *solid lines*) and semiannual (*open circles* and *dashed lines*) harmonics A_{tr} of the trend δT_{trS} (K·yr^{-1}) are shown on the lower left (Fishkova et al. 2001)

$$I_{Str} = \left\{ 1 + 0.0060 \left[1 + 1.73 \cos \frac{2\pi}{365}(t_d - 207) + 0.85 \cos \frac{4\pi}{365}(t_d - 60) \right.\right.$$
$$\left.\left. + 1.0 \cos \frac{6\pi}{365}(t_d - 70) \right] (t - 1972.5) \right\},$$

$$I_{SS} = \left\{ 1 + 0.0025 \left[1 + 0.27 \cdot \cos \frac{2\pi}{365}(t_d - 53) + 0.42 \cdot \cos \frac{4\pi}{365}(t - 90) \right.\right.$$
$$\left.\left. + 0.22 \cdot \cos \frac{6\pi}{365}(t_d - 117) \right] (F_{10.7} - 130) \right\},$$

where t_d is the day of year, t is the year's number, and $F_{10.7}$ is the index of solar activity.

4.4 Model of the 557.7-nm Atomic Oxygen Emission

The values of variations for the temperature inside the emission layer and for the layer altitude, thickness, and asymmetry have been obtained based on their average relations to intensity and to each other:

$$T_S = T_{S0}(\varphi) + \Delta T_{SA} + \Delta T_{Str} + \Delta T_{SS} \text{ (K)},$$

$$T_{S0}(\varphi) = 190 + 6 \cdot [\cos 1.8(\varphi - 40)]^{4.5} \text{ (K)},$$

$$\Delta T_{SA} = \left[2.74 A_1(\varphi) \cos \frac{2\pi}{365}(t_d - t_\varphi) + 4.67 A_2(\varphi) \cos \frac{4\pi}{365}(t_d - 120) \right.$$
$$\left. + 9.94 A_3(\varphi) \cos \frac{6\pi}{365}(t_d - 50) \right] \text{ (K)},$$

$$\Delta T_{Str} = -\left[0.07 \cdot \cos \frac{2\pi}{365}(t_d - 207) + 0.05 \cos \frac{4\pi}{365}(t_d - 60) \right.$$
$$\left. + 0.06 \cos \frac{6\pi}{365}(t_d - 70) \right] (t - 1972.5) \text{ (K)},$$

$$\Delta T_{SS} = -\left[0.34 \cos \frac{2\pi}{365}(t_d - 53) + 0.01 \cos \frac{4\pi}{365}(t - 90) \right.$$
$$\left. + 0.006 \cos \frac{6\pi}{365}(t_d - 117) \right] (F_{10.7} - 130) \text{ (K)},$$

$$Z_mS = Z_{mS0}(\varphi) + \Delta Z_{mSA} + \Delta Z_{mStr} + \Delta Z_{mSS} \text{ (km)},$$

$$Z_{mS0}(\varphi) = 103 - 6.34 [\cos 1.8(\varphi - 40)]^{4.5} \text{ (km)},$$

$$\Delta Z_{mSA} = -\left[1.76 A_1(\varphi) \cos \frac{2\pi}{365}(t_d - t_\varphi) + 0.94 A_2(\varphi) \cos \frac{4\pi}{365}(t_d - 120) \right.$$
$$\left. + 1.84 A_3(\varphi) \cos \frac{6\pi}{365}(t_d - 50) \right] \text{ (km)},$$

$$\Delta Z_{mStr} = -\left[0.034 \cos \frac{2\pi}{365}(t_d - 207) + 0.026 \cos \frac{4\pi}{365}(t_d - 60) \right.$$
$$\left. + 0.029 \cos \frac{6\pi}{365}(t_d - 70) \right] (t - 1972.5) \text{ (km)},$$

$$\Delta Z_{mSS} = -\left[0.016 \cos \frac{2\pi}{365}(t_d - 53) + 0.0052 \cos \frac{4\pi}{365}(t - 90) \right.$$
$$\left. + 0.0032 \cos \frac{6\pi}{365}(t_d - 117) \right] (F_{10.7} - 130) \text{ (km)},$$

$$W_S = W_{S0}(\varphi) + \Delta W_{SA} + \Delta W_{Str} + \Delta W_{SS} \text{ (km)},$$

$$W_{S0}(\varphi) = 10 + 1.34 [\cos 1.8(\varphi - 40)]^{4.5} \text{ (km)},$$

$$\Delta W_{SA} = -\left[0.88 A_1(\varphi) \cos \frac{2\pi}{365}(t_d - t_\varphi) + 0.47 A_2(\varphi) \cos \frac{4\pi}{365}(t_d - 120) \right.$$
$$\left. + 0.92 A_3(\varphi) \cos \frac{6\pi}{365}(t_d - 50) \right] \text{ (km)},$$

$$\Delta W_{Str} = -\left[0.017\cos\frac{2\pi}{365}(t_d-207) + 0.013\cos\frac{4\pi}{365}(t_d-60)\right.$$
$$\left. + 0.015\cos\frac{6\pi}{365}(t_d-70)\right](t-1972.5)\ (km),$$

$$\Delta W_{SS} = -\left[0.0081\cos\frac{2\pi}{365}(t_d-53) + 0.0026\cos\frac{4\pi}{365}(t-90)\right.$$
$$\left. + 0.0016\cos\frac{6\pi}{365}(t_d-117)\right](F_{10.7}-130)\ (km),$$

$$P_S = P_{S0}(\varphi) + \Delta P_{SA} + \Delta v P_{Str} + \Delta P_{SS},$$
$$P_{S0}(\varphi) = 0.85 - 0.24\left[\cos 1.8(\varphi-40)\right]^{4.5},$$

$$\Delta P_{SA} = -\left[0.0282 A_1(\varphi)\cos\frac{2\pi}{365}(t_d-t_\varphi) + 0.0150 A_2(\varphi)\cos\frac{4\pi}{365}(t_d-120)\right.$$
$$\left. + 0.0294 A_3(\varphi)\cos\frac{6\pi}{365}(t_d-50)\right],$$

$$\Delta P_{Str} = -\left[0.00054\cos\frac{2\pi}{365}(t_d-207) + 0.00042\cos\frac{4\pi}{365}(t_d-60)\right.$$
$$\left. + 0.00046\cos\frac{6\pi}{365}(t_d-70)\right](t-1972.5),$$

$$\Delta P_{SS} = -\left[0.00026\cos\frac{2\pi}{365}(t_d-53) + 0.000083\cos\frac{4\pi}{365}(t-90)\right.$$
$$\left. + 0.000051\cos\frac{6\pi}{365}(t_d-117)\right](F_{10.7}-130).$$

The obtained correlations between the 557.7-nm emission intensity and solar flux for various months of the year have been used to evaluate the seasonal variability of the temperature depending on solar activity (trend δT_F). They are related to the corresponding data for the sodium and hydroxyl emissions (Fig. 4.36). For the hydroxyl emission, the data of observations at Zvenigorod ($\varphi = 55.7°N$, $\lambda = 36.8°E$) have been used. The correlation formulas (r_F, δT_F) were constructed for individual months in contrast to the work by Neumann (1990) in which the seasonal grouping of data obtained near Wuppertal ($\varphi = 51°N$, $\lambda = 7°E$) was made only to relate them to quasi-biennial oscillations. The seasonal values of the 557.7-nm emission temperature have been estimated by the average relations between the emission temperature and intensity in view of the negative correlation between temperature and solar flux (Semenov and Shefov 1997b,c,d; Fishkova et al. 2000). For the sodium emission, this was done by the data of Fishkova et al. (2001a). For comparison, the variations of the atomic oxygen layer altitude Z(O) calculated based on the seasonal variations of the intensity, temperature, and emission layer altitude for the 557.7-nm emission are given at the top of Fig. 4.36. Examples of the altitude profile of atomic oxygen density for some points in a year are given in Fig. 2.15 (Perminov et al. 1998).

4.4 Model of the 557.7-nm Atomic Oxygen Emission

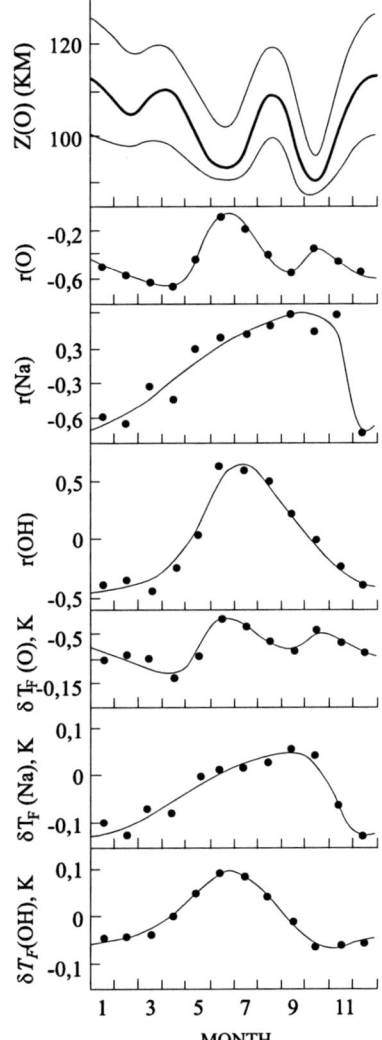

Fig. 4.36 Seasonal variations of the temperature responses to solar activity δT_F and r_F of the relation of the temperature in the OH (87 (km)), Na (92 (km)), and 557.7-nm (91–102 (km)) emission layers. For comparison, the seasonal variations of the altitude of the maximum atomic oxygen concentration layer (Perminov et al. 1998) (*thick line*) are shown. The *thin lines* specify the range of altitudes corresponding to the half-thickness of the layer (Fishkova et al. 2001b)

It can be seen that the seasonal variations of the regression coefficients, i.e., temperature response on solar activity δT_{FS} and correlation coefficients r_F for the O, Na, and OH emission layers, are different. This is obviously related to the seasonal variations of the layer altitude for atomic oxygen, which, in turn, determines the ozone density in this atmospheric region. Hence, the rate of heating of the atmosphere at the lower thermospheric altitudes, which reflects the degree of relation of the above-mentioned processes to solar flux, is largely determined by atmospheric circulation.

The seasonal variations of the long-term trend has an interesting feature. The temperature of the 557.7-nm emission layer has been estimated by the data of

Evlashin et al. (1999). Comparison of the seasonal values of the δT_{tr} (O), δT_{tr} (Na), and δT_{tr} (OH) trends for the atomic oxygen (90–102 (km)), sodium (~92 (km)) (Abastumani), and hydroxyl emissions (~87 (km)) (Zvenigorod) shows that the observed variations of the trend for the 557.7-nm emission result from the variations of the emission layer altitude that are synchronously reflected in the behavior of the δT_{tr} (O) trend (Fig. 4.37). The altitude variations of the amplitudes and phases of

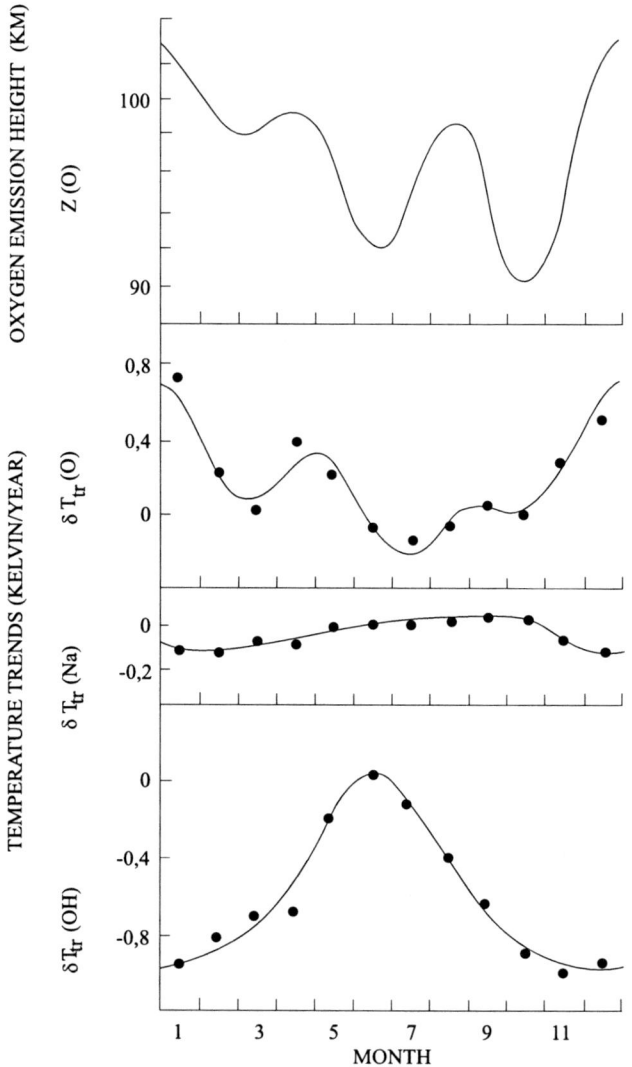

Fig. 4.37 Seasonal variations of the long-term temperature trends for the atomic oxygen, sodium, and hydroxyl emissions (*dots*). The *solid lines* are the sums of three harmonics. The variations of the altitude of the maximum atomic oxygen concentration layer (Fishkova et al. 2001b) are shown at the top

4.4 Model of the 557.7-nm Atomic Oxygen Emission

the annual and semiannual harmonics of the intensity and temperature trends are presented on the lower left of Fig. 4.35. They testify to monotonous variations of the harmonic amplitudes at altitudes of 85–100 (km) resulting in a change of sign near 92–93 (km). The phases of the harmonics also vary monotonously (Fishkova et al. 2001b).

Cyclic Aperiodic (Quasi-biennial) Variations

The notions about the character of the so-called quasi-biennial oscillations (QBOs) of the parameters of the mesopause and lower thermosphere have changed essentially in recent years. It has been found that these oscillations are caused by solar flux QBOs. This was considered in detail in Sect. 1.6.2.

It turned out that the sequence of solar flux maxima in an 11-yr cycle represents a train of oscillations with varying period and amplitude (see Figs. 4.20, 4.30–4.32) whose variations are well described by the Airy function $-Ai(-x)$ (Fadel et al. 2002; Semenov et al. 2002b).

For this case, $x = 3 - \Delta t$, where $\Delta t = t - t_o$ are the years elapsed from the beginning of the train, t_o. Since $Ai(3) \leq 0.008$, i.e., it is practically equal to zero though $Ai(x)$ asymptotically tends to zero as $x \to \infty$, it can be assumed that the beginning of the train corresponds to $x = 3$. The value of $-Ai(-1)$ is about -0.54, which seems to correspond to the point of minimum of the yearly average $F_{10.7}$ in the 11-yr cycle.

The solar activity component in the index $F_{10.7}$ is

$$\Delta F_{10.7} = -22\, Ai(3 - \Delta t).$$

The values of the Airy function can be taken from tables or calculated by approximate formulas (Abramowitz and Stegun 1964).

In view of these data, it would be more correct to refer to this type of variations as cyclic aperiodic variations (CAVs). The first broad minimum has a duration of ~ 3.8 years, the duration of the subsequent maxima is 2.8 years, decreasing to about 1.7 years in succeeding years. The lowest point of the first wide minimum falls on the minimum of the 11-yr cycle. However, the total train length is ~ 22 years; that is, it involves some years of the subsequent cycle. Therefore, some maxima of the trains of two cycles can interfere. This has the result that there is no similarity between the CAVs observed in different 11-yr cycles (Kononovich and Shefov 2003).

These CAVs for the 557.7-nm emission can be described by the formulas

$$\Delta I_{SAO}(557.7\,\text{nm}) = -0.14\, Ai\, \Delta(3 - \Delta t),$$
$$\Delta T_{SAO}(557.7\,\text{nm}) = 7\, Ai\, \Delta(3 - \Delta t), K,$$
$$\Delta Z_{mSAO}(557.7\,\text{nm}) = 0.9\, Ai\, \Delta(3 - \Delta t)\, (\text{km}),$$
$$\Delta W_{SAO}(557.7\,\text{nm}) = 0.9\, Ai\, \Delta(3 - \Delta t)\, (\text{km}).$$

The corresponding measurement data and their approximation are presented in Figs. 4.30–4.32.

Fourier analysis of the Airy function has shown (Shefov and Semenov 2006) that when the train intervals for the arguments $3 - \Delta t$ are chosen within 12–15 years, the maximum amplitudes of the harmonics correspond to periods of ~ 2.3 and 2.8 years and depend on the position of the chosen time interval within the 11-yr solar cycle.

5.5-yr Variations

This type of intensity variations has been revealed at Abastumani (Megrelishvili 1981; Megrelishvili and Fishkova 1986) (see Figs. 4.30–4.32). They can be described by the formula

$$\Delta I_{5.5} = 0.035 \cdot \cos \frac{2\pi}{5.5}(t - t_{5.5}) .$$

These variations can also be traced by the variations of the scattered light of the Earth's atmosphere at altitudes of ~ 100 (km) (Megrelishvili 1981), testifying to variations in density of both the atmosphere and its aerosol constituent.

The temperature variations can be described by the formula (Hernandez 1976)

$$\Delta T_{5.5} = 7.1 \cdot \cos \frac{2\pi}{5.5}(t - t_{5.5}) \; (K) .$$

The layer altitude and thickness can be estimated based on the correlation formulas

$$\Delta Z_{m5.5} = 0.6 \cdot \cos \frac{2\pi}{5.5}(t - t_{5.5}) \; (km),$$

$$\Delta W_{5.5} = 0.37 \cdot \cos \frac{2\pi}{5.5}(t - t_{5.5}) \; (km),$$

setting $t_{5.5} = 1959.0, 1970.0, 1981.0, 1992.0, 2003.0$. Spectral analysis of the solar flux variations in an 11-yr cycle shows that they can be described by the first harmonic whose maximum corresponds to the maximum of the cycle. In this case, the amplitude of the second harmonic (~ 5.5 years) makes 15(%) of that of the first harmonic (Kononovich 2005).

Variations Related to Solar Activity

The dependence of the intensity ΔI_F on solar flux has been derived from the data of a number of studies (Roach et al. 1953; Givishvili et al. 1996; Semenov 1996; Fishkova et al. 2001b) (see Fig. 4.26):

$$\Delta I_F = (0.0024 \pm 0.0005) \cdot (F_{10.7} - 130) .$$

To analyze the behavior of the temperature, the data reported by Hernandez (1976, 1977), Hernandez and Killeen (1988), and She et al. (1993) have been used. It should be noted that the results of long-term observations (Semenov et al. 1996, 2002a,

4.4 Model of the 557.7-nm Atomic Oxygen Emission

2005; Semenov and Shefov 1996a, 1999a; Golitsyn et al. 1996; Shefov et al. 2000a) show that in the mesopause region at ~ 90 (km), within the interval from the 19th to the 22nd cycle, temperature maxima ($\Delta T \sim 10\text{--}15$ (K)) appeared on the background of two minima below and above this altitude range during the periods of solar flux maxima, while during the periods of minima only broad minima were observed. This seems to be a characteristic feature of the altitude temperature profile of the mesopause, which was not taken into account in previous models (CIRA 1972; Barnett and Corney 1985). The corresponding measurement data are presented in Fig. 4.31. The solar flux dependence of the temperature can be described by the relation

$$\Delta T_F = -(0.12 \pm 0.015) \cdot (F_{10.7} - 130) \text{ (K)}.$$

The dependences of the emission layer altitude and thickness on solar flux have been derived by processing rocket measurements after elimination of other types of variations (see Fig. 4.32):

$$\Delta Z_{mF} = -(0.015 \pm 0.005) \cdot (F_{10.7} - 130) \text{ (km)},$$
$$\Delta W_F = -(0.016 \pm 0.006) \cdot (F_{10.7} - 130) \text{ (km)}.$$

A distinct correlation of these parameters with solar flux has been revealed by Shefov et al. (2000a).

Solar activity also affects the seasonal variability of the emission (see Fig. 4.34). Based on the data obtained at Abastumani (Fishkova et al. 2001b), it has been found that the correlation between emission intensity and solar flux maxima in spring (March and April) sharply decreases in summer and increases again at the autumn equinox. However, in the period of the autumn intensity maximum (October and November), the correlation again weakens a little (see Fig. 4.34). The quantitative characteristics of the correlations can be represented as

$$I_S(F_{10.7}, i) = [I_{MS}(i) \pm \sigma_{FS}] + [\delta I_{FS}(i) \pm \sigma_F](F_{10.7} - 130) \text{ (Rayleigh)}.$$

The correlation characteristics are given in Table 4.4. Here I_{MS} is the monthly mean intensity for 1972, σ is the root-mean-square deviation, and i is the month's number.

Long-Term Trend

After elimination of all types of variations, an average long-term change can be revealed. Based on the data of 70-yr observations at midlatitudes, a statistically significant trend has been found for intensity variations (Semenov and Shefov 1997a,d) (see Fig. 4.35):

$$\Delta I_{tr} = (0.006 \pm 0.0011)(t - 1972.5).$$

The long-term variability of the yearly average intensity of green emission was repeatedly noted (Fukuyama 1977a; Fishkova 1983; Semenov and Shefov 1997a,d).

Table 4.4 Statistical characteristics of the seasonal dependence of the monthly mean intensities of the 557.7-nm atomic oxygen emission on solar flux

Month	Altitude [O] (km)	I_{MS} (Rayleigh)	σ_{FS} (Rayleigh)	δI_{FS} (Rayleigh · $(\Delta F=1)^{-1}$)	σ_F (Rayleigh · $(\Delta F=1)^{-1}$)	δT_{FS} (K · $(\Delta F=1)^{-1}$)	r_F	σ_{rF}	t_{StF}
1	112	169	27	0.58	0.17	−0.10	0.507	0.129	2.5
2	106	235	27	0.67	0.18	−0.09	0.551	0.123	2.3
3	107	249	26	0.80	0.18	−0.10	0.613	0.109	2.5
4	110	213	27	0.97	0.19	−0.14	0.663	0.098	2.3
5	102	264	41	0.80	0.29	−0.09	0.441	0.142	2.0
6	94	364	32	0.10	0.22	−0.01	0.080	0.173	2.5
7	96	310	34	0.26	0.24	−0.03	0.185	0.168	3.0
8	107	247	28	0.48	0.20	−0.06	0.395	0.149	3.2
9	104	284	31	0.74	0.21	−0.08	0.533	0.129	1.9
10	90	377	39	0.52	0.27	−0.04	0.333	0.162	0.6
11	100	300	29	0.58	0.20	−0.06	0.456	0.138	1.7
12	111	177	19	0.45	0.13	−0.08	0.531	0.125	1.9
Yearly average	103	270	32	0.57	0.22	−0.12	0.361	0.161	2.1

4.4 Model of the 557.7-nm Atomic Oxygen Emission

The long-term emission intensity trend has clearly been traced, especially in view of the data obtained by Rayleigh in 1923–1934 (Lord and Spencer 1935; Hernandez and Silverman 1964), and the solar flux dependence of the intensity has been revealed.

To analyze the long-term variability of the green emission intensity for different seasons, the relevant data have been grouped by seasons: winter (December–January), spring (March–April), summer (June–July), and autumn (September–October). It turned out that the variations revealed in these seasonal intervals differ from the yearly average variations (Fig. 4.38, Table 4.5) (Fishkova et al. 2000). As can be seen, the most substantial differences are observed for winter and summer. For the period 1926–1992, the following relations have been obtained:

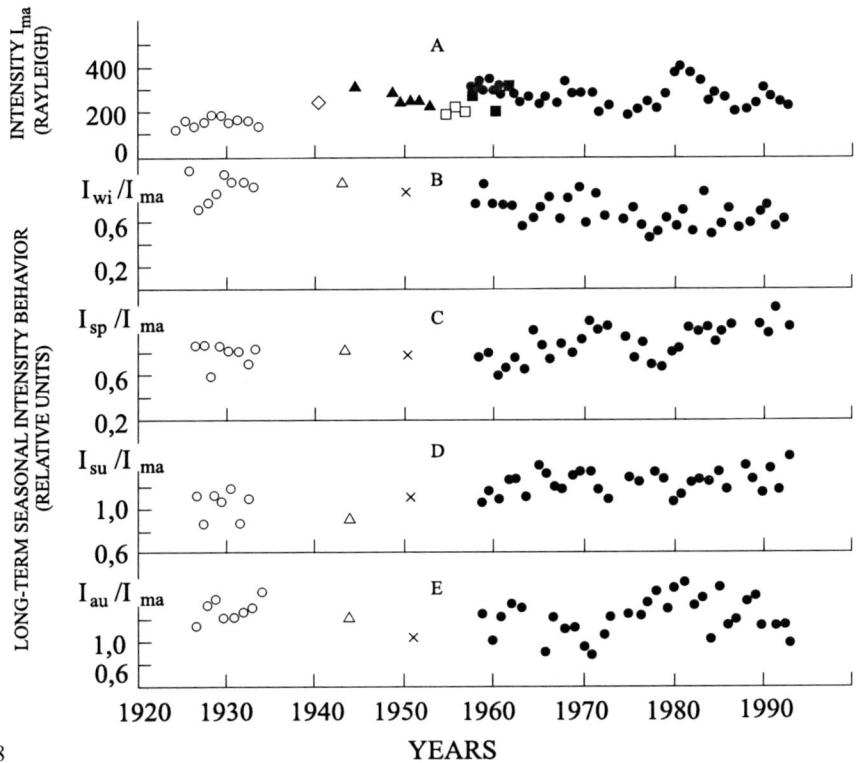

Fig. 4.38 Long-term variations of the 557.7-nm emission intensity by the data obtained at the following stations: Terling (Lord Rayleigh and Spencer Jones 1935; Hernandez and Silverman 1964) are *open circles*; Haute Provence (Dufay and Tcheng Mao-Lin 1946, 1947a,b) are *crosses*); data of Barbier et al. (1951) are *triangles*; data of Barbier (1959b) and Christophe-Glaume (1965) are *full squares*; Simeiz (Shain and Shain 1942) are *diamonds*; Cactus Peak (Roach et al. 1953) are *full triangles*; Abastumani (Fishkova et al. 2001) are *full circles*. Also presented are the yearly average variations (**A**) and the seasonal average related to yearly average variations for winter (**B**), spring (**C**), summer (**D**), and autumn (**E**) (Fishkova et al. 2000)

Table 4.5 Statistical characteristics of the seasonal dependence of the monthly average intensities of the 557.7-nm atomic oxygen emission on the long-term trend

Month	Altitude (km)	I_{MS} (Rayleigh)	σ_{trS} (Rayleigh)	δI_{trS} (Rayleigh · yr^{-1})	σ_{tr} (Rayleigh · yr^{-1})	δT_{trS} (K · yr^{-1})	r_{tr}	σ_{rtr}	t_{Str}
1	102	175	58	−3.37	0.77	0.74	−0.619	0.109	4.4
2	98	232	60	−1.67	0.79	0.22	−0.367	0.158	2.1
3	98	256	65	−0.24	0.84	0.02	−0.053	0.185	0.3
4	99	216	53	−2.20	0.69	0.40	−0.514	0.137	3.2
5	97	260	82	−2.51	1.08	0.21	−0.395	0.154	2.3
6	92	362	82	0.80	1.09	−0.07	0.128	0.171	0.7
7	93	315	73	1.76	0.98	−0.13	0.312	0.162	1.8
8	98	255	76	1.22	1.02	−0.06	0.142	0.179	0.8
9	97	289	74	0.24	0.99	0.06	0.045	0.182	0.2
10	91	377	80	−1.44	1.06	0.01	−0.249	0.174	1.4
11	93	308	75	−2.31	0.98	0.28	−0.388	0.150	2.3
12	102	181	46	−1.98	0.61	0.51	−0.500	0.131	3.3
Yearly average	97	270	42	−0.79	0.56	0.13	−0.248	0.166	1.4

$$\Delta I_{wi} = -(0.301 \pm 0.060) - (0.0054 \pm 0.0009) \cdot (t - 1972.5),$$
$$r = -0.695 \pm 0.080, \text{ winter};$$
$$\Delta I_{sp} = -(0.127 \pm 0.067) + (0.0023 \pm 0.0010) \cdot (t - 1972.5),$$
$$r = 0.354 \pm 0.138, \text{ spring};$$
$$\Delta I_{su} = (0.122 \pm 0.060) + (0.0040 \pm 0.0009) \cdot (t - 1972.5),$$
$$r = 0.600 \pm 0.104, \text{ summer};$$
$$\Delta I_{au} = (0.226 \pm 0.083) - (0.0006 \pm 0.0012) \cdot (t - 1972.5),$$
$$r = -0.079 \pm 0.153, \text{ autumn}.$$

As follows from these relations, statistically significant trends are observed for winter, spring, and summer.

For comparison, the extreme values of relative intensities for the above periods are determined by the following formulas:

$t_d = 1$ (January 1):

$$\Delta I(1) = -(0.432 \pm 0.141) - (0.0071 \pm 0.0019) \cdot (t - 1972.5), \quad r = -0.567 \pm 0.120;$$

$t_d = 60$ (March 1):

$$\Delta I(60) = -(0.082 \pm 0.213) + (0.0031 \pm 0.0029) \cdot (t - 1972.5), r = 0.200 \pm 0.178;$$

$t_d = 166$ (June 15):

$$\Delta I(166) = (0.280 \pm 0.162) + (0.0029 \pm 0.0030) \cdot (t - 1972.5), r = 0.186 \pm 0.186;$$

$t_d = 288$ (October 15):

$$\Delta I(288) = (0.400 \pm 0.193) - (0.00047 \pm 0.0036) \cdot (t - 1972.5), r = -0.026 \pm 0.192.$$

Thus, a statistically significant trend exists only for the intensity winter minimum. This suggests that dynamic processes produce large irregularities in the atmosphere which shows up in seasonal variations of the airglow emissions.

It has been shown (Semenov and Shefov 1999b) that to the period of the autumn maximum of green emission intensity there corresponds a descent of the atomic oxygen layer to ~ 92 (km). Analysis of the seasonal intensity variations shows that the point in time at which the intensity is a maximum seems to be affected by the long-term trend, namely

$$t_d = (291 \pm 5) + (0.121 \pm 0.068) \cdot (t - 1972.5), r = 0.323 \pm 0.160.$$

The behavior of the harmonic amplitudes also has some features. The amplitude of the annual harmonic depends on solar flux:

$$\Delta A_1(F_{10.7}) = -(0.00065 \pm 0.00028) \cdot (F_{10.7} - 130); \ r = -0.466 \pm 0.115; \ t_{St} = 2.3,$$

and the correlation is significant.

The effect of the long-term trend can be described by the formula

$$\Delta A_1(t) = (0.00145 \pm 0.00120) \cdot (t - 1972.5), \qquad r = 0.264 \pm 0.208 .$$

However, as can be seen, this effect is statistically nonsignificant. The amplitudes of the semiannual and 4-month harmonics and the phases of all harmonics show no effect of the trend or solar flux (Fishkova et al. 2000).

Though, as can be seen, a significant positive trend for a period of $\sim$70 years has been revealed for the emission intensity, the situation with the data about the temperature is far less certain. These data are rather few in number and cover an interval of $\sim$27 years. After their reduction to the flux $F_{10.7} = 130$, which is a mean smoothed value for a 22-yr interval and approximately corresponds to 1972, Δt do not show any trend. The now available values of ΔT_{tr} do not show a significant deviation from zero, though formally the trend is (-0.02) $(K \cdot yr^{-1})$.

As already mentioned, the long-term variations of the intensity testify to its growth for 70 years. The 11-yr mean intensity in the 1920s (Hernandez and Silverman 1964) made $\sim$60(%) of its value in 1972. For this case, the cycle average $F_{10.7}$ was 100. Estimation of the atmospheric temperature within the emission layer for the above conditions by average relations (Semenov 1997) has shown that it should be 210–215 (K) (Evlashin et al. 1999) (Fig. 4.39). The average temperature trend for the investigated period appears to be equal to (-0.2) $(K \cdot yr^{-1})$. This method can also be used to estimate the temperature for the second half of the 19th century when the observations of the 557.7-nm emission began (Semenov and Shefov 1996b).

This agrees with the active temperature measurements in the middle atmosphere of the past four or five decades that have shown its regular cooling at altitudes of 30–95 (km) (Golitsyn et al. 1996; Semenov and Lysenko 1996; Semenov et al. 1996; Semenov 1996, 1997; Semenov and Shefov 1996a, 1999a; Semenov et al. 2002a). This behavior of the temperature, according to the data for the 630-nm atomic oxygen emission (Semenov 1996) from which the effect of other types of variations has been eliminated (see Fig. 4.42), is traced up to altitudes of $\sim$270 (km).

A consequence of this was the regular decrease in atmospheric density at altitudes above 50 (km), implying its global subsidence. As a result, for the period 1923–1992, it has been obtained that

$$\Delta T_{tr} = -(0.10 \pm 0.03)(t - 1972.5) \, (K) .$$

The rocket data yield a very small regular change of altitude, $\Delta Z_{mtr} = -26$ $(m \cdot yr^{-1})$, and a positive trend of the layer thickness:

$$\Delta Z_{mtr} = -(0.026 \pm 0.010)(t - 1972.5) \, (km),$$
$$\Delta W_{tr} = (0.03 \pm 0.01)(t - 1972.5) \, (km) .$$

Comparison with the data for the hydroxyl emission (Semenov and Shefov 1996a, 1999a) shows that the variations of the layer thickness $W_{557.7}$, which reflect

4.4 Model of the 557.7-nm Atomic Oxygen Emission

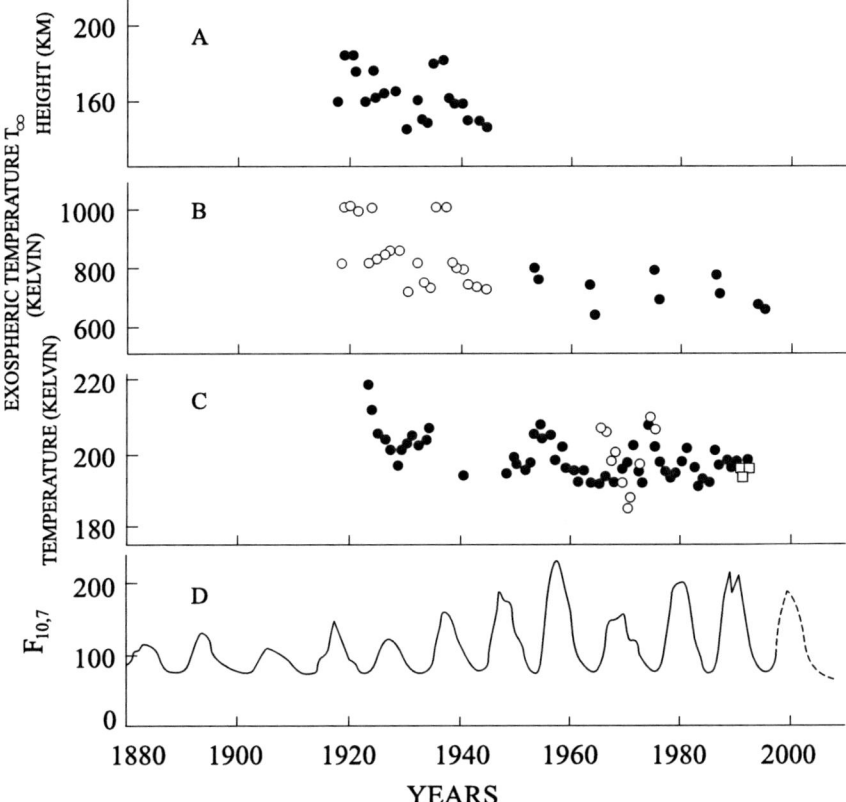

Fig. 4.39 Long-term variations of the characteristics of the upper atmosphere (Evlashin et al. 1999). (**A**) are mean auroral ray heights Z for the conditions of midnight, winter solstice ($\varphi \sim 63°$N), and minimum solar flux; (**B**) are early average temperatures of the exosphere for the conditions of midnight, midlatitudes, and minimum solar flux: data of Fig. 4.39 a (*open circles*), and of Semenov (1996) and Semenov and Lysenko (1996) (*full circles*); (**C**) are yearly average temperatures of the lower thermosphere ($\sim$100 (km)) by the measurements of temperatures (Hernandez and Killeen 1988) (*open circles*) and intensities (Givishvili et al. 1996; Semenov 1996) of the 557.7-nm atomic oxygen emission (*full circles*) and by lidar measurements (She et al. 1993) (*squares*); (**D**) yearly average solar flux

the variations of the altitude Z_m, are more realistic manifestations of the variations of the atmospheric density, which can be described by the formula

$$N = 2.44 \cdot 10^{15} \cdot \exp[-0.53 \cdot W_{557.7}] (\text{cm}^{-3}) ,$$

where $W_{557.7}$ is expressed in kilometers.

In sighting the emission layer limb, the real thickness of the layer can be obtained from the observed altitude distribution of the emission intensity by solving the inverse problem (Shefov 1978b).

The revealed long-term changes in the upper atmosphere put forward an important problem of the nature of the phenomenon under observation, since it seems that in this case a combined effect of both the anthropogenous factor associated with the increasing content of hotbed gases and the long-period solar cycles takes place. The knowledge of this effect is necessary for forecasting the further behavior of the characteristics of the Earth's atmosphere as a whole. In seeking a solution to this problem, the data of spectrophotometry of the 557.7-nm emission carried out since its detection in the second half of the 1800s, unfortunately yet not published (Semenov and Shefov 1996b), could be helpful.

Here it is necessary to mention an interesting circumstance related to the process of long-term cooling and subsidence of the upper atmosphere (Golitsyn et al. 1996; Semenov 1996; Lysenko et al. 1999; Megrelishvili and Toroshelidze 1999) which can be traced since the 1920s (Evlashin et al. 1999; Starkov et al. 2000). The yearly average trend δT_{tr} (O) for the interval 1923–1992 was $-(0.10 \pm 0.03)$ $(K \cdot yr^{-1})$. As follows from the available data, it is probable that in the second half of the 20th century the trend changed its sign as a result of a small displacement of the 557.7-nm emission layer in altitude because of the subsidence of the neutral atmosphere and also under the solar action, as already noted (Semenov 1997; Semenov and Shefov 1999a; Shefov et al. 2000a). Its yearly average value estimated only by the data of measurements performed between 1957 and 1992 is 0.10 ± 0.03 $(K \cdot yr^{-1})$. Undoubtedly, the correct estimation of the trend for these altitudes is substantially affected by the variations of the altitude of the 557.7-nm emission layer due to solar activity (Shefov et al. 2000a) which, probably, have been taken into account not completely. Besides, it is of interest that on comparison of the rate of subsidence of the neutral atmosphere, calculated by the changes of the temperature profiles, and the rate of subsidence of the atomic oxygen layer, determined by the variations of the 557.7-nm emission characteristics, it turned out that the rate of subsidence of atomic oxygen is less than that of the basic neutral constituents – oxygen and nitrogen molecules (Evlashin et al. 1999; Starkov et al. 2000). If this is actually the case, this feature implies the displacement of the atomic oxygen layer upward (~ 1.5 (km)) relative to the basic constituent layers and, apparently, as consequence, the change of the trend sign. The trend is observed to change for the last decades (Lysenko et al. 1997a,b; Lysenko and Rusina 2002a,b).

Latitudinal Variations

The latitudinal variations of the parameters of the 557.7-nm emission have already been described when we considered the behavior of the amplitudes and phases of the harmonics of seasonal variations (see Figs. 4.30–4.32):

$$\delta I_{S1}(\varphi) = (0.33 \pm 0.03)(0.33 \pm 0.03) \cdot [\sin\varphi]^{(0.60 \pm 0.01)},$$

correlation coefficient $r = 0.953 \pm 0.041$,

$$\delta I_{S2}(\varphi) = (0.38 \pm 0.03) \cdot [1 - (\sin\varphi)^{(1.1 \pm 0.08)}],$$

4.4 Model of the 557.7-nm Atomic Oxygen Emission

correlation coefficient r = 0.977 ± 0.015,

$$\delta I_{S3}(\varphi) = (0.30 \pm 0.01) \cdot [\cos 2(\varphi - 45)]^{(16\pm 1)},$$

correlation coefficient r = 0.818 ± 0.100,

$$t_{S1} = (133 \pm 7) + (2.80 \pm 0.16) \cdot \varphi,$$

where φ is the geographic latitude (degrees), the correlation coefficient r = 0.986 ± 0.009.

The phases t_{S2} and t_{S3} do not depend on latitude.

When calculating the absolute values of seasonal variations, it is necessary to consider the latitudinal dependence of the emission intensity. Following Fishkova et al. (2000), we have

$$I_{ma} = 60 + 220 \cdot [\cos 1.8(\varphi - 40)]^{4.5} \text{ (Rayleigh)}.$$

As already mentioned, the amplitudes of the variations for the emission layer temperature, altitude, and thickness are obtained by their relations with the intensity:

$$\delta T_{S1}(\varphi) = (16.0 \pm 1.4) \cdot [\sin\varphi]^{(0.60\pm 0.01)} \text{ (K)},$$
$$\delta T_{S2}(\varphi) = (18.2 \pm 1.4) \cdot [1 - (\sin\varphi)^{(1.1\pm 0.08)}] \text{ (K)},$$
$$\delta T_{S3}(\varphi) = (14.4 \pm 0.5) \cdot [\cos 2(\varphi - 45)]^{(16\pm 1)} \text{ (K)},$$
$$\delta Z_{mS1}(\varphi) = -(7.8 \pm 0.7) \cdot [\sin\varphi]^{(0.60\pm 0.01)} \text{ (km)},$$
$$\delta Z_{mS2}(\varphi) = -(9.0 \pm 0.7) \cdot [1 - (\sin\varphi)^{(1.1\pm 0.08)}] \text{ (km)},$$
$$\delta Z_{mS3}(\varphi) = -(7.1 \pm 0.2) \cdot [\cos 2(\varphi - 45)]^{(16\pm 1)} \text{ (km)},$$
$$\delta W_{S1}(\varphi) = -(4.1 \pm 0.4) \cdot [\sin\varphi]^{(0.60\pm 0.01)} \text{ (km)},$$
$$\delta W_{S2}(\varphi) = -(4.7 \pm 0.4) \cdot [1 - (\sin\varphi)^{(1.1\pm 0.08)}] \text{ (km)},$$
$$\delta W_{S3}(\varphi) = -(3.7 \pm 0.1) \cdot [\cos 2(\varphi - 45)]^{(16\pm 1)} \text{ (km)}.$$

Based on the UARS satellite measurements of the 557.7-nm emission intensity, its altitudinal–latitudinal distributions in the latitude range 40°S–40°N have been constructed. It is of interest that, according to ground-based observations, there are intensity maxima at latitudes near 30°N and also in the southern hemisphere, and the emission rate peaks at ~97 (km) (Ward 1999).

For the high latitudes, the auroral oval boundaries are described by analytic relations (Ivanov et al. 1993; Starkov 1994a,b; Semenov and Shefov 2003). The average distribution of the 557.7-nm emission intensity in the oval region at geomagnetic latitude Φ has been obtained (Monfils 1968; Silverman 1970) based on measurement data. The intensity (kilorayleighs) can be approximately calculated by the empirical formula

$$\lg I_{557.7} = -0.538 + 0.072 \cdot \exp(0.348 K_p) /$$
$$\{1 + \exp[-(\Phi - 71.4 + 1.35 \cdot K_p)/(34 - 10.4 \cdot K_p)]\}$$
$$+ (0.21 + 0.136 \cdot K_p) \exp[-(\Phi - 71.4 + 1.35 \cdot K_p)^2/(6.8 + 0.39 \cdot K_p)^2] \,.$$

However, more realistic estimates can be obtained only with a model which considers the spatial distribution of the excitation processes in the auroral zone and the effect of solar activity (Evlashin et al. 1996).

Longitudinal Variations

Used for analysis were the data of measurements performed simultaneously at different stations located in the geographic latitude range 25–55°N during the International Geophysical Year (1957–1959) (12 stations (Yao 1962)) and the International Year of Quiet Sun (1964–1965) (14 stations (Smith et al. 1968)). These intervals of years correspond to the years of the maximum and the subsequent minimum ($F_{10.7} = 74$) of the 19th solar cycle ($F_{10.7} = 225$). Based on these data, seasonal mean variations of the emission intensity have been derived for each station as for the periods of both high and low solar activity. They well agree in character with the results of long-term measurements performed at midlatitudes (Semenov and Shefov 1997a,d; Fishkova et al. 2000, 2001b; Mikhalev et al. 2001; Mikhalev and Medvedeva 2002). Since the results of the work by Fishkova et al. (2000) suggest that the seasonal change of the 557.7-nm radiation intensity depends on geographic latitude, to reduce them to unified conditions for the latitude 40°N, some corrections have been made by the relations presented elsewhere (Fishkova et al. 2000). After correction of the data for latitudinal changes, all data were reduced to the solar flux $F_{10.7} = 74$. These results are presented in Fig. 4.40.

The data obtained show appreciable stationary longitudinal variations of the 557.7-nm emission. It seems that the features of the Earth's surface relief show up in the longitudinal variability of the emission layer characteristics. Shown schematically on the top of Fig. 4.40 are the positions of the regions of continents and oceans in the longitude range for the band of latitudes $(40 \pm 10)°N$. It can be seen that the emission intensity over the Euroasian continent is greater than over the Pacific Ocean regions. On the North American continent, the observation stations were located only on the western side. Observations were not carried out over Atlantic Ocean. Because of the limited number of longitudinal measurement points, an elementary first approximation has been made that gave the following relation:

$$I_\lambda = 230 \cdot [1 + \Delta I] \cdot \{1 + \Delta I \cdot [0.16 \cdot \cos(\lambda - 10) + 0.02 \cdot \cos 2(\lambda - 170)$$
$$+ 0.01 \cdot \cos 3(\lambda - 5)]\} \text{ (Rayleigh)} \,,$$

where

$$\Delta I = 0.21 \cdot \cos \frac{2\pi}{365}(t - 230) + 0.22 \cdot \cos \frac{4\pi}{365}(t - 120) + 0.26 \cdot \cos \frac{6\pi}{365}(t - 50) \,.$$

4.4 Model of the 557.7-nm Atomic Oxygen Emission

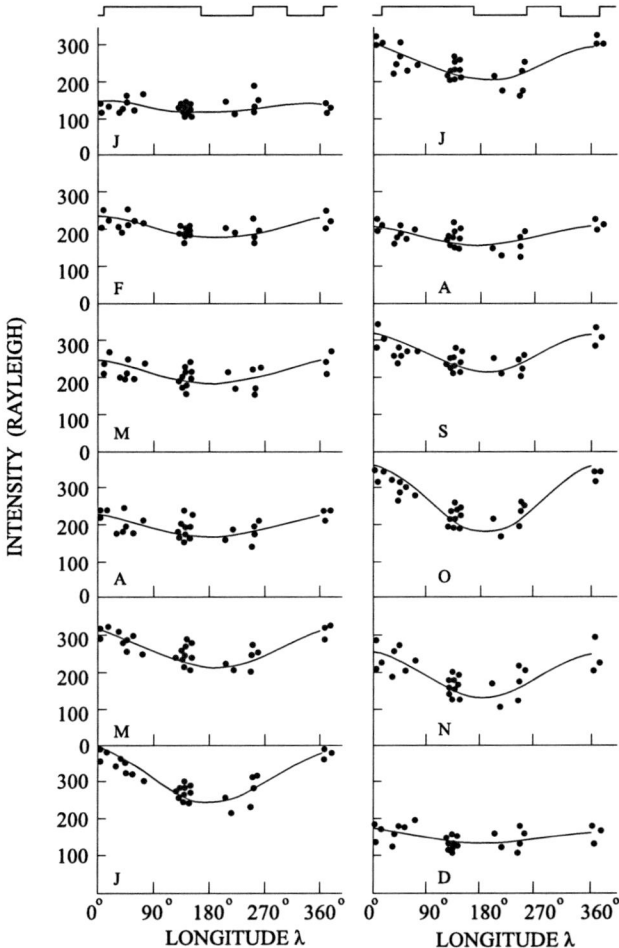

Fig. 4.40 Longitudinal variations of the intensity of the 557.7-nm atomic oxygen emission for different months of the year (Shefov and Semenov 2004b). *Dots* are measurement data; *solid lines* are approximations

As follows from the properties of the variations of emission parameters (Semenov and Shefov 1996a, 1997a,b,c,d; Shefov et al. 2000a), to an increase in intensity there corresponds a decrease in emission layer altitude and an increase in temperature (Shefov et al. 2000a).

The behavior of the intensity, temperature, and emission layer altitude completely fit to the orographic variations that have been studied by the behavior of the hydroxyl emission (Shefov et al. 1999, 2000b). This gives grounds to conclude that the altitude distribution of atomic oxygen at altitudes of 80–100 (km) is determined, with other things being equal, by the features of the Earth surface underlying relief and by the dynamics of the wind system of the lower atmosphere and therefore depends on longitude.

Besides the longitudinal variations associated with the Earth's relief, there are longitudinal variations caused by planetary waves (Shepherd et al. 1993b; Wang et al. 2000, 2002). They testify to periods of wave variations close to 16 days. These waves cover a wide range of altitudes and modulate various photochemical processes, causing their periodic changes, including the occurrence of noctilucent clouds in summer (Shefov and Semenov 2004a).

Disturbed Variations Induced by Stratospheric Warmings

The effects in the lower thermosphere during stratospheric warmings have been investigated based on the data of Fukuyama (1977b) and Fishkova (1983) (see Figs. 4.30 and 4.31). However, the intensity and temperature measurements were carried out not simultaneously and at different stations in the eastern and western hemispheres, though in close latitudes, for 10 years. Moreover, different starting points were chosen in using the epoch superposition method. In analyzing the behavior of the intensity (Fishkova 1983) by the data of Rakipova and Efimova (1975), taken for the date zero, t_{sw} was the date of the onset of stratospheric warming on the 10-mbar surface ($\sim$32 (km)), such that a steady source of maximum positive temperature change appeared in any region of the given isobaric surface. Thus, it was obtained that

$$\Delta I_{sw} = 0.04 \cdot (t - t_{sw}) \cdot \exp[-(t - t_{sw})/10] \; .$$

The relation for the temperature was obtained by the data of Hernandez (1977) (see Fig. 4.31). In contrast to the analysis of the behavior of the emission intensity, the date of the maximum of stratospheric warming, which takes place on the average 3–5 days later, was chosen for the zero reference point t_{sw}. For this case,

$$\Delta T_{sw} = -1.5 \cdot (t - t_{sw}) \cdot \exp[-(t - t_{sw})/10] \; (K) \; .$$

Here it should be stressed that correct application of the epoch superposition method is of great importance, since otherwise this can lead to erroneous conclusions (Taubenheim 1969; Taubenheim and Feister 1975).

The list of warmings in the period 1955–1986 is given by Tarasenko (1988). By overlapping the data for ΔI_{sw} and ΔT_{sw}, one can draw the conclusion that regular intensity variations ΔI_{sw} occur on the background of a planetary wave with a period of $\sim$30 days.

Variations Induced by Geomagnetic Disturbances

The effect of the composition waves propagating from the auroral zone after geomagnetic disturbances (Nasyrov 1970) on the atmosphere is similar to the variations of the hydroxyl emission (Shefov 1969a, 1973a; Semenov and Shefov 1996a, 1999a). The percentage of the cases of longitudinal orientation of the isophots of the spotty structure of 557.7-nm emission was found to increase to 50(%) (Korobeynikova and Nasyrov 1972) within 3–4 days after a disturbance event at the latitude of Ashkhabad ($\Phi = 30.6°N$).

Variations During the Periods of Meteoric Activity

During the periods of maximum intrusion of meteoric flows, a 20–30(%) increase in intensity has been noted (Shefov 1968a, 1970b). Data on variations of temperature and layer altitude and thickness are still not available.

Variations During the Periods of Seismic Disturbances

As repeatedly observed at Abastumani (Fishkova and Toroshelidze 1989; Toroshelidze 1991) and Ashkhabad (Korobeynikova et al. 1989), irregular fluctuations of the intensity of the 557.7-nm emission occurred several hours prior to an earthquake. For some local (Caucasus) earthquakes with magnitudes $M \geq 5$, the amplitude of the intensity fluctuations increased to 100(%). However, the available data are insufficient for systematization. The nature of this behavior of the intensity fluctuations is probably related to the waves accompanying the release of gases from the lithosphere before an earthquake (Akmamedov et al. 1996). The characteristic radius R (km) of the region where a foreshock shows up is determined by the relation (Toroshelidze 1991)

$$R = 1.0 \cdot \exp M.$$

Variations Induced by Internal Gravity Waves

Theoretical estimates of the possible waves propagating upward from the regions of their generation in the troposphere testify to the attenuation of the trains containing several periodic oscillations. The basic time dependence of the waves has the form (Chunchuzov 1988, 1994)

$$\delta T = \delta T_{00} \left(\frac{2\pi t}{\tau}\right)^{1/2} \exp\left(-\frac{2\pi t}{\alpha \tau}\right) \cos\left(\frac{2\pi t}{\tau} + \frac{\pi}{4}\right).$$

Here τ is the wave period, δT_{00} is the amplitude, t is the time lapsed from the beginning of the oscillation train, and α is the attenuation parameter determined by the width a of the obstacle (in interaction the jet streams with the mountain ridges or the baric formations), and by the altitude to which the wave propagates, Z, and by the zenith angle of wave propagation, θ. This is discussed in more detail in Sect. 5.2.3.

Long-term measurements of IGWs in the upper atmosphere, in particular by the characteristics of the hydroxyl emission (Krassovsky et al. 1978; Novikov 1981), gave the following universal formula for the dependence of the wave amplitude δT on period τ:

$$(\delta T/T)^2 \sim \tau^{5/3},$$

where T is the atmospheric temperature.

When the usual Fourier analysis of time series is applied, the waves periods are calculated which characterize the train. However, in this case, the amplitude depends on the degree of attenuation of the train (Shefov and Semenov 2004c).

Typical periodic intensity variations induced by IGWs are determined by the relation

$$\frac{\Delta I_{gw}}{I} = 2.4 \cdot 10^{-3} \cdot \tau_w^{1.3} \sin \frac{2\pi t}{\tau_w},$$

where τ_w is the wave period.

The amplitudes of oscillations are in average about 10(%) of the wave amplitude (Kuzmin 1975; Toroshelidze 1991). The variations due to IGWs cannot be predicted within a night, since IGWs, generated mainly by meteorological sources in the lower atmosphere, are random in character.

Little is known about the variations of the Doppler temperature under the action of IGWs (Semenov 1989). The parameter η determined by the relation (Krassovsky 1972)

$$\frac{\Delta I_{gw}}{I} = \eta \frac{\Delta T_{gw}}{T}$$

from measurements performed in winter turned out to be equal to 0.6, which is far below that for the hydroxyl emission. The average time delay θ of the intensity oscillations relative the temperature variations is $\sim$90 (s). These values of the parameters η and θ also allow the conclusion that the excited oxygen molecules produced due to the Barth process are mainly in the state $^5\Pi_g$ (Semenov 1989).

It should be noted that some types of variations that have been revealed for the intensity of the 557.7-nm emission have not yet been obtained for temperature.

Thus, the above analysis of the material of investigations of the 557.7-nm atomic oxygen emission presented in numerous publications made it possible to systematize the available data and construct empirical models for the nocturnal variations of intensity, temperature, and emission layer altitude. Once first results had been obtained, additional investigations were carried out that provided supplementary data on some types of variations and gave impetus to absolutely new approaches in their analysis.

These results were successfully used for tentative estimation of the long-term variations of the altitude distributions of ozone and atomic oxygen densities at 80–100 (km) (Semenov 1997) and their dependences on solar flux by using the theory of photochemical production of excited oxygen atoms due to the Barth process (Semenov and Shefov 1999b). The presented refined model was used to develop an empirical model of the variations of the altitude distribution of atomic oxygen,

4.5 Model of the 630-nm Atomic Oxygen Emission

which in fact determines the thermal conditions of the mesopause and lower thermosphere, at the mentioned altitudes (Sect. 7.2).

The material of long-term spectrophotometric observations carried out at various stations has been systematized to reveal regular and nonregular variations of the emission intensity, temperature, and maximum emission rate altitude of the emission layer.

Since ionized species participate in the basic photochemical processes giving rise to the 630-nm emission, the illumination of the atmosphere at the emission layer altitudes with the UV radiation of the Sun is of great importance. During twilight, the emission intensity decreases as the Sun depresses under the horizon, i.e., the effect of the solar UV radiation at thermospheric altitudes decreases, resulting in a decrease in the concentration of the reactants participating in dissociative recombination. At the same time, the inflow of photoelectrons and O^+ ions from the magnetically conjugate regions of the upper atmosphere during twilight (in winter) makes some contribution to the emission intensity.

Geomagnetic disturbances are accompanied by red auroras and occur at low and middle latitudes which are much more intense than the airglow in quiet geomagnetic conditions (Yevlashina and Yevlashin 1971; Evlashin et al. 1996).

In high-latitude auroras, the atmospheric constituents are also excited by direct impact of the secondary electrons resulting from the ionization of atmospheric neutrals by precipitating energetic electrons.

The ground-based photometric measurements presented in numerous publications refer in the main to the behavior of the emission intensity, mainly, at midlatitudes. Based on this material, information about various types of regular variations was gained (Shefov et al. 2006). The behavior of the variations resulting from various sporadic disturbances was much less investigated.

Regular measurements of the red emission intensity began during the International Geophysical Year (1957–1959). The measurement data and subsequent investigations were reported by Yao (1962) and Smith et al. (1968) and reviewed by Barbier (1961) and Fishkova (1983).

There are in general much less interferometric data on the Doppler temperature of the 630-nm emission than on its intensity. The conditions under which metastable oxygen atoms appear at thermospheric altitudes, largely in the ionospheric F_2 layer, suggest that their existence is not inconsistent with the radiation lifetime. Nevertheless, the frequency of collisions of excited oxygen atoms with the surrounding atmospheric species during the lifetime should ensure their relaxation to an equilibrium state with the temperature equal to the ambient temperature. It is well known that the collision rate for the process

$$O(^1D) + (e, O, O_2, N_2) \to O(^3P) + (e, O, O_2, N_2)$$

is determined by the relations given in Sect. 2.6 (Pavlov et al. 1999).

The lifetime of excited metastable atoms is determined by the relation

$$\tau = \frac{1}{A + \beta_e \cdot n_e + \beta_O \cdot [O] + \beta_{O_2} \cdot [O_2] + \beta_{N_2} \cdot [N_2]} \text{(s)}.$$

At the emission layer altitudes, the dominant atmospheric constituent is atomic oxygen for which $\beta_O = 2.5 \cdot 10^{-12}$ (cm$^3 \cdot$s^{-1}) (Pavlov et al. 1999). For the altitudes 180, 270, and 350 (km) to which there correspond the relative emission rates of 0.1, 1, and 0.1, respectively, for the exospheric temperature T = 800 (K) ($F_{10.7} = 130$), τ is 7, 80, and 123 (s), respectively. With the night intensity 200–300 (Rayleigh), the density of excited atoms O(^{1}D) at the maximum emission rate is $\sim 4 \cdot 10^3$ (cm^{-3}), while the density of unexcited atoms [O] is $\sim 8 \cdot 10^8$ (cm^{-3}). The fast metastable atoms O*(^{1}D) newly formed as a result of dissociative recombination (Sect. 2.1.2) collide with unexcited thermalized oxygen atoms:

$$O^*(^1D) + O(^3P) \to O^*(^3P) + O(^1D).$$

In this case, the rate of gas-kinetic collisions is $4.2 \cdot 10^{-10}$ (cm$^3 \cdot$s^{-1}) (Bates and Nicolet 1950), which is two orders of magnitude greater than the rates of the deactivation processes. The rate of collisions with excitation transfer is $\sim 10^{-10}$ (cm$^3 \cdot$s^{-1}). Therefore, the frequency of gas-kinetic collisions at the mentioned altitudes is, respectively, 0.8, 0.08, and 0.02 (s^{-1}). This implies that for fast metastable atoms the number of such collisions within the mentioned times τ is 6, 6, and 2, respectively. As a consequence, an equilibrium with the ambient temperature in the bulk of the emission layer is established since the intensity of upper part of the layer above 0.1 level of $Q_{max}(Z)$ makes several percents. Yee (1988) presents a calculated profile of the 630-nm emission for altitudes of 200, 250, and 300 (km) from which it is evident that some difference from the thermalized distribution is perceptible at intensities (related to the maximum) less than 0.05 and becomes substantial at intensities less than 0.01. Under the real conditions of interferometric measurements, these intensities correspond to the profile wings and are of the order of errors. Therefore, possible contributions of nonthermalized atoms cannot affect the measured temperature.

In the recent theoretical publications, the problem of the absence of complete thermalization of O(^{1}D) atoms formed in the upper atmosphere as a result of the dissociative recombination reaction was considered repeatedly (Kharchenko et al. 2005).

It is easy to show that, if the ratio of the content of the nonthermal component n_{nth} to the kinetic one n_{kin} is $\frac{n_{nth}}{n_{kin}} = 0.055$ and temperature $T_{nth} = 1.37 \cdot T_{rin}$ (Kharchenko et al. 2005), then the ratio of the amplitudes of the maximums of the emission Doppler profile would be equal to

$$k = \frac{I_{nth}}{I_{kin}} = \frac{n_{nth}}{n_{kin}} \cdot \sqrt{\frac{T_{kin}}{T_{nth}}} = \frac{0.055}{\sqrt{1.37}} = 0.047$$

4.5 Model of the 630-nm Atomic Oxygen Emission

because $\int_{-\infty}^{\infty} \exp\left(-\frac{x^2}{a^2}\right) dx = \sqrt{\pi} \cdot a$ (in this case $a = \Delta\lambda_D = 3.566 \cdot 10^{-3} \cdot \sqrt{\frac{T}{1000}}$ (nm) and the integral is proportional to n).

In such a case, one can present the summation of two Doppler contours with different amplitudes and different temperatures and their sum in the form of a Doppler profile with some effective temperature

$$\exp\left[-\left(\frac{\Delta\lambda}{\Delta\lambda_{Dkin}}\right)^2\right] + k \cdot \exp\left[-\left(\frac{\Delta\lambda}{\Delta\lambda_{Dnth}}\right)^2\right] = (1+k) \cdot \exp\left[-\left(\frac{\Delta\lambda}{\Delta\lambda_{Deff}}\right)^2\right].$$

After simple transformations, we obtain from this expression

$$\frac{T_{kin}}{T_{eff}} = 1 - \left(\frac{\Delta\lambda_{Dkin}}{\Delta\lambda}\right)^2 \cdot \ln\left\{\frac{1 + \frac{n_{nth}}{n_{kin}} \cdot \sqrt{\frac{T_{kin}}{T_{nth}}} \cdot \exp\left[\left(\frac{\Delta\lambda}{\Delta\lambda_{Dkin}}\right)^2 \cdot \left(1 - \frac{T_{kin}}{T_{nth}}\right)\right]}{1 + \frac{n_{nth}}{n_{kin}} \cdot \sqrt{\frac{T_{kin}}{T_{nth}}}}\right\}.$$

As far as the temperature is determined not by one part of the Doppler profile (though it is possible formally) but by a series of points of the entire contour (actually from the maximum to the intensity level not less than 5–10(%)), the average relation between the above-indicated temperatures can be calculated introducing the relation $\Delta\lambda = x \cdot \Delta\lambda_{Dkin}$ and using the following formula

$$\left\langle\frac{T_{kin}}{T_{eff}}\right\rangle = \frac{1}{m} \cdot \int_0^m \frac{T_{kin}}{T_{eff}} dx = 1 - \frac{1}{m} \cdot \int_0^m \frac{1}{x^2} \cdot \ln\left\{\frac{1 + \frac{n_{nth}}{n_{kin}} \cdot \sqrt{\frac{T_{kin}}{T_{nth}}} \cdot \exp\left[x^2 \cdot \left(1 - \frac{T_{kin}}{T_{nth}}\right)\right]}{1 + \frac{n_{nth}}{n_{kin}} \cdot \sqrt{\frac{T_{kin}}{T_{nth}}}}\right\} dx,$$

where m actually determines the number of $\Delta\lambda_{Dkin}$ in the integration interval. In reality $m \sim 2$. The application of the L'Hospital's rule confirms that the integrand has no singularities at $x = 0$. Further simplification of the analytical expression is not reasonable because it becomes less convenient for numerical calculations.

The result of the simplest way of calculation of the effective Doppler temperature using the line profile measured by a Fabry–Perot interferometer and evaluations of its difference from the kinetic temperature T by a numerical summation of two Doppler profiles are made for $T_{kin} = 1000$ (K), and the above-indicated conditions for the nonthermal component according to the recent paper by Kharchenko et al. (2005) were used. The resulting temperature (without any signs of a considerable deviation of the line profile in the wings for the values $\Delta\lambda = 2 \cdot \Delta\lambda_D$) was found less than 1020 (K). Such differences ($\sim$20 (K)) are certainly less than the measurement errors.

The calculation using the above presented analytical formula shows that for the profile levels from the maximum to 3(%) for the above-indicated conditions $\left\langle\frac{T_{eff}}{T_{kin}}\right\rangle \sim 1.015$. Even for the above-indicated nighttime conditions (for which $\frac{T_{nth}}{T_{kin}} \sim 2$) $\left\langle\frac{T_{eff}}{T_{kin}}\right\rangle \sim 1.021$.

Thus, one can conclude that the existence of the nonthermal component principally and formally quantitatively creates an increase in the measured Doppler temperature as compared to the kinetic temperature within the maximum of the 630-nm line emitting layer. However, this increase is only a few percents (or tens of K) and this value is within the measurements accuracy and moreover sinks in the variations caused by various dynamical processes. Therefore, there is almost no reasons for statements on the discrepancy between the measured Doppler and kinetic temperatures of the $O(^1D)$ atoms for altitudes of the emitting layer maximum. Thus, the measured Doppler temperature corresponds to the thermospheric temperature at the altitudes of the emission layer maximum.

The recording of the interference pattern was carried out both by the photoelectric method with scanning the spectrum or varying the pressure inside of the interferometer chamber (Hernandez and Turtle 1965) and with the help of piezoelectric elements (Hernandez 1966, 1970, 1974). However, this method involves simultaneous photoelectric measurements of the emission intensity for correcting the line profile distorted by intensity variations within the scanning time ∼15 (min). The photographic method made it possible to record the profile as a whole and thus exclude possible distortions of this type (Semenov 1975a,b, 1976, 1978; Ignatiev and Yugov 1995).

Regular temperature measurements began early in the 1960s and were conducted for more than 30 years under various heliogeophysical conditions. Their results were used in the main to study diurnal temperature variations. Long-term (12 years) measurements were performed at the Fritz Peak Observatory (39.9°N, 105.5°W) (Hernandez and Killeen 1988). More short-term observations are presented in a number of publications (Biondi and Feibelman 1968; Truttse and Yurchenko 1971; Blamont et al. 1974; Yurchenko 1975; Semenov 1976, 1978; Kondo and Tohmatsu 1976; Hernandez and Roble 1976, 1977; Thuillier et al. 1977a,b; Semenov and Shefov 1979a,b; Jacka et al. 1979; Cocks and Jacka 1979; Toroshelidze 1989, 1991; Ignatiev and Yugov 1995; Smith and Hernandez 1995; Killeen et al. 1995; Fagundes et al. 1995; Meriwether and Biondi 1995; Meriwether et al. 1996, 1997; Cierpka et al. 2003; Burns et al. 2004; Zhang and Shepherd 2004).

The essential feature of the available material is that, with few exclusions (Semenov and Shefov 1979a,b; Hernandez and Killeen 1988; Ignatiev and Yugov 1995; Smith and Hernandez 1995), there are data of long-term measurements performed at one and the same geographical place that can be systematized to reveal seasonal temperature variations. In the majority of publications, nocturnal temperature variations are considered, and attention is focused on the behavior of the wind.

Rocket and satellite measurements of the altitude profiles of the atomic oxygen emission were conducted for 1964–1995. In total there is information about 25 rocket and satellite measurement series. They were carried out practically at midlatitudes and, therefore, are rather uniform.

Satellite measurements have led Zhang and Shepherd (2004) to the conclusion that the emission rate distribution is described by the conventional Gauss function. However, investigations of this emission have shown that its emission rate altitude distribution is well approximated by an asymmetric Gauss function.

4.5 Model of the 630-nm Atomic Oxygen Emission

To describe the variability of the 630-nm emission, numerous measurement data have been used (Tarasova 1961, 1962, 1963; Tarasova and Slepova 1964; Nagata et al. 1965, 1968; Wallace and McElroy 1966; Reed and Blamont 1967; Gulledge et al. 1968; Huruhata and Nakamura 1968; Schaeffer et al. 1972; Thuillier and Blamont 1973; Sahai et al. 1975; Serafimov et al. 1977; Hays et al. 1978; Serafimov 1979; Tarasova et al. 1981; Abreu et al. 1982; Zhang and Shepherd 2004).

It should be noted, however, that not for all parameters of the 630-nm emission there was a possibility to reveal all mentioned types of variations. These are, in particular, the variations induced by seismic activity, stratospheric warmings, etc. The reason is that in some cases there are no observation data for concrete phenomena or they are available in small numbers.

Nocturnal Variations

The initiation of the 630-nm emission at night has the feature that it is governed by two processes. One of them occurs at a given geographical place and is related to dissociative recombination that is a consequence of the action of the ultraviolet radiation of the Sun. The other process is related to O^+ ions and photoelectrons coming from the conjugate region. These species are also produced due to the UV irradiation of this atmospheric region and serve as an additional source of excited oxygen atoms. All this is responsible for the peculiar character of the intensity variations at night that was revealed in the early observations (Elvey and Farnsworth 1942). The proportion between these mechanisms depends on latitude, longitude, and season. Thus, the observed emission intensity variations can be represented as

$$\Delta I(\chi_\odot, \varphi, \lambda) = \delta I(\chi_\odot, \varphi, \lambda) + \delta I(\chi_{\odot\mathrm{con}}, \varphi_{\mathrm{con}}, \lambda_{\mathrm{con}}) .$$

Their plots are given in Fig. 4.41. The solar zenith angles at a given geographical place, $\chi_\odot$, and in the conjugate region, $\chi_{\odot\mathrm{con}}$, depend on the geographic latitudes φ and φ_{con}, on the season, i.e., on the Sun declination $\delta_\odot$, and on the local solar times τ and τ_{con}. The zenith angle $\chi_\odot$ is determined by the relation

$$\cos\chi_\odot = \sin\varphi \cdot \sin\delta_\odot - \cos\varphi \cdot \cos\delta_\odot \cdot \cos\tau .$$

When considering processes in the magnetically conjugate region of the atmosphere, it is necessary to use transformations of the geographic (φ and λ) and geomagnetic (Φ and Λ) coordinates. In the central dipole approximation, which well suffices to describe processes at middle and low latitudes, the coordinate transformation is performed by a mere rotation of the axes which is given by formulas of spherical trigonometry (Sect. 1.4.2).

The coordinates of the Sun and the related quantities can be computed with a code (WinEphem 2002; Montenbruck and Pfleger 2000).

If a process occurring at an altitude Z is considered, it is necessary to use a relation which describes the spatial behavior of the corresponding magnetic line (Sect. 1.4).

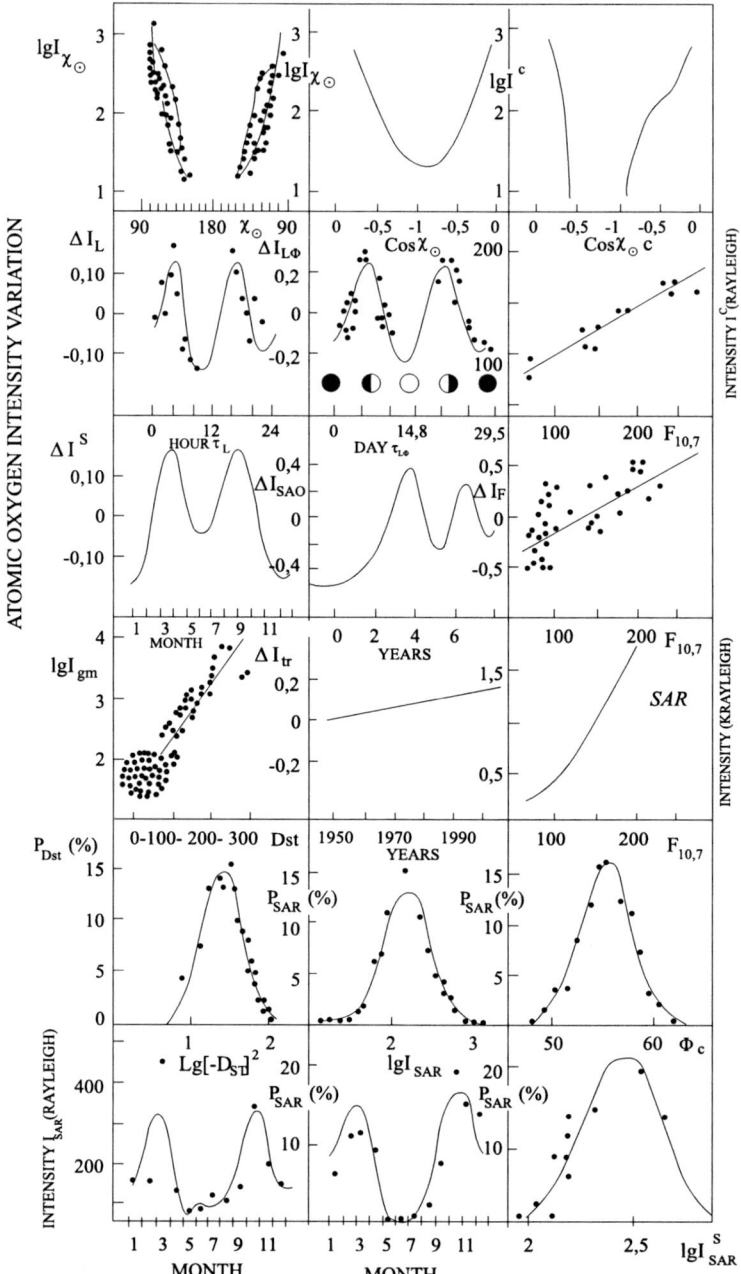

Fig. 4.41 Variations of the intensity of the 630-nm atomic oxygen emission in relation to different parameters (Shefov et al. 2006). The designations are described in the preface to Chap. 4. *Dots* are mean values of measured intensities; *solid lines* are approximations. The emission intensities under the decimal logarithm sign are expressed in Rayleighs

4.5 Model of the 630-nm Atomic Oxygen Emission

Based on the observations carried out at Zvenigorod (55.7°N, 36.8°E) (Truttse 1975), for the diurnal intensity variations for all months of year the components have been isolated that are due to the excitation of atomic oxygen as a result of dissociative recombination in the evening:

$$\log_{10} I_{\chi_\odot} = 2.84 + 1.56 \cdot \cos\left[104 \cdot (\cos\chi_\odot + 2.76)\right] + 0.1 \cdot \cos\left[209 \cdot (\cos\chi_\odot + 0.55)\right],$$

and in the morning part of night:

$$\log_{10} I_{\chi_\odot} = 2.84 + 1.56 \cdot \cos\left[104 \cdot (\cos\chi_\odot - 0.80)\right] + 0.1 \cdot \cos\left[209 \cdot (\cos\chi_\odot + 1.42)\right].$$

Here the argument is expressed in degrees.

The component of the diurnal intensity variations due to the inflow of ions and photoelectrons from the magnetically conjugate atmospheric region is described for the evening with the zenith angles $99° \leq \chi_{\odot c} \leq 110°$ by the relation

$$\log_{10} I^c_{\chi_{\odot c}} = 2.62 + 2.57 \cdot \cos\left[82 \cdot (\cos\chi_{\odot c} + 33)\right] + 0.2 \cdot \cos\left[165 \cdot (\cos\chi_{\odot c} + 0.41)\right] +$$
$$+ 0.38 \cdot \cos\left[247 \cdot (\cos\chi_{\odot c} + 1.65)\right] + 0.33 \cdot \cos\left[412 \cdot (\cos\chi_{\odot c} + 0.93)\right]$$

and for the morning with the zenith angles $115° \leq \chi_{\odot c} \leq 70°$ by

$$\log_{10} I^c_{\chi_{\odot c}} = 2.62 + 2.57 \cdot \cos\left[82 \cdot (\cos\chi_{\odot c} - 1.25)\right] + 0.2 \cdot \cos\left[165 \cdot (\cos\chi_{\odot c} + 1.59)\right] +$$
$$+ 0.38 \cdot \cos\left[247 \cdot (\cos\chi_{\odot c} + 0.4)\right] + 0.33 \cdot \cos\left[412 \cdot (\cos\chi_{\odot c} + 1)\right].$$

Analysis of the measurement data shows that the excitation of atomic oxygen at night and in the morning due to the inflow of photoelectrons and ions from the conjugate region of the upper atmosphere illuminated by the Sun, being in the other hemisphere, is effective in the main in the winter months. The contribution of this mechanism to the emission intensity depends on the level of solar activity (Fishkova 1983) and is described by the relation ($r = 0.948 \pm 0.030$)

$$\Delta I_{con} = (57 \pm 8) + (0.444 \pm 0.047) \cdot F_{10.7} \quad \text{(Rayleigh)}.$$

Since the character of the diurnal variations is determined by the solar zenith angle rather than by the local time, the temperature measurement data obtained during different seasons and at different latitudes could be combined. With the data of measurements performed during daylight hours (Cocks and Jacka 1979), regular diurnal variations have been revealed.

Here it should be noted that in a number of observations interferometric measurements were conducted for four directions (north–south and west–east) to simultaneously investigate wind motions. In this case, the zenith angles of sight were 60–70°. This, naturally, implies that the geographic coordinates of the thermospheric region at altitudes of 250–280 (km) for each direction differed from the coordinates of the observation place. Therefore, the measurement data for the western and eastern directions of sight corresponded to other local times. The difference could be as great as several tens of minutes depending on the latitude of the observation place and on the zenith angle of sight, and, naturally, it increases with latitude. The coordinates of the region of sight can be determined by the formulas

$$\sin\varphi = \sin\varphi_0 \cdot \cos\psi; \qquad \sin(\lambda_0 - \lambda) = \frac{\sin\psi}{\sqrt{1 - \sin^2\varphi_0 \cdot \cos^2\psi}},$$

where $\cos\psi = k \cdot \sin^2\zeta + \cos\zeta \cdot \sqrt{1 - k^2 \cdot \sin^2\zeta}$.

Here φ_0 and λ_0 are the geographic coordinates of the observation place; φ and λ are the coordinates of the region sighted at the zenith angle ζ; $k = \frac{1}{1+Z/R_E}$, where Z is the altitude of the sighted region and R_E is Earth's radius. Thus, for the observations in the western direction the local time is smaller and in the eastern direction is greater than in sighting toward the zenith.

The diurnal temperature variations can be determined by the formula (Fig. 4.42)

$$\Delta T_\chi = 366 \cdot \cos\left[\frac{2\pi}{360}(\chi - 26)\right] + 76 \cdot \cos\left[\frac{4\pi}{360}(\chi - 16)\right] + 21 \cdot \cos\left[\frac{6\pi}{360}(\chi - 5)\right] \text{ (K)}.$$

To describe some asymmetry of the emission parameter variations as a function of $\chi_\odot$, when calculating their evening values (local time $12 \leq \tau \leq 24$ (h)), χ is set equal to the solar zenith angle $\chi_\odot$, while for the morning part of day ($0 \leq \tau \leq 12$ (h)) it is set equal to $360 - \chi_\odot$. The daily mean temperature for quiet heliogeomagnetic conditions is 1000 (K).

The nocturnal temperature variations have the same features as the intensity variations. The pretwilight increase in temperature is caused by warming up of the atmosphere due to the inflow of superthermal electrons and ions from the magnetically conjugate atmospheric region.

Mutual relation of the diurnal variations of the maximum emission rate altitude Z_m, the thickness of the emission layer bottom part, W_l, and top part, W_u, and the layer total thickness W has revealed considerable correlations between the quantities W_l and W and the altitude Z_m:

$$W_l = 45 - 0.19 \cdot (Z_m - 250) \text{ (km)}; \qquad W = 95 - 0.19 \cdot (Z_m - 250) \text{ (km)}; r \approx -0.7.$$

However, the thickness of the layer top part, W_u, remains practically unchanged and equals 50 ± 5 (km). This has made it possible to estimate the variations of the asymmetry P:

$$\frac{1}{P} = 1 + \frac{W_l}{W_u}.$$

The solid lines in Fig. 4.43 represent the approximation by the equations

$$Z_m = 237 + 46 \cdot \cos\frac{2\pi}{360}(\chi - 182) + 1.4 \cdot \cos\frac{4\pi}{360}(\chi - 45)$$
$$+ 4.4 \cdot \cos\frac{6\pi}{360}(\chi - 105) \text{ (km)},$$

4.5 Model of the 630-nm Atomic Oxygen Emission

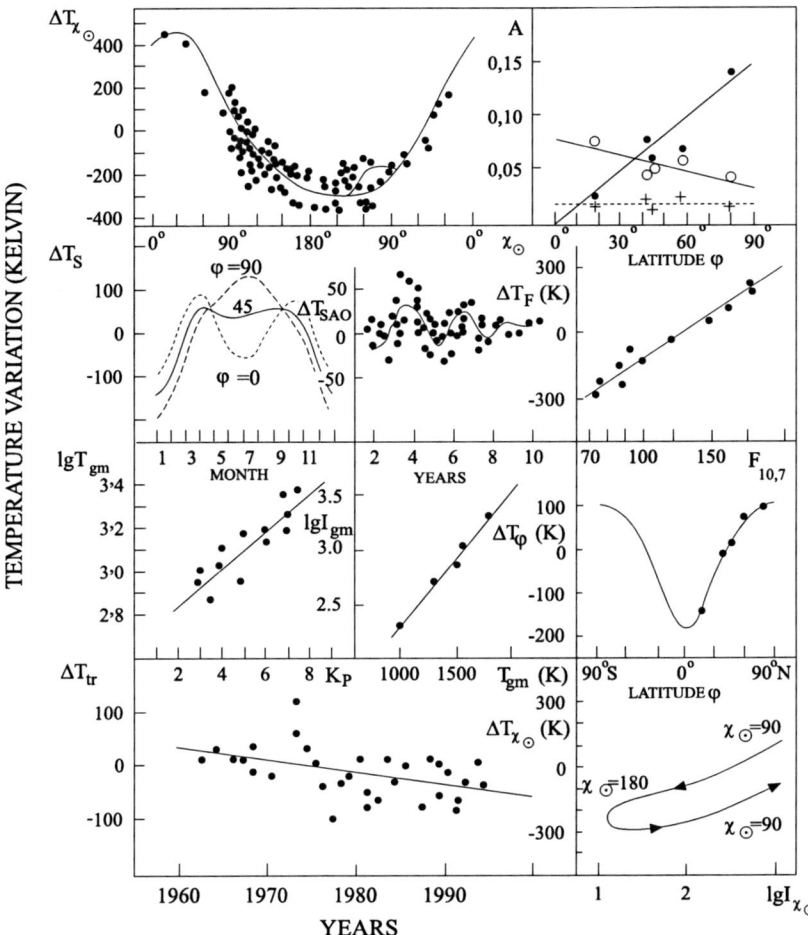

Fig. 4.42 Variations of the temperature of the 630-nm atomic oxygen emission in relation to different parameters. The designations are described in the preface to Chap. 4. *Dots* are mean measured temperatures; *solid lines* are approximations; the latitudinal variations of amplitudes of the seasonal harmonics are shown in the right upper panel: annual are *full circles*, semiannual are *hollow circles* and four-month are *crosses*; *straight lines* are regression lines

$$W_1 = 49 + 11 \cdot \cos\frac{2\pi}{360}(\chi - 5) + 2 \cdot \cos\frac{4\pi}{360}(\chi - 96)$$
$$+ 0.5 \cdot \cos\frac{6\pi}{360}(\chi - 54) \text{ (km)},$$
$$W = 100 + 16 \cdot \cos\frac{2\pi}{360}(\chi - 358) + 1.5 \cdot \cos\frac{4\pi}{360}(\chi - 163)$$
$$+ 1.5 \cdot \cos\frac{6\pi}{360}(\chi - 40) \text{ (km)},$$
$$P = 0.500 + 0.50 \cdot \cos\frac{2\pi}{360}(\chi - 180) + 0.004 \cdot \cos\frac{4\pi}{360}(\chi - 16)$$
$$+ 0.005 \cdot \cos\frac{6\pi}{360}(\chi - 105).$$

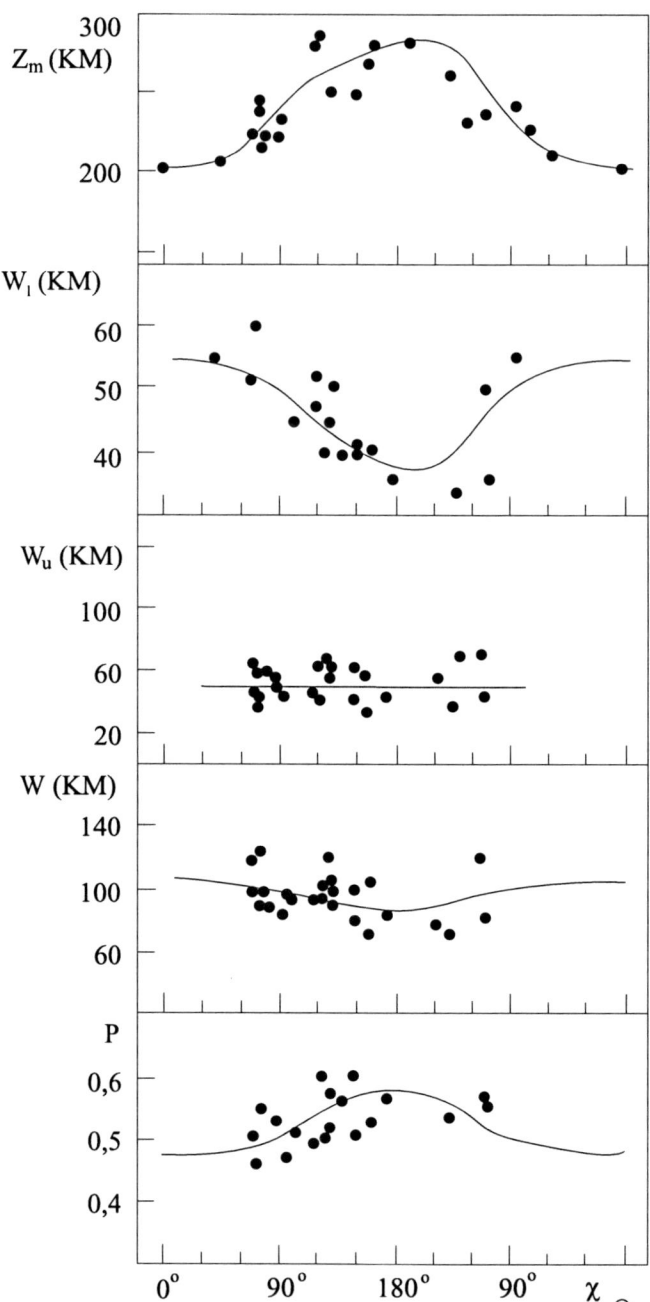

Fig. 4.43 Diurnal variations of the parameters of the 630-nm emission altitude distribution in relation to the solar zenith angle $\chi_\odot$. Z_m is the altitude of the maximum emission Q_m; W_l is the thickness of the layer bottom part at the level $Q = Q_m/2$; W_u is the thickness of the layer top part at the level $Q = Q_m/2$; W is the layer thickness, and P is the asymmetry. *Dots* are measurement data; *solid lines* are approximations

4.5 Model of the 630-nm Atomic Oxygen Emission

The relations obtained were used to calculate the altitude distributions of the emission rate for different solar zenith angles (Fig. 4.44). The emission rate variations were calculated by the intensity variations.

It can be seen that as the emission layer lowers with decreasing solar zenith angle, the bottom part of the layer expands a little, while the emission intensity increases considerably, from 100 (Rayleigh) to 25 (kilorayleigh), which is in agreement with the results of direct measurements during daylight hours (Noxon and Goody 1962).

An important feature of the concurrent intensity, temperature, and emission layer altitude variations is that the intensity variations occur synchronously with the layer altitude variations as the zenith angle $\chi_\odot$ increases or decreases, and both are ahead of the temperature rise after midnight (Fig. 4.45). This effect is also traced by the data of Zhang and Shepherd (2004). Calculations have shown that the difference in solar zenith angles at which the emission intensity and the layer altitude in the evening are equal to their respective morning values is about 30°. This difference weakly depends on both the declination of the Sun (i.e., the season) and the geographic latitude and corresponds to a time delay of about 2 (h).

Lunar Variations

The lunar intensity variations are diurnal and also depend on the phase age of the Moon.

The tidal variations in the atmosphere caused by the action of the Moon and the requirements for their analysis have been considered in Sect. 4.1.

Analysis of the measurement data published by Truttse and Belyavskaya (1975) gave the following expression for the tidal intensity variations:

$$\Delta I_L = 0.035 \cdot \cos\frac{\pi}{12}(\tau_L - 21) + 0.12 \cdot \cos\frac{\pi}{6}(\tau_L - 4) + 0.012 \cdot \cos\frac{\pi}{4}(\tau_L - 3).$$

Variations associated with the Moon's age (phase Φ) are also observed in the behavior of the emission intensity. The Moon's age can be estimated to be within several tenths of a day by the formula (Meeus 1982)

$$t_{L\Phi} = 29.53 \cdot [(t_d/365 + YYYY - 1900) \cdot 12.3685] - 1.$$

Here the square brackets denote the fractional part of the number. The coordinates of a point in a tide-deformed atmospheric region are calculated by the same formulas that are used in the calculations of diurnal lunar variations. In this case, instead of the lunar time τ_L the Moon phase age t_L is used. According to the measurements carried out at Abastumani, the intensity variations are described by the relation

$$\Delta I_{L\Phi} = 0.01 \cdot \cos\frac{2\pi}{29.53}(t_{L\Phi} - 12) + 0.21 \cdot \cos\frac{4\pi}{29.53}(t_{L\Phi} - 5)$$
$$+ 0.07 \cdot \cos\frac{6\pi}{29.53}(t_{L\Phi} - 7.6).$$

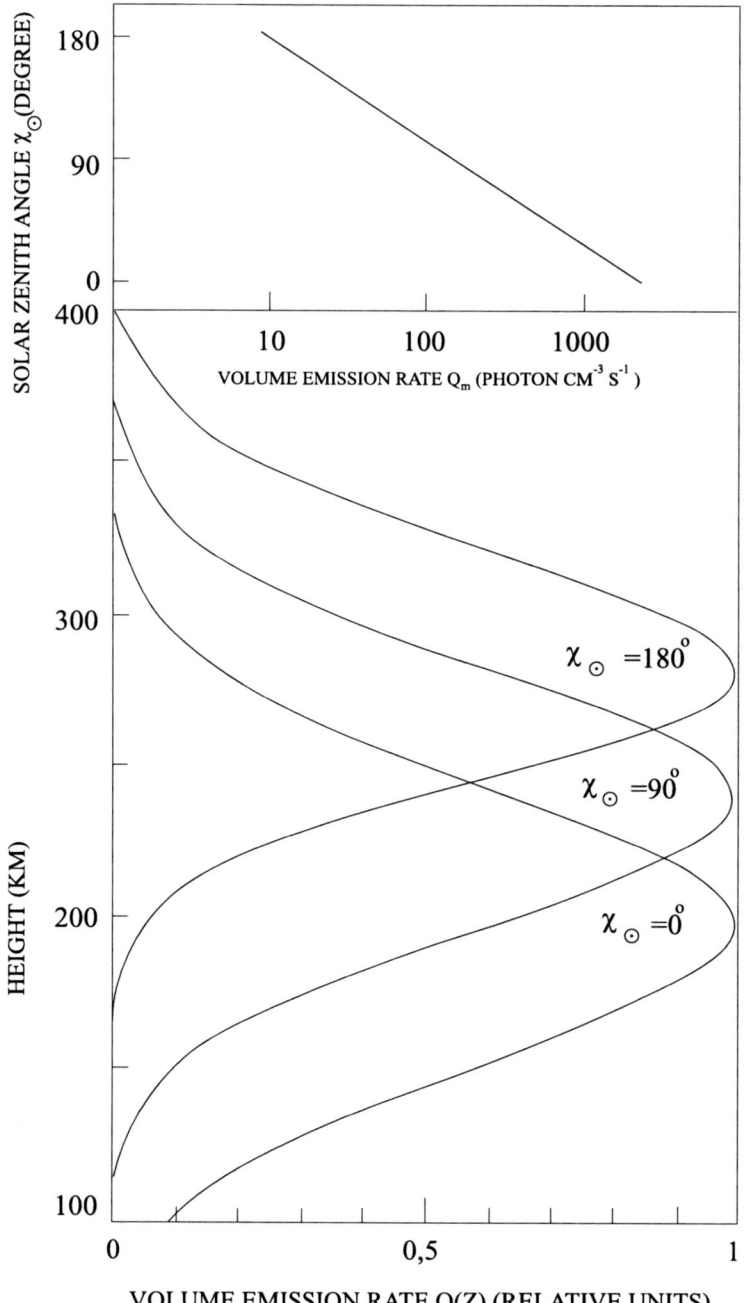

Fig. 4.44 Altitude distributions of the emission rate Q(Z) for different solar zenith angles $\chi_\odot$. The respective variations of the emission rate Q_m are shown at the top of the figure

4.5 Model of the 630-nm Atomic Oxygen Emission

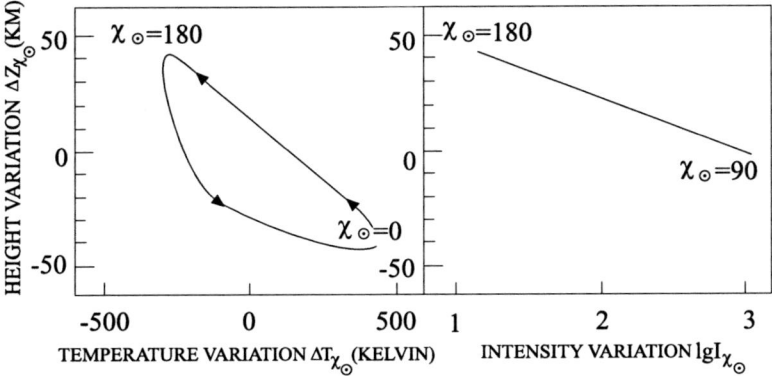

Fig. 4.45 Relation of the emission layer altitude variations to the simultaneous variations of the emission intensity and temperature for different solar zenith angles $\chi_\odot$

Seasonal Variations

The feature of the processes responsible for the initiation of the red 630-nm oxygen emission in the thermosphere determines the character of the seasonal variations of the emission intensity. Here the gradual decrease in emission intensity due to dissociative recombination and due to the contribution of the mechanism related to the magnetically conjugated region is of significance. Thus, the nightly mean intensity at midlatitudes has a maximum in summer and minima near equinoxes (Korobeynikova and Nasyrov 1974). For the time 2–3 (h) after the sunset, the intensity variations show a summer minimum and an autumn maximum (Fishkova 1983). The midnight emission intensities (6 (h) after the sunset, $\chi_\odot > 120°$), despite their small amplitudes, show semiannual variations (Truttse and Belyavskaya 1977; Fishkova 1983).

The seasonal intensity variations for given points in a day are determined by the contributions of the diurnal variations for the corresponding solar zenith angles.

Based on the data obtained at Abastumani (Fishkova 1983), they can be described by the formula

$$\Delta I_S = 0.080 \cdot \cos\frac{2\pi}{365}(t_d - 178) + 0.126 \cdot \cos\frac{4\pi}{365}(t_d - 73)$$
$$+ 0.025 \cdot \cos\frac{6\pi}{365}(t_d - 90) + 0.022 \cdot \cos\frac{8\pi}{365}(t_d - 74).$$

For 1972, the yearly average intensity I_0 was estimated to be 85 (Rayleigh).

Measurements of the emission temperature show that the character of its seasonal variations depends on latitude φ. The behavior of the variations is largely determined by the properties of the first two harmonics: annual and semiannual. Truttse and Belyavskaya (1977) present data on the temperature variations for the 630-nm emission, on the atmospheric density, and on the atomic-to-molecular oxygen density ratios. Maxima of these quantities were observed at equinoxes. The data reported by Hernandez and Roble (1977) and Hernandez and Killeen (1988) testify that the annual harmonic for the Fritz Peak Observatory (39.9°N, 254.4°E) is in agreement with the data for Zvenigorod (55.7°N, 36.8°E), namely its amplitude increases in going from the equator to the pole. The semiannual harmonic, according to the data of investigations of many emission characteristics, has a maximum amplitude at the equator and a minimum in the polar region.

The seasonal variations of the Doppler temperature derived from near-midnight measurements can be represented as (see Fig. 4.42)

$$\Delta T_S = A_1 \cdot \cos\left[\frac{2\pi}{365} \cdot (t_d - 186)\right] + A_2 \cdot \cos\left[\frac{4\pi}{365} \cdot (t_d - 93)\right]$$
$$+ A_3 \cdot \cos\left[\frac{6\pi}{365} \cdot (t_d - 70)\right] (K).$$

The harmonic amplitudes are determined by the regression relations

$$A_1 = (1.64 \pm 0.1) \cdot \varphi; \quad A_2 = (79 \pm 11) - (0.52 \pm 0.21) \cdot \varphi; \quad A_3 = 17 \pm 7.$$

The phases of the harmonics do not depend on latitude.

Variations Related to Solar Activity

Based on the data on the long-term variations of yearly average intensities (Givishvili et al. 1996), their correlation dependence on index $F_{10.7}$ (r = 0.730 ± 0.079) (see Fig. 4.41) has been obtained:

$$\Delta I_F = (0.0030 \pm 0.0010) \cdot (F_{10.7} - 130).$$

For the midnight intensities, the correlation (r = 0.782 ± 0.041) is given by the formula
$$\Delta I_F = (0.0060 \pm 0.0015) \cdot (F_{10.7} - 130).$$

The intensity (Rayleigh) of the red emission at low-latitude red auroras depends on solar flux and, simultaneously, on the magnitude of the geomagnetic disturbance characterized by geomagnetic index D_{st}. Truttse (1968a,b,1973) has derived the empirical relation

$$\log_{10} I_{gm} = (F_{10.7} - 160)/50 + (|\Phi| - 34) \cdot (-D_{st})/1460,$$

from which it follows that the necessary and sufficient condition for the occurrence of red auroras is that the indices $F_{10.7}$ and D_{st} must be simultaneously high. Since the index $F_{10.7}$ determines the thermospheric temperature and T = 379 + 3.24 $F_{10.7}$ (CIRA 1972), we have

$$\log_{10} I_{gm} = (T - 900)/170 + \||\Phi| - 34| \cdot (-D_{st})/1460.$$

4.5 Model of the 630-nm Atomic Oxygen Emission

Obviously, the further increase in temperature is related to the magnitude of the geomagnetic disturbance. This relation naturally explains the fact that even under the conditions of the solar flux maximum during the 20th cycle when the value of $F_{10.7}$ was insignificant, the red emission was not detected even in the visual observation along the emission layer from orbiting spacecrafts, since the sensitivity threshold of an eye to the 630-nm emission is over 10 (kilorayleigh).

The solar flux dependence of the temperature of the 630-nm emission was revealed at the early stage of the investigations. Based on the data of a number of observations, an approximate regression relation was obtained (see Fig. 4.42):

$$\Delta T_F = (1030 \pm 65) \cdot \log_{10} \frac{F_{10.7}}{130} \text{ (K)} .$$

As follows from the data of Killeen et al. (1995), the temperatures at high latitudes are somewhat higher than at midlatitudes.

Cyclic Aperiodic Variations Related to Solar Activity

It has been demonstrated (Shefov et al. 2006) that the intensity variations can be described by the Airy function (Abramowitz and Stegun 1964). Analysis of the frequency of observations of red auroras (Yevlashin 2005) has shown its variability during a solar cycle (ΔN_{SAO} in a year), which can be described by the Airy function:

$$\Delta N_{SAO} = -6 \cdot \text{Ai}(3 - \Delta t) ,$$

where Δt (years) is the time elapsed from the point of solar flux minimum. The maximum amplitude of the frequency variations is ~ 2.5, which corresponds to $\sim 30(\%)$ of the yearly average value.

The thermospheric temperature experiences similar variations, which can be described by the relation (see Fig. 4.42)

$$\Delta T_{SAO} = 72 \, \text{Ai}(3 - t) \text{ (K)} .$$

The maximum amplitude of the temperature variations is ~ 30 (K).

The amplitude of the intensity variations can be obtained by the relation (see Fig. 4.41)

$$\Delta I_{SAO} = 0.96 \cdot \text{Ai}(3 - t) ,$$

and the maximum amplitude is ~ 0.40.

Variations Related to Geomagnetic Activity

The effect of geomagnetic disturbances on the variations of the emission parameters depends on geomagnetic latitude. At the midlatitudes with $\Phi < 40°$, they weakly depend on the indices K_P and D_{st}. If it is required to estimate one of the indices, one can use an average relation between these indices (Jacchia 1979):

$$D_{st} = -6.5 \cdot \text{Sh}(0.535 \cdot K_p).$$

For strong disturbances ($100 \leq -D_{st} \leq 300$), the intensity variations can be determined by the formula (Truttse 1973)

$$\log_{10} I_{gm} = (1.48 \pm 0.20) + [-D_{st}/(110 \pm 10)].$$

When analyzing disturbed variations, it is difficult to distinguish between the night airglow and auroras. In this case, it is necessary to consider the latitudinal distribution of the emission 630.0 (nm). In the foregoing, the Truttse formula (Truttse 1968a,b, 1973) was given that approximates the dependence of the emission intensity on solar and geomagnetic activities.

An interesting case of a red aurora is the anomalous low-latitude red aurora that occurred on September 1–2, 1859, during which the geomagnetic index reached $D_{st} = -1760$ (nT) (Tsurutani et al. 2003), the greatest value throughout the observation period. However, this value is in doubt because it was obtained for only one place but not as a planetary average value. This was a period near the maximum of the 10th solar cycle (1856–1867) and the first maximum of the aperiodic variations. The yearly average Wolf number was $\sim$94 and the monthly mean for August $\sim$107 (maximum within the year) (Yevlashin 2005), implying that $F_{10.7}$ was $\sim$145 and 158, respectively. A similar situation took place for the low-latitude red aurora observed on October 24, 1870, at the first maximum of the aperiodic variations of the 11th cycle (1867–1878) and for the aurora observed on March 17, 1716, which, according to visual observations, was red colored and also corresponded to the first maximum of the aperiodic variations of the 3rd cycle (1712–1723) (Yevlashin 2005). For these cases, the Wolf yearly average and monthly mean numbers were 139 and 146 ($F_{10.7} \sim$ 182 and 188), respectively.

As suggested by Tsurutani et al. (2003), the geomagnetic disturbance of September 1–2, 1859, was caused by an extremely intense Sun burst (visible on its disk as a star of much greater brightness than the Sun surface), and from the available descriptions of the glow that took place at that time it can be stated that this was an intense low-latitude aurora observed at the zenith in midlatitudes ($\Phi \sim (20 \pm 10)^\circ$N). It was red, and raying forms were seen on its background. The visible glow of the raying forms near the northern horizon corresponded to latitudes $\sim 35^\circ$N. The observers characterized the glow at the zenith as bright red (Kimball 1960). Comparison with the observations of the auroras of January 21, 1957, and February 11, 1958 (Shefov and Yurchenko 1970), suggests that the intensity of this glow was some tens of megarayleighs. Analysis of red auroras (Truttse 1973) shows that as geomagnetic disturbances intensify, the latitude band of the glow becomes narrower. During red auroras the atmosphere warms up to several thousands of Kelvin due to the influx of the ring current energy (Krasovsky 1971; Lobzin and Pavlov 1998).

Following Yevlashin (2005), if we use the Truttse formula and substitute for the quantities on the right side their values characteristic of the observed anomalous aurora, namely T ~ 900 (K) (corresponding to $F_{10.7}$ ~ 158 by the model (CIRA 1972)), D_{st} ~ −1760, and Φ ~ 25°N, it appears that the intensity of the 630-nm emission should be ~100 000 (megarayleigh). In this case, the nightglow energy would be ~$3 \cdot 10^5$ (erg·cm^{-2}·s^{-1}) during 6–7 (h), which is equivalent to the brightness of the daytime solar radiation in the visible spectral region. This reasonably agrees with the estimate obtained even in view of the fact that the sensitivity of an eye in the mentioned spectral region makes ~20(%). Even if this brightness value is overestimated, a severalfold smaller illuminance is improbable as well, since such a bright glow, comparable to daylight illumination, would be noticed by the observers. This, however, was not the case. Most likely, the emission intensity was ~100 (megarayleigh) as during the aurora of February 11, 1958. Therefore, the Truttse formula must be refined for such anomalous conditions. It seems that the latitude dependence of the emission intensity should have the form of a bell-shaped function with a maximum at a latitude Φ_0 (approximately in the range 20–40°N), which, in turn, depends on D_{st}. In this case, the logarithm of the 630-nm emission intensity under strong disturbances seems to be a nonlinear function of D_{st} because the contribution of D_{st} to the emission intensity should decrease since the warming of the upper atmosphere should slowdown due to intense radiative energy removal.

First measurements of the Doppler temperature during intense auroras in the auroral zone revealed an increase in temperature to 3500 (K) (Mulyarchik 1959 1960a,b; Mulyarchik and Shcheglov 1963).

Based on the temperature measurements performed at Zvenigorod (Semenov 1978), it was found that its dependence on the index K_P for midlatitudes can be described by the formula (r = 0.977 ± 0.023) (see Fig. 4.42)

$$T_{gm} = (700 \pm 60) \cdot \exp\left(\frac{K_P}{7.24 \pm 0.9}\right) \text{ (K)}.$$

For these conditions there is a close correlation between the emission intensity and temperature, which has the form (r = 0.996 ± 0.004) (Semenov 1978)

$$I_{gm} = (7.3 \pm 1.7) \cdot \exp\left(\frac{T_{gm}}{310 \pm 15}\right) \text{ (Rayleigh)}$$

An important feature of the Doppler profiles of the 630-nm emission during auroras is the broadening of the 630-nm line profile in its bottom part. This is evidence that this emission arise at high altitudes due to a nonequilibrium process. The nonequilibrium results from incomplete relaxation of fast-excited oxygen atoms produced by dissociative recombination (Ignatiev et al. 1972, 1975, 1976, 1977, 1984; Ignatiev 1977a,b; Ignatiev and Yugov 1995; Ignatiev and Nikolashkin 2002).

Long-Term Trend

The data of long-term measurements of the 630-nm emission intensity performed since 1958 at Abastumani (φ = 41.8°N, λ = 42.8°E, Φ = 36.7°N, Λ = 120.3°E) (Fishkova 1983) and the later data (Givishvili et al. 1996) were taken as a basis to reveal a long-term trend in the intensity variations. This data set was supplemented with the data of midlatitude measurements (Barbier 1961) performed at the Haute Provence Observatory ($\varphi = 43.9°N$, $\lambda = 5.7°E$, $\Phi = 45.9°N$, $\Lambda = 86.0°E$). Upon elimination of the effect of solar activity from the series of yearly average midnight intensities, the following relation was obtained (Givishvili et al. 1996) for the interval 1953–1994 with the yearly average intensity $I_0 = 85$ (Rayleigh) for $t = 1972.5$ (see Fig. 4.41):

$$\Delta I_{tr} = (0.0025 \pm 0.0012) \cdot (t - 1972.5).$$

It is of importance that regular temperature variations have been revealed that can be eliminated in seeking long-term temperature variations. Even preliminary results indicated that there is a tendency for a long-term decrease in temperature which shows up in the behavior of some characteristics of the thermosphere (Semenov 1996; Semenov 2000). Based on the obtained parameters of the regular temperature variations, a long-term negative trend has been revealed, which is presented in Fig. 4.42. The regression relation has the form

$$\Delta T_{tr} = -(2.2 \pm 0.8) \cdot (t - 1972.5) \, (K \cdot yr^{-1})$$

for the average solar flux $F_{10.7} = 130$.

Latitude Dependence

The latitude dependence of the emission intensity is determined by the spatial distribution of the processes of dissociative recombination and by the inflow of ions and photoelectrons from the conjugate region. Moreover, the latitudinal intensity distribution is substantially affected by the midlatitude Stable Auroral Red (SAR) arcs having a longitudinal elongation. They measure in latitude about 1000 (km) (Roach and Gordon 1973).

The seasonal variations of the detection frequency of SAR arcs have maxima near equinoxes (Lobzin and Pavlov 1998). They can be described by the relation

$$P_{SAR}^S (\%) = 8.3 + 6.4 \cdot \cos \frac{2\pi}{365} (t_d - 343) + 4.0 \cdot \cos \frac{4\pi}{365} (t_d - 80)$$
$$+ 2.3 \cdot \cos \frac{6\pi}{365} (t_d - 52).$$

The corresponding intensity variations are described by the formula

$$I_{SAR}^S = 180 + 55 \cdot \cos \frac{2\pi}{365} (t_d - 343) + 80 \cdot \cos \frac{4\pi}{365} (t_d - 80)$$
$$+ 72 \cdot \cos \frac{6\pi}{365} (t_d - 52) \, (Rayleigh).$$

Mutually relating the monthly mean values of these quantities, one can see that there is a lognormal correlation between I_{SAR}^S and P_{SAR}^S (Shefov et al. 2006) (see Fig. 4.41):

4.5 Model of the 630-nm Atomic Oxygen Emission

$$P^S_{SAR}(\%) = 21 \cdot \exp\left[-\left(3.57 \cdot \log_{10}\frac{I^S_{SAR}}{280}\right)^2\right].$$

The intensity of SAR arcs depends on solar flux. Near the winter solstice in the northern hemisphere at latitudes $\Phi = 53°N$ corresponding to $L = 2.72$, the dependence of the intensity on the index $F_{10.7}$ is nonlinear (Pavlov 1998). It can be represented by the formula ($r = 0.994 \pm 0.005$)

$$I_{SAR} = (82 \pm 10) \cdot \exp[F_{10.7}/(63 \pm 4)] \text{ (Rayleigh)}.$$

SAR arcs arise due to magnetic storms, and their intensity correlates with the index D_{st}, the degree of correlation being different for different ranges of D_{st} values. For strong magnetic storms ($-200 \leq D_{st(min)} \leq -100$ (nT)), the correlation factor is -0.36 ± 0.09 (Lobzin and Pavlov 1998).

The frequency distributions depending on D_{st} are lognormal,

$$P_{SAR}(D_{st}) = 15 \cdot \exp\left[-\left(2.63 \cdot \log_{10}\left|\frac{-D_{st}}{25}\right|\right)^2\right], (\%/(\Delta\log_{10}|-D_{st}|=0.1)),$$

rather than normal, as mentioned by Ievenko and Alekseev (2004). The intensity (Rayleigh) of the 630-nm emission also obeys a lognormal distribution. The data of Ievenko and Alekseev (2004) yield

$$P_{SAR}(I) = 13 \cdot \exp\left[-\left(2.35 \cdot \log_{10}\frac{I_{SAR}}{150}\right)^2\right] (\%/(\Delta\log_{10} I_{SAR} \text{ (Rayleigh)} = 0.1)).$$

The velocity of motion V (m·s^{-1}) of an SAR arc in the equatorial (north–south) direction, according to the data by Ievenko and Alekseev (2004), also obeys a lognormal distribution:

$$P_{SAR}(V) = 22 \cdot \exp\left[-\left(6.3 \cdot \log_{10}\frac{(V_{SAR}+60)}{70}\right)^2\right] (\%/(\Delta\log_{10}(V_{SAR}+60, \text{(m/s)}) = 0.1)).$$

The maximum of the probability density distribution corresponds to the velocity $V_{SAR} = 10$ (m·s^{-1}), which agrees with the data reported by Ievenko and Alekseev (2004), but contradicts the conclusions made by these authors.

The obtained lognormal distributions of some characteristics of SAR arcs enable one, according to the results obtained by Klyatskin (1994), to conclude that these arcs are initiated due to the diffusion of radiating species in a random field of wind velocities and the auroral electron precipitation at the thermospheric altitudes.

However, the latitudinal distribution of the SAR arc intensity maximum on the meridian of Yakutsk, obtained for the period 1989–2000, is normal (Ievenko and Alekseev 2004):

$$P_{SAR}(\Phi_c) = 16 \cdot \exp\left[-\frac{(\Phi_c - 55)^2}{(3.6)^2}\right], (\%/\text{degree}).$$

In this case, the zenith of the observation station corresponded to the corrected geomagnetic latitude $\Phi_c = 57°N$. Thus, the halfwidth of an SAR arc is $\sim 6°$ or 666 (km).

Simultaneously with the analysis of seasonal variations, the latitude dependence of the mean temperature of the 630-nm emission has been revealed. It can be represented as

$$\Delta T_\varphi = -148 \cdot \cos 2\varphi - 26 \cdot \cos 4\varphi \text{ (K)}.$$

Relation to the Characteristics of the Ionospheric F2 Region

Since the 630-nm emission is initiated due to dissociative recombination, it is natural that the emission parameters are related to the characteristics of the ionospheric F2 region, such as the altitude of its maximum, $h'F2$, and the critical frequency f_0F2, which determines the electron density. For the first time, a close correlation between the ionospheric and the emission parameters has been revealed by Barbier (1959a). Numerous manifestations of this correlation have been detected in long-term observations carried out at Abastumani (Toroshelidze 1991).

This correlation is generally represented by the formula

$$I_{630} = K(f_0F2)^2 \cdot [O_2]_{200} \cdot \exp\left(-\frac{h'F2 - 200}{H}\right),$$

where $[O_2]_{200}$ is the concentration of molecular oxygen at an altitude of 200 (km) and H is the scale height in the emitting region. To describe this correlation more accurately, it is necessary to take into account the nocturnal variations of the vertical electron density profile (Serafimov and Gogoshev 1972). For this purpose, the electron density distribution was approximated by a parabolic function for the bottom part of the emission layer and by an exponential function for its top part.

Comparison of the calculations and measurements of the 630-nm emission intensity has shown good agreement for quiet geomagnetic conditions (Semenov 1976).

Orographic Variations

Measurements of the variations of the emission parameters on the leeward side of mountains were conducted first in the Urals region and then in the region of Caucasus (Semenov et al. 1981; Shefov et al. 1983; Shefov 1985; Sukhodoev et al. 1989a,b). It has been found that the temperature of the 630-nm emission was 100 and 200 (K), respectively, above that in flat regions. The spotty structure of the emission intensity distribution over Hawaiian Islands (Roach and Gordon 1973) also seems to be a manifestation of orographic disturbances. In this case, the expected increment of the temperature from its mean value can be ~ 300 (K).

The interferometric measurements performed in South America near Arequipa (Peru) (16.5°S, 288.5°E) at an altitude of ~ 5500 (m) on September 26–27, 1994, have made it possible to reveal temperature variations for the 630-nm emission at the zenith and ~400 (km) north (N), east (E), south (S), and west (W) of the observation point. The last two regions of the atmosphere were over the ocean, while the first two over the Andes mountain ridge. From the data reported by Meriwether et al. (1996,1997) on the emission intensity variations during the mentioned night (not reduced to the zenith for the measurement conditions at the zenith distance $\zeta = 60°$) and on the temperature variations, it follows that a wave with phase ~3.5 (h) propagating west–east was observed for all measurement directions. Upon elimination of this wave component, an increase in intensity (~10 (Rayleigh)) and temperature (~150 (K)) in the leeward region of the mountains was detected. Investigations of the orographic variations by the hydroxyl emission at the mesopause altitudes (~87 (km)) near the Caucasian ridge have shown that the simultaneous increase in temperature and intensity of the emission was accompanied by a decrease in altitude of the emission layer (Shefov et al. 1999, 2000b). Therefore, it seems that the data reported by Meriwether et al. (1996, 1997) can be interpreted in a similar manner.

Earthquake Effect

The response of the airglow to seismic processes was repeatedly detected during observations in the Caucasus and in Turkmenistan (Nasyrov 1978; Fishkova et al. 1985; Fishkova and Toroshelidze 1989).

During the Roodbar earthquake (M = 7) in Iran (37°N, 49.6°E) that took place on June 21, 1990, at a local solar time $\tau = 0.3$ (h), temperature variations were detected at four directions (north, east, south, and west) on sighting the upper atmosphere with a Fabry–Perot interferometer at the Vannovsky Observatory of the Physicotechnical Institute of the Turkmen Academy of Sciences near Ashkhabad (38.0°N, 58.4°E) (Akmamedov 1993). An increase in temperature by up to 300 (K) was detected at points almost equally distant north and south (940 and 870 (km), respectively) of the epicenter. A smaller increase in temperature (by ~70 (K)) was observed in the eastern, most distant (1200 (km)) region. In the western, nearest to the epicenter, region, the temperature increased by 100–150 (K). The mean atmospheric temperature the for the mentioned dates was ~1050 (K). Analysis of the nature of the phenomenon observed (Pertsev and Shalimov 1996; Akmamedov et al. 1996) allowed the authors of the cited publications to propose a possible mechanism of the temperature and intensity variations at the meridional direction. This phenomenon arise due to a combination of several processes. Before an earthquake, over its focus, there occurs a modulation of the inflow of lithospheric gases to the atmosphere. As a result, gravity waves with phases 1–2 (h) are generated. When arrived at altitudes of ~125 (km), they dissipate, and at ~200 (km) they give rise to Rayleigh–Taylor instability. This, in turn, leads to the formation of plasma bubbles, i.e., regions of lower plasma density, which propagate upward and elongate along the magnetic lines. In the bubbles, the temperature increases by ~300 (K), while the emission intensity decreases. These disturbed regions exist for ~1 (h). This pattern of the emission intensity and temperature variations induced by earthquakes is supported by measurements.

Thus, based on long-term measurements and investigations of the spatial and temporal variations of the intensity of the 630-nm atomic oxygen emission of the night and twilight airglow, which occurs at the altitudes of the ionospheric F2 layer, empirical relations have been constructed that allow one to estimate the emission intensity for various heliogeophysical conditions, such as the night time of day, lunar time, phase age of the Moon, season, year's number, and solar flux during an 11-yr cycle, and in relation to its cyclic aperiodic variations, to the magnitude of geomagnetic disturbances, and to their latitudinal manifestations in the form of subauroral red arcs at midlatitudes.

References

Abramowitz M, Stegun IA (1964) Handbook of mathematical functions with formulas, graphs and mathematical tables. Nat Bur Stand, Washington DC

Abreu VJ, Schmitt GA, Hays PB, Dachev TP (1982) Volume emission rate profiles of the 6300-A tropical nightglow obtained from the AE-E satellite: latitudinal and seasonal variations. J Geophys Res 87A:6346–6352

Agashe VV, Pawar VR, Aher GR, Nighut DN, Jahangir A (1989) Study of mesopause temperature and its behaviour from OH nightglow. Indian J Radio Space Phys 18:309–314

Akmamedov Kh (1993) Interferometric measurements of the temperature of the F_2 region of the ionosphere during the period of the Iranien earthquake of June 20, 1990. Geomagn Aeron 33:135–138

Akmamedov Kh, Pertsev NN, Romanova NN, Semenov AI, Chefranov SG, Shalimov SL, Shefov NN (1996) Possible mechanism of temperature rise in the F2-region of the ionosphere during the Iran earthquake of June 20, 1990. Geomagn Aeron 36:228–231

Apostolov EM, Letfus V (1985) Quasi-biennial oscillations of the green corona intensity. Bull Astron Inst Czechosl 36:199–205

Bakanas VV, Perminov VI (2003) Some features in the seasonal behavior of the hydroxyl emission characteristics in the upper atmosphere. Geomagn Aeron 43:363–369

Bakanas VV, Perminov VI, Semenov AI (2003) Seasonal variations of emission characteristics of the mesopause hydroxyl with different vibrational excitation. Adv Space Res 32:765–770

Baker DJ, Stair AT (1988) Rocket measurements of the altitude distribution of the hydroxyl airglow. Physica Scripta 37:611–622

Barbier D (1959a) Recherches sur la raie 6300 de luminescence atmosphèrique nocturne. Ann Géophys 15:179–217

Barbier D (1959b) Sur les variations systematiques des intensités des pricipales radiations de lumière du ciel nocturne à l'observatoire de Haute Provence. Ann Géophys 15:412–414

Barbier D (1961) Les variations d'intensité de la raie 6300 Å de la luminescence nocturne. Ann Géophys 17:3–15

Barbier D, Dufay J, Williams D (1951) Recherches sur l'émission de la raie verte de la lumière du ciel nocturne. Ann d'Astrophys 13:399–437

Barnett JJ, Corney N (1985) Middle atmosphere reference model derived from satellite data. In: Labitzke K, Barnett JJ, Edwards B (eds) Handbook for MAP, vol 16. SCOSTEP, Urbana, pp 47–85

Bates DR (1981) The green light of the night sky. Planet Space Sci 29:1061–1067

Bates DR, Nicolet M (1950) The photochemistry of atmospheric water vapour. J Geophys Res 55:301–327

Batista P, Clemesha BR, Simonich DM (1990) Seasonal variations in mesospheric sodium tidal activity. J Geophys Res 95D:7435–7442

Batista PP, Takahashi H, Clemesha BR (1994) Solar cycle and QBO effect on the mesospheric temperature and nightglow emissions at a low latitude station. Adv Space Res 14:221–224

Benedict WS, Plyler EK, Humphreys CJ (1953) The emission spectrum of OH from 1.4 to 1.7μ. J Chem Phys 21:398–402

Berg MA, Shefov NN (1962a) Emission of the hydroxyl bands and the (0,1) λ 8645 Å atmospheric band of oxygen in the nightglow. Planet Space Sci 9:167–171

Berg MA, Shefov NN (1962b) OH emission and atmospheric O_2 band λ 8645 A. In: Krassovsky VI (ed) Aurorae and airglow. N 9. USSR Acad Sci Publ House, Moscow, pp 46–52

Berg MA, Shefov NN (1963) OH emission and atmospheric O_2 band λ 8645 Å. In: Krassovsky VI (ed) Aurorae and airglow. N 10. USSR Acad Sci Publ House, Moscow, pp 10–23

Berthier P (1955) Variations saisonnières de l'intensité des bandes de molecules OH et O_2 atmosphérique dans la luminescence du ciel nocturne. Compt Rend. 240:1796–1798

Berthier P (1956) Étude spectrophotométrique de la luminescence nocturne des bandes des molecules OH et O_2 Atmosphérique. Ann Géophys 12:113–143

Biondi MA, Feibelman WA (1968) Twilight and nightglow spectral line shapes of oxygen λ 6300 and λ 5577 radiation. Planet Space Sci 16:431–443

Birnside BG, Tepley CA (1990) Airglow intensities observed in the Southern and Northern hemispheres. Planet Space Sci 38:1161–1177

Blamont JE, Luton JM, Nisbet JS (1974) Global temperature distribution from OGO-6 6300 Å airglow measurements. Radio Sci 9:247–251

Bronshten VA, Dagaev MM, Kononovich EV, Kulikovsky PG (1981) Astronomical calendar. Invariable part. Abalakin VK (ed). Nauka, Moscow

Bronshten VA, Grishin NI (1970) Noctilucent clouds. Nauka, Moscow

Burns AG, Killeen TL, Wang W, Roble RG (2004) The solar-cycle-dependent response of the thermosphere to geomagnetic storms. J Atmos Solar—Terr Phys 66:1–14

Chamberlain JW (1961) Physics of the aurora and airglow. Academic Press, New York

Chandra S, Jackman CH, Fleming EL, Russell JM III (1997) The seasonal and long term changes in mesospheric water vapor. Geophys Res Lett 24:639–642

Chapman S (1939) Notes on atmospheric sodium. Astrophys J 90:309–316

Chapman S, Lindzen RS (1970) Atmospheric tides. Reidel, Dordrecht-Holland

Christophe-Glaume J (1965) Étude de la raie 5577 Å de l'oxygène dans la luminescence atmospherique nocturne. Ann Géophys 21:1–57

Chunchuzov IP (1988) Orographic waves in the atmosphere produced by a varying wind. Izv USSR Acad Sci Atmos Oceanic Phys 24:5–12

Chunchuzov IP (1994) On possible generation mechanism for nonstationary mountain waves in the atmosphere. J Atmos Sci 15:2196–2206

Chunchuzov YeP, Shagaev MV (1983) Estimates of the coefficient of vertical turbulent diffusion in the lower thermosphere. Izv USSR Acad Sci Atmos Oceanic Phys 20:154–155

Cierpka K, Kosch MJ, Holma H, Kavanagh AJ, Hagfors TL (2003) Novel Fabry–Perot interferometer measurements of F-region ion temperature. Geophys Res Lett 30:1293, doi: 10.1029/2002GL015833

CIRA (1972). COSPAR International Reference Atmosphere. Stickland AC (ed). Akademie-Verlag, Berlin

Clayton HH (1884a) A lately discovered meteorological cycle. I. Am Meteorol J 1:130–143

Clayton HH (1884b) A lately discovered meteorological cycle. II. Am Meteorol J 1:528–534

Clemesha BR, Takahashi H (1995) Rocket-borne measurements of horizontal structure in the OH (8.3) and NaD airglow emissions. Adv Space Res 17:81–84

Clemesha BR, Simonich DM, Batista PP (1992) A long-term trend in the height of the atmospheric sodium layer: possible evidence for global change. Geophys Res Lett 19:457–460

Clemesha BR, Simonich DM, Takahashi H, Melo SML (1993) A simultaneous measurement of the vertical profiles of sodium nightglow in the upper atmosphere. Geophys Res Lett 20:1347–1350

Clough HW (1924) A systematically varying period with an average length of 28 months in weather and solar phenomena. Mon Weather Rev 52:421–439

Clough HW (1928) The 28-month period in solar activity and corresponding periods in magnetic and meteorological data. Mon Weather Rev 56:251–264

Cocks TD, Jacka F (1979) Daytime thermospheric temperatures, winds velocities and emission intensities derived from ground based observations of OI λ 630 nm airglow line profiles. J Atmos Terr Phys 41:409–415

Davis TN, Smith LL (1965) Latitudinal and seasonal variations in the night airglow. J Geophys Res 70:1127–1138

Deans AG, Shepherd GG, Evans WFJ (1976) A rocket measurement of the O_2 ($b^1\Sigma_g^+ \to X^3\Sigma_g^-$) atmospheric band nightglow altitude distribution. Geophys Res Lett 3:441–444

Dodd JA, Blumberg WAM, Lipson SJ, Lowell JR, Armstrong PS, Smith DR, Nadile RM, Wheeler NB, Huppi ER (1993) OH (v, N) column densities from high-resolution Earthlimb spectra. Geophys. Res. Lett. 20:305–308

Dodd JA, Lipson SJ, Armstrong PS, Blumberg WAM, Nadile RM, Adler-Golden SM, Marinelli WJ, Holtzclaw KW, Green BD (1994) Analysis of hydroxyl earthlimb airglow emissions: kinetic model for state-to-state dynamics of OH (v, N). J. Geophys. Res. 99:3559–3585

Dufay M (1959) Étude photoélectrique du spectre du ciel nocturne dans le proche infra-rouge. Ann Géophys 15:134–151

Dufay J, Tcheng M-L (1946) Recherches spectrophotométriques sur la lumière du ciel nocturne dans la région visible.1. Ann Géophys 2:189–230

Dufay J, Tcheng M-L (1947a) Recherches spectrophotométriques sur la lumière du ciel nocturne dans la région visible. 2. Ann Géophys 3:153–183

Dufay J, Tcheng M-L (1947b) Recherches spectrophotométriques sur la lumière du ciel nocturne dans la région visible. 3. Ann Géophys 3:282–305

Ebel A (1980) Eddy diffusion models for mesosphere and lower thermosphere. J Atmos Terr Phys 42:102–104

Elvey CT, Farnsworth AH (1942) Spectrophotometric observations of the light of the night sky. Astrophys J 96:451–467

Espy PJ, Stegman J (2002) Trends and variability of mesospheric temperature at high-latitudes. Phys Chem Earth 27:543–553

Evlashin LS, Shefov NN, Ponomarev VM (1996) Spectral energy distribution of auroral emission based on model concepts. Geomagn Aeron 36:660–666

Evlashin LS, Semenov AI, Shefov NN (1999) Long-term variations in the thermospheric temperature and density on the basis of an analysis of Störmer's aurora-height measurements. Geomagn Aeron 39:241–245

Fadel KhM, Semenov AI, Shefov NN, Sukhodoev VA, Martsvaladze NM (2002) Quasibiennial variations in the temperatures of the mesopause and lower thermosphere and solar activity. Geomagn Aeron 42:191–195

Fagundes PR, Sahai Y, Bittencourt JA, Takahashi H (1995) Observations of thermospheric neutral winds and temperatures at Cachoeira Paulista (23°S, 45°W) during a geomagnetic storms. Adv Space Res 16:27–30

Fishkova LM (1955) Intensity variations of the night sky luminosity in the near infrared region. In: Kharadze EK (ed) Bull Abastumani Astrophys Observ N 15: pp 3–23

Fishkova LM (1976) Regular nocturnal and seasonal variations of the emission intensity of OH, NaD, 5577 Å of the upper atmosphere. In: Krassovsky VI (ed) Aurorae and airglow. N 24. Soviet Radio, Moscow, pp 5–15

Fishkova LM (1978) About oscillations of intensity of night emission of the upper atmosphere in the periods of stratospheric warming up. Geomagn Aeron 18:549–550

Fishkova LM (1979) Nocturnal sodium emission in the Earth's upper atmosphere. In: Problems of the atmospheric optics. Leningrad State Univ, Leningrad, pp 154–172

Fishkova LM (1981) On variations of vibrational level populations of excited OH molecules in the upper atmosphere. In: Galperin YuI (ed) Aurorae and airglow. N 29. Radio and Svyaz, Moscow, pp 9–21

Fishkova LM (1983) The night airglow of the Earth mid-latitude upper atmosphere. In: Shefov NN (ed). Metsniereba, Tbilisi

Fishkova LM, Toroshelidze TI (1989) The reflection of seismic activity in night sky glow variations. In: Feldstein YaI, Shefov NN (eds) Aurorae and airglow. N 33. VINITI, Moscow, pp 17–23

Fishkova LM, Gokhberg MB, Pilipenko VA (1985) Relationship between night airglow and seismic activity. Ann Geophys 3:679–694

Fishkova LM, Martsvaladze NM, Shefov NN (2000) Patterns of variations in the OI 557.7-nm. Geomagn Aeron 40:782–786

Fishkova LM, Martsvaladze NM, Shefov NN (2001a) Long-term variations of the nighttime upper-atmosphere sodium emission. Geomagn Aeron 41:528–532

Fishkova LM, Martsvaladze NM, Shefov NN (2001b) Seasonal variations in the correlation of atomic oxygen 557.7-nm emission with solar activity and in long-term trend. Geomagn Aeron 41:533–539

Forbes JM, Geller MA (1972) Lunar semidiurnal variation in OI (5577 Å) nightglow. J Geophys Res 77:2942–2947

Frank LA, Signarth JB, Craven JD (1986) On the flux of small comets in to the Earth's upper atmosphere. I. Observations. Geophys Res Lett 13:303–306

Fukuyama K (1976) Airglow variations and dynamics in the lower thermosphere and upper mesosphere—I. Diurnal variations and its seasonal dependency. J Atmos Terr Phys 38:1279–1287

Fukuyama K (1977a) Airglow variations and dynamics in the lower thermosphere and upper mesophere—II. Seasonal and long-term variations. J Atmos Terr Phys 39:1–14

Fukuyama K (1977b) Airglow variations and dynamics in the lower thermosphere and upper mesosphere—III. Variations during stratospheric warming events. J Atmos Terr Phys 39:317–331

Gadsden M (1990) A secular changes in noctilucent clouds occurrence. J Atmos Terr Phys 52:247–251

Gadsden M (1998) The north-west Europe data on noctilucent clouds: a survey. J Atmos Solar—Terr Phys 60:1163–1174

Gadsden M (2002) Statistics of the annual counts of nights on which NLCs were seen. Mem Br Astron Assoc 45. Aurora section (Meeting "Mesospheric clouds", Scottand, Perth, 19–22 August, 2002)

Gann RG, Kaufman F, Biondi MA (1972) Interferometric study of the chemiluminescent excitation of sodium by active nitrogen. Chem Phys Lett 16:380–383

Gavrilieva GA, Ammosov AA (2002a) Seasonal variations of the mesopause temperature over Yakutsk (63°N, 129.5°). Geomagn Aeron 42:267–271

Gavrilieva GA, Ammosov PP (2002b) Near-mesopause temperatures registered over Yakutia. J Atmos Solar—Terr Phys 64:985–990

Givishvili GV, Leshchenko LN, Lysenko EV, Perov SP, Semenov AI, Sergeenko NP, Fishkova LM, Shefov NN (1996) Long-term trends of some characteristics of the Earth's atmosphere. I. Experimental results. Izv Atmos Oceanic Phys 32:303–312

Golitsyn GS, Semenov AI, Shefov NN, Fishkova LM, Lysenko EV, Perov SP (1996) Long-term temperature trends in the middle and upper atmosphere. Geophys Res Lett 23:1741–1744

Golitsyn GS, Semenov AI, Shefov NN (2000) Seasonal variations of the long-term temperature trend in the mesopause region. Geomagn Aeron 40:198–200

Golitsyn GS, Semenov AI, Shefov NN (2001) Thermal structure of the middle and upper atmosphere (25–110 km), as an image of its climatic change and influence of solar activity. In: Beig G (ed) Long term changes and trends in the atmosphere, vol 2. New Age Int Lim Publ, New Delhi, pp 33–42

Golitsyn GS, Semenov AI, Shefov NN, Khomich VYu (2006) The response of the middle atmosphere temperature on the solar activity during various seasons. Phys Chem Earth 31:10–15

Greer RGH, Best GT (1967) A rocket-borne photometric investigation of the oxygen lines at 5577 Å and 6300 Å, the sodium D-lines and the continuum at 5300 Å in the night airglow. Planet Space Sci 15:1857–1881

Greer RGH, Llewellyn EJ, Solheim BH, Witt G (1981) The excitation of O_2 ($b^1\Sigma_g^+$) in the nightglow. Planet Space Sci 29:383–389

Greer RGH, Murtagh DP, McDade IC, Dickinson PHG, Thomas L, Jenkins DB, Stegman J, Llewellyn EJ, Witt G, Mackinnon DJ, Williams ER (1986) ETON 1: A data base pertinent to the study of energy transfer in the oxygen nightglow. Planet Space Sci 34:771–788

Gulledge IS, Packer DM, Tilford SC, Vanderllice JT (1968) Intensity profiles of the 6300 Å and 5577 Å OI lines in the night airglow. J Geophys Res 73:5535–5547

Harris FR (1983) The atmospheric system of O_2 in nightglow. EOS Trans. AGU. 64:779

Hauchecorne A, Blix T, Gerndt R, Kokin GA, Meyer W, Shefov NN (1987) Large-scale coherence of the mesospheric and upper stratospheric temperature fluctuations. J Atmos Terr Phys 49:649–654

Hays PB, Rusch DW, Roble RG, Walker JCG (1978) The OI 6300 Å airglow. Rev Geophys Space Res 16:225–232

Helmer M, Plane JMC (1993) A study of the reaction $NaO_2 + O \rightarrow NaO + O_2$: implications for the chemistry of sodium in the upper atmosphere. J Geophys Res 98D:23207–23222

Hernandez G (1966) Analytical description of a Fabry–Perot photoelectric spectrometer. Appl Opt 5:1745–1748

Hernandez G (1970) Analytical description of a Fabry–Perot photoelectric spectrometer. 2: Numerical results. Appl Opt 9:1591–1596

Hernandez G (1974) Analytical description of a Fabry–Perot photoelectric spectrometer. 3: Off axis behavior and interference filters. Appl Opt 13:2654–2661

Hernandez G (1975) Reaction broadening of the line profiles of atomic sodium in the night airglow. Geophys Res Lett 2:103–105

Hernandez G (1976) Lower thermosphere temperatures determined from the line profiles of the OI 17,924–K (5577Å) emission in the night sky. 1. Long-term behaviour. J Geophys Res 81:5165–5172

Hernandez G (1977) Lower thermosphere temperatures determined from the line profiles of the OI 17,924–K (5577Å) emission in the night sky. 2. Interaction with the lower atmosphere during stratospheric warming. J Geophys Res 82:2127–2131

Hernandez G, Killeen TL (1988) Optical measurements of winds and kinetic temperatures in the upper atmosphere. Adv Space Res 8:149–213

Hernandez G, Roble RG (1976) Direct measurements of nighttime thermospheric winds and temperatures. I. Seasonal variations during geomagnetic quiet periods. J Geophys Res 81:2065–2074

Hernandez G, Roble RG (1977) Direct measurements of night-time thermospheric winds and temperatures. 3. Monthly variations during solar minimum. J Geophys Res 82:5505–5511

Hernandez GJ, Silverman SM (1964) A reexamination of Lord Rayleigh's data on the airglow 5577 (OI) emission. Planet Space Sci 12:97–112

Hernandez GJ, Turtle JP (1965) Nightglow 5577 Å [OI] line kinetic temperatures. Planet Space Sci 13:901–904

Hernandez G, Smith RW, Conner JF (1992) Neutral wind and temperature in the upper mesosphere above South Pole, Antarctica. Geophys Res Lett 19:53–56

Hernandez G, Wiens R, Lowe RP, Shepherd GG, Fraser GJ, Smith RW, LeBlanc L, Clark M (1995) Optical determination of the vertical wavelength of propagating upper atmosphere oscillations. Geophys Res Lett 22:2389–2392

Huruhata M, Nakamura T (1968) Rocket observations of emission heights of 6300 Å line in night airglow. In: Mitra AP, Jacchia LG, Newman WS (eds) Space Res, vol 8. North-Holland, Amsterdam, pp 699–704

Huruhata M, Nakamura T, Steiger WR (1967) A rocket observation of (OI) 5577 Å emission and continuum at 5300 Å in night airglow. Rep Ionosph Space Res Japan 21:229–232

Ievenko IB, Alekseev VN (2004) Effect of the substorm and strorm on the SAR arc dynamics: a statistical analysis. Geomagn Aeron 44:592–603

Ignatiev VM (1977a) Unusual profiles of the 5577 Å and 6300 Å emissions in aurorae. Astron Circ USSR Acad Sci N 940:2–4

Ignatiev VM (1977b) Peculiarities of contours of 5577 Å and 6300 Å lines in auroras. Geomagn Aeron 17:153–154

Ignatiev VM, Yugov VA (1995) Interferometry of the large-scale dynamics of the high-latitudinal thermosphere. Shefov NN (ed). Yakut Sci Centre Siberian Branch RAN, Yakutsk

Ignatiev VM, Nikolashkin SV (2002) The temperature of the subauroral lower thermosphere during the stratospheric warming in Januar–February and March, 2000. Geomagn Aeron 42:398–403

Ignatiev VM, Yugov VA, Alekseev KV, Atlasov KV (1972) The interferometric measurements of Doppler temperature according to λ 6300 Å width in aurora. In: Fishkova LM, Kharadze EK (eds) Bull Abastumani Astrophys Observ. N 42. pp 91–96

Ignatiev VM, Sivtseva LD, Yugov VA, Atlasov KV (1974) Regular variations of the hydroxyl rotational temperatures over Yakutsk. In: Physics of the upper atmosphere at high latitudes. N 2. Yakut Depart Siberian Branch USSR Acad Sci, Yakutsk, pp 22–31

Ignatiev VM, Yugov VA, Atlasov KV (1975) Dissociative–recombinative profiles of the 5577 Å and 6300 Å lines in aurorae. Preprint ICPhIA. Yakut Branch Sibirean Depart USSR Acad Sci, Yakutsk

Ignatiev VM, Yugov VA, Atlasov KV, Makarov GA, Borisov GV (1976) The emission 5577 Å Doppler temperature variations in aurorae. In: Krassovsky VI (ed) Aurorae and airglow. N 24. Soviet Radio, Moscow, pp 59–63

Ignatiev VM, Yugov VA, Atlasov KV (1977) Auroral atmosphere temperature measurements as based on 6300 Å emission. In: Krassovsky VI (ed) Aurorae and airglow. N 25. Soviet Radio, Moscow, pp 9–12

Ignatiev VM, Yugov VA, Atlasov KV (1984) Non-thermal profiles of oxygen atom emissions in auroras. In: Galperin YuI (ed) Aurorae and airglow. N 31. VINITI, Moscow, pp 134–140

Ivanov VE, Kirillov AS, Mal'kov MV, Sergienko TI, Starkov GV (1993) The auroral oval boundaries and planetary model of luminous intensity. Geomagn Aeron 33:630–636

Jacchia LG (1979) CIRA-1972, recent atmospheric models and improvements in progress. In: Rycroft MJ (ed) Space Res, vol 19. Pergamon Press, Oxford, pp 179–192

Jacka F, Bower AR, Wilksch PA (1979) Thermospheric temperatures and winds derived from OI λ 630 nm night airglow line profiles. J Atmos Terr Phys 41:397–407

Kalinin YuD (1952) On the certain problems of the secular variation studies of the terrestrial magnetism. In: Trans Institute Terr Magn Ionosph Radio Wave Prop USSR Acad Sci. N 8(18). Hydrometeoizdat, Leningrad, pp 5–11

Kharchenko V, Dalgarno A, Fox JL (2005) Thermospheric distribution of fast $O(^1D)$ atoms. J Geophys Res 110A:12, doi: 10.1029/2005JA011232

Killeen TL, Won YI, Niciejewski RJ, Bums AG (1995) Upper thermosphere winds and temperatures in the geomagnetic polar cap: solar cycle, geomagnetic activity, and interplanetary magnetic field dependencies. J Geophys Res 100A:21327–21342

Kimball DS (1960) A study of the aurora of 1859. Sci Rep N 6. UAG-R109. Univ Alaska, Fairbanks

Kluev OF (1985) Thermospheric temperature measurement from the emissive spectra of the AlO molecules. In: Portnyagin YuI, Chasovitin YuK (eds) Trans Institute Exper Meteorol. N 16(115). Upper atmospheric physics. Hydrometeoizdat, Moscow, pp 15–25

Klyatskin VI (1994) Statistical description of the diffusion of tracers in a random velocity field. Uspekhi Phys Nauk 164:531–544

Kondo Y, Tohmatsu T (1976) Thermospheric temperature dependence of the atomic oxygen 6300 Å emission in the twilight airglow. J Geomagn Geoelectr 28:207–218

Kononovich EV (2005) Analytical representations of mean solar activity variations during a cycle. Geomagn Aeron 45:295–302

Kononovich EV, Shefov NN (2003) Fine structure of the 11-years cycles of solar activity. Geomagn Aeron 43:156–163

Koomen M, Scolnik R, Tousey R (1956) Distribution of the night airglow (OI) 5577 Å and NaD layers measured from a rocket. J Geophys Res 61:304–306

Korobeynikova MP, Nasyrov GA (1972) Study of the nightglow emission λ 5577 Å for 1958–1967 in Ashkhabad. Ylym (Nauka), Ashkhabad

Korobeynikova MP, Nasyrov GA (1974) Structural peculiarities of the λ 6300 Å emission. In: Khrgian AKh (ed) Geophys Bull N 27. Nauka, Moscow, pp 35–39

Korobeynikova MP, Nasyrov GA (1975) Influence of the gravity wave on the behavior of the 5577 Å emission. In: Krassovsky VI (ed) Aurorae and airglow. N 23. Nauka, Moscow, pp 143–148

Korobeynikova MP, Nasyrov GA, Khamidulina VG (1966) The nightglow emission λ 5577 Å. In: Kalchaev KK, Shefov NN (eds). Tables and maps of isophotes. Ashkhabad, 1964. VINITI, Moscow

Korobeynikova MP, Nasyrov GA, Khamidulina VG (1968) The nightglow emission λ 5577 Å. In: Kalchaev KK, Shefov NN (eds). Tables and maps of isophotes. Ashkhabad, 1965–1966. VINITI, Moscow

Korobeynikova MP, Nasyrov GA, Khamidulina VG (1970) Intensity variations and dynamical characteristics of the spatial patches of the emission λ 5577 Å. In: Krassovsky VI (ed) Aurorae and airglow. N 18. Nauka, Moscow, pp 5–14

Korobeynikova MP, Nasyrov GA, Khamidulina VG (1972) The nightglow emission λ 5577 Å, λ 6300 Å. In: Savrukhin AP (ed). Tables and maps of isophotes. Ashkhabad, 1967. Ylym (Nauka), Ashkhabad

Korobeynikova MP, Nasyrov GA, Shefov NN (1979) Internal gravity wave registration in Ashkhabad and Zvenigorod. Geomagn Aeron 19:1116–1117

Korobeynikova MP, Nasyrov GA, Toroshelidze TI, Shefov NN (1983) Some results of the simultaneous studies of the internal gravity waves at some stations. In: Lysenko IA (ed) Studies of the dynamic processes in the upper atmosphere. Hydrometeoizdat, Moscow, pp 121–123

Korobeynikova MP, Chuchuzov EP, Shefov NN (1984) Horizontal eddy diffusion near the turbopause from observations of the 557.7-nm emission. Izv USSR Acad Sci Atmos Oceanic Phys 20:854–857

Korobeynikova MP, Kuliyeva RN, Goshdzhanov MI, Khamidulina VG, Shamov AA (1989) Variations of night sky emissions 557.7 nm, 630 nm and Na during earthquakes. In: Feldstein YaI, Shefov NN (eds) Aurorae and airglow. N 33. VINITI, Moscow, pp 24–27

Korsunova LP, Gorbunova TA, Bakaldina VD (1985) Solar activity influence on the turbopause variations. In: Lysenko IA (ed) Studies of the dynamic processes in the upper atmosphere. Hydrometeoizdat, Moscow, pp 175–179

Krassovsky VI (1971) The calms and the storms in the upper atmosphere (Physics of the upper atmosphere and near-Earth space). Nauka, Moscow

Krassovsky VI (1972) Infrasonic variations of the OH emission in the upper atmosphere. Ann Geophys 28:739–746

Krassovsky VI, Semenov AI (1987) On the "holes" in the spatial distribution of the OI 130 nm emission of the dayglow. Cosmic Res 25:323–324

Krassovsky VI, Shefov NN, Yarin VI (1962) Atlas of the airglow spectrum $\lambda\lambda$3000–12400 Å. Planet Space Sci 9:883–915

Krassovsky VI, Potapov BP, Semenov AI, Shagaev MV, Shefov NN, Sobolev VG, Toroshelidze TI (1977) The internal gravity waves near mesopause and hydroxyl emission. Ann Geophys 33:347–356

Krassovsky VI, Potapov BP, Semenov AI, Sobolev VG, Shagaev MV, Shefov NN (1978) Internal gravity waves near mesopause. I. Results of studies of hydroxyl emission. In: Galperin YuI (ed) Aurorae and airglow. N 26. Soviet Radio, Moscow, pp 5–29

Kropotkina EP (1976) Hydroxyl emission and the stratospheric meteorology at Abastumani, Zvenigorod, Loparskaya and Yakutsk. In: Krassovsky VI (ed) Aurorae and airglow. N 24. Soviet Radio, Moscow, pp 37–43

Kropotkina EP, Shefov NN (1977) Tidal emission variations in the mesopause. In: Krassovsky VI (ed) Aurorae and airglow. N 25. Soviet Radio, Moscow, pp 13–17

Kuzmin KI (1975) Intensity oscillations of the 5577 Å and 5893 Å emissions and geomagnetic activity. In: Krassovsky VI (ed) Aurorae and airglow. 23. Nauka, Moscow, pp 28–32

Kvifte GJ (1967) Hydroxyl rotational temperatures and intensities in the nightglow. Planet Space Sci 15:1515–1523

Labitzke K, van Loon H (1988) Associations between the 11-year solar cycle, the QBO and the atmosphere. P. I. The troposphere and stratosphere in the northern hemisphere in winter. J Atmos Terr Phys 50:197–206

Le Texier H, Solomon S, Thomas RJ, Garcia RR (1989) OH*(7–5) Meinel band dayglow and nightglow measured by the SME limb scanning near infrared spectrometer: comparison of the observed seasonal variability with two-dimensional model simulations. Ann Geophys 7:365–374

Lobzin VV, Pavlov AV (1998) Relation between emission intensity of subauroral red arcs and solar and geomagnetic activity. Geomagn Aeron 38:446–455

López-González MJ, López-Moreno JJ, Rodrigo R (1992) Atomic oxygen concentration from airglow measurements of atomic and molecular oxygen emissions in the nightglow. Planet Space Sci 40:929–940

López-González MJ, Murtagh DP, Espy PJ, López-Moreno JJ, Rodrigo R, Witt G (1996) A model study of the temporal behaviour of the emission intensity and rotational temperature of the OH Meinel bands for high-latitude summer conditions. Ann Geophys 14:59–67

Lord R, Spencer JH (1935) The light of the night sky: analyses of the intensity variations at three stations. Proc R Soc London 151A:22–55

Lowe RP (2002) Long-term trends in the temperature of the mesopause region at mid-latitudes as measured by the hydroxyl airglow. We-Heraeus seminar on trends in the upper atmosphere. Kühlungsborn, Germany. p 32

Lowe RP, Lytle EA (1973) Balloon-born spectroscopic observation of the infrared hydroxyl airglow. Appl Opt 12:579–583

Lowe RP, LeBlanc L (1993) Preliminary analysis of WINDI (UARS) hydroxyl data: apparent peak height. Abstracts. The 19th Ann. Europ. Meet. Atm. Stud. Opt. Meth. Kiruna, Sweden, August 10–14, 1992. Kiruna, pp 94–98

Lowe RP, Perminov VI (1998) Analysis of mid-latitude ground-based and WINDII/UARS observations of the hydroxyl nightglow. 32nd Scientific Assembly of COSPAR (Japan, Nagoya, 1998). Nagoya, p 131

Lowe RP, Gilbert KL, Turnbull DN (1991) High latitude summer observations of the hydroxyl airglow. Planet Space Sci 39:1263–1270

Lysenko EV, Rusina VYa (2002a) Changes in the stratospheric and mesospheric thermal conditions during the last three decades: 3. Linear trends of monthly mean temperatures. Izv Atmos Oceanic Phys 38:296–304

Lysenko EV, Rusina VYa (2002b) Changes in the stratospheric and mesospheric thermal conditions during the last three decades: 4. Trends in the height and temperature of the stratopause. Izv Atmos Oceanic Phys 38:305–311

Lysenko EV, Rusina VYa (2003) Long-term changes in the stratopause height and temperature derived from rocket measurements at various latitudes. Int J Geomagn Aeron 4:67–81

Lysenko EV, Nelidova GF, Prostova AM (1997a) Changes in the stratospheric and mesospheric thermal conditions during the last three decades: 1. The evolution of a temperature trend. Izv Atmos Oceanic Phys 33:218–225

Lysenko EV, Nelidova GF, Prostova AM (1997b) Changes in the stratospheric and mesospheric thermal conditions during the last three decades: 2. The evolution of annual and semiannual temperature oscillations. Izv Atmos Oceanic Phys 33:226–233

Lysenko EV, Perov SP, Semenov AI, Shefov NN, Sukhodoev VA, Givishvili GV, Leshchenko LN (1999) Long-term trends of the yearly mean temperature at heights from 25 to 110 km. Izv Atmos Oceanic Phys 35:393–400

Lysenko EV, Nelidova GG, Rusina VYa (2003) Annual cycles of middle atmosphere temperature trends determined from long-term rocket measurements. Int J Geomagn Aeron 4: pp 57–65

Makino T, Hagiwara Y (1971) Measurement of the altitude dependence of OH nightglow by K-10-5 rocket. Bull Inst Space Aeronaut 7:130

Marchuk GI (1982) Mathematical modeling in the surrounding medium problem. Nauka, Moscow

Matveeva OA, Semenov AI (1985) The results of hydroxyl emission observations during MAP/WINE period; stratospheric warmings (February, 1984). MAP/WINE Newsletter. N 3:4–5

McDade IC, Llewellyn EJ (1986) The excitation of $O(^1S)$ and O_2 bands in the nightglow: a brief review and preview. Can J Phys 64:1626–1630

McEven DJ, Hammel GR, Williams CW (1998) Polar airglow variations over a one–half solar cycle. In: Proc. 24th annual European meeting on atmospheric studies by optical methods (August 18–22, 1997, Andenes, Norway). Sentraltrykkeriet A/S. Bodø, pp 74–76

Meeus J (1982) Astronomical formulae for calculators, 2nd edn. Willmann-Bell Inc, Richmond Virginia

Megrelishvili TG (1981) Regularities of the variations of the scattered light and emission of the Earth twilight atmosphere. Khrgian AKh (ed). Metsniereba, Tbilisi

Megrelishvili TG, Fishkova LM (1986) The oscillations of upper atmosphere parameters with a period of about 5.5 years as given by dusk and night sky emission data. Geomagn Aeron 26:154–156

Megrelishvili TG, Toroshelidze TI (1999) Long-term trend in the mesopause density as inferred from twilight spectrophotometric observations. Geomagn Aeron 39:258–260

Meriwether JW, Biondi MA (1995) Optical interferometric observations of 630-nm intensities, thermospheric winds and temperatures near the geomagnetic equator. Adv Space Res 16:17–26

Meriwether JW, Mirick JL, Biondi MA, Herrero FA, Fesen CG (1996) Evidence of orographic wave heating in the equatorial thermosphere at solar maximum. Geophys Res Lett 23:2177–2180

Meriwether JW, Biondi MA, Herrero FA, Fesen CG, Hallenback DC (1997) Optical interferometric studies of the nighttime equatorial thermosphere: enhanced temperatures and zonal wind gradients. J Geophys Res 102A:20041–20058

Merzlyakov EG, Portnyagin YuI (1999) Long-term changes in the parameters of winds in the midlatitude lower thermosphere (90–100 km) as inferred from long-term wind measurements. Izv Atmos Oceanic Phys 35:482–493

Mikhalev AV, Medvedeva IV (2002) Seasonal behavior of emission from the upper atmosphere in 558 nm line of atomic oxygen. Atmos Oceanic Opt 15:993–997

Mikhalev AV, Medvedeva IV, Beletsky AB, Kazimirovsky ES (2001) An investigation of the upper atmospheric optical radiation in the line of atomic oxygen 557.7 nm in East Siberia. J Atmos Solar—Terr Phys 63:865–868

Miropolsky YuZ (1981) Dynamics of internal gravity waves in the ocean. Hydrometeoizdat, Leningrad

Misawa K, Takeuchi I (1976) Parallel intensity-variations of [O] 5577 Å line and $O_2(0–1)$ Atmospheric band at 8645 Å. Rep Ionosp Space Res Jpn 30:109–112

Misawa K, Takeuchi I (1977) Ground observation of the $O_2(0–1)$ atmospheric band at 8645 Å and the [O] 5577 Å line. J Geophys Res 82:2410–2412

Misawa K, Takeuchi I (1982) Nightglow intensity variations in the $O_2(0–1)$ atmospheric band, the Na D lines, the OH (6–2) band, the yellow-green continuum at 5750 Å and the oxygen green line. Ann Geophys 38:781–788

Monfils A (1968) Spectres auroraux. Space Sci Rev 8:804–845

Monin AS (1988) Theoretical principles of the geophysical hydrodynamics. Hydrometeoizdat, Leningrad

Montenbruck O, Pfleger T (2000) Astronomy on the personal computer, 4th edn. Springer, Berlin Heidelberg New York

Moreels G, Megie G, Vallance JA, Gattinger RL (1977) An oxygen–hydrogen atmospheric models and its application to the OH emission problem. J Atmos Terr Phys 39:551–570

Mulyarchik TM (1959) Interferometric measurements of the line width of the λ 5577 Å [OI] in the aurorae. Izv USSR Acad Sci Ser Geophys 1902–1903

Mulyarchik TM (1960a) Interferometric measurement of λ 6300 Å [OI] and λ 5198–5200 Å [NI] emissions from auroras. Dokl USSR Acad Sci 130:303–306

Mulyarchik TM (1960b) Interferometric measurements of the upper atmospheric temperature from the widths of the some emission lines. Izv USSR Acad Sci Ser Geophys 449–458

Mulyarchik TM, Shcheglov PV (1963) Temperature and corpuscular heating in the auroral zone. Planet Space Sci 10:215–218

Munk WH (1980) Internal wave spectra at the buoyant and internal frequencies. J Phys Oceanogr 10:1718–1728

Murtagh DP, Witt G, Stegman J, McDade IC, Llewellyn EJ, Harris F, Greer RGH (1990) An assessment of proposal $O(^1S)$ and O_2 ($b^1\Sigma_g^+$) nightglow excitation parameters. Planet Space Sci 38:43–53

Myrabø HK (1984) Temperature variation at mesopause levels during winter solstice at 78°N. Planet Space Sci 32:249–255

Myrabø HK (1987) Night airglow O_2 (0–1) atmospheric band emission during the northern polar winter. Planet Space Sci 35:1275–1279

Myrabø HK, Deehr CS, Sivjee GG (1983) Large-amplitude nightglow OH (8–3) band intensity and rotational temperature variations during 24-hour period at 78°N. J Geophys Res 88A:9255–9259

Myrabø HK, Henriksen K, Deehr CS, Romick GJ (1984) O_2 ($b^1\Sigma_g^+ - X^3\Sigma_g^-$) atmospheric band night airglow measurements in the northern polar cap region. J Geophys Res 89A:9148–9152

Nagata T, Tohmatsu T, Ogawa T (1965) Rocket measurements of the 6300 Å and 3914 Å dayglow features. Planet Space Sci 13:1273–1282

Nagata T, Tohmatsu T, Ogawa T (1968) Rocket observation of visible and ultraviolet dayglow features. J Geomagn Geoelectr 20:315–321

Nasyrov GA (1967a) Peculiarities of spatial inhomogeneities of the atomic oxygen emission λ 5577 Å in the nightglow. In: Krassovsky VI (ed) Aurorae and airglow. N 13. Nauka, Moscow, pp 5–9

Nasyrov GA (1967b) Spatial variations of nightglow in the region λ 5893 Å. In: Krassovsky VI (ed) Aurorae and airglow. N 13. Nauka, Moscow, pp 10–12

Nasyrov GA (1970) The influence of the geomagnetic activity emissions of atomic oxygen. Geomagn Aeron 10:1112–1114

Nasyrov GA (1978) On the connection of the nightglow emissions with the seismic activity. Izv Turkmenian SSR Acad Sci 119–122

Nasyrov GA (2003) Statistical regularities of variations in the sodium nightglow observed in Ashkhabad during the solar activity minimum. Geomagn Aeron 43:402–404

Neumann A (1990) QBO and solar activity effects on temperature in the mesopause region. J Atmos Terr Phys 52:165–173

Nikolashkin SV, Ignatiev VM, Yugov VA (2001) Solar activity and QBO influence on the temperature regime of the subauroral middle atmosphere. J Atmos Solar—Terr Phys 63:853–858

Novikov NN (1981) Internal gravity waves near mesopause. VI. Polar region. In: Galperin YuI (ed) Aurorae and airglow. N 29. Soviet Radio, Moscow, pp 59–67

Noxon JF, Goody RM (1962) Observation of day airglow emission. J Atmos Sci 19:342–343

Offermann D, Graef H (1992) Messungen der OH*—Temperatur. Promet 22:125–128

Offermann D, Gerndt R, Lange G, Trinks H (1983) Variations of mesopause temperatures in Europe. Adv Space Res 3:21–23

Ogawa T, Iwagami N, Nakamura M, Takano M, Tanabe H, Takeuchi A, Miyashita A, Suzuki K (1987) A simultaneous observation of the height profiles of the night airglow OI 5577 Å, O_2 Herzberg and Atmospheric bands. J Geomag Geoelectr 39:211–228

Packer DM (1961) Altitudes of the night airglow radiations. Ann Geophys 17:67–75

Pavlov AV (1996) Mechanism of the electron density depletion in the SAR arc region. Ann Geophys 14:211–212

Pavlov AV (1997) Subauroral red arcs as a conjugate phenomenon: comparison of OV1-10 satellite data with numerical calculations. Ann Geophys 15:984–998

Pavlov AV (1998) Interpreting the observations of auroral red arcs in magnetically conjugate regions. Geomagn Aeron 38:803–807

Pavlov AV, Pavlova NM, Drozdov AB (1999) Production rate of $O(^1D)$, $O(^1S)$ and $N(^2D)$ in the subauroral red arc region. Geomagn Aeron 39:201–205

Perminov VI, Semenov AI (1992) The nonequilibrium of the rotational temperature of OH bands under high level rotational excitation. Geomagn Aeron 32:306–308

Perminov VI, Semenov AI, Shefov NN, Tikhonova VV (1993) Estimates of seasonal variations in the altitude of the emitting hydroxyl layer. Geomagn Aeron 33:364–369

Perminov VI, Semenov AI, Shefov NN (1998) Deactivation of hydroxyl molecule vibrational states by atomic and molecular oxygen in the mesopause region. Geomagn Aeron 38:761–764

Perminov VI, Semenov AI, Bakanas VV, Zheleznov YuA, Khomich VYu (2004) Regular variations in the (0–1) band intensity of the oxygen emission atmospheric system. Geomagn Aeron 44:498–501

Perminov VI, Shefov NN, Semenov AI (2007) Empirical model of variations in the emission of the molecular oxygen atmospheric system. 1. Intensity. Geomagn Aeron 47:104–108

Pertsev NN, Shalimov SL (1996) The generation of atmospheric gravity waves in a seismically active region and their effect on the ionosphere. Geomagn Aeron 36:223–227

Petitdidier M, Teitelbaum H (1977) Lower thermosphere emissions and tides. Planet Space Sci 25:711–721

Plane JMC (1991) The chemistry of meteoric metals in the Earth's upper atmosphere. Int Rev Phys Chem 10:55–106

Plane JMC, Cox RM, Qian J, Pfenninger WM, Papen GC, Gardner CS, Espy PJ (1998) Mesospheric Na layer at extreme high latitudes in summer. J Geophys Res 103D:6381–6389

Plane JMC, Gardner CS, Yu J, She CY, Garcia RR, Pumphrey HC (1999) Mesospheric Na layer at 40°N: modeling and observations. J Geophys Res 104D:3773–3788

Potapov BP, Sobolev VG, Sukhodoev VA, Yarov BN (1983) Measurement of the height of layer of hydroxyl emission in Near-Moscow region. Geomagn Aeron 23:326–327

Potapov BP, Sobolev VG, Sukhodoev VA, Yarov BN (1985) On the mutual arrangement of emissional layers of hydroxyl and OI 5577 Å in the Earth upper atmosphere. Geomagn Aeron 25:685–686

Qian J, Gardner CS (1995) Simultaneous lidar measurements of mesospheric Ca, Na, and temperature profiles at Urbana, Illinois. J Geophys Res 100D:7453–7461

Rao MNM, Murty GSN, Jain VC (1982) Altitude peak of (OI) 5577 in the lower thermosphere: Chapman versus Barth mechanisms. J Atmos Terr Phys 44:559–566

Rakipova LR, Efimova LK (1975) Dynamics of the upper atmospheric layers. Hydrometeoizdat, Leningrad

Rapoport ZTs, Shefov NN (1974) Relations between variations of hydroxyl emission and radio wave absorption in D-region. Indian J Radio Space Phys 3:314–316

Rapoport ZTs, Shefov NN (1976) Connection of the disturbed variations of the hydroxyl emission and the radio wave absorption. In: Shefov NN, Savrukhin AP (eds) Studies of the upper atmospheric emission. Ylym, Ashkhabad, pp 12–16

Reed EI, Blamont JE (1967) Some results concerning the principal airglow lines as measured from the OGO-2 satellite. In: Smith–Rose RL, Bowhill SW, King JW (eds) Space Res, vol 7. North-Holland, Amsterdam, pp 337–352

Reed EI, Chandra S (1975) The global characteristics of atmospheric emissions in the lower thermosphere and their aeronomic implications. J. Geophys. Res. 80:3057–3062

Reisin ER, Scheer J (2002) Searching for trends in mesopause region airglow intensities and temperatures at EL Leoncito. Phys. Chem. Earth. 27:563–569

Rishbet H, Garriott OK (1969) Introduction to ionospheric physics. Academic Press. New York

Roach FE, Gordon JL (1973) The light of the night sky. Reidel, Dordrecht

Roach FE, Pettit HB, Williams DR, St Amand P, Davis DN (1953) A four-year study of OI 5577 Å in the nightglow. Ann d'Astrophys 16:185–205

Ross MN, Christensen AB, Meng CI, Carbary JF (1992) Structure in the UV nightglow observed from low Earth orbit. Geophys Res Lett 19:985–988

McDade IC, Llewellyn EJ (1986) The excitation of $O(^1S)$ and O_2 bands in the nightglow: a brief review and preview. Can J Phys 64:1626–1630

Reshetov BD (1973) Variability of the meteorological elements in the atmosphere. Hydrometeoizdat, Leningrad
Sahai Y, Dresñher A, Lauche H, Teixeira NR (1975) First results of 6300 Å nightglow measurements aboard a rocket launched from Natal Brazil. In: Rycroft MJ (ed) Space Res, vol 15. Akademie-Verlag, Berlin, pp 251–255
Schaeffer RC, Feldman PD, Zipf EC (1972) Dayglow [OI] $\lambda\lambda$6300 Å and 5577 Å lines in the early morning ionosphere. J Geophys Res 77:6828–6838
Scheer J, Reisin ER (1990) Rotational temperatures for OH and O_2 airglow bands measured simultaneously from El Leoncito (31°48'S). J Atmos Terr Phys 52:47–57
Scheer J, Reisin ER (1998) Extreme intensity variations of O_2b airglow induced by tidal oscillations. Adv Space Res 21:827–830
Scheer J, Reisin ER (2000) Unusually low airglow intensities in the Southern Hemisphere midlatitude mesosphere region. Earth Planets Space 52:261–266
Scheer J, Reisin ER (2002) Most prominent airglow night at El Leoncito. J Atmos Solar–Terr Phys 64/8–11:1175–1181
Scheer J, Reisin ER, Espy JP, Bittner M, Graef HH, Offermann D, Ammosov PP, Ignatiev VM (1994) Large-scale structures in hydroxyl rotational temperatures during DYANA. J Atmos Terr Phys 56:1701–1715
Scheer J, Reisin ER, Mandrini CH (2005) Solar activity signatures in mesopause region temperatures and atomic oxygen related airglow brightness at El Leoncito, Argentina. J Atmos Solar—Terr Phys 67:145–154
Schuster A (1906) On sunspot periodicities. Preliminary notice. Proc R Soc London 77A:141–145
Semenov AI (1975a) Interferometric measurements of the upper atmosphere temperature. I. Application of the cooled image converters. In: Krassovsky VI (ed) Aurorae and airglow. N 23. Nauka, Moscow, pp 64–65
Semenov AI (1975b) Doppler temperature and intensity of emission 6300 Å. Geomagn Aeron 15:876–880
Semenov AI (1976) Interferometric measurements of the upper atmosphere temperature. III. 6300 Å emission and characteristics of the atmosphere and ionosphere. In: Krassovsky VI (ed) Aurorae and airglow. N 24. Nauka, Moscow, pp 44–51
Semenov AI (1978) Interferometric measurements of the upper atmospheric temperature. V. Semiannual temperature variations according to 6300 emission. In: Krassovsky VI (ed) Aurorae and airglow. N 27. Nauka, Moscow, pp 85–86
Semenov AI (1989) The specific features of the green emission excitation process in the noctural atmosphere. In: Feldstein YaI, Shefov NN (eds) Aurorae and airglow. N 33. VINITI, Moscow, pp 74–80
Semenov AI (1996) A behavior of the lower thermosphere temperature inferred from emission measurements during the last decades. Geomagn Aeron 36:655–659
Semenov AI (1997) Long-term changes in the height profiles of ozone and atomic oxygen in the lower thermosphere. Geomagn Aeron 37:354–360
Semenov AI (2000) Long-term temperature trends for different seasons by hydroxyl emission. Phys Chem Earth Pt B 25:525–529
Semenov AI, Lysenko EV (1996) Long-term subsidence of the middle and upper atmosphere according to its ecological evolution. Environ. Radioecology and Appl. Ecology 2:3–13
Semenov AI, Shefov NN (1979a) Comparison of the atmospheric temperatures according to the CIRA-72 and nightglow data. Bulgarian Geophys Studies 5:61–66
Semenov AI, Shefov NN (1979b) Comparison of atmospheric temperatures according to CIRA-72 and nightglow data. In: Rycroft MJ (ed) Space Res, vol 19. Pergamon Press, Oxford, pp 203–206
Semenov AI, Shefov NN (1989) The effect of internal gravity waves on the dynamics and energetics of the lower thermosphere (according to characteristics of the nightglow). In: Middle atmosphere studies. Ionospheric Researches. N 47. VINITI, Moscow, pp 24–43
Semenov AI, Shefov NN (1996a) An empirical model for the variations in the hydroxyl emission. Geomagn Aeron 36:468–480

Semenov AI, Shefov NN (1996b) On the possibility of extracting valuable information on climatic changes in the Earth's atmosphere from archives of astronomical spectroscopic observations. Astron Lett 22:632–633

Semenov AI, Shefov NN (1997a) An empirical model of nocturnal variations in the 557.7-nm emission of atomic oxygen. 1. Intensity. Geomagn Aeron 37:215–221

Semenov AI, Shefov NN (1997b) An empirical model of nocturnal variations in the 557.7-nm emission of atomic oxygen. 2. Temperature. Geomagn Aeron 37:361–364

Semenov AI, Shefov NN (1997c) An empirical model of nocturnal variations in the 557.7-nm emission of atomic oxygen. 3. Emitting layer altitude. Geomagn Aeron 37:470–474

Semenov AI, Shefov NN (1997d) Empirical model of the variations of atomic oxygen emission 557.7 nm. In: Ivchenko VN (ed) Proc SPIE (23rd European Meeting on Atmospheric Studies by Optical Methods, Kiev, September 2–6, 1997), vol 3237. Int Soc Opt Engin, Bellingham, pp 113–122

Semenov AI, Shefov NN (1999a) Empirical model of hydroxyl emission variations. Int J Geomagn Aeron 1:229–242

Semenov AI, Shefov NN (1999b) Variations of the temperature and the atomic oxygen content in the mesopause and lower thermosphere region during change of the solar activity. Geomagn Aeron 39:484–487

Semenov AI, Shefov NN (2003) New knowledge of variations in the hydroxyl, sodium and atomic oxygen emissions. Geomagn Aeron 43:786–791

Semenov AI, Shefov NN (2005) Model of the vertical profile of the atomic oxygen concentration in the mesopause and lower ionosphere region. Geomagn Aeron 45:797–808

Semenov AI, Shagaev MV, Shefov NN (1981) On the effect of orographic waves on the upper atmosphere. Izv USSR Acad Sci Atmos Oceanic Phys 17:982–984

Semenov AI, Shefov NN, Fishkova LM, Lysenko EV, Perov SP, Givishvili GV, Leshchenko LN, Sergeenko NP (1996) Climatic changes in the upper and middle atmosphere. Doklady Earth Sciences 349:870–872

Semenov AI, Shefov NN, Givishvili GV, Leshchenko LN, Lysenko EV, Rusina VYa, Fishkova LM, Martsvaladze NM, Toroshelidze TI, Kashcheev BL, Oleynikov AN (2000) Seasonal peculiarities of long-term temperature trends of the middle atmosphere. Dokl Earth Sci 375:1286–1289

Semenov AI, Sukhodoev VA, Shefov NN (2002a) A model of the vertical temperature distribution in the atmosphere altitudes of 80–100 km that taking into account the solar activity and the long-term trend. Geomagn Aeron 42:239–244

Semenov AI, Bakanas VV, Perminov VI, Zheleznov YuA, Khomich VYu (2002b) The near infrared spectrum of the emission of the nighttime upper atmosphere of the Earth. Geomagn Aeron 42:390–397

Semenov AI, Shefov NN, Lysenko EV, Givishvili GV, Tikhonov AV (2002c) The seasonal peculiarities of behavior of the long-term temperature trends in the middle atmosphere at the mid-latitudes. Phys Chem Earth 27:529–534

Semenov AI, Shefov NN, Perminov VI, Khomich VYu, Fadel KhM (2005) Temperature response of the middle atmosphere on the solar activity for different seasons. Geomagn Aeron 45:236–240

Serafimov K (1979) Space research in Bulgaria. Bulgarian Acad Sci Publ House, Sofia

Serafimov KB, Gogoshev MM (1972) On the mutual relations between line 6300 Å and region F parameters. Compt Rendu Acad Bulg Sci 25:197–200

Serafimov K, Gogoshev M, Gogosheva Ts (1977) Models of the night altitudinal distribution of λ 6300 Å emission. Geomagn Aeron 17:1044–1049

Shagaev MV (1978) Characteristics of hydroxyl emission variations, pointing out the processes of its excitations. In: Krassovsky VI (ed) Aurtorae and airglow. 27. Nauka, Moscow, pp 18–25

Shain GA, Shain PF (1942) Methods of the variation investigations of the emission lines in the night sky spectrum. Dokl USSR Acad Sci 35:152–156

Shashilova NA (1983) Neutral content of the high-latitudinal atmosphere. In: Mizun YuG (ed) Ionospheric investigations. N 35. Radio and Svyaz, Moscow, pp 25–31

She CY, Yu JR, Chen H (1993) Observed thermal structure of a midlatitude mesopause. Geophys Res Lett 20:567–570
Shefov NN (1959) Intensities of some twilight and night airglow emissions. In: Krassovsky VI (ed) Spectral, electrophotometrical and radar researches of aurorae and airglow. N 1. USSR Acad Sci Publ House, Moscow, pp 25–29
Shefov NN (1960) Intensities of some night sky emissions. In: Krassovsky VI (ed) Spectral, electrophotometrical and radar researches of aurorae and airglow. N 2–3. USSR Acad Sci Publ House, Moscow, pp 57–59
Shefov NN (1961) On determination of the rotational temperature of the OH bands. In: Krassovsky VI (ed) Spectral, electrophotometrical and radar researches of aurorae and airglow. N 5. USSR Acad Sci Publ House, Moscow, pp 5–9
Shefov NN (1967a) Some properties of the hydroxyl emission. In: Krassovsky VI (ed) Aurorae and airglow. N 13. USSR Acad Sci Publ House, Moscow, pp 37–43
Shefov NN (1967b) OH emission and noctilucent clouds. In: Khvostikov IA, Witt G (eds) Noctilucent clouds. Proc Int Symp (Tallinn, 1966). VINITI, Moscow, pp 187–188
Shefov NN (1968a) Behaviour of the upper atmosphere emissions during high meteoric activity. Planet Space Sci 16:134–136
Shefov NN (1968b) Intensity and rotational temperature variations of the hydroxyl emission in the nightglow. Nature (London) 218:1238–1239
Shefov NN (1969a) Hydroxyl emission of the upper atmosphere. I. Behaviour during solar cycle, seasons and geomagnetic disturbances. Planet Space Sci 17:797–813
Shefov NN (1969b) Hydroxyl emission of the upper atmosphere. II. Effect of a sunlit atmosphere. Planet Space Sci 17:1629–1639
Shefov NN (1969c) Low-latitudinal effects of the geomagnetic disturbances. II. Hydroxyl emission as an indicator of energy introducing in the atmosphere by the corpuscular bombardment. In: Obridko VN (ed) Solar-terrestrial physics. N 1. VINITI, Moscow, pp 285–288
Shefov NN (1970a) On the correlation between the intensity emission of the atmospheric system of O_2 and the vibrational temperature of the OH bands. Astron Circ USSR Acad Sci N 589:7–8
Shefov NN (1970b) Behaviour of the upper atmosphere emissions during high meteoric activity. In: Krassovsky VI (ed) Aurorae and airglow. N 18. Nauka, Moscow, pp 21–25
Shefov NN (1971a) Hydroxyl emissions of the upper atmosphere. III. Diurnal variations. Planet Space Sci 19:129–136
Shefov NN (1971b) Hydroxyl emission of the upper atmosphere. IV correlation with the molecular oxygen emission. Planet Space Sci 19:795–796
Shefov NN (1972a) Hydroxyl emission. Ann Geophys 28:137–143
Shefov NN (1972b) Some statistical properties of the hydroxyl emission. In: Fishkova LM, Kharadze EK (eds) Bull Abastumani Astrophys Observ. N 42. pp 9–24
Shefov NN (1973a) Behaviour of the hydroxyl emission during solar cycle, seasons and geomagnetic disturbances. In: Krassovsky VI (ed) Aurorae and airglow. N 20. Nauka, Moscow, pp 23–39
Shefov NN (1973b) Relations between the hydroxyl emission of the upper atmosphere and the stratospheric warmings. Gerlands Beitr. Geophys 82:111–114
Shefov NN (1974a) Lunar tidal variations of hydroxyl emission. Indian J Radio Space Phys 3:314–316
Shefov NN (1974b) Lunar variations of hydroxyl emission. Geomagn Aeron 14:920–922
Shefov NN (1975a) Results of studies of the hydroxyl emission. In: Krassovsky VI (ed) Aurorae and airglow. N 22. Soviet Radio, Moscow, pp 71–76
Shefov NN (1975b) Emissive layer altitude of the atmospheric system of molecular oxygen. In: Krassovsky VI (eds) Aurorae and airglow. N 23. Nauka, Moscow, pp 54–58
Shefov NN (1976) Seasonal variations of the hydroxyl emission. In: Krassovsky VI (eds) Aurorae and airglow. N 24. Nauka, Moscow, pp 32–36
Shefov NN (1978a) Non-corpuscular nature of the ionospheric absorption, appearing at the middle latitudes after geomagnetic storms. In: Krassovsky VI (eds) Aurorae and airglow. N 27. Soviet Radio, Moscow, pp 36–44

Shefov NN (1978b) Altitude of the hydroxyl emission layer. In: Krassovsky VI (eds) Aurorae and airglow. N 27. Soviet Radio, Moscow, pp 45–51

Shefov NN (1985) Solar activity and near surface circulation as the commensurable sources of the thermal regime variations of the lower thermosphere. Geomagn Aeron 25:848–849

Shefov NN (1989) The recording of wave and spotted inhomogeneities of upper atmospheric emission. In: Feldstein YaI, Shefov NN (eds) Aurorae and airglow. N 33. VINITI, Moscow, pp 81–84

Shefov NN, Kropotkina EP (1975) The height variations of the λ 5577 Å emission layer. Cosmic Res 13:765–770

Shefov NN, Pertsev NN (1984) Orographic disturbances of upper atmosphere emissions. In: Taubenheim J (ed) Handbook for middle atmosphere program, vol 10. SCOSTEP, Urbana, pp 171–175

Shefov NN, Piterskaya NA (1984) Spectral and space–time characteristics of the background luminosity of the upper atmosphere. Hydroxyl emission. In: Galperin YuI (ed) Aurorae and airglow. N 31. VINITI, Moscow, pp 23–123

Shefov NN, Semenov AI (2001) An empirical model for nighttime variations in atomic sodium emission: 2. Emitting layer height. Geomagn Aeron 41:257–261

Shefov NN, Semenov AI (2002) The long-term trend of ozone at heights from 80 to 100 km at the mid-latitude mesopause for the nocturnal conditions. Phys Chem Earth 27:535–542

Shefov NN, Semenov AI (2004a) Longitudinal-temporal distribution of the occurrence frequency of noctilucent clouds. Geomagn Aeron 44:259–262

Shefov NN, Semenov AI (2004b) The longitudinal variations of the atomic oxygen emission at 557.7 nm. Geomagn and Aeron 44:620–623

Shefov NN, Semenov AI (2004c) Spectral characteristics of the IGW trains registered in the upper atmosphere. Geomagn Aeron 44:763–768

Shefov NN, Semenov AI (2006) Spectral composition of the cyclic aperiodic (quasi-biennial) variations in solar activity and the Earth's atmosphere. Geomagn Aeron 46:411–416

Shefov NN, Toroshelidze TI (1975) Upper atmosphere emission as an indicator of the dynamic processes. In: Krassovsky VI (ed) Aurorae and airglow. N 23. Nauka, Moscow, pp 42–53

Shefov NN, Truttse YuL (1969) Hydrogen and hydroxyl emissions in the nightglow. In: Ann IQSY, vol 4. MIT Press, Cambridge, MA, pp 400–406

Shefov NN, Yarin VI (1961) On the latitudinal dependence of the OH rotational temperature. In: Krassovsky VI (eds) Spectral, electrophotometrical and radar researches of aurorae and airglow. N 5. USSR Acad Sci Publ House, Moscow, pp 25–28

Shefov NN, Yarin VI (1962) Latitudinal and planetary variations of the OH airglow. In: Krassovsky VI (eds) Aurorae and airglow. N 9. USSR Acad Sci Publ House, Moscow, pp 19–23

Shefov NN, Yurchenko OT (1970) Absolute intensities of the auroral emissions in Zvenigorod. In: Krassovsky VI (ed) Aurorae and airglow. N 18. Nauka, Moscow, pp 50–96

Shefov NN, Pertsev NN, Shagaev MV, Yarov VN (1983) Orographically caused variations of upper atmospheric emissions. Izv USSR Acad Sci Atmos Oceanic Phys 19:694–698

Shefov NN, Semenov AI, Pertsev NN, Sukhodoev VA, Perminov VI (1999) Spatial distribution of IGW energy inflow into the mesopause over the lee of a mountain ridge. Geomagn Aeron 39:620–627

Shefov NN, Semenov AI, Yurchenko OT (2000a) Empirical model of variations in the atomic sodium nighttime emission: 1. Intensity. Geomagn Aeron 40:115–120

Shefov NN, Semenov AI, Pertsev NN, Sukhodoev VA (2000b) The spatial distribution of the gravity wave energy influx into the mesopause over a mountain lee. Phys Chem Earth Pt B 25:541–545

Shefov NN, Semenov AI, Yurchenko OT (2002) Empirical model of the ozone vertical distribution at the nighttime mid-latitude mesopause. Geomagn Aeron 42:383–389

Shefov NN, Semenov AI, Yurchenko OT (2006) An empirical model of the 630 nm—atomic oxygen emission variations during nighttime. 1. Intensity. Geomagn Aeron 46:236–246

Shepherd GG, Thuillier G, Gault WA, Solheim BH, Hersom C, Alunni M, Brun JF, Brune S, Chalot P, Cogger LL, Desaulniers DL, Evans WFJ, Gattinger RL, Girod F, Harvie D, Henn

RH, Kendall DJW, Llewellyn EJ, Lowe RP, Ohrt J, Pasternak F, Peillet O, Powell I, Rochon Y, Ward WE, Wiens RH, Wimperis J (1993a) WINDII, the wind imaging interferometer on the upper atmosphere research satellite. J Geophys Res 98D:10725–10750

Shepherd GG, Thuillier G, Solheim BH, Chandra S, Cogger LL, Duboin ML, Evans WFJ, Gattinger RL, Gault WA, Hersé M, Hauchecorne A, Lathuilliere C, Llewellyn EJ, Lowe RP, Teitelbaum H, Vial F (1993b) Longitudinal structure in atomic oxygen concentrations observed with WINDII on UARS. Geophys Res Lett 20:1303–1306

Silverman SM (1970) Night airglow phemenology. Space Sci Rev 11:344–379

Sipler DP, Biondi MA (1975) Evidence for chemiexcitation as the source of the sodium nighglow. Geophys Res Lett 2:106–108

Sipler DP, Biondi MA (1978) Interferometric studies of the twilight and nightglow sodium D-line profiles. Planet Space Sci 26:65–73

Siskind DE, Sharp WE (1991) A comparison of measurements of the oxygen nightglow and atomic oxygen in the lower thermosphere. Planet Space Sci 39:627–639

Sivjee GG, Hamwey RM (1987) Temperature and chemistry of the polar mesopause OH. J Geophys Res 92A:4663–4672

Sivjee GG, Walterscheid RL, Hecht JH, Hamwey RM, Schubert G, Christensen AB (1987) Effects of atmospheric disturbances on polar mesopause airglow OH emissions. J Geophys Res 92A:7651–7656

Skinner WR, Yee JH, Hays PB, Burrage MD (1998) Seasonal and local time variations in the $O(^1S)$ green line, O_2 Atmospheric band, and OH Meinel band emissions as measured by the high resolution Doppler imager. Adv Space Res 21:835–841

Smith RW, Hernandez G (1995) Upper atmospheric temperatures at South Pole. Adv Space Res 16:31–39

Smith LL, Roach FE, McKennan JM (1968) IQSY night airglow data. Rep UAG-1. Washington, DC

Starkov GV (1994a) Statistical depedences between the magnetic indices. Geomagn Aeron 34:101–103

Starkov GV (1994b) Mathematical model of auroral boundaries. Geomagn Aeron 34:331–336

Starkov GV, Yevlashin LS, Semenov AI, Shefov NN (2000) A subsidence of the thermosphere during 20th century according to the measurements of the auroral heights. Phys Chem Earth Pt B 25:547–550

States RJ, Gardner CS (1999) Structure of the mesospheric Na layer at 40°N latitude: seasonal and diurnal variations. J Geophys Res 104D:11783–11898

States RJ, Gardner CS (2000a) Thermal structure of the mesopause region (80–105 km) at 40°N latitude. Part I: seasonal variations. J Atmos Sci 57:66–77

States RJ, Gardner CS (2000b) Thermal structure of the mesopause region (80–105 km) at 40°N latitude. Part II: diurnal variations. J Atmos Sci 57:78–92

Sukhodoev VA, Yarov VI (1998) Temperature variations of the mesopause in the leeward region of the Caucasus ridge. Geomagn Aeron 38:545–548

Sukhodoev VA, Pertsev NN, Reshetov LM (1989a) Variations of characteristics of hydroxyl emission caused by orographic perturbations. In: Feldstein YaI, Shefov NN (eds) Aurorae and airglow. N 33. VINITI, Moscow, pp 61–66

Sukhodoev VA, Perminov VI, Reshetov LM, Shefov NN, Yarov VN, Smirnov AS, Nesterova TS (1989b) The orographic effect in the upper atmosphere. Izv USSR Acad Sci Atmos Oceanic Phys 25:681–685

Sukhodoev VA, Pertsev NN, Shefov NN (1992) Formation of orographic disturbances in mesopause of mountain lee. EOS Trans AGU 73: Spring Meet Suppl 223

Takahashi T, Okuda M (1974) Morphological study of the diurnal variation in the [OI] 5577 Å night airglow intensity. Sci Rep Tohoku Univ Ser 5 22:19–33

Takahashi H, Batista PP (1981) Simultaneous measurements of OH(9,4), (8,3), (7,2), (6,2) and (5,1) bands in the airglow. J Geophys Res 86A:5632–5642

Takahashi H, Clemesha BR, Sahai Y (1974) Nightglow OH (8,3) band intensities and rotational temperature at 23°S. Planet Space Sci 22:1323–1329

Takahashi H, Gobbi D, Batista PP, Melo SML, Teixeira NR, Buriti RA (1998) Dynamical influence on the equatorial airglow observed from the south american sector. Adv. Space Res. 21:817–825

Takahashi H, Sahai Y, Batista PP (1984) Tidal and solar cycle effects on the OI 5577 Å, Na D and OH (8–3) airglow emissions observed at 23°S. Planet Space Sci 32:897–902

Takahashi H, Sahai Y, Batista PP (1986) Airglow $O_2(^1\Sigma)$ atmospheric band at 8645 Å and the rotational temperature observed at 23°S. Planet Space Sci 34:301–306

Takahashi H, Sahai Y, Teixeira NR (1990) Airglow intensity and temperature response to atmospheric wave propagation in the mesopause region. Adv Space Res 10:77–81

Takahashi H, Clemesha BR, Batista PP (1995) Predominant semi-annual oscillation of the upper atmospheric airglow intensities and temperatures in the equatorial region. J Atmos Terr Phys 57:407–414

Takahashi H, Clemesha BR, Simonich DM, Melo SML, Eras A, Stegman J, Witt G (1996) Rocket measurements in the equatorial airglow: MULTIFOT 92 database. J Atmos Terr Phys 58:1943–1961

Takeuchi A, Tanaka K, Miyeshita A (1986) Atlas of zenith airglow radiations obtained at Kiso, Japan, 1979–1983. WDC C2, Tokyo

Takeuchi A, Tanaka K, Miyeshita A (1989a) Atlas of zenith airglow radiations obtained at Kiso, Japan, 1984–1988. WDC C2, Tokyo

Takeuchi A, Tanaka K, Miyeshita A (1989b) Atlas of zenith airglow radiations obtained at Kiso, Japan, 1989–1990. WDC C2, Tokyo

Taranova OG (1967) Study of space–time properties of the hydroxyl emission. In: Krassovsky VI (ed) Aurorae and airglow. N 13. USSR Acad Sci Publ House, Moscow, pp 13–21

Taranova OG, Toroshelidze TI (1970) On the measurements of the hydroxyl emission in twilight. In: Krassovsky VI (ed) Aurorae and airglow. N 18. Nauka, Moscow, pp 26–32

Tarasenko DA (1988) The structure and circularion of the stratosphere and mesosphere of the Northern hemisphere. Hydrometeoizdat, Leningrad

Tarasova TM (1961) Direct measurements of the night sky luminosity. Astron Circ USSR Acad Sci N 222:31–32

Tarasova TM (1962) Direct measurements of the night sky in the λ 8640 Å spectral region. In: Artificial satellites of the Earth. N 13. USSR Acad Sci Publ House, Moscow, pp 107–109

Tarasova TM (1963) Night sky emission line intensity distribution with respect to height. In: Priester W (ed) Space Res, vol 3. North-Holland, Amsterdam, pp 162–172

Tarasova TM (1967) On space correlation of night sky emission. In: Smith-Rose RL, Bowhill SW, King JW (eds) Space Res, vol 7. North-Holland, Amsterdam, pp 351–361

Tarasova TM (1971) Nightglow of the atmosphere on the basis of 1967–1968 rocket measurements. Preprint. Polar Geophys Institute, Apatity

Tarasova TM, Slepova VA (1964) Height distribution of radiation intensity of the night sky main emission lines. Geomagn Aeron 4:321–327

Tarasova TM, Yagodkina OI, Bogdanov NN, Yevlashin LS, Mikirov AE, Shidlovsky AA (1981) Height aurora profiles in the red and visible spectrum bands from measurements on the Franz-Josef Land. In: Isaev SI, Nadubovich YuA, Yevlashin LS (eds) Aurorae and airglow. N 28. Nauka, Moscow, pp 44–47

Taubenheim J (1969) Statistische Auswertung geophysikalischer und meteorologischer Daten. Akad Verlag, Leipzig

Taubenheim J, Feister U (1975) Ermittlung von Characteristiken stochastischer Prozesse durch Autosynchronisation. Gerlands Beitr. Geophys 84:389–398

Taylor MJ, Lowe RP, Baker DJ (1995) Hydroxyl temperature and intensity measurements during noctilucent cloud displays. Ann Geophys 13:1107–1116

Thuillier G, Blamont JE (1973) Vertical red line 6300 Å distribution and tropical nightglow morphology in quiet magnetic conditions. In: McCormac BM (ed) Physics and chemistry of upper atmosphere. Reidel, Dordrecht, pp 219–231

Thuillier G, Falin JL, Wachtel C (1977a) Experimental global model of the exospheric temperature based on measurements from the Fabry–Perot interferometer on board the OGO-6 satellite—discussion of the data and properties of the model. J Atmos Terr Phys 39:399–414

Thuillier G, Falin JL, Barlier F (1977b) Global experimental model of the exospheric temperature using optical and incoherent scatter measurements. J Atmos Terr Phys 39:1195–1202

Toroshelidze TI (1968a) Results of the spectrographic observations of the hydroxyl emission in the 10600-11200 A region in twilight. Bulletin Georgian SSR Academy of Sciences 52:N 1.57–62

Toroshelidze TI (1975) Rotational temperature variations of the hydroxyl emission. In: Krassovsky VI (ed). Aurorae and Airglow. N 3. Nauka, Moscow, 33–35

Toroshelidze TI (1989) The study of the temperature and winds on the basis of observations of the 630 nm emission in the evening sector of mean latitudes. In: Feldstein YaI, Shefov NN (eds) Aurorae and airglow. N 33. VINITI, Moscow, pp 11–16

Toroshelidze TI (1991) The analysis of the aeronomy problems on the upper atmosphere glow. In: Shefov NN (ed). Metsniereba, Tbilisi

Tousey R (1958) Rocket measurements of the night airglow. Ann Geophys 14:186–195

Turnbull DN, Lowe RP (1991) Temporal variations in the hydroxyl nightglow observed during ALOHA-90. Geophys Res Lett 18:1345–1348

Truttse YuL (1965) The spatial variations of the oxygen lines intensities. In: Krassovsky VI (ed) Aurorae and airglow. N 11. USSR Acad Sci Publ House, Moscow, pp 52–64

Truttse Yu (1968a) Upper atmosphere during geomagnetic disturbances. I. Some regular features of low-latitude auroral emissions. Planet Space Sci 16:981–992

Truttse Yu (1968b). Upper atmosphere during geomagnetic disturbances. II. Geomagnetic storms oxygen emission at 6300 Å and heating of the upper atmosphere. Planet Space Sci 16:1201–1208

Truttse Yu (1972a) Oxygen emission at 6300 Å. Ann Geophys 28:169–177

Truttse YuL (1972b) Night variations of intensity of emission 6300 Å in quiet geomagnetic conditions. Geomagn Aeron 12:561–564

Truttse YuL (1973) Upper atmosphere during geomagnetic disturbances. In: Krassovsky VI (ed) Aurorae and airglow. N 20. Nauka, Moscow, pp 5–22

Truttse YuL (1975) Oxygen emission 6300 Å during periods with the low geomagnetic activity. In: Krassovsky VI (ed) Aurorae and airglow. N 22. Nauka, Moscow, pp 60–70

Truttse YuL, Belyavskaya VD (1975) Red oxygen emission λ 6300 Å and density of the upper atmosphere. Geomagn Aeron 15:101–104

Truttse YuL, Belyavskaya VD (1977) Semiannual variations of the emission 6300 Å. Astron Circ USSR Acad Sci N 936:1–3

Truttse YuL, Gogoshev MM (1977) Red oxygen line 6300 Å and electron content in night F-region. Dokl Bulgarian Acad Sci 30:45–48

Truttse YuL, Shefov NN (1970a) Mid-latitudinal aurora on March 23–24, 1969. Astron Circ USSR Acad Sci N 562:3–5

Truttse YuL, Shefov NN (1970b) On middle-latitude aurorae of the current solar maximum. Planet Space Sci 18:1850–1854

Truttse YuL, Shefov NN (1971) Low-latitudinal effects of the geomagnetic disturbances. In: Geophys Bull N 23. Nauka, Moscow, pp 3–10

Truttse YuL, Yurchenko OT (1971) Temperature of the upper atmosphere from the 6300 Å emission data. Planet Space Sci 19:545–546

Tsurutani BT, Gonzalez WD, Lakhina GS, Alex S (2003) The extreme magnetic storm of 1–2 September 1859. J Geophys Res 108A:1268, doi: 10.1029/2002JA009504

Turnbull DN, Lowe RP (1983) Vibrational population distribution in the hydroxyl night airglow. Can. J. Phys. 61:244–250

Vassy AT, Vassy E (1976) La luminescence nocturne. In: Rawer K (ed) Handbuch der Physik. Geophysik 111/5, vol 49/5. Springer, Berlin Heidelberg New York, pp 5–116

Vincent RA, Kovalam S, Fritts DC, Isler JR (1998) Long-term MF radar observations of solar tides in the low-latitude mesosphere: interannual variability and comparisons with GSWM. J Geophys Res 103D:8667–8683

Walker JD, Reed EI (1976) Behaviour of the sodium and hydroxyl nighttime emissions during a stratospheric warmings. J Atmos Sci 33:118–130

Wallace L, McElroy MB (1966) The visual dayglow. Planet Space Sci 14:677–708

Wang DY, Ward WE, Shepherd GG, Wu DL (2000) Stationary planetary waves inferred from WINDII wind data taken within altitudes 90–120 km during 1991–96. J Atmos Sci 57:1906–1918

Wang DY, Ward WE, Solheim BH, Shepherd GG (2002) Longitudinal variations of green line emission rates observed by WINDII at altitudes 90–120 km during 1991–1996. J Atmos Solar—Terr Phys 64:1273–1286

Ward WE (1999) A simple model of diurnal variations in the mesospheric oxygen nightglow. Geophys Res Lett 26:3565–3568

Ward WE, Rochon YJ, McLandress C, Wang DY, Criswick JR, Solheim BH, Shepherd GG (1994) Correlations between the mesospheric $O(^1S)$ emission peak intensity and height, and temperature at 98 km using WINDII data. Adv Space Res 14:57–60

Watanabe T, Morioka Y, Nakamura M (1976) Altitude distribution of the O_2 and OH Meinel emissions in the nightglow. Rep Ionosph Space Res Jpn 30:41–45

Watanabe T, Nakamura M, Ogawa T (1981) Rocket measurements of O_2 Atmospheric and OH Meinel bands in the airglow. J Geophys Res 86A:5768–5774

Wiens RH, Weill G (1973) Diurnal, annual and solar cycle variations of hydroxyl and sodium intensities in the Europe–Africa sector. Planet Space Sci 21:1011–1027

WinEphem (2002) http://www.geocities.com/tmarkjames/WinEphem.html

Witt G, Stegman J, Solheim BH, Llewellyn EJ (1979) A measurement of the O_2 ($b^1\Sigma_g^+ - X^3\Sigma_g^-$) atmospheric band and the $O(^1S)$ green line in the nightglow. Planet Space Sci 27:341–350

Witt G, Stegman J, Murtagh DP, McDade IC, Greer RGH, Dickinson PHG, Jenkins DB (1984) Collisional energy transfer and the excitation of $O_2(b^1\Sigma_g^+)$ in the atmosphere. J Photochem 25:365–378

Woeikof A (1891) Cold and warm winter interchange. Meteorol Rep N 9:409–422

Woeikof A (1895) Die Schneedecke in "paaren" und "unpaaren" Wintern. Meteorol Zeits 12:77–78

Yao IG (1962) Observations of the night airglow. In: Roach FE (ed). Ann IGY, vol 24. Pergamon Press, London

Yarin VI (1961a) The OH emission according to observations in Yakutsk. In: Krassovsky VI (ed) Spectral, electrophotometrical and radar researches of aurorae and airglow. N 5. USSR Acad Sci Publ House, Moscow, pp 10–17

Yarin VI (1961b) Continuous emission and the Herzberg O_2 bands in the night airglow. In: Krassovsky VI (ed) Spectral, electrophotometrical and radar researches of aurorae and airglow. N 5. USSR Acad Sci Publ House, Moscow, pp 35–38

Yarin VI (1962a) On the dependence of intensity of OH bands on the rotational temperature. In: Krassovsky VI (ed) Aurorae and airglow. N 8. USSR Acad Sci Publ House, Moscow, pp 9–10

Yarin VI (1962b) Variations of the vibrational population rates of OH molecules. In: Krassovsky VI (ed) Aurorae and airglow. N 9. USSR Acad Sci Publ House, Moscow, pp 10–18

Yarin VI (1962c) On the molecular oxygen emissions in Yakutsk. In: Krassovsky VI (ed) Aurorae and airglow. N 9. USSR Acad Sci Publ House, Moscow, pp 34–43

Yarin VI (1970) Connection of the hydroxyl emission with the meteorological conditions above Yakutsk. In: Krassovsky VI (ed) Aurorae and airglow. N 18. USSR Acad Sci Publ House, Moscow, pp 18–20

Yee JH (1988) Non-thermal distribution of $O(^1D)$ atoms in the night-time thermosphere. Planet Space Sci 26:89–97

Yee JH, Abreu VJ (1987) Mesospheric 5577 Å green line and atmospheric motions—atmospheric explorer satellite observations. Planet Space Sci 35:1389–1395

Yee JH, Crowley G, Roble RG, Skinner WR, Burrage MD, Hays PB (1997) Global simulations and observations of $O(^1S)$, $O_2(^1\Sigma)$ and OH mesospheric nightglow emissions. J Geophys Res 102A:19949–19968

Yevlashin LS (2005) Aperiodic variations of the observation frequency of the red type-A auroras during 11-year cycle of solar activity. Geomagn Aeron 45:388–391

Yevlashina LM, Yevlashin LS (1971) Some peculiarities of the F region disturbances during red auroras of A type. In: Yevlashin LS (ed) Morphology and form of the polar ionosphere. Nauka, Leningrad, pp 137–146

Yugov VA, Nikolashkin SV, Ignatiev VM (1997) Correlation of temperature in the subauroral lower thermosphere with solar activity and phases of quasi-biennial oscillations. Geomagn Aeron 37:755–758

Yurchenko OT (1975) Interferometric measurements of the upper atmosphere temperature. II. 6300 Å emission variations. In: Krassovsky VI (ed) Aurorae and airglow. N 23. Nauka, Moscow, pp 66–68

Zhang SP, Shepherd GG (2004) Solar influence on the $O(^1D)$ dayglow emission rate: global-scale measurements by WINDII on UARS. Geophys Res Lett 31:L07804, doi: 10.1029/2002GL019447

Zhang SP, Peterson RN, Wiens RH, Shepherd GG (1993) Gravity waves from O_2 nightglow during the AIDA'89 campaign. I. Emission rate/temperature observations. J Atmos Terr Phys 55:355–375

Chapter 5
Wave Processes in the Atmosphere

All processes occurring in the upper atmosphere and the variations of the atmospheric characteristics are related to fluctuations in the energy influx. The variations of the parameters of the middle atmosphere are characterized by mesoscopic time scales from several minutes to several hours.

5.1 Internal Gravity Waves

The atmospheric waves bearing this name are caused by Archimedean buoyancy forces (which disappear in an isothermal process, $\gamma = 1$ (Golitsyn and Chunchuzov 1975)) in the atmosphere whose density varies with altitude. The periods of internal gravity waves (IGWs) are longer than the minimum Brunt–Väisälä period τ_g ($\tau_g \sim 5.2\,(\min)$ in the range 0–$100\,(\mathrm{km})$), which is determined by the formula (Hines 1974)

$$\tau_g = 2\pi \cdot \sqrt{\frac{\gamma \cdot k \cdot T}{(\gamma - 1) \cdot M \cdot m_H \cdot g^2}} = 2\pi \cdot \sqrt{\frac{\gamma \cdot H}{(\gamma - 1) \cdot g}},$$

where $\gamma = \frac{C_p}{C_v} = 1.4$ is the heat capacity ratio, k is Boltzmann's constant, T is the temperature, M is the molecular mass, m_H is the mass of a hydrogen atom, g is the free fall acceleration, and H is the scale height.

A wave with period τ propagates upward in the direction whose zenith angle θ is determined by the relation

$$\cos\theta = \frac{\tau_g}{\tau}.$$

As waves propagate upward, their energy is conserved and, hence, the amplitude increases since the density exponentially decreases with altitude:

$$A = A_0 \cdot \exp\left(\frac{Z - Z_0}{2\alpha \cdot H}\right).$$

The coefficient α takes into account the wave absorption. The limiting altitude to which the waves propagate is usually not above 100–130 (km). The vertical wavelengths λ_z vary from several kilometers to some tens of kilometers and the horizontal wavelengths λ_x from some hundreds to some thousands of kilometers.

The limiting horizontal phase velocity of IGWs can be estimated by the expression (Eckart 1960)

$$C_x = \sqrt{\frac{\gamma \cdot k \cdot T}{M \cdot m_H} \cdot \frac{1 - \left(1 - \frac{\tau_g^2}{\tau^2}\right) \cdot \frac{\tau^2}{\tau_g^2}}{1 - 1.21 \cdot \frac{\tau^2}{\tau_g^2}}} \approx 0.9 \cdot \sqrt{\frac{\gamma \cdot k \cdot T}{M \cdot m_H} \cdot \left(1 - \frac{\tau_g^2}{\tau^2}\right)}.$$

This approximate expression is valid, to within 3(%), for $\tau > 20$ (min) (Krassovsky et al. 1978). For IGWs, there is a dispersion relation according to which the wave velocity tends to zero as the wavelength decreases, while on increasing wavelength it tends to the increasing wavelength it tends to the velocity of sound, which is determined by the formula

$$C = \sqrt{\frac{\gamma \cdot k \cdot T}{M \cdot m_H}}.$$

5.1.1 Detection of IGWs in the Atmosphere

In the last decades, it became clear that internal gravity waves play an important part in the processes occurring in the upper atmosphere. Since 1960s, when during atmospheric tests of powerful atomic and thermonuclear charges grandiose disturbances were detected in the upper atmosphere, it has been commonly supposed that IGWs are generated in the main in the troposphere or at its boundary during active meteorological processes. These waves rather easily penetrate into the upper atmosphere. This is their distinction from acoustic waves which, with rare exception, do not propagate upward because of their reflection from the hot ozonosphere. However, besides the cases with explosions, it has not been revealed experimentally when and from which meteorological structures IGWs originate. However, sometimes it was supposed that they arise in the upper atmosphere during very strong geomagnetic disturbances. Nevertheless, for the overwhelming majority of detected waves, their relation to geomagnetic disturbances has not been decisively revealed.

As mentioned above, the amplitude of IGWs, as they propagate to higher and higher altitudes, increases in inverse proportion to the square root of the atmospheric density due to conservation of their energy in the case of no losses. This favored their detection in high atmospheric layers. At altitudes of 80–200 (km), IGWs are subject to spectral filtration because of the wind-induced shear, intensified turbulence, convection, and, hence, mixing of atmospheric components, and, finally, are absorbed, providing an increase in temperature in the absorption region in addition to that produced by solar radiation. Moreover, when absorbed, these waves can transfer their momentum to the air mass that absorbs them, and this is accompanied by the appearance of a wind component directed away from the IGW source. All these factors affect substantially the structure of both a neutral and an ionized atmosphere. Therefore, a detailed observation of IGWs, in particular of their amplitudes and directions of propagation, is of great importance. It is the more important as IGWs can yield valuable information on the processes responsible for their generation.

In a number of theoretical studies, it has been shown that the energetics of the upper atmosphere changes substantially during the propagation of IGWs. It seems that this fact was first noted by Hines (1968, 1974). Estimates of the energy fluxes to the upper atmosphere from the observable IGW sources (Gavrilov 1974; Chunchuzov 1978; Vincent 1984; Gavrilov 1992) have shown that they are comparable to the fluxes of solar short-wave radiation ($\sim 10\,(\mathrm{erg}\cdot\mathrm{cm}^{-2}\cdot\mathrm{s}^{-1})$), which controls the temperature of the upper atmosphere. Near the mesopause, the energy fluxes from IGWs are estimated to have values from one to some hundreds of ($\mathrm{erg}\cdot\mathrm{cm}^{-2}\cdot\mathrm{s}^{-1}$) (Hines 1968; Gavrilov 1974; Chunchuzov 1978).

Unfortunately, in contrast to numerous theoretical investigations, the methods of observations of IGWs remained few in number for a long time. It is extremely difficult to directly detect IGWs at the Earth surface because of their small amplitude at this level and the significant disturbance from random atmospheric pressure fluctuations. Besides, by performing measurements only at the Earth surface, it is impossible to trace the evolution of the parameters of IGWs and the extent of their environmental impact as they penetrate into the upper atmosphere. Wave processes in the upper atmosphere were detected by observing noctilucent clouds more than half a century ago, but their relation to IGWs was perceived only in the early 1960s. These observations could reveal, however, only very slow and short IGW (with velocities no more than $10\,(\mathrm{m}\cdot\mathrm{s}^{-1})$ and lengths of 10 (km)) (Bronshten and Grishin 1970; Fogle 1971). Now IGWs with such characteristics are detected by images of emission fields of the upper atmosphere photographed at large zenith angles (Moreels and Hersé 1977; Hapgood and Taylor 1982; Taylor and Hapgood 1990). However, these observations have the grave disadvantage that they give no way of reliably determining the directions of propagation of the waves and their periods and amplitudes. Waves with higher velocities (over $100\,(\mathrm{m}\cdot\mathrm{s}^{-1})$) and longer periods (over 100 (km)) were detected by radiophysical methods implemented simultaneously at several distant points, and the detection of radiosignals from geostationary satellites turned out especially efficient (Vasseur et al. 1972; Bertin et al. 1978). This method made it possible to determine the true velocities and directions of propagation of IGWs, since their velocities, as they moved from the bottom atmospheric layers to ionospheric altitudes, were greater than the velocities of motion of the atmosphere throughout its depth. In recent years, owing to the up-to-date television technology using CCDs, it became possible to obtain spatial intensity distributions for hydroxyl, atomic oxygen, and sodium emissions that clearly reveal a wave structure in the upper atmosphere and its dynamics (Fig. 5.1) (Gavrilieva and Ammosov 2001).

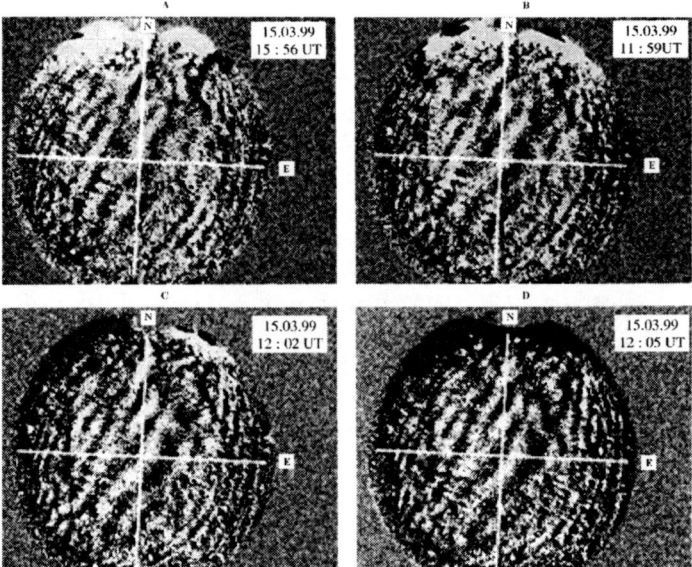

Fig. 5.1 Example of the structures of hydroxyl emission wave fields. Long waves ($\sim$26(km)) propagated southwest with a velocity of 25 (m·s^{-1}) and a "ripple" structure moved almost orthogonally (Gavrilieva and Ammosov 2001)

Most of the data on IGWs propagating to altitudes above 200(km) have been obtained by various methods of continuous radiosonde observation of moving ionospheric perturbations in the F2 region. When different ionospheric regions are observed simultaneously from three ground-based points, it is possible to determine the direction of propagation of a wave by the time lags with which the wave passed through these regions. However, because of the absorption and nonlinear processes at lower levels of the atmosphere, IGWs arrive at high altitudes considerably relaxed and distorted. Moreover, an ionospheric perturbation occurring in the geomagnetic field is an intricate selective response to IGW. Therefore, interpretation of such diverse and interdependent data is a rather difficult problem. Nevertheless, it has been established that IGWs are detected in the main at midlatitudes in winter daytime, propagating from winter polar regions. This is due to the fact that the ionized constituent of the atmosphere, being in the geomagnetic field and moving along its lines, responds only to IGWs propagating in a certain direction relative to this field. Besides, the enhanced ionization during daytime facilitates the detection of IGWs.

However, even with the above diagnostic facilities, the IGW source could not be identified with circumterrestrial meteorological structures for a long time. In few cases of detection of IGWs, their occurrence was accounted for by geomagnetic disturbances in high-altitude ($\sim$120(km)) atmospheric regions. The majority of other methods, based on measuring variations of atmospheric characteristics mainly at lower altitudes, did not provide continuous observation of IGWs. With these methods, only very restricted in space and short-term (minutes and seconds) records of these effects were possible. Among these methods are the radio and optical sounding of meteoric and artificial tracks and the rocket and satellite measurements of density, pressure, and temperature. In extensive radio observations of meteors at altitudes of 80–100 (km), it is possible to obtain data sufficient for valuable statistical inferences only for IGWs with many-hour periods. Therefore, because of the irregularity of the observation material, adequate information about the amplitudes and directions of the detected IGW has not been obtained until the end of the 1970s.

Since the early 1970s, a great amount of manifold data on characteristics of IGWs, conditions of their generation, and their effect on the dynamics and energetics of high atmospheric layers have been accumulated at A. M. Obukhov Institute of Atmospheric Physics of the Russian Academy of Sciences (IAPh). New interesting results have been obtained by the optical method of detection of the hydroxyl, atomic oxygen, and sodium emissions in the upper atmosphere. This method is based on the principle of simultaneous detection of the IGW-induced intensity and temperature variations in several regions of the night sky (actually, in a rather thin, ~10-km emission layer).

The idea of using variations of the emission intensity and medium temperature in studying acoustic gravity waves in the upper atmosphere arose at IAPh long ago. As early as in 1956, at the Seventh International Astrophysical Symposium "Les Molecules dans les Astres" (Liège, Belgium, July 12–14, 1956), Krassovsky called attention to the regularity of these variations and came out with the supposition that they can be a consequence of adiabatic processes which occur in an emission layer during the propagation of IGWs (Krassovsky 1957a,b). This was verified experimentally when the effect of IGW modulation of the hydroxyl emission intensity was discovered (Krassovsky and Shagaev 1974a,b; Shagaev 1974; Krassovsky et al. 1978; Potapov et al. 1978). Subsequently, emission intensity variations caused by IGWs propagating through the emission layer were also detected in molecular oxygen, atomic oxygen, and sodium emissions (Krassovsky et al. 1975, 1986b; Krassovsky and Shefov 1976b; Noxon 1978). These results gave impetus to numerous reports which argued in support of the new phenomenon and its importance for the study of IGWs and their effect on the characteristics of the upper atmosphere (Meriwether 1975; Armstrong 1975; Weinstock 1978; Hatfield et al. 1981; Walterscheid et al. 1987; Krassovsky et al. 1988; Semenov 1989a).

For many reasons, the hydroxyl emission turned out to be an extremely convenient and informative object in investigating various temperature and dynamic characteristics of the upper atmosphere at mesospheric altitudes. First of all, as already mentioned, this is related to the fact that this is the most powerful nightglow in the wavelength range 0.5–4.5 (μm) and, therefore, it is reliably detectable by ground-based spectrophotometric instruments. Moreover, the hydroxyl emission is localized at altitudes near the base of the thermosphere where the principal processes of IGW dissipation just occur, giving rise to temperature and dynamic variations in the atmosphere.

5.1.2 Modulation of Emission Characteristics by IGWs Propagating Through the Emission Layer

Detection and identification of the hydroxyl emission gave impetus to revealing its nature. As described in Sect. 2.2, two basic cycles of processes responsible for the excitation of hydroxyl molecules have been proposed: ozone–hydrogen and perhydroxyl. The first mechanism by which hydroxyl molecules are excited at the ninth vibrational level became universally recognized. The second mechanism could provide the excitation of molecules at levels not above the sixth one. Thus, the two mechanisms should give different types of correlation between the characteristics of OH bands from the upper and the lower vibrational levels, and these features have been actually detected in the behavior of hydroxyl emission. Their interpretation essentially depends on the proposed cycle of processes responsible for the observed behavior of the intensity variations of OH emissions from different vibrational levels.

As considered in Sect. 2.2, the realizability of the perhydroxyl mechanism is put under doubt in view of laboratory investigations which have shown that the oxygen isotope ^{18}O when reacting with perhydroxyl $H^{16}O_2$ attaches to the oxygen molecule to form $^{18}O^{16}O$ rather than to the hydroxyl molecule ^{16}OH (Kaye 1988). The formation of OH (naturally, from the ground state $v = 0$) was checked by means of the laser-excited fluorescence of molecules. Moreover, there is no information about any attempt to detect the hydroxyl emission in the infrared spectral range. Therefore, it is now yet impossible to make the decisive conclusion that the perhydroxyl reaction fails to provide excitation of hydroxyl molecules under the conditions of the upper atmosphere. This calls for further investigations of the nature of the observed features of the hydroxyl emission. It seems that here an essential role is played by the processes of vibrational relaxation of excited hydroxyl molecules by collisions with oxygen atoms and molecules, depending on the altitude in the mesopause region. It is not improbable that some features of the rotational temperature variations can be related to the interaction of excited OH molecules with hydrogen molecules (Dewangan et al. 1986). All variety and complexity of these processes, which occur under the conditions of the upper atmosphere, cannot be realized in laboratory investigations. Therefore, to solve the problem, one has to use various approaches and methods of analysis of the detected natural variations of the OH emission intensity and temperature. One of these approaches deals with the response of the OH emission layer to IGWs.

The doubtless principal advantage of the optical recording method is the possibility of detection of IGWs by simultaneously measured variations of several parameters, namely the emission intensity and the temperature of the medium in which the emission occurs, which are caused by adiabatic processes in the emission layer during the propagation of IGWs. As already mentioned, Krassovsky (1957a,b) was the first to call attention to the effect of adiabatic processes on the upper atmospheric temperature and the emissions that accompany the propagation of infrasonic oscillations. Krassovsky (1972), based on the data of observations of the hydroxyl emission intensity and temperature variations, has shown for the first time that these variations can be caused by the action of IGWs on the emission layer. On the assumption that the processes occurring in the hydroxyl emission region during the propagation of IGWs are purely adiabatic, the relative intensity variations can be related to the relative temperature fluctuations as follows:

$$\frac{\Delta I}{I} = \eta \cdot \frac{\Delta T}{T},$$

where the factor η (Krassovsky's number) is given by

$$\eta = \frac{m}{\gamma - 1} - n, \; \gamma = \frac{C_p}{C_v},$$

where C_p and C_v are the specific heats at constant pressure and at constant volume, respectively; n is the exponent in the formula for the rates of three-body reactions (ozone–hydrogen and perhydroxyl) that describes their temperature dependence; m is a coefficient depending on the degree of quenching of vibrationally excited hydroxyl (which is equal to 2 with no quenching and to 1 for complete quenching).

In Sect. 2.2.7, it was noted that in the cycle of reactions responsible for the excitation of hydroxyl, the three-body reactions mentioned above are dominant. Vibrationally excited OH molecules are produced as a result of the subsequent processes. These processes are faster than the primary processes; therefore, the detected hydroxyl emission reflects, in the main, the primary processes with an insignificant delay. The initial components of the hydroxyl excitation reactions cannot be depleted substantially within the characteristic periods (20–120 (min)) of the IGWs detected in the mesopause and lower thermosphere regions. The principal component is atomic oxygen, whose density at altitudes of about 90 (km) is estimated to be $\sim 10^{11}$ (cm^{-3}) (CIRA-1972) and the consumption no more than 10^6 (cm$^{-3} \cdot$ s^{-1}).

Thus, the initial components for the cycle of formation of excited hydroxyl are produced as a result of the reaction of three-body collisions of oxygen or hydrogen atoms with oxygen molecules or with other molecules of the atmosphere:

$$O + O_2 + M \rightarrow O_3 + M,$$
$$H + O_2 + M \rightarrow HO_2 + M,$$

where M is any component of the atmosphere. The number of molecules formed in some volume V of the emission layer, N, can be determined by the following expressions:

$$N(O_3) = \alpha_{O_3} \cdot \exp(E_i/RT) \cdot [O] \cdot [O_2] \cdot [M] \cdot V,$$
$$N(HO_2) = \alpha_{HO_2} \cdot \exp(E_i/RT) \cdot [H] \cdot [O_2] \cdot [M] \cdot V,$$

where the square brackets denote concentrations of atoms or molecules, $\alpha \cdot \exp(E_i/RT)$ is the ith reaction rate, and E_i is the activation energy of the ith reaction: $E = 1014 \pm 46$ (cal $\cdot$ mol^{-1}) for the reaction with atomic oxygen and $E = 685 \pm 128$

(cal·mol^{-1}) for the reaction with atomic hydrogen (Winick 1983); R = 1.987 (cal·deg^{-1}·mol^{-1}) is the gas constant. The concentrations of reacting components can be represented as

$$[X] = \frac{(X)}{V}; \quad [O_2] = \frac{(O_2)}{V}; \quad [M] = \frac{(M)}{V},$$

where the component symbols in parentheses designate the total number of atoms and molecules in volume V, and X stands for O or H.

For the case of an adiabatic process we have

$$V^{\gamma-1} \cdot T = C = \text{const},$$

whence

$$V = C \cdot T^{-\frac{1}{\gamma-1}}.$$

Substituting this in the above relations, we obtain

$$N = \left[\frac{\alpha(X)(O_2)(M)}{C}\right] \cdot T^{2/(\gamma-1)} \cdot \exp[E_i/RT].$$

For adiabatic processes, the terms in square brackets are invariable. Based on the foregoing, we obtain

$$\frac{\delta N}{N} = \left(\frac{2}{\gamma-1} - \frac{E_i}{RT}\right) \cdot \frac{\delta T}{T}.$$

Since the hydroxyl emission intensity is proportional to the number of molecules produced, N, we have

$$\frac{\delta I}{I} = \left(\frac{2}{\gamma-1} - \frac{E_i}{RT}\right) \cdot \frac{\delta T}{T} = \eta \cdot \frac{\delta T}{T},$$

where I is the intensity of the OH emission in the observed rotational–vibrational band. For the conditions of the mesopause (T $\sim$ 200 (K)) and for the activation energies given above, the values of the number η for the reactions forming ozone and perhydroxyl molecules can be estimated as

$$\eta_{O_3} = 2.45; \quad \eta_{HO_2} = 1.7.$$

Thus, for given temperature variations caused by IGWs propagating through the OH emission layer, the relative variations of the OH emission intensity determined in view of the ozone–hydrogen excitation mechanism are greater than that determined based on the perhydroxyl one.

It is of interest to elucidate and estimate the effect of the rates of quenching of excited hydroxyl molecules in collisions with atmospheric atoms and molecules on η. If N_{OH}-excited OH molecules are formed in a volume V in a second, the emission intensity I_{OH} in any rotational–vibrational band can be estimated as

$$I_{OH} = \frac{N_{OH} \cdot A_{OH}}{A_{OH} + [M] \cdot \beta_M},$$

5.1 Internal Gravity Waves

where A_{OH} is the probability of emission in the given band and β_M is the rate of quenching of the excited hydroxyl by the component M.

The most effective quenching should be expected if $[M] \cdot \beta_M > A_{OH}$; hence, the above expression can be written as

$$I_{OH} \approx \frac{N_{OH} \cdot A_{OH}}{[M] \cdot \beta_M}.$$

Representing the concentration of M as we earlier did for that of oxygen and hydrogen atoms, we obtain

$$I_{OH} \approx \left[\frac{N_{OH} \cdot A_{OH}}{(M) \cdot \beta_M \cdot C}\right] \cdot T^{\left[\frac{1}{\gamma-1} - \frac{E}{RT}\right]}.$$

In this case,

$$\frac{\delta I_{OH}}{I_{OH}} \approx \left(\frac{1}{\gamma-1} - \frac{E}{RT}\right) \cdot \frac{\delta T}{T} = \eta \cdot \frac{\delta T}{T},$$

that is, at high quenching rates (m = 1) and great E, the variations δI_{OH} and δT can be opposite in phase. Thus, the quenching of excited hydroxyl due to collisional processes can considerably reduce η instead of increasing it. The nature of the hydroxyl emission and the correlation between the variations of its intensity and temperature under adiabatic conditions were studied by Potapov (1975c) and Semenov (1989a,b) who considered the mechanisms of formation of excited hydroxyl molecules. The reactions with participation of ozone and perhydroxyl can proceed both with and without vibrational excitation of hydroxyl molecules. The character of these reactions is reflected in the correlation between intensity and rotational temperature under adiabatic perturbations of the state of the atmosphere during the propagation of IGWs. It has been shown that for the reaction of nonexcited ozone with atomic hydrogen, η ranges between 3.5 and 6. Subsequently, the idea of the modulation of the intensity of the upper atmospheric emissions under the action of IGWs was further developed (Weinstock 1978; Walterscheid et al. 1987; Sivjee et al. 1987; Schubert and Walterscheid 1988). By the time of publication of the work by Weinstock (1978), data of observations of quasi-periodic temperature and intensity oscillations of the $O_2(^1\Sigma)$ emission became known (Noxon 1978), which were interpreted as a consequence of the action of IGWs on the emission layer. Weinstock (1978) pointed out that when calculating the Krassovsky number η, it should be borne in mind that the fluctuations of the principal components of the reaction of excitation of $O_2(^1\Sigma)$ and the fluctuations of the atomic oxygen density are not in phase. This observation was taken into account by introducing a new term in the expression for η.

For the hydroxyl emission, we have

$$\eta = \frac{2}{\gamma-1} - \frac{E_i}{RT} - \left[1 + \frac{\gamma}{\gamma-1} \cdot \left(\frac{H}{H_X} + 1\right)\right],$$

where H_X is the scale height for hydrogen or oxygen atoms and H is the scale height for the principal component O_2. The theoretical value of η obtained by this expression agrees with that obtained by the original formula. The distinction is that the expression in square brackets can accept various values reducing η.

Attempts to develop a model which would provide an explanation of the IGW-induced fluctuations of the hydroxyl nightglow based on the consideration of the combined action of a variety of dynamic and photochemical processes are discussed elsewhere (Walterscheid et al. 1987). Proceeding from the conclusions by Krassovsky (1972) and Weinstock (1978), it was analyzed how η depends not only on the parameters that characterize the state of the atmosphere (temperature, concentrations of principal components O_2 and N_2 and minor components O, O_3, OH, and HO_2, and scale heights) and on the reaction rates, but also on the period of IGWs, their wavelengths, and the direction of their vertical propagation.

5.1.3 Choice of Measurable Parameters

The study of fast (1–3 (min)) small-scale (about 10–30 (km)) intensity variations for different emissions (OH and OI 557.7 (nm)) which arise in the mesopause and lower thermosphere region has shown that the measurement data are strongly distorted by the fluctuations of the atmospheric transparency (Taranova 1967; Fedorova 1967; Korobeynikova and Nasyrov 1972; Kuzmin 1975a).

Special investigations performed at the Zvenigorod station (Kuzmin 1975c; Krassovsky and Shefov 1976b) have shown that in poor transparency conditions the spectral distribution of the moonlight variations resembles the spectrum of the intensity variations for the 557.7-nm emission. Nevertheless, the transparency properties are practically identical for narrow spectral bands 5–10 (nm) in the range 700–1100 (nm) if they do not contain absorption bands of water vapor. Therefore, the rotational temperature of the hydroxyl emission, determined by the ratio of simultaneously measured intensities of a pair of closely spaced spectral lines, appears independent of the atmospheric transparency variations. These conditions are satisfied by neighboring lines of the P branches and the R and Q branches of many OH bands in the near-infrared spectral region.

Since the rotational temperature of the hydroxyl emission is rather close to the temperature inside the emission layer at $\sim$90 (km), its fast variations reflect the actual variations of the atmospheric temperature at these altitudes.

Nevertheless, it is well known that OH molecules with different vibrationally excited levels radiate at different altitudes. Therefore, Krassovsky's number η poses the requirement of investigating the characteristics of IGWs by recording different OH bands that characterize different photochemical processes in the atmosphere.

For the hydroxyl emission, η is 2–4 (Krassovsky and Shefov 1976b), while for the green emission $\eta \leq 1$ (Krassovsky et al. 1986b; Semenov 1989a). Therefore, the intensity of the green emission should respond to temperature variations not as dramatically as that of the hydroxyl emission. It should also be noted that this parameter is difficult to determine for the green line of atomic oxygen because of some difficulties involved in the interferometric determination of temperature by the Doppler profile of this emission. This circumstance hampers regular observations of the emission.

The hydroxyl emission enables simultaneous measurements of intensity and rotational temperature variations which makes it possible to reveal an adiabatic component (number η) in the measured variations which, if present, can serve as an argument in favor of the wave type of the disturbance observed.

The wave properties of the mesopause processes under study impose certain requirements on the method of their investigation. As shown above, the instrumentation for recording IGWs must allow for fast measuring of rotational temperature.

It is well known that the vibrational–rotational bands of hydroxyl are located in the main in the infrared spectral region. To choose the necessary lines of these bands, the wavelengths of all OH bands observable in the spectral region from 500 to 4400 (nm) were calculated by more accurate methods than those used earlier (Piterskaya 1976b). For determining the intensity ratios of the chosen lines and branches of OH bands as functions of rotational temperature, there are calculations of the relative intensities of the P and Q lines and P branches for $N = 1$–7 and $v = 1$–9 in every 5 (K) over the range 150–350 (K), which completely fits the conditions of the observations described elsewhere (Piterskaya and Shefov 1975; Piterskaya 1976a).

Practically, to determine the characteristics of IGWs, the hydroxyl (8–3), (9–4), (5–1), (6–2), (9–5), and (4–1) bands were used. The wavelengths for the center of the spectral region taken by each band were 0.73, 0.776, 0.79, 0.83, 0.99, and 1.01 (μm), respectively (Krassovsky et al. 1962).

5.1.4 Observation Conditions

The investigation of internal gravity waves was based in the main on the data of spectrographic and photometric observations of the variations of the hydroxyl emission characteristics that were conducted at the Zvenigorod station of A. M. Obukhov Institute of Atmosphere Physics of the Russian Academy of Sciences (55.7°N, 36.8°E). Regular observations have been carried out since the late 1972.

To preclude the possible effect of the diffused light from the Moon and from the light sources of the neighboring settlements and of the variability of the atmospheric transparency due to variable cloudiness, practically all observations were conducted in clear moonless nights when the Milky Way or Pleiades stars could be clearly seen. This corresponds to the transparency for stellar point sources no worse than 0.9 (Sawyer 1951).

The observation time within a night was from 3 to 6 (h).

5.1.5 Determination of IGW Characteristics by the Variations of the OH Rotational Temperature

A great deal of information is available in the literature on the variations of the rotational temperature measured by various OH bands whose amplitude for quiet geophysical conditions is generally no more than a few Kelvin's degrees. In some cases of disturbed atmospheric conditions, temperature variations of amplitude $+50\,(\text{K})$ were detected. When comparing temperature variations in various regions of the firmament, it is possible sometimes to detect similar variations, while in other cases they are hardly noticeable. For such a comparison, the results of observations in three regions of the sky which form an equilateral triangle were used. The results could be grouped by three observation patterns: there was nothing in common in the three regions, similar variations were observed only in two regions, and identical oscillations occurred in all the three regions. About a quarter of the obtained material could be assigned to the last pattern.

Common features in temperature variations detected at three azimuths are sometimes very pronounced (Krassovsky and Shagaev 1974b; Krassovsky and Shefov 1976b). Against the background of aperiodic temperature oscillations, similar periodic oscillations stand out whose amplitude is far above the random background. However, the oscillation periods in records obtained at different azimuths, as a rule, do not coincide. In that case, undoubtedly, wave-shaped propagation of temperature perturbations occurs. The velocity and direction of their propagation can be determined by the base rectangle dimensions and by the phase shift.

Figure 5.2 shows a typical periodogram of one set of values of the rotational temperature at the zenith (Krassovsky et al. 1977, 1978). The positions of maxima are different in different time intervals. If a great number of maxima are plotted on the same plot (Fig. 5.3), the mean of all amplitudes of individual monochromatic waves can be presented as a function of τ^n, where τ is the period and $n \sim 1.1$. The average periodogram resembles a periodogram obtained by ionospheric sounding (Gupta and Nagpal 1973). The only difference is that the data of ionospheric sounding give $n \sim 1.7$ and the abrupt drop of spectral densities happens near the 15-min periods. It should be noted that the average periodogram in Fig. 5.3 is also similar to the spectral density periodogram obtained from the data of radiometeoric sounding for periods over $2\,(\text{h})$. It has been revealed that $n = 0.82$ for an altitude of $90\,(\text{km})$ and 0.47 for $100\,(\text{km})$ (Spizzichino 1971).

To reveal periodic variations imperceptible on the background of the fluctuations related to irregularities of the emission layer or to other random noises, the procedure of data processing described in Sect. 3.6 is used. Fourier transformation of the autocorrelation and cross-correlation functions C_{AA}, C_{BB}, C_{CC} and C_{AB}, C_{BC}, C_{CA} computed for series of temperatures measured at points A, B, C of the sky yields three spectral density autocorrelation functions P_{AA}, P_{BB}, P_{CC} and three spectral density cross-correlation functions P_{AB}, P_{BC}, P_{CA}. The presence of a maximum corresponding to the same frequency in the three pairs of spectra suggests that there are

5.1 Internal Gravity Waves

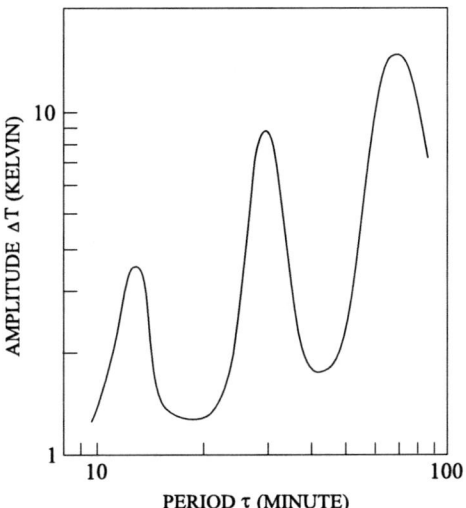

Fig. 5.2 Example of periodogram of the variations of rotational temperature for hydroxyl (Krassovsky et al. 1977, 1978)

waves or groups of waves for which the horizontal phase velocity and the direction of propagation can be determined.

The validity of this interpretation is verified by calculating the coefficients of coherence for all pairs of spectra:

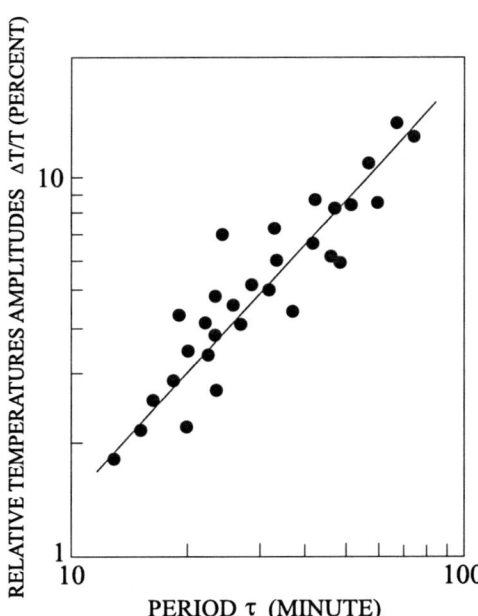

Fig. 5.3 Typical average periodogram of rotational temperature variations for hydroxyl (Krassovsky et al. 1977, 1978). *Dots* are spectral amplitude maxima; *solid line* is regression line

$$r_{AB} = \frac{P_{AB}^2}{P_{AA} \cdot P_{BB}}; \quad r_{BC} = \frac{P_{BC}^2}{P_{BB} \cdot P_{CC}}; \quad r_{CA} = \frac{P_{CA}^2}{P_{CC} \cdot P_{AA}}.$$

When all the three coefficients of coherence were close to unity, namely over 0.7, it was supposed that the same phenomenon was detected at the three points (regions of the emission layer).

Based on periodograms, all observed oscillations are subdivided into three groups: oscillations which are observed only at one of the points, at two of them, and at all the three points. We first consider only the third group. Noteworthy, in this group the three coefficients had different values, though within the chosen limits (0.7–1).

By means of ionospheric sounding, it was shown that the phase surface of the internal gravity waves that produce moving ionospheric disturbances in the F2 layer propagates downward (Georges 1968). This convincingly proved that the source of the waves is below the layer in which they are observed (Hines 1960). A similar pattern is revealed for the IGWs observed by the hydroxyl emission. Measurements of temperature fluctuations were performed at two distant points to determine the altitude of maximum hydroxyl emission (Potapov 1975a,b, 1976). It was found that the altitude at which similar fluctuations occur does not remain unchanged even within a night and that in many cases it is greater for bands from higher initial levels than that for bands from lower ones. These results agreed with the long known behavior of the hydroxyl rotational temperature: on the average, it is higher for the bands from higher initial levels (Krassovsky 1973; Berg and Shefov 1963). Subsequently, a similar conclusion was made by Suzuki and Tohmatsu (1976). Thus, by simultaneously recording two hydroxyl bands corresponding to different initial levels, it is possible to observe IGWs simultaneously at two different altitudes.

Let us use the following designations: T_1 and T_2 are the mean temperatures (K) at the lower and the upper level, respectively; A_1 and A_2 are the amplitudes of temperature variations at the lower and the upper altitude level, respectively; ΔZ is the distance between the maxima of the hydroxyl emission layers with different vibrational excitation; C_z and C_x are the vertical and the horizontal phase velocity of an IGW; Δt is the time interval between the IGW arrivals at the lower and the higher levels; τ_g is the Brunt–Väisälä period; τ is the period of the recorded oscillations; k is Boltzmann's constant; γ is the specific heat ratio ($C_p/C_v = 1.4$); M is the mean molecular mass, and g is the free fall acceleration. Let us also assume that all values of g and M, with an adequate accuracy, are constant and equivalent to their values at an altitude of about 90 (km) (CIRA-1972).

As mentioned above, the IGW amplitude increases with altitude. Thus, we have

$$\Delta Z = 2\alpha \cdot H \cdot \log_e \left(\frac{A}{A_0}\right) = \frac{2\alpha \cdot k \cdot \bar{T}}{M \cdot m_H \cdot g} \cdot \log_e \frac{A}{A_0},$$

where $\bar{T}$, the temperature mean for the lower and the upper altitude levels, can be represented as

$$T = \beta_1 T_1 = \beta_2 T_2 = \beta T,$$

5.1 Internal Gravity Waves

and then
$$\Delta Z = 2\alpha\beta k T m^{-1} g^{-1} \log_e(A_2/A_1).$$

On the other hand, we have
$$\Delta Z = C_Z \cdot \Delta t.$$

In a rough approximation, C_z, according to the dispersion relation (Lindzen 1971)
$$\frac{C_x^2}{C_z^2} = \frac{\tau^2}{\tau_g^2} \cdot \left(1 - \frac{\tau_g^2}{\tau^2}\right),$$

can be replaced by C_x. As a result, we obtain
$$\Delta Z = C_x \cdot \frac{\tau_g}{\tau} \cdot \left(1 - \frac{\tau_g^2}{\tau^2}\right)^{-\frac{1}{2}} \cdot \Delta t.$$

From these relations, it follows that
$$\alpha = \frac{C_x \cdot \tau_g \cdot M \cdot m_H \cdot g \cdot \Delta t}{2\beta \cdot k \cdot T \cdot \sqrt{1 - \frac{\tau_g^2}{\tau^2}} \cdot \log_e \frac{A}{A_0}},$$

and on substitution of the values of C_x and τ_g,
$$\alpha \approx 0.9 \cdot \pi \gamma^{1/2} \left[\frac{\gamma-1}{\gamma} + \frac{\partial H}{\partial Z}\right]^{1/2} \cdot \left[\frac{\Delta t}{\tau} \cdot \log_e \frac{A}{A_0}\right].$$

Based on experimental data (Krassovsky et al. 1978), it has been established that the expression in the second square brackets
$$\frac{\Delta t}{\tau} \cdot \log_e \frac{A}{A_0} \approx 0.142$$

within 5(%). Besides, β (depending on whether T is taken equal to the temperature of the lower or the upper hydroxyl emission layer) introduces an error of several percents. As a result it appears that $\alpha \sim 1$ within 10(%), i.e., practically within the measurement error for the IGW parameters Δt, τ, A, and A_0. Thus, the errors in α/β yield a 10(%) error to ΔZ and a 0.5(%) error to the horizontal velocities of IGWs. Comparing radiometeoric investigations of IGWs with periods over 2 (h) gave $\alpha \sim 0.5$ (Spizzichino 1971). This value of α was accounted for by the fact that the energy of IGWs in their propagation upward increases from some other sources. However, there is no decisive opinion on this point.

The rotational temperature was determined at two southern points of the sky besides the zenith. The three points formed an equilateral triangle. Assuming that the center of gravity of the lower emission layer was on the average at $\sim$90 (km) (Potapov 1975a,b, 1976) and the center of gravity of the higher level was ΔZ over, the side of the lower base triangle appears equal to 200 (km) and that of the upper one to $((90+\Delta Z)/90)\cdot 200$ (km). Under these assumptions, it is possible to calculate the wave velocities C_1 and C_2 and determine their directions at the lower and upper levels, respectively, from the point of view of the immobile observer on the Earth surface.

The above formulas can be used to calculate the horizontal phase velocity of IGWs, C_x, in the coordinate system of the moving layer, and also C_{x_1}, C_1, φ_1 and C_{x_2}, C_2, φ_2, the horizontal phase velocities, the horizontal velocities for the observer on the Earth, and the azimuths for the lower and the upper levels, respectively.

The differences $V_1 = C_1 - C_{x_1}$ and $V_2 = C_2 - C_{x_2}$ can also be determined. A table of examples of measured and calculated values of C_x, V_1, V_2 and also of C_1, C_2, and their azimuths φ_1 and φ_2 for some measurement period is presented elsewhere (Krassovsky et al. 1978). The azimuths are counted clockwise from the south direction. The tendency for velocities a little greater on the average than the limiting one (C_{x_1} and C_{x_2}) can be related not only to the measurement errors and errors in the estimation of the altitudes of the hydroxyl levels, but also to the transfer of the IGW momentum in the atmosphere and to the tail wind that favors the propagation of IGWs directed in line with this wind for greater distances. A possible reason for the latter may be the variability of the dissipative processes and of the reflectivity of the upper atmosphere.

For comparison, the wind velocities V and the azimuths of their directions φ obtained by radiometeoric methods near Moscow, at Obninsk (Portnyagin et al. 1978), have been used. They refer to the altitude range 90–100 (km). Assuming the altitude 90 (km) or $90 + \Delta Z$ (km) for radiometeoric velocities V and their azimuths φ, it is possible to calculate the projections of the meteoric velocities V_1^o and V_2^o on the directions of IGWs in the lower and the upper layers, respectively. It turned out that, within the limits of errors, the hydroxyl and radiometeoric data show quite satisfactory coincidence between V_1 and V_1^o. However, the result is different if the radiometeoric data are compared to the data on the hydroxyl emission at the altitude $(90 + \Delta Z)$ (km). The available values of V are strongly averaged over the altitudes from 80 to 100 (km). For a thinner emission layer region, true values of V can be greater. It seems that the Obninsk radiometeoric sounding data correspond to altitudes over 90 (km).

It is of importance to increase the accuracy of determination of C_1, φ_1 (or C_2, φ_2) as well as of V and φ. If V and φ are associated with a certain altitude, the true altitude of the emission layer Z can be determined from its conventional value Z_{90} of ~ 90 (km) by the relation

$$\frac{Z}{Z_{90}} = \frac{C_{x_1} + V_1^o}{C_1}.$$

When it is managed to detect simultaneously two waves with different periods, a possibility arises to calculate the velocity magnitude and direction independently for each wave.

Thus, simultaneous observations of the variations of the hydroxyl rotational temperature by the bands from high and low initial levels suggest that these variations are largely induced by IGWs propagating from the lower atmosphere to a near-mesopause region. Therefore, it seems probable that the great number of variations of the rotational temperature recorded at Zvenigorod only by one hydroxyl band are related to a similar situation (Tables 3 and 4, Krassovsky et al. 1978). With the use of a large base triangle with a side of about 200 (km), provided $\tau_g = 4.5\text{–}5\,(\text{min})$ and $\tau > 20\,(\text{min})$, only long IGWs can be reliably detected. Their horizontal phase velocity is close to the limiting one that corresponds to the average temperature of the lower and the upper layers. In many cases, the velocity of detected waves oscillates about this limit, and the deviation can be accounted for by the superposition of atmospheric circulation. These waves essentially differ from the IGWs detected in the ionospheric F2 region by radio sounding (Georges 1968). The velocities of the latter are either more than $300\,(\text{m}\cdot\text{s}^{-1})$ in the case of IGWs which occur during powerful geomagnetic storms or obviously lower $150\,(\text{m}\cdot\text{s}^{-1})$ in the case of traveling ionospheric disturbances (TIDs) in quiet geomagnetic conditions. It is well known (DeVries 1972) that during geomagnetic disturbances winds blow with very high velocities from the polar regions toward the equator. Therefore, it can be supposed that the high velocity of IGWs during geomagnetic storms is promoted by a high-velocity tail wind.

5.1.6 Localization of IGW Sources in the Troposphere

The optical method of detection of an IGW by the variations of the hydroxyl rotational temperature measured at three azimuths enables localization of the IGW source. It is well known that at great distances from the sources internal gravity waves disintegrate into a series of quasi-harmonic oscillations which propagate at different zenith angles depending on the oscillation period. The dispersion relation for IGWs has the form (Golitsyn and Romanova 1968)

$$\omega^2 = \frac{C^2}{2}\left\{k_z^2 + k_x^2 + \frac{1}{4H^2} - \left[(k_z^2 + k_x^2 + \frac{1}{4H^2})^2 - \frac{4(\gamma-1)g^2 k_x^2}{C^2}\right]^{1/2}\right\},$$

where $\omega = 2\pi/\tau$ is the oscillation frequency; C is the sound velocity; k_z and k_x are the wave numbers for the vertical and the horizontal directions, respectively, and H is the scale height.

For for IGWs at the mesopause altitudes, the following inequality is fulfilled (Hines 1960):

$$k_z^2 \gg k_x^2; \quad k_z \cdot H \gg 1.$$

In this case,

$$\tau = \frac{\tau_g}{k_x}\cdot\sqrt{k_x^2 + k_z^2 + \frac{1}{4H^2}}.$$

This relation allows one to use a procedure of calculation of a ray picture in geometrical acoustics, which, strictly speaking, is applicable for wavelengths $\lambda \ll 4\pi H \approx 100\,(\text{km})$, for deriving an analytic expression which could be used to calculate the horizontal distance L from the source of IGWs to the place of their detection. In our case, $\lambda \approx 50\,(\text{km}) < 4\pi H$. The ray slope with respect to the horizon is given by

$$\frac{dz}{dx} = \frac{C_z}{C_x} = \frac{\partial \omega/\partial k_z}{\partial \omega/\partial k_x},$$

where C_{Gz} and C_{Gx} are the group velocity components for the vertical and the horizontal directions, respectively. When calculating this angle, we take into account that $k_z < 0$ for an upward energy flow, i.e., $C_{Gz} > 0$. Then we can obtain

$$\frac{dz}{dx} = \frac{\tau_g}{\tau} \cdot \left[1 - \frac{\tau_g \cdot C_G^2}{4\pi H}\right]^{-1/2},$$

where τ is the period of the detected oscillation, which, according to our observations, is 15–120 (min) (i.e., $\tau_g^2 \ll \tau^2$). In the formulas used, the effect of the wind in the atmosphere is not taken into account, but this is quite admissible since the mean wind velocities are much lower than the velocity of the wave.

The right side of the last formula is a function of the vertical coordinate Z since the quantities τ_g and H depend on altitude. However, these dependences are very weak for altitudes below 80–100 (km); therefore, they can be neglected in some approximation. In this case, the formula becomes practical for the estimation of the horizontal distance L from the source

$$L \approx Z \cdot \frac{\tau}{\tau_g} \cdot \left[1 - \left(\frac{C_\Gamma}{C_0}\right)^2\right]^{-1/2},$$

where $C_0 = 4\pi H/\tau_g$ is the limiting velocity of IGWs ($\sim 300\,(\text{m} \cdot \text{s}^{-1})$).

The zenith angle θ at which a wave with period τ propagates upward is determined by the relation

$$\theta = \text{arctg}\left[\frac{\tau^2}{\tau_g^2} - 1\right]^{1/2}.$$

Attempts to detect IGW sources in the troposphere have been made over many years (Hines 1968). For this purpose, the ray interpolation downward along the wave phase surface was already used in the case of noctilucent clouds. Therefore, the horizontal ranges L of IGW propagation have been calculated for the conditions of an undisturbed middle atmosphere with no wind (Brodhun et al. 1974; Chubukov 1977; Chunchuzov and Shefov 1978) (Fig. 5.4). Such an interpolation results in essential limitations for waves with equal periods. The triangle vertices cannot be at the same distance from a point or a line outside the triangle. The oscillation phases at these vertices also cannot be identical. Therefore, an IGW source traceable near the Earth surface, in the case of plane waves with the same period throughout the firmament region taken by the triangle, cannot be point or linear. It necessarily has a wave surface like that of the hydroxyl emission layer in the base triangle region. All this resembles the pattern described by Lindzen (1971). For finding waves with a certain period, a coherence factor is used which a priori implies a wave train whose length is several wavelengths. Therefore, the mentioned wave surface at the tropospheric

Fig. 5.4 Distances X of horizontal propagation of gravity waves of various periods to altitudes Z equal to 80, 85, 90, 95, and 100 (km) calculated with no account of the wind (Chubukov 1977)

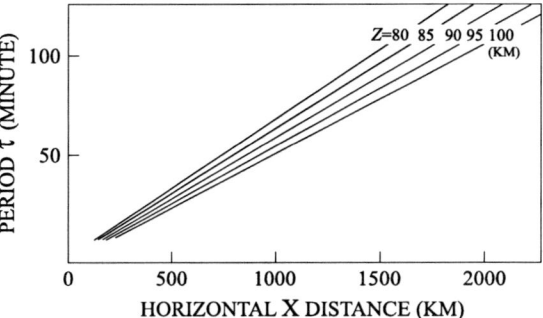

level should be greater in length than the horizontal wavelength, in particular, in the direction of IGW propagation.

However, it is not improbable that the true IGW source is located much farther than the place that is related by ray tracing to the base triangle in the hydroxyl emission layer and, moreover, is more compact. At a very large distance from such a source, the wave can be considered plane. The mean atmospheric wind in the region between the IGW source and the hydroxyl layer, depending on its direction, moves the IGW source away or brings it nearer to its position calculated for the case of no wind.

Thus, the identification of meteorological structures in which IGWs penetrating in the thermosphere are generated on the weather map is complicated due to the great length of the IGW sources and the vertical bending of the wave path because of the wind. This can well be seen by comparing the azimuths of the directions of propagation of IGWs of the same period detected by different OH emission bands (Krassovsky et al. 1978). Moreover, it is impossible to select a weather map which would correspond without an intermediate interpolation to the recorded time at which the detected ray entered the troposphere. Besides, weather maps are available in the main for the midnight of universal time and are somewhat voluntary in details.

Nevertheless, the majority of downward interpolations (if not all) point to places where active meteorological structures (fronts, occlusions, cyclones, jet streams, but not anticyclones) are located. However, it should be borne in mind that in many cases this can be accounted for by the fact that IGWs are detectable by the hydroxyl emission only on a clear sky, which, as a rule, happens in anticyclone regions. In this situation, only IGWs can be observed whose sources are located in the border regions of anticyclones containing active meteorological structures. Figures 5.5–5.7 give some examples of weather maps closely matching the times at which IGWs were detected in the hydroxyl layer. The place of observation of hydroxyl rotational temperature variations is marked with an open circle. The interpolation downward is terminated at the place marked with a full circle whose diameter is conventionally taken equal to one-third of the distance from the observation point (Krassovsky et al. 1977, 1978).

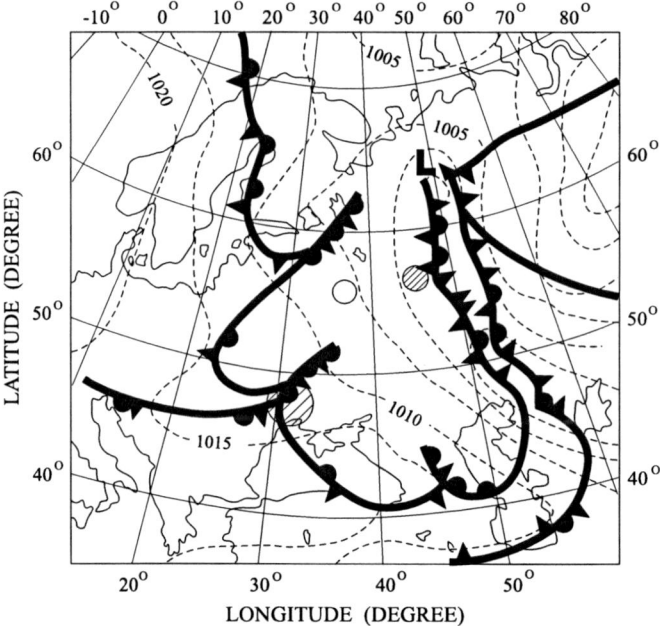

Fig. 5.5 Weather map for 03:00 a.m. Moscow time of April 21, 1974. IGWs were detected between 01:30 a.m. and 04:36 a.m. *Hollow circle* is Zvenigorod, the observation place; *shaded circles* are locations of expected IGW sources; *thick lines* are locations of meteorological fronts (Krassovsky et al. 1977, 1978), and *dashed lines* are isobars

The farther the IGW source, the greater the period of the detected wave. This, perhaps, was the reason why in simultaneous observations of IGWs at Zvenigorod and Abastumani, being 1600 (km) distant from each other, there were no maxima of the same period in the periodograms of the OH rotational temperature variations, though the expected IGW sources were close to each other. Figure 5.8 gives weather maps for the time of simultaneous detection of IGWs at different stations (Korobeynikova et al. 1979, 1983). It can be seen that the periods τ of IGWs observed from different places are noticeably different. The distances L from observation points to IGW sources evidently tend to increase with periods τ. In Fig. 5.9, measured L and τ are presented for the observations performed at Zvenigorod and Abastumani. Also given are data of observations of IGWs in the ionospheric F2 region (Hung 1977; Hung and Smith 1977). The IGW source was the Heloise tropical hurricane that occurred on September 22 and 29, 1975, over the territory of the Caribbean Sea.

It is of interest that, in contrast to traveling ionospheric disturbances (TIDs) (Gupta and Nagpal 1973), the IGWs that are detected by variations of the OH rotational temperature propagate not only in meridional but also in zonal directions. It is a common belief that the limitation of TID directions is related to the features of the interaction of atmospheric ionized species with the geomagnetic field (Gupta

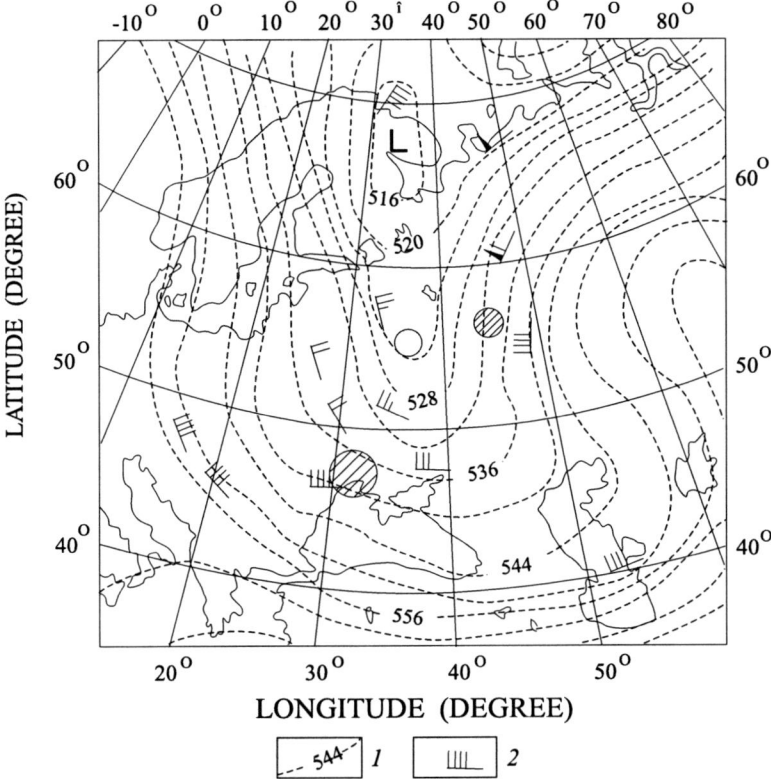

Fig. 5.6 The 500-mbar baric topography map for the weather map of Fig. 5.5 (Krassovsky et al. 1977, 1978). *Dashed lines* are altitudes of isobars in decameters. The other designations are the same as those in Fig. 5.5

and Nagpal 1973). However, for TIDs propagating in a neutral atmosphere there is no limitation. The connection of the IGWs detectable in the F2 region with poor meteorological conditions was investigated with a more correct account of the upper atmosphere circulation (Bertin et al. 1975). Nevertheless, the tropospheric origin of thermospheric IGWs does not make improbable a dependence of their amplitudes on geomagnetic activity (Shagaev 1974). However, this does not necessarily imply that in the overwhelming majority of cases the OH rotational temperature variations are produced by IGWs generated during auroras in the E region of the ionosphere. They can also be caused by the variations of the structural parameters of the upper atmosphere and its circulation (Hines 1968).

There is no evidence as yet that the TIDs related to a type D anomaly (Armstrong 1975) accompany IGWs detected by the OH emission. However, this type of TIDs is a very infrequent phenomenon, usually observable in winter during daylight hours (Heislet 1963), while IGWs are detected by the optical method mainly at night.

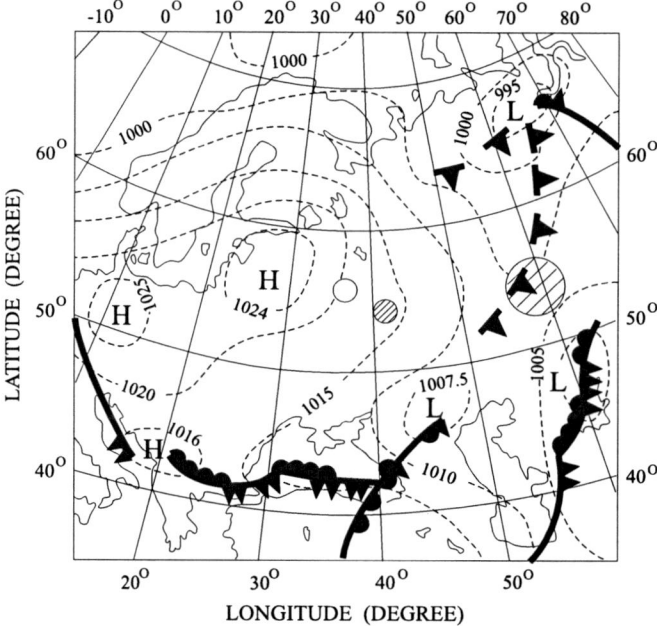

Fig. 5.7 Weather map for 03:00 a.m. Moscow time of January 7, 1976. IGWs were detected between 01:35 and 04:35 a.m. The symbols H and L mark high-pressure and low-pressure regions, respectively (Krassovsky et al. 1977, 1978). The other designations are the same as in Fig. 5.5

When propagating between the OH emission layer and the F2 region, IGWs are damped and affected by very intense atmospheric circulation. Nevertheless, the data of ionospheric sounding performed at N. V. Pushkov Institute of Terrestrial Magnetism, Ionosphere and Radiowaves Propagation, Troitsk (43 (km) from Zvenigorod), show variations with a large relative amplitude at h'F2 with the period the same as that observed in the variations of the OH rotational temperature. According to the available information, only one of two events of fluctuations of the OH rotational temperature was accompanied by TIDs (Armstrong 1975).

It was mentioned that temperature variations with different periods are detected sometimes at one or two azimuths. They, perhaps, are produced by closely located polychromatic IGW sources. In this case, the distances from the source to individual sites of the base triangle are considerably different from each other and, hence, IGWs with different periods pass through them. If this explanation is valid, the possibility arises to detect polychromatic IGW sources at the points of intersection of the circles whose centers are at the observation points of the firmament and radii correspond to the periods of the waves observed at these points. In practice, we have not points, but triangles because of the variance of the circle radii. However, in this case, the vertices of such triangles are spaced by distances substantially smaller than the wavelengths of the observed waves. Figure 5.10 exemplifies the location of polychromatic IGW sources (Krassovsky et al. 1978).

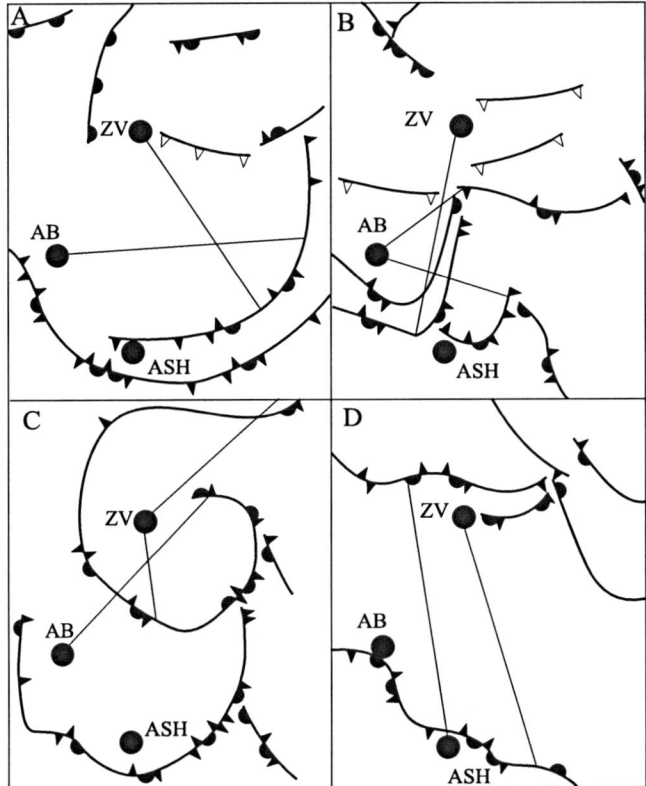

Fig. 5.8 Weather maps related to 03:00 a.m. Moscow time for dates of simultaneous detection of IGWs at several stations (Abastumani – AB, Zvenigorod – ZV, Ashkhabad – ASH) (Korobeynikova et al. 1983). April 29/30, 1973 (**A**); April 16/17, 1974 (**B**); April 26, 1974 (**C**), and October 2/3, 1975 (**D**)

A similar result can be a consequence of the Doppler effect when a monochromatic IGW source rapidly moves nearby the base triangle. Besides long waves, shorter waves sometimes appear similar to those observed in noctilucent clouds (Fogle 1971; Haurwitz 1971) or expected in slowly moving perturbances in the E region (Heislet 1963). However, waves with these periods cannot be detected in three-azimuth observations with a large base. Probably, they occur in a small region of the firmament, i.e., they can be detected with a smaller base triangle.

5.1.7 The Nature of IGW Sources

The procedure of location of IGW sources described in the previous section cannot be considered fully adequate since at the places revealed with its help there can be various active atmospheric perturbances whose character and number permanently

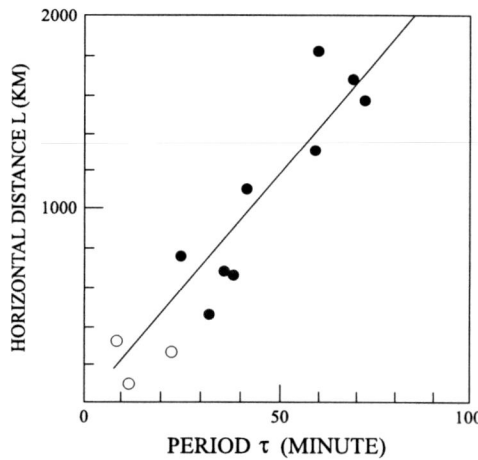

Fig. 5.9 Comparison of measured horizontal ranges L and periods τ. *Full circles* are data from Zvenigorod and Abastumani; *hollow circles* are data taken from the publications (Hung 1977; Hung and Smith 1977), and *solid line* is regression line

vary with time. This especially refers to jet streams which, as a rule, take place over such places. Therefore, to show that in most cases jet streams can be IGW sources, it was necessary, at least, to find correlations between the observed IGW parameters and the characteristics of jet streams.

The amplitude A and period τ of IGWs are the characteristics that can be most unambiguously determined in experiment. In the previous sections, it has been shown that on the average the amplitude of the detected waves was closely related to their periods. Therefore, it was accepted practical to use the ratio A/τ, which can be determined precisely enough from experimental data, as a unified characteristic of IGWs. The projection of the velocity of a jet stream on the direction of motion of an IGW

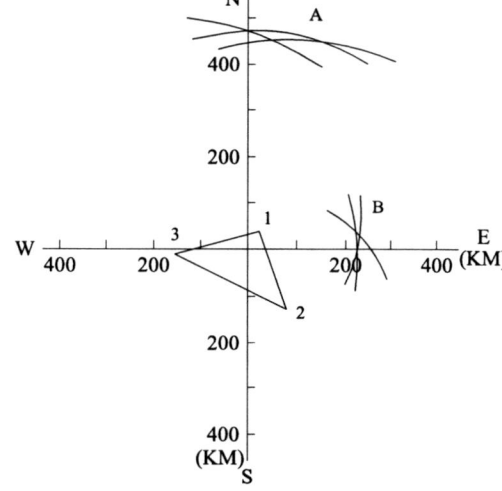

Fig. 5.10 Examples of location of nonmonochromatic IGW sources at ∼90 (km) on January 21, 1973, (**A**) and April 4, 1973 (**B**) (Krassovsky et al. 1977, 1978). 123 is base triangle

observed from the Earth surface, $V_{||}$, was taken as a characteristic of the jet stream. It should be noted, however, that this quantity is not as simple and unambiguous to determine as A/τ. The matter is that, first, the method of reverse ray tracing used to locate expected IGW sources introduces some error and, second, $V_{||}$ cannot always be determined by interpolation from baric topography maps because of lack of required data or a perturbed character of the jet stream. As a rule, jet streams are smooth near cyclones. But sometimes the jet stream at an expected place abruptly changes its velocity and direction within small distances, making the determination of $V_{||}$ impossible. Nevertheless, despite all these difficulties, a considerable body of data could be selected from the results of extensive observations of IGWs which made it possible to find a relation between A/τ and $V_{||}$ (Krassovsky et al. 1978). In these cases, $V_{||}$ was determined at that level (500, 300, 200, and 100 (mbar)) where the jet stream velocity was a maximum. Since a jet stream is rather conservative, the time lock-on to within several hours turned out reasonable. Figure 5.11 presents the relation between A/τ and $V_{||}$ obtained for monochromatic IGWs detected by the hydroxyl emission from the low vibrational levels ($v \ll 6$). It can be seen that there is a tendency for an increase in A/τ as the projection of the fair wind velocity increases and for a decrease with increasing the counter component velocity in the jet stream. This dependence is satisfactorily described by the regression equation

$$\frac{A}{\tau} \approx 0.267 \cdot \left(1 + \frac{V_{||}}{\overline{C}_x}\right),$$

where $\overline{C}_x \approx 250\,(\text{m} \cdot \text{s}^{-1})$ and the correlation coefficient is 0.81 ± 0.06.

This expression testifies that monochromatic IGW sources are linked to the coordinate system of the jet stream. It describes the Doppler effect for the observer who is in the coordinate system of the Earth. This pattern is absolutely analogous to what takes place during the observation of the radiations of moving extraterrestrial objects.

Figure 5.11 shows a similar dependence of A/τ on $V_{||}$ for polychromatic emitters with the regression line given by the expression

$$\frac{A}{\tau} \approx 0.32 \cdot \left(1 + \frac{V_{||}}{\overline{C}_x}\right),$$

where $\overline{C}_x \approx 50\,(\text{m} \cdot \text{s}^{-1})$ and the correlation factor is 0.73 ± 0.07. The expression obtained also implies that polychromatic IGW sources are linked to the coordinate system of the jet stream. However, in this case, the mean horizontal phase velocity $\overline{C}_x$ is much lower than in the case of monochromatic IGW sources.

Based on the foregoing, it can be more strongly suggested that most of the observed IGWs were generated by jet streams. Unfortunately, it was impossible to relate the IGWs and the jet stream parameters because of lack of actual data about $V_{||}$. Nevertheless, special attention should be paid to the fact that IGWs are generated only in small regions of a jet stream whose remaining part "is silent". Beneath the

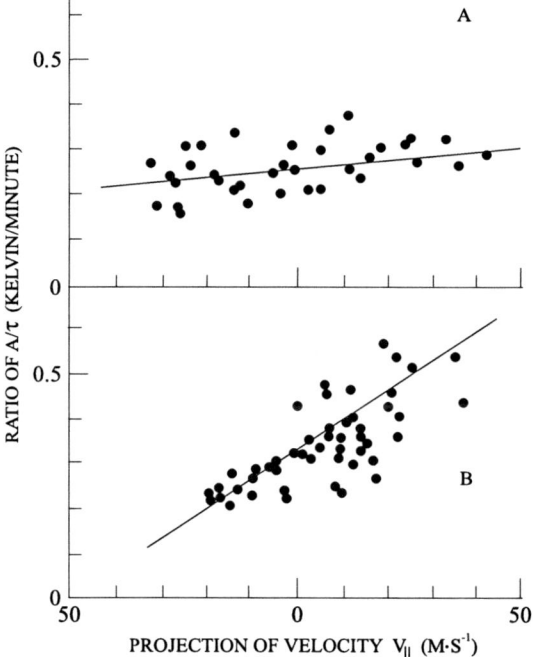

Fig. 5.11 Correlation between A/τ and $V_{\|}$ for monochromatic (**A**) and nonmonochromatic IGWs (**B**) (Krassovsky et al. 1977, 1978)

IGW-generating sites, there always are mobile atmospheric perturbances (cyclones, fronts, occlusions), which seem to create conditions favorable for IGW generation.

5.1.8 Seasonal Variability of the Spectral Distribution of IGW Amplitudes

It is, naturally, of great importance to find seasonal dependences of various characteristics of IGWs and reveal regularities in the seasonal behavior of IGWs based on these dependences. This is due to the necessity to elucidate and take into account the role of the energetic and dynamic characteristics of gravity waves, both in the overall circulation of the atmosphere and in its temperature regime (Gavrilov 1974; Chunchuzov 1981; Hirota 1984). Most of the regularities have been revealed by analyzing the radiophysical data (incoherent scattering of radiowaves, radiometeoric investigations) (Spizzichino 1969, 1971; Gavrilov and Delov 1976; Portnyagin et al. 1978; Karimov and Lukyanov 1979; Kazannikov and Portnyagin 1981a,b; Gavrilov 1987, 1996; Gavrilov and Roble 1994; Gavrilov et al. 1994) and lidar measurements (Gardner and Voelz 1985; Senft and Gardner 1991; Murayama et al. 1994; Collins et al. 1994).

However, it is well known that to estimate the effect of IGWs on the temperature regime and circulation of the upper atmosphere, data on the energy, momentum, and mass fluxes are necessary which can be calculated if data on the variations of the neutral temperature, density, and vertical velocity of the IGWs are available. These quantities can be estimated with an optical method by recording the variations of the characteristics of some airglow emissions induced by IGWs propagating through emission layers located at different altitudes.

Based on the data on variations of hydroxyl emission characteristics obtained by the optical method at Zvenigorod, an attempt has been made to reveal the seasonal variability of the spectral distribution of the relative amplitudes of IGWs, $\delta I/I$ and $\delta T/T$, on their periods (Semenov and Shefov 1989). The variations of the hydroxyl emission intensity and rotational temperature corresponding to different excitation levels were considered. Earlier it was noted that the relative amplitude of the detected waves increased with their period (τ). Bull et al. (1981) analyzed the behavior of the pressure spectral density in the surface air. They have found that $F \propto \tau^n$, where F is the energy flux, for the wave periods from 5 to 150 (min). It was noted that IGWs were regularly detected in mountain regions (n was estimated to be about 2.6). The same conclusion was made by other researchers (Tepley et al. 1981; Sukhodoev et al. 1989a,b) based on optical observations of IGWs by the Earth's airglow emissions. Vincent (1984) investigated the electron density oscillations in the D region at altitudes of about 85 (km). From the data of long-term observations it has been obtained that $F \propto \tau^{1.5}$ for the periods from 6 (min) to 24 (h).

Figures 5.12 and 5.13 present the results of a correlation analysis of the relative amplitudes (in percentage of the mean value) of the oscillations of hydroxyl emission intensity and temperature with IGW periods (15–75 (min)) for summer

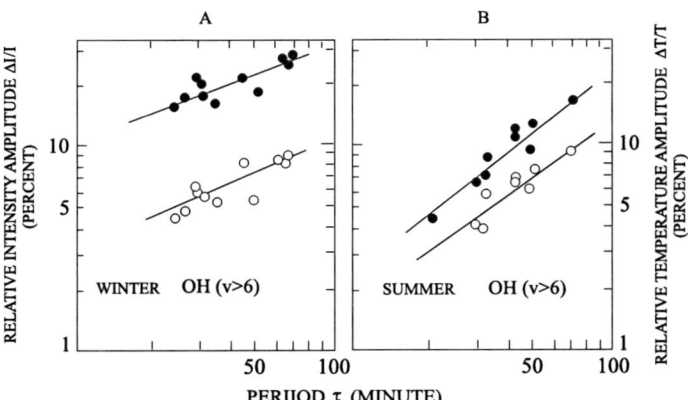

Fig. 5.12 Relative amplitudes of the variations of hydroxyl emission intensity, $\Delta I/I$, and temperature, $\Delta T/T$, for the vibrational levels v > 6 as functions of IGW periods τ in winter (**A**) and summer (**B**). R = 90 (km). *Full circles* are intensity; *hollow circles* are temperature; *solid lines* are regression lines

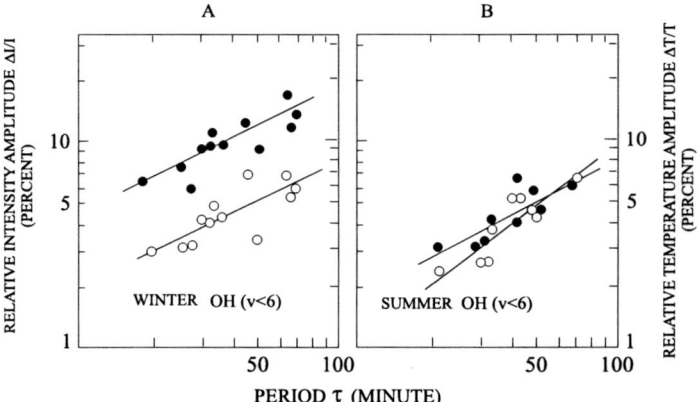

Fig. 5.13 Relative amplitudes of the variations of hydroxyl emission intensity, $\Delta I/I$, and temperature, $\Delta T/T$, for the vibrational levels $v < 6$ as functions of IGW periods τ in winter (**A**) and summer (**B**). $Z = 86$ (km). The designations are the same as in Fig. 5.12

and winter conditions. The distance between the regression lines characterizes Krassovsky's number η.

The corresponding regression equations are as follows:
for the data of Fig. 5.12:

Winter   Summer

$$\log_{10} \frac{\Delta I}{I} = 0.44 \cdot \log_{10} \tau + 0.59; \qquad \log_{10} \frac{\Delta I}{I} = 1.05 \cdot \log_{10} \tau - 0.71;$$

$$r = 0.80; \qquad r = 0.95;$$

$$\log_{10} \frac{\Delta T}{T} = 0.53 \cdot \log_{10} \tau - 0.02; \qquad \log_{10} \frac{\Delta T}{T} = 1.07 \cdot \log_{10} \tau - 0.96;$$

$$r = 0.81; \qquad r = 0.93;$$

for the data of Fig. 5.13:

Winter   Summer

$$\log_{10} \frac{\Delta I}{I} = 0.62 \cdot \log_{10} \tau + 0.02; \qquad \log_{10} \frac{\Delta I}{I} = 0.68 \cdot \log_{10} \tau - 0.42;$$

$$r = 0.73; \qquad r = 0.82;$$

$$\log_{10} \frac{\Delta T}{T} = 0.60 \cdot \log_{10} \tau - 0.31; \qquad \log_{10} \frac{\Delta T}{T} = 0.94 \cdot \log_{10} \tau - 0.91;$$

$$r = 0.74; \qquad r = 0.89;$$

The energy of IGWs is proportional to the quantity $(\Delta T/T)^2$, which in turn depends on τ^n, where n characterizes the angular coefficient of regression. As can be seen from Figs. 5.12 and 5.13, n = 1.9–2.1 for summer and 1.1–1.2 for winter, i.e., the rate of rise of the wave amplitude with period, which is characterized by the slope angles of the regression lines, shows a pronounced seasonal variability. The yearly average n = 1.55 agrees with the data obtained near the mesopause by radiomethods (Vincent 1984). It is not improbable that the decrease in energy of IGWs is due to the seasonal variability of turbulence at these altitudes (Lindzen 1971).

In the early 1980s, the Geophysical Institute of Alaska University conducted measurements of the OH emission intensity and temperature on West Spitsbergen (Myrabø et al. 1983). The observations were carried out in one region of the firmament. There was only one 24-h interval (January 1981) throughout the observation period when clouds and auroras did not impede the work. The data obtained were used to investigate the dependence of the amplitude of temperature variations on IGW period. It was found that n = 1.8, which is noticeably greater than the n value corresponding to winter conditions at Zvenigorod.

When an IGW of certain period propagates through an OH emission layer, the emission intensity and temperature vary with the same period. During the observation of IGWs by the hydroxyl emission at three azimuths at Zvenigorod, there were cases when the response of the OH emission intensity and temperature was reflected by their variations with the period of the propagating wave only at one or two azimuths. However, in some cases these periods were different for two sighted regions. These data yield noticeably greater values of n for winter: 1.8 for temperature variations and 2.6 for intensity variations. Thus, spectral differences are observed between IGWs and other oscillation processes in the hydroxyl emission which cannot be related to internal gravity waves, i.e., can be of other nature.

It is of interest that the number η is also seasonally variable. Based on measurement data, values of η for the hydroxyl (8–3) and (5–1) bands (Fig. 5.14) have been obtained. Noteworthy is that the values of η for winter are noticeably greater than for summer. The mean η (OH (8–3) band) are 3.2 for winter and 1.6 for summer. For the OH (5–1) band, they are, respectively, 2.3 and 1.3 (Krassovsky et al. 1988). The variations of the parameter η for the (9–5) and (4–1) bands are of the same character

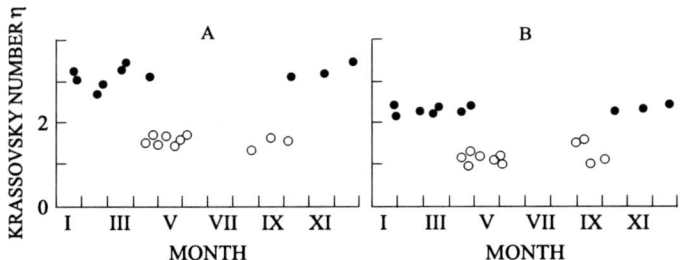

Fig. 5.14 Seasonal variations of Krassovsky number η determined by the emission in the OH (8–3) band (**A**) and in the OH (5–1) band (**B**). *Full circles* are winter; *hollow circles* are summer (Krassovsky et al. 1988)

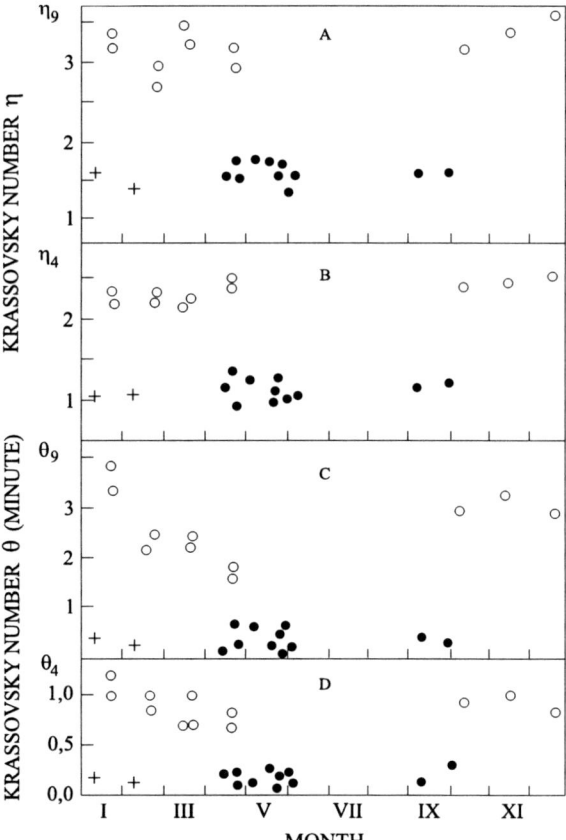

Fig. 5.15 Seasonal variations of the parameters η and θ for the OH emissions from the high (v = 9) (**A, C**) and low vibrational levels (v = 4) (**B, D**) (Shagaev 1978). *Hollow circles* are winter; *full circles* are summer; crosses are data for the period of the grandiose stratospheric warmings in January and February, 1973

(Fig. 5.15) (Shagaev 1978). It should be noted that the number θ that determines the delay of intensity variations relative to temperature variations also shows seasonal variability. Comparison of these numbers reveals a clear correlation between them (Fig. 5.16). For the observation periods near equinoxes, significant jumps of η and θ values are observed. The values of η and θ were assigned to the summer or the winter group if they were close to the values unambiguously measured in summer or in winter.

When comparing Figs. 5.14 and 5.15, we see that there is a relationship between the temperature regime of the hydroxyl layer with IGWs dissipated in it and η: in summer, at small η, the angular coefficient of the relation $T = f(\delta T/T)^2$ is positive, while in winter, as η increases, the sign of the angular coefficient changes (Fig. 5.17). It is well known that Krassovsky's number η strongly depends on the quantity m that characterizes the rate of vibrational–rotational relaxation of excited hydroxyl molecules and on whether the processes occurring in the layer during the propagation of IGWs are adiabatic (Krassovsky 1972). Therefore, the seasonal grouping of η may testify to the necessity of taking into account some

5.1 Internal Gravity Waves

Fig. 5.16 Comparison of the parameters η and θ for the OH emissions from the high (v = 9) (**A**) and low vibrational levels (v = 4) (**B**) (Shagaev 1978). The designations are the same as in Fig. 5.15

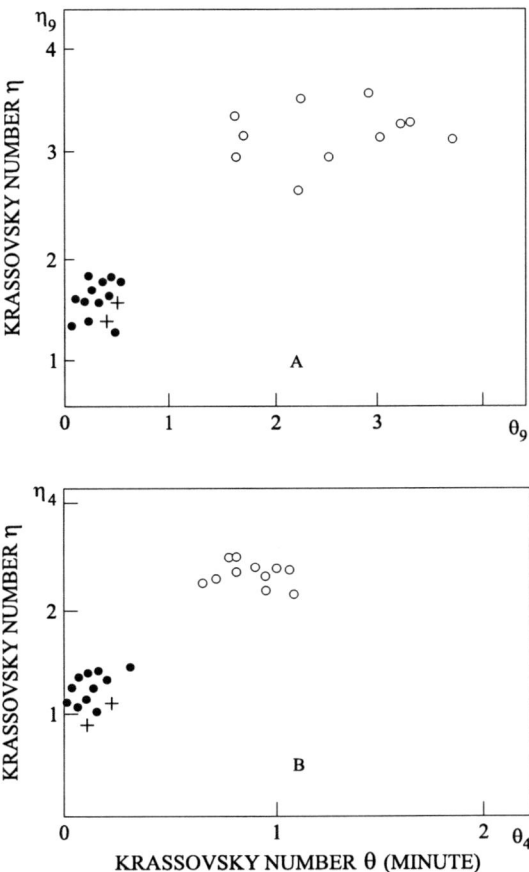

IGW-induced nonadiabatic process occurring in the layer or the seasonal variability of the processes of quenching of excited hydroxyl molecules.

5.1.9 Proportion Between the Principal Mechanisms of Hydroxyl Emission Excitation in the Mesopause

Measuring the variations of hydroxyl emission characteristics caused by IGWs propagating through the emission layer in the mesopause, it is possible to investigate not only the nature of this airglow emission but also the features of the relevant photochemical processes, namely the relative contribution of the known principal processes that result in excitation of various vibrational levels of hydroxyl molecules (Semenov 1989b).

It has been shown (Krassovsky and Shagaev 1977) that the different power dependences of the rates of photochemical reactions on temperature (taken into

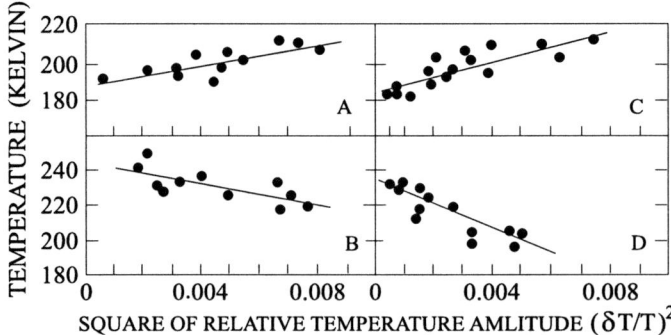

Fig. 5.17 Atmospheric temperature in the region of the hydroxyl emission layer as a function of the squared relative IGW amplitude determined for summer (**A**) and winter (**B**) by the OH (8–3) band and for summer (**C**) and winter (**D**) by the OH (5–1) band. *Dots* are measurement data; *solid lines* are regression lines (Krassovsky et al. 1988)

account by the number η) can serve as indicators of the contributions of the ozone–hydrogen and perhydroxyl reactions to the excitation of hydroxyl molecules.

When detecting IGWs by the OH emission, one more parameter is determined; this is θ, the delay of the emission intensity oscillations relative to the accompanying oscillations of the rotational temperature (Meriwether 1975; Krassovsky and Shagaev 1977). The yield of ozone and perhydroxyl molecules permanently varies. However, the emission occurs only after the appearance of excited hydroxyl, i.e., only once ozone or perhydroxyl has entered into reactions with hydrogen or oxygen atoms, respectively. The time θ characterizes this delay, which is caused by the mentioned exothermic processes.

The energy yield of the ozone–hydrogen reaction is 76.8 (kcal), which suffices to excite OH molecules from the ground electronic state ($^2\Pi$) to the ninth vibrational level (74.88 (kcal)) and of perhydroxyl molecules only to the sixth one (54.17 (kcal)). Therefore, it seems possible to estimate the contribution of the ozone process to the excitation of the lower vibrational levels by measuring simultaneously the variations of the rotational temperature and intensity of the emission from the higher and lower vibrational levels of hydroxyl molecules that are related to the action of IGWs and determining the time θ. With this purpose, data of spectrographic observations of IGWs by the hydroxyl emission in the (9–4) and (5–1) bands have been analyzed. The true delay time θ° related to the emission from the ninth and fifth levels is determined by the expressions

$$\theta_9^o = \frac{1}{A_9 + \beta_9[M]} + \frac{1}{\alpha_{O_3}[H]} \quad \text{and} \quad \theta_5^o = \frac{1}{A_5 + \beta_5[M]} + \frac{1}{\alpha_{HO_2}[O]},$$

where θ_9^o and θ_5^o are the delay times related only to the ozone–hydrogen and perhydroxyl process, respectively, i.e., with the emission from the ninth or the fifth level; $\alpha_{O_3} = 1.5 \cdot 10^{-12} \cdot \sqrt{T} \, (\text{cm}^3 \cdot \text{s}^{-1})$ and $\alpha_{HO_2} = 3 \cdot 10^{-12} \cdot \sqrt{T} \, (\text{cm}^3 \cdot \text{s}^{-1})$ are the respective reaction rates (Moreels et al. 1977); A_9 and A_5 are the Einstein coefficients for the ninth and the fifth vibrational levels of the OH molecule (Mies 1974); $\beta_9 = 8 \cdot 10^{-13} \, (\text{cm}^3 \cdot \text{s}^{-1})$ and $\beta_5 = 5 \cdot 10^{-13} \, (\text{cm}^3 \cdot \text{s}^{-1})$ are the respective rates of deactivation of OH (v = 9) and OH (v = 5) molecules in collision with molecules of the medium (Streit and Johnston 1976). Using the given values of reaction rates and typical concentrations of the atmospheric constituents for the mesopause altitudes (CIRA-1972) that correspond to the observation conditions, it can be shown that the first terms in the above relations are more than an order of magnitude smaller than the second ones; that is, we have

$$\theta_9^o \approx \frac{1}{\alpha_{O_3}[H]}, \quad \theta_5^o \approx \frac{1}{\alpha_{HO_2}[O]}.$$

The measurable intensity of the OH (9–4) band emission and, hence, delay time θ are due to the action of the ozone–hydrogen process only, while the intensity of the band (5–1) emission is determined by the population of the fifth vibrational level of OH molecules, which depends not only on the perhydroxyl process but also on the effect of the cascade transitions from the overlying levels, i.e., on the additional action of the ozone–hydrogen process. Analysis of the amplitudes of the variations of the IGW-induced formation of hydroxyl molecules at the fifth vibrational level shows that for the levels under consideration the delay times between the emission intensity and temperature variations are related as

$$\theta_5 = K_1 \cdot \theta_9^o + K_2 \cdot \theta_5^o.$$

Here θ_5 is the observed delay time resulting from the effect of both processes; k_1 and k_2 are coefficients taking into account the relative contribution of the above processes to the excitation of low vibrational levels (in this case, v = 5). The coefficient k_2 can be estimated based on the relation $k_2 \approx (I_5^o/I_5) < 1$, where I_5^o is the intensity of the emission from the fifth level caused by the perhydroxyl process, and I_5 is the observed intensity. As follows from the formulas obtained, $\theta_5^o \sim 0.3\,(\text{s})$, while the relevant formula and observations yield $\theta_9^o \approx \theta_9 \approx 200\,(\text{s})$. Thus, the terms in the last formula are incommensurable, and since the time of recording of hydroxyl emission spectra attained by now is 1 (min), the coefficient K_1 can be determined as

$$K_1 \approx \theta_5/\theta_9.$$

Thus, using measurements of θ_9 and θ_5, it is possible to trace the seasonal variability of the contribution of the ozone–hydrogen process to the excitation of hydroxyl molecules to the fifth vibrational level (Fig. 5.18), which is approximated (solid line) by the expression (with the correlation coefficient –0.89)

$$K = 0.44 - 0.19 \cdot \cos\frac{2\pi}{365} \cdot t_d,$$

where t_d is the day of year. From Fig. 5.18 it can be seen that this contribution varies within a year from ~ 0.25 to ~ 0.6, tending to increase in summer. The above range of values of the coefficient K_1 testifies to a rather significant contribution of

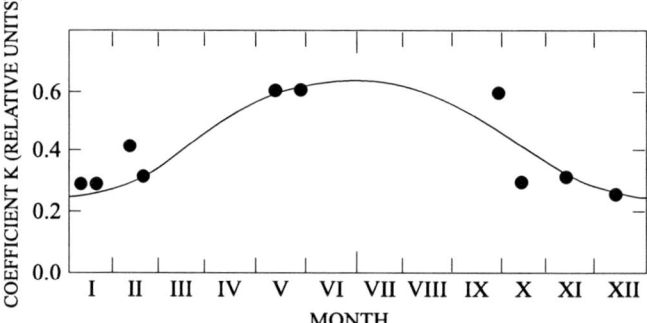

Fig. 5.18 Seasonal variations of the coefficient K that takes into account the contribution of the ozone–hydrogen process to the excitation of the fifth vibrational level of the OH molecule

the perhydroxyl process to the excitation of the lower vibrational levels of hydroxyl molecules.

Using the above expression with experimentally determined values of θ_9, it is possible to estimate the atomic hydrogen density at the hydroxyl emission altitudes for midlatitudes and night conditions. Figure 5.19 shows the resulting seasonal distribution of the atomic hydrogen density, whose analytic expression obtained by the least square technique has the form

$$\log_{10}[H] = 8.7 - 0.5 \cdot \cos\frac{2\pi}{365} \cdot t_d.$$

The obtained seasonal dependence of the atomic hydrogen density is characterized by its increase in summer up to $\sim 2 \cdot 10^9 \, (\text{cm}^{-3})$ and decrease down to $1 \cdot 10^8 \, (\text{cm}^{-3})$ in winter. This seasonal variability of the atomic hydrogen density for the thermospheric altitudes is confirmed by the data derived from long-term measurements of the intensity of hydrogen emission in the H_α line (Fishkova 1983).

Direct use of θ_5 for a similar calculation of the atomic oxygen density involves some difficulties for the reasons stated above.

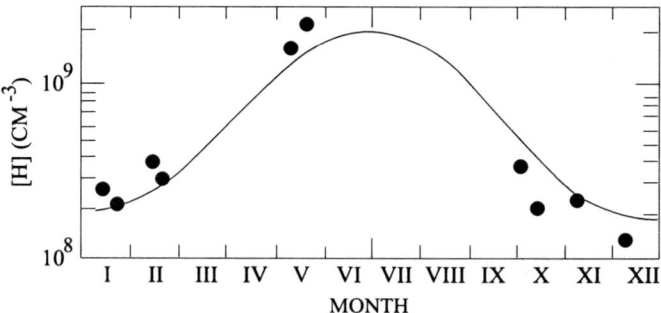

Fig. 5.19 Seasonal variations of the atomic hydrogen concentration at the OH emission altitudes in the mesopause

5.1.10 IGW-Induced Variations of the Doppler Temperature and Intensity of the 557.7-nm Emission

The effect of internal gravity waves on the variations of the 557.7-nm atomic oxygen green emission intensity was repeatedly investigated (Kuzmin 1975a,b,c,d; Noxon 1978). Unfortunately, in these investigations the behavior of the temperature in the emission layer during the propagation of IGWs was not considered. At the same time, the knowledge of the behavior of simultaneously measured IGW-induced temperature and intensity variations for the 557.7-nm emission can be useful to gain an understanding of the emission initiation mechanism. This method was repeatedly employed in studying the processes responsible for the initiation of the OH emission (Semenov and Shefov 1989; Semenov 1989a). Nevertheless, noteworthy are the requirements that should be fulfilled when using data on simultaneous intensity and temperature variations for the emission under study.

The observed diverse variations of the emission intensity and temperature can be of different nature and not necessarily related to each other. Thus, the intensity of an emission is determined not only by the temperature but also by the nonuniformly spatially distributed densities of initial reactants responsible for this emission. Many variations can be caused by the turbulence that is accompanied by upward and downward air flows and by the overall atmospheric circulation (Krassovsky and Shagaev 1974b). A method for revealing unambiguously the effects related only to IGWs is considered elsewhere (Krassovsky and Shagaev 1977; Semenov and Shefov 1989). Only the waves with the same period in temperature and intensity are attributed as IGWs by the phases of which it is possible to determine the wave direction and velocity. It has been found, however, that the intensity variations are somewhat delayed in phase relative to the temperature variations (Krassovsky et al. 1977). Besides, the ratio of the relative amplitude of intensity variations to that of temperature variations induced by IGWs (and only in this case), Krassovsky's number η, is an important aeronomic parameter which allows one to analyze the types of chemical processes giving rise to emissions. Nevertheless, this important condition was neglected by a number of researchers. Some authors interpret all variations without exclusion as being induced by IGWs (Takenuchi et al. 1981; Myrabø et al. 1983). In this case, η is determined as the ratio of the relative variations of intensity and temperature for small time intervals (from a few seconds to several minutes) and, moreover, without taking into account the phase delay between the intensity and the temperature oscillations. As a result, it appears that the values of η obtained by this method always have a significant spread from negative to positive values with some prevalence of the latter. These values were erroneously identified not once, for example, by Weinstock (1978), with the true η associated with IGWs that have rather definite positive values depending on the season.

IGWs are actually not a unique reason for the emission temperature and intensity variations since there exist variations caused by the turbulence and circulation of the upper atmosphere and also independent intensity variations caused by mixing of atoms and molecules with spatially nonuniform densities which participate in reactions giving rise to one or another type of emission. Moreover, dissipation of IGWs can be terminated by turbulence, and this may adversely affect the result of calculation of the number η.

The approach discussed was used to analyze the observations of the variations of characteristics of the 557.7-nm emission observed at Zvenigorod in winter (Krassovsky et al. 1986; Semenov 1989a).

To obtain data on η, simultaneous measurements of the emission intensity and Doppler temperature were performed at four points of the firmament with azimuths $0°$, $120°$, $180°$, and $240°$ counted clockwise from the northern direction. The emission intensity was measured by two photometers. In the first one, the optical axis was oriented with a mirror every 40 (s) to different regions of the sky at a zenith distance of $27°$. At the inlet of the second photometer (similar in optical arrangement to the first one), directed only to the zenith, a diaphragm and a short-focus lens were placed. This increased the photometer coverage to $43°$. This photometer operated synchronously with the first one, recording the green line every 40 (s). The Fabry–Perot interferometer with a cooled multistage image converter tube, capable of photographing interference rings, was used to determine the temperature by the Doppler profile of the 557.7-nm emission and the emission intensity (Semenov 1975). Observations were conducted during a night in the south direction (field of vision about $1°$) at a zenith distance of $60°$ with an exposure time of 15 (min). The time series of the Doppler temperature and intensity of the 557.7-nm emission were subjected to Fourier analysis. If the harmonics of temperature intensity variations measured for various regions had equal periods and phases differing by no more than π, it was thought that an IGW rather than some other variation was detected. Thereafter, the wave lengths, directions, and velocities were determined. A criterion for reliable determination of the wavelength of IGWs was that all amplitudes of the intensity variations coincided with the amplitude of a similar wave obtained from the data of the wide-angle photometer. This was evidence that the determined wavelength was not much less than the determined wavelength was not much less than the double photometer, equal to 150 (km) at the altitude 97 (km) where the 557.7-nm emission occurred. With this field of vision, IGWs and irregularities smaller than 150 (km), including waves of Kelvin–Helmholtz type of length less than 10–15 (km), were smoothed out. Therefore, the IGWs revealed appeared longer than 150 (km).

The data of IGW observations by the 557.7-nm [OI] emission have shown that internal gravity waves were detected only twice within several nights. It turned out that on the average Krassovsky's number $\eta = 0.6$, which is much less than for the hydroxyl emission (Krassovsky and Shagaev 1977; Krassovsky et al. 1978). The average time of delay of the intensity variations of the 557.7 (nm) emission relative to the temperature variations, θ, was 90 (s).

The theoretical dependence of the ratio of the relative amplitude of the emission intensity variations to that of the ambient temperature variations under adiabatic conditions, η, on the parameters of photochemical reactions has already been considered.

5.1 Internal Gravity Waves

According to current concepts (Trunkovsky and Semenov 1978; Bates 1978), the 557.7-nm atomic oxygen emission arise due to the chain of reactions considered in Sect. 2.5.

Following Wraight (1982), it is supposed that the molecular oxygen emissions arise due to the ($^1\Sigma-^3\Sigma$) transitions as a result of the reactions

$$O_2^* + O \rightarrow O + O_2(b^1\Sigma),$$
$$O_2(b^1\Sigma) \rightarrow O_2(X^3\Sigma) + h\nu \; (760; 864.5\,(nm)).$$

However, it was supposed (Wraight 1982) that the initial excited state of O_2^* appears from the state $^5\Pi_g$ for which the excitation reaction rate includes the factor T^{-n} with $n = 4.4$. Attention is also paid to the fact that the experimental value of η obtained by Noxon (1978a) for the (0–1) band of the O_2 Atmospheric system and used in the theory by Weinstock (1978) is obviously less than unity. If the state of O_2^* is actually $^5\Pi_g$, the theoretical η will be ~ 0.6. This is in good agreement with the experimental value of η that we have found for the 557.7-nm emission.

The intensity of the emissions of the O_2 Atmospheric system ($b^1\Sigma$) (conventionally designated by I_{760} since the probability of the (0–0) transition is ~ 20 times greater than that of the (0–1) transition) can be estimated as

$$I_{760} \sim [O_2^*] \cdot [O] \cdot \alpha_{760} \cdot W_{760} \, (\text{photon} \cdot (\text{cm}^{-2} \cdot \text{s}^{-1})),$$

where the square brackets denote the densities of the respective atoms and molecules. For the conditions of measurements at an altitude of 95 (km) (maximum of the emission intensity I_{760}), we have $[O] = 2 \cdot 10^{11}\,(\text{cm}^{-3})$ (CIRA-1972); $W_{760} = 1.1 \cdot 10^6$ (cm) (Witt et al. 1979) is the half-thickness of the emission layer, and α_{760} is the excitation reaction rate.

On the other hand, the intensity of the O green emission, also in view of the excitation reactions, is given by

$$I_{557.7} \sim [O_2^*] \cdot [O] \cdot \alpha_{557.7} \cdot W_{557.7} \, (\text{photon} \cdot (\text{cm}^{-2} \cdot \text{s}^{-1})),$$

where $\alpha_{557.7}$ is the reaction rate and $W_{557.7} = 0.9 \cdot 10^6$ (cm) is the half-thickness of the emission layer (Witt et al. 1979); for the altitude 97 (km) (maximum of the emission intensity $I_{557.7}$), we have $[O] = 3 \cdot 10^{11}\,(\text{cm}^{-3})$ (CIRA-1972). Then (as a first approximation we can assume that $[O_2^*]$ is the same in both reactions) the intensity ratio is

$$\frac{I_{760}}{I_{557.7}} = 0.8 \cdot \frac{\alpha_{760}}{\alpha_{557.7}}.$$

The time constant θ for the delay of the intensity variations with respect to the Doppler temperature variations can be estimated as

$$\theta \approx \frac{[O]}{\alpha_{557.7}} \text{ or } \alpha_{557.7} \approx \frac{[O]}{\theta}.$$

According to the results obtained above, we have θ = 90 (s). This great value of θ can be due to only the Barth reaction involved in the formation of $O(^1S)$ metastable atoms rather than due to the emission from them since the mean lifetime of $O(^1S)$ is less than 1 (s). Thus, we have $\alpha_{557.7} \approx 3.7 \cdot 10^{-14} (cm^3 \cdot s^{-1})$.

During the observations the emission I_{760} was not determined. However, according to the long-term observations carried out at midlatitudes (Christophe-Glaume 1965; Fishkova 1983) and the data reported by Shefov (1975), the intensities of the emissions under discussion for the given observation conditions can be estimated as $I_{864.5} = 420$ (Rayleigh) and $I_{557.7} = 220$ (Rayleigh). In view of the transition probabilities for the $O_2(b^1\Sigma - X^3\Sigma)$ system (Vallance Jones 1973), we obtain

$$I_{760} = I_{864.5} \cdot 20.3 = 8500 \text{ (Rayleigh)}$$

and, consequently, $\alpha_{760} \approx 1.8 \cdot 10^{-12} (cm^3 \cdot s^{-1})$. The values of α at ~200 (K) available in the literature for the reaction of formation of excited oxygen molecules show a wide range, from $8 \cdot 10^{-34} (cm^6 \cdot s^{-1})$ (McEwan and Phillips 1975) to $10^{-32} (cm^6 \cdot s^{-1})$ (Campbell and Gray 1973), and it is not clear which electronic states of O_2^* thus arise. On the assumption that the O_2^*-excited molecules produced eventually transform into $O_2(b^1\Sigma)$ molecules which emit at 760 (nm), the rate of this reaction can be estimated by the relation

$$I_{760} \approx [O]^2 \cdot [M] \cdot \alpha \cdot W_{760},$$

where $[M] = 10^3 (cm^{-3})$ (CIRA-1972). For this case we obtain $\alpha \approx 1.9 \cdot 10^{-32} (cm^6 \cdot s^{-1})$. The total yield of O_2^*, and, hence, α, should be greater since the above estimates did not take into account either the 557.7-nm emission or the Herzberg bands. However, their emission intensity is much less than that of the $O_2(b^1\Sigma - X^3\Sigma)$ system. There is no reserve for $O_2(a^1\Delta - X^3\Sigma)$ bands since $\alpha \approx 1.9 \cdot 10^{-32} (cm^6 \cdot s^{-1})$ is close to its maximum values (Campbell and Gray 1973; McDade et al. 1984). It is not improbable that these bands arise due to processes other than the Barth process.

Bates (1978), based on the data of laboratory measurements which did not specify the excited states of O_2^* molecules formed in the Bart reaction, gives some limits for $\alpha_{557.7}$:

$$\alpha_{557.7}/\alpha_M > 5.5 \, (T = 200 \, (K)), \alpha_{557.7}/\alpha_M > 33 \, (T = 300(K)).$$

Here α_M is the coefficient of deactivation of O_2^* molecules due to their collisions with O_2 and N_2 molecules. If we use, for instance, the data of McDade et al. (1984) who give $\alpha_M = 1.25 \cdot 10^{-15} (cm^3 \cdot s^{-1})$ for O_2^*, which is the progenitor of $O(^1S)$, and the estimate $\alpha_{557.7} = 3.7 \cdot 10^{-14} (cm^3 \cdot s^{-1})$, it appears that

$$\alpha_{557.7}/\alpha_M \approx 30 \, (T_{557.7} = 250 \, (K)).$$

Thus, the limiting estimates for the mentioned temperature range are valid. At the same time, it should be borne in mind that a possible excited state of O_2^* is, perhaps, $^5\Pi_g$ (Wraight 1982). However, as indicated by Wraight (1982), if O_2^* are formed in states other than $^5\Pi_g$, the coefficient n for them is much less than 4.4. In this case, the estimated η should be greater than that obtained experimentally, $\eta = 0.6$.

The analysis performed shows that the study of the IGW-induced space–time variations of the Earth's airglow emissions allows one to reveal many details of the atmospheric photochemical processes. Application of this method to the study of the temperature and intensity variations of the 557.7-nm emission has made it possible to demonstrate for the first time that O_2^* molecules in the $^5\Pi_g$ state play a substantial part in the initiation of this emission.

5.1.11 Variations of the Altitude of Hydroxyl Emission Layers with Various Vibrational Excitation

When investigating the variations of atmospheric temperature fields in the mesopause region by the variations of the hydroxyl emission rotational temperature, it is necessary to know exactly the altitudes at which the observed temperature variations, measured by the emission from OH molecules excited at various vibrational levels, take place. Systematized data on the emission layer altitude variations derived from rocket measurements were presented in Sect. 4.1 (Semenov and Shefov 1996). A method of determination of the emission layer altitude is measuring simultaneously the emission intensity variations at several points, the distance between which is comparable to the expected altitude of the layer under observation (Chamberlain 1978). This method was successfully used in determining the altitude of the O atomic oxygen green line emission at 557.7 (nm) (Barat et al. 1972).

However, the observations of fast variations of the emission intensity are complicated because of the atmospheric transparency variations whose frequency spectrum is considerably different for different regions of the firmament (Kuzmin 1975c,d). Attempts were made to avoid the effect of transparency fluctuations (Peterson and Kieffaber 1973a,b) by conducting measurements in mountains at a height of about 3000 (m) by parallactic photographic recording of the irregularities of the hydroxyl glow in the near-infrared region of the spectrum at two distant points. However, the emission layer altitude could be determined only for one night. Subsequently, the triangulation method of measuring fluctuations was modernized and realized in observations of the emission layer temperature variations (Potapov 1975a,b).

With the optical method of recording the IGW characteristics δT/T and τ, the opportunity was realized to investigate the seasonal variability of the hydroxyl emission altitudes for different vibrationally excited levels. The data of long-term spectrographic observations of the hydroxyl nightglow in the (9–4) and (5–1) bands in quiet heliogeomagnetical conditions have been analyzed. For the observation period (1981–1984), there were 27 detections of IGWs by the (9–4) band and 87 by the (5–1) band. The periods τ of the detected IGWs covered the range from 15 to 110 (min).

In the previous sections, it was noted that the relative temperature amplitudes δT/T were proportional to the periods of the observed IGWs. In this case, the relationship between these quantities was characterized by correlation coefficients of 0.7 and 0.87 for the data obtained from temperature variation measurements for the (9–4) and (5–1) bands, respectively. The correlation between the IGW temperature amplitudes and periods for these bands is given by the expressions

$$(\delta T/T)_{9-4} = 0.001 \cdot \tau + 0.024, (\delta T/T)_{5-1} = 0.0012 \cdot \tau + 0.0033.$$

Since the analysis involved data for different seasons of year, it can be supposed that the observed values of the deviations of the measured IGW relative amplitudes from the regression lines not only are of random origin but also are related to the seasonal variability of the hydroxyl emission layer altitude. This supposition rests on the fact that the observed variance (0.04) of the measured amplitudes $\delta T/T$ relative to the regression lines is greater than the error of the determination of the relative wave amplitude (0.01).

As mentioned above, the triangulation measurements of the altitude of the rotational temperature fluctuations performed during the propagation of IGWs have revealed that the emission from a higher vibrational level occurs in a higher-lying atmospheric layer. The IGW temperature amplitudes for the (9–4) band described here also appeared greater than those for the OH (5–1) band. The increase in IGW amplitude with altitude for the case of no attenuation in the range of altitudes taken by the hydroxyl emission layer (9 (km)) is determined by the expression

$$\left(\frac{\delta T}{T}\right)_{9-4} \bigg/ \left(\frac{\delta T}{T}\right)_{5-1} = \exp\left(\frac{\Delta Z}{2H}\right),$$

where ΔZ is the distance between the peaks of the altitude profiles of the OH emission layers for the (9–4) and (5–1) bands.

Thus, using the measured relative amplitudes of IGWs and taking into account that the obtained regression relations represent the dependences of the amplitudes on periods for certain yearly average altitudes of the hydroxyl emission in the (9–4) and (5–1) bands, it is possible to calculate the values of ΔZ corresponding to the deviations of the measured relative amplitudes of the temperature wave from their average values. Then, using the data of Potapov et al. (1985) on the average altitudes of the hydroxyl emission from the eighth and fifth initial vibrational levels (assuming that the emission altitudes for the eighth and ninth levels are practically close to each other) and the monthly mean variations ΔZ calculated for them, it is possible to obtain the seasonal variability of the altitudes of the hydroxyl emission in the (9–4) and (5–1) bands (Fig. 5.20A). In the same figure (B), the seasonal variability of the atmospheric temperature in the mesopause region determined by the hydroxyl emission from the ninth and fifth levels is presented. The seasonal variability of the hydroxyl emission altitude shows an interesting regularity, which is confirmed by analysis of other experimental data (Lowe and LeBlanc 1993): for winter, the regions of the OH emissions from the higher and lower vibrational levels approach each other (the mean emission altitude was ~92 (km) for the (9–4) band and ~85 (km) for the (5–1) band), while for summer, on the contrary, layering takes place (the (9–4) band is localized at ~88 (km) and the (5–1) band at ~86 (km)).

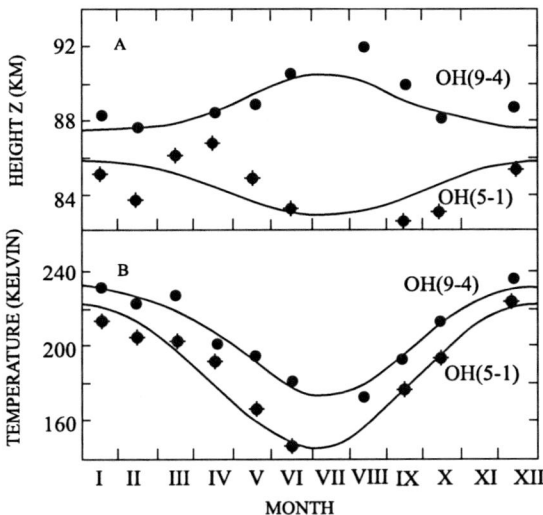

Fig. 5.20 Seasonal variations of the altitude (**A**) and temperature (**B**) of the hydroxyl emission from different vibrationally excited levels. *Dots* are OH (9–4); *dots* and *crosses* are OH (5–1); *lines* are approximations

This result is in very good agreement with the data obtained by analysis of rocket measurements (see Fig. 4.5) (Semenov and Shefov 1996).

Thus, the optical recording of IGWs offers one more method of determination and monitoring of the hydroxyl emission altitude, which, in combination with other methods, allows one to construct an empirical model of the behavior of this geophysical parameter.

5.1.12 Behavior of the Mesopause Temperature During the Propagation of IGWs in Summer and Winter

A reason for the active investigation of IGWs is that they transfer energy from the lower to the upper atmosphere. The energy fluxes carried by IGWs appear comparable to the short-wave solar fluxes that control the temperature of the upper atmosphere (Gavrilov 1974; Chunchuzov 1981). Hence, the energetics of the atmospheric layers at the lower thermosphere level varies substantially as IGWs propagate through them (Hines 1965). The IGW energy fluxes to the upper atmosphere near the mesopause are estimated as $\sim 10\,(\text{erg} \cdot (\text{cm}^{-2} \cdot \text{s}^{-1}))$ (Hines 1968; Reid 1989). At the hydroxyl airglow altitudes (90 (km)), there occurs attenuation of this energy, which is estimated by the decrement $1/Z_0 = ((1/15) - (1/20))\,(\text{km}^{-1})$ (Chunchuzov 1981). Here Z_0 is the vertical scale of energy decrease which refers only to fast IGWs and practically does not depend on their period. The physical mechanism of IGW attenuation is associated with turbulence, which, eventually, affects the thermal regime of the mesopause. Some researchers considered the effect of IGWs on the thermal conditions of the higher atmospheric layers. However, there is no consensus as to the understanding and account of this effect (Vincent 1984). Some authors argue that as IGWs dissipate, a warming of the atmosphere should be expected (Shved 1977; Kutepov and Shved 1978; Chunchuzov 1978; Kalov and Gavrilov 1985; Reid 1989), while the other advocate the reverse (Johnson 1975; Izakov 1978).

The accumulated data on IGW characteristics, obtained by optical investigations of hydroxyl emissions, allow one to trace the behavior of the atmospheric temperature in the mesopause region during the propagation of IGWs (Semenov 1988). With the data about the wave parameters at the mesopause altitude, it is possible to calculate the energy flux vertical component F_z that can affect the thermal regime of the upper atmosphere, which is of importance in gaining an insight into the energetics of the upper atmosphere magnitude. It is well known that F_z is proportional to the squared relative wave amplitude (Hines 1965; Chunchuzov 1981):

$$\langle F_z \rangle \approx \frac{\beta \cdot \rho_o \cdot g^2 \cdot \tau_g^3 \cdot C_X}{8\pi^2 \cdot \tau} \cdot \left\langle \left(\frac{\delta T}{T}\right)^2 \right\rangle.$$

Here the angle brackets imply averaging over the IGW period; ρ_o is the mean atmospheric density in the mesopause region, τ_g is the Brunt–Väisälä period, g is the free fall acceleration, C_X is the phase velocity of the wave in the horizontal direction, τ is the period of the detected waves, and β is the coefficient taking into account the IGW attenuation in the mesopause region, which is ~0.25 (Chunchuzov 1981). For the conditions near 90 (km), this expression has the form

$$\langle F_Z \rangle \approx 340 \cdot \frac{C_X}{\tau} \cdot \left\langle \left(\frac{\delta T}{T}\right)^2 \right\rangle.$$

To detect IGWs and determine their characteristics, the method of optical recording of the waves by hydroxyl emissions was used. Measurements were performed for the OH (8–3) and (5–1) bands with the emission maxima at about 90 and 85 (km), respectively (Potapov et al. 1985). Each observation night was characterized by the emission layer temperature T (an average over the time of temperature measurements) and some parameters of the detected IGWs (amplitude δT, period, phase velocity, azimuth of propagation).

The correlations of the measured temperatures with the squared relative amplitudes of IGWs detected by different hydroxyl emission bands in summer and winter are presented in Fig. 5.17. It can be seen that the average temperature in winter is greater (by ~20 (K)) than in summer for both altitude levels. The correlation between the amplitudes of IGWs propagating through the emission layer and the layer temperature is positive for summer and negative for winter, and the angular coefficients of the regression lines are smaller in absolute value for summer than for winter for both altitude levels (Semenov 1988). For the data presented in Fig. 5.17, the following regression equations have been obtained:

Altitude 90 (km), summer:

$$T = 2680 \cdot \left(\frac{\delta T}{T}\right)^2 + 189, \quad r = 0.79;$$

Altitude 90 (km), winter:

$$T = -2840 \cdot \left(\frac{\delta T}{T}\right)^2 + 243, \quad r = -0.65;$$

Altitude 85 (km), summer:

$$T = 4180 \cdot \left(\frac{\delta T}{T}\right)^2 + 184, \quad r = 0.79;$$

Altitude 85 (km), winter:

$$T = -6600 \cdot \left(\frac{\delta T}{T}\right)^2 + 233, \quad r = -0.84.$$

The variance of the average temperature for the same amplitudes of the identified waves is partially related to the number of oscillation cycles in the train and also to the existence of other IGWs. The firmament region taken by these, not identified, components can enter only partially in the field of vision of the instrumentation used. Under these conditions, reliable detection of IGWs is impossible. The variance is also caused by other wave sources, not related to the detected IGWs, that warm up some regions of the upper atmosphere, including due to the energy of the oscillation trains that have already passed through these regions before the observation.

In addition, Fig. 5.36 presents the measurements (points) performed at the Kislovodsk mountainous station of the Institute of Atmospheric Physics of the Russian Academy of Sciences when oscillations were detected in the temperature variations in the mesopause which were caused by the orographical effect in the region of the Main Caucasian ridge (Semenov and Shefov 1989; Sukhodoev et al. 1989a,b). The periods of the detected waves were in the main 7–13 (min). The good agreement of these data with the presented relation suggests that the detected temperature variations were also caused by IGWs that occurred in the atmosphere during the interaction of a wind flow with the ridge.

The temperature of any layer of the upper atmosphere depends in the main on two factors: the energy delivered to the layer, including due to IGWs, and the layer cooling due to its self-radiation leaving in the outer space. The observed behavior of the mesopause temperature is possibly related to the substantial role of the carbon dioxide emission that has a considerable impact on the thermal regime in this region of the atmosphere during the propagation of IGWs. There are data of night rocket measurements of the altitude distribution of the CO_2 emission intensity performed in winter when the hydroxyl airglow region was in a shadow and in summer when it was illuminated by the Sun (Stair et al. 1985; Ulwick et al. 1985). It has been revealed that in both cases the CO_2 emission intensity varies severalfold, and this certainly points to the time variation of the CO_2 concentration. In the hydroxyl airglow region in winter, because of the turbulence intensified by the propagating IGWs (Lindzen 1971) (the greater the wave amplitude, the more intense the turbulence), the concentration of CO_2 also increases due to its inflow from below, resulting in a more intense cooling of the layer and its predominance over the warming due to the dissipation of IGWs. In the summer atmosphere illuminated by the Sun, practically over the hydroxyl emission layer, in the range of altitudes 95–115 (km), a significant (up to 30-fold) decrease in CO_2 concentration was detected (Ulwick et al. 1985). In this case, as turbulence intensifies, CO_2 will be partially transported upward from the hydroxyl emission region, and its concentration in the layer will thus decrease. This will lead to a predominance of the warming due to the IGW energy dissipation over the cooling due to the CO_2 emission.

5.2 Orographic Disturbances

When the unsteady air flow of the terrestrial atmosphere interacts with an obstacle, disturbances occur, which give rise to various wave processes (Gossard and Hooke 1975). In the early studies of these phenomena, the standing waves that arise over mountains which are flown round by air masses were considered (Kozhevnikov 1999). Originally, it was believed that these waves can penetrate into the region of the upper atmosphere up to the mesopause, though only under favorable conditions. However, the selectivity of the reflecting layers, critical layers, and absorption mechanisms to the azimuthal direction of the wave vector results in an appreciable azimuthal anisotropy of the waves in the mountain lee region. The waves whose vectors are perpendicular to the wind velocity and also the short waves whose vectors lie in the same vertical plane with the wind velocity vector are impeded (namely, are absorbed or reflected) first (Pertsev 1989a). This is why the standing orographic waves, having a chance to penetrate into the upper atmosphere, mainly in winter, due to the features of circulation of the middle atmosphere, are not the principal cause of orographic disturbances in the downwind regions of mountain ridges both in the mesopause and lower thermosphere and at the altitudes of the thermosphere (Pertsev 1989b).

5.2.1 Orographic Disturbances in the Upper Atmosphere

Mountain ridges are best suited to study the properties of wave disturbances in the upper atmosphere that occur over the landform obstacles. The relevant airborne measurements were originally performed in the region of the Ural Mountains (64°N). This mountain ridge of height about 1000 (m) rises sharply enough above the surrounding plain on both the western and the eastern sides. Moreover, a zonal

5.2 Orographic Disturbances

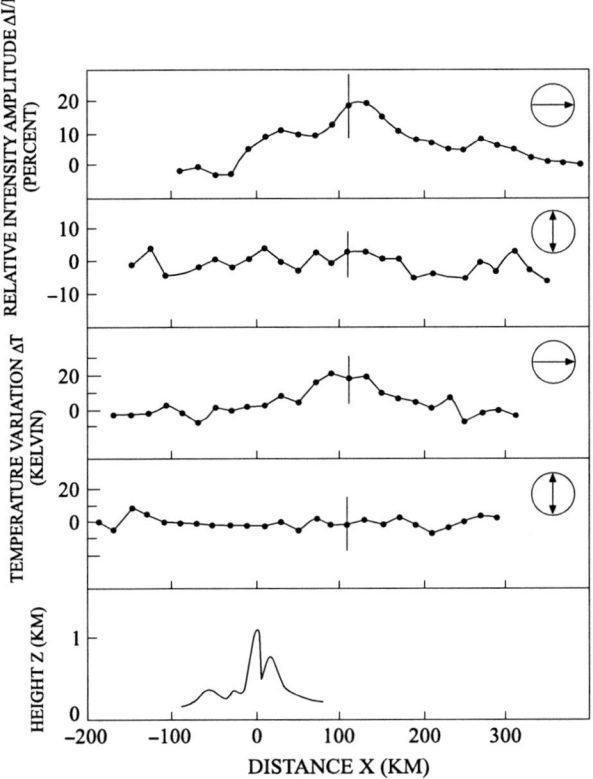

Fig. 5.21 Average variations of the hydroxyl emission rotational temperature increment $\Delta T(Z \sim 87\,(\text{km}))$ and of the relative intensity $\Delta I/I$ of the 630-nm atomic oxygen emission ($Z \sim 250\,(\text{km})$) in relation to the distance from the Ural ridge along the latitude $\varphi \sim 64°\text{N}$ at different directions of the tropospheric wind (Semenov et al. 1981; Shefov et al. 1983). The *arrows* indicate the wind direction (speed $\sim 10\,(\text{m} \cdot \text{s}^{-1})$); the *vertical lines* indicate the variance

wind practically permanently blows in this region. At altitudes of 3–4 (km) the mean wind speed is 10–20 (m · s^{-1}). Measurements of the hydroxyl emission temperature made it possible to detect its increase by about 10 (K) in the downwind region of the mountains and also a 20% relative increase in intensity of the 630-nm atomic oxygen emission; the disturbance region was 200 (km) in size (Fig. 5.21) (Semenov et al. 1981; Shefov et al. 1983; Shefov and Pertsev 1984).

Special regular measurements were started in 1985 at the Institute of Atmosphere Physics of the USSR Academy of Sciences at the mountainous station near Kislovodsk (43.7°N, 42.7°E, $Z = 2070\,(\text{m})$) to study in detail the properties of the wave processes detected in the downwind region of the Caucasian Mountains. For this purpose, two photometers were developed and fabricated. The two-channel scanning photometer (see Fig. 3.15) with interference filters [branch R (724.5 (nm)), branch Q (728.0 (nm))] was intended for determining the rotational temperature of the OH (8–3) band in eleven discrete directions in a plane perpendicular to

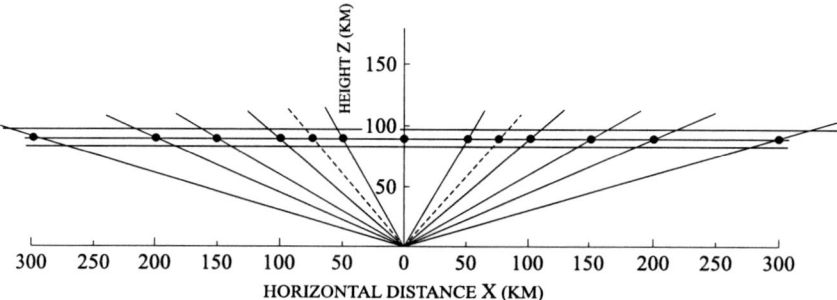

Fig. 5.22 Scheme of discrete orientation of the zenith angles of observation of the scanning photometers for the measuring of spatial distribution of the orographically disturbed hydroxyl emission temperature

the median line of the Caucasian ridge (Fig. 5.22). The chosen zenith angles provided the distribution of the sighted regions of the hydroxyl emission layer at every 50 (km) around the zenith direction. The observation place was 50 (km) north of Elbrus (5642 (m)) (Fig. 5.23). The time of measuring of the emission intensities in one region of the sky was ~40 (s); the time of one procedure of scanning the sky was 15 (min). The goal of the measurements was to determine the spatial distribution

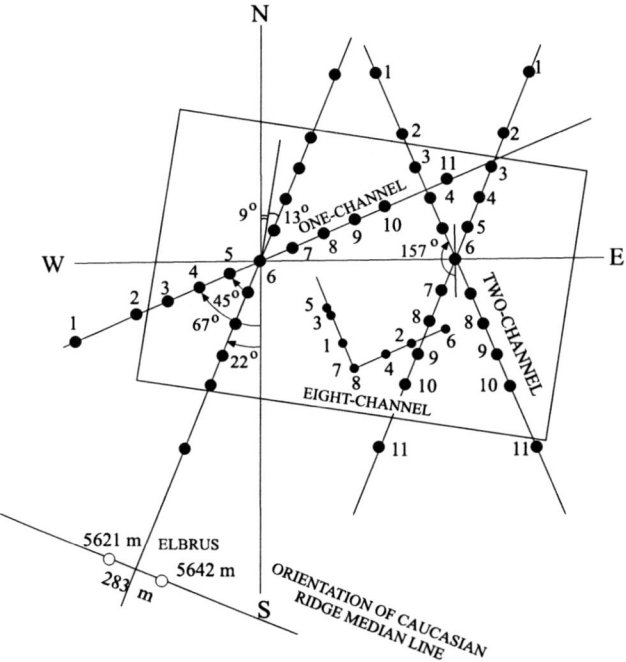

Fig. 5.23 Orientation of the observing pavilion and the planes of scanning of the photometers at Kislovodsk High-Mountainous Scientific Station

5.2 Orographic Disturbances

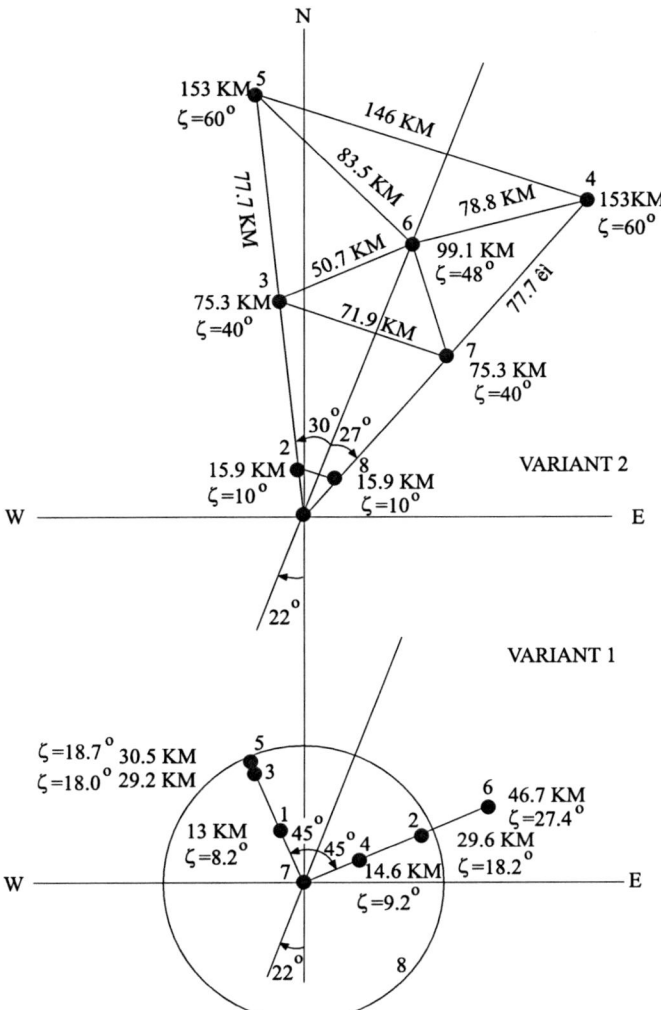

Fig. 5.24 Two orientations of the fields of vision of the eight-channel photometer

of the rotational temperature variations across the downwind region of the ridge in relation to the speed and direction of the wind over the ridge.

The second photometer had eight optical channels intended to investigate the spatial characteristics of waves in the downwind region at the hydroxyl emission layer altitudes (~ 87 (km)). The options of the used spatial structure of the measurement regions at 87 (km) are shown in Fig. 5.24. To localize the wave sources, the orientations of the fields of sight presented in Fig. 5.25 were used.

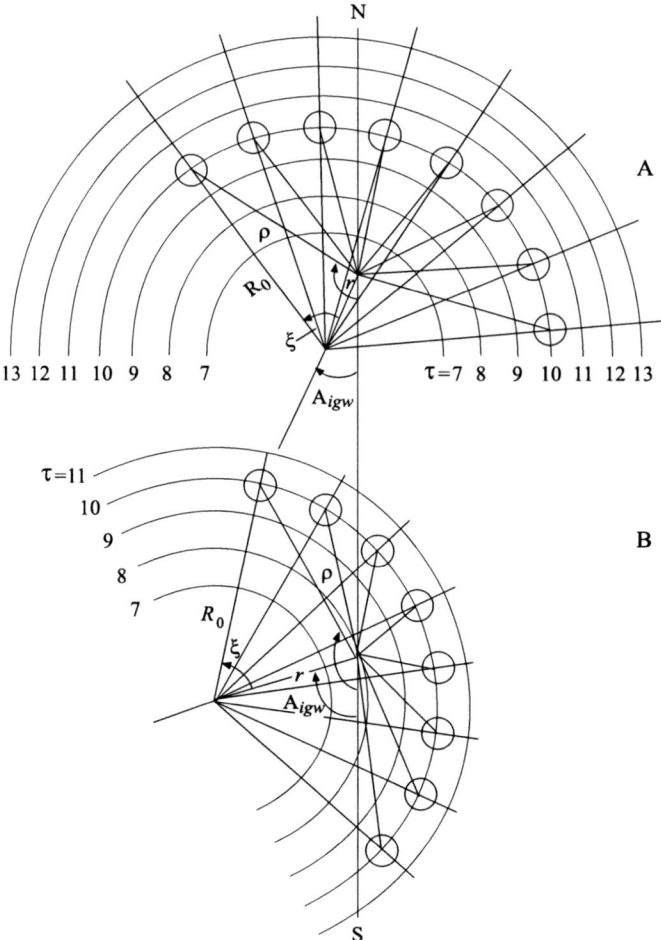

Fig. 5.25 Scheme of the orientations of the fields of observation used to localize wave sources with an eight-channel photometer. (**A**) Z = 87 (km), r = 50 (km), A_{igw} = 22° (Elbrus); (**B**) Z = 87 (km), r = 94 (km), A_{igw} = 71°

5.2.2 Measurements

Systematic measurements gave information on the processes that occur in the downwind region of the mesopause. The nightly mean rotational temperature variations ΔT_r along the vertical plane perpendicular to the median line of the Caucasian ridge have revealed a temperature maximum north of the mountains and a decrease in temperature south and north (Fig. 5.26) (Sukhodoev et al. 1989a,b). To analyze the altitude dependence of the tropospheric wind characteristics, the measured wind speeds normalized to the wind speed at the 500-mbar isobaric altitude (∼5 (km)) were compared. According to the night observations of orographic disturbances at

5.2 Orographic Disturbances

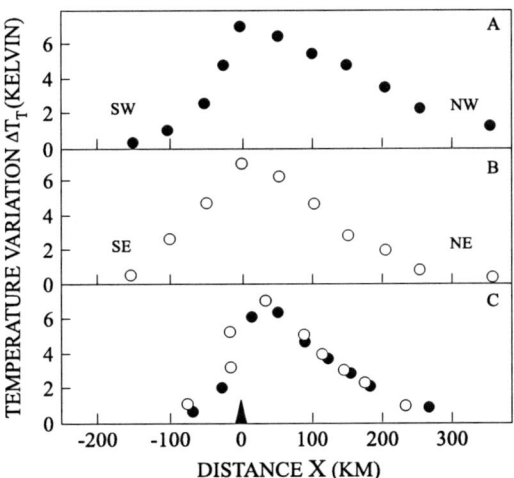

Fig. 5.26 Nightly mean distributions of the temperature of OH (8–3) derived from the observations conducted at two mutually perpendicular directions over the Caucasian ridge on September 23, 1985 (see Fig. 5.23) (Sukhodoev et al. 1989a,b). (**A**) SW–NE; (**B**) SE–NW; (**C**) projections of the distributions on the direction perpendicular to the ridge median line

three stations located near Sukhumi (43.0°N, 41.1°E), Tbilisi (41.7°N, 44.8°E), and Mineralnye Vody (44.2°N, 43.1°E), the mean wind speed linearly increases with altitude (Fig. 5.27). Relating the maximum ΔT_T and the tropospheric wind speed at various altitudes on the windward side of the mountains (Sukhumi) has shown that the maximum correlation factor corresponds to the 600-mbar isobaric altitude of ~4 (km) (see Fig. 5.28).

In view of the fact that the vibrational temperature is a manifestation of the process of vibrational relaxation depending on the atmospheric density and, hence, on the emission layer altitude, special simultaneous measurements of the distributions of the increments of rotational, ΔT_r, and vibrational temperatures, ΔT_v, across the ridge were performed. It turned out that these temperatures varied in antiphase, i.e., an increase in rotational temperature was accompanied by a decrease in vibrational temperature (Fig. 5.29) (Sukhodoev and Yarov 1998; Shefov et al. 2000a,b). This testified that the altitude of the emission layer decreased. Direct measurements performed on the UARS satellite (Lowe and Perminov 1998) showed that the altitude of the hydroxyl emission layer over the Caucasian region was lower than that over other regions (Fig. 5.30).

Also, a nonlinear dependence of ΔT_r on the wind speed at an altitude of 4 (km) has been revealed. The distance of the ΔT_T maximum from the ridge, X_M, and the width of the disturbance region, ΔX, are minimum when the prevailing wind is directed perpendicular to the median line of the ridge ($A_V^* = 0$) and increase with azimuth A_V^* (Fig. 5.31) (Sukhodoev and Yarov 1998; Shefov et al. 2000a,b).

Investigations of the generated wave spectrum have made it possible to reveal a variety of relations between the characteristics of the wave processes occurring in the region of Caucasus. The typical dependence of the spectral density on wave period is shown in Fig. 5.32. It can be seen that the spectra demonstrate permanent presence of waves with periods of 6–40 (min). It can be noted that in the range less than 5 (min), oscillations were recorded simultaneously by channels with small (4°) and large (43°) fields of vision. The large field was used for smoothing out short

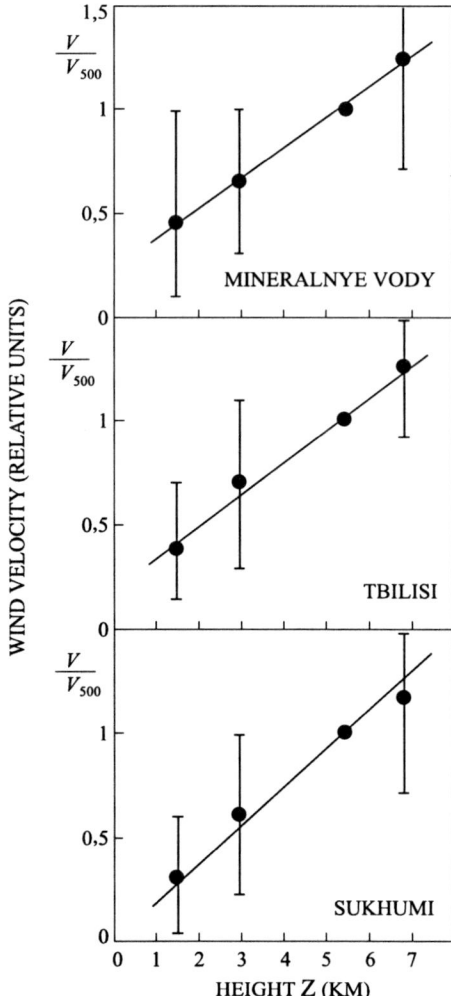

Fig. 5.27 Comparison of wind speeds at various altitudes of the troposphere according to the data obtained at Sukhumi, Tbilisi, and Mineralnye Vody. *Dots* are mean values; *solid lines* are regression lines

waves. The presence of oscillations in the range less than 5 (min), perhaps, testifies to the acoustical nature of these waves.

The oscillations present in the spectrum show a relationship between amplitude δT and period τ similar to that revealed earlier for IGWs: $(\delta T) \propto \tau^{5/3}$ (Krassovsky et al. 1978; Novikov 1981). Somewhat smaller amplitudes observed sometimes seem to be due to a limited duration of the train compared to the 2-h interval under investigation (Fig. 5.33) (Shefov and Semenov 2004b).

For the parameters of IGWs detected in the lee region of the mountains, their correlation with the wind velocity at various altitudes of the troposphere has also been revealed. As for the orographic increase in temperature in the field mesopause region, there is a correlation peak for altitudes of about 4 (km) (Figs. 5.34 and 5.35) (Sukhodoev et al. 1989a,b).

5.2 Orographic Disturbances

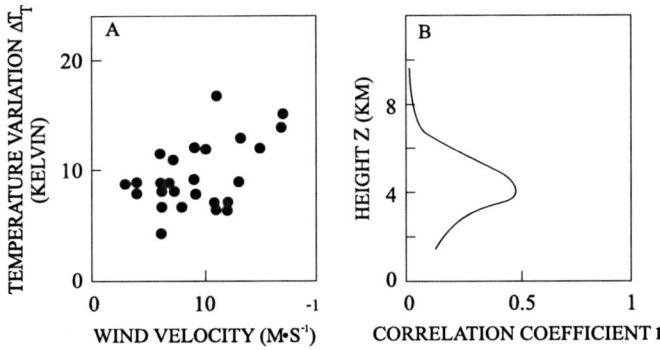

Fig. 5.28 Correlation between the amplitude of an orographic temperature increment in the mesopause and the tropospheric wind speed. (**A**) Example of the correlation field for the altitude 4 (km) (600 (mbar)), (**B**) altitude distribution of the correlation coefficient

Noteworthy, however, is the considerable nonuniformity of the wave field in the horizontal region of the mesopause covered by the eight-channel device (about 50 (km)). According to the measurements, appreciable fluctuations of the wave amplitudes and periods were observed in some regions. This seems to be due to the irregularity of the underlying surface and to the wind field that generates local wave sources. Undoubtedly, for the sighted regions corresponding to large zenith angles, an additional smoothing of the detected waves can have an effect (Shefov 1989).

Analysis of the wave characteristics obtained from the data of observations carried out at Zvenigorod gave a series of correlation formulas (Krassovsky et al. 1978). For the waves detected near the Caucasian Mountains, the following relation was obtained:

$$\frac{\delta T}{\tau} = \left(1 + \frac{V_p}{C_x}\right),$$

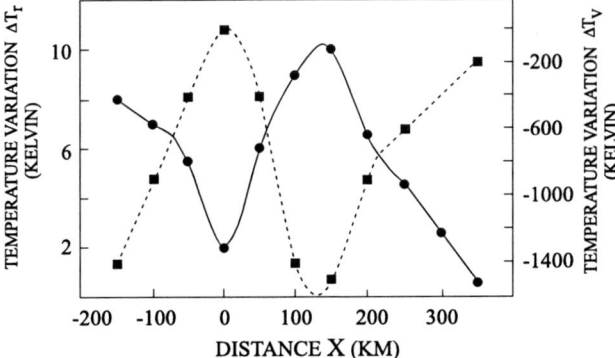

Fig. 5.29 Mean increments of the hydroxyl emission rotational temperature, ΔT_r (*dots*), and vibrational temperature, ΔT_v (*squares*), as a function of the distance X North of the Caucasian ridge (Sukhodoev and Yarov 1998; Shefov et al. 2000a,b)

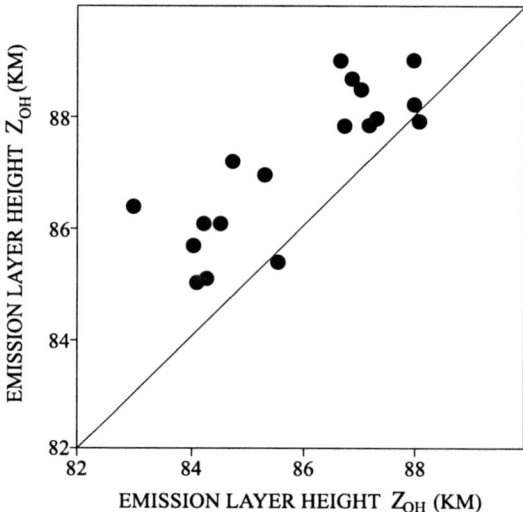

Fig. 5.30 Relation of the OH layer altitudes over the Caucasian ridge (abscissa axis) to those outside it in the latitudes range 35°–45°N (ordinate axis) measured from the WINDII/UARS satellite (Lowe and Perminov 1998)

where V_p is the projection of the wind velocity on the direction of propagation of the wave. The mean phase velocity C_x of waves of this type is about $140\,(m \cdot s^{-1})$ (Sukhodoev et al. 1989a). Since the wave periods were in the main 7–13 (min), the mean wavelength was 70–80 (km).

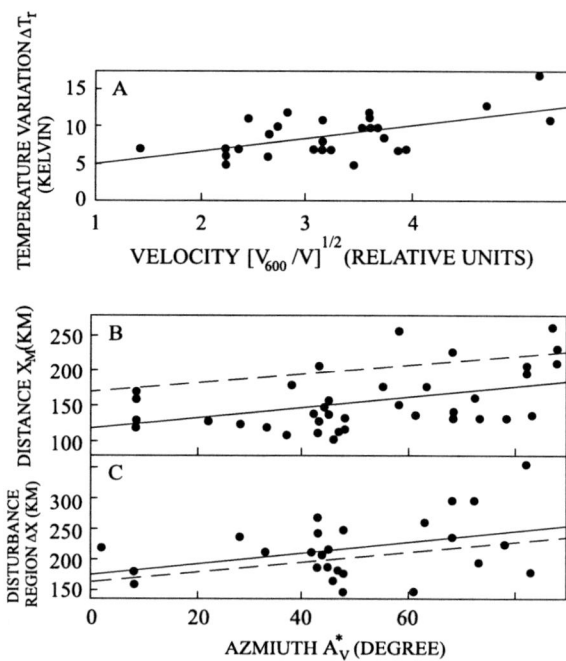

Fig. 5.31 Relation of the characteristics of an orographic disturbance: temperature ΔT_r (**A**), distances of the ΔT_r maximum from the ridge, X_M (**B**), and disturbed region halfwidth ΔX in the leeward region of the mesopause (**C**) to the parameters (speed V_{600} and azimuth A_V^*) of the prevailing tropospheric wind at the 600-mbar altitude (Sukhodoev and Yarov 1998; Shefov et al. 2000a,b). *Solid lines* are regression lines; *dashed lines* are calculation results

5.2 Orographic Disturbances

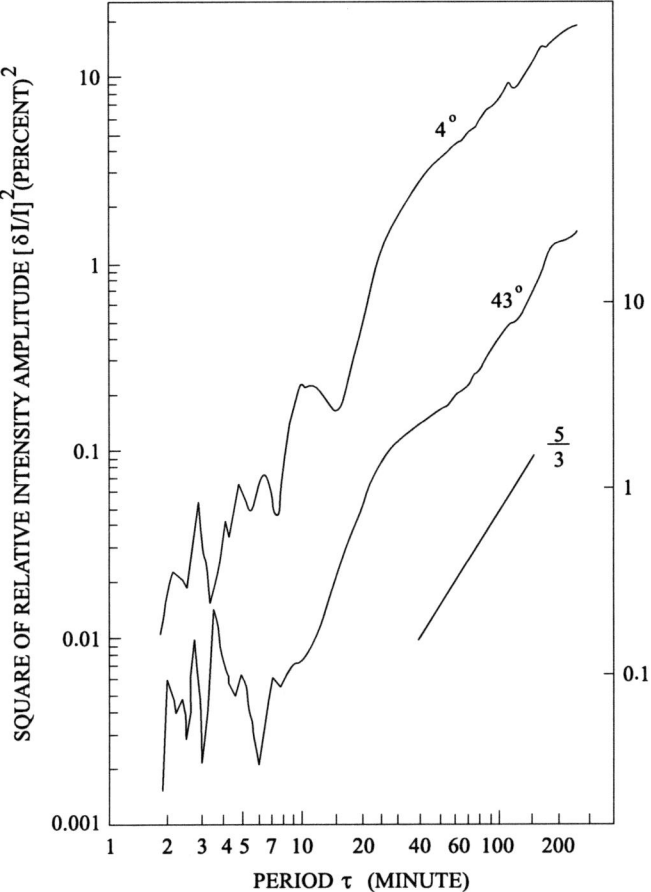

Fig. 5.32 Examples of spectra of the hydroxyl emission intensity fluctuations over the Caucasian ridge. The upper and lower curves were obtained with the field of vision equal to 4° and 43°, respectively

Therefore, all the mentioned facts allow the conclusion that the detected oscillations are fast IGWs which arise in the atmosphere during the interaction of a wind flow with a mountain ridge (Sukhodoev et al. 1989a,b). This conclusion is confirmed on relating the mean temperature T_r over the observation time to the quantity $\left(\frac{\delta T_w}{T}\right)^2$ that characterizes the vertical energy flux of the waves (Hines 1965; Chunchuzov 1978, 1981). For IGW with periods of 20–90 (min) observed in the conditions of Zvenigorod, clear correlations were obtained for summer (positive) and winter (negative) (Krassovsky et al. 1986a, 1988; Semenov and Shefov 1989). They can be described by the formula

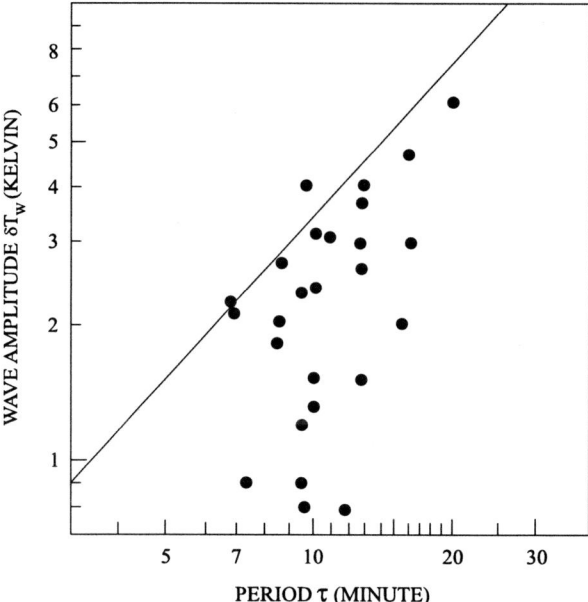

Fig. 5.33 Relation of wave amplitudes to their periods. The *straight line* corresponds to the average dependence (Krassovsky et al. 1978)

$$T = -2840 \cdot \left(\frac{\delta T_w}{T}\right)^2 + 243.$$

The measurements performed for waves with periods of 6–20 (min) at Kislovodsk completely match these correlations (Fig. 5.36). This suggests similarities between the observed wave phenomena and generality of their nature. Thus, the detected waves, by all signs, are moving internal waves.

When measurements were performed with an eight-channel photometer, oscillations were detected, as a rule, only by two (sometimes in three) recording channels. This suggests that the oscillation source is a restricted region from which a wave of period τ with a circular front starts propagating. In this case, using the method of backward ray tracing (Chunchuzov and Shefov 1978), it is possible to determine the horizontal wave propagation distance, X, as

$$X = Z \cdot \sqrt{\left(\frac{\tau}{\tau_g}\right)^2 - 1},$$

where Z is the altitude of the hydroxyl emission layer (87 (km)) and τ_g is the Brunt–Väisälä period.

Undoubtedly, the wave propagation distance should be affected by a wind flowing on its path. The relations that describe the effect of the wind speed were presented in a number of publications (Chunchuzov 1981; Kazannikov 1981). One of them has the form

5.2 Orographic Disturbances

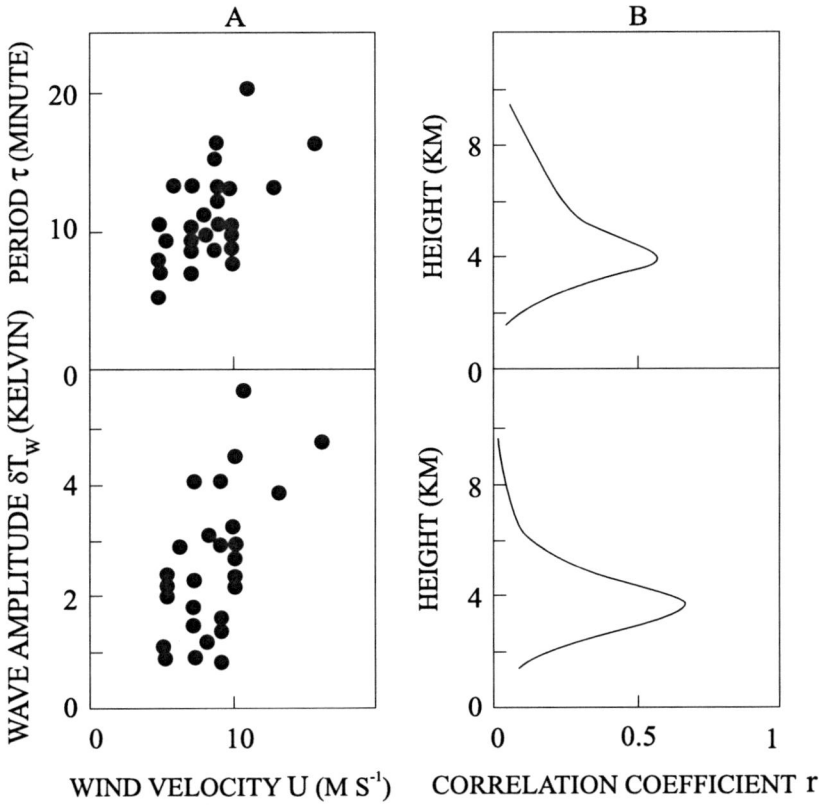

Fig. 5.34 Correlations between the periods of waves in the mesopause and their amplitudes and the speed of a wind over Sukhumi. (**A**) Examples of correlation fields for the altitude 4 (km) (600 (mbar)); (**B**) altitude distributions of correlation coefficients

$$X = Z \cdot \left(\frac{\tau}{\tau_g}\right) \cdot \left[1 - \left(\frac{V+u}{C_{max}}\right)^2\right]^{-1/2} \cdot \left(\frac{V}{V+u}\right)^2,$$

where V is the velocity of the wave, C_{max} is its maximum velocity, and u is the velocity of the wind.

Obviously, the extent of distortion of the estimated wave propagation distance is determined in large measure by the wave velocity. Since there is no information about the altitude profiles of the wind velocity and direction for concrete conditions, the wave propagation distances were estimated, to a first approximation, without taking into account the effect of the wind. It turned out that the expected sources of waves are closely associated with the Caucasian ridge and other irregularities of the Transcaucasian landform and are practically absent in the Colchis lowland (Fig. 5.37). The sources associated with certain height levels constitute a

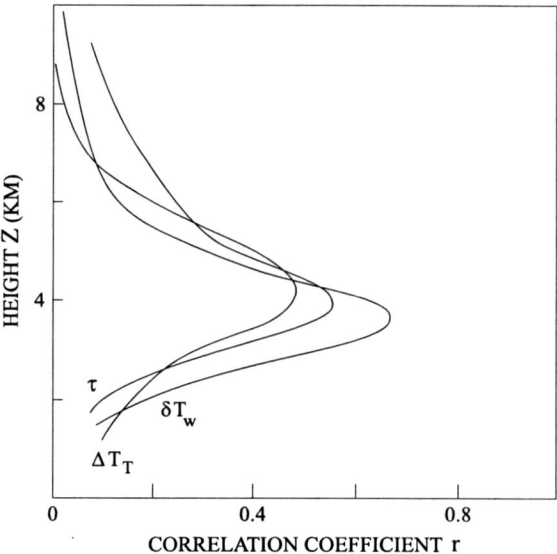

Fig. 5.35 Altitude distribution of the correlation coefficients between the orographic disturbance amplitude ΔT_v, wave amplitudes δT_w, periods τ, and the speeds of tropospheric winds over Sukhumi (Sukhodoev et al. 1989a,b)

pronounced distribution showing a high degree of correlation (~0.85) with the average profile of the ridge (Fig. 5.38) (Sukhodoev et al. 1989a,b).

Based on the correlation obtained, it is possible to derive an empirical relation between the relative number of wave sources, normalized to the full number of sources, N, per unit length of the ridge, and the height of the mountains:

$$n(Z_M) = 0.082 \cdot N \cdot (Z_M - 0.43).$$

Here Z_M is expressed in kilometers. If we describe the average smoothed profile of the ridge as

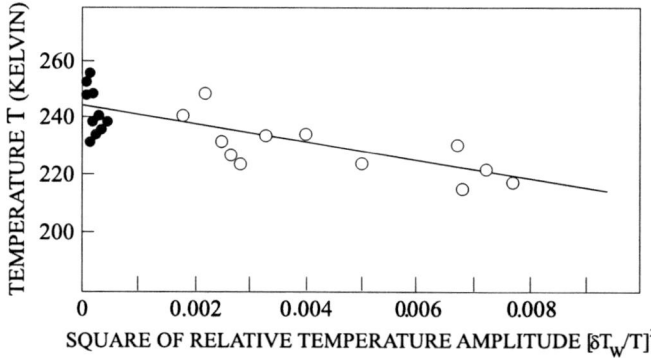

Fig. 5.36 Relation of nightly mean temperatures T and squared relative IGW amplitudes $\left(\frac{\delta T_w}{T}\right)^2$ for winter. *Hollow circles* are Zvenigorod (Semenov 1988; Semenov and Shefov 1989); *full circles* are Kislovodsk (Sukhodoev et al. 1989a)

5.2 Orographic Disturbances

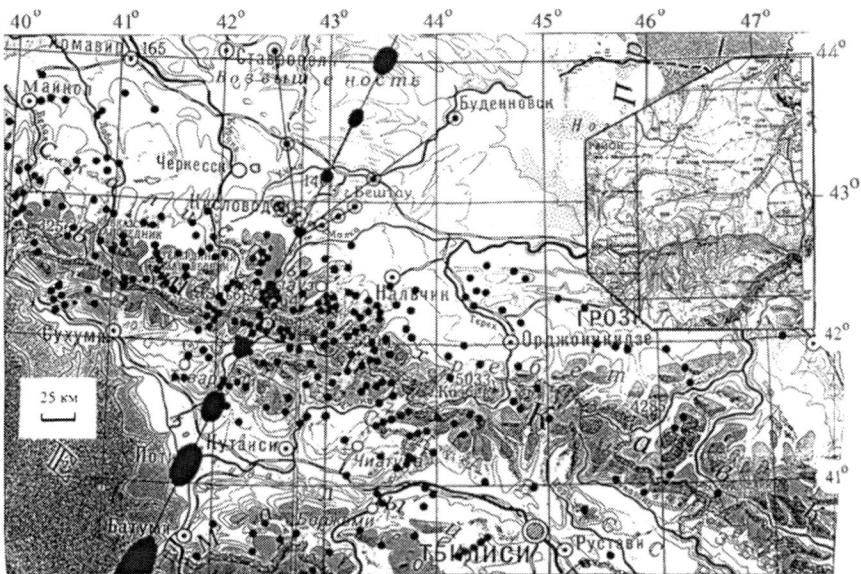

Fig. 5.37 Spatial distribution of IGW sources over the Caucasian ridge (*dots*) obtained by the method of reverse ray tracing on the base of spectral analysis of the time series data for some regions of the hydroxyl emission layer. The observation station (KHMSS), the emission layer regions observed by a two-channel photometer, and a version of observation regions with an eight-channel photometer are shown. In the inset, the region of location of KHMSS is encircled

$$Z_M = Z_M^0 \cdot \exp\left(-\frac{r^2}{r_0^2}\right),$$

where Z_M^0 is the mean height of the ridge in the central part, equal to about 3.5 (km), and take the ridge halfwidth W_M as a characteristic parameter, then we obtain $W_M = 2\sqrt{\log_e 2} \cdot r_0$. For the Caucasian ridge in its middle part, we have $W_M \sim 100$ (km), and it varies along the ridge.

5.2.3 Basic Theoretical Suppositions

The measurement data presented above testify to an energy influx to altitudes near 90 (km) over the lee region of the mountain ridge which is due to the IGWs generated over the mountains by nonstationary air flows. Based on this observation, the spatial distribution of the energy influx has been estimated theoretically (Shefov et al. 1999; Shefov et al. 2000a,b).

Following Sawyer (1959) and Chunchuzov (1978, 1981), the vertical energy flux density can be represented as

$$F_Z = p' \cdot w,$$

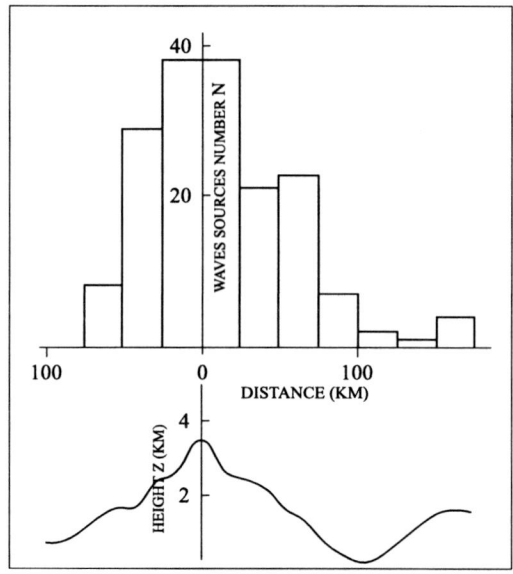

Fig. 5.38 Correlation between the number of IGW sources along the Caucasian ridge over a section of length 250 (km) at different height levels and the average ridge height profile in this section of the mountains. The minimum corresponds to the Colchis lowland. The correlation coefficient is 0.85

where p' is the amplitude of the pressure variations and w is the vertical wave speed. For a monochromatic wave propagating in a uniformly stratified atmosphere with no background wind and dissipation, p' and w are related by the expression (Chunchuzov 1988, 1994)

$$w = -\frac{k_0^2}{\omega^0 k_Z^0} \cdot \frac{p'}{\rho_0(Z)},$$

where $\rho_0(Z)$ is the atmospheric density, $\omega^0 = \omega_g \cdot \cos\theta$, $k_Z^0 = -k_0 \cdot \mathrm{tg}\theta = -k^0 \cdot \sin\theta$,

$$k_0^2 = (k^0)^2 \cdot \cos^2\theta = \left(\frac{\omega_g \cdot t}{r}\right)^2 \cdot \sin^2\theta \cdot \cos^2\theta,$$

θ is the zenith angle of the wave propagation direction; t is the time; r is the distance along the wave ray, and ω_g is the Brunt–Väisälä frequency.

Thus,

$$F_Z = \frac{(p')^2 \cdot t}{\rho_0(Z) \cdot r} \cdot \cos\theta.$$

For large distances from mountains, p' is determined by the formula (Chunchuzov, 1988, 1994)

$$p' = (2\pi)^{3/2} \frac{\rho_0(0) \cdot \exp\left(-\frac{Z}{2H}\right) \cdot Z_M^0 \cdot a^2}{\tau_g^2 \cdot r^2} \sin^2\theta \cdot \cos\theta \cdot \cos\psi$$

$$\times \Delta V_0 \cdot G \cdot \tau_0 \sqrt{\frac{2\pi \cdot t}{\tau}} \exp\left(-\frac{2\pi \cdot t}{\alpha \cdot \tau}\right) \cdot \cos\left(\frac{2\pi \cdot t}{\tau} + \frac{\pi}{4}\right),$$

5.2 Orographic Disturbances

where $\Delta V_0 \cdot G$ is the standard deviation for the vertical wind; ΔV_0 is the characteristic amplitude of the gust velocity; τ_0 is the characteristic gust duration; Z_M^0 is the height of the mountain; a is its halfwidth; τ_g is the Brunt–Väisälä period ($\tau_g = 5.2 \pm 0.7$ (min) up to altitudes of 100 (km)); ψ is the wave propagation azimuth relative to the direction of the gust; τ is the period of the wave propagating upward at the zenith angle θ; $r = \frac{Z}{\cos\theta}$ ($Z \sim 100$ (km) is the altitude of IGW dissipation in the upper atmosphere), and

$$\alpha = \frac{Z}{a \cdot \sin\theta \cdot \cos\theta}$$

(Fig. 5.39). The time t is counted from the onset of the gust over the mountain.

Next, the vertical wave energy flux density is given by (Chunchuzov 1978, 1981)

$$F_Z = \frac{4\pi^2 \cdot \rho_0(0) \cdot (Z_M^0)^2 \cdot a^4}{\tau_g^3 \cdot Z^3} \Delta V_0^2 \cdot G^2 \cdot \tau_0^2 \cdot D \cdot \sin^4\theta \cdot \cos^7\theta \cdot \cos^2\psi,$$

where

$$D = \left(\frac{2\pi \cdot t}{\tau}\right)^2 \cdot \exp\left(-\frac{2}{\alpha} \cdot \frac{2\pi \cdot t}{\tau}\right) \cdot \cos^2\left(\frac{2\pi \cdot t}{\tau} + \frac{\pi}{4}\right).$$

According to the data reported by Sato (1990), gusts of speed ΔV start arising only when the predominant wind speed V is over $V_0 \sim 5 - 7\,(\text{m} \cdot \text{s}^{-1})$. A possible approximation of this variability is

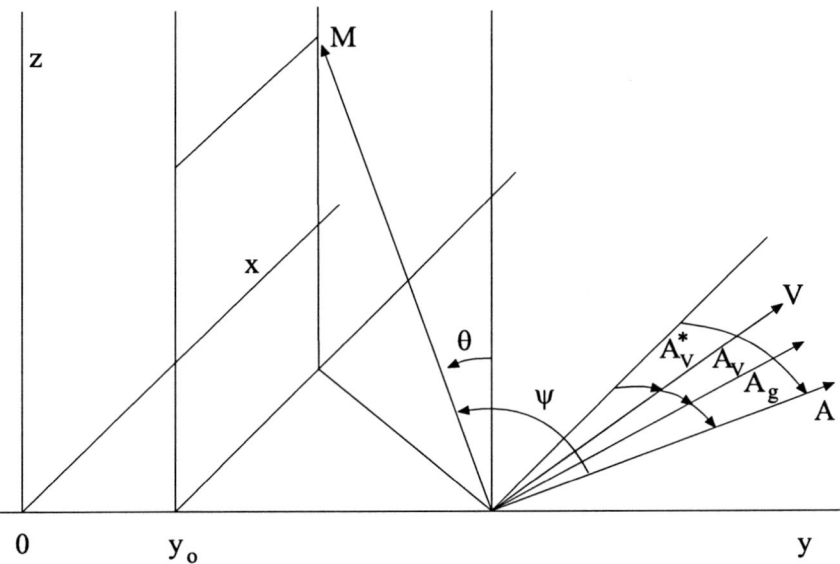

Fig. 5.39 Geometric scheme of the conditions of wave generation in the atmosphere over a mountain landform and their propagation to the leeward region of the mesopause (M)

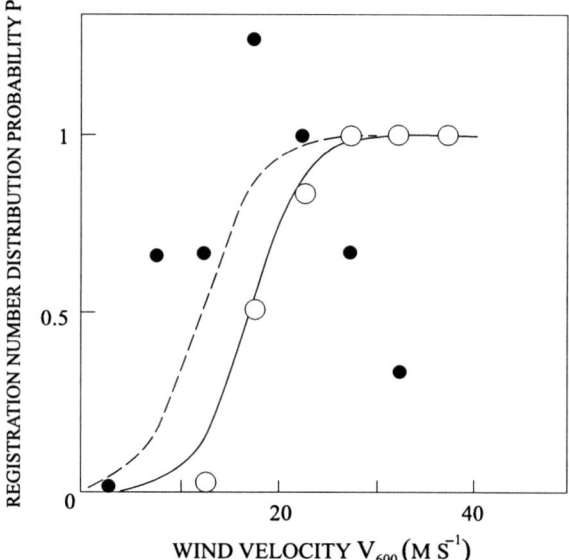

Fig. 5.40 Distribution of the number of cases where gusts with ΔV were recorded versus the wind velocity V_{600}, obtained based on the data of Sato (1990). *Full circles* are $\Delta V < 0.4\,(m \cdot s^{-1})$, *hollow circles* are $\Delta V > 0.4\,(m \cdot s^{-1})$, *solid line* is approximation taking into account only cases with $\Delta V > 0.4\,(m \cdot s^{-1})$, and *dashed line* is the same with $\Delta V < 0.4\,(m \cdot s^{-1})$ for $V < 20\,(m \cdot s^{-1})$

$$\Delta V = \frac{\alpha \cdot V + \Delta V_0}{1 + \exp\left[\frac{4(V_0 - V)}{\delta V}\right]},$$

where α is the coefficient of regression and δV is the wind speed jump. Based on the data of Sato (1990), approximate histograms have been constructed (Fig. 5.40) that represent the numbers of cases where gusts with ΔV less than $0.4\,(m \cdot s^{-1})$ (dots) and more than $0.4\,(m \cdot s^{-1})$ (open circles) versus the wind speed V_{600} at the 600-mbar isobaric level. The solid line indicates an approximation with the above formula which takes into account only ΔV values over $0.4\,(m \cdot s^{-1})$:

$$P = \frac{1}{1 + \exp\left[\frac{4(17 - V_{600})}{11}\right]}.$$

5.2.4 Analysis of the Input Parameters

Integration of a wave train with respect to time for real values $\alpha > 10$ yields

$$\bar{D} = \int_0^\infty D \, dt \approx \frac{\tau}{16\pi} \cdot \alpha^3.$$

5.2 Orographic Disturbances

Therefore, the total energy transferred in a direction θ through a unit cross-section at an altitude Z in a time equal to the train duration is determined as

$$\tilde{F}_Z = \int_0^\infty F_Z dt = \frac{\pi}{4} \cdot \frac{\rho_0(0) \cdot (Z_M^0)^2 \cdot a}{\tau_g^2 \cdot Z^2} \Delta V_0^2 \cdot G^2 \cdot \tau_0^2 \cdot \sin\theta \cdot \cos^3\theta \cdot \cos^2\psi.$$

The effective duration of a wave train, t_{eff}, can be estimated by the relation

$$t_{eff} = \tau \frac{e^2}{\alpha^2} \frac{\int_0^\infty x^2 \exp\left(-\frac{2x}{\alpha}\right)\cos^2 x\, dx}{\int_0^{2\pi} \cos^2 x\, dx},$$

whence

$$t_{eff} \approx \frac{e^2}{8\pi} \frac{Z}{a} \frac{\tau_g \cdot \left(\frac{\tau}{\tau_g}\right)^2}{\sqrt{1-\left(\frac{\tau_g}{\tau}\right)^2}}.$$

For typical values of the entering quantities, t_{eff} is $\sim 3\,(h)$, which agrees with the data reported by Chunchuzov (1978, 1981), and also corresponds to the time constant of dissipation for the turbulent energy produced in the mesopause by IGWs. Hence, the observed orographic disturbance in the mesopause over the lee region of the ridge was the result of the action of practically single wave trains generated by the assemblage of mountains. Therefore, for the given calculations, it was supposed that all wave trains from various sources over the ridge arose simultaneously.

The average squared vertical wind velocity is $\sim 120\,(cm^2 \cdot s^{-2})$ (Gage and Nastrom 1989; Van Zandt et al. 1989). Based on the above data, it can be represented as a function of the horizontal wind speed (Fig. 5.41):

$$\Delta V_0^2 = 200 \left(\frac{V_{600}}{10}\right)^{0.57} (cM^2 \cdot c^{-2}).$$

The correlation coefficient for this approximation is 0.742 ± 0.150.

The product $\Delta V_0^2 \cdot G^2$ is in fact proportional to the spectrum of the horizontal wind speed. Therefore, it is possible to use radar measurements of vertical wind velocity (Van Zandt et al. 1989; Gage and Nastrom 1989). The spectrum of vertical velocity F_w^2 is related to the spectrum of horizontal velocity F_u^2 by a formula following from a simplified theory of IGWs:

$$F_w^2 = \omega^2 \cdot \left(\frac{\tau_g}{2\pi}\right)^2 \cdot F_u^2.$$

To estimate the dispersion of the vertical wind, it is necessary to integrate F_w^2 over the frequency interval characteristic of IGWs.

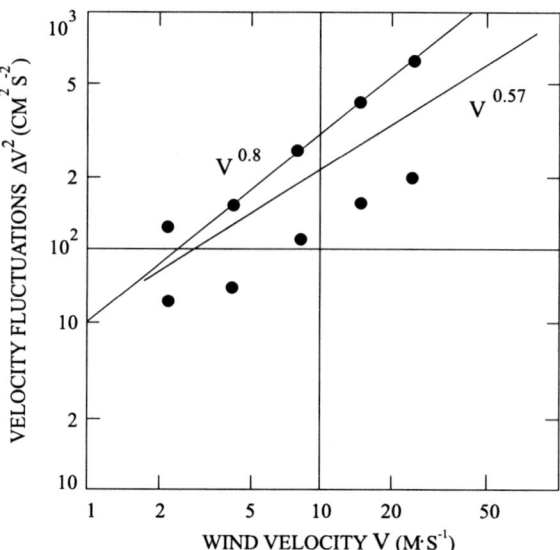

Fig. 5.41 Correlation between the squared fluctuations of vertical wind velocity, ΔV_0^2, and the wind velocity V (Gage and Nastrom 1989; Van Zandt et al. 1989). The *straight lines* indicate the obtained approximation over the top points and over the entire data set

The product $\Delta V_0^2 \cdot G^2$ represents the variance of the vertical wind velocity. Based on the data of Van Zandt et al. (1989), it can be described by the formula

$$\Delta V_0^2 \cdot G^2 = \int_{f_{min}}^{f_{max}} F_w^2(f) df,$$

where

$$F_w^2(f) = 3 \cdot 10^{-4} \cdot f^{-\frac{5}{3}} (m^2 \cdot s^{-2} \cdot Hz^{-1}), \quad f = \frac{1}{\tau} (Hz),$$

whence, in view of the relation for the horizontal wind velocity, taking into account that $1/f_{min} \sim 120\,(min)$, we obtain

$$S = \int_{f_{min}}^{f_{max}} F(f) df = 0.17 \cdot \left(\frac{V_{600}}{10}\right)^{0.57} (m^2 \cdot s^{-2}).$$

Statistical investigations of the distribution of dimensions a of a large assemblage of local mountains (Steyn and Ayotte 1985) suggest that the distribution of the number of mountains, N(a), over horizontal scales a is described by a power function:

$$N(a) \sim a^{-2.1 \pm 0.3}.$$

Verification of this relation has shown that for the Caucasian ridge the distribution

$$N(a) = N_m \cdot a^{-2}, \quad a_{min} < a < a_{max},$$

5.2 Orographic Disturbances

where $N_m = 3600$, $a_{min} \sim 0.5\,(km)$, $a_{max} \sim 10\,(km)$, is adequate.

The effective gust duration, τ_{eff}, can be estimated by the formula

$$\tau_0 = \frac{2\pi \cdot a}{V},$$

where V is the speed of the prevailing wind. Averaging over the mountain halfwidths yields

$$\tau_{eff} = \int_{a_{min}}^{a_{max}} \left(\frac{2\pi \cdot a}{V \cdot a^2}\right) da \Big/ \int_{a_{min}}^{a_{max}} \frac{da}{a^2} = \frac{2\pi \cdot a_{min}}{V} \cdot \log_e \frac{a_{max}}{a_{min}}.$$

For typical values of the input quantities, we have $\tau_{eff} \sim 10\,(min)$. This estimate agrees with the data reported elsewhere (Kolesnikova and Monin 1965; Chunchuzov 1988, 1994).

Next, the assumption has been made that the mountain heights Z_M^0 are distributed as Z_0^{-2}. Then, integrating over a and Z_M^0 and designating

$$\{E_Z\} = Q \cdot \varphi,$$

we obtain

$$Q = \frac{\pi}{4} \frac{\rho_0(0) \cdot S \cdot \tau_{eff}^2 \cdot Z_{M\,min}^0 \cdot Z_{M\,max}^0 \cdot a_{min} \cdot \log_e\left(e\frac{a_{max}}{a_{min}}\right) \cdot N_m}{\tau_g^2 \cdot Z^2},$$

$$\varphi = \left(\frac{\tau_g}{\tau}\right)^4 \cdot \sqrt{\frac{\tau^2}{\tau_g^2} - 1} \cdot \cos^2\psi,$$

where $Z_{M\,min}^0$ and $Z_{M\,max}^0$ are the minimum and maximum apex heights relative to the highest mountain; a_{min} and a_{max} are the minimum and maximum halfwidths of the mountains, respectively.

5.2.5 Calculation of the Energy Flux Spatial Distribution

The obtained relations can be used to estimate the spatial distribution of the integrated vertical energy influx at a given point of the lee region in the mesopause. Since the height of the mountain ridge ($\sim 4\,(km)$) on the average) is small compared to the altitude of the mesopause and lower thermosphere ($\sim 100\,(km)$) where IGWs dissipate, the mountain height was not taken into account in the calculations of the wave propagation distance. Therefore, the wave period, in view of the phase velocity c_{ph} (Pertsev 1989b), can be derived from the relation for the horizontal wave propagation distance (Chunchuzov and Shefov 1978) in the form

$$\tau = \tau_g \cdot \sqrt{\frac{x^2}{z^2} + \frac{y^2}{z^2}} + C(c_{ph}),$$

where

$$C(c_{ph}) = 1 - 1.22 \cdot \left(\frac{c_{ph}}{c_s}\right)^2,$$

and c_s is the velocity of sound.

The schematic of the conditions of wave generation in the atmosphere over the mountains in the region of Kislovodsk High-Mountainous Scientific Station of the IAPh, RAN (KHMSS) (43.7°N, 42.7°E) and their propagation to the mesopause region is shown in Fig. 5.39.

Under the actual conditions of wind flow, the prevailing wind having azimuth A_V^* relative to the horizontal normal to the ridge median line y produces wind fluctuations with azimuth A_V relative to the predominant wind with speed V. These fluctuations, in turn, generate gusts having azimuth A_g relative to the direction of the local wind fluctuation. Thus, the gust azimuth relative to the direction of the normal to the ridge median line is

$$A = A_V^* + A_V + A_g.$$

In the chosen coordinate system, the angle ψ between the directions of the gust vector and the vector directed from the point G where the wave is generated to the point M in the mesopause at the altitude Z where it is detected is determined by the relation

$$\cos \psi = \frac{x \cdot \cos A - y \cdot \sin A}{\sqrt{x^2 + y^2 + z^2}}.$$

It is well known (Gossard and Hooke 1975) that the penetration of IGWs into the upper atmosphere is determined in many respects by the wind velocity and direction and by the vertical temperature profile in the middle atmosphere. These factors most strongly affect the waves with periods of 6–20 (min). Based on the experimental data obtained during observations in Caucasus, it has been established that the degree of wave attenuation on the average can be empirically represented by a sigma function:

$$T(\tau) = \frac{1}{1 + \exp\left(\frac{4(\tau_{00} - \tau)}{\Delta \tau}\right)},$$

where $\tau_{00} \sim 10\,(\text{min})$, and $\Delta \tau \sim 6\,(\text{min})$ is a parameter which describes the argument range where the function changes from one limiting state to another.

As already mentioned, from these observations it follows that the strongest correlation between the temperature amplitude increment ΔT due to the orographic effect and the wind velocity is noted for the 600-mbar isobaric level ($\sim 4\,(\text{km})$) (see Fig. 5.35). Using the Sukhumi meteorological data for fair weather nights at the observation point (KHMSS) at a height of 2000 (m) and 50 (km) northward from the Elbrus Mountain during which photometer measurements were carried out, the following correlation ($r \sim 0.7$) has been revealed between the prevailing wind speed

5.2 Orographic Disturbances

$V(Z_{pr})$ and the altitude Z_{pr} (see Fig. 5.27):

$$V(Z_{pr}) = V_{600} \cdot \frac{Z_{pr}}{Z_{600}}.$$

Proceeding from the relations used, the energy influx to a chosen point (x, y_0, z) of the mesopause in the lee region of the mountains will be determined by the relation

$$[E_z(x, y_0)] = Q \cdot C(c_{ph}) \cdot \int_{-y_1}^{y_2} T(\tau) \cdot \varphi \cdot dy,$$

where Q and φ are the functions considered in the previous section. Introducing the dimensionless coordinates

$$p = \frac{x}{z} \quad \text{and} \quad q = \frac{y}{z},$$

we obtain

$$[E_z(p, q_0)] = Q \cdot C(c_{ph}) \cdot \int_{-q_1}^{q_2} T(\tau) \frac{[p \cdot \cos A - (q - q_0) \cdot \sin A]^2 \sqrt{p^2 + (q - q_0)^2}}{[p^2 + (q - q_0)^2 + C(c_{ph})]^3} dq,$$

whence the integral in the previous relation has the form

$$\Phi = f_1 \cdot \cos^2 A + f_2 \cdot \sin^2 A + f_3 \cdot \sin 2A,$$

where

$$f_1 = \int_{-q_1}^{q_2} T(\tau) \cdot p^2 \cdot \mu \cdot dq,$$

$$f_2 = \int_{-q_1}^{q_2} T(\tau) \cdot (q - q_0)^2 \cdot \mu \cdot dq,$$

$$f_3 = \int_{-q_1}^{q_2} T(\tau) \cdot p \cdot (q - q_0) \cdot \mu \cdot dq,$$

$$\mu = \frac{\sqrt{p^2 + (q - q_0)^2}}{[p^2 + (q - q_0)^2 + C(c_{ph})]^3}.$$

Evidently, the observed distribution of the temperature fluctuations that is a manifestation of the IGW energy influx is determined by statistically averaging the directions of arrival of the waves at a given point of the atmosphere or, eventually, the directions of the gusts. For describing the distribution of the probability density of random azimuths A_V and A_g, the Henyey–Greenstein function (Chamberlain 1978) usually used in light-scattering optics was chosen as a modeling function. For a plane distribution, this function has the form

$$p(\cos A) = \frac{1 - g^2}{4(1 - 2g \cdot \cos A + g^2)^{3/2}}.$$

Its conveniences are the inherent angular dependence and the presence of only one parameter g, which inherently is the mean cosine of the distribution. It can readily be seen that $g = 0$ corresponds to a uniform (isotropic) distribution of the azimuths and $g = 1$ to their coincidence with the given direction. The function $p(\cos A)$ for $g = 0.5$ is shown in Fig. 5.42.

Introducing the designations

$$B_1 = 2 \int_0^\pi p(\cos A) \cdot \cos^2 A \cdot d\cos A,$$

$$B_2 = 2 \int_0^\pi p(\cos A) \cdot \sin^2 A \cdot d\cos A,$$

$$B_3 = 2 \int_0^\pi p(\cos A) \cdot \cos A \cdot \sin A \cdot d\cos A,$$

we obtain

$$B_1 = \frac{1 + 2g^2}{3}, \qquad B_2 = \frac{2(1 - g^2)}{3},$$

$$B_3 = \pi \frac{1 - g^2}{2g} \left[\frac{1}{\sqrt{1 - g^2}} P^0_{-1/2}\left(\frac{1 + g^2}{1 - g^2}\right) - \frac{4}{3} \frac{\sqrt{1 - g^2}}{g} P^1_{1/2}\left(\frac{1 + g^2}{1 - g^2}\right) \right] =$$
$$= 0.005 + 0.468 \cdot g + 0.650 \cdot g^2 - 1.441 \cdot g^3 + 0.317 \cdot g^4,$$

where P are associated Legendre functions of the first kind.

Thus, upon integration with the assumed distributions of the local wind and gust azimuths, we have

$$\Phi = T \cdot \cos^2 A_V^* + L \cdot \sin^2 A_V^* + D \cdot \sin 2A_V^*,$$

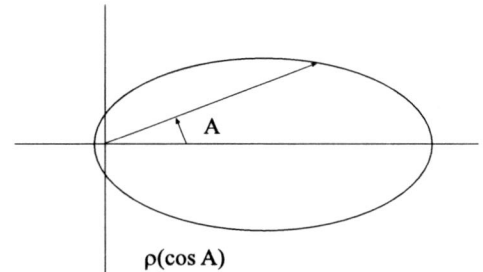

Fig. 5.42 Distribution function $p(\cos A)$ of the azimuths of fluctuations of local wind and gust directions, $g = 0.5$

where
$$T = f_1 \cdot M + f_2 \cdot N - 2f_3 \cdot P$$

is the transverse orographic function,

$$L = f_1 \cdot N + f_2 \cdot M + 2f_3 \cdot P$$

is the longitudinal orographic function, and

$$D = (f_2 - f_1) \cdot P - f_3 \cdot (M - N)$$

is the diagonal orographic function.

Here

$$M = B_{1v} \cdot B_{1g} + B_{2v} \cdot B_{2g} - 2B_{3v} \cdot B_{3g},$$
$$N = B_{2v} \cdot B_{1g} + B_{1v} \cdot B_{2g} + 2B_{3v} \cdot B_{3g},$$
$$P = (B_{1v} - B_{2v}) \cdot B_{3g} + (B_{1g} - B_{2g}) \cdot B_{3v}.$$

In these relations, the functions B_{iv} are determined by the g_v values for the local wind and the functions B_{ig} by the g_g values for the gusts.

5.2.6 Numerical Estimates of the Energy Flux Spatial Distribution

The relations obtained have been used to perform calculations for various g_v and g_g, values, namely 0, 0.25, 0.5, 0.75, and 1, taken in various combinations. It turned out that the distributions of the energy flux Φ over distance from mountain ridge p calculated for $g_v = g_g = 0.5$ show the best fit to observation data.

The behavior of the functions f_i, T, L, and D for the spatial distribution cross-section ($q_0 = 0$) is illustrated by Fig. 5.43. The computed Φ distribution has a maximum at $p \sim 2$, i.e., 170–200 (km) (Sukhodoev et al. 1992).

Comparison of the dependence of the distance X_M and the distribution peak width ΔX on the azimuth of the prevailing wind, A_V^*, shows reasonable agreement with measurements (see Fig. 5.31) (Sukhodoev and Yarov 1998; Shefov et al. 2000a,b).

It is of importance that the use of a normal distribution for the function p(cos A) gave no consent with measurements.

The dependence of the amplitude of the mesopause temperature orographic disturbance on the wind speed over the mountain ridge, which is proportional to $\sqrt{V}$ for $V > 5$–$7\,(m \cdot s^{-1})$ (Sukhodoev and Yarov 1998; Shefov et al. 2000a,b), fits well with the measurements of vertical wind fluctuations (Van Zandt et al. 1989).

The results presented allow one to construct the spatial distribution of the energy influx at altitudes of about 90 (km) over the lee region of the Caucasian Mountains for estimating the parameters of the disturbance region formed at the altitudes of the lower thermosphere (Fig. 5.44). With the obtained relations, where

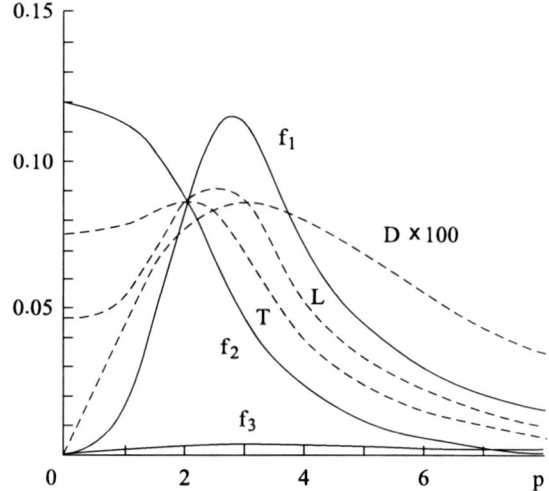

Fig. 5.43 Orographic functions f_1, f_2, and f_3 (*solid lines*) and the transverse, T, longitudinal, L, and diagonal, D, orographic functions with the parameters $g_v = g_g = 0.5$ for the spatial distribution cross-section in the KHMSS region ($y_0 = 0$) (Shefov et al. 1999, 2000a,b)

$$\rho_0(0) = 1.2 \cdot 10^{-3}\,(\text{g}\cdot\text{cm}^{-3}),\ S = 0.17 \cdot \left(\frac{V_{600}}{10}\right)^{0.57}\,(\text{m}^2\cdot\text{s}^{-2}),\ \tau_{\text{eff}} = 10\,(\text{min}),$$

$$Z_{\text{Mmax}} = 3\,(\text{km}),\ a_{\text{max}} = 10\,(\text{km}),\ a_{\text{min}} = 0.5\,(\text{km}),\ N_m = 3600,\ Z\,100\,(\text{km}),$$

$$\tau_g = 5.2\,(\text{min}),$$

it is possible to estimate the energy influx to a given point (p, q_0) of the mesopause in the downwind region of the mountains for $\sim$3 (h):

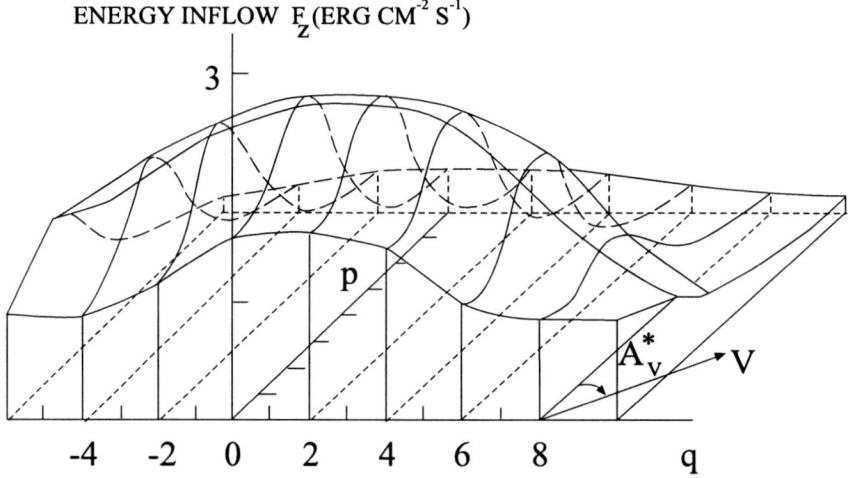

Fig. 5.44 Computed spatial distributions of the IGW energy vertical influx at altitudes of $\sim$100 (km) over the leeward region of the Caucasian Mountains at the prevailing wind velocity $V_{600} = 10\,(\text{m}\cdot\text{s}^{-1})$ and the azimuth $A_V^* = 45°$ (Shefov et al. 1999, 2000a,b)

$$[E_z(p,q_0)] = 3.2 \cdot 10^5 \,(\text{erg} \cdot \text{cm}^{-2}).$$

For the region of the temperature disturbance maximum, this yields $|F_z| \approx 3\,(\text{erg} \cdot \text{cm}^{-2} \cdot \text{s}^{-1})$, which well agrees with the estimates of the IGW energy influx (Gavrilov 1987).

5.2.7 Effect of Orographic Disturbances on the Energetics and Structure of the Upper Atmosphere

The calculations performed enable one to obtain a more detailed notion about the spatial distribution of disturbances at the altitudes of the upper atmosphere, including in the mesopause and lower thermospheric region near mountain ranges. The decrease in altitude of the hydroxyl emission layer, revealed from both ground-based (Sukhodoev and Yarov 1998) and satellite data (Lowe and Perminov 1998), suggests that at these altitudes vertical downward motions arise which accompany the IGW dissipation, and this was earlier indicated by Chunchuzov (1978, 1981). Interferometric measurements of the temperature of the 630-nm atomic oxygen emission over the Andes near Arequipa (16.5°S, 288.5°E) in Peru (Fig. 5.45) (Meriwether et al. 1996) have revealed its increase over the mountain ridge (whose height is ~5(km)) by 200–500(K) at a distance of ~430(km) from the ridge. Earlier estimates of 630-nm emission temperature variations in the F2 region obtained for Ural Mountains and Caucasus testified to an increase in temperature by 100 and 200 (K), respectively (Shefov 1985). The spotty structure of the 630-nm emission intensity over Hawaiian Islands (Fig. 5.46) (Roach and Gordon 1973) also seems to be a manifestation of the effect of orographic disturbances (Semenov et al. 1981; Shefov et al. 1983). The heights of the mountains on the islands are 3055 and 4205 (m). The expected temperature variation in this case can be ~300(K). All this allows one to suggest a possibility of detecting disturbances in other observation regions, in particular, in the north of the Scandinavian ridge over which there is an auroral zone (Shefov et al. 1999).

The characteristic spotty structure of the 557.7-nm atomic oxygen green emission testifies to the development of horizontal eddy diffusion accompanying the absorption of IGWs at the turbopause altitudes (Korobeynikova et al. 1984). For the coefficient of horizontal diffusion $K_{xx} \sim 3 \cdot 10^6\,(\text{cm}^2 \cdot \text{s}^{-1})$, the coefficient of vertical eddy diffusion $K_{zz} \sim 10^3\,(\text{cm}^2 \cdot \text{s}^{-1})$ (Lindzen 1971; Weinstock 1976; Ebel 1980; Chunchuzov and Shagaev 1983; Korobeynikova et al. 1984). Ashkhabad (38.0°N, 58.4°E) measurements were actually carried out near the Kopet-Dag mountain ridge (height 2–2.5 (km)). The average lognormal distributions of the probability density of mean intensities (~20(%) or ~50 (Rayleigh)) and sizes (~100(km)) of the spots (Semenov and Shefov 1997a) allow one to estimate the variation of the emission layer altitude in the region of a spot. The empirical model of the variations of green emission characteristics yields $\Delta I/Z_m = -4.2(\% \cdot \text{km}^{-1})$ (Semenov and Shefov 1997b,c,d, 1999). Therefore, the spot brightness $\Delta I \sim 50$ (Rayleigh) implies

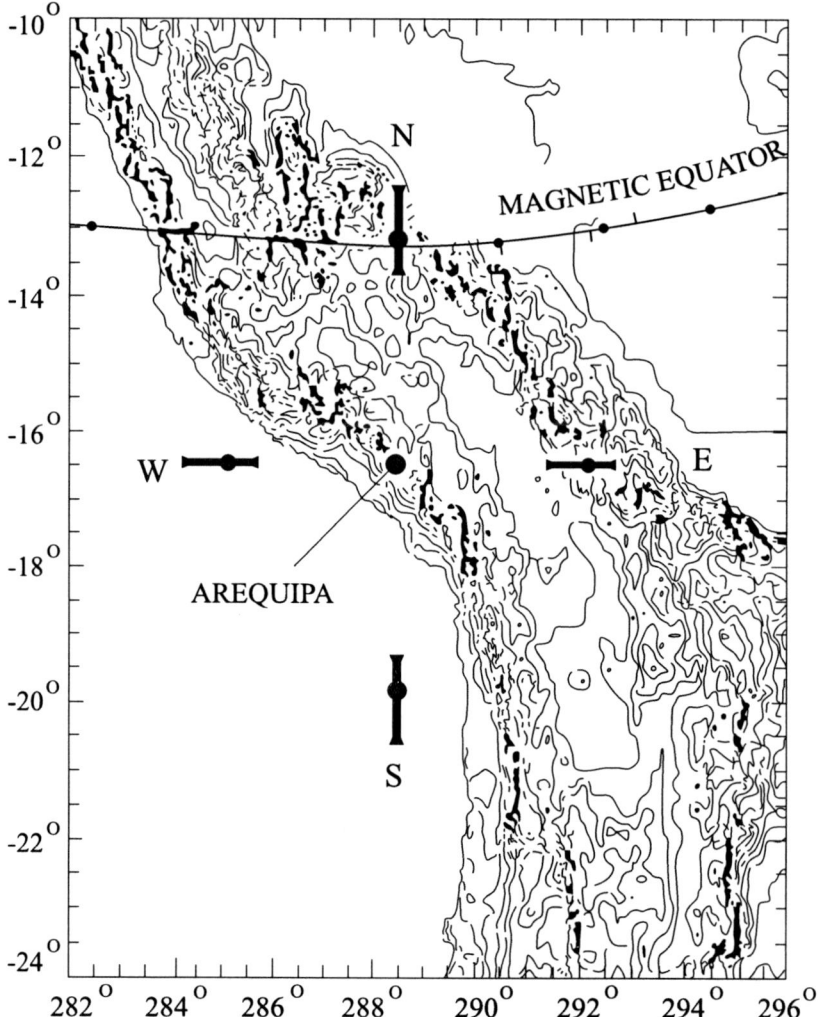

Fig. 5.45 Place of detection of a temperature orographic disturbance in the F2 region by interferometric measurements at the observatory of Arequipa, Peru (16.5°S, 288.5°E) (Meriwether et al. 1996). The *bars* indicate the positions of the upper atmospheric regions at 270 (km) over the Earth surface in which temperature was measured

the decrease of the altitude of the 557.7-nm emission layer by 4–5 (km) for an atmospheric region ∼100 (km) in size. This confirms the supposition of vertical motions in the mesopause and lower thermosphere resulting from the absorption of IGWs in the turbopause. This, in turn, implies a local deformation of the altitude distribution of the atomic oxygen density, which is responsible for many processes occurring in this upper atmospheric region. This seems to be also supported by satellite observations of the Herzberg emission bands (Ross et al. 1992) appearing in a process

5.2 Orographic Disturbances

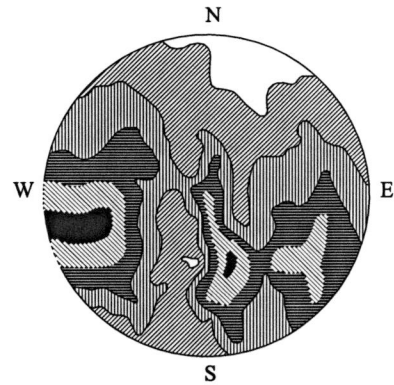

Fig. 5.46 Map of the spatial distribution of the intensity of the 630-nm atomic oxygen emission over the Earth surface, obtained from observations on Hawaiian Islands (September 11/12, 1961, 22:40 local time, meridian 195°E, Haleakala (20.7°N, 203.7°E, $Z = 3052\,(m)$) (Roach and Gordon 1973). The average stationary tropospheric wind was directed southwest

accompanying the 557.7-nm emission. Such a process manifests itself in that orographic disturbances are reflected in the meridional variation of the green line mean intensity that was revealed by the isophot maps published elsewhere (Korobeynikova et al. 1968). The results of the calculations of the sizes of disturbance regions are in agreement with measurements.

Based on the maps of 557.7-nm emission isophots mentioned above, Nasyrov (2007) has analyzed the spatial distribution of the intensity of this emission. It turned out that the character of the distribution varied depending on the direction of the wind in the troposphere relative to the median line of the ridge. In most cases, the deviations of the intensity from the mean ΔI (between 5 and 50 (Rayleigh)) increased almost monotonically in the direction of a normal to the ridge median line of that its half-space (north or south) to which the wind was directed, and the derivative of the increase was proportional to the wind speed. This independently implied that the disturbance intensity increased with wind velocity. The plot in Fig. 5.47 A relates the regression coefficient ρ of the straight lines approximating this trend of ΔI to the wind velocity V. The dots correspond to the northern region and the open circles to the southern region. The regression center line passes in fact through the origin of coordinates and is described by the relation (correlation coefficient $r = 0.77 \pm 0.06$)

$$\rho = (0.115 \pm 0.016) \cdot V.$$

Here ρ is expressed in $(Rayleigh \cdot (100\,km)^{-1})$; V is the wind speed $(m \cdot s^{-1})$. It should be noted that typical situations were those in which the wind velocity corresponded to the 500- or 300-mbar isobaric level. In some cases, high speeds were noted at a 100-mbar level at low speeds at the corresponding altitude of 5 or 9 (km), and these values well fitted to the general correlation formula. This suggests that an intense jet stream initiates gusts in the lower atmospheric layers that interact with the landform, promoting the generation of IGWs.

Based on 61 nights of observations in 1964, a clear dependence of ΔI on coordinate X with a peak within the sighted region of the 557.7-nm emission layer was revealed only in five cases. In fifteen cases, the peak was not so pronounced because of the spread of measurement points. The spread in emission intensity ΔI_{sp}

Fig. 5.47 Parameters of an orographic disturbance versus tropospheric wind speed V (Nasyrov, 2007). The rates of rise of intensity ΔI (**A**), maximum intensity ΔI_{sp} (**B**), distance X_M corresponding to ΔI_{sp} (**C**), and disturbed region halfwidth ΔX (**D**)

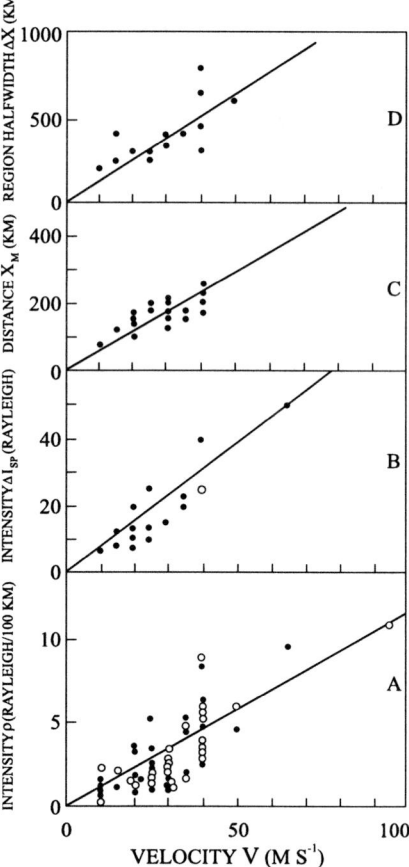

(Rayleigh) is related to the wind velocity in Fig. 5.47. The points denote individual values and the circle the value of ΔI_{sp} for the average distribution for the five mentioned cases. The regression line is described by the relation (correlation coefficient $r = 0.77 \pm 0.06$)

$$\Delta I_{sp} = (0.74 \pm 0.11) \cdot V.$$

The coordinate X_M of the peak ΔI_{sp} can be estimated only roughly since only one half (somewhat greater than the other) of the intensity distribution for the orographically disturbed region of the emission layer is known. The same refers to the width ΔX of this region at the half-intensity level. These quantities can be estimated by the relations (Fig. 5.47C and D, respectively)

$$X_M = (6 \pm 1) \cdot V; \quad r = 0.73 \pm 0.11;$$
$$\Delta X = (12 \pm 3) \cdot V; \quad r = 0.70 \pm 0.14.$$

The results obtained are essentially similar to the results of investigations of the hydroxyl emission (~ 87 (km)) near Caucasus (Sukhodoev et al. 1989a,b; Shefov

5.2 Orographic Disturbances

et al. 1999, 2000a,b). Nevertheless, since the atomic oxygen emission occurs at higher altitudes (~97 (km)), the observed disturbance region is wider and more dependent on the tropospheric wind speed. For the hydroxyl emission, the increase in temperature ΔT_{OH} at the disturbance peak is about 10 (K), which corresponds to an increase in emission intensity by ~25(%) and a decrease in emission layer altitude by ~1 (km) (Semenov and Shefov 1996). For the emission at 557.7 (nm), the intensity increase by ~10(%) (at a mean wind speed $V \sim 35 \, (m \cdot s^{-1})$ and the yearly average emission intensity for 1964 equal to 230 (Rayleigh)) corresponds to the increase in temperature $\Delta T_{557.7} \sim 5$ (K) and the decrease in altitude of maximum emission $\Delta Z_{557.7} \sim 2.5$ (km). The decrease in amplitude of the orographically induced disturbance for the atomic oxygen emission seems to be related to a decrease in amplitude of the IGWs due to their dissipation. Nevertheless, the IGWs generated by tropospheric winds flowing over a mountain landform appreciably disturb the lower thermosphere.

Thus, a mountain landform should be reflected as a stationary planetary component in the spatial variations of characteristics of the upper atmosphere and its emission layers. These effects were detected (Thuillier and Blamont 1973) though they were strongly smoothed because the measurements were performed along the limb, since the length L of a ray crossing a sighted emission layer is estimated as

$$L \approx 500 \cdot \sqrt{\frac{W}{10}} \, (km)$$

(Shefov 1978), where W is the layer thickness (km). Since for the 630-nm emission we have $W \sim 60$ (km), the ray length should be $L \sim 1220$ (km).

An interesting information can be derived from satellite measurements performed for pale stars setting beyond the horizon (Gurvich et al. 2002). The direct purpose of these measurements was to investigate the space–time distribution of the air density in the stratosphere by observations of stellar scintillations. It should, however, be noted that the glow intensity measured during the observation of feeble stars involved a considerable component from the airglow continuum. For the observations of various stars carried out sequentially at different dates during the motion of the Mir station, it turned out that the local time was the same for all cases. This made it possible to eliminate a possible effect of the variations of the continuum intensity during a night. However, for different measurements, the line of sight passed over different longitudes of the Earth surface at latitudes between 46 and 52°N. This made it possible to obtain longitudinal distributions of the emission intensity and emission layer altitude and thickness that roughly followed the location of mountain ridges (Fig. 5.48). This is evidence that orographic disturbances occur throughout the atmosphere.

The results obtained can be used to estimate the global contribution of IGWs generated near the Earth surface to the energy influx to the mesopause and lower thermospheric region (Shefov 1985).

The length of the mountains in the northern hemisphere can be estimated as $L_N \sim 6 \cdot 10^4$ (km) and in the southern region as $L_S \sim 4.5 \cdot 10^4$ (km); the planetary

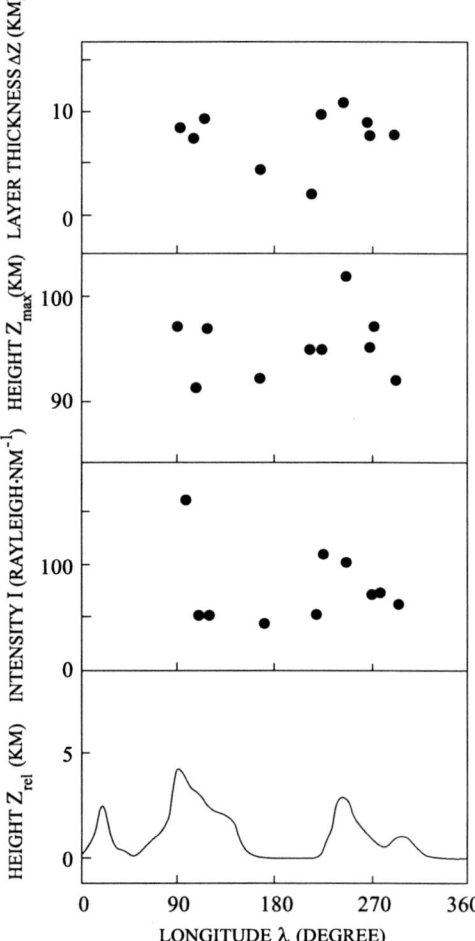

Fig. 5.48 Longitudinal distribution of the data of ten measurements of the upper atmosphere emissions in the spectral range 420–530 (nm) on the "Mir" orbital spaceship at latitudes of 46–52°N (Gurvich et al. 2002)

mountain length is $L_E \sim 1 \cdot 10^5$ (km). Naturally, individual peaks, island mountains, etc., are difficult to take into account. The average width of mountains is $W_M \sim 300$ (km), the average height $Z_M \sim 2$ (km), and the average disturbance region length $\Delta X \sim 400$ (km). Thus, the planetary mean energy influx can be estimated as 1.5–$2\,(\text{erg} \cdot \text{cm}^{-2} \cdot \text{s}^{-1})$.

The above conclusions about the disturbances in the upper atmosphere near mountain ridges can, perhaps, be extended to the IGWs generated in regions of cold fronts and cyclones over which wind flows are always present.

The variable meteorological situation over the Earth surface is responsible for the variable background component of a disturbance. This offers the possibility to reveal disturbances in the upper atmosphere based on planetary maps of the structure of the Earth surface and of the meteorological situation and take them into account in an up-to-date model of the upper atmosphere.

5.3 Planetary Waves

The Earth's atmosphere is an oscillating system with a broad spectrum of natural frequencies pertaining to oscillations induced by various physical processes. The rotating atmosphere, possessing a gyroscopic rigidity, responds to any disturbance by oscillatory motions. The waves arising for this reason are important components of atmospheric oscillations. Such planetary waves, by Rossby's definition, are atmospheric disturbances of global horizontal length. They in fact represent deviations from the symmetric motions that occur at middle and high latitudes with a time scale greater than one day (Dickinson 1969). Planetary waves have the features of global extension, large amplitudes in cold seasons of year, and very small amplitudes in summer (Gaigerov 1973).

Based on the investigations performed, it has been found that the structure of a disturbance depends on its horizontal wave number k, which is related to the wavelength as $k = \frac{2\pi}{\lambda}$. The planetary scale disturbances with small wave numbers can penetrate through the atmosphere in the form of planetary waves. The disturbances with high wave numbers retain their local eddy character and are carried in the main by the mean flow. Zonal east–west winds reliably prevent planetary waves from propagating upward. Strong zonal west–east winds, acting as a reflecting barrier, also impede vertical vertical propagation of planetary waves. However, there is no delay in wave motions near the equator. Besides, rather strong west–east winds can pass planetary waves with small wave numbers and small Rossby modes at mid-latitudes (Dickinson 1969). This creates favorable conditions for the propagation of planetary waves to the stratosphere and mesosphere in and mesosphere in winter. The amplitudes of planetary waves decrease with altitude in the region of weak zonal winds as a result of radiation processes. The interaction between vertically propagating waves and the mean flow drastically depends on the mean zonal wind profile. The waves that arrive at the critical line along which the mean zonal wind velocity coincides with their phase velocity fail to cross this line and are absorbed by the middle flow (Holton 1972). As a result, mesospheric medium- and small-scale disturbances are most pronounced at low latitudes, while at high latitudes disturbances of all scales are rather strong (Gaigerov 1973).

Thus, vertically propagating planetary waves generated in the troposphere make a contribution to the stratospheric eddy heat and momentum transfer. In relation to such disturbances, the stratosphere can be considered a region free of wave generation sources (Holton 1972).

Investigations of various processes in the mesosphere and lower thermosphere have clearly revealed variations with time scales of several days. Both stationary planetary waves associated with the features of the ground landform and traveling waves were detected.

The results of analyses of the measurements of the hydroxyl emission intensity in the OH (8–3) band performed by means of the WINDII/UARS device at latitudes between 45°S and 40°N were used to construct seasonal mean longitudinal distributions of the intensity (Rayleigh) and emission maximum altitude Z_m at latitudes $40° \pm 5°N$ (Perminov et al. 1999) (Fig. 5.49) and of the emission rate

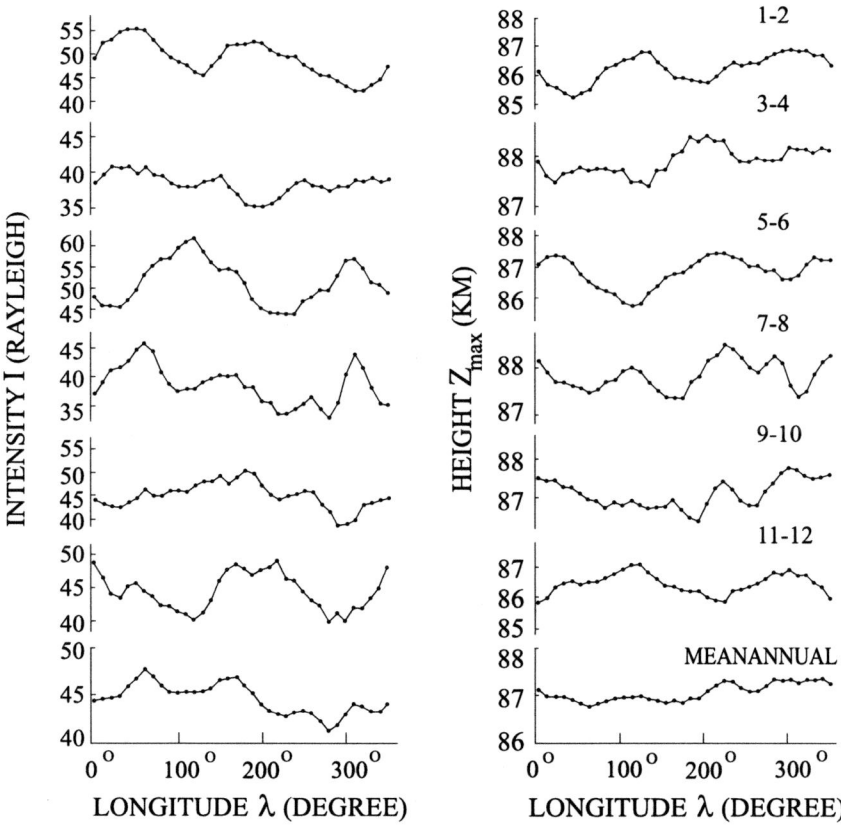

Fig. 5.49 Two-month mean and yearly average longitudinal variations of the intensity of the 734.6 (nm) OH (8–3) $P_1(N' = 2)$ line and of the emission maximum altitude Z_m obtained with WINDII/UARS in 1993 (Perminov et al. 1999)

Q_o (photon·cm^{-3}·s^{-1}) (Perminov et al. 2002) (Fig. 5.50). From these data, it follows that the longitudinal variations in intensity and layer altitude are opposite in phase, which completely agrees with other results on hydroxyl emission variations. Moreover, this feature shows up even in the yearly average distributions. Changes in phases at altitudes of 80–88 (km) testify to vertical and meridional propagation of stationary planetary waves (with a wave number of 2) from the southern to the northern hemisphere during equinoxes. This is evidence by the decrease in phase for such a wave which propagated in the direction from 80 (km) at southern latitudes to 88 (km) at northern latitudes. Similar cases were considered theoretically (Pogoreltsev and Sukhanova 1993; Pogoreltsev 1996) also revealed in the analysis of stationary planetary waves in the data on winds at altitudes of 90–120 (km) (Wang et al. 2000).

Planetary waves with periods of 2–5 days were investigated by the variations of the 557.7-nm and 630-nm atomic oxygen emissions, sodium emission, and OH

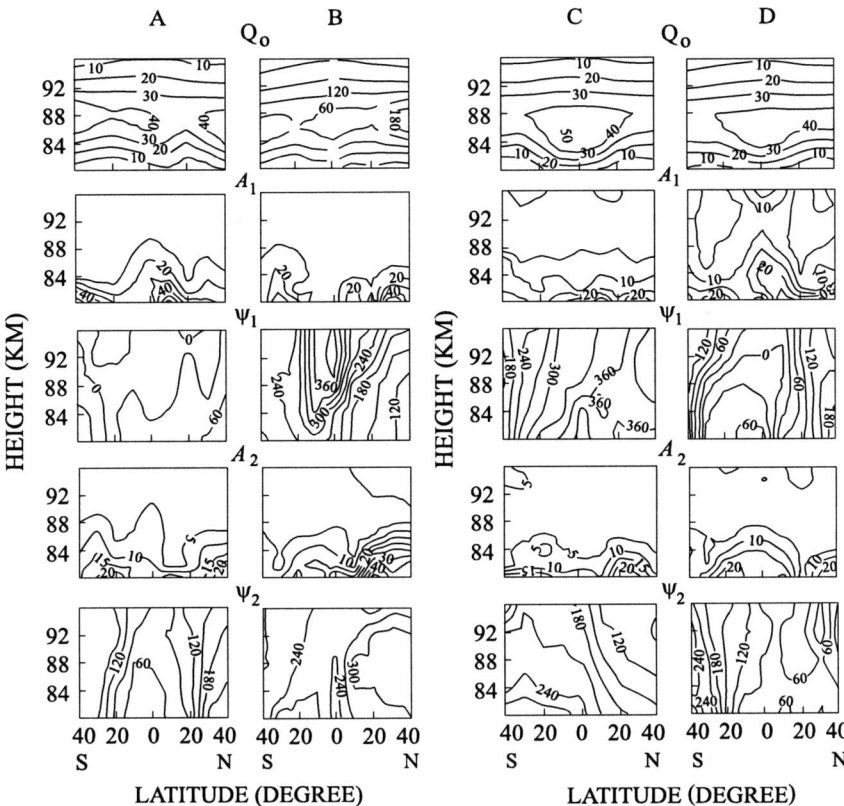

Fig. 5.50 Latitude–altitude distributions of the seasonal mean emission rate Q_0, amplitudes A_1 and A_2, and phases ψ_1 and ψ_2 of planetary waves with zonal wave numbers of 1 and 2 (Perminov et al. 2002). Solstice periods: from November 7, 1992, to February 5, 1993, (**A**) and from May 7 to August 6, 1993 (**B**); equinox periods: from February 6 to May 6, 1993, (**C**) and from August 7 to November 6, 1993 (**D**). The wave amplitudes are given in percentage of the seasonal mean values of the emission rate; the phases are indicated in degrees east of the Greenwich Meridian

(6–2) emission in the southern hemisphere near the equator (7°S, 37°W). These emissions show semiannual variabilities characteristic of the equatorial region. Quasi-two-day variations were most often detected from November to January (summer), their amplitudes being more than 30(%) for the oxygen emissions and 10(%) for the hydroxyl emission. These variations were interpreted as Rossby waves. On the other hand, the quasi-3.5-yr variations were most pronounced from April to September. Takahashi et al. (2002) interpreted these variations as ultrafast Kelvin's waves.

According to investigations by Laštovička (1997), the variations with periods of about 2, 5, 10, and 16 days are dominant in the troposphere and stratosphere and also in the mesosphere and lower thermosphere. The amplitudes of their both zonal and meridional components show a long-term trend for an increase in velocity and its increase with a decrease in solar activity (Jacobi 1998).

The variations of the intensity of the polar mesospheric summer echo (PMSE) with periods of 4–6 days are manifestations of planetary waves. Smoothed summertime variations have components with periodically varying amplitudes (Kirkwood and Réchou 1998).

The propagation of a 16-day planetary wave is related to the features of its generation. By its parameters this wave closely corresponds to the normal Rossby (1,3) mode. This is due to the latitudinal structure of waves of this type described by Hough functions $(s, n-s) = (1,1), (1,2)$, and $(2,1)$ where n is the meridional index, s is the zonal wave number, and n–s is equal to the number of zeros of the meridional component of the wave velocity between the poles (Stepanov 2000). However, the 16-day wave observable in the winter northern hemisphere can be compared to the normal (1,3) mode only in this hemisphere, and there is no correlation between the hemispheres which it could be expected for this mode. The normal-to-hemispherical mode transformation is caused first of all by the regular phase displacement to the west due to dissipation and, as a consequence, by less positive interference in the standing wave between the poles than for a Rossby wave (Stepanov 2000).

The most extensive investigations have been performed by means of radar wind measurements (Luo et al. 2002). Waves were detected in winter (October–April) when the background zonal wind had the west–east direction. The summer activity of waves was low and was restricted to a thin layer near the line of zero wind speed ($\sim$85 (km)). Smoothing the measurement data with a bandwidth filter (halfwidth 12–20 days, period 64 days) for three altitude ranges (67–73 (km), 79–85 (km), and 91–97 (km)) has revealed almost periodic variations between 1994 and 1997 whose zonal and meridional component amplitudes experienced cyclic changes (almost without failures) for three years. The amplitude was $5-7\,(m \cdot s^{-1})$ in winter and $\sim 2\,(m \cdot s^{-1})$ in summer. According to the measurements of the rotational temperature of the OH (4–2) band carried out in summer near Stockholm (59.5°N, 18.2°E) between 1992 and 1995, the amplitude of 16-day variations was $\sim$5 (K). The oscillations continued as long as about 60 days (Espy et al. 1997).

Analysis of the frequency of appearance of noctilucent clouds for a concrete longitudinal region in June/July for several years, for which most detailed observation data have been accumulated, has revealed clear 16-day variations of the probability of appearance of noctilucent clouds at latitudes near 55°N (Shefov and Semenov 2004a) (Fig. 5.58). They testify to the east–west propagation of the wave with a velocity of $\sim 17\,(m \cdot s^{-1})$, which agrees with the results reported by Luo et al. (2002).

Long-period variations (20–40 days) at the altitudes of the mesosphere and lower thermosphere (60–100 (km)) have been revealed by radar wind measurements carried out in the latitude range 30°N–70°N between 1980 and 1999 at Tromsø (70°N, 19°E), Saskatoon (52°N, 107°W), Collm (52°N, 15°E), and Yamagawa (30°N, 130°E). The wave velocities were $\sim 10\,(m \cdot s^{-1})$ in the mesosphere, $\sim 5\,(m \cdot s^{-1})$ in the lower thermosphere, and higher at mid- and low latitudes. Though the climatology of these oscillations is similar to that of the long-period normal modes of planetary waves (10–16 days), the phases, being also similar for various situations, do not fit to the phases of freely propagating waves. By relating these variations to the characteristics of solar and geomagnetic activities, the authors (Luo et al. 2001) have arrived at the conclusion that they correlate with the 27-day solar cycle associated with the synodic solar rotation period. Weak correlations with the 11-yr solar cycle in the mesosphere (positive) and in the lower thermosphere (negative) have also been revealed (Luo et al. 2001).

It should be noted that the harmonic analysis performed by Luo et al. (2001) to reveal the 27-day periodicity is absolutely inapplicable in this case since a sequence of 27-day solar cycles, each representing a synodic solar rotation period, is not a periodic process. This is due to the fact that new groups of sunspots arise as a result of a random process, and therefore the beginning of a new solar cycle is occasional. Therefore, in any case the revealed periodicity requires further analysis.

Near equinoxes strong transformations occur in the circulating atmosphere at the altitudes of the mesopause and lower thermosphere, which were detected by radar and photometer measurements. The amplitude and phase of a wave with a wave number s = 1 at latitudes 42°–62°N experienced jumps (Liu et al. 2001; Manson et al. 2002). Such a jump in the zonal direction of the wind velocity happens about a vernal equinox during about 15 days at altitudes from 105 to 75 (km). During an autumnal equinox, the process begins in the altitude range 92–105 (km) even before the middle of the summer and is stabilized a little near the summer solstice at altitudes of about 92 (km). Then the rate of lowering of the transition level gradually increases with decreasing altitude to 85 (km) and finally falls abruptly (within several days) to zero at the 75-km level. Shortly thereafter, at altitudes of 93–105 (km), another zone of zero wind velocity is formed and persists during several weeks.

These atmospheric rearrangements were already investigated for altitudes up to ∼60 (km) (Webb 1966). Therefore, extending the zone of transformation of the global circulation to the altitudes of the lower thermosphere, we have a general system of space–time variations in atmospheric dynamics.

5.4 Noctilucent Clouds

In the years when the upper atmosphere was not explored yet, noctilucent clouds were the second, after auroras, atmospheric phenomenon which testified to the existence of a terrestrial atmosphere at high altitudes. For the first time they were certainly detected in 1885, after the eruption of Krakatau volcano. A detailed history of their early investigations is described by Bronshten and Grishin (1970). Nevertheless, when examining the historical investigations of unusual atmospheric phenomena, one can suppose that noctilucent clouds occurred in the preceding years as well (Schröder 1999). Various shapes of noctilucent clouds are described by Witt (1962) as well as by Bronshten and Grishin (1970). Examples of the structure of noctilucent clouds are given in a number of publications (Witt 1962; Bronshten and Grishin 1970; Gadsden and Schröder 1989; Gadsden and Parviainen 1995).

Fig. 5.51 Example of noctilucent clouds distributed over the sky. The photo was taken at Zvenigorod on July 6, 2000

One of them is shown in Fig. 5.51. The variations of the frequency of appearance of noctilucent clouds within a night in relation to the solar zenith angle, its seasonal variations, and latitude dependence (Figs. 5.52 and 5.53) were investigated by a number of researchers (Fogle 1968; Bronshten and Grishin 1970; Villmann 1970; Gadsden 1998).

Once the altitude at which noctilucent clouds appeared had been determined ($\sim$82 (km)), it became obvious that they are a consequence of the scattering of solar

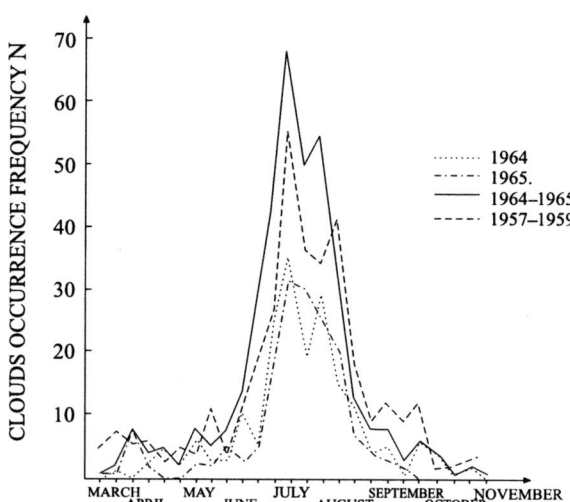

Fig. 5.52 Seasonal variations of the occurrence frequency of noctilucent clouds (Villmann 1970)

5.4 Noctilucent Clouds

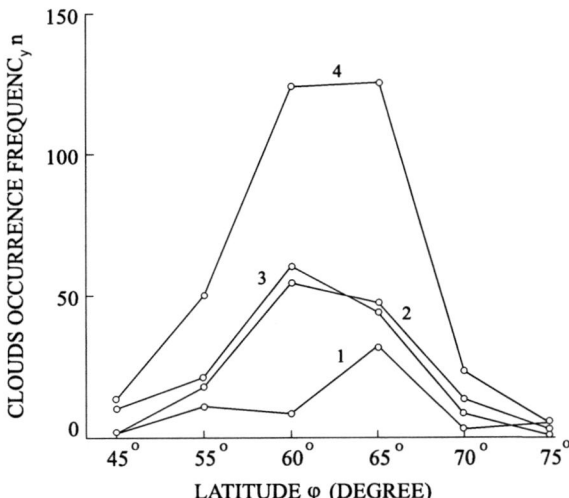

Fig. 5.53 Latitudinal distribution of the number of appearance cases of noctilucent clouds. 1 is 1957; 2 is 1958; 3 is 1959; 4 is 1957–1959 (Bronshten and Grishin 1970)

radiation by an aerosol. In this connection, much effort was directed toward developing a theory of the formation of scattering particles. There was a long-term discussion as to which is the basic agent by which sunlight is scattered – water vapor or dust, and it was understood that the prerequisite to the appearance of noctilucent clouds is the condensation of water vapor on dust particles in the mesopause region at a substantial decrease in temperature (Khvostikov 1956). Direct rocket investigations have confirmed the supposition that condensation of water vapor occurs on dust particles (Hemenwey et al. 1964). Their role in the condensation processes is essential (Gavrilov et al. 1997).

By that time, owing to investigations of the altitude distribution of the atmospheric temperature, it became known that there is a temperature minimum at the altitudes of formation of noctilucent clouds. Thus, statistical data on the frequency of observation of noctilucent clouds at latitudes of 50°–60°N that peaked in the first half of July became the direct demonstration of the seasonal variability of the temperature regime of the middle atmosphere.

Owing to the regular measurements of the hydroxyl emission, and, first of all, of the temperature in the mesopause region, performed at Zvenigorod scientific station of the Institute of Atmospheric Physics, it became possible for the first time to obtain data on the temperature variations before the appearance, during the observation, and after disappearance of noctilucent clouds (Shefov 1965, 1970, 1967, 1968). It was noticed that the appearance of noctilucent clouds was accompanied by enhanced meteor activity. These results have shown that the atmospheric temperature could decrease to 125 (K) in summer during the periods when noctilucent clouds occurred. Subsequently, measurements on Alaska carried out during observations of noctilucent clouds gave temperatures of 150–155 (K) (Taylor et al. 1995).

Long-term measurements of the hydroxyl emission temperature show that in summer (June–July), when the frequency of observations of noctilucent clouds is a maximum, the temperatures less than 150 (K) make 15–20(%) of all measurements (Shefov 1976). The frequency of appearance of noctilucent clouds for the given longitudinal region also makes 20–25(%) (Vasiliev et al. 1975).

Analysis of the long-term observations of noctilucent clouds has revealed the tendency for a long-term increase in frequency of their occurrence, N, on which background the dependence of the variations of N on the 11-yr variability of solar activity clearly shows up (Vasiliev 1970; Vasiliev and Fast 1973; Gadsden 1990, 1998). Gadsden (1998) proposed an approximation of the observed long-term variations of N by the sum of the many-year mean trend (logistic curve) and the sinusoidal component. However, this representation cannot be satisfactory, even because the solar flux variations during an 11-yr cycle are described by aperiodic functions and, moreover, their sequence is not a sinusoidal process (Kononovich 2005). Subsequently, these data were revised and the data of observations in various northwest regions of Europe were considered more correctly. This gave a significant decrease of the long-term trend in the frequency of observations of noctilucent clouds (Gadsden 2002; von Zahn 2003). Nevertheless, its dependence on the 11-yr solar flux was clearly seen in both the earlier and the subsequent data. Therefore, the data reported by Gadsden (1998, 2002) and von Zahn (2003) have been used to compare the long-term variations of the yearly average numbers of observations of noctilucent clouds, N, to the variations of the smoothed yearly average Wolf numbers W in the 20th, 21st, and 22nd 11-yr solar cycles. In doing this, the zero level of the Wolf numbers was made coincident with the level of maximum values of N corresponding to solar flux minima (Fig. 5.54A,B). Smoothed yearly average W has been calculated based on analytic approximations (Kononovich 2005). The scale of Wolf numbers in Fig. 5.54A,B is not referred to any level. Comparison of the deviations ΔN from the level of maximum N and Wolf numbers W is shown in Fig. 5.55A,B. Based on the data reported by Gadsden (1998), these deviations can be approximated by the relation

$$\Delta N = -(6.3 \pm 1) \cdot \exp\left(\frac{W}{100 \pm 20}\right); \quad r = 0.740 \pm 0.086,$$

and based on the data of Gadsden (2002) and von Zahn (2003), by the relation

$$\Delta N = -(4.5 \pm 0.9) \cdot \exp\left(\frac{W}{92 \pm 20}\right); \; r = 0.693 \pm 0.095.$$

As can be seen from Fig. 5.55B, despite some increase in dispersion of points compared to Fig. 5.55A, there is a pronounced dependence of N on solar activity. In this case, there is no regular time shift between the mean solar flux variations and the frequency of appearance of noctilucent clouds which was noted by Romejko et al. (2002, 2003).

The long-term trend of N values corresponding to solar flux minima for the interval 1965–1997 under consideration can be described as a first approximation for the first case by the formula

5.4 Noctilucent Clouds

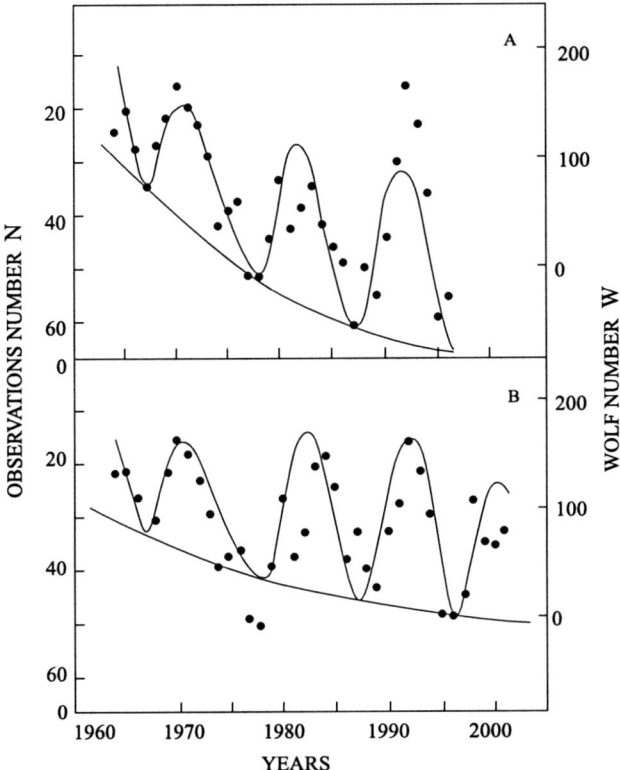

Fig. 5.54 Comparison of the yearly average numbers N of observations of noctilucent clouds according to the data of Gadsden (1998) (**A**) and Gadsden (2002) and von Zahn (2003) (**B**) (*dots*) in relation to yearly average Wolf numbers for the 20th, 21st, and 22nd 11-yr solar cycles (*solid line*). The level of values for W is the line corresponding to maximum values of N associated with solar activity minima

$$N = 21 + 1.7 \cdot (t - 1960) - 1.2 \cdot 10^{-4} \cdot (t - 1960)^{3.3},$$

and for the second one by the formula

$$N = 28 + 0.8 \cdot (t - 1960) - 0.007 \cdot (t - 1960)^2,$$

which, thus, describe the long-term behavior of the humidity at the altitudes of occurrence of noctilucent clouds since there is no temperature trend in summer (Fig. 6.10) (Golitsyn et al. 2000).

The reason for these variations is the increase in water vapor content in the mesopause region, which in turn is determined by the long-term increase in methane content in the terrestrial atmosphere (Thomas et al. 1989). The same process is responsible for the increase in atomic hydrogen in the mesopause region (Semenov 1997; Shefov and Semenov 2002; Shefov et al. 2002). Nevertheless, the long-term statistics of noctilucent clouds did not involve the observations of the atmosphere

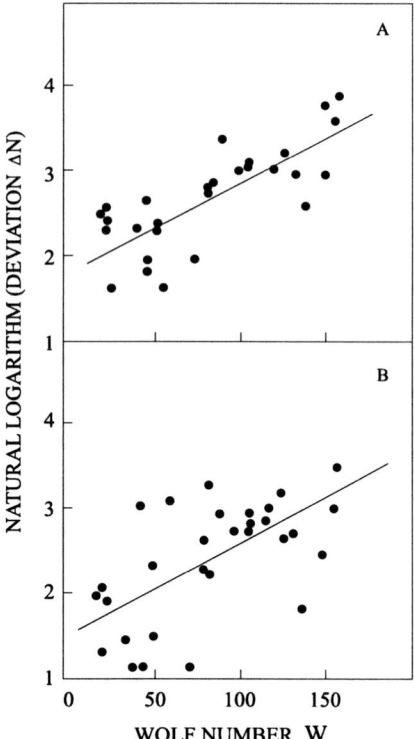

Fig. 5.55 Relation of the natural logarithms of deviations ΔN from the line corresponding to maximum values of N for minima of solar activity with numbers W for the data given in Fig. 5.54A (**A**) and Fig. 5.54B (**B**). The *straight line* is the regression line

at good transparency, but with no noctilucent clouds (Romejko et al. 2002, 2003). When these observations were taken into account, the character of the solar-flux-dependent variations has been revealed. Besides, it turned out that it is necessary to consider the seasonally mean brightness of the clouds, which was estimated with the method proposed by Grishin (1957).

Donahue et al. (1972), based on satellite photometric measurements performed in summer, detected a light-scattering layer at altitudes of ∼84.3 (km) both in the northern and in the southern polar regions of the Earth. The scattered light intensity was much greater than that of light scattered by noctilucent clouds at midlatitudes. This suggests that the atmospheric conditions at these altitudes favored water vapor condensation.

Nevertheless, one of the most important properties of noctilucent clouds was not investigated for a long time. Hines (1968) was the first to pay attention to the wave nature of the structure of noctilucent clouds. This structure arises due to internal gravity waves propagating in the mesopause region. Kropotkina and Shefov (1975) showed that lunar tides promote the increase in frequency of appearance of noctilucent clouds (Fig. 5.56). This was confirmed by Gadsden (1985).

Therefore, it is probable that the occurrence of noctilucent clouds of the first type, which represent an almost homogeneous glow (veil, following Bronshten and Grishin (1970)), is just a consequence of lunar tides which start the vapor

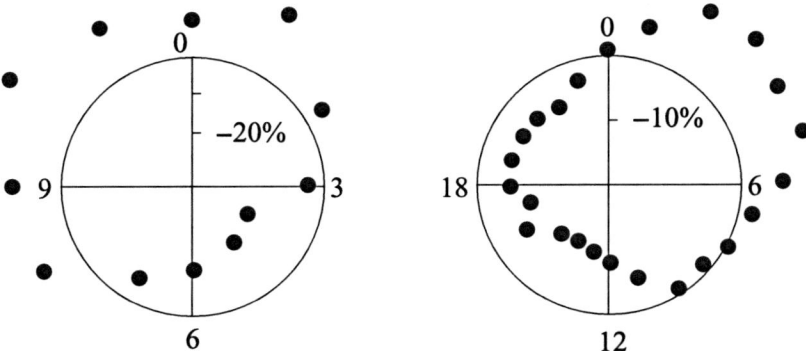

Fig. 5.56 Semidiurnal and diurnal components of the probability of appearance of noctilucent clouds caused by lunar tides about the mean value (*circles*) (Kropotkina and Shefov 1975; Krassovsky and Shefov 1976a). The *points* indicate the mean values corresponding to given hours of lunar time

condensation processes in the atmosphere when the situation becomes favorable for the formation of noctilucent clouds.

Thus, noctilucent clouds arise in the mesopause region at altitudes of 82 ± 1 (km) due to the variations in water vapor condensation processes resulting from the modulation of the atmosphere by propagating internal gravity and planetary waves. Moreover, the wave modulation of the atmosphere at the altitudes where condensation of water vapor is possible is the necessary and sufficient condition for the possibility of visual examination of light scattering in this atmospheric layer. This in turn is related to the presence of wave sources, which are active meteorological structures, such as cold fronts and cyclones. The fact that noctilucent clouds are not observed in the polar region is just related not to lack of conditions for condensation, but to lack of wave sources at high latitudes which would provide for the medium modulation to produce the necessary contrast of scattered light.

A related process to noctilucent clouds is PMSE (Polar Mesosphere Summer Echo) – radar reflections from the mesopause regions, observed in summer, which are due to the ions formed on aerosol particles at these altitudes (Lübken et al. 1996, 1998). Their variations are induced by planetary waves of various time scales (Kirkwood and Réchou 1998; Stepanov 2000).

Data of long-term observations were repeatedly used in seeking longitudinal variations of the frequency of occurrence of noctilucent clouds. However, the average distributions constructed based on data of many years did not show an appreciable nonuniformity. Analysis of the longitudinal distribution of the frequency of occurrence of noctilucent clouds performed by Pavlova (1962) has shown, after some corrections, its possible decrease in moving in the west–east direction.

Now reliable data of various types of measurements are available which point to the manifestation of the properties of the terrestrial landform by processes occurring in the upper atmosphere. First indications of this were revealed by Kessenikh and Bulatov (1944) from data obtained for the ionospheric F2 layer (increased electron density over continents compared to that over oceans). Investigations of the orographic effect in the upper atmosphere point to the appearance of a disturbed region in the lee region of the atmosphere (Sukhodoev et al. 1989a,b; Sukhodoev and Yarov 1998; Shefov et al. 2000a,b) resulting in a lowering of the hydroxyl emission layer. Recently it has been found that the variability of the ozone density in the stratosphere is a manifestation of the terrestrial landform (Kazimirovsky and Matafonov 1998a,b). Thus, the available data suggest that it is more probable that the necessary conditions for the appearance of noctilucent clouds will be provided over continents than over oceans.

Noctilucent clouds were observed in the main at latitudes $50°$–$60°$N, though sometimes they were seen to the south and to the north of this interval. However, since stationary longitudinal distributions have not been certainly revealed, search for variations was performed which could be caused by moving planetary waves. Planetary waves with various periods were detected in various processes in the middle atmosphere at altitudes of 80–100 (km) (Shepherd et al. 1993a,b; Scheer et al. 1994; Laštovička et al. 1994; Kirkwood and Stebel 2003). The amplitude of the hydroxyl emission temperature variations is ~ 5 (K) (Espy et al. 1997), implying the variation of the emission layer altitude by ~ 0.5 (km).

Figure 5.57A presents the geographic distribution of points for which the cases of observation of noctilucent are included in catalogues by Fast (1972, 1980). It can be seen that these stations are ~ 194 in number, though there is some irregularity in their longitudinal distribution (numbers of points n for each $10°$ geographic latitude interval and $20°$ longitude interval are shown in Fig. 5.57B), naturally caused in part by the presence of ocean regions. Not all stations operated simultaneously during several tens of years.

Based on the data of the catalogues composed by Fast (1972, 1980), longitude–time distributions of the number of observations of noctilucent clouds N have been constructed (Fig. 5.57C) in the main for two summer months (June–July) for each year between 1957 and 1978 when the frequency of observations peaked and slightly varied within this period (Shefov and Semenov 2004a). This has been done to reduce the effect of the seasonal variability of the probability of occurrence of noctilucent clouds. The number N is the yearly average number of observations (June–July) in the given interval at the solar flux $F_{10.7} = 124 \pm 50$.

Examination of data for each year has revealed the existence of appreciable longitudinal and time nonuniformities in the frequency of observations of noctilucent clouds. The longitudinal distribution clearly testifies to the concentration of the yearly (June–July) average probability of observations $N_{mac} = 1.6 \pm 0.7$ over continents. Few observations carried out on islands in oceans, however, suggest that events of this type infrequently occur over oceans: $N_{mao} = 0.4 \pm 0.2$ (in Fig. 5.57C the oceans are indicated with broken lines).

The available data, however, do not show a clear effect of the height of the underlying surface landform (schematically shown for latitudes $50°$–$60°$N in Fig. 5.57D). It is natural that the observation data do not involve a possible effect of cloudiness and its longitudinal distribution, which affects the number of observations of noctilucent clouds.

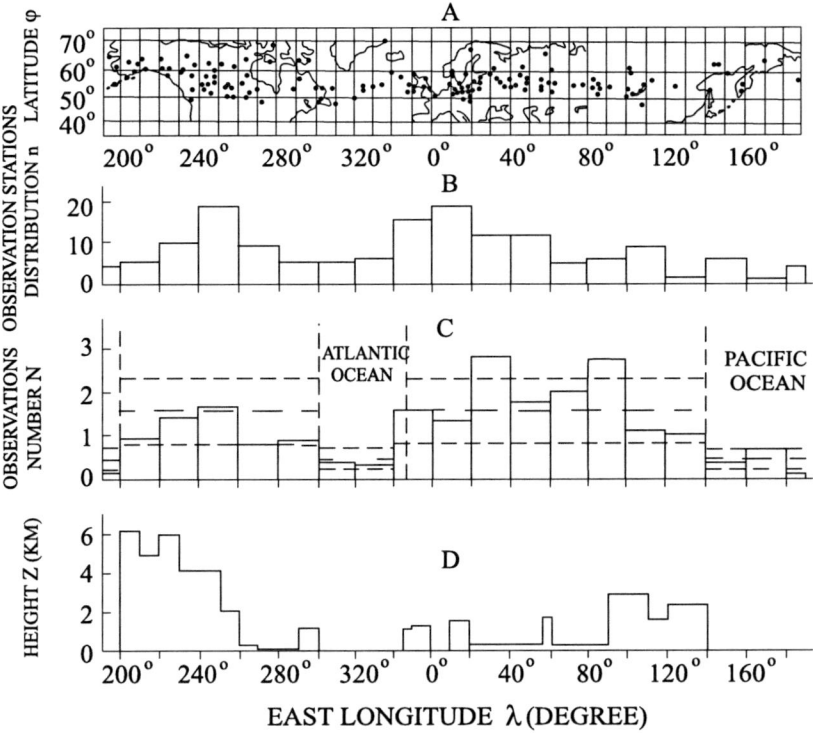

Fig. 5.57 Longitudinal distributions of the conditions of registration of noctilucent clouds (Shefov and Semenov 2004a). The geographic position of the points at which noctilucent clouds were observed (**A**); the longitudinal distribution of the number n of observations points for 20° longitudinal intervals (**B**); the yearly (June–July) average long-term distribution of the frequency of observations of noctilucent clouds, N, in the indicated longitudinal intervals (**C**); the average schematic image of change of height Z of a relief (**D**). *Horizontal dashed lines* indicate the mean values of N over continents and oceans and their variance

To analyze the longitudinal–time behavior of the number of observations of noctilucent clouds for June–July of different years, the numbers N were determined for five nights and 20° intervals of longitudes. These data show that the time variations of the frequency of occurrence of noctilucent clouds during the summer period have two or three peaks, which are detected in different time slices for different intervals of longitudes, and the peaks appear in different longitudinal zones not simultaneously, but with a certain time shift. To obtain statistically valid results, the years were selected for which the data on observation of noctilucent clouds covered a maximum number of longitudinal regions (1966, 1967, 1968, 1972, 1976). By mutual superposition of the time distributions for various longitudinal intervals on the time scale, the cases of best coincidence of their maxima have been selected. The combined distribution is shown in Fig. 5.58.

The use of the data for all years of observation made it possible to state that in summer the region where the probability of appearance of noctilucent clouds is a

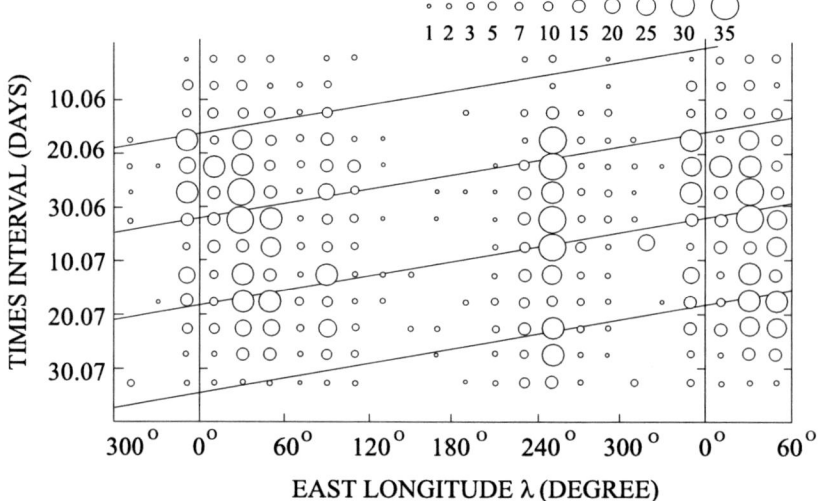

Fig. 5.58 Total longitudinal–time distribution of the frequency of observations of noctilucent clouds for 5-day time intervals in 20° longitudinal intervals (Shefov and Semenov 2004a) The size of a circle corresponds to the number of observations of noctilucent clouds indicated at the top of the figure. *Straight lines* indicate the positions of maxima of the time distributions of the number of observations and correspond to a 16-day period

maximum moves in the east–west direction with a period of 16 days (Shefov and Semenov 2004a).

Analysis of the longitudinal and time distribution of the cases of observation of noctilucent clouds in summer (June–July) for two tens of years has revealed an increase in the probability of occurrence of noctilucent clouds over continents compared to oceans. Besides, the longitudinal distribution of appearance of noctilucent clouds for the same date is determined by the periods of planetary waves which thus modulate the frequency of their appearance in a given longitudinal region. The longitudinal–time distribution of the frequency of appearance of noctilucent clouds points to the presence of planetary waves with a period of 16 days which propagate in the western direction. An interesting case of observation of noctilucent clouds took place near Krasnoyarsk in the evening before the solar eclipse of July 31, 1981, and within 6 (min) during and after the eclipse full phase (Gadsden and Schröder 1989). This points to the fact that this observation period was near the phase of maximum of a planetary wave during which the atmospheric conditions favored the formation of noctilucent clouds. Observations of noctilucent clouds also suggest that there exist waves with a period of 5 days (Kirkwood and Stebel, 2003).

This result is of importance from the viewpoint of obtaining data on the temperature variations in the mesopause under the action of wave processes. Since the data on the long-term behavior of airglow emissions in the upper atmosphere are not as numerous as the observations of noctilucent clouds, these observations are used to gain information about emission processes.

The photos of noctilucent clouds taken for many years offer the possibility, not used up to now, to investigate their wave structure. The methods developed to locate the sources of IGW generation in the troposphere in combination with aerological weather data on the temperature and the wind speed and direction at different altitudes in the troposphere could be helpful in analyzing the meteorological situation accompanying the occurrence of IGWs, just as this was done in investigations of orographic disturbances.

5.5 Spotty Structure of the Airglow Intensity

The salient feature of the airglow is its spotty structure. It was detected and first investigated in the early 1950s based on measurements of the spatial distribution of the intensity of the atomic oxygen emission at 557.7 (nm) (Roach and Pettit 1952; Huruhata 1953; Roach et al. 1958). The results have shown that the spots have varying sizes up to 1000 (km) and move with velocities of 80–100 $(m \cdot s^{-1})$.

The long-term observations carried out at the Vannovsky station near Ashkhabad gave a great body of measurement data presented in a number of publications (Truttse 1965; Korobeynikova et al. 1966, 1968, 1970; Korobeynikova and Nasyrov 1972; Kalchaev et al. 1970; Nasyrov 1967). Figure 4.27 shows an example of the spotty structure of the 557.7-nm emission.

Data of a statistical analysis of the spotty structure of the airglow are presented in Sect. 4.4. It turned out that the behavior of the spot intensity shows a certain regularity. The amplitude of spots relative to the background makes on the average 20–30(%), being sometimes greater. The spots, when moving, vary in size. These properties are due to eddy diffusion (Korobeynikova et al. 1984). First attempts of statistical systematization of the probabilities of the intensity distribution of spots, their sizes, and velocities of motion have shown that they are asymmetric and can be well approximated by a lognormal distribution (Semenov and Shefov 1989). In other words, normal distributions have logarithms of the spot intensity, size, and velocity and of the emission layer thickness.

It should be stressed that theoretical considerations of the processes of diffusion of a passive impurity in a random field of velocities lead to the conclusion that the probability densities of the rates of variation of the observable characteristics of the atmosphere can be described by lognormal distributions (Klyatskin 1994).

However, the spotty structure of the airglow is not only due to the dissipation of IGWs from atmospheric sources but also due to orographic disturbances (Shefov et al. 1999; Nasyrov 2007). This conclusion also follows from the data of observations of the spotty structure of the 630-nm atomic oxygen emission which were carried out on Hawaiian Islands (Roach and Gordon 1973) and also from the measurements of the temperature of this emission over the Andes (Meriwether et al. 1996). The spotty structure of the airglow shows up not only in the spatial distribution of its intensity observed from the Earth surface. Records of the airglow taken from satellites along the limb have shown that there are vertical turbulent motions in the emission layers (Ross et al. 1992).

5.5.1 Recording of the Wave and Spotty Irregularities of the Airglow

The use of an optical instrument having a certain angular aperture for recording various irregularities of emission layers inevitably results in distortions of the parameters of detected phenomena both due to the finite thickness of an emission layer W and an increase in length of a ray in the layer with zenith angle ζ of the line of sight and due to the finite width 2α of the instrument aperture. Thus, the sighted region of the emission layer is an ellipse with a major semiaxis $2a_0$. The degree of smoothing of a wave amplitude or glow irregularity is determined by the parameter $p = \lambda/2a_0$ (λ being the wavelength) or $p = d/2a_0$ (d being the characteristic size of the irregularity) (Shefov 1989).

For a wave described by the expression $\cos\frac{2\pi}{\lambda}L$ which propagates inside an emission layer along the direction L characterized by a zenith angle θ and azimuth $A-A_V$ relative to the azimuth of the line of sight A_V, and also for an intensity spot, the factor μ of decrease (smoothing) of the wave amplitude is determined by expression

$$\mu = \frac{\left|\int_{-\infty}^{\infty}\int_{-1}^{1}\int_{-v'}^{v'} \varphi(w,u,v)\cdot f(w,u,v)\,dv\,du\,dw\right|}{\int_{-\infty}^{\infty}\int_{-1}^{1}\int_{-v'}^{v'} \varphi(w,u,v)\,dv\,du\,dw},$$

where the function f describes the characteristics of the process to be smoothed – a wave or a spot, and φ is a smoothing function. It is determined by the form of the altitude distribution of the emission rate and by the sensitivity distribution of the instrument field of vision. Experimental data show that in some cases the distribution of the emission rate Q(Z) over altitude Z is well approximated by the Gauss formula

$$Q(Z) = \frac{1}{\sqrt{\pi}H_0}\cdot\exp\left[-\frac{(Z-Z_m)^2}{H_0^2}\right],$$

where Z_m is the altitude of maximum emission. The slight asymmetry in this case cannot have a considerable effect. It is more convenient to use the layer thickness W_0 at the level of half the value of Q(Z) instead of the parameter H_0, i.e.,

$$W_0 = 2\sqrt{\ln 2}\cdot H_0.$$

The aperture of the device can be presented analytically by the conventional formula (Miroshnikov 1977)

$$\cos^n\left(\frac{\pi}{2}\cdot\frac{\alpha'}{\alpha}\right),$$

where the parameter n was set equal to 2, close to its actual value, and the running coordinate α' for the geometrical conditions of observations in a spherical atmosphere is determined by the expression

5.5 Spotty Structure of the Airglow Intensity

$$\text{tg}\alpha' = \frac{\sqrt{v^2\left(1 - \frac{\sin^2\beta}{\cos^2\alpha}\right) + (u + \text{tg}\alpha \cdot \text{tg}\beta)^2 \cdot \cos^2\beta}}{\frac{1 - \frac{\sin^2\beta}{\cos^2\alpha}}{\text{tg}\alpha \cdot \cos\beta} + (u + \text{tg}\alpha \cdot \text{tg}\beta) \cdot \sin\beta}.$$

Here u and v are normalized variables along the major and minor axes of the sighted ellipse, respectively,

$$w = \frac{Z - Z_m}{H_0}, \quad v' = \sqrt{1 - u^2}, \quad \sin\beta = \frac{1}{1 + \frac{Z}{R_E}} \cdot \sin\zeta,$$

where R_E is the Earth's radius.

Thus, the smoothing weight function is determined by the product

$$\cos^2\left(\frac{\pi}{2} \cdot \frac{\alpha'}{\alpha}\right) \cdot \exp(-w^2).$$

However, the wave amplitude and the spot size do not remain constant on varying altitude. Since the atmospheric density decreases with as increase in altitude, the wave amplitude is proportional to (Gossard and Hooke 1975)

$$\exp\left(\frac{Z - Z_m}{2H}\right) = \exp\left(\frac{H_0}{2H}w\right),$$

where H is the scale height.

Analysis (Korobeynikova et al. 1984) of the observed spotty irregularities of the 557.7-nm emission intensity by the data of earlier measurements (Korobeynikova and Nasyrov 1972) has shown that on the average the radial distribution of the intensity in a spot of radius ρ seems to be determined by horizontal eddy diffusion and is described by Einstein's formula (Landau and Lifshits 1986)

$$\exp\left(-\frac{\rho^2}{\rho_0^2}\right) = \exp\left(-\frac{\rho^2}{4K_H \cdot t}\right),$$

where t is time.

For the lower thermospheric region (80–100 (km)), the altitude dependence of the coefficient of horizontal eddy diffusion, K_H, can be determined by the empirical relation (Korobeynikova et al. 1984)

$$K_H = \left(\frac{Z}{56} - 1\right)^2 (10^6 \text{m}^2 \cdot \text{s}^{-1}) = \left(\frac{wH}{Z_m - 56} + 1\right)^2 (\text{km}^2 \cdot \text{s}^{-1}).$$

The function f_W which describes the characteristics of a wave within the sighted part of the layer has the form

$f_W = \cos(w, u, v)$

$$= \cos\left\{\frac{\pi}{p} \cdot \frac{a}{a_0} \cdot \left[\left[\left(w \cdot \frac{H_0}{a} \cdot \text{tg}\beta + \text{tg}\alpha \cdot \text{tg}\beta - \frac{a_0}{a} \cdot \text{tg}\alpha \cdot \text{tg}\beta_0\right) \cdot \cos(A - A_V) + \right.\right.\right.$$
$$\left.\left.+ \sqrt{u^2 + v^2\left(1 - \frac{\sin^2\beta}{\cos^2\alpha}\right)} \cdot \cos\left(A - A_V - \arctan\frac{v\sqrt{1 - \frac{\sin^2\beta}{\cos^2\alpha}}}{u}\right)\right]$$
$$\left.\left.\cdot \sin\theta + w \cdot \frac{H_0}{a} \cdot \cos\theta\right]\right\},$$

where $a = r \cdot \frac{\text{tg}\alpha \cdot \cos\beta}{1 - \frac{\sin^2\beta}{\cos^2\alpha}}$, and the distance along the line of sight is determined by

$$r = R_E \left[\sqrt{\left(1 + \frac{Z}{R_E}\right)^2 - \sin^2\zeta} - \cos\zeta\right].$$

For an emission intensity spot, we have

$$f_{SP} = \exp\psi(w, u, v) = \exp\left\{\frac{1}{p^2} \cdot \frac{a^2}{a_0^2} \cdot \left[\left(w \cdot \frac{H_0}{a} \cdot \text{tg}\beta + \text{tg}\alpha \cdot \text{tg}\beta - \frac{a_0}{a} \cdot \text{tg}\alpha \cdot \text{tg}\beta_0 + u\right)^2\right.\right.$$
$$\left.\left.+ v^2 \cdot \left(1 - \frac{\sin^2\beta}{\cos^2\alpha}\right)\right] \cdot \frac{1}{\left(1 + \frac{wH_0}{Z-56}\right)^2}\right\}.$$

Here β_0 is the angle between the line of sight and the vertical at the observation place at the altitude Z_m of maximum emission.

The results of calculations by the above formulas for the conditions corresponding to the observations (zenith angles ζ) of an orographic effect near the Caucasian ridge ($2\alpha = 4°$, $Z_m = 90$ (km), $H = 5.64$ (km), $W_0 = 13$ (km), $A-A_V = 45°$, $\theta = 90°$) (Sukhodoev et al. 1989a,b) are given in Fig. 5.59. It can be seen that for the mentioned parameters, typical lengths of suppressed waves are less than 8 (km) for observations at the zenith and can reach 100 in sighting at a zenith angle of 75°. For a given smoothing coefficient μ, the wavelengths is related to the zenith angle as

$$\lambda_\zeta = \lambda_0 \cdot 10^{\frac{\zeta}{65}}.$$

For spots, the smoothing also becomes more efficient with increasing zenith angle. Comparison of the calculation results with the observation data on the amplitudes obtained with an eight-channel photometer for the same waves (Sukhodoev

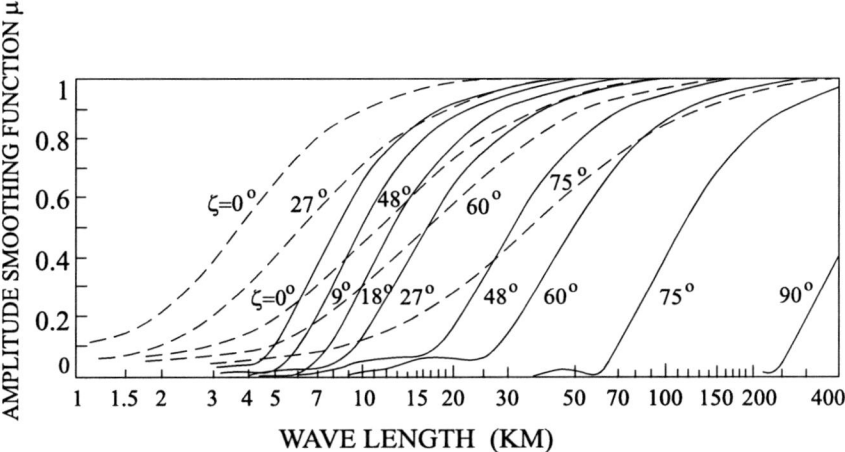

Fig. 5.59 Dependence of the function of smoothing of wave amplitude μ on wavelength λ (*solid lines*) and on spot size d (*dashed lines*) for different zenith angles ζ of the line of sight (Shefov 1989)

et al. 1989a,b) has shown good correlation ($\sim$0.8) between the calculated and measured μ values for seven sections of the emission layer.

5.6 Winds

In a consideration of characteristics of wind flows, the terminological definition of wind direction is of importance. In the meteorological literature, wind direction is reported by the direction from which it originates. This is a physically absolutely inconvenient definition. Therefore, in the present consideration of the problems of winds interacting with atmospheric constituents (no matter whether in the lower or in the upper atmosphere) by wind direction everywhere is implied the direction of the wind velocity vector, i.e., the direction in which the wind blows.

Wind motions in the upper atmosphere are a manifestation of irregular energy influx and propagation of waves of different origin. Since the energy influx depends on the altitude above the Earth surface, the wind system is responsible for the global atmospheric spatial–temporal circulation that has a strong effect on the rates of the photochemical processes occurring in the atmosphere (Krassovsky and Shefov 1978, 1980). Here it is necessary to distinguish between two types of the effect of atmospheric circulation. The first one refers to wave processes induced in the lower atmosphere – from IGWs to tidal and planetary waves transferring energy upward. These phenomena have already been considered in the previous sections. The second type is related to wind phenomena which take place directly at the altitudes of emission layers and enhance the mixing of chemically active atmospheric constituents. In these cases, measurements of emission characteristics (Doppler velocities in interferometry) allow one to determine wind speeds. The altitude distributions of wind directions, speeds, and their variations (Roper et al. 1993; Wang et al. 1997; Rajaram and Gurubaran 1998; Liu et al. 2001) testify that emission layers are present very frequently at the boundaries of the altitude distributions of wind speed jumps.

The methods of most extensive measurements of wind velocity are based on tracing the motions of ionized structures in the upper atmosphere. At the altitudes of the lower thermosphere, this is performed by measuring meteor trails with radars (Kashcheev is devoted et al. 1967; Portnyagin et al. 1978; D'yachenko et al. 1986; Fakhrutdinova 2004). Special ionospheric methods are applied at the altitudes of the ionospheric F2 layer of the thermosphere (Kazimirovsky and Kokourov 1979). At altitudes above 100 (km), artificial shone clouds were used (see Sect. 3.5.9) (Andreeva et al. 1991).

The optical technique mainly used for determining wind speed is related to the interferometric method of estimation of the radial wind velocity by the line of sight of its velocity in a emission layer. For this purpose, actually only the 557.7-nm and 630-nm atomic oxygen emissions are used. The features of these methods have been discussed in Chap. 3.

5.6.1 Relation of the Variations of Hydroxyl Airglow Characteristics to the Tropospheric Wind

We now consider manifestations of direct correlations between the airglow and the wind conditions in the lower atmosphere. In the late 1960s (Savada 1965) and subsequently (Kagan and Shkutova 1985), it was supposed that semidiurnal lunar tides in the ocean make a significant contribution to the semidiurnal lunar oscillations throughout the atmospheric depth. The Scandinavian ridge is an essential obstacle to such tidal streams from the Atlantic (Northern sea). Therefore, it can be expected that the winds originating at the mountain boundary of the sea disturb the amplitude and phase of semidiurnal oscillations of the atmosphere not only at its base but also at altitudes near the mesopause. To reveal this effect, experimental data on the temperature and wind variations near the mesopause were considered in relation to the surface wind near the Scandinavian mountains.

With that end in view, measurements of the hydroxyl rotational temperature were performed by the OH (9–3) band at the Zvenigorod station. The time of one temperature measurement was 5 (min). Observations were carried out near a new moon for several hours. From the observation data, 25 values of mesopause temperature averaged over three-hour time slices were taken. The center of these intervals fell at 22:30 (LT). The temperature values were divided into two groups according to the wind direction over the Scandinavian mountains, which was determined from synoptic maps at a level 700 (mbar) ($\sim$3 (km)) at a point with coordinates (64°N, 10°E). The first group included data about the mesopause temperature for the wind over Scandinavia directed toward Moscow. The second group of data corresponded to the wind in which there was no velocity vector component directed to Moscow. In most cases, to this group there corresponded the wind directed along the Scandinavian ridge to its northern end.

It is well known that a variation in atmospheric temperature caused by an internal gravity wave (IGW) or a tidal wave is accompanied by a corresponding variation in speed of the horizontal wind. This dependence is described by the expression (Hines 1974; Myrabø 1984)

$$\frac{\Delta T}{T} \approx \pm i \cdot \sqrt{\gamma - 1} \cdot \frac{\Delta V}{C},$$

where ΔT and ΔV are the deviations of the temperature and wind speed from their mean values, T is the mean temperature for the period of measurements, C is the sound velocity at the temperature T, γ is the heat capacity ratio, and i is a quantity which characterizes the phase shift between the temperature and the wind speed variations. For harmonic oscillations, ΔT and ΔV can be replaced by their amplitude values A_T and A_V. Near the mesopause at midlatitudes for winter the mean temperature is about 210 (K) (CIRA-1972) and the mean sound velocity is about 290 (m · s^{-1}). Thus, the previous expression can be represented as

$$A_T \approx \pm i \cdot 0.46 \cdot A_V.$$

The observed variations in mesopause temperature were related to the wind speed variations obtained from radiometeoric measurements performed at Obninsk ($\sim$70 (km) away from Zvenigorod). A correlation between these quantities has been revealed which can be described as

$$\Delta T = T - T_0 = \alpha \cdot V_{Obn}$$

or

$$A_T = \rho \cdot A_{Obn},$$

where T is the mesopause temperature measured at Zvenigorod and T_0 is its value at $V_{Obn} = 0$, where V_{Obn} is the wind speed in the meteor zone, according to Obninsk measurements, A_T and $A_{v(Obn)}$ are the amplitude values of the temperature variations (Zvenigorod) and wind variations (Obninsk), and ρ is the regression coefficient.

It is well known that the variations of hydroxyl emission rotational temperature are affected by lunar tides whose magnitude depends on lunar time (Shefov 1974a,b). All temperature variations observed at Zvenigorod corresponded to the following period of Moon's phases: the last quarter – a new moon – the first quarter. The measured values of V_{Obn} were averaged over three-hour intervals for one-day periods before and after 22:30 (LT) for each date of temperature measurements at Zvenigorod. The coefficients of series for V_{Obn} were correlated with the coefficients of temperature series separately for the mentioned two groups formed in accordance with the wind direction in the region of the Scandinavian ridge. Regression coefficients were calculated for the T and V_{Obn} time series with a time shift of 3k hours relative to the chosen midtime 22.30 (LT), with k taking integer values -8, -7, -6, ..., 6, 7, 8. For the obtained correlation formulas, the correlation coefficient r varied between $+0.3$ and $+0.8$. The regression function $\rho(\Delta t)$, representing the dependence of the regression coefficient ρ on the time shift Δt, was subjected to Fourier analysis to reveal tidal oscillations. The resulting oscillation spectrum contained a semidiurnal component. Simultaneously, harmonic analysis was applied to radiometeoric measurements (Obninsk) to reveal the amplitudes and phases of the 24-, 12-, and 8-h components of wind variations. For these variations, the semidiurnal harmonic was predominant as well. Using the harmonic analysis data on the amplitude of the semidiurnal component, $A_{v(Obn)}$, and on the regression coefficient ρ obtained by correlation analysis in the above expression, the amplitude value of the temperature of the semidiurnal lunar-tidal component has been obtained (Semenov 1987).

The same procedure was used to process data on the correlations between the variations of the radiometeoric wind V_{Obn} at Obninsk and the variations of the winds over Scandinavia, V_S, and Moscow, V_M, which were taken from synoptic bulletins for 700-mbar levels. As before, processing was conducted separately for the two conditions of wind motions over the Scandinavian ridge. The maximum values of the coefficients of correlations of the T, V_S, and V_M series with the radiometeoric wind (Obninsk), to which there corresponded certain regression coefficients ρ used to find the amplitudes of the semidiurnal harmonics of T, V_S, and V_M, varied in the range 0.6–0.9.

The results obtained show that the direction of the wind over Scandinavia noticeably affects the semidiurnal lunar amplitudes and phases of the wind measured by the radiometeoric method at Obninsk and of the mesopause temperatures (Zvenigorod). This fact supports the supposition that semidiurnal lunar tides in the ocean play a considerable part in the global atmospheric tidal oscillations. It is especially interesting in view of the fact that marine tides, collapsing at the Scandinavian coast, can influence the wind blowing over the Scandinavian mountains. Clearly, the amplitude of the wind over the mountains, V_S, in the first group of data is greater than in the second one.

5.6.2 Wind Parameters Derived from Optical Measurements

Investigations of wind velocities in the lower thermosphere (100 (km)) and in the region of the ionospheric F2 layer by means of interferometric measurements of the Doppler profiles of the 557.7-nm and 630-nm atomic oxygen emissions present a highly complicated engineering problem. It is required not only to develop and build a precision optical instrument but also to provide its stable operation and calibration and develop methods for processing records to be obtained (see Chap. 3). This is why observations of this type are carried out at a limited number of points. The relevant studies are presented in detail elsewhere (Hernandez and Roble 1976;

Yugov and Ignatiev 1983; Semenov 1982, 1985; Hernandez and Killeen 1988; Hernandez et al. 1992; Ignatiev and Yugov 1995).

Systematization of the main characteristics of diurnal wind variations derived from 630-nm emission data has made it possible to gain a general idea of the behavior of night variations under quiet geomagnetic conditions for the zonal (positive west–east direction) and the meridional (positive south–north direction) wind components. Analysis of measurement data shows that at altitudes of 200–400 (km) the wind blows in the west–east direction between 18:00 and 24:00 (LT) and in the east–west direction between 04:00 and 12:00 (LT). Thus, the wind speed is $\sim 160\,(\text{m} \cdot \text{s}^{-1})$ at altitudes of about 240 (km) between 01:00 and 02:00 (LT). Such a wind is most intense at the equator. The variations of the zonal wind speed within a day depend on latitude.

The daily variations also depend on solar flux. For instance, during a period of high solar activity at altitudes of ~ 240 (km) at latitudes near $42°$N, the west–east (W $\rightarrow$ E) zonal wind was over $100\,(\text{m} \cdot \text{s}^{-1})$ between 18:00 and 01:00 (LT). At low solar activity, it was over $100\,(\text{m} \cdot \text{s}^{-1})$ between 19:00 and 23:00 (LT) (Hernandez and Roble 1976).

The meridional wind after midnight near the equator and at midlatitudes is directed from the pole to the equator, i.e., N $\rightarrow$ S in the northern hemisphere and S $\rightarrow$ N in the southern hemisphere (Meriwether et al. 1997).

Investigations of wind motions by the 557.7-nm emission show a substantial seasonal variability of the wind direction. In the second half of night, within approximately three hours, there is a wind blowing from the east to the west (E $\rightarrow$ W) in September and from December to April. In the rest of the daytime, the wind is directed west to east (W $\rightarrow$ E) throughout the year (Phillips et al. 1994; Fauliot et al. 1995, 1997).

References

Andreeva LA, Kluev OF, Portnyagin YuI, Khananiyan AA (1991) Studies of the upper atmospheric processes by the artificial luminous cloud method. Hydrometeoizdat, Leningrad
Armstrong EB (1975) The influence of a gravity wave on the airglow hydroxyl rotational temperature at night. J Atmos Terr Phys 37:1585–1591
Barat J, Blamont JE, Petitdidier M, Sidi C, Teitelbaum M (1972) Mise on évidence expérimentale d'une structure inhomogène à petite échelle dans la couche émissive de l'oxygène atomique à 5577. Ann Geophys 28:145–148
Bates DR (1978) On the proposal of Chapman and Barth for O (^{1}S) formation in the upper atmosphere. Planet Space Sci 27:717–718
Berg MA, Shefov NN (1963) OH emission and atmospheric O_2 band λ 8645 A. In: Krassovsky VI (ed) Aurorae and airglow. N 10. USSR Acad Sci Publ House, Moscow, pp 10–23
Bertin F, Testud J, Kersley L (1975) Medium scale gravity waves in the ionospheric F-region and their possible origin in weather disturbances. Planet Space Sci 23:493–507
Brodhun D, Bull G, Neisser J (1974) On the identification of tropospheric sources of gravity waves observed in the mesosphere. Zeits Meteorol 24:299–308
Bronshten VA, Grishin NI (1970) Noctilucent clouds. Nauka, Moscow

Bull G, Dubois R, Neisser J, Stangenberg JG (1981) Untersuchengen über Schwerewellen in Gebirgsnabe. Zeits Meteorol 31:267–279

Campbell IM, Gray CN (1973) Rate constant for oxygen O (^{1}P) recombination and association with N (^{4}S). Chem Phys Lett 18:607–609

Chamberlain JW (1978) Theory of planetary atmospheres. Academic Press, New York

Christophe-Glaume J (1965) Étude de la raie 5577 A de l'oxygène dans la luminescence atmospherique nocturne. Ann Géophys 21:1–57

Chubukov VP (1977) The paths of gravity waves reaching the mesopause. In: Krassovsky VI (ed) Aurorae and airglow. N 25. Soviet Radio, Moscow, pp 18–22

Chunchuzov IP (1988) Orographic waves in the atmosphere produced by a varying wind. Izv USSR Acad Sci Atmos Oceanic Phys 24:5–12

Chunchuzov IP (1994) On possible generation mechanism for nonstationary mountain waves in the atmosphere. J Atmos Sci 15:2196–2206

Chunchuzov YeP (1978) On energy balance characteristics of the internal gravity waves observed from hydroxyl emission near the mesopause. Izv USSR Acad Sci Atmos Oceanic Phys 14:770–772

Chunchuzov YeP (1981) Internal gravity waves near mesopause. V. Use of the Earth's atmosphere oscillation theory. In: Galperin YuI (ed) Aurorae and airglow. N 29. Radio and Svyaz, Moscow, pp 44–58

Chunchuzov YeP, Shagaev MV (1983) Estimates of the coefficient of vertical turbulent diffusion in the lower thermosphere. Izv USSR Acad Sci Atmos Oceanic Phys 20:154–155

Chunchuzov YeP, Shefov NN (1978) Propagation range of internal gravity waves in the mesosphere. Izv USSR Acad Sci Atmos Oceanic Phys 14:849–850

CIRA-1972. COSPAR International Reference Atmosphere. Stickland AC (ed) Akademie-Verlag, Berlin

Collins R, Nomura A, Gardner C (1994) Gravity waves in the upper mesosphere over Antarctica: lidar observations at South Pole and Syowa. J Geophys Res 99D:5475–5485

DeVries LL (1972) Structure and motion on the thermosphere shown by density data from the low-q accelerometer calibration system (logas). In: Bowhill SA, Jaffe LD, Rycroft MJ (eds) Space Res, vol 12. Akademie-Verlag, Berlin, pp 867–879

Dewangan DP, Flower DR, Danby G (1986). Rotational excitation of OH by H_2: a comparison between theory and experiment. J Phys B 19:L747–L753

Dickinson RE (1969) Planetary waves and large-scale disturbances in the stratosphere and mesosphere. In: Sechrist CF (ed) Meteorological and chemical factors in D – region aeronomy – record of the third aeronomy conference (USA, Illinois, September 23–26, 1968). Univ Illinois, Urbana, pp 80–87

Donahue TM, Guenter B, Blamont JE (1972) Noctilucent clouds in daytime circumpolar particulate layers near the summer mesopause. J Atmos Sci 29:1205–1209

D'yachenko VA, Lysenko IA, Portnyagin YuI (1986) Climatic regime of the lower thermospheric wind. Obninsk: All-Union Sci Res Institute Hydrology-Meteorology Information, International Data Centre

Ebel A (1980) Eddy diffusion models for mesosphere and lower thermosphere. J Atmos Terr Phys 42:102–104

Eckart C (1960) Hydrodynamics of the oceans and atmospheres. Pergamon Press, Oxford

Espy PJ, Stegman J, Witt G (1997) Interannual variations of the quasi-16-day oscillation in the polar summer mesospheric temperature. J Geophys Res 102D:1983–1990

Fakhrutdinova AN (2004) A circulation of the mesosphere-lower thermosphere of the middle latitudes. Teptin GM (ed) Kazan State University, Kazan

Fast NP (1972) A catalogue of the noctilucent cloud appearances according to the world data. Tomsk Univ Publ House, Tomsk

Fast NP (1980) A catalogue of the noctilucent cloud appearances according to the world data. Tomsk Univ Publ House, Tomsk

Fauliot V, Thuillier G, Hersé M (1995) Observations of the E-region horizontal winds in the auroral zone and at mid-latitudes by a ground-based interferometer. Ann Geophys 13:1172–1186

Fauliot V, Thuillier G, Vial F (1997) Mean vertical wind in the mesosphere-lower thermosphere region (80–120 (km)) deduced from the WINDII observations on board UARS. Ann Geophys 15:1221–1231

Fedorova NI (1967) Hydroxyl emission at high latitudes. In: Krassovsky VI (ed) Aurorae and airglow. N 13. USSR Acad Sci Publ House, Moscow, pp 22–36

Fishkova LM (1983) The night airglow of the Earth mid-latitude upper atmosphere. Shefov NN (ed) Metsniereba, Tbilisi

Fogle B (1968) The climatology of noctilucent clouds according to observations made from North America during 1964–66. Meteorol Mag 97:193–204

Fogle B (1971) Noctilucent clouds – their characteristics and interpretation. In: Webb WL (ed) Thermospheric circulation. The MIT Press, Cambridge, pp 95–107

Gadsden M (1985) Observations of noctilucent clouds from north–west Europe. Ann Geophys 3:119–126

Gadsden M (1990) A secular changes in noctilucent clouds occurrence. J Atmos Terr Phys:52:247–251

Gadsden M (1998) The north–west Europe data on noctilucent clouds: a survey. J Atmos Solar-Terr Phys 60:1163–1174

Gadsden M (2002) Statistics of the annual counts of nights on which NLCs were seen. Memoris British Astron Assoc, vol 45, Aurora section (meeting "Mesospheric clouds", Perth, Scotland, 19–22 August).

Gadsden M, Schröder W (1989) Noctilucent clouds. Springer-Verlag, New York

Gadsden M, Parviainen P (1995) Observing noctilucent clouds. Kall Kwik Centre, Perth, Scotland

Gaigerov SS (1973) Investigation of the synoptic processes in the atmospheric high layers. Bugaev VA (ed) Hydrometeoizdat, Leningrad

Gage KS, Nastrom GD (1989) A simple model for the enhanced frequency spectrum of vertical base on tilting of atmospheric layers by lee waves. In: Liu CH, Edwards B (eds) Handbook for MAP, vol 28. SCOSTEP, Urbana, pp 292–298

Gardner CS, Voelz DG (1985) Lidar measurements of gravity wave saturation effects in the sodium layer. Geophys Res Lett 12:765–768

Gavrilov AA, Kaidalov OV, Kazannikov AM (1997) Mesospheric clouds and their connection to the mesopause properties and the influx of meteoric substance. Solar Syst Res 31:472–479

Gavrilov NM (1974) Thermal effect of internal gravity waves in the upper atmosphere. Izv USSR Acad Sci Atmos Oceanic Phys 10:83–84

Gavrilov NM (1987) Study of internal gravity waves in meteor zone. In: Poper RP (ed) Handbook for MAP, vol 25. SCOSTEP, Urbana, pp 153–166

Gavrilov NM (1992) Internal gravity waves in the mesopause region: hydrodynamical sources and climatological patterns. Adv Space Res 12:113–121

Gavrilov NM (1996) Intensity structure of ion-temperature perturbations in the thermosphere. Geomagn Aeronomy 36:124–128

Gavrilov NM, Delov IA (1976) Measurement of parameters of internal gravitational waves in meteor zone. Geomagn Aeronomy 16:293–297

Gavrilov NM, Roble R (1994) The effect of gravity waves on the global mean temperature and composition structure of the upper atmosphere. J Geophys Res 99D:25773–25780

Gavrilov NM, Richmond AD, Bertin F, Lafenile M (1994) Investigation of seasonal and interannual variations of internal gravity wave intensity in the thermosphere over Saint Santin. J Geophys Res 99A:6297–6306

Gavrilieva GA, Ammosov PP (2001) Observations of gravity wave propagation in the all-sky infrared airglow. Geomagn Aeronomy 41:363–369

Georges TM (1968) HF Doppler studies of traveling ionospheric disturbances. J Atmos Terr Phys 30:735–746

Golitsyn GS, Chunchuzov EP (1975) Acoustic-gravity waves in the atmosphere. In: Krassovsky VI (ed) Aurorae and airglow. N 23. Nauka, Moscow, pp 5–21

Golitsyn GS, Romanova NN (1968) Vertical propagation of sound in an atmosphere with the height variable viscosity. Izv USSR Acad Sci Atmos Oceanic Phys 4:118–120

Golitsyn GS, Semenov AI, Shefov NN (2000) Seasonal variations of the long-term temperature trend in the mesopause region. Geomagn Aeronomy 40:198–200

Gossard EE, Hooke WH (1975) Waves in the atmosphere. Elsevier Scientific Publishing Company, Amsterdam

Grishin NI (1957) Manual to observations of the noctilucent clouds. USSR Acad Sci Publ House, Moscow

Gupta AB, Nagpal OP (1973) F_2-region ionospheric response to atmospheric gravity waves. Ann Geophys 29:307–319

Gurvich AS, Vorobiev VV, Savchenko SA, Pakhomov AI, Padalka GI, Shefov NN, Semenov AI (2002) The 420–530 (nm) region nightglow of the upper atmosphere as measured onboard Mir research platform in 1999. Geomag Aeronomy 42:514–519

Hapgood MA, Taylor MJ (1982) Analysis of airglow image data. Ann Geophys 38:805–813

Hatfield R, Tuan TF, Silverman SM (1981) On the effects of atmospheric gravity waves on profiles of H, O_3 and OH emission. J Geophys Res 86A:2429–2437

Haurwitz B (1971) Wave forms in noctilucent clouds. In: Webb WL (ed) Thermospheric circulation. The MIT Press, Cambridge, pp 95–107

Heislet LH (1963) Observation of movement of perturbances in the F-region. J Atmos Terr Phys 25:72–86

Hemenwey CL, Soberman RK, Witt G (1964) Sampling of noctilucent cloud particles. Tellus 16:84–88

Hernandez G, Killeen TL (1988) Optical measurements of winds and kinetic temperatures in the upper atmosphere. Adv Space Res 8:149–213

Hernandez G, Roble RG (1976) Direct measurements of nighttime thermospheric winds and temperatures. I. Seasonal variations during geomagnetic quiet periods. J Geophys Res 81:2065–2074

Hernandez G, Smith RW, Conner JF (1992) Neutral wind and temperature in the upper mesosphere above South Pole, Antarctica. Geophys Res Lett 19:53–56

Hines CO (1960) Internal atmospheric gravity waves at ionospheric heights. Can J Phys 38:1441–1481

Hines CO (1965) Dynamical heating of the upper atmosphere. J Geophys Res 70:177–183

Hines CO (1968) A possible source of waves in noctilucent clouds. J Atmos Sci 25:937–942

Hines CO (1974) The upper atmosphere in motion. Heffernan Press, Worcester Massachusetts

Hirota I (1984) Climatology of gravity wave in the middle atmosphere. J Atmos Terr Phys 46:767–773

Holton J (1972) An introduction to dynamic meteorology. Academic Press, New York

Hung RJ (1977) Observation of upper atmospheric disturbances caused by hurricanes and tropical storms. In: Rycroft MJ, Stickland AC (eds) Space Res, vol 17. Pergamon Press, Oxford, pp 205–210

Hung RJ, Smith RE (1977) Study of stratospheric – ionospheric coupling during thunderstorms and tornados. In: Rycroft MJ, Stickland AC (eds) Space Res, vol 17. Pergamon Press, Oxford, pp 211–215

Huruhata M (1953) Photoelectric studies of the night sky light. Ann Tokyo Astron Observ 3:165–174

Ignatiev VM, Yugov VA (1995) Interferometry of the large-scale dynamics of the high-latitudinal thermosphere. Shefov NN (ed) Yakut Sci Centre Siberian Branch RAN, Yakutsk

Izakov MN (1978) On the turbulence influence on the thermal regime of the planet thermospheres. Cosmic Res 16:403–411

Jacobi Ch (1998) On the solar cycle dependence of winds and planetary waves as seen from mid–latitude D1 LF mesopause region wind measurements. Ann Geophys 16:1534–1543

Johnson FS (1975) Transport processes in the upper atmosphere. J Atmos Sci 32:1658–1662

Kagan BA, Shkutova NV (1985) On the affect of ocean tides on gravitational tides in the atmosphere. Oceanology 25:193–200

Kalchaev KK, Korobeynikova MP, Nasyrov GA, Khamidulina VG (1970) Search of the rapid spatial variations of the intensity of the oxygen green line. In: Krassovsky VI (ed) Aurorae and airglow. N 18. Nauka, Moscow, pp 15–17

Kalov YeD, Gavrilov NM (1985) Investigation of seasonal changes of gravity wave parameters in the meteor zone. Izv USSR Acad Sci Atmos Oceanic Phys 21:791–795

Karimov KA, Lukyanov AY (1979) Statistical characteristics of internal gravity waves in the meteor zone. Izv USSR Acad Sci 15:604–606

Kashcheev BL, Lebedinets BN, Lagutin MF (1967) Meteor phenomena in the Earth's atmosphere. Nauka, Moscow

Kaye JA (1988) On the possible role of the reaction $O + HO_2 \rightarrow OH + O_2$ in OH airglow. J Geophys Res 93:285–288

Kazannikov AM (1981) On the wind influence on the internal gravity wave trajectories. Geomagn Aeronomy 21:369–371

Kazannikov AM, Portnyagin YuI (1981a) On some results of study internal gravity waves by using radiometeor observation data. Izv USSR Acad Sci Atmos Oceanic Phys 17:70–72

Kazannikov AM, Portnyagin YuI (1981b) Characteristics of the small-scale movements in the meteor zone. Geomagn Aeronomy 21:371–372

Kazimirovsky ES, Kokourov VD (1979) Ionospheric movements. Erofeev NM (ed) Nauka, Novosibirsk

Kazimirovsky ES, Matafonov GK (1998a) The continental and orographic "structures" in the global distribution of the total ozone content. Dokl Earth Sci 361:544–546

Kazimirovsky ES, Matafonov GK (1998b) Continental scale and orographic "structures" in the global distribution of the total ozone content. J Atmos Solar-Terr Phys 60:993–995

Kessenikh VN, Bulatov ND (1944) Continental effect in the geographic distribution of the electronic concentration of the F_2 layer. Dokl USSR Acad Sci 45:250–254

Khvostikov IA (1956) On the nature of the noctilucent clouds. Izv USSR Acad Sci Ser Geophys N 7:869–871

Kirkwood S, Réchou A (1998) Planetary-wave modulation of PMSE. Geophys Res Lett 25:4509–4512

Kirkwood S, Stebel K (2003) Influence of planetary waves on noctilucent cloud occurrence over N W Europe. J Geophys Res 108D:8440. doi:10.1029/2002JD002356

Klyatskin VI (1994) Statistical description of the diffusion of tracers in a random velocity field. Uspekhi Phys Nauk 164:531–544

Kolesnikova VK, Monin AS (1965) Fluctuations of meteorological field spectra. Izv USSR Acad Sci Atmos Oceanic Phys 1:377–386

Kononovich EV (2005) Analytical representations of mean solar activity variations during a cycle. Geomagn Aeronomy 45:295–302

Korobeynikova MP, Nasyrov GA (1972) Study of the nightglow emission λ 5577 A for 1958–1967 in Ashkhabad. Ylym (Nauka), Ashkhabad

Korobeynikova MP, Nasyrov GA, Khamidulina VG (1966) The nightglow emission λ 5577 Å. Tables and maps of isophotes. Ashkhabad, 1964. Kalchaev KK, Shefov NN (eds) VINITI, Moscow

Korobeynikova MP, Nasyrov GA, Khamidulina VG (1968) The nightglow emission λ 5577 Å. Tables and maps of isophotes. Ashkhabad, 1965–1966. Kalchaev KK, Shefov NN (eds) VINITI, Moscow

Korobeynikova MP, Nasyrov GA, Khamidulina VG (1970) Intensity variations and dynamical characteristics of the spatial patches of the emission λ 5577 A. In: Krassovsky VI (ed) Aurorae and airglow. N 18. Nauka, Moscow, pp 5–14

Korobeynikova MP, Nasyrov GA, Toroshelidze TI, Shefov NN (1983) Some results of the simultaneous studies of the internal gravity waves at some stations. In: Lysenko IA (ed) Studies of the dynamic processes in the upper atmosphere. Hydrometeoizdat, Moscow, pp 121–123

Korobeynikova MP, Chuchuzov EP, Shefov NN (1984) Horizontal eddy diffusion near the turbopause from observations of the 557.7-nm emission. Izv USSR Acad Sci Atmos Oceanic Phys 20:854–857

Korobeynikova MP, Nasyrov GA, Shefov NN (1979) Internal gravity wave registration in Ashkhabad and Zvenigorod. Geomagn Aeronomy. 19:1116–1117

Kozhevnikov VN (1999) Atmospheric disturbances during mountain streamling. Scientific World, Moscow

Krassovsky VI (1957a) Nature of the intensity variations of the terrestrial atmosphere emission. Mém Soc Roy Sci Liège 18:58–67

Krassovsky VI (1957b) Nature of the intensity changes of the terrestrial atmosphere. Izv USSR Acad Sci Ser Geophys 664–669

Krassovsky VI (1972) Infrasonic variations of the OH emission in the upper atmosphere. Ann Geophys 28:739–746

Krassovsky VI (1973) Rotational and vibrational hydroxyl excitation in the laboratory and the night airglow. J Atmos Terr Phys 35:705–711

Krassovsky VI, Shagaev MV (1974a) Optical method of recording of acoustic gravity waves in the upper atmosphere. J Atmos Terr Phys 36:373–375

Krassovsky VI, Shagaev MV (1974b) Inhomogeneities and wavelike variations of the rotational temperature of atmospheric hydroxyl. Planet Space Sci 22:1334–1337

Krassovsky VI, Shagaev MV (1977) On the nature of hydroxyl airglow. Planet Space Sci 25:509–510

Krassovsky VI, Shefov NN (1976a) Wavelike and tidal movements near mesopause. In: Lauter EA, Taubenheim J, Böhm S (eds) Physica Solariterrestris. N 1. Akademie-Verlag, Potsdam, pp 61–66

Krassovsky VI, Shefov NN (1976b) Internal gravity wave studies by the optical method. Gerlands Beiträge Geophys 85:175–185

Krassovsky VI, Shefov NN (1978) Hydroxyl emission, waves and winds in the radiometeor region. Planet Space Sci 26:793–795

Krassovsky VI, Shefov NN (1980) On the relationship between the temperature and the circulation in the mesopause region. Geomagn Aeronomy 20:758–760

Krassovsky VI, Shefov NN, Yarin VI (1962) Atlas of the airglow spectrum $\lambda\lambda$ 3000–12400 Å. Planet Space Sci 9:883–915

Krassovsky VI, Kuzmin KI, Piterskaya NA, Semenov AI, Shagaev MM, Shefov NN, Toroshelidze TI (1975) Results of some airglow observations of internal gravitational waves. Planet Space Sci 23:896–898

Krassovsky VI, Semenov AI, Shefov NN, Yurchenko OT (1976) Predawn emission at 6300 Å and super-thermal ions from conjugate points. J Atmos Terr Phys 38:999–1001

Krassovsky VI, Potapov BP, Semenov AI, Shagaev MV, Shefov NN, Sobolev VG, Toroshelidze TI (1977) The internal gravity waves near mesopause and hydroxyl emission. Ann Geophys 33:347–356

Krassovsky VI, Potapov BP, Semenov AI, Sobolev VG, Shagaev MV, Shefov NN (1978) Internal gravity waves near mesopause. I. Results of studies of hydroxyl emission. In: Galperin YuI (ed) Aurorae and airglow. N 26. Soviet Radio, Moscow, pp 5–29

Krassovsky VI, Matveeva OA, Chunchuzov YeP (1986a) On mechanism for removing the energy of internal gravity waves from the upper atmosphere. Izv USSR Acad Sci Atmos Oceanic Phys 22:325–326

Krassovsky VI, Semenov AI, Sobolev VG, Tikhonov AV (1986b) Variations of the Doppler temperature and the 557,7 (nm) emission intensity at the passage of inner gravitational waves. Geomagn Aeronomy 26:941–945

Krassovsky VI, Matveeva OA, Semenov AI (1988) Dependence of the upper atmospheric warming from IGW amplitude. In: Lysenko IA (ed) Studies of the dynamic processes in the upper atmosphere. Hydrometeoizdat, Moscow, pp 56–59

Kropotkina EP, Shefov NN (1975) Lunar tide influence on the occurrence probability of the noctilucent clouds. Izv USSR Acad Sci Atmos Oceanic Phys 11:1179–1181

Kutepov AA, Shved GM (1978) Radiative transfer of the 15-μm CO_2 band with the breakdown of local thermodynamic equilibrium in the Earth's atmosphere. Izv USSR Acad Sci Atmos Oceanic Phys 14:28–43

Kuzmin KI (1975a) Intensity oscillations of the 5577 A and 5893 A emissions and geomagnetic activity. In: Krassovsky VI (ed) Aurorae and airglow. N 23. Nauka, Moscow, pp 28–32

Kuzmin KI (1975b) A possible correlation of internal gravity waves at height of about 90–100 (km) with atmospheric lows. Izv USSR Acad Sci Atmos Oceanic Phys 11:120–121

Kuzmin KI (1975c) Spectral analysis of variations of green emission of the night sky. Geomagn Aeronomy 15:567–568

Kuzmin KI (1975d) Fast variations of the green emission of atomic oxygen in the upper atmosphere. Izv USSR Acad Sci Atmos Oceanic Phys 11:246

Landau LD, Lifshits EM (1986) Theoretical Physics, vol 6. Hydrodynamics. Nauka, Moscow

Laštovička J (1997) Observations of tides and planetary waves in the atmosphere-ionosphere system. Adv Space Res 20:1209–1222

Laštovička J, Fišer V, Pancheva D (1994) Long-term trends in planetary wave activity (2–15 days) at 80–100 (km) inferred from radio wave absorption. J Atmos Terr Phys 56:893–899

Lindzen RS (1971) Tides and gravity waves in the upper atmosphere. In: Fiocco G (ed) Mesospheric models and related experiments. D Reidel Publishing Company, Dordrecht, pp 122–130

Liu HL, Roble RG, Taylor MJ, Pendleton WR (2001) Mesospheric planetary waves at northern hemisphere fall equinox. Geophys Res Lett 28:1903–1905

Lübken FJ, Fricke KH, Langer M (1996) Noctilucent clouds and the thermal structure near the Arctic mesopause in summer. J Geophys Res 101D:9489–9508

Lübken FJ, Rapp M, Blix T, Thrane E (1998) Microphysical and turbulent measurements of the Schmidt number in the vicinity of polar mesosphere summer echo. Geophys Res Lett 25:893–896

Lowe RP, LeBlanc L (1993) Preliminary analysis of WINDI (UARS) hydroxyl data: apparent peak height. Abstracts. The 19th Ann Europ Meet Atm Stud Opt Meth Kiruna, Sweden, Aug 10–14, 1992. Kiruna, pp 94–98

Lowe RP, Perminov VI (1998) Analysis of mid-latitude ground-based and WINDII/UARS observations of the hydroxyl nightglow. 32nd Scientific Assembly of COSPAR (Japan, Nagoya, 1998). Nagoya, p 131

Luo Y, Manson AH, Meek CE, Igarashi K, Jacobi Ch (2001) Extra long period (20–40 day) oscillations in the mesospheric and lower thermospheric winds: observations in Canada, Europe and Japan, and considerations of possible solar influences. J Atmos Solar–Terr Phys 63:835–852

Luo Y, Manson AH, Meek CE, Thayaparan T, MacDougall J, Hocking WK (2002) The 16-day wave in the mesosphere and lower thermosphere: simultaneous observations at Saskatoon (52°N, 107°W) and London (43°N, 81°W), Canada. J Atmos Solar–Terr Phys 64:1287–1307

Manson AH, Meek CE, Stegman J, Espy PJ, Roble RG, Hall CM, Hoffmann P, Jacobi Ch (2002) Springtime transitions in mesopause airglow and dynamics: photometer and MF radar observations in Scandinavian and Canadian sectors. J Atmos Solar-Terr Phys 64:1131–1146

McDade IC, Llewellyn EJ, Greer RG, Witt G (1984) Altitude dependence of the vibrational distribution of $O_2\,(c^1\Sigma_u^-)$ in the nightglow and possible effect of vibrational excitation in the formation of $O\,(^1S)$. Can J Phys 62:780–788

McEwan MJ, Phillips LF (1975) Chemistry of the atmosphere. Edward Arnold, London

Meriwether JW (1975) High latitude airglow observations of correlated short-term fluctuations in the hydroxyl Meinel 8–3 band intensity and rotational temperature. Planet Space Sci 23:1211–1221

Meriwether JW, Mirick JL, Biondi MA, Herrero FA, Fesen CG (1996) Evidence of orographic wave heating in the equatorial thermosphere at solar maximum. Geophys Res Lett 23:2177–2180

Meriwether JW, Biondi MA, Herrero FA, Fesen CG, Hallenback DC (1997) Optical interferometric studies of the nighttime equatorial thermosphere: enhanced temperatures and zonal wind gradients. J Geophys Res 102A:20041–20058

Mies FH (1974) Calculated vibrational transition probabilities of $OH(X^2\Pi)$. J Molecular Spectrosc 53:150–188

Miroshnikov MM (1977) The theoretical principles of the optical-electronic instruments. Mashinostroenie, Leningrad

Moreels G, Hersé M (1977) Photographic evidence of waves around the 85 (km) level. Planet Space Sci 25:265–273

Moreels G, Megie G, Vallance Jones A, Gattinger RL (1977) An oxygen-hydrogen atmospheric models and its application to the OH emission problem. J Atmos Terr Phys 39:551–570

Murayama Y, Tsuda T, Wilson R, Nakane H, Hayachida SA, Sugimoto N, Matsui I, Sasano Y (1994) Gravity wave in the upper stratosphere and lower mesosphere observed with Rayleigh lidar at Tsukuba, Japan. Geophys Res Lett 21:1539–1542

Myrabø HK (1984) Temperature variation at mesopause levels during winter solstice at 78°N. Planet Space Sci 32:249–255

Myrabø HK, Deehr CS, Sivjee GG (1983) Large-amplitude nightglow OH (8–3) band intensity and rotational temperature variations during 24–hour period at 78° N. J Geophys Res 88A:9255–9259

Nasyrov GA (1967) Peculiarities of spatial inhomogeneities of the atomic oxygen emission λ 5577 A in the nightglow. In: Krassovsky VI (ed) Aurorae and airglow. N 13. Nauka, Moscow, pp 5–9.

Nasyrov GA (2007) Orography-caused variations in the 557.7 (nm) atomic oxygen emission intensity. Geomagn Aeronomy 47:101–103

Novikov NN (1981) Internal gravity waves near mesopause. VI. Polar region. In: Galperin YuI (ed) Aurorae and airglow. N 29. Soviet Radio, Moscow, pp 59–67

Noxon JF (1978) Effect of internal gravity waves upon night airglow temperatures. Geophys Res Lett 5:25–27

Pavlova TD (1962) A study of the longitudinal distribution of noctilucent clouds. In: Villmann ChI (ed) Proceedings of III Conference noctilucent clouds (Tallinn, May 16–19, 1961). Institute Phys Astron Estonian SSR Acad Sci, Tallinn, pp 119–125

Perminov VI, Lowe RP, Pertsev NN (1999) Longitudinal variations in the hydroxyl nightglow. Adv Space Res 24:1609–1612

Perminov VI, Pertsev NN, Shefov NN (2002) Stationary planetary variations of the hydroxyl emission. Geomagn Aeronomy 42:610–613

Pertsev NN (1989a) Azimuth anisotropy of mountain lee waves in the upper atmosphere. Izv USSR Acad Sci Atmos Oceanic Phys 25:432–436

Pertsev NN (1989b) Heating of the upper layers of the atmospheric by dissipation of mountain lee waves. Izv USSR Acad Sci Atmos Oceanic Phys 25:562–564

Peterson AW, Kieffaber LM (1973a) Infrared photography of OH airglow structures. Nature (London) 242:321–322

Peterson AW, Kieffaber LM (1973b) Photgraphic parallax height of infrared airglow structures. Nature (London) 244:92–93

Phillips A, Manson AH, Meek CE, Llewellyn EJ (1994) A long-term comparison of middle atmosphere winds measured at Saskatoon (52°N, 107°W) by medium – frequency radar and a Fabry – Perot interferometer. J Geophys Res 99:12923–12935

Piterskaya NA (1976a) The relative intensity of the OH band lines with $\Delta v = 1$. In: Krassovsky VI (ed) Aurorae and airglow. N 25. Nauka, Moscow, pp 23–59

Piterskaya NA (1976b) The wavelengths of the OH rotational-vibrational bands. In: Krassovsky VI (ed) Aurorae and airglow. N 25. Nauka, Moscow, pp 60–70

Piterskaya NA, Shefov NN (1975) Intensity distribution of the OH rotation-vibration bands. In: Krassovsky VI (ed) Aurorae and airglow. N 23. Nauka, Moscow, pp 69–122

Pogoreltsev AI (1996) Simulation of the influence of stationary planetary waves on the zonally averaged circulation of the mesosphere/lower thermosphere region. J Atmos Terr Phys 58:902–909

Pogoreltsev AI, Sukhanova SA (1993) Simulation of global structure of stationary planetary waves in the mesosphere and lower thermosphere. J Atmos Terr Phys 55:33–40

Portnyagin YuI, Sprenger K, Lysenko IA, Schminder R, Orlyansky AD, Greiziger KM, Il'ichev YuD, Kürschner R, Schennig B (1978) The wind measurements at heights 90–100 (km) by ground-based methods. Portnyagin YuI, Sprenger K (eds) Hydrometeoizdat, Leningrad

Potapov BP (1975a) Determination of the effective height of fluctuations of hydroxyl emissions. Planet Space Sci 23:1346–1347

Potapov BP (1975b) Triangular measurements of the hydroxyl emission height. Astron Circ USSR Acad Sci N 856:5–7

Potapov BP (1975c) The nature of the hydroxyl emission and the correlation between its intensity and temperature. In: Krassovsky VI (ed) Aurorae and airglow. N 23. Nauka, Moscow, pp 36–41

Potapov BP (1976) Measurement of the altitude of the OH emission by the fluctuations of the rotational temperature. In: Krassovsky VI (ed) Aurorae and airglow. N 24. Soviet Radio, Moscow, pp 21–26

Potapov BP, Semenov AI, Sobolev VG, Shagaev MV (1978) Internal gravity waves near mesopause. II. Instruments and optical methods of measurements. In: Galperin YuI (ed) Aurorae and airglow. N 26. Soviet Radio, Moscow, pp 30–65

Potapov BP, Sobolev VG, Sukhodoev VA, Yarov V.N (1985) On the mutual arrangement of emissional layers of hydroxyl and OI 5577 A in the Earth upper atmosphere. Geomagn Aeronomy 25:685–686

Rajaram R, Gurubaran S (1998) Seasonal variabilities of low – latitude mesospheric winds. Ann Geophys 16:197–204

Reid IM (1989) Observations of gravity waves scales fluxes and saturation during MAP. In: Edwards B (ed) Handbook for MAP, vol 27. SCOSTEP, Urbana Illinois, pp 87–103

Roach FE, Gordon JL (1973) The light of the night sky. D Reidel Publishing Company, Dordrecht, Holland

Roach FE, Pettit HB (1952) Excitation patterns in the nightglow. Mém Soc Roy Sci Liège 12:13–42

Roach FE, Tansberg-Hanssen E, Megill RL (1958) The characteristic size of the airglow cells. J Atmos Terr Phys 13:113–121

Romejko VA, Pertsev NN, Dalin PA (2002) Long-term observations of noctilucent clouds in Moscow: a data-base and statistical analysis. Geomagn Aeronomy 42:670–675

Romejko VA, Dalin PA, Pertsev NN (2003) Forty years of noctilucent cloud observations near Moscow: database and simple statistics. J Geophys Res 108D:8440. doi:10.1029/2002JD002364

Roper RG, Adams GW, Brosnahan JW (1993) Tidal winds at mesopause altitudes over Arecibo (18°N, 67°W), 5–11 April 1989 (AIDA'89). J Atmos Terr Phys 55:289–312

Ross MN, Christensen AB, Meng CI, Carbary JF (1992) Structure in the UV nightglow observed from low Earth orbit. Geophys Res Lett 19:985–988

Sato K (1990) Vertical wind disturbances in the troposphere and lower stratosphere observed by the MU radar. J Atmos Sci 47:2803–2817

Savada R (1965) The possible effect of oceans on the atmospheric lunar tide. J Atmos Sci 22:636–643

Sawyer JS (1959) The introduction of effects of topography into methods of numerical forecasting. Quart J Roy Met Soc 85:31–43

Sawyer RA (1951) Experimental spectroscopy, 2nd edn. Academic Press, New York

Scheer J, Reisin ER, Espy JP, Bittner M, Graef HH, Offermann D, Ammosov PP, Ignatiev VM (1994) Large – scale structures in hydroxyl rotational temperatures during DYANA. J Atmos Terr Phys 56:1701–1715

Schröder W (1999) Were noctilucent clouds caused by Krakatoa eruption? A case study of the research problems before 1885. Bull Amer Meteorol Soc 80:2081–2085

Schubert G, Walterscheid RL (1988) Wave-driven fluctuations in OH nightglow from extended source region. J Geophys Res 93A:9903–9915

Semenov AI (1975) Interferometric measurements of the upper atmosphere temperature. I. Application of the cooled image converters. In: Krassovsky VI (ed) Aurorae and airglow. N 23. Nauka, Moscow, pp 64–65

Semenov AI (1982) Zonal winds in the ionospheric F2-region. Geomagn Aeronomy 22:497–498

Semenov AI (1985) Interferometric measurements of the temperature and wind in the F_2 – region at the different latitudes. In: Lysenko IA (ed) Studies of the dynamic processes in the upper atmosphere. Hydrometeoizdat, Moscow, pp 64–68

Semenov AI (1987) Comparison of variations of the hydroxyl emission in the atmosphere and wind characteristics. Geomagn Aeronomy 27:853–855

Semenov AI (1988) Seasonal variations of the hydroxyl rotational temperature. Geomagn Aeronomy 28:333–334

Semenov AI (1989a) The specific features of the green emission excitation process in the nocturnal atmosphere. In: Feldstein YaI, Shefov NN (eds) Aurorae and airglow. N 33. VINITI, Moscow, pp 74–80

Semenov AI (1989b) Relation between the ozone-hydrogen and superhydroxyl excitation mechanism of hydroxyl emission. Geomagn Aeronomy 29:687–689

Semenov AI (1997) Long-term changes in the height profiles of ozone and atomic oxygen in the lower thermosphere. Geomagn Aeronomy 37:354–360

Semenov AI, Shefov NN (1989) The effect of internal gravity waves on the dynamics and energetics of the lower thermosphere (according to characteristics of the nightglow). In: Kazimirovsky ES (ed) Middle atmosphere studies. Ionospheric Researches. N 47. VINITI, Moscow, pp 24–43

Semenov AI, Shefov NN (1996) An empirical model for the variations in the hydroxyl emission. Geomagn Aeronomy 36:468–480

Semenov AI, Shefov NN (1997a) An empirical model of nocturnal variations in the 557.7-nm emission of atomic oxygen. 1. Intensity. Geomagn Aeronomy 37:215–221

Semenov AI, Shefov NN (1997b) An empirical model of nocturnal variations in the 557.7-nm emission of atomic oxygen. 2. Temperature. Geomagn Aeronomy 37:361–364

Semenov AI, Shefov NN (1997c) An empirical model of nocturnal variations in the 557.7-nm emission of atomic oxygen. 3. Emitting layer altitude. Geomagn Aeronomy 37:470–474

Semenov AI, Shefov NN (1997d) Empirical model of the variations of atomic oxygen emission 557.7 (nm). In: Ivchenko VN (ed) Proceedings of SPIE (23rd Europ Meet Atmos Stud Opt Meth, Kiev, Sept 2–6, 1997), vol 3237. The Intern Soc Opt Eng, pp 113–122

Semenov AI, Shefov NN (1999) Empirical model of hydroxyl emission variations. Int J Geomagn Aeronomy 1:229–242

Semenov AI, Shagaev MV, Shefov NN (1981) On the effect of orographic waves on the upper atmosphere. Izv USSR Acad Sci Atmos Oceanic Phys 17:982–984

Senft D, Gardner C (1991) Seasonal variability of gravity wave activity and spectra in the mesopause region at Urbana. J Geophys Res 96D:17229–17264

Shagaev MV (1974) About the connection of rapid variation of rotational temperature of atmospheric hydroxyl with geomagnetic activity. Geomagn Aeronomy 14:759–760

Shagaev MV (1978) Characteristics of hydroxyl emission variations, pointing out the processes of its excitations. In: Krassovsky VI (ed) Aurorae and airglow. N 27. Nauka, Moscow, pp 18–25

Shefov NN (1965) The upper atmosphere emissions and the noctilucent clouds. In: Krassovsky VI (ed) Aurorae and airglow. N 11. Nauka, Moscow, pp 48–51

Shefov NN (1967) OH emission and noctilucent clouds. In: Khvostikov IA, Witt G (eds) Noctilucent clouds. Proc Intern Symp (Tallinn, 1966). VINITI, Moscow, pp 187–188

Shefov NN (1968) Behaviour of the upper atmosphere emissions during high meteoric activity. Planet Space Sci 16:134–136

Shefov NN (1970) Behaviour of the upper atmosphere emissions during high meteoric activity. In: Krassovsky VI (ed) Aurorae and airglow. N 18. Nauka, Moscow, pp 21–25

Shefov NN (1974a) Lunar variations of hydroxyl emission. Geomagn Aeronomy 14:920–922

Shefov NN (1974b) Lunar tidal variations of hydroxyl emission. Indian J Radio Space Phys 3:313–314

Shefov NN (1975) Emissive layer altitude of the atmospheric system of molecular oxygen. In: Krassovsky VI (ed) Aurorae and airglow. N 23. Nauka, Moscow, pp 54–58

Shefov NN (1976) Seasonal variations of the hydroxyl emission. In: Krassovsky VI (ed) Aurorae and airglow. N 24. Soviet Radio, Moscow, pp 32–36

Shefov NN (1978) Altiude of the hydroxyl emission layer. In: Krassovsky VI (ed) Aurorae and airglow. N 27. Soviet Radio, Moscow, pp 45–51

Shefov NN (1985) Solar activity and near surface circulation as the commensurable sources of the thermal regime variations of the lower thermosphere. Geomagn Aeronomy 25:848–849

Shefov NN (1989) The recording of wave and spotted inhomogeneities of upper atmospheric emission. In: Feldstein YaI, Shefov NN (eds) Aurorae and airglow. N 33. VINITI, Moscow, pp 81–84

Shefov NN, Pertsev NN (1984) Orographic disturbances of upper atmosphere emissions. In: Taubenheim J (ed) Handbook for Middle Atmosphere Program. N 10. SCOSTEP, Urbana, pp 171–175

Shefov NN, Semenov AI (2002) The long-term trend of ozone at heights from 80 to 100 (km) at the mid-latitude mesopause for the nocturnal conditions. Phys Chem Earth 27:535–542

Shefov NN, Semenov AI (2004a) Longitudinal-temporal distribution of the occurrence frequency of noctilucent clouds. Geomagn Aeronomy 44:259–262

Shefov NN, Semenov AI (2004b) Spectral characteristics of the IGW trains registered in the upper atmosphere. Geomagn Aeronomy 44:763–768

Shefov NN, Pertsev NN, Shagaev MV, Yarov VN (1983) Orographically caused variations of upper atmospheric emissions. Izv USSR Acad Sci Atmos Oceanic Phys 19:694–698

Shefov NN, Semenov AI, Pertsev NN, Sukhodoev VA, Perminov VI (1999) Spatial distribution of IGW energy inflow into the mesopause over the lee of a mountain ridge. Geomagn Aeronomy 39:620–627

Shefov NN, Semenov AI, Pertsev NN, Sukhodoev VA (2000a) The spatial distribution of the gravity wave energy influx into the mesopause over a mountain lee. Phys Chem Earth Pt B 25:541–545

Shefov NN, Semenov AI, Pertsev NN (2000b) Dependencies of the amplitude of the temperature enhancement maximum and atomic oxygen concentrations in the mesopause region on seasons and solar activity level. Phys Chem Earth Pt B 25:537–539

Shefov NN, Semenov AI, Yurchenko OT (2002) Empirical model of the ozone vertical distribution at the nighttime mid-latitude mesopause. Geomagn Aeronomy 42:383–389

Shepherd GG, Thuillier G, Gault WA, Solheim BH, Hersom C, Alunni M, Brun JF, Brune S, Chalot P, Cogger LL, Desaulniers DL, Evans WFJ, Gattinger RL, Girod F, Harvie D, Henn RH, Kendall DJW, Llewellyn EJ, Lowe RP, Ohrt J, Pasternak F, Peillet O, Powell I, Rochon Y, Ward WE, Wiens RH, Wimperis J (1993a) WINDII the wind imaging interferometer on the upper atmosphere research satellite. J Geophys Res 98D:10725–10750

Shepherd GG, Thuillier G, Solheim BH, Chandra S, Cogger LL, Duboin ML, Evans WFJ, Gattinger RL, Gault WA, Hersé M, Hauchecorne A, Lathuilliere C, Llewellyn EJ, Lowe RP, Teitelbaum H, Vial F (1993b) Longitudinal structure in atomic oxygen concentrations observed with WINDII on UARS. Geophys Res Lett 20:1303–1306

Shved GM (1977) Thermal atmospheric regime at the mid-latitudes in the mesopause vicinity (70–100 (km)). In: Invest geomagnetism aeronomy solar phys. N 43. Nauka, Moscow, pp 182–191

Sivjee GG, Walterscheid RL, Hecht JH, Hamwey RM, Schubert G, Christensen AB (1987) Effects of atmospheric disturbances on polar mesopause airglow OH emissions. J Geophys Res 92A:7651–7656

Spizzichino A (1969) Etude experimentale des vents dans la haute atmosphère. Ann Geophys 25:5–28

Spizzichino A (1971) Meteor trail radar winds over Europe. In: Webb WL (ed) Thermospheric circulation. The MIT Press, Cambridge, pp 117–180

Stair AT, Sharma RD, Nadile RM, Baker DJ, Grieder WF (1985) Observations of limb radiance with cryogenic spectral infrared rocket experiment. J Geophys Res 90A:9763–9775

Stepanov BE (2000) Propagation of a 16-day planetary wave. Izv Atmos Oceanic Phys 36:691–697

Steyn DG, Ayotte KW (1985) Application of two-dimensional spectra to mesoscale modeling. J Atmos Sci 42:2884–2887

Streit GE, Johnston HS (1976) Reaction and quenching of vibrationally excited hydroxyl radicals. J Chem Phys 64:95–103

Sukhodoev VA, Yarov VI (1998) Temperature variations of the mesopause in the leeward region of the Caucasus ridge. Geomagn Aeronomy 38:545–548

Sukhodoev VA, Pertsev NN, Reshetov LM (1989a) Variations of characteristics of hydroxyl emission caused by orographic perturbations. In: Feldstein YaI, Shefov NN (eds) Aurorae and airglow. N 33. VINITI, Moscow, pp 61–66

Sukhodoev VA, Perminov VI, Reshetov LM, Shefov NN, Yarov VN, Smirnov AS, Nesterova TS (1989b) The orographic effect in the upper atmosphere. Izv USSR Acad Sci Atmos Oceanic Phys 25:681–685

Sukhodoev VA, Pertsev NN, Shefov NN (1992) Formation of orographic disturbances in mesopause of mountain lee. EOS Trans AGU, 73:Spring Meet Suppl, 223

Suzuki K, Tohmatsu T (1976) An interpretation of the rotational temperature of the airglow hydroxyl emissions. Planet Space Sci 24:665–671

Takahashi H, Buriti RA, Gobbi D, Batista PP (2002) Equatorial planetary wave signatures observed in mesospheric airglow emissions. J Atmos Solar-Terr Phys 64:1263–1272

Takenuchi I, Nisawa K, Kato J, Acyama I (1981) Seasonal variations of the correlation among nightglow radiation and emission mechanism of OH nightglow emission. J Atmos Terr Phys 43:157–164

Taranova OG (1967) Study of space-time properties of the hydroxyl emission. In: Krassovsky VI (ed) Aurorae and airglow. N 13. USSR Acad Sci Publ House, Moscow, pp 13–21

Taylor MJ, Hapgood MA (1990) On the origin of ripple-type wave structure in the OH nightglow emission. Planet Space Sci 38:1421–1430

Taylor MJ, Lowe RP, Baker DJ (1995) Hydroxyl temperature and intensity measurements during noctilucent cloud displays. Ann Geophys 13:1107–1116

Tepley CA, Burnside RG, Meriwether JW (1981) Horizontal thermal structure of the mesosphere from observations of OH (8–3) band emission. Planet Space Sci 29:1241–1249

Thomas G. Olivero JJ, Jensen EJ, Schröder W, Toon OB (1989) Relation between increasing methane and the presence of ice clouds at the mesopause. Nature, London 338:490–492

Thuillier G, Blamont JE (1973) Vertical red line 6300 Å distribution and tropical nightglow morphology in quiet magnetic conditions. In: McCormac BM (ed) Physics and chemistry of upper atmosphere. D Reidel Publishing Company, Dordrecht, pp 219–231

Trunkovsky EM, Semenov AI (1978) Interferometric measurements of the upper atmosphere temperature. IV. Analysis of photographic interferograms. In: Krassovsky VI (ed) Aurorae and airglow. N 27. Soviet Radio, Moscow, pp 66–84

Truttse YuL (1965) The spatial variations of the oxygen lines intensities. In: Krassovsky VI (ed) Aurorae and airglow. N 11. USSR Acad Sci Publ House, Moscow, pp 52–64

Ulwick JC, Baker KD, Stair AT, Frings W, Hennig R, Grossmann KU, Hegblom ER (1985) Rocketborne measurements of atmospheric fluxes. J Atmos Terr Phys 47:123–131

Vallance Jones A (1973) The infrared spectrum of the airglow. Space Sci Rev 15:355–400

Vasseur G, Reddy CA, Testud J (1972) Observations of waves and travelling disturbances. In: Bowhill SA, Jaffe LD, Rycroft MJ (eds) Space Research, Vol 12. Akademie-Verlag, Berlin, pp 1109–1131

Van Zandt TE, Nastrom GD, Green JL, Gage KS (1989) The spectrum of vertical velocity from radar observations. In: Liu CH, Edwards B (eds) Handbook for MAP, Vol 28. SCOSTEP, Urbana, pp 377–383

Vasiliev NV, Fast NP (1973) On the connection of the mesospheric clouds with several cosmophysical phenomena. Astron Geodesy, Vol 241. Tomsk Univ Publ House, Tomsk, pp 64–72

Vasiliev OB (1970) Frequency spectrum of the noctilucent cloud displays and their connection with solar activity. In: Ikaunieks JJ (ed) Physics of the mesospheric (noctilucent) clouds (Riga, November, 20–23, 1968). Zinatne, Riga, pp 121–135

Vasiliev OB, Villmann ChI, Melnikova IN, Chubey MS (1975) Technique of statistical analysis of observations of mesospheric clouds applied by the network of stations of the Hydrometeorological Service. In: Vasiliev OB (ed) Meteorol Invest. N 22. Nauka, Moscow, pp 110–118

Villmann ChI (1970) On the space-time regularities of the noctilucent cloud displays. In: Ikaunieks JJ (ed) Physics of the mesospheric (noctilucent) clouds (Riga, November, 20–23, 1968). Zinatne, Riga, pp 103–113

Vincent RA (1984) Gravity wave motions in the mesosphere. J Atmos Terr Phys 46:119–128 von Zahn U (2003) Are noctilucent clouds truly a "miner's canary" for global change? EOS Trans AGU 84: 261, 264

Walterscheid RL, Schubert G, Straus JM (1987) A dynamical chemical model of wave-driven fluctuations in the OH nightglow. J Geophys Res 92A:1241–1254

Wang DY, McLandress C, Fleming EL, Ward WE, Solheim B, Shepherd GG (1997) Empirical model of 90–120 km horizontal winds from wind – imaging interferometer green line measurements in 1992–1993. J Geophys Res 102D:6729–6745

Wang DY, Ward WE, Shepherd GG, Wu DL (2000) Stationary planetary waves inferred from WINDII wind data taken within altitudes 90–120 km during 1991–96. J Atmos Sci 57:1906–1918

Webb WL (1966) Structure of the stratosphere and mesosphere. Academic Press, New York

Weinstock J (1976) Nonlinear theory of acoustic-gravity waves. 1. Saturation and enhanced diffusion. J Geophys Res 81:633–652

Weinstock J (1978) Theory of the interaction of gravity waves with O_2 (^1S) airglow. J Geophys Res 83:5175–5185

Winick JR (1983) Photochemical processes in the mesosphere and lower thermosphere. In: Carovillano RI, Forbes JM (eds) Solar-Terrestrial Physics. D Reidel Publ Co, Hingham Mass, pp 677–732

Witt G (1962) Height, structure, and displacements of noctilucent clouds. Tellus 14:1–18

Witt G, Stegman J, Solheim BH, Llewellyn EJ (1979) A measurement of the $O_2(b^1\Sigma_g^+ - X^3\Sigma_g^-)$ atmospheric band and the $O(^1S)$ green line in the nightglow. Planet Space Sci 27:341–350

Wraight PC (1982) Association of atomic oxygen and airglow excitation mechanism. Planet Space Sci 30:251–259

Yugov VA, Ignatiev VM (1983) The neutral wind measurement of the ionospheric F-region. In: Lysenko IA (ed) Studies of the dynamic processes in the upper atmosphere. Hydrometeoizdat, Moscow, pp 121–123

Chapter 6
Climatic Changes in the Upper Atmosphere

The current concept is that the Earth's climate changes on the global scale. According to the data of the World Meteorological Organization (WMO/UNEP 1990), the average global air temperature in the ground layer, being a very sensitive measure of climatic changes, has increased by 0.3–0.6 (K) over the last 100 years. Moreover, the rate of warming over the last three decades significantly exceeds its value averaged over the century. There are grounds to believe that the economic activity of mankind plays an important part in the global climate change.

In connection with this problem, investigations of the state of upper atmospheric layers have been the particular concern of experts in the last few years. This is caused, on the one hand, by the possible catastrophic consequences of anthropogenic change of the chemical composition of the atmosphere as a whole, exemplified by the ozone layer depletion (Dütsch and Staehelin 1989; Stolarski et al. 1992). On the other hand, according to theoretical research (see, e.g., (Roble and Dickinson 1989; Rind et al. 1990)), global changes of the thermal and dynamic regimes resulting from an increasing inflow of greenhouse gases are more pronounced in the upper atmospheric layers compared to the surface ones.

Analysis of aerological data obtained in the last three decades confirms this theoretical notion and demonstrates that the positive trend of the upper tropospheric temperature and the negative trend of the middle stratospheric temperature are approximately an order of magnitude greater than the temperature trend in the ground atmospheric layer (Angel 1988; Miller et al. 1992). Even greater negative temperature trend – also by an order of magnitude – was retrieved from time series of rocket, satellite, lidar, and radiophysical measurements performed in the upper stratosphere and mesosphere (Kokin et al. 1990; Taubenheim et al. 1990; Angel 1991; Aikin et al. 1991; Hauchecorne et al. 1991). However, unlike radiosonde measurements carried out in the mode of global monitoring since the 1950s, measurements in the middle atmosphere are less representative and cover a much shorter time interval. Thus, climatological data of six US rocket stations for altitudes of 25–55 (km) were generalized over an 18-yr period (Angel 1991), and the data of five Russian rocket stations for altitudes of 25–80 (km) cover periods from 18 to 27 years (Kokin et al. 1990). The periods of regular satellite temperature measurements at altitudes up to 55 (km) (Aikin et al. 1991) and lidar temperature measurements at altitudes of 33–75 (km) (Hauchecorne et al. 1991) are about half the above ones. Observations of the D layer of the ionosphere carried out in the last three decades (Taubenheim et al. 1990) and of the sodium layer ($\sim$92 (km)) between 1972 and 1992 (Clemesha et al. 1992) indirectly testify to cooling of the mesosphere, though the authors deny a decrease of the sodium layer altitude in their subsequent papers (Clemesha et al. 2003, 2004).

The estimated linear trend of the mesospheric temperature ranges from -2 (K) (Angel 1991) to -15 (K) per decade (Clancy and Rush 1989; Beig et al. 2003). This wide spread in estimates is due to different factors among which are (1) statistical nonuniformity of time series of data, (2) errors of the measurement methods used, (3) disadvantages of methods of statistical analysis, and (4) limited observation periods. Each of these factors should be borne in mind when analyzing individual observations. However, qualitative agreement between the trend estimates obtained by all the mentioned observation methods indicates that the first two factors are not dominant, and the quantitative discrepancies are caused mostly by the different statistical analysis methods employed and, which is especially important, by limited lengths of time series.

This is especially true for the parameters of the upper atmospheric layers, since they depend on the cyclic variability of the solar flux much stronger than the corresponding parameters of the ground layer. In this regard, optical and radiophysical investigations of the upper atmosphere over several decades supplemented with comparatively short series of rocket, satellite, and lidar measurements of the temperature in the middle atmosphere allow reliable statements on long-term changes of the parameters in the upper atmosphere.

In estimating the linear trends of the atmospheric temperature at different altitudes, it is of importance that the length of measurement series be over at least one or two decades. Therefore, the problem of detecting long-term changes of any characteristic of a geophysical process calls for preliminary elimination of periodic and random variations of all types from the data to be analyzed. Analysis of hydroxyl radiation temperature data that provide the basis for estimation of long-term trends has revealed that the trend itself shows noticeable seasonal, latitudinal, and altitudinal variability; it also depends on the measurement period. Violation of these conditions is a reason for lacking an unambiguous idea of the character of long-term changes of the atmospheric state in the mesopause and lower thermosphere.

Hence, before analyzing average long-term changes, one should reduce the available time series of data to uniform heliogeophysical conditions. If data under consideration involve an effect of geomagnetic activity, they should be corrected or excluded from the analysis. Unfortunately, it is not always that the authors of publications on long-term trends mention whether these self-evident requirements were fulfilled. In addition, the trend value must be accompanied by the indication of the relevant measurement period.

Attempts undertaken to revise both temperature measurements in the middle atmosphere and long-term trends retrieved on their basis (Nielsen et al. 2002; Sigernes et al. 2003a,b; Laštovička, 2005) were mostly oriented on the adjustment of trend values calculated for the preceding time intervals to those derived from measurements performed in the last few years. Of course, this approach cannot be satisfactory, because there is some evidence for a long-term trend change for several decades. It seems likely that the reason for this process is not only the impact of anthropogenic factors on the Earth's atmosphere but also the effect of the secular variability of solar activity. Therefore, when analyzing the accumulated data, one must take into account the features of the trend behavior.

Since the upper atmosphere is sensitive to disturbances of its stationary state (temperature, composition, etc.), investigations of the evolution of its state can be the decisive factor for elucidation of a tendency for the Earth's climate evolution and of the role of the anthropogenic factor in this process.

The longest and most systematic data have been provided by ionospheric measurements of the electron concentration n_e in the E (110–120 (km)) and F2 layers (240–350 (km)) and by spectrophotometric measurements of the characteristics of emissions in the upper atmosphere. Analysis of almost half-secular ionospheric observations has revealed long-term trends of the electron concentration n_e in these layers. Long-term data on zonal and meridional wind velocity components in the lower thermosphere (Jacobi 1998; Jacobi and Kürschner 2002) as well as on the planetary wave amplitudes based on radiophysical wind research (Laštovička et al. 1994; Laštovička 1997a,b, 2002) also indicate the presence of systematic long-term changes of the mean characteristics. The Earth's airglow has been studied comprehensively in the Soviet Union since 1948. By the present time, a great body of information on emissions from hydroxyl ($\sim$87 (km)), sodium at 589.3 (nm) ($\sim$92 (km)), and atomic oxygen at 557.7 (nm) ($\sim$97 (km)) and 630 (nm) ($\sim$270 (km)) has been accumulated (Fishkova 1983; Semenov and Shefov 1996; Shefov 1969). It should be emphasized that to estimate long-term trends from the data collected from the late 1950s till the early 1990s, the assumption of their linearity in the period under consideration was naturally made, because there was no clear evidence for their nonlinear changes during this half-secular period.

The available data of measurements of the intensities and temperatures of the hydroxyl (OH) and 557.7-nm atomic oxygen emissions carried out in Japan, England, France, Germany, and Sweden allowed only their dependence on solar activity to be established, since the length of the time series of data was not over 20 years. A set of irregular measurements of the hydroxyl rotational temperature performed at several equatorial stations does not allow an independent conclusion on its long-term variations, since in the most cases, the observation period was a few years and never longer than the 11-yr solar cycle; therefore, these measurements can be used only in combination with other data.

Undoubtedly, the problem of long-term systematic changes of the thermal state of the middle and upper atmosphere still remains urgent. Since the period under consideration involves not only the increase of the content of greenhouse gases but also the secular increase of the average solar flux, whose maximum was observed in the late 1990s, it seems likely that the tendency for a reduction of solar activity in the subsequent years can have the result that small mean temperature changes will be detected for time intervals shorter than 20 years.

In this chapter, an attempt is made to summarize the data of investigations of the temperature regime in the middle atmosphere performed by different methods in Russia in combination with the available data of similar measurements performed in other countries. On this basis, the behavior of atmospheric temperature at different altitudes in different seasons, its dependence on solar activity, and its long-term trends have been elucidated.

6.1 The Temperature Trend in the Middle Atmosphere

Rocket Measurements

Regular investigations of the temperature regime in the middle atmosphere with an M-100B two-step meteorological rocket that transported main and auxiliary equipment to altitudes up to 90 (km) were started in the Soviet Union in the early 1960s and continued until the late 1990s. Measurements were performed as the equipment was moved downward by parachute from an altitude of 85 (km). The atmospheric temperature was measured with resistance thermometers. To reduce aerodynamic heating of the thermometers, the head part of the meteorological rocket was stabilized relative to the total velocity vector, and a special high-altitude parachute was used to measure the wind speed and direction. When processing the rocket data, a number of corrections were introduced including aerodynamic and radiative corrections together with the correction for the thermometer self-radiation and Joule heating by the measuring current. The method of data processing is described in detail elsewhere (Izakov et al. 1967; Lysenko 1981; Lysenko et al. 1982; Schmidlin 1986).

The resultant error in measuring the atmospheric temperature at altitudes up to 40 (km) is almost completely determined by the instrumental errors of the measuring temperature converter and radio telemetry channel. The rms error was 2.7 (K). At altitudes above 50 (km), the random component of the resultant error was almost completely determined by the corresponding errors of temperature corrections, among which the aerodynamic error was dominant (Lysenko 1981). The rms error increased with altitude and was equal to 6 (K) for the 60–75 (km) layer (Lysenko et al. 1982).

Measurements of Airglow Emissions

The airglow characteristics (intensity and temperature) were measured by the method of optical spectrometry. The results obtained in Russia were mainly based on observations carried out at the stations of Zvenigorod (55.7°N, 36.8°E), Abastumani (41.8°N, 42.8°E), and Yakutsk (62.0°N, 129.7°E). In addition, the observation data obtained at the foreign stations of Delaware (42.8°N, 81.4°W), Quebec (46.9°N, 71.1°W), Maynooth (53.2°N, 6.4°W), Stockholm (59.5°N, 18.2°E), Wuppertal (51°N, 7°E), Fritz Peak (39.9°N, 105.5°W), and Fort Collins (40.6°N, 105.0°W) were used. It should be noted that the time series of Russian spectrophotometric data are several times longer than the foreign ones. In Russia, the relevant observations have been performed from the mid-1950s until now, while in abroad, they are irregular. Therefore, the conclusions about long-term changes of the temperature regime in the upper atmospheric layers based only on the data of foreign stations are sometimes inconsistent with the results of investigations in Russia, because they refer to different periods and dates. At the same time, when used in combination, these data completely confirm the results obtained in Russia.

At Abastumani, photoelectric measurements of (7–3), (8–4), (3–0), and (4–1) hydroxyl band intensities in the spectral range 900–1040 (nm) have been carried out since 1948, and the emission lines of atomic oxygen (at 557.7 (nm) and 630.0 (nm)) and sodium (at 589.3 (nm)) have been measured since 1958. Hydroxyl emission spectra were measured between 1957 and 1972 to determine the hydroxyl rotational temperature.

At the Zvenigorod station, the hydroxyl band intensities and rotational temperatures have been measured from the mid-1950s up to now in the spectral range 580–1150 (nm) with spectrographs initially equipped with image converter tubes providing for photographic recording of spectra. In the last decade, they have been equipped with charge-coupled photodetectors (Semenov et al. 2002b).

The rotational temperature of hydroxyl emission, carrying information on the ambient temperature at OH airglow altitudes, is determined from the intensity distribution over the detected OH rotational bands (Shefov 1961). The measurement technique has already been considered in detail in Sect. 2.2.5.

The rms error in reconstructing absolute values of the measured emission intensities was about 5(%). The error in determining the temperature at different stations was within 1–2 (K).

Different OH bands were used to measure the hydroxyl emission at different stations in different years. It is well known, however, that there is a systematic difference (of a few degrees of Kelvin) in rotational temperatures between bands from different initial vibrational energy levels (Shefov 1961, 1976; Berg and Shefov 1963). According to subsequent investigations (Semenov et al. 2002; Bakanas and Perminov 2003; Bakanas et al. 2003), this difference also depends on season. Therefore, to compare the temperature data obtained at different stations, all of them were reduced to the same OH (5–2) band.

Certainly, the necessary condition for complete comparability of measurements performed at different stations is the use of the intensity factors that allow one to calculate OH rotational temperatures without regard of the vibrational level number. Unfortunately, these obvious requirements are ignored, to say the least. This problem has also been discussed in Sect. 2.2.5.

6.1.1 Long-Term Yearly Average Temperature Trends

Rocket Data. Temperature of the Stratosphere and Mesosphere

The main results by which one can judge on long-term temperature changes in the stratosphere and mesosphere have been obtained at the rocket stations of

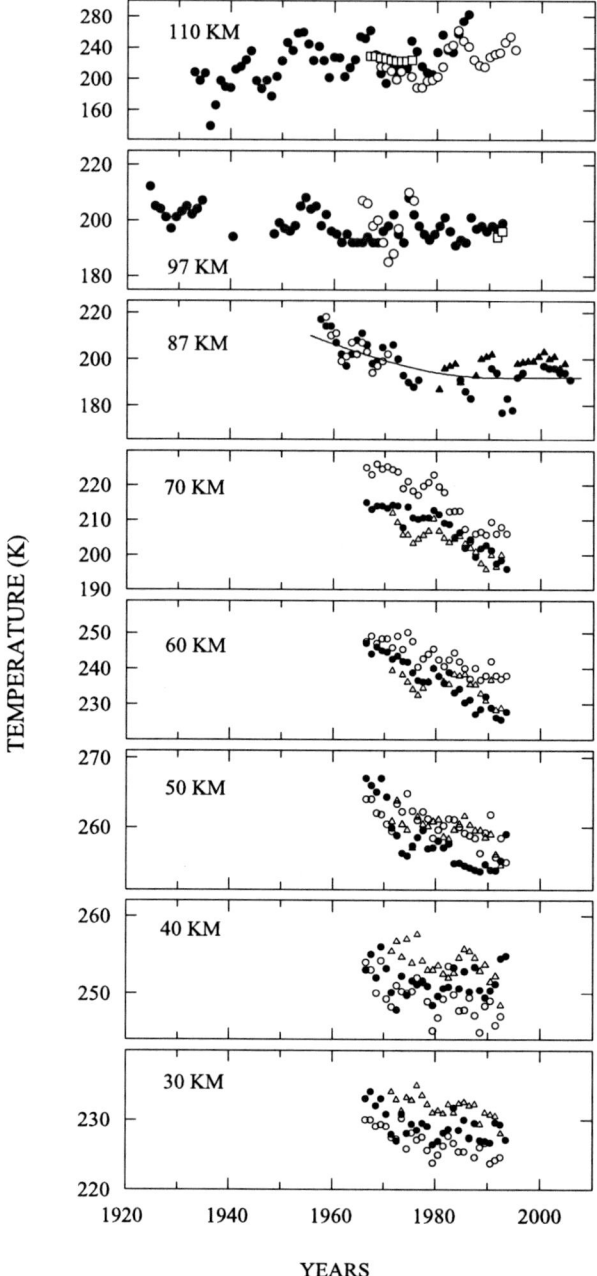

Fig. 6.1 Long-term variations of the yearly average atmospheric temperature at altitudes from 30 to 70 (km) retrieved from rocket measurements at high (*hollow circles*), middle (*full circles*), and low latitudes (*triangles*) (Lysenko 1981; Kokin et al. 1990; Kokin and Lysenko 1994); temperature variations at 87 (km) were retrieved from hydroxyl emission data of Zvenigorod

6.1 The Temperature Trend in the Middle Atmosphere

Heiss Island (80.6°N, 58.0°E) since 1964, Volgograd (48.7°N, 45.8°E) since 1965, Balkhash (46.8°N, 74.6°E) since 1973, Thumba (8.5°N, 76.8°E) since 1971, and Molodezhnaya (67.7°S, 45.8°E) since 1969. Weekly sounding with meteorological rockets was performed until 1993. The temperature was measured in the altitude range from 21–23 to 75–80 (km).

Figure 6.1 shows time series of yearly average atmospheric temperatures in 10-km layers centered at altitudes of 30, 40, 50, 60, and 70 (km) retrieved from measurements at high-latitude, midlatitude (Volgograd), and tropical stations (Kokin and Lysenko 1994; Golitsyn et al. 1996). It can clearly be seen that the yearly average temperatures decrease with time for all altitude layers throughout the observation period. In the mesosphere, a negative trend was most pronounced, and the temperature in the stratosphere also decreased by no less than 1 (K) per decade.

It is obvious that representative quantitative trend estimates cannot be obtained without taking into account long-period oscillations comparable in duration to solar cycles. To this end, the time series of atmospheric temperatures measured at different altitudes were analyzed using a specially developed adaptive system of statistical analysis (ASSA) (Rosenfeld 1986) based on the standard system of spectral analysis. Its special feature is that the initial set of harmonics expected from observations (basic functions) was specified a priori and then revised based on the results of spectral and correlation analysis of residual noise. Significant regular components present in the residual noise strongly worsen the accuracy of trend evaluation. Therefore, a reliable conclusion about the trend can be drawn only after determination of the entire spectrum of regular oscillations. To this end, power density spectra were calculated using the Hemming window together with the spectra of maximum residual noise entropy. Pronounced spectral maxima were checked for regularity, that is, harmonics with the corresponding periods were included in the set of basic functions. If the amplitudes of these harmonics appeared high enough and the corresponding maxima in the residual noise spectrum disappeared completely or partly, this gave grounds to consider the harmonics significant.

The monthly mean temperatures at altitudes from 25 to 75 (km) were analyzed with a step of 5 (km). Preliminary investigations (Kokin et al. 1990) demonstrated that to estimate a trend correctly, harmonics with periods of 0.5, 1.0, 2.0, 5.5, and 11 years should be considered. As to quasi-biennial oscillations, the residual noise spectrum had maxima corresponding to periods from 17 to 36 months. Therefore, the harmonic with a period of 24 months was also added to the basic functions (Kokin et al. 1990). Subsequently, based on the observations carried out at the

---

Fig. 6.1 (continued) (*full circles*), Abastumani (*hollow circles*) (Golitsyn et al. 1996; Semenov and Lysenko 1996; Givishvili et al., 1996), and Wuppertal (*triangles*) (Offermann and Graef 1992; Bittner et al. 2002; Offermann et al. 2004); temperature variations at 97 (km) were retrieved from 557.7-nm atomic oxygen emission data: interferometric measurements (*hollow circles*) (Hernandez and Killeen 1988), lidar measurements (*squares*) (She et al. 1993), and estimates from the emission intensity (*full circles*) (Evlashin et al. 1999; Starkov et al. 2000); and temperature variations at 110 (km) were retrieved from ionospheric (*full circles*) (Givishvili and Leshchenko 1995, 2000) and incoherent scattering measurements (*hollow circles*) (Alcaydé et al. 1979)

stations of Heiss Island and Volgograd for more than 25 years, calculations were performed with the 22-yr harmonic added to the basic functions. Analysis of the calculation results has shown that the difference between the trend values obtained with and without the 22-yr harmonic was within the rms error of the trend estimate.

Vertical profiles of the temperature trends retrieved from rocket measurements for different latitudes are shown in Fig. 6.2. It can be seen that negative temperature trends took place almost at all altitude levels from 25 to 75 (km). The only exception is the level at 45 (km) for which, according to the data of four stations (except for Heiss Island), the trend was absent. The data of all five stations showed that the upper stratosphere cooled with a rate of $-(0.1-0.2)$ $(K \cdot yr^{-1})$. In the mesosphere at midlatitudes of the eastern part of the northern hemisphere, the maximum negative temperature trend equal to 0.8 $(K \cdot yr^{-1})$ was observed at altitudes of 55–60 (km); at altitudes of 70–75 (km), the trend was -0.6 $(K \cdot yr^{-1})$. According to the data of

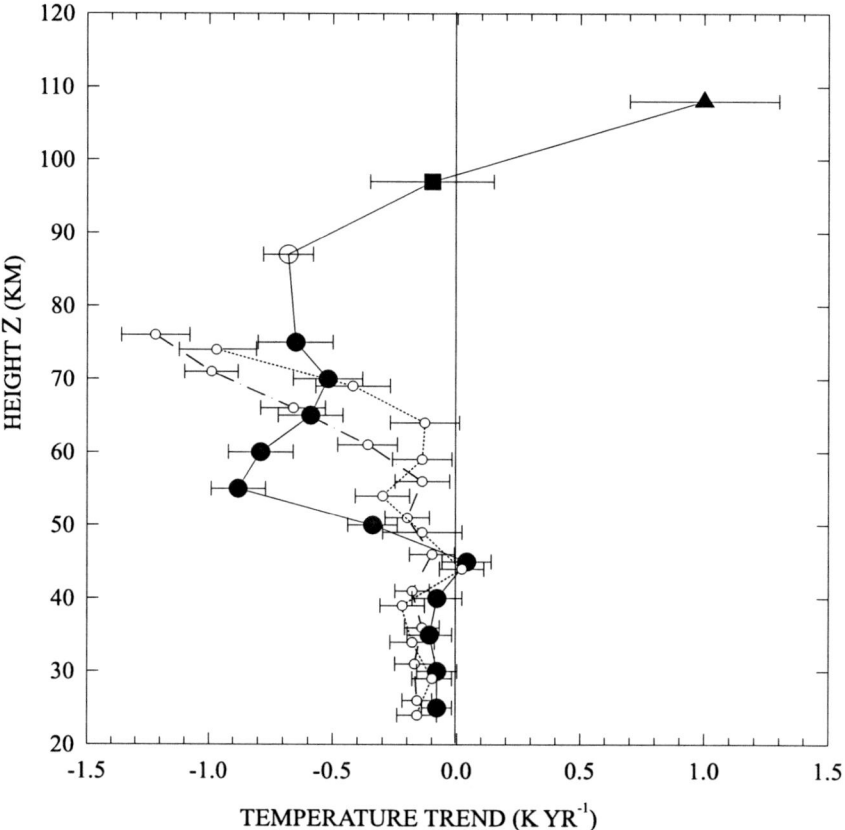

Fig. 6.2 Vertical profiles of the yearly average temperature trends retrieved from rocket (25–75 (km)) (small *open circles* and *full circles*); optical (87 and 97 (km)) (*open circle* and *full square*), and ionospheric measurements (108 (km)) (*full triangle*) at high (*dotted curve*), middle (*solid curve*), and low latitudes (*dashed curve*) (Golitsyn et al. 1996)

the Molodezhnaya and Heiss Island high-latitude stations, the trend values for the lower mesosphere and the upper stratosphere were approximately identical. Above 55 (km), the trend sharply increased and reached -1.1 (K·!yr^{-1}) at altitudes of 70–75 (km). Nearly the same vertical profile of the temperature trend was observed at the Thumba tropical station, but the sharp increase of the negative temperature trend occurred above 65 (km). It reached -1.0 (K·yr^{-1}) at an altitude of 75 (km).

The temperature trends were also analyzed (Schmidlin 1996) based on rocket measurements in America. The results of this work are in agreement with the rocket measurements performed in Russia for all latitudes at least up to 40 (km). Some disagreement takes place for equatorial measurements at an altitude of 50 (km), where, according to Thumba measurement data, a slight positive trend ($+0.02$ (K·yr^{-1})) was detected, whereas American researchers indicated a negative trend (-0.23 (K·yr^{-1})) at the Kwajalein station (9°N). The trend values for altitudes above 50 (km) were lacking in Rosenfeld's work (1986) because of the uncertainty of corrections that should be introduced to compare the temperatures calculated for different years.

Measurements of the Airglow Characteristics

Results of long-term measurements of the yearly average integral intensities of the hydroxyl emission and atomic oxygen green emission are shown in Fig. 6.3. To analyze the behavior of the 557.7-nm emission intensity, in addition to Abastumani measurements, data of observations carried out at other midlatitude stations, such as Terling (52.0°N, 1.0°W) (Lord Rayleigh and Spencer Jones 1935; Hernandez and Silverman 1964), Simeiz (44.4°N, 34.0°E) (Shain and Shain 1942), Haute Provence (43.9°N, 5.7°E) (Dufay and Tcheng Mao-Lin 1946, 1947a,b), and Cactus Peak (36.1°N, 117.8°W) (Roach et al. 1953) were used. Despite significant quasi-regular year-to-year variations, an increase in the OH emission intensity is observed throughout the observation period. The trend value was estimated by processing Abastumani data with the use of the ASSA code. Analysis of monthly mean values showed that the OH emission intensity increased between January 1948 and December 1992 with an average rate of $1.46 \pm 0.09(\% \cdot \text{yr}^{-1})$ with the confidence probability $P = 0.95$. In data processing, the 22-, 11-, and 5.5-yr long-period variations with amplitudes of 0.05–0.06 (megarayleigh) were taken into account. An earlier analysis of the variations of the 557.7-nm emission characteristics (Shefov and Kropotkina 1975) revealed a clearly defined solar flux dependence of the emission intensity and emission layer altitude as well as seasonal and lunar–tidal variations (Semenov and Shefov 1997a,b,c,d). Processing of the 557.7-nm emission data with the ASSA code revealed 22- and 11-yr harmonic oscillations with amplitudes of about 40 (Rayleigh) and 5.5-yr oscillations with amplitude of about 10 (Rayleigh). The positive trend amplitude averaged over the observation period was $0.6 \pm 0.09(\% \cdot \text{yr}^{-1})$ (Semenov and Lysenko 1996).

As mentioned above, the rotational temperature can be determined from the intensity distribution over the rotational–vibrational OH bands. Since the emission

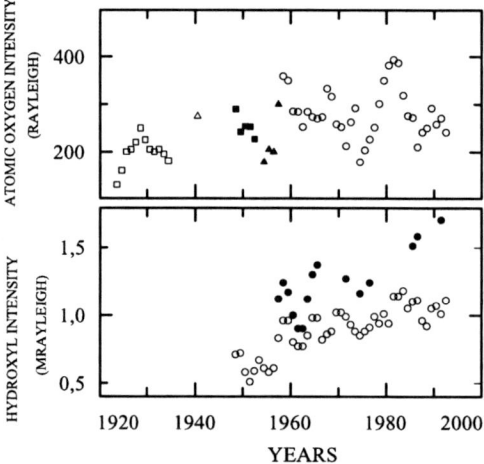

Fig. 6.3 Long-term variations of the yearly average 557.7-nm and hydroxyl emission intensities. The 557.7-nm emission intensity was retrieved from measurements performed at Abastumani (*open circles*), Haute Provence (*full triangles*), Cactus Peak (*full squares*), Simeiz (*open triangle*), and Terling (*open squares*). The hydroxyl emission intensity was retrieved from measurements performed at Abastumani (*open circles*) and Zvenigorod (*full circles*) (Golitsyn et al. 1996; Lysenko et al. 1999)

layer has a maximum at $\sim$87 (km) (Shefov 1978; Baker and Stair 1988; Semenov and Shefov 1996), the observed temperature variations should be attributed to the OH emission layer with a halfwidth of $\sim$8.5 (km) centered at 87 (km). Figure 6.1 shows these temperature data for Abastumani (1957–1972) and Zvenigorod (1957–1995). Over the indicated periods, the temperature has a negative trend for both stations equal to $-(0.7 \pm 0.1)$ (K·yr^{-1}) (Givishvili et al. 1996) (see Fig. 6.1). Offermann and Graef (1992), based on their own data, stated that the temperature trend in the mesopause region was $+1$ (K·yr^{-1}). However, being superimposed on a much longer time series of analogous data, they fitted well with that series and thus confirmed the trend value reported by Givishvili et al. (1996).

After observations carried out between 1995 and 2005, the opportunity arose to trace the temperature behavior in the mesopause region. In addition, recent special-purpose investigations of the proportion between the rotational temperatures for OH bands related to different vibrationally excited levels (Bakanas and Perminov 2003; Bakanas et al. 2003) have revealed that there are seasonal differences between these temperatures, but their yearly average values differ insignificantly (see Fig. 4.1). Therefore, for the period 1984–1986, when temperatures were determined by the OH (8–3) band, the temperature measurements were corrected to fit the OH fifth vibrational level, since the temperatures presented in Fig. 6.1 were calculated for the OH (5–2) band (Semenov and Shefov 1996). Full triangles in the same figure indicate the temperature data obtained at the Wuppertal station (Offermann and Graef 1992; Bittner et al. 2002; Offermann et al. 2004). The data set shown in this figure enabled a noticeable decrease in the negative temperature trend in the mesopause region over the last few years (from the late 1980s) to be detected. This suggests a noticeable nonlinearity in the long-term temperature behavior at these altitudes. Thus, the temperature trend over the period under consideration can be subdivided into two portions: an almost linear temperature decrease with a rate of about -0.6 (K·yr^{-1}) in 1955–1985 and a trend close to zero in 1985–2005. Unfortunately, it is too early to discuss the more recent temperature trend, because to this end, observations must be continued to encompass at least one or two solar cycles. The solid curve in Fig. 6.1 that approximates the long-term temperature variations between 1955 and 2005 is described by the relation

$$T = 198 - 0.6 \cdot (t - 1972) + 0.012 \cdot (t - 1972)^2, (K).$$

Figure 6.1 also illustrates the variations of the yearly average temperature of atomic oxygen emission at 557.7 (nm) ($\sim$97 (km)) retrieved from interferometric (Hernandez and Killeen 1988) and lidar measurements (She et al. 1993, 1995) performed at the same altitudes. The temperatures estimated from the data on the 557.7-nm emission intensity (Semenov and Lysenko 1996) are also shown in the figure. A weak tendency toward a decrease in temperature at $\sim$97 (km) at midlatitudes can be seen for the entire data set (1924–1992). The linear temperature trend for this period was about -0.1 (K·yr^{-1}) (see Fig. 6.2).

The temperature in the ionospheric, E layer ($\sim$110 (km)) (Givishvili and Leshchenko 2000) was estimated from the data of vertical sounding (Givishvili and Leshchenko 1993, 1995) performed at the stations near Moscow (55.5°N), Slough (52.5°N), and Juliusruh (55.6°N) and from the data of incoherent scattering measurements (1967–1975) performed at Saint-Santin (44.6°N, 2°E) (Alcaydé et al. 1979). According to the data reported by Givishvili and Leshchenko (1995, 2000), a systematic increase in atmospheric temperature at $\sim$110 (km) with an average rate of about $+1$ (K·yr^{-1}) has been observed since 1931 till now (see Fig. 6.2).

Vertical Profiles of the Yearly Average Temperature at Altitudes of 25–110 (km)

Figure 6.4 shows the altitude–time profiles of the yearly average temperature in the middle atmosphere at midlatitudes of the northern hemisphere derived by Lysenko et al. (1999) and Semenov (2000) from the data presented elsewhere (Golitsyn et al. 1996; Semenov and Lysenko 1996; Givishvili et al. 1996; Semenov et al. 1996). It also shows a time series of the yearly average radio-frequency emission flux $F_{10.7}$. The individual profiles shown in the figure are shifted by 10 (K). They were reconstructed from the data of rocket (25, 30, 35, ..., 75 (km)), optical (87 and 97 (km)), and ionospheric measurements (110 (km)) performed in 1955–1995 with their subsequent interpolation and extrapolation. The retrieved vertical profiles were smoothed using the method of spline interpolation. An analysis of this family of profiles has revealed a number of new and important features. First, the yearly average vertical temperature profiles undergo systematic year-to-year changes. Second, a temperature maximum whose amplitude ($\Delta T \approx 20\text{--}25$ (K)) correlates (r = 0.88 ± 0.04) with the solar flux (Semenov and Shefov 1999b) is

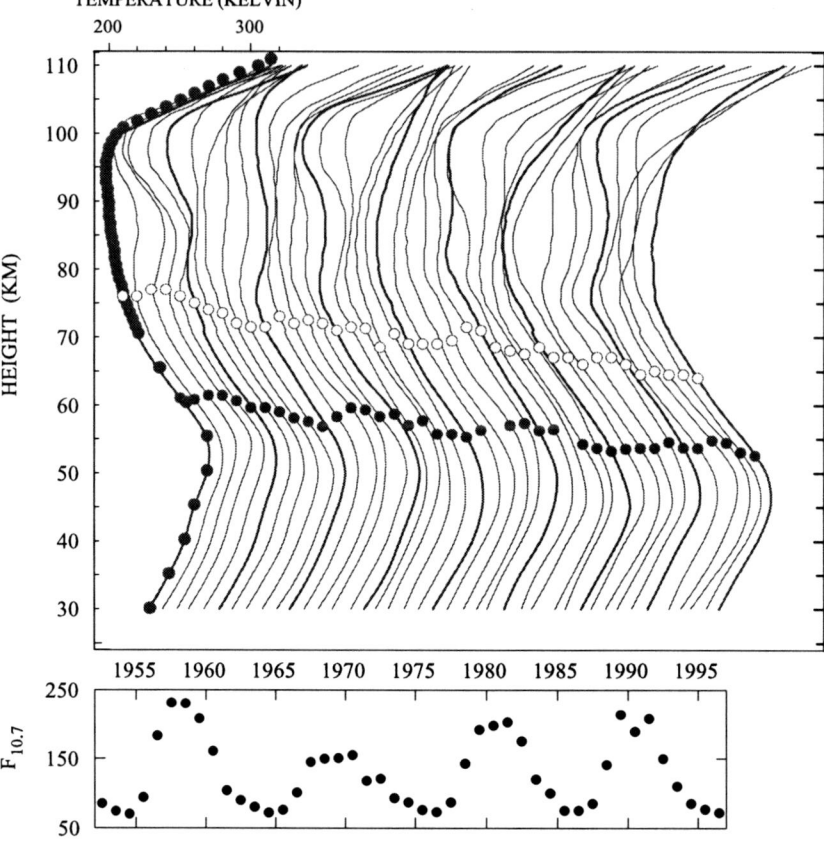

Fig. 6.4 Vertical yearly average temperature profiles (25–110 (km)) at midlatitudes retrieved for 1955–1995 from measurement data (Alcaydé et al. 1979; Hernandez and Killeen 1988; Kokin et al. 1990; Offermann and Graef 1992; She et al. 1993; Kokin and Lysenko 1994; Givishvili et al. 1996; Semenov et al. 1996; Golitsyn et al. 1996; Givishvili and Leshchenko 1999; Evlashin et al. 1999). Each profile is shifted from the preceding one by 10 (K). *Bold curves* illustrate the temperature profiles measured every 5 years, *open* and *full circles* indicate altitudes at which the temperature was equal to 210 and 250 (K), respectively. The temperature scale is indicated at the top left of the figure. The variations of the solar activity (yearly average index $F_{10.7}$) (Lysenko et al. 1999; Semenov 2000) are shown below

observed at altitudes of 85–95 (km). Since the vertical temperature profile was constructed using spline interpolation of measurement data for altitudes of 70, 75, 87, 97, and 110 (km), the altitude of this maximum needs further refinement. Nevertheless, it should be emphasized that the temperature at altitudes of ∼92 (km) retrieved from sodium emission data (Shefov and Semenov 2001) is in agreement with the data shown in Fig. 6.4 within ±4 (K).

Figure 6.5 shows vertical temperature profiles for periods of maximum (*solid curves*) and minimum solar fluxes (*dotted curves*) (Lysenko et al. 1999). A significant difference between them can be seen. The inset, which presents all vertical temperature profiles for the entire observation period, gives an idea of the dynamic

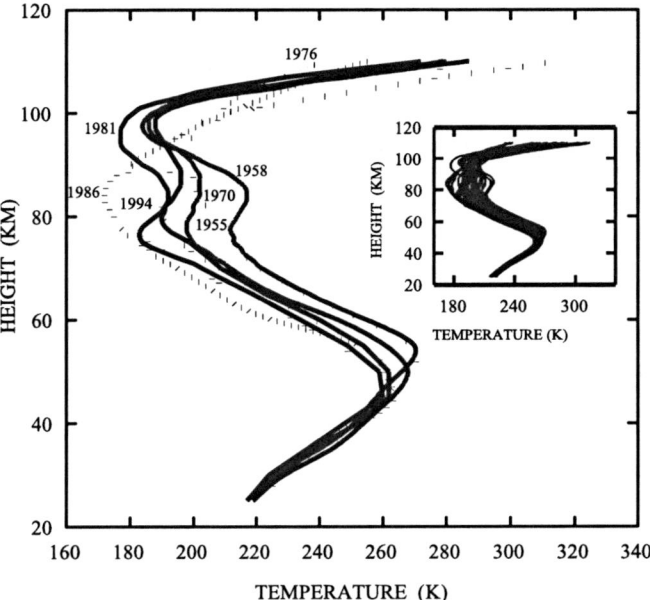

Fig. 6.5 Examples of the yearly average vertical temperature profiles in years of maximum (1958, 1970, 1981, 1989) (*solid curves*) and minimum solar activity (1955, 1962, 1976, 1986, 1994) (*dashed curves*). The inset shows all temperature profiles from Fig. 6.4 (Lysenko et al. 1999)

character of vertical temperature variations at altitudes of 70–100 (km). It can be seen that the behavior of vertical temperature profiles differs from that predicted by the existing models, which do not take into account the effect of the solar flux and long-term trends.

Analysis of the features of the temperature regime (see Figs. 6.4 and 6.6) shows that constant-temperature altitude levels slightly ascend in the range 30–45 (km) with rates up to 50 $(m \cdot yr^{-1})$; they systematically descend in the altitude range 45–75 (km) with a rate reaching -250 $(m \cdot yr^{-1})$ near 75 (km). Above 80 (km), these levels undergo systematic variations that are manifestations of the effect of the solar flux and long-term trend. For the period between 1955 and 1995, the altitude of the minimum temperature at solar minima decreased from 95 to 83 (km), and the minimum temperature decreased from 197 to 177 (K) with a rate of $-(0.5 \pm 0.1)$ $(K \cdot yr^{-1})$.

An important feature of vertical temperature profiles for altitudes of 80–100 (km) is a maximum observed systematically in periods of high solar activity during several 11-yr cycles. This maximum was also revealed in the data of short-term lidar measurements carried out in 1990–1993 at the Fort Collins Observatory (40.6°N, 105°W) (She et al. 1993, 1995; Yu and She 1995). In the subsequent measurement series dated up to the end of 1997 (minimum solar activity) (She et al. 1998), the authors of the publications noted a gradual decrease in the amplitude of the temperature maximum at altitudes of 80–100 (km). However, they did not relate the temperature maximum in this altitude range to the solar flux variations that took place

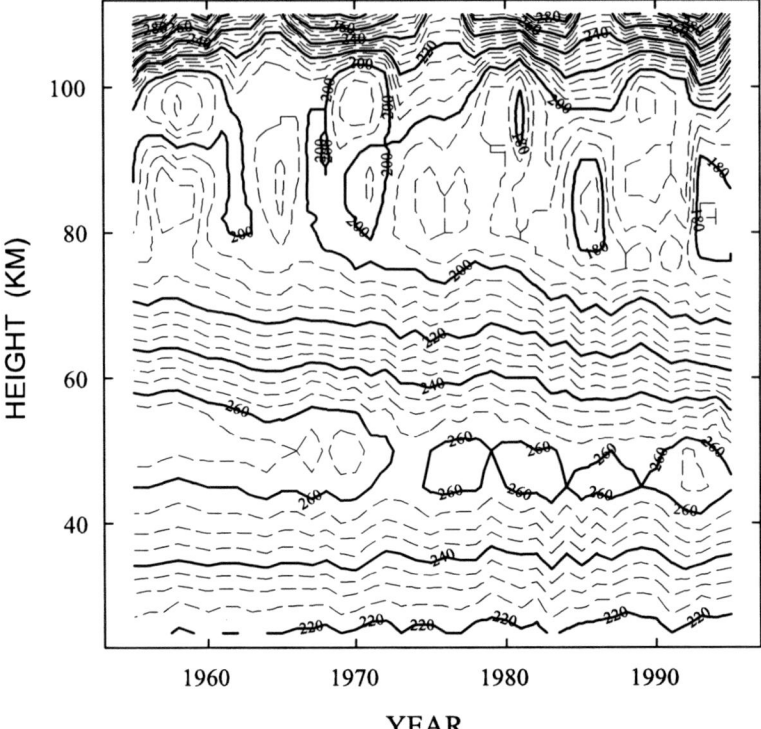

Fig. 6.6 Altitude–time distribution of the yearly average temperature at midlatitudes over a 40-yr period. Isotherms are drawn with a step of 5 (K) (Lysenko et al. 1999)

during the observation period and suggested the influence of the Pinatubo volcano eruption (July 1991).

In the last few years, one more temperature maximum with amplitude of 10–30 (K) was detected at altitudes of 60–70 (km) (Meriwether and Gardner 2000); its behavior is poorly known yet.

As can be seen from Fig. 6.7, the temperature difference in the regions of minima of the vertical temperature profiles (below T_L and above T_U of the maximum observed around 90 (km)) varies with time in the periods of increased solar activity, correlates with the solar flux, has a long-term trend, and changes its sign in the period under consideration. Its behavior can be described as

$$T_L - T_U = -0.72 \cdot (t - 1972.5) + 0.1(F_{10.7} - 13), (K) .$$

In addition, long-term variations of the temperature minimum at altitudes of 70–110 (km) in the periods of solar minimum between 1955 and 1995 also testify to a monotonic decrease of its altitude Z_M and temperature T_M (Fig. 6.8) (Semenov and Shefov 1999b).

6.1 The Temperature Trend in the Middle Atmosphere

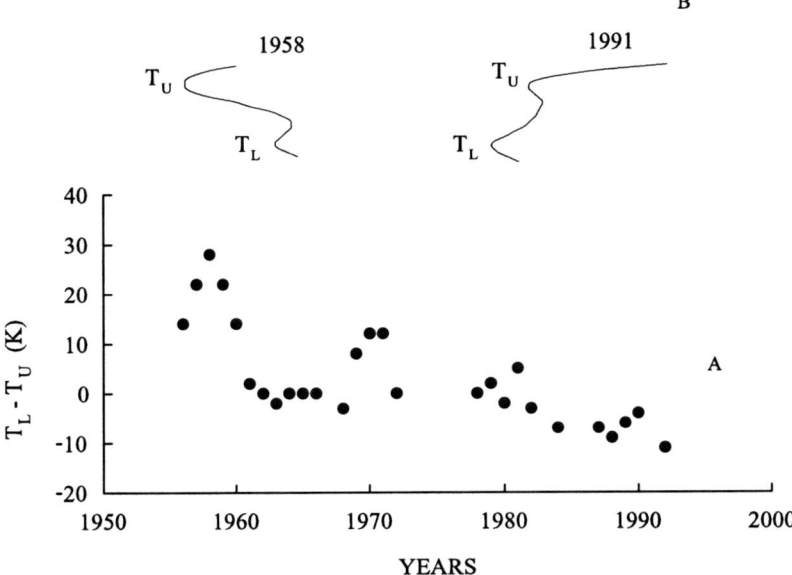

Fig. 6.7 Time dependences of the temperature difference between the lower and upper temperature minima (**A**) and examples (**B**) of vertical temperature profiles for years of solar maxima (the 19th and 22nd cycles) (Semenov and Shefov 1999)

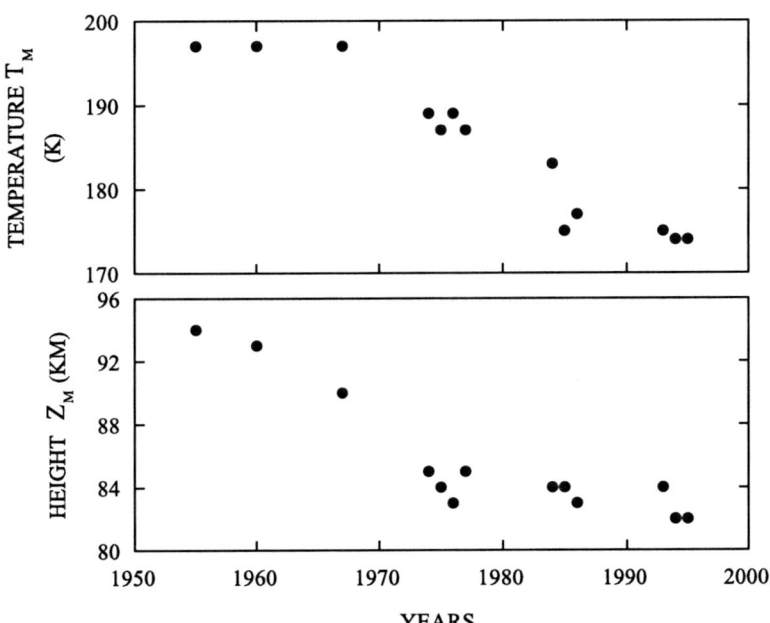

Fig. 6.8 Long-term variations of the altitude and temperature of the temperature minimum at 70–110 (km) at solar minima in 1955–1995 (Semenov and Shefov 1999)

The temperature maximum can be accounted for the influence of the atomic oxygen recombination process (Fomichev and Shved 1994) mainly due to variations of the altitude of the layer with a maximum atomic oxygen density caused by changes of solar activity (Semenov and Shefov 1999b). This is supported by the data on seasonal variations of the temperature of the upper minimum in the mesopause region (Yu and She 1995) and on vertical profiles of atomic oxygen density (Perminov et al. 1998). It should be emphasized that the seasonal and daily variations of the altitude of the temperature maximum (She et al. 1995; Meriwether and Gardner 2000) are in agreement with the variations of the atomic oxygen layer altitude Z_O and density $[O(Z_O)]$ (Fig. 6.9) (Semenov and Shefov 1997c, 1999b; Golitsyn et al. 2001). Their average dependences on solar flux have the form

$$Z_O = 98.5 - 0.029 \cdot (F_{10.7} - 130), (\text{km}),$$
$$[O(Z_O)] = [8 - 0.011 \cdot (F_{10.7} - 130)] \cdot 10^{11}, (\text{cm}^{-3}).$$

Considering the altitude variations of the temperature in relation to those of the atomic oxygen density, one can see a strong correlation between them ($r = 0.96 \pm 0.033$), namely, the increase in temperature is described by the expression

$$\Delta T(Z) = (38 \pm 5) \cdot \log_e \left[(0.85 \pm 0.06) \cdot \frac{[O(Z)]_{\text{hsa}}}{[O(Z)]_{\text{lsa}}} \right],$$

where $[O(Z)]_{\text{hsa}}$ and $[O(Z)]_{\text{lsa}}$ denote the atomic oxygen densities below 95 (km) for high (hsa) and low solar activity (lsa), respectively (Semenov and Shefov 1999b).

It is of importance that the correlation between the temperature change and the atomic oxygen density ratio does not result from the calculations based on the photochemical process of excitation of the 557.7-nm emission and vertical temperature profiles. The difference between the vertical density profiles $[O(Z)]_{\text{hsa}}$ and $[O(Z)]_{\text{lsa}}$ follows from different green emission layer altitudes that depend on solar activity.

Obviously, a substantial effect of the vertical wind velocity component at these altitudes should also be taken into account. Its average value is about 2 $(\text{cm} \cdot \text{s}^{-1})$ and depends significantly on latitude (Portnyagin et al. 1995).

Therefore, it seems likely that the seasonal variations of the minimum temperature altitude ($\sim$100 (km) in winter and $\sim$88 (km) in summer) (Berger and von Zahn 1999) are due to variations of the atomic oxygen layer altitude.

Satellite data on the behavior of the altitude of 557.7-nm green atomic oxygen emission maximum (Fauliot et al. 1997) that is a manifestation of the planetary circulation are in good agreement with the predictions of an empirical model which describes the variability of the parameters of this emission (Semenov and Shefov 1997a,b,c,d, 1999b). The exothermic reactions with participation of the HO_X and O_X atmospheric constituents that proceed at these altitudes can also contribute to the energy balance (She et al. 1995). The difference between the energy influxes in the middle atmosphere for high and low solar activity clearly follows from the data given in Fig. 6.5.

6.1 The Temperature Trend in the Middle Atmosphere

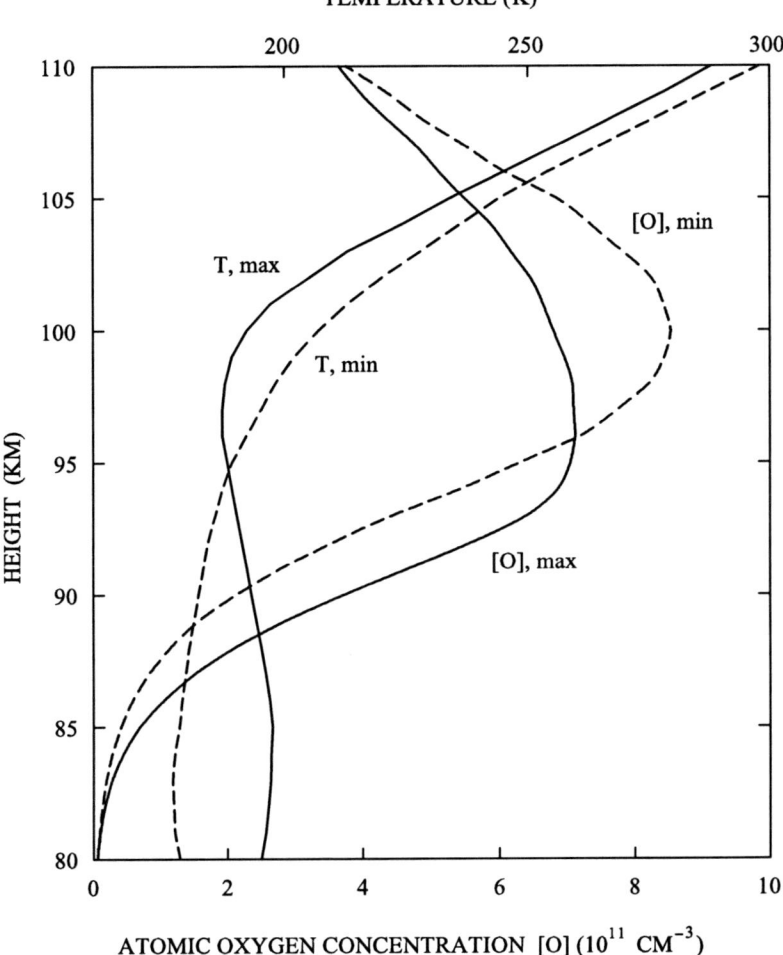

Fig. 6.9 Vertical profiles of the yearly average temperature and atomic oxygen concentration for years of solar maximum (1989, $F_{10.7} = 214$; *solid curves*) and minimum (1995, $F_{10.7} = 78$; *dashed curves*). The atomic oxygen concentrations are calculated with help of an empirical model of 557.7-nm emission variations (Semenov and Shefov 1999; Golitsyn et al. 2001)

6.1.2 Seasonal Behavior of Long-Term Temperature Trends for the Middle Atmosphere

Analysis of long-term monthly mean temperature variations at different altitudes revealed considerable differences between the trends in different seasons. The trends differ most significantly in winter and in summer. By way of example, Fig. 6.10 shows time series of the monthly mean temperatures in December and June (months of the winter and summer solstice) retrieved from rocket, spectrophotometric, and

Fig. 6.10 Long-term variations of the monthly average winter (December) and summer (June) temperatures at different altitudes retrieved from the data of rocket, spectrophotometric, and radiophysical measurements. *Closed symbols* are for December, and *open symbols* are for June. *Circles* for an altitude of 87 (km) are for Zvenigorod, *squares* for Wuppertal, *triangles* for Yakutsk, *inverted triangles* for Maynooth, and *diamonds* for Quebec and Delaware. *Solid lines* are the regression lines (Semenov et al. 2000)

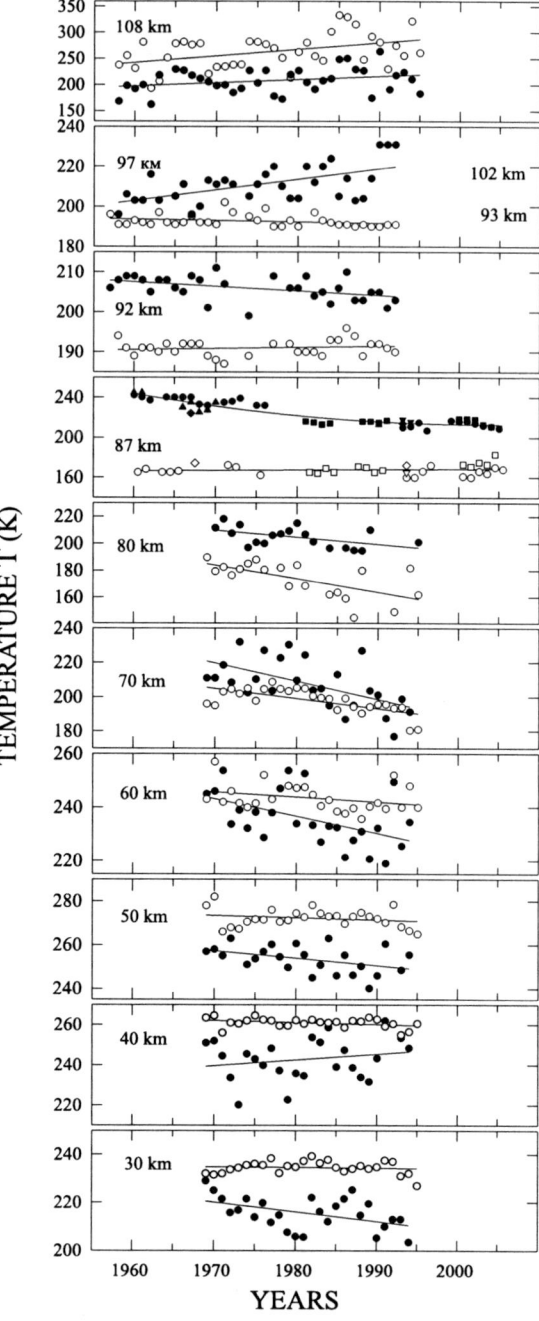

radiophysical data. The rocket data (25–80 (km)) had been obtained at Volgograd (48.7°N) between 1969 and 1995. The spectrophotometric data on the temperature regime at an altitude of 87 (km) were retrieved from the hydroxyl emission measurements performed at Zvenigorod (55.7°N) in 1960–1998, Wuppertal (51°N) in 1980–1998 (Offermann and Graef 1992; Offermann et al. 2002, 2004), Maynooth (53.2°N) in 1993 (Mulligan et al. 1995), Quebec (46.8°N) in 1967 (Lowe 1969; Snelling and Hampson 1969), Delaware (42.8°N) in 1993 (She and Lowe 1998), and Yakutsk (62°N) in 1960–1990 (Yarin 1961, 1962; Ignatiev et al. 1974; Atlasov et al. 1975; Scheer et al. 1994; Ammosov and Gavrilieva 1996). The temperature around 92 (km) was retrieved from an analysis of the behavior of the atmospheric sodium emission. Long-term (1957–1992) series of the measured sodium emission intensity are available for Abastumani (41.8°N) (Fishkova 1979, 1983; Fishkova et al. 2001a), and the long-term (1972–1987) series of the sodium layer altitude were retrieved from lidar measurements performed in Brazil (23.2°S) (Clemesha et al. 1992). According to lidar data (Qian and Gardner 1995; States and Gardner 1999), there is a strong correlation between the sodium density in the layer and the temperature at its density maximum. Long-term measurements of the seasonal sodium emission intensity variations (Fishkova 1979, 1983; Fishkova et al. 2001a; Shefov et al. 2000; Shefov and Semenov 2001) made it possible to reduce the data to identical heliogeophysical conditions and to derive correlation relations between the sodium emission intensity I_{Na} (in Rayleigh) and the atmospheric temperature around 92 (km) (Fishkova et al. 2001a; Shefov et al. 2000):

$$T(92(km)) = (185 \pm 0.8) + (0.20 \pm 0.01) \cdot I_{Na}, (K),$$

with the correlation coefficient $r = 0.952 \pm 0.020$. Here it should be noted that the temperatures at an altitude of $\sim$92 (km) obtained by this method are in good agreement ($\sigma \sim 4(K)$) with the yearly average temperatures for 1955–1995 (Lysenko et al. 1999; Semenov and Shefov 1999b).

Similarly, based on the long-term observations of the 557.7-nm atomic oxygen emission intensity (Fishkova et al. 2000, 2001b) and on its correlation with the atmospheric temperature at the emission layer altitude ($\sim$97 (km)), the behavior of the temperature in different seasons was estimated using a procedure described elsewhere (Evlashin et al. 1999; Fishkova et al. 2001b; Starkov et al. 2000).

To estimate the temperature of the lower thermosphere (105–110 (km)), results of vertical sounding (VS) of the ionosphere – the critical frequencies f_oE of the midday E layer corresponding to the stationary daytime conditions in the ionosphere (Givishvili and Leshchenko 2000) – were used. To eliminate the effect of solar flux on the results, the data were reduced to the fixed solar flux $F_{10.7} = 130$. In this case, according to an empirical model of the ionosphere (Fatkullin et al. 1981), the maximum of the midlatitude E layer is located at $\sim$109 (km) in winter and at $\sim$107 (km) in summer. Thus, the retrieved temperatures refer to altitudes of 107–109 (km) and correspond to the daily mean temperatures since, following Hedin (1983), the daily temperature oscillations at these altitudes, equal to 1–2 (K), can be neglected. The error in estimating the monthly mean temperature was not over 7 (K). In this analysis, data of measurements performed at three stations – Slough (52.5°N) in 1958–1987, Juliusruh (55.6°N), and Moscow (55.5°N) in 1958–1994 – were used.

The temperature trends for different seasons were estimated quantitatively with smoothing long-term oscillations, whose periods were comparable with the 11-yr solar cycle. The method of statistical processing of time series is described in detail elsewhere (Givishvili et al. 1996). Figure 6.10 shows an example of long-term temperature variations at different altitudes in winter (December) and in summer (June). From the figure, it can be seen that their behavior changes significantly. This is especially clearly seen in Fig. 6.11 that illustrates the seasonal variations of the temperature trend at different altitudes in the region of midlatitudes. Full circles show the trend values calculated by statistical processing for each month, and solid curves approximate the seasonal trend variations $\delta T_{tr}(t_d, z)$ retrieved on the basis of the sum of harmonics with periods of 12, 6, 4, 3, 2.4, and 2 months using the expression

$$\delta T_{tr}(t_d, Z) = \delta T_{trMA}(Z) + A_1(Z) \cdot \cos \frac{2\pi}{365}(t_d - t_1)$$
$$+ A_2(Z) \cdot \cos \frac{4\pi}{365}(t_d - t_2) + A_3(Z) \cdot \cos \frac{6\pi}{365}(t_d - t_3)$$
$$+ A_4(Z) \cdot \cos \frac{8\pi}{365}(t_d - t_4) + A_5(Z) \cdot \cos \frac{10\pi}{365}(t_d - t_5)$$
$$+ A_6(Z) \cdot \cos \frac{12\pi}{365}(t_d - t_6),$$

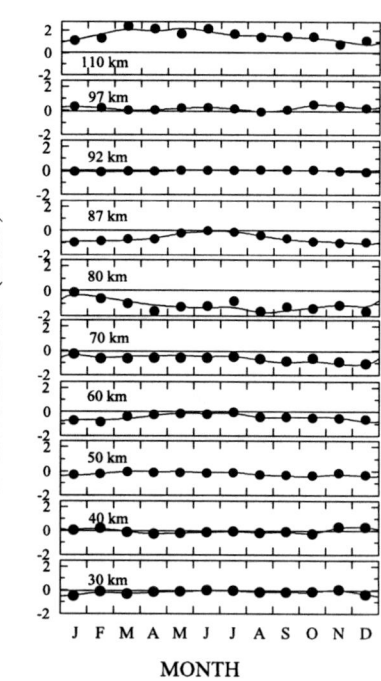

Fig. 6.11 Seasonal behavior of the long-term temperature trends ($K\,yr^{-1}$) in the atmosphere at different altitudes for midlatitudes. *Full circles* show the measured monthly mean temperatures, and *solid curves* show their approximations by the sum of harmonics (Semenov et al. 2002a)

6.1 The Temperature Trend in the Middle Atmosphere

Table 6.1 Amplitudes and phases of harmonics with periods of 12, 6, 4, 3, 2.4, and 2 months approximating the seasonal variations of temperature trends at different altitudes

Z (km)	δT_{trMA} $(K \cdot yr^{-1})$	Harmonic amplitudes $(K \cdot yr^{-1})$						Harmonic phases (days)					
		A_1	A_2	A_3	A_4	A_5	A_6	t_1	t_2	t_3	t_4	t_5	t_6
110	1.56	0.53	0.20	0.10	0.08	0.07	0.04	136	88	45	50	6	10
105	1.02	0.32	0.07	0.09	0.14	0.04	0.02	136	115	63	0	12	15
100	0.44	0.09	0.13	0.08	0.09	0.01	0.001	314	144	44	12	11	0
97	0.19	0.11	0.12	0.12	0.06	0.01	0.01	336	148	44	21	14	23
95	0.05	0.17	0.12	0.09	0.05	0.02	0.01	327	160	48	29	10	19
92	−0.03	0.09	0.01	0.02	0.01	0.01	0.01	214	170	37	35	8	15
90	−0.14	0.34	0.07	0.03	0.01	0.02	0.004	188	170	50	13	1	13
87	−0.63	0.44	0.13	0.05	0.03	0.01	0.02	173	178	47	30	6	21
85	−0.53	0.52	0.12	0.07	0.04	0.06	0.03	169	47	4	40	27	15
82	−0.81	0.45	0.14	0.12	0.08	0.04	0.04	100	29	27	27	20	15
80	−1.13	0.48	0.15	0.15	0.12	0.06	0.06	40	24	42	21	23	15
75	−0.90	0.40	0.12	0.15	0.13	0.06	0.05	68	32	42	23	11	10
70	−0.67	0.32	0.09	0.08	0.13	0.07	0.04	115	28	31	18	10	12
65	−0.54	0.26	0.08	0.04	0.07	0.03	0.05	157	30	50	16	19	14
60	−0.45	0.28	0.10	0.06	0.06	0.02	0.03	175	136	50	25	35	16
55	−0.39	0.30	0.12	0.11	0.02	0.04	0.03	173	131	91	40	35	18
50	−0.22	0.19	0.03	0.06	0.01	0.01	0.03	125	119	75	50	30	16
45	−0.01	0.15	0.05	0.06	0.05	0.01	0.01	0	5	62	52	56	20
40	−0.04	0.19	0.12	0.05	0.07	0.06	0.02	356	182	78	57	42	13
35	−0.12	0.12	0.04	0.04	0.06	0.05	0.03	306	143	69	50	37	10
30	−0.19	0.12	0.06	0.09	0.06	0.05	0.01	202	127	62	45	37	20
25	−0.22	0.13	0.04	0.03	0.03	0.02	0.01	173	86	65	48	36	25

where $\delta T_{tr}(t_d, z)$ is the yearly average trend value at altitude Z; A_1, A_2, A_3, A_4, A_5, and A_6 are the harmonic amplitudes; t_1, t_2, t_3, t_4, t_5, and t_6 are their phases, t_d is the day of the year. The harmonic amplitudes and phases are given in Table 6.1.

The vertical profiles of monthly mean temperature trends shown in Fig. 6.12 were drawn based on the results presented in Fig. 6.11. To the right of the figure, the yearly average trend is shown. Designations in Fig. 6.12 are the same as in Fig. 6.11: full circles denote the trend values calculated from the regression equation, and solid curves show their approximations, as a rule, by the sum of harmonics with periods of 12, 6, and 4 months. It should be noted that the sums of six harmonics were used to approximate the vertical winter temperature profiles (November–February). The vertical temperature profiles shown in Fig. 6.3 were drawn with allowance for the seasonal variations of the altitude of the ionospheric E layer (Fatkullin et al. 1981) and of the 557.7-nm emission layer (Semenov and Shefov 1997c,d). The seasonal and long-term variations of the sodium layer altitude were small (∼1 (km)) (Clemesha et al. 1992, 2003, 2004; Qian and Gardner 1995; States and Gardner 1999; Shefov and Semenov 2001) and hence were not taken into account in constructing the vertical profiles.

As can be seen from Figs. 6.11 and 6.12, the long-term temperature trend for different seasons has different values at different altitudes in the middle atmosphere.

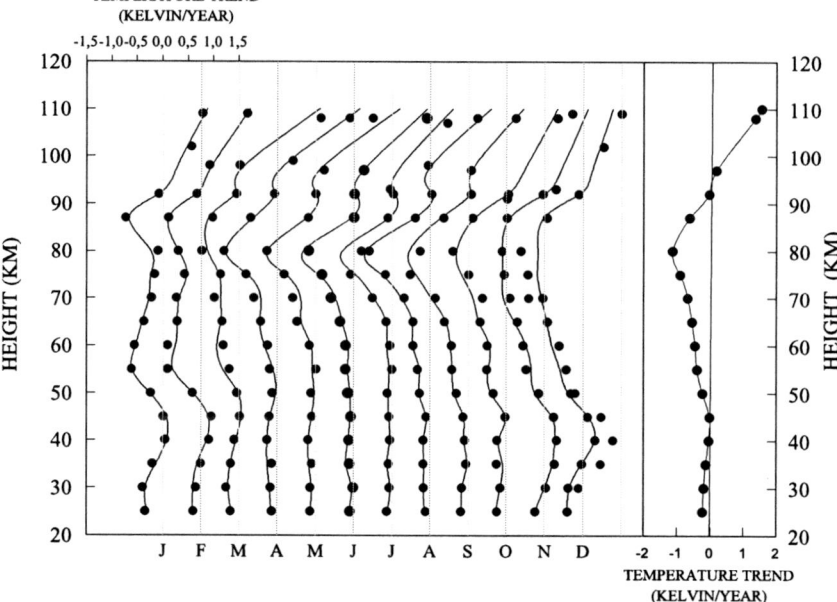

Fig. 6.12 Vertical profiles of the long-term temperature trends ($K \cdot yr^{-1}$) for different months of the year. *Full circles* show the measured temperatures, and *solid curves* show their approximations by the sum of harmonics. *Thin straight lines* above the serial month numbers show the zero temperature trends for each profile. The trend scale is at the top of the figure. The vertical profile of the yearly average temperature trend (Semenov et al. 2002a) is shown to the right of the figure

In summer, the stratospheric trend remains almost unchanged with an increase in altitude; it is equal to approximately -0.1 ($K \cdot yr^{-1}$). In the mesosphere, it sharply increases to a maximum negative value of approximately -1.1 ($K \cdot yr^{-1}$) in the layer 75–80 (km) and then its absolute value noticeably decreases. In the mesopause region, the trend becomes positive, increasing to 1.1 ($K \cdot yr^{-1}$) at an altitude of 107 (km). In winter, the negative trend $-(0.3–0.5)$ ($K \cdot yr^{-1}$) decreases in the middle stratosphere, passes through zero, and becomes positive in the upper stratosphere. The maximum positive trend changing from 0.3 ($K \cdot yr^{-1}$) in February and November to 0.7 ($K \cdot yr^{-1}$) in December is observed in the layer of 40–45 (km). Near the stratopause, the trend vanishes, and in the lower mesosphere, the negative trend rapidly increases with altitude. In the middle and upper mesosphere, the trend increases much more slowly: from -0.7 ($K \cdot yr^{-1}$) at 60 (km) to -0.9 ($K \cdot yr^{-1}$) around 87 (km). At the altitude of the E-layer maximum, the temperature trend has a positive value of $1–2$ ($K \cdot yr^{-1}$) throughout the year.

As to the positive temperature trend in the winter upper stratosphere, we can state the following: The observed similarity of rocket data and predictions of a numerical simulation of climate in the middle atmosphere with the doubled carbon dioxide content (Rind et al. 1990) suggests that the increase in the temperature of the upper atmosphere in winter is a consequence of the change of the atmospheric dynamics because of an increase in the concentration of greenhouse gases. The results of the simulation demonstrate the trend contrast in adjacent autumn–winter months, namely, an increase in temperature in November, January, and February and a temperature decrease in December (Rind et al. 1990). This simulation result agrees in general with the observations; however, negative trend values were measured in January rather than in December (see Fig. 6.12). In addition, according to the data of Rind et al. (1990), the winter trend positive anomaly should become more pronounced in going from midlatitudes to North Pole. Observations, however, demonstrate that the period of this seasonal anomaly and the absolute value of the positive trend are much greater at midlatitudes (Volgograd) than at high latitudes (Heiss Island) (Lysenko et al. 2003; Lysenko and Rusina 2003).

Positive trends in the upper stratosphere were detected at another midlatitude station (Balkhash, 46.8°N) (Kokin and Lysenko 1994). Analysis of the data of rocket sounding performed at Riori (39°N) for a 25-yr period of regular observations (Keckhut and Kodera 1999) has demonstrated that in December, the temperature trend in a layer of 34–47 (km) acquires positive values with a maximum of $\sim$0.5 $(K \cdot yr^{-1})$ around 41 (km). It seems that the anomalous winter temperature trend in the upper stratosphere has the greatest absolute value and the greatest seasonal duration in the 50–60°N latitude belt, decreasing toward North Pole and toward the equator. The temperature trend in the middle and upper stratosphere over the Thumba tropical station (8°N) is negative throughout a year (Lysenko et al. 2003; Lysenko and Rusina 2003).

Positive values of the temperature trend in the lower thermosphere (95–110 (km)) are stably observed throughout a year and cannot be explained by changes of the chemical composition of the atmosphere. In any case, a numerical simulation of climatic changes in the upper atmosphere resulting from an increase in the content of greenhouse gases indicates cooling of this layer and of the thermosphere as a whole (Roble and Dickinson 1989). A possible reason for positive trend values is deformation of the vertical temperature profile owing to a subsidence of the thermosphere. If we assume that the subsidence occurs in such a manner that the same temperature value corresponds to the same density in the thermosphere, the product of the rate of subsidence by the vertical temperature gradient will yield a positive temperature increment with time, that is, the deformation effect which can partly compensate for and at some altitudes even exceed the negative temperature trend in the thermosphere. Rough estimation of the deformation effect by the yearly average temperature vertical profile predicted with the MSIS model and by the rate of subsidence of the upper atmosphere (Semenov and Lysenko 1996) has shown that its maximum value is about 3 $(K \cdot yr^{-1})$ at an altitude of 135 (km). Assuming, with no regard of the deformation effect, that the negative temperature trend linearly increases from -0.7 $(K \cdot yr^{-1})$ in the mesopause to about -2.5 $(K \cdot yr^{-1})$ at $\sim$300 (km) (Semenov and Lysenko 1996; Semenov 1996), we obtain that in reality, with allowance for the subsidence of the thermosphere, a positive trend should be observed at altitudes of 100–170 (km) with a maximum of $\sim$2 $(K \cdot yr^{-1})$ at an altitude of 130 (km). The observed seasonal behavior of the trend at these altitudes is, perhaps, related to seasonal changes of the atomic oxygen layer altitude (Semenov 1997).

Attention should be drawn to the presence of a transition zone at altitudes of 92–93 (km) near the mesopause. The temperature trends are, as a rule, negative in all seasons below this zone and positive above it. The altitude of this transition zone remains almost unchanged throughout a year. In summer, this zone with almost zero temperature trend extends downward to altitudes of 87 (km) and at high latitudes (70°N) even to 82 (km) (Lübken et al. 1996) – the altitude of noctilucent clouds. The altitude of maximum negative trend undergoes annual variations, namely, it is maximum in winter ($\sim$87 (km)) and minimum in summer ($\sim$78 (km)).

The winter temperature trend retrieved from 20-yr (1980–2001) measurements carried out on Spitsbergen Island is reported by Nielsen et al. (2002). However, winter measurements for each year were performed in different periods and reflected random perturbations during these periods (Sigernes et al. 2003a,b). To eliminate the influence of these possible perturbations and to reveal a tendency in the behavior of winter mesopause temperature from its annual course, these data were reanalyzed by Semenov and Shefov (2006). To this end, the figures published by Sigernes et al. (2003b) were reduced to a common timescale (Fig. 6.13). It is clear that the data of temperature measurements must be reduced to identical heliogeophysical conditions to exclude the effect of all regular and irregular variations. Only after this procedure, the data can be used to reveal long-term changes. Sigernes et al. (2003b) used the data of measurements performed during a polar night in November–February ($305 \leq t_d \leq 60$). However, they did not indicate the local time of measurements that lasted 3 hours a day. In winter, the solar zenith angle changes by $\sim$10° between 18:00 and 00:00 at this latitude. Direct measurements for 24 (h) demonstrated large random variations caused by wave processes. However, the daily mean variations retrieved from continuous airglow measurements for 19 days showed that possible average changes of the OH temperature were insignificant, no more than a few degrees of Kelvin (Myrabø et al. 1983; Myrabø 1984). Nevertheless, seasonal variations are noticeable throughout the winter period. Naturally, various random changes occur on their background.

It was originally necessary to exclude the influence of solar activity. According to Semenov et al. (2005), the response of temperature of the hydroxyl emission on solar activity depends on the months of a year. Therefore, on the basis of the data presented in figure of Sigernes et al. (2003b), the mean monthly values (for $F_{10.7} = 130$) for the specified years have been originally obtained. Then they have been used for obtaining the empirical relationships on the basis of the mean monthly values for separate months of each year.

$$T(\text{November}) = 206 + 0.12 \cdot (F_{10.7} - 130) \quad (r = 0.48)$$
$$T(\text{December}) = 210 + 0.09 \cdot (F_{10.7} - 130) \quad (r = 0.64)$$
$$T(\text{January}) = 207 + 0.06 \cdot (F_{10.7} - 130) \quad (r = 0.65)$$
$$T(\text{February}) = 201 + 0.03 \cdot (F_{10.7} - 130) \quad (r = 0.40)$$

Unfortunately, because of a small number of the used data, the presented dependences not always possess sufficient reliability. In this case, it was necessary to exclude a small amount of the data owing to their significant deviation from set of

6.1 The Temperature Trend in the Middle Atmosphere

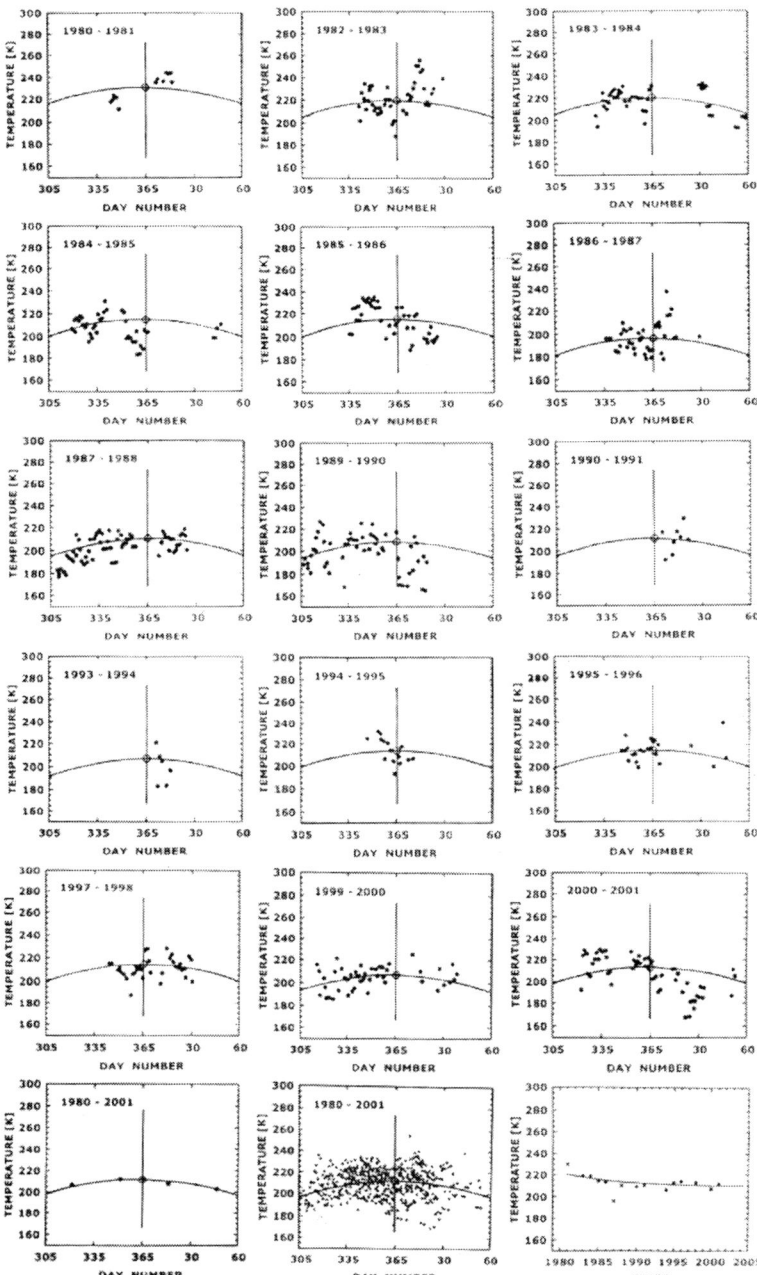

Fig. 6.13 Daily mean temperature of the OH (6–2) emission retrieved from observations at the auroral station Adventdalen, Spitsbergen (Svalbard) (see text) shown by *full circles* (Sigernes et al. 2003a,b). All data are shown on the *lower left*. The *solid curve* here shows the seasonal mean variations. This curve is also shown on the plots for each winter season. The *large open circle* shows the winter temperature (January 01). Long-term changes of these values are shown on the lower right (Semenov et al. 2006)

other values forming closer group. Nonetheless, the calculated values of the coefficients of regression (the response of temperature to solar activity) satisfactorily agree with the results of researches of the response of temperature on the solar activity, obtained from the analysis of the rocket measurements in high latitudes on Heiss Island.

The same relationships enable one to obtain the estimation of the regularity of average seasonal variations of temperature during the winter period for latitude of station Spitsbergen for $F_{10.7} = 130$. It has been known (Semenov and Shefov 1996) that the seasonal variations of temperature of the hydroxyl emission are well described practically by the first harmonic. On this basis, the approximation for a considered case also can be presented as (the plot is located in the bottom left part of Fig. 6.13)

$$T_i^0(t_d) = T_i^{00} + \Delta T^0 \cdot \cos \frac{2\pi}{365} t_d,$$

where i is a number of year, t_d is the day of year, ΔT° is an amplitude of the seasonal variations, T_i^{00} is mean annual value of a temperature of the given year (i). Numerical values of parameters are equal: $T_i^{00} = 180(K)$, $\Delta T^\circ = 30(K)$ for $F_{10.7} = 130$.

After introduction of a reduction on solar activity, the given works (Sigernes et al. 2003b) for different years are shown in Fig. 6.13.

For obtaining the average variations by other methods, the data of measurements of temperature of the winter periods of each year have been combined on a graph by consecutive shift along the axis of ordinates concerning the measurement data for chosen in our case of the winter period of 1983–1984. Such procedure has allowed to provide the minimal dispersion between the data of various years. Results of the made systematization are shown in the bottom middle part of Fig. 6.13. Average distribution of probability of a deviation of the measured values of temperature from the mean winter variations is well represented by Gaussian distribution with a dispersion 21 (K). By this is meant that the other essential regular variations are excluded, and deviations are caused by the casual reasons.

In this connection, the average seasonal temperature variations were first retrieved from the results presented by Sigernes et al. (2003b). All data were brought into coincidence in one plot by shifting along the ordinate to provide a minimum variance. The so revealed behavior of the winter mesopause temperature at high latitudes of the northern hemisphere is in good agreement with that at high latitudes of the southern hemisphere (Davis, 68.6°S, 78.0°E) (Burns et al. 2001). The average regularity of change of temperature revealed in such a way is presented in Fig. 6.13 as a solid line. The average temperature variation law, illustrated by the straight line in the figure, was applied to the data of each year (the vertical straight line corresponds to $t_d = 0$), denoted by large open circles, which were retrieved for each year. Thus, these temperature values involve only long-term variations shown on the lower right of Fig. 6.13. From these data, it follows that from the end of 80th years, the temperature trend slowly decreased. Such behavior of a trend well corresponds to the long-term temperature trend for winter conditions of the middle latitudes (Fig. 6.10).

It should be noted that the solar flux effect is estimated to be $\sim 3\ (K \cdot (100(sfu))^{-1})$, which is in agreement with the respective data for winter hydroxyl emission altitudes (Semenov et al. 2005). This is especially important because the polar atmosphere at these altitudes remains unilluminated by the Sun for a long time (Spitsbergen).

The observed long-term behavior of the mesopause temperature at midlatitudes in summer allows the conclusion that during this period, favorable conditions for the formation of noctilucent clouds (NC) (T < 150(K)) are realized only in 15–20(%) of all cases. This is in good agreement with the average frequency of NC observation (15–20(%)) per month in the chosen region (55°N, 90–120°E) derived by analyzing the data collected by Fast (1972). At the same time, the constancy of the summer mesopause temperature (Lübken et al. 1996; Golitsyn et al. 2000; Semenov 2000) and the long-term increase in frequency of NC occurrence (Gadsden 1990) might be evidence of an increase in humidity of the upper atmospheric layers. This is confirmed by the increased contents of atomic hydrogen, methane, and water vapor in the atmosphere (Semenov 1997; Chandra et al. 1997). Analysis of the data of Gadsden (1998, 2002) and von Zahn (2003) considered in Sect. 5.4 (see Fig. 5.54) has led us to conclude that the average long-term changes of the frequency of occurrence of noctilucent clouds in the periods of solar minima indicate an increase in atmospheric humidity because the summer temperature remains at the same level for several decades (Golitsyn et al. 2000).

The space–time behavior of temperature trends in the atmosphere at altitudes of 25–110 (km) is illustrated in detail in Fig. 6.4. Clearly seen are positive winter temperature trends in the upper stratosphere, deep negative trends in the upper mesosphere throughout a year, a region of zero temperature trend whose maximum width is $\sim$10 (km) in summer, and a stable positive trend in the lower thermosphere.

6.1.3 Latitudinal Variations of the Temperature Trend

By the present time, the long-term temperature measurements in the middle atmosphere (25–75 (km)) performed between 1964 and 1994 have been presented in a number of publications (Kokin et al. 1990; Keckhut et al. 1995; Givishvili et al. 1996; Golitsyn et al. 1996; Taubenheim et al. 1997; Keckhut and Kodera 1999; Keckhut et al. 1999). Their analysis have shown that significant climatic changes took place in the given region of the atmosphere over the last 30 years that were indicated as trends of the thermal regimes of the stratosphere and mesosphere.

Here we should immediately emphasize that the necessary condition for a comparison of the temperature trends obtained by different researchers at different atmospheric altitudes is to reduce the relevant measurement data to the same time intervals, latitudes, and seasons. If this condition is not fulfilled, for instance, as in the works by Beig (2000), Beig and Fadnavis (2001), Beig et al. (2003), and Laštovička (2005), any discussion of the inferences from measurement data seems to be senseless, the more so that Laštovička (2005) gives no bibliographic data on the cited works. The importance of this approach stems from the fact that the data of long-term continuous temperature measurements in the mesopause region demonstrate a nonlinear character of both the long-term winter temperature variations and the yearly average temperature variations, which was especially pronounced between 1990 and 2000.

Whereas a linear approximation of the temperature variations was still correct in analyzing the data of 1957–1995, since there were no grounds to suggest changes of the long-term temperature trends, the nonlinear character of temperature variations was evident already for the period 1957–2005. This is seen from Fig. 6.1 that shows the yearly average temperature variations derived without elimination of the effect of solar activity and from Fig. 6.10 where the winter temperatures are presented that were derived in view of this effect.

This temperature behavior corresponds to a decrease in absolute values of the negative trend which can subsequently lead even to the change of the trend sign. This conclusion can be made in view of the observed secular variation of the solar activity (Dergachev and Raspopov 2000; Bashkirtsev and Mashnich 2003; Komitov and Kaftan 2003).

Subsequently, seasonal and altitudinal features of the long-term variations in the middle atmosphere temperature were also examined (Semenov et al. 2000, 2002a,c). The dependence of the temperature at different altitudes on solar flux was derived by Semenov et al. (2005). The above-mentioned results call for the development of a new model of the middle atmosphere to take into account the revealed long-term variations of its parameters on preset time intervals (Semenov et al. 2004). The existing CIRA and MSIS models do not consider long-term variations.

The approximation of latitudinal, seasonal, and altitudinal variations of the long-term temperature trend performed by Perminov and Semenov (2007) could be a component of a new model of the middle atmosphere suitable for a description of the behavior of the middle atmosphere over the past decades. To this end, long-term temperature measurements at altitudes of 25–75 (km) carried out at the stations of rocket sounding of the atmosphere of the USSR and RF Meteorological Committee at low, middle, and high latitudes between 1964 and 1994 were used.

The data to be analyzed were borrowed from the results of rocket sounding of the atmosphere (Bulletin 1964–1994) for Heiss Island (80.6°N, 58°E), Molodyezhnaya (67.7°S, 45.8°E), Volgograd (48.7°N, 45.8°E), and Thumba (8.5°N, 76.8°E). The database included more than 4500 measurements performed between 1964 and 1994. As a rule, the measurements were carried out every week around local midnight using M-100 meteorological rockets. The random measurement error was 2–3 (K) at altitudes of 25–45 (km) and about 6 (K) at 60–75 (km). The vertical temperature profiles were presented in Bulletin with an altitude step from 1 to 5 (km). In the present study, the monthly mean temperatures at altitudes of 25, 30, 35, ..., 75 (km) were used.

As shown in Sect. 7.4 (Figs. 7.10 and 7.11) (Semenov et al. 2005), the temperature regime of the middle atmosphere responds noticeably to changes in solar activity. Thus, the solar activity index should be taken into account in studying long-term temperature changes. Analysis of the data obtained shows a clear latitude dependence of the long-term variations of yearly average temperature. Though it is

6.1 The Temperature Trend in the Middle Atmosphere

doubtless that the points of temperature measurements are few in number, there is an opportunity to evaluate the tendency of latitudinal variations. It should be noted that measurement points were localized in a rather narrow (45–77°E) latitude belt. Therefore, the longitudinal variations of the parameters under investigation in this latitude belt cannot be expected significant. In addition, it was assumed that the transitions from one hemisphere to another near the equator and from eastern to western hemisphere in the polar regions are also smooth. On this basis, the arrangement of the monthly mean temperatures and temperature trend in a great circle of the meridian ($\lambda \sim (60 \pm 15)°E$) for latitudes ranging from 0° to 360° enables the average tendencies for the latitudinal variations to be estimated.

It is quite clear that to determine the latitudinal laws of temperature trend variations, its seasonal and vertical variations must simultaneously be taken into account. Therefore, to determine a temperature trend at altitude Z (km), the approximation

$$T(Z, t_d, \varphi, t, F_{10.7}) = T_0(Z, t_d, \varphi) + \Delta T(Z, t_d, \varphi) \cdot (t - 1972) + \delta T_F(Z) \cdot (F_{10.7} - 130)$$

was used, where $T(Z, t_d, \varphi, t, F_{10.7})$ is the monthly mean temperature (in (K)), $T_0(Z, t_d, \varphi)$ is the vertical temperature profile in 1972 with the solar activity index $F_{10.7} = 130$, $\Delta T(Z, t_d, \varphi)$ is the linear trend (in ($K \cdot yr^{-1}$)), t denotes the year, δT_F is the temperature response to a change of the solar activity index (in ($K \cdot sfu^{-1}$)). The least squares method was used to calculate the long-term linear temperature trends for four stations of rocket sounding located at altitudes of 25–75 (km) with a step of 5 (km) for all months. Figure 6.14 shows the calculated trend values. From the figure, it can be seen that the temperature trend depends on the latitude, season, and altitude. Based on these results, analytical approximations of these dependences were obtained.

To calculate the latitudinal component, the data for each month and altitude were approximated by the least squares method using the following analytic expression:

$$\Delta T(Z, t_d, \varphi) = \Delta T_0(Z, t_d) - A(Z, t_d) \cdot \sin \frac{2\pi}{360}(\varphi - \varphi_0) \, .$$

Here φ is the latitude (in deg) which in this case ranges from 0° to 360° and is counted from the equator toward North Pole along the meridian, t_d is the serial number of the day in the year, $\Delta T_0(Z, t_d)$ is the average meridional temperature trend (in ($K \cdot yr^{-1}$)), $A(Z, t_d)$ (in ($K \cdot yr^{-1}$)) and φ_0 (in deg) are the amplitude and phase of the harmonic of the meridional temperature trend. Thus, this data presentation could not cause longitudinal changes of the temperature trend. In this case, the number of approximation points in the above-mentioned latitude belt was increased and the continuity condition was satisfied. As an example, Fig. 6.15 shows latitudinal approximations for January and July at an altitude of 50 (km). Also as an example, Fig. 6.16 shows seasonal values of the average solstice and equinox meridional trend and amplitude of the first meridional harmonic for all examined altitudes. The phase $\varphi_0 = 0$ is reduced to 90 in all cases. As a result, the harmonic component changed sign from plus to minus when going from the northern to southern hemisphere, and the above formula becomes

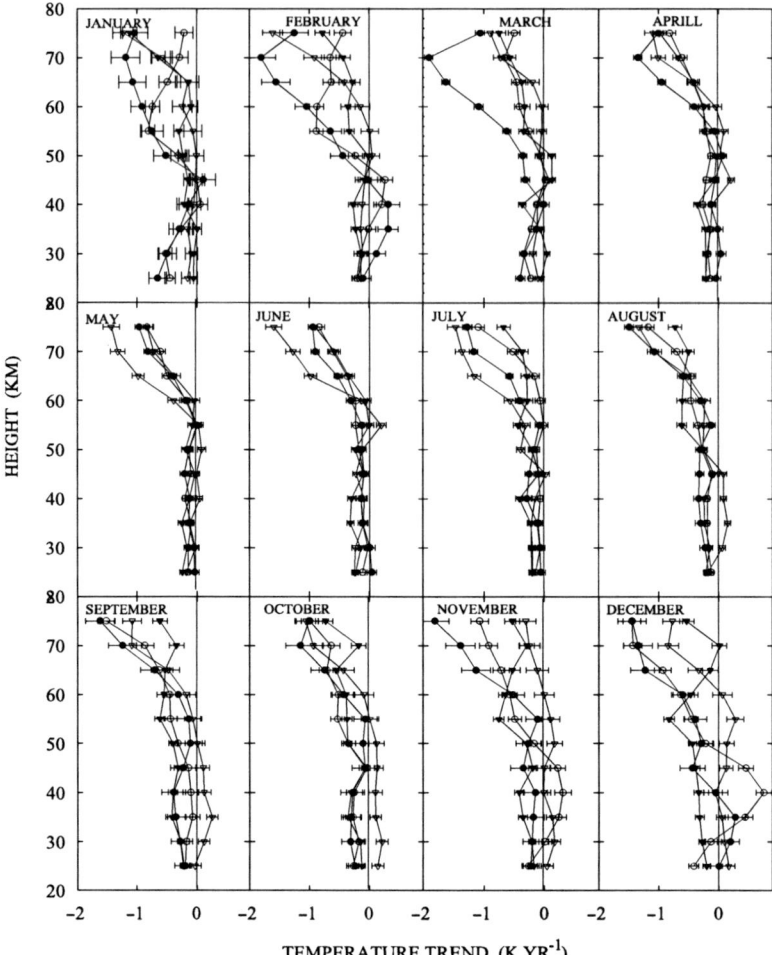

Fig. 6.14 Monthly mean vertical profiles of long-term temperature trends retrieved from observations at high (Heiss Island, *full circles*; Molodyezhnaya, *hollow triangles*), middle (Volgograd, *open circles*), and low latitudes (Thumba, *full triangles*). *Horizontal bars* show standard deviations from the monthly mean values

$$\Delta T(Z, t_d, \varphi) = \Delta T_0(Z, t_d) - A(Z, t_d) \cdot \sin \frac{2\pi}{360} \varphi.$$

To describe the seasonal behavior of the average meridional temperature trend $\Delta T_0(Z, t_d)$ and meridional harmonic amplitude $A(Z, t_d)$, they were approximated for each altitude by the sum of six harmonics:

6.1 The Temperature Trend in the Middle Atmosphere

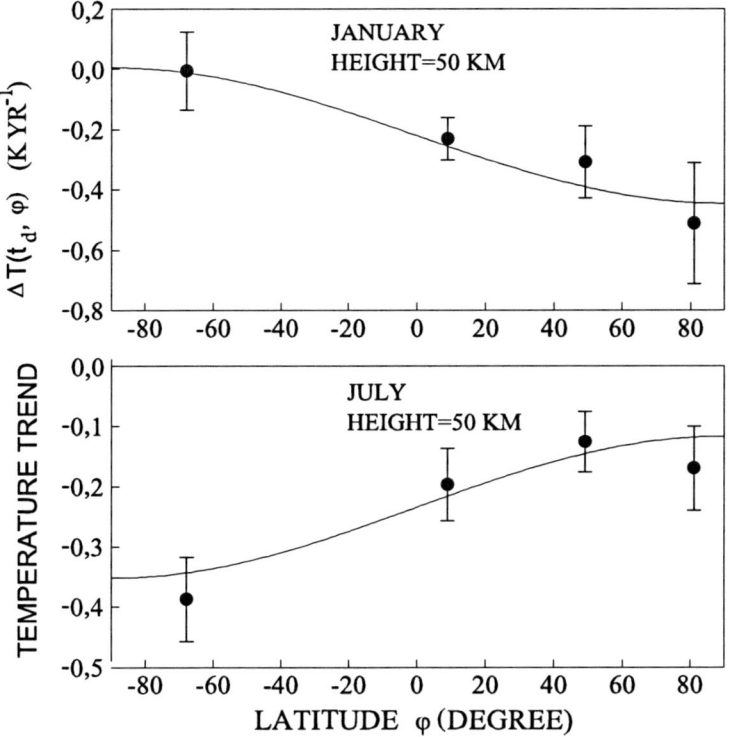

Fig. 6.15 Examples of approximations of the temperature trend latitudinal variations by a harmonic function

$$\Delta T_0(Z, t_d) = B_{00}(Z) + \sum_{n=1}^{6} B_{0n}(Z) \cdot \cos\frac{2\pi}{365.25}(t_d - t_{0n}),$$

$$A(Z, t_d) = B_{10}(Z) + \sum_{n=1}^{6} B_{1n}(Z) \cdot \cos\frac{2\pi}{365.25}(t_d - t_{1n}).$$

Then the amplitudes $B_{kn}(z)$ (in $(K \cdot yr^{-1})$) and phases t_{kn} (day of the year) ($k = 0, 1$) were approximated as functions of altitude Z by the sum of five harmonics:

$$B_{kn}(Z) = C_{0kn}(Z) + \sum_{m=1}^{6} C_{mkn}(Z) \cdot \cos\frac{2\pi m}{75}(Z - Z_{mkn}),$$

$$t_{kn}(Z) = D_{0kn}(Z) + \sum_{m=1}^{6} D_{mkn}(Z) \cdot \cos\frac{2\pi m}{75}(Z - Z_{mkn}),$$

where $n = 0, \ldots, 6$ and $m = 1, \ldots, 5$ are the serial numbers of the corresponding harmonics. The coefficients C_{mkn}, D_{mkn} (in $(K \cdot yr^{-1})$), and Z_{mkn} (in (km)) are given

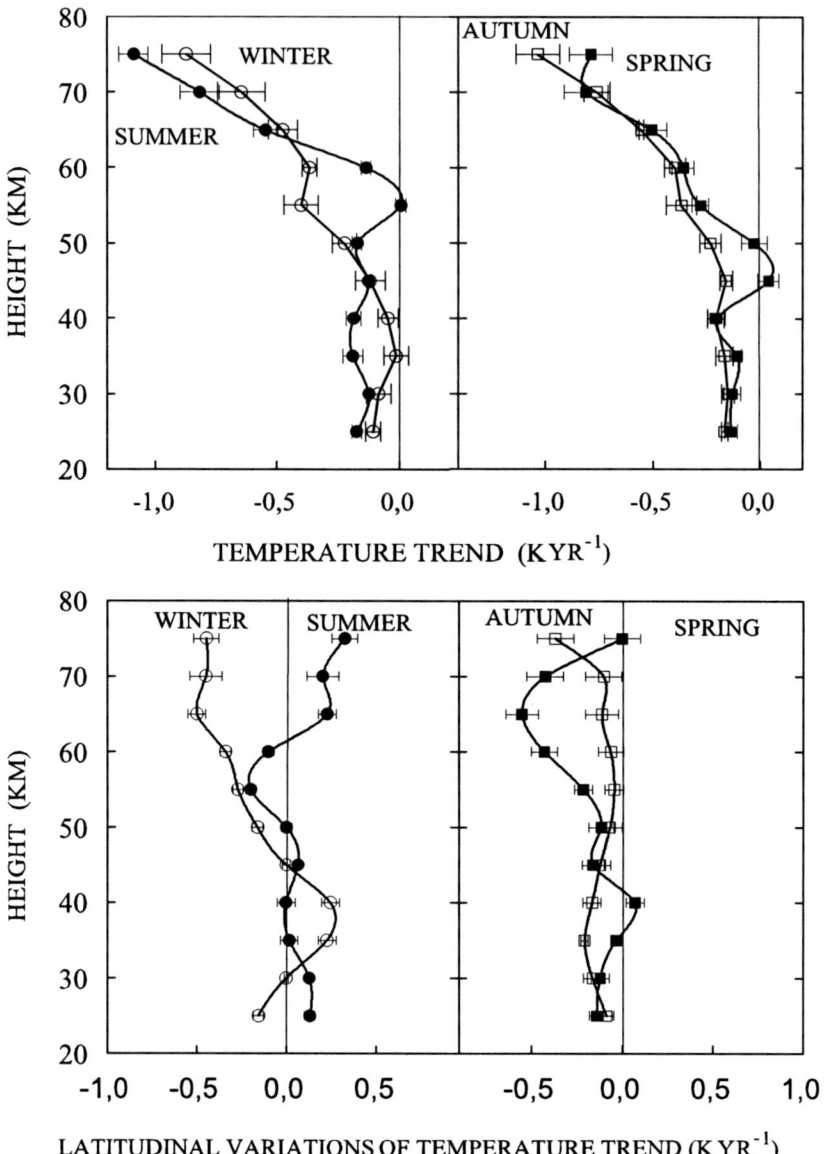

Fig. 6.16 Vertical changes of the temperature trend averaged over the meridian (60° E) (**top panels**) and amplitudes (**bottom panels**) of its latitudinal variations in different seasons

in Table 6.2. Figure 6.17 illustrates an example of the approximation by the sum of five harmonics.

Analyzing the results presented in Table 6.2 and in the figures, it is interesting to note that from 1964 to 1994, the middle atmosphere cooled everywhere with

6.1 The Temperature Trend in the Middle Atmosphere

Table 6.2 Approximation coefficients for the latitudinal, seasonal, and altitude variations of the long-term temperature trend in the middle atmosphere (25–75 (km))

B_{kn} (K·yr⁻¹) t_{kn} (day)		C_{0kn} (K·yr⁻¹) or D_{0kn} (day of year)	C_{1kn} (K·yr⁻¹) or D_{1kn} (day of year)	Z_{1kn} (km)	C_{2kn} (K·yr⁻¹) or D_{2kn} (day of year)	Z_{2nk} (km)	C_{3kn} (K·yr⁻¹) or D_{3kn} (day of year)	Z_{3kn} (km)	C_{4kn} (K·yr⁻¹) or D_{4kn} (day of year)	Z_{4kn} (km)	C_{5kn} (K·yr⁻¹) or D_{5kn} (day of year)	Z_{5kn} (km)
ΔT_0 (K·yr⁻¹)	B_{00}	−0.30	0.39	29.8	0.30	16.4	0.20	15.1	0.11	14.3	0.06	14.9
	B_{02}	−0.27	0.64	50.2	0.49	31.1	0.3	0.2	0.17	2.9	0.05	5.6
	t_{02}	−19.3	253.8	56.8	176.5	33.1	135.8	1.6	67.1	4.0	35.2	5.7
	B_{04}	−0.27	0.58	50.0	0.48	31.3	0.3	24.8	0.16	2.8	0.04	4.8
	t_{04}	82.7	136.5	10.0	113.5	12.5	70.6	12.4	42.2	12.6	11.4	12.6
	B_{06}	−0.3	0.59	50.1	0.48	31.5	0.32	0.3	0.18	3.6	0.05	5.8
	t_{06}	10.0	4.8	64.7	8.8	23.5	7.6	12.6	6.9	0.5	9.4	14.1
A (K·yr⁻¹)	B_{11}	−0.12	0.1	42.9	0.32	0.7	0.17	1.5	0.16	3.4	0.07	4.3
	t_{11}	653.9	1167.3	12.5	920.1	12.5	614.0	12.4	296.7	12.8	100.6	12.9
	B_{13}	0.72	1.19	12.4	0.92	12.7	0.62	12.8	0.33	12.9	0.11	13.0
	t_{13}	−277.4	495.2	51.1	424.7	31.5	279.6	0.3	151.1	3.2	60.7	5.3
	B_{15}	−0.09	0.3	49.8	0.3	32.1	0.19	0.7	0.1	2.8	0.04	5.5
	t_{15}	−356.2	636.9	49.3	507.5	30.5	334.4	24.2	168.9	2.0	43.5	3.9

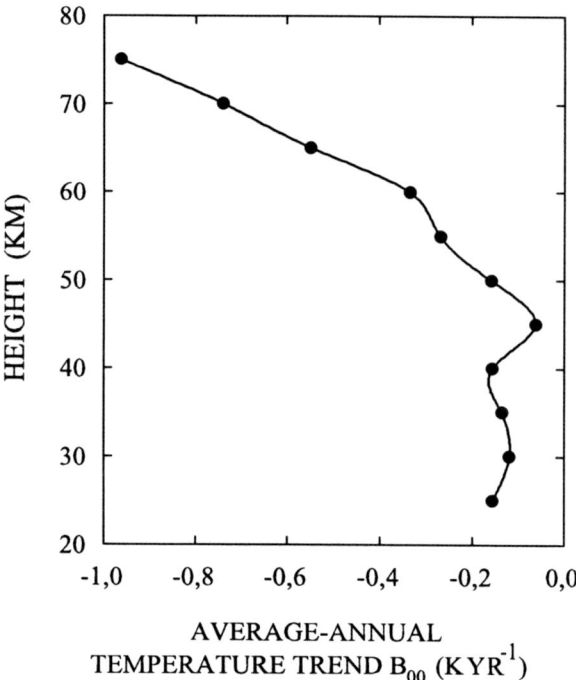

Fig. 6.17 Yearly average temperature trend at midlatitudes (*full circles*) as a function of the altitude. The *solid curve* shows the approximation

an average rate of -0.35 $(K \cdot yr^{-1})$ or, in other words, its temperature decreased by more than 10 (K) for 30 years. Vertical changes of the cooling rate are characterized by intensification of this process with increasing altitude: the trend is 0.1–0.2 $(K \cdot yr^{-1})$ at altitudes of 25–45 (km) and reaches -0.95 $(K \cdot yr^{-1})$ around 75 (km) (see Fig. 6.17). However, the trend undergoes strong seasonal changes around 55 (km). Thus, the amplitude of its annual change (0.4 $(K \cdot yr^{-1})$) is comparable to the yearly average temperature. As a result, the long-term trend at this altitude undergoes significant changes within a year: from 0 to 0.8 $(K \cdot yr^{-1})$.

The above features of the long-term temperature trend in the middle atmosphere refer to its average meridional values. Its latitudinal changes are illustrated in Figs. 6.15 and 6.16. As mentioned above, the latitudinal behavior of the temperature trend is characterized by a maximum over North or South Pole. The amplitudes of the meridional harmonic are shown at the bottom of Fig. 6.16.

An important result is that in winter the maximum absolute values of the negative temperature trends are observed at high latitudes, while in summer their absolute values are minimal during the observation period. This explains well the almost complete absence of temperature trends in the polar zone in summer as well as the negative trend values of about -1 $(K \cdot yr^{-1})$ in winter in the mesopause region retrieved from the measurements performed at the station Spitsbergen Island in the Arctic (Sigernes et al. 2003b) refined by Semenov et al. (2006) (see Fig. 6.13) and of $-(1.0 \pm 0.5)(K \cdot yr^{-1})$ retrieved from the measurements performed at the Davis station in the Antarctic (French and Burns 2004).

Analysis of the latitudinal changes of the temperature trend as functions of the altitude has shown that its greatest changes, as a rule, are observed in the mesosphere (60–75 (km)) with its typical behavior in winter shown at the top of Fig. 6.15 and in summer with the latitudinal variations illustrated at the top of Fig. 6.16. In the lower stratosphere (30–40 (km)), the seasonal behavior of the temperature trend latitudinal variations (shown at the bottom of Fig. 6.15) is opposite to that in the mesosphere.

6.2 Temperature Trend in the Thermosphere

Ionospheric Measurements

Among the methods of diagnostics of the ionosphere, the method of panoramic (multifrequency) radar or vertical sounding (VS), which provides information on the vertical profile of free electron density (n_e) at altitudes Z = 100–300 (km), is frequently used. Among its numerous advantages, the remarkable uniformity of VS data should be mentioned, which is especially important in analyzing long series of observations. This is explained by the fact that the error in measuring ionospheric plasma frequencies (f_0) used to calculate n_e values (n_e, in cm^{-3}), is proportional to f_0^2, in (MHz^2) depends not so much on the parameters of the employed instruments as on the state of the environment being monitored (Manual 1978). Moreover, the limiting accuracy of determining f_0 is ± 0.05 (MHz) for the E layer and ± 0.1 (MHz) for the F2 layer, given that the instrument sensitivity varies over a wide range. It is equivalent to an error of about 2–5(%) in determining electron density. The trend values for these altitudes are presented elsewhere (Ulich and Turunen 1997; Bremer 1998).

Besides this method, the analysis of the long-term variations of the characteristics of the radiowaves propagation is widely used for diagnostics of the ionosphere at the D-layer altitudes (Taubenheim et al. 1990, 1991, 1997; Bremer 1992; Laštovička 1997b). Investigations of the radiowave absorption on various paths allow the variations in the temperature regime of this atmospheric region to be estimated.

Emission Measurements

The main emission at altitudes of 240–270 (km) is the atomic oxygen emission at 630 (nm). Interferometric methods of temperature measurements have already been considered in detail in the preceding sections. The long-term variations of the Doppler temperature are presented in Fig. 4.42. The plots demonstrate that the temperature decreased during the 30-yr period between 1963 and 1994 with a negative trend of $-(2.2 \pm 0.8)$ (K · yr^{-1}). This implies that the temperature of the thermosphere at an altitude of $\sim$250 (km) decreased by 65–70 (K).

6.3 Trends of the Atmospheric Density and Composition

6.3.1 Long-Term Variations in Concentration of Neutral and Ionized Atmospheric Constituents

The data on the intensities of the hydroxyl, 557.7-nm atomic oxygen, and 589.3-nm sodium emissions that have been accumulated by the present time (Semenov and Shefov 1996, 1997a,c,d; Semenov 1997; Semenov and Shefov 1999a; Shefov et al. 2000, 2002) can be used to retrieve the ozone and atomic oxygen densities at altitudes of 80–100 (km). Systematization of long-term investigations of these emissions provided the basis for the development of empirical models of their characteristics, including the intensity, temperature, and emission layer altitude. Based on theoretical studies of the mechanisms of these emissions, the behavior of the vertical ozone and atomic oxygen profiles was investigated under various heliogeophysical conditions (Semenov 1997). It is well known (Semenov 1997) that the ozone density is related to the atomic hydrogen density in the mesopause region and that the main source of hydrogen in the upper atmosphere is methane (Brasseur and Solomon 2005). Over the last decades, the methane concentration in the lower atmosphere increased with a trend of $\sim 1.5(\% \cdot \text{yr}^{-1})$, which resulted in an increase of its concentration over the last four decades by a factor of 1.4 (Thomas et al. 1989).

The data of observations of the H_α emission that characterize the atomic hydrogen content at high altitudes (Fishkova 1983) testified to the dynamism of the processes of atomic hydrogen generation and consumption in the mesosphere.

Rocket measurements of the atomic hydrogen density in the mesopause region are few in number (Meier and Prinz 1970; Anderson et al. 1980; Sharp and Kita 1987; Ulwick et al. 1987). Nevertheless, their analysis has revealed that the hydrogen density is significantly affected by the long-term trend and solar activity. Following Le Texier et al. (1987), the hydrogen density undergoes seasonal variations of $\sim 20(\%)$ that cannot be retrieved from the available data. Figure 7.9 shows the variations of the relative atomic hydrogen concentration with solar activity and with time from which the effect of trend was eliminated and which were reduced to $F_{10.7} = 130$ (Semenov 1997; Shefov and Semenov 2002). The positive trend of atomic hydrogen density so obtained is in good agreement with the trend of methane (WMO/UNEP 1990; Brasseur and Solomon 2005). The yearly average hydrogen density at an altitude of 87 (km) under standard heliogeophysical conditions (Semenov and Shefov 1996; Semenov 1997) in 1972 was $[H]_0 = 1.4 \cdot 10^8$ (cm^{-3}) (Semenov 1997).

To elucidate the behavior of the long-term changes of vertical ozone and atomic oxygen profiles, the latter were calculated (Semenov 1997) for the yearly average conditions in 1955, 1972, and 1995. These conditions were reduced to the same solar flux $F_{10.7} = 130$. Used for the vertical profiles of atomic hydrogen were those predicted by the model of Keneshea et al. (1979). Figure 6.18 shows the calculated vertical profiles of atomic oxygen and ozone concentrations. The retrieved vertical profiles and absolute densities are in good agreement with the typical data on

6.3 Trends of the Atmospheric Density and Composition

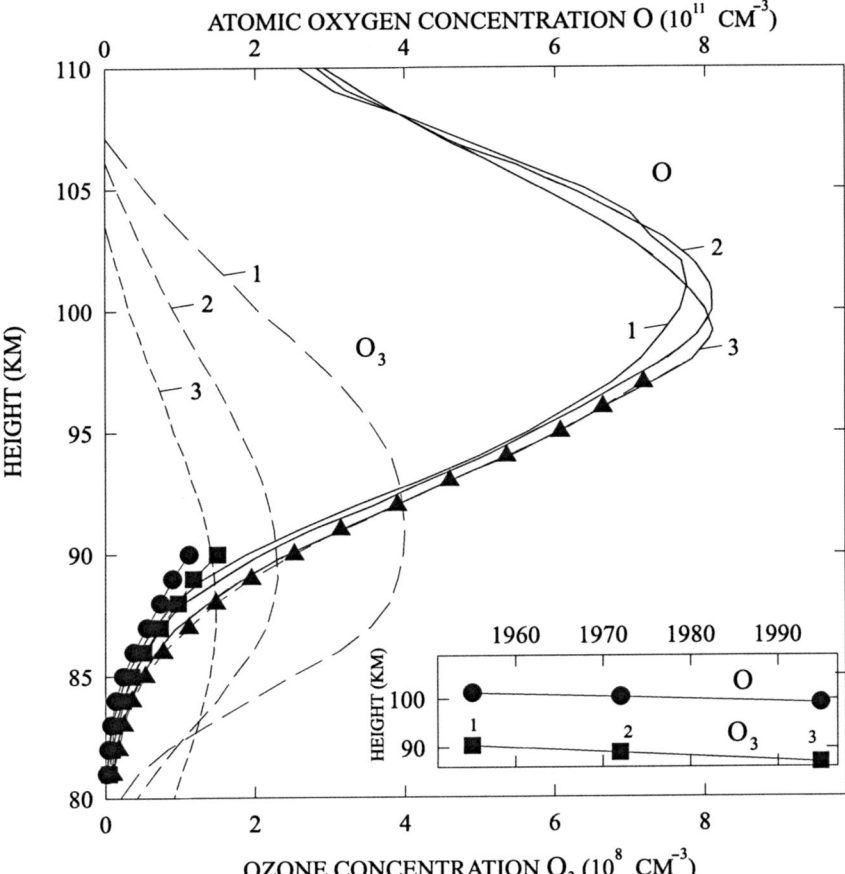

Fig. 6.18 Yearly average vertical profiles of the ozone (*dashed curves*) and atomic oxygen concentrations (*solid curves*) at night in 1955 (*curves* 1), 1972 (*curves* 2), and 1995 (*curves* 3) (Semenov 1997) calculated based on empirical models of the OH and 557.7-nm emission variations (Semenov and Shefov 1996, 1997a,c,d), respectively. *Full circles, squares,* and *triangles* indicate the atomic oxygen concentration calculated from the OH emission data for 1955, 1972, and 1995, respectively. Long-term variations of the O and O_3 maximum altitudes retrieved from the respective vertical concentration profiles (Semenov 1997) are shown in the inset on the lower right of the figure

ozone and atomic oxygen reported in many publications (Witt et al. 1979; McDade and Llewellyn 1988; Murphree et al. 1984; Vaughan 1982). The long-term average content of atomic oxygen at altitude of 90–110 (km) slightly varied during the observation period, mainly due to the subsidence of the layer lower boundary, whereas the ozone content at altitudes of 80–100 (km) decreased by a factor of 2–2.5. Obviously, this was related to the increase in atomic hydrogen content responsible for the decrease in ozone concentration due to the ozone–hydrogen reaction. The positive trend of the OH emission intensity (see Fig. 6.3) confirms this conclusion. It should be noted that the ozone concentration decreases mainly at hydroxyl emission layer altitudes; in turn, hydroxyl is localized near the lower boundary of the atomic oxygen layer. Therefore, the variations of the vertical atomic oxygen profiles at altitudes of 85–95 (km) affect significantly the behavior of the hydroxyl emission (Perminov et al. 1998).

Analysis of the results obtained at seven vertical sounding stations of Western Europe and CIS countries shows the presence of noticeable regional features of long-term n_e trends at altitudes of the E-layer maximum ($\sim$106 (km)). Between the early 1960s and the present time, the n_e trends that are not connected with changes of heliogeophysical conditions have been significant and positive for eastern longitudes and insignificant for western longitudes. At the same time, between the early 1930s and the late 1950s, the n_e trend was significant and negative for both segments of longitudes (Givishvili and Leshchenko 1993).

Analysis of the data of vertical sounding has revealed the presence of a statistically significant long-term decrease in n_e in the region of the F2-layer maximum under both quiet and perturbed heliogeophysical conditions (negative ionospheric storms) (Sergeenko and Kuleshova 1994).

The results of the above analysis of rocket, spectrophotometric, and ionospheric data demonstrate that long-term changes of the temperature and dynamic regimes of the upper atmosphere and its structural characteristics took place throughout the atmospheric depth. Moreover, the absolute values of statistically significant trend estimates for some middle and upper atmospheric parameters were substantially greater than the values of climatic trends revealed in the troposphere and in the atmospheric surface layer. This fact is a serious argument in support of the necessity to stimulate and extend the use of the existing methods and means of monitoring of the upper atmospheric layers to obtain uniform data on the variations of different characteristics of the upper atmosphere whose period is over 50 years, including the variations caused by anthropogenic factors.

The long-term cooling of the middle atmosphere was accompanied by a reduction of the total density of the lower thermosphere, resulting in the global subsidence of the upper atmosphere. As a result, the boundary conditions at altitudes of about 100 (km), conventionally set constant in the construction of models of the upper atmosphere, have changed substantially. Therefore, because of the presence of the atmospheric temperature trend at different altitudes, the up-to-date models of the neutral atmosphere and ionosphere must consider the changes of the temperature profile throughout the atmospheric depth. Otherwise, without taking into account the long-term temperature variations, discrepancy will arise between measurements and results of model calculations for the heliogeophysical conditions corresponding to the measurements.

Moreover, the existing models do not consider the actual dependence of the temperature in the stratosphere, mesosphere, and lower thermosphere on solar activity.

The indicated tendencies for temperature and density changes in the middle and upper atmosphere raise an important problem of their further development in the next decades. Attempts of systematization and analysis of various series of long-term rocket, emission, and lidar measurements have been undertaken in the last few years to reveal a tendency of the changes in the temperature regime over the last decades of the 20th century (Lysenko et al. 1997a,b; Lysenko and Rusina 2002a,b; Beig et al. 2003). The result obtained testifies to a decrease in absolute values of the negative temperature trends in the stratosphere and lower mesosphere.

6.3.2 Estimation of the Long-Term Variations of the Density of the Upper Atmosphere by the Evolution of the Parameters of Satellite Orbits

A method of revealing the long-term variations in the state of the upper atmosphere is based on the determination of the atmospheric density at different altitudes. To this end, a number of studies were performed in which space and time density variations were investigated invoking data on the parameters of orbits of artificial satellites launched in the early 1960s, and the existing models of the upper atmosphere were refined (Keating et al. 2000; Emmert et al. 2004; Volkov and Suevalov 2005). The long-term variations of the atmospheric density at altitudes of 400–1100 (km) were investigated for the period covering the 20th–22nd solar cycles based on the measured parameters of artificial satellite orbits. To reveal and analyze the long-period density variations in the upper atmosphere, the measured parameters of artificial satellites (TLE of the NORAD system) orbited no less than 10 years were used.

The determination of long-period variations of the density of the upper atmosphere from the deceleration of an artificial satellite can be carried out reliably enough only by comparing atmospheric density or deceleration data obtained under identical or close conditions. To provide such conditions, it seems optimal to compare the deceleration (decrease of the revolution period) or the density retrieved from the deceleration of the same artificial satellite (whenever possible, with a simple aerodynamic shape) during periods of minimum solar activity. To compare data obtained for one artificial satellite to investigate the relative density variations with time, it is not necessary to know precisely the ballistic factor of the satellite. It is also important that the ballistic factor of an artificial satellite having a simple aerodynamic shape does not change throughout the flight.

Based on the data on the deceleration of 27 long-lived satellites orbited between 1967 and 2002, it has been found that the linear density trend at altitudes of ~ 400 (km) was $-(3.1 \pm 0.9)(\%)$ for 10 years. It has also been revealed that the trend value increased with altitude (Emmert et al. 2004):

$$\frac{\Delta \rho(Z)}{\rho(480\,\mathrm{km})} = -3.1 - 0.004 \cdot (Z - 480), (\% \cdot (10\,\mathrm{yr})^{-1}).$$

A small decrease in absolute value of the density trend with increasing solar activity has also been established:

$$\frac{\Delta\rho(Z)}{\rho(480\,\text{km})} = -3 + 0.017 \cdot (F_{10.7} - 130), \, (\% \cdot (10\,\text{yr})^{-1}).$$

Nevertheless, the data obtained for different satellites and different time intervals yield different trend values. If we choose a 20-yr interval in the 20th–22nd 11-yr solar cycles, the average density trend at altitudes of 400–1100 (km) estimated from the data of 22 satellites (Volkov and Suevalov 2005) will be $-(5 \pm 1)(\% \cdot (10\,\text{yr})^{-1})$.

Thus, the above analysis of the deceleration data for 27 artificial satellites obtained between 1974 and 1997 allows the conclusion that the atmospheric density had a stable tendency to decrease: no one of the considered data sets showed an increase in density.

6.4 Long-Term Subsidence of the Middle and Upper Atmosphere

The long-term systematic decrease of the temperature of the middle and upper atmosphere during the period under consideration (40–50 years) should be accompanied by a decrease of the total atmospheric density at different altitudes. The temperature profiles presented in Fig. 6.4 show a decrease of altitudes with constant-temperature values (for example, at altitudes of 50–75 (km) with temperatures of 210 (K) (*open circles*) and 250 (K) (*full circles*)), thereby indicating a gradual subsidence of the middle atmosphere throughout the observation period. The subsidence is understood as a decrease of the altitudes of atmospheric layers of certain density. For more vivid presentation of the tendencies for changes of altitude levels with constant-temperature values in the altitude range under consideration, the same data are presented in Fig. 6.6, allowing the long-term behavior of atmospheric temperature at different altitudes to be revealed with higher vertical resolution. It can well be seen that the altitudes of levels with constant temperatures increase up to ~ 40 (km), while a pronounced tendency for their decrease is observed above ~ 50 (km).

The vertical profiles of the rate of subsidence for such layers with allowance for the negative long-term temperature trends in the middle and upper atmosphere were calculated (Semenov and Lysenko 1996; Lysenko et al. 1999; Semenov 1996) based on the following solution of the barometric equation:

$$\log_e N(Z) = \log_e N(Z_o) - \log_e \frac{T(Z)}{T(Z_o)} - \frac{M_o m_H g_0}{k} \int_{Z_o}^{Z} \frac{M_r(Z)dZ}{\left(1+\frac{Z}{R_E}\right)^2 T(Z)},$$

where $N(Z)$ and $N(Z_0)$ are the concentrations of the atmospheric neutrals at altitudes Z and $Z_0 = 0$, respectively; $T(Z)$ and $T(Z_0)$ are the temperatures at the same altitudes; m_H is the mass of a hydrogen atom; M_0 and $M(Z)$ are the molecular masses at altitudes Z_O and Z, respectively; $M_r(Z) = M(Z)/M_0$; g_0 is the free fall acceleration at altitude Z_0; k is Boltzmann's constant; and R_E is the Earth's radius. The multiplier $M_O \cdot m_H \cdot g_O/k = 34.162\,(\text{K} \cdot \text{km}^{-1})$.

6.4 Long-Term Subsidence of the Middle and Upper Atmosphere

Based on this formula, the change ΔZ of the altitude of the layer, which characterizes the corresponding change of its density $\Delta\rho(Z)$ or of its temperature $\Delta T(Z)$ retrieved from the emission parameters, can be estimated by the relation

$$\Delta Z = -H(Z) \cdot \frac{\Delta\rho(Z)}{\rho(Z)} = -\frac{H(Z)}{T(Z)} \cdot \Delta T(Z), \text{ (km)}.$$

Here $H(Z)$ is the scale height. Thus, the rate of subsidence can be estimated as

$$r(Z) = 1000 \cdot \frac{\Delta Z}{\Delta t}, (m \cdot yr^{-1}),$$

where Δt denotes the period (years) during which the data were obtained.

Results of calculations have shown that the average rates of subsidence of the atmosphere at altitudes of 70 and 100 (km) caused by a substantial change of the temperature of the middle atmosphere over a 40-yr period are about -50 and -120 $(m \cdot yr^{-1})$, respectively (Evlashin et al. 1999; Semenov et al. 2000) (Fig. 6.19).

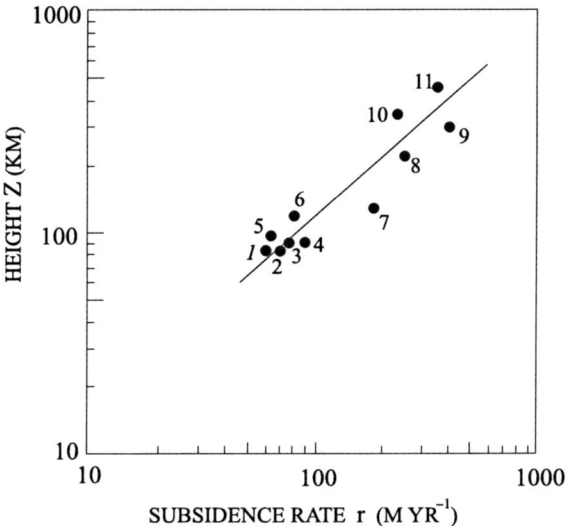

Fig. 6.19 Subsidence rate of the upper atmosphere versus altitude (Semenov et al. 2000) determined by measurements in the ionospheric D layer (Taubenheim et al. 1997) (1); by data on twilight scattering in the mesopause region at 85 (km) (Megrelishvili and Toroshelidze 1999) (2); by the altitude of radar meteor trail returns (Semenov et al. 2000) (3); by the ozone layer altitude in the mesosphere (Semenov 1997; Shefov and Semenov 2002) (4); by the atomic oxygen layer altitude determined based on the 557.7-nm emission measurements (Semenov 1997) (5); by mass-spectrometer measurements of atomic oxygen density at 120 and 130 (km) (Pokhunkov et al. 2003) (6 and 7); by 630-nm atomic oxygen emission measurements (8); by ionospheric measurements of the F2-layer altitude (Ulich and Turunen 1997) and (Bremer 1998) (9) and (10), respectively, and by satellite deceleration measurements (Keating et al. 2000; Emmert et al. 2004; Volkov and Suevalov 2005) (11). The *straight line* is the regression line

For higher altitudes (270–350 (km)), analysis of rocket measurements of vertical profiles for the 630-nm atomic oxygen emission layer (8) revealed a long-term decrease of the altitude of maximum emission, which was noted earlier (Semenov 1996; Semenov and Lysenko 1996). From the temperature variations of different types for this emission (see Fig. 4.42), it can be inferred that the temperature trend is -2.2 $(K \cdot yr^{-1})$. The altitudes of the ionospheric F2 layer are given in accordance with the measurements reported by Ulich and Turunen (1997) (9) and Bremer (1992, 1998) (10).

Analysis of long-term measurements of the deceleration of 27 satellites (Keating et al. 2000; Emmert et al. 2004; Volkov and Suevalov 2005) (11) enabled the rate of subsidence to be estimated for altitudes of ~500 (km). The straight line in the figure is the regression line.

Approximation of this correlation yields the following relationship between the rate of subsidence and the altitude:

$$r(Z) = -90 \cdot \left(\frac{Z}{100}\right)^{1.1}, (m \cdot yr^{-1}),$$

with the correlation coefficient $r = 0.87 \pm 0.08$.

Of special interest are the results obtained for the thermospheric altitudes based on investigations of auroral arcs. Long-term measurements were performed by Störmer (1955) in the south of Norway between 1918 and 1944. Analysis of these data revealed that a long-term decrease of the altitudes took place even in the first half of the 20th century (Yevlashin et al. 1998; Evlashin et al. 1999). Independent measurements of vertical profiles of the relative frequency of observation of the

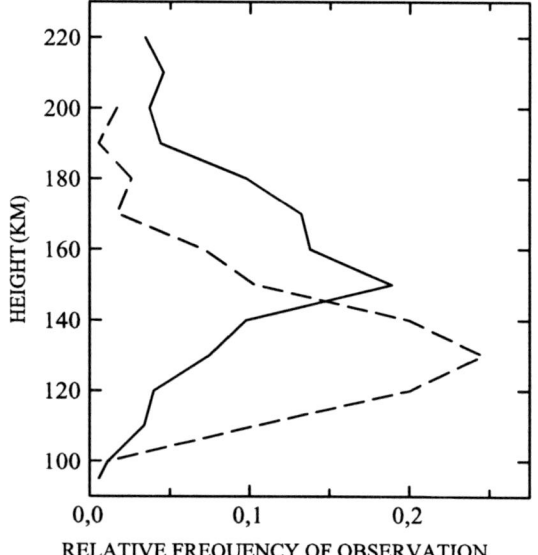

Fig. 6.20 Vertical profiles of the relative frequency of observation n of the lower edge of auroras on the daytime side of the auroral oval for different periods of in the years of solar maximum. The *solid curve* is for 1958/1959, and the *dashed curve* is for 1988/1989 (Starkov and Shefov 2001)

lower edge of auroral arcs on the daytime side of the auroral oval in the years of solar maxima 1958/1959 on Spitsbergen Island (~79°N) (Starkov 1968; Starkov and Shefov 2001) and 1988/1989 on Heiss Island (~81°N) (Starkov and Shefov 2001) gave radically different profiles (Fig. 6.20). However, an important feature of these data is that, unlike other data, they were obtained in winter at high latitudes. The rates of altitude variation estimated from these measurements are -880 (m·yr^{-1}) at ~175 (km) and -450 (m·yr^{-1}) at ~145 (km). These direct results clearly testify to the subsidence of the atmosphere at high latitudes that occurred at that period. Unfortunately, it is difficult now to reduce these data to the midlatitude and yearly average conditions.

The estimated rate of subsidence naturally refers to the observation period. It seems likely that the process of evolution of the thermal regime began to change in character by the late 1990s, which is reflected in variations of temperature trends. Under the influence of long-term variations in solar activity, the subsidence of the atmosphere can be changed by its elevation.

References

Aikin AC, Chanin ML, Nash J, Kendig DJ (1991) Temperature trends in the lower mesosphere. Geophys Res Lett 18:416–419

Alcaydé D, Fontanari J, Kockarts G, Bauer P, Bernard R (1979) Temperature, molecular nitrogen concentration and turbulence in the lower thermosphere inferred from incoherent scatter data. Ann Géophys 35:41–51

Ammosov PP, Gavrilieva GA (1996) The energy contribution of long-period waves to the mesopause. Geomagn Aeronomy 36:570–573

Anderson DE, Feldman PD, Gentieu EP, Meier RR (1980) The UV dayglow 2, L_α and L_β emissions and the H distribution in the mesosphere and thermosphere. Geophys Res Lett 7:529–532

Angel JK (1988) Variations and trends in tropospheric and stratospheric global temperatures, 1958–1987. J Clim 1:1296–1313

Angel JK (1991) Stratospheric temperature change as a function of height and sunspot number during 1972–89 based on rocketsonde and radiosonde data. J Clim 4:1170–1180

Atlasov KV, Borisov GB, Ignatiev VM, Sivtseva LV, Yugov VA (1975) The rotational temperatures of the hydroxyl bands (9,3) and (6,1) measured over Yakutsk during 1965–1970. In: Physics of the upper atmosphere at the high latitudes. N 3. Yakut Depart Siberian Branch USSR Acad Sci, Yakutsk, pp 174–186

Bakanas VV, Perminov VI (2003) Some features in the seasonal behavior of the hydroxyl emission characteristics in the upper atmosphere. Geomagn Aeronomy 43:363–369

Bakanas VV, Perminov VI, Semenov AI (2003) Seasonal variations of emission characteristics of the mesopause hydroxyl with different vibrational excitation. Adv Space Res 32:765–770

Baker DJ, Stair AT (1988) Rocket measurements of the altitude distribution of the hydroxyl airglow. Phys Scr 37:611–622

Bashkirtsev VS, Mashnich GP (2003) Will be face global warming in the nearest future? Geomagn Aeronomy 43:124–127

Beig G (2000) The relative importance of solar activity and anthropogenic influences on the ion composition, temperature, and associated neutrals of the middle atmosphere. J Geophys Res 105D:19841–19856

Beig G, Fadnavis S (2001) In search of greenhouse signals in the equatorial middle atmosphere. Geophys Res Lett 28:4603–4606

Beig G, Keckhut P, Lowe RP, Roble R, Mlynczak MG, Scheer J, Fomichev V, Offermann D, French WJR, Shepherd MG, Semenov AI, Remsberg E, She CY, Luebken FJ, Bremer J, Clemesha BR, Stegman J, Sigernes F, Fadnavis S (2003) Review of mesospheric temperature trends. Rev Geophys 41:1015, doi: 10.1029/2002RG000121

Berg MA, Shefov NN (1963) The hydroxyl emission with different vibrational excitation. In: Krasovsky VI (ed) Aurorae and Airglow. N 10. USSR Acad Sci Publ House, Moscow, pp 19–23

Berger U, von Zahn U (1999) The two-level structure of the mesopause: A model study. J Geophys Res 104D:22083–22093

Bittner M, Offermann D, Graef HH, Donner M, Hamilton K (2002) An 18-year time series of OH rotational temperatures and middle atmosphere decadal variations. J Atmos Sol Terr Phys 64:1147–1166

Brasseur G, Solomon S (2005) Aeronomy of the middle atmosphere, 3rd edn. Springer-Verlag, Dordrecht, Holland

Bremer J (1992) Ionospheric trend in mid-latitudes as a possible indicator of the atmospheric greenhouse effect. J Atmos Terr Phys 52:1505–1511

Bremer J (1998) Trends in the ionospheric E and F regions over Europe. Ann Geophys 16:986–996

Bull Results Rocket Sounding Atmosphere (1964–1994). Hydrometeoizdat, Moscow

Burns GB, French WJR, Greet PA, Williams PFB, Finlayson K, Lowe RP (2001) The mesopause region above Davis, Antarctica (68.6°S, 78.0°E). In: Beig G (ed) Long term changes and trends in the atmosphere, Vol 2. New Age International Limit Publications, Delhi, pp 43–52

Chandra S, Jackman CH, Fleming EL, Russell JM III (1997) The seasonal and long term changes in mesospheric water vapor. Geophys Res Lett 24:639–642

Clancy RT, Rush DW (1989) Climatology and trends of mesospheric (58–90 (km)) temperature based upon 1982–1986 SME limb scattering profiles. J Geophys Res 94D:3377–3393

Clemesha BR, Batista PP, Simonich DM (2003) Long-term variations in the centroid height of the atmospheric sodium layer. Adv Space Res 32:1707–1711

Clemesha BR, Batista PP, Simonich DM (2004) Negligible long-term temperature trend in the upper atmosphere at 23°S. J Geophys Res 109D05302: doi: 10.1029/2003JD004343. 1–8

Clemesha BR, Simonich D, Batista PP (1992) A long-term trend in the height of the atmospheric sodium layer: possible evidence for global change. Geophys Res Lett 19:457–460

Dergachev VA, Raspopov OM (2000) Long-term processes on the Sun controlling trends in the solar irradiance and the Earth's surface temperature. Geomagn Aeronomy 40:279–283

Dufay J, Tcheng Mao-Lin (1946) Recherches spectrophotométriques sur la lumière du ciel nocturne dans la région visible. 1. Ann Géophys 2:189–230

Dufay J, Tcheng Mao-Lin (1947a) Recherches spectrophotométriques sur la lumière du ciel nocturne dans la région visible. 2. Ann Géophys 3:153–183

Dufay J, Tcheng Mao-Lin (1947b) Recherches spectrophotométriques sur la lumière du ciel nocturne dans la région visible. 3. Ann Géophys 3:282–305

Dütsch HV, Staehelin J (1989) Discussion of the 60 year total ozone record at Arosa based on measurements of the vertical distribution and meteorological parameter. Planet Space Sci 37:1587–1599

Emmert JT, Picone JM, Lean JL, Knowles SH (2004) Global change in the thermosphere: compelling evidence of a secular decrease in density. J Geophys Res 109A: A02301, doi: 10.1029/2003 JA010176. 1–12

Evlashin LS, Semenov AI, Shefov NN (1999) Long-term variations in the thermospheric temperature and density on the basis of an analysis of Störmer's aurora-height measurements. Geomagn Aeronomy 39:241–245

Fast NP (1972) A catalogue of the noctilucent cloud appearances according to the world data. Tomsk University Publications House, Tomsk

Fatkullin MN, Zelenova TI, Kozlov VK, Legenka AD, Soboleva TN (1981) Empirical models of the mid-latitudinal ionosphere. Nauka, Moscow

Fauliot V, Thuillier G, Vial F (1997) Mean vertical wind in the mesosphere-lower thermosphere region (80–120 (km)) deduced from the WINDII observations on board UARS. Ann Geophys 15:1221–1231

Fishkova LM (1979) Nocturnal sodium emission in the Earth's upper atmosphere. In: Problems of the atmospheric optics. Leningrad State University, Leningrad, pp 154–172

Fishkova LM (1983) The night airglow of the Earth mid-latitude upper atmosphere. Shefov NN (ed). Metsniereba, Tbilisi

Fishkova LM, Martsvaladze NM, Shefov NN (2000) Patterns of variations in the OI 557.7-nm. Geomagn Aeronomy 40:782–786

Fishkova LM, Martsvaladze NM, Shefov NN (2001a) Long-term variations of the nighttime upper-atmosphere sodium emission. Geomagn Aeronomy 41:528–532

Fishkova LM, Martsvaladze NM, Shefov NN (2001b) Seasonal variations in the correlation of atomic oxygen 557.7-nm emission with solar activity and in long-term trend. Geomagn Aeronomy 41:533–539

Fomichev VI, Shved GM (1994) On the closeness of the middle atmosphere to the state of radiative equilibrium on estimation of net dynamical heating. J Atmos Terr Phys 56:479–485

French WJR, Burns GB (2004) The influence of large–scale oscillations on long–term trend assessment in hydroxyl temperatures over Davis, Antarctica. J Atmos Sol Terr Phys 66:493–506

Gadsden M (1990) A secular changes in noctilucent clouds occurrence. J Atmos Terr Phys 52:247–251

Gadsden M (1998) The north-west Europe data on noctilucent clouds: a survey. J Atmos Sol Terr Phys 60:1163–1174

Gadsden M (2002) Statistics of the annual counts of nights on which NLCs were seen. In: Mem British Astron Assoc, Aurora section (Meeting ≪ Mesospheric clouds ≫, Scottand, Perth, 19–22) Vol 45

Givishvili GV, Leshchenko LN (1993) Long-term trends of the mid-latitude ionosphere and thermospheric properties. Dokl Earth Sci 333: 86–89

Givishvili GV, Leshchenko LN (1995) Climatic trend dynamics of the midlatitude ionospheric E region. Geomagn Aeronomy 35:434–440

Givishvili GV, Leshchenko LN (1999) Long-term temperature variations of the mid-latitudinal region of the E-ionosphere. Dokl Earth Sci 368:682–684

Givishvili GV, Leshchenko LN (2000) Long-term temperature variations of the mid-latitudinal lower thermosphere. Dokl Earth Sci 371A:539–541

Givishvili GV, Leshchenko LN, Lysenko EV, Perov SP, Semenov AI, Sergeenko NP, Fishkova LM, Shefov NN (1996) Long-term trends of some characteristics of the Earth's atmosphere. I. Experimental results. Izv Russ Acad Sci Atmos Ocean Phys 32:303–312

Golitsyn GS, Semenov AI, Shefov NN, Fishkova LM, Lysenko EV, Perov SP (1996) Long-term temperature trends in the middle and upper atmosphere. Geophys Res Lett 23:1741–1744

Golitsyn GS, Semenov AI, Shefov NN (2000) Seasonal variations of the long-term temperature trend in the mesopause region. Geomagn Aeronomy 40:198–200

Golitsyn GS, Semenov AI, Shefov NN (2001) Thermal structure of the middle and upper atmosphere (25–110 (km)), as an image of its climatic change and influence of solar activity. In: Beig G (ed) Long term changes and trends in the atmosphere, Vol 2. New Age International Limit Publications, New Delhi, pp 33–42

Hauchecorne A, Chanin ML, Keckhut P (1991) Climatology and trends of the middle atmospheric temperature (33–87) as seen by Rayleigh lidar over the south of France. J Geophys Res 96D:15297–15309

Hedin AE (1983) A revised thermospheric model based on mass spectrometer and incoherent scatter data: MSIS-83. J Geophys Res 88A:10170–10188

Hernandez GJ, Silverman SM (1964) A reexamination of Lord Rayleigh's data on the airglow 5577 (OI) emission. Planet Space Sci 12:97–112

Hernandez G, Killeen TL (1988) Optical measurements of winds and kinetic temperatures in the upper atmosphere. Adv Space Res 8:149–213

Ignatiev VM, Sivtseva LD, Yugov VA, Atlasov KV (1974) Regular variations of the hydroxyl rotational temperatures over Yakutsk. In: Physics of the upper atmosphere at high latitudes. N 2. Yakut Depart Siberian Branch USSR Acad Sci, Yakutsk, pp 22–31

Izakov MN, Kokin GA, Perov SP (1967) A measurement method of the pressure and temperature during the rocket meteorological sounding. Meteorol Hydrol 12:70–86

Jacobi Ch (1998) On the solar cycle dependence of winds and planetary waves as seen from mid-latitude D1 LF mesopause region wind measurements. Ann Geophys 16:1534–1543

Jacobi Ch, Kürschner D (2002) A possible connection of mid – latitude mesosphere/lower thermosphere zonal winds and the southern oscillation. Phys Chem Earth 27:571–577

Keating GM, Tolson RH, Bradford MS (2000) Evidence of long term global decline in the Earth's thermospheric densities apparently related to anthropogenic effects. Geophys Res Lett 27:1523–1526

Keckhut P, Hauchecorne A, Chanin ML (1995) Midlatitude long-term variability of the middle atmosphere: trends and cyclic and episodic changes. J Geophys Res 100D:18887–18897

Keckhut P, Kodera K (1999) Long-term changes of the upper stratosphere as seen by japanese rocketsondes at Ryori (39°N, 141°E). Ann Geophys 17: 1210–1217

Keckhut P, Schmidlin FJ, Hauchecorne A, Chanin ML (1999) Stratospheric and mesospheric cooling trend estimates from US rocketsondes at low latitude stations (8°S–34°N), taking into account instrumental changes and natural variability. J Atmos Sol Terr Phys 61(6):447–459

Keneshea TJ, Zimmermann SP, Philbrick CR (1979) A dynamic model of the mesosphere and lower thermosphere. Planet Space Sci 27:385–401

Kokin GA, Lysenko EV (1994) On temperature trends of the atmosphere from rocket and radiosonde data. J Atmos Terr Phys 56:1035–1040

Kokin GA, Lysenko EV, Rosenfeld SKh (1990)Temperature changes in the stratosphere and mesosphere in 19651968 based on rocket sounding data. Izv Acad Sci USSR Atmos Oceanic Phys 26:518–523

Komitov BP, Kaftan VI (2003) Solar activity variations for the last millennia. Will the next long-period solar minimum be formed. Geomagn Aeronomy 43:553–561

Laštovička J (1997a) Observations of tides and planetary waves in the atmosphere-ionosphere system. Adv Space Res 20:1209–1222

Laštovička J (1997b) Long-term trends in the upper middle atmosphere as detected by ionospheric measurements. Adv Space Res 20:2065–2073

Laštovička J (2002) Long-term changes and trends in the lower ionosphere. Phys Chem Earth 27:497–507

Laštovička J (2005) Progress in trend studies: highlights of the TREND2004 Workshop. Adv Space Res 35:1359–1365

Laštovička J, Fišer V, Pancheva D (1994) Long-term trends in planetary wave activity (2–15 days) at 80–100 (km) inferred from radio wave absorption. J Atmos Terr Phys 56:893–899

Le Texier H, Solomon S, Garcia RR (1987) Seasonal variability of the OH Meinel band. Planet Space Sci 35:977–989

Lord Rayleigh, Spencer Jones H (1935) The light of the night sky: analyses of the intensity variations at three stations. Proc R Soc Lond 151A: 22–55

Lowe RP (1969) Interferometric spectra of the Earth's airglow (1,2 to 1,6 (μm)). Philos Trans R Soc Lond A264:163–169

Lübken FJ, Fricke KH, Langer M (1996) Noctilucent clouds and the thermal structure near the Arctic mesopause in summer. J Geophys Res 101D:9489–9508

Lysenko EV (1981) An error of the measurement method of the atmospheric temperature with aid the rocket resistence thermometer. Trans Central Aerolog Observ 144:28–44

Lysenko EV, Rosenfeld SKh, Speransky KE (1982) Experimental determination of the error characteristics of the rocket meteorological measurements. Meteorol Hydrol 10:46–53

Lysenko EV, Nelidova GF, Prostova AM (1997a) Changes in the stratospheric and mesospheric thermal conditions during the last three decades: 1. the evolution of a temperature trend. Izv Atmos Ocean Phys 33:218–225

Lysenko EV, Nelidova GF, Prostova AM (1997b) Changes in the stratospheric and mesospheric thermal conditions during the last three decades: 2. The evolution of annual and semiannual temperature oscillations. Izv Atmos Ocean Phys 33:226–233

Lysenko EV, Perov SP, Semenov AI, Shefov NN , Sukhodoev VA, Givishvili GV, Leshchenko LN (1999) Long-term trends of the yearly mean temperature at heights from 25 to 110 (km). Izv Atmos Ocean Phys 35:393–400

Lysenko EV, Rusina VYa (2002a) Changes in the stratospheric and mesospheric thermal conditions during the last three decades: 3. Linear trends of monthly mean temperatures. Izv Atmos Ocean Phys 38:296–304

Lysenko EV, Rusina VYa (2002b) Changes in the stratospheric and mesospheric thermal conditions during the last three decades: 4. trends in the height and temperature of the stratopause. Izv Atmos Ocean Phys 38:305–311

Lysenko EV, Nelidova GG, Rusina VYa (2003) Annual cycles of middle atmosphere temperature trends determined from long-term rocket measurements. Int J Geomagn Aeronomy 4(1):57–65

Lysenko EV, Rusina VYa (2003) Long-term changes in the stratopause height and temperature derived from rocket measurements at various latitudes. Int J Geomagn Aeronomy 4:67–81

Manual (1978) Manual of URSI for the interpretation and treatment of ionograms. Nauka, Moscow

McDade IC, Llewellyn EJ (1988) Mesospheric oxygen atom densities inferred from night-time OH Meinel band emission rates. Planet Space Sci 36:897–905

Megrelishvili TG, Toroshelidze TI (1999) Long-term trend in the mesopause density as inferred from twilight spectrophotometric observations. Geomagn Aeronomy 39:258–260

Meier RR, Prinz DK (1970) Absorption of the solar Lyman alpha line by geocoronal atomic hydrogen. J Geophys Res 75:6969–6979

Meriwether JW, Gardner CS (2000) A review of the mesosphere inversion layer phenomenon. J Geophys Res 105D:12405–12416

Miller A, Nagatani JRM, Tiao GC, Niu XF, Reinsel GC, Wuebbles D, Grant K (1992) Comparisons of observed ozone and temperature trends in their lower stratosphere. Geophys Res Lett 19:929–932

Mulligan FJ, Horgan DF, Galligan IG, Griffin EM (1995) Mesopause temperatures and integrated band brightnesses calculated from airglow on emission recorded at Maynooth (53.2°N, 6.4°W) during 1993. J Atmos Terr Phys 57:1623–1637

Murphree JS, Elphinstone RD, Cogger LL (1984) Dynamics of the lower thermosphere consistent with satellite observations of 5577 Å airglow: I. Method of analysis. Can J Phys 62:370–381

Myrabø HK (1984) Temperature variation at mesopause levels during winter solstice at 78°N. Planet Space Sci 32:249–255

Myrabø HK, Deehr CS, Sivjee GG (1983) Large-amplitude nightglow OH (8–3) band intensity and rotational temperature variations during 24–hour period at 78°N. J Geophys Res 88A:9255–9259

Nielsen KP, Sigernes F, Raustein E, Deehr CS (2002) The 20-year change of the Svalbard OH–temperatures. Phys Chem Earth 27:555–561

Offermann D, Graef H (1992) Messungen der OH* – Temperature. Promet 22:125–128

Offermann D, Donner M, Semenov AI (2002) Hydroxyl temperatures: variability and trends. In: We-Heraeus Seminar on trends in the upper atmosphere. Kühlungsborn, Germany, p 38

Offermann D, Donner M, Knieling P, Naujokat B (2004) Middle atmosphere temperature changes and the duration of summer. J Atmos Sol Terr Phys 66:437–450

Perminov VI, Semenov AI (2007) Model of the latitudinal, seasonal and altitudinal changes of the long-term temperature trend of the middle atmosphere. Geomagn Aeronomy 47:682–690

Perminov VI, Semenov AI, Shefov NN (1998) Deactivation of hydroxyl molecule vibrational states by atomic and molecular oxygen in the mesopause region. Geomagn Aeronomy 38:761–764

Pokhunkov AA, Rybin VV, Tulinov GF (2003) Atomic oxygen trend in the midlatitude and equatorial thermosphere. Geomagn Aeronomy 43:641–646

Portnyagin YuI, Forbes JF, Solovjeva TV, Miyahara S, DeLuca C (1995) Momentum and heat sources at the mesosphere and lower thermosphere regions 70–110 (km). J Atmos Terr Phys 57:967–977

Qian J, Gardner CS (1995) Simultaneous lidar measurements of mesospheric Ca, Na, and temperature profiles at Urbana, Illinois. J Geophys Res 100D:7453–7461

Rind D, Suozzo R, Balachandran NK, Prather MJ (1990) Climate changes and the middle atmosphere. Pt 1: the doubled CO_2 climate. J Atmos Sci 47:475–494

Roach FE, Pettit HB, Williams DR, St Amand P, Davis DN (1953) A four-year study of OI 5577 Å in the nightglow. Ann d'Astrophys 16: 185–205

Roble RG, Dickinson RF (1989) How will changes in carbon dioxide and methane modify the mean structure of the mesosphere and thermosphere? Geophys Res Lett 16:1441–1444

Rosenfeld SKh (1986) An analysis of the regular components of the observation series, prescribed on the nonuniform sequence of data. Meteorol Hydrol 3: 5–14

Scheer J, Reisin ER, Espy JP, Bittner M, Graef HH, Offermann D, Ammosov PP, Ignatiev VM (1994) Large – scale structures in hydroxyl rotational temperatures during DYANA. J Atmos Terr Phys 56:1701–1715

Schmidlin FJ (1986) Rocket techniques used to measure the middle atmosphere. In: Goldberg RA (ed) Handbook for MAP, Vol 19. SCOSTEP, Urbana, pp 1–28

Schmidlin FJ (1996) Rocketsonde temperatures trend long-term stratospheric behavior. Paper C2-1-0002. 31 Science Assembly COSPAR. Birmingham, July 14–21

Semenov AI (1996) A behavior of the lower thermosphere temperature inferred from emission measurements during the last decades. Geomagn Aeronomy 36:655–659

Semenov AI (1997) Long-term changes in the height profiles of ozone and atomic oxygen in the lower thermosphere. Geomagn Aeronomy 37:354–360

Semenov AI (2000) Long-term temperature trends for different seasons by hydroxyl emission. Phys Chem Earth Pt B 25:525–529

Semenov AI, Lysenko EV (1996) Long-term subsidence of the middle and upper atmosphere according to its ecological evolution. Environ Radioecol Appl Ecol 2:3–13

Semenov AI, Shefov NN (1996) An empirical model for the variations in the hydroxyl emission. Geomagn Aeronomy 36:468–480

Semenov AI, Shefov NN (1997a) An empirical model of nocturnal variations in the 557.7-nm emission of atomic oxygen. 1. Intensity. Geomagn Aeronomy 37:215–221

Semenov AI, Shefov NN (1997b) An empirical model of nocturnal variations in the 557.7-nm emission of atomic oxygen. 2. Temperature. Geomagn Aeronomy 37:361–364

Semenov AI, Shefov NN (1997c) An empirical model of nocturnal variations in the 557.7-nm emission of atomic oxygen. 3. Emitting layer altitude. Geomagn Aeronomy 37:470–474

Semenov AI, Shefov NN (1997d) Empirical model of the variations of atomic oxygen emission 557.7 (nm). In: Ivchenko VN (ed) Proceedings of SPIE (23rd Eur Meet Atmos Stud Opt Meth, Kiev, September 2–6, 1997), Vol 3237. The Int Soc Opt Eng, Bellingham, pp 113–122

Semenov AI, Shefov NN (1999a) Empirical model of hydroxyl emission variations. Int J Geomagn Aeronomy 1:229–242

Semenov AI, Shefov NN (1999b) Variations of the temperature and the atomic oxygen content in the mesopause and lower thermosphere region during change of the solar activity. Geomagn Aeronomy 39: 484–487

Semenov AI, Shefov NN (2006) Long-term variations of a temperature and structure of the middle atmosphere within last century. In: Proc29th Ann Apatity Seminar "Physics of Auroral Phenomena"(February 27–March 3, 2006). Kola Sci Center Russian Acad Sci Polar Geophys Institute RAS, Apatity, pp 250–253

Semenov AI, Shefov NN, Fishkova LM, Lysenko EV, Perov SP, Givishvili GV, Leshchenko LN, Sergeenko NP (1996) Climatic changes in the upper and middle atmosphere. Dokl Earth Sci 349: 870–872

Semenov AI, Shefov NN, Givishvili GV, Leshchenko LN, Lysenko EV, Rusina VYa, Fishkova LM, Martsvaladze NM, Toroshelidze TI, Kashcheev BL, Oleynikov AN (2000) Seasonal peculiarities of long-term temperature trends of the middle atmosphere. Dokl Earth Sci 375: 1286–1289

Semenov AI, Sukhodoev VA, Shefov NN (2002a) A model of the vertical temperature distribution in the atmosphere altitudes of 80–100 (km) that taking into account the solar activity and the long-term trend. Geomagn Aeronomy 42: 239–244

Semenov AI, Bakanas VV, Perminov VI, Zheleznov YuA, Khomich. VYu (2002b) The near infrared spectrum of the emission of the nighttime upper atmosphere of the Earth. Geomagn Aeronomy 42:390–397

Semenov AI, Shefov NN, Lysenko EV, Givishvili GV, Tikhonov AV (2002c) The seasonal peculiarities of behavior of the long-term temperature trends in the middle atmosphere at the mid-latitudes. Phys Chem Earth 27:529–534

Semenov AI, Pertsev NN, Shefov NN, Perminov VI, Bakanas VV (2004) Calculation of the vertical profiles of the atmospheric temperature and number density at altitudes of 30–110 (km). Geomagn Aeronomy 44:773–778

Semenov AI, Shefov NN, Perminov VI, Khomich VYu, Fadel KhM (2005) Temperature response of the middle atmosphere on the solar activity for different seasons. Geomagn Aeronomy 45:236–240

Semenov AI, Shefov NN, Sukhodoev VA (2006) Re-analysis of the long-term hydroxyl rotational temperature trend according to measurements at Spitsbergen. In: Proceedings of 29th Ann Apatity Seminar "Physics of Auroral Phenomena" (February 27–March 3, 2006). Kola Sci Center Russian Acad Sci Polar Geophys Institute RAS, Apatity, pp 245–249

Sergeenko NP, Kuleshova VP (1994) Climatic variations of the disturbance properties in the ionosphere and upper atmosphere. Dokl Earth Sci 334:534–536

Shain GA, Shain PF (1942) Methods of the variation investigations of the emission lines in the night sky spectrum. Dokl USSR Acad Sci 35:152–156

Sharp WE, Kita D (1987) In situ measurement of atomic hydrogen in the upper atmosphere. J Geophys Res 92D:4319–4324

She CY, Lowe RP (1998) Seasonal temperature variations in the mesopause region at mid–latitude: comparison of lidar and hydroxyl rotational temperatures using WINDII/UARS OH height profiles. J Atmos Sol Terr Phys 60:1573–1583

She CY, Yu JR, Chen H (1993) Observed thermal structure of a midlatitude mesopause. Geophys Res Lett 20:567–570

She CY, Yu JR, Krueger DA, Roble R, Keckhut P, Hauchecorne A, Chanin ML (1995) Vertical structure of the midlatitude temperature from stratosphere to mesopause (30–105 (km)). Adv Space Res 20:377–380

She CY, Thiel SW, Krueger DA (1998) Observed episodic warming at 86 and 100 (km) between 1990 and 1997: Effects of Mount Pinatubo eruption. Geophys Res Lett 25:497–500

Shefov NN (1961) On determination of the rotational temperature of the OH bands. In: Krasovsky VI (ed) Spectral, electrophotometrical and radar researches of aurorae and airglow. N 5. USSR Acad Sci Publ House, Moscow, pp 5–9

Shefov NN (1969) Hydroxyl emission of the upper atmosphere. I. Behaviour during solar cycle, seasons and geomagnetic disturbances. Planet Space Sci 17:797–813

Shefov NN (1976) Seasonal variations of the hydroxyl emission. In: Krassovsky VI (ed) Aurorae and Airglow. N 24. Soviet Radio, Moscow, pp 32–36

Shefov NN (1978) Altitude of the hydroxyl emission layer. In: Krassovsky VI (ed) Aurorae and Airglow. N 27. Soviet Radio, Moscow, pp 45–51

Shefov NN, Kropotkina EP (1975) The height variations of the $\lambda 5577$ A emission layer. Cosmic Res 13:765–770

Shefov NN, Semenov AI (2001) An empirical model for nighttime variations in atomic sodium emission: 2. Emitting layer height. Geomagn Aeronomy 41:257–261

Shefov NN, Semenov AI (2002) The long-term trend of ozone at heights from 80 to 100 (km) at the mid-latitude mesopause for the nocturnal conditions. Phys Chem Earth 27:535–542

Shefov NN, Semenov AI, Yurchenko OT (2000) Empirical model of variations in the atomic sodium nighttime emission: 1. Intensity. Geomagn Aeronomy 40:115–120

Shefov NN, Semenov AI, Yurchenko OT (2002) Empirical model of the ozone vertical distribution at the nighttime mid-latitude mesopause. Geomagn Aeronomy 42:383–389

Sigernes F, Nielsen KP, Deehr CS, Svenøe T, Shumilov N, Havnes O (2003a) The hydroxyl rotational temperature record from the auroral station in Adventdalen, Svalbard (78°N, 15°E):

preliminary results. In: Kultima J (ed) Sodankylä Geophys Observ Publ, Vol 92. Oulu University Press, Oulu, pp 61–67

Sigernes F, Nielsen KP, Deehr CS, Svenøe T, Shumilov N, Havnes O (2003b) Hydroxyl rotational temperature record from the auroral station in Adventalen, Svalbard (78°N, 15°E). J Geophys Res 108A: 1342, doi: 10.1029/2001JA009023. SIA1–SIA3

Snelling D, Hampson J (1969) Water vapor concentration and neutral reactions in the mesosphere and stratosphere. In: Sechrist CF (ed) Aeronomy Rep. N 32. University of Illinois, Urbana, pp 223–234

Starkov GV (1968) Aurora altitudes in the polar cap. Geomagn Aeronomy 8:36–41

Starkov GV, Shefov NN (2001) Long-term changes of the auroral heights in dayside of the auroral oval. Geomagn Aeronomy 41:763–765

Starkov GV, Yevlashin LS, Semenov AI, Shefov NN (2000) A subsidence of the thermosphere during 20th century according to the measurements of the auroral heights. Phys Chem Earth Pt B 25:547–550

States RJ, Gardner CS (1999) Structure of the mesospheric Na layer at 40°N latitude: seasonal and diurnal variations. J Geophys Res 104D:11783–11898

Stolarski R, Bojkov R, Bishop L, Zerefos C, Staehelin J, Zavodny J (1992) Measured trends in stratospheric ozone. Science 256:342–349

Störmer C (1955) The polar aurora. Clarendon Press, Oxford

Taubenheim J, von Cossart G, Entzian G (1990) Evidence of CO_2 – induced progressive cooling of the middle atmosphere derived from radio observations. Adv Space Res 10:171–174

Taubenheim J, Berendorf K, Krueger W, Entzian G (1991) Trends und langfristige Variationen der mittleren Atmosphäre in Abhängigkeit von der Höhe. Rep Potsdam Institute Climate Impact Res 1:247–251

Taubenheim J, Entzian G, Berendorf K (1997) Long-term decrease of mesospheric temperature, 1963–1995, inferred from radiowave reflection heights. Adv Space Res 20:2059–2063

Thomas G, Olivero JJ, Jensen EJ, Schröder W, Toon OB (1989) Relation between increasing methane and the presence of ice clouds at the mesopause. Nature (London) 338:490–492

Ulich T, Turunen E (1997) Evidence for long-term cooling of the upper atmosphere in ionosonde data. Geophys Res Lett 24:1103–1106

Ulwick JC, Baker KD, Baker DJ, Steed AJ, Pendleton WR, Grossmann K, Brueckelmann HG (1987) Mesospheric minor species determinations from rocket and ground-based i.r. measurements. Geophys Res Lett 49:855–862

Vaughan G (1982) Diurnal variation of mesospheric ozone. Nature (London) 296:133–135

Volkov II, Suevalov VV (2005) Estimate of the long-term density variations in the upper atmosphere of the Earth at minimums of solar activity from evolution of the orbital parameters of the Earth's artificial satellites. Solar System Res 39:177–183

von Zahn U (2003) Are noctilucent clouds truly a « miner's canary » for global change? EOS Trans AGU. 84: pp 261, 264

Witt G, Stegman J, Solheim BH, Llewellyn EJ (1979) A measurement of the $O_2(b^1\Sigma_g^+ - X^3\Sigma_g^-)$ atmospheric band and the $O(^1S)$ green line in the nightglow. Planet Space Sci 27:341–350

WMO/UNEP (1990), Scientific assessment of climate change. Intergovern Panel Climate Change, Geneva

Yarin VI (1961) The OH emission according to observations in Yakutsk. In: Krassovsky VI (ed) Spectral, electrophotometrical and radar researches of aurorae and airglow. N 5. USSR Acad Sci Publ House, Moscow, pp 10–17

Yarin VI (1962) On the dependence of intensity of OH bands on the rotational temperature. In: Krassovsky VI (ed) Aurorae and Airglow. N 8. USSR Acad Sci Publ House, Moscow, pp 9–10

Yevlashin LS, Semenov AI, Shefov NN (1998) Long-term variations of the thermospheric temperature and density on the base of auroral characteristics. In: Proceedings of 21st Ann Seminar. Phys Auroral phenomena. Kola Sci Centre, Apatity, pp 117–120

Yu JR, She CY (1995) Climatology of a midlatitude mesopause region observed by a lidar at Fort Collins, Colorado (40.6°N,105°W). J Geophys Res 100D: 7441–7452

Chapter 7
Models of Vertical Profiles of Some Characteristics of the Upper Atmosphere

An important feature of the airglow arising from the upper atmosphere is that it is a sensitive indicator of the concentration of small atmospheric constituents as well as of the temperature and dynamic regimes at various altitude levels. This gives considerable opportunity for the development of models which would describe the varying characteristics of not only the airglow emissions but also the atmospheric constituents responsible for their occurrence. The most informative is the hydroxyl emission. Nevertheless, some general properties of the hydroxyl, sodium, 864.5-nm molecular oxygen, and 557.7-nm atomic oxygen emissions enable one to construct models of vertical profiles of most photochemically active constituents of the mesopause and lower thermosphere, such as atomic oxygen, ozone, and atomic hydrogen. The emission of the atmospheric continuum caused by the photochemical reactions of nitric oxide with atomic oxygen and ozone has already been considered in the preceding sections. Unfortunately, the available data are lacking to develop such models. This in particular is due to the fact that to record the airglow continuum calls for special conditions of clear atmospheric weather and no artificial illumination which are difficult to satisfy, especially nowadays. It is also difficult to measure the infrared airglow produced by NO molecules from the Earth surface.

7.1 Ozone in the Mesopause Region

The importance of the ozone layer in the mesopause region became obvious once the hydroxyl emission had been detected and its photochemical origin had been elucidated. Mesospheric ozone was studied in the UV range of the spectrum at night based on the light absorption by stars (Roble and Hays 1974) and in the daytime based on the 1.27-μm emission of molecular oxygen (Thomas 1990a) using ground-based microwave radiometers (Wilson and Schwartz 1981) and satellites (Ricaud et al. 1996). Based on these limb measurements, daily and seasonal variations of the vertical ozone profiles have been revealed. Therefore, the attempts to retrieve the ozone characteristics from the OH emission data (Hecht et al. 2000) with model descriptions of some input parameters that disregarded the heliogeophysical conditions at which the measurements were performed cannot be considered satisfactory.

An empirical model of variations of the vertical ozone concentration profiles at night at altitudes of 80–100 (km) was constructed (Shefov et al. 2002; Shefov and Semenov 2002) based on empirical models of variations of the hydroxyl, atomic oxygen (557.7 (nm)), and sodium (589.0–589.6 (nm)) emissions as well as on the vertical temperature profiles measured in the middle atmosphere at altitudes of 30–110 (km). In addition, the important photochemical processes with participation of the ozone molecule were also included.

The photochemistry of ozone is a subject of long-term research (Nicolet 1971). Therefore, to analyze the hydroxyl (Semenov 1997) and sodium emissions (Shefov et al. 2000a; Shefov and Semenov 2001; Fishkova et al. 2001a), the recent data on the photochemical processes giving rise to these emissions (Chikashi et al. 1989; Plane 1991; Herschbach et al. 1992; Helmer and Plane 1993; McNeil et al. 1995; Clemesha et al. 1995) were used. All of them have been considered in Sect. 2.3.1.

In view of these photochemical processes, the relation (Semenov 1997)

$$[O_3] = \frac{Q_{OH}(Z) \cdot \{A_9 + \beta_{N_2} \cdot [N_2] + \beta_{O_2} \cdot [O_2]\}}{A_9 \cdot \gamma_9 \cdot \alpha_{OH} \cdot [H] \cdot B}$$

was derived, where

$$B = 1 - \frac{Q_{OH}(Z) \cdot \beta_O}{A_9 \cdot \gamma_9 \cdot \{\alpha_{O_3}(N_2) \cdot [N_2] + \alpha_{O_3}(O_2) \cdot [O_2]\} \cdot [O_2]}.$$

The vertical profiles of the atomic oxygen density in the lower thermosphere reflected in the behavior of the emission at 557.7 (nm) undergo considerable variations depending on the time of day, season, and solar activity and a long-term trend (Perminov et al. 1998; Semenov and Shefov 1999b; Shefov et al. 2000a; Fishkova et al. 2000) that was disregarded in the conventional models. As demonstrated by Semenov (1997), the Barth mechanism of excitation of the 557.7-nm emission allows vertical profiles of the atomic oxygen density to be reconstructed from rocket measurements with good accuracy.

As mentioned, empirical models of the variations of sodium (Shefov et al. 2000a; Shefov and Semenov 2001; Fishkova et al. 2001a), hydroxyl (Semenov and Shefov 1996), and atomic oxygen emissions (Semenov and Shefov 1999b; Shefov et al. 2000a; Fishkova et al. 2000, 2001b) and of the atomic hydrogen density (Semenov 1997; Shefov et al. 2000a; Semenov and Shefov 1997c,d) were used to calculate the vertical profiles of the ozone density $[O_3(Z)]$. Data on the behavior of atomic hydrogen are required to retrieve the ozone density from hydroxyl emission measurements. Direct rocket measurements of hydrogen density at altitudes of 80–100 (km) are few in number. Nevertheless, the use of these measurements has made it possible to reveal a very close positive correlation between hydrogen density and solar activity and a positive long-term trend for 1959–1984 (Sect. 7.3). This is also confirmed by a long-term increase of the methane (the main supplier of hydrogen in the upper atmosphere) content in the ground atmospheric layer

(Nicolet 1971; Smith et al. 2000). The trends for a relative increase in atomic hydrogen density in the mesopause region and in methane density in the lower atmosphere appear identical. Measurements of the water vapor content in the stratosphere (Smith et al. 2000) and mesosphere (~80 (km)) (Chandra et al. 1997) and of their variations also indicate the presence of a significant positive trend. However, these circumstances were disregarded in the existing MSIS-E-1990 and CIRA-1986 models and in the work by Thomas (1990b). Therefore, the vertical hydrogen profiles were corrected for the solar flux ($F_{10.7}$) and for the long-term trend (Semenov 1997; Shefov et al. 2000a).

Figure 7.1 presents the calculated monthly mean vertical profiles of the ozone density for low (1976 and 1986) and high solar activities (1980 and 1991). To the right of the figure, the yearly average profiles are shown. These pairs of years were chosen because the solar fluxes (both monthly mean and yearly average ones) were almost identical for each specified pair of years; at the same time, there was a gap of 10–11 years between two observations. This allows the presence of trends and the effect of solar activity to be revealed independently. Comparing the data, one can see that the ozone density (at night) decreases with increasing solar activity. This is a consequence of the positive correlation between the atomic hydrogen density and the solar flux (Semenov 1997; Shefov et al. 2000a), since atomic hydrogen is

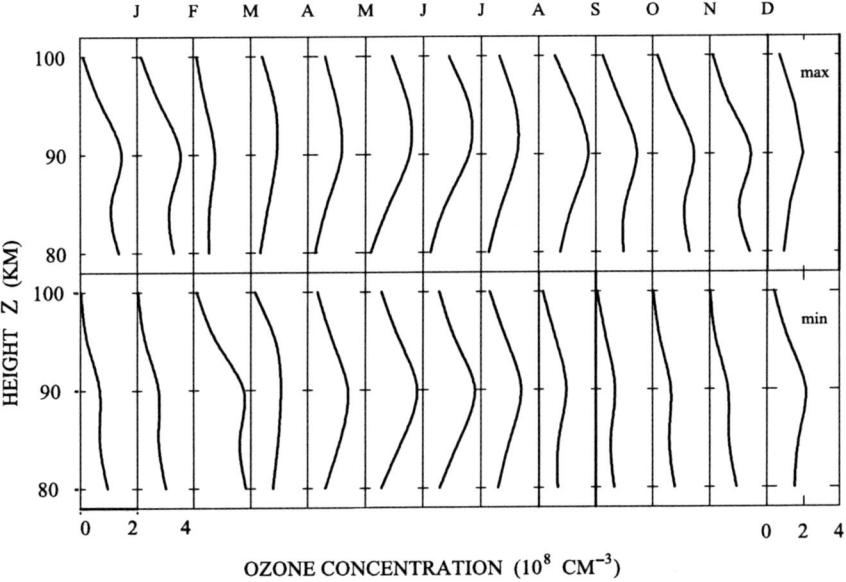

Fig. 7.1 Monthly mean vertical profiles of the ozone density for high ($F_{10.7} \sim 203$ in 1980 and 1991) and low solar activity ($F_{10.7} \sim 74$ in 1976 and 1986). The yearly average profiles are shown to the right of the figure. The graduation points of the upper scale designating months are simultaneously the reference points of the $[O_3(Z)]$ profile for the corresponding month. The lower scale is for January (Shefov et al. 2002; Shefov and Semenov 2002)

the main component of catalytic ozone decomposition in the course of the ozone–hydrogen reaction.

The average vertical profile $[O_3(Z)]$ for $F_{10.7} = 130$ can be approximated by an asymmetric Gauss function, as already done for the emission intensity profiles (Chap. 4). For altitudes in the range 80–100 (km) and $z \geq 90$ (km), we have

$$[O_3(Z)]_0 = 2 \cdot 10^8 \cdot \exp\left[-\frac{\log_e 2 \cdot (Z-90)^2}{P^2 \cdot W^2}\right], \quad (cm^{-3}),$$

and for $z \leq 90$ (km),

$$[O_3(Z)]_0 = 2 \cdot 10^8 \cdot \exp\left[-\frac{\log_e 2 \cdot (Z-90)^2}{(1-P)^2 \cdot W^2}\right], \quad (cm^{-3}),$$

with the asymmetry $P = 0.353$ and profile halfwidth $W = 22.5$ (km).

It turns out that the seasonal variations of the ozone density at different altitudes are different (Fig. 7.2) and the ozone column content N_{O_3} (cm^{-2}) at altitudes of 80–100 (km) also undergoes seasonal variations. Solid curves here indicate variations for high solar activity ($F_{10.7} = 203$), and dashed curves are for low solar activity ($F_{10.7} = 74$).

Figure 7.3 presents vertical profiles of amplitudes A and phases t (days) of the harmonics of seasonal variations reduced to $F_{10.7} = 130$ that enter in the approximation

$$\Sigma = 1 + A_1 \cdot \cos\left[\frac{2\pi(t_d - t_1)}{365}\right] + A_2 \cdot \cos\left[\frac{4\pi(t_d - t_2)}{365}\right] + A_3 \cdot \cos\left[\frac{6\pi(t_d - t_3)}{365}\right],$$

where t_d is the serial number of the day of year. It can be seen that the annual harmonic is dominant at midlatitudes.

Based on the data obtained, the effect of solar activity on the harmonics of seasonal variations can be estimated. As a first approximation, from the available data we obtain

$$A_i(Z, F_{10.1}) = A_{0i} \cdot [1 + \delta A_{trSi} \cdot (F_{10.1} - 130)],$$

where

$$\delta A_{trS0} = \frac{Z-90}{2400}, \quad \delta A_{trS1} = -10^{-5} \cdot (F_{10.7} - 130) \cdot \exp\left[\frac{123-Z}{6.3}\right],$$

$$\delta A_{trS2} = -10^{-5} \cdot (F_{10.7} - 130) \cdot \exp\left[\frac{148-Z}{10}\right], \quad \delta A_{trS3} = 0,$$

$$t_1 = 182 - \frac{364}{1 + \exp\left(\frac{Z-80}{2}\right)}, \quad t_2 = \frac{365}{1 + \exp\left(\frac{100-Z}{5}\right)}, \quad t_3 = 60 + \frac{80}{1 + \exp\left(\frac{Z-80}{5}\right)}.$$

After reduction of $[O_3(Z)]$ to the constant flux $F_{10.7} = 200$ for the solar activity maximum and $F_{10.7} = 74$ for the solar activity minimum, it turned out that the difference between long-term trends for individual months of the year and their seasonal

7.1 Ozone in the Mesopause Region

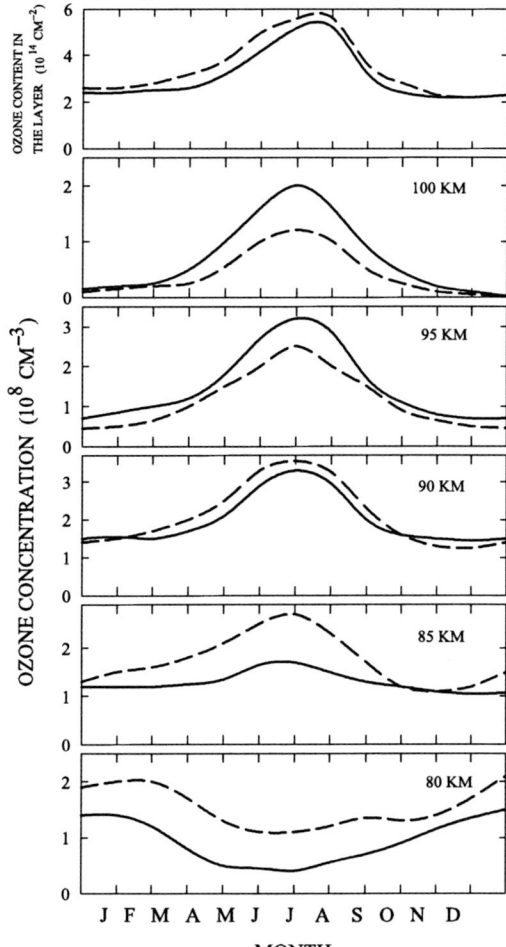

Fig. 7.2 Examples of seasonal variations of the ozone density at high ($F_{10.7} \sim 203$, *solid curves*) and low solar activity ($F_{10.7} \sim 74$, *dashed curves*) and of the ozone molecular density in the 80–100 (km) layer (Shefov et al. 2002; Shefov and Semenov 2002)

variations was insignificant. Therefore, the data were averaged, and on their basis, the change of the trend for different altitude levels was retrieved in the form

$$\delta[O_3(Z)]_{tr} = -0.0016 \cdot (Z - 85) \, .$$

Thus, the ozone density at altitudes of 80–100 (km) can be described by the relation

$$[O_3(Z)] = [O_3(Z)]_0 \cdot \{1 + \delta[O_3(Z)]_{tr} \cdot (t - 1972.5)\} \cdot \{1 + (F_{10.7} - 130)\} \cdot \Sigma \, .$$

The total ozone density in the 80–100 (km) layer is described by the relation

$$N_{O_3} = 3.4 \cdot 10^{14} \cdot \left[1 + 0.39 \cdot \cos\frac{2\pi(t_d - 182)}{365} + 0.18 \cdot \cos\frac{4\pi(t_d - 7)}{365} + 0.05 \cdot \cos\frac{6\pi(t_d - 70)}{365}\right]$$

$$\times [1 - 0.001 \cdot (F_{10.7} - 130)] \cdot [1 - 0.0073 \cdot (t - 1972.5)], \quad (cm^{-2}) \, .$$

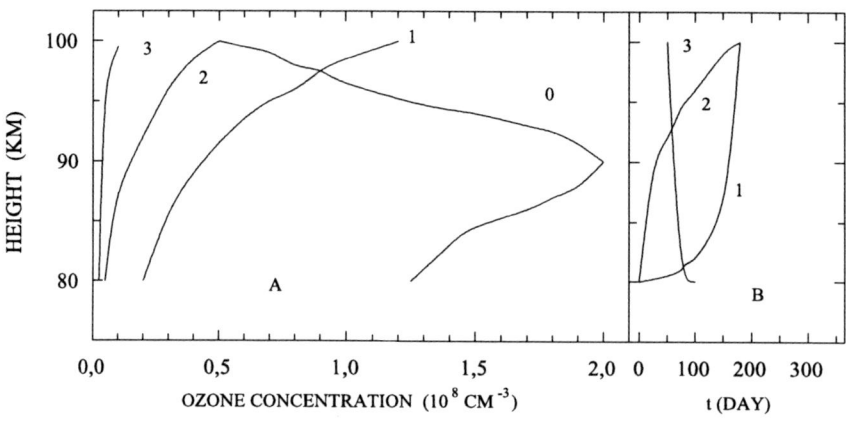

VERTICAL VARIATIONS OF AMPLITUDES A AND PHASES t

Fig. 7.3 Altitude variations of the ozone density (*curve* 0), amplitudes A of annual (*curve* 1), semiannual (*curve* 2), and 4-month harmonics (*curve* 3), and phases t of the harmonics of seasonal variations ($F_{10.7} = 130$) (Shefov et al. 2002; Shefov and Semenov 2002)

The harmonic amplitudes weakly depend on solar activity.

The determination of the ozone density from the sodium emission data at altitudes around 92 (km) (where the ozone density and the emission rate peak) has revealed its similar dependences on time and solar activity.

Used long-term yearly average variations of the atmospheric characteristics illustrate the dependence from the behavior of the solar activity (Fig. 7.4) (Shefov and Semenov 2002).

The obtained empirical dependence of ozone density variations at altitudes of 80–100 (km) is of special interest in solving the problem on the thermal regime in this atmospheric region, especially in elucidating its long-term behavior. In the middle atmosphere at the altitudes of the stratosphere and mesopause, the prevailing contribution to the energy influx due to the solar UV radiation absorption is provided by ozone despite its low relative content. It has already been demonstrated (Semenov 1997) that the ozone density at altitudes of 85–100 (km) at moderate solar activity decreased threefold for the period from 1955 to 1995. The results presented here demonstrate that the greatest change (three times and more) occurs in the top part of the layer.

As follows from the ozone photochemistry in the mesopause region, the basic process of ozone destruction at night is the ozone–hydrogen reaction responsible for the initiation of the hydroxyl emission. Atomic hydrogen serves as a catalyst for the relevant sequence of reactions. In the daytime, the ozone density decreases by a factor of $\sim$1.5 at altitudes under consideration due to the photodissociation reaction (Nicolet 1971)

$$O_3 + h\nu(\lambda \leq 1180\,\text{nm}) \rightarrow O_2(X^3\Sigma_g^-) + O(^3P), \quad j = 10^{-2}(s^{-1}).$$

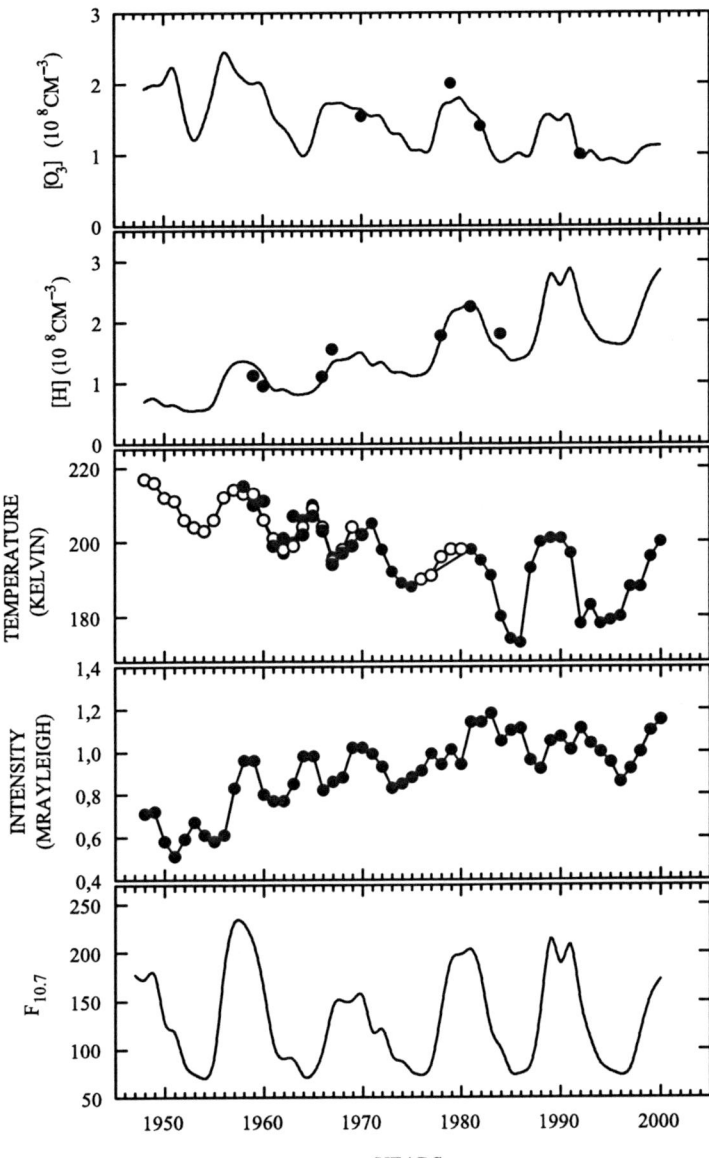

Fig. 7.4 Long-term yearly average variations of the atmospheric characteristics (Shefov and Semenov 2002) including (from below upwards) the behavior of the solar flux, integral hydroxyl emission intensity (*full circles*), temperature at 87 (km) (*full circles* indicate the measurement data (Semenov 2000) and *open circles* the results of interpolation and extrapolation using the empirical model (Semenov and Shefov 1996)), atomic hydrogen density (*solid curve*) calculated from the data (Semenov 1997; Shefov and Semenov 2002) with *full circles* illustrating rocket measurements reduced to 95 (km) based on the model (Keneshea et al. 1979), calculated ozone density at 95 (km) (*solid curve*) with *full circles* illustrating the satellite and rocket data (Roble and Hays 1974; Vaughan 1982) and the daytime data (Thomas et al. 1983) reduced to the night conditions at 95 (km) based on (Kichhoff 1986; Takahashi et al. 1996)

The absorption of solar UV radiation causes a proportional decrease of the energy influx. The relative decrease of the ozone density $N(O_3)$ at altitudes of 80–100 (km) is $\sim 30(\%)$ over the 40-yr period. This was sufficient to provide the observed cooling of the atmosphere in this altitude range (Golitsyn et al. 1996).

The problem of heating and cooling in the altitude range under consideration was discussed repeatedly (Mlynczak and Solomon 1993; Akmaev and Fomichev 2000; Volodin 2000). Nevertheless, the effect of long-term temperature trends has been taken into account only in the last few years (Gruzdev and Brasseur 2005). They are manifestations of the strong relation between the cooling of the mesosphere and the ozone concentration variations. The data presented above allow the temperature trend to be estimated based on the calculations of the heat influx at the altitudes in point (Brasseur and Solomon 2005). The decrease in ozone density results in a decrease in total heat influx by $\sim 20(\%)$ for the 40-yr period. This is equivalent to the decrease in temperature around 90 (km) with a rate of about -1 ($K \cdot yr^{-1}$). It seems that the increased molecular concentration of CO_2 and the additional heating of this atmospheric region by infrared radiation (Ogibalov et al. 2000) partly compensate for the cooling; as a result, the temperature trend at altitudes of ~ 87 (km) is -0.68 ($K \cdot yr^{-1}$) (Golitsyn et al. 1996).

Thus, the long-term decrease in ozone density in this region of the atmosphere is a consequence of the increase in methane content in the lower atmosphere that determines the atomic hydrogen density in the mesopause region. The decrease in ozone density is a reason for the long-term decrease in energy influx at altitudes of 80–100 (km) that is responsible for the significant long-term negative temperature trend.

7.2 Atomic Oxygen

Atomic oxygen at altitudes of 80–110 (km) is an important constituent of the middle atmosphere. The dissociation of molecular oxygen upon exposure to ultraviolet radiation of the Sun and the subsequent absorbed energy conversion in the process of atomic oxygen recombination determine the thermal and dynamic regimes in this region of the middle atmosphere. The space–time behavior of atomic oxygen affects significantly the atmospheric characteristics in this altitude range due to the interaction of atomic oxygen with carbon dioxide (CO_2) molecules. Laboratory measurements (Khvorostovskaya et al. 2002) performed for the reaction of deactivation of the vibrationally excited carbon dioxide molecule

$$CO_2(01^10) + O \rightarrow CO_2(00^00) + O$$

gave the rates of this reaction for temperatures inherent in the mesopause region. It has been found that the reaction rates are different for temperatures below 260 (K) and above 300 (K): $k^O_{01^10,00^00} = 1.56 \cdot 10^{-12} (cm^3 \cdot c^{-1})$ and $k^O_{01^10,00^00} = 1.40 \cdot 10^{-12} (cm^3 \cdot c^{-1})$, respectively. The reaction rate varies almost linearly between

7.2 Atomic Oxygen

these values. Therefore, it can be well approximated by one formula for temperatures typical of the upper mesosphere and lower thermosphere:

$$k^O_{01^10,00^00} = \left\{1.40 + 0.16 \Big/ \left[1 + \exp\left(\frac{T-280}{10}\right)\right]\right\} \cdot 10^{-12} (\text{cm}^3 \cdot \text{c}^{-1}).$$

The measured temperature profiles at the indicated altitudes are in close correlation with the altitude distributions of the atomic oxygen density calculated from the 557.7-nm emission data (Semenov and Shefov 1999b). This testifies to a significant effect of the interaction between atomic oxygen and CO_2 molecules. Obviously, the reverse process of excitation of CO_2 molecules on collisions with oxygen atoms also occurs. Its role in cooling of the medium becomes significant at altitudes above 100 (km) where the optical thickness of the atmosphere for the 15-μm emission is less than unity.

Attempts to calculate the vertical profiles of the atomic oxygen density and to construct models of its variations were undertaken repeatedly (Thomas and Young 1981; McDade et al. 1986; Siskind and Sharp 1991; López-González et al. 1992; Melo et al. 1996, 2001). Nevertheless, the vertical atomic oxygen profiles predicted by the CIRA-1996 model (Llewellyn and McDade 1996) were constructed with no regard for their actual significant variations with heliogeophysical conditions. The base for their construction was measurements of the intensity of the 1.27-μm emission arising in the daytime from the (0–0) band of the O_2 Infrared Atmospheric system due to photolysis of ozone molecules. Ozone, in turn, is produced due to the three-body recombination reaction with participation of oxygen atoms. It is natural that in calculations of these multistage photochemical processes, some average model temperature and neutral composition of the upper atmosphere were used with no regard for the actual heliogeophysical conditions. Semenov and Shefov (1997c,d) analyzed measurements of the 557.7-nm emission intensity with the use of simplified photochemical relations disregarding deactivation of excited oxygen atoms by atomic oxygen. The vertical profiles of atomic oxygen density were approximated by the Chapmen function, which is not as good for this purpose as the asymmetric Gauss function.

Long-term measurements of the characteristics of the 557.7-nm atomic oxygen, sodium, and hydroxyl emissions allowed their regular intensity and temperature variations to be determined at the emission layer altitudes (Semenov and Shefov 1996, 1997a,c,d; Semenov 1997; Fishkova et al. 2000, 2001b; Shefov et al. 2000a; Shefov and Semenov 2001, 2004b). The rocket, spectrophotometric, and radiophysical temperature measurements in the middle atmosphere at altitudes of 25–110 (km) (Semenov et al. 1996, 2000, 2002a,b,c; Golitsyn et al. 1996, 2001; Lysenko et al. 1999; Semenov and Shefov 1999a,b; Semenov 2000) provided the basis for the construction of monthly mean vertical temperature profiles for different months of the year at midlatitudes for a more than 40-yr period. These data were used to construct a model of monthly mean vertical profiles of the atomic oxygen density in the mesopause and lower thermosphere as functions of the heliogeophysical conditions.

The spectrophotometric data on the temperature variations of different types in the mesopause and lower atmosphere accumulated by the present time together with the rocket and radiophysical data enabled an empirical model of the temperature and neutral composition of the atmosphere at altitudes of 30–110 (km) to be constructed. This model predicted the seasonal and latitudinal variations of the above-indicated parameters together with their dependence on solar activity and their long-term trend for all indicated altitudes of the middle atmosphere (Semenov et al. 2004). These properties could not be considered with the preceding models of the atmosphere because they, as a rule, had been constructed based on data obtained within rather short periods (10–15 years).

7.2.1 Input Experimental Data for the Model of Atomic Oxygen

Regular experimental data on the characteristics of hydroxyl, sodium, and atomic oxygen emissions were obtained during several decades at Zvenigorod Observatory of A.M. Obukhov Institute of Atmospheric Physics of the Russian Academy of Sciences (55.7°N, 36.8°E), Abastumani Astrophysical Observatory of the Georgian Academy of Sciences (41.8°N, 42.8°E), Institute of Cosmophysical Research and Aeronomy of the Yakutsk Science Center of the Siberian Division of the Russian Academy of Sciences (62.0°N, 129.7°E) as well as during shorter periods at a number of observation stations located at different latitudes and listed in the publications by Semenov and Shefov (1996, 1997a,c,d, 1999a). These authors performed a statistical systematization of the measurement data and proposed empirical approximate formulas to describe the time and latitudinal intensity and temperature variations within emission layers and the layer altitudes. Data on the characteristics of layers and their altitudes were retrieved by analyzing the published rocket measurements. More recent studies made it possible to refine some variation types, such as cyclic aperiodic (quasi-biennial) variations (Fadel et al. 2002) and long-term temperature trends at different altitudes of the middle atmosphere (Semenov et al. 1996, 2000, 2002a,b; Golitsyn et al. 1996, 2000, 2001; Lysenko et al. 1999; Shefov et al. 2000b; Shefov and Semenov 2002; Semenov and Shefov 2003).

7.2.2 Photochemical Basis of the Model

The photochemical mechanism of excitation of the atomic oxygen emission at 557.7 (nm) has been considered in detail in Sect. 2.5. It is based on the Barth mechanism involving a sequence of chemical reactions resulting in the formation of excited metastable oxygen molecules. It engenders some systems of molecular emission bands in the emission spectrum of the night upper atmosphere (Barth and Hildebrandt 1961) and incorporates a sequence of reactions (Bates 1988). The

7.2 Atomic Oxygen

required relationships between the 557.7-nm emission intensity and the atomic oxygen density are presented in this chapter.

7.2.3 Empirical Regularities of 557.7-nm Emission Variations

Empirical regularities of variations of the emission layer parameters and time variations of the 557.7-nm emission intensity and temperature within the emission layer retrieved from Semenov and Shefov (1997a,b,c,d) and Semenov (1997) have been presented in Sect. 4.4. The seasonal variations of the intensity, temperature, and altitude of maximum emission are in close correlation (Figs. 4.30–4.33). The solar activity also affects systematically the layer parameters (Fig. 7.5).

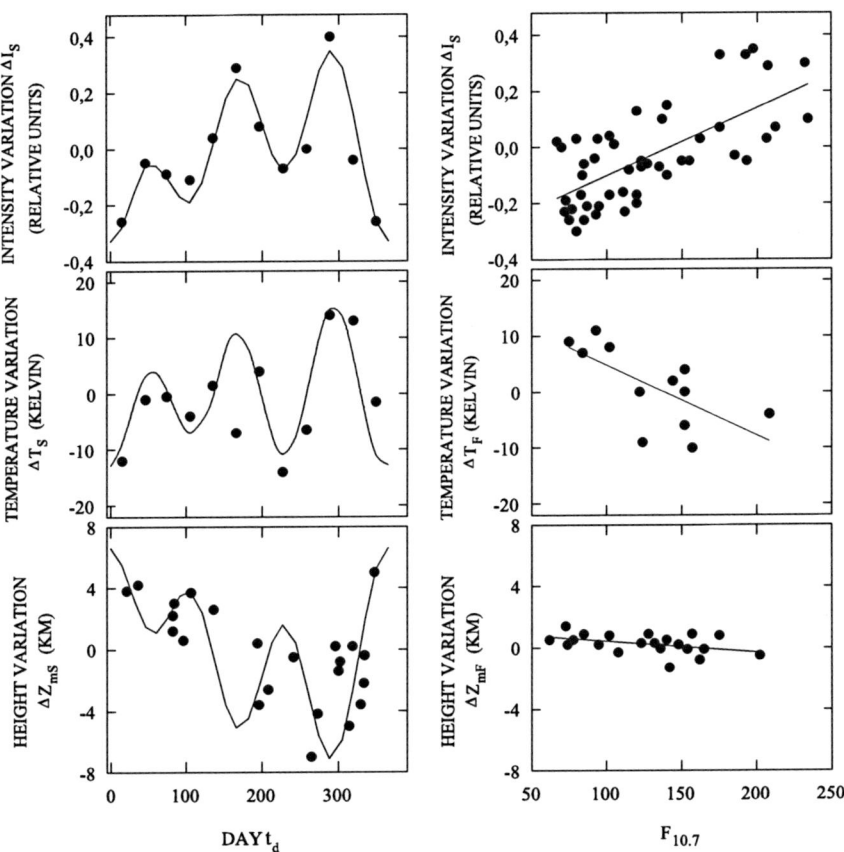

Fig. 7.5 Variations of the intensity, temperature, and maximum emission altitude of the 557.7-nm atomic oxygen emission about their yearly average values as functions of the season and solar activity

7.2.4 Model of Variations of the Atomic Oxygen Density in the Mesopause and Lower Thermosphere

As described in the previous sections, the use of relations that describe the behavior of the emission parameters and of the atmospheric temperature and neutral constituent densities calculated for a number of heliogeophysical conditions based on long-term measurements of the middle atmosphere temperature and 557.7-nm emission characteristics has enabled a model of variations of the atomic oxygen density to be constructed for different altitudes of the mesopause and lower thermosphere.

Figures 7.6 and 7.7 show the [O] vertical profiles for different months the of year and low ($F_{10.7} = 75$ in 1976 and 1986) and high solar activities ($F_{10.7} = 203$ in 1980 and 1991). It should be noted that, as follows from the seasonal behavior of the variations of the 557.7-nm emission intensity, the position of the atomic oxygen layer maximum also changes considerably. This was also noted by Perminov et al. (1998). This effect is especially pronounced in summer and autumn (October–November).

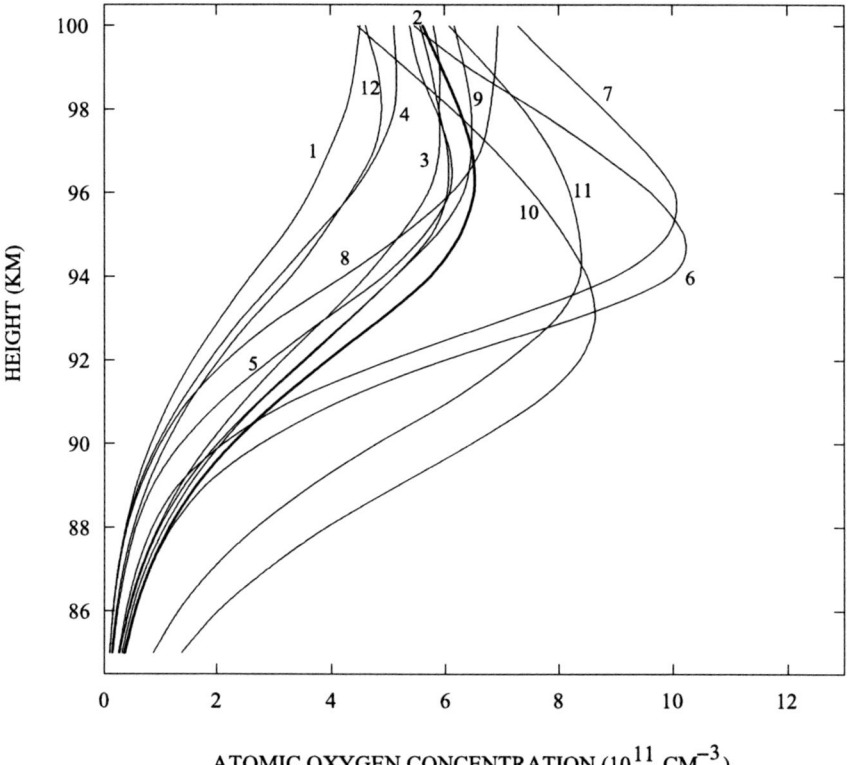

Fig. 7.6 Model vertical profiles of atomic oxygen density ([O], 10^{11} (cm^{-3})) for different months of year and solar activity minimum ($F_{10.7} = 75$). Figures adjacent to the curves indicate the serial month number, and the thick curve shows the yearly average profile

7.2 Atomic Oxygen

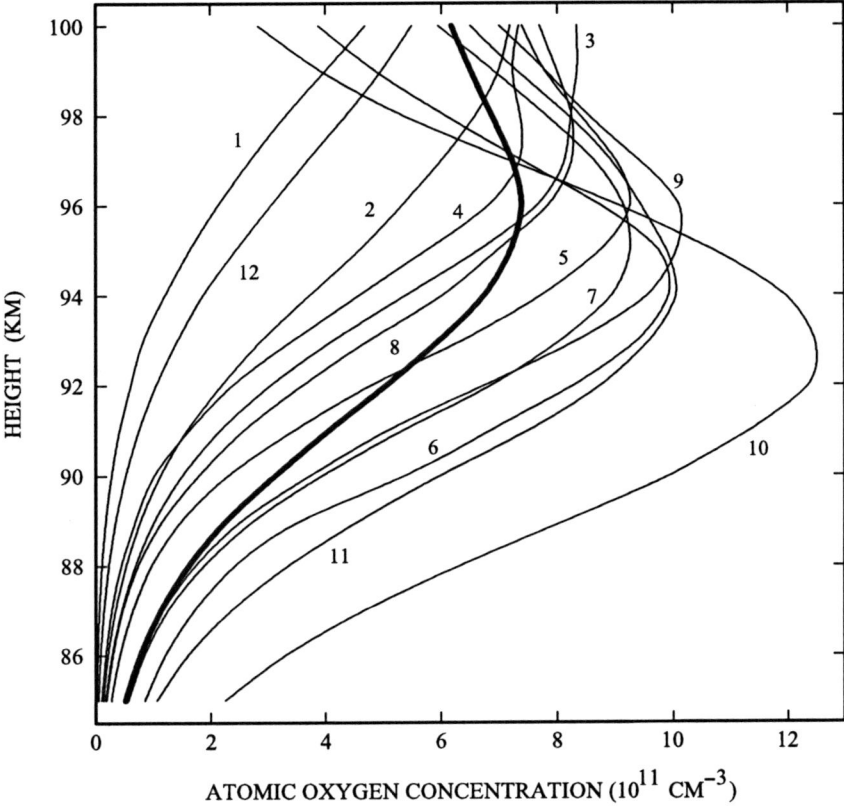

Fig. 7.7 Model vertical profiles of the atomic oxygen density for different months of the year (*curves* 1–12) and solar maximum ($F_{10.7} = 203$). Designations are the same as in Fig. 7.6

An increase in solar activity results in an increase in atomic oxygen density at the altitude of maximum emission and in a descent of the lower boundary of the emission layer. This was also pointed out by Semenov and Shefov (1999b).

Measurements of vertical profiles of the atomic oxygen density in the mesopause and lower thermosphere were repeatedly performed in the preceding years by different methods, including rocket measurements of the 130-nm atomic oxygen emission from a special source scattered in the atmosphere (Henderson 1974; Dickinson et al. 1976; Trinks et al. 1978; Howlett et al. 1980; Sharp 1980; Offermann et al. 1981; Greer et al. 1986). The results obtained showed a significant spread in absolute values of the density at the altitude of maximum emission, which also varied. The results of model calculations for the 557.7-nm emission presented above revealed a negative correlation between the altitude of maximum emission and the atomic oxygen density. A negative correlation is clearly traced between the 557.7-nm emission intensity and the altitude of maximum emission for both the seasonal variations and the variations associated with the solar cycle (Semenov and Shefov 1997a;

Semenov 1997; Shefov et al. 2000a) (see Figs. 4.30 and 4.32). The rocket data have large variance, apparently due to random atmospheric perturbations; nevertheless, they show the same tendency for correlation.

Among rocket measurements of the atomic oxygen density there are simultaneous measurements of vertical profiles of the 557.7-nm atomic oxygen emission intensity, and this offers an opportunity to compare model calculations and measurement data. However, to calculate the atomic oxygen density in the emission layer, not only the longitudinal variations of the 557.7-nm emission characteristics (Shefov and Semenov 2004b) but also the effect of the running planetary waves (Shefov and Semenov 2004a) responsible for descents and ascents of the emission layer should be taken into account. The layer ascent is accompanied by a decrease in temperature in the mesopause, which in summer can lead to the formation of noctilucent clouds (Fast 1980). This situation was observed, for instance, during rocket measurements on June 29, 1974 (Trinks et al. 1978), and on July 11, 1977 (Thomas and Young 1981). At the same time, on July 5, 1956 (Heppner and Meredith 1958), noctilucent clouds were observed only in the eastern hemisphere (Fast 1972). Direct investigations of the longitudinal variations of the 557.7-nm emission intensity associated with the running planetary waves are presented in the paper by Shepherd et al. (1993). They demonstrate that, for instance, in the winter of 1992 (January) the longitudinal mean intensity was $\sim$160–190 (Rayleigh) at latitudes of 40–42°N, and the observed amplitude of intensity variations was $\sim$50(%). These results well explain the difference between the measurement and model data under the indicated heliogeophysical conditions because the model cannot consider randomly arising perturbations. Therefore, it is difficult to estimate the effect of running planetary waves on rocket measurements. Sometimes this can be done for summer observations by using data on noctilucent clouds.

The calculated atomic oxygen densities at the altitude of maximum emission and the layer altitudes $Z_m([O])$ for different months of the year and solar fluxes presented in Figs. 7.6 and 7.7 are compared in Fig. 7.8 with direct rocket measurements of [O] performed on April 1, 1974, and January 4, 1976 (Dickinson et al. 1976), June 29, 1974 (Trinks et al. 1978), March 25, 1972, and March 30, 1972 (Offermann et al. 1981), July 11, 1977 (Thomas and Young 1981), March 23, 1982 (Greer et al. 1986), and with the [O] values retrieved from the 557.7-nm emission characteristics measured on July 5, 1956 (Heppner and Meredith 1958), May 22, 1969 (Dandekar and Turtle 1971), February 3, 1973 (Kulkarni 1976), and December 19, 1981 (López-Moreno et al. 1984). As can be seen from this figure, the absolute values of the model and experimental atomic oxygen densities are in agreement. This was also noted in the work by Semenov (1997), where it was demonstrated that the empirical model of the variations of the 557.7-nm emission and the photochemical model of its initiation used for calculations predicted vertical distributions of the atomic oxygen density which were in good agreement with that measured by Greer et al. (1986).

From the results of comparison, it also follows that the measurements demonstrate a large spread in atomic oxygen concentration values similar to the concentration seasonal variations predicted by the empirical model, showing, however, a

7.2 Atomic Oxygen

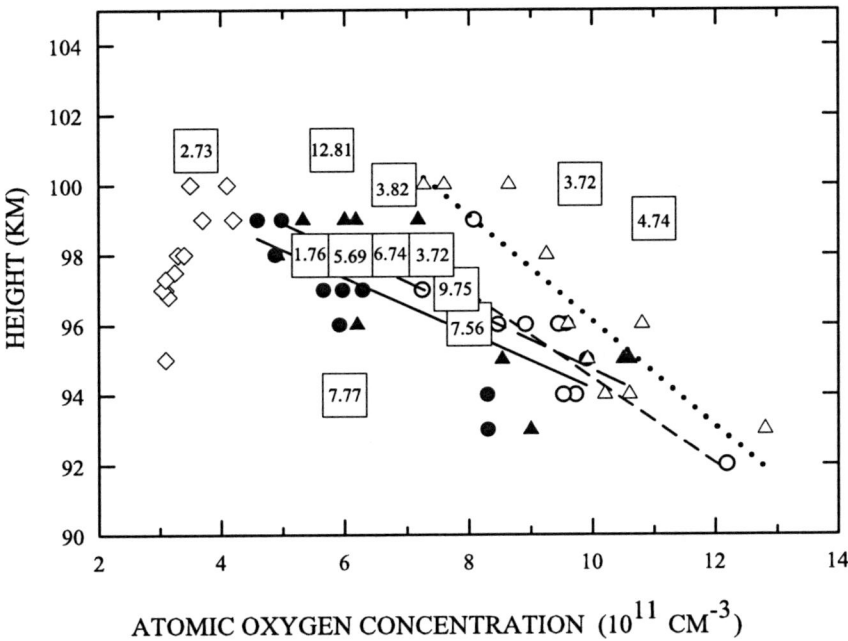

Fig. 7.8 Dependence of the density [O] on the altitude of the atomic oxygen layer maximum according to the data shown in Figs. 7.6 and 7.7 in 1976 (1) and 1986 (2) (*solar minima*), 1980 (3) and 1991 (*solar maxima*); rocket measurements of the atomic oxygen density (5) (*figures enclosed in squares near the symbols designate the month and year of measurements; see the text*); densities calculated for different months of year (at 45°N latitude) using the CIRA-1996 model (6) (Llewellyn and McDade 1996); and regression lines for the data of 1976 (7), 1986 (8), 1980 (9), and 1991 (10)

general tendency to vary with altitude. This testifies to the reliability of the dependence of maximum atomic oxygen concentration on the layer altitude established from our data. It should be added that the concentration values predicted by our model correspond to different solar activities, which is manifested through different slopes of the regression lines. Figure 7.8 also shows the altitude dependence of the maximum atomic oxygen concentration calculated with the use of the CIRA-1996 model (diamonds) (Llewellyn and McDade 1996). It can be seen that this model does not reproduce the observed time variations of the atomic oxygen concentration and its dependence on solar activity and gives a smaller absolute concentration.

Thus, the empirical model of variations of the vertical atomic oxygen concentration profile at altitudes of 80–100 (km) developed on the basis of long-term measurements of the parameters of the 557.7-nm atomic oxygen emission layer at night takes into account their various regular time variations, dependence on solar activity, latitude, and longitude as well as on the behavior of the long-term temperature trend in the twentieth century. For the twenty-first century, knowledge of the tendency for the long-term temperature trend behavior in the middle atmosphere is required.

7.3 Atomic Hydrogen

The atomic hydrogen content in the region of the mesopause and lower thermosphere affects significantly the energy balance in this atmospheric region. As considered in Sect. 2.2, the ozone–hydrogen reaction gives rise to the hydroxyl emission for which atomic hydrogen plays the role of a catalyst.

Nowadays, it is well known that the main supplier of hydrogen in the upper atmosphere is methane (Brasseur and Solomon 2005). Over the last decades, a significant increase in the methane content with a trend of $\sim 1.5(\% \cdot yr^{-1})$ has been observed in the lower atmosphere, which resulted in an increase in the methane concentration by a factor of 1.4 over the last four decades (Thomas et al. 1989).

Observations of the H_α emission in the night upper atmosphere characterizing the atomic hydrogen content at high altitudes (Fishkova 1983) testify to the dynamism of processes of its formation and consumption in the mesosphere (Semenov 1997). Rocket data on the atomic hydrogen density in the mesopause region are few in number (Meier and Prinz 1970; Anderson et al. 1980; Sharp and Kita 1987; Ulwick et al. 1987). They refer to only seven experiments between 1959 and 1984 in which the absorption of the L_α solar radiation by atmospheric hydrogen was measured. Nevertheless, these measurements cover a wide range of solar fluxes. Therefore, it was possible to separate the effects of long-term changes and solar activity. The seasonal variations reported by Le Texier et al. (1987) make $\sim 20(\%)$ and could not be retrieved from the available data. The variations of the relative atomic hydrogen density (revealed upon elimination of the trend) versus solar flux and, reduced to the solar flux $F_{10.7} = 130$, versus time are presented in Fig. 7.9 for a number of years (Semenov 1997; Shefov and Semenov 2002). They can be approximated by the expressions

$$\frac{\Delta_F[H]}{[H]_0} = (0.0048 \pm 0.0007) \cdot (F_{10.7} - 130), \qquad r = 0.948 \pm 0.042,$$

$$\frac{\Delta_{tr}[H]}{[H]_0} = (0.0245 \pm 0.0048) \cdot (t - 1972.5), \qquad r = 0.915 \pm 0.067.$$

The positive trend of the atomic hydrogen content so obtained is in good agreement with the methane trend (Brasseur and Solomon 2005; Thomas et al. 1989) also shown in Fig. 7.9. The yearly average value of the hydrogen density under standard heliogeophysical conditions, already indicated in Chap. 4, was $[H]_0 = 1.4 \cdot 10^8 (cm^{-3})$ at $Z = 87$ (km) in 1972. Thus, the atomic hydrogen concentration is determined by the expression

$$[H] = [H]_0 \cdot \left(1 + \frac{\Delta_F[H]}{[H]_0}\right) \cdot \left(1 + \frac{\Delta_{tr}[H]}{[H]_0}\right), \quad (cm^{-3}).$$

Based on the data reported by Keneshea et al. (1979) and Sharp and Kita (1987), the vertical profile of the atomic hydrogen density in the mesopause region can be described by an asymmetric Gaussian distribution with the parameters $Z_m = 84$ (km),

7.3 Atomic Hydrogen

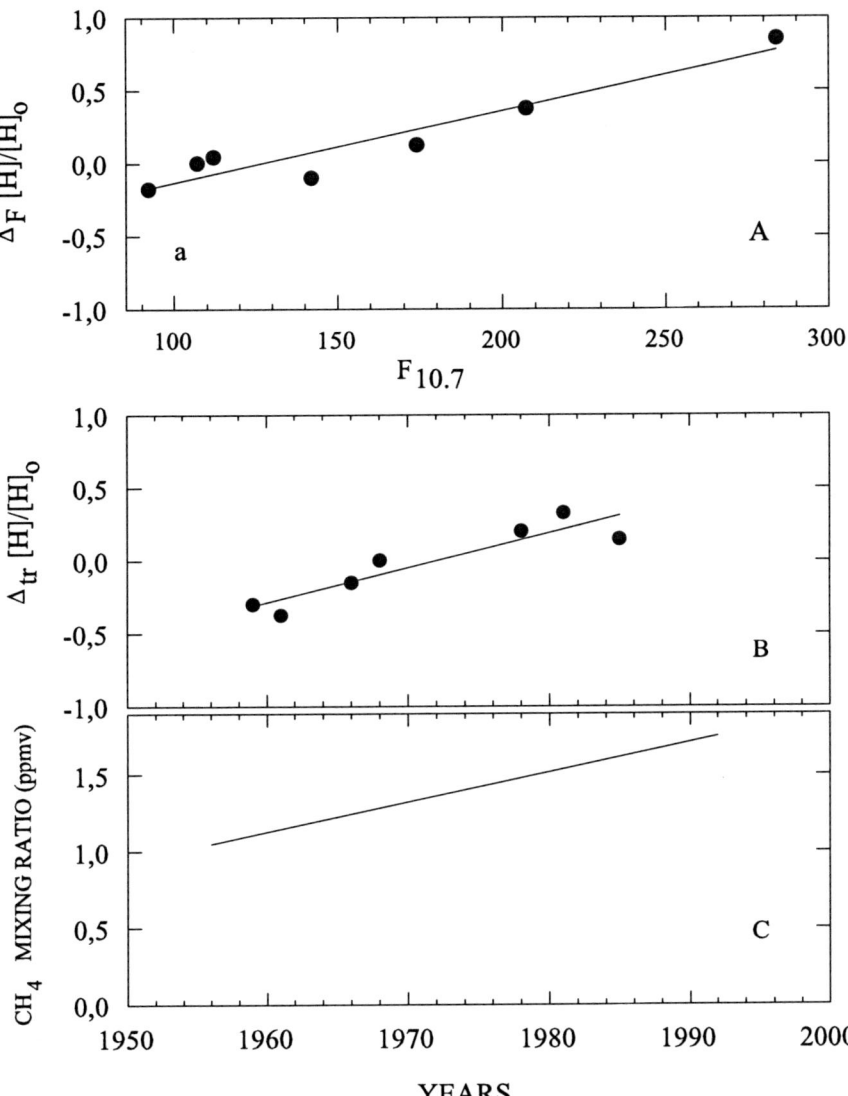

Fig. 7.9 Variations of the relative atomic hydrogen density in the mesopause region versus solar flux (**A**) and time in 1959–1984 (**B**) retrieved from rocket measurements (Semenov 1997; Shefov and Semenov 2002). Straight lines are regression lines. For comparison, the average trend of methane content in the atmosphere is shown (**C**) (Rees and Fuller-Rowell 1990)

$[H(Z_m)] = 2 \cdot 10^8 (cm^{-3})$, W = 14 (km), and P = 0.56 (Semenov 1997). Unfortunately, data on the tendency of long-term changes of the altitude of this atomic hydrogen layer in the mesopause region are lacking.

These data were used to retrieve long-term variations of the atomic hydrogen density between 1950 and 2000. They are presented in Fig. 7.4 (atomic hydrogen density). Solid curves in the figure indicate the variations of the atomic hydrogen density calculated by the data reported by Semenov (1997) and Shefov and Semenov (2002); full circles show the rocket data reduced to an altitude of 95 (km) using the model by Keneshea et al. (1979).

7.4 Temperature of the Middle Atmosphere

The response of the monthly mean temperature of the middle atmosphere to solar activity was analyzed (Semenov et al. 2005; Pertsev et al. 2005) based on the long-term rocket and spectrophotometric data on its airglow emissions obtained during

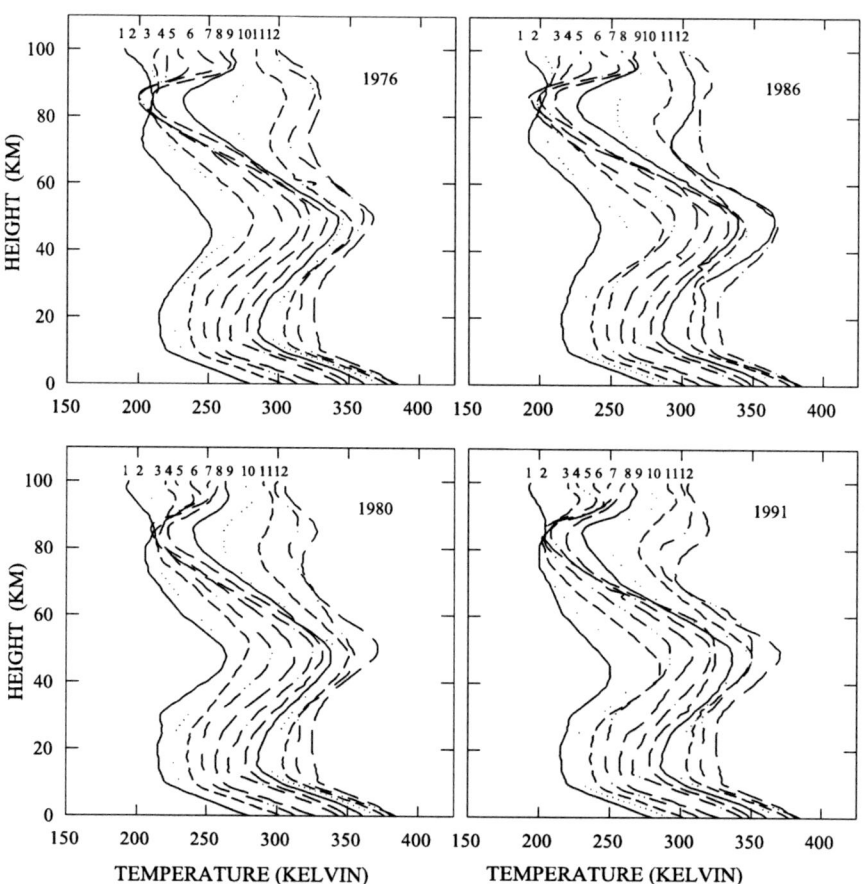

Fig. 7.10 Vertical profiles of the monthly mean temperature of the middle atmosphere in years of solar minimum (1976 and 1986) and maximum (1980 and 1991). The temperature profile for each month (*figures near the curves specify the serial month number and year*) is shifted from the preceding one by 10 (K) along the horizontal axis. The temperature scale is for the January temperature profile

7.4 Temperature of the Middle Atmosphere

several 11-yr solar cycles (Fig. 7.10) The results have already been discussed in Sect. 1.6.4.4.

The vertical profiles of the of response variation of the middle atmosphere temperature with solar activity testify to their clearly expressed nonlinearity. The substantial seasonal variability of the solar activity effect in the mesopause region is obviously caused by the different seasonal vertical profiles of the temperature.

Thus, the vertical profiles of the temperature response to solar activity at altitudes of 30–100 (km) demonstrate that the maximum seasonal variations are observed at altitudes of 80–95 (km) (about $-(5 \pm 1.7)$ $(K \cdot (100\,sfu)^{-1})$ in winter and about $+(8 \pm 1.7)$ $(K \cdot (100\,sfu)^{-1})$ in summer), and the minimum ones are observed at altitudes of 55–70 (km) (about $+(2 \pm 0.4)$ $(K \cdot (100\,sfu)^{-1})$ in winter and about $-(1 \pm 0.4)$ $(K \cdot (100\,sfu)^{-1})$ in summer). The altitude of the zero temperature response to solar activity in the middle atmosphere varies between 55 and 70 (km) throughout the year ($\sim$70 (km) in spring–summer and $\sim$55 (km) in autumn–winter). This is especially clearly seen from Fig. 7.11. The proposed empirical model of the response of the middle atmosphere temperature to solar activity at different altitudes and in different seasons can be used to estimate the temperature variations at various altitudes depending on solar activity and season.

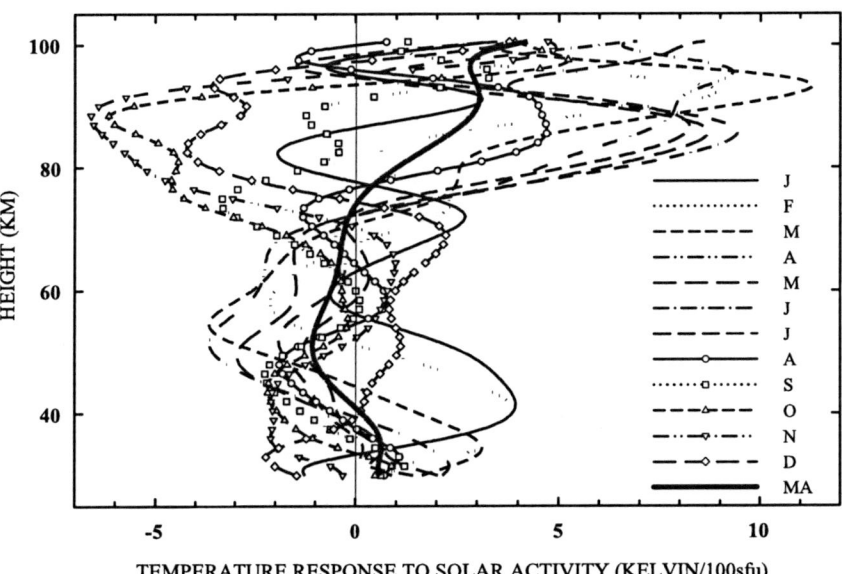

Fig. 7.11 Model vertical profiles of $\delta T/\delta F_{10.7}$ for different months of the year. Here curves 1–12 show monthly mean temperature profiles, and curve 13 shows the yearly average temperature profile

References

Akmaev RA, Fomichev VI (2000) A model estimate of cooling in the mesosphere and lower thermosphere due to the CO_2 increase over the last 3–4 decades. Geophys Res Lett 27:2113–2116

Anderson DE, Feldman PD, Gentieu EP, Meier RR (1980) The UV dayglow 2, L_α and L_β emissions and the H distribution in the mesosphere and thermosphere. Geophys Res Lett 7:529–532

Barth CA, Hildebrandt AF (1961) The 5577 Å airglow emission mechanism. J Geophys Res 66:985–986

Bates DR (1988) Excitation of 557.7 nm OI line in nightglow. Planet Space Sci 36:883–889

Brasseur G, Solomon S (2005) Aeronomy of the middle atmosphere, 3rd edn. Springer-Verlag, Dordrecht, Holland

Chandra S, Jackman CH, Fleming EL, Russell JMIII (1997) The seasonal and long term changes in mesospheric water vapor. Geophys Res Lett 24:639–642

Chikashi Y, Masahara F, Hirota E (1989) Detection of the NaO radical by microwave spectroscopy. J Chem Phys 90:3033–3037

CIRA-1986 (1990) COSPAR International Reference Atmosphere: 1986. Part II: Middle atmosphere models. In: Rees D, Barnett JJ, Labitzke K (eds) Advances in Space Research, Vol 10. pp 1–525

Clemesha BR, Simonich DM, Takahashi H, Melo SML, Plane JMC (1995) Experimental evidence for photochemical control of the atmospheric sodium layer. J Geophys Res 100(D9):18909–18916

Dandekar BS, Turtle JP (1971) Atomic oxygen concentration from the measurement of the [OI] 5577 A emission of the airglow. Planet Space Sci 19:949–957

Dickinson PHG, Twiddy ND, Young RA (1976) Atomic oxygen concentration in the lower ionosphere. In: Rycroft MJ (ed) Space Research, Vol 16. Akademie-Verlag, Berlin, pp 301–305

Fadel KhM, Semenov AI, Shefov NN, Sukhodoev VA, Martsvaladze NM (2002) Quasibiennial variations in the temperatures of the mesopause and lower thermosphere and solar activity. Geomagn Aeron 42:191–195

Fast NP (1972) A catalogue of the noctilucent cloud appearances according to the world data. Tomsk University Publishing House, Tomsk

Fast NP (1980) A catalogue of the noctilucent cloud appearances according to the world data. Tomsk University Publishing House, Tomsk

Fishkova LM (1983) The night airglow of the earth mid-latitude upper atmosphere. Shefov NN (ed) Metsniereba, Tbilisi

Fishkova LM, Martsvaladze NM, Shefov NN (2000) Patterns of variations in the OI 557.7-nm. Geomagn Aeron 40:782–786

Fishkova LM, Martsvaladze NM, Shefov NN (2001a) Long-term variations of the nighttime upper-atmosphere sodium emission. Geomagn Aeron 41:528–532

Fishkova LM, Martsvaladze NM, Shefov NN (2001b) Seasonal variations in the correlation of atomic oxygen 557.7-nm emission with solar activity and in long-term trend. Geomagn Aeron 41:533–539

Golitsyn GS, Semenov AI, Shefov NN, Fishkova LM, Lysenko EV, Perov SP (1996) Long-term temperature trends in the middle and upper atmosphere. Geophys Res Lett 23:1741–1744

Golitsyn GS, Semenov AI, Shefov NN (2000) Seasonal variations of the long-term temperature trend in the mesopause region. Geomagn Aeron 40:198–200

Golitsyn GS, Semenov AI, Shefov NN (2001) Thermal structure of the middle and upper atmosphere (25–110 km), as an image of its climatic change and influence of solar activity. In: Beig G (ed) Long term changes and trends in the atmosphere, Vol 2. New Age International Limited Publishers, New Delhi, pp 33–42

Greer RGH, Murtagh DP, McDade IC, Dickinson PHG, Thomas L, Jenkins DB, Stegman J, Llewellyn EJ, Witt G, Mackinnon DJ, Williams ER (1986) ETON 1: A data base pertinent to the study of energy transfer in the oxygen nightglow. Planet Space Sci 34:771–788

Gruzdev AN, Brasseur GP (2005) Long-term changes in the mesosphere calculated by two dimensional model. J Geophys Res 110, D03304:1–18. doi:10.1029/2003JD004410

Hecht JH, Collins S, Kruschwitz C, Kelley MC, Roble RG, Walterscheid RL (2000) The excitation of the Na airglow from Coqui Dos rocket and ground-based observations. Geophys Res Lett 27:453–456

Helmer M, Plane JMC (1993) A study of the reaction $NaO_2 + O \rightarrow NaO + O_2$: implications for the chemistry of sodium in the upper atmosphere. J Geophys Res 98D:23207–23222

Henderson WR (1974) Atomic oxygen profile measurements. J Geophys Res 79:3819–3826

Heppner JP, Meredith LH (1958) Nightglow emission altitudes from rocket measurements. J Geophys Res 63:51–65

Herschbach DR, Kolb CE, Worsnop DR, Shi X (1992) Excitation mechanism of the mesospheric sodium nightglow. Nature (London) 356:414–416

Howlett LC, Baker KD, Megill LR, Shaw AW, Pendleton WR (1980) Measurement of a structured profile of atomic oxygen in the mesosphere and lower thermosphere. J Geophys Res 85A:1291–1296

Keneshea TJ, Zimmermann SP, Philbrick CR (1979) A dynamic model of the mesosphere and lower thermosphere. Planet Space Sci 27:385–401

Khvorostovskaya LE, Potekhin IYu, Shved GM, Ogibalov BP, Uzyukova TV (2002) Measurement of the rate constant for quenching CO_2 (01^10) by atomic oxygen at low temperatures: reassessment of the rate of cooling by the CO_2 15 µm emission in the lower thermosphere. Izvestiya Atmos Oceanic Phys 38:613–624

Kirchhoff VWJH (1986) Theory of the atmospheric sodium layer: a review. Can J Phys 64:1664–1672

Kulkarni PV (1976) Rocket study of 5577 Å OI emission at night over the magnetic equator. J Geophys Res 81:3740–3744

Le Texier H, Solomon S, Garcia RR (1987) Seasonal variability of the OH Meinel band. Planet Space Sci 35:977–989

Llewellyn EJ, McDade IC (1996) A reference model for atomic oxygen in the terrestrial atmosphere. Adv Space Res 18:209–226

López-González MJ, López-Moreno JJ, Rodrigo R (1992) Atomic oxygen concentration from airglow measurements of atomic and molecular oxygen emissions in the nightglow. Planet Space Sci 40:929–940

López-Moreno JJ, Vidal S, Rodrigo R, Llewellyn EJ (1984) Rocket-borne photometric measurements of $O_2(^1\Delta_g)$, green line and OH Meinel bands in the nightglow. Ann Geophys 2:61–66

Lysenko EV, Perov SP, Semenov AI, Shefov NN, Sukhodoev VA, Givishvili GV, Leshchenko LN (1999) Long-term trends of the yearly mean temperature at heights from 25 to 110 km. Izvestiya Atmos Oceanic Phys 35:393–400

McDade IC, Murtagh DP, Greer RGH, Dickinson PHG, Witt G, Stegman J, Llewellyn EJ, Thomas L, Jenkins DB (1986) ETON 2: Quenching parameters for the proposed precursors of $O_2(b^1\Sigma_g^+)$ and $O(^1S)$ in the terrestrial nightglow. Planet Space Sci 34:789–800

McNeil WJ, Murad E, Lai ST (1995) Comprehensive model for the atmospheric sodium layer. J Geophys Res 100D:16847–16855

Meier RR, Prinz DK (1970) Absorption of the solar Lyman alpha line by geocoronal atomic hydrogen. J Geophys Res 75:6969–6979

Melo SML, Takahashi H, Clemesha BR, Batista PP, Simonich DM (1996) Atomic oxygen concentration from rocket airglow observations in the equatorial region. J Atmos Terr Phys 58:1935–1942

Melo SML, McDade IC, Takahashi H (2001) Atomic oxygen density profiles from ground-based nightglow measurements at 23°S. J Geophys Res 106D:15377–15384

Mlynczak MG, Solomon S (1993) A detailed evaluation of the heating efficiency in the middle atmosphere. J Geophys Res 98D:10517–10541

Nicolet M (1971) Aeronomic reactions of hydrogen and ozone. In: Fiocco G (ed) Mesospheric model and related experiments. D. Reidel Publishing Company, Dordrecht, pp 1–51

Offermann D, Friedrich V, Ross P, von Zahn U (1981) Neutral gas composition measurements between 80 and 120 km. Planet Space Sci 29:747–764

Ogibalov BP, Fomichev BI, Kutepov AA (2000) Radiative heating effected by infrared CO_2 bands in the middle and upper atmosphere. Izvestiya Atmos Oceanic Phys 36:454–464

Perminov VI, Semenov AI, Shefov NN (1998) Deactivation of hydroxyl molecule vibrational states by atomic and molecular oxygen in the mesopause region. Geomagn Aeron 38:761–764

Pertsev NN, Semenov AI, Shefov NN (2005) Long-term variations of temperature and neutral density of the mid-latitude middle atmosphere by rocket and optical data. In: Warmbein B (ed) Proceedings 17th European Space Agency Symposium on European rocket and balloon programmes and related research (Norway, Sandefjord, 30 May–2 June). ESA Publications, Noordwijk, pp 245–250

Plane JMC (1991) The chemistry of meteoric metals in the earth's upper atmosphere. Int Rev Phys Chem 10:55–106

Rees D, Fuller-Rowell TJ (1990) Numerical simulations of the seasonal/latitudinal variations of atomic oxygen and nitric oxide in the lower thermosphere and mesosphere. Adv Space Res 10:83–102

Ricaud P, de la Noë J, Connor BJ, Froidevaux L, Waters JW, Harwood RS, MacKenzie IA, Peckham GE (1996) Diurnal variability of mesospheric ozone as measured by the UARS microwave limb sounder instrument: theoretical and ground-based variations. J Geophys Res 101D:10077–10089

Roble RG, Hays PB (1974) On determining the ozone number density distribution from OAO-2 stellar occultation measurements. Planet Space Sci 22:1337–1340

Semenov AI (1997) Long-term changes in the height profiles of ozone and atomic oxygen in the lower thermosphere. Geomagn Aeron 37:354–360

Semenov AI (2000) Long-term temperature trends for different seasons by hydroxyl emission. Phys Chem Earth 25:525–529

Semenov AI, Shefov NN (1996) An empirical model for the variations in the hydroxyl emission. Geomagn Aeron 36:468–480

Semenov AI, Shefov NN (1997a) An empirical model of nocturnal variations in the 557.7-nm emission of atomic oxygen. 1. Intensity. Geomagn Aeron 37:215–221

Semenov AI, Shefov NN (1997b) An empirical model of nocturnal variations in the 557.7-nm emission of atomic oxygen. 2. Temperature. Geomagn Aeron 37:361–364

Semenov AI, Shefov NN (1997c) An empirical model of nocturnal variations in the 557.7-nm emission of atomic oxygen. 3. Emitting layer altitude. Geomagn Aeron 37:470–474

Semenov AI, Shefov NN (1997d) Empirical model of the variations of atomic oxygen emission 557.7-nm. In: Ivchenko VN (ed) Proceedings SPIE (23rd European Meeting on Atmospheric Studies by Optical Methods, Kiev, September 2–6, 1997), Vol 3237. The International Society for Optical Engineering, Bellingham, pp 113–122

Semenov AI, Shefov NN (1999a) Empirical model of hydroxyl emission variations. Intern J Geomagn Aeron 1:229–242

Semenov AI, Shefov NN (1999b) Variations of the temperature and the atomic oxygen content in the mesopause and lower thermosphere region during change of the solar activity. Geomagn Aeron 39:484–487

Semenov AI, Shefov NN (2003) New knowledge of variations in the hydroxyl, sodium and atomic oxygen emissions. Geomagn Aeron 43:786–791

Semenov AI, Shefov NN, Fishkova LM, Lysenko EV, Perov SP, Givishvili GV, Leshchenko LN, Sergeenko NP (1996) Climatic changes in the upper and middle atmosphere. Doklady Earth Sci 349:870–872

Semenov AI, Shefov NN, Givishvili GV, Leshchenko LN, Lysenko EV, Rusina VYa, Fishkova LM, Martsvaladze NM, Toroshelidze TI, Kashcheev BL, Oleynikov AN (2000) Seasonal peculiarities of long-term temperature trends of the middle atmosphere. Doklady Earth Sci 375:1286–1289

Semenov AI, Sukhodoev VA, Shefov NN (2002a) A model of the vertical temperature distribution in the atmosphere altitudes of 80–100 km that taking into account the solar activity and the long-term trend. Geomagn Aeron 42:239–244

Semenov AI, Bakanas VV, Perminov VI, Zheleznov YuA, Khomich. VYu (2002b) The near infrared spectrum of the emission of the nighttime upper atmosphere of the earth. Geomagn Aeron 42:390–397

Semenov AI, Shefov NN, Lysenko EV, Givishvili GV, Tikhonov AV (2002c) The seasonal peculiarities of behavior of the long-term temperature trends in the middle atmosphere at the mid-latitudes. Phys Chem Earth 27:529–534

Semenov AI, Pertsev NN, Shefov NN, Perminov VI, Bakanas VV (2004) Calculation of the vertical profiles of the atmospheric temperature and number density at altitudes of 30–110 km. Geomagn Aeron 44:773–778

Semenov AI, Shefov NN, Perminov VI, Khomich VYu, Fadel KhM (2005) Temperature response of the middle atmosphere on the solar activity for different seasons. Geomagn Aeron 45:236–240

Sharp WE (1980) Absolute concentrations of $O(^3P)$ in the lower thermosphere at night. Geophys Res Lett 7:485–488

Sharp WE, Kita D (1987) In situ measurement of atomic hydrogen in the upper atmosphere. J Geophys Res 92D:4319–4324

Shefov NN, Semenov AI (2001) An empirical model for nighttime variations in atomic sodium emission: 2. Emitting layer height. Geomagn Aeron 41:257–261

Shefov NN, Semenov AI (2002) The long-term trend of ozone at heights from 80 to 100 km at the mid-latitude mesopause for the nocturnal conditions. Phys Chem Earth 27:535–542

Shefov NN, Semenov AI (2004a) Longitudinal-temporal distribution of the occurrence frequency of noctilucent clouds. Geomagn Aeron 44:259–262

Shefov NN, Semenov AI (2004b) The longitudinal variations of the atomic oxygen emission at 557.7 nm. Geomagn Aeron 44:620–623

Shefov NN, Semenov AI, Yurchenko OT (2000a) Empirical model of variations in the atomic sodium nighttime emission: 1. Intensity. Geomagn Aeron 40:115–120

Shefov NN, Semenov AI, Pertsev NN (2000b) Dependencies of the amplitude of the temperature enhancement maximum and atomic oxygen concentrations in the mesopause region on seasons and solar activity level. Phys Chem Earth Pt B 25:537–539

Shefov NN, Semenov AI, Yurchenko OT (2002) Empirical model of the ozone vertical distribution at the nighttime mid-latitude mesopause. Geomagn Aeron 42:383–389

Shepherd GG, Thuillier G, Solheim BH, Chandra S, Cogger LL, Duboin ML, Evans WFJ, Gattinger RL, Gault WA, Hersé M, Hauchecorne A, Lathuilliere C, Llewellyn EJ, Lowe RP, Teitelbaum H, Vial F (1993) Longitudinal structure in atomic oxygen concentrations observed with WINDII on UARS. Geophys Res Lett 20:1303–1306

Siskind DE, Sharp WE (1991) A comparison of measurements of the oxygen nightglow and atomic oxygen in the lower thermosphere. Planet Space Sci 39:627–639

Smith CA, Toumi R, Haigh JD (2000) Seasonal trends in stratospheric water vapour. Geophys Res Lett 27:1687–1690

Takahashi H, Melo SML, Clemesha BR, Simonich DM (1996) Atomic hydrogen and ozone concentrations derived from simultaneous lidar and rocket airglow measurements in the equatorial region. J Geophys Res 101D:4033–4040

Thomas RJ (1990a) Seasonal ozone variations in the upper mesosphere. J Geophys Res 95D:7395–7401

Thomas RJ (1990b) Atomic hydrogen and atomic oxygen density in the mesopause region: global and seasonal variations deduced from Solar Mesospheric Explorer near-infrared emission. J Geophys Res 95D:16457–16476

Thomas RJ, Young RA (1981) Measurement of atomic oxygen and related airglow in the lower thermosphere. J Geophys Res 86C:7389–7393

Thomas RJ, Barth CA, Rottman GJ, Rusch DW, Mount GH, Lawrence GM, Sanders RW, Thomas GE, Clemens LE (1983) Ozone density distribution in the mesosphere (50–90) km measured by the SME limb scanning near infrared spectrometer. Geophys Res Lett 10:245–248

Thomas G, Olivero JJ, Jensen EJ, Schröder W, Toon OB (1989) Relation between increasing methane and the presence of ice clouds at the mesopause. Nature (London) 338:490–492

Trinks H, Offermann D, von Zahn U, Steinhauer C (1978) Neutral composition measurements between 90- and 220-km altitude by rocket-borne mass spectrometer. J Geophys Res 83A:2169–2176

Ulwick JC, Baker KD, Baker DJ, Steed AJ, Pendleton WR, Grossmann K, Brueckelmann HG (1987) Mesospheric minor species determinations from rocket and ground-based i.r. measurements. Geophys Res Lett 49:855–862

Vaughan G (1982) Diurnal variation of mesospheric ozone. Nature (London) 296:133–135

Volodin EM (2000) Sensitivity of the stratosphere and the mesosphere to observed changes in ozone and carbon dioxide concentrations as simulated by the Institute of Numerical Mathematics atmospheric general circulation model. Izvestiya Atmos Oceanic Phys 36:566–573

Wilson WJ, Schwartz RR (1981) Diurnal variations of mesospheric ozone using millimeter measurements. J Geophys Res 86D:7535–7538

Index

activity, 2, 7, 26–29, 61–64, 68, 70–74, 77–86, 88–100, 104, 107–118, 121, 133, 149, 169–171, 218, 222, 224–226, 229, 230, 245, 303, 372, 403, 406, 410, 415, 421, 431, 432, 434, 444, 445, 453, 457, 459, 460, 473, 484, 487, 488, 490, 491, 493, 495, 502, 504, 513, 515, 526, 531–533, 535–537, 539, 541–544, 547–549, 571, 628, 631–634, 647, 651, 653, 656–658, 661–664, 672–674, 676, 684, 686, 688, 689, 696, 698, 699, 703, 705, 706, 708–710, 712–716, 720–723, 725, 726, 728–730, 732, 733
Airy function, 71, 74–77, 79–84, 294, 296, 372, 432, 443, 444, 457, 458, 493, 494, 523
albedo, 22, 23, 97, 125, 396
alkali metals, 182, 347, 349
aperture, 4, 272, 276–278, 280, 282, 284, 285, 292, 293, 296, 337–339, 374, 382, 384, 391, 480, 640
atomic oxygen, 553, 555, 557, 559, 561, 584, 585, 587, 589, 595, 619–621, 623, 626, 639, 644, 646, 653, 654, 656, 657, 659
azimuth A, 42, 334, 371

bands, 556, 560–562, 564, 566, 569, 579, 582, 588–590, 592, 620, 654
Barth mechanism, 712, 720
brightness of, 17–20, 24, 25, 114, 137, 270–272, 312, 317, 348, 349, 390, 525, 634

carbon dioxide CO_2, 9
carbon dioxide CO_2, 243
Carrington cycle, 415, 444
characteristic curve, 305–307, 350–353

coefficient of absorption, 179, 228, 465
collision frequency, 126
continuum emission, 232, 346
correlation between intensity, 559
cross-section, 96, 100, 121, 123, 124, 134, 169, 170, 172, 194, 210, 216, 226, 229, 241, 270, 272, 318, 338, 449, 611, 617, 618
cyclic aperiodic variations, 70, 73, 75, 79, 493, 530

deactivation, 1, 99, 120, 125, 128, 133, 135, 142, 159, 161–163, 168, 175, 192, 193, 201, 203, 204, 209, 215, 216, 234, 236, 240, 242, 468, 510, 540, 583, 588, 718, 719
declination, 53, 57, 100, 385, 426–428, 451, 470, 481, 484, 513, 519
density, 6, 10, 41, 43, 60, 64, 88, 89, 104, 110, 116, 120, 132, 164, 169–171, 177, 179, 184, 200, 205, 206, 214, 215, 222, 226, 228–230, 234–237, 241, 242, 244, 269, 305, 306, 314, 345–347, 368, 371, 372, 392, 403, 410, 423, 436, 449–451, 456, 466, 468, 474, 490, 491, 494, 500, 501, 510, 521, 522, 527–529, 532, 538, 539, 547, 551, 552, 555, 557, 559, 562, 577, 584, 592, 599, 607–609, 615, 619, 620, 623, 636, 641, 648, 667, 676, 679, 683, 695, 696, 698–701, 704, 707, 709, 710, 712–719, 721–728, 731–734
determination of the, 586, 589, 590
developer, 303, 304, 306
diffusion, 1, 46, 60, 64, 65, 70, 71, 79, 85, 112, 179, 182, 183, 197, 236, 237, 324, 325, 347–349, 410, 477, 478, 480, 527, 531, 532, 535, 536, 619, 639, 641, 648, 651, 659

735

dipole coordinates, 109
dissociation, 1, 10, 26, 60, 89, 168–172, 202, 234, 434, 718
dissociative recombination, 60, 127, 206, 208, 209, 510, 513, 515, 525
Doppler, 5, 126, 127, 153, 203, 210, 218, 221, 228, 293, 294, 296, 298, 337, 354, 355, 360, 388, 394, 396, 397, 401, 404, 405, 408, 510–512, 525, 545, 561, 573, 575, 585–587, 643, 646, 649, 652, 695
Doppler temperature, 449, 473–475, 482, 508, 509, 511, 512, 522, 525, 535, 541
dynamical, 114, 512, 536, 651, 659, 705

Earth, ix, 2, 7–9, 15–17, 20, 23, 30–36, 39, 40, 42–45, 47, 49, 51, 57, 61, 62, 86–88, 94–96, 98–103, 105, 110, 112, 114–116, 125, 127, 142, 147, 157, 159, 198, 209, 269, 271, 273, 274, 282, 284, 285, 289, 293, 298, 303, 313, 317, 318, 329–331, 340–342, 344–346, 350, 351, 365, 382, 392, 394, 402–405, 407, 409, 410, 427, 430, 465, 505, 532, 533, 536, 538, 540–542, 544–546, 553, 566, 568, 575, 577, 589, 620, 621, 623–625, 634, 639, 641, 643, 648, 649, 651, 652, 655, 657, 705–711, 732, 733
eccentricity e, 342
ecliptic, 25, 26, 36, 38–40, 42, 53, 54, 103, 427
electronic, 8, 122, 125, 133–135, 159, 185, 186, 201, 203, 229, 237, 281, 318, 407, 416, 582, 588, 651, 653
emission layer, 33, 41, 42, 44–48, 51, 52, 60, 116, 126, 132, 136, 148, 155–157, 164, 166, 173, 176, 178, 179, 195, 200, 204–206, 232, 238, 243, 244, 281, 298, 300, 301, 329–331, 334, 349, 367, 374, 392, 394, 395, 397, 416, 418–423, 425, 426, 429, 445, 446, 448, 450, 451, 466, 468–478, 480, 482, 483, 485, 490–492, 495, 500–505, 508–510, 512, 516, 519, 521, 528, 544, 555, 556, 558, 565, 566, 568, 569, 572, 579, 582, 590, 594, 596, 597, 599, 604, 607, 619–621, 623, 636, 640, 644, 669, 670, 676, 679, 681, 696, 698, 702, 709, 719–721, 723–725
entrance pupil, 284, 285, 289, 292, 312, 391
equation of time, 53
equator, 24, 25, 36, 42, 101, 230, 427, 428, 431, 436, 439, 442, 443, 474, 522, 527, 538, 546, 567, 625, 627, 647, 683, 689, 731
equatorial coordinates, 42, 54

excitation, 1, 10, 29, 60, 121–124, 128, 134, 157, 159, 164, 166–168, 175, 176, 178, 179, 182, 196, 206, 208–211, 215, 217, 218, 220, 221, 227, 239–244, 282, 401, 409, 416, 450, 465, 466, 504, 510, 515, 530, 533, 534, 538–542, 545, 548, 556–559, 564, 577, 581–584, 587, 648, 652, 653, 656, 659, 676, 703, 704, 712, 719, 720, 731
exosphere, 222, 501

fiber optics, 281, 339
field of view, 281, 282
fluorescence, 556
forbidden transitions, 207, 465
formulas, 566, 568, 583, 601, 642, 646
Fourier analysis, 75, 81, 83, 296, 359, 362, 365, 366, 372, 377, 432, 456, 494, 508, 586, 646

galactic coordinates, 36, 37
galactic cosmic rays, 95, 108
generation of IGW, 369, 381, 621
geocorona, 29, 61, 88, 222, 226, 228
geographic coordinates, 30, 32, 39, 52, 53, 101, 105, 107, 340, 343, 344
geomagnetic, 2, 3, 10, 28, 34, 35, 55–57, 60, 61, 73, 86, 87, 100–107, 109, 111, 113, 115, 117, 127, 207, 209, 212, 226, 230, 241, 340, 349, 365, 415, 432, 436–441, 443, 444, 465, 475, 503, 506, 509, 513, 516, 522–524, 528, 530–532, 534, 535, 537–539, 543, 547, 571, 656, 662, 709

horizontal, 36, 60, 83, 87, 100, 102, 103, 107, 298, 324, 329, 331, 343, 374, 394, 407, 446, 476–478, 531, 619, 728
hour angle t, 36
hydrogen, 4, 5, 9, 10, 26, 27, 29, 60, 61, 88, 123, 163–168, 172, 217, 222, 223, 225–230, 245, 294, 409, 416, 435, 538, 551, 556–560, 582–584, 633, 654, 656, 687, 696, 697, 700, 707, 709, 711–713, 716–718, 726–728, 731, 733

image, 4, 5, 269, 273, 275, 278, 280, 281, 284, 285, 294, 296, 297, 302–306, 312–318, 321–323, 325–328, 336, 337, 339, 346, 350, 351, 353, 358, 360–362, 365, 383, 388, 392, 393, 407–409, 477, 533, 541, 586, 637, 650, 655, 665, 705, 730
inclination I, 100
incoherent scattering, 5, 576, 667, 671
Infrared Atmospheric system, 465

inhomogeneities, 654, 657
instrumental, x, 195, 294, 296, 301, 351, 358, 359, 362, 367, 383, 388, 393, 664, 706
interferometer, 586, 648, 654, 657, 659
internal gravity waves, 109, 166, 177, 314, 326, 331, 348, 351, 366, 373, 406, 426, 470, 473, 536, 538, 551, 552, 561, 564, 567, 579, 585, 586, 634, 648, 649, 651–654, 656
ionization, 26, 28, 29, 95, 121–123, 177, 210, 220, 349, 509, 554
ionospheric absorption, 410, 438, 443

jet stream, 381, 574, 575, 621

kinetic, 98, 99, 103, 112, 119, 134, 404, 449, 510–512, 534, 650, 705

Latitude, 526, 627
limb measurements, 52, 711
longitude, 24, 25, 30, 34, 36–38, 53–55, 58, 61, 342–345, 415, 421, 427, 430, 504, 506, 513, 725
lunar variations, 415, 454, 484, 519

magnetic field, 1, 35, 61, 68, 70, 79, 84, 86, 87, 99–101, 103, 105, 112–114, 127, 349, 535
magnetosphere, 1, 10, 86, 87, 97, 103, 107, 109, 110, 114, 115, 117
maximum, 562, 564, 575, 587, 588, 598, 599, 602, 605, 613, 617, 619, 622, 623, 625, 626, 632–634, 637, 638, 640, 642, 646, 653, 657
measurements, 553, 555, 561, 576, 579, 583, 584, 586–596, 598, 599, 601, 604, 611, 614, 617, 619–621, 623–625, 628, 629, 631, 632, 634, 635, 639, 641, 643–646, 649, 650, 653–655, 658, 659
middle atmosphere, 5, 6, 79, 82, 89, 90, 92, 94, 95, 99, 110, 112, 116, 135, 170, 241, 347, 348, 375, 409, 434, 435, 533, 537, 542, 544, 551, 568, 594, 614, 631, 636, 650, 654, 661, 662, 664, 671, 676, 681, 682, 687, 688, 692–694, 698, 700, 701, 703–710, 712, 716, 718–720, 722, 725, 728–733
molecular oxygen, 555, 587, 656
Moon, 19–23, 57–59, 110, 113, 114, 117, 271, 347, 372, 390, 391, 393, 415, 427–430, 454, 471, 484, 519, 530, 561, 645

natural, x, 3, 86, 89, 94, 114, 154, 178, 182, 184, 269, 285, 293, 308, 338, 361, 362, 365, 383, 390, 392, 394, 528, 556, 625, 634, 636, 706, 719
neutrino, 78, 109, 113, 115
night sky, 4, 6–8, 18, 110, 114, 115, 118, 127, 271, 404–406, 408, 530, 532–534, 536–538, 540, 542, 543, 546, 555, 650, 653, 655, 706, 709
nitric oxide NO, 9
nitrogen dioxide NO_2, 9
noctilucent clouds, 95, 97, 116, 154, 435, 436, 447, 506, 533, 543, 544, 553, 568, 573, 628–639, 649–652, 654–657, 659, 684, 687, 705, 710, 724, 733

orographic disturbances, 340, 528, 545, 594, 598, 619, 621, 623, 639, 658
oxygen, 3, 4, 9, 10, 27, 48, 60, 87–89, 96, 99, 104, 115, 122, 126, 127, 132–134, 136, 137, 142, 147, 161–173, 175, 185, 186, 189, 190, 192, 194, 197, 200, 202–204, 206–210, 219, 222, 225–229, 233, 238, 241–245, 293, 301, 303, 328, 329, 336, 340, 346, 347, 349, 354, 388, 394, 397, 403, 404, 408–410, 415, 416, 434, 450, 465–468, 470, 473–477, 481–484, 487, 488, 490–492, 496, 498–502, 505, 508–510, 512–515, 517, 521, 522, 525, 528, 530, 531, 533–535, 537–545, 547, 548, 663, 665, 667, 669, 671, 676, 677, 679, 683, 695–698, 701, 702, 705, 707, 708, 711, 712, 718–725, 730–733
ozone O_3, 9
ozone–hydrogen reaction, 697, 714, 716, 726

photodissociation, 60, 114, 121, 164, 168–172, 234, 236, 240, 716
photography, 6, 8, 277, 278, 282, 285, 305, 318, 402, 410, 411, 654
photoionization, 10, 27–29, 60, 61, 121, 122, 176, 177, 179, 182, 194, 212, 215, 216, 220, 240, 349
photometry, 116, 284, 352, 359
planetary waves, 95, 97, 113, 114, 116, 430, 449, 484, 506, 548, 625–629, 635, 636, 638, 643, 650, 651, 653, 654, 659, 706, 724

radiation flux, 271, 353
Rayleigh scattering, 108, 123, 125, 239
refraction, 39–42, 108, 112, 295, 337, 361
relaxation, 70, 79, 125, 126, 128, 132–136, 154–156, 159, 168, 197, 373, 419, 439, 440, 509, 525, 556, 580, 599

rotational, 16, 60, 110, 125, 128–130, 134–142, 148, 150–157, 159, 160, 163, 165, 166, 188, 196, 197, 199, 200, 243, 282, 287, 290, 292, 331, 346, 351, 352, 369, 388, 392, 393, 401, 406, 408, 409, 411, 416, 418–423, 429, 431, 436, 439, 442, 445, 446, 448, 449, 466, 468, 535, 537, 539–541, 543–546, 548, 556, 558–567, 569–572, 577, 580, 582, 589, 590, 595, 597–599, 601, 628, 644, 645, 647, 652–656, 658, 663, 665, 669, 670, 703–705, 707–710

scanning, 294, 297, 318, 336, 353, 354, 360–362, 388, 402, 405, 512, 537, 595, 596, 734
sky, 230, 280–282, 298, 339, 359, 367, 368, 371, 382, 390, 394, 401, 562, 565, 569, 586, 596, 630
Solar, 656
solar, ix, 1, 2, 7, 9, 10, 19, 20, 23–30, 43, 44, 52–54, 60–64, 68, 70–100, 103, 104, 107, 109–118, 121, 124, 125, 133, 149, 157, 164, 169–171, 179, 183, 200, 202, 207, 209, 212–214, 216–219, 221, 222, 224–227, 229, 230, 239, 242, 245, 339, 340, 365, 372, 373, 382, 403, 405, 406, 410, 415, 421, 426, 427, 431–435, 444, 445, 450, 451, 453–463, 469, 470, 472, 473, 477, 480–482, 484–488, 490, 491, 493–497, 499–502, 504, 508, 509, 513, 515, 516, 518–527, 529–535, 537–539, 542–544, 546–549, 628, 632–634, 647, 651, 653, 657, 658, 662–664, 667, 669, 671–677, 679, 680, 684, 686–689, 696, 698–700, 702–710, 712–718, 720–733
solar flux, 632, 636, 647
spectrum, ix, 3–10, 26–29, 86, 95, 98, 111–113, 116, 117, 124, 137, 148, 179, 183, 213, 214, 217, 218, 230, 243, 271, 273, 277, 278, 280, 281, 284, 294, 297, 302, 305, 312, 313, 328, 336, 339, 340, 350, 351, 353, 354, 357, 359–362, 365, 366, 369, 370, 383, 390–392, 406, 409, 414, 416, 424, 473, 512, 531, 536, 542, 546, 560, 589, 599, 600, 611, 625, 646, 649, 652, 658, 667, 709, 711, 720, 733
spots, 61, 62, 70, 321, 322, 434, 444, 476–478, 619, 639, 642
stellar sky, 23, 113
steric factor, 121
stratospheric warming, 404, 415, 446, 464, 506, 513, 533–535, 538, 543, 547

Sun, ix, 1, 2, 17, 19, 20, 25–27, 36, 43, 53, 54, 56, 61, 62, 65, 70, 71, 74, 78, 79, 84, 86, 94, 103, 105, 111–113, 115, 117, 124, 125, 168, 169, 178, 183, 185, 207, 209, 216, 218, 220, 271, 289, 340, 342, 347, 349, 350, 391, 415, 416, 426, 432, 434, 444, 445, 451, 470, 481, 504, 509, 513, 515, 519, 524, 593, 594, 687, 704, 718

temperature, x, 1, 2, 6, 19, 22, 23, 26, 41, 60, 61, 63, 74, 80, 82, 89–95, 99, 108, 110, 111, 113–116, 119–121, 123, 125, 126, 128, 132, 134, 136, 142, 148, 150, 153–157, 159, 161, 163, 164, 166, 167, 172, 175–178, 196, 197, 199, 200, 203, 206–208, 210, 213, 215, 218, 222, 226, 230, 235, 242, 244, 269, 270, 272, 279, 282, 287, 288, 290–294, 296–298, 301, 304, 312, 314–317, 320, 321, 327–329, 331, 337, 340, 346–348, 351–353, 358, 360, 363, 365, 368, 369, 371–373, 388, 390–393, 397, 401–405, 407–409, 411, 413, 414, 416–424, 426, 429–431, 434–437, 439, 442, 444–446, 448, 449, 453, 456–461, 463, 464, 466–468, 470, 472–475, 478, 482, 484, 485, 489–495, 500, 506–512, 515–517, 519, 521, 523, 525, 526, 528–535, 537, 539–543, 545–549, 551–553, 555–567, 569–572, 576–583, 585–593, 595–603, 614, 615, 617, 619, 620, 623, 628, 631, 633, 636, 638, 639, 644–647, 649, 650, 652, 654–658, 661–677, 679–696, 698–712, 717–722, 724, 725, 728–733
the emission layer, 555–557, 559, 560, 562, 564, 566, 581, 585, 587, 589, 592, 599, 607, 619, 622, 636, 639, 640, 643
thermosphere, 48, 60, 61, 73, 80, 87, 89, 95, 110, 114–117, 169, 215, 227, 229, 235, 244, 245, 347–349, 401, 403–405, 408, 410, 432, 434, 456, 465, 466, 493, 501, 506, 509, 521, 526, 531–535, 538, 540–542, 544, 545, 548, 549, 555, 557, 560, 569, 591, 594, 613, 617, 620, 625, 628, 629, 644, 646, 648, 649, 653, 654, 656, 662, 663, 679, 683, 687, 695, 698, 703–708, 710–712, 719, 722, 723, 726, 730–733
tides, 46, 109, 113, 114, 426, 428, 531, 540, 547, 634, 635, 644–646, 650, 653
time, x, 2, 3, 10, 19, 28, 30, 36, 42, 52–55, 57–59, 61, 64, 65, 68, 71–84, 87, 89, 93, 94, 99, 101–103, 107–109, 112, 114–116, 120, 126, 128, 130, 134, 135,

138, 142, 148, 159, 182, 189, 197, 203, 207, 230, 245, 269, 270, 272, 273, 281, 290, 291, 296, 302, 304–308, 312, 318, 321, 322, 325, 327, 329, 332, 333, 336, 342, 344, 346, 348, 351, 361, 365–377, 381, 382, 385, 391, 393, 394, 398, 401, 409, 414–416, 426, 428, 432, 434, 441, 445, 448, 451, 457, 463, 465, 466, 470, 473, 476, 478, 480–482, 494, 499, 507, 508, 516, 519, 521, 523, 524, 529, 530, 545, 551, 554, 557, 559, 561, 562, 564, 569, 570, 572, 573, 575, 582, 583, 585–587, 589, 592, 594, 596, 603, 607, 609–611, 623, 625, 629, 631, 632, 635–638, 644–646
translational, 125–128, 134
trend, 89, 90, 113, 115, 154, 291, 369, 370, 400, 403, 409, 415, 434, 435, 453, 455, 461–463, 473, 488, 491, 492, 495, 497, 499, 500, 502, 526, 533, 538, 544, 621, 628, 632, 633, 650, 657, 661–663, 667–671, 673, 674, 680–684, 686–700, 702, 704–709, 712, 713, 715, 718, 720, 725–727, 730, 733
troposphere, 2, 94, 96, 113, 348, 401, 406, 507, 537, 552, 568, 569, 600, 621, 625, 627, 639, 655, 698
turbopause, 97, 98, 112, 179, 348, 477, 478, 536, 619, 620, 651
twilight conditions, 28, 183

vertical, ix, 43, 46, 70, 100, 104, 115, 117, 172, 182, 197, 288, 321, 323, 324, 336, 342, 394, 413, 420, 447, 449, 476–478, 528, 531, 534, 542, 544, 552, 560, 564, 567–569, 577, 591, 592, 594, 595, 598, 603, 607–609, 611–614, 617–620, 625, 626, 639, 642, 648, 649, 657, 658, 669, 671–676, 679, 681–683, 686, 688–690, 695–698, 700, 702, 704, 708, 709, 711–714, 719, 722–726, 729, 733
vibrational, 60, 123, 125, 128–143, 148, 150–153, 155–168, 185, 188, 192, 195, 197, 237, 238, 240–243, 286, 290, 349, 401, 416–424, 433, 439, 445, 446, 448, 467, 468, 530, 532, 540, 543, 548, 556, 558, 559, 561, 564, 575, 577, 578, 580–584, 589, 590, 599, 601, 652–654, 665, 670, 703, 704, 707, 732

wave train, 76, 373, 374, 376, 377, 381, 568, 610, 611
wavelength, 552, 586
wind, 60, 82, 86, 87, 95–98, 100, 103, 104, 108, 110, 115–117, 183, 294, 298, 299, 346–348, 373, 374, 388, 394, 397, 398, 401, 402, 404, 408, 410, 415, 432, 435, 446, 451, 506, 527, 534, 538, 548, 552, 553, 566–569, 575, 593–595, 597–605, 608–614, 616–618, 621–625, 628, 629, 639, 643–651, 653–655, 657, 659, 663, 664, 676, 704, 706
Wolf number, 524
Wolf number W, 61

zenith angle, 28–30, 32, 33, 39, 42–45, 53, 96, 121, 125, 157, 179, 200, 281, 282, 288, 298–300, 329–331, 336, 343, 347, 374, 376, 385, 386, 395, 398, 415, 426, 427, 451, 469, 470, 473, 480, 481, 513, 515, 516, 519–521, 551, 568, 608, 609, 630, 640, 642, 684
Zodiacal Light, 3, 20, 25, 26, 36, 42, 109, 111, 117, 230

Printed in the United States
116960LV00001BA/13/P